Heavy Metal Toxicity and Neurodegeneration

Heavy Metal Toxicity and Neurodegeneration

Edited by

Prasann Kumar
Department of Agronomy, School of Agriculture, Lovely Professional University,
Phagwara, Punjab, India

Neha Gogia
Yale University, New Haven, CT, United States

ELSEVIER

ACADEMIC PRESS
An imprint of Elsevier

Academic Press is an imprint of Elsevier
125 London Wall, London EC2Y 5AS, United Kingdom
50 Hampshire Street, 5th Floor, Cambridge, MA 02139, United States

ISBN: 978-0-443-36575-1

For Information on all Academic Press publications
visit our website at https://www.elsevier.com/books-and-journals

Publisher: Stacy Masucci
Acquisitions Editor: Megan McManus
Editorial Project Manager: Pat Gonzalez
Production Project Manager: Vijayaraj Purushothaman
Cover Designer: Greg Harris

Typeset by MPS Limited, Chennai, India

Contents

List of contributors

Nilabja Banerjee Department of Agronomy, School of Agriculture, Lovely Professional University, Phagwara, Punjab, India

Bushra Bashir School of Pharmaceutical Sciences, Lovely Professional University, Phagwara, Punjab, India

Krupanshi Bharadava Biophysics & Structural Biology, Research & Development Cell, Parul University, Vadodara, Gujarat, India

Debjani Choudhury Department of Plant Pathology, School of Agriculture, Lovely Professional University, Phagwara, Punjab, India

Ashok S. Dambale Department of Agronomy, School of Agriculture, Lovely Professional University, Phagwara, Punjab, India

Shipa Rani Dey Faculty of Agricultural Sciences, GLA University, Mathura, Uttar Pradesh, India

Neha Gogia Yale University, New Haven, CT, United States

B. Gopika Department of Agronomy, School of Agriculture, Lovely Professional University, Phagwara, Punjab, India

K. Jaisuriyan Department of Agronomy, School of Agriculture, Lovely Professional University, Phagwara, Punjab, India

Sapna Jarial Department of Agricultural Economics and Extension, School of Agriculture, Lovely Professional University, Phagwara, Punjab, India

Prabhjot Singh Jassal Department of Biotechnology, School of Bioengineering and Biosciences, Lovely Professional University, Phagwara, Punjab, India

Radhey Shyam Kaushal Biophysics & Structural Biology, Research & Development Cell, Parul University, Vadodara, Gujarat, India; Department of Life Sciences—Biotechnology, Parul Institute of Applied Sciences, Parul University, Vadodara, Gujarat, India

Mridu Kulwant Department of Environmental Studies, The Maharaja Sayajirao University of Baroda, Vadodara, India

Prasann Kumar Department of Agronomy, School of Agriculture, Lovely Professional University, Phagwara, Punjab, India

Sanjay Kumar Structural and Molecular Biology, Center for Discovery and Innovation Building, The City College of New York, New York, NY, United States

Makey Sri Chakra Naga Vardhan Department of Agronomy, School of Agriculture, Lovely Professional University, Phagwara, Punjab, India

Divya Patel Department of Environmental Studies, The Maharaja Sayajirao University of Baroda, Vadodara, India; National Sugar Institute, Kanpur, Uttar Pradesh, India

Mastu Patel Department of Agronomy, School of Agriculture, Lovely Professional University, Phagwara, Punjab, India

Amit Raj Department of Agronomy, School of Agriculture, Lovely Professional University, Phagwara, Punjab, India

Patan Ruksar Department of Agronomy, School of Agriculture, Lovely Professional University, Phagwara, Punjab, India

Monika Sharma Department of Agronomy, School of Agriculture, Lovely Professional University, Phagwara, Punjab, India

Shivalika Sood Department of Agricultural Economics and Extension, School of Agriculture, Lovely Professional University, Phagwara, Punjab, India

Venkata Suneel Kumar Velugolu Department of Agronomy, School of Agriculture, Lovely Professional University, Phagwara, Punjab, India

Sukriti Vishwas Chitkara College of Pharmacy, Chitkara University, Rajpura, Punjab, India

Tusha Yadav Department of Biotechnology, School of Bioengineering and Biosciences, Lovely Professional University, Phagwara, Punjab, India

Preface

It is with great pleasure and gratitude that we present *Heavy Metal Toxicity and Neurodegeneration*, a book that delves into the critical intersection of environmental toxins and their impacts on neurological health. The idea of this book was born out of both curiosity and concern. Curiosity about the complexity in functioning of the human brain and concern over the invisible toxins that threaten human health. Over the years, we all have witnessed a continuous increase in the devastating effects of neurodegenerative diseases, not just in clinical settings but also among friends, families, and our loved ones. Seeing once-dynamic, energetic, vibrant minds fade into helplessness, and decline led us to questions: *What if some of neurodegenerative cases were preventable? What if something in our environment was accelerating the progression of these diseases? What if the air we breathe, water we drink, and food we eat are slowly altering our minds? How many cases of neurodegeneration could be deaccelerated or may be prevented if we address environmental toxins?* The deeper we explored, the clearer one of the potential links became, heavy metal toxicity is a silent yet another significant contributor toward neurodegeneration.

Heavy metals are a group of dense, metallic elements that are poisonous at higher concentrations. Some of the most common heavy metals that pose health risks to humans include mercury, lead, arsenic, and cadmium. These infiltrate our (human) bodies through contaminated water, air pollution, industrial exposure, and even everyday consumer products. They accumulate in the brain and nervous system, impair the functioning of vital cellular processes, fuel oxidative stress, neuroinflammation, mitochondrial dysfunction (the hallmarks of many neurodegenerative diseases), and potentially causing damage to the neurons. Despite substantial scientific evidence, the public awareness regarding this issue remains alarmingly low. This book is for scientists, researchers, students, medical professionals, environmental health advocates, and for anyone who seeks answers to these questions and is concerned about their brain health.

This book aims to bridge the gap between the indisciplinary fields of agriculture, toxicology, and neuroscience, and highlighting the role of heavy metals in neurodegeneration and strategies for reducing their effects. Inside, you will discover a comprehensive analysis of the mechanisms underlying heavy metals neurotoxicity, insights from recent research, case studies that show real-world consequences, and potential strategies for detoxification and prevention. Instead of just addressing the issue, this book offers hope through solutions and protective strategies. It is a collective effort, bringing together the contributions of distinguished scientists and researchers committed to advancing our understanding of the damaging effects of heavy metals on the brain.

As editors, we have had the privilege of working with exceptional scholars whose insights and thought-provoking perspectives have shaped this book into a comprehensive resource. We would like to express our sincere gratitude to each of the chapter authors for their hard work and dedication for this project. Our sincere thanks also extend to Elsevier for their unwavering support in making this work a reality.

We invite you to read this book with curiosity and an open mind. The burden of neurodegenerative diseases is immense on the society. Understanding and addressing the contributing factors such as heavy metal toxicity is a vital step toward their prevention. This book is not merely a book that involves scientific discussion, it serves as a call for both awareness and action. Our aspiration is that this book will not only educate but also inspire change, whether it is in research, public policy, clinical practice, or individual health choices or decisions.

Our brain is the most valuable organ. Safeguarding it is not just a medical obligation but a duty and shared responsibility for us all.

Editors
Dr. Prasann Kumar
Dr. Neha Gogia

Chapter 1

Biological pathways of heavy metal toxicity

Mridu Kulwant[1], and Divya Patel[1,2]

[1]*Department of Environmental Studies, The Maharaja Sayajirao University of Baroda, Vadodara, India,* [2]*National Sugar Institute, Kanpur, Uttar Pradesh, India*

1.1 Introduction

Natural elements with high atomic weight and density and metallic characteristics are known as heavy metals. Their chemical form and concentration determine their bioavailability and toxicity, and they are present in the Earth's crust. Lead, mercury, cadmium, and arsenic are a few examples of hazardous substances, even in trace amounts (Abd Elnabi et al., 2023). The toxicity, environmental behavior, chemical characteristics, exposure source, and biological function of heavy metals are the basis for their classification. Even in low amounts, toxic heavy metals, including lead (Pb), mercury (Hg), and cadmium (Cd), have detrimental biological consequences. Iron (Fe), manganese (Mn), copper, and zinc (Zn) are essential elements that can be harmful (Patel et al., 2023; Tchounwou, Yedjou, Patlolla, & Sutton, 2012). Metalloids, bioaccumulate metals, transition and posttransition metals, persistent and nondegradable metals, and exposure sources are all examples of environmental behavior (Ali, Khan, & Ilahi, 2019). In the periodic table, posttransition metals appear after transition, and transition metals partially fill d-orbitals. Both metallic and nonmetallic characteristics are present in metalloids. Anthropogenic metals from human activity, naturally occurring metals in geological formations, and critical metals needed for metabolic functions, redox reactions, and enzyme function are some examples of exposure sources (Ali & Khan, 2017). Nonessential metals are frequently poisonous and have no recognized biological function. Their behavior, toxicity, and management techniques in biological and environmental systems are better understood thanks to this categorization (Saad, El-Sikaily, & Kassem, 2016).

Their natural and artificial sources can determine the degree, frequency, and manner of exposure to heavy metals. The Earth's crust is one of the natural sources; when it weathers, heavy metals, including lead, cadmium, and arsenic, are released into soil and water. Volcanic action also releases these metals as gases or particles (Angon, Islam, Kc, Anjum, & Poudel, 2024). Natural biological cycles, forest fires, hydrothermal vents, and sediment erosion are examples of biological sources. While forest fires release stored metals into the sky, microorganisms and plants may mobilize heavy metals in soil and sediment through metabolic activities. While sediment erosion carries metals to aquatic systems, hydrothermal vents discharge iron and manganese into ocean waters (Wang & Azam, 2024).

The discharge of heavy metals into the environment, which causes local and global pollution, is mainly caused by human activity. Heavy metals, including arsenic, mercury, lead, cadmium, and nickel, are released into the air and soil by industrial processes like mining, metal processing, smelting, electronics manufacturing, agriculture, and energy generation. Phosphate fertilizers, pesticides, and tainted water are used in agriculture, while burning coal emits lead, arsenic, and mercury into the atmosphere (Briffa, Sinagra, & Blundell, 2020). Heavy metals are by-products of fuel consumption and oil refinement. Wastewater and sewage sludge, building materials, and inappropriate waste disposal are examples of urban and residential sources. Lead, lithium, and mercury are among the elements released during the disposal of batteries and e-waste. Heavy metals seep into groundwater and soil from landfills (Masindi, Mkhonza, & Tekere, 2021). Aviation fuel releases lead into the environment, and car exhaust and brake pads release cadmium and lead into the atmosphere. The adverse effects of heavy metals on the environment can be lessened with

Heavy Metal Toxicity and Neurodegeneration. DOI: https://doi.org/10.1016/B978-0-443-36575-1.00031-5

appropriate waste management and transportation procedures. Heavy metals can infiltrate biological systems through consumption, inhalation, and contact with tainted food, water, or soil. Determining these paths and sources is essential to creating focused mitigation plans (Wang, Ting, & Zhao, 2024).

1.1.1 Importance of studying toxicity pathways

Concern over the relationship between environmental factors and human health is developing. Pollution and chemical exposure have a substantial adverse influence on public health. This highlights the necessity of integrated approaches to address these issues (Sundas, Contreras, Mujahid, Beneyto, & Vehi, 2024). Mortality rates rise due to respiratory and cardiovascular disorders brought on by air pollution, mainly PM (Nawaz et al., 2023). Food and water are contaminated by chemicals, including pesticides and endocrine-disrupting compounds, which can cause cancer and reproductive problems, among other chronic health problems (Gonsioroski, Mourikes, & Flaws, 2020). Climate change raises the occurrence of outbreaks and vector-borne illnesses, which pose health problems.

1.1.2 Relevance to regulatory frameworks and public health policies

Environmental safety, occupational health, and public health are all impacted by heavy metal toxicity. To reduce exposure and lessen health hazards, regulatory frameworks and regulations have been developed as a result of an understanding of biological processes such as oxidative stress, enzyme disruption, genotoxicity, and carcinogenicity; some of these are listed in Table 1.1 (Tchounwou et al. 2012).

1.1.3 Occupational health and safety policies

The Mines Act of 1952 tracks exposure to lead, arsenic, and mercury in mining operations, the Factories Act of 1948 controls occupational exposure to heavy metals, and the Personal Protective Equipment Regulations guarantee worker safety in dangerous settings (Sudhakar et al., 2024). While blood lead level monitoring systems are in place for children and industrial workers, the National Programme on Prevention and Control of Fluorosis & Heavy Metal Toxicity concentrates on screening and treatment. Programs for risk communication and public awareness are being implemented to inform people about the dangers of heavy metals in food, water, and workplace environments (Egan, Tsai, & Chuke, 2019). Lack of compliance, industrial pollution from small-scale companies, and bioaccumulation in the food chain due to persistent heavy metals in crops, soil, and water are some issues facing heavy metal regulation. More vigorous enforcement of industrial emissions and waste disposal regulations, routine biomonitoring of heavy metal levels in high-risk populations, creating plant and microbial bioremediation techniques, and incorporating computational modeling for heavy metal toxicity and exposure risks are some of the policy recommendations. Regulatory frameworks and public health policies are essential to protect human health and the environment against heavy metal toxicity. Strict enforcement, ongoing monitoring, and creative remediation techniques are needed (Bala et al., 2022; Sonone, Jadhav, Sankhla, & Kumar, 2020).

1.1.4 Objectives of the chapter

This chapter provides an overview of heavy metal toxicity, as well as information on sources of exposure, essential and nonessential heavy metals, and biological systems.

1.2 Overview of heavy metals and their toxic effects

The presence of toxic heavy metals can negatively affect human health and the environment when they are present in high concentrations. Heavy metals that contribute to the spread of poisonous diseases include lead (Pb), mercury (Hg), cadmium (Cd), arsenic (As), chromium, nickel (Ni), copper (Cu), zinc (Zn), aluminum (Al), and selenium (Se). Many neurological effects can be caused by lead, including neurological damage, developmental disorders, kidney damage, and high blood pressure (Witkowska, Słowik, & Chilicka, 2021).

1.2.1 Common toxic heavy metals

Damage to the nervous system, cognitive impairment, kidney failure, and kidney damage can be caused by mercury. In addition to kidney damage, cadmium can also lead to lung disease and bone demineralization. Skin lesions (The term "skin lesion" refers to an area of skin that differs from the surrounding tissue in texture, color, and shape), cancers, cardiovascular diseases, and neurotoxicity are among the effects of arsenic. There is a possibility that chromium can cause lung cancer, respiratory problems, and skin irritations (Xie, Tong, Chu, Feng, & Geng, 2022). In addition to causing skin allergies and respiratory issues, nickel can also be carcinogenic. The liver, gastrointestinal tract, and nervous system can be damaged from copper exposure. In addition to nausea and stomach cramps, zinc can reduce our immune system badly. Aluminum may be linked to Alzheimer's disease and might be neurotoxic. Nervous system disorders, gastrointestinal disorders, and hair and nail loss can all be brought on by selenium (Genchi, Carocci, Lauria, Sinicropi, & Catalano, 2020).

TABLE 1.1 List of prominent standard guidelines for heavy metals.

Sr. no.	Different guidelines (worldwide)	
1.	WHO guidelines	
	Lead (Pb)	≤10 μg/L in drinking water
	Arsenic (As)	≤10 μg/L in drinking water
	Cadmium (Cd)	≤3 μg/kg body weight per week
	Mercury (Hg)	≤6 μg/kg body weight per week
2.	USEPA	
	CWA	Regulates heavy metal discharges into water bodies.
	SDWA	Sets MCLs for toxic metals in water.
	RCRA	Manages hazardous waste containing heavy metals.
3.	European Union REACH Regulation	
	Registration, Evaluation, Authorisation, and Restriction of Chemicals (REACH)	Limits heavy metal use in industries.
	RoHS	Controls the levels of cadmium (Cd), lead (Pb), mercury (Hg), and hexavalent chromium (Cr^{6+}) in electrical and electronic devices.
4.	BIS—Drinking Water Quality	
	Lead (Pb)	≤10 μg/L
	Arsenic (As)	≤10 μg/L
	Cadmium (Cd)	≤3 μg/L
5.	CPCB guidelines	
6.	The legislation establishes industrial discharge limits for heavy metals into water bodies and regulates emissions from tanneries, mining, battery manufacturing, and other industries.	
7.	Hazardous and Other Wastes (Management and Transboundary Movement) Rules, 2016	
8.	FSSAI Regulations	

BIS, Bureau of Indian Standards; *CPCB*, Central Pollution Control Board; *CWA*, Clean Water Act; *FSSAI*, Food Safety and Standards Authority of India; *MCLs*, maximum contaminant levels; *RCRA*, Resource Conservation and Recovery Act; *RoHS*, Restriction of Hazardous Substances; *SDWA*, Safe Drinking Water Act; *USEPA*, United States Environmental Protection Agency; *WHO*, World Health Organization.

1.2.2 Mechanisms of toxicity

One of the leading causes of heavy metal toxicity is oxidative stress, which damages proteins, lipids, and DNA and causes cell death and some illnesses. Antioxidants like vitamins C and E, polyphenol-rich diets, and cholesterol treatment are some ways to fight this (Patel & Kulwant, 2024).

1.2.3 Oxidative stress induction

One of the main toxicity mechanisms caused by heavy metals is oxidative stress, which arises from an imbalance between the body's capacity to detoxify these intermediates through antioxidant defenses and the creation of reactive oxygen species (ROS). Heavy metals producing ROS induce oxidative stress that can harm DNA, lipids, and proteins, resulting in cellular malfunction and death. Heavy metal-mediated ROS generation: metals like iron and copper engage in Fenton and Haber–Weiss reactions, which produce hydroxyl radicals and ROS (Abdal Dayem et al., 2017).

Example:

$$Fe^{2+} + H_2O_2 \rightarrow Fe^{3+} + OH^- + OH$$

In addition to disrupting mitochondrial electron transport, which results in electron leakage and excessive ROS generation, heavy metals such as chromium undergo redox cycling, which produces ROS as by-products (Juan, Pérez de la Lastra, Plou, & Pérez-Lebeña, 2021). Lipid peroxidation: besides oxidizing polyunsaturated fatty acids

(PUFAs), which damage cell membranes and compromise cellular integrity, ROS produce malondialdehyde (MDA), a hazardous byproduct that further damages proteins and DNA. Example: lipid peroxidation is encouraged by lead (Pb) and cadmium (Cd), which can cause kidney damage and neurotoxicity (Su et al., 2019). Protein oxidation and enzyme inhibition: proteins' sulfhydryl groups are bound by heavy metals, which results in structural degradation and the loss of enzymes, misfolding, aggregation, and functional impairment. Example: thiol-containing enzymes are inhibited by arsenic (As), which interferes with cellular metabolism (Shen, Li, Cullen, Weinfeld, & Le, 2013). DNA damage and genotoxicity: ROS harm DNA, resulting in mutations, chromosomal abnormalities, single- and double-stranded DNA damage, and base alterations such as 8-oxo-dG production. Example: hexavalent chromium (Cr^{6+}) contributes to carcinogenesis by causing breaks in DNA strands (Srinivas, Tan, Vellayappan, & Jeyasekharan, 2019). Depletion of antioxidant cellular antioxidants such as GSH, SOD, CAT, and GPx are depleted by heavy metals and are responsible for detoxifying and neutralizing ROS. Example: it is well known that cadmium (Cd) lowers GSH levels, which increases oxidative stress and damages the kidneys (Birben, Sahiner, Sackesen, Erzurum, & Kalayci, 2012).

1.2.4 Disruption of enzymatic processes

Poisonous heavy metals interfere with cellular enzymatic functions by attaching themselves to vital biomolecules, leading to oxidative damage, metabolic imbalances, and cellular toxicity. Heavy metals can replace essential metal cofactors like Zn^{2+}, Fe^{2+}, Cu^{2+}, and Mg^{2+} that are necessary for enzymatic activity when they attach to the active sites of enzymes (Jan et al., 2015). Lead modifies neuronal signal transmission by replacing calcium in enzymes such as protein kinase C. In metalloproteins, cadmium takes the role of zinc, preventing them from functioning. Essential enzymes are rendered inactive by the strong connections that heavy metals like arsenic, mercury, and cadmium form with -SH groups (Bressler & Goldstein, 1991). Cellular respiration is inhibited by arsenic's binding to lipoic acid, a necessary cofactor in pyruvate dehydrogenase. Enzyme cysteine residues are bound by mercury, which results in protein misfolding and dysfunction. Heavy metals cause oxidative stress and cellular damage by interfering with antioxidant defense enzymes. Lead and cadmium, for instance, deactivate Catalase and glutathione peroxidase (CATx), increasing the build up of hydrogen peroxide, and substitute zinc (Zn^{2+}) in superoxide dismutase (SOD), decreasing its capacity to neutralize superoxide radicals (Balali-Mood, Naseri, Tahergorabi, Khazdair, & Sadeghi, 2021).

1.2.5 Genotoxicity and carcinogenicity

Chronic exposure to heavy metals increases the likelihood of mutations, genomic instability, and unchecked cell proliferation. Heavy metals also cause genotoxicity and carcinogenicity through DNA damage, epigenetic changes, and interference with repair systems. Lead, arsenic, and chromium are heavy metals that can harm DNA in several ways. ROS are involved in oxidative damage, which results in base changes and breaks in DNA strands (Koyama, Kamogashira, & Yamasoba, 2024). Some different classes of carcinogenic heavy metals are listed in Table 1.2. While cadmium binds to DNA, breaking its structure and causing strand breaks, chromium forms stable adducts by direct DNA contact. Mercury and arsenic also cause DNA strands to become alkylated and cross-linked. Because they prevent nucleotide excision repair and base excision repair, certain metals, including arsenic and cadmium, can interfere with DNA repair processes, increasing the likelihood of mutations and creating

TABLE 1.2 Heavy metals classified as human carcinogens.

Heavy metal	Cancer	Types of mechanism	References
Arsenic (As^{3+})	Skin, lung, bladder	DNA damage, oxidative stress, epigenetic changes	Muenyi, Ljungman, and States (2015)
Hexavalent Chromium (Cr^{6+})	Lung cancer	DNA adduct formation, ROS generation	Nigam, Priya, Bajpai, and Kumar (2014)
Cadmium (Cd^{2+})	Lung, prostate, and kidney	DNA repair inhibition, epigenetic changes	Rasin, Basheer, Haribabu, Santibanez, and Garrote (2025)
Nickel (Ni^{2+})	Lung, nasal	Epigenetic silencing of tumor suppressors	Sun, Shamy, and Costa (2013)
Lead (Pb^{2+})	Possible carcinogen	Genotoxicity, oxidative stress	Liu et al. (2018)

genomic instability (Chatterjee & Walker, 2017). Chromosome abnormalities, including clastogenic effects, aneuploidy, and the production of micronuclei, are indicators of genotoxicity in human cells and can be brought on by heavy metals like lead and arsenic (Kou, Wu, Ren, & Yang, 2020). Through DNA methylation alterations, histone modifications, and miRNA dysregulation, heavy metals like arsenic and cadmium can influence the expression of genes, impact oncogenes and tumor suppressor genes, and promote tumor development through global DNA hypomethylation and histone modifications (Kou et al., 2020). Heavy metals can cause uncontrolled cell proliferation by interfering with tumor suppressor genes like p53 and Rb. Hexavalent chromium, for instance, mutates p53, which accelerates cancer development. Furthermore, some metals stimulate excessive cell division and raise the risk of tumors by acting as mitogens. HIF-1α is activated by nickel and cobalt, which promotes angiogenesis and tumor development (Haidar, Fatema, Shoily, & Sajib, 2023).

Heavy metals cause genotoxicity by damaging DNA, preventing repair processes, and encouraging chromosomal instability. They also cause cancer through oxidative stress, epigenetic changes, and disruption of tumor suppressor genes. Prolonged exposure to heavy metals greatly increases cancer risk; hence, regulatory control and mitigation measures are crucial (Torgovnick & Schumacher, 2015).

1.2.6 Dose–response relationships

A key idea in toxicology, risk assessment, and regulatory standards is the dose–response relationship, which describes how exposure levels affect heavy metal toxicity (Shinde et al., 2024). The dose–response relationship explains the link between exposure to certain chemicals. Lead and children's cognitive impairment are two examples of toxicity that rise proportionately with dose, according to linear dose–response ratios. According to threshold dose–response relationships, hazardous effects, such as kidney damage and cadmium, must occur at a specific dose. Low dosages may have diverse or opposing effects, as demonstrated by nonmonotonic dose–response relationships. For example, arsenic can cause adaptive cellular responses at low doses but becomes highly poisonous and carcinogenic at greater concentrations (Moffett, Mumtaz, Sullivan, & Whittaker, 2022). Some defined prominent parameters have been tabulated in Table 1.3.

Heavy metal dose–response curves demonstrate that there is no safe threshold for exposure to some metals and that exposure to them raises the risk of cancer. There is no known safe threshold for the neurodevelopmental consequences of lead exposure, which causes neurotoxicity in children. Kidney damage only occurs when persistent exposure to cadmium is above a threshold level. To define safety limits, determine maximum contaminant levels, and direct risk assessment for occupational exposure in sectors such as manufacturing, electronics, and mining, regulatory bodies rely on dose–response data. Establishing toxicity thresholds, regulatory restrictions, and public health initiatives to lower heavy metal exposure all depend on an understanding of dose–response relationships (Moulis, Bulat, & Buha Djordjevic, 2020).

1.2.7 Bioaccumulation and biomagnification

As heavy metals like lead, mercury, cadmium, and arsenic are not biodegradable, they continue to exist in the environment. Through food, water, and air, they infiltrate organisms and cause bioaccumulation and biomagnification, which endangers ecosystems and human health (Mitra et al., 2022). The term "bioaccumulation" describes how harmful chemicals, such as mercury (Hg), gradually build up in fish tissues, reaching large amounts in long-lived species like tuna (Kumar et al., 2023). Predators like sharks and eagles have greater concentrations of metals in the food chain due to biomagnification than do primary eaters like tiny fish (Saidon, Szabó, Budai, & Lehel, 2024). While plants acquire heavy metals from polluted soil through phytoaccumulation, aquatic creatures absorb them from the water through their gills and body surfaces (Kamal, Ghaly, Mahmoud, & Côté, 2004). Long-term retention of heavy metals might result from their accumulation in tissues like the kidneys and bones. Bone storage of lead and cadmium results in kidney damage and calcium replacement. Long-term toxicity can result from the inadequate excretion of some heavy metals (Jomova, Alomar, Nepovimova, Kuca, & Valko, 2025).

1.3 Molecular and cellular pathways of heavy metal toxicity

1.3.1 Oxidative stress and antioxidant defense

By producing ROS, harming cellular macromolecules, causing DNA damage, lipid peroxidation, protein malfunction, and cell death, heavy metals such as lead, mercury, cadmium, and arsenic induce oxidative stress and drive cells to activate antioxidant defense systems (Ghosh, Das, Chaffee, Roy, & Sen, 2018). DNA mutations and oxidative damage can be brought on by heavy metals such as lead, mercury, cadmium, and arsenic. Lead causes lipid peroxidation, neurotoxicity, mitochondrial dysfunction, and inhibition of antioxidant enzymes. Mercury causes protein oxidation, Fenton-like reactions, DNA strand breakage, and glutathione depletion (Haidar et al., 2023). Superoxide anion, hydrogen peroxide, hydroxyl radicals, and singlet oxygen are examples of ROS. Singlet oxygen assaults lipids and proteins, hydrogen peroxide produces hydroxyl radicals, which in turn cause serious damage, and

TABLE 1.3 The dose–response relationship: key parameters.

S. No.	Parameter	Definition	Example (heavy metal effects)	References
1.	NOAEL (No Observed Adverse Effect Level)	Professional opinions are based on study design, drug indication, expected pharmacology, and off-target effects. NOAEL does not have a standard definition.	**Cadmium NOAEL: ~1 μg/kg/day** (kidney effects)	Charkiewicz, Omeljaniuk, Nowak, Garley, and Nikliński (2023), Dorato and Engelhardt (2005)
2.	LOAEL (Lowest Observed Adverse Effect Level)	Lowest dose where toxicity is observed	**Arsenic LOAEL: ~0.5 μg/kg/day** (skin lesions, cancer)	Abd Elnabi et al. (2023), Charkiewicz et al. (2023)
3.	LD50 (Lethal Dose for 50% of Population)	Dose causing death in 50% of test subjects	**Mercury LD50:** ~1 mg/kg (acute toxicity)	Chinedu, Arome, and Ameh (2013), Erhirhie, Ihekwereme, and Ilodigwe (2018)
4.	Benchmark Dose (BMD)	Dose causing a predefined effect level	Used in **risk assessment** for setting safety limits	Committee et al. (2022), Vieira Silva, Ringblom, Moldeus, Törnqvist, and Öberg (2021)

superoxide anion damages proteins and DNA (Jakubczyk et al., 2020). ROS can harm cells through oxidative stress, which includes protein oxidation, DNA damage, and lipid peroxidation. Cell death, increased permeability, and membrane instability can result from these damages. Strand breakage, base changes, and cross-linking may all cause DNA damage, which can result in genotoxicity, mutations, and the emergence of cancer. ROS oxidation can also result in protein oxidation and enzyme inhibition, which can lead to aggregation and loss of function (Jomova et al., 2023). To counteract ROS and stop oxidative damage, cells employ both enzymatic and nonenzymatic antioxidants. Superoxide is broken down into hydrogen peroxide (H_2O_2) by enzymatic antioxidants such as superoxide dismutase (SOD), H_2O_2 is broken down into water and oxygen by catalase, and H_2O_2 and lipid peroxides are reduced by glutathione peroxidase (GPx). Antioxidants that are not enzymatic, such as glutathione (GSH), neutralize ROS, detoxify heavy metals, and shield lipid membranes from peroxidation. Increasing consumption of fruits, vegetables, and polyphenols, employing chelation treatment, creating antioxidants based on nanoparticles, and enforcing stringent environmental rules are some methods to lessen oxidative stress caused by heavy metals. Reducing metal exposure and strengthening antioxidant defenses are essential for toxicity mitigation (Irato & Santovito, 2021; Kozlov, Javadov, & Sommer, 2024).

1.3.2 Protein and enzyme interference

Lead, mercury, cadmium, and arsenic are examples of heavy metals that damage proteins, enzymes, and vital biological processes, resulting in toxicity and cellular malfunction. Cellular dysfunction results from their interference with signaling networks, enzyme activity, protein structure, and metabolic processes (Jaishankar, Tseten, Anbalagan, Mathew, & Beeregowda, 2014). The structure and function of proteins are changed when heavy metals such as As, Cd, and Hg attach to their thiol groups. Mercury interferes with cellular metabolism by deactivating enzymes that contain thiols. Additionally, heavy metals cause malfunction by substituting vital elements like Zn, Fe, and Ca in enzymes. Lead competes with calcium in brain signaling proteins, causing neurotransmission to be disrupted, and Cd substitutes zinc in metalloproteins, reducing their function (Tamás, Sharma, Ibstedt, Jacobson, & Christen, 2014).

As enzyme inhibitors, heavy metals obstruct metabolic processes and alter proteins through oxidation. Lead (Pb) inhibits delta-aminolevulinic acid dehydratase (ALAD), which hinders heme synthesis and results in anemia, whereas arsenic (As) inhibits pyruvate dehydrogenase, which interferes with the creation of ATP. Moreover, proteins oxidized by heavy metals lose their functionality. Mercury (Hg) causes apoptosis by changing the potential of the mitochondrial membrane. Ca^{2+}-dependent enzymes and signaling proteins are disrupted by heavy metals, which interfere with calcium signaling. Lead (Pb) interferes with neurotransmission by breaking down calmodulin and protein kinase C, and cadmium interferes with cell signaling and bone metabolism by imitating calcium (Renu et al., 2021). Protein misfolding and ER stress brought on by heavy metals result in apoptosis and neurodegenerative illnesses. They damage glutathione-dependent enzymes and interfere with detoxification processes. Metal excretion is decreased by arsenic and cadmium because they interfere with protein folding chaperones. Glutathione is depleted by mercury, which increases oxidative stress. Neurodegenerative disorders and apoptosis are brought on by prolonged ER stress (Poothong, Jang, & Kaufman, 2021).

Heavy metals impair cellular processes by interacting with proteins, enzymes, and signaling pathways and cause oxidative stress, metabolic malfunction, and apoptosis. Heat shock proteins help in protein refolding and stress response, whereas metallothioneins guard against protein damage. Protein oxidation is stopped by antioxidant enzymes, which also neutralize ROS. Damaged and misfolded proteins are eliminated by autophagy and proteasomal breakdown. Comprehending these pathways is essential for formulating regulatory policies and treatment approaches (Hu et al., 2022).

1.4 Organ-specific pathways of toxicity

1.4.1 Neurotoxicity (brain and nervous system)

The degree of targeted toxicity caused by heavy metal accumulation in organs varies according to dosage, exposure time, and metal bioavailability. Lead, mercury, and arsenic are examples of heavy metals that can be neurotoxic to the brain and neurological system. In contrast to mercury, which binds to cysteine like methionine, lead imitates calcium and enters neurons through calcium channels. These metals cause motor dysfunction and cognitive impairments by interfering with neurotransmitters. Arsenic damages mitochondrial activity and ATP synthesis in neurons, whereas mercury produces ROS that lead to inflammation and neuronal death (Wijayawardena, Mallavarapu, & Naidu, 2016). Cognitive impairment, memory loss, neurodevelopmental problems (like autism and ADHD), and neurodegenerative illnesses (like Parkinson's and Alzheimer's) are examples of toxicity outcomes (Hajri et al., 2022).

1.4.2 Nephrotoxicity (kidneys)

The build up of metals, including cadmium (Cd), lead (Pb), and mercury (Hg), causes nephrotoxicity in the kidneys.

These metals cause ion balance abnormalities, mitochondrial damage, and renal tubular damage. Chronic kidney disease, renal failure, and electrolyte abnormalities are among the harmful consequences that follow (Barbier, Jacquillet, Tauc, Cougnon, & Poujeol, 2005).

1.4.3 Hepatotoxicity (liver)

Mercury, cadmium, and arsenic are metals that can cause fatty liver disease and fibrosis, lipid peroxidation, and liver enzyme dysfunction. Mercury increases ROS, causing fatty liver disease and fibrosis; cadmium interferes with glutathione metabolism; and Arsenic inhibits detoxification (Rana, Tangpong, & Rahman, 2018).

1.4.4 Pulmonary toxicity (lungs)

Lead, cadmium, and arsenic are metals that can raise the risk of lung cancer and induce inflammation, oxidative stress, and lung fibrosis. Lead builds up in lung tissues and causes chronic obstructive pulmonary disease, whereas cadmium reduces lung surfactant generation, affecting lung function. Lung cancer, chronic bronchitis, and lung fibrosis are among the harmful effects (Paithankar, Saini, Dwivedi, Sharma, & Chowdhuri, 2021).

1.4.5 Cardiovascular toxicity (heart and blood vessels)

Metals such as lead, cadmium, and arsenic can lead to endothelial dysfunction, hypertension, and poisoning of the heart muscle. Lead causes vasoconstriction by interfering with the synthesis of nitric oxide, and cadmium causes atherosclerosis by increasing oxidative stress. Arsenic disrupts ion channels, resulting in heart cell damage and cardiac arrhythmias(Alissa & Ferns, 2011).

1.4.6 Gastrointestinal toxicity (digestive system)

Metals, including arsenic (As), lead (Pb), and mercury (Hg), can harm mucous membranes, disrupt gut flora, block enzymes, and induce malabsorption of nutrients. Lead inhibits digestive enzymes, which affects nutritional absorption and raises the risk of diabetes, while arsenic destroys the intestinal lining. Dysbiosis of the gut microbiota and stomach ulcers are examples of toxicity effects (Jomova et al., 2025).

1.4.7 Reproductive and develop21wq44mental toxicity

Metals like lead (Pb), mercury (Hg), and cadmium (Cd) alter the balance of hormones, influencing the amounts of testosterone and estrogen. Lead lowers fertility by interfering with spermatogenesis. Lead results in low birth weight and early delivery, whereas mercury harms embryonic brain development. Hormonal abnormalities, birth malformations, and infertility are all consequences of toxicity (Rahman & Singh, 2019).

While detoxification assistance, such as glutathione boosters and probiotic treatment, aids in metal excretion, chelation therapy, dietary antioxidants, such as vitamin C, vitamin E, selenium, and polyphenols, and environmental laws such as the WHO, EPA, and BIS, restrict industrial heavy metal emissions (Sears, 2013). The disruption of cellular pathways, enzyme functioning, and oxidative equilibrium caused by heavy metals results in organ-specific toxicity, which facilitates the creation of efficient detoxification plans and regulatory regulations (Koyama et al., 2024).

1.5 Heavy metals detoxification and treatment

It is well-known that some heavy metals are potentially toxic in high concentrations to living beings, whether plants, animals, or microbes, while some are nutritionally essential elements that play key roles in different physiological and biological processes, such as cobalt, zinc, iron, chromium, copper, molybdenum, manganese, and selenium (Mitra et al., 2022). Certain types of heavy metals also have applications in medical fields. For example, arsenic compounds have been widely used in medical sciences to treat illnesses, including trypanosomes, syphilis, and amebic diarrhea (Paul, Galván, Yoshinaga-Sakurai, Rosen, & Yoshinaga, 2023). In addition to its well-known gastroprotective benefits and effectiveness in treating *Helicobacter pylori* infection, bismuth compounds have been utilized widely as medicines, particularly for treating stomach cancer. They also have broad antimicrobial, antileishmanial, and anticancer characteristics (Keogan & Griffith, 2014). Copper, gold, zinc, and silver are extensively used as antiinflammatory (Thunus, Dauphin, Moiny, Deby, & Deby-Dupont, 1995). Suppose these heavy metals enter the body through inhalation, injection, or skin contact. In that case, they can cause several health problems, such as neurological disorders, kidney and liver damage, reproductive issues, anemia, cancer risk, respiratory problems, skin lesions, developmental delays (in children), and cognitive impairments, depending on the specific metal and exposure level. (Mitra et al., 2022). Removing the exposure source and preventing the patient from coming into contact with it is the first step in treating heavy metal poisoning. First-line therapy focuses on treating symptoms, providing support, and ensuring that the kidney, liver, respiratory, and cardiovascular systems are monitored. Chelating agent administration is the most widely used treatment (Kim, Kim, & Kumar, 2019). Fig. 1.1 lists a few recognized medicinal

FIGURE 1.1 Some reported natural resources for heavy metal poisoning treatments are listed.

herbs and other natural seeds helpful in treating heavy metal intoxication. The efficacy and mechanism of several therapeutic strategies, including levodopa, para-aminosalicylic acid, and ethylenediaminetetraacetic acid (EDTA), have been established. Antiinflammatory substances, natural and synthetic antioxidants, glutamate protectors, and ATP/ADP ratio protectors are other therapeutic strategies that have been investigated to treat heavy metal toxicity. These strategies have been utilized to reduce Mn-induced neurotoxicity (Marreilha Dos Santos, Andrade, & Aschner, 2017). According to a prior study, curcumin plus arsenic may significantly shield rats' brains against oxidative stress and dopaminergic alterations brought on by arsenic (Yadav et al., 2010). Almond and walnut eating at a 400 mg/kg/day dose significantly reduced cadmium-induced depression, anxiety, and memory loss. Neurochemical abnormalities in rats have normalized after supplementing with certain nuts. Some studies show that long-term supplementation with almonds and walnuts supplies essential nutrients that can help overcome dietary deficiencies and reduce heavy-metal poisoning (Batool et al., 2019).

Commonly used chelating drugs for chronic arsenic toxicity include penicillamine, dimercaptosuccinic acid (DMSA), and 2,3-dimercapto-1-propanesulfonic acid (DMPS). It was observed that when DMSA and long carbon chain analogues like mono cyclohexyl and mono isoamyl DMSA were used concurrently, the arsenic load was reduced more effectively than when DMSA was used alone (Mishra, Mehta, & Flora, 2008). Furthermore, chelation therapy combined with adjuvant antioxidant therapy is effective in treating metal toxicity. In lead poisoning, it was discovered that combining DMSA and α-lipoic acid improved oxidative damage. Additionally, the chelating activity of zinc and $CaNa_2$ EDTA was enhanced. Zinc and selenium have been demonstrated to guard against nonessential metals (Kim et al., 2019).

1.6 Biomarkers of heavy metal toxicity

Biomarkers are valuable tools for assessing the interactions of biological systems with heavy metals or other potential contaminants. As such, they aid in determining how organisms react to heavy metals (Kim et al., 2019).

1.6.1 Biomarkers of exposure

These show whether heavy metals are present in the body and in what quantities. Usually, they are measured in biological samples such as tissues, blood, urine, hair, and nails (Vinothkannan, Emmanuel Charles, Rajaram, Al-Sadoon, & Gulnaz, 2023). Biomarkers' exposures to heavy metals such as lead (Olympio, Salles, Akiba, & Luz, 2023),

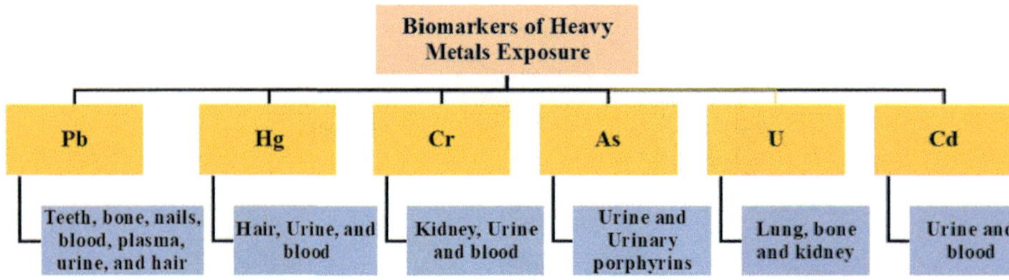

FIGURE 1.2 Biomarkers of heavy metal exposure.

TABLE 1.4 Biomarkers aid in extrapolating data from accessible sites to processes in inaccessible sites.

S. No.	Category	Descriptions
1.	Molecular lesions	DNA adducts, glycosylated hemoglobin, protein adducts, and amino acid conjugates by electrophilic chemicals, chromosomal aberrations, and micronuclei induction by clastogens
2.	Metabolized exogenous agents	phenol (benzene), DDE (DDT), cotinine (nicotine), retinol (β-carotene), benzo(a)pyrene dihydrodiolepoxide (benzo(a)pyrene), acrolein (cyclophosphamide);
3.	Endogenously produced biomolecules	acetylcholinesterase (organophosphate pesticide), gamma-glutamyltransferase (liver toxins). Porphyrin ratios (Pb and other metals) and disease markers, e.g., alpha-fetoprotein (liver cancer), SGOT (myocardial infarction), and creatinine kinase (muscle trauma);
4.	Cellular/tissue changes	sperm motility, sperm count, macrophage activity, red blood cell counts, and lymphocyte ratios
5.	Unchanged exogenous agents	asbestos fibers, PCBs, aflatoxins, some solvents, and heavy metals.

chromium, cadmium (Ventura et al., 2021), arsenic (Hughes, 2006), uranium (Guéguen et al., 2016), and mercury are shown in Fig. 1.2.

1.6.2 Biomarkers effect

Biomarkers' impact is linked to a physiological system instead of exposure to a specific xenobiotic. Accordingly, they are described as biochemical, physiological, behavioral, or other measurable changes in an organism that, depending on their severity, may be linked to a known or suspected illness or health issue. Kakkar et al. separated the data into five categories using biomarkers, which aid in extrapolating information from accessible areas to processes in inaccessible sites (Table 1.4) (Kakkar & Jaffery, 2005).

1.7 Case studies of heavy metal toxicity

Exposure to heavy metals through ingestion, inhalation, or skin contamination can cause poisoning. Mining, burning fossil fuels, industrial processes, agriculture, and the consumption of plants, fish, and meat that have acquired heavy metals from contaminated soil, streams, and seawater are some ways humans might come into contact with these metals. Multiple heavy metal toxicities can be more severe and result in significant morbidity. Toxic levels of heavy metals are present in specific traditional medical formulations (Rajkumar, Lee, & Gupta, 2020). Children are more susceptible than adults. Many case studies have been conducted globally to assess the toxicity and health hazards of ingesting heavy metals (Mitra et al., 2022).

In a family case report that Yildirim presented, five family members, ages 20–54, were poisoned by liquid mercury (Yıldırım et al., 2012). Case 1 arrived at their clinic complaining of fever, coughing, and nocturnal sweats. On day 4 of hospitalization, the patient passed away due to acute respiratory distress syndrome, pneumonia, anemia, and neutropenia. 0.4×10^3 mm^{-3} (normal range: 4.3–10.3 × 103 mm^{-3}) was her white blood cell (WBC) count, and 10.8 g/dL (normal range: 11.5–16.0 g/dL) was her hemoglobin level. In Case 2, bicytopenia was evident; the platelet count (PLT) count was 88 × 103 mm^{-3} (normal range: 150–400 × 103 mm^{-3}) and the leukocyte count was 1.3 × 103 mm^{-3} (normal range: 4.3–10.3 × 103 mm^{-3}). There was toxic peripheral neuropathy in cases two and three. In example number three, the PLT count was 123 × 103 mm^{-3} (typical range: 150–400 × 103 mm^{-3}). Aside from elevated blood mercury levels, Cases 4 and 5 showed no beneficial results. Both patients complained of headaches and exhaustion.

Methylmercury poisoning in humans caused by consuming fish and shellfish contaminated by methylmercury released in wastewater from a chemical factory is known as Minamata disease. In May 1956, Minamata disease was publicly "discovered" for the first time in Minamata City, southwest of Kyushu Island, Japan. Marine products from Minamata Bay were found to contain high levels of mercury contamination (5.61–35.7 ppm). Additionally, high levels of mercury were discovered in the hair of patients, their relatives, and people who lived along the Shiranui Sea coast (Harada, 1995). Among the heavy metals, arsenic has been examined extensively worldwide, particularly in developing nations such as Bangladesh, Nepal, Vietnam, Pakistan, and India (Shaji et al., 2021). Arsenic's poisonous and detrimental effects have been documented in numerous earlier research studies. Belkin and associates studied chronic arsenic toxicity in rural China caused by home coal combustion. Based on the case study, they found that prolonged exposure to arsenic significantly negatively impacts residents in China's southwest Guizhou Province (Belkin, Zheng, Zhou, & Finkelman, 2008). Numerous earlier research studies confirmed this claim because arsenic poisoning is also a significant problem in Bangladesh (Ahmad, Khan, & Haque, 2018). After surveying the villages of Bangladesh's Laxmipur, Barisal, and Madaripur districts, (Milton, Hasan, Rahman, and Rahman (2001) discovered conclusive proof that consuming inorganic arsenic in drinking water causes pulmonary side effects, such as coughing, chest pain, and dyspnea. One of Thailand's most significant environmental issues is the cadmium pollution of rice fields in the Mae Sot area of Tak Province (Somporn, Chaiwong, Hasan, & Chaiwong, 2023). Thus Songprasert and colleagues studied cadmium exposure in the Mae Sot region of Tak Province, Thailand. According to that study, the Mae Sot community lost more healthy lives than those who lived in areas with less or no cadmium (Songprasert et al., 2015).

1.8 Conclusion

Toxicity from heavy metals is a global health and environmental issue. Heavy metal biomarkers, their biological route, and their toxicity have all been adequately demonstrated. The rapid, uncontrolled use of insecticides, medicines, and other substances that contain heavy metals has made their exposure and release intolerable. Certain heavy metals have medicinal uses, and essential trace heavy metals are required for regular cellular and enzymatic functions. However, high concentrations of these and other nonessential metals and metalloids can harm biomolecules and reduce cellular function in living things, including plants. Heavy metal poisoning has been treated with a variety of therapeutic approaches. Earlier researchers have published a few case studies for certain heavy metals, such as uranium. Still, numerous case studies have been conducted worldwide on the acute and chronic toxicity of other heavy metals such as lead, arsenic, cadmium, and mercury chromium. Therefore more research is recommended in cases of uranium toxicity where drinking water has significant levels of uranium. Furthermore, it is advised that cases involving certain heavy metals, such as lead, arsenic, cadmium, and mercury, be regularly studied.

Acknowledgment

The authors are thankful to the prestigious National Sugar Institute, Kanpur, and the Maharaja Sayajirao University of Baroda, Vadodara, Gujarat, as well as to other people involved in this study for their direct and indirect contributions.

References

Abd Elnabi, M. K., Elkaliny, N. E., Elyazied, M. M., Azab, S. H., Elkhalifa, S. A., Elmasry, S., et al. (2023). Toxicity of heavy metals and recent advances in their removal: A review. *Toxics, 11*, 580.

Abdal Dayem, A., Hossain, M. K., Lee, S. B., Kim, K., Saha, S. K., Yang, G. M., et al. (2017). The role of reactive oxygen species (ROS) in the biological activities of metallic nanoparticles. *International Journal of Molecular Sciences, 18*, 120.

Ahmad, S. A., Khan, M. H., & Haque, M. (2018). Arsenic contamination in groundwater in Bangladesh: Implications and challenges for healthcare policy. *Risk Management and Healthcare Policy, 11*, 251–261.

Ali, H., Khan, E., & Ilahi, I. (2019). Environmental chemistry and ecotoxicology of hazardous heavy metals: Environmental persistence, toxicity, and bioaccumulation. *Journal of Chemistry, 2019*, 6730305.

Ali, H., & Khan, E. (2017). What are heavy metals? Long-standing controversy over the scientific use of the term 'heavy metals'–proposal of a comprehensive definition. *Toxicological & Environmental Chemistry, 100*, 1–25.

Alissa, E. M., & Ferns, G. A. (2011). Heavy metal poisoning and cardiovascular disease. *Journal of Toxicology, 2011*, 870125.

Angon, P. B., Islam, M. S., Kc, S., Das, A., Anjum, N., Poudel, A., et al. (2024). Sources, effects and present perspectives of heavy metals contamination: Soil, plants and human food chain. *Heliyon, 10*, e28357.

Bala, S., Garg, D., Thirumalesh, B. V., Sharma, M., Sridhar, K., Inbaraj, B. S., et al. (2022). Recent strategies for bioremediation of emerging pollutants: A review for a green and sustainable environment. *Toxics, 10*, 484.

Balali-Mood, M., Naseri, K., Tahergorabi, Z., Khazdair, M. R., & Sadeghi, M. (2021). Toxic mechanisms of five heavy metals:

Mercury, lead, chromium, cadmium, and arsenic. *Frontiers in Pharmacology, 12*, 643972.

Barbier, O., Jacquillet, G., Tauc, M., Cougnon, M., & Poujeol, P. (2005). Effect of heavy metals on, and handling by, the kidney. *Nephron Physiology, 99*, p105–p110.

Batool, Z., Agha, F., Tabassum, S., Batool, T. S., Siddiqui, R. A., & Haider, S. (2019). Prevention of cadmium-induced neurotoxicity in rats by essential nutrients present in nuts. *Acta Neurobiologiae Experimentalis, 79*, 169–183.

Belkin, H. E., Zheng, B., Zhou, D., & Finkelman, R. B. (2008). Chapter seventeen—Chronic arsenic poisoning from domestic combustion of coal in rural china: a case study of the relationship between earth materials and human health. In De Vivo B, Belkin HE, & Lima A (Eds.). *Environmental geochemistry* (pp. 401–420). Amsterdam: Elsevier.

Birben, E., Sahiner, U. M., Sackesen, C., Erzurum, S., & Kalayci, O. (2012). Oxidative stress and antioxidant defense. *World Allergy Organization Journal, 5*, 9–19.

Bressler, J. P., & Goldstein, G. W. (1991). Mechanisms of lead neurotoxicity. *Biochemical Pharmacology, 41*, 479–484.

Briffa, J., Sinagra, E., & Blundell, R. (2020). Heavy metal pollution in the environment and their toxicological effects on humans. *Heliyon, 6*, e04691.

Charkiewicz, A. E., Omeljaniuk, W. J., Nowak, K., Garley, M., & Nikliński, J. (2023). Cadmium toxicity and health effects—A brief summary. *Molecules (Basel, Switzerland), 28*, 6620.

Chatterjee, N., & Walker, G. C. (2017). Mechanisms of DNA damage, repair, and mutagenesis. *Environmental and Molecular Mutagenesis, 58*, 235–263.

Chinedu, E., Arome, D., & Ameh, F. S. (2013). A new method for determining acute toxicity in animal models. *Toxicology International, 20*, 224–226.

Committee, E. S., More, S. J., Bampidis, V., Benford, D., Bragard, C., Halldorsson, T. I., et al. (2022). Guidance on the use of the benchmark dose approach in risk assessment. *EFSA Journal, 20*, e07584.

Dorato, M. A., & Engelhardt, J. A. (2005). The no-observed-adverse-effect-level in drug safety evaluations: Use, issues, and definition(s). *Regulatory Toxicology and Pharmacology: RTP, 42*, 265–274.

Egan, K. B., Tsai, R. J., & Chuke, S. O. (2019). Integrating childhood and adult blood lead surveillance to improve identification and intervention efforts. *Journal of Public Health Management and Practice, 25*(Suppl 1 LEAD POISONING PREVENTION), S98–S104.

Erhirhie, E. O., Ihekwereme, C. P., & Ilodigwe, E. E. (2018). Advances in acute toxicity testing: Strengths, weaknesses and regulatory acceptance. *Interdisciplinary Toxicology, 11*, 5–12.

Genchi, G., Carocci, A., Lauria, G., Sinicropi, M. S., & Catalano, A. (2020). Nickel: Human health and environmental toxicology. *International Journal of Environmental Research and Public Health, 17*, 679.

Ghosh, N., Das, A., Chaffee, S., Roy, S., & Sen, C. K. (2018). Chapter 4 – Reactive oxygen species, oxidative damage and cell death. In S. Chatterjee, W. Jungraithmayr, & D. Bagchi (Eds.). *Immunity and inflammation in health and disease* (pp. 45–55). Academic Press.

Gonsioroski, A., Mourikes, V. E., & Flaws, J. A. (2020). Endocrine disruptors in water and their effects on the reproductive system. *International Journal of Molecular Sciences, 21*, 1929.

Guéguen, Y., Roy, L., Hornhardt, S., Badie, C., Hall, J., Baatout, S., et al. (2016). Biomarkers for uranium risk assessment for the development of the CURE (Concerted Uranium Research in Europe) molecular epidemiological protocol. *Radiation Research, 187*, 107–127.

Haidar, Z., Fatema, K., Shoily, S. S., & Sajib, A. A. (2023). Disease-associated metabolic pathways affected by heavy metals and metalloid. *Toxicology Reports, 10*, 554–570.

Hajri, M., Abbes, Z., Yahia, H. B., Jelili, S., Halayem, S., Mrabet, A., et al. (2022). Cognitive deficits in children with autism spectrum disorders: Toward an integrative approach combining social and non-social cognition. *Frontiers in Psychiatry, 13*, 917121.

Harada, M. (1995). Minamata disease: Methylmercury poisoning in japan caused by environmental pollution. *Critical Reviews in Toxicology, 25*, 1–24.

Hu, C., Yang, J., Qi, Z., Wu, H., Wang, B., Zou, F., et al. (2020). Heat shock proteins: Biological functions, pathological roles, and therapeutic opportunities. *MedComm, 2022*(3), e161.

Hughes, M. F. (2006). Biomarkers of exposure: A case study with inorganic arsenic. *Environmental Health Perspectives, 114*, 1790–1796.

Irato, P., & Santovito, G. (2021). Enzymatic and non-enzymatic molecules with antioxidant function. *Antioxidants, 10*, 579.

Jaishankar, M., Tseten, T., Anbalagan, N., Mathew, B. B., & Beeregowda, K. N. (2014). Toxicity, mechanism and health effects of some heavy metals. *Interdiscip Toxicol, 7*, 60–72.

Jakubczyk, K., Dec, K., Kałduńska, J., Kawczuga, D., Kochman, J., & Janda, K. (2020). Reactive oxygen species – Sources, functions, oxidative damage. *Polski Merkuriusz Lekarski: Organ Polskiego Towarzystwa Lekarskiego, 48*, 124–127.

Jan, A. T., Azam, M., Siddiqui, K., Ali, A., Choi, I., & Haq, Q. M. (2015). Heavy metals and human health: Mechanistic insight into toxicity and counter defense system of antioxidants. *International Journal of Molecular Sciences, 16*, 29592–29630.

Jomova, K., Alomar, S. Y., Nepovimova, E., Kuca, K., & Valko, M. (2025). Heavy metals: Toxicity and human health effects. *Archives of Toxicology, 99*, 153–209.

Jomova, K., Raptova, R., Alomar, S. Y., Alwasel, S. H., Nepovimova, E., Kuca, K., et al. (2023). Reactive oxygen species, toxicity, oxidative stress, and antioxidants: Chronic diseases and aging. *Archives of Toxicology, 97*, 2499–2574.

Juan, C. A., Pérez de la Lastra, J. M., Plou, F. J., & Pérez-Lebeña, E. (2021). The chemistry of reactive oxygen species (ROS) revisited: Outlining their role in biological macromolecules (DNA, lipids and proteins) and induced pathologies. *International Journal of Molecular Sciences, 22*, 4642.

Kakkar, P., & Jaffery, F. N. (2005). Biological markers for metal toxicity. *Environmental Toxicology and Pharmacology, 19*, 335–349.

Kamal, M., Ghaly, A. E., Mahmoud, N., & Côté, R. (2004). Phytoaccumulation of heavy metals by aquatic plants. *Environment International, 29*, 1029–1039.

Keogan, D. M., & Griffith, D. M. (2014). Current and potential applications of bismuth-based drugs. *Molecules (Basel, Switzerland), 19*, 15258–15297.

Kim, J.-J., Kim, Y.-S., & Kumar, V. (2019). Heavy metal toxicity: An update of chelating therapeutic strategies. *Journal of Trace Elements in Medicine and Biology, 54*, 226–231.

Kou, F., Wu, L., Ren, X., & Yang, L. (2020). Chromosome abnormalities: New insights into their clinical significance in cancer. *Molecular Therapy Oncolytics, 17*, 562–570.

Koyama, H., Kamogashira, T., & Yamasoba, T. (2024). Heavy metal exposure: Molecular pathways, clinical implications, and protective strategies. *Antioxidants (Basel), 13*, 76.

Kozlov, A. V., Javadov, S., & Sommer, N. (2024). Cellular ROS and antioxidants: Physiological and pathological role. *Antioxidants, 13*, 602.

Kumar, V., Umesh, M., Shanmugam, M. K., Chakraborty, P., Duhan, L., Gummadi, S. N., et al. (2023). A retrospection on mercury contamination, bioaccumulation, and toxicity in diverse environments: Current insights and future prospects. *Sustainability, 15*, 13292.

Liu, X., Wu, J., Shi, W., Shi, W., Liu, H., & Wu, X. (2018). Lead induces genotoxicity via oxidative stress and promoter methylation of DNA repair genes in human lymphoblastoid TK6 cells. *Medical Science Monitor: International Medical Journal of Experimental and Clinical Research, 24*, 4295–4304.

Marreilha Dos Santos, A. P., Andrade, V., & Aschner, M. (2017). Neuroprotective and therapeutic strategies for manganese-induced neurotoxicity. *Clinical Pharmacology and Translational Medicine, 1*, 54–62.

Masindi, V., Mkhonza, P., & Tekere, M. (2021). Sources of heavy metals pollution. In *Remediation of heavy metals. environmental chemistry for a sustainable world (vol. 70)*. Cham: Springer, 419–454. Retrieved from https://doi.org/10.1007/978-3-030-80334-6_17.

Milton, A. H., Hasan, Z., Rahman, A., & Rahman, M. (2001). Chronic arsenic poisoning and respiratory effects in Bangladesh. *Journal of Occupational Health, 43*, 136–140.

Mishra, D., Mehta, A., & Flora, S. J. S. (2008). Reversal of arsenic-induced hepatic apoptosis with combined administration of DMSA and its analogues in guinea pigs: Role of glutathione and linked enzymes. *Chemical Research in Toxicology, 21*, 400–407.

Mitra, S., Chakraborty, A., Tareq, A., Emran, T., Nainu, F., Khusro, A., et al. (2022). Impact of heavy metals on the environment and human health: Novel therapeutic insights to counter the toxicity. *Journal of King Saud University – Science, 34*, 101865.

Moffett, D. B., Mumtaz, M. M., Sullivan, D. W., & Whittaker, M. H. (2022). Chapter 13 – General considerations of dose-effect and dose-response relationships*. In G. F. Nordberg, & M. Costa (Eds.). *Handbook on the toxicology of metals (fifth edition)* (pp. 299–317). Academic Press. Retrieved from https://doi.org/10.1016/B978-0-444-59453-2.00010-X. ISBN 9780444594532 In press.

Moulis, J.-M., Bulat, Z., & Buha Djordjevic, A. (2020). Threshold in the toxicology of metals: Challenges and pitfalls of the concept. *Current Opinion in Toxicology, 19*, 28–33.

Muenyi, C. S., Ljungman, M., & States, J. C. (2015). Arsenic disruption of DNA damage responses-potential role in carcinogenesis and chemotherapy. *Biomolecules, 5*, 2184–2193.

Nawaz, M. O., Henze, D. K., Anenberg, S. C., Braun, C., Miller, J., & Pronk, E. (2023). A source apportionment and emission scenario assessment of PM2.5- and O3-related health impacts in G20 countries. *GeoHealth, 7* e2022GH000713.

Nigam, A., Priya, S., Bajpai, P., & Kumar, S. (2014). Cytogenomics of hexavalent chromium (Cr^{6+}) exposed cells: A comprehensive review. *The Indian Journal of Medical Research, 139*, 349–370.

Olympio, K. P. K., Salles, F. J., Akiba, N., & Luz, M. S. (2023). Biomarkers of lead exposure: Platforms and analysis. In V. B. Patel, V. R. Preedy, & R. Rajendram (Eds.). *Biomarkers in toxicology* (pp. 489–513). Cham: Springer International Publishing.

Paithankar, J. G., Saini, S., Dwivedi, S., Sharma, A., & Chowdhuri, D. K. (2021). Heavy metal associated health hazards: An interplay of oxidative stress and signal transduction. *Chemosphere, 262*, 128350.

Patel, D., Kulwant, M., Shirin, S., Varshney, R., Pandey, G., & Yadav, A. K. (2023). Fluoride removal from aqueous solution using iron-based materials: Preparation, characterization, and applications. In A. K. Yadav, S. Shirin, & V. P. Singh (Eds.). *Advanced treatment technologies for fluoride removal in water: Water purification* (pp. 71–92). Cham: Springer Nature Switzerland.

Patel, D., & Kulwant, M. (2024). Chapter 10 – Bioremediation of xenobiotic contamination in soil. In P. Kumar, A. L. Srivastav, V. Chaudhary, E. D. van Hullebusch, & R. Busquets (Eds.). *Bioremediation of emerging contaminants from soils* (pp. 203–225). Elsevier.

Paul, N. P., Galván, A. E., Yoshinaga-Sakurai, K., Rosen, B. P., & Yoshinaga, M. (2023). Arsenic in medicine: Past, present and future. *Biometals: An International Journal on the Role of Metal Ions in Biology, Biochemistry, and Medicine, 36*, 283–301.

Poothong, J., Jang, I., & Kaufman, R. J. (2021). Defects in protein folding and/or quality control cause protein aggregation in the endoplasmic reticulum. *Progress in Molecular and Subcellular Biology, 59*, 115–143.

Rahman, Z., & Singh, V. P. (2019). The relative impact of toxic heavy metals (THMs) (arsenic (As), cadmium (Cd), chromium (Cr)(VI), mercury (Hg), and lead (Pb)) on the total environment: An overview. *Environmental Monitoring and Assessment, 191*, 419.

Rana, M. N., Tangpong, J., & Rahman, M. M. (2018). Toxicodynamics of lead, cadmium, mercury and arsenic-induced kidney toxicity and treatment strategy: A mini review. *Toxicology Reports, 5*, 704–713.

Rasin, P., VA, A., Basheer, S. M., Haribabu, J., Santibanez, J. F., Garrote, C. A., et al. (2025). Exposure to cadmium and its impacts on human health: A short review. *Journal of Hazardous Materials Advances, 17*, 100608.

Renu, K., Chakraborty, R., Myakala, H., Koti, R., Famurewa, A. C., Madhyastha, H., et al. (2021). Molecular mechanism of heavy metals (lead, chromium, arsenic, mercury, nickel and cadmium)-induced hepatotoxicity – A review. *Chemosphere, 271*, 129735.

Rajkumar, V., Lee, V.R., & Gupta, V. (2020). *Heavy metal toxicity*. In: StatPearls [Internet]. Treasure Island (FL): StatPearls Publishing.

Saad, A., El-Sikaily, A., & Kassem, H. (2016). Essential, non-essential metals and human health. *Blue Biotechnology Journal, 3*, 87–135.

Saidon, N. B., Szabó, R., Budai, P., & Lehel, J. (2024). Trophic transfer and biomagnification potential of environmental contaminants (heavy metals) in aquatic ecosystems. *Environmental Pollution (Barking, Essex: 1987), 340*, 122815.

Sears, M. E. (2013). Chelation: Harnessing and enhancing heavy metal detoxification—A review. *TheScientificWorldJournal, 2013*, 219840.

Shaji, E., Santosh, M., Sarath, K. V., Prakash, P., Deepchand, V., & Divya, B. V. (2021). Arsenic contamination of groundwater: A global synopsis with focus on the Indian Peninsula. *Geoscience Frontiers, 12*, 101079.

Shen, S., Li, X.-F., Cullen, W. R., Weinfeld, M., & Le, X. C. (2013). Arsenic binding to proteins. *Chemical Reviews, 113*, 7769–7792.

Shinde, S., Kaur, S., Tekade, M., Pawar, B., Gupta, T., Vasdev, N., et al. (2024). Chapter 6 - Toxicological risk assessment and risk management. In *Public health and toxicology issues in drug research*. Academic Press, 143–166. Retrieved from https://doi.org/10.1016/B978-0-443-15842-1.00022-3.

Somporn, R., Chaiwong, N., Hasan, M. M., & Chaiwong, S. (2023). Cadmium exposure and renal function biomarkers: A 10-year follow-up of Thailand's Tak Province population. *Sustainability, 15*, 11291.

Songprasert, N., Sukaew, T., Kusreesakul, K., Swaddiwudhipong, W., Padungtod, C., & Bundhamcharoen, K. (2015). Additional burden of diseases associated with cadmium exposure: A case study of cadmium contaminated rice fields in Mae Sot District, Tak Province, Thailand. *International Journal of Environmental Research and Public Health, 12*, 9199–9217.

Sonone, S. S., Jadhav, S., Sankhla, M. S., & Kumar, R. (2020). Water contamination by heavy metals and their toxic effect on aquaculture and human health through food chain. *Letters in Applied NanoBioScience, 10*(2), 2148–2166.

Srinivas, U. S., Tan, B. W. Q., Vellayappan, B. A., & Jeyasekharan, A. D. (2019). ROS and the DNA damage response in cancer. *Redox Biology, 25*, 101084.

Su, L. J., Zhang, J. H., Gomez, H., Murugan, R., Hong, X., Xu, D., et al. (2019). Reactive oxygen species-induced lipid peroxidation in apoptosis, autophagy, and ferroptosis. *Oxidative Medicine and Cellular Longevity, 2019*, 5080843.

Sudhakar, P. P., Dave, H., & Kulwant, M. (2024). Chapter 3 – Electronic waste management in developing economies: Challenges and approaches. In R. Singh, S. Mohapatra, & M.-C. Jong (Eds.). *Solid waste management for resource-efficient systems* (pp. 43–63). Elsevier. Retrieved from https://doi.org/10.1016/B978-0-443-23775-1.00013-8. ISBN 9780443237751.

Sun, H., Shamy, M., & Costa, M. (2013). Nickel and epigenetic gene silencing. *Genes (Basel), 4*, 583–595.

Sundas, A., Contreras, I., Mujahid, O., Beneyto, A., & Vehi, J. (2024). The effects of environmental factors on general human health: A scoping review. *Healthcare, 12*, 2123.

Tamás, M. J., Sharma, S. K., Ibstedt, S., Jacobson, T., & Christen, P. (2014). Heavy metals and metalloids as a cause for protein misfolding and aggregation. *Biomolecules, 4*, 252–267.

Tchounwou, P. B., Yedjou, C. G., Patlolla, A. K., & Sutton, D. J. (2012). Heavy metal toxicity and the environment. *Experientia Supplementum, 101*, 133–164.

Thunus, L., Dauphin, J. F., Moiny, G., Deby, C., & Deby-Dupont, G. (1995). Anti-inflammatory properties of copper, gold and silver, individually and as mixtures. *Analyst, 120*, 967–973.

Torgovnick, A., & Schumacher, B. (2015). DNA repair mechanisms in cancer development and therapy. *Frontiers in Genetics, 6*, 157.

Ventura, C., Gomes, B. C., Oberemm, A., Louro, H., Huuskonen, P., Mustieles, V., et al. (2021). Biomarkers of effect as determined in human biomonitoring studies on hexavalent chromium and cadmium in the period 2008–2020. *Environmental Research, 197*, 110998.

Vieira Silva, A., Ringblom, J., Moldeus, P., Törnqvist, E., & Öberg, M. (2021). Benchmark dose-response analyses for multiple endpoints in drug safety evaluation. *Toxicology and Applied Pharmacology, 433*, 115732.

Vinothkannan, A., Emmanuel Charles, P., Rajaram, R., Al-Sadoon, M. K., & Gulnaz, A. (2023). Survey to identify the metal accumulation pathway in humans using hair and nail as biomarkers from fisherfolk population. *Chemosphere, 319*, 138020.

Wang, B., Ting, Z. J., & Zhao, M. (2024). Sustainable aviation fuels: Key opportunities and challenges in lowering carbon emissions for aviation industry. *Carbon Capture Science & Technology, 13*, 100263.

Wang, J., & Azam, W. (2024). Natural resource scarcity, fossil fuel energy consumption, and total greenhouse gas emissions in top emitting countries. *Geoscience Frontiers, 15*, 101757.

Wijayawardena, M. A. A., Megharaj, M., & Naidu, R. (2016). Exposure, toxicity, health impacts, and bioavailability of heavy metal mixtures. *Advances in Agronomy, 138*, 175–234. Retrieved from https://doi.org/10.1016/bs.agron.2016.03.002.

Witkowska, D., Słowik, J., & Chilicka, K. (2021). Heavy metals and human health: Possible exposure pathways and the competition for protein binding sites. *Molecules (Basel, Switzerland), 26*, 6060.

Xie, Z., Tong, S., Chu, X., Feng, T., & Geng, M. (2022). Chronic kidney disease and cognitive impairment: The kidney-brain axis. *Kidney Diseases (Basel, Switzerland), 8*, 275–285.

Yadav, R. S., Shukla, R. K., Sankhwar, M. L., Patel, D. K., Ansari, R. W., Pant, A. B., et al. (2010). Neuroprotective effect of curcumin in arsenic-induced neurotoxicity in rats. *Neurotoxicology, 31*, 533–539.

Yıldırım, R., Erdem, F., Gündoğdu, M., Bilen, Y., Koca, E., Yıllıkoğlu, Y., et al. (2012). Mercury toxicity: A family case report. *Turkish Journal of Hematology, 29*, 76–79.

Chapter 2

Mechanisms of neurodegeneration induced by heavy metals

Prasann Kumar
Department of Agronomy, School of Agriculture, Lovely Professional University, Phagwara, Punjab, India

2.1 Introduction

Neurodegenerative diseases (NDs) are among the diseases that induce the progressive degeneration of the nervous system's structure and function. These diseases are commonly defined by progressive neurodegeneration accompanied by a decline in cognitive, motor, or functional skills. Alzheimer's disease (AD), Parkinson's disease (PD), and amyotrophic lateral sclerosis (ALS) represent some of the most common NDs (Dooka et al., 2024; Karri et al., 2018; Yu et al., 2024). In addition, other diseases exist that represent massive mobility challenges to the individual and the healthcare system, such as Huntington's disease (HD) and multiple sclerosis (MS).

AD is the most common subtype of dementia. It is marked by the accumulation of amyloid-beta plaques and tau protein tangles in the brain, leading to neuronal dysfunction and death. These pathological signatures can disrupt synaptic signals, turning off cognitive faculties such as memory, reasoning, and other mental abilities. AD most commonly occurs among adults 65 or older (though less common early-onset forms can happen). It is a complex disease, and many genetic, environmental, and lifestyle factors contribute to its onset. The apolipoprotein E (APOE) ε4 allele represents one of the most substantial genetic risk factors, but its pathogenic role once the disease has been initiated is complex. The exact avenues by which amyloid plaque and tau tangle accumulate still need complete elucidation, neuroinflammation, oxidative damage, and metal homeostasis have recently been implicated in AD pathogenesis (Mohankumar et al., 2024; Raj, Kaur, Gupta, & Singh, 2021).

PD is a progressive neurological condition that affects movement control due to the degeneration of dopaminergic neurons in the substantial nigra part of the brain. PD has its defining motor signs of tremor, bradykinesia, rigidity, and postural instability. Cognitive impairment and other non-motor symptoms, including mood disorders and autonomic dysfunction, are critical contributors to disease progression and part of the disease itself. PD is presumed to be induced by genetic and environmental factors (physiological phenomenon). However, the precise mechanism of PD is still unknown. Mutations in many genes associated with familial PD have been identified, including SNCA, LRRK2, and PARK7. However, exposure to environmental pollutants such as pesticides or heavy metals could also increase the chances of developing the disease. PD also necessitates the intracytoplasmic development of Lewy bodies, aggregates of the intracerebral amino acid, alpha-synuclein, that aggregate and disturb typical cellular homeostasis, contributing to neurodegeneration (Bjørklund, Đorđević, Hamdan, Wallace, & Peana, 2024; Kumar et al., 2025).

ALS is a progressive neurodegenerative disorder that targets upper and lower motor neurons and ultimately causes muscle weakness, atrophy, and paralysis. ALS is most often marked by a progressive loss of voluntary control of muscles, ultimately threatening respiratory function and precipitating death, usually within 3–5 years of diagnosis. ALS is predominantly sporadic, although 5%–10% of cases are familial, and mutations in the C9ORF72, SOD1, TARDBP, and FUS genes have been implicated. ALS has no known causal agent, but it is believed to relate to genetic mutations, protein misfolding, mitochondrial dysfunction, and neuroinflammation. ALS shares standard features with other NDs, such as PD and AD, whereby aggregation of proteins, oxidative damage, and neuroinflammatory pathways may all play a role in its pathogenesis (Famurewa et al., 2022; Gomes da Silva et al., 2024).

Heavy Metal Toxicity and Neurodegeneration. DOI: https://doi.org/10.1016/B978-0-443-36575-1.00008-X

There are other NDs, like HD and MS, which are associated with the burden of these diseases on individuals' quality of life (QoL), similar to AD, PD, and ALS. HD is a trinucleotide repeat expansion in the HTT gene responsible for the genetic disorder of HD, which causes progressive motor, cognitive, and psychiatric symptoms. MS is an autoimmune-mediated disease that results in inflammation and neurodegeneration due to the attack of the myelin sheath, which is a protective layer surrounding the neurons, by the immune system. While these diseases are distinct with their unique pathophysiological features, they have common pathways, including neuronal injury, cellular stress, and dysregulation of essential molecular mechanisms necessary for neuronal survival. Substantial efforts are underway to better understand these diseases' complex etiology and overlapping pathophysiological mechanisms and to explore potential therapeutic targets and interventions to slow or arrest disease progression.

Environmental exposure to heavy metals such as lead (Pb), mercury (Hg), arsenic (As), and cadmium (Cd) has been increasingly considered a risk factor for the initiation and/or progression of several NDs, including AD, PD, and HD. These metals, which are common in industrial processes, contaminate water, soil, air, and even specific food sources, and they could accumulate in the brain and directly and indirectly damage neural structures. Long-term exposure to these harmful elements may alter cellular functions, generate oxidative stress, and derail blood–brain barrier (BBB) permeability, allowing toxic substances to infiltrate the central nervous system (CNS). Therefore, the physiological vulnerability to progressive neurodegenerative changes is enhanced as a function of age, with heavy metals being a significant environmental risk factor.

The way heavy metals impact the onset of ND is complex—the induction of oxidative stress is one of the most researched pathways (Fu, Li, Song, Cen, & Wu, 2023; Kwakye et al., 2019). Heavy metals like mercury and lead can produce ROS via redox cycling, leading to cellular damage, lipid peroxidation, and mitochondrial dysfunction. These ROS induce damage to proteins, lipids, and nucleic acids, resulting in cell death, especially in neurons, which are very susceptible to oxidative stress. Mitochondria, essential in generating energy for neurons, are highly vulnerable to the harmful effects of heavy metals. Metal overload may cause a hindrance in mitochondrial activity that harms adenosine triphosphate (ATP) generation and leads to oxidative stress and neurodegeneration.

Aside from being involved in oxidative stress, heavy metals also interfere with several signaling pathways that modulate cell survival and apoptosis. Cadmium, for instance, may interact with specific proteins involved in defense mechanisms within cells, especially antioxidant enzymes, which diminish their activity. Moreover, arsenic also alters neuroinflammatory pathways by activating microglia and astrocytes, which release proinflammatory cytokines and amplify neuronal damage. Neuroinflammation represents a prominent feature of a wide variety of NDs, which is further exacerbated by long-term exposure to heavy metals, producing a chronic inflammatory milieu that increases the rate of disease progression (Althobaiti, 2025; López-Granero, Aschner, & Sánchez-Santed, 2023).

A second key mechanism is metal homeostasis impairment and protein aggregation. However, the brain is susceptible to metal imbalance, and disrupted metal ion homeostasis generates toxic protein aggregates. AD has been linked to an imbalance of metals in the body, with the build-up of Aβ plaques and tau tangles in the brain correlating with aberrant levels of metal ions, including copper, iron, and zinc. These metals may be displaced by heavy metals such as lead and mercury, which may, in turn, promote the aggregation of Aβ and tau through direct interactions with these proteins, increasing their propensity to aggregate. This protein accumulation results in the dysregulation of neuronal function and the defining characteristics of neurodegenerative plaques.

Overall, environmental exposure to heavy metals is a crucial factor in the pathogenesis of NDs via multiple networked mechanisms, such as oxidative stress, mitochondrial dysfunction, inflammation, and protein aggregation. Knowledge of these routes is essential in the design of targeted therapeutic strategies that will help to decrease or inhibit the toxic consequences of heavy metal exposure and prevent or slow the progression of neurodegenerative disorders.

The rationale of this chapter, in the context of the need to understand the critical mechanisms by which various heavy metals mediate neurodegeneration, would allow us to advance our understanding of the pathophysiology of NDs. However, extensive studies have now established links between environmental exposure to heavy metals and neurotoxic conditions such as AD, PD, and HD, but However, extensive studies have now established links between environmental exposure to heavy metals and neurotoxic conditions such as AD, PD, and HD, the detailed molecular and cellular mechanisms remain incompletely understood. The purpose of this chapter is to review current knowledge on the mechanisms that link candidates, such as lead, mercury, arsenic, and cadmium, to cellular systems modification, oxidative stress generation, mitochondrial dysfunction, and, when relevant, the stimulus of the inflammatory response, which feeds back on neuronal degeneration. Our analysis of these processes in granular detail seeks to provide insights into potential new therapeutic strategies that will alleviate the deleterious impact of metal toxicity on brain health (Lin, Wang, Huang, Xu, & Yang, 2024; Vellingiri et al., 2022; Wang & Matsushita, 2021).

TABLE 2.1 A comprehensive table detailing the neurological effects of heavy metals (Dong, Fu, & Jiang, 2024; Samuel, Datta, Khandge, & Selvarajan, 2021).

Heavy metal	Common sources of exposure	Routes of entry into the body	Bioaccumulation Sites	Neurological effects	Cellular and molecular mechanisms of toxicity	Associated neurodegenerative diseases	Biomarkers for detection	Potential detoxification strategies
Lead (Pb)	Contaminated water, old paints, industrial emissions, batteries	Ingestion (water, food), inhalation (dust, fumes), dermal absorption	Brain (hippocampus, cortex), bones	Cognitive decline, attention deficits, behavioral disorders	Oxidative stress, mitochondrial dysfunction, disruption of neurotransmitter signaling, impaired synaptic plasticity	AD, PD, ADHD	BLL, ALA in urine	Chelation therapy (EDTA, DMSA), dietary antioxidants (vitamin C, Zinc)
Mercury (Hg)	Fish (methylmercury), dental amalgams, industrial emissions, vaccines (thimerosal)	Inhalation (vapors), ingestion (fish, water), dermal absorption	Brain (cerebellum, cortex), kidneys	Tremors, memory loss, sensory impairment, motor dysfunction	Increased ROS, BBB disruption, inhibition of antioxidant enzymes, neuronal apoptosis	ALS, PD, minamata disease	Blood mercury levels, hair, and urine mercury concentrations	Chelation therapy (DMSA, DMPS), Selenium supplementation, glutathione enhancement
Arsenic (As)	Contaminated groundwater, pesticides, industrial smelting, rice and seafood	Ingestion (contaminated food, water), inhalation (industrial exposure)	Brain, liver, kidneys, skin	Memory impairment, neuroinflammation, peripheral neuropathy	Interferes with neurotransmission, induces DNA damage, increases oxidative stress	AD, cognitive dysfunction, neuropathy	Urinary arsenic levels, hair, and nail arsenic content	Activated charcoal, Selenium supplementation, *N*-acetylcysteine
Cadmium (Cd)	Industrial pollution, cigarette smoke, contaminated food (rice, shellfish)	Inhalation (smoke, industrial fumes), ingestion (contaminated food)	Brain, kidneys, liver	Motor dysfunction, cognitive impairment, depression, anxiety	Disrupts calcium homeostasis induces lipid peroxidation, interferes with dopamine metabolism	PD, ALS, cognitive decline	Blood and urine cadmium levels, metallothionein expression	Chelation therapy (DMSA), zinc and selenium supplementation
Aluminum (Al)	Processed foods, cookware, cosmetics, drinking water, antiperspirants	Ingestion (food, water), inhalation (industrial exposure)	Brain (hippocampus, cortex), bones	Memory loss, confusion, neuroinflammation, and motor dysfunction	Promotes amyloid-beta aggregation, inhibits proteasomal degradation, induces ROS	AD, PD, cognitive dysfunction	Serum aluminum levels, brain MRI (aluminum deposition)	Silicon-rich mineral water, deferoxamine therapy

(Continued)

TABLE 2.1 (Continued)

Heavy metal	Common sources of exposure	Routes of entry into the body	Bioaccumulation Sites	Neurological effects	Cellular and molecular mechanisms of toxicity	Associated neurodegenerative diseases	Biomarkers for detection	Potential detoxification strategies
Manganese (Mn)	Welding fumes, contaminated drinking water, pesticides, and mining	Inhalation (industrial fumes), ingestion (water, food)	Basal ganglia, liver, pancreas	Parkinsonian-like symptoms, tremors, and mood disturbances	Mitochondrial dysfunction, dopaminergic neuron damage, impaired astrocyte function	Manganism (Parkinsonism-like disorder), PD	Blood and urine manganese levels, MRI of the basal ganglia	Iron and calcium supplementation, chelation therapy (EDTA)
Nickel (Ni)	Stainless steel, batteries, coins, and industrial pollution	Inhalation (industrial fumes), ingestion (food, water), dermal absorption	Brain, liver, lungs	Cognitive impairment, depression, anxiety	Induces oxidative stress, disrupts calcium signaling, neuroinflammation	Cognitive dysfunction, depression	Blood and urine nickel levels	Antioxidant therapy (vitamin C, selenium), chelation therapy
Chromium (Cr)	Industrial waste, contaminated water, leather tanning, and welding	Inhalation (dust, fumes), ingestion (contaminated food, water)	Brain, liver, kidneys	Neuroinflammation, impaired motor coordination, memory deficits	DNA damage, oxidative stress, disruption of dopamine metabolism	PD, cognitive decline	Blood chromium levels, urinary chromium excretion	Chelation therapy (DMSA, EDTA), antioxidant supplementation

AD, Alzheimer's disease; *ALA*, δ-aminolevulinic acid; *ALS*, amyotrophic lateral sclerosis; *BBB*, blood–brain barrier; *BLL*, blood lead levels; *PD*, Parkinson's disease; *ROS*, reactive oxygen species.
Source: Dong, Q., Fu, H., Jiang, H. (2024). The role of exosome-shuttled miRNAs in heavy metal-induced peripheral tissues and neuroinflammation in Alzheimer's disease. *Biomedicine Pharmacotherapy*, *176*, 116880. https://doi.org/10.1016/j.biopha.2024.116880; Samuel, M. S., Datta, S., Khandge, R. S., Selvarajan, E. (2021). A state of the art review on characterization of heavy metal binding metallothioneins proteins and their widespread applications. *Science of the Total Environment*, *775*, 145829. https://doi.org/10.1016/j.scitotenv.2021.145829.

The importance of this subject in current medicine and environmental health deserves to be emphasized. As environmental pollution caused by metalliferous mining, industrialization, and urbanization continues to increase, heavy metal exposure risks to public health remain of concern. Understanding the neurotoxic effects of these metals is essential to prevention, improving clinical outcomes in affected populations, and the policies that minimize exposure. In addition, this knowledge is crucial for developing biomarkers for diagnosing and monitoring NDs related to environmental pollutants (Table 2.1).

2.2 Heavy metals and their sources

Neurotoxic effects of lead (Pb), mercury (Hg), arsenic (As), cadmium (Cd), and aluminum (Al) are extensively characterized, and all of the above contribute to neurodegeneration via multiple mechanisms, including oxidative stress, mitochondrial dysfunction, inflammation, and neurotransmitter imbalance. Lead (Pb) is one of the most studied neurotoxic metals linked to cognitive function, memory, and learning ability impairments. Lead is a neurotoxin, which, after chronic exposure, can cause the death of neuronal cells, disturb calcium homeostasis, and disrupt synaptic transmission (found in old paint, contaminated water, and industrial emissions). Lead poisoning is even more harmful in children, whose underdeveloped nervous systems are easily compromised, resulting in reduced IQ, attention deficits, and behavioral disorders. Among adults, long-term exposure to lead has been associated with a higher risk of NDs, including AD and Parkinson's. Mercury (Hg) is a potent neurotoxin in elemental, inorganic, and organic forms (Adegbola & Adetutu, 2024; Sun, Johnson, & Aluko, 2025). Its organic form, methylmercury, has the highest bioavailability and passage across the BBB. Mercury is an unavoidable contaminant, appearing in seafood, industrial waste, and dental amalgams, that causes neuronal degeneration through oxidative stress, impairment of neurotransmission, and apoptosis in brain cells. It attacks primarily motor function, coordination, and thinking ability; it has been linked in humans to diseases like Minamata disease and to possibly hastening the progression of AD and PD. Arsenic (As), a metalloid in drinking water, pesticides, and industrial pollutants, acts through interference with ATP production, generation of ROS, and disruption of the neurotransmitter signaling system. Chronic arsenic exposure has been linked with cognitive impairment, neuroinflammation, and amyloid-beta deposition, a distinctive feature of AD. It is also associated with a greater risk of depression, anxiety, and other neuropsychiatric disorders from its potential impact on the biological pathways of dopamine and serotonin. Cadmium (Cd) is a heavy metal and environmental toxicant found in cigarette smoke, industrial waste, and contaminated food. It has a long biological half-life that allows for accumulation in the brain (Li, Ren, Crabbe, Wang, & Ma, 2021; Moyano et al., 2022). It plays a role in neurodegeneration by inducing oxidative stress, damaging the BBB, and disrupting calcium signaling, critical for the communication between neurons. Research has indicated that exposure to cadmium can adversely affect learning and memory, promote neuroinflammation, and play a role in the progression of neurodegenerative pathologies such as AD and PD. Cadmium also disrupts the homeostasis of essential trace metals such as zinc, which are necessary for regulating cognitive function and neuronal health. Aluminum (Al) is not often regarded as a heavy metal. Still, it has been related to the etiology of NDs owing to its deposition and oxidative stress in the brain. Aluminum exposure, found in processed foods, drinking water, antacids, and certain cosmetics, has been linked to cognitive impairment, increased amyloid plaque formation, and tau protein aggregation, the main pathological features of AD. Aluminum is thought to exert neurotoxic effects by disrupting iron metabolism, mediating higher oxidative damage and neuronal death. Moreover, aluminum impairs synaptic plasticity, influencing cognitive behavior such as learning and memory (Deng, Ren, Hou, Deng, & Cheng, 2023; Lasisi-Sholola, Hammed, Ajike, Akhigbe, & Afolabi, 2024; Pota et al., 2024). These heavy metals act jointly as neurodegenerative agents by overlapping mechanisms, such as oxidative damage, impairment of mitochondrial function, neuroinflammation, and alteration of neurotransmission. These findings underscore the critical need for preventive measures, such as reducing environmental exposure, dietary interventions high in antioxidants, and pharmacological interventions.

However, different forms of exposure to these toxic substances are responsible for adverse effects on human health and the environment. Exposure to toxicants mainly occurs via occupational exposure to chemicals, heavy metals, etc. Mining, construction, agriculture, and manufacturing workers are prone to direct contact and prolonged exposure to toxic and harmful materials. For instance, the miners risk inhaling delicate particulate matter that contains lead, arsenic, or silica, which can lead to serious respiratory and systemic health issues. Similarly, agricultural workers are repeatedly exposed to pesticides and herbicides that can have both acute and chronic health effects, such as neurological disease and cancers. Factory workers who were exposed to chemicals such as benzene, asbestos, and formaldehyde will likely suffer long-term health problems, from respiratory illness to more carcinogenic effects. For the most part, individual risks can be minimized through using PPE (personal protective equipment) and compliance with safety regulations. However, enforcement is weak in many places, with few safeguards, so occupational hazards are ever present. The need for stringent rules and effective monitoring is crucial to ensure the safety of workers and the public (Cai et al., 2023; Karri, Schuhmacher, & Kumar, 2016; Wu et al., 2023).

Environmental contamination is another significant source of exposure to it across all populations, especially those who do not work. Air pollution: Nitrogen oxides, sulfur dioxide, volatile organic compounds (VOCs), and particulate matter released via industrial emissions, vehicle exhaust, and biomass burning are linked with higher rates of respiratory illnesses, cardiovascular diseases, increase in cancer risk, and premature death. Pollutants from industrial or agricultural runoff or improper waste disposal would contaminate water sources and run into drinking water sources, including heavy metals (lead, mercury, cadmium), pesticides, and pharmaceuticals. Long-term exposure to polluted water may contribute to developmental disorders, kidney impairments, or endocrine disruption. Soil pollution is another crucial environmental problem caused by industrial waste disposal, pesticide use, and airborne deposition of heavy metals. The most contaminated soil not only restricts the agricultural productive potential of the space but also represents a direct threat to the health of humans via dermal contact, inhalation of dust, and ingestion of soil, particularly in children who might play in polluted places. Climate change aggravates the consequences of environmental pollution because it modifies the spatial dynamics of pollutants and amplifies the volume, frequency, and intensity of exposure risks (Paithankar, Saini, Dwivedi, Sharma, & Chowdhuri, 2021; Popov Aleksandrov et al., 2021; Tang et al., 2023).

Food contamination, particularly from seafood, crops, and processed food products, is another regular avenue of exposure. Heavy metals, including mercury and lead, are neurotoxins that bioaccumulate through food chains, eventually contaminating seafood and endangering human health, particularly in vulnerable groups such as pregnant women and young children. That means leaching pollutants such as PCBs and dioxins from industrial facilities are quickly taken up by fish and shellfish and can lead to chronic health effects, such as immune suppression or hormonal disruptions. Practising so-called modern farming like the Dutch Flanders Fields and the Yangtze River delta, where human participation in food production is very high, also leads to excessive residual pesticides in crops, which finally enter the human food supply. The plants, which are ultimately part of your food, too, can take up the heavy metals above. (Processed and packaged foods also pose a danger, as they tend to be laced with additives, preservatives, and the packaging material bisphenol A [BPA] that has been shown to throw hormonal balance off-kilter.) Food safety is also worsened by the presence of naturally occurring toxins called mycotoxins, which are produced by mildew or fungi on improperly stored grains and nuts and are associated with liver damage and increased cancer risk. Regulating food safety standards, monitoring contaminants, and encouraging sustainable agricultural practices can help reduce the incidence of food contamination (Anyanwu, Orish, Ezejiofor, Nwaogazie, & Orisakwe, 2020; Rehman et al., 2021).

Due to high exposure levels, industrial activities pose significant risks to employees and the general population. Mining operations release heavy metals, from mercury to arsenic, into the environment, contaminating air, water, and soil. Smelting and metal refining processes compound heavy metal pollution in surrounding ecosystems. Manufacturing industries, such as chemical plants and electronics production, produce toxic waste containing persistent organic pollutants, solvents, and heavy metals. These pollutants harm the people who work in the factories and end up being discharged into nearby water bodies, threatening local aquatic life and communities dependent on these water sources. The heavy reliance on synthetic fertilizers and pesticides in agri-business practices causes groundwater pollution and bioaccumulation of toxic chemicals throughout the food chain. Industrial livestock farming also emits antibiotic-resistant bacteria and nitrates that can leach into drinking water aquifers, endangering human and environmental health. The use of industrial sites, which are increasingly encroaching on residential areas, poses a threat to exposure, especially concerning vulnerable individuals such as children and older people. This highlights the urgent need for demanding industrial regulations and robust environmental surveillance to prevent exposure risks (Bhattacharya, Bose, & Munshi, 2024; Li, Xia, Zorec, Parpura, & Verkhratsky, 2021; Stachowicz, 2023).

Familiar environmental sources of exposure contribute to this cumulative burden of toxicants in our daily lives, often taking the form of household products. Exposure to lead is another common hazard, most often from lead-based paints in older houses, and it can have serious health consequences, particularly in children, ranging from cognitive decline to developmental and behavioral problems. Even outdated plumbing systems with lead pipes can release lead into the drinking water supply, increasing the risk of chronic lead poisoning. Many of the everyday products we use to clean our homes, as well as air fresheners and personal care products, can be sources of VOCs and endocrine-disrupting chemicals (EDCs) things like phthalates and parabens, which are linked to hormonal imbalances and reproductive health issues. Culinary processes like using PFC-coated nonstick cookware can directly lead to one of these toxic exposures, as the compounds persist in the human body and the surrounding environment. The improper disposal of electronic waste, batteries, and household chemicals leads to localized soil and water contamination and increased exposure risks.

Using pesticides and insecticides indoors can lead to indoor air pollution and direct dermal exposure through contact with insecticide-treated surfaces, particularly for toddlers who crawl on treated surfaces. However, others are aware that tighter stances, including regulation of toxic agents and the availability of safer ones, are needed to reduce home exposures and protect public health.

Multilingual exposure sources lead to toxic burdens from household products, industry, food, environment, and occupational exposures. The risks begin very early on and accumulate over a lifetime; they have serious health consequences that include neurological disorders, respiratory disease, cancers, and endocrine disruption. As health professionals, environmental scientists, policymakers, and the general public interested in environmental health, you are crucial in advocating for change. There are various means of this, from regulations, industry practices, public education, and encouraging safer substitutes to mitigate these harms (Hussien et al., 2018; Kim, Ha, Hwang, & Kim, 2025; Nguyen & Kim, 2022). Improving human health and ecosystems through ameliorating exposure by implementing enhanced environmental monitoring systems, improving the enforcement of relevant safety standards, and enabling scientific studies on exposure mitigation are critical to reducing the overall burden of toxic chemicals.

2.3 Mechanisms of neurodegeneration induced by heavy metals

When there are more ROS and RNS, oxidative stress can develop. It is an imbalance between ROS, nitrogen atoms, and body antioxidants to neutralize these substances. ROS and RNS molecules are highly reactive and come from oxygen and nitrogen. They include the superoxide anion ($O_2^{\bullet-}$), hydroxyl radical (•OH), hydrogen peroxide (H_2O_2), nitric oxide (NO•), and peroxynitrite ($ONOO^-$). The leading site of this occurrence within an organism's cells is the mitochondria where oxygen reacts by phosphorylation to create superoxide radicals. ROS and RNS are essential in cell signaling and immune defense. Still, too much accumulation due to environmental pressure, like pollution, radiation, toxins, or chronic inflammation, can lead to oxidative damage. This will upset the balance of cells to bring about various pathological conditions in humans like neurodegenerative disorders, cardiovascular diseases, cancers, and aging. One of the significant effects of oxidative stress is its harmful impact on lipid membranes, proteins, DNA, and other macromolecules. When ROS initiates lipid peroxidation, it causes the oxidation degradation of polyunsaturated fatty acids, generating lipid peroxides and reactive aldehydic metabolites. Such cellular ships sink as they plow further damage into neighboring cells' hulls. This process undermines membrane-attached receptors' structure and function, fluidity in general, and linked signal pathways. The consequences are ultimately cellular malfunction, which eventually leads to cellular orientation. Likewise, proteins may be damaged by oxidative modification, changing their structure and function. It is known that ROS and RNS-B—carbonylation, nitration, oxidation of amino acid residues (especially cysteine, tyrosine, and methionine), etc.—lead to protein misfolding and aggregation as well as degradation. If not acted upon, these jobs will speak again: their environs are less forgiving than critical remarks (Aschner et al., 2024; Oyeyemi, Owonikoko, Okoro, Adagbonyi, & Ajeigbe, 2024; Perera, Mauran, & Wickramasinghe, 2023). What they have done to us! Oxidative damage affects enzymatic activity, interferes with cellular signaling pathways, and helps create decomposed proteins—a diagnostic feature in various NDs such as AD or PD. Furthermore, the consequences of oxidative stress on DNA damage are pretty serious and directly contribute to the properties of carcinogenesis. Meanwhile, ROS can engender strand breaks, base modifications, and DNA cross-linking, while 8-hydroxy-2′-deoxyguanosine, or 8-OHdG, has become a well-known biomarker for oxidative DNA damage. Continued damage to the DNA activates cell cycle arrest mechanisms and cell repair processes. If the damage is beyond repair, various cells undergo apoptosis. However, chronic oxidative stress overwhelms the cells' ability to repair themselves; this brings about mutagenesis and carcinogenesis. Cells develop sophisticated defences against oxidative stress, including enzymatic and nonenzymatic antioxidants. The key enzymatic antioxidants are superoxide dismutase, which breaks down superoxide radicals to hydrogen peroxide; catalase, which splits the latter into water and oxygen; and glutathione peroxidase (GPx), using glutathione (GSH) as a substrate, to reduce peroxides. Nonenzymatic antioxidants, Including vitamin E, selenium, and various natural plant compounds, support essential functions of the human body. Vitamin C, along with Coenzyme Q10, actually helps recycle vitamin E after it loses one of its electrons while reacting with an oxidant to make oxidized vitamin E (Table 2.2).

The nuclear factor erythroid 2-related factor 2 (Nrf2) signaling system is vital for controlling the expression of antioxidant enzymes and other protective genes, thereby ensuring cellular resistance to oxidation. Nonetheless, prolonged oxidative stress can cause any secondary defences against free radicals to break down, increasing cell injury and disease progression. Understanding the mechanisms underlying oxidative stress and its effects on cell components is necessary to target therapeutic strategies to minimize the damage.

Mitochondria are crucial in energy production, maintaining calcium ion concentrations (intracellular calcium homeostasis), and apoptosis. If mitochondrial function

TABLE 2.2 Mechanisms of heavy metal-induced neurotoxicity.

Heavy metal	Primary sites of neurotoxicity	Key molecular and cellular targets	Oxidative stress mechanisms	Mitochondrial dysfunction pathways	Neuroinflammation and immune response	Neurotransmitter system disruptions	BBB alterations	Protein aggregation and misfolding effects	Epigenetic and genetic modifications
Lead (Pb)	Hippocampus, cortex, basal ganglia	Neurons, astrocytes, and microglia	Increases ROS and lipid peroxidation, depletes glutathione (GSH), inhibits antioxidant enzymes (SOD, catalase)	It impairs the ETC, reduces ATP production, and increases mitochondrial permeability.	Activates microglia and astrocytes and upregulates proinflammatory cytokines (TNF-α, IL-6)	Inhibits NMDA receptors, alters dopamine and acetylcholine signaling	Disrupts endothelial tight junctions, increases permeability	Induces tau hyperphosphorylation, promotes beta-amyloid aggregation	DNA methylation changes, histone modifications, miRNA dysregulation
Mercury (Hg)	Cerebellum, cortex, hippocampus	Neurons, oligodendrocytes, astrocytes	Induces lipid peroxidation, depletes GSH, disrupts thiol homeostasis	Inhibits ETC complexes (I, III, IV), promotes mitochondrial swelling, enhances cytochrome c release	Increases TNF-α, IL-1β, NF-κB activation, microglial overactivation	Reduces dopamine and serotonin levels, impairs glutamate uptake	It damages endothelial cells, increases BBB permeability	Promotes alpha-synuclein misfolding, triggers APP cleavage	Alters DNA repair genes cause hypomethylation of neural genes
Arsenic (As)	Hippocampus, striatum, cortex	Neurons, astrocytes, oligodendrocytes	Generates superoxide radicals, inhibits antioxidant enzymes, disrupts NADPH homeostasis	Induces mitochondrial fission, decreases ATP synthesis, increases apoptosis via caspase-3	Activates NF-κB elevates IL-6 and TNF-α, disrupts microglial signaling	Impairs GABAergic and cholinergic neurotransmission alters glutamate receptor expression	Induces tight junction breakdown, enhances metal ion transport	Triggers neurofibrillary tangle formation, enhances tau phosphorylation	Inhibits DNA repair mechanisms, induces global DNA hypomethylation
Cadmium (Cd)	Cerebral cortex, hippocampus, basal ganglia	Neurons, microglia, endothelial cells	Increases superoxide and hydroxyl radicals, depletes GSH, inhibits SOD	Suppresses ETC complex I and IV, reduces mitochondrial membrane potential, enhances apoptosis	Increases IL-1β, IL-18, upregulates inflammasomes (NLRP3), promotes microgliosis	Alters dopamine, serotonin, and glutamate transporters	Disrupts endothelial cell tight junctions and induces vascular inflammation	Induces misfolded protein aggregates, disrupts proteasomal degradation	Increases DNA strand breaks, modifies histone acetylation patterns
Aluminum (Al)	Hippocampus, cortex, cerebellum	Neurons, microglia, oligodendrocytes	Catalyzes ROS formation inhibits catalase and peroxidase activity and induces lipid peroxidation.	It Impairs mitochondrial biogenesis, disrupts calcium homeostasis, and promotes apoptosis.	Activates NF-κB, IL-6, TNF-α, enhances microglial reactivity	Reduces acetylcholine synthesis, inhibits NMDA receptor function	Enhances BBB permeability, induces tight junction protein loss	Increases amyloid-beta aggregation, disrupts tau homeostasis	Induces hypermethylation of neuroprotective genes, modifies noncoding RNA expression
Manganese (Mn)	Basal ganglia, substantia nigra, striatum	Dopaminergic neurons, astrocytes, microglia	Generates ROS via Fenton-like reactions, disrupts glutathione homeostasis	Inhibits mitochondrial Mn-SOD, disrupts complex I function, impairs ATP production	Upregulates TNF-α, IFN-γ, activates NF-κB pathways	Increases dopamine oxidation, alters glutamate excitotoxicity	Induces endothelial stress, disrupts BBB integrity	Enhances alpha-synuclein aggregation, promotes tau misfolding	Alters histone methylation modifies DNA methylation in dopamine-related genes
Nickel (Ni)	Hippocampus, cortex, olfactory bulb	Neurons, astrocytes, endothelial cells	It disrupts iron homeostasis, increases free radical generation, and inhibits antioxidant enzymes.	Reduces mitochondrial membrane potential, promotes cytochrome c release	Activates TNF-α, IL-8, triggers astrogliosis	Impairs cholinergic and dopaminergic neurotransmission	Enhances BBB disruption via endothelial inflammation	Induces protein misfolding, promotes tau hyperphosphorylation	Increases DNA strand breaks, alters histone modifications
Chromium (Cr)	Cortex, striatum, hippocampus	Neurons, microglia, endothelial cells	Induces hydroxyl radical formation, disrupts NADH/NADPH balance	Alters mitochondrial permeability inhibiting complex III and IV activity	Elevates IL-1β, TNF-α, increases microglial activation	Reduces dopamine receptor expression, impairs glutamate clearance	Enhances BBB permeability, disrupts tight junction integrity	Induces tau and beta-amyloid aggregation, interferes with protein clearance pathways	Increases DNA adduct formation, modifies epigenetic regulatory genes

APP, Amyloid precursor protein; *ATP*, adenosine triphosphate; *BBB*, blood–brain barrier; *ETC*, electron transport chain; *ROS*, reactive oxygen species.
Source: Esan, O. O., Igado, O. O., Femi-Akinlosotu, O. M., Oyagbemi, A. A., Omobowale, T. O., Oladele, O. A., Nwulia, E. (2024). Alchornea laxiflora (Benth.) Pax K. Hoffman extract protects against lead-induced neurodegeneration in cockerel chickens. *IBRO Neuroscience Reports, 17*, 65–72. https://doi.org/10.1016/j.ibneur.2024.06.004; Karri, V., Schuhmacher, M., Kumar, V. (2020). A systems toxicology approach to compare the heavy metal mixtures (Pb, As, MeHg) impact in neurodegenerative diseases. *Food and Chemical Toxicology, 139*, 111257. https://doi.org/10.1016/j.fct.2020.111257; Park, K., Kwak, I.-S. (2023). Growth retardation and suppression of ubiquitin-dependent catabolic processes in the brackish water clam Corbicula japonica in response to salinity changes and bioaccumulation of toxic heavy metals. *Environmental Pollution, 337*, 122554. https://doi.org/10.1016/j.envpol.2023.122554.

becomes dysfunctional, various diseases may arise for which there is no precedent for recovery. Both biochemical and structural alterations accompany the procedure through which mitochondrial dysfunction occurs. This decreases bioenergy, increases mitochondrial permeability, and upsets the balance of calcium ion distribution, thus being a crucial factor in cell survival and death pathways. The chief causes of mitochondrial damage are mutations in mitochondria caused by genetic defects, oxidative stress, and environmental toxins. ROS produced within the mitochondria alter mitochondrial DNA (mtDNA), lipids, and proteins by oxidation-rendering them dysfunctional. Breakdown in the performance of the mitochondrial electron transport chain (ETC) will stop ATP generation, increase the generation of secondary ROS, and create a vicious circle of mitochondrial impairment. Damage to mitochondria also prevents the tricarboxylic acid cycle from going through its steps smoothly, disrupting the production of critically short-lived metabolic intermediates (Haidar, Fatema, Shoily, & Sajib, 2023; Joneidi et al., 2019; Khan & Parvez, 2015).

If mitochondria do not work right, ATP production will fail; this is important for the expected balance between cells. These membranes must carry out oxidation and join phosphate residues for energy production. However, disturbances in the ETC complexes, especially I and III, lead to nonstandard H+ formation for ATP synthase. In addition, the increase in mitochondrial permeability transition pore (mPTP) permeability caused by oxidative stress brings about energy shortages. When the mPTP opens, an influx of calcium, depletion of ATP, and, eventually, oxidative stress depress the mitochondrial membrane potential. Individuals are faced with inner life's termination stage. This releases proapoptotic factors such as cytochrome *c* into the cytosol to trigger apoptotic cascades. Long-term ATP depletion will also lead to necrotic cell death, not least in situations like ischemia-reperfusion injury or NDs.

ATP production (ATP synthesis, in the case of mitochondrial impairment) is vitally important to the homeostasis of cells. ATP is not put out in the requisite quantities when things go awry. The liver's oxidative phosphorylation system produces most ATP and works mainly on the inner mitochondrial membrane. However, defects in ETC complexes, mostly I and III, lead to insufficient formation of the proton gradient needed by ATP synthetases. Higher levels of mitochondrial permeability will result in more severe energy failures. Once the mPTP has opened up, caused perhaps by too much calcium or due to oxidative stress or ATN depletion, the $\Delta\Psi m$ is lost across the mitochondrial membrane. This lets out a variety of proapoptotic factors, such as cytochrome *c*, into the cytosolic medium. Chronic lack of ATP will also give rise to necrotic cell death, particularly in cases of ischemia-reperfusion injury or ND (Arun et al., 2021; Hussien et al., 2018; Nguyen & Kim, 2022; Stachowicz, 2023; Tripathi, Kharkwal, Mishra, & Singh, 2024).

The presence of calcium ions (Ca^{2+}) is critical for mitochondrial function. These act as some signal molecule or one kind of object that regulates ATP production and cell survival. However, the balance of calcium in mitochondria is thrown off by disorder in their function, and thereby pathological consequences follow rapidly. Under normal conditions, the concentration of calcium ions in the cytoplasm is regulated within a specific range by these organelles: they trap it there temporarily so to speak in their matrix. However, increased calcium levels, especially as it enters the matrix from outside in conditions like excess pressure due to ion channels or endoplasmic reticulum (ER). A conventional sign that the mitochondria are overloaded with calcium, opening the mPTP, producing excess ROS, and an acute fall off in $\Delta\Psi m$. So it must be considered that when people want to die young by eating too much cytochrome *c* and inadvertently letting out death-bringing factors like caspase into their bodies to wreak havoc on their DNA this is very unfortunate. Besides, calcium imbalance will make further disease pathologic by starting an inflammatory cascade locally and throughout the system.

Few pathologies do not show mitochondrial dysfunction, which manifests in bioenergetic failure, increased mitochondrial permeability, and disordered calcium signaling. Understanding the molecular mechanisms underlying mitochondrial impairment is crucial in developing target-oriented therapeutic interventions. Strategies that favor improving mitochondrial function, such as antioxidants, mitochondrial-targeted drugs, and gene therapy, offer hope for ending mitochondrial dysfunction-induced disease. The following steps shall be to hone in on enemy targets at the molecular level and work out precision medicine strategies for restoring mitochondrial health and overcoming disease (Nguyen & Kim, 2022; Oyeyemi et al., 2024; Park & Kwak, 2023; Rehman et al., 2021).

Neuroinflammation is a complex physiological response of the CNS to injury, infection, or neurodegenerative processes. It is primarily mediated by the activation of resident immune cells, including microglia and astrocytes, which play a crucial role in maintaining CNS homeostasis and responding to pathological conditions. While neuroinflammation serves as a protective mechanism under acute conditions, chronic or dysregulated neuroinflammatory responses contribute to the progression of NDs such as AD, PD, MS, and ALS. The key processes involved in neuroinflammation include the activation of microglia and astrocytes, the release of proinflammatory cytokines, and the disruption of the BBB, leading to neurovascular damage.

Microglia are the primary immune cells of the CNS, functioning as resident macrophages responsible for immune surveillance and responding to neuronal injury. In a healthy brain, microglia are ramified and characterized by long, branched processes that continuously monitor their microenvironment. Upon encountering pathological stimuli such as pathogens, damaged neurons, or aggregated proteins, microglia undergo morphological and functional changes, shifting to an activated state. Activated microglia secrete various proinflammatory mediators, including tumor necrosis factor-alpha (TNF-α), interleukin-6 (IL-6), and interleukin-1 beta (IL-1β), which can either support neuroprotection or exacerbate neuronal damage depending on the context of activation (Esan et al., 2024; Khan & Parvez, 2015; Li, Xia, et al., 2021; Perera et al., 2023). Similarly, astrocytes, the most abundant glial cells in the CNS, play a vital role in neuroinflammation. Under normal physiological conditions, astrocytes contribute to synaptic regulation, metabolic support, and maintenance of the BBB. However, in response to CNS insults, astrocytes become reactive, undergoing hypertrophy and upregulating the expression of glial fibrillary acidic protein. Reactive astrocytes contribute to neuroinflammation by producing cytokines, chemokines, and ROS, which can influence microglial activity and neuronal survival. The dual role of astrocytes in neuroinflammation, where they can either promote repair or drive neurodegeneration, is dictated by the balance between proinflammatory (A1) and antiinflammatory (A2) astrocytic phenotypes. One hallmark of neuroinflammation is the excessive production and release of proinflammatory cytokines, which mediate communication between glial cells and neurons. TNF-α, IL-6, and IL-1β are among the most studied cytokines in neuroinflammatory responses.

- **TNF-α:** TNF-α is a pleiotropic cytokine central to neuroinflammation. It is primarily secreted by activated microglia and astrocytes in response to injury, infection, or neurodegeneration. TNF-α exerts its effects through TNF receptors (TNFR1 and TNFR2), with TNFR1 predominantly mediating apoptotic and inflammatory pathways, while TNFR2 is associated with neuroprotection and regeneration. Elevated levels of TNF-α in the CNS have been implicated in synaptic dysfunction, neuronal apoptosis, and the progression of NDs.
- **IL-6:** IL-6 is another critical cytokine involved in neuroinflammation, with proinflammatory and antiinflammatory properties. Under pathological conditions, IL-6 is released by activated microglia and astrocytes, contributing to immune cell recruitment and modulation of neuronal survival. While acute IL-6 signaling is essential for tissue repair, chronic elevation of IL-6 has been associated with NDs by exacerbating oxidative stress, mitochondrial dysfunction, and excitotoxicity.
- **IL-1β:** IL-1β is a potent inflammatory cytokine produced by microglia in response to injury and infection. It activates intracellular signaling cascades that enhance the expression of adhesion molecules, chemokines, and secondary inflammatory mediators, amplifying neuroinflammation. IL-1β has been implicated in disrupting synaptic plasticity, impairing learning and memory, and promoting neurotoxicity in diseases such as AD and MS.

The interplay between these cytokines orchestrates the neuroinflammatory milieu, determining whether the response leads to neuroprotection or neurodegeneration. Chronic activation of proinflammatory pathways can contribute to sustained neuroinflammation, leading to progressive neuronal damage and functional impairment. The BBB is a highly selective barrier composed of endothelial cells, pericytes, and astrocytic end feet that regulate the passage of molecules between the bloodstream and the CNS. It plays a crucial role in maintaining brain homeostasis by preventing the entry of toxins, pathogens, and peripheral immune cells into the brain parenchyma (Aschner et al., 2024; Hussien et al., 2018; Khan & Parvez, 2015; Park & Kwak, 2023). However, neuroinflammatory processes can compromise BBB integrity, leading to neurovascular damage and increased permeability. Microglial activation and cytokine release contribute significantly to BBB dysfunction. Proinflammatory cytokines such as TNF-α and IL-1β disrupt tight junction proteins (e.g., occludin, claudins, and zonula occludens-1), weakening the endothelial barrier and allowing the infiltration of peripheral immune cells, such as monocytes and T-cells, into the CNS. This infiltration further exacerbates neuroinflammation by triggering additional immune responses and oxidative stress. Astrocytes, crucial for maintaining BBB function, undergo structural and functional changes during neuroinflammation. Reactive astrocytes release matrix metalloproteinases, which degrade the extracellular matrix and further compromise BBB integrity. Moreover, increased expression of vascular adhesion molecules, such as intercellular adhesion molecule-1 and vascular cell adhesion molecule-1, facilitates leukocyte transmigration into the CNS, aggravating neuroinflammatory damage (Bhattacharya et al., 2024; Khan & Parvez, 2015; Park & Kwak, 2023; Perera et al., 2023).

Neurovascular dysfunction resulting from BBB disruption is a key contributor to NDs. In conditions like AD, impaired BBB function leads to the accumulation of toxic proteins such as amyloid-beta (Aβ), exacerbating neuronal loss. In MS, BBB breakdown facilitates the entry of autoreactive T-cells, promoting demyelination and neurodegeneration. The extent of BBB disruption and neurovascular damage is a critical determinant of disease progression and severity. Neuroinflammation is a multifaceted process that involves the activation of microglia and astrocytes, the release of proinflammatory cytokines, and the disruption of the BBB, leading to neurovascular damage. While acute neuroinflammatory responses are necessary for tissue repair and immune defense, chronic neuroinflammation is implicated in the pathogenesis of various NDs (Aschner et al., 2024; Nguyen & Kim, 2022; Park & Kwak, 2023). Understanding the molecular mechanisms governing neuroinflammation is essential for developing therapeutic strategies to modulate glial activation, cytokine signaling, and BBB integrity to mitigate neurodegenerative damage. Future research should focus on identifying novel antiinflammatory targets and therapeutic interventions to counteract the detrimental effects of neuroinflammation while preserving its protective functions in CNS homeostasis.

The disruption of cellular signaling pathways plays a crucial role in the pathophysiology of various neurological disorders, mainly through mechanisms such as calcium dysregulation, impairment of neurotrophic factors like brain-derived neurotrophic factor (BDNF) and nerve growth factor (NGF), and alterations in synaptic plasticity. Cellular signaling pathways are essential for maintaining normal neuronal function, intracellular communication, and overall homeostasis in the nervous system. Any perturbation in these intricate signaling cascades can lead to significant deficits in cognition, synaptic integrity, and neuronal survival, contributing to neurodegenerative and neuropsychiatric conditions such as AD, PD, HD, schizophrenia, and depression. Among the various disruptions, calcium dysregulation is a central contributor, given calcium's fundamental role as a second messenger in neuronal signaling, synaptic transmission, and gene expression (Bhattacharya et al., 2024; Hussien et al., 2018; Joneidi et al., 2019; Karri et al., 2020).

Calcium homeostasis in neurons is tightly regulated by voltage-gated calcium channels (VGCCs), ligand-gated ion channels (such as *N*-methyl-*D*-aspartate receptors [NMDARs]), ER-associated calcium stores, and mitochondrial calcium buffering. Perturbations in these calcium-handling mechanisms can result in excessive calcium influx or inefficient calcium sequestration, leading to excitotoxicity and cell death. For instance, hyperactivation of NMDARs can lead to excessive calcium influx, overwhelming mitochondrial buffering capacity and triggering apoptosis via cytochrome *c* release and caspase activation. In NDs, such as AD, calcium dysregulation is a hallmark feature, with amyloid-beta oligomers inducing hyperactivation of NMDARs and leading to calcium overload. Furthermore, mutations in presenilin proteins, which are components of the γ-secretase complex involved in amyloid precursor protein processing, have been linked to aberrant calcium release from the ER, further contributing to neuronal toxicity. Similarly, in PD, mitochondrial dysfunction exacerbates calcium dysregulation by impairing ATP production required for calcium homeostasis, rendering dopaminergic neurons particularly vulnerable to degeneration.

Neurotrophic factors, particularly BDNF and NGF, are pivotal in neuronal survival, differentiation, and synaptic plasticity. The impact of cellular signaling disruption on these neurotrophic factors is profound, as it can lead to reduced expression, impaired receptor signaling, and inefficient downstream pathway activation. BDNF, which exerts its effects primarily through the TrkB receptor, is crucial for synaptic plasticity, long-term potentiation (LTP), and memory formation. Reduced BDNF levels have been implicated in NDs and psychiatric disorders, where compromised TrkB signaling results in synaptic deficits and neuronal atrophy. In AD, decreased BDNF expression correlates with hippocampal atrophy and cognitive decline, and amyloid-beta aggregates have been shown to disrupt BDNF-TrkB signaling, exacerbating synaptic dysfunction. Similarly, in major depressive disorder, impaired BDNF signaling contributes to reduced neurogenesis and synaptic connectivity, underlying the pathophysiology of depression. NGF primarily binds to the TrkA receptor and is critical for maintaining and surviving cholinergic neurons in the basal forebrain. Reduced NGF levels have been associated with AD, leading to cholinergic deficits and cognitive impairments (Aschner et al., 2024; Esan et al., 2024; Stachowicz, 2023). Additionally, pro-NGF, the precursor form of NGF, can preferentially bind to the p75 neurotrophin receptor (p75NTR), triggering apoptotic pathways instead of promoting neuronal survival, further exacerbating neurodegeneration.

Synaptic plasticity, the ability of synapses to strengthen or weaken over time in response to activity, is essential for learning, memory, and adaptive behaviors. The disruption of cellular signaling pathways can significantly alter synaptic plasticity, leading to cognitive impairments and neurological disorders. LTP and long-term depression (LTD) are two key mechanisms of synaptic plasticity, which rely on precise calcium signaling and neurotrophic factor regulation. In AD, synaptic dysfunction is one of the earliest pathological features, with amyloid-beta oligomers disrupting LTP by impairing NMDAR-mediated calcium influx and downstream signaling pathways such as the CaMKII and CREB pathways. Additionally, tau hyperphosphorylation, another hallmark of AD, disrupts cytoskeletal integrity and dendritic spine morphology, further contributing to synaptic deficits. In schizophrenia, alterations in synaptic plasticity have been linked to dysregulated glutamatergic transmission, particularly involving NMDAR hypofunction. This impairment in NMDAR activity leads to deficits in LTP and altered excitatory-inhibitory balance, contributing to cognitive and behavioral abnormalities observed in patients. Furthermore, synaptic plasticity critically depends on neurotrophic factor support, with reduced BDNF levels impairing synapses' structural and functional integrity. In depression, chronic stress and glucocorticoid dysregulation lead to reduced BDNF expression, resulting in synaptic atrophy in key brain regions such as the hippocampus and prefrontal cortex (Li, Ren, et al., 2021; Li, Xia, et al., 2021; Tang et al., 2023).

Mitochondrial dysfunction is another crucial factor contributing to the disruption of cellular signaling pathways, particularly in the context of calcium dysregulation and synaptic plasticity. Mitochondria are key regulators of intracellular calcium homeostasis, ATP production, and ROS generation. Dysfunctional mitochondria can lead to impaired calcium buffering, increased oxidative stress, and energy deficits, negatively impacting synaptic function and neuronal survival. In PD, mitochondrial dysfunction is a primary pathological feature, with mutations in genes such as PINK1 and Parkin leading to defective mitophagy and accumulated damaged mitochondria (Bhattacharya et al., 2024; Deng et al., 2023; Popov Aleksandrov et al., 2021). This results in impaired ATP production and excessive ROS generation, exacerbating calcium dysregulation and synaptic loss. Similarly, in HD, mitochondrial dysfunction contributes to neuronal excitotoxicity, with mutant huntingtin protein impairing mitochondrial calcium uptake and leading to synaptic dysfunction. Given the intricate interplay between mitochondrial function, calcium signaling, and synaptic plasticity, therapeutic strategies targeting mitochondrial health hold promise for mitigating neuronal damage in NDs.

Inflammatory processes further exacerbate the disruption of cellular signaling pathways, mainly through the activation of microglia and astrocytes. Neuroinflammation, characterized by excessive production of proinflammatory cytokines, such as TNF-α, IL-1β, and IL-6, can impair synaptic plasticity and neurotrophic factor signaling. Chronic neuroinflammation has been implicated in AD, PD, and depression, where activated microglia release excessive glutamate, leading to excitotoxicity and synaptic degeneration. Additionally, inflammatory cytokines can downregulate BDNF expression and TrkB signaling, further compromising synaptic function. In depression, elevated levels of inflammatory markers have been correlated with reduced synaptic plasticity and impaired hippocampal neurogenesis, suggesting a direct link between inflammation and synaptic dysfunction. Astrocytes, which play a crucial role in maintaining synaptic homeostasis, can also contribute to neuroinflammation by releasing glutamate and proinflammatory mediators, exacerbating neuronal damage (Cai et al., 2023; Pota et al., 2024; Rehman et al., 2021; Stachowicz, 2023).

Given the extensive impact of cellular signaling disruptions on calcium homeostasis, neurotrophic factor regulation, and synaptic plasticity, therapeutic interventions targeting these pathways hold promise for treating neurological and psychiatric disorders. Pharmacological strategies aimed at restoring calcium homeostasis, such as NMDAR modulators and calcium channel blockers, have been explored in preclinical and clinical studies. For instance, memantine, an NMDAR antagonist, has been approved for treating AD, as it helps reduce excitotoxicity by preventing excessive calcium influx. Additionally, BDNF-mimetic compounds and TrkB agonists are being investigated as potential therapeutics for NDs and depression. Enhancing synaptic plasticity through neurotrophic factor supplementation, exercise, and dietary interventions has also shown promise in preclinical models (Karri et al., 2016; Lasisi-Sholola et al., 2024; Rehman et al., 2021; Sun et al., 2025). Furthermore, mitochondrial-targeted therapies, such as antioxidants and mitophagy enhancers, are being explored for their potential to restore cellular signaling homeostasis and prevent neuronal degeneration.

The disruption of cellular signaling pathways, mainly through calcium dysregulation, impaired neurotrophic factor signaling, and altered synaptic plasticity, plays a critical role in the pathophysiology of various neurological disorders. Calcium homeostasis is essential for neuronal function, and its dysregulation can lead to excitotoxicity and cell death. Neurotrophic factors such as BDNF and NGF are crucial for synaptic maintenance and neuronal survival, and their impaired signaling contributes to neurodegenerative and psychiatric conditions. These disruptions significantly affect synaptic plasticity, a fundamental mechanism for learning and memory, leading to cognitive deficits and neurological dysfunction (Li, Xia, et al., 2021; Nguyen & Kim, 2022; Paithankar et al., 2021; Wu et al., 2023). Understanding these intricate signaling mechanisms provides valuable insights into potential therapeutic targets for mitigating neurodegeneration and improving brain health. Future research focusing on restoring cellular signaling homeostasis through pharmacological and nonpharmacological approaches holds promise for advancing treatments for neurodegenerative and psychiatric disorders.

The precise regulation of protein degradation is crucial for maintaining cellular homeostasis, and its disruption is a hallmark of numerous neurodegenerative disorders. Two primary systems orchestrate protein degradation in eukaryotic cells: the ubiquitin–proteasome system (UPS) and the autophagy-lysosomal pathway. These mechanisms are essential for eliminating misfolded, damaged, or aggregated proteins that, if accumulated, can lead to cellular toxicity and functional decline. Dysfunctions in these pathways contribute to pathological conditions such as AD, PD, and ALS, where protein aggregation and cellular stress lead to progressive neuronal degeneration. Understanding the roles of these systems and their failures in neurodegeneration provides insights into potential therapeutic strategies.

The UPS is a highly selective proteolytic system that degrades short-lived and misfolded proteins through ubiquitination and subsequent proteasomal degradation. This system relies on a cascade of enzymatic actions involving ubiquitin-activating enzymes (E1s), ubiquitin-conjugating enzymes (E2s), and ubiquitin ligases (E3s), which tag proteins for degradation via the 26S proteasome. Impairments in any component of this cascade can accumulate aberrant proteins, resulting in neurotoxicity and cell death. In NDs, UPS dysfunction is frequently observed. For instance, in PD, mutations in Parkin, an E3 ubiquitin ligase, impair the degradation of misfolded or damaged proteins, promoting the accumulation of α-synuclein aggregates characteristic of Lewy bodies. Similarly, in AD, reduced proteasomal activity has been linked to the accumulation of hyperphosphorylated tau and amyloid-beta peptides, exacerbating neurodegeneration. Studies have shown that proteasomal function declines with aging, which may partly explain the late-onset nature of many NDs (Bjørklund et al., 2024; Fu et al., 2023; Lin et al., 2024; Samuel et al., 2021).

The involvement of deubiquitinating enzymes (DUBs) further complicates the UPS's role in neurodegeneration. These enzymes regulate protein turnover by removing ubiquitin chains, thereby rescuing proteins from degradation. Dysregulation of DUBs, such as ubiquitin carboxyl-terminal hydrolase L1 (UCH-L1), has been implicated in neurodegenerative conditions. Inhibition or dysregulation of DUBs can either enhance or impair protein degradation, leading to excessive protein accumulation or inappropriate degradation of essential proteins. Autophagy is a crucial intracellular degradation process that eliminates long-lived proteins, damaged organelles, and toxic aggregates. It has three main pathways: macroautophagy, microautophagy, and chaperone-mediated autophagy (CMA). Macroautophagy plays a predominant role in neuroprotection by engulfing and degrading large protein aggregates that the proteasome cannot process.

In NDs, impaired autophagic flux has been linked to the accumulation of toxic proteins. For example, in AD, studies have demonstrated a failure in the clearance of autophagic vesicles, leading to the accumulation of dysfunctional organelles and amyloid-beta aggregates in neurons. Similarly, in Huntington's disease, mutant huntingtin (mHTT) protein disrupts autophagy by sequestering critical autophagy-related proteins, leading to aggregate formation and neuronal toxicity. CMA, which selectively degrades proteins containing a KFERQ-like motif by targeting them to lysosomes, is also impaired in NDs. In PD, CMA dysfunction has been observed with α-synuclein mutants that resist lysosomal degradation, promoting aggregate formation and dopaminergic neuronal death. Furthermore, genetic or pharmacological activation of autophagy has been shown to mitigate neurodegenerative phenotypes in animal models, highlighting its potential as a therapeutic target. Another key regulator of autophagy is the mammalian target of the rapamycin (mTOR) signaling pathway. Under nutrient-rich conditions, mTOR inhibits autophagy, while under stress or starvation, its inhibition triggers autophagic degradation. Dysregulation of mTOR signaling in NDs results in impaired autophagy, exacerbating protein accumulation. Pharmacological approaches targeting mTOR, such as rapamycin and its analogues, have demonstrated promise in restoring autophagic activity and ameliorating disease symptoms in preclinical models (Dong et al., 2024; Dooka et al., 2024; Famurewa et al., 2022; Lin et al., 2024).

Protein misfolding and aggregation are standard pathological features of NDs. Many neurotoxic proteins, including amyloid-beta, tau, and α-synuclein, adopt β-sheet-rich structures prone to aggregation. These aggregates disrupt cellular homeostasis by sequestering essential proteins, impairing organellar function, and inducing oxidative stress and inflammation.

In AD, amyloid-beta peptides aggregate to form extracellular plaques, while hyperphosphorylated tau forms intracellular neurofibrillary tangles (NFT). The imbalance between amyloid-beta production and clearance leads to its accumulation, triggering synaptic dysfunction and neuronal death. Moreover, amyloid-beta oligomers interact with tau, exacerbating its aggregation and further promoting neuronal toxicity. Similarly, in PD, misfolded α-synuclein forms insoluble aggregates known as Lewy bodies, contributing to dopaminergic neuron loss. The propagation of α-synuclein aggregates between neurons via prion-like mechanisms exacerbates neurodegeneration. This phenomenon has also been observed with tau in AD and with TDP-43 in ALS, suggesting a shared mechanism across neurodegenerative disorders. Molecular chaperones, such as heat shock proteins, play a critical role in preventing protein aggregation by assisting in protein folding and directing misfolded proteins to degradation pathways. However, in neurodegenerative conditions, chaperone systems become overwhelmed, leading to aggregation and toxicity. Pharmacological chaperones and proteostasis regulators have emerged as potential therapeutic strategies to enhance protein quality control and mitigate aggregation-related toxicity. Given the critical role of impaired protein degradation in neurodegeneration, therapeutic interventions aimed at enhancing UPS function, autophagy, and proteostasis are actively being explored.

1. **Proteasome activators:** Small molecules that enhance proteasome activity, such as 20S proteasome activators, have shown promise in reducing protein aggregates and improving neuronal survival.

2. **Autophagy inducers:** Pharmacological agents like rapamycin, trehalose, and lithium promote autophagy and facilitate aggregate clearance. Genetic approaches to modulate autophagy-related genes also hold therapeutic potential.
3. **Chaperone-based therapies:** Enhancing molecular chaperone activity using small molecules or gene therapy strategies can mitigate protein misfolding and aggregation.
4. **Gene-editing technologies:** CRISPR-based approaches targeting genes involved in protein degradation and aggregation could provide precision medicine strategies for neurodegenerative disorders.
5. **Targeting prion-like propagation:** Strategies that inhibit the cell-to-cell spread of misfolded proteins could slow disease progression in synucleinopathies and tauopathies.

Impairment of protein degradation systems plays a central role in the pathogenesis of NDs. Dysfunctions in the UPS, autophagy deficits, and protein misfolding contribute to neuronal toxicity and disease progression. Targeting these pathways offers a promising avenue for therapeutic intervention. Advances in molecular and pharmacological strategies to restore proteostasis and enhance protein clearance mechanisms may pave the way for future effective treatments for neurodegenerative disorders.

2.4 Mechanisms of metal accumulation in the brain

The BBB is a highly specialized, selective barrier that regulates the transport of substances between the systemic circulation and the CNS. Composed of endothelial cells with tight junctions, pericytes, and astrocytic end feet, the BBB prevents the entry of neurotoxic substances while facilitating the transport of essential nutrients. However, heavy metals such as lead (Pb), mercury (Hg), cadmium (Cd), and arsenic (As) have been shown to breach this protective barrier, leading to neurotoxicity and various neurological disorders. Heavy metals can penetrate the BBB through multiple mechanisms, including passive diffusion, ion channels, and transporter-mediated uptake. Lipophilic compounds, such as organic mercury (methylmercury), readily diffuse across the BBB due to their high solubility in lipid membranes. In contrast, metals like lead and cadmium exploit ion channels and transporters to gain entry into the CNS. Lead mimics calcium ions and is transported via VGCCs, whereas cadmium competes with zinc transporters to cross the BBB. Additionally, in its trivalent form (As^{3+}), arsenic interacts with aquaglyceroporins, facilitating its passage into neural tissues. Various mechanisms mediate metal ions' transport across the BBB, including carrier-mediated transport, endocytosis, and ion channels. Divalent metal transporter 1 (DMT1) plays a crucial role in facilitating the entry of iron (Fe^{2+}), lead (Pb^{2+}), and manganese (Mn^{2+}) into neural tissues. Similarly, transferrin receptors (TfR) mediate the uptake of iron-bound transferrin, which can also be hijacked by metal ions such as aluminum and cadmium. Organic cation transporters and ATP-binding cassette (ABC) transporters further contribute to the movement of metals across the BBB, either by facilitating uptake or actively effluxing toxic metals out of the brain. Exposure to heavy metals disrupts BBB integrity by inducing oxidative stress, inflammation, and endothelial dysfunction (Adegbola & Adetutu, 2024; Fu et al., 2023; Raj et al., 2021; Wang & Matsushita, 2021). Lead exposure, for instance, has been shown to downregulate tight junction proteins such as occludin and claudin-5, thereby increasing BBB permeability. Similarly, mercury exposure triggers neuroinflammation by activating microglia and astrocytes, releasing proinflammatory cytokines such as TNF-α, IL-6, and IL-1β. Arsenic and cadmium generate ROS, leading to lipid peroxidation and damage to endothelial cells. Collectively, these disruptions compromise the BBB, allowing further infiltration of neurotoxic substances and exacerbating neuronal injury. Metallothioneins (MTs) are low-molecular-weight, cysteine-rich proteins that play a vital role in metal homeostasis and detoxification. These proteins bind heavy metals such as zinc (Zn), copper (Cu), and cadmium (Cd) through thiol groups, thereby reducing metal-induced oxidative damage. However, excessive heavy metal exposure can lead to the overexpression of MTs, altering their normal function. Other chelating proteins, such as ferritin and transferrin, are involved in iron homeostasis, whereas ceruloplasmin regulates copper metabolism. Dysregulation of these proteins can contribute to neurodegeneration and cognitive impairment. Heavy metal exposure disturbs metal homeostasis, leading to neuronal toxicity and dysfunction. For example, lead disrupts calcium signaling by interfering with VGCCs, impairing synaptic transmission and neurodevelopmental deficits. Similarly, mercury forms strong bonds with sulfhydryl groups in proteins, disrupting enzymatic activity and mitochondrial function. Cadmium exposure leads to zinc depletion, affecting synaptic plasticity and cognitive functions. The disruption of these essential mental homeostatic processes contributes to NDs such as AD, PD, and ALS. Iron, copper, and zinc are essential trace metals in various neurological processes, including neurotransmitter synthesis, myelination, and synaptic function. However, imbalances in these metals contribute to oxidative stress and neurotoxicity. As observed in NDs, excess iron accumulation leads to Fenton reactions that generate hydroxyl radicals, causing oxidative damage to neuronal cells. Copper imbalance is implicated in Wilson's disease and AD, where copper-induced oxidative stress promotes amyloid-beta aggregation. Zinc plays a dual role in neuroprotection and neurotoxicity, while essential for synaptic function, excessive zinc can lead to excitotoxicity and neuronal apoptosis. Heavy metals compete with essential metals for binding sites on biological molecules, displacing crucial metal ions (Althobaiti, 2025; Gomes da Silva et al., 2024; Mohankumar et al., 2024; Wang & Matsushita, 2021). For instance, lead displaces calcium in calcium-binding proteins, disrupting neuronal signaling and synaptic plasticity. Similarly, cadmium replaces zinc in metalloproteins, impairing their enzymatic activity and structural integrity. Mercury's high affinity for thiol groups in proteins leads to the displacement of essential sulfur-containing amino acids, affecting cellular metabolism and redox balance. Metalloproteins, which rely on metal cofactors for their structural and functional stability, are particularly susceptible to heavy metal toxicity. For example, superoxide dismutase (SOD), a critical antioxidant enzyme, requires zinc and copper for its activity. Heavy metal exposure can alter the binding affinity of these cofactors, leading to reduced enzymatic efficiency and increased oxidative stress. Another example is cytochrome *c* oxidase, a key enzyme in the mitochondrial ETC, which requires copper for its function. Mercury and lead interfere with this enzyme, impairing mitochondrial respiration and ATP production. The disruption of metalloproteins contributes to neuronal cell death and the progression of neurodegenerative disorders. The BBB serves as a crucial defense mechanism against neurotoxic insults, but its integrity is compromised by heavy metal exposure. The ability of heavy metals to cross the BBB through various transport mechanisms leads to neuronal accumulation and toxicity. Metal-binding proteins such as MTs, transferrin, and ceruloplasmin play a critical role in regulating metal homeostasis, but their function is disrupted by excessive metal exposure. Additionally, heavy metals displace essential metal cofactors from biological systems, leading to metalloprotein dysfunction and neuronal toxicity (Table 2.3).

These mechanisms collectively contribute to neurodegeneration and highlight the urgent need for strategies to mitigate heavy metal exposure and its impact on brain health. Future research should focus on developing neuroprotective agents that enhance BBB integrity, restore metal homeostasis, and prevent metalloprotein dysfunction to combat heavy metal-induced neurotoxicity.

2.5 Genetic and epigenetic factors in metal-induced neurodegeneration

Genetic susceptibility plays a crucial role in determining an individual's response to heavy metal exposure, influencing both the accumulation of metals in the body and the severity of toxicity they induce. Inherited genetic factors contribute significantly to metal accumulation by regulating transport proteins, detoxification enzymes, and cellular defense mechanisms. Genetic variations, including single nucleotide polymorphisms, can alter the function of metal transporters such as DMT1, ZIP (Zrt/Irt-like protein), and ABC transporters, which are essential in regulating the uptake and efflux of metals like lead, cadmium, mercury, and arsenic. Mutations in genes encoding these proteins can enhance metal retention in target tissues, including the brain, liver, and kidneys, thereby increasing toxicity and the risk of associated diseases (Dooka et al., 2024; Fu et al., 2023; Karri et al., 2018; Kwakye et al., 2019). For example, polymorphisms in the MT gene family, which encodes small cysteine-rich proteins that bind and sequester toxic metals, have been linked to variations in heavy metal tolerance among individuals. Some people with specific MT gene variants may exhibit greater resilience to metal toxicity, while others may be more susceptible due to reduced metal-binding efficiency.

Furthermore, inherited genetic factors play a pivotal role in determining an individual's vulnerability to metal-induced neurodegeneration. Specific genetic polymorphisms have been linked to neurological disorders associated with metal exposure, such as AD, PD, and ALS. The APOE gene, particularly the APOE ε4 allele, has been implicated in an increased risk of developing AD in individuals exposed to high levels of aluminum, lead, and mercury. This allele is associated with reduced clearance of amyloid-beta plaques, exacerbated by metal-induced oxidative stress and neuroinflammation. Similarly, polymorphisms in genes such as PARK2, SNCA, and LRRK2, which are involved in the pathogenesis of PD, have been linked to increased sensitivity to manganese and other neurotoxic metals. Variants in these genes can lead to impaired metal homeostasis, mitochondrial dysfunction, and dopaminergic neuronal damage, all of which contribute to the progression of neurodegenerative disorders. Moreover, genetic predisposition to ALS has been associated with mutations in the SOD1 (superoxide dismutase 1) gene, which encodes an enzyme that neutralizes ROS. Heavy metals such as lead, mercury, and arsenic can disrupt SOD1 activity, increasing oxidative stress and motor neuron degeneration in individuals carrying SOD1 mutations. These findings underscore the importance of genetic susceptibility in modulating the toxic effects of heavy metals, highlighting the need for personalized approaches to mitigate metal-induced health risks (Fu et al., 2023; Gomes da Silva et al., 2024; Mohankumar et al., 2024; Samuel et al., 2021; Vellingiri et al., 2022).

Beyond genetic inheritance, epigenetic modifications play a crucial role in mediating the effects of heavy metal exposure on gene expression and disease susceptibility. Epigenetic modifications, including DNA methylation, histone modifications, and noncoding RNA regulation, do

TABLE 2.3 Comprehensive table comparing heavy metal accumulation in different neurodegenerative diseases.

Heavy metal	Neurodegenerative disease	Brain regions affected	Levels of accumulation in the brain	Mechanisms of toxicity	Key cellular targets	Biomarkers for detection	Evidence from postmortem studies	Genetic and epigenetic susceptibility factors	Therapeutic approaches for metal detoxification
Lead (Pb)	AD	Hippocampus, cerebral cortex	Moderate to high	Oxidative stress, neuroinflammation, BBB disruption	Neurons, astrocytes	Elevated Pb in CSF/ blood, increased ROS markers	Pb detected in AD patient brains with neuronal loss	Polymorphisms in metal transporters (e.g., HFE gene)	Chelation therapy (EDTA), antioxidants (vitamins C and E)
Mercury (Hg)	PD	Substantia nigra, striatum	High	Mitochondrial dysfunction, dopaminergic neuron toxicity	Dopaminergic neurons, microglia	Elevated Hg levels in blood, increased lipid peroxidation	Postmortem brains show Hg deposits in substantia nigra	Genetic variations in MT1 and MT2 (metallothionein genes)	Selenium-based chelators, NAC
Arsenic (As)	ALS	Motor cortex, spinal cord	Moderate	Protein aggregation, DNA damage, inflammatory response	Motor neurons, oligodendrocytes	High As levels in CSF, increased oxidative DNA damage markers	ALS patient brains exhibit arsenic accumulation	Epigenetic modifications in oxidative stress response genes	Dietary Se supplementation, chelation (DMPS, DMSA)
Cadmium (Cd)	HD	Striatum, cortex	Low to moderate	Disrupts calcium signaling, induces apoptosis	Neurons, astrocytes	Cd detected in HD patient CSF, increased inflammatory cytokines	Elevated Cd levels in HD-affected basal ganglia	Alterations in calcium homeostasis genes	Chelation (EDTA), zinc supplementation
Aluminum (Al)	MS	White matter, cerebral cortex	Moderate	Myelin damage, neuroinflammation, protein misfolding	Oligodendrocytes, microglia	Elevated Al in CSF, Increased autoantibodies	Al aggregates found in demyelinated MS plaques	Epigenetic modifications in myelin-related genes	Desferrioxamine therapy, curcumin-based detox strategies
Manganese (Mn)	Parkinsonism	Basal ganglia, globus pallidus	High	Mitochondrial dysfunction, dopamine depletion	Dopaminergic neurons, astrocytes	High Mn levels in blood, increased neuroinflammatory markers	Elevated Mn in PD-affected basal ganglia	SLC30A10 mutations linked to Mn toxicity	Iron chelators, dietary Mn restriction

AD, Alzheimer's disease; *ALS*, amyotrophic lateral sclerosis; *HD*, Huntington's disease; *MS*, multiple sclerosis; *NAC*, N-acetylcysteine; *PD*, Parkinson's disease.
Source: Adegbola, P. I., Adetutu, A. (2024). Genetic and epigenetic modulations in toxicity: The two-sided roles of heavy metals and polycyclic aromatic hydrocarbons from the environment. *Toxicology Reports, 12*, 502–519. https://doi.org/10.1016/j.toxrep.2024.04.010; Dong, Q., Fu, H., Jiang, H. (2024). The role of exosome-shuttled miRNAs in heavy metal-induced peripheral tissues and neuroinflammation in Alzheimer's disease. *Biomedicine Pharmacotherapy, 176*, 116880. https://doi.org/10.1016/j.biopha.2024.116880; Kumar, A., Abhinav, S., Rachamalla, M., Ali, M., Navin, S., Kumar, D., Ghosh, A. K. (2025). Chapter 29 – Treatment of heavy metal induced neurotoxicity via advanced drug delivery systems. In P. A. Chawla, R. Loebenberg, K. Dua, V. Parikh, V. Chawla (Eds.), *Novel drug delivery systems in the management of disorders* (pp. 393–407). Academic Press. https://doi.org/10.1016/B978-0-443-13474-6.00025-1; Kwakye, G. F., Jiménez, J. A., Thomas, M. G., Kingsley, B. A., McIlvin, M., Saito, M. A., Korley, E. M. (2019). Heterozygous huntingtin promotes cadmium neurotoxicity and neurodegeneration in striatal cells via altered metal transport and protein kinase C delta dependent oxidative stress and apoptosis signaling mechanisms. *Neurotoxicology, 70*, 48–61. https://doi.org/10.1016/j.neuro.2018.10.012; Lin, J., Wang, L., Huang, M., Xu, G., Yang, M. (2024). Metabolic changes induced by heavy metal copper exposure in human ovarian granulosa cells. *Ecotoxicology and Environmental Safety, 285*, 117078. https://doi.org/10.1016/j.ecoenv.2024.117078.

not alter the DNA sequence but can significantly impact gene activity and cellular responses to environmental stressors, including toxic metals. DNA methylation, a process in which methyl groups are added to cytosine residues in CpG dinucleotides, is a key mechanism for regulating gene expression. Heavy metals such as arsenic, cadmium, and lead have been shown to disrupt DNA methylation patterns, leading to aberrant gene activation or silencing. For instance, arsenic exposure has been linked to global DNA hypomethylation, which can promote genomic instability and increase the risk of cancer (Dong et al., 2024; Dooka et al., 2024; Famurewa et al., 2022; Lin et al., 2024; Raj et al., 2021). Conversely, hypermethylation of tumor suppressor genes, such as p16INK4a and RASSF1A, has been observed in individuals exposed to cadmium, leading to impaired cell cycle regulation and increased carcinogenesis. Similarly, lead exposure has been associated with altered methylation of genes involved in neurodevelopment, potentially contributing to cognitive impairments and behavioral disorders in children.

Histone modifications represent another layer of epigenetic regulation influenced by heavy metal exposure. Histones are proteins that package DNA into chromatin, and their chemical modifications, such as acetylation, methylation, phosphorylation, and ubiquitination, play a vital role in modulating gene expression. Heavy metals can interfere with histone-modifying enzymes, disrupting chromatin structure and transcriptional regulation. For example, cadmium has been shown to inhibit histone deacetylases (HDACs), leading to increased histone acetylation and aberrant gene activation. This disruption can contribute to inflammatory responses, cellular stress, and the progression of diseases such as cancer and neurodegeneration. In contrast, arsenic exposure has been linked to histone H3K9 and H3K27 methylation changes associated with gene repression and epigenetic silencing of DNA repair genes. These histone modifications can exacerbate DNA damage, impair cellular repair mechanisms, and promote malignant transformation. Furthermore, alterations in histone methylation have been observed in individuals exposed to mercury, affecting genes involved in immune function and neurological health (Fu et al., 2023; Karri et al., 2018; Kumar et al., 2025; Samuel et al., 2021; Wang & Matsushita, 2021).

Noncoding RNAs, including microRNAs (miRNAs) and long noncoding RNAs (lncRNAs), play an emerging role in mediating the effects of heavy metals on gene expression. miRNAs are small RNA molecules that regulate gene expression by binding to messenger RNAs and preventing their translation into proteins. Heavy metal exposure has been shown to dysregulate miRNA expression, leading to cellular pathways related to apoptosis, inflammation, and oxidative stress changes. For instance, arsenic exposure has been linked to altered expression of miR-21, a miRNA associated with cancer progression and immune regulation. Similarly, cadmium exposure has been found to upregulate miR-146a, which modulates inflammatory responses and has been implicated in autoimmune diseases (Kwakye et al., 2019; López-Granero et al., 2023; Mohankumar et al., 2024; Samuel et al., 2021; Vellingiri et al., 2022). Mercury exposure has been associated with changes in miR-124 expression, critical in neuronal development and function. Dysregulation of these miRNAs can contribute to the toxic effects of heavy metals by influencing key biological processes and exacerbating disease susceptibility. Noncoding RNAs (lncRNAs) also play a role in heavy metal toxicity by modulating chromatin structure, transcription, and posttranscriptional regulation. Recent studies have shown that exposure to lead and cadmium can alter lncRNA expression patterns, affecting genes involved in cell proliferation, differentiation, and apoptosis. These findings suggest that noncoding RNAs serve as critical mediators of metal-induced epigenetic changes and may represent potential biomarkers for assessing metal toxicity and disease risk (Table 2.4).

The interplay between genetic susceptibility and epigenetic modifications highlights the complexity of heavy metal toxicity and its impact on human health. Individuals with genetic predispositions to metal accumulation or NDs may be more vulnerable to epigenetic alterations induced by heavy metal exposure. For example, individuals carrying the APOE ε4 allele may experience more significant epigenetic dysregulation in response to lead or mercury exposure, further increasing their risk of AD. Similarly, genetic variants in DNA methyltransferases and histone-modifying enzymes may influence an individual's susceptibility to metal-induced epigenetic changes (Dong et al., 2024; Dooka et al., 2024; Famurewa et al., 2022; Lin et al., 2024; Raj et al., 2021). Understanding these interactions is essential for developing targeted strategies to mitigate heavy metal toxicity and prevent related diseases. Future research should focus on identifying genetic and epigenetic biomarkers to predict individual susceptibility to heavy metals, enabling personalized interventions and risk assessment strategies. Additionally, exploring the potential of epigenetic therapies, such as DNA methylation and histone-modifying enzyme inhibitors, may offer novel approaches for counteracting metal-induced toxic effects. By integrating genetic and epigenetic insights, researchers can advance the development of precision medicine approaches to address the growing health challenges of heavy metal exposure.

2.6 Neurodegeneration in specific diseases

2.6.1 Alzheimer's disease

AD is a continuously progressing neurodegenerative condition marked by the degeneration of cognitive function,

TABLE 2.4 Genetic and epigenetic factors in metal-induced neurodegeneration.

Heavy metal	Neurodegenerative disease	Brain regions affected	Levels of accumulation in the brain	Mechanisms of toxicity	Key cellular targets	Biomarkers for detection	Evidence from postmortem studies	Genetic and epigenetic susceptibility factors	Therapeutic approaches for metal detoxification
Lead (Pb)	AD	Hippocampus, cerebral cortex	Moderate to high	Oxidative stress, neuroinflammation, BBB disruption	Neurons, astrocytes	Elevated Pb in CSF/ blood, increased ROS markers	Pb detected in AD patient brains with neuronal loss	Polymorphisms in metal transporters (e.g., HFE gene)	Chelation therapy (EDTA), antioxidants (vitamins C and E), neuroprotective agents (curcumin, resveratrol)
Mercury (Hg)	PD	Substantia nigra, striatum	High	Mitochondrial dysfunction, dopaminergic neuron toxicity	Dopaminergic neurons, microglia	Elevated Hg levels in blood, increased lipid peroxidation	Postmortem brains show Hg deposits in the substantia nigra	Genetic variations in MT1 and MT2 (metallothionein genes)	Selenium-based chelators, NAC, glutathione enhancement
Arsenic (As)	ALS	Motor cortex, spinal cord	Moderate	Protein aggregation, DNA damage, inflammatory response	Motor neurons, oligodendrocytes	High As levels in CSF increased oxidative DNA damage markers	ALS patient brains exhibit arsenic accumulation	Epigenetic modifications in oxidative stress response genes	Dietary Se supplementation, chelation (DMPS, DMSA), antiinflammatory agents
Cadmium (Cd)	HD	Striatum, cortex	Low to moderate	Disrupts calcium signaling, induces apoptosis	Neurons, astrocytes	Cd detected in HD patient CSF, Increased inflammatory cytokines	Elevated Cd levels in HD-affected basal ganglia	Alterations in calcium homeostasis genes	Chelation (EDTA), zinc supplementation, neurotrophic factors
Aluminum (Al)	MS	White matter, cerebral cortex	Moderate	Myelin damage, neuroinflammation, protein misfolding	Oligodendrocytes, microglia	Elevated Al in CSF, increased autoantibodies	Al aggregates found in demyelinated MS plaques	Epigenetic modifications in myelin-related genes	Desferrioxamine therapy, curcumin-based detox strategies, antiinflammatory diet
Manganese (Mn)	Parkinsonism	Basal ganglia, globus pallidus	High	Mitochondrial dysfunction, dopamine depletion	Dopaminergic neurons, astrocytes	High Mn levels in blood, increased neuroinflammatory markers	Elevated Mn in PD-affected basal ganglia	SLC30A10 mutations linked to Mn toxicity	Iron chelators, dietary Mn restriction, neuroprotective peptides

AD, Alzheimer's disease; *ALS*, amyotrophic lateral sclerosis; *HD*, Huntington's disease; *MS*, multiple sclerosis; *PD*, Parkinson's disease.
Source: Bjørklund, G., Đorđević, A. B., Hamdan, H., Wallace, D. R., Peana, M. (2024). Metal-induced autoimmunity in neurological disorders: A review of current understanding and future directions. *Autoimmunity Reviews, 23*(3), 103509. https://doi.org/10.1016/j.autrev.2023.103509; Fu, X., Li, H., Song, L., Cen, M., Wu, J. (2023). Association of urinary heavy metals co-exposure and adult depression: Modification of physical activity. *NeuroToxicology, 95*, 117–126. https://doi.org/10.1016/j.neuro.2023.01.008; Samuel, M. S., Datta, S., Khandge, R. S., Selvarajan, E. (2021). A state of the art review on characterization of heavy metal binding metallothioneins proteins and their widespread applications. *Science of The Total Environment, 775*, 145829. https://doi.org/10.1016/j.scitotenv.2021.145829; Vellingiri, B., Suriyanarayanan, A., Selvaraj, P., Abraham, K. S., Pasha, M. Y., Winster, H., Gopalakrishnan, A. V., G, S., Reddy, J. K., Ayyadurai, N., Kumar, N., Giridharan, B., P, S., Rao, K. R. S. S., Nachimuthu, S. K., Narayanasamy, A., Mahalaxmi, I., Venkatesan, D. (2022). Role of heavy metals (copper (Cu), arsenic (As), cadmium (Cd), iron (Fe) and lithium (Li)) induced neurotoxicity. *Chemosphere, 301*, 134625. https://doi.org/10.1016/j.chemosphere.2022.134625.

memory loss, and the loss of neurons on a substantial scale with two significant pathological hallmarks, extracellular amyloid-beta (Aβ) plaques and intracellular NFT composed primarily of hyperphosphorylated tau protein. There is more and more substantial evidence that metallic ions, especially aluminum (Al), iron (Fe), cupra (Cu), and zinc (Zn), play a critical role in Aβ tau aggregate neuritic toxic damage. Aluminum, a known neurotoxin, has been linked to Alzheimer's disease due to its role in promoting tau hyperphosphorylation, disrupting protein balance (proteostasis), and increasing oxidative stress. These effects contribute to synaptic dysfunction and the loss of neurons commonly seen in Alzheimer-type conditions. Al disturbs binary homeostasis, leading to iron dyshomeostasis and the subsequent neuroinflammation, lipid peroxidation, and damage to mitochondria and neurons. It has been shown that Al can bind to tau protein, modifying its conformation and increasing its tendency to form NFTs, which is this disease's characteristic! (AD pathology) What's more, and worse still, aluminum aggravates amyloid pathology by speeding up Aβ aggregation and deposition, thus increasing neuroinflammation and oxidative damage. Yet further, Iron, another key metal in the pathology of AD, produces oxidative stress through the Fenton reaction. But the damaged part above the iron level of the AD brain, particularly in the hippocampus and cortex, corresponds to an increasing burden among amyloid plaques and tau pathology. Iron increases the oligomerisation and toxicity of Aβ through redox cycling, thereby aggravating synaptic dysfunction and neuronal loss (Dong et al., 2024; Gomes da Silva et al., 2024; Kumar et al., 2025; López-Granero et al., 2023; Sun et al., 2025). Copper and zinc, essential metal ions for normal brain function, also play a double role in AD. While these metals are necessary in your brain for synaptic communication or to operate enzymes, their excessive activity can lead to AD pathogenesis. Copper interacts with amyloid-β peptides in such a way as to alter the dynamics of their aggregation and promote toxic oligomer formation. It can also catalyze the production of hydrogen peroxide, leading to oxidative stress and neuronal damage Conversely, zinc can modulate both tau phosphorylation and tau aggregation by interacting with kinases and phosphatases that regulate tau homeostasisIt's a proven fact that too much zinc increases Aβ aggregation, which produces these dense plaques and even more significant synaptic damage All metals participate in a myriad of different ways in the synaptic dysfunction and neuron death of AD pathology Aβ and tau pathologies each disturb synaptic integrity by disrupting neurotransmitter release, receptor traffic, ependritic hilar stability Aβ oligomers interact with synaptic receptors such as NMDA and AMPA, being responsible for excitotoxicity and calcium dyshomeostasis, both of which erode synaptic plasticity. What is potentially more serious is that tau pathology contributes to synaptic breakdown or malfunction because it destroys the integrity of synaptic architecture. It becomes clear why people with type AD think poorly and have memory deficits: tau proteins detach from microtubules—that lead to its cytoskeletal instability, destroying dendritic spines vital for learning memory functions. Furthermore, synaptic junctions are lost key elements of cognitive ability (Althobaiti, 2025; Dong et al., 2024; Gomes da Silva et al., 2024; Li, Ren, et al., 2021). Moreover, aggregates composed in this way travel the length of axons from one neuron to another, thus contributing to dementia throughout the brain as well as hastening already extensive damage done to neurons. As a result of amyloid and tau pathology, together with synaptic dysfunction and oxidative stress, the pathology is thus amplified to the point where neurons are being destroyed faster than they can be replaced. Consequently, a vicious cycle is created. Oxidative stress is caused by the release of proinflammatory cytokines (activation of microglia is another example) or reactive nitrogen species. This, in turn, amplifies the excitation of neurons. The interplay of metal imbalance, amyloid and tau toxicity, mitochondrial dysfunction, and neuroinflammation ultimately leads to widespread neuronal death and characteristic cognitive decline in AD. Understanding the role of metals in AD pathogenesis provides potential therapeutic targets, such as metal chelation therapy and an outcome that preserves synaptic function and medicines that might slow or halt the progression of the disease (Gomes da Silva et al., 2024; Kumar et al., 2025; Kwakye et al., 2019; López-Granero et al., 2023; Vellingiri et al., 2022).

2.6.2 Parkinson's disease

PD is a progressive neurodegenerative disorder characterized by the loss of dopaminergic neurons in the substantia nigra, leading to motor and nonmotor symptoms. One of the significant environmental factors implicated in PD pathogenesis is metal toxicity, particularly manganese (Mn) and iron (Fe) overload, which significantly contribute to oxidative stress, mitochondrial dysfunction, and neuroinflammation. Manganese exposure is known to induce a Parkinsonian-like syndrome called manganism, which shares similarities with PD but has distinct pathological features. Mn accumulation in the basal ganglia disrupts dopamine homeostasis, impairs synaptic transmission, and increases oxidative stress, leading to neuronal damage. Mn interacts with various cellular components, including mitochondria and glial cells, exacerbating neurodegeneration. Excessive Mn exposure leads to increased ROS production, depletion of antioxidants like glutathione, and alterations in calcium homeostasis, all contributing to neuronal death. Iron overload, on the other hand, is a well-recognized feature of PD, as increased Fe levels are consistently observed in the substantia nigra of PD patients. Fe catalyzes the Fenton reaction, leading to excessive ROS generation and lipid peroxidation, which damages mitochondrial membranes and disrupts the ETC function. The interplay between Mn and Fe in PD pathology is complex, as both metals can synergistically enhance oxidative stress, leading to dopaminergic neuronal loss (Bjørklund et al., 2024; Dong et al., 2024; Kwakye et al., 2019; Raj et al., 2021; Sun et al., 2025). The mechanisms underlying metal-induced mitochondrial dysfunction in PD involve impaired bioenergetics, decreased ATP production, and mtDNA damage. Manganese disrupts mitochondrial dynamics by impairing mitochondrial fusion and fission, leading to bioenergetic failure and neuronal death. Mn exposure also inhibits mitochondrial complex I and III activity, exacerbating ROS accumulation and reducing ATP synthesis. Similarly, iron-induced mitochondrial dysfunction involves ferroptosis, an iron-dependent form of programmed cell death characterized by lipid peroxidation and glutathione depletion. Iron overload in PD enhances mitochondrial oxidative stress by increasing iron–sulfur cluster instability, leading to electron leakage and ETC dysfunction. mPTP opening, another critical event in metal-induced toxicity, results in cytochrome *c* release and apoptosis of dopaminergic neurons. In addition to mitochondrial dysfunction, Mn and Fe contribute to neuroinflammation, a hallmark of PD pathology. Mn exposure triggers microglial activation, leading to the release of proinflammatory cytokines such as TNF-α, IL-6, and IL-1β, which amplify neurotoxicity. Mn-induced neuroinflammation involves the activation of nuclear factor-kappa B (NF-κB) and mitogen-activated protein kinase (MAPK) signaling pathways, exacerbating dopaminergic neuron damage. Iron also plays a critical role in neuroinflammation by promoting microglial activation and ferroptosis-related neuronal injury. Excess Fe enhances lipid peroxidation-mediated inflammation, further accelerating PD progression. The combined effects of Mn and Fe in PD highlight the importance of metal homeostasis in neuronal survival and disease progression. Strategies aimed at chelating excess Mn and Fe, restoring mitochondrial function, and modulating neuroinflammation hold therapeutic potential for mitigating metal-induced neurodegeneration in PD. Understanding the intricate relationship between metal toxicity, mitochondrial dysfunction, and neuroinflammation is crucial for developing novel interventions to slow or prevent PD progression (Famurewa et al., 2022; Kumar et al., 2025; Lin et al., 2024; Pota et al., 2024; Sun et al., 2025; Wang & Matsushita, 2021).

2.6.3 Amyotrophic lateral sclerosis

ALS is a progressive neurodegenerative disorder characterized by the selective loss of motor neurons in the brain and spinal cord, leading to muscle weakness, atrophy, and eventual paralysis. Emerging research has pointed to the potential role of heavy metals in the pathogenesis of ALS, mainly through their interaction with the glutamate system and their accumulation in motor neurons, which triggers neuroinflammatory responses. The glutamate system plays a crucial role in neurotransmission, but excessive glutamate can lead to excitotoxicity, a well-documented mechanism implicated in ALS. Heavy metals such as lead, mercury, arsenic, and cadmium have been shown to disrupt glutamate homeostasis by interfering with glutamate transporters and receptors, leading to increased extracellular glutamate levels and subsequent excitotoxic damage. Aluminum, for example, has been found to impair astrocytic glutamate uptake by inhibiting excitatory amino acid transporters (EAATs), particularly EAAT2, which is primarily responsible for removing excess glutamate from synaptic clefts. Similarly, lead exposure has been associated with enhanced glutamate receptor activity, particularly NMDA receptors, leading to calcium influx and neuronal toxicity. This dysregulation exacerbates oxidative stress and mitochondrial dysfunction, two hallmarks of ALS pathology. Furthermore, the accumulation of heavy metals in motor neurons contributes to their degeneration through direct cytotoxic effects and the initiation of neuroinflammatory responses (Bjørklund et al., 2024; Dong et al., 2024; Fu et al., 2023; Li, Ren, et al., 2021; Raj et al., 2021). Motor neurons, due to their high metabolic demand and limited capacity for metal detoxification, are particularly vulnerable to heavy metal accumulation. Studies have found elevated levels of metals such as iron, copper, and zinc in the spinal cord tissues of ALS patients. While essential trace metals like copper and zinc play vital roles in enzymatic functions, their dyshomeostasis can contribute to ALS progression. Mutations in the copper–zinc superoxide dismutase 1 (SOD1) gene, one of the major genetic contributors to ALS, lead to toxic gain-of-function properties that promote oxidative damage and protein aggregation. Additionally, excessive iron accumulation has been observed in the motor cortex and spinal cord of ALS patients, promoting the Fenton reaction, which generates ROS and exacerbates neurodegeneration. The presence of heavy metals in motor neurons also triggers neuroinflammatory responses by activating microglia and astrocytes, which release proinflammatory cytokines such as TNF-α, IL-1β, and IL-6. These cytokines amplify oxidative stress and neuronal injury, creating a vicious cycle of inflammation and neurodegeneration. Mercury, for instance, has been shown to activate microglia and increase nitric oxide production, leading to further neuronal damage. The interaction between heavy metals and neuroinflammation in ALS suggests a potential link between environmental exposures and disease onset, supporting the intriguing hypothesis that ALS results from a complex interplay of genetic and environmental factors. Understanding the mechanistic underpinnings of heavy metal toxicity in ALS could pave the way for novel neuroprotective interventions that target metal homeostasis and neuroinflammatory pathways (Bjørklund et al., 2024; Famurewa et al., 2022; Li, Ren, et al., 2021; Lin et al., 2024; López-Granero et al., 2023; Samuel et al., 2021).

2.6.4 Other disorders

Early-onset NDs and the impact of environmental pollutants on neurodevelopmental disorders such as autism represent pressing concerns in modern medical and environmental research, as emerging evidence suggests strong links between environmental toxins and neurological health. Early-onset NDs, including conditions like juvenile Parkinson's, Huntington's disease, and certain forms of Alzheimer's, manifest at an unusually young age due to a combination of genetic predispositions, environmental exposures, and metabolic dysfunctions that accelerate neuronal damage. While genetic mutations such as those in the HTT gene for Huntington's disease or the PARK2 gene for Parkinson's contribute significantly to early-onset neurodegeneration, increasing research indicates that environmental toxins, heavy metals, and pesticides exacerbate these conditions by inducing oxidative stress, mitochondrial dysfunction, and chronic neuroinflammation, leading to accelerated neuronal death and cognitive decline. Similarly, the impact of environmental pollutants on neurodevelopmental disorders like autism spectrum disorder (ASD) has garnered significant attention, as prenatal and early childhood exposure to heavy metals like lead and mercury, air pollutants, and EDCs such as BPA have been implicated in altering brain development and synaptic connectivity. Epidemiological studies have demonstrated that children exposed to high levels of air pollution, excellent particulate matter (PM2.5), polycyclic aromatic hydrocarbons, and VOCs have an increased risk of developing autism due to the disruption of neurodevelopmental pathways, impaired neuronal migration, and increased inflammation in the CNS. Additionally, exposure to organophosphate pesticides such as chlorpyrifos during pregnancy has been linked to the altered fetal brain structure, reduced cognitive function, and increased risk of ASD by interfering with acetylcholine signaling, a crucial neurotransmitter involved in neural development and synaptic plasticity. Another concerning environmental factor is exposure to heavy metals such as lead, mercury, and arsenic, which have been shown to impair neuronal differentiation, induce neuroinflammation, and contribute to synaptic dysfunction, all hallmarks of neurodevelopmental disorders. Mercury, in particular, disrupts glutamate homeostasis and increases oxidative stress, leading to excitotoxicity that affects cognitive function, language development, and social behaviors associated with ASD. Similarly, arsenic exposure has been linked to epigenetic modifications that alter gene expression related to brain development, increasing the risk of intellectual disabilities and behavioral impairments. Beyond individual toxicants, the cumulative effect of multiple environmental stressors exacerbates neurodevelopmental risks, as children with genetic susceptibility to autism may experience heightened vulnerability when exposed to environmental pollutants, a phenomenon known as the gene–environment interaction model. Furthermore, maternal exposure to EDCs such as phthalates and polychlorinated biphenyls (PCBs) has been associated with altered levels of neurotrophic factors essential for normal brain growth, leading to an increased prevalence of neurodevelopmental disorders (Adegbola & Adetutu, 2024; Althobaiti, 2025; Fu et al., 2023; Gomes da Silva et al., 2024; Raj et al., 2021; Wang & Matsushita, 2021). Given the rising incidence of early-onset NDs and autism, understanding the interplay between genetics and environmental factors is crucial for developing preventive strategies, such as reducing prenatal exposure to neurotoxic chemicals, advocating for stricter environmental regulations, and promoting early interventions that mitigate neurodevelopmental damage. Public health initiatives aimed at reducing air pollution, monitoring industrial waste disposal, and educating populations on the risks of household toxins could play a critical role in protecting vulnerable individuals from developing neurological conditions. Advances in biomonitoring, epigenetic research, and neurotoxicology pave the way for a better understanding of how environmental pollutants impact brain health, emphasizing the urgent need for global policies prioritizing neuroprotection and long-term cognitive well-being (Fu et al., 2023; Gomes da Silva et al., 2024; Lin et al., 2024; Moyano et al., 2022; Samuel et al., 2021).

2.7 Therapeutic approaches to mitigate metal-induced neurodegeneration

Chelation therapy involves using chelating agents to remove heavy metals from the body, particularly in cases of metal-induced neurotoxicity. The most commonly used chelators include ethylenediaminetetraacetic acid (EDTA), dimercaptosuccinic acid (DMSA), and British Anti-Lewisite (BAL). These agents bind to toxic metal ions, facilitating their excretion through urine or feces, thereby reducing their bioavailability and potential toxicity to neuronal cells. EDTA has been widely used in cases of lead poisoning, as it forms stable complexes with lead ions, significantly reducing their concentration in blood and tissues. However, EDTA is not selective and can also chelate essential metal ions such as calcium and zinc, potentially leading to deficiencies and adverse side effects. On the other hand, DMSA has a higher specificity for lead, mercury, and arsenic and is often preferred due to its oral bioavailability and relatively safer profile. BAL, initially developed as an antidote to arsenic poisoning, is less commonly used today due to its potential for severe side effects (Bjørklund et al., 2024; Dong et al., 2024; Dooka et al., 2024; López-Granero et al., 2023).

Despite the efficacy of chelators in reducing heavy metal loads, challenges remain in their application for neurotoxicity treatment. One major limitation is the inability of many chelating agents to cross the BBB efficiently, limiting their effectiveness in removing metals already accumulated in the CNS. Additionally, chelation therapy can lead to metal redistribution, where mobilized toxic metals are temporarily transported through the bloodstream before complete excretion, potentially exacerbating toxicity in sensitive tissues. Therefore adjunct therapies, such as antioxidants and antiinflammatory agents, are often considered to mitigate the potential oxidative damage caused by mobilized heavy metals.

Oxidative stress and neuroinflammation are major contributors to metal-induced neurotoxicity. Heavy metals, such as lead, mercury, and cadmium, can generate ROS that lead to oxidative damage of neuronal membranes, proteins, and DNA. This oxidative stress triggers inflammatory responses, further exacerbating neuronal injury and dysfunction. Consequently, antioxidants and antiinflammatory agents have been explored as therapeutic strategies to counteract these damaging effects. Several natural compounds have shown promise in mitigating oxidative stress and neuroinflammation. Curcumin, a polyphenol derived from turmeric, has demonstrated potent antioxidant and antiinflammatory properties. It scavenges free radicals, inhibits proinflammatory cytokines, and modulates signaling pathways involved in neuroprotection. However, its clinical application is limited by poor bioavailability and rapid metabolism. Strategies such as nanoparticle-based formulations and combination therapy with bioenhancers like piperine are being explored to overcome these limitations (Kumar et al., 2025; Kwakye et al., 2019; Lin et al., 2024; Samuel et al., 2021; Yu et al., 2024).

Resveratrol, another natural compound found in grapes and berries, has been shown to activate sirtuins, which play a critical role in cellular defense mechanisms against oxidative stress. Additionally, resveratrol modulates inflammatory pathways by inhibiting nuclear factor-kappa B (NF-\u03baB) signaling, a key regulator of inflammation. Its neuroprotective effects have been demonstrated in various preclinical models of neurodegeneration, making it a promising candidate for therapeutic intervention. Synthetic antioxidants, such as *N*-acetylcysteine (NAC), have also been investigated for their potential to replenish intracellular glutathione levels and reduce oxidative damage. NAC has shown neuroprotective effects in conditions such as Parkinson's and Alzheimer's by modulating mitochondrial function and reducing inflammation. However, the clinical translation of antioxidant therapies remains challenging due to variability in individual responses and the complexity of oxidative stress pathways in NDs (Althobaiti, 2025; Famurewa et al., 2022; Kumar et al., 2025; Mohankumar et al., 2024).

Mitochondria play a crucial role in neuronal energy metabolism and cell survival. Heavy metal toxicity disrupts mitochondrial function by impairing ETC activity, reducing ATP production, and increasing oxidative stress. As a result, mitochondrial dysfunction is a key feature of NDs associated with metal toxicity. Therapeutic strategies aimed at protecting mitochondria are gaining attention as potential interventions for neurodegeneration. Coenzyme Q10 (CoQ10) is a well-known mitochondrial protectant that is an antioxidant and an electron carrier in the mitochondrial respiratory chain. CoQ10 supplementation has enhanced mitochondrial function, reduced oxidative damage, and improved neuronal survival in various neurodegenerative models. However, its bioavailability is limited, prompting the development of nanoformulations and liposomal delivery systems to enhance its therapeutic potential.

Peroxisome proliferator-activated receptor (PPAR) agonists, such as pioglitazone and rosiglitazone, have also been investigated for their neuroprotective effects. These agents modulate mitochondrial biogenesis, improve glucose metabolism, and reduce neuroinflammation. Studies have shown that PPAR agonists can protect against neurodegenerative processes in conditions like AD and PD. However, their long-term safety and efficacy need further clinical validation. Emerging mitochondrial-targeted antioxidants, such as MitoQ and SS-31, are designed to selectively accumulate within mitochondria, providing enhanced protection against oxidative stress. These compounds have shown promising results in preclinical studies and are being explored for their potential in treating neurodegenerative disorders associated with metal toxicity. Advancements in gene therapy and nanomedicine offer promising avenues for the targeted delivery of therapeutic agents in neurodegeneration treatment. Gene therapy involves using viral or nonviral vectors to introduce genetic material that can modulate disease-related pathways. At the same time, nanomedicine utilizes nanotechnology-based approaches for precision drug delivery (Anyanwu et al., 2020; Esan et al., 2024; Paithankar et al., 2021; Perera et al., 2023; Rehman et al., 2021; Stachowicz, 2023).

Gene therapy can potentially address metal-induced neurotoxicity by targeting genes involved in metal homeostasis, oxidative stress response, and neuronal repair. For instance, overexpression of MTs, metal-binding proteins, has been explored to sequester toxic metals and reduce their neurotoxic effects. Gene-editing technologies such as

CRISPR-Cas9 hold promise for correcting genetic defects associated with impaired metal metabolism and neurodegeneration. Nanomedicine has revolutionized drug delivery by enabling the transport of therapeutic agents across the BBB, a significant challenge in neurodegenerative treatment. Nanoparticles, liposomes, and dendrimers are designed to encapsulate antioxidants, chelators, and anti-inflammatory agents, enhancing their bioavailability and targeted delivery to affected brain regions. For example, nanoparticle-based formulations of curcumin and resveratrol have improved neuroprotective effects by enhancing their stability and penetration into neuronal tissues.

Moreover, functionalised nanoparticles can be engineered to respond to specific pathological conditions, such as oxidative stress or inflammation, enabling controlled release of therapeutic agents. This approach enhances treatment efficacy while minimizing systemic side effects. Ongoing research in nanomedicine aims to develop multifunctional platforms that combine imaging and therapy, allowing real-time monitoring of disease progression and treatment response. While gene therapy and nanomedicine hold great potential, challenges such as immune responses, off-target effects, and long-term safety must be addressed before widespread clinical application. Regulatory frameworks and ethical considerations also play a crucial role in developing and deploying these advanced therapeutic strategies (Arun et al., 2021; Bhattacharya et al., 2024; Haidar et al., 2023; Karri et al., 2020; Kim et al., 2025; Nguyen & Kim, 2022; Rehman et al., 2021).

The treatment of metal-induced neurotoxicity and NDs requires a multifaceted approach that integrates chelation therapy, antioxidants, mitochondrial protectants, gene therapy, and nanomedicine. Chelation therapy remains a cornerstone for metal detoxification, but its limitations necessitate complementary strategies such as antioxidants and mitochondrial protectants to mitigate oxidative stress and cellular damage. Emerging technologies in gene therapy and nanomedicine offer promising solutions for targeted intervention, enhancing treatment efficacy, and overcoming challenges related to drug delivery. Continued research and clinical trials are essential to translate these innovative approaches into effective therapies for neurodegeneration, ultimately improving patient outcomes and quality of life.

2.8 Environmental and public health implications

Heavy metal contamination has emerged as a significant environmental and public health concern, mainly due to its association with NDs and other chronic health conditions. Epidemiological studies have consistently linked exposure to heavy metals such as lead, mercury, arsenic, and cadmium to an increased risk of neurodegenerative disorders, including AD, PD, and ALS. These metals disrupt neural homeostasis through mechanisms involving oxidative stress, mitochondrial dysfunction, and protein aggregation, ultimately leading to neuronal damage and cognitive decline. Longitudinal studies have shown that populations residing near industrial areas, mining sites, or regions with high levels of environmental pollution are at a disproportionately higher risk of developing such conditions. Geographic and demographic factors play a crucial role in determining the severity of heavy metal toxicity, as regions with high levels of industrial emissions or poor regulatory enforcement often exhibit higher incidences of neurological disorders. Vulnerable populations, such as children, older people, and individuals with preexisting conditions, are particularly susceptible to the neurotoxic effects of heavy metals due to their prolonged exposure and reduced detoxification capacity. Additionally, socioeconomic disparities exacerbate the impact of metal toxicity, as lower income communities often have limited access to clean water, proper nutrition, and healthcare services, further increasing their risk of neurodegeneration (Anyanwu et al., 2020; Khan & Parvez, 2015; Kim et al., 2025; Li, Xia, et al., 2021; Oyeyemi et al., 2024).

In response to the growing evidence of heavy metal-induced neurotoxicity, numerous regulatory measures and public health interventions have been implemented to mitigate exposure and safeguard human health. Governments and international organizations, recognizing the global nature of the issue, have established stringent policies to reduce environmental heavy metal contamination through industrial regulations, emissions control, and waste management practices. The European Union's REACH regulation and the United States Environmental Protection Agency guidelines have set strict limits on permissible heavy metal concentrations in air, water, and soil to minimize human exposure. International collaborations, such as the Minamata Convention on Mercury, have been instrumental in curbing mercury emissions from industrial sources and promoting safer alternatives. Despite these efforts, enforcing such regulations remains challenging in many developing countries, where inadequate monitoring systems and limited financial resources hinder effective implementation. To address these gaps, targeted prevention strategies have been developed for high-risk populations, including dietary interventions, regular health screenings, and community-based education programs. Nutritional strategies, such as increased intake of antioxidants and essential minerals like zinc and selenium, have been shown to mitigate the effects of heavy metal toxicity by enhancing detoxification pathways and reducing oxidative damage in neural tissues.

Public health initiatives and educational programs are pivotal in raising awareness about the dangers of heavy metal exposure and promoting safer environmental practices. Community engagement efforts, including workshops, public campaigns, and school-based interventions, effectively disseminate information on sources of heavy metal exposure and preventive measures and empower individuals to take action. Governments and nongovernmental organizations have collaborated to implement programs that provide access to clean drinking water, promote sustainable agricultural practices, and advocate for stricter environmental regulations. The role of healthcare professionals, who are well-equipped to identify early signs of heavy metal toxicity and recommend appropriate interventions, is equally crucial. Their involvement should instill confidence in the audience about the healthcare system's ability to address heavy metal toxicity. Advances in biomonitoring techniques have also facilitated the early detection of heavy metal accumulation in biological samples, allowing for timely intervention and treatment. Furthermore, emerging technologies such as phytoremediation and bioremediation are being explored as sustainable solutions for reducing environmental heavy metal contamination. Integrating scientific research, policy development, and public engagement can enable a comprehensive approach to minimize heavy metal exposure and its associated health risks. Continuing investment in research and innovation, robust regulatory frameworks, and proactive community involvement will mitigate heavy metal contamination's environmental and public health implications and safeguard future generations from the devastating effects of NDs.

2.9 Conclusion

Heavy metal–induced neurodegeneration is a complex and multifaceted process involving several key mechanisms that contribute to neuronal damage and progressive cognitive decline. The primary pathways implicated in this process include oxidative stress, mitochondrial dysfunction, protein misfolding, neuroinflammation, and disruptions in neurotransmitter homeostasis. Heavy metals such as lead, mercury, cadmium, and arsenic generate ROS, leading to lipid peroxidation, DNA damage, and neuronal apoptosis. Mitochondrial dysfunction is another crucial mechanism, as heavy metals impair the ETC, reduce ATP production, and trigger the release of proapoptotic factors, exacerbating neuronal death. Additionally, heavy metals interfere with protein homeostasis by promoting protein misfolding and aggregation, which are characteristic of NDs like AD and PD. Neuroinflammation, a key driver of neurodegeneration, is also heightened by heavy metals, as they activate microglia and astrocytes, leading to chronic inflammation and increased production of proinflammatory cytokines. Furthermore, heavy metals disrupt neurotransmitter systems by altering synaptic transmission, impairing dopamine, glutamate, and acetylcholine signaling, contributing to cognitive dysfunction, motor impairment, and emotional disturbances. Given the increasing recognition of these pathways, future research must focus on further elucidating the molecular and cellular interactions between heavy metals and neurodegeneration. Understanding these relationships will provide critical insights into how environmental and occupational exposure to heavy metals contributes to NDs. Research should prioritize identifying biomarkers for early detection and developing neuroprotective strategies that mitigate heavy metal toxicity. There is also a pressing need for interdisciplinary studies integrating toxicology, neurology, and environmental sciences. The importance of these multidisciplinary studies cannot be overstated, as they are crucial in comprehensively assessing the impact of chronic heavy metal exposure on brain health. Moreover, emerging therapeutic approaches such as chelation therapy, antioxidant supplementation, and gene editing hold promise in mitigating the effects of heavy metals on the nervous system. Advanced nanotechnology-based drug delivery systems may also enhance the efficacy of neuroprotective agents while minimizing systemic toxicity. From an environmental management perspective, stricter regulations and policies should be implemented to limit heavy metal contamination in air, water, soil, and food sources. Public health initiatives should focus on raising awareness of the risks associated with heavy metal exposure and promoting preventive measures, particularly in vulnerable populations. Furthermore, sustainable remediation strategies such as phytoremediation, biochar application, and microbial bioremediation should be explored to reduce heavy metal contamination in affected ecosystems. The implications of heavy metal–induced neurodegeneration for neurotoxicology and public health are profound, necessitating urgent action to address this growing global burden. Heavy metal exposure is particularly concerning in industrialized and developing regions with prevalent environmental pollution and occupational hazards. Governments and healthcare organizations must work together to implement policies that protect individuals from excessive heavy metal exposure while supporting research initiatives that explore novel therapeutic interventions. Additionally, global efforts should promote eco-friendly industrial practices, enhance waste management systems, and develop sustainable agricultural practices that prevent heavy metal accumulation in food crops. By integrating scientific advancements with proactive policy-making, society can effectively reduce the incidence of heavy metal-induced neurodegeneration and improve overall brain health. In conclusion, the intricate interplay between heavy metals and NDs underscores the importance of continued research, therapeutic innovation, and environmental sustainability. The need for interdisciplinary studies is clear, and it is through collaboration and integration of different fields that we can make significant strides in understanding and addressing heavy metal–induced neurodegeneration.

References

Adegbola, P. I., & Adetutu, A. (2024). Genetic and epigenetic modulations in toxicity: The two-sided roles of heavy metals and polycyclic aromatic hydrocarbons from the environment. *Toxicology Reports, 12*, 502–519. Retrieved from https://doi.org/10.1016/j.toxrep.2024.04.010.

Althobaiti, N. A. (2025). Heavy metals exposure and Alzheimer's disease: Underlying mechanisms and advancing therapeutic approaches. *Behavioural Brain Research, 476*, 115212. Retrieved from https://doi.org/10.1016/j.bbr.2024.115212.

Anyanwu, B. O., Orish, C. N., Ezejiofor, A. N., Nwaogazie, I. L., & Orisakwe, O. E. (2020). Neuroprotective effect of Costus afer on low dose heavy metal mixture (lead, cadmium and mercury) induced neurotoxicity via antioxidant, anti-inflammatory activities. *Toxicology Reports, 7*, 1032–1038. Retrieved from https://doi.org/10.1016/j.toxrep.2020.08.008.

Arun, K. B., Madhavan, A., Sindhu, R., Emmanual, S., Binod, P., Pugazhendhi, A., ... Pandey, A. (2021). Probiotics and gut microbiome – Prospects and challenges in remediating heavy metal toxicity. *Journal of Hazardous Materials, 420*, 126676. Retrieved from https://doi.org/10.1016/j.jhazmat.2021.126676.

Aschner, M., Skalny, A. V., Paoliello, M. M. B., Tinkova, M. N., Martins, A. C., Santamaria, A., ... Tinkov, A. A. (2024). Retinal toxicity of heavy metals and its involvement in retinal pathology. *Food and Chemical Toxicology, 188*, 114685. Retrieved from https://doi.org/10.1016/j.fct.2024.114685.

Bhattacharya, S., Bose, M., & Munshi, C. B. (2024). Paradigms in heavy metal induced autophagy: A toxicological approach. In *Reference Module in Biomedical Sciences*. Elsevier. Retrieved from https://doi.org/10.1016/B978-0-323-95488-4.00136-4.

Bjørklund, G., Đorđević, A. B., Hamdan, H., Wallace, D. R., & Peana, M. (2024). Metal-induced autoimmunity in neurological disorders: A review of current understanding and future directions. *Autoimmunity Reviews, 23*(3), 103509. Retrieved from https://doi.org/10.1016/j.autrev.2023.103509.

Cai, H., Bao, Y., Cheng, H., Ge, X., Zhang, M., Feng, X., ... Yang, X. (2023). Zinc homeostasis may reverse the synergistic neurotoxicity of heavy metal mixtures in Caenorhabditis elegans. *Science of the Total Environment, 868*, 161699. Retrieved from https://doi.org/10.1016/j.scitotenv.2023.161699.

Deng, S., Ren, B., Hou, B., Deng, R., & Cheng, S. (2023). Antimony-complexed heavy metal wastewater in antimony mining areas: Source, risk and treatment. *Environmental Technology & Innovation, 32*, 103355. Retrieved from https://doi.org/10.1016/j.eti.2023.103355.

Dong, Q., Fu, H., & Jiang, H. (2024). The role of exosome-shuttled miRNAs in heavy metal-induced peripheral tissues and neuroinflammation in Alzheimer's disease. *Biomedicine & Pharmacotherapy, 176*, 116880. Retrieved from https://doi.org/10.1016/j.biopha.2024.116880.

Dooka, D. B., Anyachor, C. P., Orish, C. N., Ezejiofor, A. N., Obasi, C. N., Frazzoli, C., & Orisakwe, O. E. (2024). Understanding the impact of plant-derived food grade substances PDFGS on metal-induced neurodegeneration: A systematic review of preclinical evidence. *Journal of Trace Elements and Minerals, 7*, 100110. Retrieved from https://doi.org/10.1016/j.jtemin.2023.100110.

Esan, O. O., Igado, O. O., Femi-Akinlosotu, O. M., Oyagbemi, A. A., Omobowale, T. O., Oladele, O. A., & Nwulia, E. (2024). Alchornea laxiflora (Benth.) Pax & K. Hoffman extract protects against lead-induced neurodegeneration in cockerel chickens. *IBRO Neuroscience Reports, 17*, 65–72. Retrieved from https://doi.org/10.1016/j.ibneur.2024.06.004.

Famurewa, A. C., Renu, K., Eladl, M. A., Chakraborty, R., Myakala, H., El-Sherbiny, M., ... Valsala Gopalakrishnan, A. (2022). Hesperidin and hesperetin against heavy metal toxicity: Insight on the molecular mechanism of mitigation. *Biomedicine & Pharmacotherapy, 149*, 112914. Retrieved from https://doi.org/10.1016/j.biopha.2022.112914.

Fu, X., Li, H., Song, L., Cen, M., & Wu, J. (2023). Association of urinary heavy metals co-exposure and adult depression: Modification of physical activity. *Neurotoxicology, 95*, 117–126. Retrieved from https://doi.org/10.1016/j.neuro.2023.01.008.

Gomes da Silva, J., Chagas, C. A., Giselly dos Santos Souza, T., Chagas de Araújo, M., Alexandre de Araújo, L. C., Melo Santos, A. M., ... Betânia Melo de Oliveira, M. (2024). RETRACTED: Using structural equation modeling to assess the genotoxic and mutagenic effects of heavy metal contamination in the freshwater ecosystems: A study involving Oreochromis niloticus in an urban river. *Science of the Total Environment, 913*, 169529. Retrieved from https://doi.org/10.1016/j.scitotenv.2023.169529.

Haidar, Z., Fatema, K., Shoily, S. S., & Sajib, A. A. (2023). Disease-associated metabolic pathways affected by heavy metals and metalloid. *Toxicology Reports, 10*, 554–570. Retrieved from https://doi.org/10.1016/j.toxrep.2023.04.010.

Hussien, H. M., Abd-Elmegied, A., Ghareeb, D. A., Hafez, H. S., Ahmed, H. E. A., & El-moneam, N. A. (2018). Neuroprotective effect of berberine against environmental heavy metals-induced neurotoxicity and Alzheimer's-like disease in rats. *Food and Chemical Toxicology, 111*, 432–444. Retrieved from https://doi.org/10.1016/j.fct.2017.11.025.

Joneidi, Z., Mortazavi, Y., Memari, F., Roointan, A., Chahardouli, B., & Rostami, S. (2019). The impact of genetic variation on metabolism of heavy metals: Genetic predisposition? *Biomedicine & Pharmacotherapy, 113*, 108642. Retrieved from https://doi.org/10.1016/j.biopha.2019.108642.

Karri, V., Ramos, D., Martinez, J. B., Odena, A., Oliveira, E., Coort, S. L., ... Kumar, V. (2018). Differential protein expression of hippocampal cells associated with heavy metals (Pb, As, and MeHg) neurotoxicity: Deepening into the molecular mechanism of neurodegenerative diseases. *Journal of Proteomics, 187*, 106–125. Retrieved from https://doi.org/10.1016/j.jprot.2018.06.020.

Karri, V., Schuhmacher, M., & Kumar, V. (2016). Heavy metals (Pb, Cd, As and MeHg) as risk factors for cognitive dysfunction: A general review of metal mixture mechanism in brain. *Environmental Toxicology and Pharmacology, 48*, 203–213. Retrieved from https://doi.org/10.1016/j.etap.2016.09.016.

Karri, V., Schuhmacher, M., & Kumar, V. (2020). A systems toxicology approach to compare the heavy metal mixtures (Pb, As, MeHg) impact in neurodegenerative diseases. *Food and Chemical Toxicology, 139*, 111257. Retrieved from https://doi.org/10.1016/j.fct.2020.111257.

Khan, M. H. A., & Parvez, S. (2015). Hesperidin ameliorates heavy metal induced toxicity mediated by oxidative stress in brain of Wistar rats. *Journal of Trace Elements in Medicine and Biology, 31*, 53–60. Retrieved from https://doi.org/10.1016/j.jtemb.2015.03.002.

Kim, B., Ha, Y., Hwang, J., & Kim, H.-J. (2025). Association between chronic ambient heavy metal exposure and mental health in Korean adult patients with asthma and the general population. *Chemosphere, 370*, 144002. Retrieved from https://doi.org/10.1016/j.chemosphere.2024.144002.

Kumar, A., Abhinav, S., Rachamalla, M., Ali, M., Navin, S., Kumar, D., & Ghosh, A. K. (2025). Chapter 29 – Treatment of heavy metal induced neurotoxicity via advanced drug delivery systems. In P. A. Chawla, R. Loebenberg, K. Dua, V. Parikh, & V. Chawla (Eds.). *Novel drug delivery systems in the management of disorders* (pp. 393–407). Academic Press. Retrieved from https://doi.org/10.1016/B978-0-443-13474-6.00025-1.

Kwakye, G. F., Jiménez, J. A., Thomas, M. G., Kingsley, B. A., McIlvin, M., Saito, M. A., & Korley, E. M. (2019). Heterozygous huntingtin promotes cadmium neurotoxicity and neurodegeneration in striatal cells via altered metal transport and protein kinase C delta dependent oxidative stress and apoptosis signaling mechanisms. *Neurotoxicology, 70*, 48–61. Retrieved from https://doi.org/10.1016/j.neuro.2018.10.012.

Lasisi-Sholola, A. S., Hammed, S. O., Ajike, R. A., Akhigbe, R. E., & Afolabi, O. A. (2024). Estrogen replacement therapy reverses spatial memory loss and pyramidal cell neurodegeneration in the prefrontal cortex of lead-exposed ovariectomized Wistar rats. *Current Research in Toxicology, 7*, 100200. Retrieved from https://doi.org/10.1016/j.crtox.2024.100200.

Li, B., Xia, M., Zorec, R., Parpura, V., & Verkhratsky, A. (2021). Astrocytes in heavy metal neurotoxicity and neurodegeneration. *Brain Research, 1752*, 147234. Retrieved from https://doi.org/10.1016/j.brainres.2020.147234.

Li, X., Ren, Z., Crabbe, M. J. C., Wang, L., & Ma, W. (2021). Genetic modifications of metallothionein enhance the tolerance and bioaccumulation of heavy metals in Escherichia coli. *Ecotoxicology and Environmental Safety, 222*, 112512. Retrieved from https://doi.org/10.1016/j.ecoenv.2021.112512.

Lin, J., Wang, L., Huang, M., Xu, G., & Yang, M. (2024). Metabolic changes induced by heavy metal copper exposure in human ovarian granulosa cells. *Ecotoxicology and Environmental Safety, 285*, 117078. Retrieved from https://doi.org/10.1016/j.ecoenv.2024.117078.

López-Granero, C., Aschner, M., & Sánchez-Santed, F. (2023). Chapter 22 – Environmental toxicants (OPs and heavy metals) in the diet: What are their repercussions on behavioral/neurological systems? In C. R. Martin, V. B. Patel, & V. R. Preedy (Eds.). *Diet and nutrition in neurological disorders* (pp. 411–427). Academic Press. Retrieved from https://doi.org/10.1016/B978-0-323-89834-8.00038-6.

Mohankumar, T., Venugopal, D., Palaniyappan, J., Beerappa, R., Duraisamy, E., & Velu, S. (2024). Environmental exposure to heavy metals in ambient air and its human health implications. In P. K. Shit, D. K. Datta, B. Bera, A. Islam, & P. P. Adhikary (Eds.). *Spatial modeling of environmental pollution and ecological risk: Advances in pollution research* (pp. 41–69). Woodhead Publishing. Retrieved from https://doi.org/10.1016/B978-0-323-95282-8.00028-6.

Moyano, P., Vicente-Zurdo, D., Blázquez-Barbadillo, C., Menéndez, J. C., González, J. F., Rosales-Conrado, N., & Pino, J.del (2022). Neuroprotective mechanisms of multitarget 7-aminophenanthridin-6(5H)-one derivatives against metal-induced amyloid proteins generation and aggregation. *Food and Chemical Toxicology, 167*, 113264. Retrieved from https://doi.org/10.1016/j.fct.2022.113264.

Nguyen, H. D., & Kim, M.-S. (2022). Exposure to a mixture of heavy metals induces cognitive impairment: Genes and microRNAs involved. *Toxicology, 471*, 153164. Retrieved from https://doi.org/10.1016/j.tox.2022.153164.

Oyeyemi, A. W., Owonikoko, W. M., Okoro, T. D., Adagbonyi, O., & Ajeigbe, K. O. (2024). Water contamination: A culprit of serum heavy metals concentration, oxidative stress and health risk among residents of a Nigerian crude oil-producing community. *Toxicology Reports, 12*, 375–388. Retrieved from https://doi.org/10.1016/j.toxrep.2024.03.006.

Paithankar, J. G., Saini, S., Dwivedi, S., Sharma, A., & Chowdhuri, D. K. (2021). Heavy metal associated health hazards: An interplay of oxidative stress and signal transduction. *Chemosphere, 262*, 128350. Retrieved from https://doi.org/10.1016/j.chemosphere.2020.128350.

Park, K., & Kwak, I.-S. (2023). Growth retardation and suppression of ubiquitin-dependent catabolic processes in the brackish water clam Corbicula japonica in response to salinity changes and bioaccumulation of toxic heavy metals. *Environmental Pollution, 337*, 122554. Retrieved from https://doi.org/10.1016/j.envpol.2023.122554.

Perera, I. C., Mauran, S., & Wickramasinghe, P. M. (2023). 19 - Bioremediation of heavy metals using yeast. In A Daverey, K. Dutta, S Joshi, & T Gea (Eds.). *Advances in yeast biotechnology for biofuels and sustainability: Value-added products and environmental remediation applications* (pp. 475–501). Elsevier. Retrieved from https://doi.org/10.1016/B978-0-323-95449-5.00004-7.

Popov Aleksandrov, A., Mirkov, I., Tucovic, D., Kulas, J., Zeljkovic, M., Popovic, D., ... Kataranovski, M. (2021). Immunomodulation by heavy metals as a contributing factor to inflammatory diseases and autoimmune reactions: Cadmium as an example. *Immunology Letters, 240*, 106–122. Retrieved from https://doi.org/10.1016/j.imlet.2021.10.003.

Pota, K., Nilewar, S., Mantsorov, C., Zumwalt, L., Nguyen, N., Bowers, C. J., ... Green, K. N. (2024). Impact of indole inclusion in the design of multi-tactical metal-binding tetra-aza macrocycles that target the molecular features of neurodegeneration. *ACS Chemical Neuroscience, 15*(20), 3776–3790. Retrieved from https://doi.org/10.1021/acschemneuro.4c00530.

Raj, K., Kaur, P., Gupta, G. D., & Singh, S. (2021). Metals associated neurodegeneration in Parkinson's disease: Insight to physiological, pathological mechanisms and management. *Neuroscience Letters, 753*, 135873. Retrieved from https://doi.org/10.1016/j.neulet.2021.135873.

Rehman, A. U., Nazir, S., Irshad, R., Tahir, K., ur Rehman, K., Islam, R. U., & Wahab, Z. (2021). Toxicity of heavy metals in plants and animals and their uptake by magnetic iron oxide nanoparticles. *Journal of Molecular Liquids, 321*, 114455. Retrieved from https://doi.org/10.1016/j.molliq.2020.114455.

Samuel, M. S., Datta, S., Khandge, R. S., & Selvarajan, E. (2021). A state of the art review on characterization of heavy metal binding metallothioneins proteins and their widespread applications. *Science of the Total Environment, 775*, 145829. Retrieved from https://doi.org/10.1016/j.scitotenv.2021.145829.

Stachowicz, K. (2023). Regulation of COX-2 expression by selected trace elements and heavy metals: Health implications, and changes in neuronal plasticity. A review. *Journal of Trace Elements in Medicine and Biology, 79*, 127226. Retrieved from https://doi.org/10.1016/j.jtemb.2023.127226.

Sun, H., Johnson, D. S., & Aluko, R. E. (2025). Nutrient and heavy metals composition of dried fish varieties from Bangladesh. *LWT, 215*, 117233. Retrieved from https://doi.org/10.1016/j.lwt.2024.117233.

Tang, J., Zheng, F., Liu, X., Li, Y., Guo, Z., Lin, X., ... Li, H. (2023). Cobalt induces neurodegeneration through FTO-triggered autophagy impairment by targeting TSC1 in an m6A-YTHDF2-dependent manner. *Journal of Hazardous Materials, 453*, 131354. Retrieved from https://doi.org/10.1016/j.jhazmat.2023.131354.

Tripathi, S., Kharkwal, G., Mishra, R., & Singh, G. (2024). Nuclear factor erythroid 2-related factor 2 (Nrf2) signaling in heavy metals-induced oxidative stress. *Heliyon, 10*(18), e37545. Retrieved from https://doi.org/10.1016/j.heliyon.2024.e37545.

Vellingiri, B., Suriyanarayanan, A., Selvaraj, P., Abraham, K. S., Pasha, M. Y., Winster, H., ... Venkatesan, D. (2022). Role of heavy metals (copper (Cu), arsenic (As), cadmium (Cd), iron (Fe) and lithium (Li)) induced neurotoxicity. *Chemosphere, 301*, 134625. Retrieved from https://doi.org/10.1016/j.chemosphere.2022.134625.

Wang, H., & Matsushita, M. T. (2021). Heavy metals and adult neurogenesis. *Current Opinion in Toxicology, 26*, 14–21. Retrieved from https://doi.org/10.1016/j.cotox.2021.03.006.

Wu, L., Cui, F., Zhang, S., Ding, X., Gao, W., Chen, L., ... Niu, P. (2023). Associations between multiple heavy metals exposure and neural damage biomarkers in welders: A cross-sectional study. *Science of the Total Environment, 869*, 161812. Retrieved from https://doi.org/10.1016/j.scitotenv.2023.161812.

Yu, G., Wu, L., Su, Q., Ji, X., Zhou, J., Wu, S., ... Li, H. (2024). Neurotoxic effects of heavy metal pollutants in the environment: Focusing on epigenetic mechanisms. *Environmental Pollution, 345*, 123563. Retrieved from https://doi.org/10.1016/j.envpol.2024.123563.

Chapter 3

Lead toxicity and its impact on cognitive function

K. Jaisuriyan, and Prasann Kumar
Department of Agronomy, School of Agriculture, Lovely Professional University, Phagwara, Punjab, India

3.1 Introduction

Lead is one of the toxic heavy metals present widely in the environment. It causes pollution by threatening the well-being of human society. Lead has a wide range of utilizations, from caulking materials to raw materials for mechanical spares, making it a pervasive material and a reason for industrial and economic advancement. Specifically, lead materials are used predominantly as an input in the storage of batteries, electronic devices, circuits, etc., resulting in the bloom of the electric and telecommunication industries. In contrast to its usefulness, their intervention with the ecosystem and the human has recently increased (Chandwani, Kayasth, Naik, & Amaresan, 2023). This creates a research thirst to study the consequences, which endorses the severity of lead exposure that inflames the human body by affecting the regular functionalities of vital organs such as heart, brain, liver, and kidneys. Lead exposure mainly creates an imbalance in the nervous system, which is the principal target, and impacts its toxicity in the human body (Santana et al., 2023). Lead toxicity exposure at the juvenile stage through the intervention with contaminated soil and environment causes chronic health impacts on nervous and metabolic functions throughout life (Hernández, Zhang, Filippelli, & Boehm, 2023). Few studies reported that lead exposure of less than 3.5 μg/dL in the blood causes impairment in cognitive functionalities, which is even lesser than the prescribed reference values by disease control and prevention authorities that warn the highly toxic nature of lead with no safe limit of exposure (Axelrad, Coffman, Kirrane, & Klemick, 2022; Mei, Liu, Zhao, He, & Gu, 2023). In addition, several studies affirm the worst impacts of lead toxicity on the human central nervous system. Reduction in intelligence quotient results in learning disabilities, which are the predominant effects of lead exposure in juvenile conditions. In addition, this kind of toxicant exposure inflames the nervous system and causes vital disorders such as Alzheimer's disease (AD), Parkinson's disease, hyper and hypoactivity ailments, and autism (Bandaru, Murumulla, & Challa, 2023; Kargar-Shouroki, Mehri, & Sepahi-Zoeram, 2023). Further toxic concentration disrupts regular routines and mechanisms by inducing oxidative stress, and disorganizing neural networks constitutes synaptic deterioration (Cai et al., 2023; Mei et al., 2023).

3.2 Intelligence quotient reduction by elevated lead concentration

Certain studies ensure the interdependence between the IQ reduction associated with lead toxicity in the juvenile stage (Mei et al., 2023; Pocock, Smith, & Baghurst, 1994). The length of exposure period plays a vital role in the increment or reduction of the IQ, which is supported by the metaanalysis performed among the groups with an exposure period of less than 4.5 years have more excellent quotient averages when compared with the groups with an exposure period higher than 4.5 years (Heidari et al., 2022). Increased blood lead concentration levels lead to numerous neurological dysfunctions around understanding spatial variations, recalling memory contexts, adaptbility to unfamiliar circumstances, and abstraction. They found an extended reaction time for these tasks and lower IQ levels among the lead-exposed children, which endorses that lead toxicity decreases the quotient values and negatively impacts regular cognitive functionalities (Halabicky et al., 2022) (Fig. 3.1).

In addition, maternal blood finds that a rise in lead concentration at 1 g/dL decreases the fetus's language development factor by 1.5 points, associated with negative consequences on neurodevelopment (Farías et al., 2022). Further, higher lead concentration promotes the causes of dyslexia, which is confirmed by the comparison between

Heavy Metal Toxicity and Neurodegeneration. DOI: https://doi.org/10.1016/B978-0-443-36575-1.00012-1

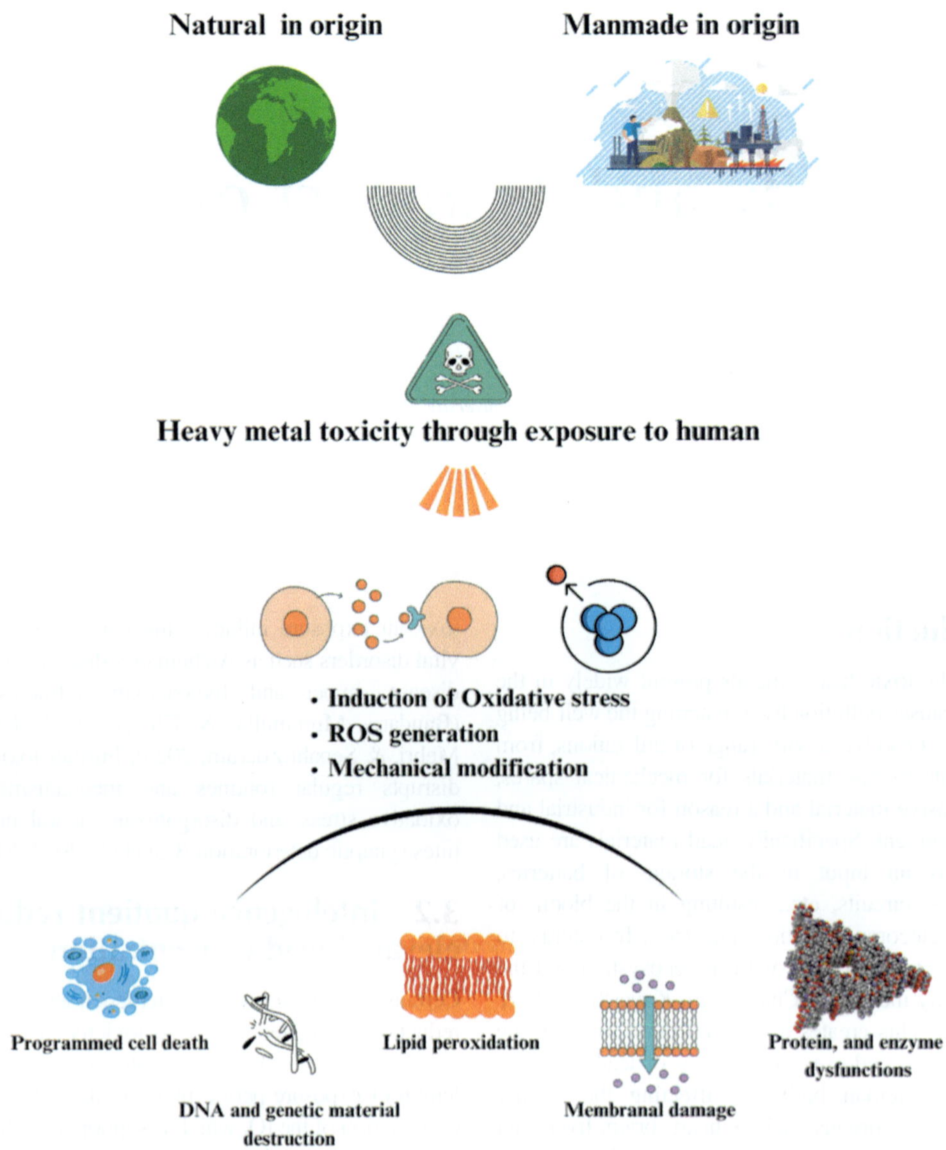

FIGURE 3.1 Sources of heavy metals and their toxicity-associated impacts on human health while exposure and its effects on the cellular functionalities.

dyslexia-infected and noninfected groups (Huang et al., 2022). Thus these conditions revealed a positive correlation between lead toxicity IQ impairments and cognitive response functionalities (Mei et al., 2023).

3.3 Neurodegenerative diseases under lead-exposed conditions

3.3.1 Alzheimer's disease

AD is one of the neurodegenerative diseases typified by the accumulation of β-amyloids and tau species characterized by hyperphosphorylation, found majorly in senior citizens (Wang, Guan et al., 2023; Wang, Hu, Liu, & Guo, 2023). Lead exposure induces the deposition of β-amyloids, supported by the study imposed on rats by external lead dosage, resulting in the layering of β-amyloids in the cerebral blood vessels (Zhou et al., 2022). In addition, lead exposure in cerebrospinal fluid production sites and microvessels of rats resulted in the generation of interleukin (IL)-1β and tumor necrosis factor-alpha, which are the inflammatory agents that play a crucial role in the neurodegenerative process, especially in Alzheimer-infected individuals (Zhang et al., 2022)—further evidenced by the generation of those inflammatory agents at the hippocampus region of gestating Wistar rats followed by juvenile rats in forebrain sites treated with lead acetate at the concentration of 15 mg/kg (Balali-Mood, Naseri,

Tahergorabi, Khazdair, & Sadeghi, 2021). Hence, by the development of factors contributing to Alzheimer's toxicity, it can be concluded that lead exposure and increased blood lead levels induce AD (Mei et al., 2023).

3.3.2 Parkinson's disease

Parkinson's disease is another principal neurodegenerative ailment categorized by motor retardation and idle tremors caused by several factors like inheritance, senescence, and prevailing external conditions (Vrijsen, Houdou, Cascalho, Eggermont, & Vangheluwe, 2023). Prolonged ingestion and exposure to metal contamination increase the vulnerability to Parkinson's disease (Gonzalez-Alcocer et al., 2023). An analytical study from various databases of gene-expressing variabilities like GSE145361, GSE72774, and GSE72776 correlates the relationship between Parkinson's disease and chronic lead exposure (Paul, Horvath, Del Rosario, Bronstein, & Ritz, 2021).

3.3.3 Autism

Generally, autism spectrum disorder is associated with aberrant and dysfunctional social behavior, a communication disorder. The cause of autism is less known, but studies support the view of lead exposure-induced autism. In vivo studies using BTBR mice species under lead-exposed conditions resulted in abnormal activities of mice, like ritualistic behavior of burying marbles, less interaction and gathering into groups, establishing the correlation between lead toxicity and symptoms denoting autism (Almutairi et al., 2022). *Drosophila* accumulated with lead in their brain exhibits symptoms that resemble autism along with oxidative stress. The symptoms are often associated with a reduction in climbing ability, unusual personal grooming, high detachment, and a fall in memory and sex expression, which indisputably elucidates the lead-induced oxidative stress, which further results in autism spectrum disorder (Shilpa, Anupama, Antony, & Gurushankara, 2021). In addition, the mean blood lead levels of disordered individuals were found to be higher than those of healthy individuals, which defines the risk and probability of autism under increased lead concentrations (Omotosho, Akinade, Lagunju, & Yakubu, 2021).

Various animal, human, and culture models confirmed the nexus between the lead levels and neurodegenerative disorders, including AD, Parkinson's diseases, autism, and attention-deficit hyperactivity disorders (Moore et al., 2022), which is often distinguished by induced oxidative stress, organ inflammation (Mao et al., 2020), neural apoptosis, irregularities, aberrations (Chen et al., 2022; Liu et al., 2020), myelination of cerebellar fibers (Nam, Seo, Nahm, & Chang, 2019), and generation of neurotoxicants like quinolinic acids (Mei et al., 2023) described in Fig. 3.2.

Another fact is that Lead is an epigenetic toxic metal that toxifies the frameworks of DNA methylation, histone modifications, and noncoding RNAs (ncRNAs) that govern the symptoms of oxidative stress, organ inflammation, neural apoptosis, irregularities, aberrations, etc., eventually causing neurodegenerative diseases. This chapter summarizes lead toxicity and its impacts on cognitive functions and health by overviewing the various epigenetic techniques resulting in neurodegeneration, including DNA methylation, histone modification, and ncRNAs (Tables 3.1 and 3.2).

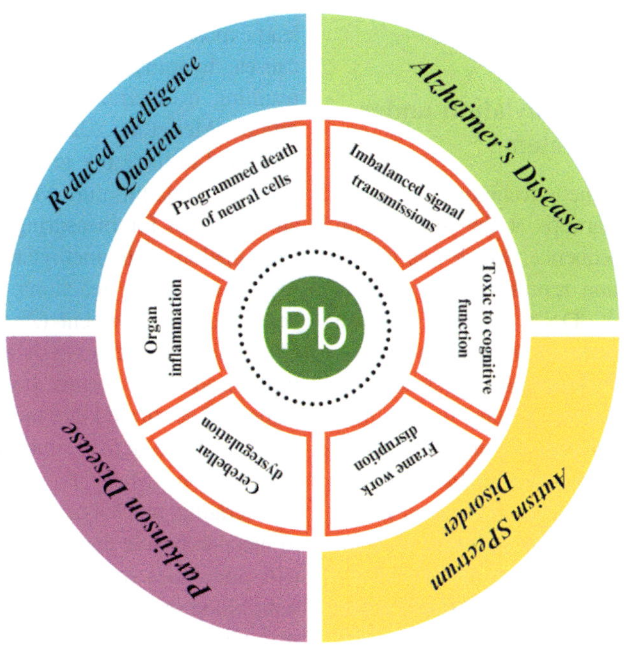

FIGURE 3.2 Lead-induced toxic consequences followed by the cognitive dysfunctional disorders.

TABLE 3.1 Role of various proteins, enzymes, and RNAs in inducing cognitive diseases.

Enzymes, proteins, and RNAs	Mechanisms	Effects of lead exposure	Results on cognitive health	References
Dnmt1	DNA methylation	Decreases	Alzheimer's disease	Yang et al. (2020)
Dnmt3a	DNA methylation	Increases	Alzheimer's disease	Yang et al. (2020)
MeCP2	DNA methylation	Decreases	Alzheimer's disease	Yang et al. (2020)
SGPD	DNA methylation	Increases	Parkinson's disease	Mei et al. (2023)
PEG	DNA methylation	Increases	Parkinson's disease	Mei et al. (2023)
H3K9ac	Histone modification	Decreases	Alzheimer's disease	Zhou et al. (2020)
H3K5me	Histone modification	Decreases	Alzheimer's disease	Mei et al. (2023)
has-miR-3651	Noncoding RNAs	Decreases	Alzheimer's disease	Mei et al. (2023)
has-miR-150-5p	Noncoding RNAs	Decreases	Alzheimer's disease	Mei et al. (2023)
has-miR-664b-3p	Noncoding RNAs	Decreases	Alzheimer's disease	Mei et al. (2023)
has-miR-627	Noncoding RNAs	Increases	Alzheimer's disease	Mei et al. (2023)
lncRNA120992	Noncoding RNAs	Increases	Alzheimer's disease	Nan et al. (2018)
miR-106b	Noncoding RNAs	Decreases	Alzheimer's disease	Xue et al. (2020)
miR-203	Noncoding RNAs	Increases	Parkinson's disease	Mei et al. (2023)
miR-137	Noncoding RNAs	Decreases	Neurodegenerative activities	Gu et al. (2019)

3.4 DNA methylation mechanisms and adjunct regulatory proteins under lead toxicity

3.4.1 Significance of DNA methylation under lead toxicity on the nervous system

DNA methylation is a central epigenetic technique characterized by the inclusion of methyl groups at the fifth carbon position with no change in the sequence, which retards the functionalities through transcription repressors and inhibits the proteins binding with DNA. DNA methyltransferase enzyme (DNMT), found in the mammalian body, is responsible for the catalysis of 5-methylcytosine, which aids in transporting the *S*-adenosylmethionine groups to the fifth position of cytosine. This DNA methylation is a pivotal factor in the processes of neuron production, specialization, synapse generation, acquisition, and retention (Xie, Xie, Wei, Li, & Lin, 2023), which is highly influenced by exposure to lead, causing malfunctioning in DNA methylation (Kebbe et al., 2023). DNA hypomethylation impacts neuron production and cognitive function, increasing lead levels during the perinatal phase of male mice, whose toxicity can potentially target DNA methylation of cortical neurons (Dou et al., 2019). Studies on C57BL/6J mice revealed the potential activity of lead exposure towards DNA hypomethylation. Lead exposure causes hypomethylation in multipotent migratory cells, resulting in aberrated communication compared to control groups (Mei et al., 2023). A total of 356 methylated sites are identified in the labor groups exposed to lead occupationally using Illumina Infinium Methylation EPIC Bead chip (850K), an ideal tool for measuring DNA methylation. In addition, examination of the methylated sites established their engagements in redox reactions, glutathione biosynthesis, and cognitive establishment (Zhang, He, Feng, & Shao, 2019). On the other hand, analysis of circulatory blood with various blood lead levels (below 100 and above 100 μg/L) using the same tool confirmed that the exposure induces methylation of genes that are actively taking part in impulses and cellular proliferation (Meng et al., 2022). P16, involved in cell proliferating mechanisms, is the principal gene susceptible to metal exposure and is amplified during neurodegeneration. p16 undergoes more significant methylation in the groups with blood lead levels of 51–100 μg/dL, which is higher the rate when compared with the groups of low blood lead levels (6–11 μg/dL), which evidences the trigger of the p16 gene to

TABLE 3.2 Effects of lead toxicity on the cognitive functions and mechanisms through confirmative in vitro and in vivo studies (Mei et al., 2023).

Observation	Treatment	Exposed concentration	Exposure duration	Targeted function	Mechanisms	Results	References
C57BL/ 6 J mice	Lead acetate	0, 2.1, and 32 ppm	Pre- and postnatal period	DNA methylation	Occurrence of DNA Hypomethylation	Cognitive dysfunction	Dou et al. (2019)
C57BL/ 6 J mice	Lead acetate	300 ppm	10 gestational days	DNA methylation	Hypomethylation of the gene Chd7	Autism induction	Hill et al. (2015)
Human	Occupational Blood lead levels (BLL)	Low BLL (< 2.469 μg/L) and high BLL (> 371.42 μg/L)	< 5 years (low BLL) and > 9 years (high BLL)	DNA methylation	Overexpression of NiNJ2 in high-BLL groups	Neural inflammation	Zhang et al. (2019)
SD rats	Lead acetate	0.05%, 0.1%, and 0.2%	Pre- and postnatal period	DNA methylation	Activation of DNMTs and an increase in the levels of MeCP2	Infection at the Hippocampus CA1 region	Wang, Wu et al. (2022), Wang, Yang et al. (2022)
C57/ B16 mice	Lead acetate	0 or 100 ppm	Before breeding and throughout lactation	Histone methylation	Alteration in H3K9/14AC levels	Cognitive impairment	Schneider, Anderson, Kidd, Sobolewski, and Cory-Slechta (2016)
PC12 cells	Lead acetate	10 μM	24 hours	Histone acetylation	Decrease in H3K9Me3 concentrations	Memory functional disruption	Wu et al. (2018)
SD rats	Lead acetate	250 ppm	2 months	Histone acetylation	Accelerated HDAC2 levels	Memory functional disruption	Wu et al. (2018)
PC12 cells	Lead acetate	10 μM	24 hours	Histone acetylation	Reduced concentrations of Ac-H3K9	Memory functional disruption	Zhou et al. (2020)
SD rats	Lead acetate	150 ppm	3 months	Histone acetylation	Accelerated HDAC2 levels	Memory functional disruption	Zhou et al. (2020)
SD rats	Lead acetate	250 ppm	Postnatal period	ncRNA	Decreased miR-137 and EZH2 activities	Memory dysfunctions	Gu et al. (2019)
C57BL/ 6 mice	Lead acetate	0.2%	3 months	ncRNA	Increased miR-138-5P concentrations	Impaired cognitive functions	Liu et al., (2021)
C57BL/ 6 J mice	Lead acetate	0.2%	20 postnatal days	ncRNA	Increased miR132 and miR-106b levels along with tau gene expressions and binding of AβPP mRNA	Alzheimer's disease emergence	Masoud et al. (2016)
HT22 and PC12 cells	Lead cells	0, 5, 10, 25, 50, 100 μM	24 or 48 hours	ncRNA	Increased miR-106b-5P and decreased XIAP protein levels	Alzheimer's disease emergence	Xue et al. (2020)
C57BL/ 6 mice	Lead acetate	0.2%	3 months	ncRNA	miR-124-3P activities reduced BACEI concentrations	Alzheimer's disease emergence	Wang, Wu et al. (2022), Wang, Yang et al. (2022)

methylate when subjected to lead exposure (Kovatsi et al., 2010; Mei et al., 2023). 5-Methylcytosine, 5-hydroxymethylcytosine, and 7-methylguanine are the key forms of DNA methylation in which 5-methylcytosine is major, generated by the inclusion of methyl groups in cytosine's fifth carbon position aided by DNMT (Joshi, Liu, Breslin SJ, & Zhang, 2022; Tian et al., 2023) and highly vulnerable to lead levels (Lin et al., 2021). SH-SY5Y cell model dosed with various lead levels resulted in modifications of 5-methylcytosine ranges associated with enzymatic activity towards epigenetic alterations. In certain conditions, the levels of 5-methylcytosine tend to decrease even after exposure termination and culture break often engaged with nerve cells distinction, and this reduction gives a credible mechanism for the chronic impact of the neurodegenerative capability of lead exposure (Lin et al., 2021). All these outcomes put forward the latent ability of lead toxicity to influence DNA methylation, which results in neurodegeneration (Mei et al., 2023).

3.4.2 Importance of DNA methylation complexes accompanied by lead toxicity

DNMT positively correlates with regulating the DNA methylation process in mammals. In addition, demethylases and methyl binders are also listed for their synergistic relation towards DNA methylation (Kim, Delgado, & Ko, 2023). Reconfiguration of these enzyme activities directly influences the DNA methylation process. Such outcomes are obtained due to lead exposure, which alters the expression range of methyl transferase, including the group of enzymes DNMT3a, DNMT3b, DNMT1, and $MeCP_2$ called Rett syndrome protein (Yang et al., 2020) that are tasked with neuron maturity along with intraneuronal signaling. These activities ultimately aid in building the nervous system and maintaining its functionalities (Xu et al., 2020). Methylcytosine phosphate accommodates guanine-binding protein no-2 (MeCP2), catalyzes the binding of methylated CpG dinucleotides involved in the transcription process of genes that engage with topographical memory, neuron transmissions, distinction, refined function, etc. (Vuu, Roberts, & Rastegar, 2023). SD rat model exposed to 0.2% lead concentration triggered the activities of DNMT and MeCP2, resulting in the refined methylation extents of specific genes associated with synaptic flexibility and organ injury at CA_1 pyramidal cells of hippocampus region through accommodation of pathogens (Wang, Wu et al., 2022; Wang, Yang et al., 2022). At the same time, the degrees of DNA methylase along with DNMT1, DNMT3a, and $MeCP_2$ mRNA are lowered when the SH-SY5Y cells are treated with the lead levels between 5 and 100 μM, contributing towards the lessened potential of the cells coupled with reduced DNA methylation (Mei et al., 2023). However, in contrast, lead exposure to mice at its juvenile stages shrunk the concentrations of DNMT1 and $MeCP_2$ in the brain cerebrum, while DNMT3a was left unaffected (Mei et al., 2023). This might be due to the adaptation mechanisms of DNA methylation towards lead exposure during the prenatal or juvenile stage. Hippocampus analysis of female long Evan mice propagated with the various lead acetate concentrations brought about aberrated modifications in DNMT1, DNMT3a, and $MeCP_2$, which affirms the lead toxicity's contribution to neurodegeneration through DNA methylation (Mei et al., 2023). Although there is an uncertainty and broader research gap, DNA methylation aligned with the oscillating protein concentrations of DNMT1, DNMT3a, and MeCP2 participates in central nervous system toxicity concerning lead exposure (Fig. 3.1).

3.4.3 Histone revisions under lead provoked neurodegeneration

Histone proteins are crucial in lending structural anchoring towards chromosomal functionalities in somatic cells of eukaryotes and certain prokaryotes under the *Archaea* domain. Nucleosomes are produced by the characterization of DNAs, surrounded by histone proteins. They are the principal protein complexes found in chromatin and perform gene activation through epigenetic scaffolds that form nucleosomes (Knapp, Naik, Ray, van Haaften, & Bicknell, 2023). In addition, they are a significant component in building chromatin, followed by the regulation of genes. Protein methylation incorporates lysine and arginine, phosphorylation entails serine and threonine, and further acetylation, crotonylation, borylation, and ubiquitin conjugation are classified as posttranslational modifications stated under histone revisions (Zhang et al., 2021). Premature exposure to lead toxicity pushes individuals to undergo DNA methylation and modification of histone proteins simultaneously, ultimately affecting the epigenetic mechanisms (Suseelan & Pinna, 2023). HEK293T human cells, when exposed to lead toxicity, result in reduced DNA methylation, MeCPG, and H3K9me3 histone complex levels. These complexes are actively engaged in gene expression and regulation, which is influenced by histone deacetylases (HDACs) that also aid in DNA methylation (Sánchez, Lin, Xie, Freeman, & Yuan, 2022). The actions of DNA methylation and HDACs are prevented by the generation of inhibiting factors which counteract the effects and aid in refining neurodegeneration through epigenetic approaches (Wang, Zhang, & Xu, 2020).

3.4.4 Methylation of histone proteins

The methylation of H3 and H4 histone complexes characterizes posttranscriptional alterations. Methyl groups of histone complexes attach to the lysine residues with the substrate *S*-adenosylmethionine and the enzyme Histone methyl transferase. The methylation of lysine can be categorized widely as monomethylated, demethylated, and trimethylated as per the number of methyl groups attached and results in the expression of genes (Al Ojaimi et al., 2022). Lead acetate treatments at the rate of 100 ppm by oral hydration were administered to female C57/Bl6 mice during their gestation phase. Their progenies are examined for histone modifications at frontal lobes and memory center, resulting in alteration in the histone methylation complexes, that is, H3K9/14ac, and H3K9me3 levels, followed by variations in sex expressions according to the target brain site (Mei et al., 2023). On the other hand, undifferentiated SH-SY5Y cells exposed to lead levels of 15 and 50 ppb with the exposure period of 96 hours slightly influenced the survival and morphological expressions with varied H3K27me3 and H3K9me3 levels of the respective cells. The treated cells are allowed to differentiate till the formation of neuron-like structures under lead-free conditions yielded a further drop in the H3K27me3 range in the cells accommodated with 15 ppb lead levels. In contrast, the cells dosed with 50 ppb showed gradual restoration (Lin et al., 2021). Several studies report the direct correlation between lead exposure and modification of histones. Female rates under 125 ppm lead levels obtained memory dysfunction or chronic memory impairment along with subdued H3K27me3 ranges at the hippocampus region (Xiao et al., 2020), followed by mice accommodated with 5 μM of lead brought about 30% loss in the H3K27me3 levels at the same hippocampus region associated with the minimal to zero activity of EZH2 enzyme involved in catalysing the methylation of H3K27me3 formation (Gu et al., 2019).

3.4.5 Action of acetylation

Acetylation of histone is another histone modification which is characterized by the binding of acetyl groups with histone under the supervision of histone acetyltransferase (HAT), followed by HDAC, which performs reverse acetylation (Shvedunova & Akhtar, 2022). Exposure to lead levels reduces the activity of HAT enzymes with decreased Ac-H3K9 levels and boosts the HDAC2 activity controlled through trichostatin A administration or shHDAC2 gene manipulation. This strategy helps in elevating the Ac-H3K9 levels even under lead toxicity, which clearly defines that toxicity triggers HDAC2 and harms neural health. Further, HDAC2 and HDAC1 physically bind together to produce HDAC1/2 complexes that warp the DNA tightly and prevent gene expressions, resulting in retard neuron growth and loss of memory. The suppression of HDAC1/2 complexes gives positive outreach of better neuron development with substantial memory even under lead-exposed conditions (Mei et al., 2023; Wu et al., 2018). PC12 cells, when treated with lead at the rate of 5 and 25 mg/L, hiked their H3 histone protein acetylation by HAT p300 and uplifted in gene transcription. In contrast, H3K9ac showed a downward increment under protracted exposure to lead, where H3ac tends to increase. This contradiction is due to the specific alteration, susceptibility of histone complexes to lead concentrations, and internal and external conditions (Mei et al., 2023; Zhou et al., 2020). Like this phenomenon, H3K27me3 to H3K9Ac and H3K4me2 ratios in the cerebral mantle are inversely related to each other, observed in mice under perpetual lead exposure (Eid, Bihaqi, Renehan, & Zawia, 2016; Mei et al., 2023). HDAC activities under lead-exposed neurodegeneration are crucial and obvious. Heavy metal toxicities, lead and cadmium, separately cut down the mass of dendrites, resulting in weakened recalling and grasping ability, specifically at the CA1 and DG locations in the hippocampus of SD rats, confirmed through experiments like Morris's water maze, Y-maze and Golgi-Cox staining techniques. HDAC2 functions in the neurodegeneration of PC12 cells under metal exposure such as lead and cadmium, while trichostatin A, along with shHDAC2, smoothens the adverse impacts created by HDAC2 even under metal-exposed conditions, which is found crucial in inducing neurodegeneration under such conditions (Zhou et al., 2020). Lead toxicity targets rats' PC12 cells and hippocampus regions with a rise in HDAC4 concentrations, ultimately resulting in deformation. Suppression of HDAC4 using inhibition factors like LMK-235 inhibiting agent, shHDAC4 plasmids affects the health synergistically amid exposure to lead (Gu et al., 2020). In another way, the sirtuin 1 (SIRT1) protein from the sirtuin group of nicotinamide adenine dinucleotide participates in protecting cognitive functions and health (Dang et al., 2022). Levels of sirtuin 1 protein reduced in male mice exposed to lead (0.2% lead acetate) for 3-month period, confirmed through the analysis of their levels in cortex and hippocampus regions at the intervals of 1, 4, and 16 months from the lead administration (Liu et al., 2021). All this evidence proves the interference of lead in cognitive function through a molecular focus.

3.5 Noncoding RNA approaches under lead exposure

ncRNAs are abundant RNA transcripts in humans' genetic framework. Its contribution exceeds 90% of overall transcripts. Recent research focuses on the role of these ncRNA in maintaining neuron functions under infectious or metal-exposed conditions. Among the ncRNAs, microRNA (miRNA), long ncRNA (lncRNA), and circular RNA (circRNA) are found to have greater involvement towards the regulation of cognitive health under AD and Parkinson's diseased states (Gada et al., 2022).

3.6 Significance of microRNA in lead-exposed circumstances

miRNAs are comparatively minor, built with 18–24 nucleotides, and involved in regulating cellular biological operations. They are well known for their contribution towards gene expressions, especially during posttranscriptional stages. Their contribution at the cellular level involves distinction, interaction, embryo formation, energy production, and organogenesis (Elangovan et al., 2023). Latest findings report the involvement of miRNA in AD and Parkinson's disease occurrence due to Lead exposure (Wang, Guan et al., 2023; Wang, Hu et al., 2023). Alzheimer's is more prone to the malfunctioning of genes during prenatal and early postnatal phases (Kumar et al., 2023). Specific research grasps attention by examining the association between miRNA and AD. miRNAs in the blood are susceptible to lead levels and are efficient enough for the induction of lead triggered Alzheimer's (LTA) (Wang, Wu et al., 2022; Wang, Yang et al., 2022). Blood-derived miRNA study on AD and LTA through segregated groups elucidates the link between has-miR-627 miRNA and LTA where other miRNAs like hsa-miR-3651, hsa-miR-150-5p, and hsa-miR-664b-3p do not have any correlations. In contrast, miR-3651 and miR-664b-3p miRNA groups are linked with LTA, as pointed out in brain studies of the same groups (Mei et al., 2023). Lead exposure at early stages targets miRNAs like miR-106b, miR-29a, miR-29b, and miR-132, which causes neurodegeneration, depletes cognitive functions, and increases AD induction factors. Lead exposure through mammary feeding to mice for the first 20 days after birth elevated the levels of such miRNAs during the lateral stages of growth (20, 180, and 700 days from birth). Among the miRNAs, miR-106b regulates amyloid beta (Aβ) protein precursor whereas miR-29b functions through binding with specificity protein-1, a transcription factor that highly influences the incidence of neural poisoning under lead-exposed conditions (Masoud, Bihaqi, Machan, Zawia, & Renehan, 2016; Mei et al., 2023). Such miRNAs control the genetic mechanisms and functionalities and produce neurotoxicants, which push towards lead-mediated neurodegeneration. A few culture experiments affirm the role of miR-124-3p/β-site APP-cleaving enzyme 1 (BACE1) in lead-mediated Alzheimer's through amyloidal mechanisms. SH-SY5Y cells under lead toxicity yielded a more significant number of BACE1 mRNAs resulting in increased Aβ formation that directly weakens cognitive health (Wang, Wu et al., 2022; Wang, Yang et al., 2022). A wide range of studies validates the role of miR-34b miRNA in governing the expression of synapse-associated membrane protein 2 (VAMP2) gene, enrolled in standard communicative functions of the brain, which is affected negatively under lead exposure, leading to neural dysfunctions and disorders. The suppression of VAMP2 by overexpression of miR-34b is recorded in both the culture and live study models administered with lead, reducing neurotransmitters, which affects the emergence of nervous system disorders (Yang et al., 2022). Increased concentrations of lead give rise to more miR-130 miRNA, which retards the SNAP-25 functions associated with the weakening of neural connections and the reduced size of the cortex involved in regulating attentive activities. Further suppression of miRNA brings back the functionalities of the SNAP-25 gene and the reallocation of synaptic connections in mice, even under toxicity (Wang, Guan et al., 2023; Wang, Hu et al., 2023). Lead exposure of HT22 cells leads to cell death through oxidative stress induced by the generation of ROS. This is due to the disruption of SLC7A11 protein molecules involved in antioxidant production like cysteine, glutathione, and malondialdehyde, which further creates an imbalance in these antioxidants, resulting in oxidative cell injury. In addition, the overexpression of miR-378a-3p under lead exposure mitigates the SLC7A11 protein and culminates in cell death through oxidative stress. Further, controlled activation of miR-378a-3p retains the levels of SLC7A11 protein. It maintains the healthy state of the cells regardless of lead concentrations which demystify the relation between lead-mediated neuro dysfunctionalities and miR-378a-3p expressions (Wang, Wu et al., 2022; Wang, Yang et al., 2022). Besides, culture models of HT22 and PC12 cells administered with lead levels upgrade miR-106b-5p miRNA proportions. This specific miRNA contributes to the XIAP levels, where this miRNA attaches to XIAP, which is observed through a luciferase reporter assay, which saves premature cell death. The more the lead volumes, the more the miR-106b-5p miRNA proportions, and the less the XIAP levels force cells to die eventually under lead toxicity. Like the previous studies, blocking the activities of miR-106b-5p miRNA balances the XIAP levels and optimizes cellular functions (Xue et al., 2020).

3.7 Importance of long noncoding RNA under lead-mediated neurodegeneration

lncRNA is a ncRNA that governs the evolution of creatures, embryonic progress, energy production and cancer cells development; it contains more than 200 nucleotides (Li & Wang, 2023) which establishes a strong relationship between nerve cell distinction and cognitive improvement (Chen et al., 2021). Lead administration in animals for the first 3 weeks after their birth cuts the intensities of about 69 lncRNA in the CA1 part of the hippocampus region from the later stage of the eighth week since lead administration. lncRNAs like AABR07065531.5 and LOC102550577 establish connections with messenger RNAs and thus contribute to cognitive functions whose interactions are highly varied under lead toxicity, which increases the probability of neurodegeneration (Singh et al., 2022). Likewise, a lncRNA, LncRNAL20992, in PC12 cells gets triggered when the cells are exposed to lead treatments causing nervous system damage through programmed cell death and retarded cell division. In addition, this specific lncRNA finds association with neural improvement factors such as mitochondrial apoptosis-inducing factor (AIFM1), chaperone protein (HSC70), glucose regulating protein (GRP78) and Lamin A/C (LMNA) (Hribljan, Salamon, Đemaili, Alić, & Mitrečić, 2018; Mei et al., 2023), which are highly susceptible under lead exposure through the activity of LncRNAL20992 that elucidates the correlation between lncRNA and lead toxicity (Mei et al., 2023; Nan et al., 2018).

3.8 Circular RNA contributes to cognitive health under lead exposure

circRNAs are the ncRNAs, have a high capability of withstanding destruction, and tolerate RNA exonucleases by their single-stranded covalent circular skeleton (Ren, Guan, Zhou, Cai, & Shan, 2023), and supervise cognitive health and disorders (Chen et al., 2021) under lead toxicity (Wang et al., 2020). This circRNA is an influential factor in disposing of the amyloid beta peptides, and infectious agents of AD were studied in mice (Mei et al., 2023; Zhang et al., 2018) (Tables 3.1 and 3.2).

The lead toxicity and circRNA are related through the amyloid beta peptide disposals, the persistence of which generates lead-mediated AD (Zhou et al., 2022). Ubiquitin protein ligase A (UBE2A), a type of protein engaged in the disposal of amyloid beta peptides, is suppressed by the greater availability of miR-7 through its linkage with UBE2A. These linkage mechanisms should be optimum for the disposal process, but when there is a greater availability of miR-7, it yields more linkages and results in the functional reduction of UBE2A (Mei et al., 2023). When a circRNA called ciRS-7 binds with miR-7, it prevents the regular routines of miR-7 through minimal linkage with UBE2A, creating a balanced stature in the disposal of amyloid beta peptides. Under lead toxic conditions, the amplitude of ciRS-7 rises with higher inhibition in the linkage mechanism between miR-7 and UBE2A protein. This condition again creates an imbalance in the disposing process and favors the occurrence of AD, where lead toxicity becomes a principal factor of contribution (Mei et al., 2023). Available reports define the role of circRNA in maintaining cognitive health under lead exposure, but further investigations will enlighten the hidden mechanisms of circRNA under such circumstances (Jiang, Wang, Wang, Feng, & Wang, 2022; Xu et al., 2022).

3.9 Conclusion

To wrap up, lead toxicity has the potential to induce nervous function disorders, such as Alzheimer's, Parkinson's, and autism, that are often linked to malfunctioning of DNA methylation, histone alterations, and expressions of ncRNAs, with multifaceted mechanisms along with variations in genetic expressions and functions. The competitiveness of lncRNA and circRNA towards binding with miRNA and the significance of lncRNA and miRNA in regulating DNA methylation and histone methylation are interrelated and thus constitutes for the complex interactions. It is noteworthy to report their tie under cognitive state influenced by lead toxicity through peculiar techniques like bioinformatics, next-generation sequencing, etc.

References

Al Ojaimi, M., Banimortada, B. J., Othman, A., Riedhammer, K. M., Almannai, M., & El-Hattab, A. W. (2022). Disorders of histone methylation: Molecular basis and clinical syndromes. *Clinical Genetics, 102*(3), 169–181.

Almutairi, M. M., Nadeem, A., Ansari, M. A., Bakheet, S. A., Attia, S. M., Albekairi, T. H., ... Ahmad, S. F. (2022). Lead (Pb) exposure exacerbates behavioral and immune abnormalities by upregulating Th17 and NF-κB-related signaling in BTBR T+ Itpr3tf/J autistic mouse model. *Neurotoxicology, 91*, 340–348.

Axelrad, D. A., Coffman, E., Kirrane, E. F., & Klemick, H. (2022). The relationship between childhood blood lead levels below 5 μg/dL and childhood intelligence quotient (IQ): Protocol for a systematic review and meta-analysis. *Environment International, 169*, 107475.

Balali-Mood, M., Naseri, K., Tahergorabi, Z., Khazdair, M. R., & Sadeghi, M. (2021). Toxic mechanisms of five heavy metals: Mercury, lead, chromium, cadmium, and arsenic. *Frontiers in Pharmacology, 12*, 643972.

Bandaru, L. J. M., Murumulla, L., & Challa, S. (2023). Exposure of combination of environmental pollutant, lead (Pb) and β-amyloid peptides causes mitochondrial dysfunction and oxidative stress in human neuronal cells. *Journal of Bioenergetics and Biomembranes, 55*(1), 79–89.

Cai, H., Bao, Y., Cheng, H., Ge, X., Zhang, M., Feng, X., ... Yang, X. (2023). Zinc homeostasis may reverse the synergistic neurotoxicity of heavy metal mixtures in *Caenorhabditis elegans*. *Science of The Total Environment, 868*, 161699.

Chandwani, S., Kayasth, R., Naik, H., & Amaresan, N. (2023). Current status and future prospect of managing lead (Pb) stress through microbes for sustainable agriculture. *Environmental Monitoring and Assessment, 195*(4), 479.

Chen, L., Liu, Y., Jia, P., Zhang, H., Yin, Z., Hu, D., ... Ge, Y. (2022). Acute lead acetate induces neurotoxicity through decreased synaptic plasticity-related protein expression and disordered dendritic formation in nerve cells. *Environmental Science and Pollution Research, 29*(39), 58927–58935.

Chen, M., Lai, X., Wang, X., Ying, J., Zhang, L., Zhou, B., ... Hua, F. (2021). Long non-coding RNAs and circular RNAs: Insights into microglia and astrocyte mediated neurological diseases. *Frontiers in Molecular Neuroscience, 14*, 745066.

Dang, R., Wang, M., Li, X., Wang, H., Liu, L., Wu, Q., ... Xie, P. (2022). Edaravone ameliorates depressive and anxiety-like behaviors via Sirt1/Nrf2/HO-1/Gpx4 pathway. *Journal of Neuroinflammation, 19*(1), 41.

Dou, J. F., Farooqui, Z., Faulk, C. D., Barks, A. K., Jones, T., Dolinoy, D. C., & Bakulski, K. M. (2019). Perinatal lead (Pb) exposure and cortical neuron-specific DNA methylation in male mice. *Genes, 10*(4), 274.

Eid, A., Bihaqi, S. W., Renehan, W. E., & Zawia, N. H. (2016). Developmental lead exposure and lifespan alterations in epigenetic regulators and their correspondence to biomarkers of Alzheimer's disease. *Alzheimer's & Dementia: Diagnosis, Assessment & Disease Monitoring, 2*, 123–131.

Elangovan, A., Venkatesan, D., Selvaraj, P., Pasha, M. Y., Babu, H. W. S., Iyer, M., ... Vellingiri, B. (2023). miRNA in Parkinson's disease: From pathogenesis to theranostic approaches. *Journal of Cellular Physiology, 238*(2), 329–354.

Farías, P., Hernández-Bonilla, D., Moreno-Macías, H., Montes-López, S., Schnaas, L., Texcalac-Sangrador, J. L., Ríos, C., & Riojas-Rodríguez, H (2022). Prenatal co-exposure to manganese, mercury, and lead, and neurodevelopment in children during the first year of life. *International Journal of Environmental Research and Public Health, 19*(20), 13020. Retrieved from https://doi.org/10.3390/ijerph192013020.

Gada, Y., Pandey, A., Jadhav, N., Ajgaonkar, S., Mehta, D., & Nair, S. (2022). New vistas in microRNA regulatory interactome in neuropathic pain. *Frontiers in Pharmacology, 12*, 778014.

Gonzalez-Alcocer, A., Duarte-Jurado, A. P., Soto-Dominguez, A., Loera-Arias, M. D. J., Villarreal-Silva, E. E., Saucedo-Cardenas, O., ... Rodriguez-Rocha, H. (2023). Unscrambling the role of redox-active biometals in dopaminergic neuronal death and promising metal chelation-based therapy for Parkinson's disease. *International Journal of Molecular Sciences, 24*(2), 1256.

Gu, X., Huang, X., Li, D., Bi, N., Yu, X., & Wang, H. L. (2020). Nuclear accumulation of histone deacetylase 4 (HDAC4) by PP1-mediated dephosphorylation exerts neurotoxicity in Pb-exposed neural cells. *Neurotoxicology, 81*, 395–405.

Gu, X., Xu, Y., Xue, W. Z., Wu, Y., Ye, Z., Xiao, G., & Wang, H. L. (2019). Interplay of miR-137 and EZH2 contributes to the genome-wide redistribution of H3K27me3 underlying the Pb-induced memory impairment. *Cell Death & Disease, 10*(9), 671.

Halabicky, O. M., Ji, X., Gur, R. E., Gur, R. C., Yan, C., Chen, A., & Liu, J. (2022). Childhood lead exposure and sex-based neurobehavioral functioning in adolescence. *Neurotoxicology, 93*, 1–8.

Heidari, S., Mostafaei, S., Razazian, N., Rajati, M., Saeedi, A., & Rajati, F. (2022). The effect of lead exposure on IQ test scores in children under 12 years: A systematic review and meta-analysis of case-control studies. *Systematic Reviews, 11*(1), 106.

Hernández, M., Zhang, Y., Filippelli, G. M., & Boehm, S. L., II (2023). Early-life low-level lead exposure alters anxiety-like behavior, voluntary alcohol consumption and AC5 protein content in adult male and female C57BL/6 J mice. *Neurotoxicology and Teratology, 95*, 107149.

Hill, D. S., Cabrera, R., Wallis Schultz, D., Zhu, H., Lu, W., Finnell, R. H., & Wlodarczyk, B. J. (2015). Autism-like behavior and epigenetic changes associated with autism as consequences of in utero exposure to environmental pollutants in a mouse model. *Behavioural Neurology, 2015*(1), 426263.

Hribljan, V., Salamon, I., Đemaili, A., Alić, I., & Mitrečić, D. (2018). Transplantation of neural stem cells in the mouse model of ischemic brain stroke and expression of genes involved in programmed cell death. *Croatian Medical Journal, 59*(5), 203–212.

Huang, A., Zhang, J., Wu, K., Liu, C., Huang, Q., Zhang, X., ... Huang, Y. (2022). Exposure to multiple metals and the risk of dyslexia—A case control study in Shantou, China. *Environmental Pollution, 307*, 119518.

Jiang, M., Wang, Y., Wang, J., Feng, S., & Wang, X. (2022). The etiological roles of miRNAs, lncRNAs, and circRNAs in neuropathic pain: A narrative review. *Journal of Clinical Laboratory Analysis, 36*(8), e24592.

Joshi, K., Liu, S., Breslin SJ, P., & Zhang, J. (2022). Mechanisms that regulate the activities of TET proteins. *Cellular and Molecular Life Sciences, 79*(7), 363.

Kargar-Shouroki, F., Mehri, H., & Sepahi-Zoeram, F. (2023). Biochemical and hematological effects of lead exposure in Iranian battery workers. *International Journal of Occupational Safety and Ergonomics, 29*(2), 661–667.

Kebbe, M., Naud, P., Assous, I., Gagnon, E., McCall, A., Villeneuve, L., ... Calderone, A. (2023). Distinct protein kinase C isoforms drive the cell cycle re-entry of two separate populations of neonatal rat ventricular cardiomyocytes. *American Journal of Physiology-Cell Physiology, 325*(2), C406–C419.

Kim, M., Delgado, E., & Ko, S. (2023). DNA methylation in cell plasticity and malignant transformation in liver diseases. *Pharmacology & Therapeutics, 241*, 108334 Yang, L., Li, X., Jiang.

Knapp, K., Naik, N., Ray, S., van Haaften, G., & Bicknell, L. S. (2023). Histones: Coming of age in Mendelian genetic disorders. *Journal of Medical Genetics, 60*(3), 1–10.

Kovatsi, L., Georgiou, E., Ioannou, A., Haitoglou, C., Tzimagiorgis, G., Tsoukali, H., & Kouidou, S. (2010). p16 promoter methylation in Pb^{2+}-exposed individuals. *Clinical Toxicology, 48*(2), 124–128.

Kumar, R., Fatima, F., Yadav, G., Singh, S., Haldar, S., Alexiou, A., & Ashraf, G. M. (2023). Epigenetic modifications by estrogen and androgen in Alzheimer's disease. *CNS & Neurological Disorders-Drug Targets (Formerly Current Drug Targets-CNS & Neurological Disorders), 22*(1), 6–17.

Li, K., & Wang, Z. (2023). lncRNA NEAT1: Key player in neurodegenerative diseases. *Ageing Research Reviews, 86*, 101878.

Lin, L. F., Xie, J., Sánchez, O. F., Bryan, C., Freeman, J. L., & Yuan, C. (2021). Low dose lead exposure induces alterations on heterochromatin hallmarks persisting through SH-SY5Y cell differentiation. *Chemosphere, 264*, 128486.

Liu, C. M., Tian, Z. K., Zhang, Y. J., Ming, Q. L., Ma, J. Q., & Ji, L. P. (2020). Effects of gastrodin against lead-induced brain injury in mice associated with the Wnt/Nrf2 pathway. *Nutrients, 12*(6), 1805.

Liu, R., Wang, Y., Bai, L., Wang, R., Wu, Y., Liu, M., ... Huang, H. (2021). Time-course miRNA alterations and SIRT1 inhibition triggered by adolescent lead exposure in mice. *Toxicology Research, 10*(4), 667–676.

Mao, G., Liu, H., Ding, Y., Zhang, W., Chen, H., Zhao, T., ... Yang, L. (2020). Evaluation of combined developmental neurological toxicity of di (n-butyl) phthalates and lead using immature mice. *Environmental Science and Pollution Research, 27*, 9318–9326.

Masoud, A. M., Bihaqi, S. W., Machan, J. T., Zawia, N. H., & Renehan, W. E. (2016). Early-life exposure to lead (Pb) alters the expression of

microRNA that target proteins associated with Alzheimer's disease. *Journal of Alzheimer's Disease, 51*(4), 1257–1264.
Mei, Z., Liu, G., Zhao, B., He, Z., & Gu, S. (2023). Emerging roles of epigenetics in lead-induced neurotoxicity. *Environment International, 181*, 108253. Retrieved from https://doi.org/10.1016/j.envint.2023.108253.
Meng, Y., Zhou, M., Wang, T., Zhang, G., Tu, Y., Gong, S., ... Xia, Z. L. (2022). Occupational lead exposure on genome-wide DNA methylation and DNA damage. *Environmental Pollution, 304*, 119252.
Moore, S., Paalanen, L., Melymuk, L., Katsonouri, A., Kolossa-Gehring, M., & Tolonen, H. (2022). The association between ADHD and environmental chemicals—A scoping review. *International Journal of Environmental Research and Public Health, 19*(5), 2849.
Nam, S. M., Seo, J. S., Nahm, S. S., & Chang, B. J. (2019). Effects of ascorbic acid on osteopontin expression and axonal myelination in the developing cerebellum of lead-exposed rat pups. *International Journal of Environmental Research and Public Health, 16*(6), 983.
Nan, A., Jia, Y., Li, X., Liu, M., Zhang, N., Chen, L., ... Jiang, Y. (2018). Editor's highlight: lncRNAL20992 regulates apoptotic proteins to promote lead-induced neuronal apoptosis. *Toxicological Sciences, 161*(1), 115–124.
Omotosho, I. O., Akinade, A. O., Lagunju, I. A., & Yakubu, M. A. (2021). Oxidative stress indices in ASD children in Sub-Sahara Africa. *Journal of Neurodevelopmental Disorders, 13*, 1–8.
Paul, K. C., Horvath, S., Del Rosario, I., Bronstein, J. M., & Ritz, B. (2021). DNA methylation biomarker for cumulative lead exposure is associated with Parkinson's disease. *Clinical Epigenetics, 13*, 1–4.
Pocock, S. J., Smith, M., & Baghurst, P. (1994). Environmental lead and children's intelligence: A systematic review of epidemiological evidence. *BMJ (Clinical Research ed.), 309*(6963), 1189–1197.
Ren, B., Guan, M. X., Zhou, T., Cai, X., & Shan, G. (2023). Emerging functions of mitochondria-encoded noncoding RNAs. *Trends in Genetics, 39*(2), 125–139.
Sánchez, O. F., Lin, L. F., Xie, J., Freeman, J. L., & Yuan, C. (2022). Lead exposure induces dysregulation of constitutive heterochromatin hallmarks in live cells. *Current Research in Toxicology, 3*, 100061.
Santana, A. B., Spelta, L. E. W., Sobalvarro, J. V. M., Podestá, M. H. M. C., Garcia, R. C. T., Dos Reis, T. M., & Torres, L. H. (2023). Gestational lead exposure and its effects on fetal/infant development—A systematic review. *Reproductive Toxicology, 117*, 108342.
Schneider, J. S., Anderson, D. W., Kidd, S. K., Sobolewski, M., & Cory-Slechta, D. A. (2016). Sex-dependent effects of lead and prenatal stress on post-translational histone modifications in frontal cortex and hippocampus in the early postnatal brain. *Neurotoxicology, 54*, 65–71.
Shilpa, O., Anupama, K. P., Antony, A., & Gurushankara, H. P. (2021). Lead (Pb)-induced oxidative stress mediates sex-specific autistic-like behaviour in Drosophila melanogaster. *Molecular Neurobiology, 58*, 6378–6393.
Shvedunova, M., & Akhtar, A. (2022). Modulation of cellular processes by histone and non-histone protein acetylation. *Nature Reviews. Molecular Cell Biology, 23*(5), 329–349.
Singh, G., Singh, V., Kim, T., Ertel, A., Fu, W., & Schneider, J. S. (2022). Altered genome-wide hippocampal gene expression profiles following early life lead exposure and their potential for reversal by environmental enrichment. *Scientific Reports, 12*(1), 11937.
Suseelan, S., & Pinna, G. (2023). Heterogeneity in major depressive disorder: The need for biomarker-based personalized treatments. *Advances in Clinical Chemistry, 112*, 1–67.
Tian, Y., Xiao, H., Yang, Y., Zhang, P., Yuan, J., Zhang, W., ... Zhang, X. (2023). Crosstalk between 5-methylcytosine and N6-methyladenosine machinery defines disease progression, therapeutic response and pharmacogenomic landscape in hepatocellular carcinoma. *Molecular Cancer, 22*(1), 5.
Vrijsen, S., Houdou, M., Cascalho, A., Eggermont, J., & Vangheluwe, P. (2023). Polyamines in Parkinson's disease: Balancing between neurotoxicity and neuroprotection. *Annual Review of Biochemistry, 92*(1), 435–464.
Vuu, Y. M., Roberts, C. T., & Rastegar, M. (2023). MeCP2 is an epigenetic factor that links DNA methylation with brain metabolism. *International Journal of Molecular Sciences, 24*(4), 4218.
Wang, R., Wu, Z., Liu, R., Bai, L., Lin, Y., Ba, Y., & Huang, H. (2022). Age-related miRNAs dysregulation and abnormal BACE1 expression following Pb exposure in adolescent mice. *Environmental Toxicology, 37*(8), 1902–1913.
Wang, R., Yang, M., Wu, Y., Liu, R., Liu, M., Li, Q., ... Huang, H. (2022). SIRT1 modifies DNA methylation linked to synaptic deficits induced by Pb in vitro and in vivo. *International Journal of Biological Macromolecules, 217*, 219–228.
Wang, T., Guan, R. L., Zou, Y. F., Zheng, G., Shen, X. F., Cao, Z. P., ... Luo, W. J. (2023). MiR-130/SNAP-25 axis regulates presynaptic alteration in anterior cingulate cortex involved in lead induced attention deficits. *Journal of Hazardous Materials, 443*, 130249.
Wang, T., Zhang, J., & Xu, Y. (2020). Epigenetic basis of lead-induced neurological disorders. *International Journal of Environmental Research and Public Health, 17*(13), 4878.
Wang, Y., Hu, H., Liu, X., & Guo, X. (2023). Hypoglycemic medicines in the treatment of Alzheimer's disease: Pathophysiological links between AD and glucose metabolism. *Frontiers in Pharmacology, 14*, 1138499.
Wu, Y., Xu, Y., Huang, X., Ye, D., Han, M., & Wang, H. L. (2018). Regulatory roles of histone deacetylases 1 and 2 in Pb-induced neurotoxicity. *Toxicological Sciences, 162*(2), 688–701.
Xiao, J., Wang, T., Xu, Y., Gu, X., Li, D., Niu, K., ... Wang, H. L. (2020). Long-term probiotic intervention mitigates memory dysfunction through a novel H3K27me3-based mechanism in lead-exposed rats. *Translational Psychiatry, 10*(1), 25.
Xie, J., Xie, L., Wei, H., Li, X. J., & Lin, L. (2023). Dynamic regulation of DNA methylation and brain functions. *Biology, 12*(2), 152.
Xu, L., Huo, X., Liu, Y., Zhang, Y., Qin, Q., & Xu, X. (2020). Hearing loss risk and DNA methylation signatures in preschool children following lead and cadmium exposure from an electronic waste recycling area. *Chemosphere, 246*, 125829.
Xu, Z., Chen, Y., Ma, L., Chen, Y., Liu, J., Guo, Y., ... Shu, Y. (2022). Role of exosomal non-coding RNAs from tumor cells and tumor-associated macrophages in the tumor microenvironment. *Molecular Therapy, 30*(10), 3133–3154.
Xue, C., Kang, B., Su, P., Wang, D., Zhao, F., Zhang, J., ... Cao, Z. (2020). MicroRNA-106b-5p participates in lead (Pb^{2+})-induced cell viability inhibition by targeting XIAP in HT-22 and PC12 cells. *Toxicology in Vitro, 66*, 104876.
Yang, C., Kang, B., Cao, Z., Zhang, J., Zhao, F., Wang, D., ... Chen, J. (2022). Early-life Pb exposure might exert synapse-toxic effects via

inhibiting synapse-associated membrane protein 2 (VAMP2) mediated by upregulation of miR-34b. *Journal of Alzheimer's Disease, 87*(2), 619–633.

Yang, L., Li, X., Jiang, A., Li, X., Chang, W., Chen, J., & Ye, F. (2020). Metformin alleviates lead-induced mitochondrial fragmentation via AMPK/Nrf2 activation in SH-SY5Y cells. *Redox Biology, 36*, 101626.

Zhang, T., Pang, P., Fang, Z., Guo, Y., Li, H., Li, X., ... Lu, Y. (2018). Expression of BC1 impairs spatial learning and memory in Alzheimer's disease via APP translation. *Molecular Neurobiology, 55*, 6007–6020.

Zhang, X. X., He, Z., Feng, B., & Shao, H. (2019). An epigenome-wide DNA methylation study of workers with occupational exposure to lead. *Journal of Applied Toxicology, 39*(9), 1311–1319.

Zhang, Y., Liu, D., Yao, X., Wen, J., Wang, Y., & Zhang, Y. (2022). DMTHB ameliorates memory impairment in Alzheimer's disease mice through regulation of neuroinflammation. *Neuroscience Letters, 785*, 136770.

Zhang, Y., Sun, Z., Jia, J., Du, T., Zhang, N., Tang, Y., & Fang, D. (2021). Overview of histone modification. In D. Fang, & J. Han (Eds.). *Histone mutations and cancer. Advances in experimental medicine and biology (vol. 1283)* (pp. 1–16). Springer. Retrieved from https://doi.org/10.1007/978-981-15-8104-5_1.

Zhou, C. C., Wang, X. J., Li, Z. C., Lu, W. J., Zhang, Y. T., Shen, F. M., & Li, D. J. (2022). Lead exposure in developmental ages promotes Aβ accumulation by disturbing Aβ transportation in blood-cerebrospinal fluid barrier/blood–brain barriers and impairing Aβ clearance in the liver. *Biological Trace Element Research, 200*(8), 3702–3711. Retrieved from https://doi.org/10.1007/s12011-021-02969-8.

Zhou, R., Zhao, J., Li, D., Chen, Y., Xiao, Y., Fan, A., ... Wang, H. L. (2020). Combined exposure of lead and cadmium leads to aggravated neurotoxicity through regulating the expression of histone deacetylase 2. *Chemosphere, 252*, 126589.

Chapter 4

Mercury poisoning: effects on the central nervous system

Neha Gogia[1], and Prasann Kumar[2]
[1]*Yale University, New Haven, CT, United States,* [2]*Department of Agronomy, School of Agriculture, Lovely Professional University, Phagwara, Punjab, India*

4.1 Introduction

Mercury poisoning is a significant public health concern due to its toxic effects on the human body, particularly the central nervous system (CNS). Mercury has three primary forms—elemental (metallic), inorganic, and organic, with distinct toxicity profiles and exposure routes. Elemental mercury (Hg^0) is commonly used in thermometers, barometers, and dental amalgams (Devi, Dogra, Kumar, & Singh, 2024; Jagadesh et al., 2024; Raj, Goutam, Kumar, & Singh, 2024). It can be hazardous when inhaled as a vapor, leading to neurological and respiratory damage. Inorganic mercury compounds (Hg^{2+}) in industrial chemicals and skin-lightening products can cause kidney and gastrointestinal toxicity. Organic mercury, particularly methylmercury and ethylmercury, is the most neurotoxic form, primarily accumulating in aquatic food chains and leading to severe neurological disorders in humans. The impact of mercury poisoning has been widely studied, with historical cases highlighting its devastating consequences, particularly in communities exposed to contaminated food and water sources. Among the most infamous cases of mercury poisoning is the Minamata disease outbreak in Japan, which became a turning point in understanding and regulating mercury pollution. This tragic incident occurred in the mid-20th century when a chemical company, Chisso Corporation, discharged industrial waste containing methylmercury into Minamata Bay. Local communities, heavily reliant on seafood, unknowingly consumed contaminated fish and shellfish, leading to widespread neurological disorders. The symptoms included ataxia, sensory disturbances, muscle weakness, vision impairment, and, in severe cases, congenital deformities in newborns (Kumar & Choudhury, 2024a, 2024b; Srivastav et al., 2024). Minamata disease not only affected thousands of people but also underscored the long-term environmental persistence of mercury, influencing global policies on industrial waste management and environmental protection. Another significant case was the Iraq grain disaster in the early 1970s, where wheat and barley seeds treated with methylmercury fungicides were imported as part of an agricultural aid program. Due to miscommunication, many rural populations consumed these seeds as food rather than planting them, leading to mass mercury poisoning. Symptoms ranged from mild paresthesia to severe neurotoxicity, coma, and death. This tragedy reinforced the dangers of mercury in agriculture and the need for strict regulatory control over its use in food production. Beyond these major outbreaks, mercury exposure has remained a persistent issue in many parts of the world, particularly in communities engaged in small-scale gold mining. Artisanal and small-scale gold miners (ASGM) often use mercury to extract gold from ore, releasing mercury vapor into the air and contaminating local water sources (Kumar & Dey, 2024; Kumar & Singh, 2024; Kumar, Dey, & Choudhury, 2024). The inhalation of mercury vapor and the bioaccumulation of methylmercury in fish consumed by miners and their families contribute to widespread neurological and systemic health issues. Similarly, mercury exposure from industrial emissions, coal-fired power plants, and waste incineration continues to be a leading environmental concern, with mercury deposits accumulating in ecosystems and entering the human food chain. Dental amalgams, which contain mercury, have been another source of exposure, particularly in individuals with multiple fillings. While mercury exposure from dental amalgams is considered low compared to other sources, chronic exposure remains a concern, especially in susceptible populations

Heavy Metal Toxicity and Neurodegeneration. DOI: https://doi.org/10.1016/B978-0-443-36575-1.00014-5

such as pregnant women and individuals with preexisting neurological conditions. Another common but often overlooked source of mercury exposure is the use of certain skin-lightening creams and cosmetics, particularly in regions where unregulated beauty products are prevalent. These products may contain inorganic mercury, which can be absorbed through the skin, leading to systemic toxicity and kidney damage. Additionally, vaccines historically contained thimerosal, an ethylmercury-based preservative, which raised concerns about potential health risks, particularly in children. However, extensive scientific research has shown no significant link between thimerosal in vaccines and neurological disorders such as autism, leading to its phased reduction in pediatric vaccines in many countries as a precautionary measure. The widespread use of mercury in traditional medicine and cultural practices also contributes to exposure risks in some communities. Certain ayurvedic and unani medicines have been found to contain mercury, which, when ingested, can lead to chronic poisoning. Similarly, some religious or spiritual practices involve burning mercury-containing substances, exposing individuals to toxic fumes. In the modern era, seafood consumption remains one of the most common sources of mercury exposure, particularly for populations that rely heavily on fish as a dietary staple (Kumar & Choudhury, 2024c, 2024d; Sharma, Devi, & Kumar, 2023). Predatory fish such as tuna, swordfish, and sharks accumulate high levels of methylmercury due to their position in the food chain. Pregnant women, nursing mothers, and young children are especially vulnerable to mercury toxicity from seafood, as it can affect fetal and early childhood brain development. Regulatory agencies, such as the US Environmental Protection Agency (EPA) and the Food and Drug Administration (FDA), have issued guidelines on fish consumption to minimize mercury exposure, emphasizing the importance of choosing low-mercury fish such as salmon, tilapia, and sardines while avoiding high-mercury species. The severity of mercury poisoning depends on the form of mercury, the route of exposure, and the duration of exposure. Acute mercury poisoning, often resulting from occupational accidents or industrial spills, can cause immediate symptoms such as coughing, chest pain, nausea, and neurological disturbances. Chronic exposure, on the other hand, leads to insidious neurological deterioration, manifesting as memory loss, tremors, mood changes, and coordination difficulties. In children, even low levels of mercury exposure can lead to developmental delays, cognitive deficits, and behavioral abnormalities. The recognition of mercury as a global environmental toxin has led to several international efforts to mitigate its impact. The Minamata Convention on Mercury, adopted in 2013, is a landmark global treaty aimed at reducing mercury emissions, phasing out mercury-containing products, and promoting safer alternatives. The treaty has spurred efforts to eliminate mercury in artisanal gold mining, restrict mercury in industrial processes, and implement mercury waste management strategies. While significant progress has been made in regulating mercury emissions and reducing exposure, challenges remain, particularly in developing countries where enforcement of regulations is limited and economic reliance on mercury-intensive industries persists. The continuous monitoring of mercury pollution, advancement in remediation technologies, and global cooperation remain critical in addressing the ongoing threat of mercury poisoning (Dey, Sharma, & Kumar, 2023; Reddy, Kumar, & Dwivedi, 2022; Sharma, Dey, & Kumar, 2023). The public must be educated on the risks of mercury exposure, the importance of dietary choices in minimizing mercury intake, and the need for safer industrial and agricultural practices. As scientific research continues to unveil new insights into mercury toxicity, the development of innovative treatment strategies, such as targeted chelation therapies and neuroprotective interventions, will be essential in mitigating the long-term effects of mercury poisoning. Ultimately, the battle against mercury toxicity is not just a scientific or medical challenge but also an environmental and socioeconomic issue that requires a multidisciplinary approach for effective prevention, management, and remediation.

4.2 Forms of mercury and their neurotoxicity

Mercury exists in three primary forms—elemental (Hg^0), inorganic (Hg^{2+}), and organic (methylmercury and ethylmercury), each with distinct routes of exposure, toxicokinetics, and effects on the human body. Among these, the CNS is particularly vulnerable to mercury toxicity due to its high lipid content, complex neural architecture, and dependence on neurotransmitter balance for optimal function. The neurotoxicity of mercury varies significantly across its different forms, with elemental and organic mercury posing the greatest threat to the CNS. In contrast, inorganic mercury exhibits more pronounced effects on peripheral organs such as the kidneys. Understanding the mechanisms of toxicity associated with these mercury species is essential for developing effective prevention and treatment strategies for mercury poisoning. This underscores the urgent need for regulatory measures to control mercury exposure and protect public health. Elemental mercury (Hg^0) is the most volatile form of mercury and is commonly found in thermometers, barometers, fluorescent lamps, dental amalgams, and some industrial processes. While it is relatively inert in liquid form, exposure occurs primarily through inhalation of mercury vapor, which is easily absorbed by the lungs, leading to systemic distribution and neurotoxic effects.

Unlike inorganic mercury compounds, which cannot cross the blood–brain barrier (BBB), inhaled elemental mercury is highly lipophilic, allowing it to penetrate the CNS and accumulate in brain tissues (Dey, Devi, & Kumar, 2023; Kumar & Mistri, 2020; Sharma, Devi, Dey, & Kumar, 2023; Sinam, Kumar, & Singh, 2023).

Once inside the CNS, elemental mercury undergoes oxidation to Hg^{2+}, which becomes trapped in the brain, causing prolonged toxic effects. The neurotoxicity of elemental mercury primarily results from oxidative stress, disruption of neurotransmitter systems, and interference with neuronal function. Elemental mercury can induce significant alterations in the levels of dopamine, serotonin, and glutamate, leading to cognitive impairments, mood disturbances, and motor dysfunction. Additionally, mercury disrupts calcium homeostasis in neurons, leading to excitotoxicity, neuronal apoptosis, and demyelination. Chronic exposure to elemental mercury vapour has been linked to a range of neuropsychiatric symptoms, including anxiety, depression, memory deficits, tremors, and irritability a condition historically referred to as "mad hatter syndrome," named after 19th-century hat makers who suffered from mercury toxicity due to prolonged exposure to mercuric nitrate used in felting processes. Severe cases of elemental mercury poisoning may result in tremors, ataxia, muscle weakness, and cognitive decline, symptoms reminiscent of neurodegenerative disorders such as Parkinson's and Alzheimer's diseases (Devi, Dey, Saini, et al., 2023; Dey et al., 2023; Goud, Singh, & Kumar, 2022; Kumar, Sharma, Saini, & Dey, 2021; Sharma, Devi, Kumar, Dey, & Dwivedi, 2023). Occupational exposure to elemental mercury remains a concern in industries that handle mercury-containing equipment, gold mining, and dental practices where amalgam fillings are used. Additionally, improper disposal or accidental breakage of mercury-containing items, such as thermometers or fluorescent bulbs, can lead to environmental contamination and indoor air exposure. Mercury vapour exposure is particularly hazardous in poorly ventilated spaces, where prolonged inhalation can lead to high systemic mercury levels.

Inorganic mercury (Hg^{2+}) is found in mercury salts such as mercuric chloride ($HgCl_2$) and mercuric sulfide (HgS). Unlike elemental and organic mercury, inorganic mercury has a limited ability to cross the BBB due to its hydrophilic nature. As a result, its neurotoxic effects are less pronounced in the CNS but more prominent in peripheral organs, particularly the kidneys. However, chronic exposure to inorganic mercury can still result in significant neurological symptoms, mainly due to its indirect effects on the nervous system. One of the essential mechanisms of inorganic mercury toxicity is oxidative stress. Inorganic mercury interacts with sulfhydryl (–SH) groups in proteins, leading to enzyme inhibition, mitochondrial dysfunction, and lipid peroxidation. This oxidative stress can contribute to neuroinflammation, peripheral neuropathy, and immunological disturbances. Furthermore, inorganic mercury disrupts calcium signaling and cellular homeostasis, leading to cell death in multiple organ systems. Peripheral neurotoxicity caused by inorganic mercury often manifests as paresthesia, muscle weakness, and coordination deficits. Patients with chronic exposure may develop tremors, mood swings, and memory impairment, although these effects are generally less severe than those observed with elemental or organic mercury exposure. Additionally, inorganic mercury has been associated with autoimmune dysfunctions, including nephrotic syndrome and autoimmune glomerulonephritis, further exacerbating systemic health problems (Dey, Kumar, & Singh, 2023; Karam, Kumar, Anand, & Choudhury, 2023; Kumar et al., 2020; Kumar, 2023; Sharma & Kumar, 2023).

Sources of inorganic mercury exposure include industrial emissions, contaminated food and water, certain skin-lightening creams, and some traditional medicines. Inorganic mercury is a byproduct of mercury mining and refining processes, making workers particularly vulnerable. Dental professionals and individuals with multiple mercury-containing dental amalgams may also experience minor exposure, although the clinical significance remains debated. Another significant source of inorganic mercury exposure is illegal gold mining operations, where mercury compounds extract gold from ore. Mercury runoff from these processes contaminates soil and water, threatening human populations and ecosystems. Chronic exposure in affected communities has led to increased cases of mercury-induced kidney and neurological disorders. Although inorganic mercury has limited CNS penetration, some studies suggest that it can contribute to neurodegenerative processes when combined with other environmental toxins or predisposing genetic factors. When exposure is sustained over long periods, it may exacerbate neuroinflammatory conditions and lead to cumulative neurotoxicity. Organic mercury compounds, particularly methylmercury (MeHg) and ethylmercury, pose the most significant risk to the CNS due to their high bioavailability, long biological half-life, and potent neurotoxicity. Methylmercury is formed primarily through microbial methylation of inorganic mercury in aquatic environments, leading to its accumulation in fish and seafood. This form of mercury readily crosses the BBB and placental barrier, making it particularly dangerous for pregnant women and developing fetuses. The primary mechanism of methylmercury toxicity involves interacting with sulfhydryl groups in proteins, disrupting essential cellular functions (Aley, Singh, & Kumar, 2022; Hidangmayum, Dwivedi, Kumar, & Upadhyay, 2023; Kumar & Kumar, 2023; Kumar, Pathak, Kumar, & Dwivedi, 2018). Methylmercury binds to neuronal proteins and enzymes, leading to mitochondrial damage, oxidative stress, and impaired neurotransmitter function. One of the most critical effects of methylmercury is its disruption of glutamate homeostasis, resulting in excitotoxicity and neuronal death. This is particularly detrimental to developing brains, where neuronal plasticity and synaptic connections are still forming.

Severe methylmercury poisoning leads to devastating neurological conditions, as seen in historical cases such as Minamata disease. Chronic exposure to methylmercury results in sensory deficits, ataxia, muscle weakness, speech impairment, and cognitive decline. In infants exposed to high levels of methylmercury in utero, profound developmental delays, microcephaly, cerebral palsy, and intellectual disabilities can occur. Even low-level exposure has been associated with subtle cognitive impairments, reduced IQ, and behavioral changes in children. Ethylmercury, another organic mercury compound, is commonly found in the preservative thimerosal, which has been used in some vaccines and pharmaceutical products. However, ethylmercury is rapidly metabolized and excreted compared to methylmercury, reducing its potential for bioaccumulation and neurotoxicity. Although concerns were raised regarding thimerosal and its possible link to autism, extensive research has found no credible evidence supporting this association, leading to its continued use in multidose vaccine preparations in some regions. Dietary exposure to methylmercury remains the most significant concern, particularly for populations consuming large amounts of predatory fish such as tuna, shark, swordfish, and king mackerel. Regulatory agencies, including the World Health Organization (WHO) and the US FDA, have issued guidelines recommending limited consumption of high-mercury fish, particularly for pregnant women, nursing mothers, and young children (Das et al., 2022; Devi & Kumar, 2023a, 2023b; Kumari, Singh, & Kumar, 2022; Kumari, Kumar, & Singh, 2023).

The bioaccumulation of methylmercury in aquatic food chains underscores the importance of environmental monitoring and pollution control efforts. Mercury emissions from coal-fired power plants, waste incineration, and industrial discharge contribute to methylmercury contamination in water bodies, posing a long-term risk to human and environmental health. International agreements such as the Minamata Convention aim to reduce mercury emissions and limit the use of mercury in industrial processes to mitigate these risks. The long-term neurological effects of methylmercury exposure remain a significant concern, as even subclinical exposure can lead to subtle cognitive and motor impairments over time. Emerging research suggests potential links between chronic low-level mercury exposure and neurodegenerative disorders such as Alzheimer's and Parkinson's diseases, highlighting the need for continued investigation into the long-term consequences of mercury toxicity (Kumar & Dwivedi, 2020; Kumar, Mandal, & Dwivedi, 2011a; Kumar, Latha, & Singh, 2023; Kumar, Saini, & Sharma, 2023; Pankaj, Kumar, Nigam, & Mishra, 2012a; Saini, Sharma, & Kumar, 2023). The neurotoxic effects of mercury vary depending on its chemical form, route of exposure, and duration of exposure. Elemental mercury primarily affects the CNS through inhalation and subsequent oxidation in the brain, leading to chronic neurological dysfunction. Inorganic mercury, while less neurotoxic, can contribute to peripheral neuropathy and systemic toxicity. Organic mercury, particularly methylmercury, poses the most significant threat due to its ability to bioaccumulate and cross critical biological barriers, leading to severe and irreversible CNS damage. Understanding the mechanisms of mercury toxicity and implementing regulatory measures to reduce exposure are crucial steps in protecting human health from this pervasive environmental toxin (Table 4.1).

4.3 Mechanisms of mercury-induced neurotoxicity

Mercury toxicity has devastating effects on the CNS through multiple interrelated mechanisms, including oxidative stress, mitochondrial dysfunction, neurotransmitter disruption, neuronal apoptosis, demyelination, and the breakdown of the BBB. These mechanisms collectively contribute to progressive neurological deterioration, cognitive impairment, motor dysfunction, and increased susceptibility to neurodegenerative diseases. Mercury's neurotoxicity is primarily mediated by its strong affinity for thiol (–SH) groups in proteins, which disrupts enzymatic functions, impairs cellular homeostasis, and leads to widespread neuronal damage. Among the different forms of mercury, methylmercury (MeHg) and elemental mercury vapour (Hg^0) pose the greatest threat to the CNS due to their ability to penetrate the BBB and accumulate in brain tissues. Understanding mercury-induced neurotoxicity's biochemical and molecular basis is critical for developing strategies to mitigate its harmful effects and improve neuroprotection. One of the primary mechanisms of mercury-induced neurotoxicity is oxidative stress, which results from an imbalance between reactive oxygen species (ROS) production and the brain's antioxidant defense systems. Mercury interacts with cellular components, particularly thiol-containing proteins, leading to the inhibition of antioxidant enzymes such as superoxide dismutase (SOD), catalase, and glutathione peroxidase. This inhibition reduces the brain's ability to neutralize free radicals, causing an accumulation of ROS, which in turn damages lipids, proteins, and DNA. Lipid peroxidation is particularly concerning in the brain due to its high content of polyunsaturated fatty acids, which are highly susceptible to oxidative damage (Devi, Dey, & Kumar, 2023; Kumar, Mandal, & Dwivedi, 2013; Kumar, Harshavardhan, et al., 2018; Kumar, Pathak, Amarnath, et al., 2018; Pathak,

TABLE 4.1 Classification of mercury and its toxicological properties (Kumar et al., 2021; Naik & Kumar, 2020; Pankaj, Kumar, Nigam, & Mishra, 2012b).

Mercury form	Chemical structure	Common sources	Toxicokinetic	Associated health effects
Elemental mercury (Hg^0)	Hg (liquid at room temperature)	-Thermometers, barometers, dental amalgams, artisanal gold mining, industrial emissions	-Absorbed mainly via inhalation (80%) but poorly through the skin and GI tract.-Easily crosses the blood–brain barrier and accumulates in the CNS.-Excreted through urine and feces.	-Tremors, mood swings, memory loss, and cognitive dysfunction (CNS toxicity).-Respiratory failure from high acute exposure.-Kidney damage at prolonged exposure.
Inorganic mercury (Hg^+, Hg^{2+})	$HgCl_2$ (mercuric chloride), Hg_2Cl_2 (mercurous chloride)	-Industrial waste, chloralkali industry, batteries, fungicides, cosmetics (skin-lightening creams), some traditional medicines	-Poor GI absorption (~7%–15%).-Does not cross the blood-brain barrier easily.-Accumulates in the kidneys, leading to nephrotoxicity.	-Severe kidney damage (nephrotoxicity).-Corrosive gastrointestinal effects (stomach pain, vomiting, diarrhea).-Dermatitis and hypersensitivity reactions.
Organic mercury (methylmercury, ethylmercury, phenylmercury)	CH_3Hg^+ (methylmercury), $C_2H_5Hg^+$ (ethylmercury)	-Contaminated seafood (methylmercury from bioaccumulation in fish), vaccines (ethylmercury from thimerosal preservative), fungicides (phenylmercury)	-Highly absorbed (> 90%) via the GI tract.-Readily crosses the placenta and blood–brain barrier, accumulating in fetal tissues and brain.-Excreted very slowly (half-life ~50 days).	-Severe neurotoxicity: cognitive impairment, developmental delays, motor dysfunction.-Prenatal exposure is linked to IQ deficits, cerebral palsy, and autism spectrum disorders.-Ataxia, hearing loss, visual disturbances (Minamata disease).

CNS, Central nervous system.

Source: Based on Devi, P., Dey, S. R., Saini, L., Kumar, P., Panigrahi, S., Dwivedi, P. (2023). Toward sustainable agriculture: Strategies involving phytoprotectants against reactive oxygen species. In *Reactive oxygen species: Prospects in plant metabolism* (pp. 229–247). Phagwara, India: Springer Nature, Department of Agronomy, School of Agriculture, Lovely Professional University. https://doi.org/10.1007/978-981-19-9794-5_13; Kumar, V., Dwivedi, P., Kumar, P., Singh, B. N., Pandey, D. K., Kumar, V., Bose, B. (2021). Mitigation of heat stress responses in crops using nitrate primed seeds. *South African Journal of Botany, 140*, 25–36. https://doi.org/10.1016/j.sajb.2021.03.024; Naik, M., Kumar, P. (2020). Role of growth regulators and microbes for metal detoxification in plants and soil. *Plant Archives, 20*, 2820–2824; Pankaj, P. K., Kumar, P., Nigam, R. C., Mishra, P. K. (2012b). Monitoring and surveillance of synthetic pyrethroids and organophosphate in different brands of soft drinks. *Journal of Chemical and Pharmaceutical Research, 4*, 3939–3943.

Kumar, Mishra, & Kumar, 2017). The peroxidation of neuronal membranes leads to loss of membrane integrity, impaired signal transduction, and increased neuronal vulnerability to apoptosis. Mitochondria, the primary energy-producing organelles in cells, are significant targets of mercury toxicity. Mercury disrupts mitochondrial function by inhibiting electron transport chain complexes, leading to reduced ATP production and increased production of ROS. The inhibition of mitochondrial complexes I and III results in electron leakage, promoting the formation of superoxide radicals that further damage mitochondrial DNA (mtDNA). Unlike nuclear DNA, mtDNA lacks robust repair mechanisms, making it highly vulnerable to oxidative damage. This leads to mitochondrial dysfunction, energy deficits, and neuronal death. Additionally, mercury-induced mitochondrial dysfunction triggers the release of cytochrome c and activation of caspase-dependent apoptotic pathways, contributing to widespread neurodegeneration.

In addition to oxidative stress and mitochondrial dysfunction, mercury toxicity severely disrupts neurotransmitter systems, particularly those involving dopamine, glutamate, and gamma-aminobutyric acid (GABA). Dopaminergic neurons are particularly susceptible to mercury toxicity due to their high metabolic activity and reliance on oxidative phosphorylation. Mercury-induced oxidative stress reduces dopamine synthesis by inhibiting tyrosine hydroxylase, the rate-limiting enzyme in dopamine production. Furthermore, mercury disrupts dopamine storage and release by impairing vesicular monoamine transporter function, reducing synaptic dopamine availability. These disruptions contribute to movement disorders, cognitive deficits, and psychiatric symptoms similar to those observed in Parkinson's disease. Mercury exposure has also been linked to alterations in glutamatergic neurotransmission, which plays a critical role in excitatory signaling, synaptic plasticity, and memory formation. Mercury disrupts glutamate homeostasis by inhibiting glutamate uptake by astrocytes, leading to excessive extracellular glutamate accumulation. This results in excitotoxicity, a pathological process in which excessive activation of *N*-methyl-*D*-aspartate (NMDA) receptors leads to calcium influx, mitochondrial dysfunction, and neuronal death. Excitotoxicity is a key contributor to neurodegenerative diseases such as Alzheimer's and Huntington's disease, highlighting the long-term consequences of mercury exposure on brain health (Kumar & Naik, 2020; Kumar & Pathak, 2019; Kumar, Rameshwar, & Kumar, 2019; Kumar, Kumar, et al., 2018; Kumar, Yumnam, et al., 2018). Conversely, mercury also interferes with GABAergic neurotransmission, essential for inhibitory signaling in the CNS. Mercury reduces GABA receptor function, leading to increased neuronal excitability and an imbalance between excitatory and inhibitory neurotransmission. This imbalance manifests as anxiety, seizures, and motor dysfunction, further exacerbating the neurological consequences of mercury exposure.

Neuronal apoptosis and demyelination are critical pathological outcomes of mercury toxicity that contribute to progressive neurodegeneration. Mercury-induced oxidative stress and mitochondrial dysfunction activate intrinsic apoptotic pathways, leading to programmed cell death in neurons and activating caspase enzymes, particularly caspase-3 and caspase-9, resulting in DNA fragmentation, chromatin condensation, and cellular disintegration. Prolonged mercury exposure leads to widespread neuronal loss, particularly in cognition, motor control, and sensory processing regions. The cerebellum, hippocampus, and cerebral cortex are particularly vulnerable to mercury-induced apoptosis, which explains the cognitive impairments, motor deficits, and sensory disturbances observed in mercury poisoning cases. In addition to neuronal apoptosis, mercury toxicity damages oligodendrocytes, the cells responsible for producing myelin, the protective sheath that insulates axons and facilitates rapid nerve conduction. Mercury disrupts the synthesis and maintenance of myelin proteins, leading to demyelination and impaired neuronal communication. Demyelination results in conduction block, reduced signal transmission speed, and increased susceptibility to neurodegenerative disorders such as multiple sclerosis (MS). Losing myelin integrity also exposes axons to further oxidative damage, exacerbating neuroinflammation and neuronal degeneration (Kumar, Singh, & Dwivedi, 2017; Kumar, Devi, & Dey, 2021; Kumar, Goud, Devi, & Koul, 2022; Kumar, Koul, & Sharma, 2022).

Another critical aspect of mercury-induced neurotoxicity is its ability to compromise the integrity of the BBB, a selective barrier that protects the brain from harmful substances. The BBB is composed of endothelial cells, astrocytes, and pericytes, which work together to regulate the passage of molecules into the CNS. Mercury disrupts BBB integrity by inducing oxidative stress, inflammation, and endothelial dysfunction. It damages tight junction proteins such as occludin and claudin, increasing barrier permeability. This increased permeability allows neurotoxic substances, immune cells, and inflammatory mediators to infiltrate the brain, exacerbating neuronal damage and inflammation. The disruption of the BBB also facilitates the accumulation of mercury within the CNS, creating a vicious cycle of toxicity and neurodegeneration. Mercury-induced BBB dysfunction has been implicated in the pathogenesis of neurodegenerative diseases such as Alzheimer's and Parkinson's disease, as increased BBB permeability is a hallmark feature of these disorders (Chakraborty et al., 2021; Dwivedi & Kumar, 2011; Kumar, Mandal, & Dwivedi, 2011b; Kumar, Goud, et al., 2022; Upadhyay et al., 2023). Furthermore, mercury exposure has been linked to neuroinflammation, characterized by the activation of microglia and astrocytes, the resident immune cells of the CNS. Chronic microglial activation results in the release of proinflammatory cytokines such as tumor necrosis factor-alpha (TNF-α), interleukin-1β (IL-1β), and interleukin-6 (IL-6), which amplify neuronal damage and contribute to neurodegenerative processes.

The long-term neurological consequences of mercury exposure extend beyond immediate neurotoxicity, with evidence suggesting a link between chronic mercury exposure and an increased risk of neurodegenerative diseases. Mercury's ability to induce oxidative stress, disrupt neurotransmitter systems, promote apoptosis, and compromise BBB integrity creates an environment conducive to developing conditions such as Alzheimer's disease, Parkinson's disease, and amyotrophic lateral sclerosis (ALS). Studies have shown that mercury exposure leads to the accumulation of amyloid-beta plaques and tau protein hyperphosphorylation, both of which are hallmarks of Alzheimer's disease. Similarly, mercury's detrimental effects on dopaminergic neurons mirror the pathophysiology of Parkinson's disease, suggesting a potential role for mercury in the etiology of these disorders (Chakraborty et al., 2021; Devi *et al.* 2023; Kumar et al., 2011b; Kumar et al., 2017; Kumar et al., 2019). The neurotoxic effects of mercury are particularly concerning for vulnerable populations, including pregnant women, infants, and individuals with preexisting neurological conditions. Prenatal mercury exposure has been associated with developmental delays, intellectual disabilities, and motor impairments in children. Mercury readily crosses the placenta and accumulates in fetal brain tissues, disrupting neurodevelopment and increasing the risk of neurodevelopmental disorders such as autism spectrum disorder (ASD) and attention-deficit/hyperactivity disorder (ADHD).

The complexity of mercury-induced neurotoxicity underscores the need for continued research and regulatory measures to reduce mercury exposure and mitigate its harmful effects. Public health initiatives aimed at reducing mercury emissions, limiting industrial mercury use, and promoting safer alternatives to mercury-containing products are crucial for minimizing the risk of mercury poisoning. Additionally, dietary guidelines recommending reduced consumption of high-mercury fish, particularly for pregnant women and young children, can help prevent methylmercury-related neurotoxicity. Advances in neuroprotective strategies, including antioxidant therapy, mitochondrial-targeted interventions, and pharmacological agents that enhance mercury excretion, hold promise for mitigating the impact of mercury on the CNS. Chelation therapy using agents such as dimercaptosuccinic acid (DMSA) and dimercaprol (BAL) has been used to facilitate mercury elimination from the body. However, its efficacy in treating chronic mercury neurotoxicity remains an area of ongoing investigation. Mercury exerts profound neurotoxic effects through oxidative stress, mitochondrial dysfunction, neurotransmitter disruption, neuronal apoptosis, demyelination, and BBB breakdown. These mechanisms collectively contribute to cognitive impairment, motor dysfunction, and an increased risk of neurodegenerative diseases. Understanding the intricate pathways of mercury-induced neurotoxicity is essential for developing effective prevention and treatment strategies, ultimately safeguarding public health from the devastating impact of mercury poisoning (Table 4.2).

4.4 Neurodevelopmental and cognitive impacts

The effects of environmental, genetic, and chemical factors on fetal and childhood brain development have been a subject of extensive scientific inquiry, with growing evidence suggesting that exposure to neurotoxic agents during critical developmental windows can lead to long-term cognitive, behavioral, and neurological impairments. The fetal brain is particularly vulnerable to harmful substances due to its rapid growth and differentiation, where exposure to heavy metals, pesticides, endocrine disruptors, and air pollutants can result in structural and functional anomalies. One of the most concerning neurotoxic agents is lead, which, even at low levels, has been linked to reductions in IQ, attentional deficits, and behavioral problems in children. Similarly, prenatal exposure to mercury, often through maternal consumption of contaminated seafood, can impair neuronal migration, synapse formation, and neurotransmitter regulation, leading to deficits in motor skills, language acquisition, and executive function. Additionally, maternal alcohol exposure, as seen in fetal alcohol spectrum disorders, results in significant neurodevelopmental disorders characterized by reduced brain volume, poor impulse control, learning disabilities, and social difficulties (Kumar & Pathak, 2019; Kumar et al., 2017; Kumar et al., 2022; Kumar, Goud, Devi, Dey, & Dwivedi, 2022; Pathak et al., 2017). Similarly, prenatal exposure to high levels of maternal stress, which leads to increased cortisol levels, can alter fetal brain programming by affecting the development of the hypothalamic–pituitary–adrenal (HPA) axis, ultimately predisposing children to anxiety, depression, and cognitive dysfunction. Furthermore, insufficient maternal nutrition, particularly deficiencies in folate, iodine, and omega-3 fatty acids, has been linked to compromised fetal brain development, increasing the risk of neural tube defects, intellectual disabilities, and poor academic performance later in life. The role of air pollution, excellent particulate matter (PM2.5), has also emerged as a critical factor influencing neurodevelopment, with studies showing that prenatal and early-life exposure

TABLE 4.2 Sources of mercury exposure, neurotoxic effects, and prevention strategies.

Source of exposure	Primary route of exposure	Most affected populations	Key neurotoxic effects	Prevention strategies
Industrial emissions (coal-fired power plants, cement production, metal smelting, chloralkali industry)	Inhalation of mercury vapor from air pollution	-Industrial workers-Communities near factories	-Cognitive decline, memory loss, mood disorders-Chronic exposure: tremors, insomnia, neuropathy	-Strict emissions control and mercury-free industrial processes-Use of air filtration systems and masks for workers
Environmental pollution (contaminated water bodies, mercury spills, soil contamination)	Inhalation (mercury vapor), dermal absorption, ingestion of contaminated water or crops	-People living near polluted areas-Farmers using contaminated water	-Developmental delays, reduced motor coordination-Neuroinflammation and oxidative stress	-Soil and water remediation technologies-Regulations on industrial waste disposal
Dietary exposure (methylmercury in seafood) (large predatory fish: shark, swordfish, tuna)	Ingestion (bioaccumulation in fish)	-Pregnant women and fetuses-Children-Populations relying on fish-based diets	-Prenatal exposure: low IQ, autism spectrum disorders, cerebral palsy-Visual and auditory impairments	-Seafood consumption guidelines (FDA and EPA limits)-Preference for low-mercury fish (e.g., salmon, shrimp, tilapia)
Dental amalgams (silver-mercury fillings)	Inhalation of mercury vapor (slow release over time)	-Individuals with multiple amalgam fillings-Dentists and dental assistants	-Chronic exposure: subtle cognitive deficits, anxiety, mood changes	-Use of mercury-free dental materials-Ventilation systems in dental clinics
Occupational exposure (miners, factory workers, healthcare workers, researchers handling mercury)	Inhalation, dermal absorption	-Gold miners (artisanal and small-scale mining)-Chemical and factory workers-Laboratory personnel	-Severe neurotoxicity: tremors, ataxia, personality changes-Kidney damage from inorganic mercury	-Protective equipment (masks, gloves)-Safer alternatives for gold extraction (cyanidation vs mercury amalgamation)
Traditional and herbal medicines (ayurvedic, unani, Chinese remedies containing mercury)	Ingestion of mercury-containing formulations	-Consumers of traditional medicine	-CNS toxicity, kidney damage, tremors	-Strict regulations on mercury content in herbal products-Public awareness campaigns
Mercury-containing household Items (thermometers, barometers, fluorescent lamps, batteries)	Inhalation (if broken), dermal absorption	-General public-Children (accidental exposure)	-Acute exposure: dizziness, nausea, headache-Chronic exposure: cognitive and motor dysfunction	-Safe disposal and recycling of mercury-containing items-Ban on mercury thermometers in many countries

CNS, Central nervous system; *EPA*, Environmental Protection Agency; *FDA*, Food and Drug Administration.

Source: Based on Dwivedi, P., Kumar, P. (2011). Anti-diabetic medicinal plants and their conservation: Waging green war on diabetes. *Medicinal Plants, 3*, 181–189. https://doi.org/10.5958/j.0975-4261.3.3.031; Kumar, P., Goud, E. L., Devi, P., Koul, B. (2022). Metal pollutants in the environment. In *Environmental microbiology: Emerging technologies* (pp. 291–323). Jalandhar, Punjab, India: De Gruyter, Department of Agronomy, School of Agriculture, Lovely Professional University. https://doi.org/10.1515/9783110727227-012; Kumar, P., Singh, B. N., Dwivedi, P. (2017). Plant growth regulators, plant adaptability and plant productivity: A review on abscisic acid (ABA) signaling in plants under emerging environmental stresses. In *Sustaining future food security in changing environments* (pp. 81–97). Varanasi, India: Nova Science Publishers, Inc., Department of Plant Physiology, Institute of Agricultural Sciences, Banaras Hindu University; Kumar, P., Mandal, B., Dwivedi, P. (2013). Combating heavy metal toxicity from hazardous waste sites by harnessing scavenging activity of some vegetable plants. *Vegetos, 26*, 416–425. https://doi.org/10.5958/j.2229-4473.26.2.106; Kumar, P., Kumar, S., Harshavardhan, M., Naik, M., Yumnam, J., Kumar, P. S., Purnima (2018). Evaluation of plant height and leaf length of sorghum grown under different sources of nutrition. *Annals of Biology, 34*, 284–286.

to traffic-related air pollution can lead to reductions in cortical thickness, disrupted white matter integrity, and increased risk of ADHD and ASD.

Beyond environmental toxins, childhood brain development can also be impaired by early-life stress, infections, and inflammation, which can alter synaptic plasticity, neurotransmitter balance, and myelination processes. Early childhood experiences play a crucial role in shaping cognitive and behavioral trajectories, where adverse childhood experiences, such as neglect, abuse, and household dysfunction, have been strongly associated with alterations in brain structure and function. Studies using neuroimaging techniques have revealed that children exposed to chronic stress exhibit reductions in hippocampal volume, which is critical for memory formation, along with hyperactivity of the amygdala, leading to heightened emotional reactivity and anxiety disorders. These neural changes underpin impairments in learning, memory, and behavior, as seen in conditions like PTSD, oppositional defiant disorder, and conduct disorder. Moreover, early childhood infections, such as bacterial meningitis, viral encephalitis, and severe cases of COVID-19, have been linked to neuroinflammation and cognitive deficits due to the disruption of the BBB and subsequent neuronal damage. Neuroinflammatory processes triggered by infections can lead to prolonged microglial activation, which, if excessive, can cause synaptic pruning abnormalities and long-term impairments in learning and executive function. Furthermore, metabolic disorders such as congenital hypothyroidism and phenylketonuria (PKU) significantly impact brain development when left untreated, with hypothyroidism leading to severe intellectual disabilities and PKU causing progressive cognitive decline due to the accumulation of neurotoxic phenylalanine (Kumar & Naik, 2020; Kumar et al., 2011b; Kumar, Devi, et al., 2021; Kumar, Goud, Devi, Dey, et al., 2022; Kumar, Harshavardhan, et al., 2018).

Genetic predispositions interacting with environmental factors also strongly influence learning, memory, and behavior impairment. Disorders such as dyslexia, ADHD, and specific learning disabilities have been linked to genetic variations affecting neuronal connectivity, neurotransmitter pathways, and cortical development. For instance, mutations in the FOXP2 gene, which plays a crucial role in speech and language processing, have been associated with developmental language disorders characterized by impaired verbal comprehension and articulation difficulties. ADHD, a condition marked by inattention, hyperactivity, and impulsivity, has been linked to polymorphisms in dopamine receptor genes (DRD4 and DRD5), suggesting a strong neurochemical basis for attention regulation. Moreover, studies have shown that children with ADHD exhibit reduced activation in the prefrontal cortex, leading to difficulties in working memory, impulse control, and goal-directed behavior. Similarly, dyslexia, a neurodevelopmental disorder affecting reading and phonological processing, has been associated with abnormalities in the left temporoparietal cortex, a region critical for language decoding and word recognition. These cognitive impairments have far-reaching implications for academic achievement, social interactions, and emotional well-being, often requiring targeted interventions such as behavioral therapy, pharmacological treatments, and specialized educational support (Chakraborty et al., 2021; Kumar et al., 2017, 2019, 2022; Kumar, Harshavardhan, et al., 2018).

ASDs, a group of neurodevelopmental conditions characterized by social communication deficits, repetitive behaviors, and restricted interests, have been increasingly linked to both genetic and environmental factors. Research has identified over 1000 risk genes associated with ASD, including mutations in the SHANK3, CHD8, and NRXN1 genes involved in synaptic formation, neuronal migration, and neurotransmission. However, genetic susceptibility alone does not account for the rising prevalence of ASD, with growing evidence implicating environmental factors such as prenatal exposure to valproic acid, maternal infections, air pollution, and endocrine-disrupting chemicals in the etiology of autism. For example, maternal immune activation caused by infections during pregnancy has been shown to increase proinflammatory cytokines such as interleukin-6 (IL-6) and tumor necrosis factor-alpha (TNF-α), leading to altered brain development and ASD-like behaviors in animal models. Additionally, studies have linked ASD to disruptions in the gut microbiome, suggesting that microbiota–gut–brain axis dysregulation may contribute to neurodevelopmental abnormalities through mechanisms involving short-chain fatty acids, immune modulation, and vagus nerve signaling. Furthermore, children with ASD often exhibit atypical neural connectivity patterns, with hyperconnectivity in local brain regions but reduced long-range connectivity, leading to deficits in information integration, sensory processing, and social cognition.

Beyond neurodevelopmental disorders, emerging research has also drawn links between early-life neurotoxic exposures and an increased risk of neurodegenerative diseases in adulthood. Conditions such as Alzheimer's disease, Parkinson's disease, and ALS have been hypothesized to have developmental origins, where early-life disruptions in neurodevelopment predispose individuals to accelerated cognitive decline and neuronal degeneration later in life (Kumar et al., 2019; Kumar, Devi, et al., 2021; Kumar, Harshavardhan, et al., 2018). For instance, lead exposure during childhood has been associated with an increased risk of Alzheimer's disease due to its ability to induce oxidative stress, disrupt mitochondrial function, and promote amyloid-beta plaque accumulation. Similarly, pesticide exposure, particularly organophosphates, has been implicated in Parkinson's disease due to its neurotoxic effects on dopaminergic neurons in the substantia nigra, a hallmark of Parkinsonian pathology. Moreover, studies suggest that neuroinflammation during fetal or early childhood development may create a "preconditioned" brain state, making individuals more susceptible to neurodegenerative processes in later life. This concept, often referred to as the "developmental origins of neurodegeneration," posits that early insults to the developing brain can lead to latent vulnerabilities that manifest decades later as cognitive decline, movement disorders, or dementia.

Overall, the intricate interplay between genetic, environmental, and epigenetic factors shapes the trajectory of fetal and childhood brain development, with profound implications for cognitive function, behavior, and long-term neurological health. The growing body of research underscores the importance of mitigating early-life exposures to neurotoxic substances, optimizing maternal and child nutrition, and implementing early interventions to support neurodevelopmental resilience. As scientific understanding of brain plasticity and neurodevelopmental disorders advances, targeted therapies, precision medicine, and preventive strategies will be crucial in improving cognitive outcomes and reducing the burden of neurological and neuropsychiatric disorders across the lifespan (Table 4.3).

4.5 Clinical manifestations of mercury toxicity in the central nervous system

Sensory disturbances, motor dysfunction, and psychological or psychiatric symptoms represent critical manifestations of various neurological disorders, often arising from underlying structural, metabolic, autoimmune, or neurodegenerative processes. Sensory disturbances, including paresthesia, vision impairment, and hearing loss, are commonly associated with neuropathic conditions, metabolic imbalances, toxic exposures, and neuroinflammatory diseases. Paresthesia, characterized by abnormal sensations such as tingling, numbness, burning, or prickling, typically results from peripheral nerve dysfunction, CNS involvement, or disruptions in sensory pathways. It is frequently observed in conditions such as diabetic neuropathy, where chronic hyperglycemia leads to nerve damage through oxidative stress, microvascular dysfunction, and

TABLE 4.3 Neurotoxic mechanisms of mercury and their clinical consequences.

Neurotoxic mechanism	Description	Targeted brain regions	Clinical consequences
Oxidative stress	Mercury induces excessive reactive oxygen species (ROS) and inhibits antioxidant defense (e.g., glutathione depletion), leading to neuronal damage.	-Cerebral cortex (cognitive function)-Hippocampus (memory formation)-Cerebellum (motor control)	-Memory loss, cognitive impairment-Increased risk of neurodegenerative diseases (e.g., Alzheimer's, Parkinson's)-Motor dysfunction (ataxia, tremors)
Mitochondrial dysfunction	Mercury disrupts mitochondrial enzymes, impairs ATP production, and leads to cellular energy deficits, causing neuronal apoptosis.	-Hippocampus-Basal ganglia (movement regulation)-Frontal cortex (executive functions)	-Fatigue, muscle weakness-Neurodegeneration, reduced learning ability-Motor disorders similar to Parkinson's disease
Neurotransmitter disruption	Mercury alters dopamine, serotonin, and glutamate levels, affecting neural signaling and plasticity.	-Striatum (dopaminergic system)-Prefrontal cortex (executive function, mood)	-Mood swings, anxiety, depression -Impulsivity, attention deficits (linked to ADHD-like symptoms)-Movement disorders (similar to Parkinson's)
Neuroinflammation	Mercury activates microglia and astrocytes, leading to chronic inflammation, neuronal death, and synaptic dysfunction.	-Cerebral cortex-Hippocampus-Amygdala (emotional regulation)	-Increased risk of neurodevelopmental disorders (e.g., autism spectrum disorder, schizophrenia)-Cognitive decline, emotional instability

Source: Based on Chakraborty, S., Kumar, P., Sanyal, R., Mane, A. B., Arvind Prasanth, D., Patil, M., Dey, A. (2021). Unravelling the regulatory role of miRNAs in secondary metabolite production in medicinal crops. *Plant Gene, 27,* 100303. https://doi.org/10.1016/j.plgene.2021.100303; Devi, P., Dey, S. R., Kumar, P. (2023). Integration of water resources management in rural areas. In *Water resources management for rural development: Challenges and mitigation* (pp. 139–152). Punjab, Phagwara, India: Elsevier, Department of Agronomy, School of Agriculture, Lovely Professional University. https://doi.org/10.1016/B978-0-443-18778-0.00013-1; Kumar, P., Mandal, B., Dwivedi, P. (2011). Heavy metal scavenging capacity of Mentha spicata and Allium cepa. *Medicinal Plants, 3,* 315–318. https://doi.org/10.5958/j.0975-4261.3.4.053; Kumar, P., Yumnam, J., Kumar, P.S., Misao, L., Jyoti, N., Naik, M.Harshavardhan, M. (2018). Cadmium induced changes in germination of maize seed treated with mycorrhiza. *Annals of Agri-bio Research, 23,* 169–170; Kumar, P., Pathak, S. (2019). Responsiveness index of sorghum (Sorghum bicolor (L.) Moench) grown under cadmium contaminated soil treated with putrescine and mycorrhiza. *Bangladesh Journal of Botany, 48,* 139–143.

inflammatory processes. Similarly, MS, an autoimmune demyelinating disorder of the CNS, often presents with paresthesia as a result of demyelination in the somatosensory pathways, leading to impaired nerve conduction and aberrant sensory processing. Heavy metal toxicity, mainly due to lead, arsenic, or mercury exposure, can also cause sensory disturbances by inducing neurotoxicity, mitochondrial dysfunction, and alterations in neurotransmitter balance, resulting in both acute and chronic parenthetic symptoms. Additionally, chemotherapy-induced peripheral neuropathy (CIPN), a common adverse effect of chemotherapeutic agents such as platinum-based drugs, taxanes, and vinca alkaloids, leads to sensory abnormalities due to axonal degeneration, mitochondrial dysfunction, and inflammatory responses.

Vision disturbances, another common sensory dysfunction, may result from retinal disorders, optic nerve damage, cortical processing deficits, or systemic diseases affecting ocular circulation. Optic neuritis, frequently associated with MS and other autoimmune conditions, leads to acute vision loss due to inflammatory demyelination of the optic nerve, often accompanied by pain during eye movement and color vision deficits. Retinal degenerative disorders such as age-related macular degeneration and diabetic retinopathy contribute to progressive vision impairment due to photoreceptor loss, vascular abnormalities, and chronic inflammation. Furthermore, mitochondrial diseases such as Leber's hereditary optic neuropathy lead to sudden vision loss due to mitochondrial dysfunction-induced retinal ganglion cell apoptosis. Toxic exposures, including methanol poisoning, can also cause vision disturbances by inducing optic neuropathy through direct mitochondrial toxicity and metabolic acidosis (Kumar & Naik, 2020, Kumar et al., 2019, Kumar, Devi, et al., 2021, Kumar et al., 2022, Kumar, Goud, Devi, Dey, et al., 2022, Kumar, Pathak, Amarnath, et al., 2018).

Hearing loss, another form of sensory impairment, may be conductive, sensorineural, or mixed, depending on the underlying pathology. Sensorineural hearing loss (SNHL) often results from damage to the cochlear hair cells, auditory nerve, or central auditory processing centers in the brainstem and cortex. Ototoxicity induced by aminoglycoside antibiotics (e.g., gentamicin), platinum-based chemotherapy (e.g., cisplatin), or loop diuretics (e.g., furosemide) can lead to irreversible SNHL due to direct damage to the inner ear structures, particularly the cochlear hair cells and spiral ganglion neurons. Autoimmune inner ear disease and neuroinflammatory disorders such as neurosyphilis, Lyme disease, and viral infections (e.g., cytomegalovirus, measles, or mumps) can also contribute to progressive hearing loss through immune-mediated cochlear damage and inflammation. In some neurodegenerative disorders, such as Parkinson's and Alzheimer's, central auditory processing deficits may lead to difficulties in sound localization, speech discrimination, and aural perception, even without peripheral hearing loss. Motor dysfunction, encompassing ataxia, tremors, and spasticity, is another significant consequence of neurological disease arising from dysfunction in the cerebellum, basal ganglia, corticospinal tracts, or peripheral nervous system. Ataxia, characterized by impaired coordination, balance, and voluntary movement control, results from damage to the cerebellum or its associated pathways. It is frequently observed in hereditary spinocerebellar ataxias, where genetic mutations lead to progressive cerebellar degeneration and neuronal loss, resulting in gait instability, dysmetria, and dysarthria. Acquired ataxias can arise from alcohol-induced neurotoxicity, thiamine deficiency (Wernicke's encephalopathy), autoimmune cerebellar ataxia, and paraneoplastic syndromes, where immune-mediated attacks on cerebellar Purkinje cells lead to progressive motor dysfunction. Ataxia can also result from toxic exposures, such as mercury poisoning, which disrupts cerebellar function by inducing neuronal apoptosis, oxidative stress, and neurotransmitter dysregulation (Chakraborty et al., 2021; Devi et al., 2023; Kumar & Pathak, 2019; Kumar et al., 2011b; Kumar, Devi, et al., 2021; Kumar et al., 2022).

Tremors, rhythmic involuntary movements, can be classified into resting, postural, kinetic, or intentional tremors, depending on their characteristics and underlying neural circuits involved. Essential tremor, the most common movement disorder, primarily affects the hands, head, and voice and is believed to result from dysfunction in the cerebellothalamocortical pathways. Parkinsonian tremors, typically seen in Parkinson's disease, are resting tremors that emerge due to dopaminergic neuron loss in the substantia nigra and alterations in basal ganglia circuitry. Cerebellar tremors, often seen in MS, stroke, or cerebellar degeneration, are intentional tremors that worsen with goal-directed movement due to impaired error correction mechanisms within the cerebellum. Additionally, drug-induced tremors can occur due to the use of medications such as lithium, valproate, or stimulants that alter neurotransmitter dynamics.

Spasticity, characterized by increased muscle tone and hyperactive stretch reflexes, arises from damage to the upper motor neurons or corticospinal pathways. It is commonly observed in conditions such as cerebral palsy, MS, stroke, and spinal cord injury. In MS, demyelination of motor pathways leads to disrupted inhibitory signaling, resulting in excessive muscle contractions, stiffness, and impaired voluntary movement control. In stroke patients, corticospinal tract damage results in loss of descending inhibitory control, causing exaggerated reflex responses and difficulty in motor execution. Spasticity can also be seen in hereditary spastic paraplegia, where genetic mutations lead to progressive degeneration of corticospinal motor neurons, causing lower limb spasticity and gait disturbances.

Psychological and psychiatric symptoms, including anxiety, depression, and personality changes, frequently accompany neurological disorders, often arising due to alterations in neurotransmitter systems, structural brain changes, and chronic stress responses. Anxiety disorders, characterized by excessive worry, hypervigilance, and autonomic hyperactivity, are commonly associated with dysfunction in the amygdala, prefrontal cortex, and limbic circuits. Neurotransmitter imbalances involving serotonin, GABA, and norepinephrine contribute to heightened stress reactivity and impaired emotional regulation. Anxiety symptoms are frequently observed in neurodegenerative diseases such as Alzheimer's disease, Parkinson's disease, and Huntington's disease, where progressive neural degeneration affects emotional processing centers. Additionally, neuroinflammatory conditions, autoimmune encephalitis, and postviral syndromes such as COVID-19 have been linked to heightened anxiety symptoms due to cytokine-mediated alterations in brain function (Devi et al., 2023; Kumar & Pathak, 2019; Kumar et al., 2013; Kumar, Devi, et al., 2021; Kumar et al., 2022; Kumar, Goud, Devi, Dey, et al., 2022; Kumar, Harshavardhan, et al., 2018; Kumar, Kumar, et al., 2018).

Depression, another prevalent psychiatric symptom in neurological disorders, is strongly linked to dysregulation of the monoaminergic system, neuroinflammation, and structural changes in the hippocampus and prefrontal cortex. Chronic neuroinflammatory states, as seen in autoimmune diseases, traumatic brain injury (TBI), and chronic neurodegenerative disorders, contribute to depressive symptoms by increasing proinflammatory cytokines such as interleukin-1 beta (IL-1β), tumor necrosis factor-alpha (TNF-α), and interleukin-6 (IL-6), which interfere with neurogenesis, synaptic plasticity, and neurotransmitter function. Depression is particularly prevalent in Parkinson's disease due to the loss of dopaminergic and serotonergic neurons, leading to anhedonia, fatigue, and reduced motivation. Additionally, poststroke depression is a well-documented phenomenon, often resulting from ischemic damage to mood-regulating brain regions such as the prefrontal cortex and basal ganglia.

Personality changes, including impulsivity, emotional instability, aggression, or apathy, are frequently observed in frontal lobe disorders, dementia, and neuropsychiatric conditions. Frontotemporal dementia, a disorder characterized by progressive degeneration of the frontal and temporal lobes, leads to significant alterations in personality, including disinhibition, compulsive behaviors, and emotional blunting. Similarly, TBI affecting the prefrontal cortex can result in impulsivity, poor judgment, and emotional dysregulation. Autoimmune neuropsychiatric disorders, such as anti-NMDA receptor encephalitis, can lead to abrupt personality changes, paranoia, hallucinations, and cognitive decline due to immune-mediated disruption of synaptic function. Overall, sensory disturbances, motor dysfunction, and psychiatric symptoms frequently overlap in neurological diseases, highlighting the complex interplay between neural circuits, neurotransmitter systems, immune mechanisms, and environmental factors. Understanding these interrelated symptoms is crucial for accurate diagnosis, effective treatment strategies, and improving patient outcomes in neurodegenerative, autoimmune, toxicological, and metabolic disorders (Table 4.4).

4.6 Diagnosis and biomarkers of mercury exposure

Mercury exposure remains a significant public health concern due to its neurotoxic effects, and accurate diagnosis is crucial for clinical management and epidemiological studies. Biomarkers play a vital role in assessing exposure levels, determining the extent of toxicity, and guiding medical interventions. The primary biomarkers for mercury exposure include blood, urine, and hair mercury levels, each providing specific insights into the type, duration, and severity of exposure. Blood mercury levels primarily reflect recent exposure to methylmercury, the organic form of mercury commonly found in seafood. Methylmercury has a high affinity for red blood cells, and blood mercury concentrations tend to correlate with dietary intake, making it a valuable biomarker for ongoing exposure. Generally, blood mercury levels above 5 μg/L are considered elevated, with toxic effects more likely to occur as concentrations increase. However, since blood mercury levels decline relatively quickly after exposure ceases, they may not be ideal for assessing long-term mercury accumulation. On the other hand, urine mercury levels are handy for evaluating exposure to elemental mercury vapour and inorganic mercury compounds, which are commonly encountered in occupational settings, dental amalgams, and specific industrial processes (Chakraborty et al., 2021; Kumar et al., 2011b; Kumar et al., 2017; Kumar et al., 2019; Kumar, Harshavardhan, et al., 2018; Pathak et al., 2017). Unlike methylmercury, which is mainly excreted via feces, inorganic mercury is predominantly eliminated through the kidneys, making urinary mercury a reliable biomarker for cumulative exposure. Normal urine mercury levels are generally below 4 μg/g creatinine, with levels above 10 μg/g creatinine often indicating excessive exposure. Chronic exposure to elemental mercury vapour, such as in industrial workers or individuals with high dental amalgam burdens, can lead to significantly elevated urine mercury levels, often accompanied by symptoms such as tremors, memory disturbances, and mood disorders. Since urinary mercury

TABLE 4.4 Neurological disorders linked to mercury exposure.

Neurological disorder	Associated mercury form	Clinical presentation	Pathophysiology
Minamata disease	Methylmercury (organic mercury from contaminated seafood)	-Sensory disturbances (numbness, paresthesia)-Ataxia, muscle weakness-Vision and hearing impairments -Cognitive and motor dysfunction	-Bioaccumulation of methylmercury in the CNS, leading to neuronal degeneration in the cerebral and cerebellar cortex-Severe oxidative stress and mitochondrial dysfunction
ASD	Methylmercury, ethylmercury (via maternal seafood consumption, vaccines)	-Impaired social communication-Repetitive behaviors, sensory sensitivities-Cognitive and speech delays	-Neuroinflammation and glial activation-Excitotoxicity due to glutamate imbalance-Epigenetic modifications affecting neurodevelopment
MS	Elemental and inorganic mercury (occupational exposure, dental amalgams)	-Muscle weakness, spasticity-Visual disturbances (optic neuritis)-Cognitive impairment, fatigue	-Demyelination of neurons due to mercury-induced immune dysregulation-Increased autoimmunity and chronic neuroinflammation
Neurodevelopmental delays	Methylmercury (prenatal exposure)	-Learning disabilities reduced IQ-Attention deficits, hyperactivity-Delayed motor development	Disruption of neuronal migration and synaptogenesis-Altered dopamine and serotonin pathways affecting cognition and behavior

ASD, Autism spectrum disorder; *CNS*, central nervous system; *MS*, multiple sclerosis.
Source: based on Devi, P., Dey, S. R., Kumar, P. (2023). Integration of water resources management in rural areas. In *Water resources management for rural development: Challenges and mitigation* (pp. 139–152). Punjab, Phagwara, India: Elsevier, Department of Agronomy, School of Agriculture, Lovely Professional University. https://doi.org/10.1016/B978-0-443-18778-0.00013-1; Kumar, P., Koul, B., Sharma, M. (2022). Phytoremediation of heavy metals. In *Heavy metals in plants: Physiological to molecular approach* (pp. 369–388). Punjab, Jalandhar, India: CRC Press, Department of Agronomy, School of Agriculture, Lovely Professional University. https://doi.org/10.1201/9781003110576-17; Kumar, P., Mandal, B., Dwivedi, P. (2013). Combating heavy metal toxicity from hazardous waste sites by harnessing scavenging activity of some vegetable plants. *Vegetos, 26*, 416–425. https://doi.org/10.5958/j.2229-4473.26.2.106; Kumar, P., Yumnam, J., Kumar, P. S., Misao, L., Jyoti, N., Naik, M., Harshavardhan, M. (2018). Cadmium induced changes in germination of maize seed treated with mycorrhiza. *Annals of Agri-bio Research, 23*, 169–170; Kumar, P., Harshavardhan, M., Kumar, P.S., Yumnam, J., Jyoti, N., Naik, M., Kumar, S. (2018). Effect on chlorophyll a/b ratio in cadmium contaminated maize leaves treated with putrescine and mycorrhiza. *Annals of Biology, 34*, 281–283; Kumar, P., Pathak, S. (2019). Responsiveness index of sorghum (Sorghum bicolor (L.) Moench) grown under cadmium contaminated soil treated with putrescine and mycorrhiza. *Bangladesh Journal of Botany, 48*, 139–143.

excretion reflects the body's burden of inorganic mercury, it remains an essential tool for monitoring individuals at risk of long-term exposure.

Hair mercury levels serve as a valuable biomarker for assessing chronic exposure to methylmercury, particularly in populations with high fish consumption. Since methylmercury readily binds to sulfhydryl groups in keratin, mercury becomes incorporated into the hair shaft as it grows, providing a historical exposure record. Hair samples can be segmented to estimate past exposure levels over time, making them particularly useful for retrospective exposure assessment. Normal hair mercury levels are typically below 1 μg/g, while levels exceeding 5 μg/g suggest significant exposure. Epidemiological studies have demonstrated strong correlations between maternal hair mercury levels and adverse neurodevelopmental outcomes in children, reinforcing the utility of hair analysis in assessing prenatal exposure risks. However, external contamination from environmental mercury sources can sometimes lead to misleading results, necessitating careful sample collection and preparation. Beyond traditional biomarkers, imaging techniques such as magnetic resonance imaging (MRI) and positron emission tomography (PET) scans have emerged as valuable tools for detecting the neurological effects of mercury toxicity. MRI can reveal structural and functional brain abnormalities associated with chronic mercury exposure, including cortical atrophy, white matter lesions, and altered connectivity patterns. Diffusion tensor imaging (DTI), a specialized form of MRI, has been beneficial in detecting mercury-induced changes in white matter integrity, as seen in chronic occupational exposure or methylmercury poisoning cases. Studies have reported significant reductions in fractional anisotropy in key brain regions involved in cognitive and motor functions, highlighting the impact of mercury on neural connectivity. PET scans, which assess metabolic activity in the brain, have provided insights into mercury-induced neurotoxicity by detecting changes in glucose metabolism and neurotransmitter function (Devi et al., 2023; Kumar et al., 2017; Kumar et al., 2019;

Kumar, Goud, Devi, Dey, et al., 2022; Kumar, Kumar, et al., 2018; Kumar, Pathak, Amarnath, et al., 2018). Mercury exposure has been associated with decreased dopamine receptor availability and reduced metabolic activity in regions such as the basal ganglia and prefrontal cortex, correlating with symptoms of tremors, cognitive dysfunction, and mood disturbances. While these imaging techniques are primarily used in research settings, they hold promise for advancing clinical diagnosis and understanding the pathophysiology of mercury toxicity. Neurophysiological tests and cognitive assessments play a crucial role in evaluating the functional impact of mercury exposure on the nervous system. Electrophysiological tests such as electroencephalography (EEG) and nerve conduction studies (NCSs) have demonstrated alterations in neural activity among individuals with mercury toxicity. EEG studies have reported increased slow-wave activity and reduced coherence between brain regions, suggesting disruptions in neural synchronization. NCS findings have revealed peripheral neuropathy in cases of chronic mercury exposure, characterized by reduced nerve conduction velocities and prolonged latency periods, which can manifest as sensory disturbances, muscle weakness, and coordination deficits. Cognitive assessments further aid in detecting neuropsychological impairments associated with mercury exposure. Standardized tests such as the Wechsler Adult Intelligence Scale, Rey Auditory Verbal Learning Test, and Wisconsin Card Sorting Test have identified deficits in attention, memory, executive function, and psychomotor speed among exposed individuals. Methylmercury exposure, particularly in utero and during early childhood, has been linked to lower IQ scores, impaired language development, and reduced visuospatial skills, emphasizing the vulnerability of the developing brain to mercury toxicity. These diagnostic approaches and biomarkers collectively provide a comprehensive framework for assessing mercury exposure and its health effects. While blood, urine, and hair mercury levels serve as primary biomarkers for exposure assessment, imaging techniques and neurophysiological tests offer deeper insights into the neurological consequences of mercury toxicity. Given the widespread environmental presence of mercury and its potential for bioaccumulation, continued advancements in diagnostic methodologies are essential for early detection, risk assessment, and the development of targeted interventions to mitigate the impact of mercury exposure on human health (Kumar & Pathak, 2019; Kumar et al., 2011b; Kumar et al., 2013; Kumar *et al.* 2022; Pathak et al., 2017).

4.7 Treatment and management strategies

Chelation therapy, supportive therapies, and dietary and lifestyle modifications play crucial roles in the management of heavy metal toxicity, particularly in cases of lead, mercury, arsenic, and cadmium poisoning, which pose significant health risks due to their capacity to accumulate in tissues and interfere with essential biological processes. Chelation therapy, which involves the administration of chelating agents such as 2,3-dimercaptopropane-1-sulfonic acid (DMPS) and meso-2,3-DMSA, is a cornerstone intervention in detoxification protocols, acting by binding to metal ions and forming stable, water-soluble complexes that can be excreted through the kidneys. It is a water-soluble analogue of dimercaprol, has demonstrated efficacy in enhancing the urinary excretion of mercury, arsenic, and lead, thereby reducing systemic toxicity. Similarly, an orally bioavailable chelator with a high affinity for lead and mercury, has been widely utilized due to its favorable safety profile and ability to cross the BBB to some extent, facilitating the removal of neurotoxic metals. These chelating agents operate via thiol (–SH) groups, interacting with metal cations through coordinate covalent bonding, displacing toxic metals from cellular and extracellular compartments. However, chelation therapy is not without risks, as the indiscriminate binding of essential trace elements such as zinc, copper, and selenium can lead to micronutrient depletion, necessitating careful monitoring and supplementation during treatment. Additionally, the administration of chelators must be tailored to the specific metal burden and patient physiology, as excessive mobilization of metals can transiently increase toxicity by redistributing them to sensitive tissues before excretion (Kumar & Pathak, 2019; Kumar et al., 2011b; Kumar, Devi, et al., 2021; Kumar, Harshavardhan, et al., 2018; Kumar, Yumnam, et al., 2018).

Supportive therapies, including antioxidants and neuroprotective agents, mitigate oxidative damage and cellular dysfunction associated with heavy metal exposure. Heavy metals induce oxidative stress through the generation of ROS, disruption of mitochondrial function, and inhibition of antioxidant enzyme systems such as SOD, catalase, and glutathione peroxidase. This oxidative imbalance results in lipid peroxidation, protein denaturation, and DNA damage, contributing to neurodegenerative processes, cardiovascular disease, and impaired immune function. To counteract these effects, antioxidant supplementation, including glutathione, *N*-acetylcysteine (NAC), alpha-lipoic acid (ALA), vitamin C, and vitamin E, has been employed to enhance detoxification pathways and protect cellular integrity. Glutathione, a tripeptide composed of glutamate, cysteine, and glycine, plays a pivotal role in metal detoxification by conjugating heavy metals and facilitating their biliary and renal excretion. NAC, a precursor to glutathione, has been extensively studied for its ability to replenish intracellular glutathione levels, scavenge free radicals, and chelate metals such as mercury and lead. ALA, a dithiol-containing compound, exhibits dual antioxidant and metal-chelating properties. It is particularly valuable in preventing heavy metal-induced neurotoxicity by crossing the BBB and reducing metal accumulation in neural tissues. Vitamins C and E act synergistically to neutralize lipid peroxides and regenerate oxidized antioxidants, stabilizing cellular membranes and reducing systemic inflammation. Furthermore, selenium and zinc supplementation have been shown to counteract heavy metal toxicity by competitively inhibiting metal absorption and restoring the activity of antioxidant enzymes often impaired by toxic metal exposure (Kumar et al., 2011b; Kumar et al., 2013; Kumar et al., 2017; Kumar, Kumar, et al., 2018; Pathak et al., 2017).

Neuroprotective agents are particularly relevant in mercury and lead toxicity, as these metals exert profound neurotoxic effects by disrupting neurotransmitter systems, inducing neuroinflammation, and impairing synaptic plasticity. Acetyl-L-carnitine, a mitochondrial cofactor, has demonstrated neuroprotective properties by enhancing neuronal energy metabolism, reducing oxidative stress, and preventing excitotoxicity. Curcumin, the bioactive compound in turmeric, has been investigated for its antiinflammatory and metal-chelating effects, offering potential benefits in mitigating neurodegenerative damage induced by heavy metals (Devi et al., 2023; Dwivedi & Kumar, 2011; Kumar & Pathak, 2019; Kumar et al., 2022; Kumar, Yumnam, et al., 2018). Polyphenols such as resveratrol and quercetin exhibit neuroprotective and anti-inflammatory activities by modulating intracellular signaling pathways, reducing glial cell activation, and scavenging free radicals. Melatonin, a neurohormone with potent antioxidant properties, has been shown to mitigate heavy metal-induced neurotoxicity by regulating mitochondrial function, modulating inflammatory cytokines, and protecting against excitotoxicity. Collectively, these neuroprotective strategies serve to ameliorate cognitive deficits, prevent neuronal apoptosis, and enhance the efficacy of chelation therapy by minimizing oxidative and inflammatory damage.

Dietary and lifestyle modifications are fundamental components of an integrative approach to heavy metal detoxification, as dietary intake and environmental exposures significantly influence metal bioaccumulation and toxicity. A nutrient-dense diet rich in organic fruits, vegetables, whole grains, and lean proteins provides essential vitamins, minerals, and phytonutrients that support detoxification pathways and enhance metal excretion. Cruciferous vegetables such as broccoli, Brussels sprouts, and kale contain glucosinolates that induce phase II detoxification enzymes, facilitating the elimination of heavy metals. Sulfur-containing foods, including garlic and onions, provide cysteine, a precursor to glutathione, thereby enhancing the body's intrinsic metal-binding capacity. Fiber-rich foods, such as flaxseeds, chia seeds, and legumes, aid in the elimination of heavy metals by binding to them in the gastrointestinal tract and reducing enterohepatic recirculation. Additionally, fermented foods and probiotics, such as yoghurt, kefir, and kimchi, contribute to gut microbiome balance and reduce heavy metal absorption by binding metals in the intestinal lumen. Adequate hydration is critical to supporting renal excretion of chelated metals, with a focus on filtered water to minimize additional exposure to contaminants (Dwivedi & Kumar, 2011; Kumar et al., 2011b; Kumar et al., 2017; Kumar et al., 2019; Kumar, Kumar, et al., 2018).

Lifestyle interventions to reduce environmental exposure to heavy metals are equally crucial in preventing bioaccumulation and chronic toxicity. Avoiding contaminated water sources, particularly those with high levels of arsenic or lead, is essential, and using high-quality water filtration systems can significantly reduce metal intake. Reducing occupational and household exposure to heavy metals through proper ventilation and protective equipment and avoiding contaminated consumer products minimizes the risk of chronic accumulation. Personal care products, including cosmetics, toothpaste, and deodorants, should be scrutinized for heavy metal contaminants such as lead and aluminum, as dermal absorption contributes to systemic toxicity. Additionally, minimizing the consumption of high-mercury fish, such as tuna, swordfish, and king mackerel, while opting for low-mercury alternatives like salmon and sardines is a practical dietary strategy to limit mercury exposure. Smoking cessation is critical, as tobacco smoke is a significant source of cadmium exposure, and reducing alcohol consumption supports liver detoxification and prevents additional oxidative stress.

Physical activity and sauna therapy have been explored as adjunctive detoxification strategies, as sweating facilitates the excretion of heavy metals through the skin. Regular aerobic exercise enhances circulation, promotes lymphatic drainage, and supports hepatic and renal detoxification. Infrared sauna therapy, in particular, has been shown to mobilize and eliminate heavy metals such as arsenic, lead, and mercury through perspiration while reducing oxidative stress and inflammation. Stress management techniques, including meditation, yoga, and deep breathing exercises, support detoxification by modulating the HPA axis and reducing cortisol-mediated oxidative damage. Sleep optimization is another critical factor, as the glymphatic system, responsible for clearing metabolic waste from the brain, is most active during deep sleep and contributes to removing neurotoxic heavy metals.

A comprehensive approach to heavy metal detoxification involves a combination of chelation therapy, supportive antioxidant and neuroprotective interventions, and dietary and lifestyle modifications to enhance metal excretion, mitigate oxidative damage, and prevent reexposure.

While DMPS and DMSA remain primary pharmacological agents for chelation therapy, their efficacy is enhanced when combined with antioxidant support and neuroprotective compounds that counteract heavy metal–induced damage. Dietary strategies prioritizing sulfur-rich, fiber-dense, and polyphenol-containing foods provide essential detox cofactors. At the same time, lifestyle modifications that minimize environmental exposure and promote metabolic waste clearance contribute to long-term heavy metal burden reduction. As heavy metal toxicity remains a pervasive public health concern, integrating these evidence-based approaches provides a holistic framework for enhancing detoxification, improving neurological and systemic health outcomes, and reducing the risk of chronic disease associated with heavy metal bioaccumulation (Kumar et al., 2013, Kumar et al., 2019, Kumar, Devi, et al., 2021, Kumar, Goud, Devi, Dey, et al., 2022, Kumar, Harshavardhan, et al., 2018).

4.8 Preventive measures and policy interventions

Regulations on industrial emissions and mercury use, public health initiatives concerning seafood consumption, and advances in environmental mercury remediation collectively represent critical components of global efforts to mitigate mercury pollution and its associated health risks. Mercury, a persistent and bioaccumulated toxic metal, originates from both natural and anthropogenic sources, with industrial activities such as coal combustion, mining, cement production, and waste incineration contributing significantly to atmospheric mercury emissions. To address this issue, international frameworks, such as the Minamata Convention on Mercury, adopted in 2013 under the United Nations Environment Programme, establish legally binding measures to reduce mercury emissions, phase out mercury-containing products, and promote safer alternatives. Countries that have ratified the Minamata Convention commit to implementing stringent regulatory measures, including emission limits on coal-fired power plants, restrictions on mercury use in artisanal and ASGM, and the prohibition of mercury in specific products such as thermometers, fluorescent lamps, and batteries. In addition to international agreements, national and regional regulatory bodies, including the US EPA, the European Union (EU), and China's Ministry of Ecology and Environment, enforce industry-specific mercury emission standards through policies such as the Mercury and Air Toxics Standards in the United States and the Industrial Emissions Directive in the EU. These regulations mandate the installation of advanced pollution control technologies, such as activated carbon injection, flue gas desulfurisation, and electrostatic precipitators, to capture mercury emissions before they enter the atmosphere. Despite regulatory progress, challenges remain, particularly in developing nations where industrial mercury emissions persist due to weak enforcement, inadequate infrastructure, and economic reliance on mercury-intensive industries. Addressing these challenges requires a combination of policy enforcement, technological innovation, and international cooperation to support developing countries in transitioning to mercury-free industrial processes.

Public health initiatives aimed at reducing mercury exposure focus primarily on seafood consumption guidelines, as methylmercury, the most toxic form of mercury, bioaccumulates in aquatic food chains and poses significant health risks to humans, particularly vulnerable populations such as pregnant women, infants, and individuals with preexisting neurological conditions. The US FDA and the EPA have established seafood consumption advisories that categorize fish species based on their mercury content, encouraging the consumption of low-mercury fish such as salmon, sardines, trout, and tilapia while advising against high-mercury species such as shark, swordfish, king mackerel, and tilefish. Similarly, the European Food Safety Authority and the WHO provide dietary recommendations to balance the nutritional benefits of seafood consumption, particularly omega-3 fatty acids, with the risks of mercury exposure. Public awareness campaigns, labeling requirements, and consumer education programs have been implemented to empower individuals to make informed dietary choices. Yet, disparities in awareness and adherence persist, particularly in regions with high seafood dependency where alternative protein sources are limited. In addition to seafood advisories, public health initiatives address mercury exposure from other sources, including dental amalgams, vaccines containing thimerosal, and traditional medicines that may contain mercury compounds. While the use of mercury in dental fillings has declined due to the availability of safer alternatives such as composite resins and glass ionomers, concerns remain regarding legacy mercury exposure from older amalgams, prompting discussions on best practices for removal and disposal. Similarly, efforts to phase out thimerosal-containing vaccines in favor of single-dose formulations aim to reduce unnecessary mercury exposure without compromising immunization programs, particularly in developing nations where multidose vaccines remain essential due to cost and storage considerations. Community-based interventions, occupational safety programs, and biomonitoring initiatives further contribute to public health strategies by identifying at-risk populations, promoting safe work practices in mercury-related industries, and facilitating early detection of mercury toxicity through blood, hair, and urine testing (Devi et al., 2023; Dwivedi & Kumar, 2011; Kumar & Pathak, 2019; Kumar et al., 2011b; Kumar, Devi, et al., 2021; Kumar, Kumar, et al., 2018; Kumar, Pathak, Amarnath, et al., 2018).

Advances in environmental mercury remediation have emerged as essential strategies for mitigating mercury contamination in air, water, and soil, with innovative technologies focusing on source control and remediation of existing pollution. In air pollution control, advancements in sorbent-based capture systems, such as impregnated activated carbon and novel metal-based sorbents, have improved mercury removal efficiency from industrial flue gases. These technologies work by adsorbing mercury onto high-surface-area materials, which can then be safely disposed of or regenerated for reuse. In addition, emerging catalytic oxidation techniques, including manganese-based catalysts and plasma-assisted oxidation, enhance mercury capture by converting elemental mercury into oxidized forms more readily removed by scrubbers. In aquatic environments, remediation efforts focus on eliminating methylmercury from contaminated water bodies through advanced filtration and bioremediation approaches. Constructed wetlands, which utilize mercury-accumulating plants and microbial consortia, have shown promise in reducing methylmercury levels in freshwater ecosystems by facilitating microbial demethylation and sequestration in plant biomass. Furthermore, phytoremediation strategies involving genetically engineered hyperaccumulator plants are being explored to enhance mercury uptake and stabilization in soil and sediment. Another promising approach is using biochar amendments in contaminated sites, as biochar, derived from the pyrolysis of organic biomass, has demonstrated strong mercury-binding capabilities, reducing bioavailability and preventing leaching into groundwater. In situ chemical stabilization techniques, such as applying activated carbon and sulfur-based amendments, further contribute to soil remediation by immobilizing mercury in less bioavailable forms, thereby minimizing ecological and human health risks. Additionally, nanotechnology-based remediation strategies, including engineered nanoparticles such as nanoscale zero-valent iron and functionalised silica nanoparticles, are being investigated for their ability to selectively capture and transform mercury into stable, nontoxic forms. While these remediation technologies hold great potential, challenges such as scalability, cost-effectiveness, and long-term stability remain key areas of research and development (Dwivedi & Kumar, 2011; Kumar et al., 2017; Kumar et al., 2022; Kumar, Harshavardhan, et al., 2018; Kumar, Kumar, et al., 2018; Kumar, Yumnam, et al., 2018; Pathak et al., 2017).

Despite significant progress in mercury regulations, public health initiatives, and remediation technologies, ongoing efforts are required to address the global burden of mercury pollution and its far-reaching environmental and health consequences. Strengthening international cooperation through knowledge-sharing platforms, capacity-building initiatives, and technology transfer programs can facilitate the adoption of best practices in mercury management across diverse geopolitical and economic contexts. Furthermore, continued research into alternative industrial processes, sustainable mining practices, and circular economy approaches can reduce global reliance on mercury while promoting safer and more sustainable economic development. As scientific understanding of mercury toxicity evolves, integrating multidisciplinary approaches that combine policy, public health, and technological innovation will be essential to achieving long-term reductions in mercury exposure and protecting human and environmental health.

4.9 Future directions and research perspectives

Emerging therapeutic strategies, long-term neurodevelopmental studies, and innovations in mercury detoxification are critical areas of research addressing mercury exposure's complex and persistent health effects. As mercury toxicity poses significant risks to human health, particularly neurodevelopmental disorders, cardiovascular dysfunction, and immune system dysregulation, researchers are exploring novel therapeutic approaches to mitigate its harmful effects, enhance detoxification, and protect vulnerable populations from long-term neurological damage. Among the most promising emerging therapeutic strategies are chelation therapies that utilize advanced agents such as meso-2,3-DMSA, 2,3-dimercapto-1-propanesulfonic acid (DMPS), and ALA to bind and facilitate the excretion of mercury from the body. While traditional chelators like ethylenediaminetetraacetic acid (EDTA) have been widely used, newer compounds with improved specificity and lower toxicity are under investigation to enhance efficacy while minimizing side effects such as mineral depletion and oxidative stress. In addition to chelation, supportive therapies that include antioxidants, neuroprotective agents, and mitochondrial stabilizers are crucial in mitigating mercury-induced cellular damage. Compounds such as glutathione, NAC, selenium, and vitamin C are being studied to counteract oxidative stress, reduce neuroinflammation, and enhance cellular detoxification pathways. Furthermore, emerging research on plant-derived polyphenols, including curcumin, resveratrol, and quercetin, suggests that these natural compounds may offer neuroprotective benefits by modulating inflammatory responses, enhancing antioxidant capacity, and promoting mercury excretion through bile and urine. In parallel, advances in gene therapy and epigenetic modulation are opening new avenues for targeting mercury-induced genetic and epigenetic alterations that contribute to neurodevelopmental and neurodegenerative disorders. Experimental studies indicate that interventions to restore DNA methylation patterns, histone modifications, and microRNA expression profiles could potentially reverse some of the long-term neurological consequences of mercury exposure. While these therapeutic strategies hold great promise, challenges remain in optimizing treatment protocols, ensuring patient safety, and understanding individual variations in mercury metabolism and response to therapy.

Long-term neurodevelopmental studies are essential for understanding the impact of prenatal and early-life mercury exposure on cognitive function, behavior, and neurological health across the lifespan. Epidemiological research has consistently shown that mercury, particularly methylmercury found in contaminated seafood, can cross the placenta and accumulate in the developing fetal brain, disrupting neuronal migration, synaptogenesis, and neurotransmitter regulation. Large-scale cohort studies, such as the Seychelles Child Development Study and the Faroe Islands Study, have provided valuable insights into the dose–response relationship between maternal seafood consumption, mercury exposure, and child neurodevelopmental outcomes. These studies have highlighted associations between prenatal methylmercury exposure and deficits in language skills, memory, attention, and motor coordination, even at low exposure levels. However, the complexity of these findings is underscored by the potential neuroprotective effects of omega-3 fatty acids and other nutrients present in fish, necessitating careful risk–benefit assessments when formulating dietary recommendations for pregnant women. Advances in neuroimaging techniques, including functional MRI, DTI, and PET, are allowing researchers to visualize structural and functional changes in the brains of individuals exposed to mercury. These imaging modalities have revealed alterations in white matter integrity, reduced cortical thickness, and disrupted connectivity in key neural circuits involved in executive function, memory, and emotional regulation. Furthermore, studies employing EEG and magnetoencephalography provide insights into how mercury exposure affects neural oscillations, synaptic plasticity, and cortical excitability, which may underlie cognitive and behavioral impairments observed in exposed populations. In addition to human studies, animal models of mercury neurotoxicity are shedding light on the molecular mechanisms underlying mercury-induced neurodevelopmental deficits. Rodent and primate studies have demonstrated that mercury exposure disrupts calcium homeostasis, impairs glutamatergic and GABAergic neurotransmission, and triggers neuroinflammatory responses mediated by microglial activation and cytokine release. These findings are crucial for identifying potential therapeutic targets and developing interventions to mitigate the long-term neurological consequences of mercury exposure.

Innovations in mercury detoxification are revolutionizing approaches to reducing mercury body burden, preventing bioaccumulation, and mitigating environmental exposure risks. Traditional detoxification strategies have relied heavily on pharmacological interventions such as chelation therapy; however, recent advancements in nanotechnology, bioengineered materials, and microbiome-based detoxification are expanding the toolkit for mercury remediation. One of the most exciting developments is using functionalised nanoparticles, including silica-based, polymeric, and carbon-based nanomaterials, for targeted mercury sequestration in biological systems. These nanoparticles can be engineered to selectively bind mercury ions, facilitate their excretion, and minimize systemic toxicity. Additionally, nano chelators, which combine chelating ligands with nanocarriers, offer enhanced stability, improved biodistribution, and prolonged circulation times, thereby increasing their efficacy in mercury removal. Another promising innovation is the application of synthetic biology and microbiome engineering to enhance the body's natural detoxification pathways. Certain gut bacteria, such as members of the Lactobacillus and Bifidobacterium genera, have been found to possess mercury-reducing and methylmercury-demethylating capabilities, suggesting that probiotic-based interventions could play a role in reducing mercury bioavailability and toxicity. Genetic engineering approaches are being explored to enhance these microbial detoxification capacities, potentially leading to the development of next-generation probiotics designed to mitigate mercury exposure in high-risk populations. Furthermore, advances in organoid and lab-on-a-chip technologies enable researchers to model mercury toxicity in complex biological systems, facilitating the development of personalized detoxification strategies based on an individual's genetic and metabolic profile.

In addition to biomedical approaches, environmental innovations in mercury detoxification are critical for reducing human exposure at the source. Novel bioremediation strategies, including phytoremediation, microbial consortia engineering, and biochar-based remediation, are optimized for mercury-contaminated soil and water systems. Hyperaccumulator plants such as Brassica juncea (Indian mustard) and aquatic macrophytes like Eichhornia crassipes (water hyacinth) are genetically modified to enhance their mercury uptake and sequestration capabilities. Similarly, microbial consortia that include mercury-resistant bacteria and fungi are being deployed in bioreactors and constructed wetlands to accelerate the breakdown and immobilization of mercury compounds. Another emerging approach is electrochemical detoxification, which utilizes electrochemical reduction and oxidation processes to convert toxic mercury species into inert or recoverable forms. These electrochemical systems, integrated with advanced membrane filtration and catalytic degradation techniques, are being developed for industrial wastewater treatment, offering scalable and cost-effective solutions for mercury removal. Additionally, advancements in atmospheric mercury capture technologies, such as sorbent-based filters and catalytic oxidation systems, are helping to reduce mercury emissions from coal-fired power plants, waste incinerators, and other industrial sources. While these innovations represent significant progress, challenges such as regulatory approval, large-scale implementation, and long-term sustainability remain key focus areas for researchers and policymakers.

As the understanding of mercury toxicity continues to evolve, integrating emerging therapeutic strategies, long-term neurodevelopmental studies, and cutting-edge detoxification technologies will be essential for reducing the global burden of mercury-related diseases. Collaborative efforts between clinicians, neuroscientists, toxicologists, and environmental scientists are necessary to translate laboratory discoveries into effective public health interventions and clinical treatments. Moreover, personalized medicine approaches that consider genetic polymorphisms in mercury metabolism, individual detoxification capacities, and coexposures to other environmental toxins may pave the way for tailored therapeutic regimens. Future research directions should optimize combination therapies that synergistically enhance mercury excretion, protect neuronal function, and restore cellular homeostasis. By leveraging multidisciplinary innovations and fostering global cooperation, the scientific community can develop comprehensive strategies to combat mercury toxicity and safeguard human health and the environment for future generations.

4.10 Conclusion

The key findings from emerging therapeutic strategies, long-term neurodevelopmental studies, and innovations in mercury detoxification underscore the urgency of addressing mercury toxicity as a global health and environmental concern. Research has demonstrated that mercury, particularly in its organic form as methylmercury, poses significant risks to neurodevelopment, cardiovascular health, and immune function. Chelation therapies such as DMSA and DMPS have shown promise in facilitating mercury excretion, but concerns over their long-term safety and efficacy remain. Adjunctive therapies, including antioxidants like glutathione, selenium, and NAC, protect against mercury-induced oxidative stress, neuroinflammation, and mitochondrial dysfunction. Advances in neuroimaging and biomolecular studies have revealed structural and functional changes in the brains of mercury-exposed individuals, emphasizing the need for continued monitoring of affected populations, particularly children and pregnant women. Longitudinal studies have consistently linked prenatal and early-life mercury exposure to cognitive deficits, attention disorders, and motor impairments, reinforcing the necessity of strict dietary guidelines and seafood consumption advisories. Innovations in detoxification strategies, including nanotechnology-based chelators, microbiome-based interventions, and bioremediation approaches, offer promising avenues for reducing mercury bioaccumulation and mitigating environmental contamination. Despite these scientific advancements, the persistence of mercury pollution in air, water, and soil highlights the need for comprehensive global management strategies.

Addressing mercury toxicity requires a coordinated global response integrating regulatory enforcement, public health initiatives, and technological innovations. While international agreements such as the Minamata Convention on Mercury have established frameworks to reduce mercury emissions and prevent human exposure, gaps in implementation and compliance remain significant challenges. Industrial emissions from coal combustion, ASGM, and improper waste disposal continue to be significant sources of mercury pollution, necessitating stricter regulations and enforcement mechanisms. Public health campaigns must also be intensified to raise awareness about mercury exposure risks, particularly in vulnerable populations such as pregnant women, children, and communities reliant on fish-based diets. Improved seafood consumption guidelines that balance the nutritional benefits of fish with mercury exposure risks are essential for mitigating health impacts. Furthermore, investment in environmentally sustainable technologies, such as mercury-free industrial processes, enhanced emission control systems, and novel remediation techniques, is crucial for minimizing mercury pollution's long-term ecological and human health consequences.

A call for interdisciplinary research and policy collaboration is imperative to advance mercury management strategies and develop effective interventions. Collaborative efforts between neuroscientists, toxicologists, environmental scientists, public health officials, and policymakers are necessary to bridge the gap between research findings and regulatory actions. Integrated approaches combining biomedical, environmental, and policy solutions will be essential for tackling mercury toxicity at individual and systemic levels. Funding for large-scale longitudinal studies, clinical trials for novel therapeutic interventions, and the development of next-generation remediation technologies should be prioritized. International cooperation in data sharing, monitoring, and enforcing mercury regulations will also enhance global efforts to reduce mercury exposure and safeguard public health. By fostering interdisciplinary collaboration and adopting a holistic approach, the scientific and policy communities can develop sustainable strategies to mitigate mercury's impact on human health and the environment for future generations.

References

Aley, P., Singh, J., & Kumar, P. (2022). Chapter 23 – Adapting the changing environment: Microbial way of life. In A. Kumar, J. Singh, & L. F. R. Ferreira (Eds.). *Microbiome under changing climate* (pp. 507–525). Woodhead Publishing. Retrieved from https://doi.org/10.1016/B978-0-323-90571-8.00023-7.

Chakraborty, S., Kumar, P., Sanyal, R., Mane, A. B., Arvind Prasanth, D., Patil, M., & Dey, A. (2021). Unravelling the regulatory role of miRNAs in secondary metabolite production in medicinal crops. *Plant Gene, 27*, 100303. Retrieved from https://doi.org/10.1016/j.plgene.2021.100303.

Das, T., Saha, S. C., Sunita, K., Majumder, M., Ghorai, M., Mane, A. B., ... Dey, A. (2022). Promising botanical-derived monoamine oxidase (MAO) inhibitors: Pharmacological aspects and structure-activity studies. *South African Journal of Botany, 146*, 127–145. Retrieved from https://doi.org/10.1016/j.sajb.2021.09.019.

Devi, P., Dey, S. R., & Kumar, P. (2023). Integration of water resources management in rural areas. In *Water resources management for rural development: Challenges and mitigation*. Punjab, Phagwara, India: Elsevier, Department of Agronomy, School of Agriculture, Lovely Professional University, 139–152. Retrieved from https://doi.org/10.1016/B978-0-443-18778-0.00013-1.

Devi, P., Dey, S. R., Saini, L., Kumar, P., Panigrahi, S., & Dwivedi, P. (2023). Toward sustainable agriculture: Strategies involving phytoprotectants against reactive oxygen species. In *Reactive oxygen species: Prospects in plant metabolism*. Phagwara, India: Springer Nature, Department of Agronomy, School of Agriculture, Lovely Professional University, 229–247. Retrieved from https://doi.org/10.1007/978-981-19-9794-5_13.

Devi, P., Dogra, P., Kumar, P., & Singh, J. (2024). Global status of biomass energy programmes-challenges and roadmap. In *Biomass energy for sustainable development*. Phagwara, India: CRC Press, Lovely Professional University, 394–414. Retrieved from https://doi.org/10.1201/9781003406501-20.

Devi, P., & Kumar, P. (2023a). Effect of microbial oxidative enzymes on soil health. In *Microbial oxidative enzymes: Biotechnological applications*. Phagwara, Punjab, 144411, India: De Gruyter, Department of Agronomy, School of Agriculture, Lovely Professional University, 37–61. Retrieved from https://doi.org/10.1515/9783111062235-003.

Devi, P., & Kumar, P. (2023b). Role of oxidative enzymes in the alleviation of oxidative stress in plant system. In *Microbial oxidative enzymes: Biotechnological applications*. Phagwara, Punjab, India: De Gruyter, Department of Agronomy, School of Agriculture, Lovely Professional University, 109–128. Retrieved from https://doi.org/10.1515/9783111062235-006.

Dey, S. R., Devi, P., & Kumar, P. (2023). Organic contaminants in aquatic environments: Sources and impact assessment. In *Metal organic frameworks for wastewater contaminant removal*. Punjab, India: Wiley, Department of Agronomy, School of Agriculture, Lovely Professional University, 299–317.

Dey, S. R., Kumar, P., & Singh, J. (2023). Growth of mustard under open field conditions with salicylic acid and antimicrobial thiourea. *Biopesticides International, 19*, 183–192. Retrieved from https://doi.org/10.59467/BI.2023.19.183.

Dey, S. R., Sharma, M., & Kumar, P. (2023). Effects and responses of chromium on plants. In *Environmental science and engineering*. Punjab, Phagwara, India: Springer Science and Business Media Deutschland GmbH, Department of Agronomy, School of Agriculture, Lovely Professional University, 385–427. Retrieved from https://doi.org/10.1007/978-3-031-44029-8_14.

Dwivedi, P., & Kumar, P. (2011). Anti-diabetic medicinal plants and their conservation: Waging green war on diabetes. *Medicinal Plants, 3*, 181–189. Retrieved from https://doi.org/10.5958/j.0975-4261.3.3.031.

Goud, E. L., Singh, J., & Kumar, P. (2022). Chapter 19 - Climate change and their impact on global food production. In A. Kumar, J. Singh, & L. F. R. Ferreira (Eds.). *Microbiome under changing climate* (pp. 415–436). Woodhead Publishing. Retrieved from https://doi.org/10.1016/B978-0-323-90571-8.00019-5.

Hidangmayum, A., Dwivedi, P., Kumar, P., & Upadhyay, S. K. (2023). Seed priming and foliar application of chitosan ameliorate drought stress responses in mungbean genotypes through modulation of morpho-physiological attributes and increased antioxidative defense mechanism. *Journal of Plant Growth Regulation, 42*, 6137–6154. Retrieved from https://doi.org/10.1007/s00344-022-10792-1.

Jagadesh, M., Dash, M., Kumari, A., Singh, S. K., Verma, K. K., Kumar, P., ... Sharma, S. K. (2024). Revealing the hidden world of soil microbes: Metagenomic insights into plant, bacteria, and fungi interactions for sustainable agriculture and ecosystem restoration. *Microbiological Research, 285*. Retrieved from https://doi.org/10.1016/j.micres.2024.127764.

Karam, N., Kumar, P., Anand, Y. R., & Choudhury, D. (2023). Management of rice blast—A review of the strategies available. *Biopesticides International, 19*, 107–116. Retrieved from https://doi.org/10.59467/BI.2023.19.107.

Kumar, D., Rameshwar, S. D., & Kumar, P. (2019). Effect of intergated application of inorganic and organic fertilizers on the roots of chickpea. *Plant Archives, 19*, 857–860.

Kumar, P. (2023). Degradation of xenobiotics by bacteria and fungi: An overview. *Biopesticides International, 19*, 117–132. Retrieved from https://doi.org/10.59467/BI.2023.19.117.

Kumar, P., & Choudhury, D. (2024a). Microbial native soil bacteria against cadmium toxicity. In *Cadmium toxicity in water*. Punjab, Phagwara, India: Springer Water. Springer Nature, Department of Agronomy, School of Agriculture, Lovely Professional University, 193–226. Retrieved from https://doi.org/10.1007/978-3-031-54005-9_9.

Kumar, P., & Choudhury, D. (2024b). Toxicity of rhizospheric cadmium contaminated soil and its phytoremediation. In *Cadmium toxicity in water*. Punjab, Phagwara, India: Springer Water. Springer Nature, Department of Agronomy, School of Agriculture, Lovely Professional University, 227–264. Retrieved from https://doi.org/10.1007/978-3-031-54005-9_10.

Kumar, P., & Choudhury, D. (2024c). Bioremediation of wastewater using hydroponics. In *Hydroponics and environmental bioremediation: Wastewater treatment*. Punjab, Phagwara, India: Springer Water. Springer Nature, Department of Agronomy, School of Agriculture, Lovely Professional University, 27–64. Retrieved from https://doi.org/10.1007/978-3-031-53258-0_2.

Kumar, P., & Choudhury, D. (2024d). Hydroponic removal of organic contaminants from water. In *Hydroponics and environmental bioremediation: Wastewater treatment*. Punjab, Phagwara, India: Springer Water. Springer Nature, Department of Agronomy, School of Agriculture, Lovely Professional University, 143–164. Retrieved from https://doi.org/10.1007/978-3-031-53258-0_6.

Kumar, P., Devi, P., & Dey, S. R. (2021). Fungal volatile compounds: A source of novel in plant protection agents. In *Volatiles and metabolites of microbes*. Elsevier, 83–104.

Kumar, P., & Dey, S. R. (2024). Hydroponics phytoremediation: An overview. In *Hydroponics and environmental bioremediation: Wastewater treatment*. Punjab, Phagwara, India: Springer Water. Springer Nature, Department of Agronomy, School of Agriculture, Lovely Professional University, 361–396. Retrieved from https://doi.org/10.1007/978-3-031-53258-0_13.

Kumar, P., Dey, S. R., & Choudhury, D. (2024). Effectiveness of cadmium on biochemical shift of pea plant treated with mycorrhiza and putrescine. *Nature Environment and Pollution Technology, 23*, 275–285. Retrieved from https://doi.org/10.46488/NEPT.2024.v23i01.023.

Kumar, P., & Dwivedi, P. (2020). Lignin estimation in sorghum leaves grown under hazardous waste site. *Plant Archives, 20*, 2558–2561.

Kumar, P., Goud, E. L., Devi, P., Dey, S. R., & Dwivedi, P. (2022). Heavy metals: Transport in plants and their physiological and toxicological effects. In *Plant metal and metalloid transporters*. Punjab, Jalandhar, India: Springer Nature, Department of Agronomy, School of Agriculture, Lovely Professional University, 23–54. Retrieved from https://doi.org/10.1007/978-981-19-6103-8_2.

Kumar, P., Goud, E. L., Devi, P., & Koul, B. (2022). Metal pollutants in the environment. In *Environmental microbiology: Emerging technologies*. Jalandhar, Punjab, India: De Gruyter, Department of Agronomy, School of Agriculture, Lovely Professional University, 291–323. Retrieved from https://doi.org/10.1515/9783110727227-012.

Kumar, P., Koul, B., & Sharma, M. (2022). Phytoremediation of heavy metals. In *Heavy metals in plants: Physiological to molecular approach*. Punjab, Jalandhar, India: CRC Press, Department of Agronomy, School of Agriculture, Lovely Professional University, 369–388. Retrieved from https://doi.org/10.1201/9781003110576-17.

Kumar, P., Kumar, T., Singh, S., Tuteja, N., Prasad, R., & Singh, J. (2020). Potassium: A key modulator for cell homeostasis. *Journal of Biotechnology, 324*, 198–210.

Kumar, P., Latha, D. H., & Singh, J. (2023). Rhizoremediation of organic emerging soil contaminants: Green technology. In *Clean technologies toward a sustainable future: Physicochemical, biochemical and biotechnological approaches*. Phagwara, Punjab, India: IWA Publishing, Department of Agronomy, School of Agriculture, Lovely Professional University, 211–229. Retrieved from https://doi.org/10.2166/9781789063783_0211.

Kumar, P., Mandal, B., & Dwivedi, P. (2011a). Heavy metals scavenging of soils and sludges by ornamental plants. *Journal of Applied Horticulture, 13*, 144–146.

Kumar, P., Mandal, B., & Dwivedi, P. (2011b). Heavy metal scavenging capacity of *Mentha spicata* and *Allium cepa*. *Medicinal Plants, 3*, 315–318. Retrieved from https://doi.org/10.5958/j.0975-4261.3.4.053.

Kumar, P., Mandal, B., & Dwivedi, P. (2013). Combating heavy metal toxicity from hazardous waste sites by harnessing scavenging activity of some vegetable plants. *Vegetos, 26*, 416–425. Retrieved from https://doi.org/10.5958/j.2229-4473.26.2.106.

Kumar, P., & Mistri, T. K. (2020). Transcription factors in SOX family: Potent regulators for cancer initiation and development in the human body. *Seminars in Cancer Biology, 67*, 105–113. Retrieved from https://doi.org/10.1016/j.semcancer.2019.06.016.

Kumar, P., & Naik, M. (2020). Biotic symbiosis and plant growth regulators as a strategy against cadmium and lead stress in chickpea. *Plant Archives, 20*, 2495–2500.

Kumar, P., & Pathak, S. (2019). Responsiveness index of sorghum (*Sorghum bicolor* (L.) Moench) grown under cadmium contaminated soil treated with putrescine and mycorrhiza. *Bangladesh Journal of Botany, 48*, 139–143.

Kumar, P., Saini, L., & Sharma, M. (2023). Assessment of the phytochemical constituents and metabolites in the medicinal plants and herbal medicine used in the treatment and management of respiratory diseases. In *Reference series in phytochemistry*. Phagwara, India: Springer Science and Business Media B.V., Department of Agronomy, School of Agronomy, Lovely Professional University. Retrieved from https://doi.org/10.1007/978-3-031-21973-3_19-1.

Kumar, P., Sharma, K., Saini, L., & Dey, S. R. (2021). Chapter 8 – Role and behavior of microbial volatile organic compounds in mitigating stress. In A. Kumar, J. Singh, & J. B. T. Samuel (Eds.). *Volatiles and metabolites of microbes* (pp. 143–161). Academic Press. Retrieved from https://doi.org/10.1016/B978-0-12-824523-1.00010-9.

Kumar, P., Singh, B. N., & Dwivedi, P. (2017). Plant growth regulators, plant adaptability and plant productivity: A review on abscisic acid (ABA) signaling in plants under emerging environmental stresses. In *Sustaining future food security in changing environments*. Varanasi, India: Nova Science Publishers, Inc., Department of Plant Physiology, Institute of Agricultural Sciences, Banaras Hindu University, 81–97.

Kumar, P., & Singh, J. (2024). Hydroponic: An eco-friendly future. In *Hydroponics and environmental bioremediation: Wastewater treatment*. Punjab, Phagwara, India: Springer Water. Springer Nature, Department of Agronomy, School of Agriculture, Lovely Professional University, 223–267. Retrieved from https://doi.org/10.1007/978-3-031-53258-0_9.

Kumar, P., Pathak, S., Kumar, M., & Dwivedi, P. (2018). Role of secondary metabolites for the mitigation of cadmium toxicity in sorghum grown under mycorrhizal inoculated hazardous waste site. In N. Kumar (Ed.). *Biotechnological approaches for medicinal and aromatic plants: Conservation, genetic improvement and utilization* (pp. 199–212). Singapore: Springer. Retrieved from https://doi.org/10.1007/978-981-13-0535-1_8.

Kumar, V., Dwivedi, P., Kumar, P., Singh, B. N., Pandey, D. K., Kumar, V., & Bose, B. (2021). Mitigation of heat stress responses in crops using nitrate primed seeds. *South African Journal of Botany, 140*, 25–36. Retrieved from https://doi.org/10.1016/j.sajb.2021.03.024.

Kumar, V. A., & Kumar, P. (2023). Revolutionizing agriculture with satellite technology for farmers: A review. *Biopesticides International, 19*, 97–106. Retrieved from https://doi.org/10.59467/BI.2023.19.97.

Kumar, P., Harshavardhan, M., Kumar, P. S., Yumnam, J., Jyoti, N., Naik, M., ... Kumar, S. (2018). Effect on chlorophyll a/b ratio in cadmium contaminated maize leaves treated with putrescine and mycorrhiza. *Annals of Biology, 34*, 281–283.

Kumar, P., Kumar, S., Harshavardhan, M., Naik, M., Yumnam, J., Kumar, P. S., ... Purnima (2018). Evaluation of plant height and leaf length of sorghum grown under different sources of nutrition. *Annals of Biology, 34*, 284–286.

Kumar, P., Pathak, S., Amarnath, K. S., Veerendra Brahma Teja, P., Dileep, B., Kumar, K., ... Siddique, A. (2018). Effect of growth regulator on morpho-physiological attributes of chilli: A case study. *Plant Archives, 18*, 1771–1776.

Kumar, P., Yumnam, J., Kumar, P. S., Misao, L., Jyoti, N., Naik, M., ... Harshavardhan, M. (2018). Cadmium induced changes in germination of maize seed treated with mycorrhiza. *Annals of Agri-bio Research, 23*, 169–170.

Kumari, P., Kumar, P., & Singh, J. (2023). Toxic contamination of aquatic habitat and its remediation. In *Spatial modeling of environmental pollution and ecological risk*. Punjab, Phagwara, India: Elsevier, Department of Agronomy, School of Agriculture, Lovely Professional University, 209–220. Retrieved from https://doi.org/10.1016/B978-0-323-95282-8.00021-3.

Kumari, P., Singh, J., & Kumar, P. (2022). Chapter 21 - Impact of bioenergy for the diminution of an ascending global variability and change in the climate. In A. Kumar, J. Singh, & L. F. R. Ferreira (Eds.). *Microbiome under changing climate* (pp. 469–487). Woodhead Publishing. Retrieved from https://doi.org/10.1016/B978-0-323-90571-8.00021-3.

Naik, M., & Kumar, P. (2020). Role of growth regulators and microbes for metal detoxification in plants and soil. *Plant Archives, 20*, 2820–2824.

Pankaj, P. K., Kumar, P., Nigam, R. C., & Mishra, P. K. (2012a). To studies organochlorine insecticides in different brands of cold drinks. *Journal of Chemical and Pharmaceutical Research, 4*, 3934–3938.

Pankaj, P. K., Kumar, P., Nigam, R. C., & Mishra, P. K. (2012b). Monitoring and surveillance of synthetic pyrethroids and organophosphate in different brands of soft drinks. *Journal of Chemical and Pharmaceutical Research, 4*, 3939–3943.

Pathak, S., Kumar, P., Mishra, P. K., & Kumar, M. (2017). Mycorrhiza assisted approach for bioremediation with special reference to biosorption. *Pollution Research, 36*, 329–332.

Raj, A., Goutam, E., Kumar, P., & Singh, J. (2024). Polyphenol oxidase: An enzyme of plants. In *Polyphenol oxidases: Function, wastewater remediation, and biosensors*. Phagwara, Punjab, India: De Gruyter, Department of Agronomy, School of Agriculture, Lovely Professional University, 105–128. Retrieved from https://doi.org/10.1515/9783111033525-004.

Reddy, S. S., Kumar, P., & Dwivedi, P. (2022). Heavy metal transporters, phytoremediation potential, and biofortification. In *Plant metal and metalloid transporters*. Springer, 387–405.

Saini, L., Sharma, K., & Kumar, P. (2023). Water efficiency in agriculture as a vital approach toward water management in rural areas. In *Water resources management for rural development: Challenges and mitigation*. Punjab, Phagwara, India: Elsevier, Department of Agronomy, School of Agriculture, Lovely Professional University, 127–138. Retrieved from https://doi.org/10.1016/B978-0-443-18778-0.00020-9.

Sharma, K., Devi, P., Dey, S. R., & Kumar, P. (2023). Mercury phytovolatilization: an overview of the mechanism and mitigation. In *Role of green chemistry in ecosystem restoration to achieve environmental sustainability*. Punjab, Phagwara, India: Elsevier, Department of Agronomy, School of Agriculture, Lovely Professional University, 325–331. Retrieved from https://doi.org/10.1016/B978-0-443-15291-7.00031-6.

Sharma, K., Devi, P., & Kumar, P. (2023). Metal-organic frameworks for wastewater contaminants removal. In *Metal organic frameworks for wastewater contaminant removal*. Punjab, Phagwara, India: Wiley, Department of Agronomy, School of Agriculture, Lovely Professional University, 95–117.

Sharma, K., Devi, P., Kumar, P., Dey, A., & Dwivedi, P. (2023). Hazardous phytotoxic nature of reactive oxygen species in agriculture. In *Reactive oxygen species: prospects in plant metabolism*. Phagwara, India: Springer Nature, Department of Agronomy, School of Agriculture, Lovely Professional University, 135–146. Retrieved from https://doi.org/10.1007/978-981-19-9794-5_8.

Sharma, K., & Kumar, P. (2023). Microbe-mediated impact on cell membrane stability and leaf pigmentation in pearl millet under toxic mercury stress in soil. *Biopesticides International, 19*, 217–228. Retrieved from https://doi.org/10.59467/BI.2023.19.217.

Sharma, M., Dey, S. R., & Kumar, P. (2023). Antioxidant defence: A key mechanism of chromium tolerance. In *Environmental science and engineering*. Punjab, Phagwara, India: Springer Science and Business Media Deutschland GmbH, Department of Agronomy, School of Agriculture, Lovely Professional University, 91–116. Retrieved from https://doi.org/10.1007/978-3-031-44029-8_5.

Sinam, V., Kumar, P., & Singh, J. (2023). Bioremediation and ecorestoration strategies of aquatic environment. In *Spatial modeling of environmental pollution and ecological risk*. Punjab, Phagwara, India: Elsevier, Department of Agronomy, School of Agriculture, Lovely Professional University, 483–499. Retrieved from https://doi.org/10.1016/B978-0-323-95282-8.00025-0.

Srivastav, A. L., Patel, N., Rani, L., Kumar, P., Dutt, I., Maddodi, B. S., & Chaudhary, V. K. (2024). Sustainable options for fertilizer management in agriculture to prevent water contamination: A review. *Environment, Development and Sustainability, 26*, 8303–8327. Retrieved from https://doi.org/10.1007/s10668-023-03117-z.

Upadhyay, S. K., Devi, P., Kumar, V., Pathak, H. K., Kumar, P., Rajput, V. D., & Dwivedi, P. (2023). Efficient removal of total arsenic (As3+/5+) from contaminated water by novel strategies mediated iron and plant extract activated waste flowers of marigold. *Chemosphere, 313*, 137551. Retrieved from https://doi.org/10.1016/j.chemosphere.2022.137551.

Chapter 5

Arsenic exposure and neurotoxic effects

Prasann Kumar
Department of Agronomy, School of Agriculture, Lovely Professional University, Phagwara, Punjab, India

5.1 Introduction

The nervous system manifestations due to arsenic exposure are many and strikingly diverse; hence, doctors classify them under different neurological disorders depending on which segment of the nervous system gets affected. Symptoms, which are inclusive of the neuropathic signs, such as numbness, tingling, and a burning sensation in the distal extremities, may also yield peripheral nerve damage in arsenic-exposed patients. Significantly, from this nerve injury, signs progress to muscle weakness and muscle atrophy if exposure is sustained. Electrophysiological studies have shown reduced peripheral nerve conduction velocities, showing how arsenic contributes to the dysfunction of these nerves (Rodríguez, Jiménez-Capdeville, & Giordano, 2003). Following long-term ingestion of arsenic, a plethora of symptoms ensue in the central nervous system (CNS). Encephalopathy is a disease that disrupts cognitive functions and moods with some alteration of consciousness together with mental retardation. Diagnosis relies on the patient's general clinical pattern, with memory loss, concentration difficulties, and unpredictable emotional responses as frequent indicators. Medical imaging studies of the brains of victims of arsenic toxicity will reveal significant loss of brain tissues together with areas of damage in white matter (Rodríguez et al., 2003). Impairments affecting the motor system and sensory functions constituted another facet of the medical conditions arising from arsenic exposure apart from peripheral neuropathy. Arsenic patients have reported several symptoms, like decrements in movement and muscle pains, together with headaches, vertigo, disorientation, hearing difficulty, and dizziness. Functionality in everyday life is diminished by arsenic through various symptomatic attacks on the motor and sensory nervous system.

5.2 Exploring the silent epidemic, arsenic-induced neurotoxicity, and strategies for early intervention

Arsenic exposure has long been recognized as a significant environmental health hazard, with well-documented carcinogenic effects. However, the neurotoxic potential of arsenic has often been overlooked despite mounting evidence of its detrimental impact on brain health, especially in vulnerable populations. Arsenic-induced neurotoxicity is a complex, multifactorial process that involves a range of cellular and molecular mechanisms, including oxidative stress, inflammation, and disruption of essential signaling pathways. These mechanisms culminate in various neurocognitive deficits, ranging from learning and memory impairments to more severe conditions like developmental delays and neurodegenerative diseases. One of the primary sources of arsenic exposure is contaminated drinking water, particularly in regions where natural arsenic contamination of groundwater is prevalent, such as parts of South Asia and Latin America. Occupational exposure, especially in mining, smelting, and pesticide production, also contributes significantly to the burden of arsenic-induced neurotoxicity. Moreover, arsenic contamination in the food supply, notably rice and seafood, is another critical pathway for chronic low-level exposure. The long-term neurotoxic effects of arsenic, however, often manifest insidiously, making early detection and intervention particularly challenging.

The neurotoxic effects of arsenic extend to both children and adults, with fetuses and young children being especially susceptible due to the developing brain's heightened vulnerability to environmental toxins. In children, arsenic exposure has been associated with a range of cognitive impairments, including reduced intelligence quotient (IQ), attention deficits, and difficulties in language development.

Heavy Metal Toxicity and Neurodegeneration. DOI: https://doi.org/10.1016/B978-0-443-36575-1.00026-1

Studies have also highlighted the link between early-life exposure and later-life neurodegenerative conditions, such as Alzheimer's and Parkinson's disease (AD and PD). The molecular mechanisms underlying arsenic-induced neurotoxicity include generating reactive oxygen species (ROS), mitochondrial dysfunction, and alterations in the integrity of the blood–brain barrier (BBB). These processes initiate a cascade of events that ultimately lead to neuronal cell death, synaptic dysfunction, and cognitive decline. The ability of arsenic to interfere with neurotransmitter systems, particularly dopamine and acetylcholine, further complicates its neurotoxic effects, impairing both motor and mental functions. Furthermore, arsenic's epigenetic effects—such as DNA methylation and histone modifications—have been shown to alter gene expression patterns, exacerbating its neurotoxic impact and potentially contributing to the transgenerational effects observed in exposed populations (Garg & Bandyopadhyay, 2024; Huang et al., 2024; Vellingiri et al., 2022).

Given the insidious nature of arsenic-induced neurotoxicity, strategies for early intervention are critical to mitigate its long-term effects. Early identification of arsenic exposure through biomonitoring is essential, with blood, urine, and hair analysis as valuable tools for assessing exposure levels. However, these methods are not without limitations, mainly when dealing with chronic low-level exposure. Emerging neuroimaging techniques, such as MRI and PET scans, offer promising avenues for detecting early brain changes associated with arsenic exposure, including structural and functional alterations in the hippocampus and prefrontal cortex. Additionally, neurocognitive testing and behavioral assessments are crucial for identifying subtle cognitive impairments that may go unnoticed, especially in children. Once arsenic exposure is detected, interventions aimed at reducing the toxic burden on the brain become paramount. Water purification technologies, such as reverse osmosis (RO) and activated alumina filtration, can significantly reduce arsenic concentrations in drinking water, thus mitigating one of the most common routes of exposure. Nutritional interventions, including antioxidants like vitamins C and E, can help counteract oxidative stress and protect against neuronal damage. In some cases, chelation therapy may be employed to accelerate arsenic excretion, although its effectiveness in reducing neurotoxic effects remains a subject of ongoing research (Bartos et al., 2024; Dutta et al., 2025). Additionally, pharmacological approaches targeting the pathways of neuroinflammation and oxidative damage are being explored as potential therapeutic strategies to counteract arsenic-induced brain damage. Public health initiatives focusing on community awareness and education are also crucial in reducing arsenic exposure, particularly in regions where contaminated drinking water remains a significant concern. In conclusion, arsenic-induced neurotoxicity represents a silent epidemic with profound implications for public health, particularly in developing countries. While considerable progress has been made in understanding the mechanisms of arsenic toxicity, much remains to be done regarding early detection, intervention, and prevention. As research continues to evolve, the potential for novel therapeutic strategies and improved diagnostic tools offers hope for reducing the neurotoxic burden of arsenic exposure on vulnerable populations (Table 5.1).

5.3 Developmental neurotoxicity

The developing brain remains exceptionally vulnerable to toxic substances, which thus makes arsenic exposure during very sensitive phases of its development have long-lasting dire consequences. Research in animal models has shown that direct exposure to arsenic during pregnancy and the first months of life unusually disrupts normal CNS development. Histopathological analysis revealed reduced brain size with an observed decreased number of glial and neuronal cells and changes in neurotransmitter systems. Changes to mechanisms of synaptic plasticity become hindered owing to structural and functional aberrations in the nervous system itself (Rodríguez et al., 2003). Following research-based epidemiological findings, children who experienced early-life arsenic exposure developed quantifiable cognitive failures. Children living in Bangladesh who drank arsenic-tainted water performed worse in the tests of intelligence than children who were not exposed to arsenic. Researchers identified three primary cognitive areas that showed deficits after tests on verbal comprehension, working memory, and processing speed (Wasserman et al., 2004). Behavioral and cognitive development effects arise after exposure to arsenic in early life. Exposure to arsenic during the newborn period results in three hallmark behavioral alterations: increased anxiety behaviors, decreased social behaviors, and impaired learning ability (Table 5.2). Studies suggest that arsenic interferes with neural networks' control of emotion regulation and social behavior.

5.4 Mechanisms of arsenic neurotoxicity

Neuronal injury induced by arsenic might occur through a multitude of complex mechanisms that act in a sequenced manner. Neurotoxicity, triggered predominantly through arsenic, appears by generating ROS, an effect better termed oxidative stress. High levels of ROS result in oxidative

TABLE 5.1 Significant sources of arsenic exposure, including environmental, occupational, and dietary sources, estimated exposure levels, and associated risks.

Source type	Specific source	Exposure route	Estimated exposure level	Associated risks
Environmental	Contaminated drinking water	Oral (ingestion)	50–500 μg/L (depending on region)	Neurological deficits, developmental delays, and cancer risk
	Air pollution (arsenic-containing particles)	Inhalation, dermal	Low to moderate exposure levels	Respiratory issues, cardiovascular disease, and lung cancer
	Soil and dust in arsenic-contaminated areas	Dermal, ingestion	1–10 μg/g (in soil)	Skin lesions, respiratory disorders, and developmental impacts
Occupational	Mining (gold, coal, arsenic mining)	Inhalation, dermal	0.1–10 mg/m^3	Lung cancer, skin cancer, and respiratory diseases
	Smelting (copper, lead, zinc)	Inhalation, dermal	1–5 mg/m^3	Chronic respiratory issues, cancer, neurological damage
	Pesticide manufacturing and usage	Inhalation, dermal	Moderate to high exposure	Cognitive impairment, reproductive issues, and cancer risks
Dietary	Rice and rice-based products	Oral (ingestion)	0.05–0.2 μg/g (in rice)	Cancer, developmental delays, neurotoxic effects
	Seafood (fish, shellfish)	Oral (ingestion)	0.02–0.5 μg/g	Cognitive impairment, endocrine disruption
	Vegetables (especially leafy greens)	Oral (ingestion)	0.01–0.2 μg/g	Developmental effects, immune system disruption
	Drinking water (contaminated sources)	Oral (ingestion)	10–100 μg/L (WHO limit: 10 μg/L)	Cancer, neurological, and developmental effects
Other	Cigarette smoke (arsenic as a contaminant)	Inhalation	Varies by smoking intensity	Lung cancer, cardiovascular disease, and neurodevelopmental disorders

Based on, Garg and Bandyopadhyay (2025), Hashim, Ahmed, Rahiman, Manikkoth, and Leena Pramod (2022), Luo and Shu (2023), Tolins, Ruchirawat, and Landrigan (2014).

TABLE 5.2 Neurotoxic effects of arsenic exposure.

Category	Effects	Mechanism	References
Peripheral nervous system	Neuropathy, muscle weakness, burning sensation, tingling in the extremities	Reduced nerve conduction velocity, oxidative stress	Rodríguez et al. (2003)
Central nervous system	Encephalopathy, memory loss, concentration deficits, mood disorders	Neuronal degeneration, oxidative damage, and inflammation	Rodríguez et al. (2003)
Developmental neurotoxicity	Cognitive deficits, lower IQ, behavioral changes (anxiety, social deficits)	Disruption in neurodevelopment, neurotransmitter imbalance	Wasserman et al. (2004)
Neurochemical disruptions	Altered dopamine, norepinephrine, and serotonin levels	Disruption in neurotransmitter synthesis and signaling	
Molecular mechanisms	Oxidative stress, mitochondrial dysfunction, DNA damage, epigenetic changes	ROS production, impaired mitochondrial function, and histone modifications	Prakash, Soni, and Kumar (2016)

IQ, Intelligence quotient.

damage to proteins, DNA, and lipids within neuronal cells. Additionally, oxidative stress further damages cellular homeostasis and causes mitochondrial dysfunction, ultimately aggravating neuronal apoptosis. Mitochondria create cellular energy through oxidative phosphorylation to drive neuronal cellular processes, and therefore the damage to the cell mitochondria upon arsenic exposure yields an energy deficit in neurons. Apart from impairing mitochondrial membrane potential and proapoptotic factor leakage, arsenic also interferes with mitochondrial oxidative phosphorylation. These disturbances in metabolism lead to poor survival and function of neurons (Prakash et al., 2016). A growing body of evidence shows that environmental exposure to arsenic produces gene expression patterns altered via epigenetic modifications. Part of the exposed modification mechanisms included DNA methylation, histone modifications, and variations in microRNA (miRNA) expression. Such modifications disturb normal neuron development and functioning, translating to transient or persistent behavioral abnormalities. Continual arsenic exposure leads to an upregulation of neuroinflammatory actions in the brain. Immune activation of microglia and astrocytes, the building blocks of the adult CNS, results in macroglial activation with the profusion of proinflammatory cytokines. Indeed, the direct neurotoxic effect on the system may be exacerbated.

Arsenic metabolism in the human body follows a complex biochemical pathway primarily involving reduction and methylation processes. Ingested arsenate (AsV) is first reduced to arsenite (AsIII) via enzymatic reactions, primarily involving glutathione (GSH) and arsenate reductases. The more toxic arsenite is then methylated in the liver by arsenic methyltransferase (AS3MT) in a process dependent on *S*-adenosylmethionine, producing monomethyl arsenic acid (MMA) and dimethylarsinic acid (DMA). While methylation was once believed to be a detoxification mechanism, MMA and DMA can generate ROS and contribute to cellular damage. The biotransformation of arsenic significantly impacts its ability to penetrate biological barriers, particularly the BBB, which regulates the entry of substances into neural tissues. Due to its high affinity for thiol groups in proteins, arsenite disrupts endothelial tight junctions, increasing BBB permeability and allowing its accumulation in neural tissues. Additionally, methylated arsenicals have been shown to cross the BBB via aquaporin transporters and influence neuronal function, leading to neurotoxicity. Once inside the brain, arsenic induces oxidative stress and mitochondrial dysfunction, disrupting cellular homeostasis. It promotes the overproduction of ROS, leading to lipid peroxidation, protein oxidation, and DNA damage. Mitochondria, the cell's powerhouse, are particularly vulnerable; arsenic impairs electron transport chain (ETC) function, reducing ATP production and enhancing oxidative stress by inhibiting complexes I and III of the ETC. This results in mitochondrial membrane depolarization, cytochrome c release, and apoptosis, contributing to neurodegenerative processes. Its disruption of neurotransmitter systems further exacerbates the neurotoxic effects of arsenic (Yan et al., 2023; Yang et al., 2024; Zhang et al., 2023, 2024). Dopaminergic dysfunction occurs through the depletion of dopamine levels and increased oxidative degradation, leading to impaired motor function and cognitive processes. Similarly, arsenic alters serotoninergic pathways by reducing serotonin synthesis and degrading its receptors, which may contribute to mood disorders such as depression and anxiety. The glutamatergic system, crucial for synaptic plasticity and learning, is also affected, with arsenic exposure leading to excitotoxicity through excessive activation of *N*-methyl-*D*-aspartate receptors (NMDARs), calcium influx, and neuronal death. Beyond these immediate biochemical disruptions, arsenic induces epigenetic modifications that have long-term effects on gene expression. DNA methylation patterns are altered, leading to aberrant silencing or activation of key regulatory genes. Arsenic exposure also affects histone modifications, such as acetylation and methylation, which influence chromatin structure and gene transcription. Additionally, arsenic modulates miRNA expression, which can dysregulate multiple cellular pathways, including stress response, apoptosis, and neurodevelopment. Collectively, these epigenetic changes contribute to chronic neurological disorders, including cognitive decline, neurodevelopmental abnormalities, and increased susceptibility to neurodegenerative diseases such as Alzheimer's and Parkinson's (Chu et al., 2025; Gopnar et al., 2023; Ma et al., 2024). The interplay between arsenic metabolism, oxidative stress, neurotransmitter disruption, and epigenetic modifications highlights the complexity of arsenic-induced neurotoxicity and underscores the need for further research into mitigation strategies (Table 5.3).

5.5 Long-term cognitive and behavioral impacts

The neurotoxic consequences of arsenic affect primary brain functions and developmental processes, impacting cognitive and behavioral aspects of health over time. It continues to impair and affect memory and issues related to attention span, executive function, and spatial abilities,

TABLE 5.3 Molecular and cellular mechanisms of arsenic-induced neurotoxicity, including oxidative stress, apoptosis, and neurotransmitter disruption.

Mechanism	Description	Impact on neural cells
Oxidative stress	Arsenic generates ROS, leading to cellular damage.	ROS cause lipid peroxidation, DNA damage, protein oxidation, and mitochondrial dysfunction, impairing cell survival.
Apoptosis (programmed cell death)	Arsenic triggers the intrinsic and extrinsic apoptotic pathways by activating proapoptotic proteins (e.g., Bax, caspases).	This leads to neuronal cell death, contributing to neurodegeneration and cognitive impairments.
Inflammation	Arsenic activates inflammatory pathways via NF-κB and MAPK signaling, promoting neuroinflammation.	It increases the release of proinflammatory cytokines (e.g., IL-1β, TNF-α), contributing to brain injury and exacerbating neurotoxic effects.
Mitochondrial dysfunction	Arsenic disrupts mitochondrial membrane potential and oxidative phosphorylation.	Impairs ATP production, increasing ROS generation and leading to neuronal energy deficits, which can accelerate cell death.
Neurotransmitter disruption	Arsenic interferes with neurotransmitter systems, particularly dopamine, serotonin, and acetylcholine.	It alters synaptic plasticity, impairs cognitive functions, and affects mood regulation, leading to deficits in learning and memory.
DNA damage and genomic instability	Arsenic induces DNA strand breaks and mutations through direct interaction with cellular DNA.	This leads to genomic instability, activating DNA repair pathways and increasing the risk of mutations, which can contribute to cancer and neurodegeneration.
ER stress	Arsenic causes the accumulation of misfolded proteins in the ER, triggering the UPR.	Prolonged ER stress leads to cell death by activating apoptotic signaling pathways, especially in neurons.
BBB disruption	Arsenic can alter the integrity of the BBB by affecting tight junction proteins.	Increases the permeability of the BBB, allowing harmful substances to enter the brain, exacerbating neurotoxicity.
Epigenetic modifications	Arsenic induces changes in DNA methylation, histone modification, and noncoding RNA expression.	Altering gene expression patterns can have long-term effects on brain development, cognition, and behavior, with potential transgenerational impacts.
Synaptic dysfunction	Arsenic disrupts synaptic signaling by altering ion channels and receptor function.	Impairs neurotransmission, leading to synaptic plasticity, learning, and memory formation deficits.

BBB, Blood–brain barrier; *ER*, endoplasmic reticulum; *ROS*, reactive oxygen species; *UPR*, unfolded protein response.
Source: Du, Luo, Huang, and Zhang (2022), He et al. (2021), Wu et al. (2024).

TABLE 5.4 Epidemiological studies on arsenic exposure and its neurotoxic effects, including study location, population studied, arsenic levels, and significant findings.

Study location	Population studied	Arsenic levels	Major findings
Bangladesh	2600 children	0–250 μg/L (in drinking water)	Arsenic exposure linked to lower IQ scores in children; neurodevelopmental delays and cognitive impairments observed.
Taiwan	170 adults	0–300 μg/L (in drinking water)	Chronic arsenic exposure is associated with neurocognitive decline, memory impairments, and motor dysfunction in adults.
Chile	900 school-aged children	10–800 μg/L (in drinking water)	Increased risk of learning disabilities and attention deficits in children exposed to high water arsenic levels.
India (West Bengal)	1500 children and adolescents	0–500 μg/L (in drinking water)	Exposure to arsenic is associated with decreased cognitive function and behavioral changes in children and adolescents.
Argentina	250 children and adolescents	0–300 μg/L (in drinking water)	Neurodevelopmental effects include poor performance in memory and attention tasks in children exposed to arsenic.
Bangladesh (Chittagong)	500 children	50–250 μg/L (in drinking water)	Arsenic exposure correlated with delayed neurodevelopment, including lower IQ scores, impaired motor skills, and behavior issues.
The United States (New Hampshire)	300 adults, rural population	5–200 μg/L (in drinking water)	Arsenic exposure is linked to cognitive decline, particularly in older adults with impairments in memory and learning.
Nepal	350 adults	10–200 μg/L (in drinking water)	Chronic exposure to arsenic is associated with peripheral neuropathy and cognitive dysfunction in adults.
Mexico (Sonora)	1000 children and adolescents	10–500 μg/L (in drinking water)	Children exposed to high levels of arsenic showed attention deficits, reduced IQ scores, and developmental delays.
Bangladesh (Khulna)	400 children	50–400 μg/L (in drinking water)	Arsenic exposure is linked to poor motor function and cognitive delays, with effects observed more strongly in younger children.

IQ, Intelligence quotient.
Source: Banerjee and Chatterji (2024), Shayan, Barangi, Hosseinzadeh, and Mehri (2023), Wu, Kreutz, Dixon, and Tokar (2025).

often arising from long-term chronic exposure to arsenic. Such damage disrupts normal daily functioning, disrupting the degenerated quality of life. A population study points out that infants who were exposed to arsenic would have low intelligence levels and memory problems. Behavioral changes produced in humans must be considered behavioral in addition to cognitive deficits that arise from arsenic. Exposure during both very early prenatal and postnatal periods results in anxiety-like behavior with impairment in socialization and decrements in learning ability (Table 5.4). Evidence suggests that arsenic exposure disrupts regulatory neural circuits that underlie emotion regulation as well as mechanisms of social behavior.

5.6 Neurochemical and molecular alterations

The human body undergoes significant changes in molecular processes and neurochemical pathways after the uptake of arsenic. When exposed to arsenic, poor communication between the brain cells is maintained with disturbed neurotransmitters. Research shows that arsenic alters the norepinephrine concentrations, contrary to how it affects dopamine systems and serotonin synthesis, while these substances govern mood functioning and brain operations. When arsenic penetrates the body, it damages neurons' structure and functional properties by dismantling their cytoskeletal networks. Further, neurological dysfunction results from disrupted cytoskeleton structure, disrupting axonal transport and connections between the neurons, thus creating cognitive and behavioral defects.

5.7 Protective strategies and interventions

Multiple intervention measures are required to treat arsenic-induced neurotoxicity. Certain antioxidants and micronutrients are potent and possible additions to the arsenal for interventions against neurotoxic damage from exposure to arsenic. Curcumin and resveratrol supplementation reduced oxidative stress and improved cognitive outcomes in animal subjects fed with arsenic. Establishing strict policies governing water and food would become a significant barrier to ingesting toxic arsenic. The constitution of a routine monitoring and remediation system would reduce levels of exposure along with associated health risks.

5.8 Neurodevelopmental effects of arsenic exposure

5.8.1 Impact on fetal and early childhood brain development

Exposure to toxicants such as arsenic during fetal and early childhood development has profound and often irreversible effects on brain development, leading to cognitive, behavioral, and neurological deficits. The developing brain is highly susceptible to environmental toxins due to its rapid growth, high metabolic demand, and immature detoxification systems. Arsenic exposure during pregnancy occurs primarily through contaminated drinking water, food, and air, allowing it to cross the placenta and accumulate in fetal tissues, including the developing brain (Chu et al., 2025; Garg & Bandyopadhyay, 2025; He et al., 2021; Shayan et al., 2023; Zhang et al., 2024). The ability of arsenic to bypass the BBB during early development exacerbates its neurotoxic effects, as the fetal BBB is not fully developed, allowing more excellent permeability to toxic substances. One of the primary mechanisms through which arsenic disrupts neurodevelopment is oxidative stress and mitochondrial dysfunction. The developing brain requires a finely tuned balance of ROS and antioxidant defenses for proper neuronal differentiation, migration, and synaptogenesis. Arsenic-induced oxidative stress leads to excessive ROS production, lipid peroxidation, DNA damage, and mitochondrial impairment, resulting in neuronal apoptosis and structural abnormalities in critical brain regions such as the hippocampus and cortex, essential for memory, learning, and cognitive function. Another significant consequence of arsenic exposure is the disruption of neurotransmitter systems that regulate brain function. Arsenic interferes with dopamine synthesis, reducing dopaminergic activity in regions associated with attention and executive function, increasing the risk of attention deficit hyperactivity disorder (ADHD) and impaired cognitive abilities. It also affects serotonin levels, which play a crucial role in mood regulation and emotional development, increasing the likelihood of anxiety and depressive disorders later in life (Andrade, Mateus, Batoréu, Aschner, & dos Santos, 2013; Chu et al., 2025; de Castro et al., 2009; Ma et al., 2024; Ugbaja et al., 2022). Additionally, arsenic disrupts the glutamatergic system, which is fundamental to synaptic plasticity and learning, leading to memory-formation deficits and reduced cognitive flexibility. Epigenetic modifications caused by arsenic exposure further compound its detrimental effects on brain development by altering gene expression in a manner that persists into adulthood. Arsenic induces DNA methylation changes that can silence or activate critical genes involved in neurodevelopment, leading to abnormal neuronal differentiation and connectivity. Histone modifications and miRNA dysregulation further impact gene expression patterns, leading to long-term brain structure and function alterations. These epigenetic changes may also increase susceptibility to neurodevelopmental disorders such as autism spectrum disorder (ASD) and intellectual disabilities. Epidemiological studies have shown that children exposed to arsenic in utero or during early childhood exhibit lower IQ scores, impaired memory, language difficulties, and reduced processing speed, highlighting the far-reaching consequences of prenatal and early-life exposure. Behavioral deficits such as increased impulsivity, decreased attention span, and social difficulties have also been observed, further emphasizing the role of arsenic in neurodevelopmental impairments. Given the critical nature of brain development during fetal and early childhood stages, arsenic exposure represents a significant public health concern, necessitating urgent interventions to reduce exposure levels, implement early detection strategies, and develop therapeutic approaches to mitigate its long-term effects (Table 5.5).

5.8.2 Cognitive impairments and reduced intelligence quotient in children

Chronic exposure to environmental toxins such as arsenic during early childhood has been strongly linked to cognitive impairments and reduced IQ, affecting a child's ability to learn, reason, and process information effectively. The developing brain is particularly vulnerable to arsenic due to its high metabolic activity, rapid synaptogenesis, and incomplete BBB, allowing toxicants to penetrate neural tissues more quickly. One of the primary mechanisms by which arsenic affects cognitive function is through oxidative stress and mitochondrial dysfunction. Excessive production of ROS damages neuronal membranes, proteins, and DNA, leading to neuroinflammation, apoptosis, and impaired neural connectivity. Since mitochondria play a crucial role in providing energy for synaptic transmission and plasticity, arsenic-induced mitochondrial dysfunction reduces ATP production, further impairing learning and memory processes (Du et al., 2022; Garg & Bandyopadhyay, 2025; Gopnar et al., 2023; Ma et al., 2024; Ugbaja et al., 2022; Yan et al., 2023). Another key factor in cognitive decline is the disruption of neurotransmitter systems, particularly dopamine, serotonin, and glutamate, which are essential for attention, mood regulation, and cognitive flexibility. Arsenic exposure reduces dopamine levels, leading to deficits in working memory and attention, critical for problem-solving and academic

TABLE 5.5 Biomonitoring studies on arsenic levels in blood, urine, hair, and nails across different populations and exposure levels.

Study location	Population studied	Biomarker analyzed	Arsenic levels (range)	Exposure source	Major findings
Bangladesh	500 children and adults	Urine, hair	Urine: 20–200 μg/L, hair: 0.5–6.5 μg/g	Drinking water, food	Elevated urinary arsenic levels in children correlated with cognitive impairments; higher arsenic in hair linked to developmental delays.
Taiwan	400 adults	Urine, blood	Urine: 10–150 μg/L, blood: 5–50 μg/L	Drinking water	Higher arsenic levels in urine and blood are associated with neurocognitive dysfunction, including memory loss and attention deficits.
Chile	200 children	Urine, hair	Urine: 20–250 μg/L, hair: 0.3–3.0 μg/g	Drinking water	Significant increase in urinary arsenic levels associated with cognitive impairments and learning disabilities in children.
India (West Bengal)	1,000 children	Urine, blood	Urine: 5–100 μg/L, blood: 1–30 μg/L	Drinking water, food	Elevated arsenic in urine and blood is linked to poorer cognitive function and behavioral changes in children.
Mexico (Sonora)	350 children	Urine, hair	Urine: 20–300 μg/L, hair: 1.0–5.0 μg/g	Drinking water, food	Higher arsenic levels in urine and hair correlated with developmental delays, attention deficits, and lower IQ in children.
The United States (New Hampshire)	250 adults (rural population)	Urine, nails	Urine: 5–50 μg/L, nails: 0.2–1.5 μg/g	Drinking water, food	Arsenic exposure is associated with reduced cognitive function and memory impairments, more pronounced in those with elevated nail arsenic levels.
Nepal	300 adults	Urine, nails	Urine: 10–150 μg/L, nails: 0.5–3.0 μg/g	Drinking water	Chronic arsenic exposure is indicated by high urinary and nail arsenic levels associated with neurocognitive dysfunction.
Bangladesh (Khulna)	200 children	Urine, hair	Urine: 20–250 μg/L, hair: 0.3–3.0 μg/g	Drinking water	Children with elevated arsenic levels in urine and hair showed motor dysfunction and cognitive delays.
Argentina	150 adults and children	Blood, urine	Blood: 5–50 μg/L, urine: 10–100 μg/L	Drinking water, food	Increased blood and urine arsenic levels are linked to memory loss and learning difficulties in both adults and children.
Bangladesh (Chittagong)	300 adults and children	Urine, blood	Urine: 10–200 μg/L, blood: 2–30 μg/L	Drinking water, food	Elevated arsenic levels in urine and blood correlated with cognitive decline and behavioral problems in both adults and children.

IQ, Intelligence quotient.
Source: Garg and Bandyopadhyay (2025), Gopnar et al. (2023), He et al. (2021), Wu et al. (2025), Zhang et al. (2023).

performance. Serotonergic disruptions contribute to emotional instability, increased anxiety, and mood disorders, all of which can negatively impact learning abilities and social interactions. Additionally, arsenic affects the glutamatergic system by altering NMDAR function, vital for synaptic plasticity and long-term potentiation (LTP), the fundamental process underlying learning and memory consolidation. These disruptions collectively contribute to slower cognitive processing speeds, difficulty retaining information, and reduced intellectual abilities in exposed children. Epigenetic modifications exacerbate cognitive impairments by altering gene expression patterns critical for brain development and function. Arsenic exposure leads to abnormal DNA methylation, histone modifications, and miRNA dysregulation, which can silence genes necessary for neuronal differentiation, synapse formation, and neuroprotection. These epigenetic changes persist beyond early development, increasing the risk of neurodevelopmental disorders such as ADHD, ASD, and learning disabilities. Studies have consistently shown that children exposed to high levels of arsenic exhibit significantly lower IQ scores, poorer verbal and nonverbal reasoning skills, and deficits in executive function, which encompasses planning, decision-making, and cognitive flexibility. Longitudinal studies have also indicated that early-life arsenic exposure has lasting effects, with cognitive deficits persisting into adolescence and adulthood, reducing academic achievements, and socioeconomic opportunities (Chu et al., 2025; He et al., 2021; Luo & Shu, 2023; Shayan et al., 2023; Yang et al., 2024; Zhang et al., 2023, 2024). The impact of arsenic on cognitive abilities is further compounded by factors such as malnutrition, socioeconomic status, and coexposure to other environmental pollutants, which may intensify neurotoxic effects. Given the substantial evidence linking arsenic to cognitive decline in children, urgent measures must be taken to mitigate exposure through improved water filtration, dietary interventions, and early cognitive assessments to identify and support affected individuals. Public health initiatives focusing on arsenic-free water sources, nutritional supplementation to counteract oxidative stress, and educational support for at-risk children are crucial in minimizing the long-term consequences of arsenic-induced cognitive impairments.

5.8.3 Behavioral and psychological disorder

Early-life exposure to environmental toxins such as arsenic has been increasingly linked to behavioral and psychological disorders, including ADHD, ASD, and anxiety. The developing brain is susceptible to arsenic due to its incomplete BBB, making it more susceptible to neurotoxic damage that can disrupt key neurodevelopmental processes. One of the primary mechanisms by which arsenic contributes to behavioral disorders is oxidative stress and mitochondrial dysfunction, which lead to neuronal apoptosis, impaired neurotransmission, and altered neural connectivity. Excessive ROS production damages critical brain regions involved in behavior regulation, such as the prefrontal cortex, hippocampus, and amygdala, leading to deficits in impulse control, emotional regulation, and social cognition. Disruptions in neurotransmitter systems also play a crucial role in arsenic-induced behavioral and psychological disorders. Arsenic exposure reduces dopamine levels, which are essential for attention, motivation, and executive function, increasing the risk of ADHD-like symptoms such as hyperactivity, impulsivity, and inattention. Dopaminergic dysfunction in the prefrontal cortex and striatum, regions responsible for cognitive control and reward processing, is strongly associated with ADHD pathology. Similarly, arsenic alters serotoninergic signaling, which is critical for mood regulation and social behavior, contributing to increased susceptibility to anxiety and depression (Banerjee & Chatterji, 2024; Tolins et al., 2014; Ugbaja et al., 2022; Yang et al., 2024). The disruption of serotonin homeostasis impairs emotional stability and stress response, leading to heightened anxiety levels, increased aggression, and social withdrawal. Moreover, arsenic affects the glutamatergic system, which plays a vital role in synaptic plasticity, learning, and memory. Dysregulation of glutamate signaling can result in excitotoxicity, neuronal damage, and impaired synaptic connectivity, contributing to ASD-related symptoms such as sensory processing deficits, restricted interests, and impaired social interactions. Epigenetic modifications caused by arsenic exposure further exacerbate the risk of behavioral and psychological disorders by altering gene expression patterns involved in brain development and function. Arsenic-induced DNA methylation changes, histone modifications, and miRNA dysregulation can silence or overactivate key genes responsible for neuronal differentiation, synapse formation, and neuroinflammation, increasing the likelihood of neurodevelopmental disorders. Studies have demonstrated that prenatal and early childhood arsenic exposure is associated with a higher prevalence of ASD-like behaviors, including repetitive movements, impaired communication, and difficulty in social engagement. Additionally, arsenic exposure has been linked to increased anxiety-related behaviors, heightened stress responses, and depression in both animal models and human studies, suggesting a direct impact on the hypothalamic–pituitary–adrenal axis, which regulates the body's response to stress. The cumulative effects of arsenic on neurotransmission, oxidative stress, and epigenetic regulation contribute to a broad spectrum of behavioral and psychological impairments that persist into adolescence and adulthood, affecting academic performance, social relationships, and overall mental well-being. Given the strong association between arsenic exposure and neurodevelopmental disorders, urgent interventions are needed to minimize exposure, enhance early detection, and develop targeted therapeutic strategies to mitigate long-term behavioral and psychological consequences in affected individuals (Chu et al., 2025; de Castro et al., 2009; Du et al., 2022).

5.8.4 Studies linking prenatal and early-life arsenic exposure to neurodevelopmental delays

Numerous studies have established a strong link between prenatal and early-life arsenic exposure and neurodevelopmental delays, demonstrating significant deficits in cognitive function, motor skills, attention, and social behaviors in exposed children. Arsenic, a well-documented neurotoxin, readily crosses the placenta during pregnancy, exposing the developing fetal brain to toxic effects that persist into infancy and early childhood. The immature BBB in foetuses and neonates allows arsenic to accumulate in neural tissues, interfering with crucial neurodevelopmental processes such as neuronal differentiation, synaptogenesis, and myelination (Wu et al., 2024, 2025; Yan et al., 2023). One of the primary mechanisms through which arsenic induces neurodevelopmental delays is oxidative stress and mitochondrial dysfunction. Several studies have shown that prenatal arsenic exposure increases the production of ROS, leading to lipid peroxidation, protein damage, and DNA fragmentation in neuronal cells. These oxidative insults impair synaptic plasticity, disrupt energy metabolism, and lead to apoptosis in critical brain regions such as the hippocampus and prefrontal cortex, essential for memory, learning, and executive function. Mitochondrial dysfunction further exacerbates these effects by reducing ATP production, impairing the energy supply for proper neurodevelopment. Epidemiological studies have consistently reported cognitive deficits in children with prenatal and early-life arsenic exposure. Research conducted in regions with high arsenic contamination in drinking water, such as Bangladesh, India, and the United States, has demonstrated significant reductions in IQ scores, verbal comprehension, and working memory in exposed children compared to unexposed controls. Longitudinal studies indicate that these cognitive impairments persist into later childhood and adolescence, affecting academic performance and overall intellectual development (Andrade et al., 2013; He et al., 2021; Tolins et al., 2014; Zhang et al., 2023). The long-term effects of arsenic exposure cannot be underestimated, as they have a profound impact on the future of these children. Neuroimaging studies have further corroborated these findings by revealing structural and functional abnormalities in arsenic-exposed children, including reduced gray matter volume, altered white matter integrity, and diminished connectivity in brain regions associated with cognitive and motor functions. Additionally, arsenic exposure has been shown to disrupt neurotransmitter systems that are critical for neurodevelopment. Several studies have reported decreased dopamine levels in arsenic-exposed children, leading to attention deficits, hyperactivity, and impulsivity, characteristics commonly observed in ADHD. Similarly, disruptions in serotoninergic pathways contribute to emotional instability, increased anxiety, and social withdrawal, further exacerbating neurodevelopmental delays. Glutamatergic dysregulation, which affects synaptic plasticity and learning, has also been observed, suggesting a broader impact on neurodevelopmental processes (Chu et al., 2025; Shayan et al., 2023; Wu et al., 2024; Yang et al., 2024). Epigenetic studies have provided further evidence linking arsenic exposure to neurodevelopmental delays. Arsenic has been shown to alter DNA methylation patterns, histone modifications, and miRNA expression, leading to long-term changes in gene expression that affect neuronal differentiation, synapse formation, and brain plasticity. These epigenetic changes not only contribute to immediate neurodevelopmental deficits but may also increase the risk of neurodevelopmental disorders such as ASD and intellectual disabilities. The cumulative evidence from epidemiological, neuroimaging, and molecular studies underscores the urgent need for public health interventions to reduce arsenic exposure in pregnant women and young children. Early screening programs, nutritional interventions to counteract oxidative stress, and policies to improve water quality are critical steps in mitigating the long-term impact of arsenic on neurodevelopment and ensuring healthier cognitive and behavioral outcomes for future generations (Du et al., 2022; Garg & Bandyopadhyay, 2025; Ugbaja et al., 2022; Wu et al., 2025; Zhang et al., 2023).

5.9 Neurodegenerative and adult-onset disorders

Chronic exposure to arsenic has been increasingly linked to the onset and progression of neurodegenerative diseases such as AD, PD, and other related conditions. Arsenic, a well-established environmental toxin, has been shown to exert neurotoxic effects on the brain through various mechanisms, including oxidative stress, mitochondrial dysfunction, and alterations in neurotransmitter systems. These factors contribute to neuronal cell death, synaptic dysfunction, and the accumulation of abnormal protein aggregates, which are hallmark features of neurodegenerative diseases. Epidemiological studies have revealed a significant association between long-term arsenic exposure and cognitive decline, as well as the increased risk of developing AD and PD. The neurotoxic properties of arsenic stem from its ability to penetrate the BBB and disrupt neuronal homeostasis, particularly in brain regions vulnerable to neurodegeneration. In AD, for instance, arsenic exposure has been implicated in the formation and accumulation of amyloid-beta plaques, a key pathological feature of AD. Amyloid-beta is thought to disrupt neuronal communication, induce inflammation, and cause oxidative damage to neurons, leading to cognitive impairment and memory loss. Arsenic can exacerbate the aggregation of amyloid-beta by promoting oxidative stress and mitochondrial dysfunction, both of which are believed to play central roles in AD pathogenesis (Gopnar et al., 2023; Wu et al., 2024, 2025; Yan et al., 2023). Additionally, arsenic exposure may increase the production of ROS, contributing to the neuroinflammatory response and neuronal damage in AD-affected brain regions such as the hippocampus and cortex. In PD, chronic arsenic exposure has been shown to damage dopaminergic neurons in the substantia nigra, a brain region that plays a critical role in motor control. Dopamine depletion due to neuronal damage is a hallmark of PD, and arsenic's impact on dopamine metabolism and neurotransmitter release further exacerbates this condition. Studies have demonstrated that arsenic-induced mitochondrial dysfunction and oxidative stress contribute to the degeneration of dopaminergic neurons, leading to the characteristic motor symptoms of PD, such as tremors, rigidity, and bradykinesia. Moreover, arsenic exposure may increase the production of inflammatory cytokines, which can further exacerbate neurodegeneration in PD by inducing glial activation and promoting neuroinflammation. Other neurodegenerative diseases, such as amyotrophic lateral sclerosis (ALS) and Huntington's disease (HD), have also been linked to chronic arsenic exposure, albeit to a lesser extent. In ALS, arsenic may contribute to the degeneration of motor neurons, leading to muscle weakness, atrophy, and paralysis. Similarly, in HD, arsenic exposure could exacerbate the accumulation of mutant huntingtin protein aggregates, which are toxic to neurons and impair cellular function. Chronic arsenic exposure also contributes to synaptic dysfunction, a key feature of many neurodegenerative diseases (Banerjee & Chatterji, 2024; Ma et al., 2024; Shayan et al., 2023; Zhang et al., 2024). Neurons rely on the precise transmission of signals across synapses to maintain communication, and the disruption of this process can lead to cognitive and motor impairments. Arsenic-induced oxidative stress impairs synaptic plasticity, essential for learning and memory. In AD, for example, arsenic-induced synaptic dysfunction has been shown to impair LTP, a process that strengthens synapses and facilitates memory formation. The disruption of synaptic function may also contribute to cognitive decline and the progression of Alzheimer's and other neurodegenerative diseases. Mitochondrial dysfunction, induced by chronic arsenic exposure, further compounds synaptic dysfunction by depleting the energy supply necessary for synaptic vesicle release and neurotransmitter recycling. This impairment in synaptic activity leads to reduced cognitive performance and is thought to be a key factor in the development of neurodegenerative conditions (Chu et al., 2025; He et al., 2021; Luo & Shu, 2023; Shayan et al., 2023; Yan et al., 2023). Epidemiological studies have provided compelling evidence linking chronic arsenic exposure to an increased risk of neurodegeneration, particularly in regions where arsenic contamination of drinking water is prevalent. For example, in Bangladesh, where high levels of arsenic are present in groundwater, studies have shown a significant association between long-term arsenic exposure and cognitive decline, as well as an increased risk of developing neurodegenerative diseases such as AD and PD. Similarly, in parts of the United States, elevated arsenic levels in drinking water have been linked to reduced cognitive performance, particularly in older adults, suggesting that chronic exposure to low to moderate levels of arsenic may contribute to age-related cognitive decline and neurodegeneration. Longitudinal studies have further supported these findings by demonstrating that individuals exposed to arsenic over extended periods exhibit a higher incidence of cognitive impairment, dementia, and neurodegenerative diseases. These studies highlight the growing concern regarding environmental arsenic exposure as a public health issue, especially in areas with limited access to clean water. One of the key challenges in understanding the relationship between arsenic exposure and neurodegeneration is the potential interaction with genetic susceptibility factors. Not all individuals exposed to arsenic develop neurodegenerative diseases, suggesting that genetic factors may play a crucial role in modulating the neurotoxic effects of arsenic. Variations in genes involved in arsenic metabolism, oxidative stress response, and mitochondrial function may influence an individual's susceptibility to arsenic-induced neurodegeneration (Banerjee & Chatterji, 2024; de Castro et al., 2009; Du et al., 2022; Tolins et al., 2014; Zhang et al., 2023). For example, polymorphisms in genes encoding enzymes involved in the biotransformation of arsenic, such as arsenic methyltransferase (AS3MT), may alter the rate at which arsenic is detoxified in the body, affecting its accumulation in the brain and subsequent neurotoxic effects. Individuals with genetic variations that result in inefficient arsenic detoxification may be at a higher risk of developing neurodegenerative diseases, as they are more likely to experience prolonged exposure to toxic levels of arsenic in neural tissues. Additionally, genetic variations in antioxidant defense genes, such as superoxide dismutase (SOD) and catalase, may affect an individual's ability to counteract oxidative stress induced by arsenic exposure. Those with lower antioxidant capacity may be more vulnerable to the oxidative damage caused by arsenic, increasing the risk of neuronal injury and neurodegeneration. Mitochondrial DNA mutations may also play a role in susceptibility to arsenic-induced neurodegeneration, as mitochondrial dysfunction is a key mechanism by which arsenic exerts its toxic effects. Genetic variations that compromise mitochondrial function could exacerbate the effects of chronic arsenic exposure, leading to accelerated neuronal cell death and synaptic dysfunction. The interplay between arsenic exposure and genetic susceptibility is a critical area of research, as understanding these interactions could help identify individuals at higher risk for neurodegenerative diseases and lead to more targeted prevention and therapeutic strategies (Chu et al., 2025; Du et al., 2022; He et al., 2021; Shayan et al., 2023). Chronic arsenic exposure is strongly associated with the onset and progression of neurodegenerative diseases such as AD and PD. The mechanisms through which arsenic induces neuronal cell death and synaptic dysfunction include oxidative stress, mitochondrial dysfunction, and disruption of neurotransmitter systems. Epidemiological studies provide strong evidence for the link between arsenic exposure and neurodegeneration, particularly in regions with high levels of arsenic contamination. Moreover, genetic susceptibility factors play a crucial role in modulating the effects of arsenic on brain health, highlighting the need for further research to identify vulnerable populations and develop effective strategies to reduce arsenic exposure and mitigate its neurotoxic effects (Table 5.6).

5.10 Molecular and cellular effects

5.10.1 DNA damage and repair mechanisms

DNA damage and the subsequent repair mechanisms are essential in maintaining cellular integrity and genomic stability. DNA is constantly subjected to a variety of internal and external insults, such as oxidative stress, UV radiation, and chemical exposure, which can lead to different forms of damage, including base modifications, single-strand breaks, double-strand breaks (DSBs), and crosslinks. If left unaddressed, such damage can lead to mutations, genomic instability, and the development of various diseases, including cancer. To counteract these threats, cells have evolved a complex network of DNA repair mechanisms that recognize and repair damaged DNA, ensuring the fidelity of the genome. One of the most critical DNA repair pathways is base excision repair (BER), which addresses small base lesions caused by oxidation, deamination, and alkylation. In BER, the damaged base is recognized and excised by a DNA glycosylase enzyme, followed by an AP endonuclease removing the sugar-phosphate backbone (de Castro et al., 2009; Luo & Shu, 2023; Ma et al., 2024; Wu et al., 2025; Zhang et al., 2024). The resulting gap is then filled by DNA polymerase, and the strand is sealed by DNA ligase. Another vital repair pathway is nucleotide excision repair (NER), which specifically repairs bulky DNA lesions, such as those caused by UV radiation. In NER, a multiprotein complex recognizes and excises a short single-stranded segment containing the lesion. DNA polymerase then fills the gap, and the DNA is ligated. A failure in NER is often linked to disorders like xeroderma pigmentosum, where individuals are susceptible to UV radiation due to defective DNA repair. Homologous recombination (HR) and nonhomologous end joining (NHEJ) are the primary repair mechanisms for more severe forms of DNA damage, such as DSBs. DSBs are particularly dangerous because if left unrepaired or repaired incorrectly, they can result in chromosomal rearrangements, loss of genetic material, or cell death. In HR, a template from the sister chromatid is used to accurately repair the break, ensuring that no genetic information is lost. This process is error-free but occurs primarily during the S and G2 phases of the cell cycle. In contrast, NHEJ repairs DSBs by directly ligating the broken ends without a homologous template. While NHEJ is a quicker repair mechanism, it is more prone to errors, which can lead to small insertions or deletions at the repair site, potentially causing mutations. Furthermore, mismatch repair (MMR) is crucial in fixing errors during DNA replication. It detects and repairs base-base mismatches and insertion–deletion loops, preventing mutations from being passed to daughter cells. The MMR system relies on several key proteins, including MSH2, MSH6, and MLH1, which recognize mismatches and initiate the repair process (Du et al., 2022; Shayan et al., 2023; Ugbaja et al., 2022; Yan et al., 2023). Defects in MMR are linked to hereditary nonpolyposis colorectal cancer, a condition that increases the risk of various cancers. DNA repair mechanisms are intricately linked to the cellular response to DNA damage, with key regulatory pathways such as the ATM/ATR checkpoint signaling pathways and the p53 tumor suppressor gene playing critical roles. The ATM/ATR kinases sense DNA damage and initiate cell cycle checkpoints, which either allow time for repair or promote apoptosis if the damage is too severe. The tumor suppressor protein p53 is critical in this context, as it can induce cell cycle arrest, activate DNA repair, or initiate programmed cell death in response to irreparable DNA damage. Dysregulation of these pathways is a common feature in many cancers, as defective DNA repair and the inability to appropriately respond to damage can lead to uncontrolled cell proliferation. Furthermore, chronic DNA damage can also result in cellular senescence, a state in which cells permanently cease to divide. While senescence can act as a protective mechanism to prevent the propagation of damaged DNA, the

TABLE 5.6 Different strategies for preventing and mitigating arsenic neurotoxicity, including water purification methods, dietary interventions, and pharmacological treatments.

Strategy type	Strategy	Description	Effectiveness	Key references/studies
Water purification	**Filtration using activated alumina**	Activated alumina-based filters remove arsenic from drinking water by adsorbing it onto the filter material.	Effective in reducing arsenic levels to safe limits in drinking water.	Studies in Bangladesh and India show a significant reduction in exposure.
	Reverse osmosis	Water passes through a semipermeable membrane to remove arsenic and other contaminants.	Highly effective, it can reduce arsenic by up to 99%.	Commonly used in domestic water treatment in arsenic-endemic areas.
	Iron-based filtration	Iron oxide–coated sand or iron filings are used to adsorb arsenic from water.	Moderate effectiveness, suitable for household-scale purification.	Research in Asia has shown success in localized water treatment systems.
	Solar-powered distillation	It uses solar energy to evaporate water and condense it, leaving arsenic behind.	It is effective in regions with abundant sunlight but is energy-intensive.	Applied in remote arsenic-affected areas in rural countries.
Dietary interventions	**Increased intake of antioxidants**	Diets rich in antioxidants (e.g., vitamin C, vitamin E, selenium) may reduce oxidative stress induced by arsenic exposure.	Moderate effectiveness in reducing oxidative damage and mitigating cognitive effects.	Studies in Taiwan and Bangladesh suggest benefits in reducing arsenic-induced damage.
	High-protein diet (e.g., rice, legumes)	Protein-rich foods may reduce arsenic absorption in the body by binding arsenic and preventing its uptake into cells.	It may have moderate protective effects, particularly in rice-heavy diets.	Dietary studies in arsenic-affected areas, especially in South Asia.
	Fortification with micronutrients	Supplementation with essential micronutrients (zinc, iron, folic acid) to counteract arsenic toxicity and support brain function.	Moderate evidence supports cognitive benefits in children.	Studies in India and Bangladesh have shown potential benefits.
Pharmacological treatments	**Chelation therapy**	Cheating agents (e.g., dimercaprol, DMSA) bind and remove arsenic from the body.	Highly effective in cases of acute poisoning; limited evidence in chronic low-level exposure.	Used in clinical settings for arsenic poisoning but not widely adopted for chronic exposure.
	Antioxidant therapy (e.g., *N*-acetylcysteine, alpha-lipoic acid)	Pharmacological antioxidants may reduce oxidative stress caused by arsenic exposure and protect against cellular damage in neural tissues.	Promising preclinical results; moderate evidence in human trials.	Ongoing clinical trials in arsenic-endemic regions.
	Neuroprotective agents (e.g., curcumin, resveratrol)	Natural compounds with neuroprotective properties that may mitigate arsenic-induced cognitive and neural damage.	Early evidence of efficacy in preclinical models; human trials needed.	Preliminary studies in Asia and the United States suggest potential benefits.
Lifestyle and environmental	**Reduction of environmental arsenic exposure**	Implementing measures to reduce arsenic exposure in daily life, such as using clean water, avoiding arsenic-rich foods, and reducing occupational exposure.	Critical for long-term prevention of arsenic neurotoxicity.	Public health initiatives in arsenic-endemic regions.
	Public awareness and education	Community programs to educate about the risks of arsenic exposure and available protective measures such as water purification and dietary strategies.	Vital for empowering communities to reduce exposure and mitigate risks.	Successful in arsenic-affected areas, particularly in South Asia.

Source: de Castro et al. (2009), He et al. (2021), Luo and Shu (2023), Yang et al. (2024).

accumulation of senescent cells in tissues can contribute to ageing and age-related diseases. DNA damage and repair mechanisms are vital for preserving genomic integrity. Cells rely on a sophisticated array of repair pathways to detect and fix different types of DNA lesions, and defects in these pathways can lead to various diseases, including cancer and neurodegenerative disorders. The ability of cells to accurately repair DNA damage and maintain genomic stability is crucial for overall organismal health and longevity (Du et al., 2022; Gopnar et al., 2023; Wu et al., 2025; Yan et al., 2023).

5.10.2 Inflammation and immune response in the brain

Inflammation and immune responses are crucial in maintaining homeostasis and defending against various insults, such as infections, injuries, and toxins. However, when these responses become dysregulated or chronic, they can contribute to the pathogenesis of neurodegenerative diseases and other neurological disorders. The brain's immune system comprises microglia, the resident immune cells of the CNS, astrocytes, neurons, and the BBB. Under normal conditions, microglia perform surveillance functions, constantly monitoring the brain for signs of damage or infection. When they detect injury, infection, or the presence of harmful stimuli, microglia become activated and initiate an inflammatory response. This response involves releasing proinflammatory cytokines, chemokines, and ROS to recruit additional immune cells and promote tissue repair. Astrocytes, another important glial cell type, activate the inflammatory response by releasing cytokines and other signaling molecules. While these responses are necessary for protecting the brain from acute damage, prolonged or excessive activation of microglia and astrocytes can have detrimental effects. Chronic neuroinflammation has been implicated in the development and progression of a variety of neurodegenerative diseases, including AD, PD, multiple sclerosis, and ALS. In these conditions, sustained microglial activation and the release of inflammatory mediators can lead to neuronal damage, synaptic dysfunction, and cognitive and motor impairments. The inflammatory response can also disrupt the BBB, which protects the brain from harmful substances and pathogens (de Castro et al., 2009; He et al., 2021; Ma et al., 2024; Ugbaja et al., 2022; Zhang et al., 2024). When the BBB is compromised, harmful molecules, such as immune cells and circulating cytokines, can enter the brain and exacerbate the inflammatory response. This creates a vicious cycle where inflammation leads to further neuronal injury, triggering additional immune activation. One of the key factors driving neuroinflammation is the accumulation of misfolded proteins, such as amyloid-beta in AD or alpha-synuclein in PD. These proteins can act as signals to activate microglia and trigger an inflammatory cascade. In AD, amyloid plaques bind to receptors on microglia, activating them and releasing proinflammatory cytokines. This inflammatory response can contribute to neuronal damage and the formation of neurofibrillary tangles, another hallmark of the disease. In PD, alpha-synuclein aggregates form Lewy bodies within dopaminergic neurons, activating microglia and releasing toxic inflammatory molecules. These cytokines and ROS can further damage the surrounding neurons, particularly the dopaminergic neurons in the substantia nigra, which are central to motor control. Additionally, neuroinflammation can impair the function of the BBB, making it more permeable to toxic molecules and immune cells. The peripheral immune system also regulates immune responses in the brain. In autoimmune diseases like multiple sclerosis, the immune system mistakenly attacks the myelin sheath surrounding neurons, leading to demyelination and neuroinflammation. T cells and antibodies cross the BBB and attack the CNS, triggering an inflammatory cascade that contributes to the progression of the disease. The immune response in the brain is not always detrimental. Acute inflammation can promote tissue repair and facilitate the clearance of damaged cells or pathogens. However, when the inflammation becomes chronic, it can contribute to neurodegeneration. The resolution of inflammation is an active process that involves the clearance of inflammatory mediators, the restoration of tissue homeostasis, and the return of microglia to their resting state (Gopnar et al., 2023; He et al., 2021; Shayan et al., 2023; Ugbaja et al., 2022; Yan et al., 2023). Inflammatory resolution is a tightly regulated process, and defects in this resolution process may contribute to the persistence of neuroinflammation and the development of neurodegenerative diseases. Several factors, including genetic predisposition, environmental exposures, and lifestyle factors, can influence the inflammatory response in the brain. For instance, chronic stress, poor diet, and lack of physical activity have all been shown to promote neuroinflammation and increase the risk of developing neurodegenerative diseases. Conversely, antiinflammatory interventions, such as using nonsteroidal antiinflammatory drugs, exercise, and dietary interventions rich in antioxidants, have been proposed as potential therapeutic strategies to mitigate neuroinflammation and its associated risks. Overall, inflammation and immune responses in the brain are essential for protecting the CNS from damage. However, when these responses become chronic or dysregulated, they can contribute to the pathogenesis of neurodegenerative diseases. Understanding the mechanisms underlying neuroinflammation and identifying strategies to modulate the immune response in the brain hold great promise for developing therapeutic approaches to treat a wide range of neurological disorders (Fig. 5.1).

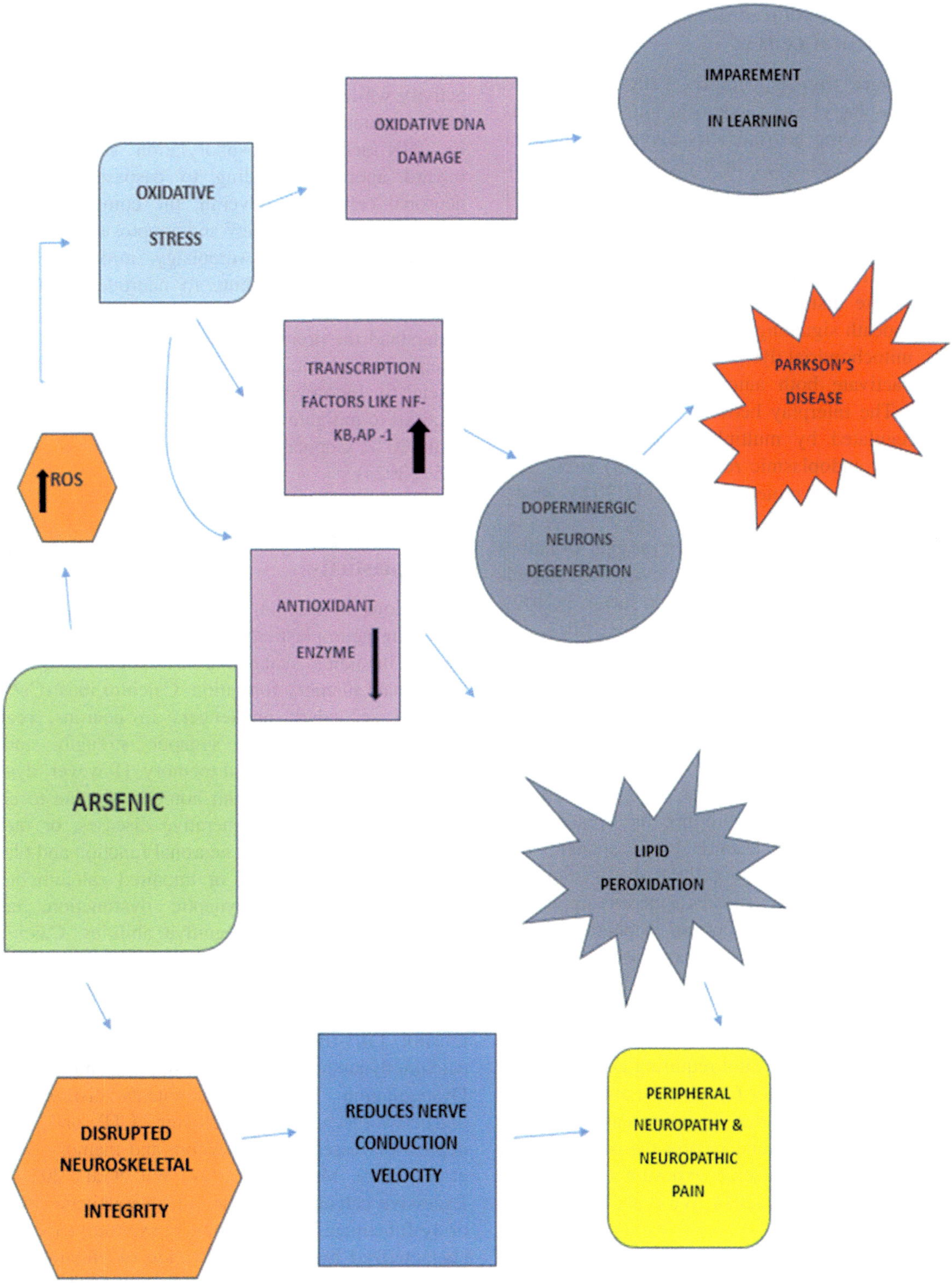

FIGURE 5.1 Mechanism of arsenic neurodegeneration.

5.10.3 Role of arsenic-induced autophagy and apoptosis in neural cells

Arsenic exposure triggers complex cellular responses, including autophagy and apoptosis, in neural cells, ultimately influencing neurotoxicity and neurodegenerative processes. Autophagy, a highly conserved catabolic process, plays a dual role in arsenic-induced neurotoxicity, acting as a protective mechanism to remove damaged cellular components and as a precursor to programmed cell death under prolonged stress conditions. Arsenic exposure disrupts the balance between prosurvival and prodeath signaling pathways, leading to oxidative stress, mitochondrial dysfunction, and DNA damage, which can activate both autophagy and apoptosis in neural cells. The interplay between these two pathways is tightly regulated by multiple molecular mediators, including ROS, endoplasmic reticulum (ER) stress, and key autophagy-related proteins (He et al., 2021; Zhang et al., 2023, 2024).

Oxidative stress is a critical upstream trigger of autophagy in arsenic-exposed neural cells. Elevated ROS levels activate nuclear factor erythroid 2-related factor 2 (Nrf2), which regulates antioxidant defense responses, but excessive oxidative stress overwhelms the cellular antioxidant capacity, leading to damage to lipids, proteins, and DNA. Mitochondrial dysfunction caused by arsenic disrupts the ETC, further increasing ROS production and inducing mitophagy, a selective autophagy targeting damaged mitochondria. However, chronic arsenic exposure can impair lysosomal function, inhibiting the clearance of damaged organelles and leading to autophagic flux blockade, which contributes to cellular dysfunction and neurotoxicity. ER stress, a critical factor linking arsenic exposure to autophagy and apoptosis in neural cells, is induced by the disruption of ER homeostasis. Arsenic disrupts ER homeostasis by causing misfolded protein accumulation, which activates the unfolded protein response (UPR). Key UPR mediators, including protein kinase R-like ER kinase, inositol-requiring enzyme 1, and activating transcription factor 6, modulate autophagy and apoptosis through downstream signaling cascades. This ER homeostasis disruption underscores the complexity of the cellular response to arsenic exposure (Banerjee & Chatterji, 2024; Tolins et al., 2014; Wu et al., 2024; Zhang et al., 2023).

The regulation of autophagy and apoptosis is also influenced by arsenic's impact on key signaling pathways such as the phosphatidylinositol 3-kinase/protein kinase B and mitogen-activated protein kinase (MAPK) pathways. Arsenic exposure inhibits Akt activation, suppressing the mammalian target of rapamycin (mTOR), a master regulator of autophagy inhibition. The suppression of mTOR facilitates autophagy initiation but may also sensitize neurons to apoptotic signals. Concurrently, arsenic activates MAPK pathways, particularly c-Jun N-terminal kinase (JNK) and p38 MAPK, which enhance autophagic activity while promoting proapoptotic signaling via Bcl-2 family proteins. The disruption of Bcl-2/Bax interactions by JNK-mediated phosphorylation shifts the balance toward apoptosis, leading to caspase activation and neuronal cell death. Overall, the complex interplay of arsenic-induced autophagy and apoptosis in neural cells is a critical study area. Autophagy, initially a protective mechanism, can contribute to neurodegeneration when dysregulated. The urgency of further research to fully understand the precise molecular mechanisms underlying arsenic neurotoxicity is clear. This understanding is crucial for the development of targeted therapeutic interventions to mitigate arsenic-induced neuronal damage (Du et al., 2022; Gopnar et al., 2023; Shayan et al., 2023; Yan et al., 2023).

5.10.4 Disruptions in calcium signaling and neuroplasticity

Disruptions in calcium signaling play a crucial role in impairing neuroplasticity, affecting various neuronal processes, including synaptic transmission, dendritic remodeling, and memory formation. Calcium ions (Ca^{2+}) serve as vital secondary messengers in neurons, regulating neurotransmitter release, synaptic strength, and gene expression for learning and memory. However, dysregulation of intracellular calcium homeostasis due to environmental toxins, neurodegenerative diseases, or metabolic stress can severely impact neuronal function and plasticity. Excessive calcium influx or impaired calcium buffering mechanisms lead to synaptic dysfunction, neuronal damage, and impaired cognitive abilities. Calcium signaling is tightly controlled by multiple channels, pumps, and binding proteins such as voltage-gated calcium channels (VGCCs), NMDARs, ryanodine receptors (RyRs), and inositol 1,4,5-trisphosphate receptors, which orchestrate calcium dynamics in the cytoplasm, ER, and mitochondria. Dysregulation of these components disrupts synaptic plasticity mechanisms, including LTP and long-term depression, essential for learning and memory (de Castro et al., 2009; Ma et al., 2024; Wu et al., 2024, 2025). Excessive calcium entry through overactivated NMDARs or dysfunctional VGCCs results in excitotoxicity, a process characterized by the overactivation of glutamatergic signaling, leading to neuronal injury and death. This excitotoxicity is a hallmark of neurodegenerative diseases such as AD, PD, and HD, where impaired calcium homeostasis contributes to synaptic loss and neuronal atrophy. Mitochondrial calcium overload further exacerbates oxidative stress and impairs ATP production, leading to neuronal energy deficits and apoptotic cell death. Additionally, ER stress due to dysregulated calcium storage and release affects protein folding and synaptic protein synthesis, which are crucial for maintaining synaptic integrity. Alterations in calcium-dependent signaling pathways, including calcium/calmodulin-dependent protein kinase II, protein kinase C, and calcineurin, interfere with synaptic remodeling and dendritic spine dynamics, further impairing neuroplasticity. Calcium-dependent activation of transcription factors such as cyclic adenosine monophosphate response element-binding protein (CREB) is essential for synaptic strengthening and neurogenesis. However, disruptions in calcium signaling lead to CREB dysregulation, affecting the expression of plasticity-related genes and synaptic proteins, thereby weakening neuronal connectivity. In neurodevelopmental disorders such as ASD and schizophrenia, aberrant calcium signaling has been implicated in altered synaptic pruning and circuit formation, resulting in cognitive and behavioral deficits. Environmental neurotoxins such as heavy metals, pesticides, and industrial pollutants also contribute to calcium dysregulation by interfering with calcium transporters and receptors, exacerbating neurotoxicity, and impairing synaptic function. Therapeutic interventions targeting calcium homeostasis, such as calcium channel blockers, NMDAR modulators, and neuroprotective agents, hold promise in mitigating calcium-induced neurotoxicity and preserving neuroplasticity. Understanding the intricate relationship between calcium signaling and neuroplasticity is crucial for developing effective strategies to prevent and treat neurological disorders associated with calcium dysregulation. Future research should focus on elucidating the precise molecular mechanisms underlying calcium-mediated neuroplasticity deficits and identifying potential therapeutic targets to restore normal neuronal function (Shayan et al., 2023; Ugbaja et al., 2022; Wu et al., 2024; Yan et al., 2023).

5.11 Detection and biomarkers of arsenic neurotoxicity

5.11.1 Biomonitoring arsenic levels in blood, urine, and hair

Biomonitoring arsenic levels in blood, urine, and hair is critical for assessing arsenic exposure, metabolism, and potential health risks in individuals and populations. Arsenic, a highly toxic metalloid, enters the human body primarily through contaminated drinking water, food, air, and occupational exposure, and its biomonitoring is essential for understanding both acute and chronic toxicity. Blood, urine, and hair are key biological matrices for detecting arsenic levels, each providing unique insights into exposure timing, extent, and duration. Blood arsenic levels reflect recent exposure and the initial distribution phase of arsenic before its metabolism and excretion, making blood valuable for detecting acute poisoning and assessing ongoing exposure in individuals with high environmental or occupational risks. Urinary arsenic is the most commonly used biomarker due to its ability to reflect recent exposure and the body's capacity to metabolize and excrete inorganic arsenic. After ingestion, inorganic arsenic undergoes methylation in the liver, producing MMA and DMA, excreted primarily through urine. The urinary arsenic profile helps distinguish between toxic inorganic and organic arsenic species derived from dietary sources such as seafood, which contain relatively nontoxic arsenobetaine and arsenocholine (Gopnar et al., 2023; Shayan et al., 2023; Wu et al., 2024; Yan et al., 2023). Advanced techniques such as high-performance liquid chromatography coupled with inductively coupled plasma mass spectrometry (ICP-MS) are used for speciation, allowing differentiation between toxic and nontoxic arsenic forms. Hair arsenic analysis serves as an essential biomarker for chronic exposure, as arsenic binds to keratin in hair and accumulates over time, providing a long-term exposure record that can span weeks to months. Hair samples are instrumental in forensic toxicology and epidemiological studies for reconstructing past exposure patterns and identifying populations at risk of chronic arsenic poisoning. However, external contamination from arsenic-containing dust, water, or personal care products can affect hair arsenic levels, requiring rigorous sample preparation and washing protocols to ensure accuracy. Integrating multimatrix biomonitoring enhances exposure assessment by correlating arsenic concentrations in blood, urine, and hair to determine the toxicokinetics and bioaccumulation of arsenic in the body. Genetic polymorphisms in arsenic metabolism, age, sex, nutritional status, and coexposure to other toxicants influence individual variability in arsenic biomonitoring results. Chronic arsenic exposure has been linked to various health effects, including skin lesions, cardiovascular diseases, neurotoxicity, and cancers of the bladder, lung, and skin, necessitating regular biomonitoring in high-risk populations (Ma et al., 2024; Wu et al., 2024; Yang et al., 2024). Public health interventions based on biomonitoring data can aid in developing regulatory policies, implementing mitigation strategies, and raising awareness about arsenic contamination sources. Future advancements in biosensor technology, metabolomics, and noninvasive sampling methods may further enhance the sensitivity, specificity, and accessibility of arsenic biomonitoring, contributing to more effective risk assessment and early detection of arsenic-related health effects. Comprehensive biomonitoring programs integrating exposure assessment with health surveillance are essential for safeguarding public health and preventing long-term arsenic toxicity in vulnerable communities.

5.11.2 Neuroimaging techniques for arsenic-induced brain changes

Neuroimaging techniques are crucial in assessing arsenic-induced brain changes, providing valuable insights into the structural, functional, and metabolic alterations associated with arsenic neurotoxicity. Chronic exposure to arsenic through contaminated drinking water, occupational hazards, or environmental pollution leads to significant neurological impairments, including cognitive decline, neuroinflammation, and neurodegenerative processes. Advanced neuroimaging modalities such as magnetic resonance imaging (MRI), functional MRI (fMRI), diffusion tensor imaging (DTI), positron emission tomography (PET), and magnetic resonance spectroscopy (MRS) enable noninvasive evaluation of arsenic-related neurotoxicity, aiding in early diagnosis, monitoring disease progression, and developing targeted therapeutic interventions. Structural MRI reveals arsenic-induced brain atrophy, particularly in regions critical for memory and cognition, such as the hippocampus, prefrontal cortex, and cerebellum, where arsenic-mediated oxidative stress and neuroinflammation lead to neuronal loss and volume reduction. DTI provides insights into white matter integrity, showing disruptions in axonal connectivity due to arsenic-induced demyelination and microstructural damage, contributing to impaired cognitive function and motor deficits. fMRI assesses alterations in neural activity and connectivity patterns, highlighting impaired synaptic transmission and dysfunctional networks in arsenic-exposed individuals, particularly in regions associated with executive function, learning, and emotional regulation (Ma et al., 2024; Shayan et al., 2023; Wu et al., 2024; Yan et al., 2023). PET imaging using radiolabeled tracers detects arsenic-induced neuroinflammation by measuring microglial activation and neurotoxic protein accumulation, hallmarks of arsenic-mediated neurodegeneration. Additionally, PET scans reveal alterations in cerebral glucose metabolism, indicating disrupted neuronal energy homeostasis caused by mitochondrial dysfunction, a key mechanism underlying arsenic neurotoxicity. MRS further enhances neurotoxicological assessments by measuring brain metabolite levels, including reductions in *N*-acetyl aspartate, a marker of neuronal integrity, increased choline levels associated with neuroinflammation, and altered glutamate/glutamine ratios reflecting excitotoxicity. These neurochemical imbalances correlate with arsenic's impact on synaptic plasticity, neurotransmitter regulation, and neurodevelopmental processes. Emerging imaging techniques such as high-resolution functional connectivity mapping and machine learning-based radiomics offer more precise characterization of arsenic-induced brain alterations, improving risk assessment and early detection of neurotoxicity. Longitudinal neuroimaging studies are essential for understanding the progression of arsenic-related neural damage and identifying potential biomarkers for early intervention. Integrating multimodal imaging approaches with neurobehavioral assessments and biomarker analysis enhances understanding of arsenic's neurotoxic mechanisms, paving the way for developing neuroprotective strategies. While neuroimaging provides a powerful tool for studying arsenic-induced brain changes, challenges such as high costs, limited accessibility, and variability in imaging biomarkers necessitate further research to standardize protocols and improve diagnostic accuracy (Luo & Shu, 2023; Ugbaja et al., 2022; Wu et al., 2025; Zhang et al., 2023, 2024). Future advancements in neuroimaging, including ultra-high-field MRI and molecular imaging techniques, hold promise for detecting subtle neurotoxic effects of arsenic at earlier stages, enabling timely therapeutic interventions. A comprehensive approach combining neuroimaging with epidemiological studies and molecular research will contribute to a better understanding of arsenic neurotoxicity, ultimately leading to improved public health policies and strategies for preventing arsenic-related neurological disorders.

5.11.3 Genetic and epigenetic markers of arsenic toxicity

Genetic and epigenetic markers of arsenic toxicity play a crucial role in understanding individual susceptibility, mechanisms of toxicity, and long-term health consequences associated with arsenic exposure. Arsenic, a widely distributed environmental toxin, exerts its toxic effects through oxidative stress, DNA damage, and disruption of key metabolic and signaling pathways, leading to a range of diseases, including cancer, neurodegeneration, and cardiovascular disorders. Genetic polymorphisms in arsenic metabolism genes, such as arsenite methyltransferase (AS3MT), glutathione *S*-transferases (GSTs), and cytochrome P450 enzymes, influence the efficiency of arsenic biotransformation and excretion, impacting individual toxicity responses. AS3MT is a critical enzyme that catalyzes the methylation of inorganic arsenic into MMA and DMA, and genetic variations in AS3MT affect arsenic methylation capacity, leading to the accumulation of toxic intermediates. Polymorphisms in GST genes, including GSTT1 and GSTM1, influence glutathione-mediated detoxification of arsenic-induced ROS, altering oxidative stress responses and cellular damage. Additionally, single-nucleotide polymorphisms in DNA repair genes, such as XRCC1 and OGG1, modulate the ability to repair arsenic-induced DNA damage, contributing to increased genomic instability and carcinogenesis. Epigenetic modifications, including DNA methylation, histone modifications, and miRNA regulation, mediate arsenic toxicity by altering gene expression patterns without changing the DNA sequence. Arsenic exposure is associated with global DNA hypomethylation and site-specific hypermethylation of tumor suppressor genes, such as p16 and RASSF1A, leading to aberrant cell cycle regulation and increased cancer risk. Hypomethylation of repetitive DNA elements, such as LINE-1 and Alu sequences, promotes genomic instability and inflammatory responses, exacerbating arsenic toxicity. Arsenic-induced histone modifications, such as acetylation and methylation of histones H3 and H4, influence chromatin accessibility and transcriptional activity of genes involved in oxidative stress, apoptosis, and inflammation. miRNAs, small noncoding RNAs that regulate gene expression posttranscriptionally, are also dysregulated by arsenic exposure, with key miRNAs such as miR-21, miR-222, and miR-29 playing roles in arsenic-mediated carcinogenesis, apoptosis, and immune dysregulation. The interplay between genetic polymorphisms and epigenetic alterations determines individual susceptibility to arsenic toxicity, with environmental and lifestyle factors further modifying risk. High-throughput omics technologies, including genome-wide association studies, epigenome-wide association studies, and transcriptomic profiling, have identified arsenic-responsive genetic and epigenetic biomarkers that can be used for early detection and risk assessment (Gopnar et al., 2023; He et al., 2021; Shayan et al., 2023; Zhang et al., 2024). Understanding these molecular signatures enables the development of precision medicine approaches for arsenic-exposed populations, including targeted interventions to mitigate toxicity effects. Future research should focus on integrating multiomics approaches to uncover novel biomarkers and elucidate the complex gene-environment interactions in arsenic toxicity. Additionally, the reversibility of arsenic-induced epigenetic changes provides opportunities for therapeutic interventions, such as dietary methyl donors, histone deacetylase inhibitors, and miRNA-based therapies, to counteract arsenic's toxic effects. Advancing knowledge on genetic and epigenetic markers of arsenic toxicity will enhance risk assessment, improve early diagnosis, and contribute to effective strategies for preventing arsenic-related diseases.

5.11.4 Behavioral and neurocognitive assessments for early detection

Behavioral and neurocognitive assessments play a crucial role in the early detection of arsenic-induced neurotoxicity, enabling timely intervention and mitigation of long-term neurological consequences. Chronic arsenic exposure, primarily through contaminated drinking water and food, has been associated with cognitive deficits, memory impairment, neurodevelopmental delays, and increased risk of neurodegenerative disorders such as AD and PD. Standardized neuropsychological testing methods, including IQ assessments, executive function tests, and memory evaluations, help identify subtle cognitive impairments in arsenic-exposed populations before significant structural brain damage occurs. Behavioral assessments, including attention and impulsivity tests, are particularly valuable in detecting arsenic-induced neurodevelopmental disruptions in children, as exposure during critical growth periods can result in impaired learning, reduced cognitive flexibility, and lower academic performance (de Castro et al., 2009; Luo & Shu, 2023; Ma et al., 2024; Wu et al., 2025; Zhang et al., 2024). The Mini-Mental State Examination and Montreal Cognitive Assessment are widely used screening tools for detecting mild cognitive impairment and early-stage dementia in arsenic-exposed adults, offering insights into memory, visuospatial skills, language abilities, and executive functioning. Additionally, computerized neurocognitive testing platforms, such as the Cambridge Neuropsychological Test Automated Battery and NIH Toolbox Cognition Battery, provide objective, quantitative measures of cognitive performance, allowing for precise longitudinal monitoring of arsenic-related cognitive decline. Functional assessments, including reaction time tests, psychomotor speed evaluations, and gait analysis, are valuable for detecting early motor impairments and coordination deficits linked to arsenic-induced neurotoxicity. Electrophysiological techniques, such as event-related potentials (ERPs) and quantitative electroencephalography, offer noninvasive biomarkers of neural dysfunction, revealing arsenic-associated changes in cortical excitability, synaptic plasticity, and neural network connectivity. For example, reduced P300 amplitude in ERP studies has been correlated with impaired cognitive processing and attentional deficits in arsenic-exposed individuals, serving as an early indicator of neurotoxicity. Advanced eye-tracking technology is also emerging as a promising tool for assessing visual attention and oculomotor control, providing sensitive measures of arsenic-related disruptions in cognitive and sensorimotor integration. In children, standardized developmental assessments, such as the Bayley Scales of Infant Development (BSID) and the Wechsler Intelligence Scale for Children, help evaluate arsenic-induced delays in language acquisition, motor coordination, and problem-solving skills. Socioemotional assessments, including behavioral checklists and parental reports, further capture arsenic-related neuropsychiatric manifestations such as increased anxiety, aggression, and social withdrawal. Integrating neurocognitive and behavioral assessments with neuroimaging techniques, such as fMRI and DTI, enhances the ability to detect arsenic-induced neural alterations, correlating cognitive deficits with structural and functional brain abnormalities. Longitudinal cohort studies tracking arsenic-exposed individuals over time provide critical insights into the progression of neurotoxic effects, allowing for the identification of early behavioral biomarkers predictive of later cognitive decline. Early detection through behavioral and neurocognitive assessments is essential for implementing targeted interventions, including cognitive training, nutritional supplementation, and environmental remediation strategies to reduce arsenic exposure. Future research should focus on developing standardized, culturally sensitive cognitive assessment protocols for arsenic-exposed populations, improving diagnostic accuracy, and facilitating early therapeutic strategies to prevent long-term neurological impairments (Gopnar et al., 2023; Shayan et al., 2023; Wu et al., 2024; Yang et al., 2024).

5.12 Strategies for prevention and mitigation

Water purification technologies to reduce arsenic exposure, nutritional and pharmacological interventions, policy regulations, and community awareness programs are essential for mitigating the adverse health effects associated with arsenic contamination. Arsenic contamination in drinking water is a significant public health concern, particularly in regions with high groundwater arsenic levels, such as South Asia, parts of the United States, and Latin America. Various water purification technologies have been developed to remove arsenic from drinking water, including adsorption, coagulation–flocculation, membrane filtration, ion exchange, and biological remediation. Adsorption-based methods utilize materials such as activated alumina, iron oxide, and graphene-based nanomaterials to bind arsenic and remove it from water. Iron-based adsorbents, particularly zero-valent iron and ferric hydroxide, have demonstrated high efficiency in arsenic removal due to their strong affinity for arsenate (AsV) and arsenite (ASI). Coagulation–flocculation methods involve the addition of chemical coagulants such as ferric chloride or aluminum sulfate, which aggregate arsenic particles and allow their removal through sedimentation (Luo & Shu, 2023; Ma et al., 2024; Shayan et al., 2023; Zhang et al., 2024). Membrane filtration techniques, including RO and nanofiltration, provide effective arsenic removal by using semipermeable membranes that selectively filter out contaminants; however, these technologies are expensive and generate brine waste, necessitating proper disposal strategies. Ion exchange resins selectively remove arsenic by replacing it with chloride or sulfate ions, but their efficiency depends on water chemistry and requires periodic regeneration. Biological remediation, including arsenic-resistant bacteria and genetically engineered microorganisms, is an emerging approach that harnesses microbial metabolic pathways to transform and sequester arsenic, offering a sustainable and low-cost solution. Integrating these technologies in household and community-scale purification systems ensures access to safe drinking water, significantly reducing arsenic exposure and associated health risks. Nutritional and pharmacological interventions are vital in mitigating arsenic toxicity and its long-term effects. Nutritional strategies focus on enhancing the body's natural detoxification pathways through dietary supplementation with essential nutrients, antioxidants, and methyl donors (Du et al., 2022; Ugbaja et al., 2022; Wu et al., 2025; Yan et al., 2023). Selenium, a vital trace element, has been shown to counteract arsenic toxicity by forming stable arsenic-selenium complexes that reduce arsenic bioavailability and enhance excretion. Zinc supplementation also offers protective effects by reducing oxidative stress and modulating arsenic-induced inflammatory responses. Methyl-donor nutrients such as folate, vitamin B12, and choline facilitate the biotransformation of inorganic arsenic into less toxic methylated species, accelerating arsenic excretion and lowering systemic toxicity. Antioxidants, including vitamins C and E, polyphenols, and flavonoids, help mitigate arsenic-induced oxidative damage by scavenging ROS and restoring cellular redox balance. Curcumin, a bioactive compound in turmeric, has shown neuroprotective and hepatoprotective effects in arsenic toxicity by modulating antioxidant defense mechanisms and reducing apoptosis. Chelation therapy, a pharmacological intervention, involves the use of chelating agents such as dimercaptosuccinic acid (DMSA), meso-2,3-DMSA (meso-DMSA), and British Anti-Lewisite (BAL) to bind arsenic and facilitate its urinary excretion. While chelation therapy effectively reduces arsenic burden in cases of acute poisoning, its efficacy in chronic arsenic exposure remains limited due to potential side effects and the risk of depleting essential trace elements. Novel pharmacological approaches are being explored to enhance treatment efficacy, including using plant-derived bioactive compounds, metal chelators with improved specificity, and gene-based therapies targeting arsenic detoxification pathways. A combination of nutritional and pharmacological interventions tailored to individual exposure levels and genetic susceptibility can significantly mitigate the toxic effects of arsenic, reducing the risk of chronic diseases such as cancer, neurodegeneration, and cardiovascular disorders (Du et al., 2022; Gopnar et al., 2023; Wu et al., 2025; Yan et al., 2023).

Policy regulations and public health initiatives are critical in addressing arsenic contamination and safeguarding public health. Governments and international organizations, including the World Health Organization (WHO) and the Environmental Protection Agency, have established regulatory standards for arsenic levels in drinking water, with the WHO-recommended limit set at 10 parts per billion (ppb). However, enforcement and compliance remain challenging in low-resource settings, where millions rely on arsenic-contaminated groundwater. National policies must prioritize implementing stringent water quality monitoring programs and ensure that arsenic levels are regularly assessed in high-risk regions. Adopting low-cost, decentralized water treatment solutions, such as community-based filtration systems and household arsenic removal units, is essential for improving access to safe water in affected areas. Public health initiatives should focus on screening and surveillance programs to identify arsenic-exposed populations and monitor health outcomes. Integrating arsenic screening into routine health check-ups enables early detection of arsenic-induced conditions, allowing timely medical intervention. Research funding should be allocated to developing innovative arsenic mitigation technologies, including cost-effective water purification methods and biomarkers for early detection of arsenic toxicity. Policymakers should also promote sustainable land use practices and industrial regulations to prevent anthropogenic arsenic contamination from mining, agriculture, and manufacturing activities. International collaboration and knowledge-sharing platforms facilitate the exchange of best practices and technological advancements in arsenic mitigation, accelerating progress toward global arsenic-free water goals (de Castro et al., 2009; He et al., 2021; Ma et al., 2024; Ugbaja et al., 2022).

Community awareness and education programs are fundamental in empowering individuals and communities to take proactive measures against arsenic exposure. Public education campaigns should focus on disseminating information about the sources of arsenic contamination, health risks, and preventive strategies. Visual aids, interactive workshops, and mobile health applications can enhance public engagement and knowledge retention. Schools and community centers should integrate arsenic awareness into educational curricula, teaching children and young adults about safe water practices and the importance of proper nutrition in mitigating arsenic toxicity. Women, often primary caregivers, should be targeted for educational interventions, as they play a key role in ensuring household water safety and nutritional well-being. Community-led participatory approaches, including citizen science initiatives, can foster local engagement in water quality monitoring, enabling residents to test and report arsenic levels in their water sources. Government and nongovernmental organizations should collaborate to provide technical support and resources for implementing community-based water purification projects (Banerjee & Chatterji, 2024; Chu et al., 2025; de Castro et al., 2009; Garg & Bandyopadhyay, 2025; Zhang et al., 2024). Training programs for healthcare professionals can improve their ability to diagnose and manage arsenic-related health conditions, ensuring that affected individuals receive appropriate medical care. Behavioral change interventions, such as incentivising safe water sources and promoting dietary modifications, can significantly reduce arsenic exposure and its associated health risks. Social media platforms and digital tools can further amplify awareness campaigns, reaching a broader audience and encouraging collective action. Strengthening community resilience through education, capacity-building, and participatory governance fosters long-term sustainability in arsenic mitigation efforts, ensuring that future generations are protected from the adverse effects of arsenic contamination. A multifaceted approach encompassing water purification technologies, nutritional and pharmacological interventions, policy regulations, and community awareness programs is essential for reducing arsenic exposure and its associated health risks. Advancements in water treatment technologies, such as adsorption, membrane filtration, and biological remediation, provide effective solutions for arsenic removal, ensuring access to safe drinking water (Gopnar et al., 2023; Ugbaja et al., 2022; Wu et al., 2024). Nutritional and pharmacological strategies, including antioxidant supplementation and chelation therapy, offer protective effects against arsenic-induced toxicity, mitigating long-term health consequences. Strong policy frameworks and regulatory enforcement are necessary to monitor arsenic levels, implement mitigation measures, and support research and development efforts. Public health initiatives, including screening programs and risk assessment strategies, enable early detection and intervention for arsenic-exposed populations. Community awareness and education programs empower individuals with the knowledge and resources to adopt preventive measures, fostering a culture of safe water practices and health-conscious behavior. Collaborative efforts among governments, researchers, healthcare professionals, and local communities are crucial in addressing arsenic contamination and ensuring global water safety. Future research should focus on developing integrated, cost-effective solutions that combine water treatment, nutritional interventions, and policy-driven approaches to achieve sustainable arsenic mitigation. By prioritizing comprehensive and coordinated action, the global community can work toward eliminating arsenic-related health hazards, protecting vulnerable populations, and securing safe water access.

5.13 Future directions and research gaps

Long-term epidemiological studies are essential for understanding the full spectrum of arsenic toxicity, particularly its chronic effects on human health, including carcinogenic, neurotoxic, cardiovascular, and developmental consequences. While acute arsenic poisoning presents with overt symptoms such as gastrointestinal distress and skin lesions, chronic exposure is insidious, manifesting over years or decades with cumulative damage to multiple organ systems. Large-scale, longitudinal studies are necessary to identify dose–response relationships, genetic susceptibility factors, and transgenerational effects of arsenic exposure. Additionally, such studies can provide critical data on regional variations in arsenic toxicity based on environmental factors, dietary intake, and socioeconomic conditions. Longitudinal cohorts with repeated biomonitoring of arsenic levels in blood, urine, hair, and nails, coupled with clinical assessments and neurocognitive evaluations, can help delineate the trajectory of arsenic-induced pathologies. Moreover, epigenetic studies in exposed populations can show how arsenic alters DNA methylation, histone modifications, and miRNA expression, potentially affecting gene regulation across generations. These studies can also inform regulatory policies by determining safe exposure limits and identifying high-risk populations that require targeted interventions. Integrating data from epidemiological studies with advances in exosomes, which assess lifetime environmental exposures, will further refine our understanding of arsenic toxicity and its interaction with other environmental and genetic risk factors. Such research is crucial for developing precision medicine approaches tailored to individuals with chronic arsenic exposure, ultimately improving health outcomes in affected populations.

Advancements in neurotoxicology research on arsenic have provided more profound insights into the molecular and cellular mechanisms underlying its deleterious effects on the nervous system. Arsenic readily crosses the BBB, accumulating in the brain and disrupting neuronal function through oxidative stress, mitochondrial dysfunction, and impaired neurotransmission. Studies have demonstrated that arsenic exposure alters synaptic plasticity, reducing key neurotransmitters such as dopamine, serotonin, and acetylcholine levels, essential for cognition, mood regulation, and motor control. Additionally, arsenic-induced neuroinflammation, characterized by increased microglial activation and elevated proinflammatory cytokines, exacerbates neuronal damage and contributes to neurodegenerative processes. Recent research has highlighted the role of arsenic in disrupting calcium signaling, which is crucial for neuronal excitability and synaptic communication. Dysregulation of calcium homeostasis can trigger apoptosis and impair neurodevelopment, particularly in prenatal and early childhood exposure cases. Animal models have further elucidated the impact of arsenic on brain morphology, showing reduced hippocampal volume, altered dendritic arborization, and impaired neurogenesis. Furthermore, advancements in omics technologies, including transcriptomics, proteomics, and metabolomics, have identified arsenic-responsive biomarkers that can serve as early indicators of neurotoxicity. These findings underscore the need for targeted therapeutic strategies aimed at mitigating arsenic-induced neurotoxicity and preserving cognitive function in exposed individuals.

The potential for novel therapeutic strategies to counteract arsenic-induced damage is a rapidly growing area of research, focusing on pharmacological interventions, dietary modifications, and regenerative medicine approaches. Antioxidant therapy has shown promise in preclinical and clinical studies, as arsenic toxicity is primarily mediated by oxidative stress and free radical formation. Compounds such as *N*-acetylcysteine, resveratrol, quercetin, and melatonin have demonstrated neuroprotective properties by scavenging ROS and enhancing the expression of antioxidant enzymes such as SOD and glutathione peroxidase (GPx). Additionally, green tea, blueberries, and curcumin polyphenols exhibit antiinflammatory and antiapoptotic effects, potentially counteracting arsenic-induced neuronal damage. Chelation therapy remains a primary pharmacological approach for reducing arsenic burden, with agents such as dimercaprol, DMSA, and DMPS (2,3-dimercapto-1-propanesulfonic acid) effectively binding arsenic and facilitating its excretion. However, long-term chelation therapy poses risks of essential metal depletion, necessitating careful monitoring and supplementation with trace elements such as zinc, selenium, and magnesium. Gene therapy and epigenetic modifications are emerging as innovative strategies to restore normal cellular function in arsenic-exposed individuals. By targeting specific pathways affected by arsenic toxicity, such as DNA repair mechanisms and histone modifications, researchers aim to develop precision therapies that reverse arsenic-induced epigenetic changes. Stem cell therapy also potentially regenerates damaged neural tissues, with mesenchymal stem cells demonstrating neuroprotective effects in arsenic-induced toxicity models. Furthermore, advances in gut microbiome research suggest that probiotics and prebiotics could mitigate arsenic toxicity by enhancing microbial arsenic metabolism and reducing systemic absorption. Integrating these therapeutic strategies into clinical practice requires rigorous clinical trials to assess their safety, efficacy, and long-term benefits.

The role of emerging technologies in arsenic detection and risk assessment is critical for implementing effective mitigation strategies and protecting public health. Traditional arsenic detection methods, such as atomic absorption spectrometry and ICP-MS, provide high sensitivity and specificity but are expensive and require specialized laboratory infrastructure. Recent advancements in biosensor technology, nanomaterials, and artificial intelligence (AI)-driven analytics have paved the way for more accessible and real-time arsenic monitoring solutions. Electrochemical biosensors utilizing gold nanoparticles, carbon nanotubes, and graphene-based materials have demonstrated rapid and ultrasensitive arsenic detection, enabling on-site water testing in affected communities. Paper-based microfluidic devices, or lab-on-a-chip systems, offer low-cost, portable alternatives for field testing, allowing individuals to assess their exposure levels with minimal technical expertise. AI-driven predictive modeling and machine learning algorithms have enhanced risk assessment capabilities by integrating environmental data, biomonitoring results, and epidemiological trends to identify high-risk populations and predict future arsenic hotspots. Remote sensing and geospatial mapping techniques further aid in monitoring arsenic contamination in groundwater, enabling targeted intervention strategies. Additionally, emerging point-of-care diagnostic tools incorporating miRNA and proteomic biomarkers offer the potential for early detection of arsenic toxicity before clinical symptoms manifest. Wearable biosensors with noninvasive detection capabilities, such as sweat or saliva analysis, represent a promising frontier in personal arsenic exposure monitoring. These technological advancements facilitate real-time risk assessment, empowering individuals and policymakers to take proactive measures in arsenic mitigation. Addressing the long-term health effects of arsenic exposure requires a multidisciplinary approach that integrates epidemiological research, neurotoxicology advancements, novel therapeutic strategies, and emerging technologies for detection and risk assessment. Long-term epidemiological studies provide invaluable data on the chronic effects of arsenic exposure, guiding public health policies and intervention strategies. Advances in neurotoxicology have elucidated the mechanisms by which arsenic disrupts neuronal function, offering potential targets for therapeutic intervention. Novel pharmacological and nutritional therapies, including antioxidant supplementation, chelation therapy, and stem cell-based approaches, hold promise for mitigating arsenic-induced damage. Meanwhile, cutting-edge detection technologies, such as biosensors, AI-driven analytics, and remote sensing, enhance risk assessment capabilities, enabling early detection and timely intervention. Collaborative efforts among researchers, healthcare professionals, policymakers, and affected communities are essential for developing sustainable solutions to arsenic toxicity. Future research should prioritize integrating multidisciplinary approaches, leveraging technological innovations and translational science to create personalized, cost-effective strategies for arsenic mitigation. By advancing our understanding of arsenic toxicity and implementing proactive measures, we can reduce the global burden of arsenic-related diseases and safeguard public health for future generations.

5.14 Conclusion

The key findings from extensive research on arsenic toxicity highlight its severe and multifaceted impact on human health, particularly its neurotoxic effects. Chronic arsenic exposure is linked to significant disruptions in neurodevelopment, neuroplasticity, and cognitive function, with both prenatal and postnatal exposure contributing to long-term deficits in memory, learning, and behavior. Studies have demonstrated that arsenic readily crosses the BBB, leading to oxidative stress, mitochondrial dysfunction, impaired neurotransmitter signaling, and neuroinflammation. Disruptions in calcium homeostasis, epigenetic modifications, and alterations in synaptic plasticity further exacerbate the detrimental effects of arsenic on the nervous system. Neuroimaging studies have provided direct evidence of arsenic-induced structural and functional changes in the brain, including reduced hippocampal volume, altered white matter integrity, and changes in cortical thickness. Additionally, biomonitoring efforts have established that arsenic accumulates in biological matrices such as blood, urine, and hair, with long-term exposure correlating with increased risks of neurodegenerative diseases, developmental disorders, and cognitive decline. Recent advances in detection technologies, including biosensors and AI-driven predictive modeling, have improved our ability to monitor and assess arsenic exposure in real time, facilitating early intervention strategies. Furthermore, nutritional and pharmacological interventions, such as antioxidant therapy, chelation therapy, and dietary supplementation with essential micronutrients, have shown promise in mitigating arsenic-induced neurotoxicity. However, despite these scientific advancements, arsenic exposure remains a pressing global health issue, necessitating stronger public health policies and regulatory frameworks to limit contamination and protect vulnerable populations. The implications of these findings for public health and regulatory policies are profound, emphasizing the urgent need for stricter arsenic standards in drinking water, food, and occupational settings. Many countries still lack comprehensive regulations addressing arsenic contamination, and enforcement mechanisms are often inadequate. The WHO has set the permissible limit of arsenic in drinking water at 10 μg/L. Yet, millions worldwide continue to consume water exceeding this threshold, particularly in arsenic-endemic regions such as Bangladesh, India, China, and parts of the United States. Strengthening water purification infrastructure, implementing large-scale arsenic remediation programs, and promoting community-based filtration technologies are crucial steps in reducing exposure. Additionally, policies should address arsenic contamination in food sources, particularly rice and other crops that accumulate arsenic from contaminated irrigation water. Regulatory agencies must enhance monitoring of arsenic levels in agricultural products and enforce safety limits to prevent dietary exposure. Occupational safety measures are also essential, particularly for workers in industries such as mining, smelting, and pesticide manufacturing, where arsenic exposure risks are high. Public health initiatives should prioritize early detection and intervention programs, including widespread biomonitoring efforts, neurocognitive assessments, and community awareness campaigns. Integrating arsenic exposure assessments into routine healthcare screenings can aid in early diagnosis and preventive care, particularly for at-risk populations such as children, pregnant women, and the elderly. Reducing the neurotoxic burden of arsenic exposure requires a multifaceted approach that combines scientific research, technological innovation, regulatory enforcement, and public health initiatives. While significant progress has been made in understanding arsenic's neurotoxic effects, there is an urgent need for global collaboration to translate these findings into actionable policies and interventions. Governments, researchers, healthcare professionals, and community organizations must work together to implement sustainable solutions that ensure safe water, food, and environment. Strengthening research efforts in neurotoxicology, advancing early detection technologies, and developing targeted therapeutic strategies will be critical in mitigating the long-term impact of arsenic exposure. Education and awareness campaigns should also be central in empowering communities to take proactive measures to reduce exposure and advocate for safer environmental practices. Protecting public health from arsenic toxicity will require sustained commitment and innovative approaches to minimize exposure, enhance early intervention, and improve long-term health outcomes for affected populations.

Acknowledgment

The author acknowledges that he used partial generative AI and AI-assisted technologies in the writing process; these technologies were only used to improve the readability and language of the work. The human author performed the editing.

References

Andrade, V., Mateus, M. L., Batoréu, M. C., Aschner, M., & dos Santos, A. P. M. (2013). Urinary delta-ALA: A potential biomarker of exposure and neurotoxic effect in rats co-treated with a mixture of lead, arsenic and manganese. *Neurotoxicology, 38*, 33–41. Retrieved from https://doi.org/10.1016/j.neuro.2013.06.003.

Banerjee, A., & Chatterji, U. (2024). Prevalence of perturbed gut microbiota in pathophysiology of arsenic-induced anxiety- and depression-like behaviour in mice. *Chemosphere, 364*, 143293. Retrieved from https://doi.org/10.1016/j.chemosphere.2024.143293.

Bartos, M., Gallegos, C. E., Mónaco, N., Lencinas, I., Dominguez, S., Bras, C., ... Gumilar, F. (2024). Developmental exposure to arsenic reduces anxiety levels and leads to a depressive-like behavior in female offspring rats: Molecular changes in the prefrontal cortex. *Neurotoxicology, 104*, 85–94. Retrieved from https://doi.org/10.1016/j.neuro.2024.07.013.

Chu, X., Li, C., Hao, Y., Shi, J., Wu, Z., Dong, C., ... Han, X. (2025). Targeting Nrf2/HO-1 signaling by crocin: Role in attenuation of arsenic trioxide-induced neurotoxicity in mice. *Journal of Ethnopharmacology, 337*, 118858. Retrieved from https://doi.org/10.1016/j.jep.2024.118858.

de Castro, M. R., Lima, J. V., Salomão de Freitas, D. P., de Souza Valente, R., Dummer, N. S., de Aguiar, R. B., ... Barros, D. M. (2009). Behavioral and neurotoxic effects of arsenic exposure in zebrafish (Danio rerio, Teleostei: Cyprinidae). *Comparative Biochemistry and Physiology Part C: Toxicology & Pharmacology, 150*, 337–342. Retrieved from https://doi.org/10.1016/j.cbpc.2009.05.017.

Du, X., Luo, L., Huang, Q., & Zhang, J. (2022). Cortex metabolome and proteome analysis reveals chronic arsenic exposure via drinking water induces developmental neurotoxicity through hnRNP L mediated mitochondrial dysfunction in male rats. *The Science of the Total Environment, 820*, 153325. Retrieved from https://doi.org/10.1016/j.scitotenv.2022.153325.

Dutta, A., Roy, R., Pandey, M., Chhetry, S., Phukan, B. C., Roy, A., ... Borah, A. (2025). Arsenic-induced mice model of Parkinson's disease: Revealing the neurotoxicity of arsenic through mitochondrial complexes inhibition and dopaminergic neurodegeneration in the substantia nigra region of brain. *Brain Research, 1851*, 149493. Retrieved from https://doi.org/10.1016/j.brainres.2025.149493.

Garg, A., & Bandyopadhyay, S. (2024). Role of an interdependent Wnt, GSK3-β/β-catenin and HB-EGF/EGFR mechanism in arsenic-induced hippocampal neurotoxicity in adult mice. *Chemosphere, 352*, 141375. Retrieved from https://doi.org/10.1016/j.chemosphere.2024.141375.

Garg, A., & Bandyopadhyay, S. (2025). A comprehensive review of arsenic-induced neurotoxicity: Exploring the role of glial cell pathways and mechanisms. *Chemosphere, 372*, 144046. Retrieved from https://doi.org/10.1016/j.chemosphere.2024.144046.

Gopnar, V. V., Rakshit, D., Bandakinda, M., Kulhari, U., Sahu, B. D., & Mishra, A. (2023). Fisetin attenuates arsenic and fluoride subacute co-exposure induced neurotoxicity via regulating TNF-α mediated activation of NLRP3 inflammasome. *Neurotoxicology, 97*, 133–149. Retrieved from https://doi.org/10.1016/j.neuro.2023.06.006.

Hashim, A., Ahmed, M. G., Rahiman, N. B. A., Manikkoth, S., & Leena Pramod, K. (2022). Evaluation of the neuroprotective activity of *P. amarus* in attenuating arsenic-induced neurotoxicity – An in vivo study. *Phytomedicine Plus, 2*, 100316. Retrieved from https://doi.org/10.1016/j.phyplu.2022.100316.

He, Z., Zhang, Y., Zhang, H., Zhou, C., Ma, Q., Deng, P., ... Yu, Z. (2021). NAC antagonizes arsenic-induced neurotoxicity through TMEM179 by inhibiting oxidative stress in oli-neu cells. *Ecotoxicology and Environmental Safety, 223*, 112554. Retrieved from https://doi.org/10.1016/j.ecoenv.2021.112554.

Huang, Z., Guo, L., Huang, J., Chen, X., Sun, J., Ye, Y., ... Sun, X. (2024). Neurotoxicity and brain metabolic dysfunction induced by long-term food-derived arsenic exposure. *Food Bioscience, 59*, 103804. Retrieved from https://doi.org/10.1016/j.fbio.2024.103804.

Luo, J., & Shu, W. (2023). 15 – Arsenic-induced developmental neurotoxicity. In S. J. S. B. T.-H.of A. T. Flora (Ed.). Oxford: Academic Press. Retrieved from https://doi.org/10.1016/B978-0-323-89847-8.00025-0 Second E.

Ma, X., Liu, Y., Ding, B., Lu, S., Ni, B., Chen, Y., ... Liu, X. (2024). Anthocyanins from blueberry ameliorated arsenic-induced memory impairment, oxidative stress, and mitochondrial-biosynthesis imbalance in rat hippocampal neurons. *Cellular Signalling, 119*, 111177. Retrieved from https://doi.org/10.1016/j.cellsig.2024.111177.

Prakash, C., Soni, M., & Kumar, V. (2016). Mitochondrial oxidative stress and dysfunction in arsenic neurotoxicity: A review. *Journal of Applied Toxicology, 36*(2), 179–188. Retrieved from https://doi.org/10.1002/jat.3256.

Rodríguez, V. M., Jiménez-Capdeville, M. E., & Giordano, M. (2003). The effects of arsenic exposure on the nervous system. *Toxicology Letters, 145*(1), 1–18. Retrieved from https://doi.org/10.1016/S0378-4274(03)00262-5.

Shayan, M., Barangi, S., Hosseinzadeh, H., & Mehri, S. (2023). The protective effect of natural or chemical compounds against arsenic-induced neurotoxicity: Cellular and molecular mechanisms. *Food and Chemical Toxicology: an International Journal Published for the British Industrial Biological Research Association, 175*, 113691. Retrieved from https://doi.org/10.1016/j.fct.2023.113691.

Tolins, M., Ruchirawat, M., & Landrigan, P. (2014). The developmental neurotoxicity of arsenic: Cognitive and behavioural consequences of early life exposure. *Annals of Global Health, 80*(4), 303–314. Retrieved from https://doi.org/10.1016/j.aogh.2014.09.005.

Ugbaja, R. N., Akinhanmi, T. F., Onunkwor, B. O., Ugwor, E. I., James, A. S., Babalola, A. A., ... Ugbaja, V. C. (2022). Flavonoid-rich fractions of *C. volubile* and *V. amygdalina* alleviates arsenic-induced neurotoxicity by improving neurosignaling and antioxidant capacity in rats' brain. *Brain Disorders, 7*, 100050. Retrieved from https://doi.org/10.1016/j.dscb.2022.100050.

Vellingiri, B., Suriyanarayanan, A., Selvaraj, P., Abraham, K. S., Pasha, M. Y., Winster, H., ... Venkatesan, D. (2022). Role of heavy metals (copper (Cu), arsenic (As), cadmium (Cd), iron (Fe) and lithium (Li)) induced neurotoxicity. *Chemosphere, 301*, 134625. Retrieved from https://doi.org/10.1016/j.chemosphere.2022.134625.

Wasserman, G. A., Liu, X., Parvez, F., Ahsan, H., Factor-Litvak, P., van Geen, A., et al. (2004). Water arsenic exposure and children's intellectual function in Araihazar, Bangladesh. *Environmental Health Perspectives, 112*(13), 1329–1333.

Wu, S., Zhong, G., Su, Q., Hu, T., Rao, G., Li, T., ... Hu, L. (2024). Arsenic induced neurotoxicity in the brain of ducks: The potential involvement of the gut-brain axis. *Journal of Trace Elements in Medicine and Biology: Organ of the Society for Minerals and Trace Elements (GMS), 81*, 127336. Retrieved from https://doi.org/10.1016/j.jtemb.2023.127336.

Wu, X., Kreutz, A., Dixon, D., & Tokar, E. J. (2025). Engineering human cerebral organoids to explore mechanisms of arsenic-induced developmental neurotoxicity. *Toxicology and Applied Pharmacology, 496*, 117230. Retrieved from https://doi.org/10.1016/j.taap.2025.117230.

Yan, X., Zhang, J., Li, J., Zhang, X., Wang, Y., Chen, X., ... Shen, L. (2023). Effects of arsenic exposure on trace element levels in the hippocampus and cortex of rats and their gender differences. *Journal of Trace Elements in Medicine and Biology: Organ of the Society for Minerals and Trace Elements (GMS), 80*, 127289. Retrieved from https://doi.org/10.1016/j.jtemb.2023.127289.

Yang, X., Chu, F., Jiao, Z., Yu, H., Yang, W., Li, Y., ... Sun, H. (2024). Ellagic acid ameliorates arsenic-induced neuronal ferroptosis and cognitive impairment via Nrf2/GPX4 signaling pathway. *Ecotoxicology and Environmental Safety, 283*, 116833. Retrieved from https://doi.org/10.1016/j.ecoenv.2024.116833.

Zhang, C., Li, Y., Yu, H., Li, T., Ye, L., Zhang, X., ... Dong, S. (2024). Co-exposure of nanoplastics and arsenic causes neurotoxicity in zebrafish (*Danio rerio*) through disrupting homeostasis of microbiota–intestine–brain axis. *The Science of the Total Environment, 912*, 169430. Retrieved from https://doi.org/10.1016/j.scitotenv.2023.169430.

Zhang, X., Yang, H., Wang, Y., Zhang, J., Zhang, H., Cao, X., ... Shen, L. (2023). Proteomic study on the mechanism of arsenic neurotoxicity in the rat cerebral cortex and the protective mechanism of dictyophora polysaccharides against arsenic neurotoxicity. *ACS Chemical Neuroscience, 14*, 2302–2319. Retrieved from https://doi.org/10.1021/acschemneuro.3c00009.

Chapter 6

Cadmium and its role in neurodegenerative diseases

Prasann Kumar
Department of Agronomy, School of Agriculture, Lovely Professional University, Phagwara, Punjab, India

6.1 Introduction

Cadmium (Cd) is a highly toxic heavy metal that poses significant environmental and public health concerns due to its nondegradable nature and persistent bioaccumulation in ecosystems. Classified as a Group 1 human carcinogen by the International Agency for Research on Cancer (IARC), cadmium has a strong association with cancer risk in humans (Subaraja, Arokiyaraj, & Mathew, 2024; Williams, Chen, Otero, & Moore, 2022). It interferes with essential metal metabolism, leading to severe health effects. Unlike other heavy metals that may have physiological roles in trace amounts, cadmium exerts solely toxic effects even at low concentrations. Its widespread distribution in the environment is driven by natural and anthropogenic activities, with mobility and bioavailability influenced by factors such as pH, soil composition, and redox potential. However, the good news is that cadmium exposure is mainly preventable. Industrial activities such as mining, metal smelting, battery production, and improper waste disposal release cadmium into the environment, while phosphate fertilizers contribute to its accumulation in agricultural soil. The food chain represents a significant pathway of cadmium exposure, mainly through leafy vegetables, rice, grains, and seafood grown or raised in contaminated environments. Cigarette smoking is another important source, as tobacco plants absorb cadmium from the soil, leading to high blood cadmium levels in smokers (Bovio et al., 2024; Hamzeh et al., 2023). Water sources become contaminated due to industrial effluents, landfill leaching, and rock dissolution, affecting human populations and aquatic ecosystems. Once absorbed into biological systems through ingestion, inhalation, or dermal exposure, cadmium binds to metallothioneins, which regulate metal homeostasis and detoxification. However, excessive cadmium exposure overwhelms these mechanisms, resulting in systemic toxicity. The kidneys are the primary target organs, where cadmium accumulation in the renal cortex leads to proteinuria, nephrotoxicity, and irreversible kidney damage, as seen in cases of Itai–Itai disease. The liver is also affected due to cadmium-induced oxidative stress, which triggers lipid peroxidation, inflammation, and hepatocellular apoptosis, contributing to liver fibrosis and a heightened risk of hepatocellular carcinoma. Inhalation of cadmium fumes or dust causes pulmonary toxicity, including chronic obstructive pulmonary disease, lung fibrosis, and an increased risk of lung cancer, particularly in occupational settings. Cadmium also disrupts calcium metabolism, leading to skeletal toxicity characterized by osteomalacia, increased fracture risk, and vitamin D metabolism interference, mimicking osteoporosis effects, particularly in postmenopausal women. Additionally, cadmium is a potent carcinogen, inducing DNA damage through oxidative stress, inhibition of DNA repair enzymes, and epigenetic modifications, increasing the risk of lung, prostate, and kidney cancers. Emerging evidence also suggests cadmium crosses the blood–brain barrier (BBB), leading to neurodegenerative effects such as neuronal apoptosis, impaired neurotransmitter function, and cognitive deficits linked to disorders like Alzheimer's and Parkinson's disease (AD and PD). Given its widespread presence, persistent toxicity, and lack of biological utility, cadmium exposure must be minimized through stringent regulatory measures, environmental remediation, and public health interventions. Reducing industrial emissions, promoting safer agricultural practices, and enhancing awareness of cadmium toxicity are critical steps toward mitigating its adverse effects on human health and the environment, empowering us all to take responsibility for our health and the health of our planet.

Heavy Metal Toxicity and Neurodegeneration. DOI: https://doi.org/10.1016/B978-0-443-36575-1.00019-4

6.2 Mechanisms of cadmium-induced neurotoxicity

6.2.1 Oxidative stress and reactive oxygen species generation

Oxidative stress is a cellular condition arising from an imbalance between the production of reactive oxygen species (ROS) and the remarkable capability of antioxidant defense systems to neutralize them, thereby preventing potential damage to biomolecules such as lipids, proteins, and nucleic acids. ROS are highly reactive molecules that include free radicals such as superoxide anion ($O_2{\bullet}^-$), hydroxyl radical (•OH), and nonradical species like hydrogen peroxide (H_2O_2) and singlet oxygen ($^1O_2$). These molecules are primarily generated as byproducts of aerobic metabolism, particularly within the mitochondria during oxidative phosphorylation, where electron leakage from the electron transport chain (ETC) results in the partial reduction of oxygen. Additionally, ROS are produced by peroxisomes, cytochrome P450 enzymes, and nicotinamide adenine dinucleotide phosphate (NADPH) oxidases, which play crucial roles in cellular signaling, immune responses, and stress adaptation. Although low-to-moderate levels of ROS are essential for redox signaling, excessive accumulation can disrupt cellular homeostasis and induce oxidative damage, leading to various pathological conditions, including NDs, cancer, cardiovascular disorders, and aging-related deterioration (Esmaeili et al., 2022; Zhang et al., 2024).

Mitochondria, the cell's powerhouses, are considered the primary source of endogenous ROS, where ETC complexes I and III contribute significantly to $O_2{\bullet}^-$ generation. This radical is rapidly converted into H_2O_2 by mitochondrial superoxide dismutase (SOD), which can diffuse across membranes and participate in Fenton or Haber–Weiss reactions, leading to the formation of the highly reactive •OH. The hydroxyl radical is destructive as it indiscriminately attacks macromolecules, causing lipid peroxidation, protein oxidation, and DNA strand breaks. Besides mitochondrial sources, peroxisomes contribute to ROS production through fatty acid β-oxidation, which generates H_2O_2 as a by-product of acyl-CoA oxidases. The NADPH oxidases (NOX) family of enzymes, membrane-bound complexes found in immune cells, endothelial cells, and other tissues, deliberately generate ROS as part of antimicrobial defense mechanisms and redox-mediated signal transduction. Similarly, cytochrome P450 enzymes, which function in xenobiotic metabolism, contribute to ROS generation through electron transfer reactions involving molecular oxygen.

Under physiological conditions, cells employ a robust network of enzymatic and nonenzymatic antioxidant defenses to counteract ROS accumulation. Enzymatic antioxidants include superoxide dismutases (SODs), catalase, glutathione peroxidases (GPxs), and peroxiredoxins (Prxs), which collectively function to neutralize superoxide and hydrogen peroxide. SOD catalyzes the dismutation of $O_2{\bullet}^-$ into H_2O_2, which is subsequently decomposed into water and oxygen by catalase or reduced by GPxs using glutathione (GSH) as an electron donor. Nonenzymatic antioxidants such as vitamins C and E, carotenoids, flavonoids, and polyphenols also play crucial roles in scavenging ROS and preventing oxidative damage. Thiol-containing molecules like GSH and thioredoxin (Trx) further regulate the intracellular redox state, maintaining the reduced environment necessary for normal cellular functions.

When the balance between ROS production and antioxidant defenses is disrupted, oxidative stress ensues, leading to cellular dysfunction and apoptosis through mitochondrial permeability transition, activation of redox-sensitive transcription factors, and induction of proinflammatory pathways (Divya et al., 2025; Li et al., 2022). Chronic oxidative stress has been implicated in the pathogenesis of various diseases, including AD, PD, atherosclerosis, and diabetes mellitus. In addition, ROS-mediated oxidative damage is a major contributor to aging, as demonstrated by the free radical theory of aging, which postulates that cumulative oxidative damage accelerates cellular senescence and tissue degeneration. Understanding ROS generation and oxidative stress mechanisms is not just crucial. Still, it is a call to action for developing therapeutic strategies that target oxidative damage and enhance cellular resilience to oxidative insults (Table 6.1).

6.2.2 Mitochondrial dysfunction and energy metabolism impairment

Mitochondrial dysfunction and energy metabolism impairment are critical factors in numerous diseases, including neurodegenerative disorders, metabolic syndromes, and aging-related conditions. Mitochondria serve as the primary energy-producing organelles in eukaryotic cells, generating adenosine triphosphate (ATP) through oxidative phosphorylation (OXPHOS) within the ETC. Dysfunction in mitochondria can arise due to genetic mutations in mitochondrial DNA (mtDNA) or nuclear-encoded mitochondrial genes, environmental toxins, oxidative stress, and impaired mitochondrial dynamics, including fission, fusion, and mitophagy. A decline in mitochondrial function leads to insufficient ATP production, increased production of ROS, and metabolic reprogramming, collectively impairing cellular homeostasis. In NDs such as PD, AD, and amyotrophic lateral sclerosis (ALS), mitochondrial dysfunction is a central feature, contributing to neuronal energy deficits and excessive oxidative stress, leading to neuroinflammation and cell death. In PD, mutations in

TABLE 6.1 Mechanisms of cadmium-induced neurotoxicity and their impact on neural function.

Mechanism	Description	Impact on neural function
Oxidative stress	Cd exposure increases ROS production, leading to oxidative damage.	Lipid peroxidation, protein oxidation, DNA damage, neuronal apoptosis.
Mitochondrial dysfunction	Cd disrupts mitochondrial respiration, reduces ATP production, and increases mitochondrial ROS.	Energy deficits, synaptic dysfunction, and increased susceptibility to neurodegeneration.
Neuroinflammation	Activates microglia and astrocytes, increasing proinflammatory cytokine release (TNF-α, IL-6).	Chronic neuroinflammation, neuronal damage, and impaired neurogenesis.
Disruption of metal homeostasis	Cd competes with essential metals (Zn, Fe, Ca), impairing enzyme activity and signaling.	Dysregulated synaptic transmission, impaired neuronal survival, and neurodegenerative processes.
ER stress	Cd interferes with protein folding and triggers the UPR.	Neuronal dysfunction, misfolded protein accumulation, and cell death.
The BBB disruption	Cd exposure weakens BBB integrity, increasing permeability to toxins.	Elevated neurotoxin entry, enhanced neuroinflammation, and neuronal injury.
Apoptosis and autophagy dysfunction	Cd activates apoptotic pathways and disrupts autophagy regulation.	Loss of neuronal cells, exacerbation of neurodegenerative processes.
Neurotransmitter dysregulation	Cd alters dopamine, glutamate, and acetylcholine levels, affecting neurotransmission.	Cognitive decline, motor dysfunction, behavioral impairments.

ATP, Adenosine triphosphate; *BBB*, blood–brain barrier; *ER*, endoplasmic reticulum; *ROS*, reactive oxygen species; *UPR*, unfolded protein response.
Source: Ahmad, Nidadavolu, Durgadoss, and Ravindranath (2014), Anjum et al. (2024), Bajpai, Bharathi, and Patel (2025).

genes such as PINK1, PARKIN, and DJ-1 impair mitophagy, leading to the accumulation of dysfunctional mitochondria, increased ROS production, and dopaminergic neuronal loss. Similarly, in AD, mitochondrial dysfunction disrupts synaptic energy demands, promotes amyloid-beta aggregation, and enhances tau pathology, exacerbating neurodegeneration. Metabolic disorders such as diabetes mellitus also exhibit mitochondrial impairment, wherein insulin resistance and altered glucose metabolism reduce mitochondrial biogenesis, leading to systemic energy deficiencies. The inability of mitochondria to efficiently oxidize glucose and fatty acids in diabetes results in lipid accumulation, oxidative stress, and systemic inflammation, contributing to complications such as neuropathy, nephropathy, and cardiovascular diseases (Anyachor et al., 2022; Mishra & Bowman, 2024; Uhegbu & Memudu, 2019; Wankhede et al., 2025). Cancer cells, in contrast, often exhibit metabolic reprogramming through the Warburg effect, wherein they rely on glycolysis for ATP generation even in the presence of oxygen, facilitating rapid proliferation. This metabolic shift is influenced by mitochondrial alterations, including mutations in ETC components, loss of mitochondrial genome integrity, and deregulation of mitochondrial apoptotic pathways, which contribute to chemoresistance and tumor progression. Aging is another significant factor associated with mitochondrial dysfunction, as the accumulation of mtDNA mutations, oxidative damage, and impaired mitophagy leads to a progressive decline in energy metabolism and increased susceptibility to age-related diseases. Cellular senescence, a hallmark of aging, is exacerbated by mitochondrial dysfunction, as decreased ATP levels and increased ROS promote chronic inflammation and tissue degeneration. Strategies to mitigate mitochondrial dysfunction and energy metabolism impairment involve targeting mitochondrial biogenesis, enhancing mitophagy, and modulating metabolic pathways through interventions such as caloric restriction, exercise, mitochondrial-targeted antioxidants, and pharmacological agents like NAD+ precursors and mitochondrial uncouplers. Emerging therapies, including mitochondrial transplantation, gene editing techniques such as CRISPR-Cas9 to correct mtDNA mutations, and mitochondrial replacement therapies, offer promising avenues for restoring mitochondrial function in genetic disorders. Understanding the interplay between mitochondrial dynamics, metabolic signaling pathways, and cellular stress responses is crucial for developing therapeutic strategies to combat diseases associated with mitochondrial dysfunction. Overall, mitochondrial dysfunction and energy metabolism impairment represent a fundamental aspect of cellular pathology, influencing a wide range of human diseases, and necessitate further research into targeted interventions to enhance mitochondrial health and improve disease outcomes (Abidi, Sharma, Randhawa, Shukla, & Acharya, 2023; Joshi et al., 2023; Mishra & Kaundal, 2023).

6.2.3 Neuroinflammation and activation of microglia and astrocytes

Neuroinflammation is a complex immune response within the central nervous system (CNS) that plays a crucial role in NDs, infections, and traumatic injuries. It is primarily mediated by microglia and astrocytes, the two major glial cell types responsible for immune surveillance and homeostasis in the CNS. Under physiological conditions, microglia function as the brain's resident immune cells, maintaining tissue integrity by phagocytosing debris, releasing neurotrophic factors, and modulating synaptic plasticity. However, in response to pathological stimuli such as infections, toxic insults, or protein aggregates, microglia become activated, undergoing morphological and functional changes characterized by increased proliferation, migration to sites of injury, and the release of proinflammatory mediators, including cytokines like tumor necrosis factor-alpha (TNF-α), interleukin-1 beta (IL-1β), and interleukin-6 (IL-6). These factors contribute to neuronal damage and synaptic dysfunction, particularly in chronic neuroinflammatory conditions associated with AD, PD, multiple sclerosis (MS), and ALS. In parallel, astrocytes, the most abundant glial cells, also undergo activation in response to CNS insults, leading to a process known as reactive astrogliosis. This transformation is marked by hypertrophy, upregulation of glial fibrillary acidic protein, and the secretion of inflammatory and neuroprotective molecules. While reactive astrocytes initially provide metabolic and structural support to neurons, prolonged activation can exacerbate neuroinflammation by amplifying the release of proinflammatory cytokines, ROS, and chemokines, further exacerbating neuronal damage (Acharyya et al., 2024; Rezaei et al., 2024; Zhang, Li, Ren, Peng, & Luo, 2023). The interplay between microglia and astrocytes is a critical determinant of neuroinflammatory outcomes, as activated microglia can stimulate astrocyte reactivity by releasing cytokines such as IL-1β and TNF-α. In contrast, reactive astrocytes can reciprocally influence microglial behavior through chemokines and extracellular matrix components. This bidirectional communication is particularly relevant in chronic neuroinflammatory diseases where sustained activation of glial cells contributes to synaptic loss, BBB dysfunction, and neurodegeneration. Emerging evidence suggests that neuroinflammation is not solely detrimental but also plays a dual role, wherein transient microglial and astrocytic activation can facilitate neuroprotection by clearing toxic aggregates, promoting neuronal repair, and restoring homeostasis. However, in disease states, chronic and excessive neuroinflammation results in a feedforward loop of cellular activation, oxidative stress, and excitotoxicity, which exacerbates neuronal injury. Therapeutic strategies targeting neuroinflammation focus on modulating microglial and astrocytic activation by suppressing excessive proinflammatory responses or enhancing their neuroprotective functions. Small molecules, such as nonsteroidal antiinflammatory drugs and minocycline, have shown potential in attenuating microglial activation. At the same time, astrocyte-modulating agents like ceftriaxone have been explored for their ability to restore glutamate homeostasis and reduce excitotoxicity. Additionally, emerging approaches involving the modulation of inflammasomes, such as the NLRP3 inflammasome, offer novel avenues for controlling neuroinflammation. Understanding the intricate mechanisms governing microglial and astrocytic activation is essential for developing targeted interventions to mitigate neuroinflammatory damage while preserving the beneficial functions of glial cells in maintaining CNS health and function (Table 6.2).

6.2.4 Disruption of blood–brain barrier integrity

The disruption of BBB integrity is a critical pathological event associated with numerous neurological disorders, including stroke, traumatic brain injury, MS, AD, and brain tumors. The BBB is a specialized, highly selective barrier composed of endothelial cells joined by tight junctions, pericytes, astrocytic end-feet, and basement membrane components that collectively regulate the passage of molecules and maintain CNS homeostasis. Any perturbation to this complex structure can increase permeability, allowing harmful substances such as toxins, pathogens, and inflammatory cells to infiltrate the brain parenchyma, exacerbating neuroinflammation, oxidative stress, and neuronal damage. The mechanisms underlying BBB disruption involve multiple interrelated factors, including neuroinflammation, oxidative stress, dysregulated transport systems, and endothelial dysfunction. Proinflammatory cytokines such as TNF-α, IL-1β, and IL-6 play crucial roles in compromising BBB integrity by altering tight junction protein expression, increasing endothelial permeability, and promoting leukocyte infiltration. In NDs like Alzheimer's, amyloid-beta accumulation induces endothelial cell apoptosis and disrupts pericyte-endothelial interactions, leading to vascular dysfunction and BBB breakdown. Similarly, hypoxia and reperfusion injury in ischemic stroke activate matrix metalloproteinases (MMPs), which degrade extracellular matrix proteins and tight junctions, resulting in vascular leakage and exacerbated neuronal injury (Albekairi et al., 2024; Gil, Kim, Choi, & Bae, 2025; Howard, David, Lux, & Tillement, 2024; Li et al., 2025). Additionally, oxidative stress from

TABLE 6.2 Link between cadmium exposure and major neurodegenerative disorders.

Neurodegenerative disorder	Mechanisms of Cd-induced toxicity	Molecular and cellular effects	Key findings from studies	Potential therapeutic strategies
AD	Oxidative stress, mitochondrial dysfunction, neuroinflammation, and amyloid-beta (Aβ) aggregation.	- Increases ROS production → lipid peroxidation, DNA damage. - Disrupts mitochondrial electron transport chain → ATP depletion. - Induces neuroinflammation via microglial activation → IL-6, TNF-α release. - Promotes tau hyperphosphorylation and Aβ plaque deposition.	- Cd detected in AD-affected brains, correlating with severity. - Epidemiological studies link Cd exposure to cognitive decline. - In vitro and in vivo studies show Cd increases Aβ production and tau aggregation.	- Antioxidant therapy (resveratrol, curcumin, *N*-acetylcysteine). - Chelation therapy to remove excess Cd. - Antiinflammatory agents to reduce neuroinflammation. - Zn and Se supplementation to counter Cd toxicity.
PD	Dopaminergic neuron toxicity, oxidative stress, mitochondrial damage, and α-synuclein aggregation.	- Disrupts dopamine homeostasis → neuronal apoptosis in the substantia nigra. - Induces mitochondrial dysfunction → ATP depletion & increased ROS. - Promotes α-synuclein misfolding & Lewy body formation. - Activates glial cells → neuroinflammation and cytokine release.	- Elevated Cd levels detected in PD patient brains, particularly in the substantia nigra. - Cd exposure in animal models leads to PD-like motor deficits. - In vitro studies show Cd exacerbates α-synuclein aggregation and neuronal apoptosis.	- Mitochondrial protectants (CoQ10, nicotinamide, melatonin). - Antioxidants to counteract oxidative stress. - Dopamine-replacement therapies (L-DOPA). - Heavy metal chelation is used to reduce the Cd burden.
ALS	Motor neuron toxicity, oxidative stress, glutamate excitotoxicity, and neuroinflammation.	- Increases oxidative damage in motor neurons → apoptosis. - Disrupts Ca^{2+} signaling → excitotoxicity and neuronal death. - Impairs glutamate clearance by astrocytes → synaptic dysfunction. - Alters metal-binding proteins (Cu/Zn-SOD) → accelerates ALS progression.	- Higher Cd levels were found in ALS patients compared to controls. - Studies link environmental Cd exposure to increased ALS risk. - Cd exposure leads to motor neuron degeneration in experimental models.	- Glutamate inhibitors (riluzole) to reduce excitotoxicity. - Antioxidant-based neuroprotection (vitamin E, edaravone). - Chelation therapy to eliminate excess Cd. - Dietary Zn and Se supplementation to counteract Cd neurotoxicity.

AD, Alzheimer's disease; *ALS*, Amyotrophic lateral sclerosis; *PD*, Parkinson's disease.
Source: Dahl and Bezprozvanny (2024), Fan et al. (2024), Isaev, Stelmashook, Genrikhs, and Onishchenko (2023).

excessive ROS production can damage endothelial cells, reduce tight junction protein stability, and initiate apoptotic signaling pathways, further contributing to BBB breakdown. Chronic systemic inflammation, often observed in conditions such as diabetes and hypertension, also impairs BBB function by inducing endothelial cell dysfunction and promoting cerebrovascular damage. Transporters, such as *P*-glycoprotein and other ATP-binding cassette transporters, are also pivotal in BBB integrity; their dysfunction can lead to the accumulation of neurotoxic substances, worsening neuronal degeneration. Moreover, mechanical injury to the brain, as seen in traumatic brain injury, directly damages endothelial cells and disrupts perivascular support structures, leading to immediate and prolonged BBB permeability changes. Emerging evidence suggests that microbiota–gut–brain axis interactions influence BBB integrity, with microbial metabolites such as short-chain fatty acids exerting protective effects, whereas dysbiosis-induced endotoxins contribute to barrier dysfunction. Therapeutic approaches to restore BBB integrity are currently being explored, including antiinflammatory agents, antioxidants, MMP inhibitors, and endothelial cell-stabilizing compounds. Nanotechnology-based drug delivery systems and targeted therapies that modulate tight junction dynamics, pericyte function, and neurovascular signaling pathways hold promise for mitigating BBB disruption in various CNS disorders. Additionally, stem cell therapies, particularly those using mesenchymal stem cells, have shown potential in restoring BBB function by promoting endothelial repair and reducing neuroinflammation. Understanding the molecular and cellular mechanisms underlying BBB disruption is crucial for developing effective therapeutic strategies to protect against neurovascular dysfunction and improve outcomes in neurological diseases. Future research should focus on identifying novel biomarkers for the early detection of BBB impairment and investigating precision medicine approaches that tailor interventions based on individual disease pathology. Given the complexity of BBB disruption in various pathologies, a multidisciplinary approach integrating molecular biology, neurovascular physiology, and advanced imaging techniques will be essential for advancing therapeutic developments and improving clinical outcomes for patients with BBB-related disorders (Hassan et al., 2024; Kaur et al., 2023; Urbano et al., 2023; Xu et al., 2024).

6.2.5 Interference with essential metal homeostasis (Zn, Fe, Cu)

Interference with essential metal homeostasis, particularly zinc (Zn), iron (Fe), and copper (Cu), has profound implications for cellular function, organismal health, and environmental interactions. These trace metals are crucial cofactors for numerous enzymatic processes, playing pivotal roles in redox reactions, DNA synthesis, signal transduction, and metabolic pathways. Maintaining metal homeostasis requires tightly regulated uptake, transport, storage, and efflux mechanisms; any disruption can lead to toxicity, deficiency, or metabolic imbalances, contributing to physiological dysfunction and disease. Zinc is integral to the structural and catalytic functions of over 300 enzymes and transcription factors, influencing immune response, oxidative stress mitigation, and cell proliferation. Zn deficiency impairs growth, neural development, and immune competence. At the same time, excess Zn disrupts iron and copper uptake by competing for transporters like ZIP (Zrt- and Irt-related proteins) and metallothioneins, leading to secondary deficiencies. Iron is central to oxygen transport, mitochondrial respiration, and electron transfer processes, being a key component of hemoglobin, cytochromes, and iron-sulfur clusters. Disruptions in Fe homeostasis, such as through overexpression of ferroportin or dysregulation of ferritin storage, can result in iron deficiency anemia, neurodegenerative disorders, or iron overload conditions like hemochromatosis, which enhances oxidative stress due to Fenton chemistry-mediated production of ROS. Copper, required for redox homeostasis, neurotransmitter synthesis, and mitochondrial function, is transported by CTR1 (copper transporter 1), ATP7A, and ATP7B, facilitating its distribution and excretion. Cu deficiency impairs energy metabolism, connective tissue integrity, and antioxidant defenses due to cuproenzymes like cytochrome *c* oxidase and superoxide dismutase dysfunction (Chang et al., 2024; Ebrahimi, Ebrahimi, Vergroesen, Aschner, & Sillanpää, 2024; Li et al., 2024; Rosas-Rodríguez & Valenzuela-Soto, 2021). At the same time, Cu overload leads to oxidative stress, hepatic damage, and Wilson's disease. Metal crosstalk mechanisms ensure coordinated balance; however, interference in one metal's availability can lead to compensatory but maladaptive responses in others, exacerbating metabolic dysfunction. The competition between Zn, Fe, and Cu occurs at multiple levels, including transporters, storage proteins, and regulatory pathways. For instance, high dietary Zn suppresses Cu absorption via metallothionein upregulation, while excessive Fe intake disrupts Zn uptake due to shared DMT1 (divalent metal transporter 1) pathways. Environmental and dietary factors, including heavy metal exposure, nutrient deficiencies, and genetic mutations in metal transporters, can disrupt this homeostasis, leading to systemic pathologies such as anaemia, neurodegeneration, or immunodeficiency. In agriculture, plant metal homeostasis determines crop yield, nutritional quality, and resistance to biotic and abiotic stressors. Deficiency or excess of Zn, Fe, or Cu affects enzymatic activity, photosynthetic efficiency, and oxidative stress resilience, impacting plant health and human nutrition. Biotechnological interventions, including biofortification, metal chelation strategies, and genetic engineering of metal transporters, aim to optimize metal uptake and utilization in plants and animals. Understanding the intricate network of Zn, Fe, and Cu homeostasis, their competitive interactions and their role in cellular physiology are crucial for advancing medical, environmental, and agricultural sciences, offering strategies for managing deficiencies, mitigating toxicity, and improving overall health outcomes. Future research should elucidate metal transport mechanisms, regulatory networks, and intermetallic dependencies to develop targeted therapeutic, nutritional, and agronomic interventions for metal homeostasis-related disorders(Iddrisu et al., 2024; Lu et al., 2024; Singh et al., 2023; Živančević et al., 2021).

6.3 Cadmium and major neurodegenerative diseases

6.3.1 Alzheimer's disease

AD is a neurodegenerative disorder characterized by progressive cognitive decline, synaptic dysfunction, and neuronal loss, with hallmark pathological features including amyloid-β (Aβ) plaque accumulation and neurofibrillary tangles composed of hyperphosphorylated tau protein. Among various environmental and genetic factors implicated in AD pathogenesis, cadmium (Cd), a heavy metal toxin, has emerged as a significant contributor to neurotoxicity through its role in Aβ aggregation and tau phosphorylation, as well as the disruption of acetylcholine metabolism. Cd exposure is associated with increased oxidative stress, mitochondrial dysfunction, and neuroinflammation, all of which contribute to AD pathology. The role of Cd in Aβ aggregation is primarily linked to its ability to promote amyloidogenic processing of amyloid precursor protein (APP) by modulating β- and γ-secretase activity. Cd exposure leads to an upregulation of β-site APP cleaving enzyme 1 (BACE1), increasing the generation of toxic Aβ oligomers. Additionally, Cd disrupts metal homeostasis in neurons by interfering with zinc and copper metabolism, which is critical for Aβ clearance and aggregation dynamics. Cd has been shown to induce conformational changes in Aβ peptides, accelerating their aggregation into neurotoxic oligomers and fibrils that promote synaptic dysfunction and neuronal death. Another crucial aspect of AD pathogenesis is tau hyperphosphorylation, which leads to neurofibrillary tangles that disrupt axonal transport and neuronal integrity. Cd has been implicated in dysregulating tau phosphorylation by activating kinases such as glycogen synthase kinase-3β (GSK-3β), cyclin-dependent kinase 5 (CDK5), and mitogen-activated protein kinases, all of which are known to facilitate tau phosphorylation. Additionally, Cd inhibits phosphatases such as protein phosphatase 2A, which dephosphorylates tau, typically leading to its pathological accumulation. The Cd-induced oxidative stress and neuroinflammation exacerbate tau pathology by triggering glial activation and cytokine release, further amplifying neurodegenerative processes. Furthermore, Cd has been shown to disrupt cholinergic neurotransmission, a key component of cognitive function, by interfering with acetylcholine metabolism (Bovio, Melchioretto, Forcella, Fusi, & Urani, 2021; Gupta, Ambasta, & Kumar, 2022; Han et al., 2025; Mincic, Antal, Filip, & Miere, 2024; Sun et al., 2025; Zaazaa et al., 2022). The cholinergic system, particularly in the basal forebrain, plays a vital role in learning, memory, and attention, and its dysfunction is a well-established feature of AD. Cd exposure has been found to inhibit choline acetyltransferase (ChAT), the enzyme responsible for synthesizing acetylcholine, leading to a reduction in acetylcholine levels and impaired neurotransmission. Moreover, Cd increases acetylcholinesterase (AChE) activity, accelerating acetylcholine degradation and further compromising cholinergic signaling. The disruption of acetylcholine metabolism due to Cd toxicity results in impaired synaptic plasticity, contributing to cognitive deficits observed in AD patients. Additionally, Cd-induced mitochondrial dysfunction and oxidative stress further deteriorate cholinergic neurons, exacerbating memory impairment and neurodegeneration. Cd's role in promoting Aβ aggregation, tau phosphorylation, and acetylcholine metabolism disruption highlights its multifaceted contribution to AD pathogenesis. Given the neurotoxic impact of Cd, further research is needed to explore potential therapeutic strategies to mitigate Cd-induced neurodegeneration, including metal chelation therapy, antioxidant interventions, and techniques to enhance Aβ clearance and tau dephosphorylation. Understanding the molecular mechanisms underlying Cd neurotoxicity may provide new avenues for AD prevention and treatment, particularly in populations exposed to environmental Cd contamination (Table 6.3).

6.3.2 Parkinson's disease

PD is a progressive neurodegenerative disorder characterized by the selective loss of dopaminergic neurons in the substantia nigra pars compacta (SNpc), leading to motor dysfunctions such sas bradykinesia, rigidity, tremors, and postural instability. Emerging evidence suggests that environmental toxins, including heavy metals like cadmium (Cd), significantly exacerbate neuronal damage, accelerating PD pathogenesis. Cd is a persistent environmental pollutant with high bioaccumulation potential, and it has been implicated in neurotoxicity through multiple mechanisms, including oxidative stress, mitochondrial dysfunction, neuroinflammation, and impaired proteostasis. One of the primary ways Cd contributes to PD is

TABLE 6.3 Epidemiological studies on cadmium exposure and neurodegenerative disease incidence.

Study	Study type and population	Exposure assessment	Neurodegenerative disease(s) examined	Key findings	Limitations
NHANES	Cross-sectional study, general US population	Urinary cadmium levels measured as a biomarker of chronic exposure	AD, PD, cognitive decline	Higher urinary cadmium levels are associated with lower cognitive function scores in older adults. Possible link between cadmium exposure and increased risk of AD and PD.	Cross-sectional nature limits causality. Potential confounding factors such as diet and lifestyle.
Rotterdam Study (the Netherlands)	Prospective cohort study, elderly population	Blood cadmium concentration measured over time	AD, dementia	Elevated blood cadmium levels correlated with an increased incidence of dementia and AD over a 10-year follow-up.	Potential confounding by other environmental toxins. Limited cadmium exposure sources were analyzed.
China Longitudinal Healthy Longevity Survey	Population-based study, elderly Chinese adults	Urinary and blood cadmium levels were measured	Cognitive impairment, AD	Higher cadmium levels are associated with poorer cognitive function, particularly in older individuals with existing comorbidities.	Lacks detailed dietary and occupational exposure history.
Swedish Twin Registry Study	Twin cohort study, elderly twins	Hair and blood cadmium levels measured	AD, PD	Increased cadmium exposure is associated with greater AD risk, with more potent effects in genetically predisposed individuals.	Difficulty in differentiating environmental from genetic contributions.
French PAQUID Study	Longitudinal cohort, elderly population	Blood and environmental cadmium exposure assessment	AD, cognitive decline	Higher cadmium exposure is linked to faster cognitive decline over time.	Limited generalizability beyond European populations.
CHMS	Cross-sectional, general Canadian population	Blood and urinary cadmium levels were analyzed	AD, PD, ALS	Cadmium exposure positively correlated with neurodegenerative disease prevalence, particularly in individuals with high environmental exposure (smokers, industrial workers).	Long-term follow-up is crucial to establish causality in the relationship between cadmium exposure and neurodegenerative diseases.
KoNEHS	Nationwide population-based study, Korean adults	Urinary cadmium concentrations measured	PD, cognitive dysfunction	Increased risk of PD in individuals with high urinary cadmium levels, independent of smoking status.	More comprehensive occupational exposure data are needed to understand the relationship between cadmium exposure and neurodegenerative diseases fully.

Taiwanese Environmental Exposure Study	Retrospective cohort, industrial workers vs general population	Blood and urinary cadmium levels in high-exposure vs low-exposure groups	ALS, Parkinson's disease	Occupational cadmium exposure significantly increased ALS risk. Higher PD prevalence in workers with chronic exposure.	Potential bias due to self-reported occupational history.
Italian Elderly Cohort Study	Longitudinal study, elderly residents in industrial areas	Blood cadmium levels and environmental exposure mapping	AD, PD	Individuals living in cadmium-contaminated areas showed a higher incidence of AD and PD compared to those in lower exposure areas.	No mechanistic data were provided.
Japanese Environmental Toxin Study	Case-control study, urban vs. rural populations	Cadmium levels in soil, water, and human biomarkers (blood, hair)	PD, cognitive decline	Higher cadmium exposure in urban populations correlated with greater neurodegenerative disease incidence. Rural areas with lower contamination showed lower prevalence rates.	Potential coexposure to other pollutants.
UK Biobank Study	Large-scale cohort, middle-aged to elderly adults	Blood cadmium levels analyzed for disease onset	AD, PD, ALS	Positive correlation between high blood cadmium and increased risk of neurodegenerative diseases, particularly in individuals with poor metal detoxification genetic profiles.	Further genetic and environmental interaction studies are needed to understand the relationship between cadmium exposure and neurodegenerative diseases.
Brazilian Amazon Cadmium Exposure Study	Community-based study, residents in mining and nonmining areas	Urinary cadmium biomonitoring	PD, cognitive impairment	A higher cadmium burden in mining-exposed communities is linked to increased neurodegenerative disease rates.	Lack of individual exposure history.

AD, Alzheimer's disease; *ALS*, Amyotrophic lateral sclerosis; *CHMS*, Canadian Health Measures Survey; *KoNEHS*, Korean National Environmental Health Survey; *NHANES*, National Health and Nutrition Examination Survey, Unites States; *PD*, Parkinson's disease.
Source: Han et al. (2025), Hassan et al. (2024), Iddrisu et al. (2024), Li et al. (2024).

by inducing dopaminergic neuron damage through excessive ROS production, leading to lipid peroxidation, protein oxidation, and DNA damage, which cumulatively cause neuronal apoptosis. Cd disrupts calcium homeostasis, exacerbates glutamate excitotoxicity, and impairs mitochondrial respiratory function by inhibiting complex I activity, resulting in ATP depletion and increased susceptibility of dopaminergic neurons to apoptosis. Furthermore, Cd exposure leads to chronic neuroinflammation, as evidenced by the activation of microglia and astrocytes, which release proinflammatory cytokines such as TNF-α, IL-1β, and IL-6, creating a neurotoxic environment that perpetuates neuronal loss. The accumulation of Cd in the brain also influences the misfolding and aggregation of α-synuclein, a key pathological hallmark of PD, which further contributes to dopaminergic neuronal degeneration. α-Synuclein, an intrinsically disordered protein, is prone to misfolding under oxidative stress conditions and metal toxicity. This leads to the formation of toxic oligomers and fibrils that aggregate into Lewy bodies, the histopathological signature of PD. Cd has been shown to directly interact with α-synuclein by binding to specific residues, altering its conformation, and accelerating its oligomerisation, which enhances its neurotoxicity. Moreover, Cd-induced oxidative stress disrupts the ubiquitin–proteasome system and autophagy-lysosomal pathways, essential for the clearance of misfolded proteins, thereby promoting the accumulation of α-synuclein aggregates (Bovio et al., 2021; Kaur et al., 2023; Rosas-Rodríguez & Valenzuela-Soto, 2021; Zaazaa et al., 2022; Živančević et al., 2021). Lewy body formation, primarily composed of aggregated α-synuclein and ubiquitinated proteins, is a critical event in PD progression, as it interferes with normal cellular functions, disrupts vesicular trafficking, impairs synaptic function, and triggers endoplasmic reticulum (ER) stress, ultimately leading to neuronal apoptosis. The interplay between Cd exposure, α-synuclein aggregation, and Lewy body formation establishes a vicious cycle of neurotoxicity that exacerbates dopaminergic neuron loss and accelerates PD progression. Additionally, Cd interferes with essential metal homeostasis by displacing physiological metals such as zinc (Zn) and iron (Fe), which are crucial for maintaining enzymatic activity and neuronal function and further contributing to neurodegeneration. Given the detrimental effects of Cd on dopaminergic neurons and its impact on α-synuclein pathology, therapeutic interventions targeting metal chelation, oxidative stress reduction, and proteostasis restoration may offer promising strategies to mitigate Cd-induced neurotoxicity and slow PD progression. Understanding the molecular mechanisms underlying Cd-induced dopaminergic neuron damage and its role in α-synuclein aggregation and Lewy body formation is critical for developing effective neuroprotective strategies for PD.

6.3.3 Amyotrophic lateral sclerosis

ALS is a fatal neurodegenerative disorder primarily affecting motor neurons in the brain and spinal cord, leading to progressive muscle weakness, atrophy, and eventual paralysis. The pathogenesis of ALS is multifactorial, with motor neuron toxicity and excitotoxicity playing central roles in disease progression. Motor neuron toxicity in ALS arises from a combination of genetic mutations, oxidative stress, mitochondrial dysfunction, neuroinflammation, and impaired protein homeostasis. Mutations in genes such as *SOD1*, *C9orf72*, *FUS*, and *TARDBP* result in toxic protein aggregates, disrupting cellular functions and inducing stress responses that eventually lead to neuronal death. Misfolded proteins contribute to ER stress and activate the unfolded protein response, which, when prolonged, triggers apoptotic pathways, exacerbating motor neuron degeneration. Additionally, mitochondrial dysfunction in ALS is characterized by impaired energy metabolism, increased ROS production, and defective calcium buffering, further amplifying neuronal vulnerability. Excitotoxicity, a critical driver of motor neuron degeneration in ALS, is primarily mediated by excessive glutamate signaling, leading to neuronal overactivation and calcium overload. Motor neurons are particularly susceptible to excitotoxicity due to their high expression of AMPA and NMDA glutamate receptors, combined with deficient glutamate transporters such as EAAT2, which fail to adequately clear extracellular glutamate. The resulting sustained excitatory stimulation leads to prolonged calcium influx, activating various cell death pathways, including mitochondrial dysfunction, oxidative stress, and activation of calpain and caspases. Furthermore, excessive glutamate signaling induces ER stress, impairing protein homeostasis and contributing to neurotoxicity (Chang et al., 2024; Gupta et al., 2022; Singh et al., 2023; Xu et al., 2024; Zaazaa et al., 2022; Živančević et al., 2021). The interplay between motor neuron toxicity and excitotoxicity is further exacerbated by neuroinflammation, where activated microglia and astrocytes release proinflammatory cytokines such as TNF-α, IL-1β, and IL-6, worsening neuronal injury. Additionally, the loss of supportive astrocyte and oligodendrocyte functions in ALS contributes to an inadequate neuronal environment, further compromising motor neuron survival. Dysfunctional RNA metabolism, impaired axonal transport, and defective autophagy also play significant roles in motor neuron toxicity, disrupting cellular homeostasis and exacerbating neurodegenerative processes. Several therapeutic strategies are being explored to target motor neuron toxicity and excitotoxicity in ALS, including glutamate receptor antagonists, calcium channel blockers, antioxidants, and neuroprotective agents to restore cellular homeostasis. Riluzole, an FDA-approved drug for ALS, primarily

TABLE 6.4 Biomarkers of cadmium-induced neurodegeneration in human populations.

Biomarker	Type	Sample source	Associated neurodegenerative disease(s)	Key findings	Clinical relevance	Limitations
Urinary cadmium (U-Cd)	Exposure biomarker	Urine	AD, PD, cognitive impairment	Higher U-Cd levels correlate with increased cognitive decline and risk of neurodegenerative diseases.	Noninvasive biomarker for chronic exposure assessment.	Variability due to kidney function differences, hydration status, and short-term exposure changes.
Blood cadmium (B-Cd)	Exposure biomarker	Blood	AD, PD, ALS, cognitive dysfunction	Elevated B-Cd is associated with neuronal damage, neuroinflammation, and oxidative stress.	Indicator of recent Cd exposure and systemic distribution.	Blood Cd levels fluctuate with acute exposure, which is unsuitable for long-term accumulation assessment.
Hair cadmium (H-Cd)	Exposure biomarker	Hair	AD, PD, ALS	Higher Cd levels in hair are linked to more excellent neurotoxic effects and disease severity.	Long-term exposure marker, easy sample collection.	Potential contamination from external sources and individual variability in hair growth rates.
Salivary cadmium (S-Cd)	Exposure biomarker	Saliva	AD, PD	Increased salivary Cd levels were observed in exposed populations with cognitive deficits.	Noninvasive, potential for point-of-care monitoring.	It lacks standardization and has lower reliability compared to blood and urine.
Cadmium in cerebrospinal fluid (CSF-Cd)	Exposure and effect biomarker	Cerebrospinal fluid	AD, PD, ALS	Higher CSF-Cd levels in neurodegeneration patients indicate Cd penetration into the brain.	Direct measure of neurotoxic metal burden in the central nervous system.	Invasive procedure limits routine use.
Lipid peroxidation markers (malondialdehyde, 4-HNE)	Oxidative stress biomarker	Blood, CSF	AD, PD	Elevated lipid peroxidation products in Cd-exposed individuals indicate neuronal oxidative damage.	Indicates Cd-induced oxidative stress and potential neurotoxicity.	Nonspecific; elevated levels may result from multiple oxidative stressors.
8-Hydroxydeoxyguanosine (8-OHdG)	Oxidative DNA damage biomarker	Urine, blood	AD, PD, ALS	Increased 8-OHdG levels in Cd-exposed populations, correlating with cognitive dysfunction.	Sensitive indicator of Cd-induced neurotoxicity and DNA damage.	Influenced by multiple environmental and genetic factors.
Inflammatory cytokines (IL-6, TNF-α, IL-1β)	Neuroinflammation biomarker	Blood, CSF	AD, PD, ALS	Higher proinflammatory cytokine levels were observed in Cd-exposed individuals with neurodegeneration.	Reflects Cd-induced neuroinflammation and immune response.	Nonspecific; other inflammatory conditions can elevate cytokine levels.
Amyloid-β (Aβ1–42) and tau protein	Neurodegenerative biomarkers	CSF, blood	AD	Cd exposure is linked to altered Aβ and tau levels, promoting AD pathology.	Potential indicator of Cd-induced AD progression.	Requires further validation for specificity to Cd exposure.

(Continued)

TABLE 6.4 (Continued)

Biomarker	Type	Sample source	Associated neurodegenerative disease(s)	Key findings	Clinical relevance	Limitations
α-Synuclein	Neurodegenerative biomarker	CSF, blood	PD	Elevated α-synuclein in Cd-exposed PD patients, suggesting Cd's role in synucleinopathy.	Biomarker for Cd-induced PD pathogenesis.	There is a need for additional large-scale studies to confirm causality.
Neurofilament light chain (NfL)	Neuronal damage biomarker	Blood, CSF	ALS, PD, AD	Higher NfL levels in Cd-exposed individuals with neurodegenerative conditions.	Marker for axonal damage and neurodegeneration.	It can be elevated in other neuroinflammatory and traumatic conditions.
GFAP	Astrocyte injury biomarker	Blood, CSF	AD, ALS	Elevated GFAP levels in Cd-exposed individuals indicate astrocyte dysfunction and neuroinflammation.	Helpful in tracking astroglial response to Cd-induced injury.	It lacks disease specificity and needs correlation with other markers.
Metallothioneins (MT-1, MT-2)	Metal detoxification biomarker	Blood, urine, CSF	AD, PD, ALS	Increased MT expression in response to Cd exposure, indicating a protective cellular response.	Reflects cellular adaptation to Cd toxicity.	It may not directly correlate with disease severity or onset.
BDNF	Neuroplasticity biomarker	Blood, CSF	AD, PD, ALS, cognitive decline	Decreased BDNF levels were observed in Cd-exposed populations with cognitive impairment.	Biomarker for neurodegeneration and impaired neuroplasticity.	Subject to variations due to age, stress, and lifestyle factors.
DAT levels	Dopaminergic dysfunction biomarker	Brain imaging, CSF	PD	Cd exposure is linked to reduced DAT availability, indicating dopaminergic neuron loss.	Potential early biomarker for Cd-induced PD.	It requires neuroimaging, which is expensive and not widely accessible.

AD, Alzheimer's disease; *ALS*, amyotrophic lateral sclerosis; *BDNF*, brain-derived neurotrophic factor; *DAT*, dopamine transporter; *GFAP*, glial fibrillary acidic protein; *PD*, Parkinson's disease.
Source: Han et al. (2025), Iddrisu et al. (2024), Mincic et al. (2024), Rosas-Rodríguez and Valenzuela-Soto (2021.

reduces glutamate release, thereby mitigating excitotoxicity, while edaravone, another approved treatment, acts as an antioxidant to counteract oxidative stress. Emerging approaches such as gene therapy, stem cell-based therapies, and neuroinflammatory modulators hold promise for addressing the complex pathophysiology of ALS. Despite ongoing research, ALS remains a devastating disorder with no cure, necessitating continued investigation into the intricate mechanisms underlying motor neuron toxicity and excitotoxicity to develop more effective therapeutic interventions that can slow disease progression and improve patient outcomes (Table 6.4).

6.3.4 Multiple sclerosis

MS is a chronic, immune-mediated neurological disorder characterized by the progressive demyelination of axons in the CNS, leading to neurodegeneration and disability. The pathophysiology of MS primarily involves immune dysregulation, wherein autoreactive T cells, B cells, and innate immune components orchestrate an inflammatory response against myelin, the insulating layer of nerve fibers. This immune attack results in demyelination, axonal damage, and neuronal loss, impairing nerve signal transmission and causing the clinical manifestations of the disease. While the precise etiology of MS remains elusive, genetic predisposition, environmental triggers, and dysregulated immune tolerance are key contributors to disease onset and progression.

An aberrant activation of adaptive and innate immune components drives the immune dysregulation in MS. In a healthy CNS, immune homeostasis prevents self-reactivity through regulatory T cells (Tregs) and other immunosuppressive mechanisms. However, in MS, dysregulated Tregs fail to control autoreactive CD4+ T cells, particularly Th1 and Th17 subsets, which play a crucial role in disease pathogenesis. These pathogenic T cells infiltrate the CNS by crossing the BBB, facilitated by the upregulation of adhesion molecules and chemokines. Upon entering the CNS, they release proinflammatory cytokines such as interferon-gamma (IFN-γ), interleukin-17 (IL-17), and TNF-α, which activate resident microglia and astrocytes, further amplifying inflammation. The loss of immune regulation also involves B cells, which contribute to MS pathology through antigen presentation, autoantibody production, and cytokine secretion. B cell–derived autoantibodies target myelin proteins, exacerbating demyelination and lesion formation.

The destruction of myelin in MS results from direct immune attack and secondary damage due to neuroinflammation. Oligodendrocytes, the myelin-producing cells of the CNS, are highly susceptible to inflammatory insults, leading to their apoptosis and impaired remyelination capacity. Activated microglia and macrophages contribute to demyelination by releasing ROS and MMPs, which degrade the myelin sheath and extracellular matrix components. Additionally, the BBB disruption facilitates peripheral immune cell infiltration, further fueling demyelination and neuroaxonal injury. Chronic inflammation in MS promotes mitochondrial dysfunction, leading to neuron energy deficits and exacerbating axonal degeneration. Neurotoxic factors, such as nitric oxide (NO) and glutamate, induce excitotoxicity, further compromising neuronal survival (Ebrahimi et al., 2024; Hassan et al., 2024; Singh et al., 2023).

Demyelination in MS leads to the formation of characteristic lesions, or plaques, in the white and gray matter of the CNS. These lesions disrupt nerve conduction, resulting in neurological deficits such as motor weakness, sensory disturbances, cognitive impairment, and vision loss. While remyelination can occur in early disease stages, it becomes increasingly inefficient as MS progresses due to oligodendrocyte progenitor cell depletion and a proinflammatory microenvironment that impairs regenerative processes. Advances in immunomodulatory therapies, such as monoclonal antibodies targeting B cells (e.g., ocrelizumab) and sphingosine-1-phosphate receptor modulators, aim to restore immune balance and reduce demyelination. However, neuroprotective and remyelination-enhancing strategies remain crucial areas of ongoing research to halt disease progression and promote CNS repair. Understanding the intricate immune dysregulation and demyelination mechanisms in MS is essential for developing novel therapeutic interventions to improve patient outcomes.

6.4 Epidemiological evidence and human studies

6.4.1 Cadmium levels in biological samples (blood, urine, hair)

Cadmium (Cd) is a toxic heavy metal that accumulates in the human body and is commonly measured in biological samples such as blood, urine, and hair to assess exposure levels and potential health risks. Blood cadmium levels are considered a reliable biomarker of recent exposure, as cadmium is primarily transported in the bloodstream bound to proteins such as metallothioneins. In occupational or environmental exposure scenarios, elevated blood cadmium concentrations indicate ongoing or acute exposure, often reflecting inhalation or ingestion of cadmium-containing substances. However, because cadmium has a biological half-life of 10–30 years in the human body, blood levels may not fully represent cumulative exposure. Urinary cadmium is widely used as a biomarker for long-term exposure, as it primarily reflects cadmium accumulation in the kidneys, the main organ for cadmium storage and toxicity. Cadmium is filtered by the glomeruli and reabsorbed in the proximal tubules, leading to renal damage over time. As renal function declines, urinary cadmium excretion increases, making it an essential marker for chronic exposure and early nephrotoxicity (Ebrahimi et al., 2024; Han et al., 2025; Kaur et al., 2023; Singh et al., 2023). Hair cadmium levels, although helpful in assessing historical exposure, are subject to external contamination and variability due to individual differences in hair growth, dietary factors, and hair treatments. Hair cadmium analysis is often used in forensic and environmental studies to evaluate long-term exposure trends but is less commonly used in clinical assessments due to inconsistencies in standardization. Studies indicate that chronic cadmium exposure is linked to various health issues, including kidney dysfunction, osteoporosis, cardiovascular diseases, and carcinogenesis. Elevated cadmium levels in biological samples are frequently observed in individuals exposed to industrial emissions, cigarette smoke, contaminated food, and polluted water. Smokers tend to have significantly higher blood cadmium concentrations than nonsmokers due to cadmium's high bioavailability in tobacco smoke. Occupational exposure in industries such as battery manufacturing, metal plating, and welding increases cadmium levels in biological fluids. Several analytical techniques, including inductively coupled plasma mass spectrometry (ICP-MS), atomic absorption spectrometry (AAS), and graphite furnace AAS (GFAAS), are commonly employed for precise quantification of cadmium in biological samples. These techniques provide high sensitivity and specificity, enabling accurate exposure assessment even at trace levels. Epidemiological studies have demonstrated strong correlations between elevated urinary cadmium levels and adverse renal effects, including proteinuria and reduced glomerular filtration rate. Additionally, cadmium exposure disrupts calcium metabolism, leading to bone demineralization and increased fracture risk. The World Health Organization (WHO) and the US Centers for Disease Control and Prevention have established reference values for cadmium exposure in different populations to guide public health interventions. Preventive measures such as dietary modifications, reduction of occupational exposure, and stringent environmental regulations are essential in minimizing cadmium-related health risks. While biological monitoring of cadmium levels remains crucial for exposure assessment, further research is needed to elucidate the molecular mechanisms underlying cadmium-induced toxicity and to develop effective therapeutic strategies for mitigating its harmful effects. Understanding the interactions between cadmium and essential biomolecules and its role in oxidative stress and epigenetic modifications will provide valuable insights into its long-term health consequences.

6.4.2 Correlation between Cd exposure and neurodegenerative disease prevalence

Cadmium (Cd), a heavy metal with no known biological function, exhibits neurotoxic properties that may significantly contribute to the onset and progression of NDs. Accumulating evidence suggests that chronic Cd exposure leads to oxidative stress, neuroinflammation, and synaptic dysfunction, which are hallmarks of disorders such as AD, PD, and ALS. The BBB is the primary protective mechanism against systemic toxins; however, prolonged Cd exposure disrupts BBB integrity, allowing Cd to infiltrate neural tissues and exert toxic effects. This disruption has been associated with increased permeability of the BBB, reduced expression of tight junction proteins (e.g., claudins and occludins), and enhanced neurovascular dysfunction. Consequently, Cd-induced oxidative stress leads to mitochondrial dysfunction, a well-recognized contributor to neuronal apoptosis and neurodegeneration. In AD pathology, Cd exposure has been linked to increased amyloid-β (Aβ) aggregation and tau hyperphosphorylation, two primary neuropathological markers. Cd disrupts metal ion homeostasis by displacing essential metals such as zinc and copper, crucial for the normal function of metalloenzymes involved in Aβ metabolism. Studies have shown that Cd enhances the activity of beta-site APP-cleaving enzyme 1 (BACE1), thereby promoting Aβ deposition. Furthermore, Cd-induced oxidative stress elevates levels of ROS, which in turn activate kinases such as GSK-3β and CDK5, leading to tau hyperphosphorylation and neurofibrillary tangle formation. These findings underscore the potential role of Cd as an environmental risk factor for AD (Han et al., 2025; Hassan et al., 2024; Li et al., 2024; Lu et al., 2024; Mincic et al., 2024).

Similarly, in PD, Cd exposure has been implicated in dopaminergic neurodegeneration within the substantia nigra pars compacta (SNpc). Cd interferes with dopamine metabolism by inhibiting tyrosine hydroxylase, the rate-limiting enzyme in dopamine synthesis, and inducing oxidative damage to dopaminergic neurons. Additionally, Cd exposure enhances α-synuclein aggregation, a critical step in PD pathogenesis. The interaction between Cd and α-synuclein may promote the formation of toxic oligomers, contributing to neuronal loss. In vitro and in vivo studies suggest that Cd exposure increases microglial activation and inflammatory cytokine release (e.g., IL-6, TNF-α), exacerbating neuroinflammation and neuronal death. Furthermore, Cd-induced mitochondrial dysfunction leads to the inhibition of complex I activity in the ETC, impairing ATP production and increasing susceptibility to neurodegenerative damage. ALS, a progressive motor neuron disease, has also been linked to environmental Cd exposure. Cadmium-induced toxicity in motor neurons is primarily mediated through excitotoxicity and oxidative stress. Cd disrupts calcium homeostasis by interfering with voltage-gated calcium channels, leading to excessive glutamate release and excitotoxic damage. The accumulation of Cd in motor neurons correlates with increased misfolding of superoxide dismutase 1 (SOD1), a key protein implicated in ALS pathogenesis. Furthermore, Cd exposure has been associated with epigenetic modifications, such as DNA methylation and histone acetylation changes, which may alter gene expression patterns in NDs. The correlation between Cd exposure and ND prevalence highlights the importance of environmental risk factors in neurotoxicity. Given the increasing industrial and environmental Cd contamination, further epidemiological and mechanistic studies are essential to establish causative links and develop targeted interventions for ND prevention.

6.4.3 Regional and occupational studies linking Cd exposure to cognitive decline

Regional and occupational studies have provided substantial evidence linking cadmium (Cd) exposure to cognitive decline, particularly among populations with prolonged environmental and occupational contact with the metal. These studies emphasize the role of Cd as a neurotoxicant, with significant implications for workers in industrial settings, residents in contaminated regions, and vulnerable populations such as children and older people. Mining, battery manufacturing, smelting, and welding workers are at heightened risk of Cd exposure due to inhaling Cd-laden dust and fumes. Long-term occupational exposure has been correlated with neurodegenerative effects, including impaired memory, reduced processing speed, and executive dysfunction. Several cohort studies have shown that chronic Cd exposure leads to altered brain metal homeostasis, oxidative stress, and neuroinflammation—key contributors to cognitive decline. The disruption of calcium signaling, mitochondrial dysfunction, and increased apoptosis in neuronal cells further exacerbate neurodegenerative conditions such as AD and PD. For example, a longitudinal study on battery industry workers found a significant decline in cognitive function, particularly in verbal learning and working memory, over 10 years, with urinary Cd levels as a predictive biomarker of neurocognitive impairment. Beyond occupational settings, Cd exposure has been extensively studied in environmentally contaminated regions where residents are exposed to the metal through polluted air, soil, and water. Agricultural communities relying on contaminated irrigation water and residents near industrial zones experience chronic low-dose Cd exposure associated with cognitive deficits in adults and children. Studies have revealed that Cd can accumulate in the brain over time, particularly in the hippocampus and cortex, regions crucial for learning and memory. In China, for example, areas with high levels of Cd-contaminated rice have reported increased incidences of cognitive decline and developmental delays among children attributed to early-life exposure (Bovio et al., 2021; Hassan et al., 2024; Kaur et al., 2023; Mincic et al., 2024; Živančević et al., 2021). Additionally, studies in Belgium's heavily industrialized areas have shown that higher blood Cd levels in elderly individuals correlate with poor performance in cognitive assessments, including tests measuring attention and verbal fluency. The mechanisms by which Cd exerts its neurotoxic effects are multifaceted. Cd exposure leads to oxidative stress by inducing ROS production, which damages neuronal membranes, disrupts synaptic plasticity, and impairs neurotransmitter function. Additionally, Cd has been shown to mimic and displace essential metals such as zinc and calcium, disrupting enzymatic activities critical for cognitive processes. Chronic exposure also triggers neuroinflammation, with increased activation of microglia and astrocytes, leading to sustained neuronal damage. Furthermore, epigenetic modifications, including DNA methylation and histone acetylation changes, have been implicated in Cd-induced cognitive impairments, suggesting that early-life exposure could have long-lasting effects on brain function. Preventive strategies are crucial given the strong correlation between regional and occupational Cd exposure and cognitive decline. To mitigate the neurotoxic effects of Cd exposure, stricter regulations on industrial emissions, safer occupational practices, and public health interventions targeting at-risk populations are necessary. Future research should focus on longitudinal studies to establish causal relationships and identify potential therapeutic interventions to counteract Cd-induced cognitive deficits (Chang et al., 2024; Mincic et al., 2024; Singh et al., 2023; Urbano et al., 2023; Xu et al., 2024; Živančević et al., 2021).

6.5 Biomarkers and diagnostic approaches

6.5.1 Biomarkers of cadmium exposure (urinary Cd, blood Cd levels)

Cadmium (Cd) is a toxic heavy metal that poses serious health risks upon exposure, primarily through occupational and environmental sources such as industrial emissions, contaminated water, and dietary intake via crops grown in contaminated soil. Biomarkers of cadmium exposure, particularly urinary cadmium (U-Cd) and blood cadmium (B-Cd) levels, serve as essential indicators for assessing exposure and potential health effects. Urinary cadmium levels are widely recognized as a biomarker for long-term exposure, reflecting the cumulative body burden due to cadmium's long biological half-life, which can exceed 10–30 years. Once absorbed, cadmium primarily accumulates in the kidneys, exerting nephrotoxic effects by inducing oxidative stress, inflammation, and cellular damage, leading to a gradual decline in renal function. Urinary cadmium levels remain elevated even after exposure ceases, making U-Cd a reliable indicator of chronic exposure and early kidney dysfunction. In contrast, blood cadmium levels are a biomarker of recent exposure, reflecting the cadmium present in circulation before it is sequestered in target organs. B-Cd levels fluctuate based on recent inhalation or ingestion of cadmium-contaminated substances, making them particularly useful for assessing acute or ongoing exposure scenarios. The relationship between U-Cd and B-Cd depends on exposure duration and individual factors such as renal function, smoking habits, and genetic predisposition, as smoking is a significant source of cadmium exposure due to tobacco plants naturally accumulating the metal from soil (Han et al., 2025; Lu et al., 2024; Rosas-Rodríguez & Valenzuela-Soto, 2021; Urbano et al., 2023; Xu et al., 2024). Studies indicate that even low-level chronic exposure to cadmium, as reflected in urinary and blood biomarkers, is associated with an increased risk of osteoporosis, cardiovascular diseases, and cancers, particularly lung and kidney cancer, due to cadmium's classification as a Group 1 carcinogen by the IARC. Biomonitoring studies utilize U-Cd and B-Cd measurements to establish reference values for exposed and nonexposed populations, a crucial step in setting regulatory frameworks for occupational and environmental health guidelines. Analytical methods for measuring cadmium biomarkers include ICP-MS, GFAAS, and anodic stripping voltammetry, each offering high sensitivity and precision for detecting cadmium at trace levels. Factors such as age, gender, nutritional status, and concurrent exposure to other nephrotoxic agents influence biomarker variability, necessitating careful interpretation of results in epidemiological studies. Preventive measures, including dietary modifications, chelation therapy, and policies to reduce cadmium emissions from industrial sources, are essential in mitigating exposure and lowering biomarker levels. Given the irreversible nature of cadmium-induced toxicity, early detection through biomarker monitoring is critical in preventing severe health outcomes and implementing timely interventions to reduce cadmium-related morbidity and mortality. In conclusion, urinary and blood cadmium biomarkers are indispensable tools in toxicological assessments, providing valuable insights into exposure dynamics, health risks, and the effectiveness of regulatory measures to reduce cadmium-related hazards to human health (Chang et al., 2024; Han et al., 2025; Hassan et al., 2024; Iddrisu et al., 2024; Rosas-Rodríguez & Valenzuela-Soto, 2021; Zaazaa et al., 2022).

6.5.2 Potential early indicators of neurotoxicity

Neurotoxicity refers to the adverse effects of chemical, biological, or physical agents on the nervous system, potentially leading to functional impairments and long-term neurological disorders. Early indicators of neurotoxicity are critical in detecting and preventing severe neurological damage, and they can manifest in behavioral, physiological, cognitive, and molecular changes before more pronounced symptoms emerge. Behavioral alterations are among the earliest warning signs of neurotoxicity, with individuals exhibiting changes in mood, anxiety levels, irritability, or depression, often before significant structural damage occurs(Han et al., 2025; Lu et al., 2024; Mincic et al., 2024; Rosas-Rodríguez & Valenzuela-Soto, 2021; Urbano et al., 2023). Motor impairments such as tremors, coordination difficulties, muscle weakness, or involuntary movements may also develop in response to neurotoxic exposure. Cognitive impairments, including memory loss, difficulty concentrating, and reduced problem-solving abilities, can serve as early markers of neurotoxic effects, particularly in cases where exposure to toxicants affects areas of the brain responsible for executive function and learning. Additionally, physiological changes such as altered sleep patterns, headaches, dizziness, and unexplained fatigue may indicate early neurotoxic effects. On a molecular level, early indicators of neurotoxicity include increased oxidative stress, neuroinflammation, mitochondrial dysfunction, and disruption of neurotransmitter balance. Oxidative stress arises from an imbalance between the production of ROS and the body's ability to neutralize them, leading to neuronal damage and apoptosis. Neuroinflammation, characterized by the activation of microglial cells and astrocytes, contributes to synaptic dysfunction and neuronal degeneration, further exacerbating neurotoxic effects. Mitochondrial dysfunction results in impaired energy production and increased vulnerability to cell death, significantly impacting neuronal health. Disruptions in neurotransmitter systems, including alterations in dopamine, serotonin, glutamate, and acetylcholine levels, can lead to imbalances that affect mood, cognition, and motor function, often serving as an early sign of neurotoxicity. Biomarkers such as elevated neurofilament proteins, tau proteins, or specific cytokines in cerebrospinal fluid and blood may provide measurable indicators of neurotoxicity before clinical symptoms become apparent. Environmental and occupational exposures to neurotoxic substances, such as heavy metals (lead, mercury, arsenic), pesticides, solvents, and industrial chemicals, are familiar sources of neurotoxic effects. Chronic exposure to these substances can lead to progressive neurological deterioration, making early detection essential in mitigating long-term consequences (Bovio et al., 2021; Chang et al., 2024; Gupta et al., 2022;

Hassan et al., 2024; Singh et al., 2023; Živančević et al., 2021). Advances in neuroimaging techniques, including functional MRI and positron emission tomography, offer valuable tools for detecting early neurotoxic changes by visualizing alterations in brain structure and function. Electrophysiological assessments, such as electroencephalography (EEG) and nerve conduction studies, can identify subclinical neurotoxic effects by detecting abnormalities in neural activity. Genetic predisposition may also play a role in susceptibility to neurotoxicity, with specific individuals being more vulnerable to the impact of toxicants due to genetic variations in detoxification pathways and neural repair mechanisms. Given the complexity of neurotoxicity, a multidisciplinary approach involving toxicologists, neurologists, and environmental health experts is necessary to develop effective screening, prevention, and intervention strategies. Public health initiatives aimed at reducing exposure to neurotoxic substances, enhancing workplace safety, and promoting awareness of early neurotoxic indicators are crucial in preventing long-term neurological damage. Future research into novel biomarkers and neuroprotective strategies will be essential in advancing early detection and intervention efforts, ultimately reducing the burden of neurotoxicity on individuals and society (Table 6.5).

6.6 Therapeutic strategies and protective interventions

6.6.1 Chelation therapy and metal-binding agents

Chelation therapy is a medical and biochemical treatment involving chelating agents to bind and remove heavy metals and other toxic substances from the body. This process is widely used in cases of heavy metal poisoning, such as lead, mercury, arsenic, and cadmium toxicity, and is also being explored for various other medical conditions. Chelating agents are organic compounds that form multiple bonds with metal ions, creating stable, water-soluble complexes that can be excreted through urine or feces. Some of the most commonly used chelating agents include ethylenediaminetetraacetic acid (EDTA), dimercaptosuccinic acid (DMSA), dimercaprol (BAL), and deferoxamine. EDTA, for instance, is widely used for lead poisoning and is administered either orally or intravenously, depending on the severity of the condition. DMSA, an FDA-approved chelating agent, is commonly used for lead and mercury detoxification, particularly in children with high lead levels. Dimercaprol, originally developed as an antidote for arsenic poisoning during World War II, is still employed for acute heavy metal poisoning, especially when other agents are ineffective. On the other hand, deferoxamine is used explicitly for iron overload conditions, such as thalassemia, where excessive iron accumulation can lead to organ damage. The effectiveness of chelation therapy depends on several factors, including the type of metal involved, the chelating agent used, the dosage, and the patient's overall health. While chelation therapy has proven effective in treating metal poisoning, it is not without risks. Common side effects include gastrointestinal discomfort, kidney stress, allergic reactions, and depletion of essential minerals such as calcium, zinc, and magnesium. This is because chelating agents are not always selective and can bind to beneficial metals required for physiological functions. Researchers have recently explored novel chelating agents and metal-binding molecules with improved specificity and reduced toxicity (Chang et al., 2024; Mincic et al., 2024; Rosas-Rodríguez & Valenzuela-Soto, 2021; Singh et al., 2023; Xu et al., 2024; Živančević et al., 2021). Natural chelators, such as phytochelatins, flavonoids, and polyphenols, have gained attention for their ability to bind heavy metals while providing antioxidant benefits. Certain plant-based compounds, including curcumin, quercetin, and catechins, exhibit metal-binding properties and are being investigated for their role in mitigating heavy metal toxicity. Additionally, engineered peptides and nanoparticles are being developed to selectively target specific metals and enhance their excretion without causing significant side effects. Chelation therapy is also being studied for its potential role in NDs such as Alzheimer's and Parkinson's, where metal accumulation in the brain is thought to contribute to disease progression. Some researchers suggest that abnormal deposition of metals like iron, copper, and aluminum may promote oxidative stress and neurotoxicity, and chelating agents might help mitigate these effects. However, clinical trials in this area remain inconclusive, and further research is needed to validate these hypotheses. While chelation therapy remains a vital treatment for heavy metal poisoning, advancements in chelating agents and metal-binding technologies continue to improve its safety and efficacy, paving the way for broader applications in medicine and environmental health (Li et al., 2024; Mincic et al., 2024; Singh et al., 2023; Urbano et al., 2023; Živančević et al., 2021).

6.6.2 Antioxidants and neuroprotective compounds (curcumin, resveratrol, *N*-acetylcysteine)

Antioxidants and neuroprotective compounds such as curcumin, resveratrol, and *N*-acetylcysteine (NAC) have gained significant attention for their potential in combating oxidative stress and NDs. Oxidative stress, caused by an imbalance between the production of ROS and the body's ability to neutralize them, plays a crucial role in the progression of neurodegenerative disorders like AD, PD, and ALS. Curcumin, a polyphenol found in turmeric, is known for its potent antioxidant, antiinflammatory, and

TABLE 6.5 Cadmium bioaccumulation pathways and organs affected by chronic exposure.

Exposure pathway	Primary route of entry	Organs affected	Mechanism of bioaccumulation	Health effects	Key findings in chronic exposure
Inhalation (airborne Cd, occupational exposure, tobacco smoke)	Lungs	Lungs, blood, kidneys, brain	Cd is inhaled as fine particulates, absorbed into alveoli, enters the bloodstream, and accumulates in target organs.	Chronic lung diseases, fibrosis, kidney damage, and neurotoxicity.	Cd from smoking and industrial exposure contributes significantly to systemic bioaccumulation and long-term organ toxicity.
Ingestion (contaminated food and water)	Gastrointestinal tract	Liver, kidneys, brain, bones	Absorbed via intestinal epithelium, it binds to metallothioneins and is transported via the blood to organs.	Nephrotoxicity, hepatotoxicity, osteoporosis, neurodegeneration.	Dietary Cd exposure is a major contributor to long-term accumulation, especially in contaminated crops or water populations.
Dermal absorption (industrial and agricultural exposure)	Skin	Blood, kidneys, liver	Limited absorption through intact skin, but significant via wounds or prolonged exposure.	Skin irritation and systemic toxicity in cases of chronic exposure.	It is a less significant route than inhalation or ingestion, but it contributes to occupational hazards.
Placental Transfer (maternal exposure during pregnancy)	Maternal circulation	Fetal brain, liver, kidneys	Crosses the placenta, bioaccumulates in fetal tissues, and disrupts development.	Low birth weight, impaired neurodevelopment, and increased risk of metabolic disorders.	Cd exposure in utero is associated with developmental delays and epigenetic modifications.
Breastfeeding (lactational transfer)	Breast milk	Infant brain, liver, kidneys	Low levels detected in breast milk; potential transfer to infants.	Neurodevelopmental toxicity, immune dysfunction.	The risk of Cd transfer is lower than placental transfer but may contribute to early-life exposure.
Hepatic accumulation (primary detoxification organ)	Portal circulation	Liver, bile, kidneys	Cd binds to metallothioneins in hepatocytes, is partially excreted in bile, but persists in tissues.	Hepatic dysfunction, oxidative stress, and inflammatory damage.	The liver not only plays a key role in Cd detoxification but also suffers significant oxidative damage from chronic exposure.
Renal accumulation (primary site of chronic Cd toxicity)	Blood filtration	Kidneys	Cd accumulates in proximal tubules, impairs reabsorption, and leads to renal dysfunction.	Chronic kidney disease (CKD), proteinuria, and electrolyte imbalances.	Cd-induced nephrotoxicity is irreversible and progresses even after exposure ceases.
Neurological accumulation (crossing the BBB)	BBB	Brain (cortex, hippocampus, basal ganglia)	Cd crosses the BBB, accumulates in neurons and glial cells, and triggers oxidative stress and inflammation.	Cognitive decline, neurodegeneration, increased risk of Alzheimer's and Parkinson's diseases.	Cd interferes with neurotransmitter function, promotes amyloid-beta accumulation, and increases neuroinflammation.
Skeletal system deposition (bone storage)	Circulatory system	Bones	Cd disrupts calcium metabolism, replaces calcium in hydroxyapatite, and weakens bone structure.	Osteoporosis, bone fractures, and increased risk of skeletal deformities.	Chronic Cd exposure leads to Itai–Itai disease, characterized by severe bone pain and fractures.
Endocrine system interference (hormonal disruption)	Circulatory system	Thyroid, adrenal glands, pancreas	Alters hormone secretion and interferes with receptor functions.	Thyroid dysfunction, insulin resistance, and metabolic disorders.	Cd exposure is linked to diabetes, hypothyroidism, and adrenal stress responses.
Reproductive system toxicity (male and female infertility)	Blood, testes, ovaries	Testes, ovaries, placenta	It accumulates in gonadal tissues and disrupts steroidogenesis and gametogenesis.	Reduced sperm count, ovulatory dysfunction, and increased risk of infertility.	Chronic Cd exposure is a known endocrine disruptor affecting reproductive health.

BBB, Blood–brain barrier.
Source: Gupta et al. (2022), Hassan et al. (2024), Kaur et al. (2023), Rosas-Rodríguez and Valenzuela-Soto (2021), and Singh et al. (2023).

neuroprotective properties. It modulates multiple molecular targets, including nuclear factor kappa B (NF-κB), brain-derived neurotrophic factor, and amyloid-beta aggregation, which are implicated in neurodegeneration. Curcumin's ability to cross the BBB makes it a promising therapeutic agent for neuroprotection, as it can directly mitigate neuroinflammation, reduce oxidative stress, and promote neuronal survival (Ebrahimi et al., 2024; Han et al., 2025; Kaur et al., 2023; Singh et al., 2023; Xu et al., 2024). However, its poor bioavailability due to rapid metabolism and low solubility has led to the development of curcumin analogues and nanoformulations to enhance its therapeutic efficacy. Resveratrol, a natural polyphenol found in grapes, red wine, and berries, exerts neuroprotective effects by activating the silent information regulator 2 (SIRT1) pathway, which plays a key role in cellular stress resistance and longevity. SIRT1 activation enhances mitochondrial function, reduces neuroinflammation, and protects against amyloid-beta toxicity, making resveratrol a potential therapeutic agent for Alzheimer's and other neurodegenerative conditions. Additionally, resveratrol possesses antioxidant properties that help neutralize free radicals and prevent neuron oxidative damage. Its role in modulating autophagy and mitochondrial biogenesis further contributes to its neuroprotective potential. However, like curcumin, resveratrol faces challenges in bioavailability, necessitating novel drug delivery systems such as liposomes and nanoparticles. NAC, a precursor to glutathione, is another potent antioxidant and neuroprotective compound extensively studied for its role in neurodegeneration. Glutathione is a critical antioxidant that protects cells from oxidative damage and maintains redox homeostasis. NAC supplementation increases intracellular glutathione levels, reducing oxidative stress and preventing neuronal damage. Additionally, NAC exerts antiinflammatory effects by inhibiting proinflammatory cytokines and modulating glutamatergic neurotransmission, which is crucial for synaptic plasticity and cognitive function. It has shown promise in clinical studies for conditions such as PD, AD, and traumatic brain injury. Unlike curcumin and resveratrol, NAC has relatively high bioavailability, making it a more feasible therapeutic option for neuroprotection. Despite their potential, the clinical translation of these antioxidants faces challenges such as bioavailability, metabolism, and dosage optimization. Combining these compounds with advanced drug delivery systems, dietary interventions, or synergistic formulations may enhance their neuroprotective effects and improve therapeutic outcomes. Future research should optimize their pharmacokinetics, identify novel derivatives, and conduct large-scale clinical trials to establish their efficacy in NDs. Given the growing burden of neurodegenerative disorders worldwide, the exploration of natural antioxidants and neuroprotective compounds remains a promising avenue for developing preventive and therapeutic strategies. Integrating these compounds into personalized medicine may offer new hope for patients suffering from debilitating neurodegenerative conditions, ultimately improving brain health and cognitive longevity (Table 6.6).

6.6.3 Dietary interventions to reduce Cd absorption (Zn, Se supplementation)

Dietary interventions play a crucial role in mitigating cadmium (Cd) absorption and toxicity, with zinc (Zn) and selenium (Se) supplementation being among the most effective strategies. Cadmium is a highly toxic heavy metal that accumulates in the body through food, water, and environmental exposure, leading to severe health effects, including kidney damage, bone demineralization, and disruptions in cellular metabolism. Due to its chemical similarity to essential metals like Zn, Cd competes for absorption and transport pathways, particularly in the intestines and kidneys. Zn supplementation is a well-recognized approach to reducing Cd absorption, as Zn and Cd share common transporters such as divalent metal transporter 1 (DMT1) and ZIP proteins. Increasing dietary Zn intake reduces Cd uptake by saturating these transporters, thereby preventing Cd from entering systemic circulation and accumulating in organs. Moreover, Zn synthesizes metallothioneins, small cysteine-rich proteins that bind and sequester Cd, facilitating its detoxification and excretion. Zn also plays a pivotal role in maintaining cellular antioxidant defense mechanisms by supporting the function of superoxide dismutase (SOD), an enzyme that neutralizes oxidative stress caused by Cd exposure. Selenium supplementation is another effective strategy for reducing Cd absorption and mitigating its toxicity. Se forms stable complexes with Cd, making it biologically inert and promoting excretion. Additionally, Se enhances the activity of glutathione peroxidase (GPx), a key antioxidant enzyme that counteracts Cd-induced oxidative stress by neutralizing lipid peroxides and hydrogen peroxide. Se also modulates the expression of stress-response genes and prevents Cd-induced apoptosis in kidney and liver cells. Research has shown that adequate Se intake can significantly reduce Cd accumulation in tissues and ameliorate its toxic effects on organ function(Chang et al., 2024; Han et al., 2025; Lu et al., 2024; Mincic et al., 2024; Rosas-Rodríguez & Valenzuela-Soto, 2021; Xu et al., 2024). Combining Zn and Se supplementation has been explored as a synergistic approach to counteracting Cd toxicity. Zn primarily reduces Cd absorption and competes for binding sites, while Se facilitates detoxification and enhances antioxidant defenses. Dietary sources of Zn include meat, shellfish, legumes, and seeds, whereas Se is abundant in Brazil nuts, seafood, eggs, and whole grains. However, excessive intake of either mineral can lead to imbalances and adverse health effects, making it essential to maintain appropriate dietary levels. Apart from Zn and Se supplementation, other nutritional factors influence Cd absorption. High dietary calcium (Ca) and iron (Fe) have been reported to reduce Cd uptake by competing for intestinal transporters. At the same time, phytates and fiber in plant-based foods can chelate Cd and limit its bioavailability. Conversely, low protein intake and deficiencies in essential minerals can increase Cd retention and exacerbate its toxic effects. Future research should focus on optimizing dietary interventions, exploring novel food-based strategies, and developing functional foods enriched with Zn and Se to protect populations at risk of Cd exposure. Personalized nutrition approaches considering genetic variations in metal metabolism may further enhance the efficacy of these interventions. Given the widespread presence of Cd in the environment and its long biological half-life, dietary strategies remain a critical component of public health efforts to minimize Cd-related health risks and promote overall well-being(Ebrahimi et al., 2024; Gupta et al., 2022; Li et al., 2025; Singh et al., 2023; Urbano et al., 2023; Xu et al., 2024).

6.6.4 Public health strategies for Cd exposure reduction

Public health strategies for reducing cadmium (Cd) exposure are essential to minimize the toxic effects of this heavy metal on human health, particularly in populations exposed to contaminated food, water, air, and occupational settings. Cd is a persistent environmental pollutant with a long biological half-life, accumulating in the kidneys, liver, and bones, leading to renal dysfunction, osteoporosis, cardiovascular diseases, and carcinogenic risks. Effective public health interventions require a multipronged approach, including regulatory policies, environmental monitoring, dietary modifications, occupational safety measures, and public awareness campaigns. Regulatory policies are crucial in controlling Cd emissions from industrial activities such as mining, smelting, battery production, and phosphate fertilizer manufacturing. Governments and international organizations like the WHO and the US Environmental Protection Agency (EPA) have established permissible limits for Cd in air, water, soil, and food to mitigate human exposure. Stricter environmental regulations on Cd emissions, proper disposal of electronic waste, and restrictions on Cd-containing products such as pigments, plastics, and rechargeable batteries are critical to reducing environmental contamination. Monitoring and remediation of Cd-contaminated sites, particularly agricultural lands, are necessary to prevent food chain contamination. Phytoremediation using hyperaccumulator plants and soil amendments such as biochar, zeolite, and organic matter can help reduce Cd bioavailability in agricultural soils, limiting crop uptake. Dietary modifications are another key strategy in reducing Cd absorption in humans. Increasing the intake of essential minerals such as zinc (Zn), selenium (Se), calcium (Ca), and iron (Fe) can help mitigate Cd uptake by competing for intestinal absorption sites. Encouraging the consumption of diverse, nutrient-rich diets and promoting food safety measures such as washing and peeling vegetables grown in Cd-contaminated areas can further reduce dietary Cd exposure. Occupational safety measures are critical for workers in high-risk industries, including mining, welding, battery manufacturing, and metal smelting. Implementing workplace air quality regulations, providing personal protective equipment (PPE), ensuring proper ventilation, and conducting regular health screenings for Cd exposure can help protect workers from chronic toxicity. Additionally, reducing Cd-containing fertilizers and monitoring heavy metal levels in irrigation water is essential for minimizing agricultural contamination. Public awareness campaigns and health education programs are vital in promoting behavioral changes to reduce Cd exposure (Bovio et al., 2021; Chang et al., 2024; Mincic et al., 2024; Sun et al., 2025; Živančević et al., 2021). Governments and health agencies should educate communities about the sources of Cd, its health risks, and preventive measures through targeted campaigns, school programs, and community workshops. Encouraging smoking cessation is also crucial, as tobacco is a significant source of Cd exposure. Public health policies should also prioritize early detection and intervention for populations at high risk of Cd toxicity. Screening programs for Cd biomarkers, such as urinary and blood Cd levels, can help identify exposure hotspots and enable timely medical intervention. Research and innovation in alternative materials and clean technologies can reduce Cd exposure. Developing safer alternatives to Cd-based industrial materials, promoting green chemistry, and investing in cleaner production techniques can help minimize environmental Cd pollution in the long run. Given the global nature of Cd contamination and its severe health consequences, a comprehensive, collaborative approach involving governments, industries, scientists, and communities is necessary to implement effective Cd exposure reduction strategies and protect public health.

6.7 Future perspectives and research directions

The need for longitudinal studies on cadmium (Cd) and neurodegeneration has become increasingly apparent due to growing evidence linking Cd exposure to various neurotoxic effects, including cognitive decline, neuroinflammation, and NDs such as AD, PD, and ALS. Cd is a known environmental pollutant with a long biological half-life, leading to its accumulation in tissues over time,

TABLE 6.6 Comparative analysis of cadmium-induced neurotoxic mechanisms in animal models and humans.

Neurotoxic mechanism	Findings in animal models	Findings in humans	Key differences and similarities
Oxidative stress and ROS generation	Increased ROS and lipid peroxidation in brain tissues (cortex, hippocampus, cerebellum); depletion of antioxidant enzymes (SOD, CAT, GPx).	Elevated oxidative stress markers in exposed individuals' blood and CSF; reduced antioxidant capacity in patients with neurodegenerative diseases.	Both exhibit oxidative damage, but human studies show chronic accumulation effects over decades, whereas animal studies focus on acute/subchronic exposure.
Mitochondrial dysfunction	Decreased ATP production, mitochondrial membrane potential loss, cytochrome *c* release, increased mitochondrial swelling, and apoptosis in rodent models.	Mitochondrial abnormalities were detected in postmortem brain tissues of Cd-exposed individuals, and impaired energy metabolism was linked to neurodegeneration.	Animal studies confirm mechanistic pathways; human evidence is mostly postmortem or inferred from indirect markers.
Neuroinflammation	Upregulation of proinflammatory cytokines (TNF-α, IL-6, IL-1β) and activation of microglia and astrocytes.	Elevated neuroinflammatory markers in Cd-exposed populations and increased glial activation are seen in neurodegenerative diseases.	Substantial similarity exists between animal and human findings, though human studies rely on biomarker correlations rather than direct mechanistic proof.
The BBB disruption	Increased permeability of the BBB, reduction in tight junction proteins (ZO-1, occludin), and enhanced Cd entry into the brain.	Evidence of BBB disruption in Cd-exposed workers, possible contribution to Alzheimer's and Parkinson's pathology.	Animal models confirm direct Cd effects on BBB integrity; human data suggest indirect contributions via chronic exposure.
Neurotransmitter dysregulation	Decreased acetylcholine, dopamine, serotonin, and glutamate levels; altered synaptic plasticity.	Altered neurotransmitter levels observed in Cd-exposed populations, higher prevalence of depression, and cognitive deficits.	Mechanistic pathways are similar, but human studies rely on functional assessments rather than direct biochemical measurements.
Amyloid-beta (Aβ) aggregation and Alzheimer's pathology	Cd increases Aβ aggregation, reduces Aβ clearance, and upregulates β-secretase, leading to plaque formation.	Elevated Cd levels in Alzheimer's patients' brains, correlation with amyloid burden.	Animal models provide causal evidence; human studies indicate associative links but lack direct proof of causation.
α-Synuclein aggregation and Parkinson's pathology	Cd exposure increases α-synuclein misfolding and aggregation in dopaminergic neurons and reduces dopamine transporter expression.	Higher Cd levels detected in Parkinson's patients' substantia nigra, correlation with disease severity.	Similar mechanistic patterns, but human studies are primarily correlational.
Motor function impairment and ALS pathology	Cd exposure leads to motor neuron degeneration, impaired neuromuscular junction function, and reduced spinal cord integrity in rodents.	Cd exposure is associated with increased ALS incidence; elevated Cd levels are found in ALS patient tissues.	Animal models confirm motor neuron toxicity, but human studies are observational and lack experimental control.
Epigenetic modifications	Cd induces DNA methylation changes, histone modifications, and miRNA dysregulation in brain tissues.	Altered DNA methylation patterns in Cd-exposed individuals, a potential link to neurodegenerative risk.	Substantial similarity in mechanisms, but human studies focus on peripheral biomarkers rather than direct neural epigenetic changes.
Neural stem cell impairment and	Reduced neurogenesis, inhibition of neuronal differentiation, and	CD exposure during pregnancy is linked to neurodevelopmental	Animal models confirm direct neurodevelopmental toxicity;

(*Continued*)

TABLE 6.6 (Continued)

Neurotoxic mechanism	Findings in animal models	Findings in humans	Key differences and similarities
neurodevelopmental effects	apoptosis of neural progenitor cells in rodent models.	delays and lower cognitive function in children.	human evidence is mostly epidemiological.
Cognitive and behavioral deficits	Memory impairment, anxiety-like behavior, and learning deficits observed in Cd-exposed rodents.	Cognitive decline, higher dementia risk, and mood disorders are reported in Cd-exposed populations.	Both show strong correlations, but human studies face confounding variables.
Neuronal apoptosis and cell death	Increased caspase-3 activation and neuronal apoptosis via rodent models' intrinsic (mitochondrial) and extrinsic pathways.	Evidence of increased neuronal apoptosis in Cd-exposed individuals but limited direct analysis in live human brains.	Mechanistically similar, but human studies are mainly postmortem analyses.
Metal dyshomeostasis and disruption of essential metal ions (Zn, Fe, Cu)	Cd competes with Zn, Fe, and Cu, disrupting enzymatic functions, metallothionein binding, and neuronal homeostasis.	Altered metal levels in Cd-exposed individuals are linked to Alzheimer's and Parkinson's pathology.	Strongly conserved mechanisms between animals and humans, supporting translational relevance.

BBB, Blood–brain barrier; *CSF*, cerebrospinal fluid; *ROS*, reactive oxygen species.
Source: Chang et al. (2024), Ebrahimi et al. (2024), Han et al. (2025), Singh et al. (2023).

including the brain. Despite numerous cross-sectional studies highlighting the potential neurotoxic effects of Cd, the lack of comprehensive, long-term studies hinders a definitive understanding of its progressive impact on the CNS. Longitudinal studies are crucial for establishing causal relationships between chronic Cd exposure and neurodegenerative outcomes by tracking cognitive decline, neurophysiological changes, and biomarker fluctuations over extended periods. Such studies would enable researchers to determine critical exposure windows, identify vulnerable populations, and elucidate the cumulative effects of Cd toxicity on neuronal function. Additionally, integrating neuroimaging techniques, advanced biomarker assessments, and genetic profiling into longitudinal research can provide deeper insights into the mechanisms by which Cd contributes to neurodegeneration. Establishing large-scale, population-based cohort studies that monitor individuals from early life to old age would be instrumental in uncovering the long-term neurological consequences of Cd exposure and guiding future public health interventions.

Emerging therapeutic targets for Cd-induced neurotoxicity are being explored as researchers strive to develop effective interventions to mitigate its harmful effects on the CNS. Given the multifaceted nature of Cd-induced neurotoxicity, therapeutic strategies must address oxidative stress, neuroinflammation, mitochondrial dysfunction, and synaptic impairment. One promising approach involves targeting oxidative stress pathways, as Cd exposure leads to excessive production of ROS and depletion of antioxidant defenses. Antioxidants such as NAC, curcumin, resveratrol, and selenium have shown neuroprotective potential by scavenging free radicals, enhancing glutathione levels, and reducing lipid peroxidation in Cd-exposed neuronal cells. Another key therapeutic target is neuroinflammation, which is exacerbated by Cd-induced activation of microglia and astrocytes, leading to the release of proinflammatory cytokines such as TNF-α and IL-6. Antiinflammatory agents, such as polyphenols, omega-3 fatty acids, and pharmacological inhibitors of NF-κB and cyclooxygenase-2 (COX-2), have shown potential in modulating neuroinflammatory pathways. Mitochondrial dysfunction is another hallmark of Cd-induced neurotoxicity, leading to ATP depletion, mitochondrial membrane potential disruption, and neuronal apoptosis. Mitochondria-targeted therapies, including coenzyme Q10, mitochondrial permeability transition pore (mPTP) inhibitors, and mitophagy-enhancing compounds, are being investigated to restore mitochondrial integrity and function in Cd-exposed neurons. Additionally, targeting metal homeostasis is a promising strategy, as Cd disrupts essential metal ion transporters and competes with zinc, calcium, and iron for binding sites. Chelation therapy using metal chelators such as *N,N'*-bis(2-mercaptoethyl) isophthalamide (NBMI) and meso-2,3-DMSA has demonstrated efficacy in reducing Cd accumulation and alleviating neurotoxicity. Advances in nanotechnology are also opening new avenues for drug delivery, allowing for the development of nanoparticle-based formulations that enhance the bioavailability and targeted delivery of neuroprotective compounds. Future therapeutic strategies should focus on combination therapies that simultaneously address multiple pathways of Cd-induced neurotoxicity, optimizing neuroprotection while minimizing adverse effects.

Policy implications and regulations on Cd exposure are critical to mitigating its adverse health effects and preventing further environmental contamination. Given the pervasive nature of Cd pollution, regulatory measures must be implemented at national and international levels to limit Cd emissions, reduce occupational and dietary exposure, and protect vulnerable populations. Existing policies established by organizations such as the WHO, the US EPA, and the European Food Safety Authority provide guidelines on permissible Cd levels in air, water, soil, and food products. However, enforcement of these regulations varies across regions, necessitating stronger compliance mechanisms and continuous monitoring. Stricter emission controls on mining, smelting, battery manufacturing, and phosphate fertilizer production are essential to reducing environmental Cd contamination. Implementing cleaner production technologies, promoting alternative materials, and enforcing proper waste disposal practices can significantly decrease industrial Cd release. In the agricultural sector, policies should focus on reducing Cd-containing fertilizers and monitoring Cd levels in irrigation water to prevent soil contamination. Strengthening food safety regulations is also imperative, as dietary intake remains a significant route of Cd exposure. Establishing maximum residue limits (MRLs) for Cd in food products, particularly rice, leafy vegetables, and seafood, can help reduce human ingestion of Cd. Public health policies should also prioritize screening programs for Cd biomarkers in high-risk populations, enabling early detection and intervention. Occupational safety regulations must be reinforced to protect workers in industries with high Cd exposure risks. Implementing workplace air quality standards, providing PPE, conducting regular health screenings, and offering hazard education programs can significantly reduce occupational Cd toxicity. Public awareness campaigns are crucial in informing communities about the sources, risks, and prevention strategies for Cd exposure. Governments and health agencies should promote educational initiatives encouraging dietary modifications, smoking cessation, and safe handling of Cd-containing products. Moreover, international cooperation is essential for addressing transboundary Cd pollution, requiring collaborative efforts among nations to establish global emission reduction targets, harmonize regulatory standards, and share best practices for Cd management. Future policy directions should also promote research and innovation in sustainable remediation technologies, such as phytoremediation, biochar application, and bioremediation, to mitigate environmental Cd contamination. Strengthening interdisciplinary collaborations between scientists, policymakers, industry stakeholders, and public health authorities will be crucial in developing comprehensive strategies to address Cd exposure and its long-term health implications. Given the emerging evidence linking Cd exposure to NDs, regulatory agencies should consider incorporating neurotoxicity assessments into Cd risk evaluations, ensuring that policies reflect the latest scientific findings. By implementing a holistic approach that integrates environmental regulations, public health interventions, and scientific advancements, policymakers can effectively mitigate Cd exposure risks and protect human health from its detrimental effects.

6.8 Conclusion

Cadmium (Cd) exposure has emerged as a significant environmental and public health concern due to its persistence in the environment and its toxic effects on multiple organ systems, particularly the CNS. Key findings from recent research indicate that Cd is a potent neurotoxin that contributes to neurodegeneration through oxidative stress, neuroinflammation, mitochondrial dysfunction, and synaptic impairment. Studies have demonstrated that Cd exposure increases the production of ROS, leading to lipid peroxidation, DNA damage, and neuronal apoptosis. Furthermore, Cd disrupts essential metal homeostasis by competing with zinc, calcium, and iron for binding sites, impairing critical enzymatic and signaling functions necessary for neuronal survival. Neuroinflammatory pathways are also activated in response to Cd exposure, with microglial and astrocyte activation leading to the release of proinflammatory cytokines such as TNF-α and IL-6, exacerbating neuronal damage. Additionally, Cd-induced mitochondrial dysfunction results in energy deficits, ATP depletion, and increased susceptibility to NDs, including AD, PD, and ALS. Epidemiological studies have further established correlations between chronic Cd exposure and cognitive decline, memory impairment, and increased risk of neurodegenerative disorders, emphasizing the urgent need for preventive measures. Given the long biological half-life of Cd and its ability to bioaccumulate in tissues over time, early intervention strategies are essential to minimize long-term neurological damage and associated health burdens.

Mitigating Cd exposure is of paramount importance in preventing NDs and preserving cognitive health, particularly in aging populations and vulnerable groups with higher exposure risks. One of the most effective approaches involves implementing stringent environmental regulations to limit Cd emissions from industrial activities such as mining, smelting, battery manufacturing, and phosphate fertilizer production. Strengthening air and water quality standards, enforcing proper waste disposal, and promoting safer alternatives to Cd-containing products can significantly reduce environmental contamination and subsequent human exposure. In the agricultural sector, minimizing Cd uptake by crops through soil amendments, biochar application, and controlled fertilizer use can help lower dietary Cd intake. Public health initiatives should raise awareness about Cd sources and promote dietary modifications that reduce Cd absorption while enhancing neuroprotection. Increasing the intake of essential minerals such as zinc (Zn), selenium (Se), calcium (Ca), and iron (Fe) can help mitigate Cd toxicity by competing for absorption sites and reducing Cd bioavailability. Occupational safety measures, including workplace air monitoring, PPE, and health surveillance programs, prevent Cd toxicity among workers in high-risk industries. Additionally, smoking cessation campaigns should be prioritized, as tobacco is a significant source of Cd exposure. Advances in therapeutic research offer promising interventions for mitigating Cd-induced neurotoxicity, including antioxidants, antiinflammatory agents, mitochondrial protectants, and chelation therapy. However, a multifaceted approach that integrates environmental policies, public health strategies, and scientific innovations is necessary to reduce Cd exposure and lower the risk of NDs. Given the increasing evidence of Cd's neurotoxic effects, future research should continue to explore longitudinal studies, novel therapeutic targets, and policy-driven solutions to protect global populations from the detrimental impact of Cd exposure on brain health.

Acknowledgment

The author used AI for language improvement only; all final edits were made by the author. The human author performed the editing.

References

Abidi, S. M. S., Sharma, C., Randhawa, S., Shukla, A. K., & Acharya, A. (2023). A review on nanotechnological perspective of "the amyloid cascade hypothesis" for neurodegenerative diseases. *International Journal of Biological Macromolecules, 253*, 126821. Retrieved from https://doi.org/10.1016/j.ijbiomac.2023.126821.

Acharyya, S., Kumar, S. H., Chouksey, A., Soni, N., Nazeer, N., & Mishra, P. K. (2024). The enigma of mitochondrial epigenetic alterations in air pollution-induced neurodegenerative diseases. *Neurotoxicology, 105*, 158–183. Retrieved from https://doi.org/10.1016/j.neuro.2024.10.002.

Ahmad, F., Nidadavolu, P., Durgadoss, L., & Ravindranath, V. (2014). Critical cysteines in Akt1 regulate its activity and proteasomal degradation: Implications for neurodegenerative diseases. *Free Radical Biology and Medicine, 74*, 118–128. Retrieved from https://doi.org/10.1016/j.freeradbiomed.2014.06.004.

Albekairi, T. H., Alanazi, M. M., Ansari, M. A., Nadeem, A., Attia, S. M., Bakheet, S. A., ... Ahmad, S. F. (2024). Cadmium exposure exacerbates immunological abnormalities in a BTBR T+ Itpr3tf/J autistic mouse model by upregulating inflammatory mediators in CD45R-expressing cells. *Journal of Neuroimmunology, 386*, 578253. Retrieved from https://doi.org/10.1016/j.jneuroim.2023.578253.

Anjum, S., Ali, H., Naseer, F., Abduh, M. S., Qadir, H., Kakar, S., ... Ahmad, T. (2024). Antioxidant activity of *Carica papaya* & *Persea americana* fruits against cadmium induced neurotoxicity, nephrotoxicity, and hepatotoxicity in rats with a computational approach. *Journal of Trace Elements in Medicine and Biology, 81*, 127324. Retrieved from https://doi.org/10.1016/j.jtemb.2023.127324.

Anyachor, C. P., Dooka, D. B., Orish, C. N., Amadi, C. N., Bocca, B., Ruggieri, F., ... Orisakwe, O. E. (2022). Mechanistic considerations and biomarkers level in nickel-induced neurodegenerative diseases: An updated systematic review. *IBRO Neuroscience Reports, 13*, 136–146. Retrieved from https://doi.org/10.1016/j.ibneur.2022.07.005.

Bajpai, A., Bharathi, V., & Patel, B. K. (2025). Therapeutic targeting of the oxidative stress generated by pathological molecular pathways in the neurodegenerative diseases, ALS and Huntington's. *European Journal of Pharmacology, 987*, 177187. Retrieved from https://doi.org/10.1016/j.ejphar.2024.177187.

Bovio, F., Melchioretto, P., Forcella, M., Fusi, P., & Urani, C. (2021). Cadmium promotes glycolysis upregulation and glutamine dependency in human neuronal cells. *Neurochemistry International, 149*, 105144. Retrieved from https://doi.org/10.1016/j.neuint.2021.105144.

Bovio, F., Perciballi, E., Melchioretto, P., Ferrari, D., Forcella, M., Fusi, P., & Urani, C. (2024). Morphological and metabolic changes in microglia exposed to cadmium: Cues on neurotoxic mechanisms. *Environmental Research, 240*, 117470. Retrieved from https://doi.org/10.1016/j.envres.2023.117470.

Chang, Y., Jiang, X., Dou, J., Xie, R., Zhao, W., Cao, Y., ... Liu, Y. (2024). Investigating the potential risk of cadmium exposure on seizure severity and anxiety-like behaviors through the ferroptosis pathway in epileptic mice: An integrated multi-omics approach. *Journal of Hazardous Materials, 480*, 135814. Retrieved from https://doi.org/10.1016/j.jhazmat.2024.135814.

Dahl, R., & Bezprozvanny, I. (2024). SERCA pump as a novel therapeutic target for treating neurodegenerative disorders. *Biochemical and Biophysical Research Communications, 734*, 150748. Retrieved from https://doi.org/10.1016/j.bbrc.2024.150748.

Divya, D., Thekkekkara, D., Sai, T. E., Ramakrishna, D., Manjula, S. N., & Goswami, S. K. (2025). Chapter 2 – Molecular etiopathogenesis of neurodegenerative diseases. In T. S. Koduru, R. A. M. Osmani, E. Singh, & S. B. T.-T. N. R. Dutta (Eds.). Academic Press. Retrieved from https://doi.org/10.1016/B978-0-443-28822-7.00002-7.

Ebrahimi, M., Ebrahimi, M., Vergroesen, J. E., Aschner, M., & Sillanpää, M. (2024). Environmental exposures to cadmium and lead as potential causes of eye diseases. *Journal of Trace Elements in Medicine and Biology, 82*, 127358. Retrieved from https://doi.org/10.1016/j.jtemb.2023.127358.

Esmaeili, Y., Yarjanli, Z., Pakniya, F., Bidram, E., Łos, M. J., Eshraghi, M., ... Zarrabi, A. (2022). Targeting autophagy, oxidative stress, and ER stress for neurodegenerative disease treatment. *Journal of Controlled Release, 345*, 147–175. Retrieved from https://doi.org/10.1016/j.jconrel.2022.03.001.

Fan, Y.-G., Wu, T.-Y., Zhao, L.-X., Jia, R.-J., Ren, H., Hou, W.-J., & Wang, Z.-Y. (2024). From zinc homeostasis to disease progression: Unveiling the neurodegenerative puzzle. *Pharmacological Research, 199*, 107039. Retrieved from https://doi.org/10.1016/j.phrs.2023.107039.

Gil, J., Kim, D., Choi, S., & Bae, O.-N. (2025). Cadmium-induced iron dysregulation contributes to functional impairment in brain endothelial cells via the ferroptosis pathway. *Toxicology and Applied Pharmacology, 495*, 117233. Retrieved from https://doi.org/10.1016/j.taap.2025.117233.

Gupta, R., Ambasta, R. K., & Kumar, P. (2022). Multifaced role of protein deacetylase sirtuins in neurodegenerative disease. *Neuroscience & Biobehavioral Reviews, 132*, 976–997. Retrieved from https://doi.org/10.1016/j.neubiorev.2021.10.047.

Hamzeh, O., Rabiei, F., Shakeri, M., Parsian, H., Saadat, P., & Rostami-Mansoor, S. (2023). Mitochondrial dysfunction and inflammasome activation in neurodegenerative diseases: Mechanisms and therapeutic implications. *Mitochondrion, 73*, 72–83. Retrieved from https://doi.org/10.1016/j.mito.2023.10.003.

Han, X., Guo, P., Fan, Q., Zhou, Q., Xu, M., Long, X., ... Tong, Q. (2025). Synergistic toxicity of cadmium and triadimefon on the microbiota and health of Rana dybowskii tadpoles. *Comparative Biochemistry and Physiology Part C: Toxicology & Pharmacology, 289*, 110092. Retrieved from https://doi.org/10.1016/j.cbpc.2024.110092.

Hassan, H. M., Gazwi, H. S. S., Hussein Refaat, I., Hussein Zaki, A., Milošević, M., Al-Rejaie, S. S., ... Yassien, E. E. (2024). Effects of Laurus extract against cadmium-induced neurotoxicity in rats. *Journal of King Saud University – Science, 36*(11), 103488. Retrieved from https://doi.org/10.1016/j.jksus.2024.103488.

Howard, J. A., David, L., Lux, F., & Tillement, O. (2024). Low-level, chronic ingestion of lead and cadmium: The unspoken danger for at-risk populations. *Journal of Hazardous Materials, 478*, 135361. Retrieved from https://doi.org/10.1016/j.jhazmat.2024.135361.

Iddrisu, L., Moses Mkulo, E., Danso, F., Alatwinusa Yohana, M., Chen, Y., Fang, Z., & Gooneratne, R. (2024). Biomarker for the toxicity evaluation of cadmium based on Glutathione: Recent Progress and challenges. *Microchemical Journal, 207*, 111935. Retrieved from https://doi.org/10.1016/j.microc.2024.111935.

Isaev, N. K., Stelmashook, E. V., Genrikhs, E. E., & Onishchenko, G. E. (2023). Interaction between mitophagy, cadmium and zinc. *Journal of Trace Elements in Medicine and Biology, 79*, 127230. Retrieved from https://doi.org/10.1016/j.jtemb.2023.127230.

Joshi, R., Missong, H., Mishra, J., Kaur, S., Saini, S., Kandimalla, R., ... Bhatti, J. S. (2023). Nanotheranostics revolutionizing neurodegenerative diseases: From precision diagnosis to targeted therapies. *Journal of Drug Delivery Science and Technology, 89*, 105067. Retrieved from https://doi.org/10.1016/j.jddst.2023.105067.

Kaur, A., Singh, N., Kaur, H., Kakoty, V., Sharma, D. S., Khursheed, R., ... Singh, S. K. (2023). Neurodegenerative diseases and brain delivery of therapeutics: Bridging the gap using dendrimers. *Journal of Drug Delivery Science and Technology, 87*, 104868. Retrieved from https://doi.org/10.1016/j.jddst.2023.104868.

Li, G., Sun, C., Zhu, L., Zeng, Y., Li, J., & Mei, Y. (2024). High cadmium exposure impairs adult hippocampal neurogenesis via disruption of store-operated calcium entry. *Ecotoxicology and Environmental Safety, 286*, 117162. Retrieved from https://doi.org/10.1016/j.ecoenv.2024.117162.

Li, Q., Wang, L., Ji, D., Yu, W., Zhang, Y., Xiang, Y., ... Deng, A. (2022). Metformin attenuates cadmium-induced degeneration of spiral ganglion neuron via restoring autophagic flux in primary culture. *Journal of Inorganic Biochemistry, 234*, 111901. Retrieved from https://doi.org/10.1016/j.jinorgbio.2022.111901.

Li, S., You, M., Chen, C., Fu, J., Xu, Y., Pi, J., & Wang, Y. (2025). Direct engulfment of synapses by overactivated microglia due to cadmium exposure and the protective role of Nrf2. *Ecotoxicology and Environmental Safety, 290*, 117620. Retrieved from https://doi.org/10.1016/j.ecoenv.2024.117620.

Lu, Y., Shen, J., Li, B., Zhao, P., Li, B., Fang, Z., ... Fan, Z. (2024). Protective effects of Weizmannia coagulans LBK against acute cadmium toxicity in mice. *Food Bioscience, 62*, 105111. Retrieved from https://doi.org/10.1016/j.fbio.2024.105111.

Mincic, A. M., Antal, M., Filip, L., & Miere, D. (2024). Modulation of gut microbiome in the treatment of neurodegenerative diseases: A systematic review. *Clinical Nutrition, 43*(7), 1832–1849. Retrieved from https://doi.org/10.1016/j.clnu.2024.05.036.

Mishra, R., & Bowman, A. B. (2024). Chapter Six - The role of stem cells in the study and treatment of neurodegenerative diseases with environmental etiology. In A. B. Bowman, M. Aschner, & L. G. B. T.-A.in N. Costa (Vol. Eds.), *Stem cells in neurotoxicology: Vol. 12*, (pp. 173–206). Academic Press. Retrieved from https://doi.org/10.1016/bs.ant.2024.07.001.

Mishra, Y., & Kaundal, R. K. (2023). Role of SIRT3 in mitochondrial biology and its therapeutic implications in neurodegenerative disorders. *Drug Discovery Today, 28*(6), 103583. Retrieved from https://doi.org/10.1016/j.drudis.2023.103583.

Rezaei, K., Mastali, G., Abbasgholinejad, E., Bafrani, M. A., Shahmohammadi, A., Sadri, Z., & Zahed, M. A. (2024). Cadmium neurotoxicity: Insights into behavioral effect and neurodegenerative diseases. *Chemosphere, 364*, 143180. Retrieved from https://doi.org/10.1016/j.chemosphere.2024.143180.

Rosas-Rodríguez, J. A., & Valenzuela-Soto, E. M. (2021). The glycine betaine role in neurodegenerative, cardiovascular, hepatic, and renal diseases: Insights into disease and dysfunction networks. *Life Sciences, 285*, 119943. Retrieved from https://doi.org/10.1016/j.lfs.2021.119943.

Singh, N., Vishwas, S., Kaur, A., Kaur, H., Kakoty, V., Khursheed, R., ... Abdel-Daim, M. M. (2023). Harnessing role of sesamol and its nanoformulations against neurodegenerative diseases. *Biomedicine & Pharmacotherapy, 167*, 115512. Retrieved from https://doi.org/10.1016/j.biopha.2023.115512.

Subaraja, M., Arokiyaraj, S., & Mathew, P. (2024). Neuroprotective effect of huperzine-A against cadmium chloride-induced Huntington's disease in Drosophila melanogaster model. *Journal of King Saud University - Science, 36*(8), 103319. Retrieved from https://doi.org/10.1016/j.jksus.2024.103319.

Sun, Y., Liu, J., Xiao, K., Chen, S., Ma, J., Cao, M., ... Du, Z. (2025). STEAP3-SLC39A8-mediated microglia ferroptosis involved in neurotoxicity in rats after exposure to lead and cadmium combined. *International Immunopharmacology, 146*, 113854. Retrieved from https://doi.org/10.1016/j.intimp.2024.113854.

Uhegbu, H., & Memudu, A. E. (2019). Prohylactic role of lawsonia inermis(henna) leaf aqueous extract on cadmium-induced neurodegenerative rat model: A new drug therapy for Alzheimer'sdisease. *IBRO Reports, 7*, 5–6. Retrieved from https://doi.org/10.1016/j.ibror.2019.09.018.

Urbano, T., Verzelloni, P., Malavolti, M., Sucato, S., Polledri, E., Agnoli, C., ... Filippini, T. (2023). Influence of dietary patterns on urinary excretion of cadmium in an Italian population: A cross-sectional study. *Journal of Trace Elements in Medicine and Biology, 80*, 127298. Retrieved from https://doi.org/10.1016/j.jtemb.2023.127298.

Wankhede, N. L., Kale, M. B., Shende, P., Ratne, N., Taksande, B. G., Upaganlawar, A. B., & Magrulkar, S. V. (2025). Chapter 1 – Etiology and treatment challenges for neurodegenerative disorders. In V. Patel, M. S. Rajput, J. S. Shah, & T. B. T.-T. T.for the C. N. S. Mehta (Eds.). Academic Press. Retrieved from https://doi.org/10.1016/B978-0-443-23841-3.00001-7.

Williams, E. T., Chen, X., Otero, P. A., & Moore, D. J. (2022). Understanding the contributions of VPS35 and the retromer in neurodegenerative disease. *Neurobiology of Disease, 170*, 105768. Retrieved from https://doi.org/10.1016/j.nbd.2022.105768.

Xu, L., Zhang, T., Zhu, B., Tao, H., Liu, Y., Liu, X., ... Meng, X. (2024). Mitochondrial quality control disorder in neurodegenerative disorders: Potential and advantages of traditional Chinese medicines. *Journal of Pharmaceutical Analysis*101146. Retrieved from https://doi.org/10.1016/j.jpha.2024.101146.

Zaazaa, A. M., Abd El-Motelp, B. A., Ali, N. A., Youssef, A. M., Sayed, M. A., & Mohamed, S. H. (2022). Stem cell-derived exosomes and copper sulfide nanoparticles attenuate the progression of neurodegenerative disorders induced by cadmium in rats. *Heliyon, 8*(1), e08622. Retrieved from https://doi.org/10.1016/j.heliyon.2021.e08622.

Zhang, Y., Zhang, C., Zhang, C., Bin, X., Jiang, J., & Huang, C. (2024). Leukocyte telomere length mediates the association between cadmium exposure and cognitive function in US older adults. *Journal of Psychiatric Research, 169*, 166–173. Retrieved from https://doi.org/10.1016/j.jpsychires.2023.11.023.

Zhang, Y.-Y., Li, X.-S., Ren, K.-D., Peng, J., & Luo, X.-J. (2023). Restoration of metal homeostasis: A potential strategy against neurodegenerative diseases. *Ageing Research Reviews, 87*, 101931. Retrieved from https://doi.org/10.1016/j.arr.2023.101931.

Živančević, K., Baralić, K., Jorgovanović, D., Buha Djordjević, A., Ćurčić, M., Antonijević Miljaković, E., ... Đukić-Ćosić, D. (2021). Elucidating the influence of environmentally relevant toxic metal mixture on molecular mechanisms involved in the development of neurodegenerative diseases: In silico toxicogenomic data-mining. *Environmental Research, 194*, 110727. Retrieved from https://doi.org/10.1016/j.envres.2021.110727.

Chapter 7

Chromium toxicity: neurological implications

Prasann Kumar
Department of Agronomy, School of Agriculture, Lovely Professional University, Phagwara, Punjab, India

7.1 Introduction

Chromium (Cr) is a transition metal that has a presence in various oxidation states, with trivalent chromium (Cr(III)) and hexavalent chromium (Cr(VI)) being the most stable and biologically significant forms. Even the trace amount of toxic Cr(III) released during the process of stainless-steel production can constitute yet another harmful form of concrete (e.g., it is carcinogenic). The bonding of chromium with other substances counts for close family relationships (Chen, Fan, et al., 2024; Mohankumar et al., 2024). Cr(VI) are no less frequently used compounds than PtV and VtC for catalysts, employed most notably in the chemical industry metallurgy, electroplating, tanning leather, pigment manufacture, and stainless-steel fabrication is where chromium is applied most extensively. Cr(VI) compounds are also commonly used to ensure products hold up over time. Despite this, the environmental significance of industrial emissions containing Cr(VI) has a good side and a bad side. These discharges pose serious risks not only to ecology but also to the health of man. Once released into the environment, Cr(VI) is highly soluble in water and readily disperses into groundwater and soil, while Cr(III) has low mobility and is less toxic. Various human activities result in chromium pollution, a common pollutant in industrialized regions (Glicklich et al., 2024; Siddig et al., 2025). Key sources of exposure to chromium include occupational settings, where people engaged in chrome plating, welding, and tanning are frequently exposed via inhalation of dust and fumes to harmful Cr(VI) material. It has been found that prolonged exposure to these kinds of attacks is likely to increase the risk of lung cancer significantly, air passage blockages, and systemic diseases for occupants of the affected premises. Another significant means by which people are exposed to chromium is through drinking water and soil that the metal has contaminated; this is particularly true in areas without adequate facilities for managing industrial waste. Wastewater from tanneries, mining sites, and leachates derived from solid waste sites contribute to groundwater becoming increasingly contaminated through Cr(VI). This represents a serious threat to the purity of drinking water. Agricultural land flooded with water containing high concentrations of chromium may result in very harmful levels of metals accumulating in the soil. Once the theory of bioaccumulation comes into play, it has practically no limits, and all kinds of plants and crops will absorb these metals through their roots and leaves. Selective sources of chromium exposure occur via dietary intake. The human diet also contains chromium, mostly in the form of Cr(III), as an essential trace element (Iyer et al., 2023; Mohebi et al., 2024). However, neither Cr(VI) nor Cr(III) can be considered suitable for human foods, so the point must be settled compared to other toxicants within the human diet. Ingesting Cr(VI) at toxic levels and for a long enough duration to have deleterious health consequences is a situation mainly brought on by industrial pollution and food processing. The toxicity of chromium, especially Cr (VI), is a critical public health issue because it can cause oxidative stress, DNA damage, and systemic poisoning. Cr (VI) is either reduced internally to Cr(III) by certain types of cell-surface receptors that have sulfur-grouped central, or it changes form entirely once it is absorbed inside a cell, producing reactive oxygen species (ROS) and intermediates, which then go on mess up homeostasis at the cellular level. One is worried about the effects of the toxic compound chromium on people, for although it has no taste or smell and colorings used to process wheat, it can mask any feeling when swallowed in unknown quantities with food. Finally, one notes that a focus in recent years

Heavy Metal Toxicity and Neurodegeneration. DOI: https://doi.org/10.1016/B978-0-443-36575-1.00025-X

has been understanding the dangers associated with having higher tensions fulfilled while wiring yourself up (i.e., electrical power). Besides being extremely dangerous for those participating in this activity, workshop buildings also leak, with gases making one wonder whether the dealers are. While a tiny fraction of the world's vehicles now come so equipped, the evidence suggests that burning natural gas for propulsion may be an even better option in the future. Some considerations include (1) potential health hazards from pollutant combustion products, such as dioxin from atomic bombs (Chen, Xu, et al., 2024; Shen et al., 2024). Nay, better give up now! It is a common saying that power is not to be safely possessed. It might evaporate at any moment, but so would electricity. Dangers that may now be realized of having a society increasingly dependent upon electrical energy include the long-term effects on human health. Exposures to Cr(VI) have challenged the assumption that this material is private property. Choral and other highly toxic forms of Cr are extracted from rusted steel on factory sites, harboring the smog-making machinery that will poison entire neighborhoods with its emissions for years. Chromium is an agent when used in glass processing. Like sodium, it can kill the human body or make it stronger (F, Cl). Chromium's role in neurotoxicity becomes all the more pressing as it resonates throughout the nervous system. Preliminary findings on nervous system pathology have only been recognized over the past decade. The nervous system has recorded a worrying signal about chromium, reminiscent of the situation a few years back when no consideration was given to that telephone conversation element. Like voltage-sensitive auditory relays in high overcast, nerves can lightly twitch or hum along. Toxic manifestations of Cr (VI) exposure on the nervous system include cognitive deficits, neuroinflammation and neuro-degenerative changes, with evidence suggesting that chromium-induced oxidative destruction contributes to neuronal damage populations exposed to Cr(VI), prolonged exposure has been shown to lead toward alterations in the levels of neurotransmitters prevailing within their bodies and changes contrasting them from normal values, impairment in synaptic function and disarray of neuronal signaling pathways. Furthermore, a connection with cognitive impairment detected reports of head pain, dizziness, and mood disturbances in workers involved in chromate-producing or using industries (Azzam et al., 2023; Chow & Foo, 2024). Extensive animal studies have shown that exposure to chromium can lead to vascular changes in the brain, disrupt the blood–brain barrier (BBB), and facilitate the accumulation of toxic metals. The evidence has been mounting for some time that chromium may play a role in neurological illness, afflicting patients with such dreaded ailments as Parkinson's disease (PD) and Alzheimer's disease (AD). The damaging effects of toxic chromium compounds, either causing greater oxidative stress by generating a large amount more ROS themselves or via indirect means to disable defense systems against oxidation dark among imaginary narratives concerned with coats of arms, are wholly equivocal with neural development: Studies have reported fifth- and sixth-grade scores going down in children who are starting low even lower if already behind on the game used by spring grade 4. Will lower-than-average intelligence have long-term consequences for children? No studies exist on the subject of this disappointing political rhetoric. However, economists have argued that low IQs are responsible for many of the world's problems. The reason remains unknown, but such speculation is worth noting. The long-term consequences of chronic, low-dose exposure are a present area of active research. For this, epidemiological studies would be desirable to establish a clearer relationship between environmental chromium exposure and neurological disorders among human populations. Rigorous regulations are needed to protect human health and the environment from chromium contamination (Gezahegn et al., 2024; Jegadheeshwari et al., 2024). Advanced remediation methods, such as bioremediation and nanotechnology-based remediation strategies, must be developed to address the issue of water circulation in industrial settings and clean up contamination in water supplies. Public health programs should monitor chromium levels in areas with high levels of exposure, improve industrial waste disposal methods, and develop public awareness of the dangers associated with chromium exposure (Table 7.1).

While chromium is crucial in various industrial applications and biological processes, its toxic effects, especially on humans, can result in serious environmental and health burdens via exposure to hexavalent forms. The effects of chromium toxicity on neural tissue prompt immediate calls for research aimed at identifying mechanisms of neurotoxicity and developing measures to prevent or ameliorate exposure.

7.2 Mechanisms of chromium toxicity

As a transition metal, chromium has various valence states and is one of the most widespread environmental pollutants. It is well known for its toxicity in many organs, including the nervous system. Neurotoxicity in chromium is due to oxidative stress, mitochondrial derangement, and interference with critical cell signaling pathways. This string of changes presents neuroinflammation, neuronal injury, and systematic cell death. Meanwhile, Cr(VI) and Cr(VI), a metabolite of Cr(VI), are more harmful than natural because they are readily soluble in water and can enter our bodies through nonspecific anionic channels. Once it has entered the cell, Cr(VI) is transformed into Cr (III). Before this process is completed, superoxide anions

TABLE 7.1 Sources of chromium exposure in the environment and its potential health impact.

Source	Chromium form	Medium of exposure	Potential health impact	Affected area	Common industries/activities	References
Industrial emissions	Hexavalent Chromium (Cr(VI))	Air	Lung cancer and respiratory issues	Near industrial sites	Chrome plating, welding, and steel production	Saha et al. (2011), Mishra and Bharagava (2016)
Contaminated water	Hexavalent Chromium (Cr(VI))	Water	Digestive system issues and cancer	Municipal water sources	Industrial discharges and wastewater runoff	Georgaki et al. (2023), Naja and Volesky (2017)
Contaminated soil	Trivalent Chromium (Cr(III))	Soil	Soil toxicity and ecosystem disruption	Areas near industrial sites	Mining and smelting operations	Lukina et al. (2016), Tchounwou et al. (2012)
Landfills/waste disposal	Hexavalent Chromium (Cr(VI))	Soil and groundwater	Groundwater contamination toxicity	Landfills, disposal sites	Waste management and landfill leaching	Saravanan and Ramesh (2024), Gupta et al. (2022)
Industrial sludge	Both Cr(VI) and Cr(III)	Soil, water, and air	Soil and water contamination	Waste disposal sites	Electroplating and industrial processes	Dhal et al. (2013), Sun et al. (2022)
Chromium-based pesticides	Chromium compounds (Cr(III))	Soil, water, and air	Toxicity to plants, animals, and humans	Agricultural fields	Use in pest control and farming practices	Kapoor et al. (2022), Saxena (2025)

and hydroxyl radicals will predominate inside the reduced cells (Nehzomi & Shirani, 2025; Olsen et al., 2024; Taiwo et al., 2024). These ROS destroy the cell's normal redox balance by overwhelming endogenous antioxidant defenses. Superoxide dismutase, catalase, and glutathione peroxidase cannot cope with them, and in consequence, the cells suffer from oxidative stress, which can lead to catastrophic structural changes in lipid molecules. These result in decreased brain cells, neuronal proteins, and DNA. Suffering from a dual blow of oxidative damage and interference in normal neurotransmitter function, neurons can no longer maintain their long-term fitness as the brain continues functioning. Next to oxidative damage, chromium toxicity is deeply implicated in functional impairments in mitochondria. Mitochondrial DNA and crucial proteins within the electron transport chain are targets for ROS-induced oxidative injury involving Cr(VI) O. As a result, ATP production is reduced. However, the initiated apoptosis will get a further boost because of the action of caspases and cytochrome *C* release, which activates apoptotic pathways again (Erkman et al., 2025; Jeffers et al., 2024; Priyadarshini et al., 2024). Another way chromium causes trouble for normal neurons is that it interferes with the calcium balance, which everyone, particularly neuroscientists, should be aware of. Chromium-induced ROS set off calcium dysregulation in cells through ER stress. These contribute strongly to neuronal excitotoxicity, synaptic dysfunction, and aberrant neurotransmitter release, a triple threat for the ailing nerve cell! Meanwhile, disruptions are further aggravated in certain signal pathways. This includes interference with the nuclear factor erythroid 2-related factor 2 (Nrf2), mitogen-activated protein kinases (MAPKs), and nuclear factor-kappa B (NF-κB). Nrf2, which is a master antioxidant regulator, is blocked by chromium toxicological stress that results in cellular defenses being down. Consequently, excessive activation of MAPKs, including such enzymes as ERK (extracellular signal-regulated kinase), JNK (c-Jun N-terminal kinase), and p38, causes cells to die. On the other hand, excessive NF-κB activation leads to an upregulation in proinflammatory mediators like TNF-α (tumor necrosis factor-alpha). There is IL-6, version 1.6 beta. Those three cytokines form a noxious environment for the neurons that will not disappear. Neuroinflammation is the main avenue by which chromium causes its neurotoxicity. Exposure to chromium will activate astrocytes and microglial cells in the central nervous system (CNS) and act as its immune response agents (defenders). This causes a flood of prostaglandins from inflammation to sweep the entire system, wreaking havoc on everything without exception; those who suffer most deeply have nerves (Bonfiglio et al., 2024; Kaur et al., 2024; Mutlu, 2024). Microglia activation especially releases an enzyme called inducible nitric oxide synthase (iNOS), which creates nitric oxide (NO). This substance will combine with superoxide anions to form peroxynitrite—bringing about lipid radicals to boot, protein nitration, and DNA fragmentation all at once. So now we see that this decline in neurons is a state of runaway inflammation. Changes in synaptic plasticity are chaos this time—all due to the firing rates; after-effects of such changes become permanent changes in memories. Furthermore, the neuroinflammation caused by chromium will compromise the integrity of the BBB, bringing nutrients and other environmental toxins into the CNS's milieu. The BBB is weakened by not just one environmental toxin that gets its foot in between cells after another collapse by paying off local guards. However, as soon as you break down brain barriers created by endothelial cells along blood vessels, everything from the surrounding environment may suddenly rush into our most important organ. This means also that any unsleeping disruptors of normal life can gain entry. Furthermore, exit routes are sealed off: waste from certain processes inside our brains is not sent out as efficiently as before. It accumulates, leading to even more damage while doing nothing but more situated damage to local neighbors. The disturbance of the BBB makes the brain more susceptible to chromium and other neurotoxicants, causing a combined toxicity that can contribute to neurodegenerative diseases, such as AD, PD, and amyotrophic lateral sclerosis (ALS). The interaction between chromium and other environmental toxins is a cause for concern (Amarloei et al., 2025; Heng et al., 2024; Rhodes et al., 2005; Wong et al., 2023). The neurotoxic examples of such risk go beyond those who bear the toxic exposure. For example, studies have shown that people exposed to both chromium and lead suffer greater oxidative stress, mitochondrial dysfunction, and neuroinflammation than those exposed to only one or the other toxic metal. This leads to faster degeneration of their neuronal tissue and poorer mental test scores. Chromium and lead both enter the whole nervous system via dopamine transporters. Chromium and arsenic produce more intense neuroinflammation in the hippocampus, leading to a corresponding decline in learning ability and reduced memory function. In addition, it has been found that chromium can add to the toxic effects of many pesticides, PAHs, and industrial solvents. Together with these findings, we can see that chromium is toxic, and its effects vary. This research implies that it will be necessary to rigorously regulate environmental exposures and develop targeted therapies for the nervous system to mitigate these deleterious effects (Fig. 7.1).

7.3 Clinical manifestations of chromium neurotoxicity

Many neurological systems have demonstrated that chromium, a heavy metal employed extensively in industry,

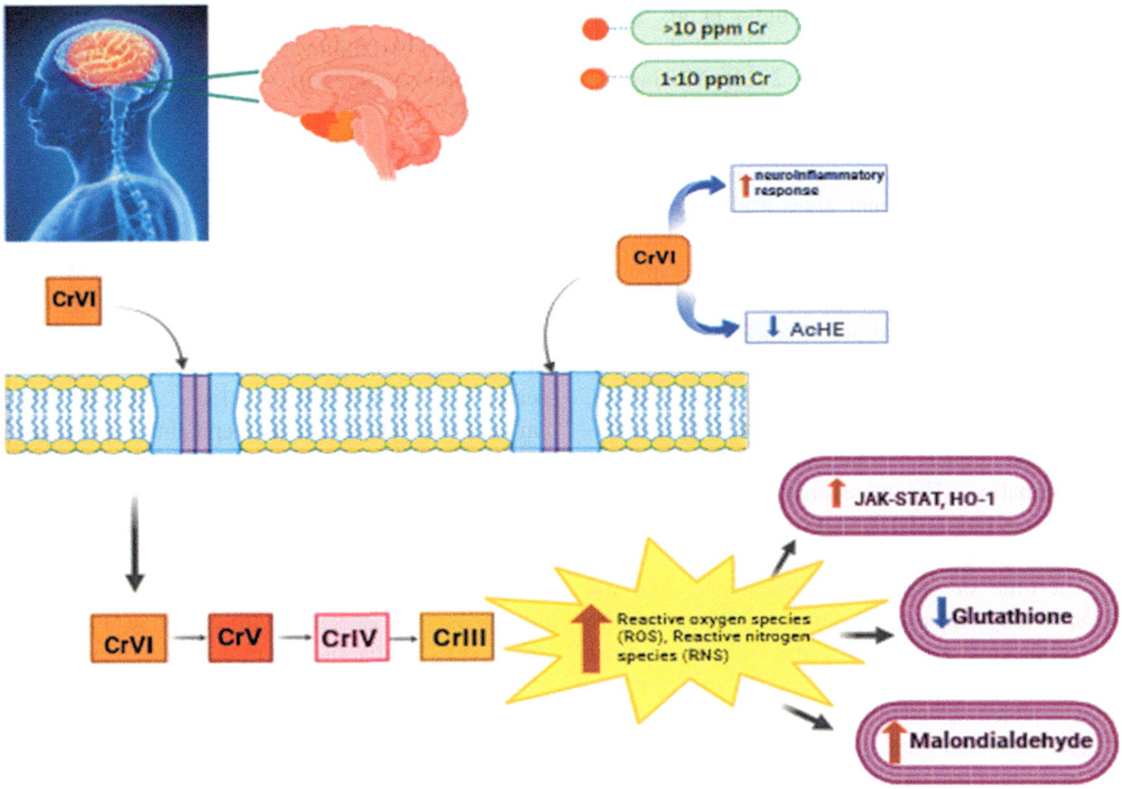

FIGURE 7.1 Neurotoxicity due to chromium toxicity.

leads to cognitive impairment and memory loss, of which dizziness is one symptom. Other possible CNS disorders caused by exposure to chromium include amnesias (both calcareous and noncalcareous), attention span disorders such as attention-deficit/hyperactivity disorder (ADD/ADHD), and motor tics. Occupations entail chromium exposure and contact with chromium-contaminated water or environmental pollution. Chromium poisoning primarily harms the CNS, with systemic effects. Chromium crosses the BBB through oxidative stress and interferes with neurophysiological processes (Azzam et al., 2023; Jeffers et al., 2024; Mohebi et al., 2024; Nehzomi & Shirani, 2025; Shen et al., 2024). This not only hinders cognitive resources but also leads to their degradation. Studies indicate that if you have been exposed to chromium for long enough, your cognitive resources will deteriorate. This disturbs your learning and processing ability and the executive functions of your mind. The inflammation returning your brain cells in response to chromium exposure stirs up microglial cells, which link to the release of proinflammatory cytokines like IL-6. In addition to TNF-alpha (tumor necrosis factor-alpha), these interleukins barge in on molecular pathways already out of whack to cause neurodegenerative cascades. That's what explains how rampant forms of neuronal destruction and synapse-synthesis errors can show up as declining lucidity in thought and variable behavior closer to mania or psychopathy. Meanwhile, chromium increases ROS production, damaging your body's natural antioxidative defense mechanisms. The resulting accumulation of ROS harms lipids, proteins, and DNA, which again reduces the integrity of neural tissue. Such oxidative harm causes slippage of the neuro depressive membrane battery through damage to mitochondria. In this way, neurons commit suicide. The gliotransmitter web, a second major node for memory processing, breaks down due to chromium and becomes something far short of human efficiency. Thus we find unusually severe problems with spatial orientation and working memory. White rats have been used to study this. As a consequence of neurotoxicity induced by chromium, attention deficits are also a common phenomenon. This is due to its disturbance of neurotransmitter equilibrium, especially those pathways that involve dopamine (DA) mechanisms, acetylcholine (Ach) whether made from implied release per se or through diffusing into synapses near producing neurons or glutamate (the most common excitatory amino acid in the brain). Such disruption in dopaminergic activity diminishes synaptic plasticity and response to rewards. Attention, persistence, and the ability to inhibit all suffering. Also, chromium exposure yields motor disturbances such as tremors, weakness in the muscles, and poor coordination, which look exactly like the symptoms of Parkinsonism (Bonfiglio et al., 2024; Jegadheeshwari et al., 2024; Mohankumar et al., 2024; Priyadarshini et al., 2024; Shen et al., 2024). Moving too frequently fails when basic ganglia stop working correctly, as neuroinflammation forges into them and oxidative stress commands more of the dopaminergic neurons to die off. So, speaking again, this account gives a clear indication that chromium may be a consistent factor in the etiology of neurodegenerative illnesses like AD and PD. In AD, chromium helps Aβ protein aggregates form more quickly and produce more. These two are characteristic pathological markers for the disease state now known as Gad (General Anxiety Disorder). Invigorated by chromium, FCT grabs onto carb holes, which speeds up Aβ aggregation. Sporadic and familial AD both share the pathology of Aβ plaque formation. The more arousing symptoms of these dementias come on later; they include trouble with memory, speaking and understanding what is happening around us in time or space. Furthermore, in AD, the combination of oily and retards signals mediating insulin cues within brain cells means both types will pass on their toxic load with additional force. Similarly, chromium exposure can induce selective degeneration of dopamine. Studies have indicated that exposure to chromium is correlated with an increased risk of Parkinsonism. This supports its role in dopaminergic neurotoxicity byzantine. For those hoping to see long-term neurological breakthroughs from our understanding of chromium poison, early intervention is critical. When you periodically check human work and life environment, testing chromium levels in the blood and urine would allow for early identification of symptoms (Gezahegn et al., 2024; Jeffers et al., 2024; Mohebi et al., 2024; Shen et al., 2024; Taiwo et al., 2024). Currently, diagnostic imaging techniques, such as magnetic resonance imaging (MRI) and positron emission tomography (PET), provide data that help monitor patients' exposure to chromium. Procedures also include using molecular and functional neuroimaging to mark sites affected by CNS inflammation. This provides a basis for prescribable therapies, interventions if necessary, and planning even more control from environmental regulations and occupational safety equipment such as dust masks, gloves, and water purification systems. Therapeutic interventions using antioxidants and antiinflammatory agents are likely key methods to counteract chromium-induced neurotoxicity. *N*-Acetylcysteine (NAC), vitamin E, polyphenols, and other antioxidant drugs are promising treatments against toxicity following chronic chromium intake. Lifestyle changes include using many antioxidants (like in capsules) and plenty of outdoor exercise. Doing math problems in your head for work stress relief with the aid of mental exercise and regular compound workouts of upper body muscles with weights can help combat all those factors that lead to certain neural degeneration! It is a testament to their increasing awareness of the neurological sequelae of chromium exposure that there is now much more emphasis on early detection and treatment measures in ADMD.

7.4 Cognitive impairment and neurodegeneration

Although chromium (Cr) is a metal of several valences, Cr (VI) and Cr(III) are the most toxic. In living systems, Cr (III) is often considered an essential trace element involved in the metabolism of glucose or lipids, but Cr(VI) can have acute and chronic toxic effects. Besides its chronic carcinogenic activity, recent epidemiological and experimental studies have stirred up doubts about the cognitive perils of chromium exposure, as well as its role in the development of neurodegenerative diseases such as AD or PD. Epidemiological evidence suggests that both occupational and environmental chromium exposure, particularly to Cr(VI), are associated with cognitive impairment up to neurodegeneration life cycle events. For example, industrial workers who inhale or contact dermally with chromium compounds show signs of cognitive dysfunction like forgetfulness and failure to think problems through. In addition, communities living near sites where water is contaminated by this heavy metal report high prevalence rates for neurological disorders. When rodents are exposed to chromium, they learn less well and have poorer memory than controls with accompanying neuroinflammatory and oxidative stress markers in areas of their brain that correspond to human cortex lude oxidant. But how does chromium lead to neurodegenerative diseases? It is a complex process that involves many factors, including oxidative stress, inflammation in the brain (neuroinflammation), disturbance of metal homeostasis, and interference with key proteins thought to underlie AD and PD pathogenesis (Azzam et al., 2023; Chen, Xu, et al., 2024; Erkman et al., 2025; Glicklich et al., 2024; Nehzomi & Shirani, 2025). One deciding pathway is chromium-generated oxidative stress, which occurs when Cr(VI), upon intracellular reduction, can produce reactive (a late form). This burden of oxidative damage spreads throughout the cell's macromolecular contents, lipids, proteins, and nucleic acids, all of which ultimately result in neuronal apoptosis and synaptic failure. Notably, oxidative stress is a hallmark of AD and PD, essential in their pathogenesis.

In the setting of AD cognition, chronic figures are involved. Industries responsible for such pollution can be organized, too. In the context of AD, chromium exposure has been related to amyloid-beta (Aβ) accumulation and tau protein hyperphosphorylation—two main pathological features of AD. Several studies have shown that chromium interferes with amyloid precursor protein processing while promoting the amyloidogenic pathway and increasing Aβ production. The chromium-induced oxidative stress worsens Aβ aggregation, which breeds toxic plaques. Furthermore, chromium exposure disturbs kinase and phosphatase activity, which leads to abnormal tau phosphorylation and the formation of neurofibrillary tangles. Pathology in PD is also significantly influenced by chromium toxicity, primarily its effects on dopaminergic neurons in the substantia nigra. Chromium exposure has been shown to induce mitochondrial dysfunction, which reduces ATP production and produces oxidative stress, selectively killing off dopaminergic neurons. In addition, chromium disturbs iron homeostasis; a standard report notes that the substantia nigra accumulates more iron. This iron dyshomeostasis results in further oxidative stress and inflammation, promoting dopaminergic neuron degeneration. This starts a vicious cycle! In turn, the activity of microglia and astrocytes increases; these cells release proinflammatory cytokines such as TNF-α and IL-6, which worsen neuronal damage and hasten PD progression (Azzam et al., 2023; Glicklich et al., 2024; Jeffers et al., 2024; Mohebi et al., 2024; Olsen et al., 2024; Shen et al., 2024).

Another possible passage for chromium injury leading to neurodegeneration is the interference of insulin signaling in the brain by it. Chromium affects insulin sensitivity, and disturbed insulin signaling has been implicated in AD and PD. Insulin resistance damages glucose metabolism in the brain, which means oxidative stress, inflammation, and synaptic dysfunction—all bad! Studies have shown that chromium exposure changes the glucose transport and insulin receptor function of neurons, hampering neurodegenerative processes. Though there is now a growing body of evidence that shows the relationship between cognitive impairment and neurodegenerative diseases. Indeed, it remains a question that needs to be solved further, even to retributive molecular mechanisms down the line. More thorough studies in a broad context and over long-time scales are essential to determine whether or not chronic exposure to chromium can heighten someone's risk of getting neurodegenerative diseases. In addition, there is a desperate need to conduct mechanistic research at the cellular and molecular levels to determine how chromium interacts with key pathways leading to neurodegeneration (Azzam et al., 2023; Erkman et al., 2025; Iyer et al., 2023; Jegadheeshwari et al., 2024; Siddig et al., 2025). Investigating related genetic and epigenetic factors may shed some light on why certain people are susceptible to the neurodegeneration that follows chromium poisoning. And such an important research question has not been effectively answered so far at all. Exploring therapeutic interventions further to reduce chromate toxicity in the nervous system is also vital. Antioxidants, metal chelators, antiinflammatory agents, etc., may offer protective effects against chromium-induced neurotoxicity. Furthermore, understanding the role of dietary chromium intake in processes leading to neurodegeneration and its combination with other trace minerals could give rise to valuable advice on eating for better health and help guide public health policy (Table 7.2).

Accumulating evidence implicates that chromium exposure contributes to both cognitive decline and an increased likelihood of getting neurodegenerative diseases such as AD and PD. These processes' mechanisms involve oxidative stress, neuroinflammation, amyloid-beta accumulation, Tau protein hyperphosphorylation, loss of dopaminergic neurons, and disrupted insulin signaling. With the widespread industrial use of chromium today and its possible environmental contamination, urgent investigation into the neurotoxic effects of chromium is needed to better protect against the dispossession of a healthy brain. It is essential to narrow these research gaps because this will be seminal for guiding regulations and ultimately protecting public health (Bonfiglio et al., 2024; Mohankumar et al., 2024; Olsen et al., 2024; Shen et al., 2024).

7.5 Oxidative stress and neuroinflammation

Chromium (Cr) is an industrial metal extensively used in numerous ways. It has been recognized as a neurotoxic agent, where its adverse effects are mainly mediated by oxidative stress and neuroinflammation. Understanding the role of oxidative stress in the mechanism of chromium neurotoxicity is essential, as it has been shown that the redox balance in neural cells is fundamentally disrupted during chromium exposure, resulting in a cascade of cell damage and dysfunction. Chromium can be found in several oxidation states; hexavalent chromium (Cr(VI)) is the most toxic due to its solubility and ability to penetrate biological membranes, such as the BBB. Upon entering the cells, Cr(VI) is reduced intracellularly to Cr(III), with excessive ROS as by-products (Erkman et al., 2025; Jegadheeshwari et al., 2024; Priyadarshini et al., 2024; Shen et al., 2024; Siddig et al., 2025). These ROS, including superoxide anions (O_2^-), hydroxyl radicals (OH^-), and hydrogen peroxide (H_2O_2), induce oxidative stress by exceeding the endogenous antioxidant defense systems, resulting in lipid peroxidation, protein oxidation, and DNA damage. The high oxygen consumption, high polyunsaturated fatty acid content, and relatively low levels of antioxidant enzymes, including superoxide dismutase (SOD), catalase (CAT), and glutathione peroxidase (GPx), make the brain particularly susceptible to oxidative stress.

Chromium is a vital neurotoxicant that can lead to the loss of the antioxidant defense system. The antioxidant system eliminates ROS in a normal physiological status to maintain cellular homeostasis. Chromium exposure, on the

TABLE 7.2 Chromium-induced cognitive decline and neurodegenerative diseases.

Aspect	Mechanism in cognitive impairment	Chromium-induced effects	Neurodegeneration in Alzheimer's	Neurodegeneration in Parkinson's	Potential outcomes	References
Oxidative stress	Impaired synaptic function and neuronal damage due to oxidative injury.	Chromium is a source of ROS, which leads to oxidative damage.	Oxidative stress promotes amyloid-beta aggregation and tau phosphorylation.	Oxidative stress causes the loss of dopaminergic neurons.	Worsening cognitive impairment and motor dysfunction.	Althobaiti (2024), Firdous et al. (2024)
Inflammation	Chronic inflammation in the brain impairs cognitive function.	Chromium exposure increases proinflammatory cytokines in the brain.	Inflammation worsens amyloid-beta plaque formation and neuronal death.	Inflammation causes neurodegeneration in the substantia nigra.	Accelerated progression of cognitive and motor deficits.	Alquezar et al. (2020), Bartman et al. (2024)
Neurotransmitter disruption	Imbalance in neurotransmitters (e.g., acetylcholine) affects learning.	Chromium may disrupt acetylcholine, dopamine, and glutamate signaling.	Altered neurotransmitter levels affect cognitive processes like memory.	Dopamine dysregulation leads to motor impairments and cognitive deficits.	The decline in memory, attention, and motor skills.	Gorini and Tonacci (2024), Guevara-Ramírez et al. (2024)
Mitochondrial dysfunction	Mitochondrial damage results in energy deficits, affecting cognition.	Chromium can impair mitochondrial function, reducing ATP production.	Mitochondrial dysfunction exacerbates neurodegeneration and amyloid toxicity.	Mitochondrial dysfunction is central to dopaminergic cell death.	Progressive cognitive impairment and motor decline due to energy deficits.	Ajibo et al. (2024), Zuo et al. (2024)
Metal ion imbalance	Disruption of metal homeostasis can affect enzyme function and protein aggregation.	Chromium interferes with the balance of essential metals like copper and zinc.	Metal dysregulation promotes amyloid-beta aggregation and tau toxicity.	Chromium-induced iron dysregulation enhances the aggregation of toxic proteins.	Worsened cognitive function and neurodegeneration due to mental imbalance.	Parui et al. (2024), Richter-Bisson et al. (2023)

other hand, causes the depletion of vital antioxidants. One is reduced glutathione (GSH), an essential thiol-containing molecule for detoxifying ROS and maintaining redox balance. Indeed, studies have demonstrated that GSH levels are lowered with the addition of chromium while oxidized glutathione (GSSG) is elevated, consequently reducing the total antioxidant capacity of neural cells. In addition, Cr-induced oxidative stress is known to inhibit the activities of antioxidant enzymes, including the superoxide dismutase (SOD) that catalyzes the dismutation of superoxide radicals into hydrogen peroxide and catalase (CAT) that subsequently decomposes hydrogen peroxide into water and oxygen (Azzam et al., 2023; Erkman et al., 2025; Iyer et al., 2023; Jegadheeshwari et al., 2024; Mohankumar et al., 2024; Taiwo et al., 2024). The impairment of these processes increases ROS levels and promotes mitochondrial dysfunction and apoptosis in neuronal cells. However, the primary ROS source, mitochondria, is a primary target of chromium toxicity. We demonstrate that chromium-induced mitochondrial membrane potential collapse and cytochrome *C* release activate caspase-dependent apoptotic cascades, leading to neuronal cell death. Additionally, oxidative stress impairs calcium homeostasis, leading to excitotoxicity and subsequent neuronal injury.

Mounting evidence indicates that oxidative stress due to high chromium concentrations induces inflammatory pathway activation in the nervous system, ultimately contributing to chronic neuroinflammation and subsequent neurodegeneration. Activated microglia and astrocytes primarily mediate neuroinflammation via the release of proinflammatory cytokines, including tumor necrosis factor-alpha (TNF-α), interleukin-1 beta (IL-1β), and interleukin-6 (IL-6). Chromium exposure activates microglial cells via TLR signaling, mechanistic NF-κB and MAPK signaling pathways that lead to an overproduction of the inflammatory mediators. At the same time, excessive production of ROS activates NOD-like receptor protein 3 (NLRP3) inflammasome, an essential part of the innate immune response responsible for the maturation and secretion of IL-1β and IL-18, thus exaggerating neuroinflammatory responses (Erkman et al., 2025; Gezahegn et al., 2024; Nehzomi & Shirani, 2025; Shen et al., 2024; Siddig et al., 2025). Moreover, chromium-mediated oxidative stress induces cyclooxygenase-2 (COX-2) and inducible nitric oxide synthase (iNOS), leading to increased production of prostaglandins and nitric oxide (NO), respectively, and aggravates neurotoxicity.

This cycle of oxidative stress and neuroinflammation forms a vicious cycle that drives further neuronal destruction. In chronic neuroinflammation, continuous ROS production further aggravates oxidative stress and mitochondrial impairment, leading to neuronal loss and cognitive deficits. Further, chromium-induced neuroinflammation is related to increased BBB permeability, permitting peripheral immune cells to invade the CNS and aggravate the inflammation. Experimental studies have shown that exposure to chromium disrupts the tight junction proteins of BBB like claudin-5 and occludin, causing degeneration of the BBB with subsequent paracellular leakage and enhanced neurotoxic effects. Furthermore, microglia and astrocytes secrete excitotoxic molecules (glutamate), leading to synaptic impairment and neuronal excitotoxicity (Erkman et al., 2025; Gezahegn et al., 2024; Iyer et al., 2023; Mohankumar et al., 2024; Priyadarshini et al., 2024).

Chronic oxidative stress and neuroinflammation have been implicated in various neurodegenerative diseases such as AD, PD, and ALS. Thus it is reasonable to speculate that chromium exposure may be involved in developing these diseases. Chromium promoted profound oxidative stress, which induced Aβ aggregation and tau hyperphosphorylation, contributing to AD. In addition, chromium exposure has been implicated in impairing insulin signaling within grease, which promotes neurodegeneration, wherein the condition is referred to as type 3 diabetes, chromium-rich diet, and PD Therefore PD can also be considered a disease in which metal-induced mitochondrial dysfunction and oxidative stress promote the selective death of dopaminergic neurons in the substantia nigra leading to movement disorders characteristic of PD, and chromium-rich diets may all worsen these (Azzam et al., 2023; Erkman et al., 2025; Olsen et al., 2024; Siddig et al., 2025; Taiwo et al., 2024). Likewise, in ALS, chromium exposure has been linked to motor neuron degeneration by way of oxidative stress and excitotoxicity mechanisms.

In addition, attempts to counteract chromium-induced neurotoxicity have been directed towards oxidative stress amelioration and inflammation pathway targeting. Natural antioxidants such as curcumin, resveratrol, and quercetin have also been reported to protect against chromium-induced oxidative damage through ROS scavenging and upregulation of endogenous antioxidant enzymes. Moreover, neuroprotective pharmacological agents involving neuroinflammatory pathways like NF-κB inhibitors and NLRP3 inflammasome blockers may be potential candidates for alleviating chromium-mediated neuroinflammation and maintaining neuronal function. Chelation therapy has also been investigated to minimize chromium accumulation in the CNS and decrease its toxic effects through agents like deferoxamine and NAC. Further studies are warranted to elucidate better the underlying molecular mechanisms responsible for chromium-induced neurotoxicity and the development of effective therapeutic strategies for preventing and treating chromium-related neurodegenerative disorders (Jeffers et al., 2024; Jegadheeshwari et al., 2024; Mohankumar et al., 2024; Mohebi et al., 2024; Shen et al., 2024).

The neurotoxic effect of chromium is mediated, at least in part, by oxidative stress and neuroinflammation. Inhibition of antioxidant defenses and increased ROS production by chromium regulate redox homeostasis and induce apoptosis via mitochondrial damage in nerve cells (neurons). Moreover, chromium-induced oxidative stress activates inflammatory pathways in the brain, leading to chronic neuroinflammation and further worsening neurodegenerative processes. This interplay between oxidative stress and neuroinflammation drives a vicious cycle that exacerbates neuronal injury and contributes to the pathogenesis of a range of neurodegenerative diseases. These mechanisms deserve further investigation to establish new therapeutic strategies to prevent chromium-induced neurotoxicity and protect neural function based on their pharmacological properties (Fig. 7.2).

7.6 Environmental and occupational exposure

Chromium (Cr), a metalloid of widespread occurrence, has several oxidation states, where Cr(III) and Cr(VI) are of primary biological and environmental significance. Cr(III) is an essential micronutrient important for glucose metabolism, and Cr(VI), which is highly toxic and carcinogenic, has significant environmental and occupational health concerns. This is the same set of sentences with different constructions, and it has also been revised for clarity, concision, or coherence. Microbial chromate reduction has gained attention for the reclamation of chromium-contaminated effluents (Bonfiglio et al., 2024; Chow & Foo, 2024; Mohebi et al., 2024; Olsen et al., 2024; Taiwo et al., 2024). Electroplating, stainless steel, leather tanneries, and pigment industries discharge substantial quantities of Cr(VI) into the air, water bodies, and soil. The industry's chromium emission to the environment is due to waste disposal, industrial by-products, and atmospheric deposition by inhalable particulate matter emitted from these industries. Besides, the chromium released to the environment persists, bioaccumulating in aquatic systems, soil, and crops and running into the food chain. The most significant sources of chromium contamination of water are chromates leached from industrial effluents, hazardous waste sites, and run-off from chromate-treated wood and pesticides. Hexavalent chromium is a relatively water-soluble metal atom which enhances mobility and bioavailability. Drinking contaminated water can lead to health problems like gastrointestinal illness, kidney

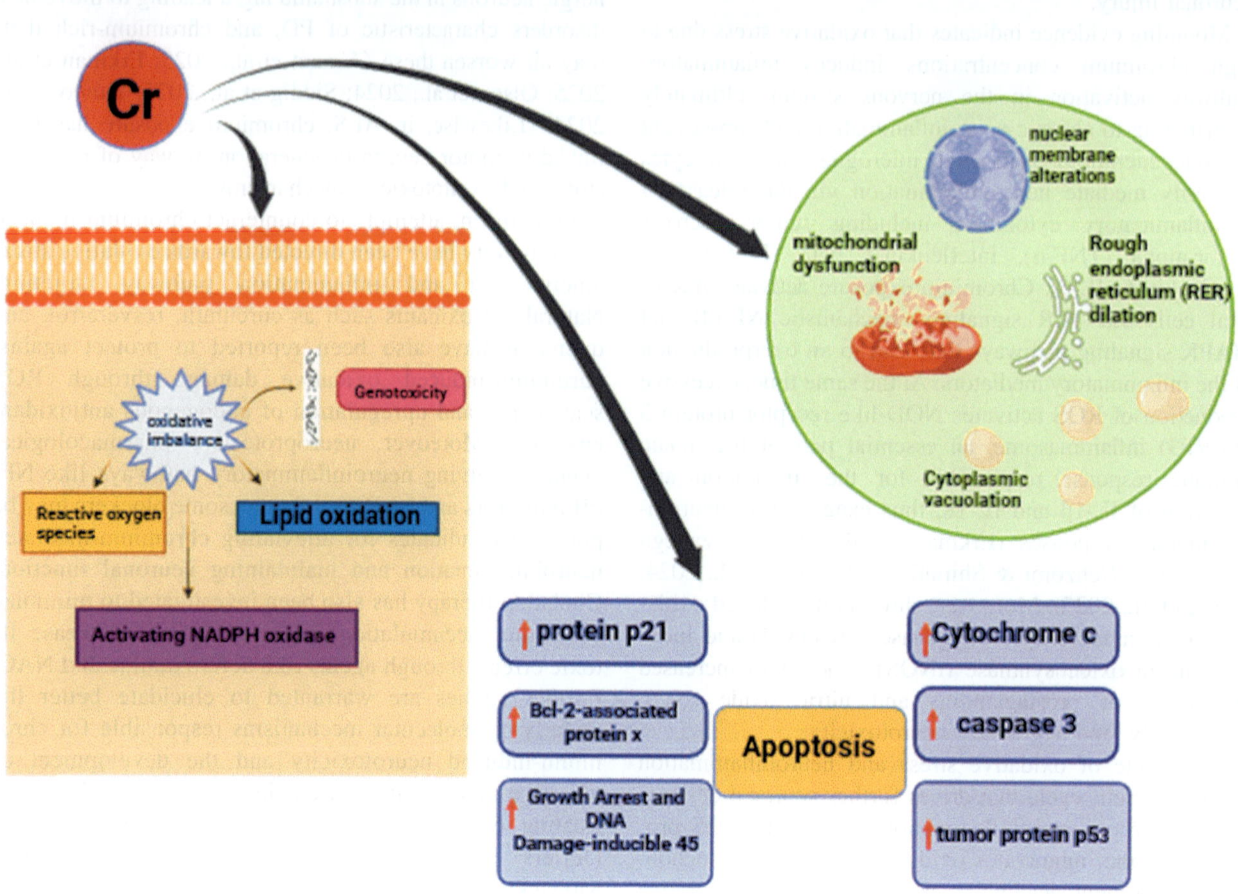

FIGURE 7.2 Mechanism of chromium toxicity.

damage and cancer from long-term exposure. The United States Environmental Protection Agency (EPA) establishes a maximum contaminant level of 0.1 mg/L for total chromium in drinking water to reduce risks to health. Soil pollution is another environmental issue plaguing areas near industrial belts. Improper disposal of industrial bone sludge and mining operations may contaminate the soil with chromium, chromate-based fertilizers and pesticides. The fact that Cr(VI) maintains a high persistence in the soil is devastating rooted plants can absorb it, thus introducing chromium to the food chain. Chromium is one of a few metals whose toxicity for plants is highly species-specific. Even so, high levels generally inhibit seed germination, root growth and plant development, leading to lower agricultural yields and premature germination of soil seeds, which disrupts the ecosystem. Exposure to chromium occurs when people consume food and water with this contaminated metal. Cr(III) is found in meat, grains and vegetables and is crucial for carbohydrate and lipid metabolism. Still, ingesting Cr(VI) contaminated food sources can have health consequences. Crop soil Irrigation with contaminated water is more susceptible to humans with crops with the potential for chromium accumulation from irrigation with contaminated water or even irrigation in contaminated soil (Bonfiglio et al., 2024; Chow & Foo, 2024; Iyer et al., 2023; Mohankumar et al., 2024; Olsen et al., 2024; Taiwo et al., 2024). Understanding these alternative exposure pathways is crucial for informing effective regulatory actions to minimize chromium contamination risks. Workers' occupational exposure mainly occurs in industries processing or working with chromium, predominantly in metal plating, welding, leather tanning, and pigment manufacturing. Chromium and chromium-bearing compounds may be inhaled, dermally absorbed, and ingested as toxic chromium species, particularly at elevated concentrations. In electroplating processes, chromium compounds (mainly Cr(VI)) are widely used for corrodible metal surfaces in corrodible substances to form corrosion-resistant coatings. Workers at these facilities are exposed to chromium by breathing in fumes and aerosols produced during plating. The chromic acid present in the plating baths is one of the primary reasons for occupational exposure and a common cause of respiratory disorders, being considered responsible for a high incidence of lung cancer, chronic bronchitis and ulceration of the nasal septum. Environmental controls are in place to reduce exposure (e.g., effective ventilation, including HEPA filters), PPE, and procedural controls. Welding fume: Chromium (Cr) vapors are generated during welding; the welders working on stainless steel and Cr-containing alloys are exposed to these fumes. Welding generates fume hazards upon inhalation due to extreme heat due to oxidized chromium (Cr(VI)) in metals. Long exposure to these fumes is linked to respiratory problems, lung cancer and skin irritation. Health Risks Certain engineering controls can be beneficial in reducing a welder's exposure to harmful situations, such as local exhaust ventilation, respiratory protection and appropriate workplace design. Chromium salts, particularly Cr(III), are essential to the leather tanning industry, which processes animal hides. When tanning agents containing chromium are inadequately treated and disposed of, chromium can oxidatively convert from Cr(III) to Cr(VI), thereby greatly enhancing toxicity and environmental pollution of waste streams. Tannery workers are typically exposed to chromium via skin contact and inhalation of the dust. The longer exposure is linked to dermatitis, respiratory issues and an increased risk of cancer (Erkman et al., 2025; Gezahegn et al., 2024; Nehzomi & Shirani, 2025; Priyadarshini et al., 2024; Shen et al., 2024; Siddig et al., 2025). Ergonomics, substitutive tanning agents, and adequate protective equipment ensure the protection of tanners in the tanning industry. Limiting exposure to chromium will require more than proper handling practices from authorities; it will also require stricter environmental regulations. Most countries have rules and regulations to minimize chromium emissions and exposure in many workplaces, reducing health risks. Most importantly, we expect industries to have strong occupational health and safety programs to protect workers from chromium exposure. Engineering controls are among the most important, including ventilation systems, properly enclosed processing areas and dust suppression methods used to limit workers' exposure to airborne chromium within the workplace. Personal protective equipment (PPE), such as protective clothing, gloves, respirators and eye protection, will reduce dermal and inhalation exposure. Routine air and biological monitoring help measure chromium exposure, which can improve preventive measures via routine monitoring and risk assessment of chromium exposure. Compliance with safety regulations is ensured through worker training, safety awareness programs, safe food handling practices, hygiene factors, and emergency response protocols. Exposure to Cr (VI) at the workplace due to some hazardous materials increases health risks. However, eliminating or modifying the processes that entail the release of Cr(VI) compounds as hazardous waste will help eliminate this serious potential health hazard. We are up until October 2023. These encompass vital regulatory guidelines such as those outlined in the Clean Water Act and Safe Drinking Water Act, which set limits on chromium concentrations in water bodies to safeguard potable water and protect aquatic biodiversity. It offers high heat resistance and is used in adhesives. The EPA oversees and enforces the Resource Conservation and Recovery Act governing the disposal and treatment of chromium (Cr) bearing hazardous waste to avoid soil and groundwater contaminants. The Occupational Safety and Health Administration (OSHA) regulates chromium in the workplace by establishing permissible exposure limits (PELs) and mandating that industries implement control measures to safeguard workers (Azzam et al., 2023; Chen, Xu, et al., 2024; Mohankumar et al., 2024; Olsen et al., 2024; Priyadarshini et al., 2024). The European Union's (EU) Regulation on Registration, Evaluation, Authorisation and Restriction of Chemicals, or REACH Regulation, mandates the registration, evaluation, and restriction of chromium compounds to promote safer alternatives in the automotive industry and the economic sustainability through the use of sustainable materials. The EPA and other agencies established air quality regulatory standards limiting chromium emissions from commercial emissions sources, thereby reducing environmental contamination and human health risks. Proteins bear the largest number of chromophores in concentrations. Consequently, industrial emissions, contaminated water and soil, and diet represent important sources of exposure, and occupations associated with metal plating, welding, and tanning of leather are specific risk factors. Comprehensive occupational safety measures, environmentally sustainable industrial practices, and regulatory compliance are some of the factors that can help mitigate chromium health hazards. Nonetheless, future research should focus on establishing novel chromium bioremediation approaches, alternatives, and monitoring strategies to safeguard human health and ecosystems in the times to come. So, with the intercession of industries, policymakers, and researchers, sustainable solutions could be developed to address the effects of chromium exposure and create healthier surroundings (Fig. 7.3).

7.7 Public health implications and preventive strategies

Chromium has many oxidation states. However, trivalent Cr(III) and hexavalent chromium or Cr(VI) are the most relevant to human health. Whereas Cr(VI) is highly toxic and regarded as a human carcinogen, Cr(III) is a trace element involved in glucose metabolism and regulating insulin activity. Occupational exposure to Cr(VI) is particularly prevalent in the leather tanning, welding, stainless steel, electroplating, and pigment sectors, where workers

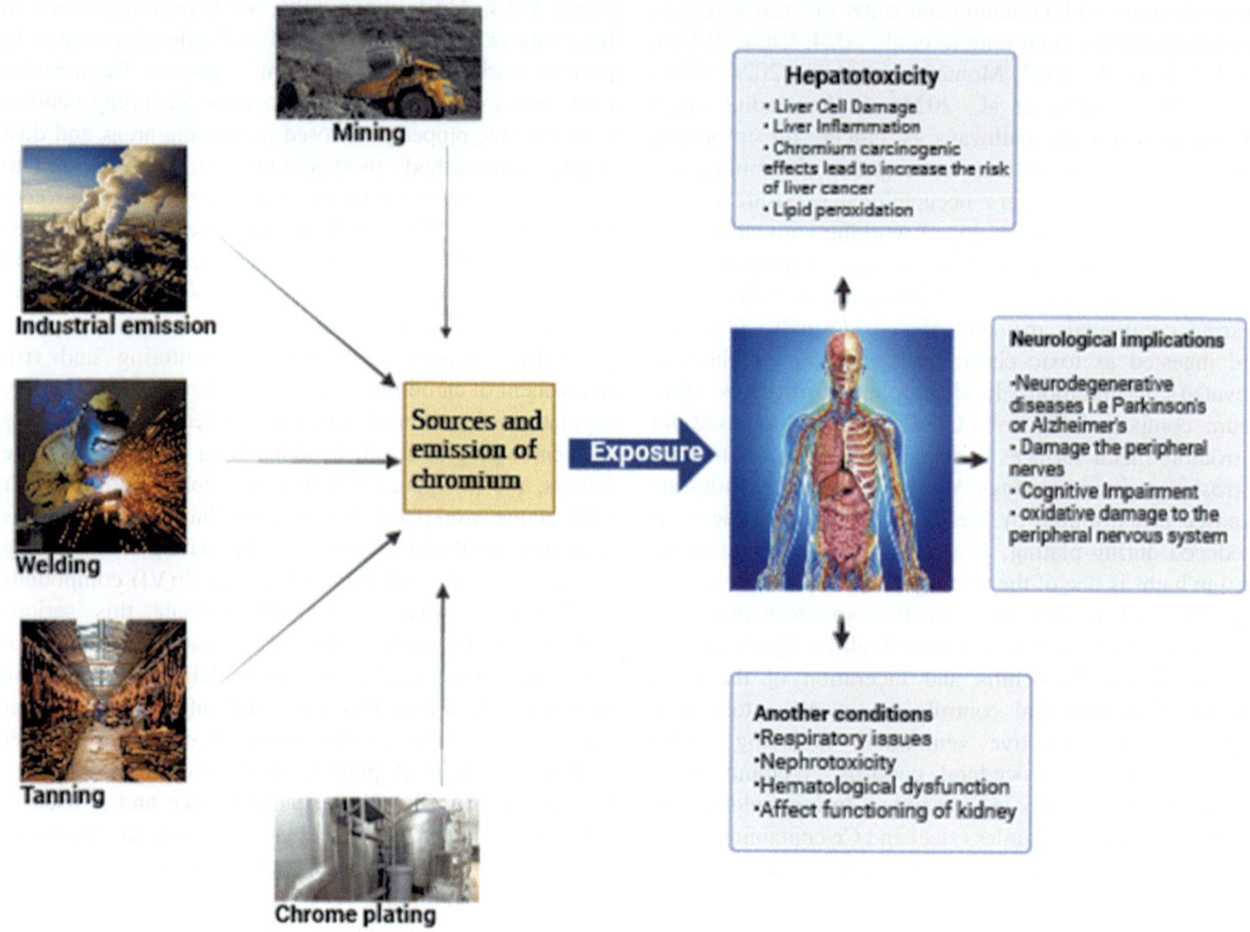

FIGURE 7.3 Sources and exposure of chromium.

face a higher risk of contact with, inhalation, and ingestion of chromium compounds. Environmental exposure to chromium via drinking water, soil, and air continues to pose a significant public health burden, particularly in communities near industrial plants, hazardous waste disposal sites, and areas with significant mining activity (Azzam et al., 2023; Chen, Xu, et al., 2024; Jegadheeshwari et al., 2024; Siddig et al., 2025; Taiwo et al., 2024).

The Cr(VI) exposure has significant toxicological effects and is a potent toxin to multiple organ systems. The respiratory system is particularly vulnerable; inhalational exposure to the toxic constituents of cigarette smoke has been linked to chronic bronchitis, nasal septum ulceration, reduced lung function, and lung cancer. Allergic dermatitis due to dermal exposure, ulcerations, and delayed wound healing systemic absorption leads to nephrotoxicity, hepatotoxicity, and hematological disturbances like anemia and leukocytosis. Neurotoxic effects of copper toxicity include cognitive defects, mood disorders, and peripheral neuropathy. Mechanisms involved in chromium toxicity include oxidative stress, DNA damage, apoptosis, and interference with signaling pathways. Cr(VI) is readily taken up by cellular membranes and is reduced intracellularly to Cr(III), which generates ROS and causes lipid peroxidation, mitochondrial dysfunction, and genomic instability, leading to its mutagenic and carcinogenic properties (Azzam et al., 2023; Iyer et al., 2023; Jegadheeshwari et al., 2024; Mohankumar et al., 2024).

Chromium toxicity presents a significant public health challenge whereby immediate and long-term policy interventions are required to minimize human exposure risk and long-term healthcare cost burden. Occupational safety regulations significantly reduce workplace exposure. Such behavior underscores the need for much lower PELs, which could be incorporated through legislation by occupational safety and health agencies, such as the US OSHA or Europe's European Chemicals Agency, that govern workplace exposures and ensure worker health. In addition to these administrative controls, engineering controls can be adopted to reduce airborne Cr(VI) concentrations, such as local exhaust ventilation, contained processes, and automated high-risk tasks. PPE (e.g., gloves, respirators, and protective clothing) minimizes direct exposure. Routine medical monitoring programs permit early identification of toxic accumulation and remediation, including biomonitoring of urine, blood, or hair chromium levels. Training workers on proper handling, disposal, and emergency response practices strengthens an occupational safety culture (Bonfiglio et al., 2024; Chen, Xu, et al., 2024; Jeffers et al., 2024; Jegadheeshwari et al., 2024; Taiwo et al., 2024).

Chromium contamination is a widespread problem affecting many communities and can be reduced with environmental remediation. Traditional remediation approaches to treating chromium contaminants include soil excavation and extraction, chemical reduction, and stabilization methods to prevent the diffusion of chromium contaminants into ground waters. Chromium-resistant bacteria and fungi isolated from heavily contaminated sites have bioremediation potential due to the microbial reduction of Cr(VI) to the less toxic Cr(III). New, powerful phytoremediation strategies based on hyperaccumulator plants offer a sustainable, efficient, and economical means of absorbing chromium from contaminated soils. As a result, strict industrial wastewater treatment methods, such as adsorption, ion exchange, membrane filtration, and electrochemical precipitation, are necessary to reduce chromium emissions in aquatic ecosystems. Such behavior must be contained with opportunities for tighter emission limits and incentives for cleaner production technologies to realize a more sustainable future and safeguard environmental and human health from continued chromium emissions (Glicklich et al., 2024; Mohebi et al., 2024; Olsen et al., 2024; Priyadarshini et al., 2024; Shen et al., 2024).

This is a critical first step toward addressing chromium and mobilizing a community to take preemptive action to protect themselves. Educating people about chromium's sources and health risks and methods for reducing exposure are the best ways to simultaneously reduce exposure for people already affected by chromium and prevent future exposure. For example, education for at-risk populations (e.g., workers, residents of contaminated environments, or policymakers) should include information on safe handling during use, disposal, and the need for periodic health screening. Building on the strengths and complementing the limitations of evidence-based and biomedical approaches, community-based participatory research approaches bring researchers, health professionals, and affected communities together to ensure that local needs are considered in any intervention. These platforms enhance knowledge transmission and behavior transformation to mitigate exposure risk and scrutiny from the media. Standard spread through social media, workshops, and public health outreach programmes improves adherence (Chow & Foo, 2024; Gezahegn et al., 2024; Iyer et al., 2023; Olsen et al., 2024; Priyadarshini et al., 2024).

Chromium toxicity is neurotoxic and can be prevented; early recognition and rapid treatment of exposure to chromium may be necessary for positive neurologic outcomes as chromium-induced neurotoxicity is an insidious process that results in progressive cognitive decline, memory impairment and motor deficits, early detection of emerging symptoms of chromium-induced neurotoxicity during this disease process is vital, leading to early medical treatment. Biomarkers of exposure and effect: Urinary chromium, oxidative stress, and inflammatory cytokines, in particular, are valuable as diagnostic tools to detect early-stage toxicity. Neuroimaging techniques, like fMRI and PET, could be used to evaluate the transformations in brain structures and functions caused by chromium, which would aid in both early diagnosis and risk assessment. Given the chronic exposure to chromium, it is worth researching the effects of chelation therapy with agents like ethylenediaminetetraacetic acid and dimercaptosuccinic acid. However, the clinical efficacy and safety of chelation therapy in the treatment of heavy metal toxicity require further verification. Antioxidants, including vitamins C and E, NAC, and polyphenols, may exert neuroprotective activity against oxidative stress and inflammation induced by chromium exposure (Azzam et al., 2023; Chen, Xu, et al., 2024; Erkman et al., 2025; Glicklich et al., 2024; Jegadheeshwari et al., 2024; Priyadarshini et al., 2024).

Health care providers are essential in identifying, managing, and preventing chromium toxicity. Doctors, toxicologists, and occupational health professionals should be trained in the clinical signs associated with chromium exposure to provide appropriate diagnostic measures and treatments. Routine screenings for high-risk subjects, including chromium-exposed workers and those living near contaminated sites, are predictive for early detection and management. Collaborating with environmental scientists and policymakers will provide a multidisciplinary approach to tackling chromium toxicity and its public health implications. Enhancing surveillance systems that detect chromium-related symptoms and developing centralized reporting databases to identify exposure trends and evaluate intervention effects would be beneficial.

Abolition of Environmental Legislation Implemented for public health awareness, regulated levels of toxic chromium in industrial plants. Considering the relatively weak regulatory frameworks in low- and middle-income countries where chromium pollution is a significant risk factor, global cooperation to tackle this contaminant is required to achieve environmental justice worldwide. Epidemiological studies designed to understand the chronic health effects of chromium exposure enhance our knowledge of its pathophysiology, characterize the health effects of environmental exposures, and have direct implications for other relevant evidence-based policy areas. This creates a positive incentive for sectors to switch to green technologies, such as chromium-free solutions for the tanning and plating sectors (Azzam et al., 2023; Jeffers et al., 2024; Mohebi et al., 2024; Nehzomi & Shirani, 2025; Shen et al., 2024).

Chromium toxicity remains a significant public health problem that may have important ramifications for both occupational and environmental health. Addressing the complex nature of chromium-induced toxicity will require a broad interdisciplinary effort encompassing occupational safety laws, environmental cleanup, public education, and early detection. Ensuring workers and communities are not exposed to chromium well calls for strong regulation, technological innovation for pollution control, and comprehensive community prevention engagement. Since chromium is known to be neurotoxic, the factors mentioned previously are critical in the case of chromium toxicity. However, with cooperation between scientists, health professionals, policymakers, and the community, sustainable solutions will emerge that will help protect human health from chromium and ensure environmental and public health in the years to come.

7.8 Future directions and research needs

The exploration of chromium neurotoxicity remains a critical and evolving area of scientific inquiry, necessitating further research into its long-term neurological effects, the mechanisms underlying chromium-induced neurodegeneration, and the development of effective preventive and therapeutic strategies. Chromium, particularly in its hexavalent form (Cr(VI)), has been implicated in various toxicological effects, including oxidative stress, DNA damage, and disruptions in cellular signaling pathways, all contributing to neurodegenerative processes. Despite substantial evidence linking chromium exposure to neurological impairments, a significant gap exists in understanding its long-term consequences on brain function and structure. Future research should focus on longitudinal studies assessing chromium exposure's cumulative and chronic effects, particularly in occupational settings and populations residing in contaminated areas. Such studies should incorporate advanced neuroimaging techniques, biomarker analyses, and cognitive assessments to establish correlations between chromium exposure levels and neurological outcomes over extended periods. Additionally, there is an urgent need to elucidate the molecular and cellular mechanisms through which chromium induces neurodegeneration. Chromium exposure has been shown to trigger oxidative stress and inflammatory responses in neural tissues, leading to mitochondrial dysfunction, apoptosis, and synaptic impairments. Investigating these pathways at a mechanistic level will provide insights into how chromium disrupts normal neuronal functions and contributes to neurodegenerative diseases such as AD, PD, and Huntington's. Understanding these mechanisms could pave the way for identifying potential molecular targets for therapeutic intervention. Given the complexity of chromium neurotoxicity, interdisciplinary collaborations between environmental scientists, toxicologists, and neurologists are essential to advance our knowledge and develop comprehensive approaches to mitigate its impact. Environmental scientists can contribute by identifying chromium contamination sources and levels, assessing human exposure risks, and creating pollution control and remediation strategies. Toxicologists can investigate the dose-dependent effects of chromium exposure on neural tissues, explore bioaccumulation patterns, and test potential detoxification strategies. On the other hand, neurologists can provide critical insights into the clinical manifestations of chromium-induced neurotoxicity, utilizing neurophysiological assessments, cognitive tests, and imaging studies to understand disease progression. Collaborative efforts should also extend to biomedical engineers, chemists, and data scientists to develop innovative detection methods, predictive models, and advanced neuroprotective strategies. Furthermore, it is imperative to translate research findings into actionable public health interventions and environmental policies to protect human health from the adverse effects of chromium exposure. Regulatory agencies and policymakers must be informed about the latest scientific evidence to establish and enforce stringent guidelines on chromium emissions, workplace exposure limits, and permissible levels in drinking water and food sources. Public health initiatives should raise awareness about chromium toxicity, promote regular health screenings for at-risk populations, and implement community-based interventions to reduce exposure. Developing safer alternatives to chromium-containing compounds in industrial applications and improving waste management practices can also significantly minimize environmental contamination. Additionally, integrating chromium neurotoxicity research into educational programs for healthcare professionals can enhance early diagnosis and management strategies for affected individuals. Ultimately, a concerted effort that bridges scientific research, interdisciplinary collaboration, and policy implementation is necessary to address the growing concerns surrounding chromium exposure and its potential long-term impact on neurological health. By prioritizing these research areas and fostering collaboration across disciplines, we can make significant strides in understanding chromium neurotoxicity, mitigating its harmful effects, and ensuring a safer and healthier future for vulnerable populations. The interplay between scientific research and policy-making should be strengthened through initiatives that promote data-sharing, interdisciplinary conferences, and collaborative grant opportunities to accelerate progress in this field. As the global industrial landscape continues to evolve, it is crucial to remain vigilant about the potential neurological risks of chromium and other environmental toxins, ensuring that scientific advancements translate into tangible benefits for public health and environmental safety.

7.9 Conclusion

Chromium, an essential trace element in its trivalent form (Cr(III)), plays a crucial role in metabolic processes. Still, its hexavalent form (Cr(VI)) is highly toxic, posing significant health risks, including severe neurological consequences. While chromium toxicity has been extensively studied for its carcinogenic and systemic effects, its impact on the CNS remains underexplored. Chronic exposure to Cr(VI) significantly affects the nervous system, as it can cross the BBB through active transport mechanisms, leading to oxidative stress, neuroinflammation, and neuronal damage. Cr(VI) acts as a potent inducer of oxidative stress, a key factor in neurodegenerative disorders, as its intracellular reduction generates ROS, leading to mitochondrial dysfunction, impaired synaptic plasticity, and neuronal apoptosis, ultimately contributing to cognitive deficits and neurodegeneration. Moreover, chromium exposure triggers neuroinflammatory responses by activating microglia and astrocytes, which release proinflammatory cytokines such as TNF-α, IL-1β, and IL-6, exacerbating neuronal damage and increasing susceptibility to neurodegenerative diseases, cognitive decline, and mood disorders. Studies suggest that Cr(VI) compromises BBB integrity, increasing vulnerability to neurotoxicants and exacerbating neurological damage. Furthermore, chromium toxicity may alter dopaminergic pathways, reducing dopamine levels and inducing Parkinsonian-like symptoms such as tremors, bradykinesia, and rigidity, indicating a potential link between chromium exposure and PD. Prenatal and early-life exposure to chromium may lead to developmental neurotoxicity, disrupting neurogenesis and synaptogenesis, resulting in cognitive impairments, attention deficits, and behavioral disorders, with potential long-term effects such as reduced IQ and an increased risk of neuropsychiatric conditions like autism spectrum disorders and attentionADD/ADHD. Prolonged exposure to Cr(VI) may contribute to chronic neurological dysfunction, with epidemiological and experimental studies suggesting cumulative neurotoxic effects that manifest later in life. Chromium-induced oxidative stress and neuroinflammation are common pathological features of neurodegenerative diseases such as AD, PD, and ALS, as it may accelerate neuronal loss, exacerbate amyloid-beta aggregation in AD, and impair dopaminergic neurons in PD. Additionally, chromium exposure has been associated with cognitive decline, memory impairment, and deficits in executive function, which are precursors to mild cognitive impairment and dementia. Chromium exposure has also been linked to psychiatric conditions such as depression, anxiety, and schizophrenia, as its disruption of serotonin and dopamine pathways may contribute to mood disorders and altered emotional regulation. Individuals exposed to high chromium levels often report chronic fatigue, muscle weakness, and coordination difficulties due to mitochondrial dysfunction and neuromuscular impairment. Despite the growing evidence of chromium's neurotoxic effects, significant knowledge gaps remain in understanding its precise mechanisms and long-term neurological consequences, necessitating further research to determine safe exposure limits, investigate genetic susceptibility, explore neuroprotective interventions, and develop sensitive biomarkers for early detection. Addressing the neurological impact of chromium toxicity requires a multidisciplinary approach involving regulatory measures, public awareness initiatives, industrial responsibility, healthcare interventions, and international research collaborations. Governments must enforce stricter regulations on industrial emissions, workplace safety, and water quality to minimize chromium exposure, while community education programs should raise awareness about exposure risks and protective measures. Industries using chromium should adopt greener technologies and implement safer waste disposal practices, and healthcare providers should incorporate neurological assessments and early interventions for at-risk populations. Global research partnerships can facilitate knowledge-sharing and accelerate the development of effective countermeasures against chromium toxicity. In conclusion, chromium toxicity poses a significant but often overlooked threat to neurological health, with Cr(VI) accumulation in the brain leading to oxidative stress, neuroinflammation, cognitive decline, and increased risk of neurodegenerative diseases and mental health disorders. Given the emerging evidence, further research is imperative to elucidate chromium's neurotoxic mechanisms and long-term effects fully. Moreover, collaborative efforts among policymakers, researchers, industries, and healthcare professionals are essential to mitigate the neurological impact of chromium toxicity and safeguard public health. Prioritizing preventive measures, stricter regulations, and heightened public awareness can reduce the burden of chromium-related neurological disorders and promote a healthier future for communities worldwide.

References

Ajibo, D. N., Orish, C. N., Ruggieri, F., Bocca, B., Battistini, B., Frazzoli, C., ... Orisakwe, O. E. (2024). An update overview on mechanistic data and biomarker levels in cobalt and chromium-induced neurodegenerative diseases. *Biological Trace Element Research, 202*(8), 3538–3564.

Alquezar, C., Felix, J. B., McCandlish, E., Buckley, B. T., Caparros-Lefebvre, D., Karch, C. M., ... Kao, A. W. (2020). Heavy metals contaminating the environment of a progressive supranuclear palsy cluster induce tau accumulation and cell death in cultured neurons. *Scientific Reports, 10*(1), 569.

Althobaiti, N. A. (2024). Heavy metals exposure and Alzheimer's disease: Underlying mechanisms and advancing therapeutic approaches. *Behavioural Brain Research, 476*, 115212. Retrieved from 10.1016/j.bbr.2024.115212.

Amarloei, A., Nourmoradi, H., Nazmara, S., Heidari, M., Mohammadi-Moghadam, F., & Mazloomi, S. (2025). Toxic heavy metals of agricultural products in developing countries and its human health risk assessment: A study from Iran. *Heliyon, 11*, e40886. Retrieved from https://doi.org/10.1016/j.heliyon.2024.e40886.

Azzam, M. A., Rizwan Khan, M., & Moustafa Youssef, H. (2023). Drinking water as a substantial source of toxic alkali, alkaline and heavy metals: Toxicity and their implications on human health. *Journal of King Saud University—Science, 35*, 102761. Retrieved from https://doi.org/10.1016/j.jksus.2023.102761.

Bartman, S., Coppotelli, G., & Ross, J. M. (2024). Mitochondrial dysfunction: A key player in brain aging and diseases. *Current Issues in Molecular Biology, 46*(3), 1987–2026.

Bonfiglio, R., Sisto, R., Casciardi, S., Palumbo, V., Scioli, M. P., Palumbo, A., ... Scimeca, M. (2024). The impact of toxic metal bioaccumulation on colorectal cancer: Unravelling the unexplored connection. *The Science of the Total Environment, 906*, 167667. Retrieved from https://doi.org/10.1016/j.scitotenv.2023.167667.

Chen, X., Xu, R., & Jiang, M. (2024). In silico prediction of carcinogenic mechanisms induced by mixture of toxic substances from E-waste dust. *Ecotoxicology and Environmental Safety, 288*, 117369. Retrieved from https://doi.org/10.1016/j.ecoenv.2024.117369.

Chen, Y., Fan, Y., Huang, Y., Liao, X., Xu, W., & Zhang, T. (2024). A comprehensive review of toxicity of coal fly ash and its leachate in the ecosystem. *Ecotoxicology and Environmental Safety, 269*, 115905. Retrieved from https://doi.org/10.1016/j.ecoenv.2023.115905.

Chow, Y. N., & Foo, K. Y. (2024). Evaluation of phytotoxicity, biochemical profiles, bioaccumulation hazard, and toxicity implications of battery manufacturing industrial discharge using a soilless cultivation system. *Process Safety and Environmental Protection, 186*, 495–506. Retrieved from https://doi.org/10.1016/j.psep.2024.03.116.

Dhal, B., Thatoi, H. N., Das, N. N., & Pandey, B. D. (2013). Chemical and microbial remediation of hexavalent chromium from contaminated soil and mining/metallurgical solid waste: A review. *Journal of Hazardous Materials, 250*, 272–291.

Erkman, F. T., Yıldırım, M., Sabuncuoğlu, S., Demir, C., Değirmenci, B., Girgin, G., ... Akcan, R. (2025). Heavy metal levels in postmortem blood samples: Unraveling links to suicide and neurological impacts. *Forensic Science International, 367*, 112390. Retrieved from https://doi.org/10.1016/j.forsciint.2025.112390.

Firdous, S. M., Khan, S. A., & Maity, A. (2024). Oxidative stress–mediated neuroinflammation in Alzheimer's disease. *Naunyn-Schmiedeberg's Archives of Pharmacology, 397*(11), 8189–8209. Retrieved from 10.1007/s00210-024-03188-3.

Georgaki, M. N., Charalambous, M., Kazakis, N., Talias, M. A., Georgakis, C., Papamitsou, T., ... Mytiglaki, C. (2023). Chromium in water and carcinogenic human health risk. *Environments, 10*(2), 33.

Gezahegn, T. F., Ambaye, A. D., Mekoyete, T. M., Malefane, M. E., Oyedotun, K. O., & Mokrani, T. (2024). Breakthroughs in nanostructured-based chemical sensors for the detection of toxic metals. *Talanta Open, 10*, 100354. Retrieved from https://doi.org/10.1016/j.talo.2024.100354.

Glicklich, D., Mustafa, M., & Wolfe, K. (2024). Toxic effects of heavy metal exposure in solid organ transplant recipients. *Transplant Reports, 9*, 100151. Retrieved from https://doi.org/10.1016/j.tpr.2024.100151.

Gorini, F., & Tonacci, A. (2024). Metal toxicity and dementia including frontotemporal dementia: Current state of knowledge. *Antioxidants, 13*(8), 938.

Guevara-Ramírez, P., Tamayo-Trujillo, R., Cadena-Ullauri, S., Ruiz-Pozo, V., Paz-Cruz, E., Annunziata, G., ... Zambrano, A. K. (2024). Heavy metals in the diet: Unraveling the molecular pathways linked to neurodegenerative disease risk. *Food and Agricultural Immunology, 35*(1), 2434457.

Gupta, M., Kumar, S., & Jat, M. K. (2022). A regional case study for flow of lead (Pb) and chromium (Cr) through solid waste management system. In *Sustainability of water resources. water science and technology library (vol. 116)*. Springer, 199–218. Retrieved from https://doi.org/10.1007/978-3-031-13467-8_13.

Heng, W., Yong, Y., Jianhang, H., & Hua, W. (2024). A novel method for effective solidifying chromium and preparing crude stainless steel from multi-metallic electroplating sludge. *Journal of Hazardous Materials, 465*, 133068. Retrieved from https://doi.org/10.1016/j.jhazmat.2023.133068.

Iyer, M., Anand, U., Thiruvenkataswamy, S., Babu, H. W. S., Narayanasamy, A., Prajapati, V. K., ... Vellingiri, B. (2023). A review of chromium (Cr) epigenetic toxicity and health hazards. *The Science of the Total Environment, 882*, 163483. Retrieved from https://doi.org/10.1016/j.scitotenv.2023.163483.

Jeffers, M., Kanniyappan, H., Cheng, K. Y., Prithweeraj, S., Perumal, G., Barba, M., ... Mathew, M. T. (2024). Toxicity risk from hip implant CoCrMo particles: The impact of dynamic flow rate on neuronal cells in microfluidic systems. *Toxicology Letters, 402*, 56–67. Retrieved from https://doi.org/10.1016/j.toxlet.2024.10.009.

Jegadheeshwari, S., Santhi, J. J., Velayutham, M., Issac, P. K., & Kesavan, M. (2024). DbGTi protein attenuates chromium (VI)-induced oxidative stress via activation of the Nrf2/HO-1 signalling pathway in zebrafish (*Danio rerio*) larval model. *International Journal of Biological Macromolecules, 280*, 136099. Retrieved from https://doi.org/10.1016/j.ijbiomac.2024.136099.

Kapoor, R. T., Mfarrej, M. F. B., Alam, P., Rinklebe, J., & Ahmad, P. (2022). Accumulation of chromium in plants and its repercussion in animals and humans. *Environmental Pollution, 301*, 119044.

Kaur, R., Garkal, A., Sarode, L., Bangar, P., Mehta, T., Singh, D. P., ... Rawal, R. (2024). Understanding arsenic toxicity: Implications for environmental exposure and human health. *Journal of Hazardous Materials Letters, 5*, 100090. Retrieved from https://doi.org/10.1016/j.hazl.2023.100090.

Lukina, A. O., Boutin, C., Rowland, O., & Carpenter, D. J. (2016). Evaluating trivalent chromium toxicity on wild terrestrial and wetland plants. *Chemosphere, 162*, 355–364.

Mishra, S., & Bharagava, R. N. (2016). Toxic and genotoxic effects of hexavalent chromium in environment and its bioremediation strategies. *Journal of Environmental Science and Health, Part C, 34*(1), 1–32.

Mohankumar, T., Venugopal, D., Palaniyappan, J., Beerappa, R., Duraisamy, E., & Velu, S. (2024). 3 – Environmental exposure to heavymetals in ambient air and its human health implications. In P. K. Shit, D. K. Datta, B. Bera, A. Islam, & P. P. Adhikary (Eds.). *Spatial modeling of environmental pollution and ecological risk: A volume in advances in pollution research* (pp. 41–69). Woodhead Publishing. Retrieved from https://doi.org/10.1016/B978-0-323-95282-8.00028-6.

Mohebi, M., Parashkoohi, M. G., & Mohammadi, A. (2024). A comparison of biofilm and suspension methods in removing heavy metals (chromium) from industrial wastewater with *Scenedesmus obliquus* and *Chlorella vulgaris*. *Results in Engineering, 24*, 103397. Retrieved from https://doi.org/10.1016/j.rineng.2024.103397.

Mutlu, T. (2024). Distribution of toxic and trace metals in fish from the Black Sea: Implications for human health risks. *Emerging Contaminants, 10*, 100295. Retrieved from https://doi.org/10.1016/j.emcon.2023.100295.

Naja, G. M., & Volesky, B. (2017). Toxicity and sources of Pb, Cd, Hg, Cr, As, and radionuclides in the environment. In *Handbook of advanced industrial and hazardous wastes management*. CRC Press, 855–903.

Nehzomi, Z. S., & Shirani, K. (2025). The gut microbiota: A key player in cadmium toxicity – Implications for disease, interventions, and combined toxicant exposures. *Journal of Trace Elements in Medicine and Biology: Organ of the Society for Minerals and Trace Elements (GMS), 88*, 127570. Retrieved from https://doi.org/10.1016/j.jtemb.2024.127570.

Olsen, T., Refsum, H., & Eiser, A. R. (2024). Hyperhomocysteinemia Is associated with a myriad of age-related illnesses: A potential role for metal toxicity. *Mayo Clinic Proceedings, 99*, 1362–1368. Retrieved from https://doi.org/10.1016/j.mayocp.2024.04.016.

Parui, R., Nongthombam, G. S., Hossain, M., Adil, L. R., Gogoi, R., Bhowmik, S., & Iyer, P. K. (2024). Impact of heavy metals on human health. In *Remediation of heavy metals: Sustainable technologies and recent advances*. Wiley, 47–81. Retrieved from https://doi.org/10.1002/9781119853589.ch4.

Priyadarshini, M., Rekha, E., Sathish, A., & Nithya, K. (2024). The relative performance of gelatin hydrogels doped with nanostructured transition-metal ferrites for chromium (VI) removal: Packed bed column studies and insights into reduction mechanisms. *Journal of Industrial and Engineering Chemistry, 137*, 144–161. Retrieved from https://doi.org/10.1016/j.jiec.2024.03.001.

Rhodes, M. C., Hébert, C. D., Herbert, R. A., Morinello, E. J., Roycroft, J. H., Travlos, G. S., ... Abdo, K. M. (2005). Absence of toxic effects in F344/N rats and B6C3F1 mice following subchronic administration of chromium picolinate monohydrate. *Food and Chemical Toxicology: An International Journal Published for the British Industrial Biological Research Association, 43*, 21–29. Retrieved from https://doi.org/10.1016/j.fct.2004.08.006.

Richter-Bisson, Z. W., Doktor, A., & Hedberg, Y. S. (2023). Serum albumin aggregation facilitated by cobalt and chromium metal ions. *ACS Applied Bio Materials, 6*(9), 3832–3841.

Saha, R., Nandi, R., & Saha, B. (2011). Sources and toxicity of hexavalent chromium. *Journal of Coordination Chemistry, 64*(10), 1782–1806.

Saravanan, G., & Ramesh, S. T. (2024). Statistical analysis of seasonal variation in the characteristics of soil like material and refuse derived fuel recovered from landfill mining. *Stochastic Environmental Research and Risk Assessment, 38*(1), 127–146.

Saxena, V. (2025). Water quality, air pollution, and climate change: Investigating the environmental impacts of industrialization and urbanization. *Water, Air, & Soil Pollution, 236*(2), 1–40.

Shen, B., Lu, R., Lv, M., Chen, J., Li, J., Long, J., ... Gong, Z. (2024). Association between the levels of toxic heavy metals and schizophrenia in the population of Guangxi, China: A case-control study. *Environmental Pollution (Barking, Essex: 1987), 363*, 125179. Retrieved from https://doi.org/10.1016/j.envpol.2024.125179.

Siddig, M. M. S., Brevik, E. C., & Sauer, D. (2025). Human health risk assessment from potentially toxic elements in the soils of Sudan: A meta-analysis. *The Science of the Total Environment, 958*, 178196. Retrieved from https://doi.org/10.1016/j.scitotenv.2024.178196.

Sun, J., Luo, Y., Ye, J., Li, C., & Shi, J. (2022). Chromium distribution, leachability and speciation in a chrome plating site. *Processes, 10*(1), 142.

Taiwo, A. M., Adenekan, A.-A. A., Olatunde, K. A., Oladoyinbo, F. O., & Oyedepo, J. A. (2024). Assessment of potentially toxic metals in selected local and foreign rice brands from Southwestern Nigeria: Implications for human health. *Journal of Food Composition and Analysis, 133*, 106428. Retrieved from https://doi.org/10.1016/j.jfca.2024.106428.

Tchounwou, P. B., Yedjou, C. G., Patlolla, A. K., & Sutton, D. J. (2012). Heavy metal toxicity and the environment. In *Molecular, Clinical and Environmental Toxicology. Experientia Supplementum (vol. 101)*. Springer, 133–164. Retrieved from https://doi.org/10.1007/978-3-7643-8340-4_6.

Wong, Y. K., Chan, W. W., Gu, D., Wong, T. W., Chan, K. J. D., Yu, J. Z., ... Lau, A. K. H. (2023). Characterization of toxic air pollutants in Hong Kong, China: Two-decadal trends and health risk assessments. *Atmospheric Environment (Oxford, England: 1994), 314*, 120129. Retrieved from https://doi.org/10.1016/j.atmosenv.2023.120129.

Zuo, M., Ye, M., Lin, H., Liao, S., Xing, X., Liu, J., ... Ren, X. (2024). Mitochondrial dysfunction in environmental toxicology: mechanisms, impacts, and health implications. *Chemical Research in Toxicology, 37*(11), 1794–1806.

Chapter 8

Interactions between heavy metals and neuroinflammatory processes

Debjani Choudhury[1], and Prasann Kumar[2]
[1]*Department of Plant Pathology, School of Agriculture, Lovely Professional University, Phagwara, Punjab, India,* [2]*Department of Agronomy, School of Agriculture, Lovely Professional University, Phagwara, Punjab, India*

8.1 Introduction

Inflammation significantly impacts numerous physiological and pathological processes. It is an adaptive response to damaging stimuli such as toxic compounds, infections, pathogens, and radiation (Medzhitov, 2008). Numerous cell types and signaling pathways are involved in this intricate process. Acute and chronic inflammation are the two categories of inflammatory processes. The body's defense mechanism, acute inflammation, is triggered by pathogens or tissue damage. Conversely, chronic inflammation is defined as persistent inflammation brought on by disruptions in the resolution program or failure to eliminate the initiating stimulus (Headland & Norling, 2015). Acute inflammatory diseases are characterized by cellular and molecular events that usually resolve independently, as opposed to autoimmune and neurodegenerative lesions resulting from a failure to resolve chronic inflammation. In chronic inflammation with unsuccessful resolution and acute inflammation with a self-limited resolution, making sense of the interaction between parenchyma cells and immune cells is a key to understanding the inflammation-repair process. In chronic inflammation with unsuccessful resolution and acute inflammation with a self-limited resolution, making sense of the interaction interlayer between parenchyma cells and immune cells is key to understanding the inflammation-repair process. Depending on stimuli, harmful factors, pathogens, etc., the change from acute to chronic inflammation may occur over weeks or even months. Chronic inflammation contributes significantly to the onset and advancement of many illnesses, such as cancer and cardiovascular diseases (Furman et al., 2019). Numerous studies have linked inflammation to the development of several neurological conditions, such as Alzheimer's disease (AD), Parkinson's disease (PD), Huntington's disease (HD), multiple sclerosis (MS), and amyotrophic lateral sclerosis (ALS). Furthermore, inflammation is a significant factor in the pathophysiological processes related to metal-induced neurological disorders (Garza-Lombó, Pappa, Panayiotidis, Gonsebatt, & Franco, 2019). Inflammation caused by metals has lately been recognized as a key pathophysiological mechanism of metals, but the pathophysiological functions of inflammation in neurological disorders and neurodegenerative diseases remain unclear. Thus this manuscript aims to compile research on inflammation caused by metals and the pathophysiological functions of inflammation in neurological disorders and neurodegenerative diseases. This chapter mainly focuses on inflammation brought on by different metals that can harm the nervous system (Fig. 8.1). To increase awareness of the connection between inflammation and neurological disorders, we briefly review how inflammation can aid in developing and progressing neurodegenerative and neurological disorders.

8.2 Neuroinflammatory diseases

In neuroinflammatory diseases, peripheral immune cells can enter the brain, and cells in the central nervous system (CNS) react to peripheral inflammatory stimuli. It is still unknown how these neurologic issues have differing degrees of recovery and aftereffects. An important mechanism shared by several neuropathological conditions is inflammation.

8.2.1 Neuroinfectious diseases

Often referred to as the first line of host defense, innate immunity quickly responds to infections, boosting adaptive

Heavy Metal Toxicity and Neurodegeneration. DOI: https://doi.org/10.1016/B978-0-443-36575-1.00007-8

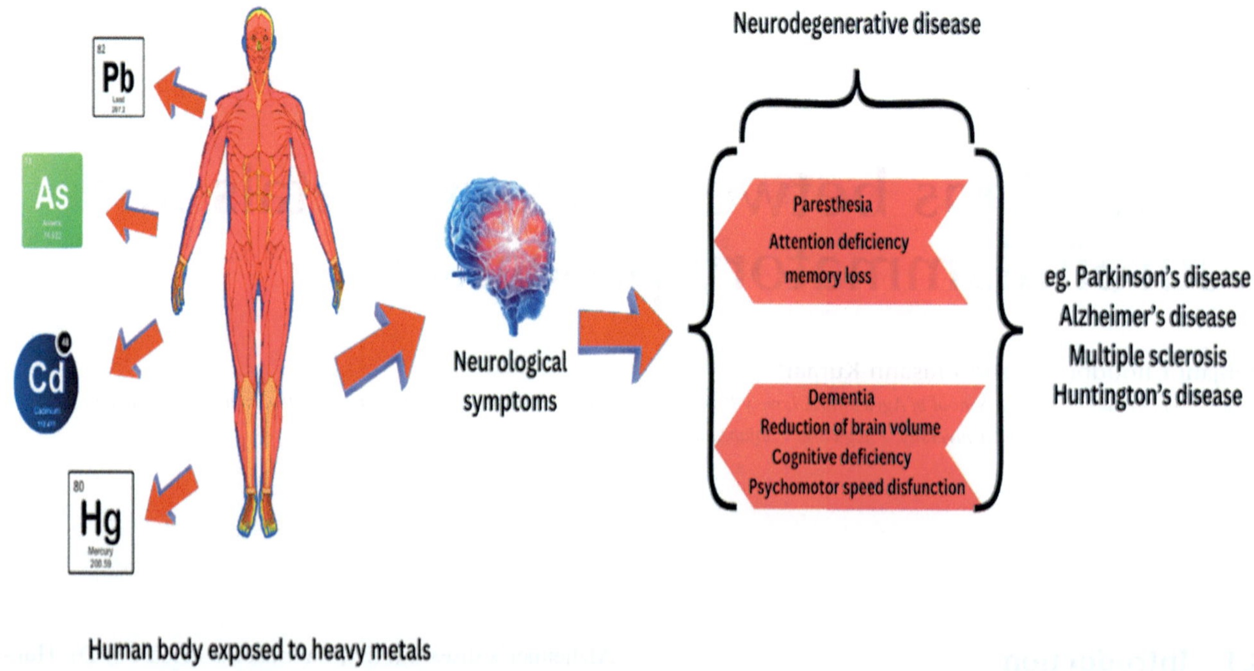

FIGURE 8.1 Heavy metals causing neurodegenerative disease.

immune responses. Specific innate immune cells and molecular pathways appear to trigger inflammatory infections. They must be functionally effective in preventing the spread of brain injury. A hallmark of CNS infections by neurotropic pathogens is neurologic dysfunction accompanied by an abrupt change in mental status brought on by inflammation (Johnson & Nath, 2018). Irreversible damage brought on by pathogens alone has been implicated in postinfectious neurologic dysfunction (Sahu & Ter, 2018; Wu et al., 2020).

Certain parasitic infections, such as Chagas disease, toxoplasmosis, human African and American trypanosomiasis, echinococcosis, cysticercosis, leishmaniasis, onchocerciasis, schistosomiasis, food-borne trematodiasis, dracunculiasis, filariasis, and soil-transmitted helminthiasis, have been shown to exhibit neurologic involvement and symptoms (Peón et al., 2017).

In recent years, infectious pathogens have been shown to play a significant role in several neurological autoimmune-related disorders. Neuroinflammation can be brought on by several systemic and organ-specific autoimmune diseases, primarily rheumatic. Examples include systemic lupus erythematosus, destructive joint diseases (Oono et al., 2018), and fibromyalgia (Abd-Ellatief, Mohamed, & Kotb, 2018). According to research on neurodegeneration, peripheral infection may have a role in the development and course of age-related neurodegeneration (Orellana et al., 2015).

Older patients seem to be more susceptible to the infection-related cognitive alterations linked to AD. While certain viral, bacterial, and fungal pathogens have also been linked to the development of disease, common infectious problems like respiratory tract infections may also cause this. It is currently unknown if these microbes are opportunistic pathogens that colonize dementia patients and exacerbate the preexisting ongoing inflammation (Castro-Gomez, Binder, & Heneka, 2019; McManus & Heneka, 2017) or if they are directly linked to the progression of AD.

8.2.2 Nonpathogen-associated central nervous system diseases

Progressively impairing cognitive and motor abilities, neurodegenerative diseases make it difficult to perform daily tasks. According to developments in genetics and animal models, the immune system plays an unexpected role in the pathophysiology and onset of diseases. There is ongoing research on the function of immune signaling pathways, growth factors, and cytokines in the pathophysiology of diseases (Hammond, Munkacsi, & Sturley, 2019).

Within hours or days, traumatic brain injury (TBI) triggers a strong immune response (Plesnila, 2016). Patients and TBI animal models have shown peripheral immune cell infiltration to the injured tissue along with activation of brain resident astrocytes and microglia; in a murine model of acute TBI, regulatory T cells have been shown to reduce neuroinflammation, T lymphocyte brain infiltration, reactive astrogliosis, interferon-γ gene expression, and temporary motor deficits (Krämer et al., 2019).

Environmental factors that impact the CNS during pregnancy and the postpartum period cause astrogliosis and microglia activation, increased proinflammatory cytokines and are closely linked to neuroinflammation (Baburamani, Supramaniam, Hagberg & Mallard, 2014; Supramaniam et al., 2013). It is a defining feature of infections as well as a side effect of noninfectious insults, such as cerebral hypoxia-ischemia (Maj, Mallard & Hagberg, 2004). Remarkably, inflammatory brain glial cells seem to play a key role in controlling the structure and function of synapses. According to research on synaptic physiology and pathophysiology, the immune system has a dynamic impact on the structure of neurodevelopmental synapses (Mottahedin et al., 2017; Riccomagno & Kolodkin, 2015). Even though brain tissue is rarely exposed to dangerous substances, it has a limited capacity for cell repair. Innate and adaptive immune molecules and receptors mediating tissue damage and repair are among the immune glial system interactions that are becoming better understood (Healy, Yaqubi, Ludwin, & Antel, 2020).

8.3 Molecular mechanisms of metal-induced inflammation

One of the chemical elements to which people are exposed daily is metal, which can impact health and cause illness (Grandjean & Landrigan, 2014). Metals are utilized extensively in many facets of life and are found naturally in the environment (Pandey, Singh, Yadav, & Mahdi, 2018). Research on the toxicity of neurotoxic metals and their involvement in different neurodegenerative diseases is becoming increasingly prevalent. The exact mechanisms by which they cause various neurodegenerative diseases are still unknown, however. Numerous metals may cause varying degrees of neuroinflammation, raising proinflammatory cytokine protein expressions and mediating neuronal toxicity (Furman et al., 2019). Here, we provide an overview of how inflammation contributes to neurological conditions brought on by metal exposure.

However, the exact mechanisms by which they cause various neurodegenerative diseases are still unknown. Numerous metals may cause varying degrees of neuroinflammation, raising proinflammatory cytokine protein expressions and mediating neuronal toxicity (Furman et al., 2019). Here, we provide an overview of how inflammation contributes to neurological conditions brought on by metal exposure.

8.4 Inflammation's role in neurological disease

The body's natural defense mechanism against damage or infection includes inflammation. Neuroinflammation, or nervous system inflammation, is a complicated phenomenon that can have beneficial and harmful effects on various neurological conditions. On the other hand, persistent neuroinflammation can be detrimental. In chronic conditions, feeding a vicious cycle of neuronal degeneration and damage. The hallmark of chronic neuroinflammation is the overactivation of glial cells, such as astrocytes. These cells release proinflammatory cytokines, chemokines, and ROS, which can exacerbate neuronal damage and cause the progressive neural deterioration observed in diseases like AD, PD, HD, and MS (Gąssowska-Dobrowolska et al., 2023). Because of this disruption, inflammation and neuronal damage are sustained as peripheral immune cells enter the CNS. Furthermore, the relationship between inflammation and neurodegenerative processes is complicated by aging and gut microbiota imbalances (Jain, Madkan, & Patil, 2023).

8.4.1 Mechanisms of neuroinflammation

The innate and adaptive reactions to dangerous conditions like infections, ischemia, stress, and trauma are referred to as neuroinflammation (Gorji, 2022). Four key characteristics are considered distinguishing characteristics in neuroinflammation: increased cytokine release, microglial cell activation, peripheral immune cell migration, and localized tissue damage (Konsman, 2022) (Fig. 8.2). Although neuroinflammation starts as a defensive response, research has shown that chronic or excessive inflammation plays a significant role in the emergence of some neurological conditions, particularly degenerative diseases (Kwon & Koh, 2020).

8.4.2 Cytokine production

Cytokines are crucial in communicating between the innate and adaptive immune systems. Glial cells, blood components, and peripheral tissues are among the cells in the CNS that are capable of producing cytokines (Seif et al., 2017). Cytokines help the cells to communicate with one another and coordinate intricate multicellular behavior (Kumari et al., 2024). Through various pathways, such as excitotoxicity, oxidative stress, HPA–axis activation, matrix metalloproteinase secretion, and apoptosis, cytokines can cause neuronal damage and death. Additionally, they may intensify the inflammatory response, resulting in additional neuronal damage (Kumari et al., 2024).

The CNS and peripheral tissues typically exhibit low cytokine levels. However, through various active transport mechanisms, peripheral cytokines and immune cells such

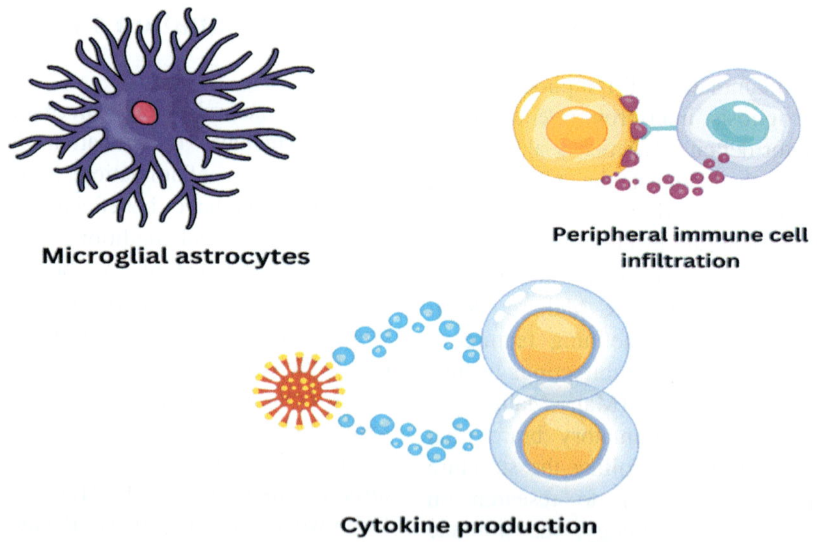

FIGURE 8.2 Mechanisms of neuroinflammation.

as T cells cross a protective blood–brain barrier (BBB) in response to immunological challenges. Peripheral cytokines also affect the CNS by diffusing across the BBB. Furthermore, cytokines are secreted by brain mast cells and microglia, which play a significant role in neuroinflammation (Lotfi et al., 2019).

8.4.3 Microglial activation

The primary line of defense in the brain is provided by microglia, resident macrophages of the CNS that phagocytose dangerous pathogens and cellular debris. Up to 20% of the glial population is made up of microglia. When the CNS is damaged or infected, microglia are typically activated and migrate toward the infection site (Qin, Ma, Chen, & Shu, 2023). They are also in charge of eliminating diseased cells and releasing various cytotoxins that defend against invasive substances. However, if microglia become overactive, the chemicals they release could end up being harmful to the healthy tissues nearby (Gao, Jiang, Tan, & Chen, 2023). Before a fully developed BBB is established, microglia enter the developing brain after emerging from the yolk sac's early erythromyeloid progenitors. Microglia support CNS homeostasis during brain development under physiological conditions by promoting the cell proliferation of neural precursors (Muzio, Viotti, & Martino, 2021). Microglia can either promote nerve growth or cause neurotoxicity, depending on how they are activated. Under normal circumstances, microglia carry out a variety of tasks that allow them to recognize and assess inflammatory cues in their surroundings, encourage synaptic plasticity and neuronal function, and eliminate cellular debris via phagocytosis (Shao, Wang, Wu, Wu, & Zhang, 2022). Maintaining general brain health and function requires these activities. Neuroinflammation, acute injuries like stroke and head trauma, as well as AD, PD, and chronic traumatic encephalopathy, can all be caused by microglia, which can become hyperactive in pathological states (Dhaiban et al., 2021). Microglial activation demonstrates that the CAN's response to both acute and chronic challenges is dynamic and adaptable. Inhibiting this aberrant microglial activation is crucial to reducing neuroinflammation and may offer a novel therapeutic strategy for several CNS disorders.

8.4.4 Peripheral immune cell infiltration

Neutrophils, macrophages, and T lymphocytes that infiltrate the peripheral immune system can also cause neuroinflammation in addition to resident immune cells. Peripheral immune cells like neutrophils, monocytes, NK cells, dendritic cells, T-cells, and even B-cells are essential for the immune response (Kanashiro et al., 2020). Neutrophils use phagocytosis to engulf microorganisms surrounding infected areas as a part of the immune response. Neutrophil migration into the CNS has been noted under certain pathological conditions, including neurodegeneration (Turner & Sharp, 2016). According to these studies, these neutrophils release matrix metalloproteinases, Proteolytic enzymes, and free radicals, aiding the BBS's disintegration (Guo, Zeng, & Gao, 2021). Second, neutrophil extracellular traps are released into the extracellular space as a result of postneutrophil activation and BBB damage, further compromising the BBB's integrity

(Zang, Chen, Zhu, Ma, & Zhai, 2022). These intricate webs are the result of neutrophil degranulation. The release of IL-17 by neutrophil migration to the cortex and hippocampus results in cytotoxicity and BBB disruption. Consequently, the damage is exacerbated, and more neutrophils are drawn to the area (Spiteri et al., 2022). The CNS contains monocytes, a type of white blood cell that can differentiate into monocyte-derived dendritic cells and monocyte-derived macrophages (Ochocka & Kaminska, 2021). The brain barrier allows monocyte-derived macrophages to enter the CNS. These infiltrating blood-borne monocytes are essential for the CNS's demyelination process and help release neurotrophic and antiinflammatory substances that control neuropathic events. Reduced acute and chronic microglia-mediated inflammation is a result of monocyte-derived macrophages. Chemokines cause NK cells to migrate toward the CNS in pathological conditions (Rivas-Fuentes, Salgado-Aguayo, Arratia-Quijada, & Gorocica-Rosete, 2021). Peripheral dendritic cells can migrate to the meninges or choroid plexus due to the release of chemokines and adhesion molecules. They can identify antigens and present them to T-cells, activating them. It is crucial to remember that there are significantly more B-cells in the CSF when there is CNS inflammation. Furthermore, B-cells considerably multiply in the perivascular space and brain parenchyma (Spiteri, Wishart, Pamphlett, Locatelli, & King, 2022). This highlights how peripheral immunoglobulin cells contribute to the complex immune responses in an inflammatory CNS.

8.4.5 Astrocytes

Astrocytes are abundant glial cells actively involved in health and illness in the CNS. New technologies have revealed the diverse range of functions of CNS astrocytes and their relationships with other cell types. Two different types of reactive astrocytes were discovered in the early phase of research into the heterogeneity of astrocyte responses: Neurotoxic astrocytes, known as A1 astrocytes, and neuroprotective astrocytes known as A2 astrocytes (Reid & Kuipers, 2021). Astrocytes preserve the CNS's homeostasis and react dynamically to illness or injury. Astrocytes become reactive during neuroinflammation, a condition marked by morphological alterations and increased glial fibrillary acidic protein. When astrocytes are in their reactive state, they can create and release various inflammatory mediators, such as cytokines and chemokines, which affect how the nearby immune cells and neurons behave. This can be protective by encouraging the removal of pathogens and debris and speeding up repair procedures (Ding et al., 2021). Astrocytes regulate several nervous system functions in both healthy and unhealthy conditions. The study of astrocytes in both states has been made possible by the realization that astrocyte reactivity is a heterogeneous process that causes a range of molecular, cellular, and functional alterations. Particular genomic programs and functions are linked to specific astrocyte subsets or activation states. Rapid advancements in the characterization of astrocyte heterogeneity and its regulation by astrocyte interactions with other cells in the CNS have been made possible by the emergence of novel genomic technologies (Khakh & Deneen, 2019). Because of their functions in CNS metabolism and neurotransmitter homeostasis, astrocytes also impact neurodegeneration. Astrocyte dysfunction can disrupt excitotoxicity by restricting glutamate uptake, interfering with the metabolic support of neurons, and changing synaptic function in neurodegenerative diseases, all of which worsen neuronal loss and cognitive decline (Satarker et al., 2022). Furthermore, astrocytes interact with microglia and other glial cells, which may intensify neuroinflammatory reactions and affect the course of the disease. Their phagocytosis ability, which is crucial for eliminating dead cells and myelin debris, can also go awry, causing persistent inflammation and impeding the body's natural healing mechanisms (Sun et al., 2023).

8.5 Environmental factors responsible for inflammation

Various factors are responsible for neuroinflammation (Fig. 8.3).

8.5.1 Traumatic brain injury

TBI causes a broad neuroinflammatory reaction that can aid in recovery if managed for a predetermined time. However, a stable low-grade irritation could result from inadequately repaired traumatized tissue, leading to chronic inflammation and ongoing damage to the surrounding tissue. Indeed, in certain TBI patients, elevated microglia activation has been discovered years after the injury. Neurodegeneration and neurological impairment are the ultimate results of chronic neuroinflammation (Djordjevic, Golam Sabbir, & Albensi, 2016). It is interesting to note that 30% of people who die suddenly from TBI have diffuse Aβ plaques in their brains (Johnson, Stewart, & Smith, 2010). Furthermore, it was discovered that TBI causes APP to build up in damaged axons (Djordjevic et al., 2016). Additionally, TBI has been linked to an increased risk of ALS and PD (Delic, Beck, Pang, & Citron, 2020; Franz et al., 2019).

8.5.2 Systemic inflammation

According to reports, people who have higher blood levels of inflammatory proteins in their midlife are more likely to develop neurodegenerative disease (Walker et al., 2019). Similarly, smaller brain volumes and aberrant white matter

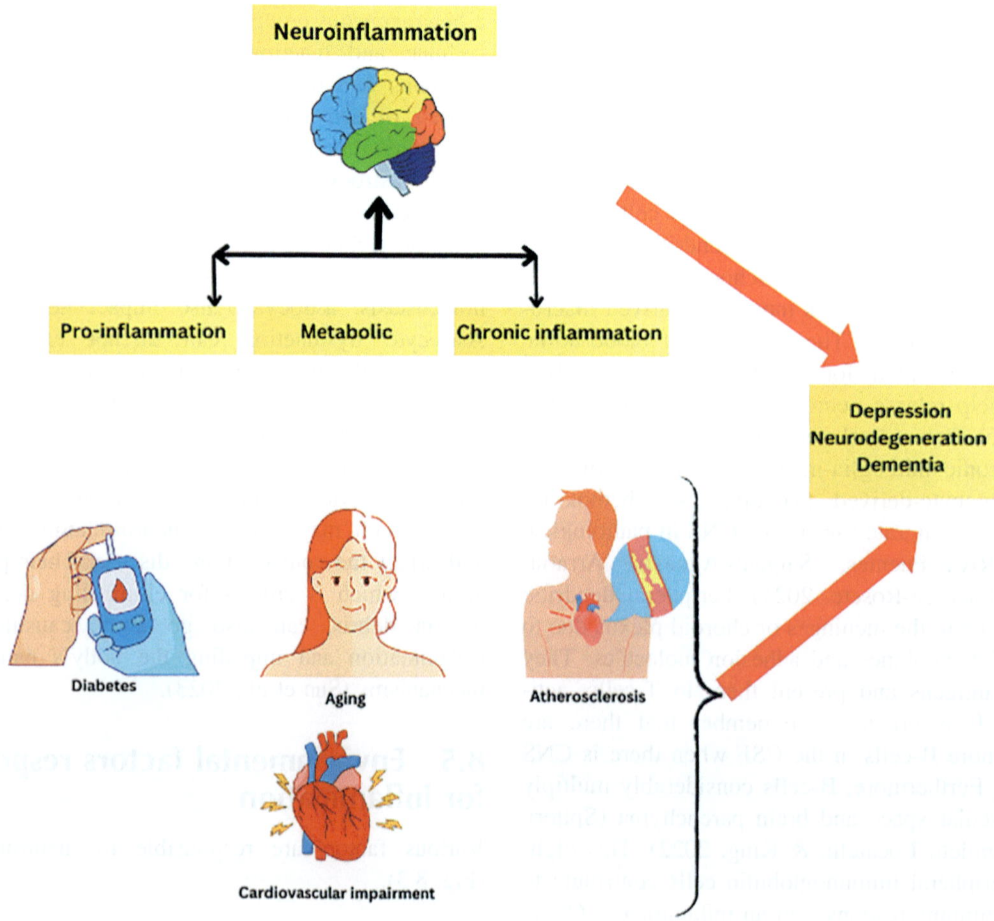

FIGURE 8.3 External factors responsible for neuroinflammation.

microstructural integrity more than 20 years later, in late life, have been linked to midlife increases in inflammatory proteins (Walker et al., 2023). All of these findings point to the possibility that systemic inflammation from the peripheral immune system, such as that brought on by infection or disruption of the gut microbiota, may accelerate the development of neurodegenerative processes decades before dementia typically manifests (Xie, Van Hoecke, & Vandenbroucke, 2022). Peripheral inflammatory molecules brought on by acute systemic viral or bacterial infections have been shown in numerous clinical and experimental investigations to impact brain function via humoral and neural pathways (Castanedo-Vazquez, Bosque-Varela, Sainz-Pelayo, & Riancho, 2019; Smeyne, Noyce, Byrne, Savica, & Marras, 2021). Numerous recent studies have documented the connection between gut microbiota and neurodegenerative disease. An imbalance in the gut microbiota causes inflammation in the intestine, which in turn triggers systemic inflammation by releasing metabolites and immune-stimulating chemicals into the bloodstream (Colombo et al., 2021). Furthermore, by attaching to particular receptors, like proinflammatory tachykinin/neurokinin receptors in the vagus nerve fiber, gut bacteria can also influence CNS inflammation (Abdel-Haq, Schlachetzki, Glass, & Mazmanian, 2019).

8.5.3 Chronic inflammation

Epidemiologically speaking, chronic conditions linked to chronic inflammation, such as diabetes, obesity, atherosclerosis, and depression, raise the risk of neurodegeneration. A chronic, low-grade state of systemic inflammation was discovered to be linked to obesity (Zatterale et al., 2020). Increased levels of free fatty acids in the bloodstream caused by dysregulated lipid metabolism in obese patients can attach to PRRs on the immune cells and trigger the release of proinflammatory cytokines such as interleukin-6 (IL-6), tumor necrosis factor α (TNF-α), adipokines, and monocyte chemoattractant protein-1 (Litwiniuk, Bik, Kalisz, & Baranowska-Bik, 2021) interrupt mitochondrial function and increase ROS production (Hoeks & Schrauwen, 2012). As a result, these inflammatory mediators can penetrate the brain, cause chronic, long-lasting neuroinflammation, and ultimately lead to neurodegeneration by disrupting the BBB (Litwiniuk et al., 2021).

8.5.4 Aging

One of the leading environmental factors contributing to many neurodegenerative diseases is aging, also associated with low-grade systemic inflammation known as "inflammaging" (Tansey et al., 2022). The repeated activation of microglia and astrocytes throughout a person's life results in oxidative stress, free radical damage, and mtDNA accumulation, increasing the likelihood that they will exhibit a primed phenotype and morphology. They show an increased inflammatory state at baseline, a more intense proinflammatory response following stimulation, and a decline in homeostasis maintenance (Lin et al., 2018). Peripheral inflammation at the systemic level causes the circulation to produce inflammatory mediators at low levels, such as TNF-α, IL-6, and C-reactive protein (CPR), which in turn fuel the inflammatory milieu in the CNS (Costantini, D'Angelo, & Reale, 2018; Tansey et al., 2022).

8.5.5 Body clock

One distinctive biological process seen in organisms at both the physiological and behavioral levels is the body clock. A common symptom of several neurodegenerative diseases is disruption of the sleep/wake cycle (Brundin, Nath, & Beckham, 2020). Modulating the immune system's time-specific activities is one of the rules of the body clock. Long-term evolution has led to the development of the circadian cycle, which is the regular variation in biological activity with the day–night cycle as its period. Research has shown that the inflammatory response of the body and immune system are directly correlated with the circadian cycle disruption (Xu, Wang, & Chen, 2023). Recent research has indicated that the circadian cycle also controls the inflammatory response in the CNS. The recovery of neural function in a focal brain ischemia model was impacted, and mice exhibited anxiety and depressive behavior as a result of the markedly elevated levels of activated microglia and proinflammatory cytokines in brain tissue observed in animal models exposed to night shift work or night light (Fonken et al., 2019; Guerrero-Vargas, Zárate-Mozo, Guzmán-Ruiz, Cárdenas-Rivera, & Escobar, 2021). Melatonin, a hormone the brain naturally produces to control the cycles of day and night, has been shown to have a direct correlation with neurodegeneration and circadian rhythms (Chen, Zhang, & Lee, 2020).

8.5.6 Cerebrovascular impairment

It has been proposed that neurodegenerative diseases are linked to cerebrovascular impairment. Additionally, peripheral immune cells are more likely to infiltrate the CNS due to cerebrovascular impairment. These cells then work in tandem with resident immune cells to cause neuroinflammation (Takata, Nakagawa, Matsumoto, & Dohgu, 2021). The BBB and bold-cerebrospinal fluid (CSF) barrier are the two crucial specialized boundaries that separate the blood from the CSF or parenchymal tissue of the CNS, shielding the latter from dangerous substances in the blood and granting immunity (Solár, Zamani, Kubíčková, Dubový, & Joukal, 2020).

8.6 Role of heavy metals in neuroinflammation

8.6.1 Aluminum (Al)

After silicon and oxygen, aluminum is the third most abundant chemical element in the Earth's crust and is thought to be the most abundant metal. Environmental Al levels have significantly increased due to the metaconsideredl's widespread use and the intensive growth of the Al industry (Brough & Jouhara, 2020). Common toxic properties of aluminum, such as prooxidant, proinflammatory, and proapoptotic activity, have been documented for a range of cell lines and tissues. More specialized neurotropic effects, such as disruption of neurotransmitter metabolism and neuronal cytoskeleton, typically mediate the neurotoxic effect of aluminium.

It was demonstrated that exposure to Al mitochondrial Mn-SOD activity, in turn, led to mitochondrial dysfunction. However, some research showed that Al increased SOD activity and gene expression (Ali, Afifi, Abdelazim, & Mosleh, 2014), which may also lead to oxidative stress by producing too much hydrogen peroxide, mainly when catalase and GPx activity are low. Al-induced downregulation of the Nrf2/Keap1 signaling pathway is probably that mediates the inhibitory effect of Al on the antioxidant system (Yu et al., 2019), whereas preventing this inhibitory effect improved the metal's prooxidant activity (Wang et al., 2017).

Al hydroxide's proinflammatory effect has been linked to inflammasome activation and microglia's subsequent release of IL-1β (Gustin et al., 2015). Apurinic/apyrimidinic endonuclease 1 may also be regarded as a negative regulator of Al neuroinflammation, according to the findings of another study (Zaky, Mohammad, Moftah, Kandeel, & Bassiouny, 2013). Notably, AI-induced inflammatory neurodegeneration may also be facilitated by the induction of systemic inflammation, which is marked by increased levels of proinflammatory cytokines

(IL-6, TNF-α) and microRNAs (miRNA-9, miRNA125b, and miRNA-146a) (Pogue, Jaber, Zhao, & Lukiw, 2017).

AL exposure has been demonstrated to raise COX-1 and COX-2 activity in the cortex and /or hippocampus in addition to the overproduction of proinflammatory cytokines (Syed, Muhammad Faisal Ikram, & Ahmed, 2015). According to Wang et al. (2015), AL-induced COX-downstream signaling entails elevated prostaglandin levels in the hippocampus as well as prostaglandin synthase and receptor expression (Wang et al., 2015).

A critical aspect of neuroinflammation is thought to be AL-induced (micro) gliosis (Blaylock, 2012). In the rat brain, exposure to AL was mainly found to cause oxidative stress-dependent glial activation (Akinrinade, Memudu, & Ogundele, 2015) with the stratum and thalamus showing the most pronounced effects (Platt, Fiddler, Riedel, & Henderson, 2001). Nanoscaled silica was also demonstrated to cause glial activation in the rat brain cortex and hippocampus as evidenced by an increase in ED1, GFAP, and nestin-positive cells (Li et al., 2009). In contrast to neurons, glial cells accumulate noticeably more iron and aluminum while also being less susceptible to oxidative stress caused by metals (Oshiro et al., 2000). These results also line up with a recent finding that a patient with cerebral amyloid angiopathy had higher levels of AL disposition in microglia and other inflammatory cells (Mold, Cottle, King, & Exley, 2019).

8.6.2 Cadmium (Cd)

One of the most hazardous industrial heavy metals, cadmium, is found in large quantities in the earth's crust and has no physiological purpose. It is frequently utilized in industries as an impurity in zinc (Zn) and lead (Pb) (Jensen & Bro-Rasmussen, 1992). Long-term exposure to Cd can have negative effects on most organ systems. It is absorbed through the inhalation of tobacco products or the consumption of contaminated food and water (Ragan, 1983). It has a long half-life in plants and animals, up to 25 years. Once absorbed, it enters the blood and builds up throughout the body, particularly in the liver and kidneys (200). Although oxidative stress, apoptosis, and mitochondrial dysfunction seem to be involved, the exact mechanisms underlying Cd-induced neurotoxicity are still unknown (Zhao et al., 2020).

Multiorgan toxicity from chronic cadmium buildup mainly affects the kidney, skeleton, liver, and nervous system (Shao et al., 2022). Among these, the nervous system is especially susceptible to cadmium poisoning. Poor performance on visuomotor tasks, altered equilibrium, and peripheral neuropathy are all risks associated with cadmium (Matsubara et al., 1990). Reduced concentration, worse cognitive function in older adults, and poor learning outcomes in children are all linked to cadmium exposure (Kippler et al., 2012). Additionally, exposure has been linked to neurodegenerative disease patterns seen in ALS, PD, and AD (Cornejo et al., 2018; Demirci, Aynalı, Demirci, Demirci, & Arıdoğan, 2017).

Cadmium has been implicated in several studies as a competitive inhibitor of voltage-gated calcium channels (VGCCs) (Kaur, Sharma, & Deshmukh, 2019; Wang, Chen et al., 2022; Wang, Cheng et al., 2022). As it competes with Ca^{2+} for space in the channel pore, cadmium mainly enters the rat cerebellar granular neuron through dihydropyridine-sensitive (L-type) VGCCs. Exposure to 100 μM cadmium prevented an increase in cytosolic calcium concentration after neuronal depolarization, and cadmium could permeate the neuron. The N-type VGCC is also implicated in cadmium-induced blockage of Ca^{2+} current in frog sympathetic neurons (Szekely et al., 2004). At voltages where Ca^{2+} channels are mainly open (0 to +30), cadmium completely and quickly blocked Ca^{2+} current, suggesting that the N-type VGCC is a pathway for cadmium entry into sympathetic neurons. The presynaptic terminal is a prominent site of cadmium uptake in neuronal cells due to the high concentration of VGCCs there (Yu et al., 2015). According to in vitro studies, a key factor involved in neurotoxicity is neuroinflammation (Huat et al., 2019). According to an in vitro study by triggering the MAPK and NF-kB pathways, Cd enhanced the release of inflammatory factors in U-87 MG human astrocytoma cells and primary human astrocytes (Phuagkhaopong et al., 2017; Yuan et al., 2015). By triggering the NF-κB pathways in glial cells, an in vitro study showed that Cd caused neuron toxicity (Yang et al., 2007). Additionally, in vivo research revealed that Cd triggered the NF-κB pathway and NLRP3 inflammasome-dependent pyroptosis in swine brains, followed by an increase in IL-1β and IL-18 release (Cai et al., 2021). Regarding cadmium toxicity, the mitochondria have become the primary targets. This is supported by experimental evidence in a rodent model where cadmium exposure to isolated mitochondria from mouse livers led to extensive organelle damage (Tang & Le, 2016). Cadmium interferes with particular protein complexes within the ETC and impairs mitochondrial function. Additionally, the permeability transition pore (PTP), a dynamic protein complex located at the interface between the inner and the outer mitochondrial compartments, opens in response to cadmium (Lee et al., 2010). Opening the PTP also signals apoptosis by releasing cytochrome *C* stores. The PTP may open because the inner mitochondrial gradient becomes weaker in a feedforward mechanism that ultimately leads to cell death. The degree to which cadmium opens the PTP is unclear. Furthermore, several studies have demonstrated that specific medications protect against Cd-induced neurotoxicity by preventing the activation of neuroinflammation (Alam et al., 2021).

According to the data mentioned previously, neuroinflammation is a key factor in Cd-induced neurotoxicity.

8.6.3 Lead (Pb)

Globally, lead, a hazardous heavy metal, has contaminated the environment and caused health issues. Human exposure to Pb happens through ingestion, inhalation, and skin absorption from contaminated soil, water, food, and consumer goods (Machoń-Grecka, Dobrakowski, Kasperczyk, Birkner, & Kasperczyk, 2022). Lead poisoning primarily affects children, and it can cause functional disorders or dysfunction in a variety of organs and systems (Reuben, 2018). A cross-sectional, retrospective study of more than 1.1 million showed that 50% of children had been detected with Pb levels in the blood (Hauptman, Niles, Gudin, & Kaufman, 2021). Studies showed that higher Pb concentration in blood levels during childhood may also reduce cortical surface area and hippocampal gray matter volume in midlife (Reuben, 2018). These findings imply that Pb results in learning, behavior, and language impairments as well as irreversible developmental harm (Allen, 2015; Ettinger et al., 2004). A growing body of research indicates that there are no safe levels of Pb and that even small amounts of the metal can cause cognitive impairment (Jakubowski, 2011). Overexposure to lead in the environment continues to be a serious global public health issue. Pb-induced neurotoxicity has been linked to several molecular processes over the last few decades, including oxidative stress, neurotransmitter metabolism disorders, and mitochondrial dysfunction (Garza, Vega, & Soto, 2006).

Neuroinflammation is strongly implicated in the pathophysiology of neurodegenerative diseases, such as AD and PD, according to a wealth of evidence (Kumawat, Kaushik, Goswami, & Basu, 2014). According to epidemiological research, the white blood cell count rises by 1.7% for every 1% increase in blood Pb levels among workers exposed to Pb occupationally (He et al., 2021).

According to in vitro research, Pb may activate primary cultures of rat microglia and astrocytes and reduce cell viability. According to microarray data, Pb raised the expression of GFAP in certain brain regions (including the hippocampus, frontal cortex, and hypothalamus) of BLAB/c mouse pups, along with an increase in the proteins IL-6, TGF-β1, and IL-18 (Kasten-Jolly, Heo, & Lawrence, 2011). According to Li et al., mice's hippocampal Pb accumulation on postnatal day 21 was boosted by maternal Pb exposure, which also resulted in EFAP, IL-1β, and IL-6 protein expression (Li et al., 2015). Additionally, Pb can trigger astrogliosis and microgliosis, which results in the dysregulation of traditional inflammatory pathways like TGF-β1, p38, p42/44 MAPK, TLR4-MyD88-NF-κB, and p38 (Liu, Chen, Zhang, Kuang, & Chen, 2015; Lu et al., 2022). According to Kanhaiya et al., microglia were activated by 10 μM Pb, which increased proinflammatory factors (Kumawat et al., 2014). Furthermore, neuroinflammation plays a crucial role in Pb-induced neurotoxicity, as evidenced by the effective prevention of Pb-induced neurotoxicity through neuroinflammation inhibition (Lu et al., 2022). Therefore, preventing Pb-induced inflammation may be a novel way to treat Pb-induced neurological disorders.

8.6.4 Mercury (Hg)

One of the most hazardous heavy metals, mercury in both organic and inorganic forms (Cesário, Mota, Caetano, Nogueira, & Canário, 2018). Mercury has been used extensively in industry, medicine, and agriculture for centuries. It is also thought to be one of the leading causes of mercury pollution in the environment (Sakamoto, Nakamura, & Murata, 2018). Furthermore, a significant amount of mercury can also be released by geothermal and volcanic activity. Human populations may be exposed to various forms of mercury through subcutaneous administration, parenteral administration, ingestion, or inhalation (Clifton II, 2007). The most well-known instance of mercury poisoning, Minamata disease, happened in Japan in the 1950s and was typified by a severe neurological condition (Eto, 2000).

Mercury has been implicated as a pathogenic factor in neurodegenerative diseases, such as AD and ALS, and has detrimental effects on several vital organs of the human body, particularly the brain (Johnson & Atchison, 2009; Xu et al., 2012). For instance, ALS was linked to mercury exposure in case–control studies, and in vivo research showed that chronic mercury exposure resulted in neurological symptoms like ataxia, tremors, and weakness in the hind limbs (Bose-O'Reilly et al., 2017; Johnson & Atchison, 2009). It has been demonstrated that several unique processes, such as mitochondrial dysfunction, oxidative stress, and neurotransmitter metabolism disorder, cause mercury-induced neurotoxicity (Carvalho et al., 2019; Liu, Yang, Xu, Xu, & Deng, 2019; Yang et al., 2020). According to an in vitro study, Hg caused a 150% increase in activated microglia, releasing proinflammatory cytokines (Liu et al., 2019). According to an in vitro study, mice exposed to mercury had higher levels of the inflammatory mediators p-NF-Kb and COX-2 in their cortex and hippocampus (Alattar et al., 2022). Furthermore, several investigations revealed that microglia were more susceptible to the harmful effects of mercury (Sharpe, Livingston, & Baskin, 2012). According to specific research, there is no correlation between exposure to mercury and serum levels of inflammatory cytokines (Monastero et al., 2017). However, more research is required to confirm the connection between neurological disorders and inflammation brought on by mercury.

In a study, it was observed that a 10-week history of severe back and abdominal pain, muscle spasms, migratory myalgias, arthralgias, paresthesias, constipation, weight loss, night sweats, blurred vision, and diffuse maculopapular rash was observed in a 19-year-old woman and her 17-year-old brother. The presence of infectious organisms was detected as unfavorable. Myokymic discharges in rhythmic to semirhythmic bursts of doublets, triplets, and quadruplets were detected by electromyography, these discharges were suggestive of neuromyotonia. Studies on toxicology revealed a high level of mercury in the urine (Shah, Butler, Mancias, Block, & Pérez, 2020). Tremors, behavioral changes, and disturbances like extreme shyness, memory loss, irritability, and persistent insomnia are among the symptoms of mercurialism, a well-defined clinical picture caused by chronic exposure to mercury vapor. Research with workers who had previously been exposed to vapHg0 revealed that symptoms may persist for years after exposure ends because of Hg^{2+} retention in the brain (Steckling et al., 2014). There is a dearth of systemic data on the effects vapHg0 on human development outside of case reports. According to Morgan et al. (2002), studies conducted on pregnant rats revealed that exposure to vapHg0 above threshold limits has a substantial negative impact on both the dams and the offspring (Morgan et al., 2002). Behavioral changes and poor coordination are two examples of these (Branco, Aschner, & Carvalho, 2021).

8.6.5 Arsenic (As)

Arsenic is the most dangerous metal toxic to human health based on its known or suspected toxicity. As a worldwide environmental contaminant, it is found in the Earth's crust everywhere (Hall, 2002). Millions of people are impacted by arsenic contamination in the environment, especially in developing nations like Bangladesh, China, India, Mexico, and Argentina (Garelick, Jones, Dybowska, & Valsami-Jones, 2008; Kaur, Kamli, & Ali, 2011). Drinking water, smoking, air, food, and cosmetics are among the exposure sources for As (Das et al., 2010; Vibol, Hashim, & Sarmani, 2015). Most of the body's organs can receive it after absorption through the skin, digestive tract, and respiratory system (Osuna-Martínez, Armienta, Bergés-Tiznado, & Páez-Osuna., 2021). The majority of the organ's normal function may be impacted by excessive As exposure (Sinha & Prasad, 2020). Additionally, several epidemiological studies have demonstrated a strong correlation between As levels in drinking water and CNS deficiencies, such as behavioral disorders, altered locomotor activity, and intellectual function (Mochizuki, 2019). Extended exposure to As has been connected to many neurological conditions, including autism, PD, and AD (Li et al., 2009; Mukherjee et al., 2003). Numerous epidemiological reports over the last twenty years have demonstrated that exposure to arsenic may change cognitive function, especially learning and memory in childhood. A recent metaanalysis examined how exposure to arsenic affected IQ tests measuring intelligence. Researchers concluded that children exposed to arsenic had an IQ drop of 0.4; although this seems to be a slight drop, these modifications may have long-term consequences (Rodríguez-Barranco et al., 2013). According to a population-based survey, the cognitively impaired (CI) group had higher hair content, and there was a dose-dependent negative correlation between hair As levels and the incidence rate of CI (Wang et al., 2021). As a result, arsenic levels in drinking water must not exceed 10 μg/L, according to WHO standards. In an epidemiological study, it was observed that children's 24-hour urine As levels increased by 5.21 μg/L in response to a daily dietary uptake of 1 ng/g As (Wang et al., 2022). The exact cause of neurotoxicity is still unknown, though. Strong evidence from fundamental molecular biology research has shown that oxidative stress, mitochondrial dysfunction, and endoplasmic reticulum stress are some of the mechanisms underlying As-induced neurotoxicity. When it comes to metal-induced neurotoxicity, such as that caused by As inflammation, it is crucials (Srivastava et al., 2009). Neuroinflammation is essential in metal-induced neurotoxicity, including that caused by As (Srivastava et al., 2009). Rat astrocytes and hippocampal microglia, for instance, release more inflammatory cytokines when exposed to As, according to an in vitro study (Mao et al., 2016; Saha, Sadhukhan, Mahalanobish, Dutta, & Sil, 2018). According to epidemiological studies, the levels of NF-Kb and TNF-α increased in blood plasma with low levels of As exposure (above 10 μg/L in the water) (Mo et al., 2011; Sinha & Prasad, 2020). Regardless of the concentration of arsenic, most epidemiological studies that examine the effects of exposure to the metal in children over several years have found that children's cognitive function is impaired. On the other hand, research assessing the effects of arsenic exposure solely during pregnancy does not consistently demonstrate cognitive impairments. In Nepal, neurodevelopment and arsenic levels in newborns were inversely correlated, according to an analysis of cord blood arsenic levels one day after birth (1.33 μg/L) and a behavioral assessment using the Brazelton scale (Parajuli, Fujiwara, Umezaki, & Watanabe, 2013). However, there were no discernible relationships between maternal and child arsenic levels and behavior, psychomotor and mental development index scores, or the mother's report of language acquisition in 18-month-old children, according to an early-life longitudinal study of prenatal and postnatal arsenic exposure (Hamadani et al., 2010). Furthermore, when age, sex, and socioeconomic status were taken into consideration, infants born to mothers with urinary arsenic levels of 80 µg/L during pregnancy did not exhibit any cognitive impairments (Tofail et al., 2009).

8.6.6 Zinc

The second most common trace element in the body, zinc, is essential for preserving regular function and serves as a structural component and cofactor for several enzymes, including those involved in DNA synthesis, RNA transcription, cell division, and activation (Prasad, 2007; Saper and Rash, 2009). Zn is found all over the body, but particularly in the brain. According to a dose–response metaanalysis, children's serum and plasma zinc levels may rise by 9% for every double of zinc intake. Comparatively, pregnant women's serum and plasma zinc levels may increase by 3% and 1%, respectively (Moran et al., 2012). About 10 times as much zinc is found in the brain as in the serum, and disorders of the CNS are closely linked to an imbalance in zinc metabolism in the brain (Frederickson & Moncrieff, 2004; Takeda, 2000). Depressive disorders have been linked to zinc deficiency (Szewczyk, Kubera, & Nowak, 2011), and neurodegenerative diseases, such as PD, have been linked to excessive zinc accumulation in the brain (Kumar et al., 2010). Clinical data revealed elevated zinc levels in PD patients' brains especially in postpartum patients (Dexter et al., 1991). Additionally, in vivo research supported the links between zinc and PD (Kumar et al., 2012). Furthermore, oligomerization of Aβ and AD was linked to zinc deficiency in both epidemiological and animal studies (Kawahara et al., 2014; Vilella et al., 2018). Obese people with inflammatory conditions have been found to have significantly lower zinc, which raises their risk of complications from obesity. According to research, zinc is essential for controlling the expression of proinflammatory cytokines via several different pathways (Fernández-Sánchez et al., 2011). White adipose tissue generates cytokines like IL-6 when inflammation occurs. CRP, a sensitive indicator of inflammation and tissue damage that also compromises vascular endothelial function, is secreted by the liver in response to IL-6 stimulation. Furthermore, TNF-α promotes IL-6 activity, which raises the synthesis of proinflammatory cytokines like adiponectin, thereby contributing to the acute-phase response (Haase & Rink, 2014). Three forms of zinc are stored in the brain: zinc ions, protein-bound zinc, and vesicular zinc (Frederickson & Moncrieff, 2004). About 85% of zinc in the CNS is closely associated with proteins that serve both structural and functional uses, including zinc-related metalloenzymes, MMPs, and zinc transcription factors. Neuronal regulation, synaptic plasticity, learning, and memory depend on zinc 2+, which is about ten times more abundant in the brain than in serum (Koh, 2001). Throughout neurodevelopment, zinc is essential for axonal growth, neuronal migration, and neuronal proliferation and differentiation (Sensi, Paoletti, Bush, & Sekler, 2009). Numerous proteins, hormones, and growth factors are synthesized with its help (Kumar, Kumar, Singh, Avasthi, & Kim, 2021). One of the earliest brain structures, the neural tube, is where stem cells and neuronal progenitor cells divide and proliferate during embryogenesis. DNA polymerase and other vital enzymes in the developing brain have been shown to require zinc (Levenson & Morris, 2011; Suh et al., 2009). Therefore a zinc deficiency during pregnancy causes aberrant fetal neuronal differentiation and hippocampus proliferation, which may affect future memory and learning functions and neurological abnormalities (Gammoh & Rink, 2017; Xue & Yong, 2020).

8.6.7 Iron (Fe)

The fourth most common element in the Earth's crust, Fe, makes up about 4.8% of the volume of crustal content and is one of the most widely distributed metals in the world. Because pure iron is silvery white and its surfaces are frequently coated in a protective layer of black ferrous ferric oxide, iron is known as a "ferrous metal" (Lieu, Heiskala, Peterson, & Yang, 2001). As a cofactor of several enzymes, Fe is essential for sustaining regular biological functions, such as DNA synthesis, mitochondrial respiration, and neurotransmitter metabolism, as a cofactor for several enzymes (Nairz & Weiss, 2020; Ward, Zucca, Duyn, Crichton, & Zecca, 2014). Fe is the most common metal in the brain and is essential for preserving the best possible biological function (Rodrigue, Daugherty, Foster, & Kennedy, 2020). There are negative consequences to both excessive accumulation and deficiency of iron. Fe deficiency anemia (IDA), which affects over 1.2 billion people and contributes substantially to the global burden of diseases, has been regarded as a worldwide public health issue (Godeau, Petit, Richard, Roquelaure, & Descatha, 2021; Pasricha, Tye-Din, Muckenthaler, & Swinkels, 2021). How the BBB, brain microvascular endothelial cells, and their capacity to transfer iron respond to systemic inflammation is still unclear. Proteins involved in iron uptake and efflux may undergo functional alterations. Reduced iron efflux through ferroportin may result from inflammation since it would raise the amount of hepcidin in the blood. Zip proteins ZIP14 and ZIP18 and inflammation, particularly IL6, which can increase Zip14 expression, may regulate iron uptake into microvascular endothelial cells (Liuzzi et al., 2005). The function of astrocytes, found on the abluminal side of the BBB, will be changed by inflammation, releasing glutamate. This excitotoxic neurotransmitter intensifies the inflammation. It is still unknown how inflammation may impact BBB permeability via tight junctions and adherens junctions, and CRP may contribute to barrier breakdown (Kuhlmann et al., 2009; Moos, Nielsen, Skjørringe, & Morgan, 2007). Iron homeostasis is altered by a persistent neuroinflammatory process in PD and AD, the two neurodegenerative illnesses. There is a difference in iron deposition in the two diseases; in PD, the increase in iron is dramatic, while in AD, only a marginal rise in iron occurs. One might assume that since inflammation and iron homeostasis are closely related, inflammatory changes involving the release of TOS, NO, and particular cytokines will have an impact on the iron proteins involved in iron homeostasis (Wessling-Resnick, 2010).

Numerous studies have shown the connection between inflammation and neurodegenerative diseases, demonstrating that controlled inflammation helps to maintain homeostasis in the CNS by removing excess Aβ, necrotic neuronal cells, and misformed and aggregated α-synuclein (Medzhitov, 2008). On the other hand, neuroinflammation can also harm the nervous system, facilitating the development and progression of neurodegenerative illnesses like AD and PD (Furman et al., 2019). Heavy metals such as Cd, Pb, Fe, and Hg are considered a common pathogenic factor, mainly due to their inability to be broken down and eliminated; these heavy metals accumulate in the brain and have detrimental neurotoxic effects (Horiguchi et al., 2010). The pathophysiological mechanism of metals in neurodegenerative diseases has drawn more attention in recent years. A few examples of mechanisms underlying metal-induced neurotoxicity are aberrant cell apoptosis, oxidative stress, neuroinflammation, and necrosis of cells (Moreno, Streifel, Sullivan, Hanneman, & Tjalkens, 2011; Xu et al., 2016; Zhao et al., 2020). The role of neuroinflammation in metal-induced neurological disorders, such as those caused by As, Fe, Zn, Mg, Cd, Pb, Hg, and Al, was validated by this review. This review emphasizes how inflammation plays a significant role in the genesis of metal neurotoxicity and neurodegenerative diseases. The mechanism of neuroinflammation in neurodegenerative diseases and metal neurotoxicity has been extensively studied in recent decades. However, more research is required to fully understand the relationship between neuroinflammation, metal neurotoxicity, neurodegenerative diseases, and the precise mechanisms of pathogenesis and etiology. Neurological disorders have a significant financial impact on patients, their families, and society as a whole. They also affect independence and quality of life, frequently necessitating full-time care in the later stages of the disease.

8.7 Conclusion

The prevalence of neurological dysfunction has increased as a result of the rise in neurodegenerative diseases brought on by an aging population. According to WHO estimates, over 55 million people have dementia, and by 2050, that figure is expected to increase to 139 million (Organization, 2021). Because more and more metals are being used in the industry, neurological disorders brought on by environmental metal pollution continue to be a global public health concern today. There is overwhelming evidence that neurodegenerative diseases are closely related to both metal overload and deficiency. Thus it is crucial to investigate the mechanisms underlying metal-induced neurological disorders and neurodegenerative diseases. Thus it is still relevant to investigate the mechanisms underlying metal-induced neurological disorders and neurodegenerative diseases. Chronic exposure to heavy metals can lead to persistent neuroinflammation, which can exacerbate diseases such as MSs, PD, and AD.

References

Abdel-Haq, R., Schlachetzki, J. C. M., Glass, C. K., & Mazmanian, S. K. (2019). Microbiome–microglia connections via the gut–brain axis. *Journal of Experimental Medicine, 216*(1), 41–59.

Abd-Ellatief, R. B., Mohamed, H. K., & Kotb, H. I. (2018). Reactive astrogliosis in an experimental model of fibromyalgia: Effect of dexmedetomidine. *Cells, Tissues, Organs, 205*(2), 105–119.

Akinrinade, I. D., Memudu, A. E., & Ogundele, O. M. (2015). Fluoride and aluminium disturb neuronal morphology, transport functions, cholinesterase, lysosomal and cell cycle activities. *Pathophysiology: The Official Journal of the International Society for Pathophysiology / ISP, 22*(2), 105–115.

Alam, S.-I., Kim, M.-W., Shah, F. A., Saeed, K., Ullah, R., & Kim, M.-O. (2021). Alpha-linolenic acid impedes cadmium-induced oxidative stress, neuroinflammation, and neurodegeneration in mouse brain. *Cells, 10*(9), 2274.

Ali, H. A., Afifi, M., Abdelazim, A. M., & Mosleh, Y. Y. (2014). Quercetin and Omega 3 ameliorate oxidative stress induced by aluminium chloride in the brain. *Journal of Molecular Neuroscience, 53*, 654–660.

Alattar, A., Alvi, A. M., Rashid, S., Hussain, N., Gul, M., Ikram, M., Khalil, A. A. K., Alshaman, R., Shah, F. A., Li, S., & Li, J. (2022). RETRACTED: Carveol ameliorates mercury-induced oxidative stress, neuroinflammation, and neurodegeneration in a mouse brain. *Neurotoxicology, 92*, 212–226. Retrieved from 10.1016/j.neuro.2022.08.006.

Allen, K. A. (2015). Is prenatal lead exposure a concern in infancy? What is the evidence? *Advances in Neonatal Care, 15*(6), 416–420.

Baburamani, A. A., Supramaniam, V. G., Hagberg, H., & Mallard, C. (2014). Microglia toxicity in preterm brain injury. *Reproductive Toxicology, 48*, 106–112.

Blaylock, R. L. (2012). Aluminum induced immunoexcitotoxicity in neurodevelopmental and neurodegenerative disorders. *Current Inorganic Chemistry (Discontinued), 2*(1), 46–53.

Bose-O'Reilly, S., Bernaudat, L., Siebert, U., Roider, G., Nowak, D., & Drasch, G. (2017). Signs and symptoms of mercury-exposed gold miners. *International Journal of Occupational Medicine and Environmental Health, 30*(2), 249–269.

Branco, V., Aschner, M., & Carvalho, C. (2021). Neurotoxicity of mercury: An old issue with contemporary significance. *Advances in Neurotoxicology, 5*, 239–262. Retrieved from 10.1016/bs.ant.2021.01.001.

Brough, D., & Jouhara, H. (2020). The aluminium industry: A review on state-of-the-art technologies, environmental impacts and possibilities for waste heat recovery. *International Journal of Thermofluids, 1*, 100007.

Brundin, P., Nath, A., & Beckham, J. D. (2020). Is COVID-19 a perfect storm for Parkinson's disease? *Trends in Neurosciences, 43*(12), 931–933.

Cai, J., Guan, H., Jiao, X., Yang, J., Chen, X., Zhang, H., Zheng, Y., et al. (2021). NLRP3 inflammasome mediated pyroptosis is involved in cadmium exposure-induced neuroinflammation through the IL-1β/IkB-α-NF-KB-NLRP3 feedback loop in Swine. *Toxicology, 453*, 152720.

Carvalho, L. V. B., Hacon, S. S., Vega, C. M., Vieira, J. A., Larentis, A. L., Mattos, R. C. O. C., Valente, D., et al. (2019). Oxidative stress levels induced by mercury exposure in amazon juvenile populations in Brazil. *International Journal of Environmental Research and Public Health, 16*(15), 2682.

Castanedo-Vazquez, D., Bosque-Varela, P., Sainz-Pelayo, A., & Riancho, J. (2019). Infectious agents and amyotrophic lateral sclerosis: Another piece of the puzzle of motor neuron degeneration. *Journal of Neurology, 266*, 27–36.

Castro-Gomez, S., Binder, J., & Heneka, M. T. (2019). Neuroinflammation as motor of Alzheimer's disease. *Der Nervenarzt, 90*, 898–906.

Cesário, R., Mota, A. M., Caetano, M., Nogueira, M., & Canário, J. (2018). Mercury and methylmercury transport and fate in the water column of tagus estuary (Portugal). *Marine Pollution Bulletin, 127*, 235–250.

Chen, D., Zhang, T., & Lee, T. H. (2020). Cellular mechanisms of melatonin: Insight from neurodegenerative diseases. *Biomolecules, 10*(8), 1158.

Clifton II, J. C. (2007). Mercury exposure and public health. *Pediatric Clinics of North America, 54*(2), 237 e1.

Colombo, A. V., Sadler, R. K., Llovera, G., Singh, V., Roth, S., Heindl, S., Sebastian Monasor, L., et al. (2021). Microbiota-derived short chain fatty acids modulate microglia and promote Aβ plaque deposition. *eLife, 10*, e59826.

Cornejo, F., Vruwink, M., Metz, C., Muñoz, P., Salgado, N., Poblete, J., ... von Bernhardi, R. (2018). Scavenger receptor-A deficiency impairs immune response of microglia and astrocytes potentiating Alzheimer's disease pathophysiology. *Brain, Behavior, and Immunity, 69*, 336–350.

Costantini, E., D'Angelo, C., & Reale, M. (2018). The role of immunosenescence in neurodegenerative diseases. *Mediators of Inflammation, 2018*(1), 6039171.

Das, A. K., Bag, S., Sahu, R., Dua, T. K., Sinha, M. K., Gangopadhyay, M., ... Dewanjee, S. (2010). Protective effect of corchorus olitorius leaves on sodium arsenite-induced toxicity in experimental rats. *Food and chemical toxicology, 48*(1), 326–335.

Delic, V., Beck, K. D., Pang, K. C. H., & Citron, B. A. (2020). Biological links between traumatic brain injury and Parkinson's disease. *Acta Neuropathologica Communications, 8*, 1–16.

Demirci, S., Aynalı, A., Demirci, K., Demirci, S., & Arıdoğan, B. C. (2017). The serum levels of resistin and its relationship with other proinflammatory cytokines in patients with Alzheimer's disease. *Clinical Psychopharmacology and Neuroscience, 15*(1), 59.

Dexter, D. T., Carayon, A., Javoy-Agid, F., Agid, Y., Wells, F. R., Daniel, S. E., ... Marsden, C. D. (1991). Alterations in the levels of iron, ferritin and other trace metals in Parkinson's disease and other neurodegenerative diseases affecting the basal ganglia. *Brain, 114*(4), 1953–1975.

Dhaiban, S., Al-Ani, M., Elemam, N. M., Al-Aawad, M. H., Al-Rawi, Z., & Maghazachi, A. A. (2021). Role of peripheral immune cells in multiple sclerosis and experimental autoimmune encephalomyelitis. *Science, 3*(1), 12.

Ding, Z.-B., Song, L.-J., Wang, Q., Kumar, G., Yan, Y.-Q., & Ma, C.-G. (2021). Astrocytes: A double-edged sword in neurodegenerative diseases. *Neural Regeneration Research, 16*(9), 1702–1710.

Djordjevic, J., Golam Sabbir, M., & Albensi, B. C. (2016). Traumatic brain injury as a risk factor for Alzheimer's disease: Is Inflammatory signaling a key player? *Current Alzheimer Research, 13*(7), 730–738.

Eto, K. (2000). Minamata disease. *Neuropathology: Official Journal of the Japanese Society of Neuropathology, 20*, S14–S19.

Ettinger, A. S., Téllez-Rojo, M. M., Amarasiriwardena, C., Bellinger, D., Peterson, K., Schwartz, J., ... Hernández-Avila, M. (2004). Effect of breast milk lead on infant blood lead levels at 1 month of age. *Environmental Health Perspectives, 112*(14), 1381–1385.

Fernández-Sánchez, A., Madrigal-Santillán, E., Bautista, M., Esquivel-Soto, J., Morales-González, Á., Esquivel-Chirino, C., Durante-Montiel, I., et al. (2011). Inflammation, oxidative stress, and obesity. *International Journal of Molecular Sciences, 12*(5), 3117–3132.

Fonken, L. K., Bedrosian, T. A., Zhang, N., Weil, Z. M., DeVries, A. C., & Nelson, R. J. (2019). Dim light at night impairs recovery from global cerebral ischemia. *Experimental Neurology, 317*, 100–109.

Franz, C. K., Joshi, D., Daley, E. L., Grant, R. A., Dalamagkas, K., Leung, A., ... Kiskinis, E. (2019). Impact of traumatic brain injury on amyotrophic lateral sclerosis: From bedside to bench. *Journal of Neurophysiology, 122*(3), 1174–1185.

Frederickson, C. J., & Moncrieff, D. W. (2004). Zinc-containing neurons. *Neuro-Signals, 3*(3), 127–139.

Furman, D., Campisi, J., Verdin, E., Carrera-Bastos, P., Targ, S., Franceschi, C., Ferrucci, L., et al. (2019). Chronic inflammation in the etiology of disease across the life span. *Nature Medicine, 25*(12), 1822–1832.

Gammoh, N. Z., & Rink, L. (2017). Zinc in infection and inflammation. *Nutrients, 9*(6), 624.

Gao, C., Jiang, J., Tan, Y., & Chen, S. (2023). Microglia in neurodegenerative diseases: Mechanism and potential therapeutic targets. *Signal Transduction and Targeted Therapy, 8*(1), 359.

Garelick, H., Jones, H., Dybowska, A., & Valsami-Jones, E. (2008). Arsenic pollution sources. In *Reviews of environmental contamination volume 197: International perspectives on arsenic pollution and remediation*. Springer, 17–60. Retrieved from https://doi.org/10.1007/978-0-387-79284-2_2.

Garza, A., Vega, R., & Soto, E. (2006). Cellular mechanisms of lead neurotoxicity. *Medical Science Monitor, 12*(3), RA57.

Garza-Lombó, C., Pappa, A., Panayiotidis, M. I., Gonsebatt, M. E., & Franco, R. (2019). Arsenic-induced neurotoxicity: A mechanistic appraisal. *JBIC Journal of Biological Inorganic Chemistry, 24*, 1305–1316.

Gąssowska-Dobrowolska, M., Chlubek, M., Kolasa, A., Tomasiak, P., Korbecki, J., Skowrońska, K., ... Baranowska-Bosiacka, I. (2023).

Microglia and astroglia—The potential role in neuroinflammation induced by pre- and neonatal exposure to lead (Pb). *International Journal of Molecular Sciences, 24*(12), 9903.

Godeau, D., Petit, A., Richard, I., Roquelaure, Y., & Descatha, A. (2021). Return-to-work, disabilities and occupational health in the age of COVID-19. *Scandinavian Journal of Work, Environment & Health, 47*(5), 408.

Gorji, A. (2022). Neuroinflammation: The pathogenic mechanism of neurological disorders. *International Journal of Molecular Sciences, 23*(10), 5744.

Grandjean, P., & Landrigan, P. J. (2014). Neurobehavioural effects of developmental toxicity. *The lancet neurology, 13*(3), 330–338.

Guerrero-Vargas, N. N., Zárate-Mozo, C., Guzmán-Ruiz, M. A., Cárdenas-Rivera, A., & Escobar, C. (2021). Time-restricted feeding prevents depressive-like and anxiety-like behaviors in male rats exposed to an experimental model of shift-work. *Journal of Neuroscience Research, 99*(2), 604–620.

Guo, Y., Zeng, H., & Gao, C. (2021). The role of neutrophil extracellular traps in central nervous system diseases and prospects for clinical application. *Oxidative Medicine and Cellular Longevity, 2021*(1), 9931742.

Gustin, A., Kirchmeyer, M., Koncina, E., Felten, P., Losciuto, S., Heurtaux, T., ... Dostert, C. (2015). NLRP3 inflammasome is expressed and functional in mouse brain microglia but not in astrocytes. *PLoS One, 10*(6), e0130624.

Haase, H., & Rink, L. (2014). Zinc signals and immune function. *Biofactors (Oxford, England), 40*(1), 27–40.

Hall, A. H. (2002). Chronic Arsenic Poisoning. *Toxicology Letters, 128*(1-3), 69–72.

Hamadani, J. D., Grantham-McGregor, S. M., Tofail, F., Nermell, B., Fängström, B., Huda, S. N., Yesmin, S., et al. (2010). Pre- and postnatal arsenic exposure and child development at 18 months of age: A cohort study in Rural Bangladesh. *International Journal of Epidemiology, 39*(5), 1206–1216.

Hammond, N., Munkacsi, A. B., & Sturley, S. L. (2019). The complexity of a monogenic neurodegenerative disease: More than two decades of therapeutic driven research into niemann-pick type C disease. *Biochimica et Biophysica Acta (BBA)—Molecular and Cell Biology of Lipids, 1864*(8), 1109–1123.

Hauptman, M., Niles, J. K., Gudin, J., & Kaufman, H. W. (2021). Individual- and community-level factors associated with detectable and elevated blood lead levels in US children: Results from a national clinical laboratory. *JAMA Pediatrics, 175*(12), 1252–1260.

He, P., Yang, C., He, D., Zhao, S., Xie, Y., Wang, H., & Ma, J. (2021). Blood lead, systemic inflammation, and blood pressure: Exploring associations and mediation effects in workers exposed to lead. *Biological Trace Element Research, 199*(7), 2573–2581.

Headland, S. E., & Norling, L. V. (2015). The resolution of inflammation: Principles and challenges. In *Seminars in immunology*. Elsevier, 149–160.

Healy, L. M., Yaqubi, M., Ludwin, S., & Antel, J. P. (2020). Species differences in immune-mediated CNS tissue injury and repair: A (neuro) inflammatory topic. *Glia, 68*(4), 811–829.

Hoeks, J., & Schrauwen, P. (2012). Muscle mitochondria and insulin resistance: A human perspective. *Trends in Endocrinology & Metabolism, 23*(9), 444–450.

Horiguchi, H., Aoshima, K., Oguma, E., Sasaki, S., Miyamoto, K., Hosoi, Y., ... Kayama, F. (2010). Latest status of cadmium accumulation and its effects on kidneys, bone, and erythropoiesis in inhabitants of the formerly cadmium-polluted jinzu river basin in Toyama, Japan, after restoration of rice paddies. *International Archives of Occupational and Environmental Health, 83*, 953–970.

Huat, T. J., Camats-Perna, J., Newcombe, E. A., Valmas, N., Kitazawa, M., & Medeiros, R. (2019). Metal toxicity links to Alzheimer's disease and neuroinflammation. *Journal of Molecular Biology, 431*(9), 1843–1868.

Jain, A., Madkan, S., & Patil, P. (2023). The role of gut microbiota in neurodegenerative diseases: Current insights and therapeutic implications. *Cureus, 15*(10).

Jakubowski, M. (2011). Low-level environmental lead exposure and intellectual impairment in children—The current concepts of risk assessment. *International Journal of Occupational Medicine and Environmental Health, 24*, 1–7.

Jensen, A., & Bro-Rasmussen, F. (1992). Environmental cadmium in Europe. *Reviews of Environmental Contamination and Toxicology, 125*, 101–181.

Johnson, F. O., & Atchison, W. D. (2009). The role of environmental mercury, lead and pesticide exposure in development of amyotrophic lateral sclerosis. *Neurotoxicology, 30*(5), 761–765.

Johnson, T. P., & Nath, A. (2018). Neurological syndromes driven by postinfectious processes or unrecognized persistent infections. *Current Opinion in Neurology, 31*(3), 318–324.

Johnson, V. E., Stewart, W., & Smith, D. H. (2010). Traumatic brain injury and amyloid-β pathology: A link to Alzheimer's disease? *Nature Reviews. Neuroscience, 11*(5), 361–370.

Kanashiro, A., Hiroki, C. H., da Fonseca, D. M., Birbrair, A., Ferreira, R. G., Bassi, G. S., Fonseca, M. D., et al. (2020). The role of neutrophils in neuro-immune modulation. *Pharmacological research, 151*, 104580.

Kasten-Jolly, J., Heo, Y., & Lawrence, D. A. (2011). Central nervous system cytokine gene expression: Modulation by lead. *Journal of Biochemical and Molecular Toxicology, 25*(1), 41–54.

Kaur, D., Sharma, V., & Deshmukh, R. (2019). Activation of microglia and astrocytes: A roadway to neuroinflammation and Alzheimer's disease. *Inflammopharmacology, 27*, 663–677.

Kaur, S., Kamli, M. R., & Ali, A. (2011). Role of arsenic and its resistance in nature. *Canadian Journal of Microbiology, 57*(10), 769–774.

Kawahara, M., Mizuno, D., Koyama, H., Konoha, K., Ohkawara, S., & Sadakane, Y. (2014). Disruption of zinc homeostasis and the pathogenesis of senile dementia. *Metallomics, 6*(2), 209–219.

Khakh, B. S., & Deneen, B. (2019). The emerging nature of astrocyte diversity. *Annual Review of Neuroscience, 42*(1), 187–207.

Kippler, M., Tofail, F., Hamadani, J. D., Gardner, R. M., Grantham-McGregor, S. M., Bottai, M., & Vahter, M. (2012). Early-life cadmium exposure and child development in 5-year-old girls and boys: A cohort study in Rural Bangladesh. *Environmental Health Perspectives, 120*(10), 1462–1468.

Koh, J.-Y. (2001). Zinc and disease of the brain. *Molecular Neurobiology, 24*, 99–106.

Konsman, J. P. (2022). Cytokines in the brain and neuroinflammation: We didn't starve the fire!. *Pharmaceuticals, 15*(2), 140.

Krämer, T. J., Hack, N., Brühl, T. J., Menzel, L., Hummel, R., Griemert, E.-V., Klein, M., et al. (2019). Depletion of regulatory T cells increases T cell brain infiltration, reactive astrogliosis, and interferon-γ gene expression in acute experimental traumatic brain injury. *Journal of Neuroinflammation, 16*, 1–14.

Kuhlmann, C. R. W., Librizzi, L., Closhen, D., Pflanzner, T., Lessmann, V., Pietrzik, C. U., ... Luhmann, H. J. (2009). Mechanisms of C-reactive protein-induced blood–brain barrier disruption. *Stroke; a Journal of Cerebral Circulation, 40*(4), 1458–1466.

Kumar, A., Ahmad, I., Shukla, S., Kumar Singh, B., Kumar Patel, D., Pandey, H. P., & Singh, C. (2010). Effect of zinc and paraquat co-exposure on neurodegeneration: Modulation of oxidative stress and expression of metallothioneins, toxicant responsive and transporter genes in rats. *Free Radical Research, 44*(8), 950–965.

Kumar, A., Singh, B. K., Ahmad, I., Shukla, S., Patel, D. K., Srivastava, G., ... Singh, C. (2012). Involvement of NADPH oxidase and glutathione in zinc-induced dopaminergic neurodegeneration in rats: Similarity with paraquat neurotoxicity. *Brain Research, 1438*, 48–64.

Kumar, V., Kumar, A., Singh, K., Avasthi, K., & Kim, J.-J. (2021). Neurobiology of zinc and its role in neurogenesis. *European Journal of Nutrition, 60*, 55–64.

Kumari, S., Dhapola, R., Sharma, P., Nagar, P., Medhi, B., & HariKrishnaReddy, D. (2024). The impact of cytokines in neuroinflammation-mediated stroke. *Cytokine & Growth Factor Reviews, 78*, 105–119. Retrieved from https://doi.org/10.1016/j.cytogfr.2024.06.002.

Kumawat, K. L., Kaushik, D. K., Goswami, P., & Basu, A. (2014). Acute exposure to lead acetate activates microglia and induces subsequent bystander neuronal death via caspase-3 activation. *Neurotoxicology, 41*, 143–153.

Kwon, H. S., & Koh, S.-H. (2020). Neuroinflammation in neurodegenerative disorders: The roles of microglia and astrocytes. *Translational neurodegeneration, 9*(1), 42.

Lee, H.-J., Suk, J.-E., Patrick, C., Bae, E.-J., Cho, J.-H., Rho, S., ... Lee, S.-J. (2010). Direct transfer of α-synuclein from neuron to astroglia causes inflammatory responses in synucleinopathies. *Journal of Biological Chemistry, 285*(12), 9262–9272.

Levenson, C. W., & Morris, D. (2011). Zinc and neurogenesis: Making new neurons from development to adulthood. *Advances in nutrition, 2*(2), 96–100.

Li, D., Lu, C., Wang, J., Hu, W., Cao, Z., Sun, D., ... Ma, X. (2009). Developmental mechanisms of arsenite toxicity in zebrafish (*Danio rerio*) embryos. *Aquatic Toxicology, 91*(3), 229–237.

Li, N., Liu, X., Zhang, P., Qiao, M., Li, H., Li, X., ... Yu, Z. (2015). The effects of early life lead exposure on the expression of interleukin (IL) 1β, IL-6, and glial fibrillary acidic protein in the hippocampus of mouse pups. *Human & Experimental Toxicology, 34*(4), 357–363.

Lieu, P. T., Heiskala, M., Peterson, P. A., & Yang, Y. (2001). The roles of iron in health and disease. *Molecular Aspects of Medicine, 22*(1–2), 1–87.

Lin, T., Liu, G. A., Perez, E., Rainer, R. D., Febo, M., Cruz-Almeida, Y., & Ebner, N. C. (2018). Systemic inflammation mediates age-related cognitive deficits. *Frontiers in Aging Neuroscience, 10*, 236.

Litwiniuk, A., Bik, W., Kalisz, M., & Baranowska-Bik, A. (2021). Inflammasome NLRP3 potentially links obesity-associated low-grade systemic inflammation and insulin resistance with Alzheimer's disease. *International Journal of Molecular Sciences, 22*(11), 5603.

Liu, J.-T., Chen, B.-Y., Zhang, J.-Q., Kuang, F., & Chen, L.-W. (2015). Lead exposure induced microgliosis and astrogliosis in hippocampus of young mice potentially by triggering TLR4–MyD88–NFκB signaling cascades. *Toxicology Letters, 239*(2), 97–107.

Liu, W., Yang, T., Xu, Z., Xu, B., & Deng, Y. (2019). Methyl-mercury induces apoptosis through ROS-mediated endoplasmic reticulum stress and mitochondrial apoptosis pathways activation in rat cortical neurons. *Free Radical Research, 53*(1), 26–44.

Liuzzi, J. P., Lichten, L. A., Rivera, S., Blanchard, R. K., Aydemir, T. B., Knutson, M. D., ... Cousins, R. J. (2005). Interleukin-6 regulates the zinc transporter Zip14 in liver and contributes to the hypozincemia of the acute-phase response. *Proceedings of the National Academy of Sciences, 102*(19), 6843–6848.

Lotfi, N., Thome, R., Rezaei, N., Zhang, G.-X., Rezaei, A., Rostami, A., & Esmaeil, N. (2019). Roles of GM-CSF in the pathogenesis of autoimmune diseases: An update. *Frontiers in Immunology, 10*, 1265.

Lu, L.-li, Zhang, Y.-wen, Li, Z.-cong, Fang, Y.-yuan, Wang, L.-lei, Zhao, Y.-song, Li, S.-jun, et al. (2022). Therapeutic effects of sodium para-aminosalicylic acid on cognitive deficits and activated ERK1/2-P90RSK/NF-KB inflammatory pathway in Pb-exposed rats. *Biological Trace Element Research, 200*(6), 2807–2815.

Machoń-Grecka, A., Dobrakowski, M., Kasperczyk, A., Birkner, E., & Kasperczyk, S. (2022). Angiogenesis and lead (Pb): Is there a connection? *Drug and Chemical Toxicology, 45*(2), 589–593.

Maj, H., Mallard, C., & Hagberg., H. (2004). Inflammatory gene profiling in the developing mouse brain after hypoxia-ischemia. *Journal of Cerebral Blood Flow & Metabolism, 24*(12), 1333–1351.

Mao, J., Yang, J., Zhang, Y., Li, T., Wang, C., Xu, L., Hu, Q., et al. (2016). Arsenic trioxide mediates HAPI microglia inflammatory response and subsequent neuron apoptosis through P38/JNK MAPK/STAT3 pathway. *Toxicology and Applied Pharmacology, 303*, 79–89.

Matsubara, E., Hirai, S., Amari, M., Shoji, M., Yamaguchi, H., Okamoto, K., ... Wakabayashi, K. (1990). A1-antichymotrypsin as a possible biochemical marker for alzheimer-type dementia. *Annals of Neurology: Official Journal of the American Neurological Association and the Child Neurology Society, 28*(4), 561–567.

McManus, R. M., & Heneka., M. T. (2017). Role of neuroinflammation in neurodegeneration: new insights. *Alzheimer's Research & Therapy, 9*, 1–7.

Medzhitov, R. (2008). Origin and physiological roles of inflammation. *Nature, 454*(7203), 428–435.

Mo, J., Xia, Y., Wade, T. J., DeMarini, D. M., Davidson, M., & Mumford, J. (2011). Altered gene expression by low-dose arsenic exposure in humans and cultured cardiomyocytes: Assessment by real-time PCR arrays. *International Journal of Environmental Research and Public Health, 8*(6), 2090–2108.

Mochizuki, H. (2019). Arsenic neurotoxicity in humans. *International Journal of Molecular Sciences, 20*(14), 3418.

Mold, M., Cottle, J., King, A., & Exley, C. (2019). Intracellular aluminium in inflammatory and glial cells in cerebral amyloid angiopathy: A case report. *International Journal of Environmental Research and Public Health, 16*(8), 1459.

Monastero, R. N., Karimi, R., Nyland, J. F., Harrington, J., Levine, K., & Meliker, J. R. (2017). Mercury exposure, serum antinuclear antibodies, and serum cytokine levels in the long island study of seafood consumption: A cross-sectional study in NY, USA. *Environmental Research, 156*, 334–340.

Moos, T., Nielsen, T. R., Skjørringe, T., & Morgan, E. H. (2007). Iron trafficking inside the brain. *Journal of Neurochemistry, 103*(5), 1730–1740.

Moran, V. H., Stammers, A.-L., Medina, M. W., Patel, S., Dykes, F., Souverein, O. W., Dullemeijer, C., et al. (2012). The relationship between zinc intake and serum/plasma zinc concentration in children:

A systematic review and dose-response meta-analysis. *Nutrients, 4*(8), 841–858.

Moreno, J. A., Streifel, K. M., Sullivan, K. A., Hanneman, W. H., & Tjalkens, R. B. (2011). Manganese-induced NF-KB activation and nitrosative stress is decreased by estrogen in juvenile mice. *Toxicological Sciences, 122*(1), 121–133.

Morgan, D. L., Chanda, S. M., Price, H. C., Fernando, R., Liu, J., Brambila, E., ... Barone, S., Jr. (2002). Disposition of inhaled mercury vapor in pregnant rats: Maternal toxicity and effects on developmental outcome. *Toxicological Sciences, 66*(2), 261–273.

Mottahedin, A., Ardalan, M., Chumak, T., Riebe, I., Ek, J., & Mallard, C. (2017). Effect of neuroinflammation on synaptic organization and function in the developing brain: Implications for neurodevelopmental and neurodegenerative disorders. *Frontiers in Cellular Neuroscience, 11*, 190.

Mukherjee, S. C., Rahman, M. M., Chowdhury, U. K., Sengupta, M. K., Lodh, D., Chanda, C. R., ... Chakraborti, D. (2003). Neuropathy in arsenic toxicity from groundwater arsenic contamination in West Bengal, India. *Journal of Environmental Science and Health, Part A, 38*(1), 165–183.

Muzio, L., Viotti, A., & Martino, G. (2021). Microglia in neuroinflammation and neurodegeneration: From understanding to therapy. *Frontiers in Neuroscience, 15*, 742065.

Nairz, M., & Weiss, G. (2020). Iron in health and disease. *Molecular Aspects of Medicine, 75*(100906), 10–1016.

Ochocka, N., & Kaminska, B. (2021). Microglia diversity in healthy and diseased brain: Insights from single-cell omics. *International journal of molecular sciences, 22*(6), 3027.

Oono, M., Fujita, Y., Uchida, N., Kawai, U., Fujita-Nakata, M., Nakanishi, M., ... Matsui, M. (2018). Rheumatoid meningitis developed in patient with stable rheumatoid arthritis and myasthenia gravis—Detailed analysis of intracranial inflammation using flow cytometry. *Journal of Neuroinflammation, 15*, 1–6.

Orellana, A. M. M., Vasconcelos, A. R., Leite, J. A., de Sá Lima, L., Andreotti, D. Z., Munhoz, C. D., ... Scavone, C. (2015). Age-related neuroinflammation and changes in AKT-GSK-3β and WNT/β-CATENIN signaling in rat hippocampus. *Aging (Albany NY), 7*(12), 1094.

Oshiro, S., Kawahara, M., Kuroda, Y., Zhang, C., Cai, Y., Kitajima, S., & Shirao, M. (2000). Glial cells contribute more to iron and aluminum accumulation but are more resistant to oxidative stress than neuronal cells. *Biochimica et Biophysica Acta (BBA)—Molecular Basis of Disease, 1502*(3), 405–414.

Osuna-Martínez, C. C., Armienta, M. A., Bergés-Tiznado, M. E., & Páez-Osuna., F. (2021). Arsenic in waters, soils, sediments, and biota from Mexico: An environmental review. *Science of The Total Environment, 752*, 142062.

Pandey, S., Singh, B., Yadav, S. K., & Mahdi, A. A. (2018). Novel biomarker for neurodegenerative diseases-motor neuron disease (MND), cerebellar ataxia (CA) and Parkinson's disease (PD). *Clinica Chimica Acta, 485*, 258–261.

Parajuli, R. P., Fujiwara, T., Umezaki, M., & Watanabe, C. (2013). Association of cord blood levels of lead, arsenic, and zinc with neurodevelopmental indicators in newborns: A birth cohort study in Chitwan Valley, Nepal. *Environmental Research, 121*, 45–51.

Pasricha, S.-R., Tye-Din, J., Muckenthaler, M. U., & Swinkels, D. W. (2021). Iron deficiency. *The Lancet, 397*(10270), 233–248.

Peón, A. N., Ledesma-Soto, Y., Olguín, J. E., Bautista-Donis, M., Sciutto, E., & Terrazas, L. I. (2017). Helminth products potently modulate experimental autoimmune encephalomyelitis by downregulating neuroinflammation and promoting a suppressive microenvironment. *Mediators of Inflammation, 2017*(1), 8494572.

Phuagkhaopong, S., Ospondpant, D., Kasemsuk, T., Sibmooh, N., Soodvilai, S., Power, C., & Vivithanaporn, P. (2017). Cadmium-induced IL-6 and IL-8 expression and release from astrocytes are mediated by MAPK and NF-KB pathways. *Neurotoxicology, 60*, 82–91.

Platt, B., Fiddler, G., Riedel, G., & Henderson, Z. (2001). Aluminium toxicity in the rat brain: Histochemical and immunocytochemical evidence. *Brain Research Bulletin, 55*(2), 257–267.

Plesnila, N. (2016). The immune system in traumatic brain injury. *Current Opinion in Pharmacology, 26*, 110–117.

Pogue, A. I., Jaber, V., Zhao, Y., & Lukiw, W. J. (2017). Systemic inflammation in C57BL/6J mice receiving dietary aluminum sulfate; up-regulation of the pro-inflammatory cytokines IL-6 and TNFα, C-reactive protein (CRP) and MiRNA-146a in blood serum. *Journal of Alzheimer's Disease & Parkinsonism, 7*(6).

Prasad, A. S. (2007). Zinc: Mechanisms of host defense. *The Journal of Nutrition, 137*(5), 1345–1349.

Qin, J., Ma, Z., Chen, X., & Shu, S. (2023). Microglia activation in central nervous system disorders: A review of recent mechanistic investigations and development efforts. *Frontiers in Neurology, 14*, 1103416.

Ragan, H. A. (1983). The bioavailability of iron, lead and cadmium via gastrointestinal absorption: A review. *Science of the Total Environment, 28*(1–3), 317–326.

Reid, J. K., & Kuipers, H. F. (2021). She doesn't even go here: The role of inflammatory astrocytes in CNS disorders. *Frontiers in Cellular Neuroscience, 15*, 704884.

Reuben, A. (2018). Childhood lead exposure and adult neurodegenerative disease. *Journal of Alzheimer's Disease, 64*(1), 17–42.

Riccomagno, M. M., & Kolodkin, A. L. (2015). Sculpting neural circuits by axon and dendrite pruning. *Annual Review of Cell and Developmental Biology, 31*(1), 779–805.

Rivas-Fuentes, S., Salgado-Aguayo, A., Arratia-Quijada, J., & Gorocica-Rosete, P. (2021). Regulation and biological functions of the CX3CL1-CX3CR1 axis and its relevance in solid cancer: A mini-review. *Journal of Cancer, 12*(2), 571.

Rodrigue, K. M., Daugherty, A. M., Foster, C. M., & Kennedy, K. M. (2020). Striatal iron content is linked to reduced fronto-striatal brain function under working memory load. *Neuroimage, 210*, 116544.

Rodríguez-Barranco, M., Lacasaña, M., Aguilar-Garduño, C., Alguacil, J., Gil, F., González-Alzaga, B., & Rojas-García, A. (2013). Association of arsenic, cadmium and manganese exposure with neurodevelopment and behavioural disorders in children: A systematic review and meta-analysis. *Science of the Total Environment 454*, 562–577.

Saha, S., Sadhukhan, P., Mahalanobish, S., Dutta, S., & Sil, P. C. (2018). Ameliorative role of genistein against age-dependent chronic arsenic toxicity in murine brains via the regulation of oxidative stress and inflammatory signaling cascades. *The Journal of Nutritional Biochemistry, 55*, 26–40.

Sahu, P. S., & Ter, E. (2018). Interactions between neurotropic pathogens, neuroinflammatory pathways, and autophagic neural cell death. *Neuroimmunology and Neuroinflammation, 5* N-A.

Sakamoto, M., Nakamura, M., & Murata, K. (2018). Mercury as a global pollutant and mercury exposure assessment and health effects. *Nihon Eiseigaku Zasshi. Japanese Journal of Hygiene, 73*(3), 258–264.

Saper, R. B., & Rash, R. (2009). Zinc: An essential micronutrient. *American Family Physician, 79*(9), 768.

Satarker, S., Bojja, S. L., Gurram, P. C., Mudgal, J., Arora, D., & Nampoothiri, M. (2022). Astrocytic glutamatergic transmission and its implications in neurodegenerative disorders. *Cells, 11*(7), 1139.

Seif, F., Khoshmirsafa, M., Aazami, H., Mohsenzadegan, M., Sedighi, G., & Bahar, M. (2017). The role of JAK-STAT signaling pathway and its regulators in the fate of T helper cells. *Cell Communication and Signaling, 15*, 1–13.

Sensi, S. L., Paoletti, P., Bush, A. I., & Sekler, I. (2009). Zinc in the physiology and pathology of the CNS. *Nature Reviews Neuroscience, 10*(11), 780–791.

Shah, E., Butler, I., Mancias, P., Block, K., & Pérez, C. A. (2020). Mercury-induced neurotoxicity and neuroinflammation: A role for heavy metal intoxication in the pathogenesis of autoimmune diseases of the nervous system. *Neurology, 94*(15_supplement), 912.

Shao, F., Wang, X., Wu, H., Wu, Q., & Zhang, J. (2022). Microglia and neuroinflammation: Crucial pathological mechanisms in traumatic brain injury-induced neurodegeneration. *Frontiers in Aging Neuroscience, 14*, 825086.

Sharpe, M. A., Livingston, A. D., & Baskin, D. S. (2012). Thimerosal-derived ethylmercury is a mitochondrial toxin in human astrocytes: Possible role of fenton chemistry in the oxidation and breakage of MtDNA. *Journal of Toxicology, 2012*(1), 373678.

Sinha, D., & Prasad, P. (2020). Health effects inflicted by chronic low-level arsenic contamination in groundwater: A global public health challenge. *Journal of Applied Toxicology, 40*(1), 87–131.

Smeyne, R. J., Noyce, A. J., Byrne, M., Savica, R., & Marras, C. (2021). Infection and risk of Parkinson's disease. *Journal of Parkinson's Disease, 11*(1), 31–43.

Solár, P., Zamani, A., Kubíčková, L., Dubový, P., & Joukal, M. (2020). Choroid plexus and the blood–cerebrospinal fluid barrier in disease. *Fluids and Barriers of the CNS, 17*, 1–29.

Spiteri, A. G., Wishart, C. L., Pamphlett, R., Locatelli, G., & King, N. J. C. (2022). Microglia and monocytes in inflammatory CNS disease: Integrating phenotype and function. *Acta Neuropathologica*, 1–46.

Srivastava, S., Vladykovskaya, E. N., Haberzettl, P., Sithu, S. D., D'Souza, S. E., & States, J. C. (2009). Arsenic exacerbates atherosclerotic lesion formation and inflammation in ApoE-/- mice. *Toxicology and Applied Pharmacology, 241*(1), 90–100.

Steckling, N., Bose-O'Reilly, S., Pinheiro, P., Plass, D., Shoko, D., Drasch, G., ... Hornberg, C. (2014). The burden of chronic mercury intoxication in artisanal small-scale gold mining in Zimbabwe: Data availability and preliminary estimates. *Environmental Health, 13*, 1–20.

Suh, S. W., Seok Joon Won, A. M., Hamby, Byung Hoon Yoo, Y., Fan, C. T., Sheline, H., ... Liu, J. (2009). Decreased brain zinc availability reduces hippocampal neurogenesis in mice and rats. *Journal of cerebral blood flow & metabolism, 29*(9), 1579–1588.

Sun, M., You, H., Hu, X., Luo, Y., Zhang, Z., Song, Y., ... Lu, H. (2023). Microglia–astrocyte interaction in neural development and neural pathogenesis. *Cells, 12*(15), 1942.

Supramaniam, V., Vontell, R., Srinivasan, L., Wyatt-Ashmead, J., Hagberg, H., & Rutherford, M. (2013). Microglia activation in the extremely preterm human brain. *Pediatric Research, 73*(3), 301–309.

Syed, H., Muhammad Faisal Ikram, A. Y., & Ahmed, T. (2015). Cyclooxygenase I and II inhibitors distinctly enhance hippocampal- and cortex-dependent cognitive functions in mice. *Molecular Medicine Reports, 12*(5), 7649–7656.

Szekely, C. A., Thorne, J. E., Zandi, P. P., Ek, M., Messias, E., Breitner, J. C. S., & Goodman, S. N. (2004). Nonsteroidal anti-inflammatory drugs for the prevention of Alzheimer's disease: A systematic review. *Neuroepidemiology, 23*(4), 159–169.

Szewczyk, B., Kubera, M., & Nowak, G. (2011). The role of zinc in neurodegenerative inflammatory pathways in depression. *Progress in Neuro-Psychopharmacology and Biological Psychiatry, 35*(3), 693–701.

Takata, F., Nakagawa, S., Matsumoto, J., & Dohgu, S. (2021). Blood-brain barrier dysfunction amplifies the development of neuroinflammation: Understanding of cellular events in brain microvascular endothelial cells for prevention and treatment of BBB dysfunction. *Frontiers in Cellular Neuroscience, 15*, 661838.

Takeda, A. (2000). Movement of zinc and its functional significance in the brain. *Brain Research Reviews, 34*(3), 137–148.

Tang, Y., & Le., W. (2016). Differential Roles of M1 and M2 Microglia in Neurodegenerative Diseases. *Molecular Neurobiology, 53*, 1181–1194.

Tansey, M. G., Wallings, R. L., Houser, M. C., Herrick, M. K., Keating, C. E., & Joers, V. (2022). Inflammation and immune dysfunction in Parkinson disease. *Nature Reviews Immunology, 22*(11), 657–673.

Tofail, F., Vahter, M., Hamadani, J. D., Nermell, B., Huda, S. N., Yunus, M., ... Grantham-McGregor, S. M. (2009). Effect of arsenic exposure during pregnancy on infant development at 7 months in Rural Matlab, Bangladesh. *Environmental Health Perspectives, 117*(2), 288–293.

Turner, R. J., & Sharp, F. R. (2016). Implications of MMP9 for blood brain barrier disruption and hemorrhagic transformation following ischemic stroke. *Frontiers in Cellular Neuroscience, 10*, 180751.

Vibol, S., Hashim, J. H., & Sarmani, S. (2015). Neurobehavioral effects of arsenic exposure among secondary school children in the Kandal Province, Cambodia. *Environmental Research, 137*, 329–337.

Vilella, A., Belletti, D., Sauer, A. K., Hagmeyer, S., Sarowar, T., Masoni, M., Stasiak, N., et al. (2018). Reduced plaque size and inflammation in the APP23 mouse model for Alzheimer's disease after chronic application of polymeric nanoparticles for CNS targeted zinc delivery. *Journal of Trace Elements in Medicine and Biology, 49*, 210–221.

Walker, K. A., Gottesman, R. F., Wu, A., Knopman, D. S., Gross, A. L., Jr., Mosley, T. H., ... Windham, B. G. (2019). Systemic inflammation during midlife and cognitive change over 20 years: The ARIC study. *Neurology, 92*(11), e1256–e1267.

Walker, K. A., Ron, C. H., Folsom, A. R., Ballantyne, C. M., Knopman, D. S., Windham, B. G., ... Gottesman, R. F. (2023). Midlife systemic inflammatory markers are associated with late-life brain volume: The ARIC study. *Neurology, 101*(8), e856–e865.

Wang, B., Cheng, H., Lin, C., Zhang, X., Duan, X., Wang, Q., & Xu, D. (2022). Arsenic exposure analysis for children living in central China: From ingestion exposure to biomarkers. *Chemosphere, 287*, 132194.

Wang, C., Chen, S., Guo, H., Jiang, H., Liu, H., Fu, H., & Wang, D. (2022). Forsythoside a mitigates Alzheimer's-like pathology by inhibiting ferroptosis-mediated neuroinflammation via Nrf2/GPX4 axis activation. *International Journal of Biological Sciences, 18*(5), 2075.

Wang, H., Ye, M., Yu, L., Wang, J., Guo, Y., Lei, W., & Yang, J. (2015). Hippocampal neuronal cyclooxygenase-2 downstream signaling imbalance in a rat model of chronic aluminium gluconate administration. *Behavioral and Brain Functions, 11*, 1–12.

Wang, X., Fan, X., Yuan, S., Jiao, W., Liu, B., Cao, J., & Jiang, W. (2017). Chlorogenic acid protects against aluminium-induced cytotoxicity through chelation and antioxidant actions in primary hippocampal neuronal cells. *Food & Function, 8*(8), 2924–2934.

Wang, X., Huang, X., Zhou, L., Chen, J., Zhang, K., Xu, Z., Huang, ... Tang, B. (2021). Association of arsenic exposure and cognitive impairment: A population-based cross-sectional study in China. *Neurotoxicology, 82*, 100–107.

Ward, R. J., Zucca, F. A., Duyn, J. H., Crichton, R. R., & Zecca, L. (2014). The role of iron in brain ageing and neurodegenerative disorders. *The Lancet Neurology, 13*(10), 1045–1060.

World Organization Health. (2021). World failing to address dementia challenge. *WHO News*. https://www.who.int/news/item/02-09-2021-world-failing-to-address-dementia-challenge. [Google Scholar].

Wessling-Resnick, M. (2010). Iron homeostasis and the inflammatory response. *Annual Review of Nutrition, 30*(1), 105–122.

Wu, Y., Xu, X., Chen, Z., Duan, J., Hashimoto, K., Yang, L., ... Yang, C. (2020). Nervous system involvement after infection with COVID-19 and other coronaviruses. *Brain, Behavior, and Immunity, 87*, 18–22.

Xie, J., Van Hoecke, L., & Vandenbroucke, R. E. (2022). The impact of systemic inflammation on Alzheimer's disease pathology. *Frontiers in Immunology, 12*, 796867.

Xu, C., Wang, X., Zhu, Y., Dong, X., Liu, C., Zhang, H., ... Chen, L. (2016). Rapamycin ameliorates cadmium-induced activation of MAPK pathway and neuronal apoptosis by preventing mitochondrial ROS inactivation of PP2A. *Neuropharmacology, 105*, 270–284.

Xu, F., Farkas, S., Kortbeek, S., Zhang, F.-X., Chen, L., Zamponi, G. W., & Syed, N. I. (2012). Mercury-induced toxicity of rat cortical neurons is mediated through *N*-methyl-*D*-aspartate receptors. *Molecular Brain, 5*, 1–14.

Xu, X., Wang, J., & Chen, G. (2023). Circadian cycle and neuroinflammation. *Open Life Sciences, 18*(1), 20220712.

Xue, M., & Yong, V. W. (2020). Neuroinflammation in intracerebral haemorrhage: Immunotherapies with potential for translation. *The Lancet Neurology, 19*(12), 1023–1032.

Yang, L., Zhang, Y., Wang, F., Luo, Z., Guo, S., & Strähle, U. (2020). Toxicity of mercury: Molecular evidence. *Chemosphere, 245*, 125586.

Yang, Z., Yang, S., Qian, S. Y., Hong, J.-S., Kadiiska, M. B., Tennant, R. W., ... Liu, J. (2007). Cadmium-induced toxicity in rat primary mid-brain neuroglia cultures: Role of oxidative stress from microglia. *Toxicological Sciences, 98*(2), 488–494.

Yu, F., Gong, P., Hu, Z., Qiu, Y., Cui, Y., Gao, X., ... Li, J. (2015). Cu(II) enhances the effect of Alzheimer's amyloid-β peptide on microglial activation. *Journal of Neuroinflammation, 12*, 1–11.

Yu, H., Zhang, J., Ji, Q., Yu, K., Wang, P., Song, M., ... Li, Y. (2019). Melatonin alleviates aluminium chloride-induced immunotoxicity by inhibiting oxidative stress and apoptosis associated with the activation of Nrf2 signaling pathway. *Ecotoxicology and Environmental Safety, 173*, 131–141.

Yuan, Y., Jiang, C., Hu, F., Wang, Q., Zhang, K., Wang, Y., Gu, J., et al. (2015). The role of mitogen-activated protein kinase in cadmium-induced primary rat cerebral cortical neurons apoptosis via a mitochondrial apoptotic pathway. *Journal of Trace Elements in Medicine and Biology, 29*, 275–283.

Zaky, A., Mohammad, B., Moftah, M., Kandeel, K. M., & Bassiouny, A. R. (2013). Apurinic/apyrimidinic endonuclease 1 Is a key modulator of aluminum-induced neuroinflammation. *BMC Neuroscience, 14*, 1–12.

Zang, X., Chen, S., Zhu, J., Ma, J., & Zhai, Y. (2022). The emerging role of central and peripheral immune systems in neurodegenerative diseases. *Frontiers in Aging Neuroscience, 14*, 872134.

Zatterale, F., Longo, M., Naderi, J., Raciti, G. A., Desiderio, A., Miele, C., & Beguinot, F. (2020). Chronic adipose tissue inflammation linking obesity to insulin resistance and type 2 diabetes. *Frontiers in Physiology, 10*, 1607.

Zhao, R., Yu, Q., Hou, L., Dong, X., Zhang, H., Chen, X., Zhou, Z., et al. (2020). Cadmium induces mitochondrial ROS inactivation of XIAP pathway leading to apoptosis in neuronal cells. *The International Journal of Biochemistry & Cell Biology, 121*, 105715.

Chapter 9

Oxidative stress and heavy metal induced neuronal damage

Patan Ruksar, and Prasann Kumar

Department of Agronomy, School of Agriculture, Lovely Professional University, Phagwara, Punjab, India

9.1 Introduction

Alzheimer's disease (AD) is the most common form of dementia, accounting for 60%–80% of cases. Around 50 million people worldwide are living with dementia, and this number is expected to rise to 139 million by 2050 due to increasing life expectancy, placing a significant burden on healthcare systems, according to Fang (2023). AD is a progressive disorder that impairs memory, thinking, language, and problem-solving abilities, eventually leading to behavioral changes and the need for full-time care. The disease is mainly caused by neurofibrillary tangles (NFTs) and amyloid-beta (Aβ) plaques. NFTs, composed of twisted tau protein, accumulate inside brain cells and disrupt their function, while NPs, formed by amyloid-beta protein, interfere with neuron communication. Amyloid-beta oligomers may trigger inflammation and neuronal death, accelerating neurodegeneration and leading to cognitive decline, as Prithul (2017) explain. Heavy metals are persistent environmental elements that enter the food chain and accumulate in the human body, posing health risks, as Vanisree et al. (2022) observed. Exposure to toxic metals—such as arsenic (As), cadmium (Cd), lead (Pb), mercury (Hg), and manganese (Mn)—can damage neuron connections, leading to memory loss and neurological disorders, including AD, as reported by Tzioras, McGeachan, Durrant, and Spires-Jones (2023). Heavy metal toxicity affects vital organs, causing kidney and stomach problems, nerve damage, weakened immunity, and increased cancer risk. This study investigates the impact of heavy metal exposure on brain health and its link to AD. It examines how toxic metals—like Pb, Hg, Cd, and aluminum (Al)—accumulate in the brain, causing oxidative stress, inflammation, and protein damage. The research reviews scientific studies connecting heavy metal toxicity to AD. It explores preventive measures and treatments to mitigate their harmful effects on brain function, aiming to raise awareness and reduce the risk of heavy metal-induced neurodegenerative diseases.

9.2 Alzheimer's disease: pathophysiology and risk factors

AD is marked by three key pathological features: amyloid plaques, tau tangles, and neuroinflammation. Amyloid plaques form when beta-amyloid protein fragments accumulate between brain cells instead of being cleared, disrupting neuron-to-neuron communication and triggering chronic inflammation, ultimately leading to widespread neuron death, as noted by Sehar, Rawat, Reddy, Kopel, and Reddy (2022). As nerve cells die, brain tissue shrinks, exacerbating memory loss and cognitive decline, emphasized by Chaudhary et al. (2024). Another hallmark of the disease is tau tangles, where hyperphosphorylated tau detaches from microtubules, disrupting nutrient transport and causing cell death, as found by Hamano et al. (2021). Additionally, Ondaro et al. (2022) observed that abnormal tau spreads between neurons, accelerating brain damage and cognitive decline. Alongside these structural changes, neuroinflammation plays a crucial role in disease progression, as microglia and astrocytes become overactive in response to amyloid plaques and tau tangles, releasing inflammatory molecules that further damage neurons. Chronic neuroinflammation weakens the blood–brain barrier (BBB), increasing neuronal vulnerability and contributing to brain atrophy, memory impairment, and cognitive decline, supported by Mira, Lira, and Cerpa (2021) and Sehar et al. (2024). Table 9.1 provides a structured overview of the impact of various heavy metals on oxidative stress pathways, neuronal damage, and potential protective interventions. Each heavy metal affects specific brain regions through distinct

Heavy Metal Toxicity and Neurodegeneration. DOI: https://doi.org/10.1016/B978-0-443-36575-1.00013-3

TABLE 9.1 Oxidative stress and heavy metal-induced neuronal damage.

Heavy metal	Source of exposure	Targeted brain regions	Mechanism of oxidative stress induction	Neurotoxic effects	Molecular markers of oxidative stress	Associated neurological disorders	Protective strategies
Lead (Pb)	Industrial emissions, contaminated water, old paints, leaded gasoline	Hippocampus, cortex, cerebellum	Increased ROS, mitochondrial dysfunction, inhibition of antioxidant enzymes, disruption of calcium homeostasis	Apoptosis, neuroinflammation, cognitive deficits, and synaptic dysfunction	Increased MDA, decreased GSH, reduced SOD and CAT activity	Alzheimer's, cognitive impairment, ADHD	Antioxidants (vitamin C, E), chelation therapy (EDTA, DMSA), dietary interventions (zinc, selenium)
Mercury (Hg)	Fish consumption, industrial waste, and dental amalgams	Cerebellum, hippocampus, motor cortex	Lipid peroxidation, glutathione depletion, mitochondrial damage, neuroinflammation	Neuronal cell death, impaired neurotransmission, and motor dysfunction	Elevated 4-HNE, reduced GPx, and GSH levels	Parkinson's, ALS, Minamata disease	NAC, selenium, omega-3 fatty acids, glutathione boosters
Cadmium (Cd)	Cigarette smoke, contaminated food and water, industrial pollution	Hippocampus, cortex, basal ganglia	ROS overproduction, mitochondrial swelling, inhibition of antioxidant defenses	Neurodegeneration, synaptic impairment, learning, and memory deficits	Increased MDA, decreased SOD and CAT activity	Alzheimer's, cognitive decline, neurodevelopmental disorders	Curcumin, quercetin, lipoic acid, selenium
Arsenic (As)	Contaminated groundwater, pesticides, industrial waste	Cortex, hippocampus, basal ganglia	Disruption of mitochondrial function, inhibition of DNA repair, lipid peroxidation	Neuroinflammation, apoptosis, altered neurotransmitter levels	Increased MDA, reduced GSH, and GPx levels	Alzheimer's, cognitive decline, neuropathy	Polyphenols, chelation therapy, B vitamins
Aluminum (Al)	Processed foods, cookware, antiperspirants, industrial exposure	Hippocampus, cortex	Increased ROS, impairment of mitochondrial respiration, inhibition of antioxidant enzymes	Neuronal loss, tau hyperphosphorylation, cognitive impairment	Elevated MDA, decreased GPx, SOD, and CAT activity	Alzheimer's, Parkinson's, neurodegenerative disorders	Silica-rich water, flavonoids, curcumin, vitamin E

NAC, *N*-acetylcysteine; *ROS*, reactive oxygen species.

oxidative mechanisms, leading to various neurotoxic outcomes and associated neurological disorders. Protective strategies include antioxidants, chelation therapies, and dietary modifications to mitigate the effects of heavy metal exposure.

9.3 Genetic risk factors

AD results from a combination of genetic and environmental factors. The APOE ε4 gene is the strongest genetic predictor of late-onset AD, influencing cholesterol metabolism and amyloid-beta clearance, increasing plaque buildup and neuronal damage, as Koutsodendris, Nelson, Rao, and Huang (2022) reported. Early-onset familial AD has been associated with mutations in the APP, PSEN1, and PSEN2 genes, leading to excessive amyloid-beta production and plaque formation, contributing significantly to neurodegeneration, as highlighted. Additionally, other genetic variants, such as TREM2 and CLU, play a crucial role in neuroinflammation and amyloid metabolism, further influencing AD progression, as Lista et al. (2024) emphasized.

Beyond genetic predisposition, environmental factors, particularly heavy metal exposure, significantly contribute to AD development. Toxic metals such as Pb, Hg, Cd, and Al have been implicated in neurotoxicity through mechanisms that promote amyloid aggregation, enhance tau hyperphosphorylation, and generate oxidative stress, triggering chronic inflammation and neuronal damage, as noted by Kaur et al. (2021). Furthermore, heavy metals disrupt mitochondrial function, impairing neuronal energy production and accelerating neurodegeneration, as Vellingiri et al. (2022) demonstrated. The interaction between genetic predisposition and environmental factors further amplifies the risk of AD. Individuals carrying the APOE ε4 allele are more vulnerable to heavy metal toxicity due to impaired detoxification, leading to increased neurotoxicity. This heightened sensitivity to environmental toxins accelerates amyloid accumulation and neurodegeneration, as found by Nabi and Tabassum (2022).

To mitigate these risks, preventive strategies focus on minimizing heavy metal exposure by avoiding contaminated water, food, and industrial pollutants. Antioxidants such as vitamin E, curcumin, and metal chelators have shown potential in counteracting metal-induced neurotoxicity, as Smirnova et al. (2023) suggested. Advances in personalized medicine, including genetic screening, can aid in identifying at-risk individuals and developing targeted interventions. Understanding the interplay between genetic predisposition and environmental exposures is crucial for designing effective prevention and therapeutic strategies. A multifaceted approach that integrates genetics, lifestyle modifications, and environmental management is essential for combating AD, as emphasized by Zhang et al. (2021).

9.4 Role of oxidative stress and neurotoxicity in Alzheimer's disease progression

AD progresses through oxidative stress and neurotoxicity, leading to neuronal damage, synaptic dysfunction, and cognitive decline. Oxidative stress arises when reactive oxygen species (ROS) exceed antioxidant defenses, causing lipid peroxidation, protein and DNA modifications, and mitochondrial dysfunction. Demirci-Cekic et al. (2022) reported that excessive ROS damages cell membranes, impairing neuronal function, while researchers highlighted mitochondrial damage as a key factor in AD progression due to impaired energy metabolism—neurotoxicity results from amyloid-beta (Aβ), tau aggregates, and heavy metals. Aβ oligomers induce oxidative stress, inflammation, and excitotoxicity, leading to synaptic dysfunction and neuron loss. Tau tangles disrupt cellular transport, accelerating neurodegeneration, as demonstrated by Thal and Tomé (2022). Additionally, heavy metals exacerbate oxidative damage, while excitotoxicity from excessive glutamate receptor activation promotes neuron death, as emphasized by Sinsky, Pichlerova, and Hanes (2021). Oxidative stress and neurotoxicity reinforce each other, amplifying amyloid-beta and tau pathology. Inflammation further accelerates brain damage and cognitive decline, as Teleanu et al. (2022) noted. To mitigate these effects, potential therapeutic strategies include antioxidants such as vitamins E and C, polyphenols like curcumin and resveratrol, and glutathione precursors, which help neutralize ROS and protect neuronal function, as suggested by Pisoschi et al. (2021). Supporting mitochondrial function with Coenzyme Q10 and NAD^+ precursors may also contribute to neuronal protection. Additionally, metal chelators that bind to toxic metals could help reduce neurotoxicity, while NMDA receptor modulators such as memantine assist in regulating glutamate levels and preventing excitotoxicity. Mazur, Malik, and Bieńko (2024) highlighted that targeting these mechanisms may offer promising avenues for slowing AD progression (Table 9.2). This table highlights the key oxidative stress factors contributing to AD, their neurotoxic mechanisms, affected brain regions, molecular markers, neurodegenerative effects, cognitive decline associations, and potential therapeutic strategies. Each factor plays a crucial role in disease progression, emphasizing the need for targeted interventions.

9.5 Heavy metals and their role in neurodegeneration

Heavy metal exposure poses significant health risks, with Pb, Hg, Cd, As, and Al being substantial contaminants. These metals enter the environment through industrial pollution, household products, and contaminated food and

TABLE 9.2 Role of oxidative stress and neurotoxicity in Alzheimer's disease progression.

Oxidative stress factor	Neurotoxic mechanism	Impacted brain regions	Molecular markers of oxidative damage	Neurodegenerative effects	Link to cognitive decline	Potential therapeutic strategies
ROS	Lipid peroxidation, mitochondrial dysfunction	Hippocampus, cortex, basal forebrain	Increased MDA, 4-HNE	Apoptosis, synaptic dysfunction	Memory loss, impaired learning	Antioxidants (vitamin E, C, polyphenols)
Mitochondrial dysfunction	ATP depletion, ROS overproduction	Hippocampus, cortex	Reduced SOD, CAT, GPx activity	Neuronal energy failure, apoptosis	Executive dysfunction, cognitive decline	Mitochondrial-targeted therapies (CoQ10, MitoQ)
Lipid peroxidation	Disruption of membrane integrity, increased permeability	Hippocampus, cortex	Elevated MDA, 4-HNE	Neuronal cell death, neuroinflammation	Declining cognitive flexibility	Polyphenols, omega-3 fatty acids
Protein oxidation	Aggregation of misfolded proteins	Hippocampus, cortex	Increased oxidized proteins	Aβ accumulation, tau hyperphosphorylation	Learning impairment, dementia	Antioxidant-rich diet, proteostasis regulators
DNA oxidation	Strand breaks, impaired repair mechanisms	Hippocampus, prefrontal cortex	Increased 8-OHdG levels	Cell cycle arrest, apoptosis	Long-term memory deficits	DNA repair enhancers, NAD+ boosters
Neuroinflammation	Activation of microglia, cytokine release	Hippocampus, basal forebrain	Increased TNF-α, IL-6	Chronic inflammation, neuronal damage	Progressive cognitive impairment	Antiinflammatory agents, curcumin, resveratrol
Tau hyperphosphorylation	ROS-induced kinase activation	Cortex, hippocampus	Increased p-Tau levels	Neurofibrillary tangle formation	Executive dysfunction, memory loss	Kinase inhibitors, tau stabilizers
Aβ toxicity	Oxidative stress-mediated aggregation	Hippocampus, cortex	Increased Aβ oligomers	Synaptic loss, neuronal death	Early-stage memory deficits	Aβ-targeting immunotherapies, metal chelators

ROS, Reactive oxygen species.

water, leading to severe health effects. Lead, commonly found in batteries, paints, and pipes, affects brain development in children, causing learning difficulties and behavioral problems. In adults, prolonged exposure results in high blood pressure, kidney issues, and fertility problems, as highlighted by Khalaf et al. (2024), who emphasized its persistence as a significant public health concern.

Mercury, present in thermometers, dental fillings, and coal-burning power plants, is highly toxic to the nervous system, leading to memory loss and nerve damage. Methylmercury, an even more toxic form found in fish, poses additional risks, particularly for pregnant women and children. Chen and Dong (2022) stressed the need for strict regulation due to mercury's bioaccumulative nature and severe neurological impacts. Similarly, cadmium, found in batteries, plastics, and fertilizers, is released into the environment through industrial emissions and cigarette smoke. Long-term Cd exposure damages the kidneys, weakens bones, and increases cancer risk. Charkiewicz, Omeljaniuk, Nowak, Garley, and Nikliński (2023) noted its widespread presence and the necessity of limiting exposure to reduce health risks. Arsenic, a naturally occurring element in groundwater, is used in pesticides, wood preservatives, and industrial processes. Contaminated drinking water leads to skin problems, heart disease, nerve damage, and various cancers. Daripa, Malav, Yadav, and Chattaraj (2023) emphasized that As exposure remains a global concern, particularly in regions with high groundwater contamination. Aluminum, widely used in food packaging, cookware, and medications, poses health risks. While not as hazardous as other heavy metals, excessive aluminum exposure has been linked to neurological disorders such as AD, as well as bone issues and kidney damage. Alkaya, Demirci, and Şevik (2022) stressed the importance of reducing exposure through safer consumer products and water purification methods to mitigate long-term health risks. Heavy metals significantly threaten brain health by contributing to neurological diseases through various toxic mechanisms. Oxidative stress is critical in neurodegeneration, generating harmful free radicals that degrade brain cells, as highlighted by Sadiq (2023). Additionally, heavy metals interfere with neurotransmitter functions, which are essential for brain cell communication, with Pb and Hg causing cognitive and behavioral issues, as reported by Kaur et al. (2021). Chronic exposure further triggers brain inflammation, increases the risk of Alzheimer's, and damages mitochondria, impairing energy production, and leading to cell death. Another concerning effect is the disruption of the BBB, which allows toxic substances to accumulate in the brain, as demonstrated by Singh and Vellapandian (2023). Given these severe effects, minimizing heavy metal exposure is crucial for protecting brain function and reducing the risk of neurological disorders.

9.6 Mechanisms of heavy metal–induced neurotoxicity

Heavy metal exposure significantly threatens brain health by disrupting essential cellular functions and leading to neurodegeneration. This chapter explores key neurotoxic mechanisms, including oxidative stress, mitochondrial dysfunction, disrupted metal balance, DNA damage, and BBB permeability. Oxidative stress and mitochondrial dysfunction play a crucial role in heavy metal-induced neurotoxicity. Metals such as Pb, Hg, As, and Cd generate excessive ROS, causing oxidative stress that damages lipids, proteins, and DNA, as Mitra et al. (2022) reported. Additionally, these metals impair mitochondrial function by disrupting ATP production, reducing energy availability, and triggering cell death, ultimately contributing to neurodegenerative diseases, as highlighted by Facecchia, Fochesato, Ray, Stohs, and Pandey (2011). Another critical factor is the disruption of metal balance and neurotransmission. The brain relies on a precise balance of metal ions for neurotransmission and enzymatic activity. Heavy metals interfere with this balance by replacing essential metals such as iron (Fe), zinc (Zn), and copper (Cu), as Ijomone, Ifenatuoha, Aluko, Ijomone, and Aschner (2020) demonstrated. Lead disrupts calcium signalling in neurons, weakening synaptic communication, while mercury and cadmium impair zinc-dependent enzymes, negatively affecting synaptic function and leading to nerve cell damage, as evidenced by Jomova, Alomar, Nepovimova, Kuca, and Valko (2024).

Beyond direct cellular damage, heavy metals also induce DNA damage and epigenetic changes. These metals bind to genetic material or induce oxidative stress, resulting in mutations and genetic instability. Liu, Du, Ge, and Liu (2024) noted that heavy metals alter gene expression through epigenetic mechanisms, including DNA methylation, histone modifications, and microRNA interference. Lead exposure has been linked to increased DNA methylation in genes essential for brain development, potentially leading to cognitive and behavioral impairments, as emphasized by Dutta et al. (2018).

Furthermore, heavy metals compromise the integrity of the BBB, which serves as a protective shield preventing harmful substances from entering the brain. They induce inflammation, oxidative stress, and vascular endothelial damage, weakening this critical barrier. Romero, Abbott, and Bradbury (2023) found that Hg can cross the BBB via amino acid transporters, accumulating in brain tissues and impairing neural function. Additional toxins can penetrate the brain when the BBB is compromised, further accelerating neurodegeneration, as demonstrated by Agnihotri and Kesari (2019). Finally, the synergistic toxicity of multiple heavy metals intensifies their harmful effects. Exposure to a combination of heavy metals, such as Pb and As, amplifies oxidative stress and disrupts neurotransmitter function more severely than exposure to a single metal. Wu, Chen, and Huang (2025) reported that these interactions increase toxicity, making even low levels of exposure more hazardous than previously assumed (Table 9.3). Table 9.3 provides a detailed overview of the mechanisms through which heavy metals induce neurotoxicity, their effects on neurotransmitter systems, associated cognitive and neurological disorders, and potential mitigation strategies.

9.7 Epidemiological and experimental evidence

Epidemiological studies have established a strong link between heavy metal exposure and the development of AD. Bakulski et al. (2020) reported that exposure to metals such as Pb, Hg, As, Cd, and Al increases the risk of cognitive decline and neurodegeneration. Long-term lead exposure, particularly in occupational settings or during early childhood, accelerates cognitive ageing and contributes to amyloid plaque accumulation in the brain. Similarly, Doroszkiewicz et al. (2023) found that individuals consuming high amounts of seafood, a significant source of methylmercury, exhibit AD-like symptoms and memory impairments. Additionally, drinking water contaminated with As has been associated with brain inflammation and tau protein buildup, while cadmium exposure has been linked to memory deficits and attention disorders. Kandimalla, Vallamkondu, Corgiat, and Gill (2016) suggested that aluminum, a common element in food and water, promotes AD pathology by facilitating amyloid plaque formation and inducing oxidative stress. Animal model studies further support these findings, demonstrating the neurotoxic effects of heavy metals on brain function. Research on genetically modified mice predisposed to AD has shown that early-life lead exposure results in increased amyloid-beta production and plaque formation, even long after exposure has ceased. Lau, Ramer, and Tremblay (2023) observed that mice exposed to methylmercury develop brain inflammation, oxidative stress, tau protein accumulation, and memory deficits.

Similarly, As exposure in rats leads to hippocampal neuron damage, cognitive dysfunction, and tau aggregation. Cariccio, Samà, Bramanti, and Mazzon (2019) found that cadmium exposure impairs neuronal plasticity, disrupts brain function, and increases oxidative stress, all of which contribute to AD progression. Laboratory studies have further demonstrated the toxic effects of heavy metals on brain cells. Human neuronal cells exposed to lead show increased amyloid plaque production, mitochondrial dysfunction, and cell death, mimicking AD pathology. Mercury exposure, as reported by Bakulski et al. (2020), disrupts calcium homeostasis, promotes oxidative damage, and accelerates tau accumulation, contributing to neuronal degeneration (Table 9.4). Azar, Yousef, El-Fawal, and Abdelnaser (2021) found that arsenic alters gene expression in brain cells, leading to chronic inflammation and impairments in cellular function. Similarly, cadmium disrupts mitochondrial function, induces endoplasmic reticulum stress, and activates apoptotic pathways, further contributing to neurodegeneration, as highlighted by Branca et al. (2020).

9.8 Prevention and potential therapeutic interventions

Heavy metals such as Pb, Hg, Cd, and Al contribute to AD by inducing oxidative stress, inflammation, and protein aggregation in the brain, as Amadi, Offor, Frazzoli, and Orisakwe (2019) noted. Preventing exposure and developing treatments are crucial, with strategies including reducing environmental exposure, using chelation therapy, incorporating antioxidants, and exploring potential drug targets, as Flora, Mittal, and Mehta (2008) discussed. Governments worldwide have implemented regulations to control industrial emissions, improve water and air quality, and ensure food safety. Water purification systems, such as reverse osmosis and activated carbon filters, effectively remove heavy metals from drinking water, as Hossain et al. (2024) highlighted. Soil contamination can also be managed through phytoremediation, where plants like sunflowers absorb toxic metals, as Rizwan et al. (2016) explained.

Dietary modifications significantly minimise exposure, such as avoiding high-mercury fish, using nonaluminum cookware, and increasing fibre intake to help eliminate metal, as Mckelvey, Oken, Spills, and Advisories (2012) described. Essential minerals such as Zn and selenium (Se) aid in reducing the body's absorption of harmful metals, as noted by Nabrzyski (2006). Chelation therapy, a widely used treatment, employs agents like EDTA, DMSA, and BAL to facilitate heavy metal removal, with research suggesting its potential benefits in Alzheimer's treatment, as explored by Fulgenzi, Vietti, and Ferrero (2020) and Bjørklund et al. (2019). In addition to synthetic chelators, natural compounds such as curcumin, cilantro, and chlorella have effectively detoxified the body, as García-Niño and Pedraza-Chaverrí (2014) explained. However, caution is necessary, as excessive chelation can lead to nutrient depletion, as cautioned by Mehrandish, Rahimian, and Shahriary (2019).

Antioxidants, particularly polyphenols like resveratrol and quercetin, are crucial in reducing oxidative damage and brain inflammation, as discussed by Bertelli and Das (2009) and Calis, Mogulkoc, and Baltaci (2020). Vitamins C and E contribute to heavy metal detoxification, while Zn and magnesium (Mg) prevent metal accumulation in brain cells, as Zatta, Lucchini, van Rensburg, and Taylor (2003) described. Compounds such as Coenzyme Q10 and alpha-lipoic acid also enhance mitochondrial function and reduce cell death induced by toxic metals, as Tripathi et al. (2023) noted. Research continuously explores innovative approaches, including nanoparticle-based chelators that selectively target toxic metals while preserving essential minerals (Table 9.5). Further studies focus on the

TABLE 9.3 Extensive table on mechanisms of heavy metal-induced neurotoxicity.

Heavy metal	Primary sources of exposure	Major affected brain regions	Cellular and molecular mechanisms	Neurotransmitter alterations	Cognitive and behavioral effects	Associated neurological diseases	Potential mitigation strategies
Lead (Pb)	Paint, water pipes, industrial emissions, contaminated soil	Hippocampus, cortex, cerebellum	Oxidative stress, mitochondrial dysfunction, apoptosis, disruption of calcium homeostasis	Decreased dopamine and glutamate, altered GABAergic function	Memory impairment, reduced IQ, attention deficits, hyperactivity	Alzheimer's disease, ADHD, autism spectrum disorder	Chelation therapy (EDTA, DMSA), antioxidant supplementation (vitamin C, E), iron and calcium-rich diet
Mercury (Hg)	Fish consumption, dental amalgams, industrial emissions	Cerebellum, cortex, basal ganglia	Lipid peroxidation, neuroinflammation, disruption of microtubules, excitotoxicity	Inhibition of glutamate reuptake, altered dopamine function	Tremors, mood disorders, cognitive decline, motor dysfunction	Minamata disease, autism spectrum disorder, Parkinson's disease	Selenium supplementation, chelation therapy, dietary restrictions on seafood
Cadmium (Cd)	Industrial waste, contaminated food and water, cigarette smoke	Hippocampus, cortex	DNA damage, oxidative stress, inhibition of neuronal differentiation	Reduced acetylcholine and dopamine levels	Learning and memory deficits, anxiety, depression	Alzheimer's disease, cognitive decline	Zinc and selenium supplementation, antioxidant therapy, dietary interventions
Arsenic (As)	Contaminated drinking water, pesticides, industrial exposure	Hippocampus, cortex, cerebellum	Epigenetic modifications, mitochondrial dysfunction, ROS generation, apoptosis	Altered cholinergic and glutamatergic function	Mood disorders, learning disabilities, memory loss	Alzheimer's disease, cognitive impairment	Chelation therapy, folate and vitamin B12 supplementation, drinking purified water
Aluminum (Al)	Food additives, cookware, industrial exposure, vaccines	Hippocampus, cortex	Tau protein hyperphosphorylation, mitochondrial dysfunction, oxidative stress	Impaired cholinergic transmission	Memory impairment, confusion, cognitive decline	Alzheimer's disease	Dietary modifications, silicon-rich water, chelation therapy
Manganese (Mn)	Welding fumes, contaminated water, occupational exposure	Basal ganglia, cortex	Oxidative stress, mitochondrial dysfunction, impaired astrocytic glutamate uptake	Increased glutamate excitotoxicity, reduced dopamine release	Motor dysfunction, mood disorders, cognitive impairment	Parkinson's-like symptoms, neurodevelopmental disorders	Iron supplementation, chelation therapy (EDTA), limiting occupational exposure
Iron (Fe) overload	Excessive iron supplementation, genetic disorders (hemochromatosis)	Hippocampus, cortex	Fenton reaction-induced oxidative stress, neuroinflammation, protein aggregation	Altered dopamine and serotonin metabolism	Memory impairment, depression, motor dysfunction	Alzheimer's disease, Parkinson's disease	Iron chelators (deferoxamine), dietary modifications
Copper (Cu) overload	Wilson's disease, contaminated water, dietary sources	Basal ganglia, cortex	ROS production, mitochondrial dysfunction, protein misfolding	Altered dopamine metabolism	Mood disorders, cognitive decline, motor dysfunction	Wilson's disease, Alzheimer's disease	Chelation therapy (penicillamine, trientine), zinc supplementation

ROS, Reactive oxygen species.

TABLE 9.4 Epidemiological and experimental evidence.

Type of study	Study design	Population/model	Exposure/ intervention	Outcome measured	Key findings	Limitations
Epidemiological	Cohort study	Humans (adults)	Air pollution (PM2.5)	Cardiovascular disease incidence	Higher PM2.5 exposure linked to increased heart disease risk	Confounding variables (e.g., smoking, diet)
Epidemiological	Case-control study	Humans (cancer patients vs. controls)	Pesticide exposure	Cancer risk (e.g., leukemia)	Certain pesticides associated with elevated cancer risk	Recall bias, selection bias
Epidemiological	Cross-sectional study	Humans (general population)	Processed food consumption	Obesity prevalence	Higher intake linked to obesity and metabolic syndrome	Cannot infer causation
Epidemiological	Ecological study	Countries (aggregated data)	High sugar consumption	Diabetes prevalence	Strong correlation but no direct causation	Ecological fallacy
Epidemiological	RCT	Humans (high-risk population)	Omega-3 supplementation	Cardiovascular events	No significant reduction in heart attack rates	Compliance issues, duration limitations
Epidemiological	Mendelian randomization	Humans (genome-wide data)	Genetic predisposition to obesity	BMI and cardiovascular risk	Genetic risk linked to metabolic disorders	Population stratification issues
Experimental	Animal study (rodents)	Mice (C57BL/6)	High-fat diet	Insulin resistance, obesity	Mice developed metabolic syndrome	Species differences limit human extrapolation
Experimental	Animal study (Nonhuman primates)	Rhesus monkeys	Long-term calorie restriction	Longevity, age-related diseases	Increased lifespan, reduced diabetes risk	Expensive, ethical concerns
Experimental	In vitro study (cell culture)	Human liver cells (HepG2)	Exposure to BPA (bisphenol A)	Oxidative stress markers	Increased oxidative stress, DNA damage	Lacks systemic metabolism of whole organisms
Experimental	Gene knockout study	Mice (CRISPR gene-editing)	Deletion of obesity-related gene	Obesity resistance, metabolic rate	Mice lacking the gene had lower fat accumulation	It may not translate directly to humans
Experimental	Drug trial (phase I)	Healthy volunteers	New cancer drug	Safety, pharmacokinetics	Tolerable side effects, good absorption	Small sample size
Experimental	Drug trial (phase II)	Cancer patients	Immunotherapy drug	Tumor regression, survival rate	Promising early results	Need for larger trials
Experimental	Drug trial (phase III)	Multinational population	COVID-19 vaccine	Infection rates, immunity duration	95% efficacy in preventing symptomatic cases	Long-term immunity unknown
Experimental	Environmental exposure study	Rats	Heavy metal (lead) exposure	Neurotoxicity, cognitive decline	Increased brain inflammation	Differences in human metabolism

RCT, Randomized controlled trial.

TABLE 9.5 Prevention and potential therapeutic interventions.

Category	Approach type	Target population	Mechanism of action	Examples	Effectiveness	Limitations
Primary prevention	Lifestyle modification	General public	Reduces risk factors through behavior change	Regular exercise, healthy diet, smoking cessation	Strong evidence for reducing chronic disease risk	Requires long-term adherence
Primary prevention	Immunisation	General public, high-risk groups	Stimulates the immune system to prevent disease	COVID-19 vaccine, HPV vaccine, Flu vaccine	Highly effective in preventing infectious diseases	Vaccine hesitancy, rare adverse effects
Primary prevention	Environmental control	General public	Reducing exposure to harmful substances	Air pollution control, clean water initiatives	Reduces disease burden	High implementation costs
Primary prevention	Genetic screening	High-risk individuals	Identifies genetic predispositions to diseases	BRCA1/2 screening for breast cancer	Allows early interventions	Ethical and privacy concerns
Primary prevention	Nutritional supplementation	General public, at-risk populations	Addresses micronutrient deficiencies	Vitamin D for bone health, Iron for anaemia	Effective in specific populations	Overuse risks toxicity
Secondary prevention	Early screening & diagnosis	High-risk individuals	Detects diseases at an early, treatable stage	Mammography, Pap smear, Colonoscopy	Increases survival rates	False positives, accessibility issues
Secondary prevention	Pharmacological prophylaxis	High-risk individuals	Prevents disease onset with medications	Aspirin for cardiovascular disease prevention	Reduces incidence of strokes/heart attacks	Risk of bleeding, long-term effects
Secondary prevention	Behavioral counselling	High-risk ndividuals	Encourages risk reduction behaviors	Smoking cessation programs, Alcohol intervention	Improves long-term health outcomes	Requires sustained engagement
Secondary prevention	Wearable health monitoring	At-risk populations	Continuous monitoring for early disease signs	Smartwatches for arrhythmia detection	Improves early detection	Data privacy concerns
Tertiary prevention	Rehabilitation & physical therapy	Patients with chronic disease	Restores function, prevents disability	Stroke rehabilitation, Cardiac rehab	Improves quality of life	Requires long-term adherence
Tertiary prevention	Chronic disease management	Patients with established conditions	Controls disease progression	Insulin therapy for diabetes, Dialysis for kidney failure	Prolongs life improves function	Expensive, side effects
Tertiary prevention	Palliative care	Terminally ill patients	Improves comfort and quality of life	Pain management in cancer patients	Enhances end-of-life care	Ethical dilemmas, resource-intensive
Tertiary prevention	Mental health support	Patients with chronic illness	Addresses the psychological burden of illness	Cognitive-behavioral therapy for chronic pain	Reduces depression, improves adherence	Stigma, accessibility issues
Therapeutic interventions	Pharmacological treatments	Patients with active disease	Directly targets disease mechanisms	Antibiotics, Chemotherapy, Antiviral drugs	High efficacy in treating specific diseases	Drug resistance, side effects

(Continued)

TABLE 9.5 (Continued)

Category	Approach type	Target population	Mechanism of action	Examples	Effectiveness	Limitations
Therapeutic interventions	Regenerative medicine	Patients with tissue damage	Uses stem cells/tissue engineering for repair	Stem cell therapy for spinal injuries	Promising for tissue regeneration	Ethical concerns, expensive
Therapeutic interventions	Gene therapy	Patients with genetic disorders	Modifies genetic material to correct disease	CRISPR for genetic diseases	Potentially curative for inherited disorders	High cost, regulatory hurdles
Therapeutic interventions	Personalised medicine	Patients with specific biomarkers	Tailors treatment based on genetic/molecular profile	Targeted cancer therapy (e.g., HER2 inhibitors)	Increases treatment precision	Requires advanced diagnostics
Therapeutic interventions	Immunotherapy	Cancer, autoimmune patients	Enhances immune response against disease	CAR-T cell therapy, Checkpoint inhibitors	Effective for cancers, autoimmune diseases	Severe immune-related side effects
Therapeutic interventions	Microbiome-based therapy	Patients with dysbiosis-linked disorders	Restores healthy gut microbiota	Fecal microbiota transplantation for C. difficile infection	Promising for gut disorders	Requires more research, regulatory challenges
Therapeutic interventions	Neuromodulation	Neurological and psychiatric patients	Modifies nerve activity using devices	Deep Brain Stimulation for Parkinson's	Effective in some neuropsychiatric conditions	Invasive, costly
Therapeutic interventions	Artificial organ and bionics	Patients with organ failure	Replaces or enhances failing organs	Artificial heart, Bionic limbs	Improves function, extends life	High cost, rejection risk

interaction between heavy metals and amyloid-beta and tau proteins, which are key markers of AD, with drugs like Clioquinol and PBT2 showing potential in mitigating neurodegeneration, as highlighted by Watt, Villemagne, and Barnham (2012) and Thakur, Karwasra, and Umar (2023).

In addition to these strategies, scientists are investigating how heavy metals influence gene expression and epigenetic changes, with CRISPR technology offering potential interventions, as explored by Hamad (2022). The gut microbiome has also emerged as a critical factor in metal detoxification, with probiotics such as *Lactobacillus rhamnosus* demonstrating the ability to reduce heavy metal absorption in the intestines, potentially lowering overall toxicity levels, as noted by Hamad (2022).

9.9 Conclusion and future perspectives

AD is a serious condition that affects memory, thinking ability, and behavior, mainly in older adults. Research shows that exposure to heavy metals like mercury (Hg), lead (Pb), cadmium (Cd), and arsenic (As) plays a significant role in the development of this disease. These harmful metals can enter the brain, causing stress, inflammation, and cell damage, which eventually lead to memory loss and other symptoms of Alzheimer's. Long-term exposure to metals like Al and Cu has also been linked to the formation of protein clumps in the brain, which are common signs of the disease. Since heavy metals are found in the environment through industrial pollution, contaminated food, and water, it is essential to continue studying their impact on brain health.

Future research should focus on understanding how these metals affect the brain over time, identifying early warning signs of the disease, and finding ways to prevent or reverse the damage. To reduce the risk, governments and health organizations should control pollution, ensure safe drinking water, and spread awareness about the dangers of heavy metal exposure. On a medical level, treatments like chelation therapy and antioxidants could help remove harmful metals from the body and protect brain cells. By taking these measures, we can work toward better prevention, early diagnosis, and improved treatments for Alzheimer's, ultimately reducing its impact on individuals and society.

References

Agnihotri, S. K., & Kesari, K. K. (2019). Mechanistic effect of heavy metals in neurological disorder and brain cancer. *Networking of Mutagens in Environmental Toxicology,* 25–47.

Alkaya, G. B., Demirci, Ç., & Şevik, H. (2022). Aluminium in food and potential role on Alzheimer's disease of aluminium. *Turkish Journal of Engineering, 6*(2), 118–127.

Amadi, C. N., Offor, S. J., Frazzoli, C., & Orisakwe, O. E. (2019). Natural antidotes and management of metal toxicity. *Environmental Science and Pollution Research, 26*, 18032–18052.

Azar, J., Yousef, M. H., El-Fawal, H. A., & Abdelnaser, A. (2021). Mercury and Alzheimer's disease: A look at the links and evidence. *Metabolic Brain Disease, 36*, 361–374.

Bakulski, K. M., Seo, Y. A., Hickman, R. C., Brandt, D., Vadari, H. S., Hu, H., ... Park, S. K. (2020). Heavy metals exposure and Alzheimer's disease and related dementias. *Journal of Alzheimer's Disease, 76*(4), 1215–1242.

Bertelli, A. A., & Das, D. K. (2009). Grapes, wines, resveratrol, and heart health. *Journal of Cardiovascular Pharmacology, 54*(6), 468–476.

Bjørklund, G., Crisponi, G., Nurchi, V. M., Cappai, R., Buha Djordjevic, A., & Aaseth, J. (2019). A review on coordination properties of thiol-containing chelating agents towards mercury, cadmium, and lead. *Molecules (Basel, Switzerland), 24*(18), 3247.

Branca, J. J., Fiorillo, C., Carrino, D., Paternostro, F., Taddei, N., Gulisano, M., ... Becatti, M. (2020). Cadmium-induced oxidative stress: Focus on the central nervous system. *Antioxidants, 9*(6), 492.

Calis, Z., Mogulkoc, R., & Baltaci, A. K. (2020). The roles of flavonols/flavonoids in neurodegeneration and neuroinflammation. *Mini Reviews in Medicinal Chemistry, 20*(15), 1475–1488.

Cariccio, V. L., Samà, A., Bramanti, P., & Mazzon, E. (2019). Mercury involvement in neuronal damage and in neurodegenerative diseases. *Biological Trace Element Research, 187*(2), 341–356.

Charkiewicz, A. E., Omeljaniuk, W. J., Nowak, K., Garley, M., & Nikliński, J. (2023). Cadmium toxicity and health effects—A brief summary. *Molecules (Basel, Switzerland), 28*(18), 6620.

Chaudhary, R. K., Mateti, U. V., Khanal, P., Rawal, K. B., Jain, P., Patil, V. S., ... Patil, B. M. (2024). Alzheimer's disease: Epidemiology, neuropathology, and neurochemistry. In *Computational and experimental studies in Alzheimer's disease*. CRC Press, 1–14.

Chen, B., & Dong, S. (2022). Mercury contamination in fish and its effects on the health of pregnant women and their fetuses, and guidance for fish consumption—a narrative review. *International Journal of Environmental Research and Public Health, 19*(23), 15929.

Daripa, A., Malav, L. C., Yadav, D. K., & Chattaraj, S. (2023). Metal contamination in water resources due to various anthropogenic activities. In *Metals in water*. Elsevier, 111–127.

Demirci-Cekic, S., Özkan, G., Avan, A. N., Uzunboy, S., Çapanoğlu, E., & Apak, R. (2022). Biomarkers of oxidative stress and antioxidant defense. *Journal of Pharmaceutical and Biomedical Analysis, 209*, 114477.

Doroszkiewicz, J., Farhan, J. A., Mroczko, J., Winkel, I., Perkowski, M., & Mroczko, B. (2023). Common and trace metals in Alzheimer's and Parkinson's diseases. *International Journal of Molecular Sciences, 24*(21), 15721.

Dutta, S., Mitra, M., Agarwal, P., Mahapatra, K., De, S., Sett, U., ... Roy, S. (2018). Oxidative and genotoxic damages in plants in response to heavy metal stress and maintenance of genome stability. *Plant Signaling & Behavior, 13*(8), e1460048.

Facecchia, K., Fochesato, L. A., Ray, S. D., Stohs, S. J., & Pandey, S. (2011). Oxidative toxicity in neurodegenerative diseases: role of mitochondrial dysfunction and therapeutic strategies. *Journal of Toxicology, 2011*(1), 683728.

Fang, X. (2023). *Effects of beta-amyloid on cerebral hypoperfusion in Alzheimer's disease* (Doctoral dissertation, The University of Mississippi Medical Center).

Flora, S. J. S., Mittal, M., & Mehta, A. (2008). Heavy metal induced oxidative stress & its possible reversal by chelation therapy. *Indian Journal of Medical Research, 128*(4), 501–523.

Fulgenzi, A., Vietti, D., & Ferrero, M. E. (2020). EDTA chelation therapy in the treatment of neurodegenerative diseases: An update. *Biomedicines, 8*(8), 269.

García-Niño, W. R., & Pedraza-Chaverrí, J. (2014). Protective effect of curcumin against heavy metals-induced liver damage. *Food and Chemical Toxicology, 69*, 182–201.

Hamad, N. (2022). *Investigating the role of PRDX1 in regulating the metallothionein genes in detoxifying heavy metals* (Master's thesis). Qatar: Hamad Bin Khalifa University.

Hamano, T., Enomoto, S., Shirafuji, N., Ikawa, M., Yamamura, O., Yen, S. H., ... Nakamoto, Y. (2021). Autophagy and tau protein. *International Journal of Molecular Sciences, 22*(14), 7475.

Hossain, A., Ahmed, M. W., Rabin, M. H., Kaium, A., Razzaque, M. A., & Zamil, S. S. (2024). Heavy metal quantification in chicken meat and egg: An emerging food safety concern. *Journal of Food Composition and Analysis, 126*, 105876.

Ijomone, O. M., Ifenatuoha, C. W., Aluko, O. M., Ijomone, O. K., & Aschner, M. (2020). The aging brain: Impact of heavy metal neurotoxicity. *Critical Reviews in Toxicology, 50*(9), 801–814.

Jomova, K., Alomar, S. Y., Nepovimova, E., Kuca, K., & Valko, M. (2024). Heavy metals: Toxicity and human health effects. *Archives of Toxicology*, 1–57.

Kandimalla, R., Vallamkondu, J., Corgiat, E. B., & Gill, K. D. (2016). Understanding aspects of Aluminum exposure in Alzheimer's disease development. *Brain Pathology, 26*(2), 139–154.

Kaur, I., Behl, T., Aleya, L., Rahman, M. H., Kumar, A., Arora, S., ... Akter, R. (2021). Role of metallic pollutants in neurodegeneration: Effects of aluminum, lead, mercury, and arsenic in mediating brain impairment events and autism spectrum disorder. *Environmental Science and Pollution Research, 28*, 8989–9001.

Khalaf, E. M., Taherian, M., Almalki, S. G., Asban, P., Kareem, A. K., Alhachami, F. R., ... Mohammadi, M. J. (2024). Relationship between exposure to heavy metals on the increased health risk and carcinogenicity of urinary tract (kidney and bladder). *Reviews on Environmental Health, 39*(3), 539–549.

Koutsodendris, N., Nelson, M. R., Rao, A., & Huang, Y. (2022). Apolipoprotein E and Alzheimer's disease: Findings, hypotheses, and potential mechanisms. *Annual Review of Pathology: Mechanisms of Disease, 17*(1), 73–99.

Lau, V., Ramer, L., & Tremblay, M.È. (2023). An aging, pathology burden, and glial senescence build-up hypothesis for late onset Alzheimer's disease. *Nature Communications, 14*(1), 1670.

Lista, S., Imbimbo, B. P., Grasso, M., Fidilio, A., Emanuele, E., Minoretti, P., ... Caraci, F. (2024). Tracking neuroinflammatory biomarkers in Alzheimer's disease: A strategy for individualized therapeutic approaches? *Journal of Neuroinflammation, 21*(1), 187.

Liu, N., Du, J., Ge, J., & Liu, S. B. (2024). DNA damage-inducing endogenous and exogenous factors and research progress. *Nucleosides, Nucleotides & Nucleic Acids*, 1–33.

Mazur, T., Malik, M., & Bieńko, D. C. (2024). The impact of chelating compounds on Cu^{2+}, $Fe^{2+/3+}$, and Zn^{2+} ions in Alzheimer's disease treatment. *Journal of Inorganic Biochemistry*, 112601.

Mckelvey, W. E. N. D. Y., Oken, E., Spills, E. M., & Advisories, F. C. (2012). Mercury and public health. *Mercury in the Environment: Pattern and Process, 267*.

Mehrandish, R., Rahimian, A., & Shahriary, A. (2019). Heavy metals detoxification: A review of herbal compounds for chelation therapy in heavy metals toxicity. *Journal of Herbmed Pharmacology, 8*(2), 69–77.

Mira, R. G., Lira, M., & Cerpa, W. (2021). Traumatic brain injury: Mechanisms of glial response. *Frontiers in Physiology, 12*, 740939.

Mitra, S., Chakraborty, A. J., Tareq, A. M., Emran, T. B., Nainu, F., Khusro, A., ... Simal-Gandara, J. (2022). Impact of heavy metals on the environment and human health: novel therapeutic insights to counter the toxicity. *Journal of King Saud University—Science, 34*(3), 101865.

Nabi, M., & Tabassum, N. (2022). Role of environmental toxicants on neurodegenerative disorders. *Frontiers in Toxicology, 4*, 837579.

Nabrzyski, M. (2006). Functional role of some minerals in foods. In *Mineral Components in Foods*. CRC Press, 123–161.

Ondaro, J., Hernandez-Eguiazu, H., Garciandia-Arcelus, M., Loera-Valencia, R., Rodriguez-Gómez, L., Jiménez-Zúñiga, A., ... Gereñu, G. (2022). Defects of nutrient signaling and autophagy in neurodegeneration. *Frontiers in Cell and Developmental Biology, 10*, 836196.

Pisoschi, A. M., Pop, A., Iordache, F., Stanca, L., Predoi, G., & Serban, A. I. (2021). Oxidative stress mitigation by antioxidants—An overview on their chemistry and influences on health status. *European Journal of Medicinal Chemistry, 209*, 112891.

Prithul, J. A. (2017). *In vitro assessment of cholinesterase inhibitory and antioxidant activities of root and rhizome of* Curcuma caesia *(L.) for the treatment of neurodegenerative disorder* (Doctoral dissertation). East West University.

Rizwan, M., Ali, S., Rizvi, H., Rinklebe, J., Tsang, D. C., Meers, E., ... Ishaque, W. (2016). Phytomanagement of heavy metals in contaminated soils using sunflower: A review. *Critical Reviews in Environmental Science and Technology, 46*(18), 1498–1528.

Romero, I. A., Abbott, N. J., & Bradbury, M. W. (2023). The blood–brain barrier in normal CNS and in metal-induced neurotoxicity. In *Toxicology of metals*. CRC Press, 561–585.

Sadiq, I. Z. (2023). Free radicals and oxidative stress: Signaling mechanisms, redox basis for human diseases, and cell cycle regulation. *Current Molecular Medicine, 23*(1), 13–35.

Sehar, U., Mukherjee, U., Khan, H., Brownell, M., Malhotra, K., Culberson, J., ... Reddy, P. H. (2024). Effects of sleep deprivation on brain atrophy in individuals with mild cognitive impairment and Alzheimer's disease. *Ageing Research Reviews*, 102397.

Sehar, U., Rawat, P., Reddy, A. P., Kopel, J., & Reddy, P. H. (2022). Amyloid beta in aging and Alzheimer's disease. *International Journal of Molecular Sciences, 23*(21), 12924.

Singh, S. A., & Vellapandian, C. (2023). Structure of the blood brain barrier and its role in the transporters for the movement of substrates across the barriers. *Current Drug Metabolism, 24*(4), 250–269.

Sinsky, J., Pichlerova, K., & Hanes, J. (2021). Tau protein interaction partners and their roles in Alzheimer's disease and other tauopathies. *International Journal of Molecular Sciences, 22*(17), 9207.

Smirnova, E., Moniruzzaman, M., Chin, S., Sureshbabu, A., Karthikeyan, A., Do, K., ... Min, T. (2023). A review of the role of curcumin in metal induced toxicity. *Antioxidants, 12*(2), 243.

Teleanu, D. M., Niculescu, A. G., Lungu, I. I., Radu, C. I., Vladâcenco, O., Roza, E., ... Teleanu, R. I. (2022). An overview of oxidative stress, neuroinflammation, and neurodegenerative diseases. *International Journal of Molecular Sciences, 23*(11), 5938.

Thakur, R., Karwasra, R., & Umar, T. (2023). Understanding Alzheimer's disease and its metal chelation therapeutics: A narrative review. *Current Pharmaceutical Design, 29*(30), 2377–2386.

Thal, D. R., & Tomé, S. O. (2022). The central role of tau in Alzheimer's disease: From neurofibrillary tangle maturation to the induction of cell death. *Brain Research Bulletin, 190*, 204–217.

Tripathi, A. K., Ray, A. K., Mishra, S. K., Bishen, S. M., Mishra, H., & Khurana, A. (2023). Molecular and therapeutic insights of alpha-lipoic acid as a potential molecule for disease prevention. *Revista Brasileira de Farmacognosia, 33*(2), 272–287.

Tzioras, M., McGeachan, R. I., Durrant, C. S., & Spires-Jones, T. L. (2023). Synaptic degeneration in Alzheimer disease. *Nature Reviews Neurology, 19*(1), 19–38.

Vanisree, C. R., Sankhla, M. S., Singh, P., Jadhav, E. B., Verma, R. K., Awasthi, K. K., ... Nagar, V. (2022). Heavy metal contamination of food crops: Transportation via food chain, human consumption, toxicity and management strategies. In *Environmental impact and remediation of heavy metals*. IntechOpen.

Vellingiri, B., Suriyanarayanan, A., Selvaraj, P., Abraham, K. S., Pasha, M. Y., Winster, H., ... Venkatesan, D. (2022). Role of heavy metals (copper (Cu), arsenic (As), cadmium (Cd), iron (Fe) and lithium (Li)) induced neurotoxicity. *Chemosphere, 301*, 134625.

Watt, A. D., Villemagne, V. L., & Barnham, K. J. (2012). Metals, membranes, and amyloid-β oligomers: Key pieces in the Alzheimer's disease puzzle? *Journal of Alzheimer's disease, 33*(s1), S283–S293.

Wu, K., Chen, Y., & Huang, W. (2025). Combined molecular toxicity mechanism of heavy metals mixtures. *Toxicological Assessment of Combined Chemicals in the Environment,* 125–172.

Zatta, P., Lucchini, R., van Rensburg, S. J., & Taylor, A. (2003). The role of metals in neurodegenerative processes: Aluminum, manganese, and zinc. *Brain Research Bulletin, 62*(1), 15–28.

Zhang, X. X., Tian, Y., Wang, Z. T., Ma, Y. H., Tan, L., & Yu, J. T. (2021). The epidemiology of Alzheimer's disease modifiable risk factors and prevention. *The Journal of Prevention of Alzheimer's Disease, 8*, 313–321.

Chapter 10

Genetic susceptibility to heavy metal toxicity

Prasann Kumar
Department of Agronomy, School of Agriculture, Lovely Professional University, Phagwara, Punjab, India

10.1 Introduction

Heavy metal toxicity represents a significant environmental and public health concern due to the widespread presence of toxic metals such as lead, mercury, arsenic, cadmium, and chromium in air, water, soil, and food. These metals persist in the environment and accumulate in biological systems, leading to adverse health effects in humans, animals, and ecosystems. Unlike organic pollutants, heavy metals are nonbiodegradable, meaning they cannot be broken down into less harmful forms, making their accumulation a long-term threat (Kumar, Saini, & Sharma, 2024f, Kumar, Khokhar, & Dey, 2024e). The sources of heavy metal exposure include industrial emissions, mining activities, agricultural runoff, wastewater discharge, contaminated food and drinking water, and even household products such as cosmetics and batteries. Depending on the metal and level of exposure, heavy metals can exert toxic effects on multiple organ systems, including the nervous, renal, cardiovascular, and immune systems. Chronic exposure to heavy metals has been linked to neurological disorders, developmental abnormalities, carcinogenesis, cardiovascular diseases, and metabolic dysfunctions. Given their pervasive nature and severe health impacts, understanding the factors contributing to differential susceptibility among individuals is crucial (Devi & Kumar, 2024a; Kumar et al., 2024g).

One of the most critical factors influencing susceptibility to heavy metal toxicity is genetic variation. While environmental exposure plays a central role, not all individuals exposed to the same heavy metal levels develop the same toxicity severity. This disparity can be attributed to genetic differences that regulate heavy metal absorption, distribution, metabolism, and excretion (Devi & Kumar, 2024b; Kumar, Karam, & Choudhury, 2024d). Key genes involved in these processes include metal transporters, detoxification enzymes, antioxidant defense mechanisms, and DNA repair pathways. For example, the solute carrier (SLC) family of transporters, such as SLC11A2 and SLC39A8, play an essential role in metal uptake, with polymorphisms in these genes affecting metal absorption efficiency and retention in the body. Similarly, ATP-binding cassette (ABC) transporters facilitate the efflux of metals and their conjugated metabolites, and genetic variants in these transporters can impact an individual's ability to excrete heavy metals effectively. Furthermore, metallothioneins (MTs), a class of cysteine-rich proteins, bind to heavy metals to neutralize their toxic effects, and genetic variations in MT genes can influence individual susceptibility to metal-induced oxidative stress and toxicity (Kumar, Choudhury, Dey, & Sharma, 2024a; Kumar & Choudhury, 2024a).

Another crucial group of genes involved in heavy metal metabolism includes detoxification enzymes such as glutathione-*S*-transferases (GSTs) and cytochrome P450 enzymes (CYPs). GSTs catalyse the conjugation of metals with glutathione, facilitating their excretion from the body. Specific polymorphisms in GST genes, such as GSTM1 and GSTT1, have been associated with altered detoxification capacity, increasing susceptibility to metal toxicity. Similarly, variations in CYP genes responsible for metabolizing environmental toxins may impact the bioactivation or detoxification of heavy metal compounds. In addition to these genetic factors, oxidative stress response genes are crucial in mitigating heavy metal-induced cellular damage. Superoxide dismutases (SODs), catalases (CATs), and nuclear factor erythroid 2-related factor 2 (NRF2) regulate antioxidant defense mechanisms, and genetic polymorphisms in these genes can influence an individual's ability to counteract oxidative stress induced by heavy metals (Kumar, Srivastav, Chaudhary, van Hullebusch, & Busquets, 2024j; Pandey, Kumar, & Dey, 2024).

Heavy Metal Toxicity and Neurodegeneration. DOI: https://doi.org/10.1016/B978-0-443-36575-1.00017-0

Beyond direct genetic factors, epigenetic modifications, such as DNA methylation, histone modifications, and noncoding RNAs (ncRNAs), also contribute to differential susceptibility to heavy metal toxicity. Heavy metal exposure has been shown to induce epigenetic alterations that affect gene expression without changing the DNA sequence. For example, arsenic exposure can lead to hypermethylation of tumor suppressor genes, increasing the risk of cancer development. Similarly, lead and cadmium have been associated with changes in histone modifications, leading to altered chromatin structure and gene expression patterns. MicroRNAs (miRNAs) have also been investigated in regulating heavy metal response, with specific miRNAs acting as biomarkers for metal toxicity. These epigenetic changes may affect the individual exposed to heavy metals and the whole family. They can also be inherited across generations, potentially amplifying the long-term health consequences of metal exposure in susceptible populations (Choudhury, Kumar, Zhimo, & Sahoo, 2024; Kumar, Sharma, & Singh, 2024h).

Understanding genetic and epigenetic susceptibility to heavy metal toxicity has significant public health implications, particularly in risk assessment, disease prevention, and the development of personalized interventions. Populations with genetic predispositions to heavy metal toxicity may require stricter exposure limits and targeted preventive measures. For instance, individuals with specific GST or MT polymorphisms may be more vulnerable to occupational heavy metal exposure and should undergo regular biomonitoring and screening. Additionally, genetic testing can help identify high-risk individuals who may benefit from dietary and lifestyle modifications to enhance detoxification pathways. Personalized nutritional interventions, such as increasing dietary intake of antioxidants like vitamin C, selenium, and zinc, may mitigate heavy metal toxicity in genetically susceptible individuals by enhancing their antioxidant defense systems (Devi et al., 2024e; Dey, Sharma, & Kumar, 2024).

Furthermore, identifying genetic markers of heavy metal susceptibility can improve environmental health policies and regulations. Governments and regulatory agencies can use genetic data to establish evidence-based exposure limits that account for variations in population susceptibility. For example, permissible exposure limits for lead or arsenic in drinking water may be reevaluated based on genetic susceptibility data, ensuring better protection for vulnerable groups. Additionally, integrating genetic insights into workplace safety measures can help industries adopt better protective strategies for workers exposed to heavy metals in mining, manufacturing, and waste disposal.

Public health interventions to reduce heavy metal toxicity should also consider the gene-environment interaction. Nutritional deficiencies, coexposure to other environmental toxins, and lifestyle factors such as smoking and alcohol consumption can exacerbate heavy metal toxicity. For instance, iron deficiency has been shown to increase lead absorption in the gut, worsening lead toxicity in vulnerable populations. Similarly, coexposure to multiple heavy metals, such as arsenic and cadmium, can have synergistic toxic effects, further complicating susceptibility patterns. Addressing these interactions through comprehensive public health strategies, such as nutritional supplementation programs and stricter environmental regulations, can help mitigate heavy metal toxicity at the population level (Kumar, Sharma, & Singh, 2024i; Kumar & Choudhury, 2024b; Sharma & Kumar, 2024).

Additionally, advancing genomics and precision medicine research offer new opportunities for developing targeted therapeutic strategies for heavy metal detoxification. Gene therapy, enzyme replacement therapy, and microbiome-based interventions are emerging as potential approaches for mitigating heavy metal toxicity in genetically susceptible individuals. For example, probiotics with metal-chelating properties have shown promise in reducing heavy metal absorption in the gut, providing a natural and noninvasive detoxification strategy. Similarly, CRISPR-based gene editing techniques may offer future possibilities for correcting genetic defects that predispose individuals to heavy metal toxicity. While these approaches are still in the early stages of development, continued research in this field holds significant potential for revolutionising heavy metal toxicity management.

Heavy metal toxicity remains a pressing global health concern, with genetic factors playing a crucial role in determining individual susceptibility to metal-induced health effects. Variations in metal transporters, detoxification enzymes, antioxidant defense mechanisms, and epigenetic modifications contribute to differential responses to heavy metal exposure. Understanding these genetic factors is essential for improving risk assessment, developing personalized interventions, and formulating effective public health policies. Integrating genetic insights into environmental health strategies can help protect vulnerable populations, enhance workplace safety measures, and drive innovation in detoxification therapies. As research in genomics, epigenetics, and precision medicine continues to evolve, the ability to predict, prevent, and mitigate heavy metal toxicity will become increasingly sophisticated, paving the way for a healthier and more sustainable future (Devi, Dey, Sharma, Kumar, & Singh, 2024a; Devi, Kaur, & Kumar, 2024d; Kumar and Raut, 2024).

10.2 Key heavy metals and their toxicity mechanisms

Heavy metals are naturally occurring elements with a high atomic weight and density, at least five times greater than water. Among the most toxic heavy metals that pose a significant threat to human health and the environment are lead (Pb), mercury (Hg), arsenic (As), and cadmium (Cd). These metals are widely distributed in the environment due to natural geological processes and human activities such as mining, industrial emissions, agricultural runoff, and improper waste disposal. Chronic exposure to these heavy metals can result in severe health effects, affecting multiple organ systems and leading to long-term disorders. Understanding these metals' toxicity, sources, and mechanisms of action is critical in developing effective mitigation strategies and public health interventions. Lead (Pb) is one of the most pervasive environmental contaminants, primarily due to its extensive use in paints, batteries, plumbing materials, and gasoline additives in the past (Devi, Kapoor, & Kumar, 2024c; Kumar, 2024a; Saurabh, Meena, Kumar, Upadhyaya, & Das, 2024). Lead exposure occurs through contaminated air, water, soil, and food, with children being particularly vulnerable due to their developing nervous systems and increased absorption of lead in the gastrointestinal tract. Once inside the body, lead disrupts various physiological functions by mimicking calcium and interfering with neurotransmitter release, synaptic plasticity, and neuronal development. Chronic lead exposure has been associated with neurotoxicity, cognitive impairments, reduced IQ, behavioral disorders, and developmental delays in children. In adults, lead toxicity can lead to hypertension, kidney damage, and reproductive issues. Lead also affects hematopoiesis by inhibiting enzymes involved in heme synthesis, leading to anemia. Since lead has a long biological half-life and can accumulate in bones, long-term exposure poses a sustained risk even after exposure ceases. Effective measures to reduce lead toxicity include phasing out lead-based products, improving water filtration systems, and implementing chelation therapy in cases of severe poisoning. Mercury (Hg) is another highly toxic heavy metal found in various forms, including elemental mercury, inorganic mercury compounds, and organic mercury compounds such as methylmercury. The primary sources of mercury contamination include industrial processes, coal combustion, artisanal gold mining, and improper disposal of mercury-containing products (Kumar, 2024b, 2024c; Kumar, Dwivedi, & Upadhyay, 2024c). Mercury bioaccumulates and biomagnifies in aquatic food chains, making fish and seafood primary sources of human exposure to methylmercury. Methylmercury is a potent neurotoxin that crosses the blood–brain barrier and disrupts neuronal function, leading to cognitive deficits, motor impairments, and developmental abnormalities, particularly in fetuses and young children. Inorganic mercury compounds can cause kidney damage and gastrointestinal irritation, while elemental mercury vapor exposure can lead to tremors, mood disturbances, and lung damage. The Minamata disease disaster in Japan exemplifies the devastating effects of mercury poisoning, where industrial mercury discharge into Minamata Bay led to widespread neurological disorders among the local population. To mitigate mercury exposure, international agreements such as the Minamata Convention aim to reduce mercury emissions, promote safe disposal practices, and regulate mercury use in industries. Arsenic (As) is a naturally occurring metalloid found in groundwater, industrial effluents, and contaminated food sources such as rice and seafood. Long-term exposure to arsenic primarily occurs through drinking water contaminated with inorganic arsenic compounds, which pose severe health risks. Arsenic toxicity is associated with skin lesions, peripheral neuropathy, cardiovascular diseases, and an increased risk of various cancers, including skin, lung, and bladder cancer. Arsenic exerts its toxic effects by interfering with cellular metabolism, disrupting mitochondrial function, and generating oxidative stress. Chronic arsenic exposure has also been linked to endocrine disruption, immune suppression, and developmental toxicity. One of the most severe consequences of arsenic exposure is its impact on epigenetic regulation, leading to altered gene expression and increased susceptibility to diseases. Large-scale arsenic poisoning incidents, such as the groundwater contamination crisis in Bangladesh and West Bengal, have highlighted the urgent need for improved water quality monitoring and alternative drinking water sources. Mitigation strategies include arsenic removal technologies such as activated alumina filters, reverse osmosis, and phytoremediation using arsenic-hyperaccumulating plants. Cadmium (Cd) is another highly toxic heavy metal primarily released into the environment through industrial activities, phosphate fertilisers, battery manufacturing, and cigarette smoke (Devi, Dogra, Kumar, & Singh, 2024b; Jagadesh et al., 2024; Raj, Goutam, Kumar, & Singh, 2024). Cadmium exposure occurs through contaminated food, water, and air, with rice and leafy vegetables being significant dietary sources due to cadmium uptake from contaminated soils. Once absorbed, cadmium accumulates in the kidneys and has a long biological half-life, making it difficult for the body to eliminate. Cadmium toxicity is associated with nephrotoxicity, osteoporosis, pulmonary damage, and an increased risk of cancers, particularly lung and prostate cancer. One of the most infamous cases of cadmium poisoning is the Itai–Itai disease in Japan, where long-term exposure to cadmium-contaminated rice and water led to severe skeletal deformities, bone fractures, and renal failure among affected populations. Cadmium exerts its toxic effects by inducing oxidative stress, inhibiting DNA repair mechanisms, and disrupting calcium homeostasis, leading to bone demineralization. Reducing cadmium exposure requires stricter environmental regulations, improved agricultural practices, and dietary interventions to minimize cadmium bioaccumulation in crops. In addition to lead, mercury, arsenic, and cadmium, several other heavy metals pose significant health risks, including chromium (Cr), nickel (Ni), and copper (Cu). Chromium exists in multiple oxidation states, with hexavalent chromium [Cr(VI)] being the most toxic form. Cr(VI) is widely used in electroplating, leather tanning, and textile manufacturing, leading to occupational and environmental exposure. Cr(VI) is a known carcinogen that causes DNA damage, oxidative stress, and respiratory disorders, particularly lung cancer in industrial workers. Nickel exposure occurs through industrial emissions, stainless steel production, and contaminated water, leading to respiratory inflammation, allergic dermatitis, and an increased lung and nasal cancer risk (Karam, Kumar, Anand, & Choudhury, 2023; Kumar, 2023; Kumar and Choudhury, 2024c). Copper, an essential trace element, can become toxic at high levels, leading to liver and neurological disorders, as seen in Wilson's disease, a genetic disorder affecting copper metabolism. Heavy metal toxicity is a global public health concern that requires comprehensive strategies for monitoring, regulation, and remediation. Biomonitoring of heavy metal levels in biological samples such as blood, urine, and hair can help assess exposure levels and identify at-risk populations. Advances in genetic and epigenetic research have provided insights into individual susceptibility to heavy metal toxicity, allowing for the development of personalized risk assessments and targeted interventions. Detoxification strategies, including chelating agents such as ethylenediaminetetraacetic acid (EDTA) and dimercaptosuccinic acid (DMSA), can help remove heavy metals from the body in cases of acute poisoning. Additionally, dietary interventions such as increasing antioxidant-rich foods, fiber intake, and essential micronutrients like selenium and zinc can help mitigate the toxic effects of heavy metals. Phytoremediation, a sustainable approach using plants to absorb and detoxify heavy metals from contaminated soils and water, has gained attention as an environment-friendly remediation technique. Policies and regulations are crucial in reducing heavy metal pollution and protecting public health. International agreements, such as the Basel Convention on Hazardous Waste Management and the Stockholm Convention on Persistent Organic Pollutants, aim to regulate toxic metals' production, use, and disposal. Strengthening environmental policies, enforcing stricter industrial emission standards, and promoting cleaner production technologies are essential to mitigating heavy metal contamination (Kumar and Choudhury, 2024d; Kumar and Kumar, 2023; Srivastav et al., 2024). Public awareness campaigns, community engagement, and education on heavy metal exposure risks can empower individuals to take preventive measures and advocate for safer environmental practices. In conclusion, lead, mercury, arsenic, cadmium, and other heavy metals threaten human health and ecosystems due to their persistence, bioaccumulation, and toxic effects. Understanding the mechanisms of heavy metal toxicity, sources of exposure, and individual susceptibility factors is crucial in developing effective mitigation strategies. Comprehensive environmental monitoring, stringent regulations, and sustainable remediation approaches are essential for reducing heavy metal pollution and safeguarding public health. As research advances, innovative solutions, including genetic screening, bioremediation, and precision medicine, hold promise in addressing the challenges posed by heavy metal toxicity in the future (Table 10.1).

10.3 Genetic determinants of heavy metal susceptibility

10.3.1 Metal transporters and cellular uptake

SLC transporters and ABC transporters play crucial roles in the absorption, distribution, and excretion of heavy metals in the human body. These transporters regulate metal homeostasis by controlling the uptake and efflux of essential and toxic metal ions across cellular membranes. SLC11A2 (also known as DMT1) and SLC39A8 are particularly important in metal transport among the SLC transporters. SLC11A2 is a divalent metal transporter primarily responsible for the uptake of iron (Fe^{2+}) but also facilitates the transport of other divalent metals such as manganese (Mn^{2+}), zinc (Zn^{2+}), cobalt (Co^{2+}), copper (Cu^{2+}), lead (Pb^{2+}), and cadmium (Cd^{2+}). It is widely expressed in the intestines, liver, and brain, where it plays a key role in metal absorption and distribution. Mutations in SLC11A2 can lead to iron metabolism disorders, such as anemia or iron overload, and influence susceptibility to heavy metal toxicity. The promiscuous nature of SLC11A2 in transporting toxic metals such as lead and cadmium makes it a critical determinant of heavy metal bioavailability and toxicity in humans. SLC39A8, another key SLC transporter, belongs to the ZIP family and is primarily involved in manganese and zinc transport (Devi and Kumar, 2023a; Kumar and Dey, 2024; Kumar and Singh, 2024; Kumari, Kumar, & Singh, 2023). It is essential for normal cellular function, as manganese is a cofactor for various enzymatic processes, including antioxidant defense and neurotransmitter metabolism. However, SLC39A8 also facilitates the uptake of toxic metals such as cadmium and lead, which can lead to neurotoxicity and developmental disorders. Genetic variations in SLC39A8 have been associated with altered manganese levels in the blood, leading to neurological conditions such as dystonia and schizophrenia. Studies have also linked SLC39A8 polymorphisms to impaired heavy metal detoxification, making it an essential genetic factor in determining individual susceptibility to heavy metal exposure. While SLC

TABLE 10.1 Genetic polymorphisms associated with susceptibility to heavy metal toxicity.

Gene	Polymorphism (SNP ID)	Heavy metal affected	Functional impact	Population relevance
SLC11A2	rs17216603	Lead (Pb)	Alters iron and lead transport efficiency	European, Asian
SLC39A8	rs13107325	Cadmium (Cd), lead (Pb)	It affects zinc and cadmium uptake	Global
GSTT1	Null allele	Mercury (Hg), cadmium (Cd)	Reduces detoxification capacity	Various populations
GSTP1	rs1695	Lead (Pb), arsenic (As)	Alters glutathione conjugation efficiency	European, Asian, African
MT1A	rs8052394	Arsenic (As), cadmium (Cd)	Modifies metal binding and detoxification ability	South Asian, Chinese
MT2A	rs10636	Cadmium (Cd), zinc (Zn)	Affects metallothionein expression	Global
NRF2	rs6721961	Cadmium (Cd), arsenic (As)	Modulates oxidative stress response	East Asian, Caucasian
CYP1A1	rs1048943	Arsenic (As), mercury (Hg)	Alters xenobiotic metabolism and detoxification	European, Asian, African
CYP2E1	rs3813867	Arsenic (As), lead (Pb)	It affects the biotransformation of toxicants	Global
HMOX1	rs2071746	Lead (Pb), mercury (Hg)	Regulates heme oxygenase and metal-induced stress	European, Asian

Source: *Based on* Devi, Kumar, Sharma, and Singh (2023), Dey, Kumar, and Singh (2023b), and Sharma and Kumar (2023).

transporters primarily mediate metal uptake, ABC transporters function in the efflux and detoxification of heavy metals. ABC transporters are a large family of transmembrane proteins that use ATP hydrolysis to transport various substrates, including heavy metals, across cellular membranes. Among them, ABCG2 and ABCC2 (also known as MRP2) play pivotal roles in heavy metal detoxification. ABCG2 is widely expressed in the intestines, kidneys, liver, and blood–brain barrier, where it functions as an efflux transporter, limiting the systemic accumulation of toxic metals. It plays a key role in reducing metal toxicity by actively exporting heavy metals such as arsenic, cadmium, and lead from cells, thereby preventing intracellular accumulation and potential damage. ABCC2, another crucial efflux transporter, is primarily found in hepatocytes and renal epithelial cells, where it facilitates the biliary and renal excretion of toxic metal conjugates. ABCC2 mediates the transport of arsenic-glutathione complexes and cadmium-glutathione conjugates, playing a significant role in phase III detoxification processes. Polymorphisms in ABC transporters can significantly influence heavy metal susceptibility, as reduced transporter function can lead to increased metal retention and toxicity. Overall, the interplay between SLC and ABC transporters is crucial for maintaining metal homeostasis, and genetic variations in these transporters can significantly impact individual vulnerability to heavy metal toxicity. Understanding these mechanisms provides valuable insights for developing targeted therapies and personalized interventions for metal-related diseases (Devi and Kumar, 2023b; Kumar, Dey, & Choudhury, 2024b; Kumar and Choudhury, 2024e; Sharma, Devi, Dey, & Kumar, 2023a).

10.3.2 Detoxification enzymes and metabolic pathways

GSTs, MTs, and cytochrome P450 enzymes (CYPs) play essential roles in the detoxification and metabolism of heavy metals, protecting cells from their toxic effects. GSTs are a family of phase II detoxification enzymes that catalyse the conjugation of glutathione (GSH) to xenobiotics, including heavy metals, facilitating their excretion. These enzymes are crucial in reducing oxidative stress induced by heavy metal exposure, as they neutralize reactive oxygen species (ROS) and prevent cellular damage. GSTs contribute to the detoxification of arsenic, cadmium, and mercury by forming glutathione-metal conjugates that are transported out of cells. Genetic polymorphisms in GST genes, such as GSTM1 and GSTT1, influence individual susceptibility to heavy metal toxicity, with specific deletions linked to increased heavy metal retention and related health risks. In addition to GSTs, MTs are vital metal-binding proteins that regulate essential metals' homeostasis and protect against toxic metal exposure. MTs have a high affinity for heavy metals like cadmium, mercury, lead, and zinc due to their cysteine-rich structure, which enables them to sequester these metals and mitigate their toxic effects (Devi, Dey, & Kumar, 2023a; Kumar and Choudhury, 2024f; Saini, Sharma, & Kumar, 2023; Sinam, Kumar, & Singh, 2023). These proteins play a crucial role in detoxification by binding and sequestering metals, reducing their bioavailability, and facilitating excretion. MTs also contribute to antioxidant defense by scavenging free radicals and protecting cells from oxidative stress induced by heavy metal exposure. The expression of MT genes is highly inducible, responding to heavy metal exposure to enhance cellular protection. However, genetic variations in MT genes can impact their expression levels and metal-binding capacity, influencing an individual's susceptibility to heavy metal toxicity. Studies have shown that individuals with reduced MT expression may be at higher risk of metal-induced diseases, including neurological disorders and cancers. Meanwhile, cytochrome P450 enzymes (CYPs) primarily metabolize organic compounds. Still, they also play an indirect role in heavy metal detoxification by modulating oxidative stress and influencing metal biotransformation. CYP enzymes, particularly CYP1A1 and CYP2E1, are involved in metabolizing environmental pollutants, including those containing heavy metals, leading to the generation of reactive metabolites. These enzymes contribute to heavy metal toxicity by generating ROS, exacerbating oxidative damage, and influencing metal-induced carcinogenesis. Additionally, CYP enzymes interact with heavy metals by modulating their bioavailability and influencing their metabolism in the liver. Polymorphisms in CYP genes can affect enzyme activity, altering an individual's ability to metabolize heavy metal–containing compounds, which may increase the risk of metal-induced toxicity and disease. The interplay between GSTs, MTs, and CYPs is crucial in determining an individual's response to heavy metal exposure, as variations in these detoxification pathways influence the body's ability to manage and excrete toxic metals effectively. Understanding the genetic and biochemical regulation of these detoxification systems provides valuable insights into the mechanisms of heavy metal toxicity and potential strategies for mitigating their harmful effects through targeted interventions and personalized medicine (Dey, Sharma, & Kumar, 2023c; Hidangmayum, Dwivedi, Kumar, & Upadhyay, 2023; Kumar, Saini, & Sharma, 2023b; Sharma, Dey, & Kumar, 2023).

10.3.3 Oxidative stress response genes

Superoxide dismutases (SODs), catalases (CATs), and nuclear factor erythroid 2-related factor 2 (NRF2) play crucial roles in cellular defense against oxidative stress induced by heavy metal toxicity. Heavy metals such as lead (Pb), cadmium (Cd), mercury (Hg), and arsenic (As) generate ROS, leading to cellular damage, inflammation, and apoptosis. The enzymatic antioxidant system, including SODs and CATs, works in coordination with transcription factors like NRF2 to maintain redox homeostasis and protect cells from oxidative damage. Superoxide dismutases (SODs) are key antioxidant enzymes responsible for the dismutation of superoxide radicals (O_2^-) into molecular oxygen (O_2) and hydrogen peroxide (H_2O_2). SODs exist in three major isoforms: SOD1 (Cu/Zn-SOD) found in the cytoplasm, SOD2 (Mn-SOD) located in the mitochondria, and SOD3 (extracellular SOD) present in the extracellular space. Each isoform plays a distinct role in detoxifying superoxide radicals generated by heavy metal exposure. For example, cadmium and mercury disrupt mitochondrial function, producing excessive superoxide, which SOD2 neutralizes (Devi et al., 2023b; Dey, Devi, & Kumar, 2023a; Sharma, Devi, & Kumar, 2023b; Sharma, Devi, Kumar, Dey, & Dwivedi, 2023c). Genetic polymorphisms or deficiencies in SOD genes can impair the ability to manage oxidative stress, making individuals more susceptible to metal-induced toxicity and associated diseases such as neurodegeneration and cancer. Hydrogen peroxide generated by SOD activity is further metabolized by catalases (CATs), another essential antioxidant enzyme involved in detoxification. CATs decompose hydrogen peroxide into water (H_2O) and oxygen (O_2), preventing the accumulation of harmful hydroxyl radicals (•OH), which can damage lipids, proteins, and DNA. Heavy metal exposure, particularly arsenic and lead, has been shown to reduce CAT activity, leading to increased oxidative stress and cellular injury. The efficiency of CAT activity is critical in mitigating oxidative damage caused by prolonged heavy metal exposure. Variations in CAT gene expression can influence individual detoxification capacity, with lower CAT activity associated with higher risks of oxidative stress-related disorders. In addition to enzymatic antioxidants, nuclear factor erythroid 2-related factor 2 (NRF2) acts as a master regulator of the antioxidant response by controlling the expression of detoxification and defense genes. Under normal conditions, NRF2 is bound to Kelch-like ECH-associated protein 1 (KEAP1), which promotes its degradation (Kumar, Latha, & Singh, 2023a; Kumar et al., 2018d; Kumar and Dwivedi, 2018; Kumar and Pathak, 2019). However, during oxidative stress induced by heavy metals, NRF2 dissociates from KEAP1, translocates to the nucleus, and binds to antioxidant response elements (AREs) in the DNA, leading to the upregulation of genes involved in antioxidant defense, including SODs, CATs, GSTs, and heme oxygenase-1 (HO-1). NRF2 activation protects against heavy metal-induced toxicity by enhancing cellular resilience against oxidative stress. However, certain heavy metals, such as cadmium and arsenic, can disrupt NRF2 signaling, impairing detoxification responses and increasing cellular vulnerability. Polymorphisms in

TABLE 10.2 Major detoxification enzymes involved in heavy metal metabolism.

Enzyme	Function	Associated heavy metals	Genetic regulation
Glutathione-*S*-transferases (GSTs)	Catalyze the conjugation of heavy metals with glutathione for detoxification	Lead (Pb), mercury (Hg), cadmium (Cd), arsenic (As)	*GSTT1, GSTP1, GSTM1* polymorphisms influence detoxification efficiency
Metallothioneins (MTs)	Bind and sequester toxic metals to reduce oxidative damage	Cadmium (Cd), arsenic (As), mercury (Hg), lead (Pb), zinc (Zn)	*MT1A, MT2A* variants regulate metal-binding capacity
Cytochrome P450 enzymes (CYPs)	Involved in phase I metabolism and biotransformation of heavy metals	Arsenic (As), lead (Pb), mercury (Hg)	*CYP1A1, CYP2E1* polymorphisms affect metal metabolism
Superoxide dismutases (SODs)	Neutralize reactive oxygen species (ROS) generated by metal exposure	Lead (Pb), cadmium (Cd), mercury (Hg), arsenic (As)	*SOD2* gene polymorphisms modulate antioxidant defense
Catalases (CATs)	Decomposes hydrogen peroxide to prevent oxidative stress	Lead (Pb), mercury (Hg), cadmium (Cd)	*CAT* gene variations influence enzyme activity
Heme oxygenase-1 (HMOX1)	Protects against oxidative stress induced by heavy metal exposure	Lead (Pb), mercury (Hg), cadmium (Cd)	*HMOX1* polymorphisms regulate stress response
Nuclear factor erythroid 2-related factor 2 (NRF2)	Regulates the expression of detoxification and antioxidant genes	Arsenic (As), cadmium (Cd), lead (Pb)	*NRF2* variants affect cellular defense mechanisms

Source: Based on Kumar, Goud, Devi, and Koul (2022b), Kumar, Goud, Devi, Dey, and Dwivedi (2022a), Naik, Kumar (2020), Pankaj, Kumar, Nigam, and Mishra (2012a).

NRF2-related genes can influence susceptibility to heavy metal toxicity and oxidative stress-related diseases. Understanding the interplay between SODs, CATs, and NRF2 provides insights into therapeutic strategies for mitigating heavy metal toxicity through targeted antioxidant interventions and personalized medicine approaches (Table 10.2).

10.4 Epigenetic regulation of heavy metal toxicity

10.4.1 DNA methylation changes due to metal exposure

Heavy metal exposure induces significant DNA methylation changes, which are crucial in gene regulation, cellular function, and disease development. DNA methylation, an essential epigenetic modification, involves adding a methyl group to cytosine residues within CpG dinucleotides, typically leading to gene silencing. Exposure to toxic metals such as lead (Pb), cadmium (Cd), arsenic (As), and mercury (Hg) disrupts standard DNA methylation patterns, contributing to altered gene expression, genomic instability, and the onset of various diseases, including cancer, neurodegenerative disorders, and developmental abnormalities. Heavy metals can cause global hypomethylation and gene-specific hypermethylation, depending on their concentration, duration of exposure, and cellular context. Global DNA hypomethylation is commonly associated with heavy metal toxicity, as it leads to chromosomal instability, activation of oncogenes, and increased transcription of transposable elements (Kumar, Singh, & Dwivedi, 2017; Kumar, Mandal, & Dwivedi, 2011a; Kumar et al., 2018c; Pankaj, Kumar, Nigam, & Mishra, 2012b). For example, arsenic exposure has been widely documented to reduce global DNA methylation levels by interfering with DNA methyltransferase (DNMT) activity, thereby contributing to carcinogenesis and other chronic diseases. Similarly, lead and cadmium exposure have been linked to DNA hypomethylation in various tissues, leading to immune dysregulation and neurotoxicity. On the other hand, heavy metals can also induce site-specific DNA hypermethylation, particularly in the promoter regions of tumor suppressor genes, resulting in their silencing and promoting malignant transformation. Arsenic and cadmium have been shown to hypermethylate key regulatory genes involved in apoptosis, DNA repair, and cell cycle control, thereby increasing cancer risk. These metals exert their effects by altering the function of DNMTs, the enzymes responsible for adding methyl groups to DNA. Cadmium, for instance, inhibits the expression of DNMT1, leading to aberrant methylation profiles that contribute to cellular dysfunction. In addition to direct interactions with DNMTs, heavy metals influence DNA methylation by generating oxidative stress, which disrupts methylation homeostasis. ROS produced by metal exposure can oxidize methylcytosine to 5-hydroxymethylcytosine, an intermediate in DNA demethylation, thereby contributing to epigenetic instability. Mercury exposure, for example, induces oxidative stress that alters the DNA methylation landscape, impacting neurodevelopmental gene expression and increasing the risk of cognitive impairments (Kotia et al., 2021; Kumar, Rameshwar, & Kumar, 2019; Kumar, Koul, & Sharma, 2022c; Kumar, Mandal, & Dwivedi, 2013). Furthermore, DNA methylation changes due to heavy metal exposure are heritable, meaning they can be passed from generation to generation, leading to transgenerational effects. Studies have demonstrated that prenatal exposure to heavy metals such as lead and arsenic can lead to altered DNA methylation patterns in offspring, affecting their immune function, neurodevelopment, and metabolic health. This suggests that heavy metal–induced epigenetic changes have long-lasting consequences beyond direct exposure, underscoring the urgency of addressing this issue. Understanding the mechanisms by which heavy metals alter DNA methylation provides critical insights into developing novel therapeutic strategies, such as demethylating agents and dietary interventions, to mitigate the adverse effects of metal toxicity on human health.

10.4.2 Histone modifications and chromatin remodeling

Histone modifications and chromatin remodeling play crucial roles in regulating gene expression and maintaining genome stability, which are significantly impacted by heavy metal exposure. Histone modifications, including methylation, acetylation, phosphorylation, ubiquitination, and sumoylation, serve as epigenetic markers that influence chromatin structure and accessibility of DNA to transcriptional machinery. Heavy metals such as arsenic (As), cadmium (Cd), lead (Pb), and mercury (Hg) disrupt these modifications, leading to altered gene expression patterns, genomic instability, and increased disease susceptibility. Among these modifications, histone acetylation and methylation are the most extensively studied regarding heavy metal toxicity. Histone acetylation, regulated by histone acetyltransferases (HATs) and histone deacetylases (HDACs), generally promotes gene transcription by relaxing chromatin structure. Heavy metal exposure, particularly to cadmium and arsenic, has been shown to alter HAT and HDAC activity, leading to aberrant gene expression (Dwivedi and Kumar, 2011; Kumar, Mandal, & Dwivedi, 2011b; Kumar and Naik, 2020; Pathak, Kumar, Mishra, & Kumar, 2017). For example, arsenic exposure has been associated with global hypoacetylation of histones H3 and H4, which results in chromatin condensation and gene silencing. This effect is particularly detrimental in tumor suppressor genes, where the loss of acetylation contributes to carcinogenesis. Conversely, lead

TABLE 10.3 Epigenetic changes due to heavy metal exposure.

Epigenetic mechanism	Effect of heavy metal exposure	Associated heavy metals	Key genes/regulatory elements
DNA methylation	Alters gene expression by hypermethylation or hypomethylation	Arsenic (As), lead (Pb), cadmium (Cd), mercury (Hg)	*DNMT1, DNMT3A, MGMT, RARβ, p16*
Histone modifications	Changes in chromatin structure via acetylation, methylation, or phosphorylation	Lead (Pb), arsenic (As), cadmium (Cd)	*H3K9, H3K27, H3K4* methylation patterns
Noncoding RNAs (miRNAs, lncRNAs, circRNAs)	Regulate gene expression post-transcriptionally by targeting detoxification and oxidative stress-related genes	Mercury (Hg), arsenic (As), lead (Pb), cadmium (Cd)	*miR-21, miR-155, miR-222, lncRNA MEG3, circRNA_0025033*

Source: Based on Kumar et al. (2018a, 2018b), Kumar and Dwivedi (2020), and Pankaj et al. (2012a).

and mercury exposure can lead to hyperacetylation of histones in specific genomic regions, resulting in the aberrant activation of genes associated with inflammation and oxidative stress. Histone methylation, another key modification, is regulated by histone methyltransferases and histone demethylases and can either activate or repress transcription depending on the specific lysine (K) or arginine (R) residues modified. Heavy metals interfere with histone methylation patterns, contributing to epigenetic dysregulation. Cadmium and arsenic exposure, for example, have been shown to increase H3K9 and H3K27 methylation, both of which are associated with transcriptional repression and gene silencing. Such modifications can silence DNA repair genes, thereby increasing genomic instability and the risk of cancer development. Conversely, heavy metals can also promote the demethylation of histones in oncogenes, leading to their overexpression and increased cell proliferation. In addition to direct histone modifications, heavy metal exposure disrupts chromatin remodeling, which involves ATP-dependent chromatin-remodeling complexes that regulate nucleosome positioning and DNA accessibility. Key chromatin remodelers, such as the SWI/SNF and NuRD complexes, are highly sensitive to metal-induced oxidative stress and epigenetic changes. Arsenic and cadmium, for instance, have been shown to disrupt the activity of these complexes, leading to impaired DNA repair and increased susceptibility to mutations. Furthermore, heavy metals can alter the expression of chromatin remodelers, exacerbating epigenetic dysregulation and contributing to long-term cellular dysfunction. The impact of heavy metals on histone modifications and chromatin remodeling is particularly concerning due to its potential transgenerational effects. Studies have demonstrated that prenatal exposure to metals such as lead and arsenic can induce heritable epigenetic changes that affect gene expression patterns in offspring. Understanding these mechanisms provides critical insights into developing targeted epigenetic therapies, such as HDAC inhibitors and histone methylation modulators, to counteract heavy metal–induced toxicity and its associated health risks (Table 10.3).

10.4.3 Role of noncoding RNAs in modulating toxicity

ncRNAs are critical in modulating cellular responses to heavy metal toxicity by regulating gene expression, maintaining genome stability, and influencing stress response pathways. These ncRNAs, including microRNAs (miRNAs), long ncRNAs (lncRNAs), and circular RNAs (circRNAs), act as key epigenetic regulators that mediate the effects of toxic metal exposure at the post-transcriptional and transcriptional levels. Heavy metals such as arsenic (As), cadmium (Cd), lead (Pb), and mercury (Hg) disrupt ncRNA expression profiles, leading to altered cellular homeostasis, oxidative stress, apoptosis, and increased susceptibility to disease. Among these, miRNAs are the most well-characterized and are known to regulate gene expression by binding to the 3' untranslated region of target mRNAs, leading to degradation or translational repression. Exposure to heavy metals alters the expression of various miRNAs, contributing to toxicity-induced pathologies. For example, cadmium exposure has been shown to upregulate miR-21, a well-known oncogenic miRNA, suppressing tumor suppressor genes such as PTEN and increasing cell proliferation. Similarly, arsenic exposure induces dysregulation of miR-29, which plays a role in DNA methylation regulation by targeting DNMT3A and DNMT3B. This alteration results in widespread epigenetic changes contributing to carcinogenesis and immune dysfunction. In addition, lead and mercury have been found to modulate miRNA expression in neural tissues, affecting neurodevelopmental processes and increasing the risk of cognitive impairments. LncRNAs, another critical class of ncRNAs, serve as scaffolds, decoys, or guides for chromatin-modifying complexes, influencing gene expression at multiple levels. Heavy metal exposure disrupts the expression of lncRNAs, leading to significant physiological consequences. For instance, arsenic exposure has been shown to upregulate lncRNA MEG3, contributing to oxidative stress and apoptosis by modulating the p53 pathway (Kumar et al., 2013, Kumar et al., 2011b; Kumar, Devi, & Dey, 2021a; Kumar et al., 2021; Naik, Kumar, 2020). Similarly, cadmium exposure alters the expression of lncRNAs involved in inflammatory signaling and DNA repair, exacerbating cellular damage and increasing susceptibility to diseases such as cancer and cardiovascular disorders. Moreover, circRNAs, a recently discovered ncRNAs, play an emerging role in heavy metal toxicity by acting as miRNA sponges, thereby regulating miRNA availability and downstream target expression. Studies have demonstrated that cadmium exposure alters circRNA expression, leading to dysregulation of oxidative stress-related genes. For example, certain circRNAs have been shown to interact with miR-34a, a key regulator of apoptosis, thereby influencing cell survival and heavy metal–induced toxicity responses. The role of ncRNAs in modulating toxicity extends beyond direct gene regulation, as they also participate in intercellular communication through exosomes. Heavy metal exposure alters the ncRNA cargo within exosomes, facilitating the systemic spread of toxicity signals and contributing to distant tissue damage. This has been mainly observed in arsenic exposure, where exosomal miRNAs are implicated in inflammatory signaling and immune suppression. Given the crucial role of ncRNAs in mediating heavy metal toxicity, they are being explored as potential biomarkers for exposure assessment and disease risk prediction. Furthermore, ncRNA-based therapeutic strategies, such as miRNA mimics and inhibitors, hold promise for mitigating the adverse effects of heavy metal exposure, offering new avenues for precision medicine in toxicology (Das et al., 2022; Goud, Singh, & Kumar, 2022; Kumar and Mistri, 2020; Kumari, Singh, & Kumar, 2022; Upadhyay et al., 2023).

10.5 Gene–environment interactions in heavy metal toxicity

10.5.1 Influence of diet and nutritional status

Diet and nutritional status play a crucial role in modulating the toxicity of heavy metals by influencing their absorption, distribution, metabolism, and excretion. Nutrients such as essential metals, antioxidants, and vitamins interact with toxic heavy metals, mitigating or exacerbating their effects on human health. Deficiencies or imbalances in key dietary components can increase susceptibility to heavy metal toxicity. At the same time, adequate nutrition can enhance detoxification mechanisms and reduce the bioavailability of toxic metals such as lead (Pb), mercury (Hg), cadmium (Cd), and arsenic (As). The interplay between dietary factors and heavy metal toxicity is particularly significant in vulnerable populations, including pregnant women, children, and individuals with preexisting health conditions.

One of the key dietary factors influencing heavy metal toxicity is the intake of essential minerals, particularly calcium (Ca), iron (Fe), zinc (Zn), and selenium (Se). These elements compete with toxic metals for absorption and binding sites in biological systems. For instance, calcium and lead share similar transport pathways, and adequate calcium intake has been shown to reduce lead absorption in the intestines, thereby decreasing its accumulation in bones and soft tissues. Similarly, iron deficiency enhances cadmium absorption and leads to divalent metal transporters such as DMT1 (SLC11A2), increasing their toxic effects. Zinc also competes with cadmium for binding to MTs, reducing cadmium-induced oxidative stress and cellular damage. Selenium plays a crucial role in detoxification by binding to mercury and arsenic, forming less toxic complexes that are more easily excreted from the body.

In addition to essential metals, dietary antioxidants such as vitamins C and E, polyphenols, and glutathione precursors help mitigate heavy metal–induced oxidative stress. Heavy metals generate ROS, leading to lipid peroxidation, DNA damage, and apoptosis. Vitamin C, a potent water-soluble antioxidant, has been shown to reduce lead and arsenic toxicity by enhancing their excretion and protecting against oxidative damage. Vitamin E, a lipid-soluble antioxidant, stabilizes cell membranes and prevents heavy metal–induced lipid peroxidation. Dietary polyphenols, found in fruits, vegetables, and green tea, exhibit metal-chelating properties and enhance the activity of endogenous antioxidant enzymes such as superoxide dismutase (SOD) and catalase (CAT), thereby reducing oxidative stress induced by heavy metals (Aley, Singh, & Kumar, 2022; Chakraborty et al., 2021; Kumar, Sharma, Saini, & Dey, 2021b; Reddy, Kumar, & Dwivedi, 2022).

Protein intake and amino acids are vital in heavy metal detoxification. Sulfur-containing amino acids, such as cysteine and methionine, serve as glutathione (GSH) precursors. This key detoxifying agent binds to heavy metals and facilitates their excretion through bile and urine. Low protein intake or deficiencies in sulfur-containing amino acids can impair glutathione synthesis, increasing heavy metal retention and toxicity. Additionally, dietary fibers found in whole grains, fruits, and vegetables can bind to heavy metals in the gastrointestinal tract, reducing their absorption and enhancing their elimination through feces.

Overall, a well-balanced diet rich in essential minerals, antioxidants, and proteins can significantly reduce the toxic effects of heavy metals. Nutritional interventions, including mineral supplementation and antioxidant therapy, offer promising strategies for mitigating heavy metal toxicity and protecting at-risk populations from long-term health consequences (Table 10.4).

10.5.2 Coexposure with other environmental toxins

Coexposure to heavy metals and other environmental toxins significantly amplifies the toxic effects on human health by inducing synergistic, additive, or antagonistic interactions at molecular, cellular, and systemic levels. In real-world environmental settings, individuals are rarely exposed to a single toxicant; instead, they encounter complex mixtures of pollutants, including pesticides, polycyclic aromatic hydrocarbons (PAHs), endocrine-disrupting chemicals (EDCs), and air pollutants, which interact with heavy metals such as lead (Pb), mercury (Hg), arsenic (As), and cadmium (Cd). These interactions can exacerbate oxidative stress, disrupt metabolic and detoxification pathways, impair neurological and immune function, and increase the risk of chronic diseases such as cancer, neurodegenerative disorders, and cardiovascular disease (Chakraborty et al., 2021; Das et al., 2022; Goud et al., 2022; Kumar et al., 2020, Kumar et al., 2013; Kumar and Mistri, 2020).

One of the primary mechanisms underlying the enhanced toxicity of coexposure is the exacerbation of oxidative stress. Heavy metals and organic pollutants independently generate ROS, leading to oxidative damage to DNA, proteins, and lipids. When combined, these toxicants often act synergistically, overwhelming cellular antioxidant defenses such as superoxide dismutase (SOD), catalase (CAT), and glutathione (GSH). For example, studies have shown that coexposure to arsenic and pesticides such as organophosphates or organochlorines significantly increases lipid peroxidation and mitochondrial dysfunction, leading to apoptosis and tissue damage. Similarly, cadmium exposure, in combination with PAHs found in cigarette smoke and industrial emissions, intensifies oxidative damage, contributing to lung toxicity and carcinogenesis.

Another critical aspect of coexposure is its impact on detoxification pathways, particularly involving Phases I and II metabolic enzymes. Cytochrome P450 (CYP) enzymes play a central role in the biotransformation of xenobiotics, but their activity is often altered when heavy metals are present alongside other toxicants. For instance, mercury and cadmium inhibit CYP activity, accumulating unmetabolised toxicants and prolonged cellular damage. Additionally, coexposure to arsenic and perfluorinated compounds has been found to dysregulate GSTs, impairing the conjugation and elimination of toxic substances, thereby enhancing their bioaccumulation and toxicity. Endocrine disruption is another significant consequence of coexposure to heavy metals and other pollutants. Environmental toxins, such as bisphenol A (BPA), phthalates, and dioxins, interfere with hormonal signaling by mimicking or antagonizing natural hormones. Heavy metals such as lead and mercury further disrupt endocrine function by interfering with hormone receptors, calcium signaling, and neurotransmitter release. The combined effects of these toxins have been linked to reproductive dysfunction, developmental abnormalities, and increased risk of hormone-related cancers. Neurological toxicity is also significantly heightened by coexposure. Heavy metals such as lead and mercury are well-known neurotoxins, and their effects are amplified when combined with air pollutants like delicate particulate matter (PM2.5) or volatile organic compounds. Such interactions exacerbate neuroinflammation, blood–brain barrier permeability, and synaptic dysfunction, contributing to cognitive decline, neurodevelopmental disorders, and an increased risk of neurodegenerative diseases like Parkinson's and Alzheimer's. Coexposure to heavy metals and other environmental toxins presents a serious public health challenge, necessitating integrated risk assessment approaches and intervention strategies to mitigate combined toxic effects (Aley et al., 2022; Kumar et al., 2018b; Kumar and Mistri, 2020; Naik, Kumar, 2020; Reddy et al., 2022; Upadhyay et al., 2023).

10.5.3 Lifestyle and occupational factors

Lifestyle and occupational factors significantly influence susceptibility to heavy metal toxicity, as they determine the extent and duration of exposure and the body's ability to detoxify and eliminate toxic metals. Various lifestyle choices, such as smoking, alcohol consumption, diet, and personal hygiene, can either exacerbate or mitigate the harmful effects of heavy metals. Similarly, occupational exposure in mining, battery manufacturing, construction, agriculture, and metal smelting increases the risk of prolonged and high-dose exposure, leading to severe health consequences (Goud et al., 2022; Kumar and Dwivedi, 2020; Kumar and Mistri, 2020; Naik, Kumar, 2020; Reddy et al., 2022). Understanding these factors is essential for developing preventive strategies and reducing the burden of heavy metal-related diseases. Smoking is one of the major lifestyle factors that exacerbate heavy metal toxicity. Tobacco leaves naturally accumulate heavy metals such as cadmium (Cd), lead (Pb), and arsenic (As) from the soil. When smoked, these metals are inhaled directly into the lungs, entering the bloodstream and accumulating in various organs. Smoking has been associated with increased oxidative stress, reduced antioxidant capacity, and impaired detoxification mechanisms, making smokers more vulnerable to the harmful effects of heavy metals. Additionally, secondhand smoke exposure can contribute to the accumulation of these toxicants, increasing health risks for nonsmokers, especially children and pregnant women.

TABLE 10.4 Genetic factors interact with environmental exposures to modulate heavy metal toxicity.

Gene/genetic factor	Environmental exposure	Mechanism of interaction	Health outcome	Relevant heavy metals
SLC11A2 (DMT1)	Dietary iron deficiency	Increased intestinal absorption of heavy metals due to iron competition	Higher body burden and toxicity	Lead (Pb), cadmium (Cd)
SLC39A8 (ZIP8)	Industrial pollution, smoking	Enhanced metal uptake and transport into cells	Increased oxidative stress and cellular damage	Cadmium (Cd), lead (Pb)
GSTT1/GSTM1 (glutathione-S-transferases)	Occupational exposure, cigarette smoke	Detoxification efficiency varies based on genetic deletion	Higher susceptibility to oxidative stress and metal-induced toxicity	Mercury (Hg), arsenic (As), lead (Pb)
MT2A (metallothioneins)	Chronic heavy metal exposure	Variants alter metal-binding capacity and detoxification	Accumulation of metals in tissues, increased toxicity	Cadmium (Cd), arsenic (As)
NRF2 (Nuclear factor erythroid 2-related factor 2)	Pesticides, air pollution	Regulates antioxidant response genes, influencing defense against oxidative stress	Reduced ability to counteract metal-induced ROS damage	Arsenic (As), Lead (Pb), mercury (Hg)
CYP2E1 (cytochrome P450 enzymes)	Industrial solvents, xenobiotics	Alters the biotransformation of heavy metals and oxidative stress response	Increased risk of liver toxicity and neurotoxicity	Arsenic (As), Lead (Pb), mercury (Hg)
miR-21, miR-155 (noncoding RNAs)	Airborne pollutants, water contamination	Regulate metal detoxification and inflammation pathways	Epigenetic modifications leading to altered gene expression	Cadmium (Cd), lead (Pb), mercury (Hg)

Source: Based on Das et al. (2022), Kumar et al. (2020), Kumar, Pathak, Kumar, and Dwivedi (2018), and Kumar and Mistri (2020).

Alcohol consumption also influences heavy metal toxicity by impairing liver function, which plays a crucial role in detoxification. Chronic alcohol consumption depletes glutathione (GSH) levels, weakens antioxidant defense mechanisms, and disrupts the metabolism of essential metals such as zinc (Zn) and selenium (Se), which are needed to counteract heavy metal-induced oxidative stress. Furthermore, alcoholic beverages, particularly those stored in improperly regulated containers, may contain traces of lead or other contaminants, increasing exposure risks. Dietary habits are another critical lifestyle factor affecting heavy metal toxicity. A diet deficient in essential minerals such as calcium, iron, zinc, and selenium increases the absorption of toxic metals as the body attempts to compensate for the missing nutrients by absorbing similar elements, including heavy metals. For example, iron-deficient individuals absorb more lead, while zinc deficiency increases cadmium retention. Conversely, a diet rich in antioxidants, fiber, and essential minerals can help reduce the absorption and bioavailability of toxic metals. Consuming contaminated seafood, substantial predatory fish such as tuna and swordfish increases exposure to methylmercury (MeHg), a highly toxic form of mercury that bioaccumulates in marine ecosystems (Goud et al., 2022; Kumar et al., 2011b; Kumar et al., 2021b; Kumar et al., 2018; Kumar and Mistri, 2020; Reddy et al., 2022).

Occupational exposure is a significant concern, particularly for mining, smelting, battery manufacturing, and agriculture workers, where exposure to lead, mercury, cadmium, and arsenic is daily. Workers in these industries often face chronic exposure to heavy metals through inhalation, dermal contact, or ingestion, leading to bioaccumulation and long-term health effects. Inadequate personal protective equipment, poor ventilation, and lack of occupational safety regulations further exacerbate the risks. Agricultural workers using pesticides containing arsenic or other metal-based compounds are also at risk of exposure, leading to increased incidence of neurological, renal, and cardiovascular disorders. Overall, lifestyle choices and occupational exposure play a crucial role in determining an individual's risk of heavy metal toxicity. Implementing preventive measures such as smoking cessation, alcohol moderation, dietary improvements, and strict occupational safety protocols can significantly reduce exposure and mitigate health risks (Table 10.5).

10.6 Genetic biomarkers and personalized risk assessment

10.6.1 Genetic screening for susceptibility

Genetic screening for susceptibility to heavy metal toxicity is crucial to understanding individual differences in response to metal exposure and identifying populations at higher risk of adverse health effects. Genetic polymorphisms in key genes involved in metal transport, detoxification, and oxidative stress response influence an individual's ability to metabolize, eliminate, or tolerate heavy metals such as lead (Pb), mercury (Hg), arsenic (As), and cadmium (Cd). Advances in genomic technologies, including genome-wide association studies (GWAS) and next-generation sequencing (NGS), have enabled researchers to identify specific genetic variants that contribute to susceptibility, paving the way for personalized risk assessments and targeted interventions. One of the key genetic determinants of heavy metal susceptibility is the SLC family of transporters, which regulate the uptake and efflux of essential and toxic metals. Polymorphisms in genes such as *SLC11A2* (divalent metal transporter 1, DMT1) and *SLC39A8* (ZIP8) affect the absorption and cellular distribution of metals like lead and cadmium. Variants in *SLC11A2* have been associated with altered lead uptake, impacting neurodevelopmental outcomes in children. Similarly, mutations in *SLC39A8* influence manganese (Mn) and cadmium transport, contributing to variations in toxicity responses among exposed individuals (Goud et al., 2022; Kumar et al., 2011b; Kumar et al., 2021b; Kumar et al., 2018; Kumar and Dwivedi, 2020; Kumari et al., 2022).

ABC transporters also play a significant role in heavy metal detoxification by mediating the efflux of toxic metals from cells. *ABCC2* (multidrug resistance protein 2, MRP2) is involved in arsenic and cadmium excretion, and polymorphisms in this gene have been linked to differential susceptibility to arsenic-induced skin lesions and cancer. Genetic variations in ABCB1 (*P*-glycoprotein) also influence mercury accumulation and neurotoxicity, particularly in populations exposed to methylmercury through seafood consumption. Detoxification enzymes, including GSTs, MTs, and cytochrome P450 (CYP) enzymes, also exhibit genetic variations that modulate heavy metal toxicity. Polymorphisms in *GSTT1* and *GSTM1* affect glutathione-mediated detoxification of metals, with individuals lacking these genes showing increased susceptibility to oxidative damage and metal-induced diseases. Variants in *MT1* and *MT2* impact the binding and sequestration of metals like cadmium and zinc, influencing renal toxicity and carcinogenic potential. Additionally, mutations in *CYP2E1* alter the metabolism of organic pollutants coexposed with heavy metals, affecting overall toxicity outcomes (Goud et al., 2022; Kumar et al., 2021b; Kumar et al., 2018; Kumar and Mistri, 2020; Naik, Kumar, 2020; Pankaj et al., 2012a).

Epigenetic modifications also contribute to genetic susceptibility by regulating gene expression in response to metal exposure. DNA methylation changes in detoxification and antioxidant genes can alter expression patterns, increasing vulnerability to toxic effects. Histone modifications and ncRNAs, such as microRNAs (miRNAs), further modulate

TABLE 10.5 Potential genetic biomarkers for heavy metal toxicity risk assessment, including their reliability and application in public health strategies.

Biomarker (gene/miRNA)	Associated heavy metals	Mechanism of action	Reliability	Application in public health strategies
SLC11A2 (DMT1)	Lead (Pb), Cadmium (Cd)	Regulates metal transport across cell membranes	High (validated in multiple studies)	Screening for susceptibility in high-risk populations
SLC39A8 (ZIP8)	Cadmium (Cd), lead (Pb)	Modulates metal uptake and cellular transport	Moderate to High	Identifying individuals at risk in polluted areas
GSTT1/GSTM1 (glutathione-*S*-transferases)	Mercury (Hg), arsenic (As), lead (Pb)	Detoxification and oxidative stress regulation	High (linked to detoxification efficiency)	Assessing susceptibility for occupational exposure
MT2A (metallothioneins)	Cadmium (Cd), arsenic (As)	Metal binding and detoxification	High	Biomarker for chronic heavy metal exposure
NRF2 (NFE2L2)	Arsenic (As), lead (Pb), mercury (Hg)	Regulates antioxidant defense mechanisms	Moderate to High	Early detection of oxidative stress-related toxicity
CYP2E1 (cytochrome P450 enzymes)	Arsenic (As), lead (Pb), mercury (Hg)	Influences metal metabolism and detoxification	Moderate	Occupational risk assessment and metabolic profiling
miR-21, miR-155 (noncoding RNAs)	Cadmium (Cd), lead (Pb), mercury (Hg)	Epigenetic regulation of metal toxicity pathways	Emerging biomarker	Epigenetic-based risk assessment and intervention
DNMT1 (DNA methyltransferase 1)	Arsenic (As), lead (Pb)	Involved in DNA methylation changes due to exposure	Moderate	Epigenetic screening for long-term exposure effects

Source: Based on Goud et al. (2022), Kumar et al. (2013, 2018, 2021), and Kumari et al. (2022).

metal-induced stress responses, influencing disease risk and progression. Genetic screening for susceptibility to heavy metal toxicity has significant implications for public health, risk assessment, and precision medicine. Identifying high-risk individuals through genetic profiling allows for targeted preventive measures, such as dietary interventions, lifestyle modifications, and enhanced occupational safety protocols. As research advances, integrating genetic screening with environmental monitoring and biomarker-based approaches will provide a comprehensive strategy for mitigating heavy metal-related health risks.

10.6.2 Biomonitoring and early detection

Biomonitoring and early detection of heavy metal toxicity are crucial in assessing human exposure, evaluating health risks, and implementing timely interventions. Given the persistence and bioaccumulative nature of toxic metals such as lead (Pb), mercury (Hg), arsenic (As), cadmium (Cd), and chromium (Cr), routine biomonitoring strategies are essential to identify early biological responses before the onset of clinical symptoms. Advances in analytical techniques, biomarker discovery, and molecular diagnostics have significantly improved the ability to detect and quantify heavy metal exposure in various biological matrices, including blood, urine, hair, nails, and tissues. These assessments enable risk stratification and inform regulatory policies to reduce environmental and occupational exposures.

One of the primary approaches in biomonitoring involves measuring the total concentration of heavy metals in biological fluids. Blood and urine are the most commonly used matrices due to their ability to reflect recent and chronic exposure. For instance, blood lead levels remain the gold standard for assessing lead toxicity, particularly in children, as even low levels can cause neurodevelopmental impairment. Similarly, urinary arsenic concentration is a reliable biomarker for arsenic exposure, especially in populations consuming contaminated groundwater. Hair and nail analysis are increasingly used for long-term exposure assessment, particularly for metals like mercury and cadmium, which accumulate in keratinized tissues over extended periods (Goud et al., 2022; Kumar et al., 2011b; Kumar et al., 2021a).

Beyond total metal concentrations, biomonitoring now incorporates molecular and biochemical biomarkers that indicate early physiological changes due to heavy metal exposure. Oxidative stress biomarkers, such as malondialdehyde (MDA), 8-hydroxy-2’-deoxyguanosine (8-OHdG), and protein carbonyl levels, provide insights into metal-induced cellular damage. These biomarkers reflect lipid peroxidation, DNA damage, and protein oxidation, hallmarks of heavy metal toxicity. Additionally, alterations in antioxidant enzyme activity, including superoxide dismutases (SODs), catalases (CATs), and glutathione peroxidases (GPxs), serve as indicators of metal-induced oxidative stress responses.

Recent advancements in omics technologies, including genomics, proteomics, and metabolomics, have facilitated the identification of novel biomarkers for early detection. Gene expression profiling has revealed that exposure to heavy metals leads to differential regulation of stress response genes, such as MTs, heat shock proteins, and cytochrome P450 enzymes (CYPs). MicroRNAs (miRNAs) have also emerged as promising biomarkers, as specific miRNA signatures have been linked to arsenic-induced carcinogenesis and cadmium-related nephrotoxicity. Proteomic and metabolomic studies further provide comprehensive insights into metabolic disruptions caused by heavy metal exposure, enabling the identification of potential diagnostic and prognostic markers.

Integrating biomonitoring with advanced sensor technologies and wearable devices has revolutionised real-time exposure assessment. Portable biosensors based on nanotechnology, electrochemical detection, and immunoassays allow for rapid, on-site screening of heavy metals in biological samples. These advancements enhance early detection capabilities, particularly in high-risk populations such as industrial workers, miners, and individuals residing in polluted regions (Aley et al., 2022; Kumar et al., 2013; Kumari et al., 2022; Reddy et al., 2022).

Overall, biomonitoring and early detection strategies are critical for mitigating the health impacts of heavy metal toxicity. By combining traditional and emerging techniques, researchers and healthcare professionals can develop precision medicine approaches for risk assessment, early intervention, and personalized detoxification strategies, ultimately improving public health outcomes.

10.6.3 Potential for precision medicine approaches

The potential for precision medicine approaches in mitigating heavy metal toxicity is gaining increasing attention due to the growing understanding of genetic, epigenetic, and molecular factors influencing individual susceptibility to metal exposure. Precision medicine, which tailors prevention, diagnosis, and treatment strategies based on an individual’s genetic makeup, metabolic profile, and environmental interactions, offers a novel way to address the challenges posed by heavy metal toxicity. Given the significant variability in how individuals absorb, metabolize, and excrete toxic metals such as lead (Pb), mercury (Hg), arsenic (As), and cadmium (Cd), precision medicine enables the identification of high-risk populations and the implementation of targeted interventions to reduce health risks.

One of the key aspects of precision medicine in heavy metal toxicity lies in genetic screening for susceptibility markers. Variants in metal transport genes such as *SLC11A2* (divalent metal transporter 1, DMT1) and *SLC39A8* (ZIP8) have been linked to differential absorption and retention of metals like lead and cadmium, influencing toxicity outcomes. Similarly, polymorphisms in detoxification enzymes such as GSTs (*GSTT1* and *GSTM1*) and MTs (*MT1* and *MT2*) contribute to individual variations in heavy metal clearance and cellular defense mechanisms. By identifying individuals with genetic predispositions to impaired metal detoxification, precision medicine can enable personalized interventions such as tailored chelation therapy, dietary modifications, and targeted pharmacological treatments (Aley et al., 2022; Kumar et al., 2021; Pankaj et al., 2012a; Reddy et al., 2022).

Epigenetic modifications, including DNA methylation, histone modifications, and ncRNA regulation, further contribute to individual differences in heavy metal toxicity responses. Studies have shown that exposure to arsenic and cadmium induces aberrant DNA methylation patterns in genes involved in cellular stress response, inflammation, and carcinogenesis. By utilizing epigenetic biomarkers, precision medicine approaches can assess early molecular alterations due to heavy metal exposure, allowing for early intervention before clinical symptoms manifest. Additionally, therapeutic strategies targeting epigenetic modifications, such as DNA methylation and HDAC inhibitors, may offer new avenues for mitigating heavy metal-induced diseases.

Metabolomic and proteomic profiling have also advanced precision medicine strategies by identifying specific metabolic pathways and protein interactions altered by heavy metal exposure. Metabolomic studies have revealed disruptions in amino acid metabolism, oxidative stress pathways, and lipid profiles associated with metal toxicity. Precision medicine can leverage this knowledge to develop biomarker-based diagnostics and personalized nutritional interventions to restore metabolic balance. For instance, targeted supplementation with antioxidants such as *N*-acetylcysteine (NAC), selenium, and zinc has been shown to mitigate oxidative stress and reduce heavy metal burden in susceptible individuals (Chakraborty et al., 2021; Das et al., 2022; Kumar and Dwivedi, 2020; Kumar et al., 2021).

Moreover, advancements in nanotechnology and drug delivery systems are enhancing precision medicine approaches by enabling the development of targeted therapeutics for heavy metal detoxification. Nanoparticle-based chelators, for example, offer higher specificity and efficiency in binding toxic metals than traditional chelating agents. Personalized pharmacological strategies can optimize dosing regimens based on genetic and metabolic profiles, reducing adverse effects while maximizing detoxification efficiency (Dwivedi and Kumar, 2011; Kumar et al., 2018b; Kumar et al., 2021). Precision medicine holds great promise in transforming the management of heavy metal toxicity by integrating genetic, epigenetic, metabolomic, and proteomic data to develop personalized risk assessments and interventions. By leveraging advances in biomarker discovery, targeted therapeutics, and individualized treatment plans, precision medicine can significantly improve health outcomes and reduce the burden of heavy metal-induced diseases in vulnerable populations.

10.7 Future directions and research gaps

The advent of genetic and omics-based approaches has revolutionised the study of heavy metal toxicity by providing deeper insights into individual susceptibility, mechanistic pathways, and molecular responses to toxic metal exposure. Traditional toxicology approaches primarily relied on epidemiological studies and biochemical assessments; however, advancements in genomics, transcriptomics, proteomics, metabolomics, and epigenomics now offer a more precise understanding of how heavy metals interact with biological systems at the molecular level. These emerging methodologies enable the identification of genetic variants, molecular biomarkers, and cellular pathways that mediate metal-induced toxicity, thereby facilitating precision medicine approaches, early diagnostics, and targeted interventions. Genomic studies, particularly GWAS, have identified key genetic polymorphisms influencing heavy metals' absorption, distribution, metabolism, and excretion (ADME). For instance, polymorphisms in metal transport genes such as *SLC11A2* (divalent metal transporter 1, DMT1) and *SLC39A8* (ZIP8) have been associated with lead and cadmium uptake variations, respectively. Similarly, single-nucleotide polymorphisms (SNPs) in detoxification genes like *GSTT1*, *GSTM1*, and *MT1* impact an individual's ability to neutralize heavy metal–induced oxidative stress. Researchers can stratify populations based on susceptibility by incorporating genetic screening into environmental health assessments and developing personalized mitigation strategies. Additionally, NGS technologies have accelerated the identification of rare genetic variants linked to differential heavy metal toxicity responses, thereby advancing risk assessment models and public health recommendations.

Beyond genomics, transcriptomics has provided valuable insights into the cellular response to heavy metal exposure by analyzing gene expression changes. RNA sequencing (RNA-seq) and microarray technologies have revealed differential expression patterns of stress response genes, inflammatory markers, and apoptotic regulators in individuals exposed to lead, mercury, arsenic, and cadmium. Studies have demonstrated that arsenic exposure leads to upregulation of proinflammatory cytokines, mitochondrial dysfunction-related genes, and DNA repair mechanisms, contributing to its carcinogenic effects. Similarly, cadmium-induced nephrotoxicity has been linked to altered expression of kidney-specific transporters and oxidative stress genes. By integrating transcriptomic data with toxicological assessments, researchers can develop early biomarkers of exposure and identify potential therapeutic targets. Proteomics and metabolomics have expanded the omics toolkit for studying heavy metal toxicity by elucidating protein interactions and metabolic perturbations associated with metal exposure. Mass spectrometry–based proteomics has identified key proteins involved in metal detoxification, such as MTs, GSTs, and cytochrome P450 enzymes (CYPs). Changes in protein abundance and posttranslational modifications provide critical insights into cellular adaptive mechanisms and potential intervention points. Likewise, metabolomic profiling has revealed significant disruptions in amino acid metabolism, lipid peroxidation, and energy homeostasis in individuals exposed to toxic metals. By leveraging proteomic and metabolomic data, researchers can develop precision medicine strategies, such as personalized supplementation with antioxidants (e.g., selenium, zinc, N-acetylcysteine) to counteract heavy metal–induced oxidative stress.

Epigenomic studies have highlighted the role of DNA methylation, histone modifications, and ncRNAs in mediating heavy metal toxicity. Heavy metals such as arsenic and cadmium induce global hypomethylation and site-specific hypermethylation, leading to tumor suppressor genes and oncogene dysregulation. These epigenetic alterations contribute to the carcinogenic and neurotoxic effects of heavy metals. Histone modifications, including acetylation and methylation, are also crucial in regulating gene expression in response to metal exposure. Furthermore, microRNAs (miRNAs) and lncRNAs have emerged as key regulators of metal-induced toxicity, modulating oxidative stress pathways, inflammatory responses, and apoptotic signaling. By integrating epigenomic data with traditional toxicology assessments, researchers can develop predictive biomarkers for early detection of metal toxicity and design epigenetic therapies to mitigate adverse health effects.

Integrating genomics with environmental health studies has become critical to understanding the gene-environment interactions that shape susceptibility to heavy metal toxicity. Environmental genomics approaches, which combine genetic data with exposure assessments, have enabled researchers to identify populations at higher risk due to genetic predisposition, occupational exposure, and lifestyle factors. For example, genome-exposure interaction studies have shown that individuals with polymorphisms in *ALAD* (aminolevulinic acid dehydratase) exhibit increased sensitivity to lead poisoning due to impaired heme biosynthesis. Similarly, genetic variants in *PON1* (paraoxonase 1) influence an individual's ability to detoxify organophosphate pesticides and heavy metals. By incorporating genomic data into environmental risk assessment models, policymakers can implement targeted public health interventions, such as screening high-risk populations, regulating industrial emissions, and developing personalized detoxification strategies. Furthermore, advancements in exposomics—the comprehensive study of environmental exposures across an individual's lifespan—have facilitated the integration of omics data with ecological health studies. High-throughput analytical techniques, such as liquid chromatography-mass spectrometry (LC-MS) and high-resolution metabolomics, have enabled the identification of metal-specific exposure signatures and their long-term health effects. Exposomic data, combined with wearable biosensors and real-time environmental monitoring, can provide a more accurate assessment of cumulative heavy metal exposure, enhancing risk prediction and early intervention strategies.

From a policy perspective, integrating genomics and environmental health studies has significant implications for public health interventions and regulatory frameworks. Governments and international health organizations must adopt evidence-based policies incorporating genetic susceptibility data into risk assessment protocols. For example, regulatory agencies such as the US Environmental Protection Agency and the European Chemicals Agency have recognized the importance of integrating omics data into chemical safety evaluations. By implementing precision toxicology approaches, policymakers can establish exposure limits tailored to vulnerable populations, improve workplace safety standards, and enhance environmental regulations to minimize heavy metal contamination in air, water, and soil. Public health interventions should also prioritize biomonitoring programs that leverage omics technologies for early detection of heavy metal toxicity. National health agencies can implement large-scale population screening programs using genetic, epigenetic, and metabolomic biomarkers to identify at-risk individuals and provide targeted health recommendations. Community-based interventions, such as dietary supplementation with protective nutrients (e.g., folate, vitamin B12, and antioxidants) and lifestyle modifications (e.g., reducing tobacco and alcohol consumption, which exacerbate metal toxicity), can further mitigate adverse health effects. Public awareness campaigns should also educate communities about the risks of heavy metal exposure, emphasizing preventive measures such as using water filtration systems, consuming low-contaminant foods, and minimizing occupational exposure through protective equipment and safe work practices.

Integrating omics-based approaches into environmental health studies also presents opportunities for developing novel therapeutic interventions for heavy metal toxicity. Advances in pharmacogenomics have enabled the identification of drug–gene interactions that influence an individual's response to metal chelators such as DMSA and EDTA. Depending on an individual's genetic and metabolic profile, personalized chelation therapy can improve treatment efficacy while reducing adverse effects. Furthermore, emerging bioremediation strategies, such as using genetically engineered microorganisms and plants for phytoremediation, offer innovative solutions for reducing environmental heavy metal contamination. Emerging genetic and omics-based approaches are transforming the field of ecological toxicology by providing a comprehensive understanding of heavy metal toxicity at the molecular level. Integrating genomics with environmental health studies enables the identification of at-risk populations, facilitates early detection through biomarker discovery, and informs precision medicine strategies for mitigating metal toxicity. From a policy perspective, incorporating omics data into public health interventions and regulatory frameworks can enhance risk assessment, improve workplace safety, and promote environmental sustainability. By leveraging these scientific advancements, researchers, policymakers, and healthcare professionals can work collaboratively to reduce the global burden of heavy metal toxicity and improve public health outcomes.

10.8 Conclusion

Recent genetic and omics-based research advancements have significantly enhanced our understanding of heavy metal toxicity, revealing critical insights into genetic susceptibility, molecular responses, and potential intervention strategies. Key findings indicate that heavy metal exposure leads to oxidative stress, DNA damage, epigenetic modifications, and disruptions in essential cellular pathways. Genomic studies have identified polymorphisms in metal transporters, detoxification enzymes, and oxidative stress response genes that influence an individual's susceptibility to heavy metal–induced toxicity. For instance, variations in genes such as *SLC11A2* (DMT1), *SLC39A8* (ZIP8), and *MT1* modulate the absorption, distribution, and excretion of toxic metals like lead, cadmium, and arsenic, thereby affecting overall toxicity levels. Additionally, transcriptomic and proteomic analyses have highlighted differential gene expression patterns and protein modifications contributing to heavy metal–induced inflammation, apoptosis, and carcinogenesis. Epigenetic alterations, including DNA methylation changes, histone modifications, and ncRNA regulation, further demonstrate how heavy metals exert long-term effects on gene expression, potentially leading to chronic diseases such as cancer, neurodegeneration, and cardiovascular disorders. These molecular insights underscore the complexity of gene–environment interactions and highlight the urgent need for precision medicine approaches to mitigate heavy metal toxicity.

The implications of these findings for mitigating heavy metal toxicity are profound, suggesting a shift towards more targeted and personalized interventions. Biomonitoring using genetic, transcriptomic, proteomic, and metabolomic biomarkers allows for early detection of heavy metal exposure, enabling timely interventions before irreversible damage occurs. Identifying individuals with genetic predispositions to heavy metal toxicity can inform risk stratification and personalized preventive strategies, such as dietary modifications, antioxidant supplementation, and lifestyle adjustments to reduce exposure. Additionally, pharmacogenomic research has demonstrated that genetic variations influence the efficacy of chelation therapy, emphasizing the need for personalized treatment regimens based on an individual's genetic and metabolic profile. Environmental and policy-level interventions must also integrate omics-based findings to establish stricter regulations on industrial emissions, improve workplace safety standards, and enhance public awareness regarding the dangers of heavy metal contamination. Incorporating omics-based data into regulatory decision-making can lead to more refined exposure limits, tailored mitigation strategies, and targeted public health initiatives, ensuring a comprehensive approach to reducing heavy metal toxicity in vulnerable populations.

Despite these advancements, there remains a critical need for further research on genetic susceptibility to heavy metal toxicity to bridge existing knowledge gaps and refine risk assessment models. Large-scale GWAS and multiomics integration are necessary to uncover additional genetic variants that contribute to differential susceptibility across diverse populations. Longitudinal cohort studies should investigate the interplay between genetic factors, epigenetic modifications, and environmental exposures over time to understand cumulative toxicity effects. Moreover, advancements in single-cell omics and spatial transcriptomics offer promising avenues for exploring tissue-specific responses to heavy metals, enabling the identification of novel therapeutic targets. Future research should also explore the role of gut microbiota in modulating heavy metal absorption and detoxification, as microbiome composition may significantly influence individual susceptibility. By advancing our understanding of genetic susceptibility and integrating cutting-edge omics technologies, researchers can develop innovative prevention and intervention strategies, ultimately reducing the burden of heavy metal toxicity on public health and fostering a safer, more sustainable environment.

Acknowledgment

The author used AI tools to improve language and readability, but all final editing was done by a human.

References

Aley, P., Singh, J., & Kumar, P. (2022). Chapter 23 – Adapting the changing environment: Microbial way of life. In A. Kumar, J. Singh, & L. F. R. B. T.-M. U. C. C. Ferreira (Eds.). Woodhead Publishing. Retrieved from https://doi.org/10.1016/B978-0-323-90571-8.00023-7.

Chakraborty, S., Kumar, P., Sanyal, R., Mane, A. B., Arvind Prasanth, D., Patil, M., & Dey, A. (2021). Unravelling the regulatory role of miRNAs in secondary metabolite production in medicinal crops. *Plant Gene, 27*, 100303. Retrieved from https://doi.org/10.1016/j.plgene.2021.100303.

Choudhury, D., Kumar, P., Zhimo, V. Y., & Sahoo, J. (2024). Crop rotation patterns and soil health management. In *Bioremediation of emerging contaminants from soils: Soil health conservation for improved ecology and food security*. Punjab, Jalandhar, India: Elsevier, Department of Plant Pathology, School of Agriculture, Lovely Professional University, 565–589. Retrieved from https://doi.org/10.1016/B978-0-443-13993-2.00025-6.

Das, T., Saha, S. C., Sunita, K., Majumder, M., Ghorai, M., Mane, A. B., ... Dey, A. (2022). Promising botanical-derived monoamine oxidase (MAO) inhibitors: Pharmacological aspects and structure-activity studies. *South African Journal of Botany, 146*, 127–145. Retrieved from https://doi.org/10.1016/j.sajb.2021.09.019.

Devi, P., & Kumar, P. (2023a). Role of oxidative enzymes in the alleviation of oxidative stress in plant system. In *Microbial oxidative enzymes: Biotechnological applications*. Phagwara, Punjab: De Gruyter, Department of Agronomy, School of Agriculture, Lovely Professional University, 109–128. Retrieved from https://doi.org/10.1515/9783111062235-006 144411, India.

Devi, P., & Kumar, P. (2023b). Effect of microbial oxidative enzymes on soil health. In *Microbial oxidative enzymes: Biotechnological applications*. Phagwara, Punjab, India: De Gruyter, Department of Agronomy, School of Agriculture, Lovely Professional University, 37–61. Retrieved from https://doi.org/10.1515/9783111062235-003.

Devi, P., & Kumar, P. (2024a). Challenges in the storage of herbal medicine products and strategies for sustainable management. In *Reference series in phytochemistry*. Phagwara, India: Springer Science and Business Media B.V., Department of Agronomy, School of Agriculture, Lovely Professional University, 1739–1767. Retrieved from https://doi.org/10.1007/978-3-031-43199-9-37.

Devi, P., & Kumar, P. (2024b). Herbal medicine and pregnancy. In *Reference series in phytochemistry*. Punjab, Phagwara, India: Springer Science and Business Media B.V., Department of Agronomy, School of Agriculture, Lovely Professional University, 693–722. Retrieved from https://doi.org/10.1007/978-3-031-43199-9-25.

Devi, P., Dey, S. R., & Kumar, P. (2023a). Integration of water resources management in rural areas. In *Water resources management for rural development: Challenges and mitigation*. Punjab, Phagwara, India: Elsevier, Department of Agronomy, School of Agriculture, Lovely Professional University, 139–152. Retrieved from https://doi.org/10.1016/B978-0-443-18778-0.00013-1.

Devi, P., Dey, S. R., Saini, L., Kumar, P., Panigrahi, S., & Dwivedi, P. (2023b). Toward sustainable agriculture: strategies involving phytoprotectants against reactive oxygen species. In *Reactive oxygen species: Prospects in plant metabolism*. Phagwara, India: Springer Nature, Department of Agronomy, School of Agriculture, Lovely Professional University, 229–247. Retrieved from https://doi.org/10.1007/978-981-19-9794-5-13.

Devi, P., Dey, S. R., Sharma, K., Kumar, P., & Singh, J. (2024a). Waste management as an opportunity for the inclusion of vulnerable groups. In *Waste management in the circular economy*. Punjab, Phagwara: Springer International Publishing, Department of Agronomy, School of Agriculture, Lovely Professional University, 207–230. Retrieved from https://doi.org/10.1007/978-3-031-42426-7-11 144411, India.

Devi, P., Dogra, P., Kumar, P., & Singh, J. (2024b). Global status of biomass energy programmes-challenges and roadmap. In *Biomass energy for sustainable development*. Phagwara, India: CRC Press, Lovely Professional University, 394–414. Retrieved from https://doi.org/10.1201/9781003406501-20.

Devi, P., Kapoor, P., & Kumar, P. (2024c). Biomarkers for toxicity detection. In *Biomarkers in environmental and human health biomonitoring: An integrated perspective*. Punjab, Phagwara, India: Elsevier, Department of Agronomy, School of Agriculture, Lovely Professional University, 313–329. Retrieved from https://doi.org/10.1016/B978-0-443-13860-7.00004-5.

Devi, P., Kaur, A., & Kumar, P. (2024d). Phytoremediation of metal contaminated soil using energy crops: Soil health maintenance along with biofuel production. In *Bioremediation of emerging contaminants from soils: Soil health conservation for improved ecology and food security*. Punjab, Phagwara, India: Elsevier, Department of Agronomy, School of Agriculture, Lovely Professional University, 307–333. Retrieved from https://doi.org/10.1016/B978-0-443-13993-2.00015-3.

Devi, P., Kumar, P., Dey, S. R., Banik, D., Kumar, G., & Mehta, C. M. (2024e). Integrated omics approaches for nutrient stress management in plants. In *Current omics advancement in plant abiotic stress biology: A volume in developments in applied microbiology and biotechnology*. Punjab, Phagwara, India: Elsevier, School of Agriculture, Lovely Professional University, 93–117. Retrieved from https://doi.org/10.1016/B978-0-443-21625-1.00008-7.

Devi, P., Kumar, P., Sharma, K., & Singh, J. (2023). Chapter 10 – Is there a relationship between myconanotechnology and sustainable development? In M. Kuddus, I. Z. Ahmad, C. M. B. T.-M. Hussain, & A. of N. in B (Eds.). *Progress in biochemistry and biotechnology* (pp. 175–207). Academic Press. Retrieved from https://doi.org/10.1016/B978-0-443-15262-7.00009-7.

Dey, S. R., Devi, P., & Kumar, P. (2023a). Organic contaminants in aquatic environments: Sources and impact assessment. In *Metal organic frameworks for wastewater contaminant removal*. Punjab: Wiley, Department of Agronomy, School of Agriculture, Lovely Professional University, 299–317 144411, India.

Dey, S. R., Kumar, P., & Singh, J. (2023b). Growth of mustard under open field conditions with salicylic acid and antimicrobial thiourea. *Biopesticides International, 19*, 183–192. Retrieved from https://doi.org/10.59467/BI.2023.19.183.

Dey, S. R., Sharma, M., & Kumar, P. (2023c). Effects and responses of chromium on plants. In *Environmental science and engineering*. Punjab, Phagwara: Springer Science and Business Media Deutschland GmbH, Department of Agronomy, School of Agriculture, Lovely Professional University, 385–427. Retrieved from https://doi.org/10.1007/978-3-031-44029-8-14 144411, India.

Dey, S. R., Sharma, M., & Kumar, P. (2024). Phytoremediation of phenolic compounds from soil. In *Bioremediation of emerging contaminants from soils: Soil health conservation for improved ecology and food security*. Punjab, Phagwara, India: Elsevier, Department of Agronomy, School of Agriculture, Lovely Professional University, 335–358. Retrieved from https://doi.org/10.1016/B978-0-443-13993-2.00016-5.

Dwivedi, P., & Kumar, P. (2011). Anti-diabetic medicinal plants and their conservation: Waging green war on diabetes. *Medicinal Plants, 3*, 181–189. Retrieved from https://doi.org/10.5958/j.0975-4261.3.3.031.

Goud, E. L., Singh, J., & Kumar, P. (2022). In A. Kumar, J. Singh, & L. F. R. B. T.-M. U. C. C. Ferreira (Eds.). *Chapter 19 – Climate change and their impact on global food production* (pp. 415–436). Woodhead Publishing. Retrieved from https://doi.org/10.1016/B978-0-323-90571-8.00019-5.

Hidangmayum, A., Dwivedi, P., Kumar, P., & Upadhyay, S. K. (2023). Seed priming and foliar application of chitosan ameliorate drought stress responses in mungbean genotypes through modulation of morpho-physiological attributes and increased antioxidative defense mechanism. *Journal of Plant Growth Regulation, 42*, 6137–6154. Retrieved from https://doi.org/10.1007/s00344-022-10792-1.

Jagadesh, M., Dash, M., Kumari, A., Singh, S. K., Verma, K. K., Kumar, P., ... Sharma, S. K. (2024). Revealing the hidden world of soil microbes: Metagenomic insights into plant, bacteria, and fungi interactions for sustainable agriculture and ecosystem restoration. *Microbiological Research, 285*. Retrieved from https://doi.org/10.1016/j.micres.2024.127764.

Karam, N., Kumar, P., Anand, Y. R., & Choudhury, D. (2023). Management of rice blast—A review of the strategies available. *Biopesticides International, 19*, 107–116. Retrieved from https://doi.org/10.59467/BI.2023.19.107.

Kotia, A., Rutu, P., Singh, V., Kumar, A., Dhoke, S., Kumar, P., & Singh, D. K., (2021). Rheological analysis of rice husk-starch suspended in water for sustainable agriculture application. In *Second international conference on functional materials, manufacturing and performances, ICFMMP 2021* (pp. 1962–1966). Elsevier Ltd, School of Mechanical Engineering, Lovely Professional Unievrsity, Punjab, Phagwara, India. https://doi.org/10.1016/j.matpr.2021.09.325.

Kumar, D., Rameshwar, S. D., & Kumar, P. (2019). Effect of intergated application of inorganic and organic fertilizers on the roots of chickpea. *Plant Archives, 19*, 857–860.

Kumar, P. (2023). Degradation of xenobiotics by bacteria and fungi: An overview. *Biopesticides International, 19*, 117–132. Retrieved from https://doi.org/10.59467/BI.2023.19.117.

Kumar, P. (2024a). Biochar production from marine algae and its application in the treatment of industrial effluent. In *Marine biomass: Biorefinery, bioproducts and environmental bioremediation*. Phagwara, Punjab: De Gruyter, Department of Agronomy, School of Agriculture, Lovely Professional University, 57–93. Retrieved from https://doi.org/10.1515/9783111353951-004 144411, India.

Kumar, P. (2024b). Application of marine biomass for carbon neutrality, sustainable environment, and economy. In *Marine biomass: Biorefinery, bioproducts and environmental bioremediation*. Phagwara, Punjab: De Gruyter, Department of Agronomy, School of Agriculture, Lovely Professional University, 95–128. Retrieved from https://doi.org/10.1515/9783111353951-005 144411, India.

Kumar, P. (2024c). Soil conservation for global sustainability. In *World sustainability series*. Punjab, Phagwara: Springer Science and Business Media Deutschland GmbH, Department of Agronomy, School of Agriculture, Lovely Professional University, 103–128. Retrieved from https://doi.org/10.1007/978-3-031-57456-6_6 144411, India.

Kumar, P., & Choudhury, D. (2024a). Optimized phytoremediation process for the sustainable management radionuclides. In *Bioremediation of emerging contaminants from soils: Soil health conservation for improved ecology and food security*. Punjab, Phagwara, India: Elsevier, Department of Agronomy, School of Agriculture, Lovely Professional University, 443–464. Retrieved from https://doi.org/10.1016/B978-0-443-13993-2.00020-7.

Kumar, P., & Choudhury, D. (2024b). Role of indigenous knowledge in agricultural soil reclamation without disturbing other ecosystems. In *Bioremediation of emerging contaminants from soils: Soil health conservation for improved ecology and food security*. Punjab, Phagwara, India: Elsevier, Department of Agronomy, School of Agriculture, Lovely Professional University, 465–488. Retrieved from https://doi.org/10.1016/B978-0-443-13993-2.00021-9.

Kumar, P., & Choudhury, D. (2024c). Microbial native soil bacteria against cadmium toxicity. In *Springer water*. Punjab, Phagwara: Springer Nature, Department of Agronomy, School of Agriculture, Lovely Professional University, 193–226. Retrieved from https://doi.org/10.1007/978-3-031-54005-9-9 144411, India.

Kumar, P., & Choudhury, D. (2024d). Toxicity of rhizospheric cadmium contaminated soil and its phytoremediation. In *Springer water*. Punjab, Phagwara: Springer Nature, Department of Agronomy, School of Agriculture, Lovely Professional University, 227–264. Retrieved from https://doi.org/10.1007/978-3-031-54005-9-10 144411, India.

Kumar, P., & Choudhury, D. (2024e). Bioremediation of wastewater using hydroponics. In *Springer water*. Punjab, Phagwara: Springer Nature, Department of Agronomy, School of Agriculture, Lovely Professional University, 27–64. Retrieved from https://doi.org/10.1007/978-3-031-53258-0_2 144411, India.

Kumar, P., & Choudhury, D. (2024f). Hydroponic removal of organic contaminants from water. In *Springer water*. Punjab, Phagwara: Springer Nature, Department of Agronomy, School of Agriculture, Lovely Professional University, 143–164. Retrieved from https://doi.org/10.1007/978-3-031-53258-0-6 144411, India.

Kumar, P., & Dey, S. R. (2024). Hydroponics phytoremediation: An overview. In *Springer water*. Punjab, Phagwara: Springer Nature, Department of Agronomy, School of Agriculture, Lovely Professional University, 361–396. Retrieved from https://doi.org/10.1007/978-3-031-53258-0-13 144411, India.

Kumar, P., & Dwivedi, P. (2018). Putrescine and glomus mycorrhiza moderate cadmium actuated stress reactions in *Zea mays* l. By means of extraordinary reference to sugar and protein. *Vegetos, 31*, 74–77. Retrieved from https://doi.org/10.5958/2229-4473.2018.00076.9.

Kumar, P., & Dwivedi, P. (2020). Lignin estimation in sorghum leaves grown under hazardous waste site. *Plant Archives, 20*, 2558–2561.

Kumar, P., & Mistri, T. K. (2020). Transcription factors in SOX family: Potent regulators for cancer initiation and development in the human body. *Seminars in Cancer Biology, 67*, 105–113. Retrieved from https://doi.org/10.1016/j.semcancer.2019.06.016.

Kumar, P., & Naik, M. (2020). Biotic symbiosis and plant growth regulators as a strategy against cadmium and lead stress in chickpea. *Plant Archives, 20*, 2495–2500.

Kumar, P., & Pathak, S. (2019). Responsiveness index of sorghum (*Sorghum bicolor* (L.) Moench) grown under cadmium contaminated soil treated with putrescine and mycorrhiza. *Bangladesh Journal of Botany, 48*, 139–143.

Kumar, P., & Raut, A. M. (2024). Microbes-assisted bioaugmentation process in the reduction of emerging industrial pollutants from soil. In *Bioremediation of emerging contaminants from soils: Soil health conservation for improved ecology and food security*. Punjab, Phagwara, India: Elsevier, Department of Agronomy, School of Agriculture, Lovely Professional University, 519–540. Retrieved from https://doi.org/10.1016/B978-0-443-13993-2.00023-2.

Kumar, P., & Singh, J. (2024). Hydroponic: An eco-friendly future. In *Springer water*. Punjab, Phagwara: Springer Nature, Department of

Agronomy, School of Agriculture, Lovely Professional University, 223–267. Retrieved from https://doi.org/10.1007/978-3-031-53258-0-9 144411, India.

Kumar, P., Mandal, B., & Dwivedi, P. (2011a). Heavy metals scavenging of soils and sludges by ornamental plants. *Journal of Applied Horticulture, 13*, 144–146.

Kumar, P., Mandal, B., & Dwivedi, P. (2011b). Heavy metal scavenging capacity of *Mentha* spicata and Allium cepa. *Medicinal Plants, 3*, 315–318. Retrieved from https://doi.org/10.5958/j.0975-4261.3.4.053.

Kumar, P., Mandal, B., & Dwivedi, P. (2013). Combating heavy metal toxicity from hazardous waste sites by harnessing scavenging activity of some vegetable plants. *Vegetos, 26*, 416–425. Retrieved from https://doi.org/10.5958/j.2229-4473.26.2.106.

Kumar, P., Singh, B. N., & Dwivedi, P. (2017). Plant growth regulators, plant adaptability and plant productivity: A review on abscisic acid (ABA) signaling in plants under emerging environmental stresses. In *Sustaining future food security in changing environments*. Varanasi, India: Nova Science Publishers, Inc., Department of Plant Physiology, Institute of Agricultural Sciences, Banaras Hindu University, 81–97.

Kumar, P., Harshavardhan, M., Kumar, P. S., Yumnam, J., Jyoti, N., Naik, M., ... Kumar, S. (2018a). Effect on chlorophyll a/b ratio in cadmium contaminated maize leaves treated with putrescine and mycorrhiza. *Annals of Biology, 34*, 281–283.

Kumar, P., Kumar, S., Harshavardhan, M., Naik, M., Yumnam, J., Kumar, P. S., ... Purnima (2018b). Evaluation of plant height and leaf length of sorghum grown under different sources of nutrition. *Annals of Biology, 34*, 284–286.

Kumar, P., Pathak, S., Amarnath, K. S., Veerendra Brahma Teja, P., Dileep, B., Kumar, K., ... Siddique, A. (2018c). Effect of growth regulator on morpho-physiological attributes of chilli: A case study. *Plant Archives, 18*, 1771–1776.

Kumar, P., Yumnam, J., Kumar, P. S., Misao, L., Jyoti, N., Naik, M., ... Harshavardhan, M. (2018d). Cadmium induced changes in germination of maize seed treated with mycorrhiza. *Annals of Agri-bio Research, 23*, 169–170.

Kumar, P., Kumar, T., Singh, S., Tuteja, N., Prasad, R., & Singh, J. (2020). Potassium: A key modulator for cell homeostasis. *Journal of Biotechnology, 324*, 198–210.

Kumar, P., Devi, P., & Dey, S. R. (2021a). In A. Kumar, J. Singh, J. B. T.-V. Samuel, & M. of M (Eds.). *Chapter 6 – Fungal volatile compounds: A source of novel in plant protection agents* (pp. 83–104). Academic Press. Retrieved from https://doi.org/10.1016/B978-0-12-824523-1.00001-8.

Kumar, P., Sharma, K., Saini, L., & Dey, S. R. (2021b). In A. Kumar, J. Singh, J. B. T.-V. Samuel, & M. of M (Eds.). *Chapter 8 – Role and behavior of microbial volatile organic compounds in mitigating stress* (pp. 143–161). Academic Press. Retrieved from https://doi.org/10.1016/B978-0-12-824523-1.00010-9.

Kumar, P., Goud, E. L., Devi, P., Dey, S. R., & Dwivedi, P. (2022a). Heavy metals: Transport in plants and their physiological and toxicological effects. In *Plant metal and metalloid transporters*. Punjab, Jalandhar, India: Springer Nature, Department of Agronomy, School of Agriculture, Lovely Professional University, 23–54. Retrieved from https://doi.org/10.1007/978-981-19-6103-8-2.

Kumar, P., Goud, E. L., Devi, P., & Koul, B. (2022b). Metal pollutants in the environment. In *Environmental microbiology: Emerging technologies*. Jalandhar, Punjab: De Gruyter, Department of Agronomy, School of Agriculture, Lovely Professional University, 291–323. Retrieved from https://doi.org/10.1515/9783110727227-012 144411, India.

Kumar, P., Koul, B., & Sharma, M. (2022c). Phytoremediation of heavy metals. In *Heavy metals in plants: Physiological to molecular approach*. Punjab, Jalandhar: CRC Press, Department of Agronomy, School of Agriculture, Lovely Professional University, 369–388. Retrieved from https://doi.org/10.1201/9781003110576-17 144411, India.

Kumar, P., Latha, D. H., & Singh, J. (2023a). Rhizoremediation of organic emerging soil contaminants: Green technology. In *Clean technologies toward a sustainable future: Physicochemical, biochemical and biotechnological approaches*. Phagwara, Punjab: IWA Publishing, Department of Agronomy, School of Agriculture, Lovely Professional University, 211–229. Retrieved from https://doi.org/10.2166/9781789063783-0211 144411, India.

Kumar, P., Saini, L., & Sharma, M. (2023b). Assessment of the phytochemical constituents and metabolites in the medicinal plants and herbal medicine used in the treatment and management of respiratory diseases. In *Reference series in phytochemistry*. Phagwara, India: Springer Science and Business Media B.V., Department of Agronomy, School of Agronomy, Lovely Professional University. Retrieved from https://doi.org/10.1007/978-3-031-21973-3-19-1.

Kumar, P., Choudhury, D., Dey, S. R., & Sharma, M. (2024a). Application of genetically modified crops against metallic contaminations. In *Bioremediation of emerging contaminants from soils: Soil health conservation for improved ecology and food security*. Punjab, Phagwara, India: Elsevier, Department of Agronomy, School of Agriculture, Lovely Professional University, 541–564. Retrieved from https://doi.org/10.1016/B978-0-443-13993-2.00024-4.

Kumar, P., Dey, S. R., & Choudhury, D. (2024b). Effectiveness of cadmium on biochemical shift of pea plant treated with mycorrhiza and putrescine. *Nature Environment and Pollution Technology, 23*, 275–285. Retrieved from https://doi.org/10.46488/NEPT.2024.v23i01.023.

Kumar, P., Dwivedi, P., & Upadhyay, S. K. (2024c). Optimization of polyamine and mycorrhiza in sorghum plant for removal of hazardous cadmium. *Plant Physiology and Biochemistry PPB / Societe Francaise de Physiologie Vegetale, 214*. Retrieved from https://doi.org/10.1016/j.plaphy.2024.108846.

Kumar, P., Karam, N., & Choudhury, D. (2024d). Assessment of the phytochemical constituents and metabolites in the medicinal plants and herbal medicine used in the treatment and management of skin diseases. In *Reference series in phytochemistry*. Punjab, Phagwara, India: Springer Science and Business Media B.V., Department of Agronomy, School of Agriculture, Lovely Professional University, 533–573. Retrieved from https://doi.org/10.1007/978-3-031-43199-9-21.

Kumar, P., Khokhar, A., & Dey, S. R. (2024e). Evidence from the use of herbal medicines in the management and prevention of common eye diseases. In *Reference series in phytochemistry*. Phagwara, India: Springer Science and Business Media B.V., Department of Agronomy, School of Agriculture, Lovely Professional University, 763–795. Retrieved from https://doi.org/10.1007/978-3-031-43199-9-27.

Kumar, P., Saini, L., & Sharma, M. (2024f). Assessment of the phytochemical constituents and metabolites in the medicinal plants and herbal medicine used in the treatment and management of respiratory diseases. In *Reference series in phytochemistry*. Punjab, Phagwara, India: Springer Science and Business Media B.V., Department of Agronomy, School of Agronomy, Lovely Professional University, 481–510. Retrieved from https://doi.org/10.1007/978-3-031-43199-9_19.

Kumar, P., Saurabh, S., & Sharma, K. (2024g). Assessment of the phytochemical constituents and metabolites in medicinal plants and herbal remedies used in the treatment and management of reproductive diseases: Polycystic ovary syndrome. In *Reference series in phytochemistry*. Punjab, Phagwara, India: Springer Science and Business Media B.V., Department of Agronomy, School of Agriculture, Lovely Professional University, 511–532. Retrieved from https://doi.org/10.1007/978-3-031-43199-9-20.

Kumar, P., Sharma, K., & Singh, J. (2024h). Analysis of complex microbial communities in soil and wastewater treatment processes. In *Microbial metagenomics in effluent treatment plant*. Punjab, Phagwara, India: Elsevier, Department of Agronomy, School of Agriculture, Lovely Professional University, 25–41. Retrieved from https://doi.org/10.1016/B978-0-443-13531-6.00002-1.

Kumar, P., Sharma, K., & Singh, J. (2024i). Response of microbial community to environment changes. In *Microbial metagenomics in effluent treatment plant*. Punjab, Phagwara, India: Elsevier, Department of Agronomy, School of Agriculture, Lovely Professional University, 43–61. Retrieved from https://doi.org/10.1016/B978-0-443-13531-6.00006-9.

Kumar, P., Srivastav, A. L., Chaudhary, V., van Hullebusch, E. D., & Busquets, R. (2024j). *Bioremediation of emerging contaminants from soils: Soil health conservation for improved ecology and food security*. Phagwara, India: Elsevier, Department of Agronomy, School of Agriculture, Lovely Professional University, Punjab. Retrieved from https://doi.org/10.1016/C2022-0-02758-6.

Kumar, P., Pathak, S., Kumar, M., & Dwivedi, P. (2018). Role of secondary metabolites for the mitigation of cadmium toxicity in sorghum grown under mycorrhizal inoculated hazardous waste site. *Biotechnological Approaches for Medicinal and Aromatic Plants Conservation, Genetic Improvement and Utilization: Conservation, Genetic Improvement and Utilization,* 199–212.

Kumar, V., Dwivedi, P., Kumar, P., Singh, B. N., Pandey, D. K., Kumar, V., & Bose, B. (2021). Mitigation of heat stress responses in crops using nitrate primed seeds. *South African Journal of Botany, 140*, 25–36. Retrieved from https://doi.org/10.1016/j.sajb.2021.03.024.

Kumar, V. A., & Kumar, P. (2023). Revolutionizing agriculture with satellite technology for farmers: A review. *Biopesticides International, 19*, 97–106. Retrieved from https://doi.org/10.59467/BI.2023.19.97.

Kumari, P., Singh, J., & Kumar, P. (2022). In A. Kumar, J. Singh, & L. F. R. B. T.-M. U. C. C. Ferreira (Eds.). *Chapter 21 – Impact of bioenergy for the diminution of an ascending global variability and change in the climate* (pp. 469–487). Woodhead Publishing. Retrieved from https://doi.org/10.1016/B978-0-323-90571-8.00021-3.

Kumari, P., Kumar, P., & Singh, J. (2023). Toxic contamination of aquatic habitat and its remediation. In *Spatial modeling of environmental pollution and ecological risk*. Punjab, Phagwara, India: Elsevier, Department of Agronomy, School of Agriculture, Lovely Professional University, 209–220. Retrieved from https://doi.org/10.1016/B978-0-323-95282-8.00021-3.

Naik, M., & Kumar, P. (2020). Role of growth regulators and microbes for metal detoxification in plants and soil. *Plant Archives, 20*, 2820–2824.

Pandey, S. K., Kumar, P., & Dey, S. R. (2024). Xenobiotic stress management by microbial endophytes. In *Bioremediation of emerging contaminants from soils: Soil health conservation for improved ecology and food security*. Madhya Pradesh, Jabalpur, India: Elsevier, College of Agriculture, Powarkheda Narmadapuram, Jawaharlal Nehru Agriculture University, 241–259. Retrieved from https://doi.org/10.1016/B978-0-443-13993-2.00012-8.

Pankaj, P. K., Kumar, P., Nigam, R. C., & Mishra, P. K. (2012a). Monitoring and surveillance of synthetic pyrethroids and organophosphate in different brands of soft drinks. *Journal of Chemical and Pharmaceutical Research, 4*, 3939–3943.

Pankaj, P. K., Kumar, P., Nigam, R. C., & Mishra, P. K. (2012b). To studies organochlorine insecticides in different brands of cold drinks. *Journal of Chemical and Pharmaceutical Research, 4*, 3934–3938.

Pathak, S., Kumar, P., Mishra, P. K., & Kumar, M. (2017). Mycorrhiza assisted approach for bioremediation with special reference to biosorption. *Pollution Research, 36*, 329–332.

Raj, A., Goutam, E., Kumar, P., & Singh, J. (2024). Polyphenol oxidase: An enzyme of plants. In *Polyphenol oxidases: Function, wastewater remediation, and biosensors*. Phagwara, Punjab: De Gruyter, Department of Agronomy, School of Agriculture, Lovely Professional University, 105–128. Retrieved from https://doi.org/10.1515/9783111033525-004 144411, India.

Reddy, S. S., Kumar, P., & Dwivedi, P. (2022). Heavy metal transporters, phytoremediation potential, and biofortification. In *Plant metal and metalloid transporters*. Springer, 387–405.

Saini, L., Sharma, K., & Kumar, P. (2023). Water efficiency in agriculture as a vital approach toward water management in rural areas. In *Water resources management for rural development: Challenges and mitigation*. Punjab, Phagwara, India: Elsevier, Department of Agronomy, School of Agriculture, Lovely Professional University, 127–138. Retrieved from https://doi.org/10.1016/B978-0-443-18778-0.00020-9.

Saurabh, K., Meena, R. K., Kumar, P., Upadhyaya, A., & Das, A. (2024). Role of hyperaccumulators in the reduction of emerging industrial pollutants from soil. In *Bioremediation of emerging contaminants from soils: Soil health conservation for improved ecology and food security*. Bihar, Patna, India: Elsevier, ICAR Research Complex for Eastern Region, 489–517. Retrieved from https://doi.org/10.1016/B978-0-443-13993-2.00022-0.

Sharma, K., & Kumar, P. (2023). Microbe-mediated impact on cell membrane stability and leaf pigmentation in pearl millet under toxic mercury stress in soil. *Biopesticides International, 19*, 217–228. Retrieved from https://doi.org/10.59467/BI.2023.19.217.

Sharma, K., & Kumar, P. (2024). Environmental threats posed by xenobiotics. In *Bioremediation of emerging contaminants from soils: Soil health conservation for improved ecology and food security*. Punjab, Phagwara, India: Elsevier, Department of Agronomy, School of Agriculture, Lovely Professional University, 183–201. Retrieved from https://doi.org/10.1016/B978-0-443-13993-2.00009-8.

Sharma, K., Devi, P., Dey, S. R., & Kumar, P. (2023a). Mercury phytovolatilization: An overview of the mechanism and mitigation. In *Role of green chemistry in ecosystem restoration to achieve environmental sustainability*. Punjab, Phagwara, India: Elsevier,

Department of Agronomy, School of Agriculture, Lovely Professional University, 325–331. Retrieved from https://doi.org/10.1016/B978-0-443-15291-7.00031-6.

Sharma, K., Devi, P., & Kumar, P. (2023b). Metal-organic frameworks for wastewater contaminants removal. In *Metal organic frameworks for wastewater contaminant removal*. Punjab, Phagwara: Wiley, Department of Agronomy, School of Agriculture, Lovely Professional University, 95–117 144411, India.

Sharma, K., Devi, P., Kumar, P., Dey, A., & Dwivedi, P. (2023c). Hazardous phytotoxic nature of reactive oxygen species in agriculture. In *Reactive oxygen species: Prospects in plant metabolism*. Phagwara, India: Springer Nature, Department of Agronomy, School of Agriculture, Lovely Professional University, 135–146. Retrieved from https://doi.org/10.1007/978-981-19-9794-5-8.

Sharma, M., Dey, S. R., & Kumar, P. (2023). Antioxidant defence: A key mechanism of chromium tolerance. In *Environmental science and engineering*. Punjab, Phagwara: Springer Science and Business Media Deutschland GmbH, Department of Agronomy, School of Agriculture, Lovely Professional University, 91–116. Retrieved from https://doi.org/10.1007/978-3-031-44029-5 144411, India.

Sinam, V., Kumar, P., & Singh, J. (2023). Bioremediation and ecorestoration strategies of aquatic environment. In *Spatial modeling of environmental pollution and ecological risk*. Punjab, Phagwara, India: Elsevier, Department of Agronomy, School of Agriculture, Lovely Professional University, 483–499. Retrieved from https://doi.org/10.1016/B978-0-323-95282-8.00025-0.

Srivastav, A. L., Patel, N., Rani, L., Kumar, P., Dutt, I., Maddodi, B. S., & Chaudhary, V. K. (2024). Sustainable options for fertilizer management in agriculture to prevent water contamination: a review. *Environment, Development and Sustainability, 26*, 8303–8327. Retrieved from https://doi.org/10.1007/s10668-023-03117-z.

Upadhyay, S. K., Devi, P., Kumar, V., Pathak, H. K., Kumar, P., Rajput, V. D., & Dwivedi, P. (2023). Efficient removal of total arsenic (As3+/5+) from contaminated water by novel strategies mediated iron and plant extract activated waste flowers of marigold. *Chemosphere, 313*, 137551. Retrieved from https://doi.org/10.1016/j.chemosphere.2022.137551.

Chapter 11

Role of glutathione in heavy metal detoxification

Prasann Kumar
Department of Agronomy, School of Agriculture, Lovely Professional University, Phagwara, Punjab, India

11.1 Introduction

Heavy metal toxicity is a significant environmental and public health concern arising from the accumulation of toxic metals such as lead (Pb), mercury (Hg), cadmium (Cd), and arsenic (As) in living organisms. Various anthropogenic activities introduce these metals into the environment, including industrial processes, mining, agricultural runoff, and improper waste disposal (Adegbola & Adetutu, 2024; Ghuge, Nikalje, Kadam, Suprasanna, & Hong, 2023). Once absorbed into biological systems, heavy metals can interfere with cellular functions once absorbed into biological systems by generating reactive oxygen species (ROS), disrupting enzyme activities, and binding to vital biomolecules, leading to oxidative stress, DNA damage, and cell death. Unlike organic pollutants, heavy metals are nonbiodegradable and persist in ecosystems for extended periods, posing long-term health risks to humans, animals, and plants. The toxic effects of heavy metals vary depending on their chemical properties, exposure levels, and biological targets. For instance, lead exposure is known to cause neurotoxicity, particularly in children, leading to cognitive deficits, developmental delays, and behavioral disorders (Hamai-Amara et al., 2024; Li, Yu, He, Zhen, & She, 2025; Liu, Xiao et al., 2024). Mercury primarily affects the central nervous system and has been linked to Minamata disease, a severe neurological disorder caused by methylmercury poisoning. Cadmium, commonly found in industrial waste and contaminated food sources, accumulates in the kidneys and leads to renal dysfunction, bone demineralization, and increased cancer risk. Arsenic exposure, mainly through contaminated drinking water, has been associated with skin lesions, cardiovascular diseases, and various forms of cancer. Given the widespread prevalence and severe consequences of heavy metal toxicity, the need for effective detoxification mechanisms is paramount in mitigating their harmful effects on living organisms (Kaur, Kaur, & Sharma, 2025; Liu, Zheng et al., 2024).

Detoxification mechanisms play a crucial role in protecting cells from the toxic impact of heavy metals by facilitating their sequestration, transformation, and excretion. The human body employs various defense strategies to counteract heavy metal toxicity, including chelation, enzymatic detoxification, antioxidant defense systems, and metal-binding proteins. Chelation therapy is a common medical intervention for heavy metal poisoning, involving the use of chelating agents such as ethylenediaminetetraacetic acid (EDTA), dimercaprol (BAL), and succimer (dimercaptosuccinic acid [DMSA]), which bind to heavy metals and facilitate their excretion through urine. At the cellular level, detoxification mechanisms involve the upregulation of stress response pathways that help neutralize oxidative damage caused by heavy metal-induced ROS production. Enzymatic antioxidants such as superoxide dismutase (SOD), catalase (CAT), and glutathione peroxidase (GPx) play vital roles in scavenging free radicals and reducing oxidative stress (Faruque et al., 2024; Sarma, Hoque, Thekkangil, Venkatarayappa, & Rajagopal, 2024). Additionally, metallothioneins (MTs), a family of cysteine-rich, metal-binding proteins, help sequester heavy metals within cells, thereby preventing their interaction with essential biomolecules. MTs exhibit a high affinity for heavy metals such as cadmium, mercury, and zinc, and their expression is upregulated in response to heavy metal exposure. Another crucial detoxification mechanism is the involvement of phase I and phase II detoxification enzymes, which facilitate the biotransformation of toxic compounds into less harmful forms. Phase I enzymes, such as cytochrome P450 monooxygenases,

Heavy Metal Toxicity and Neurodegeneration. DOI: https://doi.org/10.1016/B978-0-443-36575-1.00006-6

initiate detoxification by introducing functional groups into xenobiotic compounds. In contrast, phase II enzymes, such as glutathione-S-transferases (GSTs), conjugate these modified compounds with endogenous molecules to enhance their solubility and excretion. Among the various detoxification pathways, the glutathione system plays a central role in cellular defense against heavy metal toxicity, highlighting its importance in maintaining redox homeostasis and detoxification processes (Lei et al., 2025; Singh, Kumar, Siddiqi, & Sharma, 2024).

Glutathione (GSH), a tripeptide composed of glutamate, cysteine, and glycine, is one of the biological systems' most critical antioxidants and detoxification agents. It exists in two forms: reduced glutathione (GSH) and oxidized glutathione (GSSG), with the former being the active form responsible for redox regulation and detoxification. The significance of glutathione in cellular defense mechanisms cannot be overstated, as it plays a central role in maintaining redox balance and detoxification processes. Its ability to directly neutralize ROS, chelate heavy metals, and participate in enzymatic detoxification reactions is crucial. One of the primary roles of glutathione in heavy metal detoxification is its ability to bind to toxic metal ions through the thiol (–SH) group present in its cysteine residue. This interaction forms glutathione-metal complexes, transported out of cells via adenosine triphosphate (ATP)-dependent transporters, thereby reducing intracellular metal accumulation (Yaashikaa, Kumar, Jeevanantham, & Saravanan, 2022; Zhang, Wang et al., 2024). The formation of glutathione-metal conjugates is facilitated by GSTs, a family of enzymes that catalyze the conjugation of glutathione to electrophilic compounds, including heavy metal ions. GST-mediated detoxification is particularly relevant in protecting cells from cadmium, mercury, and arsenic toxicity, as these metals have a strong affinity for thiol groups. Glutathione peroxidase (GPx) is crucial in mitigating oxidative stress by reducing hydrogen peroxide (H_2O_2) and lipid peroxides, thereby preventing cellular damage induced by heavy metal exposure. The regeneration of reduced glutathione from its oxidized form is catalyzed by glutathione reductase (GR), which maintains the redox balance within cells and ensures the continuous availability of GSH for detoxification purposes (Mandal, Bashir, & Raj, 2025; Ummalyma, Udayan, & Sreekumar, 2024; Xi, Zhang, Xu, Cheng, & Chen, 2024).

The regulation of glutathione synthesis and metabolism is not a static process; it is tightly controlled to adapt to environmental stressors, including heavy metal exposure. The rate-limiting step in glutathione biosynthesis, catalyzed by glutamate-cysteine ligase (GCL), is a prime example of the body's adaptability. This enzyme combines glutamate and cysteine to form γ-glutamylcysteine, followed by the addition of glycine by glutathione synthetase (GS) to produce GSH. The availability of cysteine, a critical determinant of glutathione synthesis, is a testament to the body's ability to respond to environmental stressors. Under conditions of heavy metal stress, cells upregulate glutathione synthesis to counteract increased oxidative damage and enhance metal detoxification capacity. The nuclear factor erythroid 2-related factor 2 (Nrf2) signaling pathway, which plays a pivotal role in the transcriptional regulation of glutathione biosynthesis genes, further underscores the body's adaptability. Upon activation by oxidative stress, Nrf2 translocates to the nucleus (Gupta et al., 2025; Xia et al., 2024). It binds to antioxidant response elements (AREs) in the promoters of genes encoding GCL, GSTs, and other detoxification enzymes, thereby enhancing cellular resilience against heavy metal toxicity. The protective effects of glutathione extend beyond direct metal chelation and antioxidant defense, as it also modulates inflammatory responses and apoptosis in response to heavy metal exposure. Chronic heavy metal toxicity has been linked to increased inflammation and apoptosis through the activation of proinflammatory cytokines and mitochondrial dysfunction. Glutathione helps mitigate these effects by suppressing inflammatory pathways and maintaining mitochondrial integrity, thereby promoting cell survival under toxic conditions.

The depletion of glutathione levels due to excessive heavy metal exposure can have severe consequences for cellular health, leading to increased oxidative damage, mitochondrial dysfunction, and impaired detoxification capacity. Various strategies have been explored to enhance glutathione levels and improve heavy metal detoxification, including dietary supplementation with glutathione precursors such as *N*-acetylcysteine (NAC) and α-lipoic acid (ALA). These compounds help replenish intracellular GSH levels and enhance the activity of detoxification enzymes. Additionally, dietary sources rich in sulfur-containing amino acids, such as garlic, onions, and cruciferous vegetables, have been shown to support glutathione synthesis and improve heavy metal excretion. Emerging research has also focused on using phytochemicals, such as curcumin, resveratrol, and sulforaphane, to activate the Nrf2 pathway and enhance glutathione-mediated detoxification. Furthermore, probiotics and prebiotics have been investigated for their role in modulating gut microbiota composition and reducing heavy metal absorption in the gastrointestinal tract, thereby complementing glutathione-based detoxification mechanisms (Bjedov et al., 2025; Heisi, Nkuna, & Matambo, 2025; Panda, Fatnani, & Parida, 2025). Given the critical role of glutathione in cellular defense, ongoing research continues to explore novel strategies for optimizing its function and enhancing resilience against heavy metal toxicity. In conclusion, heavy metal toxicity remains a pressing environmental and health issue, necessitating the activation of effective detoxification mechanisms to mitigate its adverse effects. The glutathione system is a fundamental component of cellular defense, providing antioxidant

TABLE 11.1 Molecular composition and structure of glutathione.

Component	Chemical formula	Molecular weight (g/mol)	Functional groups involved in heavy metal binding
Glutathione (GSH)	$C_{10}H_{17}N_3O_6S$	307.32	Thiol (–SH), carboxyl (–COO^-), amino (–NH_2)
Glutamate	$C_5H_9NO_4$	147.13	Carboxyl (–COO^-)
Cysteine	$C_3H_7NO_2S$	121.16	Thiol (–SH), carboxyl (–COO^-), amino (–NH_2)
Glycine	$C_2H_5NO_2$	75.07	Amino (–NH_2), carboxyl (–COO^-)

Source: Based on Ding, Ding, Liu, Liu, and Chai (2024), Liu, FAN, HAN, and WANG (2025), Song, Sun, Saud, Fahad, and Nawaz (2025).

protection, metal chelation, and enzymatic detoxification capabilities. Understanding the intricate role of glutathione in heavy metal detoxification offers valuable insights for developing therapeutic interventions to reduce heavy metal burden and promote overall health (Table 11.1).

11.2 Structure and function of glutathione

11.2.1 Chemical composition and biosynthesis

Biomolecules' chemical composition and biosynthesis are fundamental in various biological processes, influencing cellular functions, metabolism, and overall physiological responses. In glutathione, a tripeptide composed of glutamate, cysteine, and glycine, its unique chemical structure enables it to participate in redox reactions, detoxification processes, and cellular defense mechanisms. The thiol (–SH) group of cysteine is the most reactive functional group in glutathione, giving it its antioxidant properties and ability to bind to heavy metals and other toxic substances. The biosynthesis of glutathione occurs through a tightly regulated two-step enzymatic pathway, ensuring its availability under normal and stress-induced conditions. The first step in glutathione biosynthesis is catalyzed by GCL, the rate-limiting enzyme in the process (Chakravorty et al., 2023; Prasath & Mohanraj, 2024; Priyanka & Dwivedi, 2023; Yuksel, 2024). This enzyme facilitates the formation of γ-glutamylcysteine by catalyzing the bond between the γ-carboxyl group of glutamate and the amino group of cysteine. The regulation of GCL activity is critical because cysteine availability often limits glutathione synthesis due to its relatively low intracellular concentration. The second step is catalyzed by GS, which adds glycine to γ-glutamylcysteine, forming the final glutathione molecule. This reaction requires ATP as an energy source, highlighting the importance of cellular energy balance in maintaining glutathione biosynthesis. The biosynthesis of glutathione is highly responsive to cellular stress, oxidative damage, and exposure to toxins, with regulatory mechanisms ensuring adequate production under challenging conditions. One of the key regulators of glutathione biosynthesis is the nuclear factor erythroid 2-related factor 2 (Nrf2) signaling pathway, which plays a crucial role in controlling the expression of genes involved in antioxidant defense. Under normal conditions, Nrf2 is sequestered in the cytoplasm by Kelch-like ECH-associated protein 1 (Keap1), which targets it for degradation. However, in response to oxidative stress or electrophilic compounds, Keap1 undergoes conformational changes, releasing Nrf2 and allowing it to translocate into the nucleus. Once inside the nucleus, Nrf2 binds to AREs in the promoter regions of genes encoding glutathione biosynthesis enzymes, leading to their upregulation and increased glutathione production. This emphasis on the role of Nrf2 in regulating glutathione biosynthesis is essential for the audience to understand, as it provides insights into the key players in cellular defense mechanisms (Geetha et al., 2021; Ilyas, Shahid, Shafi, Aijaz, & Wasiullah, 2025; Li et al., 2024; Pang et al., 2024). In addition to transcriptional regulation, glutathione biosynthesis is influenced by nutrient availability, metabolic state, and external environmental factors. Dietary sources of cysteine, such as sulfur-containing amino acids from protein-rich foods, contribute to maintaining glutathione levels. Other compounds, such as NAC and ALA, have been shown to enhance glutathione biosynthesis by providing precursor molecules or supporting enzymatic activity. Furthermore, metabolic pathways, such as the pentose phosphate pathway, generate reducing equivalents in the form of nicotinamide adenine dinucleotide phosphate (NADPH), which is essential for regenerating reduced glutathione from its oxidized form (GSSG). Given the central role of glutathione in cellular protection, its biosynthesis is a crucial process tightly regulated by multiple signaling pathways, enzymatic mechanisms, and metabolic inputs to maintain cellular homeostasis and resilience against oxidative stress (Anas et al., 2025; Milano, Giotta, & Lambreva, 2024; Zhang, Wu et al., 2024).

11.2.2 Intracellular distribution and regulation

The intracellular distribution and regulation of glutathione (GSH) are crucial for maintaining cellular redox balance, detoxification processes, and overall cellular homeostasis. Glutathione exists in reduced (GSH) and oxidized (GSSG) forms, with a high GSH/GSSG ratio essential for cellular health. Within the cell, glutathione is predominantly found in the cytosol. Still, it is also distributed to various organelles, including the mitochondria, nucleus, and endoplasmic reticulum (ER), where it performs specific functions. Glutathione transport into these compartments is mediated by specialized transporters that ensure an optimal intracellular distribution. In mitochondria, glutathione is critical in protecting against oxidative stress, given that mitochondria are significant sites of ROS production during oxidative phosphorylation. Mitochondrial glutathione is imported from the cytosol via the mitochondrial glutathione transporter, as mitochondria lack the enzymatic machinery for glutathione biosynthesis. Once inside, it acts as a key antioxidant, maintaining mitochondrial integrity and function by neutralizing ROS and preventing lipid peroxidation (Khan, Asaf, Kang, & Lee, 2025; Rachappanavar, Gupta, Jayaprakash, & Abbas, 2024; Rottet et al., 2023; Umar, Naeem, Hussain, Ahmad, & Xu, 2025). The nucleus also holds a distinct glutathione pool, essential for protecting DNA from oxidative damage and regulating redox-sensitive transcription factors. The nuclear GSH pool is critical for modulating gene expression in response to oxidative stress, ensuring proper cellular responses to environmental and metabolic changes. The ER, however, maintains a more oxidizing environment to facilitate appropriate protein folding and disulfide bond formation in newly synthesized proteins. The lower GSH/GSSG ratio in the ER compared to the cytosol is necessary for oxidative protein folding, a process essential for secretory and membrane protein maturation. Glutathione homeostasis is regulated through biosynthesis, recycling, and degradation. The rate-limiting enzyme in GSH synthesis, GCL, is tightly regulated by intracellular cysteine availability, feedback inhibition by GSH, and transcriptional control through the nuclear factor erythroid 2-related factor 2 (Nrf2) pathway. Under oxidative stress conditions, Nrf2 activation enhances the expression of GCL and GS, leading to increased GSH production. In addition to biosynthesis, the regeneration of GSH from its oxidized form (GSSG) is catalyzed by glutathione reductase (GR) in an NADPH-dependent reaction. The availability of NADPH, primarily generated through the pentose phosphate pathway, is crucial for sustaining glutathione recycling (Giri et al., 2024; He et al., 2025; Tripathi & Poluri, 2021; Wang et al., 2025). Cellular glutathione levels are also modulated through conjugation with xenobiotics and heavy metals by GSTs, facilitating their excretion from the cell. Furthermore, autophagic and lysosomal degradation pathways regulate GSH turnover by degrading glutathione-containing proteins and releasing amino acids for resynthesis. Disruptions in glutathione distribution and regulation can lead to oxidative stress, cellular dysfunction, and disease pathogenesis, including neurodegenerative disorders, cancer, and metabolic diseases. Therefore maintaining precise control over intracellular glutathione levels and distribution is essential for cellular resilience and adaptation to stress (Table 11.2).

11.2.3 Antioxidant properties and redox balance

Oxidative stress, characterized by an imbalance between ROS production and the antioxidant defense system, plays

TABLE 11.2 Different mechanisms by which glutathione (GSH) detoxifies heavy metals.

Mechanism	Description	Example heavy metals	Key enzymes/processes involved
Direct binding	GSH directly binds to heavy metals through its thiol (–SH) group, forming metal-GSH complexes sequestered or excreted.	Cd, Pb, Hg	Nonenzymatic binding, formation of metal-GSH conjugates
Enzymatic detoxification	GSTs catalyse the conjugation of heavy metals with GSH, facilitating their transport and removal.	As, Cr	GSTs, ABC transporters
Redox modulation	GSH neutralizes ROS from heavy metal exposure, preventing cell oxidative damage.	Cu, Fe	Glutathione peroxidase (GPx), glutathione reductase (GR), superoxide dismutase (SOD)
Phytochelatin synthesis	GSH is a precursor for phytochelatins, peptides that chelate and sequester heavy metals in plants.	Cd, Zn	Phytochelatin synthase
Regeneration and recycling	After detoxification, GSH is oxidized to glutathione disulfide (GSSG) and is regenerated back to GSH by glutathione reductase, maintaining redox balance.	All heavy metals	Glutathione reductase (GR)

GSTs, Glutathione-*S*-transferases; *ROS*, reactive oxygen species
Source: Based on Chukwu and Gulser (2025), Li et al. (2022), Wang et al. (2023).

a pivotal role in various physiological and pathological processes. ROS, including superoxide anions (O_2^-), hydrogen peroxide (H_2O_2), and hydroxyl radicals (OH•), are natural by-products of cellular metabolism, primarily generated in mitochondria during oxidative phosphorylation. While low levels of ROS are essential for cellular signaling and immune defense, excessive ROS accumulation leads to oxidative damage of lipids, proteins, and nucleic acids, contributing to aging, neurodegeneration, cardiovascular diseases, and cancer. To counteract oxidative stress, organisms have evolved intricate antioxidant defense mechanisms comprising enzymatic and nonenzymatic antioxidants that help maintain redox homeostasis.

Enzymatic antioxidants include superoxide dismutase (SOD), catalase (CAT), and glutathione peroxidase (GPx), which act in concert to neutralize ROS. SOD catalyzes the conversion of superoxide anions into hydrogen peroxide, which is subsequently detoxified by catalase or reduced to water by GPx using glutathione (GSH) as a cofactor. Glutathione, a tripeptide composed of glutamate, cysteine, and glycine, is a crucial intracellular antioxidant that scavenges free radicals and maintains the reduced cellular environment. Nonenzymatic antioxidants, such as vitamin C (ascorbic acid), vitamin E (tocopherol), polyphenols, flavonoids, and carotenoids, further contribute to cellular redox balance by directly scavenging free radicals or regenerating oxidized antioxidants (Guan et al., 2024; Gupta, Siwach, & Mishra, 2024; Li & Zhao, 2025; Rath & Das, 2025).

The redox balance is regulated through intricate signaling pathways, including the nuclear factor erythroid 2-related factor 2 (Nrf2) pathway, which governs the expression of antioxidant and detoxification genes. Under oxidative stress, Nrf2 dissociates from its cytoplasmic inhibitor Kelch-like ECH-associated protein 1 (Keap1). It translocates to the nucleus, where it binds to the ARE in the promoter region of target genes. This upregulation enhances the cellular capacity to neutralize ROS, thereby preventing oxidative damage. Additionally, redox-sensitive transcription factors, such as nuclear factor-kappa B (NF-κB) and activator protein 1 (AP-1), play significant roles in inflammation and cellular adaptation to oxidative stress.

Disruptions in redox homeostasis are implicated in various pathological conditions, including neurodegenerative diseases (Alzheimer's and Parkinson's), cardiovascular disorders, diabetes, and cancer. Chronic oxidative stress leads to mitochondrial dysfunction, impaired DNA repair mechanisms, and aberrant cell signaling, promoting cellular senescence and apoptosis. In cancer, ROS can induce genetic mutations and activate oncogenic signaling pathways, contributing to tumor progression. However, cancer cells also exploit redox adaptations to enhance survival, necessitating targeted redox-based therapeutic strategies (Khan et al., 2025; LIU et al., 2025; Pang et al., 2024; Priyanka & Dwivedi, 2023; Zhang, Wu et al., 2024).

Dietary and pharmacological interventions to enhance antioxidant defenses have gained attention in mitigating oxidative stress-related diseases. Natural antioxidants derived from fruits, vegetables, and medicinal plants exhibit protective effects by scavenging free radicals and modulating redox-sensitive pathways. Synthetic antioxidants, including NAC and edaravone, are explored for their therapeutic potential in neuroprotection and cardiovascular health. However, an excessive intake of exogenous antioxidants may disrupt redox signaling, underscoring the importance of a balanced approach. Antioxidant mechanisms are fundamental in maintaining redox homeostasis, preventing oxidative damage, and preserving cellular function. A deeper understanding of redox biology and antioxidant strategies holds promise for therapeutic advancements in combating oxidative stress-associated diseases (Geetha et al., 2021; He et al., 2025; Khan et al., 2025; Milano et al., 2024; Tripathi & Poluri, 2021).

11.3 Mechanisms of glutathione-mediated heavy metal detoxification

11.3.1 Direct binding and chelation of heavy metals

Heavy metal contamination is a critical environmental and health concern due to the toxic effects of metals such as cadmium (Cd), mercury (Hg), lead (Pb), and arsenic (As). These metals persist in ecosystems, bioaccumulate in living organisms, and disrupt physiological processes by binding directly to biomolecules or interfering with essential metal homeostasis. The direct binding of heavy metals to proteins, enzymes, and DNA alters their structural integrity and function, leading to oxidative stress, cellular dysfunction, and apoptosis. One primary mechanism through which heavy metals exert their toxicity is interacting with thiol (–SH) groups in cysteine residues of proteins and glutathione, a key intracellular antioxidant (Giri et al., 2024; He et al., 2025; Pang et al., 2024; Umar et al., 2025; Yuksel, 2024). This interaction disrupts enzymatic activity, impairs redox balance, and weakens cellular defense mechanisms. Additionally, heavy metals can replace essential metals such as zinc (Zn), calcium (Ca), and magnesium (Mg) in metalloproteins, leading to the loss of critical biological functions. For instance, Cd competes with Zn for binding sites in MTs, small cysteine-rich proteins involved in metal detoxification and storage, resulting in the inactivation of Zn-dependent enzymes and disruption of cellular signaling pathways. Similarly, Pb interferes with calcium-dependent processes by mimicking Ca^{2+}, affecting neurotransmission, bone metabolism, and muscle contraction. Arsenic, existing in both trivalent

(As^{3+}) and pentavalent (As^{5+}) forms, inhibits pyruvate dehydrogenase and other essential metabolic enzymes, leading to mitochondrial dysfunction and energy depletion. Chelation therapy is an effective strategy for counteracting heavy metal toxicity by using chelating agents that form stable complexes with metals, facilitating their excretion. Chelators, such as EDTA, DMSA, and dimercaptopropane sulfonate (DMPS), have been widely used for the removal of Pb, Hg, and As from biological systems. These chelators contain multiple functional groups, including carboxyl (–COOH), hydroxyl (–OH), and thiol (–SH), which coordinate metal ions through electrostatic and covalent interactions, reducing their bioavailability and toxicity. Phytochelatins (PCs) and MTs, naturally occurring peptides and proteins in plants and animals, also play a crucial role in metal sequestration by chelating heavy metals through their thiol groups, thus preventing cellular damage. In addition to synthetic chelators, natural compounds such as polyphenols, flavonoids, and alginates derived from plant and marine sources exhibit chelating properties and antioxidant activity, providing potential therapeutic and environmental remediation applications. Developing nanomaterial-based chelators, including functionalised nanoparticles and metal-organic frameworks, has expanded the possibilities for efficient and selective heavy metal removal from contaminated environments (Li & Zhao, 2025; Milano et al., 2024; Rachappanavar et al., 2024; Tripathi & Poluri, 2021; Umar et al., 2025). Furthermore, bioremediation approaches utilizing metal-resistant bacteria, fungi, and algae have gained attention for their ability to biosorb and bioaccumulate heavy metals through cell wall binding and intracellular sequestration. Genetic engineering of microbial and plant systems to enhance metal-binding capacities offers a promising avenue for phytoremediation and biotechnological applications. However, chelation therapy and bioremediation effectiveness depend on metal speciation, bioavailability, and environmental conditions. While chelation provides immediate detoxification benefits, improper or excessive use of chelators may lead to the depletion of essential metals and unintended side effects. Therefore a balanced approach integrating chemical, biological, and nanotechnological strategies is critical for sustainable heavy metal detoxification and environmental remediation. Understanding the molecular mechanisms of metal binding and chelation not only aids in mitigating toxicity but also provides insights into designing targeted therapeutic interventions and developing innovative materials for environmental cleanup.

11.3.2 Glutathione conjugation via glutathione-*S*-transferases

Glutathione conjugation, catalyzed by GSTs, is a crucial detoxification mechanism that protects cells from oxidative stress, xenobiotics, and electrophilic compounds, including carcinogens, environmental pollutants, and heavy metals. GSTs are a superfamily of phase II detoxification enzymes that facilitate the conjugation of reduced glutathione (GSH), a tripeptide composed of glutamate, cysteine, and glycine, to electrophilic substrates, rendering them more water-soluble and facilitating their excretion. This enzymatic reaction involves the nucleophilic attack of the thiol (–SH) group of GSH on electrophilic centers of toxic compounds, neutralizing their reactivity and preventing cellular damage. GSTs are widely distributed in various tissues, particularly in the liver, kidneys, and intestines, where they play a significant role in detoxification processes. The GST superfamily is classified into several cytosolic isoforms, including Alpha (GSTA), Mu (GSTM), Pi (GSTP), Theta (GSTT), Zeta (GSTZ), and Omega (GSTO), each exhibiting distinct substrate specificity and tissue distribution. In addition to detoxification, GSTs contribute to cellular homeostasis by regulating redox balance, modulating signaling pathways, and interacting with proteins involved in apoptosis and cell survival (Anas et al., 2025; Khan et al., 2025; Rachappanavar et al., 2024; Rottet et al., 2023; Umar et al., 2025; Zhang, Wu et al., 2024). One of the key physiological roles of GST-mediated glutathione conjugation is the detoxification of ROS-induced lipid peroxidation products, such as 4-hydroxynonenal (4-HNE) and malondialdehyde (MDA), which can otherwise form harmful protein and DNA adducts, leading to oxidative damage and carcinogenesis. Furthermore, GSTs play a crucial role in drug metabolism, influencing the efficacy and resistance of chemotherapeutic agents. Overexpression of GSTP1 has been linked to resistance in various cancers by conjugating anticancer drugs such as cisplatin, chlorambucil, and doxorubicin, leading to their inactivation and increased drug efflux. This has made GSTs a potential target for overcoming chemoresistance through the development of GST inhibitors. Besides xenobiotic detoxification, GSTs also participate in heavy metal detoxification by facilitating the conjugation of GSH to toxic metals such as cadmium (Cd), mercury (Hg), and arsenic (As), forming metal-glutathione complexes that are transported out of the cell via ATP-binding cassette (ABC) transporters. This process protects cellular proteins from direct metal binding, preventing enzyme inhibition and oxidative stress-induced damage. The regulation of GST expression is controlled by transcription factors such as nuclear factor erythroid 2-related factor 2 (Nrf2), which activates the ARE in response to oxidative and electrophilic stress. Nrf2-mediated upregulation of GSTs enhances cellular defense mechanisms against oxidative damage and contributes to chemoprevention. However, dysregulation of GSTs has been implicated in various diseases, including cancer, neurodegenerative disorders, and metabolic syndromes (Guan et al., 2024; He et al., 2025; Li et al., 2022; Zhang, Wu et al., 2024).

Polymorphisms in GST genes, such as GSTM1 and GSTT1 null variants, are associated with altered detoxification capacity and susceptibility to diseases, including lung cancer, asthma, and cardiovascular disorders. The therapeutic potential of modulating GST activity is explored through small-molecule inhibitors, natural compounds, and gene therapy approaches to enhance drug efficacy and reduce toxicity. Additionally, dietary components such as cruciferous vegetables, rich in isothiocyanates, have been shown to induce GST expression, highlighting the role of diet in modulating detoxification pathways. Understanding the molecular mechanisms of glutathione conjugation via GSTs provides valuable insights into cellular defense mechanisms, disease susceptibility, and strategies for enhancing detoxification capacity through pharmacological and nutritional interventions.

11.3.3 Role in metal sequestration and transport

Metal sequestration and transport, the silent guardians of biological systems, environmental remediation, and industrial applications, are essential in maintaining metal homeostasis, detoxification, and bioavailability. These mechanisms are not just about ensuring the proper distribution of crucial metals like iron, zinc, copper, and manganese. They are about preventing the potential harm these metals can cause. Organisms have evolved sophisticated mechanisms, including MTs, chelating molecules, and specialized transport proteins, to regulate metal uptake, storage, and efflux. MTs, low-molecular-weight, cysteine-rich proteins, bind heavy metals through thiol groups, providing a detoxification mechanism while maintaining intracellular metal balance. Similarly, PCs and organic acids such as citrate and malate assist in metal chelation, facilitating metal storage and transport within plant cells. In bacteria and fungi, siderophores are secreted to scavenge and solubilize iron from the environment, playing a vital role in microbial metal homeostasis. In plants, metal sequestration is crucial for nutrient acquisition, where transporter proteins like ZIP (ZRT/IRT-like proteins) and NRAMP (natural resistance-associated macrophage protein) facilitate metal uptake and distribution. Hyperaccumulator plants have evolved specialized metal transport mechanisms that allow them to thrive in metal-rich soils by actively sequestering heavy metals in vacuoles, thus contributing to phytoremediation strategies (Chukwu & Gulser, 2025; Guan et al., 2024; He et al., 2025; Khan et al., 2025; Tripathi & Poluri, 2021). Plant metal transport is mediated by the xylem and phloem, ensuring efficient metal distribution to different organs while preventing toxic accumulation in sensitive tissues. In environmental applications, metal sequestration is integral to bioremediation and wastewater treatment, where biosorbents, biochar, and genetically engineered microorganisms are used to capture and immobilize heavy metals from contaminated sites. Microbial biofilms play a key role in metal sequestration by producing extracellular polymeric substances that bind and precipitate metals, reducing their mobility and toxicity in the environment. Metal sequestration is employed in industrial processes, such as metallurgy, electroplating, and mining waste management, to recover valuable metals and prevent environmental contamination. Synthetic chelating agents like EDTA and DTPA are widely used for metal extraction and stabilization, though their environmental impact has led to increasing interest in biodegradable alternatives. In human health, metal sequestration is relevant in conditions such as heavy metal poisoning, where chelating agents like dimercaprol and DMSA are administered to bind and facilitate the excretion of toxic metals like lead, mercury, and arsenic (Li & Zhao, 2025; Pang et al., 2024; Rottet et al., 2023; Umar et al., 2025; Wang et al., 2025, 2023). Metal transport proteins such as transferrin and ceruloplasmin play essential roles in human physiology by regulating iron and copper homeostasis, preventing oxidative stress, and ensuring proper enzymatic functions. Disruptions in metal sequestration and transport mechanisms are implicated in neurodegenerative diseases such as Alzheimer's and Parkinson's, where metal accumulation contributes to oxidative damage and protein aggregation. Advances in nanotechnology have enabled the development of engineered nanoparticles and functionalised materials for targeted metal sequestration and transport, offering innovative solutions for environmental and biomedical applications. Understanding the molecular and biochemical basis of metal sequestration and transport is crucial for developing sustainable strategies for metal management across diverse fields, from agriculture and medicine to industrial processing and ecological restoration, highlighting its vital role in preventing metal toxicity in biological systems (Table 11.3).

11.4 Heavy metal–induced oxidative stress and GSH response

11.4.1 Reactive oxygen species generation and oxidative damage

ROS generation and oxidative damage play crucial roles in cellular metabolism, signaling, and pathological conditions, influencing diverse biological processes across various organisms. ROS, including superoxide anions (O_2^-), hydrogen peroxide (H_2O_2), and hydroxyl radicals (OH•), are primarily generated as by-products of aerobic respiration within the mitochondria, where the electron transport chain (ETC) incompletely reduces oxygen molecules, leading to superoxide formation. Enzymatic sources such as NADPH oxidases (NOX), xanthine oxidase, and cytochrome P450 enzymes produce ROS in various

TABLE 11.3 Different glutathione-*S*-transferase (GST) isoforms and their roles in heavy metal detoxification.

GST isoform	Classification	Role in metal detoxification	Example heavy metals detoxified
GSTA (Alpha)	Cytosolic GST	catalyzes the conjugation of GSH with metal-induced oxidative stress by-products and xenobiotics.	Cd, As, Hg
GSTP (Pi)	Cytosolic GST	Plays a key role in cellular defense against oxidative stress and apoptosis caused by heavy metals.	Pb, Hg, Cr
GSTM (Mu)	Cytosolic GST	Involved in detoxification of lipid peroxidation by-products induced by heavy metal exposure.	Cd, Pb, Zn
GSTT (Theta)	Cytosolic GST	Involved in the metabolism of small electrophilic molecules, including metal-related toxicants.	As, Cr
GSTO (Omega)	Cytosolic GST	It has a role in arsenic biotransformation by catalyzing arsenate reduction to arsenite.	As
MGST (Microsomal GST)	Membrane-bound GST	Protects against metal-induced oxidative stress in cellular membranes.	Cu, Fe

Source: Based on Giri et al. (2024), Gupta et al. (2024), Milano et al. (2024), Umar et al. (2025), and Zhang, Wu et al. (2024).

cellular compartments. While low-to-moderate ROS levels are essential for redox signaling, immune defense, and cellular homeostasis, excessive ROS accumulation triggers oxidative stress, disrupting biomolecules such as lipids, proteins, and nucleic acids. Lipid peroxidation, a significant consequence of oxidative damage, leads to membrane destablization, loss of fluidity, and the formation of toxic aldehydes like MDA and 4-hydroxynonenal (4-HNE), which further propagate oxidative damage through secondary reactions with cellular components (Anas et al., 2025; He et al., 2025; Li et al., 2022; Milano et al., 2024; Rottet et al., 2023; Tripathi & Poluri, 2021). Proteins exposed to oxidative stress undergo modifications such as carbonylation, nitration, and cross-linking, impairing enzymatic activity, altering structural integrity, and promoting aggregation, a hallmark of neurodegenerative disorders like AD and PD. Oxidative damage to DNA results in strand breaks, base modifications, and mutations, contributing to genomic instability and carcinogenesis. The balance between ROS generation and antioxidant defense mechanisms determines cellular fate, with antioxidant enzymes like superoxide dismutase (SOD), catalase, and glutathione peroxidase (GPx) playing vital roles in neutralizing ROS and preventing oxidative stress. Nonenzymatic antioxidants such as glutathione, vitamin C, E, and polyphenols provide crucial redox buffering capacity. Under prolonged oxidative stress, mitochondrial dysfunction exacerbates ROS production, triggering apoptosis via cytochrome *c* release and caspase activation. Chronic oxidative damage is implicated in aging, cardiovascular diseases, diabetes, cancer, and inflammatory disorders, where persistent oxidative stress drives disease progression by activating proinflammatory signaling pathways like NF-κB and MAPK, promoting cytokine release, endothelial dysfunction, and tissue damage. In plants, ROS generation occurs in chloroplasts, peroxisomes, and plasma membrane-associated NADPH oxidases, playing dual roles in stress responses and developmental processes. Abiotic stresses such as drought, salinity, and heavy metal toxicity enhance ROS production, leading to cellular damage and growth inhibition, whereas controlled ROS signaling mediates defense responses, programmed cell death, and stomatal regulation. ROS-induced oxidative damage is counteracted by enzymatic antioxidants like ascorbate peroxidase (APX) and peroxiredoxins, along with nonenzymatic antioxidants like flavonoids and carotenoids. Nanotechnology, pharmacology, and molecular biology advances have enabled the development of targeted antioxidants, gene editing strategies, and synthetic ROS scavengers to mitigate oxidative damage in medical and agricultural applications. Understanding ROS dynamics and oxidative damage mechanisms is essential for developing therapeutic interventions, improving crop resilience, and addressing environmental stressors, highlighting the complex interplay between ROS production, cellular defense mechanisms, and disease pathogenesis across biological systems (Chukwu & Gulser, 2025; Giri et al., 2024; Rachappanavar et al., 2024; Rath & Das, 2025; Umar et al., 2025).

11.4.2 Glutathione-dependent antioxidant defenses

Glutathione-dependent antioxidant defenses play a pivotal role in maintaining cellular redox homeostasis and protecting against oxidative stress by neutralizing ROS and detoxifying harmful electrophilic compounds. Glutathione (GSH), a tripeptide composed of glutamate, cysteine, and glycine, is the most abundant intracellular thiol-based antioxidant, functioning as a key component in enzymatic and nonenzymatic detoxification pathways. The GSH redox cycle is primarily regulated by glutathione peroxidases (GPx), glutathione reductase (GR), and GSTs, which collectively mitigate oxidative damage by reducing hydroperoxides, recycling oxidized glutathione (GSSG) back to its reduced form, and facilitating the conjugation of toxic xenobiotics for cellular excretion. GPx catalyzes the reduction of hydrogen peroxide (H_2O_2) and lipid peroxides using GSH as an electron donor, converting it into GSSG, which is subsequently regenerated to GSH by GR using NADPH as a reducing equivalent. This cyclic process ensures a continuous supply of reduced GSH, essential for maintaining the redox environment within the cell. GSTs, a superfamily of phase II detoxification enzymes, utilize GSH to neutralize electrophilic compounds, including lipid peroxidation by-products like 4-hydroxynonenal (4-HNE), environmental toxins, and carcinogens, thereby protecting macromolecules from oxidative modification (Gupta et al., 2024; Milano et al., 2024; Rachappanavar et al., 2024; Tripathi & Poluri, 2021; Umar et al., 2025). Apart from enzymatic roles, GSH also participates in direct ROS scavenging, metal ion chelation, and thiol-based redox signaling, modulating key transcription factors such as nuclear factor erythroid 2-related factor 2 (Nrf2), which regulates the expression of ARE-driven genes. The Nrf2-Keap1 pathway plays a crucial role in cellular defense by sensing oxidative stress and upregulating the synthesis of GSH and detoxifying enzymes, thereby enhancing resilience against oxidative damage. The intracellular GSH/GSSG ratio is a critical indicator of oxidative stress, with a shift toward GSSG accumulation reflecting cellular redox imbalance and increased vulnerability to oxidative injury. In mitochondria, GSH is indispensable for maintaining redox equilibrium, preventing lipid peroxidation, and regulating apoptotic pathways by modulating cytochrome *c* release and caspase activation. In neurological systems, GSH depletion is linked to neurodegenerative disorders such as AD, PD, and amyotrophic lateral sclerosis, where oxidative stress exacerbates protein aggregation, mitochondrial dysfunction, and neuronal loss. In cardiovascular systems, GSH-dependent defenses protect against ischemia-reperfusion injury, atherosclerosis, and endothelial dysfunction by mitigating oxidative damage to lipids and proteins (Gupta et al., 2024; Khan et al., 2025; Wang et al., 2025, 2023; Zhang, Wu et al., 2024). In cancer biology, while GSH provides cytoprotective effects against oxidative stress-induced DNA damage, its overexpression in tumor cells contributes to chemoresistance by neutralizing reactive intermediates from anticancer drugs, highlighting the dual role of GSH in cellular survival and therapy resistance. In plants, glutathione functions as a crucial antioxidant, particularly under abiotic stress conditions such as heavy metal toxicity, drought, and salinity, where it aids in redox buffering, metal chelation, and activation of stress-responsive pathways. Advances in redox biology, synthetic antioxidants, and genetic engineering have led to novel strategies to enhance GSH biosynthesis, modulate redox signaling, and develop GSH-based therapeutics for oxidative stress-related diseases. Understanding the intricate mechanisms of glutathione-dependent defenses is critical for developing interventions to mitigate oxidative damage across biological and environmental systems.

11.4.3 Effects of heavy metal exposure on GSH homeostasis

Heavy metal exposure significantly disrupts glutathione (GSH) homeostasis, leading to oxidative stress, cellular toxicity, and metabolic dysfunction in biological systems. GSH plays a crucial role in detoxifying heavy metals by acting as a primary antioxidant, a metal chelator, and a cofactor for enzymatic detoxification pathways. However, exposure to heavy metals such as cadmium (Cd), lead (Pb), mercury (Hg), arsenic (As), and chromium (Cr) depletes intracellular GSH levels through excessive oxidation, conjugation, and impaired synthesis, thereby compromising cellular defense mechanisms. Cadmium, a nonredox-active metal, disrupts GSH homeostasis by forming Cd-GSH complexes and depleting cellular GSH pools, leading to increased ROS production and mitochondrial dysfunction. Additionally, Cd inhibits glutathione reductase (GR) and GCL, key enzymes involved in GSH recycling and biosynthesis, thereby reducing the availability of reduced GSH and impairing redox balance. Lead exposure induces oxidative stress by interfering with GSH-dependent antioxidant defenses, promoting lipid peroxidation, and depleting GSH through direct binding and inhibiting GSTs, which play a critical role in detoxification (Anas et al., 2025; Giri et al., 2024; Guan et al., 2024; Li et al., 2022; Milano et al., 2024; Rachappanavar et al., 2024). Mercury, particularly in its methylmercury (MeHg) form, exerts neurotoxic effects by binding to thiol groups in GSH, leading to GSH depletion, protein oxidation, and neuronal damage. MeHg exposure also disrupts mitochondrial function by increasing hydrogen peroxide (H_2O_2) production, impairing ETC activity, and triggering apoptosis. Arsenic exposure depletes GSH by promoting excessive GSH conjugation via arsenic methylation, which is crucial for arsenic detoxification. However, chronic arsenic exposure leads to a paradoxical increase in oxidative stress due to GSH exhaustion, resulting in DNA damage, protein oxidation, and metabolic dysfunction. Chromium, particularly hexavalent chromium (Cr^{6+}), induces oxidative stress by generating free radicals and disrupting GSH homeostasis through direct interactions with thiol groups, leading to DNA strand breaks, lipid peroxidation, and apoptosis. The depletion of GSH during heavy metal exposure compromises direct antioxidant defense and weakens cellular signaling pathways, such as the Nrf2-Keap1 pathway, which regulates the expression of detoxification and antioxidant genes. Suppression of Nrf2 activity by heavy metals results in reduced expression of glutathione peroxidase (GPx), glutathione reductase (GR), and GSTs, further exacerbating oxidative damage. In addition to enzymatic inhibition, heavy metals impair GSH synthesis by reducing cysteine availability, a rate-limiting precursor for GSH biosynthesis (He et al., 2025; Li & Zhao, 2025; Milano et al., 2024; Umar et al., 2025; Wang et al., 2023). The disruption of GSH homeostasis has widespread implications in neurotoxicity, hepatotoxicity, nephrotoxicity, and cardiovascular diseases, where heavy metal-induced oxidative stress contributes to chronic inflammation, mitochondrial dysfunction, and cell death. Strategies to restore GSH homeostasis in heavy metal toxicity include supplementation with NAC, sulfur-containing antioxidants, and phytochemicals that upregulate GSH biosynthesis and recycling pathways. Understanding the intricate mechanisms by which heavy metals disrupt GSH metabolism is crucial for developing therapeutic interventions to mitigate oxidative damage and improve cellular resilience against metal-induced toxicity across biological and environmental systems (Table 11.4).

11.5 Glutathione in phytoremediation and environmental detoxification

11.5.1 Role in plant defense against heavy metals

Glutathione (GSH) plays a critical role in plant defense against heavy metal toxicity by functioning as a key antioxidant, metal chelator, and regulator of stress-responsive pathways that mitigates oxidative damage and maintains cellular homeostasis. Plants exposed to heavy metals such as cadmium (Cd), lead (Pb), mercury (Hg), arsenic (As), and chromium (Cr) experience excessive ROS production, which leads to oxidative stress, lipid peroxidation, protein oxidation, and DNA damage. GSH serves as a first line of defense by directly scavenging ROS, maintaining redox balance, and synthesizing PCs, crucial for metal sequestration and detoxification (Chukwu & Gulser, 2025; Giri et al., 2024; Li & Zhao, 2025; Rottet et al., 2023; Zhang, Wu et al., 2024). PCs are enzymatically synthesized from GSH by PC synthase (PCS) and form stable complexes with heavy metal ions, reducing their bioavailability and preventing their interaction with vital cellular components. These GSH-derived PCs play an essential role in metal homeostasis by facilitating metal transport into vacuoles, where metals are sequestered and detoxified, thereby minimizing their toxic effects in the cytoplasm. In addition to metal chelation, GSH also regulates the expression and activity of key antioxidant enzymes, including glutathione peroxidase (GPx), glutathione reductase (GR), GSTs, superoxide dismutase (SOD), and catalase (CAT), which collectively reduce oxidative stress and protect against cellular damage. The GSH/GSSG ratio is a crucial indicator of redox status in plants under heavy metal stress, shifting toward oxidized glutathione (GSSG), reflecting increased oxidative pressure and cellular damage. To counteract this imbalance, GR plays a vital role in recycling GSSG back to GSH using NADPH as a reducing equivalent, ensuring a continuous supply of reduced GSH to sustain antioxidant defense (Gupta et al., 2024; He et al., 2025; Li & Zhao, 2025; Tripathi & Poluri, 2021; Umar et al., 2025; Zhang, Wu et al., 2024). Moreover, GSH modulates the activity of key transcription factors such as nuclear factor erythroid 2-related factor 2 (Nrf2) and abscisic acid (ABA)-responsive elements, which regulate stress-responsive gene expression and enhance the plant's ability to cope with heavy metal–induced stress. In root systems, GSH influences metal uptake and translocation by modulating the expression of metal transporter proteins such as heavy metal ATPases and NRAMPs, which play crucial roles in controlling metal accumulation and distribution within plant tissues. The exogenous application of GSH or its precursors, such as NAC, has enhanced heavy metal tolerance in plants by upregulating antioxidant defense mechanisms, improving metal sequestration, and maintaining redox homeostasis. Furthermore, plants with elevated GSH levels exhibit improved growth and biomass accumulation under metal stress conditions, highlighting the importance of GSH metabolism in stress adaptation. In phytoremediation strategies, GSH-rich plants with enhanced metal tolerance are employed for the remediation of contaminated soils, as they can efficiently accumulate and detoxify heavy metals without significant growth inhibition. The role of GSH in metal defense is also closely linked to sulfur metabolism, as cysteine, the rate-limiting precursor of GSH, is synthesized through the sulfate assimilation pathway, which is upregulated in response to heavy metal exposure. Understanding the intricate role of GSH in plant defense mechanisms against heavy metal toxicity provides valuable insights into improving crop resilience, developing phytoremediation strategies, and enhancing sustainable agricultural practices in metal-contaminated environments (Chukwu & Gulser, 2025; Giri et al., 2024; Guan et al., 2024; Rottet et al., 2023; Umar et al., 2025; Zhang, Wu et al., 2024).

TABLE 11.4 Impact of heavy metal exposure on glutathione homeostasis.

Heavy metal	Effect on GSH levels	GSH/GSSG ratio changes	Upregulated enzymes	Consequences
Cadmium (Cd)	Decreases due to direct binding and oxidative stress	Decreases (higher GSSG due to oxidation)	GR, GSTs, PCS	Increased ROS, cellular damage, apoptosis
Lead (Pb)	Depletes GSH via conjugation and oxidative stress	Decreases	Glutathione peroxidase (GPx), GSTs	Lipid peroxidation, neurotoxicity, mitochondrial dysfunction
Mercury (Hg)	Binds directly to thiol groups, reducing free GSH	Decreases	GR, GPx, GSTs	Protein dysfunction, oxidative stress, neurotoxicity
Arsenic (As)	Forms As-GSH complexes, lowering free GSH	Decreases	GSTO, glutathione synthetase	Disrupts redox homeostasis, impairs cellular metabolism
Chromium (Cr)	Generates ROS, leading to GSH oxidation	Decreases	GPx, GR, GSTs	DNA damage, oxidative stress, apoptosis
Zinc (Zn)	It can upregulate GSH synthesis in low doses but deplete GSH at toxic levels	Variable, depending on the dose	γ-Glutamylcysteine synthetase (GCL), GR	Beneficial at low levels but toxic at high levels
Copper (Cu)	Induces oxidative stress, depleting GSH	Decreases	Superoxide dismutase (SOD), GPx, GSTs	Mitochondrial dysfunction, oxidative DNA damage

GR, Glutathione reductase; *GSTO*, glutathione-*S*-transferase omega; *GSTs*, glutathione-*S*-transferases; *PCS*, phytochelatin synthase; *ROS*, reactive oxygen species.

11.5.2 Glutathione enhancement strategies for improved phytoremediation

Enhancing glutathione (GSH) levels in plants is promising for improving phytoremediation efficiency in heavy metal-contaminated environments. GSH is crucial in metal detoxification, chelation, sequestration, and antioxidative defense. Various approaches have been explored to increase endogenous GSH content and optimize its functional role in phytoremediation, including genetic engineering, exogenous application of GSH precursors, and modulation of sulfur metabolism. Genetic engineering strategies involve overexpressing key enzymes involved in GSH biosynthesis and recycling, such as γ-glutamylcysteine synthetase (γ-ECS) and GS, which catalyze the sequential synthesis of GSH from its precursor amino acids glutamate, cysteine, and glycine (Anas et al., 2025; Chukwu & Gulser, 2025; Guan et al., 2024; He et al., 2025). Transgenic plants with enhanced γ-ECS and GS expression exhibit higher GSH accumulation, improved tolerance to heavy metals, and increased metal sequestration in vacuoles through the synthesis of PCs, which are derived from GSH and serve as essential chelators for metal detoxification. Additionally, overexpression of glutathione reductase (GR), an enzyme responsible for maintaining the GSH/GSSG ratio, enhances redox homeostasis, thereby reducing oxidative damage induced by heavy metal stress and improving plant survival in contaminated soils. Another effective strategy involves the exogenous application of GSH or its precursors, such as NAC, cysteine, and sulfur-containing compounds, which can stimulate endogenous GSH biosynthesis and strengthen antioxidant defense mechanisms. Studies have shown that foliar spraying or soil application of NAC leads to elevated GSH levels, enhanced PC production, and increased metal uptake in hyperaccumulator plants, thereby improving their efficiency in extracting heavy metals from contaminated environments. Sulfur supplementation, particularly in sulfate or elemental sulfur, also plays a key role in boosting GSH synthesis, as sulfur is an essential component of cysteine, the rate-limiting precursor in GSH biosynthesis. The upregulation of sulfate transporters (SULTRs) and activation of sulfur assimilation pathways have been shown to enhance GSH production and improve plant tolerance to cadmium (Cd), lead (Pb), and arsenic (As) toxicity (Guan et al., 2024; Khan et al., 2025; Li et al., 2022; Pang et al., 2024; Wang et al., 2025; Zhang, Wu et al., 2024). Furthermore, microbial-assisted phytoremediation strategies involving plant growth-promoting rhizobacteria (PGPR) and mycorrhizal fungi have been explored to enhance GSH-mediated detoxification processes. Certain rhizobacteria, such as *Pseudomonas* and *Bacillus* species, promote sulfur metabolism, induce antioxidant enzyme activity, and increase GSH levels in plant roots, improving metal uptake and tolerance. Mycorrhizal associations also enhance GSH-dependent detoxification by improving nutrient uptake, reducing metal bioavailability in the rhizosphere, and increasing plant biomass accumulation in contaminated soils. In addition to biochemical and microbial interventions, phytohormone signaling pathways, such as those regulated by ABA, salicylic acid (SA), and jasmonic acid (JA), have been implicated in modulating GSH homeostasis and enhancing metal stress tolerance. Plants treated with ABA or SA exhibit increased GSH biosynthesis and improved metal sequestration, further supporting the role of hormonal regulation in GSH enhancement strategies. Optimizing GSH metabolism through genetic, biochemical, microbial, and hormonal approaches offers a powerful strategy for improving phytoremediation efficiency, reducing heavy metal toxicity, and promoting sustainable environmental cleanup efforts (Chukwu & Gulser, 2025; Milano et al., 2024; Rachappanavar et al., 2024; Rath & Das, 2025; Tripathi & Poluri, 2021; Umar et al., 2025).

11.5.3 Genetic engineering for increased GSH production in hyperaccumulators

Genetic engineering has emerged as a powerful approach to enhance glutathione (GSH) production in hyperaccumulator plants, thereby improving their heavy metal detoxification, sequestration, and tolerance capacity. GSH plays a critical role in metal homeostasis by serving as a precursor for PCs, which chelate and transport heavy metals into vacuoles for sequestration and by mitigating oxidative stress through its antioxidant properties. Increasing GSH biosynthesis in hyperaccumulators through genetic modifications can significantly enhance their phytoremediation potential. One of the primary targets for genetic engineering is the overexpression of key enzymes involved in the GSH biosynthetic pathway, particularly γ-ECS and GS, which catalyse the stepwise synthesis of GSH from glutamate, cysteine, and glycine. Overexpression of γ-ECS, the rate-limiting enzyme in GSH biosynthesis, has significantly increased GSH levels in transgenic plants, improving metal tolerance and accumulation. Similarly, coexpression of both γ-ECS and GS further enhances GSH biosynthesis and reinforces metal sequestration mechanisms (Chukwu & Gulser, 2025; Giri et al., 2024; Guan et al., 2024; He et al., 2025; Khan et al., 2025). In addition to increasing GSH synthesis, genetic modifications targeting glutathione reductase (GR), an enzyme that maintains the reduced GSH pool by recycling oxidized glutathione (GSSG), have proven effective in sustaining high intracellular GSH levels under metal stress. Transgenic plants overexpressing GR exhibit an increased GSH/GSSG ratio, improving redox homeostasis and enhancing resistance to metal-induced oxidative stress. Furthermore, engineering the expression of PCS, which catalyzes the polymerization of GSH into metal-binding PCs, has been demonstrated to enhance heavy metal sequestration in hyperaccumulators such as Thlaspi caerulescens and Arabidopsis halleri. In addition to manipulating the GSH biosynthetic pathway, genetic engineering approaches have focused on improving sulfur assimilation, as cysteine availability is a major limiting factor in GSH production. Overexpression of SULTRs and key sulfur metabolism enzymes such as ATP sulfurylase (APS) and *O*-acetyl serine (thiol)lyase (OASTL) has been shown to enhance sulfate uptake and increase cysteine biosynthesis, ultimately leading to elevated GSH levels and improved metal tolerance (Li et al., 2022; Milano et al., 2024; Rachappanavar et al., 2024; Umar et al., 2025; Wang et al., 2023). Another promising strategy involves introducing genes that modulate transcription factors and signaling pathways associated with stress tolerance and GSH metabolism. For example, overexpression of transcription factors such as AtWRKY13 and AtbZIP23 has enhanced GSH biosynthesis and conferred increased resistance to cadmium (Cd) and arsenic (As) toxicity. Genetic modifications that enhance the expression of GSH-dependent detoxification enzymes, such as GSTs, have been explored to improve metal chelation and transport. Combining these approaches with synthetic biology tools, such as CRISPR/Cas9 genome editing, allows for precise modifications in hyperaccumulator plants to optimize GSH biosynthesis and metal detoxification pathways. Integrating these genetic engineering strategies holds significant potential for developing hyperaccumulators with superior phytoremediation capabilities, enabling efficient removal of toxic metals from contaminated environments while minimizing the impact on plant growth and productivity (Table 11.5).

11.6 Glutathione-based therapeutic approaches in animals and humans

11.6.1 GSH modulation for heavy metal toxicity treatment

Glutathione (GSH) modulation has emerged as a crucial strategy for mitigating heavy metal toxicity in plants, given its multifaceted role in metal chelation, sequestration, antioxidative defense, and detoxification. As a key tripeptide antioxidant, GSH directly detoxifies heavy metals by chelating metal ions, forming glutathione–metal complexes, and facilitating their transport into vacuoles through PC synthesis. The upregulation of GSH biosynthesis, recycling, and transport mechanisms is essential for enhancing metal tolerance and reducing oxidative stress induced by toxic metal exposure (Anas et al., 2025; Guan et al., 2024; Rachappanavar et al., 2024; Tripathi & Poluri, 2021). One approach to GSH modulation involves increasing the biosynthesis of its precursor amino acids—glutamate, cysteine, and glycine—through metabolic engineering or external supplementation. Cysteine availability is often the limiting factor in GSH synthesis, and increasing its production via overexpression of enzymes such as APS, OASTL, and SULTRs has been shown to enhance GSH accumulation and improve metal stress resilience. Additionally, overexpression of γ-ECS and GS, the two key enzymes responsible for GSH biosynthesis, leads to elevated intracellular GSH levels, facilitating improved metal chelation and sequestration. Beyond biosynthesis, maintaining an optimal GSH/GSSG ratio is critical for redox homeostasis and effective detoxification. The enzyme glutathione reductase (GR) plays a vital role in converting oxidized glutathione (GSSG) back to its reduced form (GSH), thereby sustaining cellular redox balance under metal-induced oxidative stress conditions. Transgenic plants with enhanced GR expression demonstrate increased resistance to heavy metals such as cadmium (Cd), lead (Pb), and arsenic (As), as they maintain higher levels of reduced GSH and mitigate ROS accumulation. Similarly, upregulating GSTs, which catalyse the conjugation of GSH with metal ions and xenobiotics, facilitates the detoxification and sequestration of heavy metals in vacuoles, thereby reducing their cytoplasmic toxicity. The exogenous applications of GSH or its precursors, such as NAC, have been explored as an alternative to enhance GSH-mediated detoxification pathways. Studies indicate that NAC supplementation increases GSH levels, enhances PC synthesis, and boosts the plant's ability to tolerate and accumulate heavy metals in contaminated environments. Furthermore, phytohormones such as SA, ABA, and JA have been implicated in modulating GSH homeostasis and enhancing metal stress tolerance (Chukwu & Gulser, 2025; Khan et al., 2025; Li & Zhao, 2025; Pang et al., 2024; Tripathi & Poluri, 2021). ABA and SA treatments have been shown to induce GSH biosynthesis, upregulate stress-responsive genes, and improve metal chelation efficiency. Additionally, microbial-assisted strategies using PGPR and mycorrhizal fungi further enhance GSH modulation by promoting sulfur assimilation, increasing antioxidant enzyme activity, and improving overall plant resilience to heavy metal toxicity. Advances in synthetic biology and genome editing technologies, such as CRISPR/Cas9, offer new possibilities for fine-tuning GSH biosynthetic and regulatory pathways to develop hyperaccumulator plants with superior metal detoxification capabilities. By integrating genetic, biochemical, and microbial strategies, GSH modulation is a powerful approach to mitigating heavy metal toxicity, improving phytoremediation efficiency, and promoting sustainable environmental cleanup.

11.6.2 Dietary and pharmacological interventions (e.g., *N*-acetylcysteine, glutathione supplements)

Dietary and pharmacological interventions are crucial in maintaining cellular health, preventing oxidative stress,

TABLE 11.5 Plant species that utilize glutathione (GSH) for heavy metal phytoremediation and their effectiveness.

Plant species	Heavy metals remediated	Mechanism involving glutathione	Effectiveness	References/notable studies
Brassica juncea (Indian mustard)	Cd, Pb, Hg, As, Zn	GSH-mediated metal binding, phytochelatin synthesis	High bioaccumulation capacity, adequate for soil and water remediation	Widely studied for hyperaccumulation potential
Arabidopsis thaliana	Cd, As, Cr	GSH synthesis and upregulation of GSTs	Model plant for understanding detox pathways	Commonly used in genetic studies
Typha latifolia (Cattail)	Cd, Pb, Hg	GSH involvement in sequestration and oxidative stress reduction	Effective in wetland remediation	Found in contaminated aquatic environments
Helianthus annuus (Sunflower)	Pb, Cd, Zn	GSH and phytochelatin involvement in chelation	Effective in field-based phytoremediation trials	Used in soil phytoremediation studies
Populus spp. (Poplar trees)	Hg, Cd, Zn	Enhanced GSH synthesis and glutathione reductase activity	Moderate to high accumulation	Used in phytostabilisation and phytoextraction
Salix spp. (Willow trees)	Cd, Zn, Cu	GSH-dependent sequestration and redox modulation	High tolerance and biomass production	Applied in riparian buffer zones
Medicago sativa (Alfalfa)	As, Cd	Increased GST and phytochelatin activity	Moderate accumulation, high tolerance	Studied for agricultural phytoremediation
Pteris vittata (Brake fern)	As	GSH involvement in arsenic detoxification and transport	Hyperaccumulator, highly effective	Well-documented arsenic accumulator
Lemna minor (Duckweed)	Pb, Cd, Hg	GSH-mediated oxidative stress reduction and sequestration	High efficiency in water remediation	Used in wastewater treatment studies

Source: Based on Guan et al. (2024), Tripathi and Poluri (2021), and Wang et al. (2025).

and supporting detoxification processes, mainly through compounds like NAC and glutathione supplements. NAC, a precursor to glutathione, has been widely studied for its ability to replenish intracellular glutathione levels, acting as a potent antioxidant that mitigates damage caused by ROS. As oxidative stress is implicated in a wide range of diseases, including neurodegenerative disorders, cardiovascular diseases, and liver conditions, NAC supplementation has gained attention as a therapeutic intervention. It is particularly beneficial in cases of acetaminophen toxicity, where it restores depleted glutathione stores in the liver, thereby preventing hepatocellular damage. Additionally, NAC has been explored for its mucolytic properties, helping to reduce the viscosity of mucus in respiratory conditions such as chronic obstructive pulmonary disease and cystic fibrosis. In psychiatric and neurological conditions, NAC shows promise in mitigating symptoms of obsessive-compulsive disorder, schizophrenia, and bipolar disorder by modulating glutamatergic neurotransmission and reducing neuroinflammation. Glutathione, a tripeptide composed of glutamine, cysteine, and glycine, is a key antioxidant in cellular detoxification, immune function, and mitochondrial protection. While the body naturally synthesises glutathione, factors such as aging, chronic illnesses, environmental toxins, and poor dietary habits can lead to its depletion, making supplementation a valuable strategy (Anas et al., 2025; Giri et al., 2024; Gupta et al., 2024; Li & Zhao, 2025; Rachappanavar et al., 2024; Wang et al., 2023). Oral glutathione supplements, however, have faced challenges regarding bioavailability, as the digestive process can break down glutathione before it reaches the cells. Alternative forms, such as liposomal glutathione and sublingual administration, have been developed to enhance absorption. Dietary precursors can also support endogenous glutathione production, including whey protein, cruciferous vegetables, and selenium-rich foods. In clinical applications, glutathione has been investigated for its role in conditions such as PD, AD, chronic liver disease, and immune dysfunction. Intravenous glutathione administration has shown efficacy in reducing oxidative damage in neurodegenerative disorders while also being used in integrative medicine for its purported skin-lightening effects by inhibiting melanin synthesis. NAC and glutathione supplementation are also explored in metabolic and inflammatory disorders, with emerging research suggesting their potential benefits in insulin resistance, obesity-related inflammation, and cardiovascular health. By modulating inflammatory pathways and reducing oxidative stress, NAC may improve endothelial function and reduce the risk of atherosclerosis. At the same time, glutathione's role in mitochondrial health is crucial for energy metabolism and overall cellular resilience. Furthermore, these compounds have gained attention in longevity and aging, with research indicating their potential to counteract age-related declines in antioxidant capacity. Despite their benefits, supplementation must be approached with caution, as excessive intake of NAC may lead to gastrointestinal discomfort, and high doses of glutathione could potentially disrupt normal redox signaling. Personalized approaches based on individual health status, genetic predispositions, and dietary habits are essential for optimizing the efficacy of these interventions. Future research continues to explore novel formulations and combinatory approaches that enhance the therapeutic impact of NAC and glutathione, ensuring better absorption, targeted delivery, and synergistic effects with other antioxidants and micronutrients.

11.6.3 Genetic and epigenetic regulation of GSH metabolism in detoxification

Glutathione (GSH) metabolism plays a crucial role in detoxification, redox balance, and cellular homeostasis, and its regulation is influenced by genetic and epigenetic mechanisms that determine its synthesis, recycling, and utilization in response to oxidative stress and xenobiotic exposure. The genetic regulation of GSH metabolism primarily involves key enzymes such as GCL, the rate-limiting enzyme in GSH synthesis, as well as GSTs, glutathione reductase (GSR), and glutathione peroxidases (GPXs), which facilitate detoxification and antioxidant defense. Variations in genes encoding these enzymes can significantly impact GSH levels and function, influencing susceptibility to diseases related to oxidative stress and toxin accumulation. Polymorphisms in GST genes, such as GSTM1, GSTT1, and GSTP1, have been associated with altered detoxification capacity, with deletions in GSTM1 and GSTT1 leading to reduced enzyme activity and increased vulnerability to carcinogens, pollutants, and drugs. Similarly, genetic variants in GCLC and GCLM, the catalytic and modifier subunits of GCL, respectively, can affect the rate of GSH biosynthesis, influencing cellular resilience to oxidative stress. Beyond genetic polymorphisms, epigenetic regulation plays a pivotal role in modulating GSH metabolism through DNA methylation, histone modifications, and noncoding RNAs that influence gene expression without altering the underlying DNA sequence (Chukwu & Gulser, 2025; Guan et al., 2024; He et al., 2025; Khan et al., 2025; Wang et al., 2023; Zhang, Wu et al., 2024). DNA methylation in the promoter regions of GSH-related genes can suppress their transcription, leading to decreased GSH synthesis and impaired detoxification capacity. For example, hypermethylation of the GCLC gene promoter has been observed in chronic oxidative stress, metabolic disorders, and aging conditions, contributing to reduced GSH levels and increased cellular vulnerability to damage. Histone modifications, such as acetylation and methylation, further regulate chromatin accessibility, thereby influencing the expression of genes involved in GSH metabolism. Inflammatory and environmental stressors can induce epigenetic modifications that suppress antioxidant defense mechanisms, exacerbating oxidative damage and disease progression. Additionally, microRNAs (miRNAs) have emerged as key regulators of GSH homeostasis, with specific miRNAs targeting mRNAs of GSH-related enzymes to modulate their expression posttranscriptionally. Dysregulation of miRNAs involved in GSH metabolism has been linked to various diseases, including cancer, neurodegenerative disorders, and cardiovascular conditions, where oxidative stress and impaired detoxification contribute to pathogenesis. Moreover, environmental factors such as diet, toxins, and lifestyle choices interact with genetic and epigenetic mechanisms to influence GSH metabolism. Nutrients such as sulfur-containing amino acids (cysteine and methionine), selenium, and polyphenols can enhance GSH synthesis and modulate epigenetic marks that regulate detoxification pathways. Environmental exposure, heavy metals, and chronic stress can induce adverse epigenetic changes that compromise GSH production and antioxidant defenses. Understanding the interplay between genetic predisposition and epigenetic regulation in GSH metabolism provides valuable insights for personalized medicine approaches to optimize detoxification and mitigate oxidative damage. The potential of GSH metabolism research to lead to more effective therapies underscores the potential impact of the audience's work on patient care. Future research continues to explore novel strategies to manipulate genetic and epigenetic regulators of GSH metabolism and develop more effective therapies for diseases linked to oxidative stress and detoxification impairments (Table 11.6).

11.7 Challenges and future perspectives

11.7.1 Limitations in GSH-based detoxification strategies

Despite the well-established role of glutathione (GSH) in detoxification and cellular protection, several limitations exist in GSH-based detoxification strategies, including challenges related to bioavailability, genetic variability, epigenetic modifications, metabolic constraints, and unintended physiological consequences. One of the primary limitations of GSH supplementation is its poor bioavailability, as orally administered GSH undergoes extensive degradation in the gastrointestinal tract, limiting its systemic uptake and intracellular availability. Alternative formulations such as liposomal GSH, NAC as a precursor, and sublingual or intravenous administration have been developed to enhance bioavailability; however, their long-term efficacy and safety remain under investigation. Furthermore, individual genetic variations significantly influence GSH metabolism, with polymorphisms in key genes such as GCL (GCLC, GCLM), GSTs, and glutathione reductase (GSR) affecting both GSH synthesis and detoxification efficiency. For instance, GSTM1 and GSTT1 deletions, commonly found in specific populations, impair conjugation and clearance of xenobiotics, increasing susceptibility to oxidative stress-related diseases. At the same time, GCLC polymorphisms may lead to suboptimal GSH biosynthesis, reducing cellular resilience. Additionally, epigenetic modifications such as DNA methylation and histone modifications further regulate GSH-related gene expression, potentially diminishing GSH production and detoxification efficiency under chronic oxidative stress conditions or exposure to environmental toxins. Another major constraint in GSH-based detoxification is the metabolic burden associated with its synthesis and recycling, as maintaining optimal GSH levels requires a steady supply of precursors such as cysteine, glutamate, and glycine, along with adequate enzymatic activity. Nutrient deficiencies, chronic diseases, and age-related declines in GSH synthesis exacerbate depletion, making exogenous support less effective in individuals with impaired metabolic pathways (Chukwu & Gulser, 2025; Guan et al., 2024; Rottet et al., 2023; Tripathi & Poluri, 2021; Wang et al., 2025). Moreover, while GSH plays a crucial role in neutralizing ROS and facilitating xenobiotic metabolism, excessive or unregulated GSH supplementation may have paradoxical effects, including the potential to disrupt physiological redox signaling, alter immune responses, and interfere with apoptotic processes necessary for cellular homeostasis. For example, in certain cancers, elevated GSH levels can contribute to chemotherapy resistance by neutralizing drug-induced oxidative stress, highlighting a context-dependent limitation of GSH-based interventions. Additionally, while GSH conjugation is essential for detoxification, it may lead to the formation of secondary metabolites with unintended biological activity, potentially contributing to cellular stress rather than alleviating it. For instance, the interaction of GSH with heavy metals can lead to the formation of persistent metal-glutathione complexes, which may prolong metal retention in tissues rather than promote efficient elimination. Furthermore, clinical applications of GSH supplementation and pharmacological interventions remain hindered by a lack of standardised dosing regimens, inconsistent clinical trial outcomes, and limited understanding of long-term effects. Addressing these limitations requires a more personalized approach that considers genetic and epigenetic profiles, dietary and metabolic status, and the specific detoxification demands of individuals. Future research should focus on optimizing delivery methods, exploring novel GSH analogues or mimetics, and integrating multitargeted strategies that enhance endogenous GSH production while minimizing

TABLE 11.6 Outlining dietary and pharmacological strategies to boost glutathione (GSH) levels for heavy metal detoxification in humans and animals.

Strategy type	Examples	Mechanism of action	Effectiveness	Notes
Dietary sources of GSH precursors	Garlic, onions, cruciferous vegetables (broccoli, kale, Brussels sprouts)	Rich in sulfur-containing compounds that support GSH synthesis	High	Supports natural detox pathways
Amino acid supplementation	NAC, glycine, L-glutamine	Provides cysteine and other precursors for GSH synthesis	High	NAC is widely used in clinical settings for detoxification
Antioxidant-rich foods	Berries, green tea, turmeric, citrus fruits	Reduce oxidative stress, preserving intracellular GSH levels	Moderate to high	Polyphenols and flavonoids enhance GSH recycling
Selenium supplementation	Brazil nuts, fish, meat, selenium supplements	Cofactor for glutathione peroxidase (GPx), promoting GSH activity	Moderate	Essential for antioxidant defense
Vitamin support	Vitamin C, vitamin E, B-complex vitamins	Enhances GSH recycling and redox balance	Moderate to high	Synergistic effects with GSH metabolism
Phytochemicals and herbal extracts	Curcumin (turmeric), resveratrol, milk thistle (silymarin)	Induce GSH synthesis and upregulate detox enzymes	Moderate	Silymarin is known for liver protection and GSH enhancement
Exercise and lifestyle factors	Moderate aerobic exercise, stress management	Increases GSH synthesis via hormetic response	Moderate to high	Overtraining can deplete GSH
Pharmacological agents	NAC, glutathione IV, liposomal GSH	Directly increases intracellular GSH levels	High	NAC is FDA-approved for acetaminophen toxicity and detox
Probiotics and gut health	Lactobacillus, bifidobacterium species	Supports GSH production and enhances metal detox pathways	Moderate	The gut microbiome plays a role in GSH metabolism

NAC, N-Acetylcysteine.
Source: Based on Chukwu and Gulser (2025), Tripathi and Poluri (2021), Umar et al. (2025), Wang et al. (2025), and Zhang, Wu et al. (2024).

adverse effects, ultimately improving the efficacy and safety of GSH-based detoxification strategies in clinical and therapeutic settings (Guan et al., 2024; Li et al., 2022; Rottet et al., 2023; Tripathi & Poluri, 2021; Zhang et al., 2024).

11.7.2 Potential biotechnological advancements

Biotechnological advancements can revolutionize multiple industries, from healthcare and agriculture to environmental sustainability and industrial production. One of the most promising areas of biotechnology is genetic engineering, which continues to evolve with the advent of CRISPR–Cas9 and other genome-editing technologies. These tools allow scientists to modify DNA with unprecedented precision, opening doors to developing disease-resistant crops, gene therapies for genetic disorders, and enhanced biofuel production. Genetically modified crops with improved yield, pest resistance, and climate resilience could help address global food security challenges in agriculture. Synthetic biology, another cutting-edge field, enables the design and construction of entirely new biological systems, leading to the development of engineered microorganisms capable of producing valuable compounds such as pharmaceuticals, biofuels, and biodegradable plastics (Guan et al., 2024; Li et al., 2022; Pang et al., 2024; Umar et al., 2025; Wang et al., 2025). Meanwhile, advancements in tissue engineering and regenerative medicine are paving the way for lab-grown organs and personalized medicine, which could transform healthcare by providing patients with tailor-made treatments and reducing dependency on organ transplants. Integrating artificial intelligence and machine learning into biotechnological research is accelerating drug discovery, enabling the rapid analysis of complex biological data, and optimizing bioprocessing techniques. Environmental biotechnology is also witnessing significant progress, with bioremediation techniques leveraging genetically engineered microbes to clean up oil spills, degrade plastic waste, and remove heavy metals from contaminated water and soil. Developing bio-based alternatives to traditional petroleum-based materials, such as bioplastics and bio-based textiles, is gaining traction as industries seek sustainable solutions to reduce environmental impact. Advances in metagenomics and microbiome research are unlocking new insights into the role of microbial communities in human health, agriculture, and ecosystem stability, leading to novel probiotics, improved soil health management strategies, and enhanced wastewater treatment methods. In medical biotechnology, stem cell research pushes the boundaries of regenerative therapies, offering potential cures for degenerative diseases like Parkinson's and Alzheimer's. Personalized medicine, driven by advancements in genomics and proteomics, enables more targeted and effective treatments based on an individual's genetic makeup. The advent of lab-grown meat and cellular agriculture presents an innovative approach to addressing global protein demands while reducing the environmental footprint of traditional livestock farming. Biopharmaceuticals, including monoclonal antibodies and RNA-based therapies, continue to expand treatment options for various diseases, including cancer, autoimmune disorders, and infectious diseases (Li & Zhao, 2025; Rachappanavar et al., 2024; Umar et al., 2025; Wang et al., 2023; Zhang, Wu et al., 2024). The convergence of nanotechnology and biotechnology leads to the development of innovative drug delivery systems, biosensors for early disease detection, and nano-engineered materials for improved medical implants. Advances in bioinformatics and computational biology are accelerating research in evolutionary biology, synthetic biology, and systems biology, providing deeper insights into the fundamental mechanisms of life. With the rise of foundries and automated high-throughput screening techniques, researchers can rapidly design, build, and test biological systems at an unprecedented scale. As ethical and regulatory frameworks evolve to keep pace with these advancements, biotechnology is poised to drive transformative changes across diverse fields, offering innovative solutions to some of the world's most pressing challenges, from food security and disease eradication to environmental conservation and industrial sustainability.

11.7.3 Translational research for medical and environmental applications

Translational research is a crucial bridge between scientific discoveries and their practical applications in medicine and environmental sustainability, pivotal in transforming laboratory findings into real-world solutions. In the medical field, translational research accelerates the development of novel therapies, diagnostics, and medical devices by integrating knowledge from molecular biology, genetics, and clinical sciences. Advances in genomics and precision medicine have enabled the identification of genetic markers associated with various diseases, allowing for targeted drug development and personalized treatment approaches that improve patient outcomes. The emergence of CRISPR-based gene editing technologies has opened new possibilities for correcting genetic disorders, developing cancer immunotherapies, and engineering disease-resistant cells. Stem cell research and regenerative medicine also benefit from translational efforts, with breakthroughs in tissue engineering leading to the development of lab-grown organs, bioengineered skin grafts, and innovative treatments for degenerative diseases such as PD and AD. In parallel, translational research is revolutionising the fight against infectious diseases, with rapid vaccine development platforms, monoclonal antibody therapies, and RNA-based technologies paving the way for next-generation treatments against emerging pathogens (Gupta et al., 2024; Li & Zhao, 2025; Umar et al., 2025; Wang et al., 2025, 2023; Zhang, Wu et al., 2024). Applying artificial intelligence and machine learning in biomedical research further enhances drug discovery, optimizes clinical trial designs, and enables predictive analytics for disease progression and treatment response. Beyond healthcare, translational research is vital in environmental biotechnology, where scientific advancements are being harnessed to address pressing ecological challenges. Microbial bioremediation strategies leverage genetically engineered bacteria and fungi to break down pollutants, remove heavy metals from contaminated sites, and degrade plastic waste, offering sustainable alternatives to traditional remediation techniques. Innovations in synthetic biology have enabled the engineering of microorganisms capable of converting waste into biofuels, biodegradable plastics, and other valuable bio-based products, reducing reliance on fossil fuels and promoting a circular bioeconomy. Translational research in agriculture also drives the development of biofortified crops with enhanced nutritional content, stress-resistant plant varieties, and microbial consortia that improve soil health and crop productivity while reducing the need for chemical fertilisers and pesticides (Giri et al., 2024; Gupta et al., 2024; Li & Zhao, 2025; Rottet et al., 2023). Integrating biotechnology with environmental monitoring systems is advancing the detection of pollutants, tracking of ecosystem changes, and real-time assessment of air and water quality using biosensors and lab-on-a-chip technologies. Furthermore, the convergence of nanotechnology and biotechnology leads to innovative materials, nanocarriers for targeted drug delivery, and nanoscale sensors for early disease detection and environmental surveillance. As translational research continues to evolve, collaborations between academia, industry, and government agencies are essential to overcoming regulatory, ethical, and logistical challenges that can hinder the transition from discovery to application. Establishing foundries, high-throughput screening platforms, and automated systems for large-scale bioprocessing further accelerates translating scientific innovations into commercially viable products and therapies. By fostering interdisciplinary approaches and embracing cutting-edge technologies, translational research holds immense potential to drive transformative advancements in medical and environmental fields, address global health challenges, promote sustainability, and shape a future where scientific discoveries directly contribute to improving human well-being and ecological resilience.

11.8 Conclusion

Glutathione (GSH) plays a critical role in heavy metal detoxification by acting as a major intracellular antioxidant and a key component in metal homeostasis. Recent studies have elucidated the molecular mechanisms through which glutathione participates in heavy metal detoxification, primarily through direct chelation, enzymatic conjugation via GSTs, and regulation of cellular redox balance. Glutathione interacts with toxic metals such as cadmium (Cd), lead (Pb), arsenic (As), and mercury (Hg), forming metal-GSH complexes that are subsequently transported out of cells via ABC transporters, thereby reducing intracellular toxicity. Additionally, glutathione plays a crucial role in mitigating oxidative stress–induced by heavy metal exposure, as these metals disrupt cellular redox homeostasis by generating ROS, leading to lipid peroxidation, protein oxidation, and DNA damage. Activating the glutathione redox cycle, involving glutathione reductase (GR) and glutathione peroxidase (GPx), helps maintain cellular defense mechanisms against oxidative damage. Moreover, emerging research suggests that glutathione biosynthesis is upregulated in response to heavy metal stress, with GCL and GS playing pivotal roles in adaptive cellular responses. Phytoremediation strategies leveraging glutathione-producing plants, such as hyperaccumulators, have shown promise in enhancing heavy metal tolerance and accumulation through increased GSH synthesis and metal sequestration. Advances in genetic and metabolic engineering have further demonstrated that overexpression of glutathione biosynthesis genes can significantly enhance heavy metal resistance in plants and microorganisms, offering potential applications in bioremediation. Future research directions in glutathione-based heavy metal detoxification should focus on understanding the interplay between glutathione metabolism and metal stress signaling pathways at the molecular and systems biology levels. Investigating the role of glutathione in epigenetic modifications and gene regulation under heavy metal stress may reveal novel targets for enhancing detoxification mechanisms. The development of synthetic biology approaches, including designing genetically engineered microbes with enhanced glutathione biosynthesis and metal-binding capabilities, holds great potential for environmental cleanup efforts. Additionally, exploring the application of nanotechnology in glutathione-mediated detoxification, such as the use of glutathione-functionalized nanoparticles for targeted metal sequestration, represents an emerging area of research. Further elucidation of the interactions between glutathione and metal-responsive transcription factors, including metal-regulatory transcription factor 1 (MTF-1) and nuclear factor erythroid 2-related factor 2 (Nrf2), could provide insights into regulatory networks governing metal detoxification and oxidative stress responses. Expanding research on dietary and pharmacological interventions aimed at boosting glutathione levels in organisms exposed to heavy metal toxicity may lead to novel therapeutic strategies for mitigating metal-induced health effects.

Integrating high-throughput omics technologies, including metabolomics, proteomics, and transcriptomics, with machine learning-based predictive models could enhance our understanding of glutathione-mediated detoxification pathways and facilitate the identification of biomarkers for heavy metal toxicity assessment. Overall, continued advancements in glutathione research have the potential to drive innovations in environmental bioremediation, sustainable agriculture, and medical interventions for heavy metal poisoning, ultimately contributing to improved ecosystem health and human well-being.

Acknowledgment

The author used AI to enhance language and readability, with final editing done by a human.

References

Adegbola, P. I., & Adetutu, A. (2024). Genetic and epigenetic modulations in toxicity: The two-sided roles of heavy metals and polycyclic aromatic hydrocarbons from the environment. *Toxicology Reports, 12*, 502–519. Retrieved from https://doi.org/10.1016/j.toxrep.2024.04.010.

Anas, M., Khattak, W. A., Fahad, S., Alrawiq, N., Alrawiq, H. S., Abdelsalam, N. R., ... Quraishi, U. M. (2025). Mitigating heavy metal pollution in agriculture: A multi-omics and nanotechnology approach to safeguard global wheat production. *Journal of Hazardous Materials Advances, 17*, 100584. Retrieved from https://doi.org/10.1016/j.hazadv.2024.100584.

Bjedov, D., Turić, N., Mikuška, A., Vignjević, G., Kovačić, L. S., Pavičić, A. M., ... Velki, M. (2025). The diving beetle, Cybister lateralimarginalis (De Geer, 1774), as a bioindicator for subcellular changes affected by heavy metal(loid) pollution in freshwater ecosystems. *Aquatic Toxicology (Amsterdam, Netherlands), 279*, 107258. Retrieved from https://doi.org/10.1016/j.aquatox.2025.107258.

Chakravorty, M., Nanda, M., Bisht, B., Sharma, R., Kumar, S., Mishra, A., ... Kumar, V. (2023). Heavy metal tolerance in microalgae: Detoxification mechanisms and applications. *Aquatic Toxicology (Amsterdam, Netherlands), 260*, 106555. Retrieved from https://doi.org/10.1016/j.aquatox.2023.106555.

Chukwu, E. C., & Gulser, C. (2025). Morphological, physiological, and anatomical effects of heavy metals on soil and plant health and possible remediation technologies. *Soil Security, 18*, 100178. Retrieved from https://doi.org/10.1016/j.soisec.2025.100178.

Ding, C., Ding, Z., Liu, Q., Liu, W., & Chai, L. (2024). Advances in mechanism for the microbial transformation of heavy metals: Implications for bioremediation strategies. *Chemical Communications, 60*, 12315–12332. Retrieved from https://doi.org/10.1039/d4cc03722g.

Faruque, M. O., Uddin, S., Hossain, M. M., Hossain, S. M. Z., Shafiquzzaman, M., & Razzak, S. A. (2024). A comprehensive review on microalgae-driven heavy metals removal from industrial wastewater using living and nonliving microalgae. *Journal of Hazardous Materials Advances, 16*, 100492. Retrieved from https://doi.org/10.1016/j.hazadv.2024.100492.

Geetha, N., Bhavya, G., Abhijith, P., Shekhar, R., Dayananda, K., & Jogaiah, S. (2021). Insights into nanomycoremediation: Secretomics and mycogenic biopolymer nanocomposites for heavy metal detoxification. *Journal of Hazardous Materials, 409*, 124541. Retrieved from https://doi.org/10.1016/j.jhazmat.2020.124541.

Ghuge, S. A., Nikalje, G. C., Kadam, U. S., Suprasanna, P., & Hong, J. C. (2023). Comprehensive mechanisms of heavy metal toxicity in plants, detoxification, and remediation. *Journal of Hazardous Materials, 450*, 131039. Retrieved from https://doi.org/10.1016/j.jhazmat.2023.131039.

Giri, S. S., Kim, H. J., Jung, W. J., Bin Lee, S., Joo, S. J., Gupta, S. K., & Park, S. C. (2024). Probiotics in addressing heavy metal toxicities in fish farming: Current progress and perspective. *Ecotoxicology and Environmental Safety, 282*, 116755. Retrieved from https://doi.org/10.1016/j.ecoenv.2024.116755.

Guan, J., Zhang, Y., Li, D., Shan, Q., Hu, Z., Chai, T., ... Qiao, K. (2024). Synergistic role of phenylpropanoid biosynthesis and citrate cycle pathways in heavy metal detoxification through secretion of organic acids. *Journal of Hazardous Materials, 476*, 135106. Retrieved from https://doi.org/10.1016/j.jhazmat.2024.135106.

Gupta, A., Siwach, C., & Mishra, V. K. (2024). Revolutionizing remediation: Unveiling the power of Lysinibacillus sp. in tackling heavy metal stress. *Sustainable Chemistry for the Environment, 8*, 100156. Retrieved from https://doi.org/10.1016/j.scenv.2024.100156.

Gupta, S. K., Gupta, R., Nage, S., Kumar, N., Kumari, R., Gupta, A., ... Krishnani, K. K. (2025). Heavy metal accumulation and biomarker responses of Nile tilapia from a coal void reservoir cage aquaculture system. *Journal of Hazardous Materials Advances*100631. Retrieved from https://doi.org/10.1016/j.hazadv.2025.100631.

Hamai-Amara, H., Abou-Saleh, H., A. Al-Ghouti, M., Crovella, S., Saadaoui, I., & Soubra, L. (2024). Microalgae potential to protect from heavy metals-induced carcinogenicity. *Algal Research, 78*, 103411. Retrieved from https://doi.org/10.1016/j.algal.2024.103411.

He, Z., Gao, J., Chen, X., Ru, Y., Zhang, D., & Pan, X. (2025). Efficient recovery of heavy metals and selenium from wastewater using granular sludge: The crucial role of glutathione (GSH). *Water Research, 270*, 122826. Retrieved from https://doi.org/10.1016/j.watres.2024.122826.

Heisi, H. D., Nkuna, R., & Matambo, T. (2025). Rhizosphere microbial community structure and PICRUSt2 predicted metagenomes function in heavy metal contaminated sites: A case study of the Blesbokspruit wetland. *The Science of the Total Environment, 959*, 178147. Retrieved from https://doi.org/10.1016/j.scitotenv.2024.178147.

Ilyas, T., Shahid, M., Shafi, Z., Aijaz, S. A., & Wasiullah (2025). Molecular mechanisms of methyl jasmonate (MeJAs)-mediated detoxification of heavy metals (HMs) in agricultural crops: An interactive review. *South African Journal of Botany, 177*, 139–159. Retrieved from https://doi.org/10.1016/j.sajb.2024.11.031.

Kaur, R., Kaur, H., & Sharma, A. (2025). Uptake and toxicity of heavy metals: The protective frontiers of metal binding proteins. *Journal of Geochemical Exploration, 271*, 107673. Retrieved from https://doi.org/10.1016/j.gexplo.2025.107673.

Khan, I., Asaf, S., Kang, S.-M., & Lee, I.-J. (2025). Physiological mechanisms of heavy metal detoxification in tomato plants mediated by endophytic fungi under nickel and cadmium stress. *Plant Physiology and Biochemistry: PPB / Societe Francaise de Physiologie Vegetale, 221*, 109589. Retrieved from https://doi.org/10.1016/j.plaphy.2025.109589.

Lei, L., Yang, C., Du, J., Liu, Z., Wang, Y., Wang, H., ... Xu, B. (2025). Functional analysis of AccCPR in *Apis cerana cerana* under pesticide and heavy metal stress. *Pesticide Biochemistry and Physiology, 208*, 106296. Retrieved from https://doi.org/10.1016/j.pestbp.2025.106296.

Li, X., Sun, M., Zhang, L., Finlay, R. D., Liu, R., & Lian, B. (2022). Widespread bacterial responses and their mechanism of bacterial metallogenic detoxification under high concentrations of heavy metals. *Ecotoxicology and Environmental Safety, 246*, 114193. Retrieved from https://doi.org/10.1016/j.ecoenv.2022.114193.

Li, X., Yu, Y., He, R., Zhen, Q., & She, D. (2025). Synergistic effects of aged lignin-based biochar and selenium fertilization on heavy metal remediation in agricultural soils. *Industrial Crops and Products, 225*, 120464. Retrieved from https://doi.org/10.1016/j.indcrop.2025.120464.

Li, Y., Peng, J., Cheng, Z., Zhang, K., Gu, H., Feng, J., & Liu, Y. (2024). Excessive heavy metal enrichment disturbs liver functions through the gut microbe in the great Himalayan leaf-nosed bat (*Hipposideros armiger*). *Ecotoxicology and Environmental Safety, 282*, 116758. Retrieved from https://doi.org/10.1016/j.ecoenv.2024.116758.

Li, Z.-T., & Zhao, H.-P. (2025). Sulfate-driven microbial collaboration for synergistic remediation of chloroethene-heavy metal pollution. *Water Research, 268*, 122738. Retrieved from https://doi.org/10.1016/j.watres.2024.122738.

Liu, L., FAN, X., HAN, Y., & WANG, H. (2025). Enhanced removal of heavy metals by wetland plant-microbiome symbiont: Prospect of potential strategies and mechanisms for environmental heavy metal regulation. *Pedosphere, 35*, 116–136. Retrieved from https://doi.org/10.1016/j.pedsph.2024.10.007.

Liu, L., Xiao, C., Gao, Y., Jiang, T., Xu, K., Chen, J., ... Lu, L. (2024). Inoculation of multi-metal-resistant *Bacillus* sp. to a hyperaccumulator plant Sedum alfredii for facilitating phytoextraction of heavy metals from contaminated soil. *Chemosphere, 366*, 143464. Retrieved from https://doi.org/10.1016/j.chemosphere.2024.143464.

Liu, Q., Zheng, X., Du, R., Shao, Y., Wen, Q., Shen, X., ... Hu, Y. (2024). Enrichment characteristics of Cd and Hg and regulation of heavy metal transporter signaling in Pleurotus ostreatus. *The Science of the Total Environment, 955*, 176909. Retrieved from https://doi.org/10.1016/j.scitotenv.2024.176909.

Mandal, R. R., Bashir, Z., & Raj, D. (2025). Microbe-assisted phytoremediation for sustainable management of heavy metal in wastewater – A green approach to escalate the remediation of heavy metals. *Journal of Environmental Management, 375*, 124199. Retrieved from https://doi.org/10.1016/j.jenvman.2025.124199.

Milano, F., Giotta, L., & Lambreva, M. D. (2024). Perspectives on nanomaterial-empowered bioremediation of heavy metals by photosynthetic microorganisms. *Plant Physiology and Biochemistry: PPB / Societe Francaise de Physiologie Vegetale, 216*, 109090. Retrieved from https://doi.org/10.1016/j.plaphy.2024.109090.

Panda, A., Fatnani, D., & Parida, A. K. (2025). Uptake, impact, adaptive mechanisms, and phytoremediation of heavy metals by plants: Role of transporters in heavy metal sequestration. *Plant Physiology and Biochemistry: PPB / Societe Francaise de Physiologie Vegetale, 221*, 109578. Retrieved from https://doi.org/10.1016/j.plaphy.2025.109578.

Pang, B., Zuo, D., Yang, T., Yu, J., Zhou, L., Hou, Y., ... Zhu, B. (2024). BcaSOD1 enhances cadmium tolerance in transgenic *Arabidopsis* by regulating the expression of genes related to heavy metal detoxification and arginine synthesis. *Plant Physiology and Biochemistry: PPB / Societe Francaise de Physiologie Vegetale, 206*, 108299. Retrieved from https://doi.org/10.1016/j.plaphy.2023.108299.

Prasath, R. V. A., & Mohanraj, R. (2024). In situ bioaccumulation of metals by *Prosopis juliflora* and its detoxification potential at the metal contaminated sites. *The Science of the Total Environment, 951*, 175715. Retrieved from https://doi.org/10.1016/j.scitotenv.2024.175715.

Priyanka, & Dwivedi, S. K. (2023). Fungi mediated detoxification of heavy metals: Insights on mechanisms, influencing factors and recent developments. *The Journal of Water Process Engineering, 53*, 103800. Retrieved from https://doi.org/10.1016/j.jwpe.2023.103800.

Rachappanavar, V., Gupta, S. K., Jayaprakash, G. K., & Abbas, M. (2024). Silicon mediated heavy metal stress amelioration in fruit crops. *Heliyon, 10*, e37425. Retrieved from https://doi.org/10.1016/j.heliyon.2024.e37425.

Rath, S., & Das, S. (2025). Stress response proteins within biofilm matrixome protect the cell membrane against heavy metals-induced oxidative damage in a marine bacterium *Bacillus stercoris* GST-03. *International Journal of Biological Macromolecules, 293*, 139397. Retrieved from https://doi.org/10.1016/j.ijbiomac.2024.139397.

Rottet, S., Iqbal, S., Xifaras, R., Singer, M. T., Scott, C., Deplazes, E., & Callaghan, R. (2023). Biochemical interactions between the Atm1-like transporter from *Novosphingobium aromaticivorans* and heavy metals. *Archives of Biochemistry and Biophysics, 744*, 109696. Retrieved from https://doi.org/10.1016/j.abb.2023.109696.

Sarma, U., Hoque, M. E., Thekkangil, A., Venkatarayappa, N., & Rajagopal, S. (2024). Microalgae in removing heavy metals from wastewater – An advanced green technology for urban wastewater treatment. *Journal of Hazardous Materials Advances, 15*, 100444. Retrieved from https://doi.org/10.1016/j.hazadv.2024.100444.

Singh, L. K., Kumar, A., Siddiqi, N. J., & Sharma, B. (2024). Heavy metals altered the xenobiotic metabolism of rats by targeting the GST enzyme: An in vitro and in silico study. *Toxicology, 509*, 153946. Retrieved from https://doi.org/10.1016/j.tox.2024.153946.

Song, J., Sun, Z., Saud, S., Fahad, S., & Nawaz, T. (2025). Exploring the deleterious effects of heavy metal cadmium on antioxidant defense and photosynthetic pathways in higher plants. *Plant Stress, 15*, 100716. Retrieved from https://doi.org/10.1016/j.stress.2024.100716.

Tripathi, S., & Poluri, K. M. (2021). Heavy metal detoxification mechanisms by microalgae: Insights from transcriptomics analysis. *Environmental Pollution (Barking, Essex: 1987), 285*, 117443. Retrieved from https://doi.org/10.1016/j.envpol.2021.117443.

Umar, A. W., Naeem, M., Hussain, H., Ahmad, N., & Xu, M. (2025). Starvation from within: How heavy metals compete with essential nutrients, disrupt metabolism, and impair plant growth. *Plant Science (Shannon, Ireland), 353*, 112412. Retrieved from https://doi.org/10.1016/j.plantsci.2025.112412.

Ummalyma, S. B., Udayan, A., & Sreekumar, N. (2024). Phycoremediation of emerging contaminants and heavy metals from industrial wastewater: A sustainable green approach for bioeconomy. *Bioresource Technology Reports, 27*, 101920. Retrieved from https://doi.org/10.1016/j.biteb.2024.101920.

Wang, J., Tian, Q., Kang, J., Zhou, H., Yu, X., Qiu, G., & Shen, L. (2025). Mechanistic insight of fungal–microalgal pellets in photobioreactor for heavy-metal wastewater bioremediation. *Bioresource Technology, 416*, 131794. Retrieved from https://doi.org/10.1016/j.biortech.2024.131794.

Wang, S., Yao, H., Li, L., Du, H., Guo, P., Wang, D., ... Ma, M. (2023). Differentially-expressed genes related to glutathione metabolism and heavy metal transport reveals an adaptive, genotype-specific mechanism to Hg^{2+} exposure in rice (*Oryza sativa* L.). *Environmental Pollution (Barking, Essex: 1987), 324*, 121340. Retrieved from https://doi.org/10.1016/j.envpol.2023.121340.

Xi, Y., Zhang, L., Xu, Y., Cheng, W., & Chen, C. (2024). Physiological and biochemical responses of perennial ryegrass mixed planting with legumes under heavy metal pollution. *Phyton-International Journal of Experimental Botany, 93*, 1749–1765. Retrieved from https://doi.org/10.32604/phyton.2024.051793.

Xia, W., Ghouri, F., Zhong, M., Bukhari, S. A. H., Ali, S., & Shahid, M. Q. (2024). Rice and heavy metals: A review of cadmium impact and potential remediation techniques. *The Science of the Total Environment, 957*, 177403. Retrieved from https://doi.org/10.1016/j.scitotenv.2024.177403.

Yaashikaa, P. R., Kumar, P. S., Jeevanantham, S., & Saravanan, R. (2022). A review on bioremediation approach for heavy metal detoxification and accumulation in plants. *Environmental Pollution (Barking, Essex: 1987), 301*, 119035. Retrieved from https://doi.org/10.1016/j.envpol.2022.119035.

Yuksel, E. A. (2024). L-glutamic acid and L-aspartic acid supplementation mitigate heavy metal-induced stress in *Phaseolus vulgaris* L. *Phyton-International Journal of Experimental Botany, 93*, 2189–2207. Retrieved from https://doi.org/10.32604/phyton.2024.055053.

Zhang, L., Wang, J., Shao, R., Chuai, X., Wang, S., & Yue, Z. (2024). Detoxification and removal of heavy metal by an acid-tolerant microalgae, *Graesiella* sp. MA1. *The Journal of Water Process Engineering, 64*, 105579. Retrieved from https://doi.org/10.1016/j.jwpe.2024.105579.

Zhang, Z., Wu, H., Zhang, A., Tan, M., Yan, S., & Jiang, D. (2024). Transfer of heavy metals along the food chain: A review on the pest control performance of insect natural enemies under heavy metal stress. *Journal of Hazardous Materials, 478*, 135587. Retrieved from https://doi.org/10.1016/j.jhazmat.2024.135587.

Chapter 12

Heavy metals and Alzheimer's disease: a review

Bushra Bashir[1], Sapna Jarial[2], Shivalika Sood[2], and Sukriti Vishwas[3]

[1]*School of Pharmaceutical Sciences, Lovely Professional University, Phagwara, Punjab, India,* [2]*Department of Agricultural Economics and Extension, School of Agriculture, Lovely Professional University, Phagwara, Punjab, India,* [3]*Chitkara College of Pharmacy, Chitkara University, Rajpura, Punjab, India*

12.1 Introduction

Alzheimer's disease (AD) is a progressive, irreversible neurodegenerative disorder primarily affecting the cerebral cortex and hippocampus. The disease initially manifests in the frontal and temporal lobes before gradually spreading to other regions of the neocortex at variable rates among individuals. AD is characterized by the extracellular accumulation of insoluble amyloid-beta (Aβ) plaques and the intracellular aggregation of hyperphosphorylated tau protein into neurofibrillary tangles (Rani, Ghosh, Ahmad, & Mondal, 2023). Aβ is generated through the proteolytic cleavage of amyloid precursor protein (APP) by β-secretases and γ-secretases, including presenilin 1 (PS1) and presenilin 2 (PS2). The symptomatic phase of AD lasts an average of 8–10 years, but preclinical and prodromal stages extend for over two decades. The sporadic form, which accounts for most cases, typically has an onset around 80 years of age and is primarily associated with impaired Aβ clearance (Vazzana, Macedo, Santini, Faggio, & Souto, 2014). Comorbidities such as cerebrovascular disease and hippocampal sclerosis frequently complicate diagnosis. Although a family history of AD is common in sporadic cases, less than 1% of patients exhibit the autosomal dominant inherited form of AD, which has an earlier onset (~45 years) due to pathogenic mutations in APP, PS1, or PS2, leading to excessive or aberrant Aβ production. Despite genetic differences, sporadic and familial forms share similar clinical progression and biomarker profiles (Paola., Curti., & Martina, 2023). AD exhibits molecular and pathological similarities to other neurodegenerative disorders, including Parkinson's disease and frontotemporal dementias. This raises the question of whether AD represents an inevitable aspect of aging or a distinct pathological entity. Ongoing research continues to explore the epidemiology, molecular mechanisms, diagnostic advancements, prevention strategies, and current therapeutic approaches for AD (Birla, Minocha, Kumar, Misra, & Singh, 2020).

Heavy metals may also influence the formation of Aβ plaques, a key feature of AD, beyond their role in oxidative stress. Previous studies suggest that metals like copper and zinc can directly bind to Aβ, facilitating its aggregation and the creation of harmful plaques. Heavy metals and Aβ may interact to disrupt normal neuronal function and encourage neuronal degeneration. Furthermore, heavy metals may interfere with the normal development of APP and increase the production of Aβ and its deposition in the brain.

While much evidence linking heavy metals to AD comes from animal models and laboratory studies, there is growing concern about environmental exposure to these metals, particularly in areas with high levels of industrial pollution or heavy metals in consumer products. Epidemiological studies have shown random results; some of the studies suggested the correlation between heavy metal exposure and an enhanced risk of AD, while other studies have not found a definitive link between heavy metals and AD. However, these studies often face challenges such as small sample sizes, potential confounding factors, and the difficulty of accurately measuring long-term metal exposure.

The crosstalk between heavy metal exposure and the pathophysiology of AD remains complex and not fully understood; compelling evidence suggests these metals may play a significant role in the pathogenesis of AD. Further research is crucial on the precise mechanisms

Heavy Metal Toxicity and Neurodegeneration. DOI: https://doi.org/10.1016/B978-0-443-36575-1.00020-0

through which heavy metals correlate with AD and possible avenues to reduce their neurotoxicity in the cerebral context and hippocampus area of the brain. These results could spur the creation of new prevention strategies and treatment protocols for AD.

12.2 Pathogenesis of Alzheimer's disease

Several factors contribute to the disease's progression. Although the precise etiology of AD is unknown, significant research has identified several essential pathophysiologies responsible for the disease's onset and progression.

Aβ plaque accumulation, tau protein–based neurofibrillary tangle development, oxidative stress, neuroinflammation, mitochondrial dysfunction, endoplasmic reticulum stress, and increased neuronal cell death are the main factors underlying the pathophysiology of AD. These processes frequently work together to hasten synapse dysfunction, neuronal death, and cognitive decline (Vishwas et al., 2022).

One of the main characteristics of AD is the buildup of Aβ plaques in the brain. APP, a membrane-bound protein present in neurons, is the source of the protein fragment known as Aβ. Under normal conditions, enzymes, including α-secretase, β-secretase, and γ-secretase, break down APP into smaller pieces. However, APP's enzymatic cleavage in AD is dysregulated, resulting in the synthesis of Aβ peptides, particularly Aβ42, which is prone to aggregation (Calissano, Matrone, & Amadoro, 2009).

Aβ aggregates into oligomers and fibrils that form insoluble plaques in the extracellular space, primarily in the brain's cortical regions, such as the hippocampus and cerebral cortex. These plaques stimulate microglial cells and interfere with synaptic plasticity and neural communication. Proinflammatory cytokines are released when microglia are activated, and these, together with Aβ, are part of the neuroinflammatory processes in AD. Although their precise function in neurodegeneration is still up for question, amyloid plaques are thought to set off a series of pathogenic processes, such as oxidative stress and neuronal death (Karch & Goate, 2015).

Another characteristic of AD is the buildup of neurofibrillary tangles, which are mainly made of hyperphosphorylated tau protein. Tau is a microtubule-associated protein vital for stabilizing microtubules, which are structural elements necessary for cellular material transit and neuronal shape maintenance (Lue et al., 1999). Due to aberrant hyperphosphorylation, Tau loses its capacity to connect to microtubules in AD. Therefore tau creates twisted filaments that group inside neurons to form neurofibrillary tangles, especially in memory-related cerebral cortex and hippocampus regions. The severity of AD and its clinical symptoms are closely associated with the brain's levels of tau tangles and amyloid plaques. One of the most significant clinical characteristics of AD is thought to be the buildup of tau tangles, and knowing about tauopathies—diseases characterized by aberrant tau aggregation—is essential to comprehending AD pathogenesis.

Increased amounts of reactive nitrogen species and reactive oxygen species (ROS) in the brain are caused by oxidative stress, and the body's capacity to counteract them with antioxidants. The brain is particularly vulnerable to oxidative stress because of its high metabolic activity and comparatively low levels of antioxidant defences. The presence of Aβ plaques in AD exacerbates oxidative stress by producing ROS as they aggregate. These free radicals can seriously harm lipids, proteins, and DNA, among other cellular constituents, which can impair cellular processes and result in neuronal damage (Vishwas, Awasthi, Corrie, Singh, & Gulati, 2020). Oxidative stress impairs energy production, mitochondrial function, and the intricate operation of the electron transport chain system. This results in mitochondrial malfunction and an additional rise in ROS production. It creates a vicious cycle of mitochondrial DNA damage and oxidative damage that contributes to the gradual loss of neurons observed in AD.

Additionally, neuroinflammation is one of the most crucial areas for the brain's cholinergic neuron degeneration in AD. TNF-alpha, IL-1beta, IL-6, IL-8, and PfkB are among the inflammatory mediators that cause the increased amount of accumulated tau tangles and Aβ plaques. The buildup of amyloid plaques in AD activates the brain's indigenous immune cells, known as microglia. To remove the Aβ, activated microglia emit chemokines, ROS, and proinflammatory cytokines. However, this inflammatory reaction frequently persists and causes more brain damage (Bashir et al., 2024; Zhang, Sui, Wang, & Sun, 2022).

As the principal organelles in cells that produce energy, mitochondria are essential for preserving cellular homeostasis. Mitochondrial malfunction is one of the first indications of cholinergic neuron loss in AD. The apoptosis of neurons in AD is significantly influenced by mitochondrial dysfunction. The brain's inability to generate energy is caused by mitochondrial malfunction, which includes damaged mitochondrial DNA, aberrant citric acid cycle, impaired calcium regulation, and aberrant electron transport chain function. The amyloid-beta formation, elevated ROS levels, oxidative damage, and neuroinflammation are the causes of this dysfunction (Wilhelmus, Nijland, Drukarch, De Vries, & Van Horssen, 2012).

The vital metals copper, zinc, iron, and calcium are also crucial in the pathophysiology of AD. Numerous biological activities depend on these metals, including synthesizing neurotransmitters, cell signaling, and enzyme activity. However, the concentration of these metals also changes in AD, which could help determine how severe the illness is and how quickly it progresses.

For example, too much iron can increase the production of ROS, which can catalyze oxidative damage through Fenton chemistry. Similarly, as their dysregulation increases amyloid aggregation, copper and zinc play an equal role in developing Aβ plaque. Calcium dysregulation is another significant contributing element to AD, which affects apoptotic signaling and mitochondrial activity by triggering alterations in synaptic processes.

The abovementioned pathological processes are ultimately secondary to synaptic loss and neuronal death, which are the hallmarks of AD cognitive decline. Because synapses—where neurons communicate with one another—are crucial for memory storage, memory loss and other cognitive dysfunctions of AD in these patients are linked to synaptic loss.

Neuroinflammation, oxidative stress, and the toxic action of Aβ and tau lead to synaptic deterioration. Fig. 12.1 illustrates the pathophysiology of AD.

12.3 Heavy metals and their impact on Alzheimer's disease progression

AD is a progressive disease that manifests as a decline in cognition, a reduction in memory function, and an impairment of daily life activities. Notable aspects of the pathophysiology of AD are set, namely, accumulation of Aβ plaques, formation of tau tangles, oxidative stress, and neuroinflammation. There is, however, new evidence suggesting the involvement of other factors like lifestyle, dietary habits, and even exposure to heavy metals as significantly influencing the onset and severity of AD.

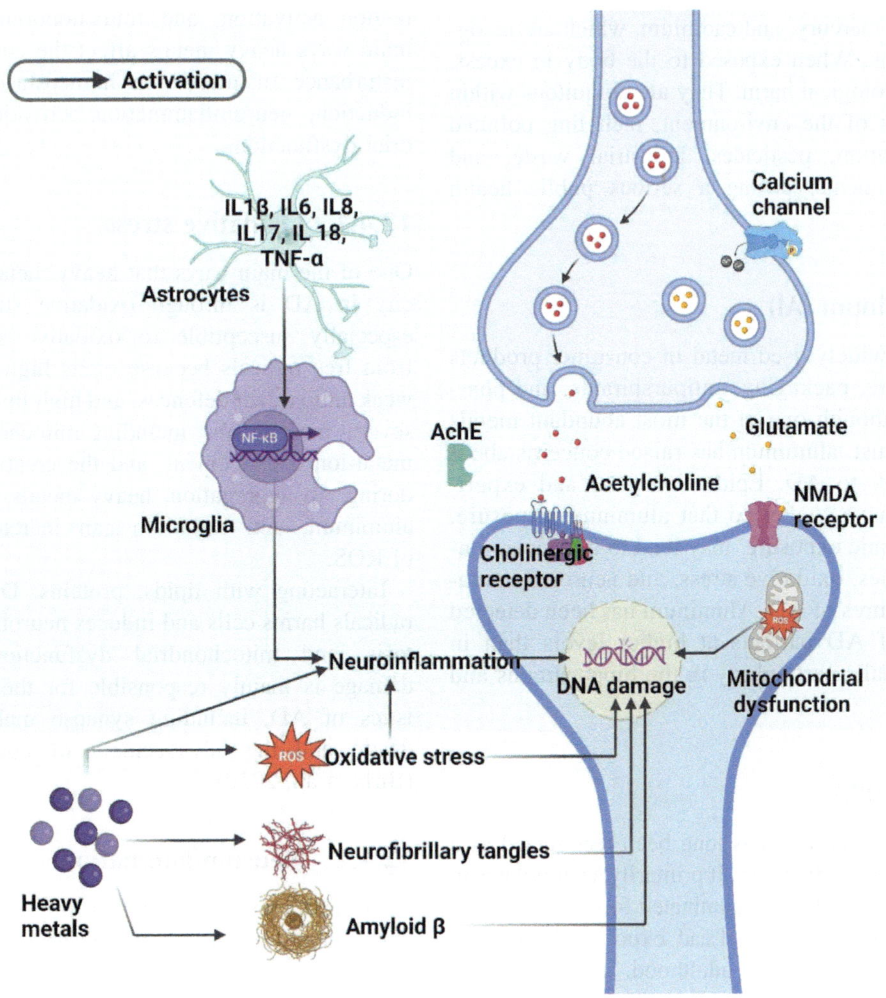

FIGURE 12.1 Pathophysiology of Alzheimer's disease.

Aluminum (Al), lead (Pb), mercury (Hg), and cadmium (Cd) are considered heavy metals that have recently been linked to the development of neurodegenerative conditions, including AD.

This chapter analyzes the linkage between heavy metals and the progression of AD and the possible mechanisms through which these metals can act as neurotoxic. Moreover, we shall examine the potential of these metals to accelerate the pathological process of AD. We will consider the neurotoxicological mechanisms of heavy metals and their effects on the essential pathological features of AD and, thereby, the clinical course of the disease.

12.3.1 Overview of heavy metals associated with Alzheimer's disease

Heavy metals are poisonous at low levels and accumulate within the body over time. They include metals like aluminum, lead, mercury, and cadmium, which are recognized neurotoxins. When exposed to the body in excess, they lead to neurological harm. They are ubiquitous within different sources of the environment, including polluted water, air pollution, pesticides, industrial waste, and working environments, posing a serious public health threat.

12.3.2 Aluminum (Al)

Aluminum is a widely used metal in consumer products such as cookware, packaging, antiperspirants, and pharmaceuticals. Although one of the most abundant metals in the Earth's crust, aluminum has raised concerns about its potential link to AD. Epidemiological and experimental studies have suggested that aluminum exposure, particularly chronic exposure, may lead to the accumulation of Aβ plaques, oxidative stress, and neuroinflammation, all key features of AD. Aluminum has been detected in the brains of AD patients at higher levels than in healthy individuals, particularly in the hippocampus and cortex.

12.3.3 Lead (Pb)

Lead is a heavy metal that has long been recognized as a significant environmental toxin. It primarily accumulates in the body through ingesting contaminated food or water and inhaling lead-containing dust. Lead exposure, especially during early development and adulthood, has been linked to cognitive decline and an increased risk of neurodegenerative diseases. In AD, lead exposure exacerbates the accumulation of amyloid plaques and induces oxidative stress, further impairing cognitive function.

12.3.4 Mercury (Hg)

Mercury is another toxic heavy metal that can accumulate in the body through dietary sources (mainly fish), industrial exposure, and dental amalgams. Mercury exposure has been associated with neurological impairments, including memory and attention deficits, which overlap with symptoms of AD. In AD, mercury has been shown to enhance Aβ aggregation, induce neuroinflammation, and cause mitochondrial dysfunction, all of which contribute to the progression of the disease.

12.4 Mechanisms of heavy metal–induced neurotoxicity

Heavy metals have a variety of neurotoxic effects that can harm neurons in several ways. The main ways heavy metals affect the course of AD are the disturbance of metal ion homeostasis, oxidative stress induction, neuroinflammation activation, and mitochondrial dysfunction. The main ways heavy metals affect the course of AD are the disturbance of metal ion homeostasis, oxidative stress induction, neuroinflammation activation, and mitochondrial dysfunction.

12.4.1 Oxidative stress

One of the main ways that heavy metals cause neurotoxicity in AD is through oxidative stress. The brain is especially susceptible to oxidative stress and damage from free radicals because of its high metabolic activity, weak antioxidant defences, and high lipid content. Through several mechanisms, including mitochondrial dysfunction, metal-ion displacement, and the creation of free radicals during Aβ aggregation, heavy metals like mercury, lead, aluminum, and cadmium can increase the production of ROS.

Interacting with lipids, proteins, DNA, ROS, or free radicals harms cells and induces neuroinflammation, apoptosis, and mitochondrial dysfunction. This oxidative damage is mainly responsible for the clinical characteristics of AD, including synapse malfunction, neuronal death, and the advancement of cognitive impairment (Birla et al., 2020).

12.4.2 Neuroinflammation

Neuroinflammation is another important mechanism via which heavy metals aid in AD development. The central nervous system's resident immune cells, known as microglial cells, are activated when metals like aluminum, mercury, and cadmium build up in the brain. To eliminate the harmful substances, activated microglia emit chemokines, ROS, and proinflammatory cytokines. However, if this inflammatory reaction persists, it may cause long-term neural damage.

Chronic neuroinflammation worsens the pathogenic characteristics of AD, such as the buildup of tau tangles and Aβ plaques and the impairment of synaptic function. Additionally, neuroinflammation hinders synaptic plasticity and neurogenesis, two processes necessary for cognitive function (Huat et al., 2019).

12.4.3 Mitochondrial dysfunction

In neurons, mitochondria play a crucial role in maintaining cellular homeostasis and providing energy for cellular functions. Heavy metal–induced mitochondrial membrane damage and malfunction result in reduced energy production and ROS accumulation. Mitochondrial malfunction also impacts calcium homeostasis, intensifying oxidative stress and neuronal damage.

Reduced ATP synthesis from the defective mitochondria in AD contributes to synapse failure, neuronal death, and cognitive decline. Also, faulty mitochondria cannot remove damaged or malfunctioning cellular components, which causes toxic compounds to accumulate and speeds up neurodegenerative processes (Cheng et al., 2021).

12.5 Accumulation and effects of heavy metals in the brain

The brain's high lipid content and the blood–brain barrier's (BBB) selective permeability make it highly susceptible to heavy metal buildup. As AD worsens with age, the BBB becomes more permeable, making it easier for heavy metals to enter the brain. Both experimental and human research have shown that heavy metals, including aluminum, lead, mercury, and cadmium, can accumulate in different parts of the brain, especially in areas related to memory and cognition, like the cortex and hippocampus.

After entering the brain, heavy metals can build up in blood vessels, glial cells, and neurons, interfering with regular biological functions. Their long-term presence in the brain contributes to neuronal toxicity and accelerates the pathological features of AD, including Aβ plaque formation, tau hyperphosphorylation, oxidative stress, and neuroinflammation.

12.6 Conclusion and future directions

AD is a complicated neurodegenerative disorder caused by Aβ buildup, tau hyperphosphorylation, oxidative stress, neuroinflammation, and mitochondrial dysfunction. Heavy metals, such as aluminum, lead, mercury, and cadmium, are thought to exacerbate these pathogenic pathways by affecting metal ion homeostasis, generating oxidative stress, and promoting chronic neuroinflammation. While preclinical studies support their neurotoxic role, epidemiological findings remain inconclusive due to variations in exposure assessments and confounding factors. Further research is required to establish a definitive causal relationship between heavy metal exposure and AD progression.

Future studies should focus on elucidating the precise molecular mechanisms of heavy metal–induced neurotoxicity using advanced omics technologies. Large-scale, longitudinal epidemiological investigations are essential to clarify exposure risks and identify vulnerable populations. Additionally, therapeutic strategies such as metal chelation, antioxidants, and neuroprotective agents require further exploration, with nanotechnology-based drug delivery systems offering the potential for targeted intervention.

From a public health perspective, stricter regulatory policies to limit environmental heavy metal exposure and public awareness initiatives are critical for mitigating neurotoxic risks. Integrating molecular research, epidemiology, and preventive strategies will be pivotal in developing effective interventions for AD and broader neurodegenerative disease prevention.

References

Bashir, B., Mittal, S., Muthukumar, A., Vishwas, S., Pandey, N. K., Gulati, M., ... Dureja, H. (2024). Harnessing the neuroprotective effect of oral administration of benfotiamine in MPTP induced Parkinson's disease in rats. *European Journal of Pharmacology, 962*, 176234.

Birla, H., Minocha, T., Kumar, G., Misra, A., & Singh, S. K. (2020). Role of oxidative stress and metal toxicity in the progression of Alzheimer's disease. *Current Neuropharmacology, 18*, 552–562. Retrieved from https://doi.org/10.2174/1570159X18666200122122512.

Calissano, P., Matrone, C., & Amadoro, G. (2009). Apoptosis and in vitro Alzheimer disease neuronal models. *Communicative & Integrative Biology, 2*, 163–169. Retrieved from https://doi.org/10.4161/cib.7704.

Cheng, H., Yang, B., Ke, T., Li, S., Yang, X., Aschner, M., & Chen, P. (2021). Mechanisms of metal-induced mitochondrial dysfunction in neurological disorders. *Toxics, 9*, 142. Retrieved from https://doi.org/10.3390/TOXICS9060142.

Huat, T. J., Camats-Perna, J., Newcombe, E. A., Valmas, N., Kitazawa, M., & Medeiros, R. (2019). Metal toxicity links to Alzheimer's disease and neuroinflammation. *Journal of Molecular Biology, 431*, 1843. Retrieved from https://doi.org/10.1016/J.JMB.2019.01.018.

Karch, C. M., & Goate, A. M. (2015). Alzheimer's disease risk genes and mechanisms of disease pathogenesis. *Biological Psychiatry, 77*, 43–51. Retrieved from https://doi.org/10.1016/j.biopsych.2014.05.006.

Lue, L.-F., Kuo, Y.-M., Roher, A. E., Brachova, L., Shen, Y., Sue, L., ... Rogers, J. (1999). Soluble amyloid B peptide concentration as a predictor of synaptic change in Alzheimer's disease. *The American Journal of Pathology, 155*, 853–862. Retrieved from https://doi.org/10.1016/S0002-9440(10)65184-X.

Paola., R.,P., Curti., C., Martina, A., (2023). Multicomponent composition comprising epigallocatechin gallate and saffron dry extract, and its use in the prevention and treatment of Parkinson's disease. IT202200000842A1. https://patentimages.storage.googleapis.com/0f/6d/1a/869e48cd2a28cb/IT202200000842A1.pdf Accessed date 25/05/2025

Rani, L., Ghosh, B., Ahmad, M. H., & Mondal, A. C. (2023). Evaluation of potential neuroprotective effects of vanillin against MPP+/MPTP-induced dysregulation of dopaminergic regulatory mechanisms in SH-SY5Y cells and a mouse model of Parkinson's disease. *Molecular Neurobiology, 60*(8), 4693–4715.

Vazzana, M., Macedo, A. S., Santini, A., Faggio, C., & Souto, E. B. (2014). Novel neuroprotective formulations based on St. John's wort extract. *Journal of Food Research, 3*, 3. Retrieved from https://doi.org/10.5539/jfr.v3n4p3.

Vishwas, S., Awasthi, A., Corrie, L., Singh, S. K., & Gulati, M. (2020). Multiple target-based combination therapy of galantamine, memantine and lycopene for the possible treatment of Alzheimer's disease. *Medical Hypotheses, 143*, 109879.

Vishwas, S., Singh, S. K., Gulati, M., Awasthi, A., Khursheed, R., Corrie, L., ... Porwal, O. (2022). Harnessing the therapeutic potential of fisetin and its nanoparticles: Journey so far and road ahead. *Chemico-Biological Interactions, 356*, 109869.

Wilhelmus, M. M. M., Nijland, P. G., Drukarch, B., De Vries, H. E., & Van Horssen, J. (2012). Involvement and interplay of Parkin, PINK1, and DJ1 in neurodegenerative and neuroinflammatory disorders. *Free Radical Biology & Medicine, 53*, 983–992.

Zhang, L., Sui, S., Wang, S., & Sun, J. (2022). Neuroprotective effect of corosolic acid against cerebral ischemia-reperfusion injury in experimental rats. *Journal of Oleo Science, 71*, 1501–1510. Retrieved from https://doi.org/10.5650/jos.ess22130.

Chapter 13

Parkinson's disease and heavy metal exposure

Amit Raj, and Prasann Kumar
Department of Agronomy, School of Agriculture, Lovely Professional University, Phagwara, Punjab, India

13.1 Introduction

Parkinson's disease (PD) remains a chronic movement disorder of considerable proportion and affects people living with the disease at an average age of 60 years. As mentioned earlier, PD causes motor and nonmotor symptoms such as tremors, akinesia or slow movement, rigidity of muscles, and postural imbalance or instability (Jagadeesan et al., 2017). In addition to these motor disorders, there are positive cognitive changes, mood disorders, and autoimmune disorders, which demonstrate severe damage to the central nervous system (CNS) of the patient. The etiology of PD is highly associated with a progressive nature of dopaminergic neurons in the substantia nigra pars compacta and thus leading to a significant diminution in the levels of dopamine in the striatum. The dopaminergic loss impairs the functioning of the basal ganglia circuitry and results in the characteristic motor disturbances that are characteristic of PD (Galvan, Devergnas, & Wichmann, 2015). The cause of PD involves genetic susceptibility and environmental factors, since the risk factors are multiple. Although there are hereditary forms of PD with an incidence range of 10%–15% of all cases, the majority of them are identified as sporadic, suggesting that the environment plays a crucial role in this ailment. For this reason, several environmental agents, such as pesticides, solvents, and heavy metals, have been mentioned in various epidemiological studies as possible risk factors for the development of PD. These external agents are believed to cause neurotoxicity, resulting in mood changes and leading to the confirmed pathways of oxidative stress, mitochondrial dysfunction, or protein misfolding. Among them, exposure to heavy metals has lately received a lot of attention in the context of PD. These include manganese (Mn), lead (Pb), mercury (Hg), iron (Fe), and other metals that are putative to be involved in the development of PD (Gianì et al., 2021). Long-term exposure to these metals will lead to the penetration of these metals into the brain and cause neurotoxicity. For example, Manganese, commonly found in industries such as welding, causes Parkinsonian syndromes, which is referred to as manganism. Oxidative stress mediated by manganese's interference with mitochondrial function is typical for causing neuronal death of dopaminergic neurons, thus causing neurotoxicity (Raj, Kaur, Gupta, & Singh, 2021).

Lead has also been examined in terms of its connection with PD development. Research has evidenced that higher levels of blood lead are associated with PD. Lead has toxic effects on the neurons in several ways, which include the suppression of antioxidant enzymes, derangement of calcium signaling, and modulation of neurotransmitter release (Mitra, Sharma, Purohit, & Sharma, 2017). These negative actions can lead to neuronal death and further deterioration of neuronal functions. Mercury, a toxic environmental chemical in water supply systems, has been studied due to its links to PD. Research has also shown a correlation between mercury levels in the population and PD, especially among individuals who consume seafood with high levels of this metal. The neurotoxicity of mercury is the linking of Hg with proteins that contain a terminal sulfhydryl group, which causes a change in the shape of the protein or aggregation of the protein molecules (Bjorklund et al., 2018). Also, mercury can cause oxidative stress and impact the mitochondria, which is one of the reasons that distinguishes the neurodegeneration of dopaminergic neurons. Unfortunately, iron at high concentrations is known to precipitate and form reactive oxygen species (ROS) through Fenton reactions and can cause oxidative degradation of the neurons. Specifically, biochemical, loss-of-function mutations of the parkin gene in mouse and human PD patients and postmortem analyses of PD substantia nigra samples have supported the oxidation hypothesis of PD and have shown that iron concentrations

Heavy Metal Toxicity and Neurodegeneration. DOI: https://doi.org/10.1016/B978-0-443-36575-1.00010-8

were raised in the substantia nigra of PD patients (You et al., 2015). Furthermore, the accumulation of iron may also promote the formation of Lewy bodies (LBs), the intracellular aggregates of α-synuclein, which is a primary marker for PD and a significant factor contributing to neuronal death. Indeed, the ways through which heavy metals are involved in the development of PD are not entirely understood and may include a multitude of factors. It should be noted that oxidative stress became a prominent issue encompassing an optimum oxidative state when an organism is experiencing an excess of prooxidants and fewer antioxidants. The heavy metals can cause direct ROS formation or affect the mitochondria's respiratory chain, leading to increased oxidative stress (Sun et al., 2022). This oxidant environment may harm lipids, proteins, and nucleics, which is the ultimate consequence of brain cell death.

Another mechanism or pathway that has been considered to play a strategic role in metal-induced neurodegeneration is oxidative stress. Because they are involved in energy production, mitochondria are seen as critical to neuron survival. Heavy metals can interfere with matrix metalloproteinases, impair electron transport chain (ETC) enzymes, and cause the release of proteins that promote apoptosis. This can cause energy depletion and subsequent cell death, contributing to the continuous decline in dopaminergic neurons (Koyama, Kamogashira, & Yamasoba, 2024). Failure of protein folding and protein aggregation are other avenues through which heavy metals can impact the development of PD. Some metals, such as mercury and iron, can interact with α-synuclein, leading to its misfolding and subsequent formation of PD's LBs characteristic. These protein aggregates can cause dysfunction of cellular homeostasis, impair proteasomal and lysosomal degradation pathways, and induce inflammation, which is known to cause the development of neuronal injury (Gandhi et al., 2019). As stated earlier, inflammation is pertinent in the neurodegenerative processes of exposure to metals. While esophageal squamous cell carcinoma cells were confirmed to play more predominant roles in tumorigenicity and medication resistance, heavy metals have also been demonstrated to stimulate microglial cells, the inherent immune cells of the CNS, to produce proinflammatory cytokines and chemokines (Battaglini et al., 2024). Long-term inflammation in the brain poses a dangerous environment for neurons since it fosters oxidative stress to enhance neurodegenerative pathways.

One of the examples of the influence of environmental factors on the determination of the PD is the situation in the Faroe Islands. The people of these islands subsist on pilot whale meat and oil have been proven to be suffering from the effects of taking in high levels of mercury. Several researchers have established that this population is at a higher risk of PD, meaning that people with mercury exposure in their diets are more prone to PD. This

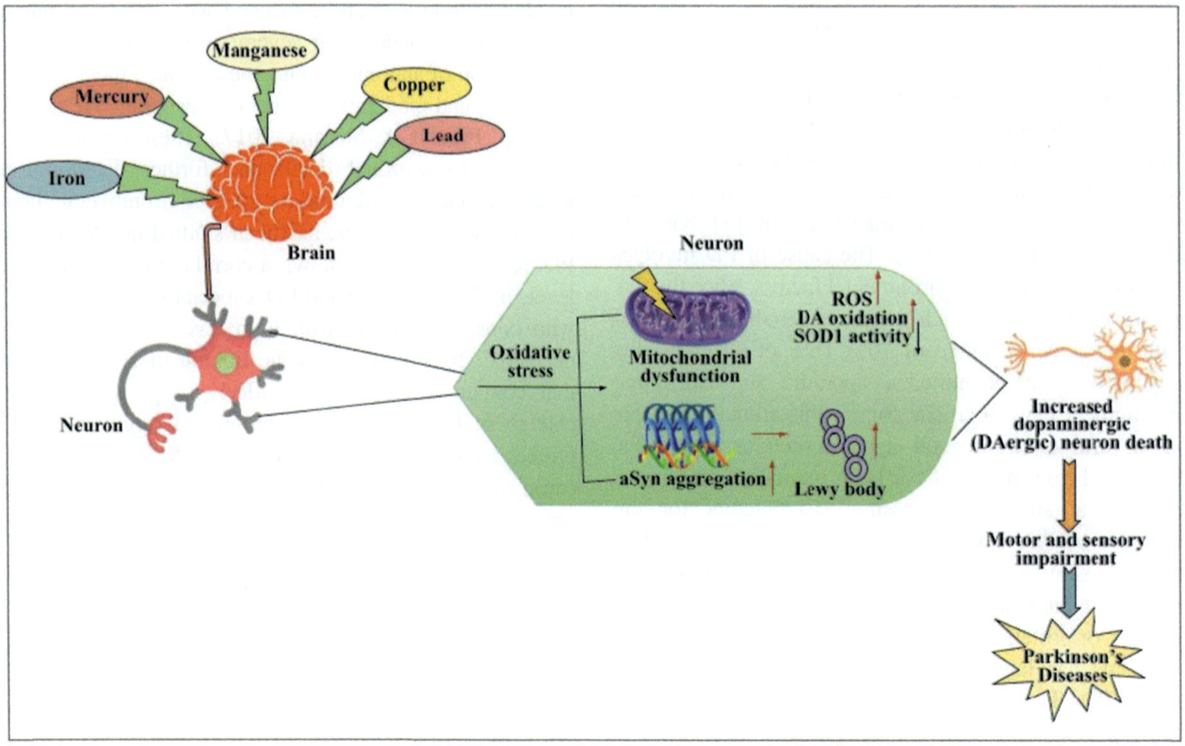

FIGURE 13.1 Correlation between heavy metal toxicity and parkinson's disease: a mechanistic study of oxidative stress pathways.

approach points to the interrelation between the factors of one's environment and disease occurrence while emphasizing the issue of diet when developing PD. The cause of Parkinson's disease is complex, and several factors play a role in its development. These include epidemiological influences, such as environmental exposures and the presence of heavy metals, which are considered important contributors to the onset of the disease (Bjorklund et al., 2018). Therefore it is essential to understand how these metals influence neurodegeneration in the course of chronic disease to be able to design efficient preventive or therapeutic measures. Future works should aim to identify more clearly the exact mechanisms through which metals may be toxic to the brain and ways that may be used to prevent such toxicity (Fig. 13.1).

13.2 Heavy metals: sources, exposure pathways, and brain accumulation

Heavy metals are metallic elements with high density and atomic number found in the environment and workplace for several reasons, such as industrialization and human intervention. Some of these metals are essential in the body in small concentrations. Still, contemplation of high concentrations leads to dangerous health risks, especially when these metals build up in the brain (Sharma et al., 2020). This discussion is delineated in the following sections.

13.2.1 Common sources of heavy metal exposure

13.2.1.1 Environmental sources

- **Industrial emissions:** Industries like mining, smelting, or manufacturing emit large amounts of fluids that contain lead, given, mercury, and cadmium, which pollute the soil and water (Raj, Kumar, & Vaghasiya, 2021).
- **Agricultural practices:** Phosphates used in fertilizers in agricultural practices may cause cadmium to enter plants and, in turn, the food chain.
- **Consumer products:** One example of a lead exposure source is objects made using lead, such as paints, some cosmetics, and deformed food containers.

13.2.1.2 Occupational sources

- **Mining and smelting:** Employees in these industries can be exposed to lead, mercury, dust, and fumes.
- **Battery manufacturing and recycling:** The management of one's material risks of lead and cadmium exposure.
- **Welding and Metal Fabrication:** Hazardous substances exploited in these processes, such as manganese, can cause the possibility of inhaling metal fumes with neurological impacts (Lucchini, Kotelchuck, & Grandjean, 2015).

13.2.2 Mechanisms of heavy metal uptake, transport, and distribution in the body

13.2.2.1 Routes of entry

- **Inhalation:** Fine particles of heavy metals or fumes of the products containing the metals penetrate the respiratory system and the blood.
- **Oral:** Ingestion of contaminated foods or water makes heavy metals enter the gastrointestinal system through which they can be transported into the bloodstream.
- **Skin penetration:** Foliar exposure to heavy metal compounds in the workplace is one such way that skin penetration is caused (Fu & Xi, 2020).

13.2.2.2 Transport and distribution

- **Transportation with proteins:** When they reach the bloodstream, heavy metals may form complexes with proteins such as albumin or metallothionein (MT) to aid their transfer to several tissues.
- **Mimicry of essential elements:** Nature has provided certain heavy metals to mimic some essential minerals used in the human body's transport system. For instance, lead can structure itself to resemble calcium, thus allowing it to form part of bones and teeth (Witkowska, Słowik, & Chilicka, 2021).
- **Hazardous biological barriers**: Certain heavy metals can cross protective barriers. An example is methylmercury, which can penetrate the placental barrier, thus exposing the fetus to dangers (Zafar, Javed, Akram, & Naqvi, 2024).

13.2.3 Brain accumulation and regional vulnerability

13.2.3.1 Blood–brain barrier penetration

- **Disruption of blood–brain barrier (BBB) integrity:** Heavy metals such as lead affect the integrity of the BBB, making it porous and allowing the passage of the metals and other neurotoxic substances.
- **Transportation of methylmercury:** Methylmercury can easily cross the BBB because it conjugates with cysteine to form a complex that can be transported along the carriers that specialize in transporting methionine (Novo et al., 2021).

13.2.3.2 Regional brain vulnerability

- **Hippocampus:** It is also associated with learning and memory loss due to the damage to the hippocampus, which is an area that plays a crucial role in such usages. These aspects are closely related to the basal ganglia, in which manganese accumulates and leads to damage in motor control in beings, thus resembling PD (Aarsland et al., 2021).
- **Cerebellum:** Mercury affects this preferred area of the human brain, interfering with regular coordination and

balance since this is the brain primarily involved in controlling movements (Chamoli & Karn, 2024).

13.3 Neurotoxic mechanisms of heavy metal–induced neurodegeneration

Lead, mercury, cadmium, arsenic, and other such chemicals can be classified as environmental neurotoxins that significantly impact diseases that affect the nervous system. The deleterious effects on neurons are mainly oxidative stress, impaired mitochondrial function, and neuroinflammation (Picca et al., 2020). Awareness of such mechanisms is essential for planning for ways and means of eliminating the attacks that bring about neurological disorders (Fig. 13.2).

13.3.1 Oxidative stress and the generation of reactive oxygen species

Oxidative stress is a pathological feature involved in the development of neurotoxicity induced by heavy metals. These alter the balance of cells, which in turn is capable of causing the generation of ROS. These reactive molecules are generated through numerous metabolic reactions and exert a lot of toxicity on the cellular structures, such as lipids, proteins, and DNA. For example, it has been evidenced that cadmium increases the generation of ROS in neuronal cells, producing lipid peroxidation and consequent cell death (Nabi & Tabassum, 2022). This study of cadmium effect on rat hippocampus cultures showed that cadmium exposure induced apoptosis by generating ROS. Thus it has elucidated the prospect of antagonizing cadmium-induced oxidative stress by using antioxidants, including *N*-acetylcysteine (NAC), as a therapeutic target (Kaushal, Chandrashekar, & Juncos, 2019). Also, lead exposure increases oxidative stress in the brain, not excluding the cortex and the hippocampus. This oxidative imbalance initiates signaling cascades, potentially leading to neuronal cell death and impaired cognitive function. Heavy metal toxicity is diverse due to the combined effects of oxidative stress and other forms of neurotoxicity.

- **Imbalance in redox homeostasis:** Metals like iron and copper can participate in redox cycling, leading to ROS generation through Fenton and Haber–Weiss reactions. High levels of ROS affect cellular components such as lipids, proteins, and DNA in neurons and, therefore, cause injury (Hassan, Noreen, Rehman, Kamal, & da Rocha, 2022).
- **Inhibition of antioxidant enzymes:** Lead and mercury inhibit antioxidant enzymes such as superoxide dismutase and glutathione peroxidase; this means that ROS cannot be quenched or erased, hence damaging the cell (Baş, Kalender, Karaboduk, & Apaydın, 2015). It further augments oxidative reactions that damage neurons in the brain. This prolongs oxidative stress and leads to the destruction of neurons.

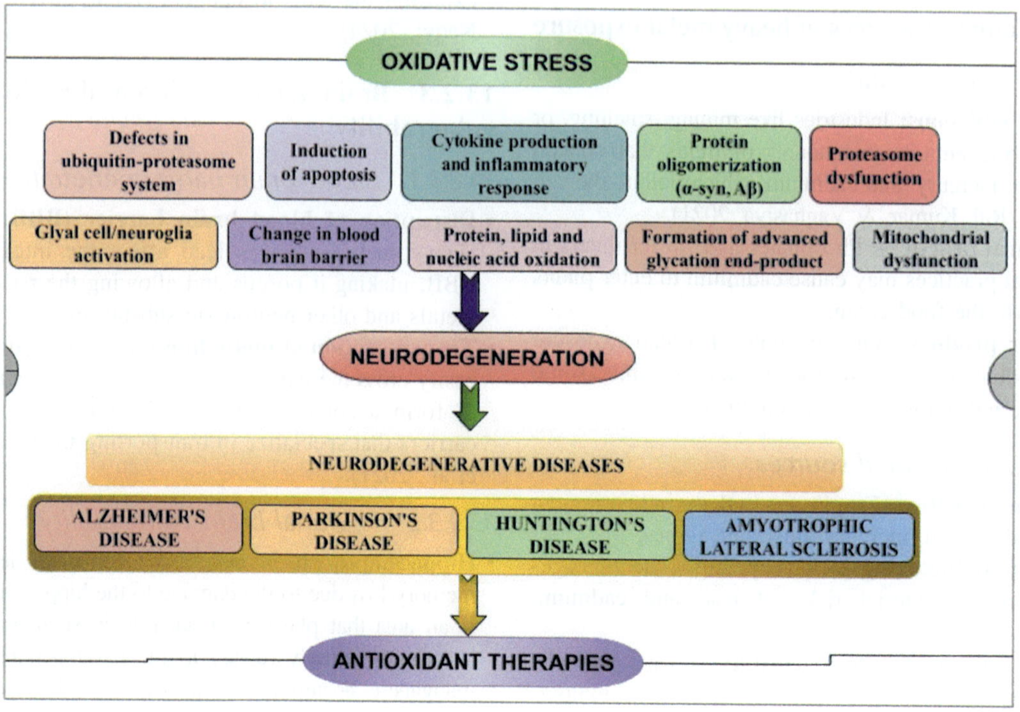

FIGURE 13.2 Oxidative stress and neurodegeneration. Oxidative stress may be involved in neurodegeneration via numerous enteropathogenic mechanisms in neurodegenerative diseases like Alzheimer's disease, Parkinson's disease, Huntington's disease, and amyotrophic lateral sclerosis. Many antioxidant therapies have been tested in recent years to reduce oxidative damage in neurodegenerative diseases.

- **NADPH oxidase:** Cadmium exposure can activate this enzyme complex, which generates superoxide radicals. Upregulation of NADPH oxidase increases ROS generation, contributing to oxidative stress and neurodegeneration (Branca et al., 2020).

13.3.2 Mitochondrial dysfunction and impairment of energy production

Mitochondria of the cells are responsible for energy production and are differentially involved in neurodegenerative diseases. Heavy metals can interfere directly with energy metabolism, thereby reducing the generation of ATP by the mitochondria while at the same time enhancing the production of radicals (Sun et al., 2022). This may harm neuronal survival, especially in high-energy-demanding brain areas. Earlier studies have illustrated that in mercury toxicity, there was modulation of membrane potential and reduced function of the respiratory chain complexes, chiefly complex I. This disruption not only slows down ATP generation but also leads to increased emission of electrons that, in turn, cause superoxide radicals. The study using neuronal cell lines provided the information that mercury impacts mitochondrial dysfunction, leading to reduced cell viability and increased apoptotic proteins, indicating that the mitochondria play crucial roles in neuronal cells (Cariccio, Samà, Bramanti, & Mazzon, 2019). Moreover, it emerged that lead could alter the cellular process as it interferes with the balance between the mitochondria, whereby the cells tend to have a higher fission rate than fusion. This dysregulation of mitochondrial morphology can hinder the transport of fundamental proteins and affect calcium balance, contributing to neuronal deterioration. The studies have indicated that antioxidant approaches utilized in limiting mitochondrial dysfunction using neuroprotective agents such as coenzyme Q10 are beneficial in reducing heavy metal neurotoxicity (Agnihotri & Aruoma, 2020).

- **Inhibition of complexes in the electron translocation chain:** Both mercury and arsenic are in vitro inhibitors of mitochondrial proteins and will hence disrupt the ETC. This inhibition slows down the electron transport process and increases electron leakage, thus increasing the generation of ROS and oxidative stress (Pace, Dagda, & Angermann, 2017).
- **Initiation of the cell death process:** This may be explained by the fact that lead exposure induces the opening of the mPTP, leading to a loss of mitochondrial membrane potential and the release of proapoptotic factors (Liu et al., 2016).
- **Impact on mitochondrial biogenesis:** Cadmium has been noted to affect both fusion and fission processes, leading to a breakdown of mitochondria into small fragments and, therefore, has an impaired function. Such modifications can affect neuronal viability and become the cause of certain neurodegenerative disorders (Chi, Chang, & Sang, 2018).

13.3.3 Neuroinflammation and the activation of microglia and astrocytes

Neuroinflammation can also be considered an essential aspect through which heavy metals act on the nervous tissue. In this regard, heavy metals can stimulate the activity of glial cells, among them microglia and astrocytes, which bring about inflammation that again worsens the condition of the neurons. This activation is manifested in the liberation of proinflammatory cytokines, chemokines, and reactive nitrogen species, which help to fuel neurotoxic changes. For instance, it is revealed that lead exposure activates microglia, leading to the release of inflammatory mediators, including tumor necrosis factor-alpha (TNF-α) and interleukin-1 beta (IL-1β). This inflammatory response can result in alteration of neurotransmission and potentially participate in the development of the diseases like Alzheimer’s (Wang, Tan, Yu, & Tan, 2015). In the latest studies, the application of inhibitors to modulate microglial activation has been shown as a practical measure capable of regulating neuroinflammation. Therefore it can potentially enhance cognitive efficacy in lead-exposed animals. In the same way, mercury has been found to stimulate the so-called astrocytes, which are essential in supporting and regulating neurons. Astrocytes put up a reaction when exposed to mercury that results in a disruption of the bipolar neuroprotective and neurotoxic signals. It was also found that mercury-treated astrocytes produced high levels of glutamate, a molecule acting as a neurotransmitter if present in excessive quantities, which can cause excitotoxicity and death of neurons.

- **Inflammation:** Heavy metals stimulate microglial activation, the immune cells in the CNS. TNF-α and IL-1β, which are produced by activated microglia, are toxic to neurons, and this can further worsen the condition (Wang et al., 2015).
- **Astrocyte pathology:** Astrocytes are crucial for maintaining neuronal health and balance. Manganese can penetrate the BBB and deposit in astrocytes, which in turn affects their functionality and causes the release of some inflammatory biomolecules. This activation of astrocytes leads to a neurotoxic state that consequently causes neurodegeneration (Doroszkiewicz et al., 2023).
- **Disruption of the BBB:** Heavy metals are becoming known for disrupting the BBB's tightness. This allows immune peripheral cells to enter the CNS and enhances inflammation. It also extends the extent of neuronal damage and death (Novo et al., 2021).

13.4 Dopaminergic neurotransmission and heavy metal toxicity

Dopaminergic neurons, which are predominantly found in the substantia nigra pars compacta, are fundamental for movement, thinking processes, and feelings. The proper functioning of these neurons is achieved by a well-coordinated process, which involves the synthesis, metabolism, and storage of dopamine in vesicles. These include lead (Pb), mercury (Hg), cadmium (Cd), and manganese (Mn), which escape the abovementioned cellular processes (Owen et al., 2019). It poses neurotoxicity effects in the development of PD, and this will cause both motor and nonmotor symptoms. This analysis explores how heavy metals disrupt dopamine biochemistry and uses battery manufacturing workers as an example to illustrate these effects.

13.4.1 Impact on dopamine synthesis, metabolism, and transport

- **Synthesis inhibition:** Dopamine production is initiated by converting the amino acid tyrosine to L-3,4-dihydroxyphenylalanine (L-DOPA through the action of the enzyme tyrosine hydroxylase (Du et al., 2022)). Although Pb and Cd have no stimulatory effects on TH activity, the two can suppress the activity by interacting with essential SH groups on the enzyme. This binding changes the conformation of the enzyme and its capability to catalyze the reaction; in this respect, there is less production of L-DOPA. The lack of substrate for further conversion by the aromatic L-amino acid decarboxylase also manifests, leading to general dopamine deficiency.
- **Metabolic disruption:** After synthesis, dopamine is metabolized by monoamine oxidase in the formation of dihydroxyphenylacetic acid, which is in turn transformed into homovanillic acid (Delcambre, Nonnenmacher, & Hiller, 2016). Metal ions can inhibit or distort these metabolic reactions. For instance, mercury may elicit oxidative stress, which, in one way or another, may impair the function of MAO, exposing the brain to toxic intermediate metabolites. Also, the metals might enhance dopamine oxidation and produce ROS and poisonous quinones hazardous to the neurons (Pavlin, Repič, Vianello, & Mavri, 2016).
- **Transport and storage interference:** Dopaminergic signaling must involve the removal of dopamine from the synaptic cleft and packaging in synaptic vesicles. The dopamine transporter (DAT) is engaged in the reabsorption process of dopamine, while the vesicular monoamine transporter 2 (VMAT2) carries dopamine into vesicles for discharge. These transporters are known to be interfered with by heavy metals. For example, it may affect DAT expression or decrease its functionality depending on the types of membrane microdomains involved in the functioning of the transporter. Also, manganese inhibits the functioning of VMAT2, which means that dopamine remains to be sequestered within the vesicles inadequately (Rasmi et al., 2024). The result is a higher dopamine concentration in the cytosol, which autooxidates and produces ROS through a vicious cycle.

13.4.2 Disruption of dopaminergic pathways and motor symptom development

- **Neuronal vulnerability to oxidative damage:** The dopaminergic neurons of the substantia nigra itself are more prone to oxidative stress partly due to qualitatively toxic metabolites of dopamine. These agents contribute to this vulnerability by catalyzing Fenton-like reactions and negatively affecting the mitochondria. Overproduction of ROS affects lipids, proteins, and nucleic acids, thus leading to mitochondrial dysfunction that plays a central role in the development of PD (Martemucci et al., 2022). This leads to the continuous oxidative damage of these neurons and, as a result, the amount of dopamine required for appropriate motor coordination decreases due to neuronal death.
- **Basal ganglia circuitry alterations:** The nigrostriatal pathway is not sustainable, and the dopaminergic neurons in the basal ganglia network are impaired. This neural network receives and sends activator and inhibitory signals to enable properly coordinated, goal-oriented movements. The toxic effects of heavy metals on dopaminergic neurons shift the net basal ganglia output in favor of inhibition; this results clinically in bradykinesia, rigidity, and tremor. However, this disruption is not only from dopamine but also from the heavy metals that generate the wrong signaling of receptors and even low DATs, which contribute to the greater instability of the motor control system (Kumar, Goyal, & Singh, 2022).
- **Emerging occupational example—battery manufacturing workers:** One case that revealed these mechanisms includes working in battery manufacturing industries in the past. It has been established that employees in these settings are often exposed to blends of heavy metals especially lead and cadmium stemming from the handling of components used in batteries (Wang & Wang, 2025). In a controlled study, these workers analyzed the amount of blood and urine metals, and neuroimaging examinations related to the striatal dopamine system were conducted. It was revealed that a nonsignificant level of heavy metal was capable of causing a decrease in DAT availability in the striatum and VMAT2 in peripheral blood mononuclear cells. These workers had negligible motor dysfunction in fine motor coordination and dexterity, even though they did not show apparent Parkinsonian symptoms. The present occupational group adds new human data from battery

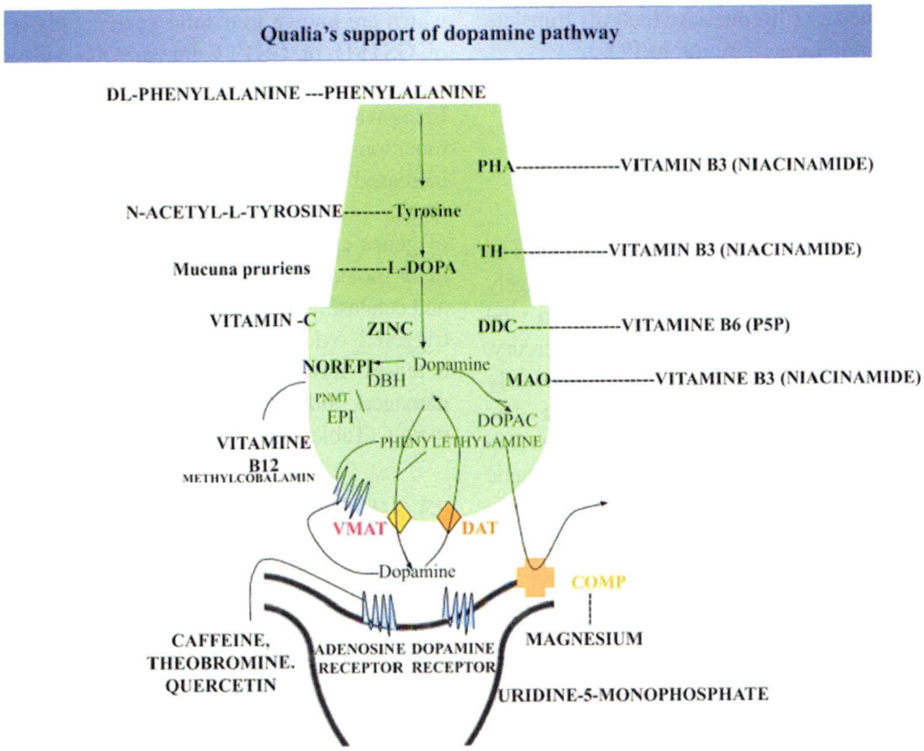

FIGURE 13.3 Schematic diagram of dopaminergic neurotransmission and sites of standard drug treatments and novel neurotransmitter disorders.

manufacturing industries to suggest that low-grade exposure to heavy metals might interfere with dopamine neurotransmission and set off neurodegenerative processes resembling PD (Fig. 13.3).

13.4.3 Potential role of heavy metals in nonmotor symptoms of Parkinson's disease

- **Cognitive impairment and mood disorders:** Apart from motor function, dopamine is involved in cognitive and mood-related functions. Therefore it was also evident that heavy metal toxicity projected its influence on the dopaminergic pathways at the cortical and limbic sites. For instance, Pb and Hg are known to impair dopaminergic transmission in the prefrontal cortex and hippocampus, which are relevant to working memory and executive functions. This disruption might be one of the reasons that lead to cognitive decline and mood changes in these patients, like depression and anxiety (Jaroudi et al., 2017).
- **Autonomic dysfunction and sleep disturbances:** Dopaminergic neurons also regulate the autonomic functions and the sleep–wake cycle. It can have clinical effects on neurotransmission in brainstem areas involved in autonomic regulation, causing symptoms such as orthostatic hypotension and profound effects on gastrointestinal motility. Moreover, as mentioned earlier, dopamine imbalance in the hypothalamic region may also disrupt circadian rhythms, leading to sleep disorders, which are nonmotor symptoms of PD (Liu, Niu, Liu, Cheng, & Le, 2021).
- **Neuroinflammation as a contributing factor:** This chapter shall establish that metal ions are robust stimulators of neuroinflammation processes. The reactive metal ions are also known to induce oxidative stress in microglial cells, which are the brain's immune cells. This leads to the production of proinflammatory cytokines that aggravate neuronal failure and impaired synaptic transmission. So, the later stage of PD offers not only increased vulnerability of dopaminergic neurons but also an impact on nondopaminergic signaling that may explain various nonmotor manifestations of PD such as cognitive impairment, mood disorders, and autonomic dysfunction (Jaroudi et al., 2017).
- **Interplay with other neurotransmitter systems:** However, it must be noted that dopamine does not stand alone in its functioning. Its interplay with other neurotransmitter systems, such as serotonin, norepinephrine, and acetylcholine, is critical for maintaining neural balance (Teleanu et al., 2022). Heavy metals can manipulate these systems since the latter affects dopamine release in both lithium-sensitive and lithium-resistant rats. For instance, there is evidence that deals with the fact that mercury affects serotonergic activity and has an impact on mood disorders. Likewise, there is downregulation and upregulation of norepinephrine, which

causes attention issues and tiredness, which all translate to increasing the nonmotor symptoms in PD.

13.5 Genetic susceptibility, individual variability, and heavy metal toxicity in Parkinson's disease

People have different reactions or effects when subjected to heavy metal elements. Unlike the exposures that averagely vary among individuals, genetic polymorphism plays a vital role in the ADME of toxic metals, including lead, mercury, cadmium, and manganese. Such causes can significantly affect cellular protection against heavy metals and can, therefore, change the propensity of an individual to develop neurodegenerative diseases such as PD. There are genetic predispositions, as well as environmental conditions, that are responsible for the heavy metal toxicity and its contribution toward neurodegenerative diseases, including PD. Some genetic polypeptide variations in an individual can affect the ability to metabolize and detoxify heavy metals and, consequently, the predisposition to the disease (Jan et al., 2015). It expounds on genetic factors concerning heavy metal metabolism, gene–environment interactions, and the importance of understanding such factors in developing risk profiling and surgical prevention measures.

13.5.1 Influence of genetic polymorphisms on heavy metal metabolism and toxicity

Some metals, even when present in small quantities, are toxic to the body; hence, the variation in the genetic makeup of metal transporters, detoxification enzymes, and antioxidant systems will define the ability of the body to manage the heavy metals. For example, variations in the genetic code of DMT1, the divalent metal transporter 1 (also known as SLC11A2), determine how effectively the metals crucial for living, such as iron and manganese, as well as the hazardous metals similar to them, enter a cell (Mani, Dsouza, & Dsouza, 2021). Forcing the variants that promote overexpression or increased activity of DMT1 will increase the accumulation of toxic metals intracellularly, leading to oxidative stress and the dysfunction of mitochondria in neurons. Likewise, the polymorphism of the genes that are mainly involved in the metabolism of lead-containing compounds like delta-aminolevulinic acid dehydratase (ALAD) will vary among the respective persons. The most significant allele is the ALAD2, which affects the affinity of the ALAD enzyme to Pb and may change the kinetics of lead and elevate its neurotoxicity (Broberg & Pawlas, 2022). Therefore the usual dosage of lead can be dangerous for people carrying this variant and cause increased concentration of lead in neural tissue and impair dopaminergic neurons. Detoxification pathways are also critical. Some of the enzymes involved in the bioactivation of these toxic metals are glutathione-*S*-transferases (GSTs), which are known to exhibit genetic polymorphism. The lack of GSTM1 or GSTT1 enzymes is evidence of the loss of capability of the human body to detoxify metals (Pljesa-Ercegovac et al., 2018). For this reason, people with these mutations have increased accumulation of metals in the cells, increased oxygen radical production, and cell damage results—supplementary information. The polymorphic status of genes encoding the enzymes antecedent to major antioxidant molecules, which include superoxide dismutase (SOD) and catalase, also contributes to the cellular reaction to metal-triggered oxidative stress. In conditions where antioxidant defenses have been reduced, neurons are likely to be damaged, suffering the cytotoxic effects of accumulated metals (Table 13.1).

13.5.2 Interaction between genetic factors and environmental exposure in Parkinson's disease risk

The onset of PD is genetically predisposed and the result of exposure to some environmental influences. It is pertinent to note that heavy metals prove toxic, especially to those people with specific genetic polymorphisms. For instance, polymorphisms in the HFE gene, which is involved in iron homeostasis, contribute to increased iron overload in the brain. When such individuals are exposed to other heavy metals, the combined effect is severe oxidative stress and damage to the mitochondria, leading to the loss of dopaminergic neurons. Besides, PD might be associated with genetic factors that cause mitochondrial dysfunction in interaction with heavy metal exposure. Mitochondrial genes are involved in energy production and program cell death. At the same time, mutations might find it difficult to defend a neuron against toxic metals' influence insofar as energy production is concerned (Kausar et al., 2018). Therefore it can be inferred that some genetic defects in mitochondrial pathways combined with metal exposure will enhance neurodegeneration. A few studies have been conducted to explain the gene–environment interactions noted next. For instance, in populations with higher industrial pollutants, GST null allele carriers have substantiated more oxidative stress indicators, biomarkers, and early neuropathological symptoms when exposed to higher levels of Pb and Cd. Therefore it goes to show that to evaluate the risk of developing PD, one has to look at one's genetic stock (Fig. 13.4).

13.5.3 Personalized risk assessment and targeted prevention strategies

Due to the importance of genetic variability in heavy metal metabolism and PD risk, there is a need to devise individual risk assessment programs that will help in the early detection of those who are most vulnerable to heavy metal toxicity. New molecular techniques make it possible to identify genetic markers that dictate one's susceptibility to

TABLE 13.1 The challenges in assessing heavy metal exposure and establishing causality in Parkinson's disease.

Challenge	Description	Impact on research	Example	Potential solutions	References
Measurement accuracy	Difficulty in quantifying heavy metal levels in tissues and fluids reliably.	Inconsistent data quality	Variability in lab results	Standardized measurement protocols	Mrštná, Krčmová, and Švec (2023)
Biological sample variability	Differences in individual metabolism affect heavy metal accumulation.	Confounding results	Variation among patients	Larger sample sizes for diverse cohorts	Balaram (2021)
Temporal exposure assessment	Challenges in determining the timing of exposure relative to disease onset.	Misinterpretation of causal links	Acute vs. chronic exposure effects	Longitudinal studies	Sigurnjak Bureš et al. (2021)
Multiple exposures	Difficulty isolating the effects of heavy metals from other environmental toxins.	Confounding variables	Coexposure to pesticides and solvents	Comprehensive exposure assessments	Belvisi et al. (2020)
Lack of biomarkers	Insufficient biomarkers for specific heavy metal exposure linked to PD.	Limited diagnostic capabilities	Unclear exposure history	Development of novel biomarkers	Broberg and Pawlas (2022)
Epidemiological data limitations	Availability of robust epidemiological data can be scarce.	Hinders large-scale studies	Inconsistent reporting in studies	Enhanced data sharing among researchers	Jaroudi et al. (2017)
Genetic variability	Genetic factors influence individual responses to heavy metal exposure.	Confounding interpretation	Different susceptibilities to PD	Stratified analyses based on genotype	Kausar, Wang, and Cui (2018)
Environmental sampling issues	Challenges in accurately sampling and analyzing environmental sources.	Misleading environmental data	Inaccurate pollution levels	Improved sampling techniques	Mrštná et al. (2023)
Control group selection	Difficulty in selecting appropriate control groups for studies.	Bias in results	Selecting nonexposed individuals	Rigorous control selection criteria	Balaram (2021)
Recall bias	Participants may not accurately recall past exposures, affecting data quality.	Reduced reliability of self-reports	Unreliable personal exposure histories	Use of objective exposure assessments	Sigurnjak Bureš et al. (2021)
Confounding factors	Other health conditions may confound results related to heavy metal exposure.	Difficulty in establishing causality	Coexisting neurological disorders	Multivariable regression models	Belvisi et al. (2020)
Environmental changes over time	Changes in pollution levels can complicate historical exposure assessments.	Misleading historical comparisons	Changes in industrial activity	Long-term environmental monitoring	Broberg and Pawlas (2022)
Ethical concerns	Ethical issues in conducting exposure studies may limit research.	Restrictions on study design	Difficulty in obtaining informed consent	Development of ethical research frameworks	Jaroudi et al. (2017)
Funding limitations	Inadequate funding restricts the scope of research initiatives.	Limited research capacity	Small pilot studies only	Advocacy for increased research funding	Mrštná et al. (2023)
Public awareness challenges	Lack of public awareness about heavy metal risks complicates data collection.	Underreporting of exposures	Low participation in studies	Community engagement and education	Balaram (2021)
Regulatory framework gaps	Inconsistent regulations regarding heavy metal exposure complicate assessments.	Unclear guidelines for researchers	Variability in permissible levels	Advocacy for better regulatory policies	Sigurnjak Bureš et al. (2021)
Longitudinal study challenges	Difficulty in maintaining cohort participation over long study periods.	Attrition bias	Loss of follow-up data	Incentives for participant retention	Belvisi et al. (2020)
Data integration issues	Challenges in integrating data from multiple sources and studies.	Fragmented understanding	Inconsistent data formats	Development of centralized databases	Broberg and Pawlas (2022)
Statistical complexity	Advanced statistical methods required to analyze complex exposure data.	Potential for misinterpretation	Difficulty in modeling interactions	Training for researchers in statistical methods	Jaroudi et al. (2017)
Interdisciplinary collaboration	There is a need for collaboration among various fields to assess exposure comprehensively.	Limited perspectives	Isolation of research disciplines	Establishment of interdisciplinary teams	Kausar et al. (2018)

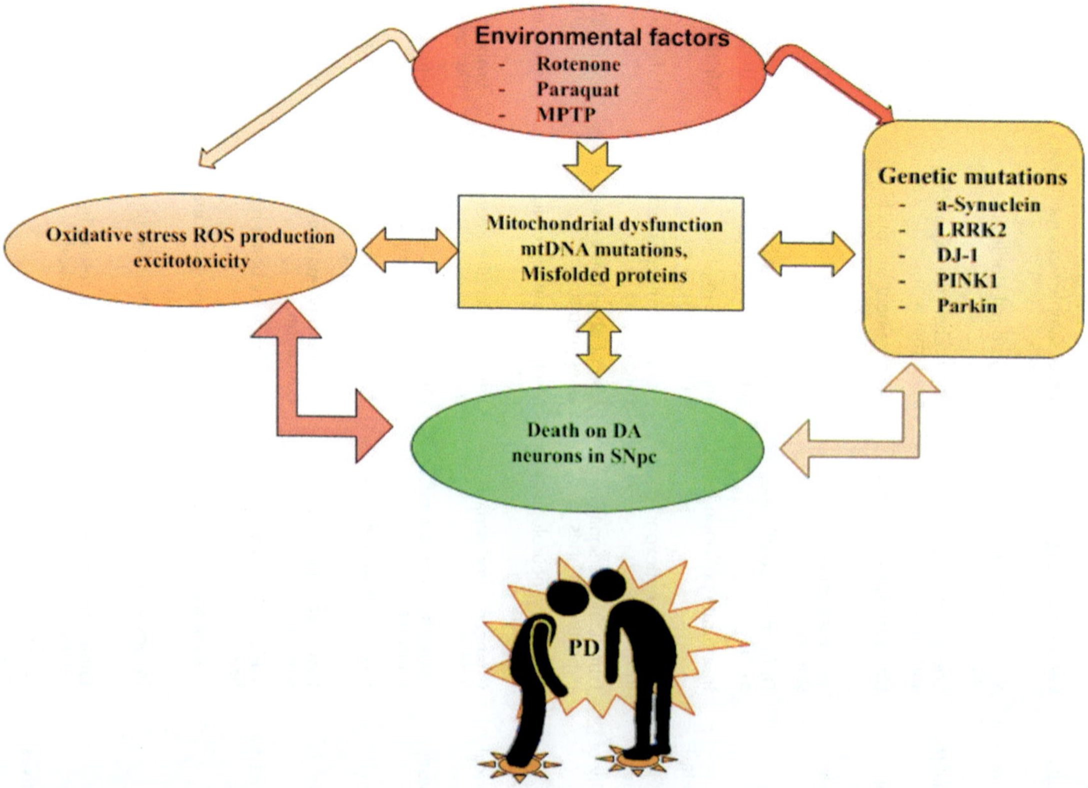

FIGURE 13.4 Potential etiological factors of Parkinson's disease. Current evidence suggests that environmental factors and genetic risk factors provoking oxidative stress, excitotoxicity, and mitochondrial dysfunction in the brain can lead to the degeneration of the midbrain dopaminergic system, resulting in Parkinson's disease.

environmental hazards, thus preventing and treating such cases when P-gal polymorphisms in MT, GST, and DRD2 can be identified in a person. As for the identified high-risk people, several measures should be implemented to prevent HM harm, including changing their habits and diets and frequently checking levels of heavy metals. The addition of other nutraceuticals that have been known to have detoxifying elements might also be helpful. For instance, curcumin and green tea extracts have been equally beneficial in boosting detoxification and dealing with oxidative stress caused by heavy metals (Bjørklund et al., 2022). Some personalized actions could include advising a person on diets that supply specific sources laden with antioxidants or supplements based on one's genotype. This chapter also seeks to establish that public health can help initiate and develop a mitigation regiment to reduce the effects of heavy metals. For instance, those communities exposed to a high risk of heavy metal contamination might require informative sheets stressing how to minimize contact with these hazardous compounds in everyday life, especially if they have an inherited genetic predisposition. Screening programs that include the subject's external environment and genetic factors may offer further information for targeted solutions. Further, it is also recommended that research investigation should advance the effects of other environmental toxicities, except for heavy metals such as pesticides and airborne pollutants, on neurodegenerative disease predisposing factors (Table 13.2).

13.6 Challenges in assessing heavy metal exposure and establishing causality in Parkinson's disease

This is because it is impossible to directly connect heavy metal exposure with PD immediately. Multiple obstacles impede the straightforward interpretation of epidemiological and experimental data. These difficulties are conceptual and methodological, owing to the measurement of heavy metals in biological samples, differentiation between multiple sources of exposure, and intrinsic problems in epidemiological studies. It is only possible, therefore, to gain insight into these difficulties to better address the challenges in designing more substantial analyses and implementing effective public health policies.

13.6.1 Limitations in accurately measuring heavy metal levels in biological samples

Another crucial issue in environmental toxicology is determining the concentration of dangerous metal ions in various biological samples. Blood, urine, hair, and, to an extent,

TABLE 13.2 Protective mechanisms and potential therapeutic interventions in Parkinson's disease.

Aspect	Description	Impact on disease progression	Examples	Future directions	References
Antioxidant defense	Endogenous antioxidants neutralize reactive oxygen species (ROS).	Reduces oxidative stress	Glutathione, superoxide dismutase	Enhancing antioxidant production	Chatterjee, Kumari, Rath, Priyadarshanee, and Das (2020)
Metal chelation	Cellular mechanisms that bind and eliminate heavy metals.	Mitigates metal toxicity	Metallothioneins	Developing targeted chelation agents	Tripathi, Kharkwal, Mishra, and Singh (2024)
Autophagy	Cellular process for degrading and recycling damaged proteins.	Promotes cellular health	ATGs	Enhancing autophagy in neurons	Sun et al. (2022)
Heat shock proteins	Proteins that help refold misfolded proteins and prevent aggregation.	Protects against protein aggregation	HSP70, HSP90	Investigating HSP induction therapies	Bjørklund, Mutter, and Aaseth (2017)
Neuroinflammation regulation	Mechanisms that control microglial activation and cytokine production.	Reduces neuroinflammation	IL-10, TGF-β	Targeting inflammatory pathways	Pfaff, Beltz, King, and Ercal (2020)
Mitochondrial protection	Mechanisms that maintain mitochondrial function and integrity.	Preserves energy production	Mitochondrial biogenesis factors	Developing mitochondrial protectants	Pljesa-Ercegovac et al. (2018)
Calcium homeostasis	Regulation of intracellular calcium levels to prevent toxicity.	Prevents excitotoxicity	Calcium-binding proteins	Novel calcium modulators	Sales et al. (2019)
Neurotrophic factors	Growth factors that support neuron survival and function.	Enhances neuronal resilience	NGF, BDNF	Exploring neurotrophic factor therapies	Carocci, Catalano, Sinicropi, and Genchi (2018)
Gene therapy	Techniques to modify gene expression to enhance protective mechanisms.	Provides long-lasting benefits	Viral vectors for gene delivery	Advancing gene editing technologies	Chawla, Rani, and Mishra (2021)
Dietary interventions	Nutritional strategies to support cellular defense mechanisms.	Supports overall brain health	Antioxidant-rich diets	Clinical trials on specific diets	Abolli et al. (2024)
Pharmacological agents	Drugs aimed at enhancing protective pathways or reducing toxicity.	May slow disease progression	*N*-Acetylcysteine, resveratrol	Screening for new drug candidates	Chatterjee et al. (2020)
Stem cell therapy	Use of stem cells to replace damaged neurons and support recovery.	Potential for regeneration	iPSCs	Investigating safety and efficacy	Tripathi et al. (2024)
Herbal remedies	Natural compounds with neuroprotective properties.	May offer additional protective effects	Curcumin, ginkgo biloba	Clinical validation of herbal efficacy	Sun et al. (2022)
Chelation therapy	Medical treatment to remove heavy metals from the body.	Reduces heavy metal burden	EDTA, DMSA	Evaluating long-term safety and efficacy	Bjørklund et al. (2017)
Lifestyle modifications	Lifestyle changes that promote overall brain health.	Enhances resilience to disease	Exercise, stress reduction	Community health initiatives	Pfaff et al. (2020)
Clinical trials	Research studies to test the efficacy of new interventions.	Provides evidence for therapeutic efficacy	Phase I, II, and III trials	Expanded recruitment for diverse populations	Pljesa-Ercegovac et al. (2018)
Biomarker development	Identifying biomarkers for early detection and monitoring therapy.	Improves diagnosis and treatment tracking	Alpha-synuclein levels	Establishing robust biomarker panels	Sales et al. (2019)
Interdisciplinary research	Collaboration across fields to enhance understanding of PD mechanisms.	Broader insights into disease pathways	Neurology, toxicology, and immunology	Creating research consortia	Carocci et al. (2018)
Public health initiatives	Programs aimed at raising awareness and prevention of heavy metal exposure.	Reduces environmental risk factors	Community screenings	Policy advocacy for ecological health	Chawla et al. (2021)
Future research directions	Identification of new areas for exploration in PD therapeutics.	Expands understanding of disease mechanisms	Novel drug targets, novel interventions	Focus on personalized medicine strategies	Abolli et al. (2024)

ATGs, Autophagy-related genes; *DMSA*, dimercaptosuccinic; *EDTA*, ethylenediaminetetraacetic; *iPSCs*, induced pluripotent stem cells; *PD*, Parkinson's disease.

cerebrospinal fluid are used as a surrogate for metals, each of which has drawbacks (Mrštná et al., 2023). For example, blood tests detecting levels of certain metals narrow down to recent exposure, but they do not show the total amount in the body. Some chemicals, such as lead (Pb), have a long biological half-life in bone, and bone biopsies are not easily and routinely accessible for investigation. This is mainly because urinary excretion depends on hydration-restricted and renal functions, making it difficult to assess the results. Besides, there are variations in the collection, handling, and storing of the samples; this can pose some problems. The most advanced methods, such as inductively coupled plasma mass spectrometry, can still be prone to matrix interferences or contamination and, therefore, can give erroneous results (Balaram, 2021). There have been improvements in analytical detection limits, but with these improvements, issues with the standard and validation of the methods by different laboratories arise. Therefore variations in assay sensitivity and specificity may fail to perform comparable exposure assessments due to differences in variability across the studies. These analytical challenges explain why researchers now need to establish other, more accurate biomarkers that can be used to determine both RTZ and AZT exposure to heavy metals in the human body.

13.6.2 Difficulties in disentangling the effects of multiple environmental exposures

Exposures to such chemicals in real-life situations seldom involve singular toxins. Thus people are subjected to multiple toxic metals, other pollutants, and chemical compounds, including pesticides, solvents, and dust particulates. This mixed-exposure setting draws a lot of confusion regarding the revelation of the effects of any one factor. Some heavy metals may enhance, reduce, or have an additive effect on another, which makes the biological effect even more complex. In other cases, the realization of more than one toxic effect at the same time may lead to their additive effect, where the product of the degree of toxicity of the two goods during their simultaneous consumption is greater than the toxicity coefficients of each product separately (Sigurnjak Bureš et al., 2021). Such interactions can conceal or mimic the biophysical signals associated with a single heavy metal, which may complicate the ability to establish neurodegenerative changes resulting from a particular exposure. Environmental exposures may also cocontribute to toxicity, that is, they may lead to the exact mechanisms of toxicity, such as oxidative stress or neuroinflammation, that make it difficult to determine the role each toxin plays in the course of the disease. Some inadequacies observed in epidemiological research include confounding factors that may arise due to the interaction of the abovementioned multiple exposures. However, identifying each component in the concocted environment is a process that requires a lot of time and resources, and it is not feasible when conducting population studies. This is made worse by the fact that another variable that also interacts with the total toxic burden includes lifestyle variables, including diet, smoking, and occupation.

13.6.3 Epidemiological challenges in establishing definitive causal links

Epidemiological observation plays a crucial role in investigating what has been described as the intervention exposure and disease risk front, but the studies all have vulnerabilities that impede causality assessment. A limitation that cannot be ignored is rate reversals, where exposure biomarkers are changed by the condition or a treatment implemented for the disease. For instance, PD patients may switch their diet or the type of medication they use; these factors can influence levels of such metals in biological tissues after diagnosis. This is the fact that the latency periods make matters worse. PD is a progressive disorder and commonly presents many years after exposure to the toxicant. Such a long latency period also exacerbates the reconstruction of the history of exposure and to distinguish when and how the total dose of the exposure affected the disease development. Further, the symptoms of PD are diverse and variable, and PD has multiple potential risk factors, which include both heredity and environmental factors, such as diet and exposure to toxins (Belvisi et al., 2020). It is hard to segregate the impact of exposure to heavy metals from other influential factors in developing the disease. Another epidemiological risk factor is selection bias. Questions about exposure and the corresponding evaluations in research are conducted using retrospective data or case-control investigations, which may indicate recall bias or errors. Using future prospective cohort studies that is more appropriate for creating temporal association is expensive and time-consuming. Even when prospective designs are used, the low prevalence of PD compared to the normal population makes it extremely difficult to secure sufficient statistical significance. Besides, the original differences within the individuals making up the population act as a contributory factor as well. Some genes that have significant implications for heavy metal toxicity include the genes involved in the transport of metal ions, metabolism of toxicants, and antioxidant enzymes. For this reason, plain epidemiological research may overstate or understate the effect of heavy metal exposure in terms of the risk of developing PD due to the violated genetic variation (Table 13.1).

13.7 Protective mechanisms and potential therapeutic interventions in Parkinson's disease

PD is a neurodegenerative and chronic sickness that is associated with the loss of dopaminergic neurons in the substantia nigra and complications in signal transmission in the brain's motor circuits. There is growing awareness that environmental factors may help define the etiology of PD, and heavy metals are likely to be involved. However, the body itself has an intrinsic cellular protective mechanism against mobile heavy metals, and there are strategies for neuroprotective ways. These projects, developments, and therapies are described in detail in this article, and the article also looks at potential trends for the future of this remedy.

13.7.1 Endogenous cellular defense mechanisms

- **MTs and metal-binding proteins:** MTs refer to low molecular weight metal-binding proteins containing thermally stable cysteine, which chelate nonessential toxic metals, including lead (Pb), cadmium (Cd), and manganese (Mn), thus inhibiting their toxicity in terms of production of ROS (Chatterjee et al., 2020). These proteins play an intracellular role in chelating metal ions in the cell and preventing oxidative stress. It is often overexpressed during the uptake of metal and thus acts to defend the organism. Likewise, other metal-chelating proteins like Ferritin also play a part in the sequestration of metals and the elimination of dangerous redox reactions.
- **Antioxidant defense systems:** Nuclear factor erythroid 2–related factor 2 or Nrf2 is the master regulator of the antioxidant response in cells. Nrf-2 is maintained mainly in the cytoplasm, but under oxidative stress, such as the effects caused by heavy metals, it translocates to the nucleus (Tripathi et al., 2024). Nrf2 then enters the nucleus, where it interacts with antioxidant response elements (AREs) and activates the expression of a group of genes, including GCL, SOD, and catalase. These enzymes, in combination, counteract ROS and balance the redox state.
- **Mitochondrial quality control and autophagy:** Mitochondria are heavily affected by the toxicity of heavy metals. In response, neurons initiate mitochondrial quality control processes such as mitophagy, an autophagic pathway responsible for the selective degradation of dysfunctional mitochondria (Sun et al., 2022). Through proteins, including PINK1 and Parkin, this process maintains the cellular energy ratio and prevents aggresomes with damaged organelles that would spread oxidative stress from forming. Further, it was observed that Hsp is involved during mental stress, and they help fold proper protein or degrade or refold the improper one.

13.7.2 Investigating potential therapeutic strategies for chelation and neuroprotection

- **Chelation therapy:** This therapy minimizes the availability of heavy metals in the system so that they pose no neurological hazards to the infant. Organic chelators previously used to treat heavy metal poisoning are ethylenediaminetetraacetic and dimercaptosuccinic. Of course, their action is universal, and the action on metal preparations also leads to the exhaustion of traces of these elements and undesirable side effects (Bjørklund et al., 2017). However, the new trend of research is based on developing more selective chelators that work on neurotoxic metals and avoid essential metals. For instance, chelators in nanoparticles increase BBB permeability and release the bound metals controlled, therebye exerting minimal harm on the neuronal systems.
- **Neuroprotective interventions:** Apart from chelation, neuroprotective strategies are designed to strengthen confessional organizations. Other supplements, including NAC and Coenzyme Q10, have been studied for their potential to aid intracellular glutathione content restoration and maintain mitochondrial stability (Pfaff et al., 2020). On the one hand, NAC can be seen as a precursor of GSH, but, on the other hand, it has direct antioxidant properties to ROS. Contributions on experimentally induced compounds are also given; these compounds activate Nrf2, increasing the transcription of various antioxidant and detoxification enzymes and offering multiple protections against metals. Furthermore, regulating inflammation is another treatment approach that could be employed. Studies reveal that heavy metals act as triggers for stimulating the production of neurotoxic cytokines from microglia. Antimicroglial, antiinflammatory, and drugs targeting NLRP3 inflammasome have been considered to reduce neuroinflammation and defend dopaminergic neurons.
- **Emerging gene therapy and personalized approaches:** Due to the genetic variability in susceptibility to the toxicity of heavy metals, gene therapy can, therefore, be considered as part of precision medicine. For instance, approaches that can increase MT levels or stimulate the activity of certain antioxidant enzymes in subjects prone to neuronal damage would provide long-term neuroprotection. However, incorporating genetic testing with exposure data increases the likelihood of early detection of the most vulnerable people who may become victims of such disasters. An example that best fits these concepts is that they apply to e-waste recycling workers in Southeast Asia. They are particularly exposed to a mixture of several heavy metals such as lead, cadmium, and mercury, which are received from dismantling various electronics products. These workers have lately been followed in detailed biomonitoring; the concentration of metals in blood and hair was evaluated, and neuropsychological performance and neuroimaging of dopaminergic activity were assessed. Indeed, workers with high-risk polymorphisms in these genes, including GSTM1 and DMT1, had higher blood metal concentrations and oxidative stress markers. These participants also manifested mild motor impairments and decreased

DAT binding potential in the striatum on PET scans (Pljesa-Ercegovac et al., 2018). As a result, another study in this group was carried out in the form of an intervention trial. Learners were given an armor of a new compound. This chelator gets to the brain and can bind with toxic metals in the body and the antioxidant NAC to produce glutathione. The changes in blood metal concentrations, antioxidant status, and motor function of treated persons were less than the control group during the 6 months. This example demonstrates the effectiveness of chelation and the importance of treatment modifications in genetically vulnerable groups.

13.7.3 Discussion of future research directions in this area

Several prospects of research activity can be highlighted that can help develop more effective treatments and elaborate new ways to protect the neuronal tissues against the toxic effects of heavy metals in the context of PD.

- **Targeted drug development**: New studies should aim to identify new chelating compounds that are superior in efficacy and toxicity profiles to the ones used in the current targeted drug development studies. Pharmacokinetics and pharmacodynamics analysis of the new compounds and their BBB permeability will play a vital role in developing effective remedies for neurodegenerative diseases related to heavy metals (Sales et al., 2019).
- **Combination therapies:** It is noticeable that further improvements might be achieved with combination therapies, including chelation agents and neuroprotective compounds. For instance, deferiprone and antioxidants could effectively reduce oxidative stress while eliminating toxic metals (Carocci et al., 2018). The trials on the safety and efficacy of such combinations will then be necessary.
- **Pharmacogenomic predictive markers:** First, finding an individual's genetic makeup that determines their heterogeneous susceptibility to toxicity and therapeutic response to certain heavy metals will form the basis for pharmacogenomic medicine (Chawla et al., 2021). The implication is that if genetic variation in heavy metals were known, interventions that target the patient could be made based on the patient's genotypic variation and thus get better results.
- **Prospective research:** Longitudinal research designs that focus on assessing the impact of specific metals and gene polymorphism on neurodegenerative disorders, the development of which over time will be significant in establishing cause-and-effect relationships as well as influencing and being influenced by the environment (Abolli et al., 2024). Such studies can offer a further understanding of prospective time frames of such susceptibility, which is valuable knowledge to prevent experimentation at such a time. That is why initiatives of the corresponding type in the field of public health should also become one of the research directions. Awareness through education to prevent contact with these environmental toxins and avoid behaviors that come with contact with such are crucial in managing this crisis on neurological health.

13.8 Public health implications and prevention strategies in Parkinson's disease

Heavy metals have been put into controversy as some of the environmental toxicants that play a part in the development of PD. Overall, it can be deduced that long-term toxic metals, lead, mercury, cadmium, and manganese, play a pivotal role in the pathogenesis of PD. This reality creates a lot of burden for public health, so prevention interventions must address priorities that can target high-risk people and settings. In this instance, we discuss how excessive exposure to heavy metals can be expressed, how health programms for mitigating the risks can be developed, and the significance of environmental management and protection. A new example regarding the communities adjacent to e-waste recycling centers embodies this.

13.8.1 Identifying vulnerable populations and high-risk environments

The first of the measures in preventing PD related to the presence of heavy metals is to identify the factors that may make people susceptible to it. The interaction of various factors within the biological and environmental factors mainly determines predisposing factors and disease risk factors. For example, heavy metals have severe effects on children and older people because their physiological systems are still developing, and the other is in continuous deterioration. These are polymorphisms in metal transporters and detoxifying enzymes, though individual genetically inheritable risk factors are also worth considering (Marras, Canning, & Goldman, 2019). Furthermore, socioeconomically average off-developed communities stay close to industrial facilities and means of production where environmental protection might not be well enhanced. Hence, residents are at a higher risk of suffering from toxic metals. High-risk areas usually entail fluctuating industrial emissions, inadequate waste handling and management, crowded human settlements, and much traffic. Other hotspots include regions close to mines, battery manufacturing factories, and electrical and electronic equipment recycling centers. For instance, residents of communities living close to e-waste recycling centers take time to be exposed to several metals, such as lead, cadmium, and mercury, which are released during the disintegration and recycling of electronic components.

This is accompanied by low remediation rates and inadequate policies enforcing a safer area, which puts these individuals at risk.

13.8.2 Developing public health interventions to reduce heavy metal exposure

Prevention of PD due to heavy metal exposure requires a multilevel public health intervention. Interventions should be designed at the community and individual levels to reduce exposure and prevent associated health risks.

- **Biomonitoring and early detection:** Organizing the common biomonitoring of high-risk groups in the communities is recommended to identify admittance to heavy metals. For example, claims management is achieved by screening through the blood, hair, or nails to determine which patients have increased levels of metals in their system. Such monitoring programs not only help in the early identification of cases but also help in the epidemiologic trends of exposure (Ben-Shlomo et al., 2024). For example, people living close to e-waste recycling centers have been recruited in biomonitoring campaigns that assess blood lead and cadmium levels, which, in some cases, are associated with minor changes in neurobehavior.
- **Community education and behavioral interventions:** The education component that needs to be implemented by the public health agencies is information campaigns about the risks associated with heavy metals. These campaigns can offer advice on how to minimize dietary habits, including proper hand washing and consuming foods that may help detoxify the internal system from potentially hazardous materials in the dust (Anyanwu, Ezejiofor, Igweze, & Orisakwe, 2018). More educational programs for high-exposure workers engaged in e-waste recycling can also be conducted to increase awareness of the usage of PPE and safety precautions.
- **Regulatory and policy measures:** The topic of regulating the discharge of heavy metals into the industrial environment. Governments should ensure that industries that use toxic metals adhere to strict emission standards and that regular checks on the effects of these industries on the environment are conducted. Some measures may include updating perishable exposure limit guidelines, the requirement of lower emission technologies, and encouraging the application of superior practices in industrial processes (Mitra et al., 2022). In regions with high traffic and industrialization of the community's automobile density, authorities may opt to implement areas with low emissions thresholds.
- **Workplace interventions:** Public health interventions should comprise integrated workplace safety programs for workplaces, especially heavy metal-producing plants. These should involve engineering controls (e.g., improved ventilation systems), administrative controls (e.g., frequent safety training and exposure monitoring), and personal protective practices (Gorini and Tonacci, 2024). Employers would be advantaged by integrating health surveillance systems that continuously monitor workers' mental burdens and neurobehavioral function so that interventions can be made on a timely basis when exposure is higher than safe levels.

13.8.3 Promoting environmental remediation and pollution control measures

Control activities, environmental remediation efforts, and pollution control are the cornerstones of long-term prevention. Remediation is designed to minimize the heavy metal burden of contaminated environments and maintain the low risk of long-term exposure.

- Soil and water remediation: Methods like phytoremediation, where plants remove or immobilize heavy metals from the soil, are a good solution for polluted urban and industrial sites. Chemical remediation methods like precipitation or adsorption of heavy metals from water sources can also significantly decrease the bioavailability of metals (Kumar, Koul, & Sharma, 2022). Local governments in risk areas must invest in high-tech water treatment plants with heavy metal extraction capabilities to provide clean drinking water to the populace.
- Technological innovations in pollution control: Industrial process control technology advancements can also contribute significantly to pollution mitigation. For example, applying sophisticated scrubbers and filtering devices in manufacturing plants can capture metal particulates before they are released into the atmosphere. In e-waste recycling, the availability of central processing facilities with state-of-the-art emissions control equipment minimizes the environmental dispersal of poisonous metals (Andeobu, Wibowo, & Grandhi, 2021). Supplementing the tenets of a circular economy, whereby waste products are recovered and recycled in a systematic clean process, also minimizes the overall toxic load.
- **E-waste recycling communities:** An ongoing project in one Southeast Asian city demonstrates the in-the-field effects of these remediation interventions. In this city, residents near the informal e-waste recycling shops had higher levels of lead and cadmium in their blood. In reaction, the local government launched a remediation program in collaboration with international environmental agencies. This program entailed building regulated, centralized e-waste processing plants with sophisticated pollution abatement technologies as well as e-waste pollution remediation by phytoremediation and soil replacement (Andeobu et al., 2021). At the same time, community health education programs were launched, and ongoing biomonitoring was undertaken. Two years later, follow-up surveys showed a substantial decrease in

residents' heavy metal levels and early neurobehavioral markers. This multipronged strategy shows that concerted prevention and remediation efforts can significantly reduce environmental exposure and potentially reduce the incidence of PD.

13.8.4 Future directions

Future research needs to address the development of sensitive, standardised biomarkers that quantitatively react to short-term exposure and cumulative heavy metal burden. An interdisciplinary investigation involving environmental science, toxicology, genomics, and neuroepidemiology will be critical in further elucidating exposure-response associations (Schymanski et al., 2019). Furthermore, applying advanced data analysis and geospatial mapping can help identify future high-risk sites for preemptive public health interventions. More coordination among government agencies, industry, and academia is needed to facilitate the translation of research findings into effective policy interventions. Cooperation should also seek to establish community backing so that remediation efforts are culturally compatible and sustainable in the long term. The public health importance of heavy metal exposure in PD cannot be overstated. At-risk populations, children, older people, and residents living in the vicinity of industrial or informal recycling plants subject themselves repeatedly exposed to high levels of risk. Successful prevention strategies, including multimodal, large-scale biomonitoring programs, public education, stringent regulatory policies, and healthy workplace intervention, each have a contributing measure. Environmental clean-up efforts and pollution controls through treatments of water and soil, and newer industrial filtration processes are critical to lowering levels of heavy metals in high-exposure environments. The recent case of Southeast Asian e-waste recycling populations demonstrates the efficacy and practicability of multidisciplinary public health interventions. Through the integration of remediation, regulation, and community-based education, heavy metal exposure can be managed, and the load of neurodegenerative diseases such as PD can be thereby reduced. In the future, biomarker technology innovation and multidisciplinary studies will further clarify these intricate interactions (Gorini & Tonacci, 2024). Data-driven, individualized public health interventions based on genetic and environmental information can improve prevention approaches. Through such partnerships, we can strive to protect public health, especially among vulnerable populations, and reduce the long-term neurological consequences of heavy metal exposure.

13.9 Conclusion

PD continues to be a multifactorial neurodegenerative disorder involving genetic susceptibility and environmental exposures, including heavy metals. Key findings emphasize the impact of heavy metals in aggravating neurodegeneration and propose a range of protective mechanisms and treatments. Nevertheless, there are considerable outstanding issues, mainly the exact mechanisms through which heavy metals participate in PD pathogenesis, interactions between genetic susceptibility and environmental exposure, and the long-term consequences of exposure across different populations. To close these gaps, we need multidisciplinary research approaches that combine the strengths of public health, neurology, genetics, and toxicology. Multidisciplinary approaches can better illuminate PD by examining how the various risk factors interact across multiple levels of biology, from population health outcomes down to molecular pathways. This integrated approach will be needed to formulate comprehensive disease progression models and identify points of potential intervention. Furthermore, translational research must be conducted to bridge the gap between laboratory research and clinical application. While preclinical research can delineate therapeutic targets, successful translation to clinical practice requires robust testing in well-executed clinical trials. These involve investigating novel drug candidates, validating the effectiveness of chelation therapies, and investigating lifestyle interventions that reverse the risk of PD. Furthermore, the use of genetic screening to predict susceptibility to environmental toxins can inform personalized prevention. Priorities should also include long-term epidemiological studies that investigate heavy metal exposure's lifetime cumulative impact and identify critical susceptibility windows. Such studies can provide valuable information on intervention timing and inform the adaptation of public health programs to reduce exposure in susceptible populations. In brief, building our understanding of PD is an interdisciplinary challenge, and the most important thing we can do to have an impact is with translational research. By prioritizing these fields of research, we can optimize our prevention measures, early diagnosis, and treatments, and improve the lives of patients with this complex and challenging disease.

References

Aarsland, D., Batzu, L., Halliday, G. M., Geurtsen, G. J., Ballard, C., Ray Chaudhuri, K., & Weintraub, D. (2021). Parkinson's disease-associated cognitive impairment. *Nature Reviews Disease Primers, 7*(1), 47. Retrieved from https://doi.org/10.1038/s41572-021-00280-3.

Abolli, S., Dehghani, S., Atlasi, R., Maleki, Z., Yunesian, M., Tabatabaei-Malazy, O., & Soleimani, Z. (2024). Arsenic and type 2 diabetes: Revealing the environmental exposure relationship through effective factors-a systematic review. *Results in Engineering, 22*, 102054. Retrieved from https://doi.org/10.1016/j.rineng.2024.102054.

Agnihotri, A., & Aruoma, O. I. (2020). Alzheimer's disease and Parkinson's disease: A nutritional toxicology perspective of the impact of oxidative stress, mitochondrial dysfunction, nutrigenomics and environmental chemicals. *Journal of the American College of Nutrition, 39*(1), 16–27. Retrieved from https://doi.org/10.1080/07315724.2019.1683379.

Andeobu, L., Wibowo, S., & Grandhi, S. (2021). A systematic review of e-waste generation and environmental management of Asia Pacific countries. *International Journal of Environmental Research and Public Health, 18*(17), 9051. Retrieved from https://doi.org/10.3390/ijerph18179051.

Anyanwu, B. O., Ezejiofor, A. N., Igweze, Z. N., & Orisakwe, O. E. (2018). Heavy metal mixture exposure and effects in developing nations: An update. *Toxics, 6*(4), 65. Retrieved from https://doi.org/10.3390/toxics6040065.

Balaram, V. (2021). Strategies to overcome interferences in elemental and isotopic geochemical analysis by quadrupole inductively coupled plasma mass spectrometry: A critical evaluation of the recent developments. *Rapid Communications in Mass Spectrometry, 35*(10), e9065. Retrieved from https://doi.org/10.1002/rcm.9065.

Baş, H., Kalender, S., Karaboduk, H., & Apaydın, F. (2015). The effects on antioxidant enzyme systems in rat brain tissues of lead nitrate and mercury chloride. *Gazi University Journal of Science, 28*(2), 169–174.

Battaglini, M., Marino, A., Montorsi, M., Carmignani, A., Ceccarelli, M. C., & Ciofani, G. (2024). Nanomaterials as microglia modulators in the treatment of central nervous system disorders. *Advanced Healthcare Materials, 13*(12), 2304180. Retrieved from https://doi.org/10.1002/adhm.202304180.

Belvisi, D., Pellicciari, R., Fabbrini, A., Costanzo, M., Pietracupa, S., De Lucia, M., & Defazio, G. (2020). Risk factors of Parkinson disease: Simultaneous assessment, interactions, and etiologic subtypes. *Neurology, 95*(18), e2500–e2508. Retrieved from https://doi.org/10.1212/WNL.0000000000010813.

Ben-Shlomo, Y., Darweesh, S., Llibre-Guerra, J., Marras, C., San Luciano, M., & Tanner, C. (2024). The epidemiology of Parkinson's disease. *The Lancet, 403*(10423), 283–292. Retrieved from https://doi.org/10.1016/S0140-6736(23)01419-8.

Bjørklund, G., Mutter, J., & Aaseth, J. (2017). Metal chelators and neurotoxicity: Lead, mercury, and arsenic. *Archives of Toxicology, 91*(12), 3787–3797. Retrieved from https://doi.org/10.1007/s00204-017-2100-0.

Bjørklund, G., Rahaman, M. S., Shanaida, M., Lysiuk, R., Oliynyk, P., Lenchyk, L., & Peana, M. (2022). Natural dietary compounds in the treatment of arsenic toxicity. *Molecules (Basel, Switzerland), 27*(15), 4871. Retrieved from https://doi.org/10.3390/molecules27154871.

Bjorklund, G., Stejskal, V., Urbina, M. A., Dadar, M., Chirumbolo, S., & Mutter, J. (2018). Metals and Parkinson's disease: Mechanisms and biochemical processes. *Current Medicinal Chemistry, 25*(19), 2198–2214. Retrieved from https://doi.org/10.2174/0929867325666171129124616.

Branca, J. J., Fiorillo, C., Carrino, D., Paternostro, F., Taddei, N., Gulisano, M., & Becatti, M. (2020). Cadmium-induced oxidative stress: Focus on the central nervous system. *Antioxidants, 9*(6), 492. Retrieved from https://doi.org/10.3390/antiox9060492.

Broberg, K., & Pawlas, N. (2022). Gene–environment interactions and metal toxicity. In *Handbook on the toxicology of metals*. Academic Press, 349–368. Retrieved from https://doi.org/10.1016/B978-0-12-823292-7.00010-3.

Cariccio, V. L., Samà, A., Bramanti, P., & Mazzon, E. (2019). Mercury involvement in neuronal damage and in neurodegenerative diseases. *Biological Trace Element Research, 187*(2), 341–356. Retrieved from https://doi.org/10.1007/s12011-018-1380-4.

Carocci, A., Catalano, A., Sinicropi, M. S., & Genchi, G. (2018). Oxidative stress and neurodegeneration: The involvement of iron. *Biometals: an International Journal on the Role of Metal Ions in Biology, Biochemistry, and Medicine, 31*, 715–735. Retrieved from https://doi.org/10.1007/s10534-018-0126-2.

Chamoli, A., & Karn, S. K. (2024). The effects of mercury exposure on neurological and cognitive dysfunction in human: A review. In *Mercury toxicity mitigation: Sustainable nexus approach. Earth and Environmental Sciences Library*. Cham: Springer, 117–135. Retrieved from https://doi.org/10.1007/978-3-031-48817-7_5.

Chatterjee, S., Kumari, S., Rath, S., Priyadarshanee, M., & Das, S. (2020). Diversity, structure and regulation of microbial metallothionein: Metal resistance and possible applications in sequestration of toxic metals. *Metallomics, 12*(11), 1637–1655. Retrieved from https://doi.org/10.1039/d0mt00140f.

Chawla, R., Rani, V., & Mishra, M. (2021). Integrated role of nanotechnology and pharmacogenetics in diagnosis and treatment of. *Pharmacogenetics, 11*.

Chi, H., Chang, H. Y., & Sang, T. K. (2018). Neuronal cell death mechanisms in major neurodegenerative diseases. *International Journal of Molecular Sciences, 19*(10), 3082. Retrieved from https://doi.org/10.3390/ijms19103082.

Delcambre, S., Nonnenmacher, Y., & Hiller, K. (2016). Dopamine metabolism and reactive oxygen species production. In *Mitochondrial mechanisms of degeneration and repair in Parkinson's disease*. Cham: Springer, 25–47. Retrieved from https://doi.org/10.1007/978-3-319-42139-1_2.

Doroszkiewicz, J., Farhan, J. A., Mroczko, J., Winkel, I., Perkowski, M., & Mroczko, B. (2023). Common and trace metals in Alzheimer's and Parkinson's diseases. *International Journal of Molecular Sciences, 24*(21), 15721. Retrieved from https://doi.org/10.3390/ijms242115721.

Du, D., Su, Y., Shang, Q., Chen, C., Tang, W., Zhang, L., & Liu, W. (2022). Biomimetic synthesis of L-DOPA inspired by tyrosine hydroxylase. *Journal of Inorganic Biochemistry, 234*, 111878. Retrieved from https://doi.org/10.1016/j.jinorgbio.2022.111878.

Fu, Z., & Xi, S. (2020). The effects of heavy metals on human metabolism. *Toxicology Mechanisms and Methods, 30*(3), 167–176. Retrieved from https://doi.org/10.1080/15376516.2019.1701594.

Galvan, A., Devergnas, A., & Wichmann, T. (2015). Alterations in neuronal activity in basal ganglia-thalamocortical circuits in the parkinsonian state. *Frontiers in Neuroanatomy, 9*, 5. Retrieved from https://doi.org/10.3389/fnana.2015.00005.

Gandhi, J., Antonelli, A. C., Afridi, A., Vatsia, S., Joshi, G., Romanov, V., & Khan, S. A. (2019). Protein misfolding and aggregation in neurodegenerative diseases: A review of pathogeneses, novel detection strategies, and potential therapeutics. *Reviews in the Neurosciences, 30*(4), 339–358. Retrieved from https://doi.org/10.1515/revneuro-2016-0035.

Gianì, F., Masto, R., Trovato, M. A., Malandrino, P., Russo, M., Pellegriti, G., & Vigneri, R. (2021). Heavy metals in the environment and thyroid cancer. *Cancers, 13*(16), 4052. Retrieved from https://doi.org/10.3390/cancers13164052.

Gorini, F., & Tonacci, A. (2024). Metal toxicity and dementia including frontotemporal dementia: current state of knowledge. *Antioxidants, 13*(8), 938. Retrieved from https://doi.org/10.3390/antiox13080938.

Hassan, W., Noreen, H., Rehman, S., Kamal, M. A., & da Rocha, J. B. (2022). Association of oxidative stress with neurological disorders. *Current Neuropharmacology, 20*(6), 1046–1072. Retrieved from https://doi.org/10.2174/1570159X19666211111141246.

Jagadeesan, A. J., Murugesan, R., Devi, S. V., Meera, M., Madhumala, G., Padmaja, M. V., & Pathak, S. (2017). Current trends in etiology, prognosis and therapeutic aspects of Parkinson's disease: A review.

Acta Bio Medica: Atenei Parmensis, 88(3), 249. Retrieved from https://doi.org/10.23750/abm.v%vi%i.6063.

Jan, A. T., Azam, M., Siddiqui, K., Ali, A., Choi, I., & Haq, Q. M. R. (2015). Heavy metals and human health: Mechanistic insight into toxicity and counter defense system of antioxidants. *International Journal of Molecular Sciences, 16*(12), 29592–29630. Retrieved from https://doi.org/10.3390/ijms161226183.

Jaroudi, W., Garami, J., Garrido, S., Hornberger, M., Keri, S., & Moustafa, A. A. (2017). Factors underlying cognitive decline in old age and Alzheimer's disease: The role of the hippocampus. *Reviews in the Neurosciences, 28*(7), 705–714. Retrieved from https://doi.org/10.1515/revneuro-2016-0086.

Kausar, S., Wang, F., & Cui, H. (2018). The role of mitochondria in reactive oxygen species generation and its implications for neurodegenerative diseases. *Cells, 7*(12), 274. Retrieved from https://doi.org/10.3390/cells7120274.

Kaushal, G. P., Chandrashekar, K., & Juncos, L. A. (2019). Molecular interactions between reactive oxygen species and autophagy in kidney disease. *International Journal of Molecular Sciences, 20*(15), 3791. Retrieved from https://doi.org/10.3390/ijms20153791.

Koyama, H., Kamogashira, T., & Yamasoba, T. (2024). Heavy metal exposure: Molecular pathways, clinical implications, and protective strategies. *Antioxidants, 13*(1), 76. Retrieved from https://doi.org/10.3390/antiox13010076.

Kumar, P., Koul, B., & Sharma, M. (2022). Phytoremediation of heavy metals. In *Heavy metals in plants*. CRC Press, 369–387.

Kumar, S., Goyal, L., & Singh, S. (2022). Tremor and rigidity in patients with Parkinson's disease: Emphasis on epidemiology, pathophysiology and contributing factors. *CNS & Neurological Disorders-Drug Targets-CNS & Neurological Disorders, 21*(7), 596–609. Retrieved from https://doi.org/10.2174/1871527320666211006142100.

Liu, G., Wang, Z. K., Wang, Z. Y., Yang, D. B., Liu, Z. P., & Wang, L. (2016). Mitochondrial permeability transition and its regulatory components are implicated in apoptosis of primary cultures of rat proximal tubular cells exposed to lead. *Archives of Toxicology, 90*, 1193–1209. Retrieved from https://doi.org/10.1007/s00204-015-1547-0.

Liu, Y., Niu, L., Liu, X., Cheng, C., & Le, W. (2021). Recent progress in non-motor features of Parkinson's disease with a focus on circadian rhythm dysregulation. *Neuroscience Bulletin, 37*, 1010–1024. Retrieved from https://doi.org/10.1007/s12264-021-00711-x.

Lucchini, R. G., Kotelchuck, D., & Grandjean, P. (2015). Principles for prevention of the toxic effects of metals. In *Handbook on the toxicology of metals*. Academic Press, 507–528. Retrieved from https://doi.org/10.1016/B978-0-444-59453-2.00024-X.

Mani, M. S., Dsouza, V. L., & Dsouza, H. S. (2021). Evaluation of divalent metal transporter 1 (DMT1)(rs224589) polymorphism on blood lead levels of occupationally exposed individuals. *Toxicology Letters, 353*, 13–19. Retrieved from https://doi.org/10.1016/j.toxlet.2021.10.002.

Marras, C., Canning, C. G., & Goldman, S. M. (2019). Environment, lifestyle, and Parkinson's disease: Implications for prevention in the next decade. *Movement Disorders, 34*(6), 801–811. Retrieved from https://doi.org/10.1002/mds.27720.

Martemucci, G., Costagliola, C., Mariano, M., D'andrea, L., Napolitano, P., & D'Alessandro, A. G. (2022). Free radical properties, source and targets, antioxidant consumption and health. *Oxygen, 2*(2), 48–78. Retrieved from https://doi.org/10.3390/oxygen2020006.

Mitra, P., Sharma, S., Purohit, P., & Sharma, P. (2017). Clinical and molecular aspects of lead toxicity: An update. *Critical Reviews in Clinical Laboratory Sciences, 54*(7-8), 506–528. Retrieved from https://doi.org/10.1080/10408363.2017.1408562.

Mitra, S., Chakraborty, A. J., Tareq, A. M., Emran, T. B., Nainu, F., Khusro, A., & Simal-Gandara, J. (2022). Impact of heavy metals on the environment and human health: Novel therapeutic insights to counter the toxicity. *Journal of King Saud University—Science, 34*(3), 101865. Retrieved from https://doi.org/10.1016/j.jksus.2022.101865.

Mrštná, K., Krčmová, L. K., & Švec, F. (2023). Advances in kynurenine analysis. *Clinica Chimica Acta, 547*, 117441. Retrieved from https://doi.org/10.1016/j.cca.2023.117441.

Nabi, M., & Tabassum, N. (2022). Role of environmental toxicants on neurodegenerative disorders. *Frontiers in Toxicology, 4*, 837579. Retrieved from https://doi.org/10.3389/ftox.2022.837579.

Novo, J. P., Martins, B., Raposo, R. S., Pereira, F. C., Oriá, R. B., Malva, J. O., & Fontes-Ribeiro, C. (2021). Cellular and molecular mechanisms mediating methylmercury neurotoxicity and neuroinflammation. *International Journal of Molecular Sciences, 22*(6), 3101. Retrieved from https://doi.org/10.3390/ijms22063101.

Owen, M. C., Gnutt, D., Gao, M., Wärmländer, S. K., Jarvet, J., Gräslund, A., & Strodel, B. (2019). Effects of in vivo conditions on amyloid aggregation. *Chemical Society Reviews, 48*(14), 3946–3996.

Pace, C., Dagda, R., & Angermann, J. (2017). Antioxidants protect against arsenic induced mitochondrial cardio-toxicity. *Toxics, 5*(4), 38. Retrieved from https://doi.org/10.3390/toxics5040038.

Pavlin, M., Repič, M., Vianello, R., & Mavri, J. (2016). The chemistry of neurodegeneration: Kinetic data and their implications. *Molecular Neurobiology, 53*, 3400–3415. Retrieved from https://doi.org/10.1007/s12035-015-9284-1.

Pfaff, A. R., Beltz, J., King, E., & Ercal, N. (2020). Medicinal thiols: Current status and new perspectives. *Mini Reviews in Medicinal Chemistry, 20*(6), 513–529. Retrieved from https://doi.org/10.2174/1389557519666191119144100.

Picca, A., Calvani, R., Coelho-Junior, H. J., Landi, F., Bernabei, R., & Marzetti, E. (2020). Mitochondrial dysfunction, oxidative stress, and neuroinflammation: Intertwined roads to neurodegeneration. *Antioxidants, 9*(8), 647. Retrieved from https://doi.org/10.3390/antiox9080647.

Pljesa-Ercegovac, M., Savic-Radojevic, A., Matic, M., Coric, V., Djukic, T., Radic, T., & Simic, T. (2018). Glutathione transferases: Potential targets to overcome chemoresistance in solid tumors. *International Journal of Molecular Sciences, 19*(12), 3785. Retrieved from https://doi.org/10.3390/ijms19123785.

Raj, A., Kumar, A., & Vaghasiya, T. P. (2021). Multidisciplinary approaches to clean up heavy metal contaminated soil: A review. *Bulletin of Environment, Pharmacology and Life Sciences, 10*(6), 53–65.

Raj, K., Kaur, P., Gupta, G. D., & Singh, S. (2021). Metals associated neurodegeneration in Parkinson's disease: Insight to physiological, pathological mechanisms and management. *Neuroscience Letters, 753*, 135873. Retrieved from https://doi.org/10.1016/j.neulet.2021.135873.

Rasmi, Y., Shokati, A., Hatamkhani, S., Farnamian, Y., Naderi, R., & Jalali, L. (2024). Assessment of the relationship between the dopaminergic pathway and severe acute respiratory syndrome coronavirus 2 infection, with related neuropathological features, and potential therapeutic approaches in COVID-19 infection. *Reviews in Medical Virology, 34*(1), e2506. Retrieved from https://doi.org/10.1002/rmv.2506.

Sales, T. A., Prandi, I. G., de Castro, A. A., Leal, D. H., da Cunha, E. F., Kuca, K., & Ramalho, T. C. (2019). Recent developments in metal-based drugs and chelating agents for neurodegenerative diseases treatments. *International Journal of Molecular Sciences, 20*(8), 1829. Retrieved from https://doi.org/10.3390/ijms20081829.

Schymanski, E. L., Baker, N. C., Williams, A. J., Singh, R. R., Trezzi, J. P., Wilmes, P., & Balling, R. (2019). Connecting environmental exposure and neurodegeneration using cheminformatics and high resolution mass spectrometry: Potential and challenges. *Environmental Science: Processes & Impacts, 21*(9), 1426–1445. Retrieved from https://doi.org/10.1039/C9EM00068B.

Sharma, S., Wakode, S., Sharma, A., Nair, N., Dhobi, M., Wani, M. A., & Pottoo, F. H. (2020). Effect of environmental toxicants on neuronal functions. *Environmental Science and Pollution Research, 27*, 44906–44921. Retrieved from https://doi.org/10.1007/s11356-020-10950-6.

Sigurnjak Bureš, M., Cvetnić, M., Miloloža, M., Kučić Grgić, D., Markić, M., Kušić, H., & Ukić, Š. (2021). Modeling the toxicity of pollutants mixtures for risk assessment: A review. *Environmental Chemistry Letters, 19*, 1629–1655. Retrieved from https://doi.org/10.1007/s10311-020-01107-5.

Sun, Q., Li, Y., Shi, L., Hussain, R., Mehmood, K., Tang, Z., & Zhang, H. (2022). Heavy metals induced mitochondrial dysfunction in animals: Molecular mechanism of toxicity. *Toxicology, 469*, 153136. Retrieved from https://doi.org/10.1016/j.tox.2022.153136.

Teleanu, R. I., Niculescu, A. G., Roza, E., Vladâcenco, O., Grumezescu, A. M., & Teleanu, D. M. (2022). Neurotransmitters—Key factors in neurological and neurodegenerative disorders of the central nervous system. *International Journal of Molecular Sciences, 23*(11), 5954. Retrieved from https://doi.org/10.3390/ijms23115954.

Tripathi, S., Kharkwal, G., Mishra, R., & Singh, G. (2024). Nuclear factor erythroid 2-related factor 2 (Nrf2) signaling in heavy metals-induced oxidative stress. *Heliyon*. Retrieved from https://doi.org/10.1016/j.heliyon.2024.e37545.

Wang, M. H. S., & Wang, L. K. (2025). Recycling and disposal of hazardous solid wastes containing heavy metals and other toxic substances. In *Control of heavy metals in the environment*. Boca Raton, FL: CRC Press, 303–324.

Wang, W. Y., Tan, M. S., Yu, J. T., & Tan, L. (2015). Role of pro-inflammatory cytokines released from microglia in Alzheimer's disease. *Annals of Translational Medicine, 3*(10), 136. Retrieved from https://doi.org/10.3978/j.issn.2305-5839.2015.03.49.

Witkowska, D., Słowik, J., & Chilicka, K. (2021). Heavy metals and human health: Possible exposure pathways and the competition for protein binding sites. *Molecules (Basel, Switzerland), 26*(19), 6060. Retrieved from https://doi.org/10.3390/molecules26196060.

You, L. H., Li, F., Wang, L., Zhao, S. E., Wang, S. M., Zhang, L. L., & Chang, Y. Z. (2015). Brain iron accumulation exacerbates the pathogenesis of MPTP-induced Parkinson's disease. *Neuroscience, 284*, 234–246. Retrieved from https://doi.org/10.1016/j.neuroscience.2014.09.071.

Zafar, A., Javed, S., Akram, N., & Naqvi, S. A. R. (2024). Health risks of mercury. In *Mercury toxicity mitigation: Sustainable nexus approach*. Cham: Springer Nature Switzerland, 67–92. Retrieved from https://doi.org/10.1007/978-3-031-48817-7_3.

Chapter 14

Multiple sclerosis: the impact of heavy metals

Prabhjot Singh Jassal, and Tusha Yadav
Department of Biotechnology, School of Bioengineering and Biosciences, Lovely Professional University, Phagwara, Punjab, India

14.1 Novelty statement

This chapter uniquely quantifies the intricate association between heavy metal exposure and the pathogenesis of multiple sclerosis (MS), highlighting its central role in environmental factors along with genetic predisposition. Unlike previous research and studies that widely involved environmental toxins with neurodegenerative diseases, this study digs into the deep specific metals, which are lead, mercury, and aluminum, illustrating their different mechanisms of action on neural networks, including oxidative stress, immune system dysregulation, and disruption of the blood–brain barrier. Additionally, it emphasizes the bidirectional impact of metal imbalances, where the accumulation of heavy metals exacerbates the MS symptoms. In contrast, a deficiency in essential metals like zinc and copper disrupts neurological functions. This study also shows an upgraded review of genetic and environmental interactions, proposing the variations in genetics that may stimulate the neurotoxic effects of heavy metals, accelerating MS progression. Furthermore, it also contributes to preventive therapies and strategies that help to mitigate the progression of MS and detoxify the body from the harmful effects of heavy metals. These therapies include chelation therapy, antioxidants, and consumption of dietary supplements and epigenetic modulations. Importantly, this study underscores the necessity for strict environmental policies and protective safety measures for human health to minimize the effects of heavy metals and offers a fresh perspective on MS etiology, paving the way for novel strategies and preventive therapies tailored to individuals with both genetic susceptibility and environmental exposure to heavy metals.

14.2 Introduction: understanding multiple sclerosis and environmental influences

MS is a genetic autoimmune disorder of the central nervous system (CNS) that results in the demyelination of neurons, gliosis, and axonal damage. Its pathogenesis included the complex interlinkage between genetic receptive and surrounding factors and inflammatory processes, which play a significant role. While the leading cause of MS remains hidden, considerable evidence proposes that both intrinsic (genetic predisposition) and extrinsic (environmental exposures) components continue its growth and development (Compston & Coles, 2008). This disease primarily affects teenagers and young adults and is characterized by relapsing-remitting courses in its starting times, which may turn into a progressive phase. Clinically, MS displays a range of neurological disorders, which may include sensory disruptions, motor impairments, and cognitive dysfunctions (Goodin, 2014).

14.2.1 Environmental set off in multiple sclerosis development

Surrounding conditions have been integrated into the pathogenesis of MS, which serves as activators in genetically receptive individuals. Epidemiology studies have disclosed geographical changes in MS prevalence, with higher occurrences in temperate areas. This highlights the central role of latitude, vitamin D deficiency, and sun exposure in regulating disease risk (Ascherio & Munger, 2007). Some viral infections, such as the Epstein–Barr virus (EBV), have been significantly associated with MS

Heavy Metal Toxicity and Neurodegeneration. DOI: https://doi.org/10.1016/B978-0-443-36575-1.00028-5

onset (Levin et al., 2010). Moreover, certain environmental factors like smoking and obesity have been known to cause MS and are more susceptible. Smoking increases the level of MS risk and elevates its progression and development by inflammatory pathways that deal with immune regulation (Goodin, 2016). Similarly, obesity during the phase of adolescence has been linked to MS, which may be because of inflammation and alterations in immune responses (Tarlinton et al., 2019). Infections beyond EBV have been implicated, including human herpesvirus-6 (HHV-6), which is linked with MS pathogenesis by molecular mimicry and immune activation (Tiwari et al., 2018).

14.2.2 Heavy metals and multiple sclerosis

An area to be focused on besides the environmental factors, exposure to heavy metals like mercury, lead, cadmium, and arsenic is well known to be a potential contributor to MS. Heavy metals are neurotoxic, which can cause blood–brain barrier, developing oxidative stress, and cause immune dysregulation (Stankiewicz et al., 2005). These functions and regulations are closely related to the pathological features of MS, proposing that exposure to heavy metals may act as a metabolic intermediate in disease etiology.

14.2.3 The interplay between genetics and environmental

The genetic inclination is significant in MS, with specific HLA-DRB1 alleles displaying a high connection with disease susceptibility. However, the concord rate of MS among monozygotic twins is approximately 30%, which shows the crucial role of environmental factors in disease indication (Oksenberg et al., 2014). Understanding how surrounding factors activate and interact with genetic risk loci could pave the way for novel preventative and therapeutic strategies. Additionally, upcoming studies suggest that heavy metal exposure like lead and cadmium may result in MS pathogenesis by mechanisms involving oxidative stress, Neuroinflammation, and damage to the blood–brain barrier (Pamphlett & Bishop, 2023). Heavy metals have also been collected in the CNS, which may lead to demyelination and neuronal damage, which process MS development (Ferraldeschi et al., 2017). A recent study states that MS patients in Poland found an association between increased blood cadmium and lead levels and the severity of MS indication, which shows that heavy metal exposure encourages MS progression (Knyszyńska et al., 2022). These metals influence mitochondrial dysfunction and impair immune regulation, which may aggravate autoimmune responses in genetically susceptible (Pamphlett & Bishop, 2023).

14.3 Heavy metals: sources and pathways of exposure

Heavy metals like mercury (Hg), cadmium (Cd), lead (Pb), and arsenic (As) occur naturally in environments that have been heavily integrated a lot with human activities. While some of these elements are necessary in small amounts for biological functions and regulations, their excessive gathering leads to toxicity. These metals have become pervasive in the surroundings due to human activities like industrialization and urbanization, posing significant harm to human health and the ecosystem (Tchounwou et al., 2012).

14.3.1 Primary sources of heavy metal contamination

Industrial emissions and urbanization: Industries like mining, metal smelting, and manufacturing-based industries release heavy metals into the air, soil, and water. For instance, lead and cadmium are the significant elements emitted by burning coal from power plants and vehicle exhaust, contributing to atmospheric deposition (Alloway, 2013). Urbanization aggravates these emissions with improper waste disposal, which increases environmental contamination.

Agricultural practices: Heavy metals primarily infiltrate and pollute the food chain through farming activities. The frequent use of phosphate fertilizers, which mainly contain cadmium, contributes to soil pollution—untreated and dirty wastewater metal irrigation like arsenic into agricultural fields (Rattan et al., 2005).

Water pollution and contamination: Aquatic ecosystems are consequential pathways for heavy metal accumulation. Mercury contamination from artisanal gold mining and industrial discharges of their contaminants into lakes and rivers has resulted in bioaccumulation in aquatic species, affecting humans through seafood uptake (Clarkson & Magos, 2006).

Household and consumer products: Daily products like cosmetics, paints, and batteries can become potential sources of exposure to heavy metals. Some examples are lead-based paints and mercury-containing skin lighting products, mostly in developing nations (Gordon et al., 2017).

14.3.2 Pathways of human exposure to heavy metals

Inhalation: Airborne particles containing heavy metals, especially emissions from industries and vehicle exhaust, are inhaled and absorbed by humans directly into the bloodstream. Cadmium and lead are often present in delicate particulate matter (PM2.5) in the atmosphere, posing the risk of lung disease and problems to urban populations (Hu et al., 2006). Moreover, exposure to heavy metals like lead, cadmium, and mercury, primarily present in airborne pollutants, is correlated with an increased risk of MS by neurotoxic effects and damage to immune function (Armon-Omer et al., 2024). A prevalence study revealed that, urinary increased levels of cadmium and mercury in MS patients compared with the control ones, with the cadmium levels relating to the disease progression (Another study illustrates that exposure to lead, regions with historical mining activity, which can show the transgenerational epigenetic effect which increased the MS susceptibility to MS by misbalancing myelin energy production and contributes to detoxification (Scholkmann, 2023). Moreover, chronic exposure to fine particulate matter (PM2.5), which usually involves heavy metals, correlates with increased proinflammatory markers in CNS, implicating air pollution as a risk for MS (Abbaszadeh et al., 2021). The interaction between these surrounding toxins and genetic predisposition underscores the need for further research into mitigating strategies.

Ingestion: Contaminated food and water are the primary sources of accumulation of heavy metals. Rice, which is grown in arsenic-rich soils, and fish contaminated with methylmercury are common examples. Children can be at risk for consuming contaminated food, which causes ingestion of lead-based paint chips (Agency for Toxic Substances and Disease Registry [ATSDR], 2019). Additionally, heavy metal exposure through contaminated food and water has been known to contribute to MS progression. Arsenic, lead, and mercury are well known to contaminate water sources and agricultural products, causing neurological effects. A study conducted in Taiwan found a relationship between soil lead levels and MS incidence among males, and long-term environmental factors contribute to disease onset (Tsai & Lee, 2013). Another US study states that increased lead and cadmium levels in sewage sludge and healthy water are related to the higher prevalence of neurodegenerative diseases, including MS (Newell et al., 2024). Ingestion of seafood contaminated with methylmercury or crops grown in areas of high metal-polluted soil can cause bioaccumulation in human beings, aggravate oxidative stress, and cause irregular immune responses (Armon-Omer et al., 2024). Furthermore, these studies also revealed high urinary cadmium and mercury levels in MS patients compared to control ones, which strengthens the link between dietary exposure to heavy metals and disease severity (Armon-Omer et al., 2024).

Dermal absorption: Dermal contact with polluted soil or water can lead to exposure to heavy metals. Occupational exposure in industries such as tanning and metalworking results in dermal absorption of chromium and cadmium (Hansen et al., 2013). Furthermore, dermal absorption of heavy metals has been widely known to contribute to MS pathogenesis. Occupational exposure to toxic metals like lead, cadmium, and chromium in industrial-based sectors like metalworking, battery manufacturing, and chemical processing has been connected to increased systemic absorption and accumulation in the body (Dehghanifiroozabadi et al., 2019). Research revealed higher levels of lead in blood in MS patients compared to control individuals, with a 1.17-fold increased risk of MS for 1 μg/L higher in blood lead levels (Dehghanifiroozabadi et al., 2019). Dermal contact with polluted water, soil, or industrial poor material can allow heavy metals to penetrate the skin barrier and enter the blood circulation system, contributing to oxidative stress and dysregulation, two key mechanisms in MS progression (Aliomrani et al., 2017). Polymorphisms in oxidative stress-related genes, such as GSTM 1, have been correlated with an ability to deduce the effect of heavy metals, making genetically susceptible more dangerous to MS when exposed to toxins (Aliomrani et al., 2017). These highlight the need for more stringent workplace and safety measures regulations.

14.3.3 Implications of heavy metal accumulation

The bioaccumulative nature of heavy metals perseveres in the biological environment and ecosystem over time. This persistence aggravates their toxicity, which can lead to severe health effects like nephrotoxicity, neurotoxicity, and immune dysfunction. Chronic exposure to heavy metals is a critical public health concern in many regions with minimum environmental regulations. Moreover, the bioaccumulative feature of these heavy metals aggravates their long-term impact on neurological health, which involves the role of MS pathogenesis. Cadmium and arsenic have been known to cause immunotoxicity that may activate specific immune responses, which increases susceptibility to autoimmune disorders like MS (Aliomrani et al., 2016a; Aliomrani et al., 2016b).

14.4 Mechanism of neurological damage

Heavy metals like mercury, lead, arsenic, and manganese are significant risks to human health, especially the nervous system. Their neurotoxic effects are primarily mediated through oxidative stress, immune system disruptions, and neuroinflammation.

14.4.1 Oxidative stress

This situation arises when there is a disparity between the body's antioxidant defense system and reactive oxygen species (ROS) production. Generally, heavy metals contribute to this imbalance by creating ROS, which has the potential to cause damage to cellular components such as proteins, lipids, and DNA. For instance, introduction to arsenic has been shown to generate oxidative stress by raising the level of ROS production, leading to neuronal damage and causing neurodegenerative disorders (Jomova et al., 2011). Likewise, mercury exposure inhibits thioredoxin reductase, a major enzyme for maintaining cellular redox, exacerbating oxidative damage in neural tissues (Carvalho, Chew, Hashemy, Lu, & Holmgren, 2008a; Carvalho et al., 2008b). Oxidative stress plays a crucial role in neurological diseases, including MS, by aggravating neuronal disruption and immune response. Heavy metals like lead, cadmium, and arsenic are the potential inducers of oxidative stress, increasing ROS production, which imbalances antioxidant defense mechanisms. In MS, this disrupts demyelination, mitochondrial dysfunction, and neuroinflammation, which increases the risk of disease. Studies illustrate that exposure increases lipid peroxidation and depletion of antioxidants like glutathione in MS patients, escalating oxidative injury to neurons and oligodendrocytes. Cadmium is related to mitochondrial respiration, producing excessive ROS and apoptosis. Likewise, exposure to mercury damages thiol-dependent antioxidant enzymes, weakening the body's cellular defense mechanism against oxidative stress and the changes of neurodegenerative disorders. Research also demonstrates that chronic exposure to heavy metals can activate epigenetic alterations that increase inflammatory responses and aggravate autoimmunity in MS. Therapeutic findings that target oxidative stress—like dietary antioxidants, chelation therapy, and environmental regulations of heavy metal pollutants—are crucial in alleviating the long-term effect of heavy metal that induces oxidative damage in MS patients.

14.4.2 Neuroinflammation

Persistent exposure to heavy metals can induce neuroinflammatory responses, as shown by the activation of glial cells and the release of proinflammatory cytokines. This inflammation causes neuronal damage and can lead to other major neurodegenerative disorders. For example, manganese exposure has been related to a rise in the level of inflammatory markers expression in the brain, leading to neuroinflammation, with studies showing the increased level of inflammatory cytokines in the CNS following lead exposure (Liu et al., 2012). Heavy metals like mercury, cadmium, and arsenic activate glial cells, primarily microglia and astrocytes, leading to sustained generation of proinflammatory cytokines such as tumor necrosis factor-alpha, interleukin-6 (IL-6), and interleukin-1 beta (Jomova et al., 2011). This persevering neuroinflammatory state contributes to blood–brain barrier disruption, oxidative stress, and demyelination, which are the significant hallmarks of MS. Research revealed that mercury exposure can intensify microbial activity, aggravating inflammation and increasing the risk of autoimmune diseases (Carvalho et al., 2008a). Similarly, cadmium has been linked to increased IL-6 production and expression, which can amplify MS progression (Liu et al., 2012). Moreover, long-time exposure to arsenic triggers inflammatory signaling pathways, which accelerate neurological toxicity and impair remyelination processes in the CNS (Jomova et al., 2011).

14.4.3 Immune system disruption

Exposure to heavy metals can damage the immune system's normal functioning, which leads to altered immune responses that can harm neural tissues. Arsenic exposure shows impaired immune function by affecting cytokine production and misbalancing the immune cell populations, which can aggravate neural damage (Dangleben, Skibola, & Smith, 2013). Mercury exposure has been associated with autoimmunity, in which the immune system attacks the body's tissues, contributing to neurological disorders (Pollard et al., 1997). These metals interact with the body's immune homeostasis by encouraging chronic inflammation of cytokines, changing immune cell production, and inducing autoimmunity. Cadmium research found that it can alter the imbalance between proinflammatory and antiinflammatory cytokines, which enhances immune activation and causes damage to neural tissues (Dangleben et al., 2013). Additionally, lead exposure has been related to a skewed T helper cell response, favouring the proinflammatory Th1/Th7 environmental immune system implicated in MS pathophysiology. Mercury, which is known for its strong immunotoxin properties, produces autoantibodies, which trigger immune attacks against the nervous system (Pollard et al., 1997). Exposure to arsenic can damage the regulatory T-cell network, which reduces the body's immune responses, thereby aggravating MS development (Dangleben et al., 2013). Furthermore, these metals also have some significant epigenetic modulations that increase the chances of autoimmune disease occurring over time.

14.4.4 Combined effects and implications

Heavy metal's toxic effects on neural tissues can often result in interconnected mechanisms. Oxidative stress can trigger neuroinflammatory responses, and disruption in the immune system can aggravate oxidative inflammation and damage. This interplay creates a vicious cycle that raises neural injury and increases the risk of developing neurodegenerative diseases. Understanding these mechanisms is essential for developing strategies to alleviate the neurotoxic effects of heavy metal exposure and creating public health awareness while making policies to reduce environmental contamination. Heavy metals have toxic neuronal effects on tissues through interconnected mechanisms that significantly contribute to MS pathogenesis. Oxidative stress, induced by metal, acts as a neuroinflammation catalyst, intensifying immune dysfunction and neuronal disruption. This interaction creates a dangerous cycle that aggravates neural damage and elevates the likelihood of developing neural diseases which include MS. Further studies have illustrated exposure to air pollutants and contaminants that contain heavy metals and particulate matter, which can trigger ROS generation, leading to oxidative stress and sudden inflammatory responses in CNS (Ehsanifar & Montazeri, 2022).

14.5 Heavy metals and multiple sclerosis pathogenesis

MS is a complex autoimmune disorder characterized by demyelination within the CNS. Both genetic predispositions and environmental exposures contribute to its progression. Exposure to heavy metals like lead, mercury, and aluminum is associated with MS pathogenesis among ecological factors.

14.5.1 Lead exposure and multiple sclerosis

This metal has been known to trigger neurotoxic effects in the CNS. A study by Napier et al. (2016) found that individuals with MS have more lead exposure compared to controls, suggesting a potential linkage between lead exposure and the rising risk of MS. Lead can potentially disrupt the blood–brain barrier and triggers oxidative stress, causing the neural damage which leads to induce the autoimmune response against CNS components.

14.5.2 Mercury exposure and autoimmunity

Exposure to mercury has been linked to triggering autoimmune reactions, which are central to MS pathology. Studies state that mercury can induce the production of autoantibodies against myelin essential protein, a crucial component of the myelin sheath, contributing to demyelination (Napier et al., 2016). Mercury can accumulate in the CNS, and its tendency is to induce neuronal damage and potentially trigger an autoimmune response against the CNS.

14.5.3 Aluminum and neuroinflammation

Aluminum exposure plays a role in MS pathogenesis. A study published in *Scientific Reports* found that increased levels of aluminum in the brains of individuals with MS suggest that aluminum accumulation may contribute to neuroinflammatory processes and demyelination (Exley, Clarkson, & Shutt, 2022). The mechanisms remain under study, but aluminum's known proinflammatory proteins could exacerbate CNS inflammation characteristics of MS (Fig. 14.1).

14.5.4 Gene–environment interactions

Specific genetic polymorphisms may increase an individual's vulnerability to the neurotoxic effects of heavy metals. Changes in genes are related to the detoxification process, which could alter the impact of heavy metal exposure on the CNS, thereby affecting MS risk and progression (Napier et al., 2016). Polymorphism in genes that express metallothionein, glutathione-*S*-transferases, and other detoxifying enzymes has been incriminated in altering the body's ability to break down and detoxify toxic metals, potentially leading to greater bioaccumulation and amplifying neurotoxicity (Knyszyńska et al., 2022). Studies revealed that individuals with genetic variations can exhibit higher levels of heavy metals in the bloodstream, which can potentially cause neurodegenerative diseases through oxidative stress or inflammatory pathways (Mohamed et al., 2022). Lead and cadmium levels in MS patients are related to a trend toward severe impairments, suggesting an interaction between the environment that modulates disease severity (Knyszyńska et al., 2022). Another study analyzing heavy metal accumulation in MS patients, who have a high level of Lead (Pb), mercury (Mg), and cadmium (Cd), compared to controls, suggested that genetic factors that trigger metal metabolism may contribute to disease severity (Mohamed et al., 2022). Some deficiencies of certain essential metals like Zinc (Zn) and Copper (Cu), which may play a crucial role in neuroprotection, have been observed in MS patients, which leads them to a misbalancing the metal toxicity and immune dysfunctions (Mohamed et al., 2022).

As Table 14.1 summarizes, heavy metals like lead, mercury, and aluminum contribute to MS pathogenesis through mechanisms. These findings illustrate the need for further research and preventive therapies to mitigate the impact of heavy metal exposure on MS patients.

14.6 Therapeutic approaches and detoxification strategies

MS is a chronic neurological disorder that is associated with immune-mediated demyelination within the CNS. Upcoming studies suggest that exposure to heavy metals, such as lead, mercury, and aluminum, may contribute to MS pathogenesis (Exley et al., 2022). Therapeutic

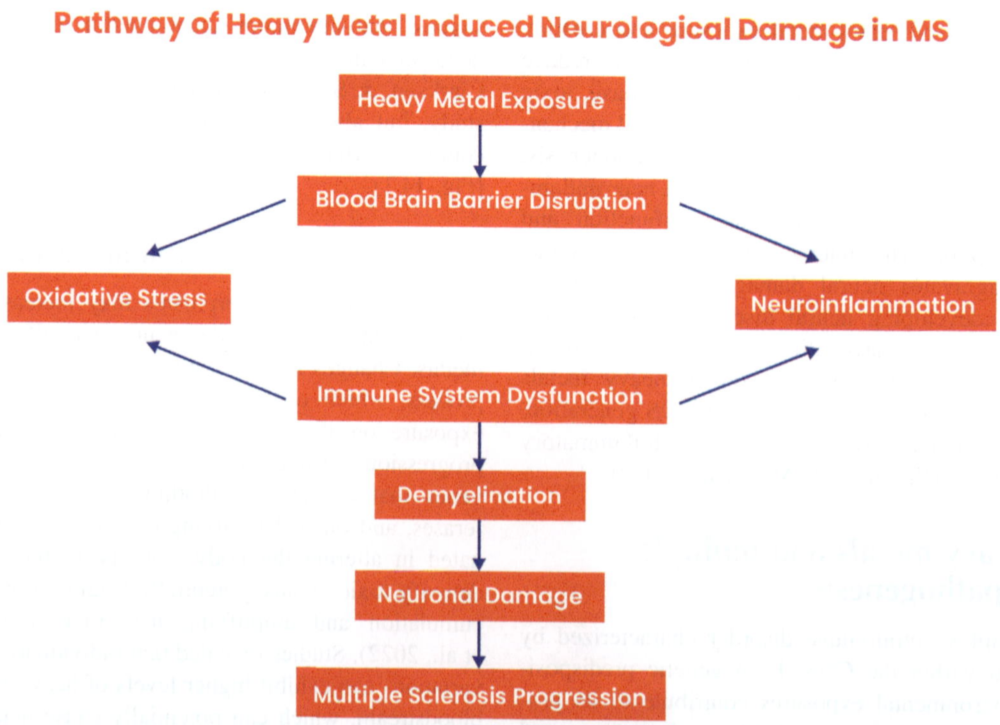

FIGURE 14.1 Pathway of heavy metal-induced neurological damage in multiple sclerosis (MS).

TABLE 14.1 Heavy metals and their impact on multiple sclerosis (MS).

Heavy metals	Impact on MS	Supporting studies	Potential mechanism
Lead (Pb)	It triggers neurotoxicity in the central nervous system, damaging the brain barrier and induces oxidative stress.	Napier et al. (2016) found higher traces of lead exposure in MS as compared to controls, which links to a higher risk of MS.	Blood–brain barrier disruption, immune trigger against the neural cells, and the oxidative demand
Mercury (Hg)	The linkage between autoimmune reactions and the production of autoantibodies against myelin and enhances demyelination recently	Research shows that mercury accumulation in the CNS can trigger neuronal disruption and immune dysfunction	Induces autoimmunity, which leads to start attacking the body's cells
Aluminum (Al)	Accumulation in the brain leads to neuroinflammation and demyelination	Scientists revealed that higher aluminum levels in patients lead to neuroinflammation	This activates the inflammatory responses in the CNS, which contribute to MS development

approaches aimed at alleviating heavy metal toxicity have garnered interest as potential strategies for managing MS.

14.6.1 Chelation therapy

Chelation therapy involves the administration of agents that bind to heavy metals, facilitating their excretion from the body. A study revealed that MS and other related disorders detailed a patient with MS who exhibited a rise in the level of aluminum, mercury, and lead, which directs a chelation challenge with calcium disodium ethylenediaminetetraacetic acid (EDTA). Succeeding chelation therapy results in the reduction of heavy metal levels and an improvement in clinical symptoms, which is a potential therapeutic benefit in MS patients with documented heavy metal accumulation (Ghoreishi et al., 2015). Another study illustrates that dimercaptosuccinic acid (DMSA) and EDTA effectively eliminate lead and mercury metals, thereby reducing oxidative stress damage (Bridges & Zalups, 2017). Emerging studies suggest that chelation therapy improves neurological function, eliminating neuroinflammation and oxidative stress, two key mechanisms that accelerate MS progression (Flora, 2017). Chelating agents like DMSA and meso-2,3-DMSA have been found to reduce metal-induced neurological toxicity by lowering ROS levels, which stabilizes mitochondrial regulations (Patrick, 2006). Studies also show the correlation between chemotherapy therapy and antioxidants like selenium and alpha-lipoic acid, which increase the detoxification pathways and further reduce the toxicity levels in the CNS. Specific experiments have shown the evidence in neurodegenerative diseases suggests that chelation therapy may be particularly beneficial in conditions where the accumulation of heavy metal toxins causes disruption and diseases like Parkinson's and Alzheimer's disease (Flora, 2017). Clinical evidence has demonstrated that chelators can reduce inflammatory cytokine levels, improving the blood–brain barrier and supporting remyelination in demyelination disorders like MS.

14.6.2 Antioxidant supplementation

Heavy metals show their neurotoxic effects through this mechanism of oxidative stress. Antioxidant supplementation has been known to counteract this oxidative damage. Compounds like *N*-acetylcysteine, alpha-lipoic acid, and vitamins C and E have been illustrated as potential compounds to reduce this oxidative stress proposed by heavy metals (Jomova et al., 2011). These antioxidants help reduce heavy metals' neurotoxic effects and offer adjunctive benefits (Méndez-Armenta & Ríos, 2007). Some trials have shown that glutathione precursors aid in lowering the level of heavy metals, potentially offering neuroprotective effects in MS (Yousefi et al., 2014). Additionally, recent studies illustrate that curcumin, resveratrol, and selenium possess potent antioxidant properties that can counteract metal-induced oxidative stress, which reduces the risk of neurological diseases (Singh et al., 2019). Selenium plays a significant role in the activity of glutathione peroxidases, an enzyme that balances the production of ROS generated by heavy metal toxicity (Burk & Hill, 2015a; Burk & Hill, 2015b). Studies also show that (CoQ10) coenzyme supplementation may improve mitochondrial functions, which reduces oxidative damage in MS patients with a history of metal exposure (Higdon & Frei, 2018). The combination of polyphenols and flavonoids found in green tea and berries has shown the ability to chelate heavy metal burden and inhibit oxidative stress, suggesting a dietary approach to reduce metal toxicity in MS (Fig. 14.2).

14.6.3 Dietary interventions

Dietary strategies help to reduce the burden of heavy metals and amplify detoxification pathways, which have been inspected. Consumption of foods rich in sulfur-containing compounds, like garlic and cruciferous vegetables, boosts the natural detoxification of body processes (Branca et al., 2018a; Branca et al., 2018b). Dietary fibers can also accumulate heavy metals in the gastrointestinal tract and their absorption (Attar et al., 2012). Emerging studies show that specific dietary components may influence the accumulation and excretion of metals in individuals with neurodegenerative conditions like MS.

Polyphenol-rich foods, which include berries, green tea, and turmeric, illustrate the ability to celebrate heavy metals and mitigate oxidative stress (Singh et al., 2020). Flavonoids, which are usually found in citrus fruits and dark chocolate, have demonstrated neuroprotective properties, enhancing oxidative defense and reducing inflammation caused by metal toxicity. Collectively, omega-3 fatty acids, found in fish and flaxseed, have been known as protective agents against heavy metal-induced neuroinflammation, offering better for MS patients (Calder, 2021).

Probiotics have been known to reduce the burden of heavy metals' bioavailability. For example, strains of *Lactobacillus* and *Bifidobacterium* have been found to bind metals such as cadmium and lead, which prevents their accumulation and absorption in the gut (Zhai et al., 2017).

14.6.4 Challenges and considerations

Even with these therapeutic strategies, challenges persist. MS's heterogeneity and multifunctional nature of pathogenesis complicate the evaluation of heavy metals' toxicity role in disease progression (Napier et al., 2016). Chelation carries some major health risks, including the depletion of potential minerals, which also results in adverse reactions (Tsai & Lee, 2013). Given the complexity of MS and its potential link with heavy metals, various therapeutic strategies and approaches have been discovered to mitigate the neurotoxic effects of these metals. In contrast, others focus on reducing oxidative stress, which removes accumulated metals and focuses on reducing oxidative stress, altering immune response. A comparative analysis of these approaches, including the benefits and challenges, is summarized in Table 14.2. This highlights the recent therapeutic landscapes and underscores the need for further research to establish their long-term efficacy and safety in MS patients.

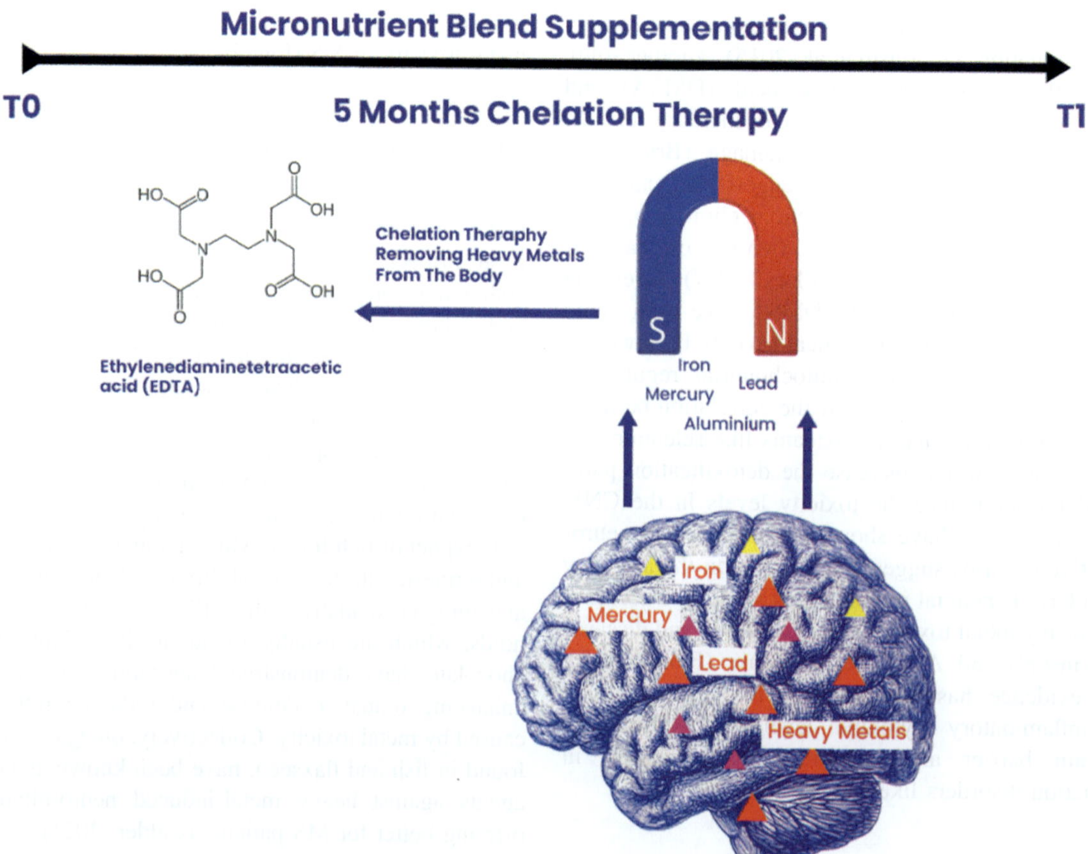

FIGURE 14.2 Chelation therapy removing heavy metals with an antioxidant supplement.

14.7 Future research and preventive strategies

14.7.1 Directions for future studies and public health measures to mitigate risks

MS is a chronic and complex neurological disorder influenced by genetic and environmental factors, including exposure to heavy metals. Increased levels of cadmium and arsenic have been shown in MS-suffering patients, proposing a potential link between the development of the disease and exposure to metals (Sarihi et al., 2021). Lead and mercury can cause neurotoxicity through mechanisms like oxidative stress, neuroinflammation, and immune system dysregulation, warranting further investigation into their role in MS pathogenesis (Napier et al., 2016)

14.7.2 Future research directions

14.7.2.1 Longitudinal studies on heavy metal exposure and MS progression

Long-term trials to monitor heavy metals in individuals over time can help establish an association between exposure and MS progression. Such studies provide insight into how chronic exposure influences disease onset and progression (Tsai & Lee, 2013). Recent longitudinal studies have demonstrated the impact of heavy environmental metals on MS progression while focusing on their cumulative effects on immune function and neurodegeneration. A Continuous-Time Markov model study analyzed ecological exposure data and its influence on MS patients and found that air pollutants containing heavy metals like cadmium correlate with increased disease activity and functional decline over time (Vazifehdan et al., 2024).

14.7.2.2 Genetic susceptibility and environmental interactions

Investigating how genetic susceptibility interacts with environmental exposures like heavy metals, the recognized environment is at higher risk. Recent studies have shown the potential relationship between environmental exposure to mercury, lead, and solvents and single-nucleotide polymorphism in MS patients. Stratify individuals based on susceptibility (Napier et al., 2016).

TABLE 14.2 Therapeutic approaches for heavy metal detox in multiple sclerosis (MS).

Therapeutic approach	Description	Potential benefits	Challenges
Chelation therapy	Use of chemical agents like EDTA and DMSA to bind heavy metals and facilitate their excretion.	Use of chemical agents like EDTA and DMSA to bind heavy metals and facilitate their excretion	It may lead to the depletion of essential minerals; it requires careful monitoring to avoid adverse effects.
Antioxidant supplementation	Intake compounds such as N-acetylcysteine, alpha-lipoic acid, and vitamins C and E to mitigate oxidative stress.	Protects neurons from heavy metal-induced damage, improving antioxidant defenses.	Efficacy in MS-specific populations still requires further research; dosage optimization is needed.
Dietary interventions	Incorporation of sulfur-rich foods (e.g., garlic, cruciferous vegetables) and dietary fibers to reduce heavy metal absorption.	It helps prevent further accumulation of heavy metals and supports natural detoxification pathways.	Effectiveness in reducing heavy metal burden is variable and requires consistent dietary adherence.
Pharmacological agents	Potential use of drugs that modulate immune response or aid in detoxification to alleviate MS symptoms.	Provides targeted therapeutic intervention to modulate immune activity and reduce symptoms	Limited research on direct effectiveness for heavy metal toxicity in MS; potential side effects.
Epigenetic modifications	Regular exercise, hydration, and avoidance of environmental toxins support overall neurological health.	It enhances overall well-being, promotes neuroprotection, and reduces the risk of exposure.	Difficult to measure impact directly on MS progression; requires lifestyle commitment.

DMSA, Dimercaptosuccinic acid; *EDTA*, ethylenediaminetetraacetic acid.

14.7.2.3 Monotonous studies on heavy metals that induce neurotoxicity

Understanding the mechanism and actions of heavy metals by which they develop neurodegeneration in MS. Heavy toxic metals contribute to blood–brain barrier dysfunction, allowing circulating poisonous materials to enter astrocytes and damage neurons (Hagemeyer et al., 2023).

14.7.2.4 Development of advanced detoxification therapies

Therapies like chelation treatments have several limitations, including the depletion of essential minerals inside our bodies. Studies of detoxification methods, like using antioxidants, have shown outstanding results in treating heavy metals' toxic behavior. Studies highlight that compounds like N-acetylcysteine and alpha-lipoic acid can mitigate oxidative stress and improve detoxification pathways in MS (Zhang et al., 2023).

14.7.2.5 Biomarker identification for early detection

Recognized biomarkers that show early exposure to heavy metals can upgrade prompt intervention, which slows MS development. Diagnostic tools like screening of heavy metals and cerebrospinal fluid help in early risk assessment (Méndez-Armenta & Ríos, 2007).

Recent Research shows that oxidative stress biomarkers like increased lipid peroxidation products (malondialdehyde) and decreased antioxidant enzyme activity (superoxide dismutase and glutathione peroxidase) can indicate heavy metal neural damage, particularly relevant to MS (Calderón-Garcidueñas et al., 2020). The recognition of heavy metals has induced epigenetic modifications, inducing changes in DNA methylation and histone acetylation. These modifications are a potential tool for early detection of toxic metal exposure and its impact on MS Progression (Santiago et al., 2017).

14.7.3 Preventive strategies

14.7.3.1 Public health policies to reduce environmental exposure

Executing proper rules and regulations to limit industrial emissions and contamination can decrease the generality of heavy metals in the environment, reducing public exposure. Countries with strict environmental policies and regulations have seen lower levels of heavy metal contamination in soil and water (Tsai & Lee, 2013).

14.7.3.2 Community education programs

Raising awareness about sources of exposure to heavy metals and encouraging behaviors to minimize contact can empower individuals to protect their health. Public health campaigns targeting at-risk communities can enhance knowledge about different sources of contamination, including industrial contaminants and household waste (Napier et al., 2016).

14.7.3.3 Regular screening in high-risk populations

Regular testing of heavy metal levels in communities with known risks can lead to early detection and prevention. High-risk groups, like workers in industries and residents of polluted and contaminated areas, can benefit from periodic screenings to check the accumulation of heavy metals (Hagemeyer et al., 2023).

14.7.3.4 Nutritional interventions

Promoting diets rich in antioxidants may help reduce heavy metal exposure. Some recommend the consumption of green leafy vegetables and cilantro, which naturally detoxify. Avoiding processed food containing harmful artificial preservatives linked to heavy metal contamination can be beneficial (Armistead & O'Farrell, 2025).

14.7.3.5 Professional safety measures

Promoting the safety of industrial workers who have potential exposure to heavy metals, who must adhere to safety protocols, and who must have protective equipment can prevent profession-related cases of MS. Regulations like installing proper ventilation and ducts, regular testing of heavy metals levels, and the use of personal protective equipment have been an alternative and efficient way to reduce workplace exposure (Zhang et al., 2023). By showing concerns about this research and preventive therapies, public health officials will work together to eliminate the risk which is associated with exposure to MS.

14.8 Conclusion

MS is a chronic, complex autoimmune disease affected by both genetic susceptibility and environmental factors, including exposure to heavy metals, which plays a significant role. Immersive studies have illustrated that heavy metals like lead, mercury, arsenic, and cadmium contribute to the pathogenesis of MS by activating oxidative stress, neuroinflammation, and immune system dysfunction. These heavy metals damage essential neurological functions by accumulating in the CNS, exhausting the blood–brain barrier, and interfering with the immune system's ability to distinguish between the body's self and foreign cells. Insufficient essential elements and trace metals like zinc and copper aggravate metal disparity, exacerbating MS development. The supply of heavy metal contamination is extensive, ranging from industrial emissions and agriculture practices to household waste and water pollution. Pathways include ingestion, inhalation, and dermal absorption of heavy metals, which significantly affect human health, especially for individuals who live in highly polluted environments. The coaction between environmental factors and genetic vulnerability suggests that several individuals may have a higher risk of developing MS due to certain genetic factors; mental exposure also interplays on disease onset and progression. Efforts to eliminate these heavy metals exposure can be solved by specific therapeutic strategies directed at detoxification and symptom management in MS patients. Chelation therapy, antioxidant supplementation, pharmacological approaches, and dietary interventions have been explored to reduce the toxicity burden of heavy metals in MS. Obstacles persist as chelation therapy carries the risk of depletion, and dietary and pharmacological interventions require further trials and research. The urge for future trials and research remains paramount in beginning a clearer understanding of the role of heavy metals in MS. Biomarker identification for early detection and genetic and environmental interaction studies are necessary to improve strategies. Preventive techniques involving stricter ecological rules, public health policies, and occupational safety protocols must be prioritized to limit heavy metal exposure in high-risk populations. Addressing heavy metal toxicity as a factor in MS requires a multidisciplinary approach and incorporation of medical, environmental, and policy-based strategies. A comprehensive understanding of metals that induce toxicity will lead to the development of more effective therapies and potentially alter the disease strategy for MS patients. Many ongoing research and protective public measures are critical in reducing the environmental burden of heavy metals and minimizing their impact on diseases like MS.

References

Abbaszadeh, S., et al. (2021). Air pollution and multiple sclerosis: A comprehensive review. *Neurological Sciences*.

Aliomrani, M., et al. (2016a). Blood concentrations of cadmium and lead in multiple sclerosis patients from Iran. *Iranian Journal of Pharmaceutical Research*.

Aliomrani, M., et al. (2016b). Glutathione *S*-transferase M1 and T1 null genotypes and multiple sclerosis risk. *Gene, 577*(1), 56–61.

Aliomrani, M., et al. (2017). Correlation between heavy metal exposure and GSTM1 polymorphism in Iranian multiple sclerosis patients. *Neurological Sciences*.

Alloway, B. J. (2013). *Heavy metals in soils: Trace metals and metalloids in soils and their bioavailability*. Springer.

Armistead, A., & O'Farrell, S. (2025). *How to detox from heavy metals, according to experts. New York Post*.

Armon-Omer, A., et al. (2024). Association between multiple sclerosis and urinary levels of toxic metals and organophosphates: A cross-sectional study in Israel. *Multiple Sclerosis and Related Disorders*.

Ascherio, A., & Munger, K. L. (2007). Environmental risk factors for multiple sclerosis. Part II: Noninfectious factors. *Annals of Neurology, 61*(6), 504–513.

Attar, A., et al. (2012). The role of dietary fibre in reducing heavy metal absorption. *Food & Chemical Toxicology.*

Branca, J. J. V., et al. (2018a). Cadmium-induced neurotoxicity: Still much ado. *Neural Regeneration Research, 13*(11), 1879–1880.

Branca, J. J. V., et al. (2018b). Role of dietary sulfur compounds in detoxification pathways. *Nutrients.*

Bridges, C. C., & Zalups, R. K. (2017). The chemical forms of mercury in human exposure and their implications for chelation therapy. *Toxicology Letters.*

Burk, R. F., & Hill, K. E. (2015a). Regulation of selenium metabolism and transport. *Annual Review of Nutrition.*

Burk, R. F., & Hill, K. E. (2015b). Selenium metabolism and its role in metal detoxification. *Annual Review of Nutrition.*

Calder, P. C. (2021). Omega-3 fatty acids and neuroinflammation: Relevance for neurodegenerative diseases. *Brain, Behavior, and Immunity.*

Calderón-Garcidueñas, L., et al. (2020). Heavy metal exposure, oxidative stress, and neurodegeneration: A review. *Neurotoxicology.*

Carvalho, C. M., et al. (2008b). Mercury and selenium interaction in mammalian cells: Effects on thioredoxin reductase and glutathione peroxidase. *Environmental Research.*

Carvalho, C. M., Chew, E. H., Hashemy, S. I., Lu, J., & Holmgren, A. (2008a). [3] Inhibition of the human thioredoxin system. A molecular mechanism of mercury toxicity. *The Journal of Biological Chemistry, 283*(18), 11913–11923.

Clarkson, T. W., & Magos, L. (2006). The toxicology of mercury and its chemical compounds. *Critical Reviews in Toxicology, 36*(8), 609–662.

Compston, A., & Coles, A. (2008). Multiple sclerosis. *The Lancet, 372*(9648), 1502–1571.

Dangleben, N. L., Skibola, C. F., & Smith, M. T. (2013). Arsenic immunotoxicity: A review. *Environmental Health, 12*, 73.

Dehghanifiroozabadi, M., et al. (2019). Blood lead levels and multiple sclerosis: A case-control study. *Multiple Sclerosis and Related Disorders.*

Ehsanifar, M., & Montazeri, Z. (2022). CNS demyelination diseases following exposure to urban air pollution. *Journal of Biomedical Research & Environmental Sciences, 3*(2), 205–209.

Exley, C., Clarkson, E., & Shutt, J. L. (2022). Potentially toxic elements in the brains of people with multiple sclerosis. *Scientific Reports, 12*(1), 27169.

Ferraldeschi, M., Romano, S., Buscarinu, M. C., Fornasiero, A., Mechelli, R., Cerasoli, B., ... Alimonti, A. (2017). Chemical Elements and Oxidative Status in Neuroinflammation. *Biometals in Neurodegenerative Diseases* (pp. 67–81). Academic Press.

Flora, S. J. S. (2017). Chelation in metal intoxication. *International Journal of Environmental Research and Public Health.*

Ghoreishi, S. M., et al. (2015). Heavy metals as environmental risk factors in patients with multiple sclerosis. *Journal of Clinical Neuroscience, 22*(2), 292–295.

Goodin, D. S. (2014). The epidemiology of multiple sclerosis: Insights to disease pathogenesis. In *Handbook of clinical neurology*231–266 122.

Goodin, D. S. (2016). The epidemiology of multiple sclerosis: Insights to a causal cascade. In *Handbook of clinical neurology.*

Gordon, B., et al. (2017). *Human exposure to lead: A major public health concern. World Health Organization.*

Hagemeyer, N., et al. (2023). The toxic metal hypothesis for neurological disorders. *Frontiers in Neurology.*

Hansen, H. R., et al. (2013). Dermal exposure to chromium and its toxicological relevance. *Journal of Applied Toxicology, 33*(9), 749–766.

Higdon, J. V., & Frei, B. (2018). Coenzyme Q10 and its potential benefits in neurodegenerative diseases. *The Journal of Nutritional Biochemistry.*

Hu, H., et al. (2006). The impact of occupational exposure to lead on the endocrine system and reproductive function. *Environmental Health Perspectives, 114*(2), 190–196.

Jomova, K., et al. (2011). Arsenic-induced oxidative stress and its role in carcinogenesis and neurodegenerative disorders. *Journal of Applied Toxicology.*

Jomova, K., Jenisova, Z., Feszterova, M., Baros, S., Liska, J., Hudecova, D., ... Valko, M. (2011). Arsenic: Toxicity, oxidative stress and human disease. *Journal of Applied Toxicology, 31*(2), 95–107.

Knyszyńska, A., et al. (2022). Searching for the relationship between the concentration of heavy metals in the blood and the clinical course of multiple sclerosis: A cross-sectional study in Poland. *International Journal of Environmental Research and Public Health.*

Levin, L. I., et al. (2010). Multiple sclerosis and Epstein-Barr virus. *Archives of Neurology, 67*(1), 21–27.

Liu, C., et al. (2012). The role of neuroinflammation in lead-induced neurotoxicity. *Toxicology and Applied Pharmacology.*

Liu, M. C., Liu, X. Q., Wang, W., Shen, X. F., Che, H. L., Guo, Y. M., & Zhang, T. (2012). Chronic lead exposure induces behavioral impairments and downregulates insulin-like growth factor 2 expression in the hippocampus of adult mice. *Neuroscience, 218*, 1–8.

Mohamed, S. R., et al. (2022). Complex profile of altered heavy metals accumulation in multiple sclerosis, a relationship with copper and zinc homeostasis. *Open Access Macedonian Journal of Medical Sciences.*

Méndez-Armenta, M., & Ríos, C. (2007). Cadmium neurotoxicity. *Environmental Toxicology and Pharmacology, 23*(3), 350–358.

Napier, M. D., Poole, C., Satten, G. A., Ashley-Koch, A., Marrie, R. A., & Williamson, D. M. (2016). Heavy metals, organic solvents and multiple sclerosis: An exploratory look at gene-environment interactions. *Archives of Environmental & Occupational Health, 71*(1), 26–34.

Newell, M., et al. (2024). Prevalence rates of neurodegenerative diseases versus human exposures to heavy metals across the United States. *Science of the Total Environment.*

Oksenberg, J. R., et al. (2014). Genetics of multiple sclerosis: From genes to pathogenesis. Nature Reviews. *Neurology, 10*(9), 517–526.

Pamphlett, R., & Bishop, D. P. (2023). The toxic metal hypothesis for neurological disorders. *Frontiers in Neurology.*

Patrick, L. (2006). Lead toxicity, a review of the literature. Part 1: Exposure, evaluation, and treatment. *Alternative Medicine Review, 11*(1), 1–22.

Pollard, K. M., Pearson, D. L., Hultman, P., Hildebrandt, B., Kono, D. H., & Hood, J. M. (1997). Xenobiotic acceleration of idiopathic systemic autoimmunity in lupus-prone bxsb mice. *Environmental Health Perspectives, 105*(9), 1066–1069.

Rattan, R. K., et al. (2005). Long-term impact of irrigation with sewage effluents on heavy metal content in soils, crops, and groundwater. *Agriculture, Ecosystems & Environment, 109*(3-4), 310–322.

Santiago, J. A., et al. (2017). Epigenetic modifications in response to environmental toxins and their impact on neurodegenerative diseases. *Neurobiology of Disease.*

Sarihi, S., et al. (2021). Toxic heavy metal concentrations in multiple sclerosis patients: A systematic review and meta-analysis. *EXCLI Journal.*

Scholkmann, F. (2023). Multiple sclerosis, extra-mitochondrial energy production in myelin and transgenerational effects of lead exposure: Possible connections. *Medical Hypotheses.*

Singh, A., et al. (2019). Curcumin and resveratrol as potential antioxidants against heavy metal-induced neurotoxicity. *Brain Research Bulletin.*

Singh, A., et al. (2020). Role of polyphenols in metal chelation and neuroprotection. *Journal of Nutritional Biochemistry.*

Stankiewicz, J., et al. (2005). Mechanisms of blood-brain barrier dysfunction in multiple sclerosis. *Journal of Neuroimmunology, 165*(1-2), 59–68.

Tarlinton, R. E., Khaibullin, T., Granatov, E., Martynova, E., Rizvanov, A., & Khaiboullina, S. (2019). The interaction between viral and environmental risk factors in the pathogenesis of multiple sclerosis. *International Journal of Molecular Sciences, 20*(2), 303.

Tchounwou, P. B., et al. (2012). Heavy metal toxicity and the environment. *EXS, 101*, 133–164.

Tiwari, S., et al. (2018). Signaling pathways and therapeutic perspectives related to environmental factors associated with multiple sclerosis. *Journal of Neuroscience Research.*

Tsai, C.-P., & Lee, C. T.-C. (2013). Multiple sclerosis incidence associated with the soil lead and arsenic concentrations in Taiwan. *PLoS One, 8*(6), e65911.

Vazifehdan, M., et al. (2024). Continuous markov models for analyzing the effect of environmental personal exposure on multiple sclerosis progression. *Studies in Health Technology and Informatics.*

Yousefi, B., et al. (2014). Arsenic trioxide augments oxidative stress and apoptosis in multiple sclerosis. *Biological Trace Element Research, 157*(1), 31–38.

Zhai, Q., et al. (2017). Probiotic-mediated detoxification of heavy metals. *Microbiology Research.*

Zhang, Y., et al. (2023). Heavy metal exposure: Molecular pathways, clinical implications, and therapeutic strategies. *Antioxidants.*

Chapter 15

Heavy metals and amyotrophic lateral sclerosis

Amit Raj, and Prasann Kumar

Department of Agronomy, School of Agriculture, Lovely Professional University, Phagwara, Punjab, India

15.1 Introduction

The motor neurons of the upper brain are both outwardly and internally affected by amyotrophic lateral sclerosis (ALS), often known as Lou Gehrig's disease, a chronic neurodegenerative and neuromuscular illness (Volonté, Apolloni, Parisi, & Amadio, 2016). The regulation of intentional skeletal muscle contraction is the domain of these motor neurons. ALS causes the progressive breakdown and death of these neurons, rendering the victim unable to control their muscles, becoming paralyzed, and finally, having to take their last breath. It falls under the group of motor neuron diseases and is usually diagnosed in adults between 40 and 70 years. The clinical manifestations of ALS are diverse, and the patients present generally progressive weakness of the limbs (limb onset ALS) or face, jaw, and neck muscles (bulbar onset ALS). Its initial signs may be twitching or muscle spasms (fasciculations), cramps, tightness or rigidity, and spasticity. In the later stages, patients are unable to do fine motor activities, become unable to walk, and develop speech problems (Masrori & Van Damme, 2020). If the disease advances, muscles shrink, weight loss occurs, and the patient loses the ability to move around. ALS affects motor neurons and their control of muscles but does not affect cognitive function in most patients. However, some patients will show frontotemporal dementia, evidenced by personality changes and poor executive function. Sensory, bowel, and bladder control are usually maintained in this condition. The primary cause of ALS remains unrecognized, but it is suggested to be a genetic and environmental disease. About 1 in 10 are familial ALS (fALS) because of gene anomalies, including SOD1, C9orf72, and TARDBP (Mathis, Goizet, Soulages, Vallat, & Le Masson, 2019). The rest is called sporadic ALS (sALS), for which oxidative stress, mitochondrial dysfunction, and neuroinflammation have been suggested. At present, there is no cure for ALS, and therapy is directed towards making the patients as comfortable as possible and enhancing their quality of life. ALS is a polygenic neurodegenerative disease that occurs due to the loss of motor neurons. To this date, its exact etiology is not entirely clear, and it is believable that many genes/environmental and lifestyle factors may play their part in the onset/progression of the disease.

It is estimated that in 10% of ALS cases, the disease is inherited and classified as fALS. SOD1, C9orf72, FUS, and TARDBP are the most well-known; the ALS2/IF platform C9orf72 hexanucleotide repeat expansion is the most familiar and affects nearly 40% of the familial cases. These mutations interfere with cellular functions such as protein balance, RNA regulation, and even cellular energy production using mitochondria. As such, when motor neurons are affected by these mutations, they cannot be reversed, and they die. It is less clearly defined than the other subtypes. Still, attitudes and lifestyle, genetic prompts, and environmental effects play an essential role in sALS, in which 90% of patients are affected. Thus the risk factors are exposure to heavy metals, pesticides, and other neurotoxins, extended activity, and head injuries. Occupational exposure involving farming, military, and competitive sporting activities has also been cited. New data point to oxidative stress, excitotoxicity due to glutamate imbalance neuroinflammation and mitochondrial damage as key mechanisms for ALS. Also, TDP-43 and SOD1 proteins are misfolded; their presence also affects neuronal toxicity through other abnormal cellular pathways (Provenzano, Torazza, Bonifacino, Bonanno, & Milanese, 2023). Age and sex are biochemical predictors; ALS occurs between the ages of 40 and 70 and is slightly more common in male individuals. Several published works have also pointed out that certain risk factors, like smoking and specific dietary patterns, may also cause one to be predisposed to develop ALS. Despite significant

Heavy Metal Toxicity and Neurodegeneration. DOI: https://doi.org/10.1016/B978-0-443-36575-1.00002-9

advancements in research to unpack the multifactorial nature of ALS, the interrelation between genetic susceptibility factors and exogenous environmental agents is still an enticing field, thus allowing future advancement in untapped therapeutic approaches.

Environmentally, the health risk associated with exposure to heavy metals has been a significant factor of concern because of its implication in the causal mechanism of neurological disorders. Lead, mercury, cadmium, and arsenic are naturally occurring metals ubiquitously released by anthropogenic sites, including industries, mining, contaminated water, pesticides, and the food chain (Fig. 15.1). These metals are nonbiodegradable and can accumulate in the human body, with dangerous effects on the nervous system after long-term exposure. Heavy metals are well established to have neurotoxic levels and affect the blood–brain barrier (BBB), affecting the nervous system in numerous ways. These alter nitric oxide/superoxide anions, which in turn produce oxidative stress, causing damage to lipids, proteins, and DNA. This oxidative damage alters key functions such as mitochondrial function and calcium regulation, leading to neuronal apoptosis. Furthermore, heavy metals disrupt neurotransmitter systems, affect synaptic plasticity, and create neuroinflammation that results from activation of the microglia and astrocytes. Selective studies indicate that heavy metal toxicity is related to neurodegenerative diseases, including AD, PD, and ALS (Ijomone, Ifenatuoha, Aluko, Ijomone, & Aschner, 2020). For example, lead and mercury demonstrated toxic effects in animals that recapitulate amyloid-beta plaques and tau protein tangles in AD, and manganese exposure causes a reduction in dopaminergic neuronal density similar to that in PD. The growing incidence of neurological disorders underlines the importance of controlling exposure to toxicity in heavy metals. The analysis of own and previously published data presents the possibility of studying the effect of heavy metals at the molecular and cellular levels, which helps to understand their contribution to the development of diseases and emphasizes the need for measures to reduce exposure and its neurotoxic effects.

15.2 Heavy metals and neurotoxicity

Heavy metals are another environmental component that has been studied. They are strongly linked to neurological illnesses like ALS. Through bioaccumulation and disruption of inherent critical neuronal processes, toxicology-grade metals such as lead, mercury, arsenic, and cadmium are recognized to induce motor neuron degeneration. The specific heavy metals associated with ALS, the harmful neurological pathways via which they function, and their role in damaging the BBB, which increases the brain's vulnerability to injury, are discussed in further detail in the following subsections (Kakaroubas, Brennan, Keon, & Saksena, 2019).

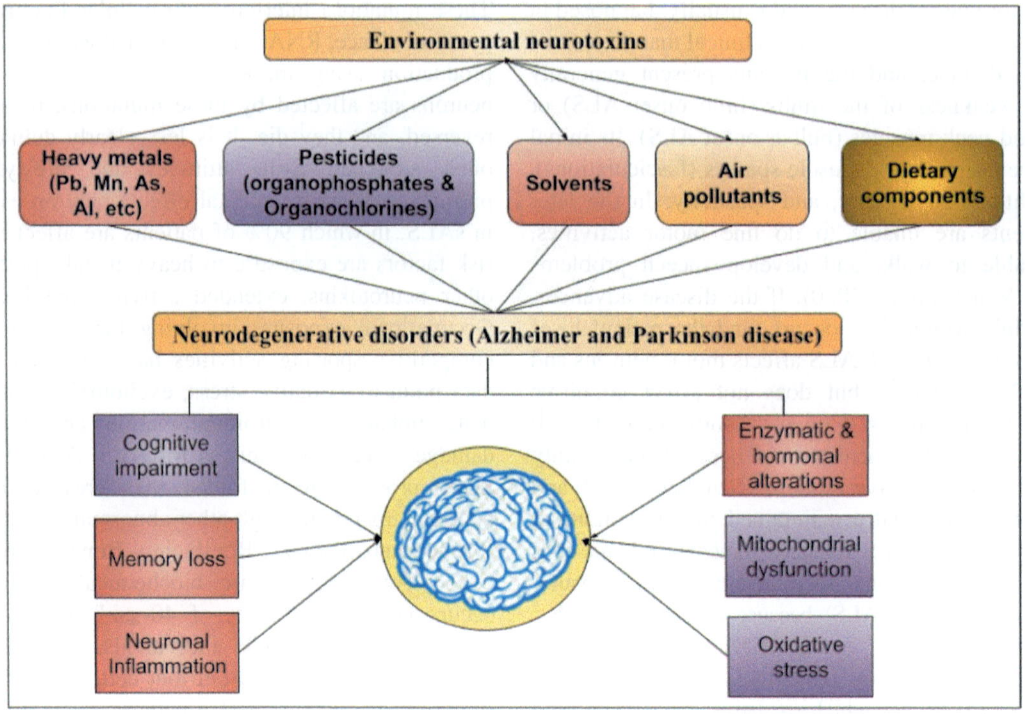

FIGURE 15.1 The influence of environmental toxicants on the onset and progression of neurodegenerative disorders.

15.2.1 Heavy metals implicated in amyotrophic lateral sclerosis

Heavy metals are neurotoxic elements that can accumulate in the nervous system and disrupt motor neuron function. Key metals associated with ALS include:

- **Lead (Pb):** Lead exposure is commonly classified as a neurotoxic hazard. It inhibits critical calcium-dependent signaling mechanisms motor neurons require and impairs neurotransmitter release. This is mainly evidenced by deficits in motor function and muscle atrophy with motor neurons (Ferraiuolo, Kirby, Grierson, Sendtner, & Shaw, 2011). Lead can be present in emissions, industrial fumes, paints, and most water sources.
- **Mercury (Hg):** Inorganic mercury that may be released from industrial processes, such as elemental mercury vapor and organic mercury like methylmercury found in fish, severely impacts the brain. Ingested mercury forms weaker complexes with cysteine and other agents that displace mercury in the brain and bind to sulfhydryl groups, inactivating proteins and enzymes vital for the neuron's survival (Kang, Wang, Guo, & Yang, 2024). There is evidence of its participation in motor dysfunction and inflammation in the brain, both of which are present in ALS.
- **Arsenic (As):** Another study showed that drinking water containing arsenic affects neurodegeneration. Arsenic inhibits energy metabolism, changes the gene profiles, and provokes oxidative stress in motor neurons. Long-term arsenic ingestion results in some characteristics akin to ALS, including muscle wasting and diminished strength (Thakur, Rachamalla, Niyogi, Datusalia, & Flora, 2021).
- **Cadmium (Cd):** Cadmium is a toxic heavy metal in industrial effluents, cigarette smoke, and polluted soils. It disrupts calcium homeostasis, impairs mitochondria, and fosters oxidative stress and positive charges. Exposure to cadmium leads to neuronal toxicity and a predisposition of motor neurons to die, causing symptoms similar to ALS (Sighencea, Popescu, & Trifu, 2024).

15.2.2 Mechanisms of neurotoxic effects

Heavy metals cause neurotoxicity through multiple mechanisms, often acting synergistically to amplify motor neuron damage in ALS:

- **Oxidative stress:** Heavy metals are great antagonists of oxidative stress. They produce putative reactive oxygen species (ROS) either through direct redox cycling or indirectly by the inactivation of antioxidant enzymes, including superoxide dismutase (SOD). ROS affects lipids, proteins, and DNA, then impairs neurons: ROS can injure cellular lipids, proteins, and DNA, therefore affecting the function of neurons. For instance, mercury interferes with the action of a vital antioxidant, glutathione, and thus increases oxidative stress to motor neurons. In addition to generating ROS, cadmium inhibits the activity of antioxidant enzymes, including catalase and superoxide dismutase (Fig. 15.2).
- **Mitochondrial dysfunction**: Metals revenue with mitotic intervention and function, which is of paramount importance for neurons. They impair the electron transport chain's ability to operate, reduce the amount of ATP produced, and raise the production of ROS. For example, mercury and arsenic affect the mitochondrial membrane potential and lead to the release of cytochrome *c*, which starts apoptosis. In addition, lead alters calcium handling into and out of mitochondria, causing energy failure and subsequent neuronal loss (Sun et al., 2022). Since motor neurons have high energy requirements, the disruption of mitochondrial components primarily contributes to their degeneration.
- **Excitotoxicity:** The mechanism of neuronal pathology in ALS also involves excitotoxicity due to overstimulation of the receptors after the release of the excitatory neurotransmitter glutamate. Heavy metals have been found to enhance excitotoxicity by enhancing the release of glutamate, blocking its reabsorption or increasing the sensitivity of the glutamate receptors. For example, lead inactivates glutamate transporters, causing increased intracellular concentrations of synaptic glutamate and the hyperactivation of NMDA (Liu et al., 2017). This leads to calcium overload, activation of calcium-dependent proteins, and, finally, causes neuronal cell death. Mercury has also been shown to affect the glutamate transporters, thus increasing excitotoxicity.

15.2.3 Role of heavy metals in disrupting the blood–brain barrier and increasing brain vulnerability

A selective structure called the BBB shields the central nervous system (CNS) from dangerous chemicals, including heavy metals. However, heavy metal exposure can weaken the BBB, making the brain more susceptible to neurotoxic insults:

- **Structural damage to the BBB:** The direct toxic attributes of lead, mercury, or arsenic are known to compromise the endothelial cells and the tight junction proteins responsible for preserving the BBB. For example, lead causes decreased occludin and claudin-5 levels, implying that lead raises the permeability of the BBB. This lets more dangerous compounds, other heavy metals, get to the CNS to extend the damage to neurons.
- **Induction of neuroinflammation:** Heavy metals trigger several glial cells in the CNS, such as microglia and astrocytes. These cells produce proinflammatory cytokines, such as TNF-α and IL-6, that compromise the stability of the BBB and sustain inflammation (Yang et al., 2022). For example, cadmium has been shown to

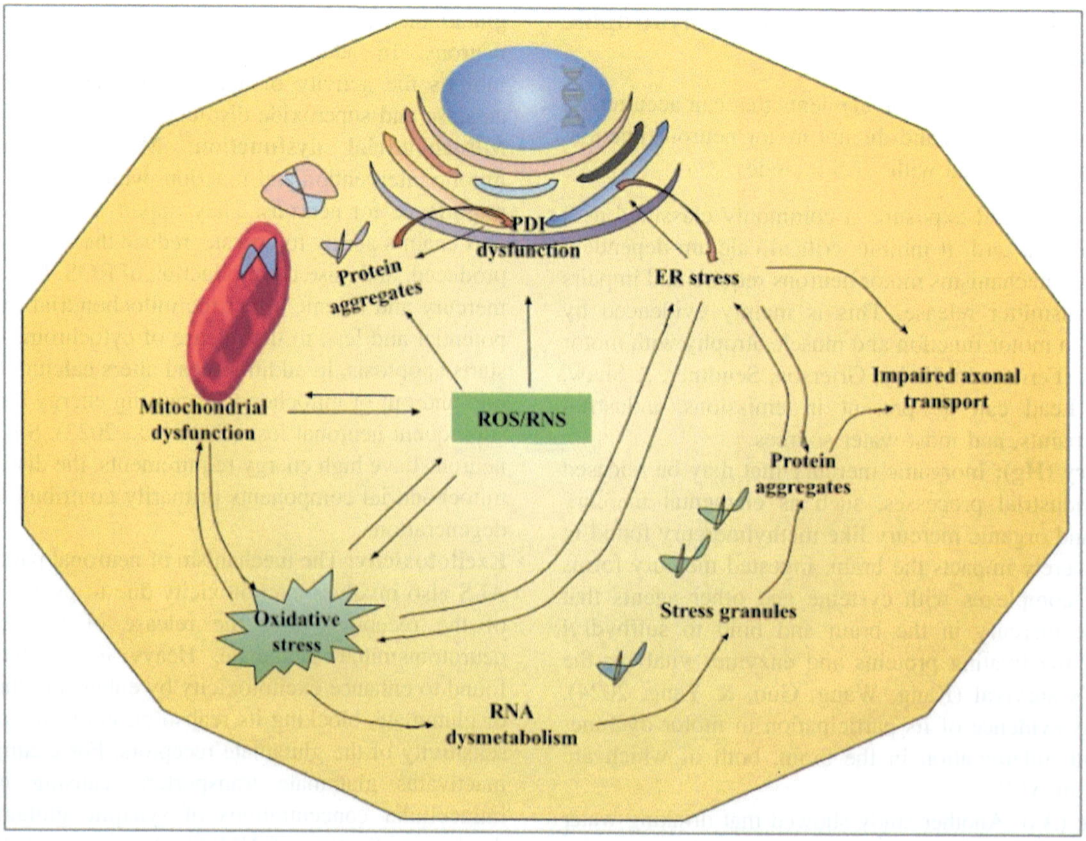

FIGURE 15.2 Mitochondrial dysfunction and oxidative stress are interconnected and form the foundation of redox imbalance observed in amyotrophic lateral sclerosis.

upregulate the inflammatory molecules, leading to BBB breakdown and neurotoxicity.

- **Facilitated transport across the BBB:** Some metal ions, such as Hg and As, can pass through the BBB through carrier-mediated processes. For instance, methylmercury forms a complex with cysteine and substitutes for methionine, so it enters the CNS through transporters of amino acids. These metals enter the body, and once inside, they build up in neuron tissue, where they produce their destructive effects.
- **Increased CNS susceptibility:** When the BBB integrity becomes compromised, not only do heavy metals accumulate in the brain, but the brain also becomes more vulnerable to other toxins, pathogens, and inflammation. However, this fosters a situation where neurotoxic effects are self-enhancing, undermining the already diminished prospects of maintaining the motor neurons' survival rate.

15.3 Epidemiological evidence

Added epidemiological studies have been made about the provision of heavy metals and neurodegenerative disorders such as ALS. While these works provide valuable data concerning potential risk factors to the environment, they also complicate the possibility of determining causality. This part discusses the epidemiological information, evaluates various study methods, and discusses the challenges of associating ALS with the environment.

15.3.1 Epidemiological studies on heavy metal exposure and amyotrophic lateral sclerosis risk

Numerous epidemiological studies have investigated the potential role of heavy metals such as lead (Pb), mercury (Hg), arsenic (As), and cadmium (Cd) in increasing ALS risk.

- **Case-control studies:** Selective exposure analysis compares the level of heavy metal exposure of ALS patients (cases) to that of patients without the disease (controls). For instance, some investigations pointed to a statistically higher blood or bone lead level in ALS patients compared with respective controls, suggesting that lead increases the risk of the disease. Likewise, high levels of mercury content have been found in ALS individuals, especially those with occupational or diet-induced exposure to this metal (Farace et al., 2020).
- **Cohort studies:** Case-control studies start with patients and healthy individuals diagnosed with ALS and compare differences in heavy metal exposure. Several

pharmaceutical-based cohort investigations have found that working with metals, such as welders or workers in smelting industries, increases one's chances of developing ALS. However, these studies are comparatively scarce owing to the protracted duration of the latency period of ALS and the difficulties inherent in assessing exposure longitudinally (Toader et al., 2023).

- **Ecological studies:** Explosive population-based epidemiology reviews the association of diseases with the environment without considering their causative agents. For instance, investigations that have compared regions with high concentrations of arsenic in the water supply have found ALS groupings. However, these conclusions are usually tainted by other regional disparities, including genetic makeup or exposure to other poisons.

15.3.2 Strengths and limitations of study designs

All the epidemiological study types have advantages and disadvantages when assessing the link between the exposition to heavy metals and ALS.

- **Ecological studies can be used to understand various population-level data and the geographical distribution of ALS cases. They can select prominent areas for further investigation**. The absence of patient-level variables could help with risk adjustment. Regularities are programmed patterns of interactions, which means that they cannot create unique reasons for their cause and effect, as could be observed from other environmental or genetic factors, which could give similar results.
- **Case-control studies:** These are more efficient regarding time and expense than cohort studies. Enable the elucidation of specific risks in interviews, biomonitoring studies, or occupational questionnaires. Many disadvantages are distinguishable, including the retrospective nature of the analysis that allows the introduction of the recall bias: participants can recall previous exposure inaccurately. It can also create selection bias; cases and controls may differ in ways other than exposure to the particular factor under study.
- **Cohort studies:** Designed as a prospective approach, this helps to avoid recall bias, and temporal associations can be determined. Ensure proper exposure data throughout the treatment period that decreases the mistake level. It takes much time in many cases, especially where diseases have prolonged incubation or dormancy periods, such as ALS. Early attrition and incomplete exposure data owing to long study periods.

15.3.3 Challenges in establishing causal relationships

However, as epidemiological research expands, qualifying heavy metal exposure to ALS as a cause and effect remains an issue due to the combined influence of environmental genetics and lifestyle.

- **Multifactorial etiology of ALS:** ALS is the quintessential multicausal disorder with the components of gene-environment interaction leading to aging at the molecular level (Tandon, Singh, & Chaturvedi, 2021). This makes it hard to understand the part played by heavy metals in developing diseases, a subject that has led to several hypotheses being postulated. For example, synergism of the genetic factors, such as SOD1 or C9orf72 mutations, and environmental factors, like heavy metals, is poorly understood.
- **Exposure assessment limitations:** Determining the extent of exposure to heavy metals has been an enormous challenge. Serological and/or hair analysis yields data on recent exposure, but it rarely indicates the prolonged or long-term exposure that could be associated with ALS. Likewise, using self-reported occupational histories or regional datasets comes with some inconsistencies.
- **Confounding factors:** It should be noted, however, that heavy metal exposure is usually bound with other environmental or lifestyle factors like smoking, diet, and pesticides. Sparser separation of these confounders is almost always possible, but it takes complex statistical techniques and study designs.
- **Latency period:** sALS often presents many years after exposure, making it challenging to decipher temporal associations. Long latency means that strategies require prospective investigations with a long follow-up period. The nature of the findings is logistically and financially challenging.
- **Genetic and environmental interactions:** Heredity may predispose one to the toxic effects of heavy metals. For instance, low antioxidant enzyme activities can predispose an individual to oxidative injury from heavy metals. Such gene-environment interactions are complex to model in epidemiological investigations and need joint consideration of genetic, environmental, and mechanistic information.

Heterogeneity in study populations: Random/example errors are factors such as genetic difference, sampling area,and different levels of SES and/or exposure levels in a study. For instance, occupational studies entail a pervasive and systematic investigation of unique occupations and fields; such studies may only cover male probabilities and omit other probabilities in women or other categories. Solving these issues requires enhanced exposure assessment methods, enhanced study subject numbers and heterogeneity, and a systematic analytic consideration of genetic, molecular, and environmental data. Epidemiological studies in this area are still relevant in understanding the contribution of heavy metals in ALS and promoting population-level prevention measures against the sources of exposure (Tables 15.1 and 15.2).

TABLE 15.1 Epidemiological evidence: heavy metal exposure and amyotrophic lateral sclerosis (ALS) risk.

Aspect	Description	Study design	Strengths	Limitations	Challenges	References
Association evidence	Studies investigating the link between heavy metals (e.g., lead, mercury, arsenic) and ALS risk	Ecological, case control	Initial insight into potential risk factors	Lack of individual exposure data	Confounding factors like genetics	Masrori and Van Damme (2020)
Ecological studies	Studies using population-level data to identify exposure-disease patterns	Ecological	Cost-effective, broad trends	Lack of individual-level exposure	Ecological fallacy	Tandon et al. (2021)
Case-control studies	Comparing heavy metal exposure in ALS patients and healthy controls	Case control	Suitable for rare diseases, detailed data collection	Recall bias, limited temporal relationships	Exposure misclassification	Landrigan, Lucchini, Kotelchuck, and Grandjean (2022)
Cohort studies	Tracking individuals over time to observe heavy metal exposure and ALS onset	Cohort	Temporal data, more substantial causality evidence	Expensive, time-consuming	Loss to follow-up, long latency of ALS	Toader et al. (2023)
Lead exposure	Investigating the role of lead in ALS risk	All study types	Historical occupational exposure data available	Difficulty in quantifying cumulative exposure	Individual susceptibility variations	Mathis et al. (2019)
Mercury exposure	Examining mercury as a neurotoxin potentially linked to ALS	Case control, cohort	Focused studies available	Small sample sizes	Uncertain dose–response relationships	Farace et al. (2020)
Arsenic exposure	Exploring arsenic's neurotoxic effects and ALS association	Ecological, cohort	Widely studied in contaminated regions	Interaction with other environmental factors	Biomarker variability	Tandon et al. (2021)
Biomarker studies	Using biomarkers like blood, hair, or urine to measure metal exposure	Case control, cohort	Objective exposure data	Biomarker half-life variability	Temporal mismatches between exposure and disease	Landrigan et al. (2022)
Genetic interactions	Investigating how genetics influence heavy metal susceptibility	Case-control genetic studies	Identifying gene-environment interactions	Small sample sizes	Complex interplay between genes and metals	Toader et al. (2023)
Smoking and metals	Examining smoking as a confounding factor in heavy metal studies	Cohort, case control	Identifies lifestyle factors	Self-reported data may be unreliable	Differentiating direct metal effects	Mathis et al. (2019)

Occupational exposure	Assessing ALS risk in metal-exposed occupations	Cohort, case control	Reliable occupational exposure data	Limited to specific industries	Generalizability to the broader population	Farace et al. (2020)
Environmental exposure	Studying ALS risk from environmental contamination sources (e.g., water, soil)	Ecological, cohort	Broad coverage of affected regions	Lack of individual exposure verification	Multiple coexposures	Tandon et al. (2021)
Dose–response relationship	Assessing how varying levels of heavy metals influence ALS risk	Case control, cohort	Provides quantitative risk insights	Uncertainty at low exposure levels	Establishing thresholds	Landrigan et al. (2022)
Longitudinal studies	Following individuals over decades to monitor exposure and ALS outcomes	Cohort	Strong temporal evidence	Long duration, high costs	Long latency of ALS symptoms	Toader et al. (2023)
Geographic patterns	Mapping ALS incidence in regions with high heavy metal exposure	Ecological	Highlights high-risk areas	Lack of individual-level data	Spatial confounding	Mathis et al. (2019)
Clinical observations	Observing ALS progression in patients with high metal exposure	Case studies	Insight into disease mechanisms	Lack of controls, limited generalizability	Individual variability	Farace et al. (2020)
Metaanalyses	Combining data from multiple studies on heavy metals and ALS	All study types	Improved statistical power	Publication bias, heterogeneity	Inconsistent study methodologies	Tandon et al. (2021)
Exposure measurement techniques	Methods used to quantify heavy metal exposure (e.g., air, water, biomarkers)	All study types	Advances in technology enable better measurements	Costly and complex equipment	Lack of standardization	Landrigan et al. (2022)
Reverse causation risks	Considering whether ALS changes behavior and increases metal exposure	Case control	Challenges existing causal assumptions	Difficult to rule out	Need for robust temporal data	Toader et al. (2023)
Establishing causality	Evaluating whether heavy metals are a cause or correlate of ALS	Cohort, experimental	Temporal relationships support causality	Ethical concerns in experimental designs	Multicausality in complex diseases	Mathis et al. (2019)

TABLE 15.2 Various biomarkers, their strengths, limitations, and relevance to assessing heavy metal exposure and its potential impact on amyotrophic lateral sclerosis (ALS) progression.

Biomarker type	Description	Strengths	Limitations	Relevance to ALS	References
Blood biomarkers	Measure heavy metal levels directly in the bloodstream	Provides current exposure levels	Short half-life for some metals	Useful for acute exposure assessment	Wu et al. (2016)
Urine biomarkers	Reflects excreted metals and metabolic byproducts	Noninvasive collection tracks recent exposure	Variability in excretion rates	It may not reflect cumulative exposure	Huang et al. (2014)
Hair biomarkers	Measures metals deposited in hair over time	Reflects long-term exposure	Susceptible to external contamination	Helpful for chronic exposure studies	Pal et al. (2022)
CSF	Assesses biomarkers directly linked to the central nervous system impact	Closely reflects nervous system status	Invasive procedure, limited availability	Potential for tracking neurological damage	Apontes (2022)
Nails	Tracks heavy metal deposition over months	Represents long-term exposure	Susceptible to environmental contamination	Emerging biomarkers for chronic exposures	Shabir et al. (2023)
Saliva	Noninvasive method for detecting some metals	An easy and quick sampling	Limited detection for some metals	May have applications for rapid screening	Sakowski, Koubek, Chen, Goutman, and Feldman (2024)
Bone	Reflects long-term storage of heavy metals like lead	Indicates cumulative exposure	Highly invasive, not routinely used	A strong indicator of historical lead exposure	Oliveira et al. (2021)
Teeth	Reflects exposure during development phases	Retains lifetime exposure history	Sample acquisition challenges	Useful for pediatric and early exposure studies	Steinacker, Barschke, and Otto (2019)
Plasma protein levels	Measures metal-binding proteins (e.g., metallothionein)	Indicates the body's response to metal exposure	May not differentiate between exposure and effect	Insight into biological response to metals	Rahmanian, Shokrzadeh, and Eskandani (2021)
Oxidative stress markers	Measures oxidative damage caused by metal exposure	Links metal exposure to cellular damage	Complex to interpret	Relevant for ALS pathology involving oxidative stress	Farace et al. (2020)
DNA damage indicators	Evaluate the genotoxic effects of heavy metals	Links metals to genetic damage	Requires advanced techniques	Highlight the role of metal-induced mutations in ALS	Wu et al. (2016)

Metal concentration ratios	Analyses ratios of different metals to assess imbalance	Identifies specific patterns of metal toxicity	May miss single-metal toxicity	Differentiates complex exposure profiles	Huang et al. (2014)
Immune markers	Measures immune response linked to heavy metal exposure	Reflects systemic inflammation	Not specific to metals	It helps connect exposure to neuroinflammation	Pal et al. (2022)
Neurofilament levels	Indicates nerve damage due to heavy metal toxicity	Correlates with neurodegeneration	Limited specificity to metals	Tracks ALS progression linked to exposure	Apontes (2022)
Metalloprotein analysis	Examines proteins that interact with metals	Highlights disruptions in metal-protein balance	Requires specialized techniques	Indicates biological effects of metal toxicity	Shabir et al. (2023)
Epigenetic modifications	Detects changes in DNA methylation linked to metal exposure	Links exposure to gene expression changes	Emerging area, complex interpretation	Investigates epigenetic links to ALS	Sakowski et al. (2024)
Mitochondrial markers	Evaluates mitochondrial damage due to heavy metals	Connects exposure to cellular energy disruption	Requires advanced technologies	Reflects ALS-related mitochondrial dysfunction	Oliveira et al. (2021)
Trace elements in CSF	Measures trace metals directly affecting the brain	Provides direct insight into neurotoxicity	Invasive collection process	Tracks metals involved in neurodegeneration	Steinacker et al. (2019)
Lipid peroxidation products	Indicates oxidative damage to lipids from heavy metals	Links exposure to oxidative stress	Lack of specificity	Links to oxidative damage in ALS pathology	Rahmanian et al. (2021)
Glutathione Levels	Reflects antioxidant response to heavy metal exposure	Indicates cellular response to oxidative stress	Sensitive to storage and handling conditions	Relevant for assessing metal-induced oxidative stress	Farace et al. (2020)

CSF, Cerebrospinal fluid.

15.4 Biomarkers of exposure and effect

Biomarkers are usually essential in interpreting the relationship between environmental stimuli, such as heavy metals, and their impact on neurodegenerative illnesses like ALS. Biomarkers also help to determine the amounts of exposure, understand biological effects, and estimate the further evolution of the illness. This section explains how biomarkers help determine heavy metal exposure and its effects on the nervous system, identifies various biomarkers, and appraises their advantages and disadvantages in estimating ALS risk and trends (Sakowski et al., 2024).

15.4.1 Biomarkers for assessing heavy metal exposure

Biomarkers of exposure are scores that quantify an individual's interaction with heavy metals like lead (Pb), mercury (Hg), arsenic (As), and cadmium (Cd). These biomarkers are typically derived from biological samples and reflect the concentration of metals in the body:

- **Blood analysis:** Blood is one of the most used indices for evaluating heavy metal amounts in the human body. For instance, blood lead levels are established as a valid biomarker of recent lead exposure and are incorporated broadly into environmental health research (Oliveira et al., 2021). The total mercury in the blood corresponds to short-term exposure, especially to organic species such as methylmercury. Blood biomarkers are desirable because samples are standardised and reflect exposure during sampling.
- **Urine analysis:** Urine is also slightly used to assess exposure to heavy metals because many of these are eliminated, such as arsenic and cadmium. The level of arsenic in the urinary sample reflects recent exposure to the metalloid, particularly in areas with affected groundwater. It is important to note that speciation analysis can differentiate between inorganic and organic species of arsenic, of which speciation information can be used to identify the source of exposure. Urinary cadmium is frequently utilized to measure integrated exposure since it indicates the amount of cadmium accumulated in the kidneys (Wu et al., 2016).
- **Hair and nail analysis**: Hair and nails integrate exposure throughout hair growth and are thus representative of long-term exposure to heavy metals. For instance, hair concentrations of methylmercury reflect the population's dietary intake of fish contaminated with this substance over several months. Hair analysis does not infringe on the body's integrity and is well suited for applications in which past exposure data ae needed; however, hair analysis is vulnerable to external influence (Huang et al., 2014).
- **Cerebrospinal fluid (CSF) analysis:** CSF could be a biomarker to evaluate the immediate consequence of cerebral exposure to heavy metals. CSF metal levels, such as arsenic or lead, should be raised to indicate neurotoxicity and the severity of CNS exposure. CSF biomarkers are mainly used in neurodegenerative diseases such as ALS, in which the CNS is always engaged (Pal et al., 2022).

15.4.2 Biomarkers of biological effect

Biomarkers of effect quantify the biological alteration from exposure to heavy metals. These biomarkers reflect the mechanisms through which heavy metals exert their neurotoxic effects and provide insights into disease risk and progression:

- **Oxidative stress biomarkers:** One key way HBTC is generated is by heavy metals-induced oxidative stress. Biomarkers of lipid peroxidation can be malondialdehyde (MDA), DNA damage can be estimated using 8-hydroxy-2′-deoxyguanosine (8-OHdG), and protein oxidation can be measured for protein carbonyls. For instance, a higher concentration of MDA has been noted in clients with lead and arsenic poisoning, indicating higher oxidative stress in motor neurons (Demirci-Cekic et al., 2022).
- **Inflammatory biomarkers:** Some heavy metals can trigger inflammation and promote neurodegeneration. Other adjuvant production is usually increased, for instance, IL-6, TNF-α, and CRP, in people with neurotoxic metal exposure. These proteins are upregulated in ALS patients, and they have a strong inverse relationship with TDP43—a known biomarker of ALS and the extent of the disease—suggesting that they could be promising biomarkers for metal-induced neuroinflammation (Steinacker et al., 2019).
- **DNA damage biomarkers:** Cadmium and arsenic are described as causing DNA damage by generating ROS or directly interacting with the DNA. γ-H2AX, a biomarker that can detect DNA double-strand breaks and micronuclei formation, is well established for genotoxicity detection. Biomarkers of DNA damage are of special significance in assessing the mutagenic activity of heavy metals and their involvement in the development of ALS (Rahmanian et al., 2021).
- **Neurodegeneration-specific biomarkers:** Biomarkers associated with neuronal damage are seen more frequently in ALS investigations today; these include NfL and tau proteins. Higher NfL in CSF or blood reflects axonal injury and positively correlates with disease progression in ALS. These biomarkers afford a connection between heavy metal exposure, neuronal toxicity, and ALS progression (Apontes, 2022).

15.4.3 Strengths and limitations of biomarkers for amyotrophic lateral sclerosis risk and progression

The utility of biomarkers in assessing heavy metal exposure and its impact on ALS is influenced by their strengths and limitations:

- **Blood biomarkers:** These are useful for assaying recently aggravated exposure to metals such as lead and mercury. This may allow the development of study-standardized implementation and data collection or analysis, helping to compare the results of different studies. However, they are not suitable for evaluating total or past exposure. They are not a representative measure of tissue-specific accumulation, especially in the CNS.
- **Urine biomarkers:** These are good for metals excreted through the renal system, such as arsenic and cadmium. They can be measured without causing harm to the participants, and it is relatively easy to get data from a large sample population. Plasma and urine levels are dependent on the hydration state. Depending on the timing and/or sample location, they can give a false-positive/negative result.
- **Hair and nail biomarkers:** Document an earlier history of exposure, especially to metallic substances such as mercury. They are medically less invasive and can be quickly done retrospectively on a set of patients. Potentially impacted by outside sources, especially environmental influences. Of course, the reason for the growth and normalization of these types of metals makes it challenging to analyze the resulting variability of growth rates and deposition of metals.
- **CSF biomarkers:** They directly contribute to CNS exposure and neurotoxicity. They are applicable when seeking information on the neurodegenerative impact of heavy metals. Invasive collection methods cannot be used for feasible studies, especially when the sample size is enormous (Shabir et al., 2023). Apa, there is less information concerning reference values and variability for ALS patient populations.
- **Biomarkers of effect:** Acquire knowledge of the molecular and cellular changes in heavy metal-induced neurotoxicity. Give early signals of biological transformations, which indeed help in risk determination. They may not be specific for work, and therefore, exposure to heavy metals may be confounded by oxidative stress and inflammation common to multiple neurotoxicity. Highly technical in analysis and variable from one case to another, complex analytical methods options are beneficial clinically.

15.5 Environmental considerations

Recent surveys have revealed evidence of electromagnetic pollution, which raises the Prospect threat to human health, especially in neurodegenerative diseases such as ALS. Little has been done to address the sources, exposure routes, and ways to prevent the increased global prevalence of toxic heavy metals. This section also discusses the sources of heavy metal pollution, how people can be exposed to the metals and the need for environmental legislation and health reforms.

15.5.1 Sources of heavy metal contamination

Lead (Pb), mercury (Hg), cadmium (Cd), and arsenic (As) are some of the typical metal toxins that find their way into the environment through various natural and artificial activities. The most significant sources include:

- **Industrial emissions:** Metallurgical, manufacturing, and fossil fuel-burning industries emit large amounts of heavy metals into the environment. For instance, lead from the degradation of lead-acid batteries or mercury from burning coal is, in considerable measure, responsible for pollution. On becoming airborne, these metals stick to the top layers of soil and water since they are nonbiodegradable (Raj, Kumar, & Vaghasiya, 2021).
- **Mining and ore processing:** Mining processes disrupt immense amounts of soil, which puts heavy metals into surrounding areas. As dumping sites or waste products from mines, these materials possess high levels of arsenic, cadmium, and other toxic elements. The discharge of this material, either freely or inadequately contained, poses a threat of seeping into water sources or endangering agricultural lands. For example, arsenic poisoning of the water supply is well-known in areas where mining plays a significant role.
- **Agricultural practices:** Agricultural activities are another source of heavy metal contamination. Phosphate fertilisers usually contain minute quantities of cadmium, and repeated application results in the build-up in the soil. Also, over the years, pesticides such as arsenic-containing pesticides and fungicides with mercury as the active ingredient have polluted agricultural land and produce (Raj, Dubey, Malla, & Kumar, 2023). The situation only gets worse when such water is used for irrigation, as this spreads metals to the food chain (Fig. 15.3).
- **Natural sources:** Heavy metals can also be mobilized naturally by volcanic activities, weathering rocks containing heavy metals, and leaching from mineral deposits. Although these sources are less prominent than anthropogenic ones, they can lead to localized pollution (Kumar, Goud, Devi, Dey, & Dwivedi, 2022a, 2022b).

15.5.2 Pathways of human exposure

Forces human encounters with heavy metals through different environmental domains, including the gaseous,

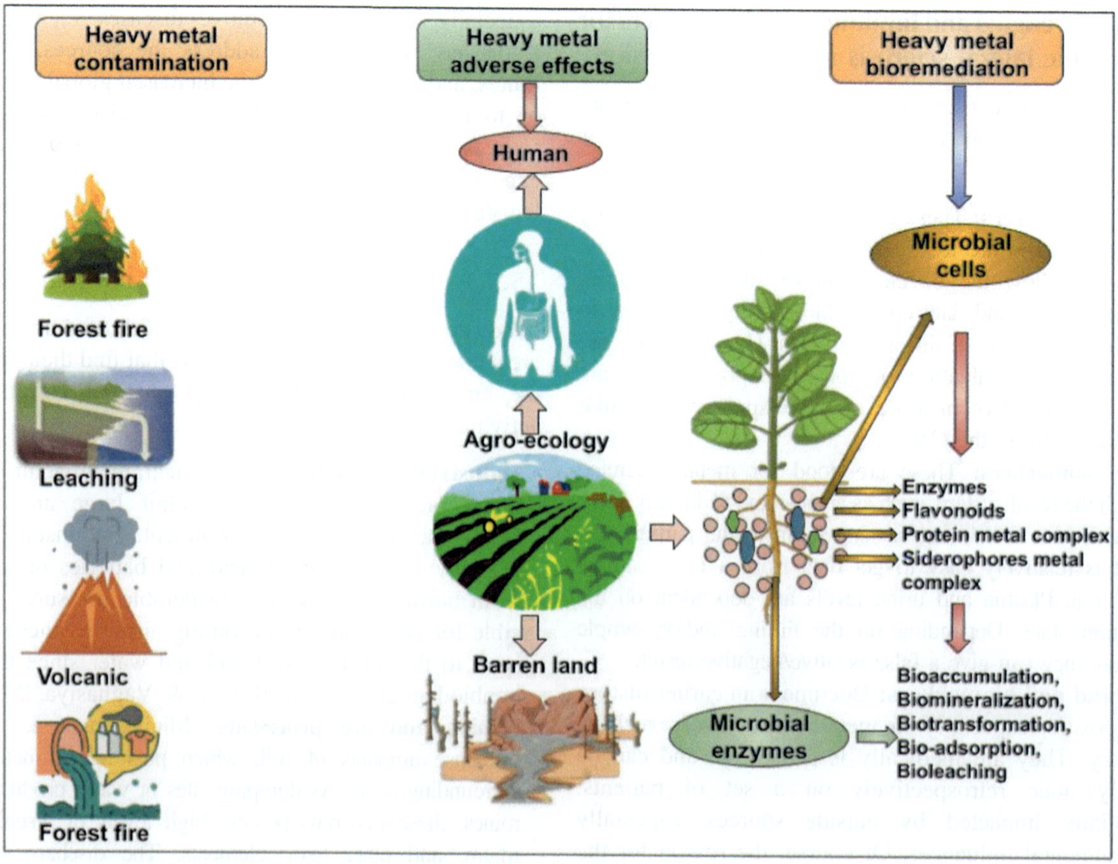

FIGURE 15.3 Major sources of heavy metal contamination and its effect at different trophic levels.

aqueous, and ingesting food products. Each pathway plays a crucial role in determining the extent and type of exposure:

- **Airborne exposure:** One of the primary ways heavy metal enters the human body is through breathing in particulates. Tiny particles of lead, mercury, or cadmium emitted into the atmosphere from industries or automobile emissions tend to remain suspended for a long time, a factor that raises the likelihood of respiratory contact (Kumar, Goud, Devi, Dey, & Dwivedi, 2022a, 2022b). Ingested, they can pass into the lungs, where they may be swallowed, or they can enter directly into the bloodstream and find their way into such parts of the human organism as the brain. For instance, inhaling lead in urban air quality was confirmed to cause neurological impacts.
- **Waterborne exposure:** Water pollution is another common source of heavy metals, most of which are contaminated water sources. Metals like arsenic, cadmium, and mercury get into the surface and groundwater through industrial effluents, agricultural leaching, and leaching from mining areas (Kumar, Choudhury, Dey, & Sharma, 2024). Waterborne arsenic toxicity is a recognized phenomenon in many parts of the world, especially in the South Asian subcontinent. Over time, the accumulation of heavy metals in the human body harms the nervous system, kidneys, and other organs.
- **Foodborne exposure:** Ingestion of contaminated food is one of the major routes through which heavy metals are taken in the body. Vegetables cultivated in metal-contaminated soils or water containing high levels of heavy metals may contain dangerous concentrations in their edible tissues (Kumar et al., 2024). For example, the arsenic accumulation in rice grains that are grown from arsenic-contaminated soils acts as the cause of arsenic exposure. Like methylmercury, mercury also biomagnifies in aquatic food chains, and tuna and swordfish contain high levels of this toxin. These metals can be absorbed through foods, which is very dangerous in the long term due to the possibility of bioaccumulation and neurotoxic effects.
- **Occupational exposure:** Employers, miners, smelters, and battery makers are most prone to exposure to heavy metals. Primary exposure to metal-containing dirt, smoke, or infected body parts often severely affects human health due to insufficient protective equipment or measures (Landrigan et al., 2022).

15.5.3 Importance of environmental regulations and public health interventions

Addressing heavy metal contamination requires a multi-faceted approach involving stringent environmental regulations, public health interventions, and community awareness:

- **Environmental regulations:** Governments and international organizations are responsible for controlling heavy metal emissions or contamination. Measures like checking emissions from industries, withdrawing leaded gasoline, and disposing of environmental waste have greatly minimized pollution worldwide (Saini, Kumar, & Singh, 2024). For example, the Minamata Convention on Mercury has been signed to reduce mercury emissions worldwide by adopting cleaner industrial processes and eliminating mercury-containing products.
- **Monitoring and remediation efforts:** The health of ecosystems through periodic sampling of air, water, and soil samples for heavy metals should be monitored continuously to contain this problem early enough. Society has progressively become capable of identifying depressions and hot and contaminated areas through recent geochemical mapping and remote sounding. Contaminative mitigation strategies used in remediating the polluted environment include phytoremediation or using plants to remove metals or chemically stabilize contaminated soil (Kumar & Choudhury, 2024).
- **Public health interventions:** It is well understood that there are barriers to the effective action of public health policies and campaigns to prevent human exposure to heavy metals. These include passing information to communities on the effects of consuming food or water containing metals, encouraging nonheme iron sources to reduce the accumulation of metals in the body and delivering safe water for drinking. For many communities exposed to arsenic-contaminated water, measures, including arsenic removal filters or the provision of safe water sources, have played a role in keeping arsenic effects low (Raman et al., 2022).
- **Occupational safety measures:** Safety measures must be adhered to within work settings that have already identified high risks of exposure. Workers should wear protective gear, especially after work duties, check their health, and install better air circulation methods to minimize exposure.
- **Global collaboration and research:** Cleaning up heavy metals is not a one-country affair. It calls for international effort, especially in managing impacts that cross national boundaries. The priority research activity in this regard is to study the long-term effects of such exposure on human bodies and the environment. Effective and efficient metal detection, analysis, and removal strategies are still significant research and policy issues.

15.6 Public health implications

ALS connection with heavy metals also poses shocking health implications; therefore focusing on metal exposure is critical. Lead (Pb), mercury (Hg), cadmium (Cd), and arsenic (As) are well-established neurotoxicants; current literature points to oxidative stress, mitochondrial dysfunction, and neuronal damage. It is essential to know such implications to minimize risk exposure, especially for the most vulnerable persons, and to formulate health policies. This section considers the general effects of the problem and points out the necessity of further research and the importance of prevention.

15.6.1 Heavy metals and amyotrophic lateral sclerosis: implications for public health

Heavy metal toxicity is a danger to life; this is because it has been mentioned to be associated with ALS, a chronic illness. The following aspects highlight its public health significance:

- **Rising ALS incidence:** ALS is a highly communicated neurodegenerative ailment with a rising occurrence rate worldwide. Due to the plausible involvement of environmental agents such as heavy metals in the etiology of ALS, awareness of ALS as a public health problem is critical. Their chronic effects of lead, cadmium, or arsenic, even at "low" doses, increase neurotoxicity, raising concerns for affected communities due to environmental pollution (Rehman, Rehman, & Akash, 2021).
- **Impact on vulnerable populations:** People of specific ages, genders, races, and living areas are most affected by toxic heavy metals. These include industrial workers, those working in mining or smelting areas directly affecting the communities, and farmers in touch with polluted water or soil. Further, high risk is compounded by social inequality as such communities are exposed to contaminated water, poor diet and sanitation, and negligible medical care (Ball, Teo, Chandra, & Chapman, 2019).
- **Cumulative and synergistic effects:** Heavy metals are fixed in the environment, and their concentration builds up in human tissues over time. Also, direct blends of different metals are often used, which could have multiplicative toxicity on the nervous system. For example, studies indicate that the interaction between lead and cadmium makes neurons more vulnerable to oxidative stress and diseases such as ALS. These findings underscore the importance of tightly constrained public health measures to minimize exposure.

15.6.2 Need for research on causal mechanisms and vulnerable populations

However, little is still known about the mechanisms through which heavy metal induces neurodegeneration of motor neurons or which ALS patients may be heavily exposed to these metals. Addressing these gaps is crucial for developing targeted public health strategies:

- **Elucidating mechanisms of neurotoxicity:** Various toxic heavy metals are known to interfere with cellular health through oxidative stress anabolism, dysfunction of mitochondria, and excitotoxicity. Nevertheless, the roles of these processes in the context of ALS are not fully understood, although their contribution is identifiable. Further investigations about toxic heavy metals and how these affect genetic and epigenetic modifications related to ALS are required. For instance, understanding oxidative stress induced by heavy metals in SOD1 mutant models may be invaluable to elucidating the disease mechanism (Goutman, Savelieff, Jang, Hur, & Feldman, 2023).
- **Epidemiological studies:** Case-control studies require larger samples to determine the associations between exposure to the specified heavy metals and ALS risk. Some CC and CS have shown positive associations; however, most CC studies were limited by small samples, recall and selection biases, and methodological flaws. More rigorous research that follows patients over a long period can provide more information on how exposed people are to these metals, how long such exposure takes, and the impact of multiple metal toxicity.
- **Identifying vulnerable populations:** Research is needed to determine who is at the greatest risk of coming into contact with these heavy metals. This encompasses evaluating workplace risks within sectors such as mining and metallurgical fields, as well as environments contaminated with water and/or soil. Genetic predisposition studies may also assist in identifying other high-risk consumers of heavy metal neurotoxins.
- **Biomarker development:** Identifying new, accurate, and valid biomarkers of heavy metal exposure and its early neurotoxicity is relevant. Urine oxidative stress, inflammation, and neuronal damage biomarkers may help initiate early interventions and shed more light on the metal–ALS connection (Benatar et al., 2016).

15.6.3 Preventive measures to protect public health

Due to these concerns, emergency measures are necessary to prevent heavy metals from threatening the general public's health. Key strategies include:

- **Environmental regulations:** Increasing measures restricting industrial discharges, mining wastewater, and agricultural effluents are essential. Strategies like the Minamata Convention on Mercury and the phase-out of leaded gasoline help lower exposure levels. Governments need to step up the regulation of heavy metal emissions, especially in industries that are most prone to them (Bartley & Fairbrother, 2025).
- **Water and soil remediation:** Water and soil pollution caused by heavy metal release are a significant method of human interaction with a contaminated environment. Phytoremediation, including the use of plants to remove metals, soil washing, and water treatment using arsenic removal filters, helps control contaminant concentration. More of these interventions should be implemented in areas with high environmental metal loads (Devi, Kaur, & Kumar, 2024).

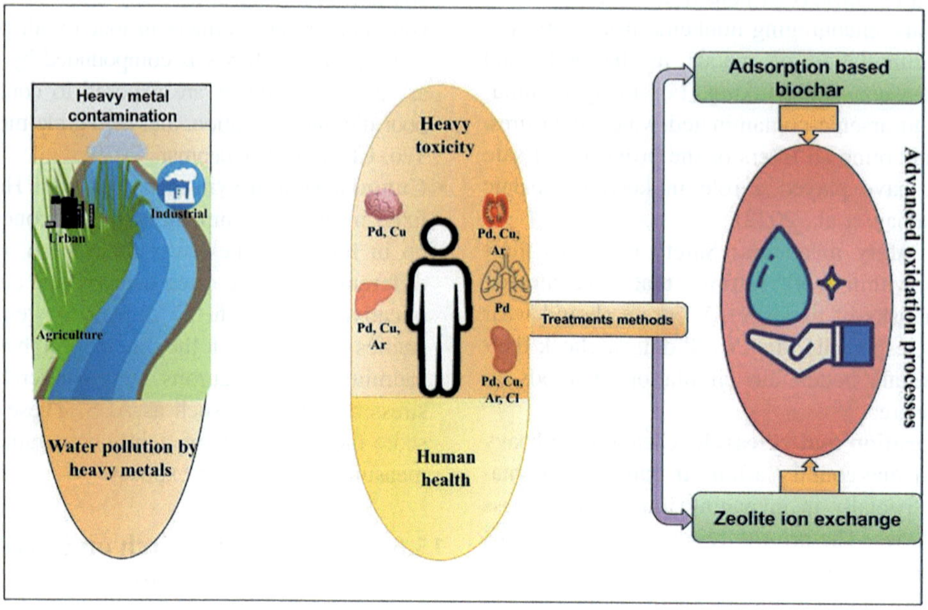

FIGURE 15.4 A comprehensive examination of heavy metal pollution sources, their correlated health hazards, and cost-effective, efficient strategies for the remediation of heavy metals from wastewater.

- **Public health education:** It is imperative to increase awareness of the dangers of high metal content to mobilize people to take protective measures. Educational campaigns usually discuss sources of contamination, like contaminated fish or uncontrolled cosmetics, and encourage safe practices that will help avoid metal accumulation in the bodies.
- **Occupational safety:** Employees in high-risk activities should be issued with PPEs, tested, and trained on proper food handling practices. The risk of occupational exposure can be reduced even further by increasing workplace ventilation and developing exposure limits.
- **Nutritional interventions:** Dietary interventions might reduce heavy metals' sev/dev/toxic impact. Oxidative damage due to metals like lead and cadmium may be prevented by diets containing antioxidants like vitamins C and E. Dental amalgam—which contains both mercury and arsenic heavy metals—and selenium and zinc intake can compete for the absorption sites.
- **Community engagement:** Implementing public health interventions in a particular community always entails various decisions being made by the people involved, making it very important to involve the community in all decision-making processes (Leask et al., 2019). Informed communities are better placed to implement preventive measures when contamination is noticed, report contamination cases, and demand cleaner places (Fig. 15.4).

15.7 Future directions and research needs

As heavy metals are linked with neurodegenerative diseases such as ALS, there is a critical medical need for more efficient identification of the causal pathways, biomarkers, and amelioration methods. Although some advancements have been made in the present decade, much work remains to be done to explain the relationship between the environment and ALS. Here, the authors define the direction for further research, stress the need for an interdisciplinary approach, and underline the need to apply the scientific findings into practice and improve population health.

15.7.1 Key areas for future research

To address the multifaceted relationship between heavy metals and ALS, targeted research efforts are needed in the following areas:

- **Large-scale epidemiological studies:** Recent epidemiological investigations comparing the level of concentrations of heavy metals to that of ALS are still lacking in adequate sample size, geographical representativeness, and confounding factors. However, this would require prospective studies with a large number of populations taken from different centers where the protocols are already uniform and an ideal chance to determine which group is at the highest risk. Evaluations based on exposure to such heavy metals during the follow-up duration could provide valuable information about the dose–response relationship of those metals and ALS development or course (Bryan et al., 2016).
- **Mechanistic investigations:** There is a need to understand the molecular and cellular aspects of heavy metal toxicity to the nervous system in detail. Future investigations should clarify the roles of heavy metals in essential pathological processes, including oxidative stress, mitochondrial dysfunction, protein misfolding, and neuroinflammation. For instance, examining how parts per million of heavy metals influence ALS-related genes like SOD1, TDP-43, and C9orf72 may reveal other contributions to disease pathogenesis. Second, more studies should be conducted to determine how HMs combine and act when several are present simultaneously interacting with genetic factors (Yang, Chen, & Wang, 2015).
- **Biomarker development:** Accurate biomarkers are needed to diagnose reexposure to heavy metals, assess neurotoxicity, and identify persons at risk for ALS. This research implies that the following steps should be taken to further biomarker detection in blood, urine, CSF, and hair. Proteomics, metabolomics, and transcriptomics could help faster tailor biomarkers associated with heavy metal intake and ALS pathological conditions (Zhang, Xu, & Zheng, 2020). Marker indices of oxidative stress and inflammation and indicators of neuronal damage may also be used as practical diagnostic markers reflecting the disease progression and the effectiveness of the corresponding therapy.

15.7.2 Need for interdisciplinary collaborations

Addressing the complexities of the heavy metal–ALS connection requires interdisciplinary collaboration among scientists, clinicians, and policymakers. Key collaborative efforts should include:

Environmental scientists and toxicologists: Environmental scientists and toxicologists play a pivotal role in identifying sources of heavy metal contamination, assessing exposure pathways, and quantifying environmental burdens. Collaborations with neurologists can help translate these findings into understanding how environmental exposures affect neural health. For instance, studying the geospatial distribution of heavy metal contamination in ALS hotspots could provide valuable epidemiological insights (Vasta et al., 2023).

Neurologists and clinicians: Neurologists and clinicians are essential for linking environmental exposure to ALS diagnosis and progression. They can contribute by conducting detailed clinical evaluations, collecting patient histories, and correlating biomarkers with disease outcomes. Collaborative clinical trials incorporating environmental exposure assessments could also help identify subgroups of ALS patients who may benefit from targeted interventions (Farace et al., 2020).

Data scientists and computational biologists: Integrating big data analytics and machine learning techniques is crucial for uncovering hidden patterns in complex datasets. Data scientists can analyze large-scale environmental, genetic, and clinical datasets to identify correlations and predict ALS risk factors. Computational modeling can also simulate the impact of heavy metal exposure on neural pathways, providing insights into potential therapeutic targets (Geraci et al., 2024).

PolicyMakers and public health experts: Bridging the gap between scientific research and practical policy implementation requires active engagement with public health experts and policymakers. Collaborations should aim to develop evidence-based regulations for reducing heavy metal emissions and improving environmental monitoring.

Community involvement: Engaging local communities in research can enhance data collection, promote awareness, and ensure that interventions address community-specific needs. For example, participatory research programs involving affected populations could provide valuable data on occupational and dietary exposure patterns.

15.7.3 Translating research into public health interventions and policies

Education and dissemination of research evidence to form operational health programs in communities that are at a high risk of developing ALS due to the effects of heavy metals are vital. This requires a multipronged approach that includes:

- **Development of exposure guidelines:** This evidence must be used to set acceptable levels of heavy metals in the air, water, soil, and the food that is polluted by them. These guidelines must take into consideration additivity and interactions, as well as differences in vulnerability among populations. For example, an increased concentration of arsenic in drinking water would reduce exposure in affected areas.
- **Environmental remediation efforts:** Practical research applications entail the need to make real investments in technologies for cleaning up contaminated environments. Current methods of cleaning up contaminated soil and water include phytoremediation and bioremediation, as well as more advanced types of filtration. Further, studies about the cost-effective and efficient method for removing a large number of contaminants should be focused.
- **Targeted interventions for at-risk populations:** Special attention should be paid to protecting industrial workers, farmers, and inhabitants of areas abounding with contamination. These may include supplementary dining, periodic checkups, and clean water availability to prevent cases of waterborne diseases. Occupational measures should also include protection from hazards at the workplace with safeguards that apply the legal requirements and accessories.
- **Public awareness campaigns:** Conventional risk management involves creating awareness of the dangers associated with exposure to such metals and measures that can be taken to minimize various risks to the individual and the entire community. Campaigns should also pass on messages about sources of heavy metals in foods we consume, safe food preparation and water, and regular health checkups.
- **Global and regional policy frameworks:** International cooperation is important for combating cross-border pollution and developing best practices for contaminant metals. Regional frameworks should relate to specific issues of the particular region, for instance, the problems of the SSA regarding informal mining or arsenic poisoning in the SA region (Gelaye, 2024). Policy leaders should support research funding and activities involving or embracing a combination of ecological, clinical, and population-based approaches.

15.8 Conclusion

The complexity of the relations between metals and ALS indicates an important health issue that calls for a solution. Despite extensive advances in characterising the neuronal toxicity of heavy metals, including lead, mercury, arsenic, and cadmium, proof of the etiology of ALS resulting from these environmental agents has not been provided. The present state of knowledge points to the involvement of heavy metals in exerting toxic effects through increased oxidative stress, impaired mitochondrial function, glutamate-mediated excitotoxicity, and inflammation, features that are prominent in ALS. However, given the array of interactions, this study establishes a correlation between genetic predisposition, coexposures, and individual susceptibility to SARS-CoV-2; more studies are still needed to disentangle the exact relationship. Heavy metal pollution is a concern worldwide due to industrial discharges, mining, agriculture, and polluted water supply. These environmental toxins not only significantly contribute to the increasing causal factors of ALS but also pose severe threats to other related health conditions. Their nonbiodegradability, tendencies towards accumulation in ecosystems, and the possibility of long-term exposure make the required preventative measures more rigorous. Preventative measures can only begin by ensuring environmental compliance, applying and rolling out technologies for cleaning up hazardous substances, and raising awareness of the implications of inhaling contaminated air, drinking water, and eating tainted food. Local nutritional interventions, safe working practices, and campaigns aimed at high-risk groups must form part of the overarching solution to heavy metal exposure. As this issue binder together various aspects of the patient's life, interdisciplinary work is crucial. Environmental scientists, toxicologists, neurologists, data scientists, and policymakers must collaborate to support research advancements and inform data for policy application. Large-scale cross-sectional and intervention epidemiological studies complement mechanistic in vitro and in vivo cellular and molecular research on heavy metals to determine causation and vulnerable human populations. Furthermore, biomarker development has the potential to provide a means of detecting the early burden of heavy metal exposure affecting ALS progression and prognosis. This means that dedication and synergy in meeting the challenge of transboundary contamination and developing standards of acceptable management of heavy metals need international cooperation. National and regional programmes to fit particular exposure causes and modes, including small-scale mining or waterborne pollutants, are therefore required in addition to global efforts. Communal educational goals and involvement are as important as individual educational goals to ensure that individuals take preventive measures to observe cleaner environments. Altogether, confronting the heavy metal–ALS connection is an issue of scientific investigation and global stewardship. Through the enhanced cross-sectoral partnership, encouraging a more comprehensive approach, and giving attention to turning preventive measures into priority and promoting groundbreaking research, the adverse effects of high concentrations of heavy metals on ALS and people's health, in general, could be addressed. All of these endeavors will not only enhance the quality of those who have ALS but also pave the way for enhancing the general health of communities globally.

References

Apontes, P. (2022). New opportunities for treatment of neurodegenerative disease through the modulation of TDP-43. In *TDP-43 and neurodegeneration*. Academic Press, 183–250.

Ball, N., Teo, W. P., Chandra, S., & Chapman, J. (2019). Parkinson's disease and the environment. *Frontiers in Neurology, 10*, 421551.

Bartley, T., & Fairbrother, M. (2025). Tackling toxins: Case studies of industrial pollutants and implications for climate policy. *Regulation & Governance, 19*(2), 329–348.

Benatar, M., Boylan, K., Jeromin, A., Rutkove, S. B., Berry, J., Atassi, N., & Bruijn, L. (2016). ALS biomarkers for therapy development: State of the field and future directions. *Muscle & Nerve, 53*(2), 169–182.

Bryan, L., Kaye, W., Antao, V., Mehta, P., Muravov, O., & Horton, D. K. (2016). Preliminary results of national amyotrophic lateral sclerosis (ALS) registry risk factor survey data. *PLoS One, 11*(4), e0153683.

Demirci-Cekic, S., Özkan, G., Avan, A. N., Uzunboy, S., Çapanoğlu, E., & Apak, R. (2022). Biomarkers of oxidative stress and antioxidant defense. *Journal of Pharmaceutical and Biomedical Analysis, 209*, 114477.

Devi, P., Kaur, A., & Kumar, P. (2024). Phytoremediation of metal contaminated soil using energy crops: Soil health maintenance along with biofuel production. In *Bioremediation of emerging contaminants from soils*. Elsevier, 307–333.

Farače, C., Fenu, G., Lintas, S., Oggiano, R., Pisano, A., Sabalic, A., ... Madeddu, R. (2020). Amyotrophic lateral sclerosis and lead: A systematic update. *Neurotoxicology, 81*, 80–88.

Ferraiuolo, L., Kirby, J., Grierson, A. J., Sendtner, M., & Shaw, P. J. (2011). Molecular pathways of motor neuron injury in amyotrophic lateral sclerosis. *Nature Reviews Neurology, 7*(11), 616–630.

Gelaye, Y. (2024). Public health and economic burden of heavy metals in Ethiopia. *Heliyon, 10*(19), e39022. Retrieved from https://doi.org/10.1016/j.heliyon.2024.e39022. (Accessed 24 may 2025).

Geraci, J., Bhargava, R., Qorri, B., Leonchyk, P., Cook, D., Cook, M., ... Pani, L. (2024). Machine learning hypothesis-generation for patient stratification and target discovery in rare disease: Our experience with open science in ALS. *Frontiers in Computational Neuroscience, 17*, 1199736.

Goutman, S. A., Savelieff, M. G., Jang, D. G., Hur, J., & Feldman, E. L. (2023). The amyotrophic lateral sclerosis exposome: Recent advances and future directions. *Nature Reviews Neurology, 19*(10), 617–634.

Huang, M., Chen, X., Shao, D., Zhao, Y., Wang, W., & Wong, M. H. (2014). Risk assessment of arsenic and other metals via atmospheric particles, and effects of atmospheric exposure and other demographic factors on their accumulations in human scalp hair in urban area of Guangzhou, China. *Ecotoxicology and Environmental Safety, 102*, 84–92.

Ijomone, O. M., Ifenatuoha, C. W., Aluko, O. M., Ijomone, O. K., & Aschner, M. (2020). The aging brain: Impact of heavy metal neurotoxicity. *Critical Reviews in Toxicology, 50*(9), 801–814.

Kakaroubas, N., Brennan, S., Keon, M., & Saksena, N. K. (2019). Pathomechanisms of blood-brain barrier disruption in ALS. *Neuroscience Journal, 2019*(1), 2537698.

Kang, B., Wang, J., Guo, S., & Yang, L. (2024). Mercury-induced toxicity: Mechanisms, molecular pathways, and gene regulation. *Science of the Total Environment,* 173577.

Kumar, P., & Choudhury, D. (2024). Toxicity of rhizospheric cadmium contaminated soil and its phytoremediation. In *Cadmium toxicity in water: Challenges and solutions*. Cham: Springer Nature Switzerland, 227–264.

Kumar, P., Choudhury, D., Dey, S. R., & Sharma, M. (2024). Application of genetically modified crops against metallic contaminations. In *Bioremediation of emerging contaminants from soils*. Elsevier, 541–564.

Kumar, P., Goud, E. L., Devi, P., Dey, S. R., & Dwivedi, P. (2022a). Heavy metals: Transport in plants and their physiological and toxicological effects. *Plant metal and metalloid transporters* (pp. 23–54). Singapore: Springer Nature Singapore.

Kumar, P., Goud, E. L., Devi, P., Dey, S. R., & Dwivedi, P. (2022b). Heavy Metals: Transport in Plants and Their Physiological and Toxicological Effects. In K. Kumar, & S. Srivastava (Eds.). *Plant Metal and Metalloid Transporters*. Singapore: Springer. Retrieved from https://doi.org/10.1007/978-981-19-6103-8_2.

Landrigan, P. J., Lucchini, R. G., Kotelchuck, D., & Grandjean, P. (2022). Principles for prevention of the toxic effects of metals. In *Handbook on the toxicology of metals*. Academic Press, 685–703.

Leask, C. F., Sandlund, M., Skelton, D. A., Altenburg, T. M., Cardon, G., Chinapaw, M. J., ... GrandStandSafe Step and Teenage Girls on the Move Research Groups. (2019). Framework, principles and recommendations for utilising participatory methodologies in the co-creation and evaluation of public health interventions. *Research Involvement and Engagement, 5*, 1–16.

Liu, W., Xu, Z., Yang, T., Xu, B., Deng, Y., & Feng, S. (2017). Memantine, a low-affinity NMDA receptor antagonist, protects against methylmercury-induced cytotoxicity of rat primary cultured cortical neurons, involvement of Ca^{2+} dyshomeostasis antagonism, and indirect antioxidation effects. *Molecular Neurobiology, 54*, 5034–5050.

Masrori, P., & Van Damme, P. (2020). Amyotrophic lateral sclerosis: A clinical review. *European Journal of Neurology, 27*(10), 1918–1929.

Mathis, S., Goizet, C., Soulages, A., Vallat, J. M., & Le Masson, G. (2019). Genetics of amyotrophic lateral sclerosis: A review. *Journal of the Neurological Sciences, 399*, 217–226.

Oliveira, A. S. D., Costa, E. A., Pereira, E. C., Freitas, M. A. S. D., Freire, B. M., Batista, B. L., ... Olympio, K. P. K. (2021). The applicability of fingernail lead and cadmium levels as subchronic exposure biomarkers for preschool children. *Science of the Total Environment, 758*, 143583.

Pal, A., Cerchiaro, G., Rani, I., Ventriglia, M., Rongioletti, M., Longobardi, A., & Squitti, R. (2022). Iron in Alzheimer's disease: From physiology to disease disabilities. *Biomolecules, 12*(9), 1248.

Provenzano, F., Torazza, C., Bonifacino, T., Bonanno, G., & Milanese, M. (2023). The key role of astrocytes in amyotrophic lateral sclerosis and their commitment to glutamate excitotoxicity. *International Journal of Molecular Sciences, 24*(20), 15430.

Rahmanian, N., Shokrzadeh, M., & Eskandani, M. (2021). Recent advances in γH2AX biomarker-based genotoxicity assays: A marker of DNA damage and repair. *DNA Repair, 108*, 103243.

Raj, A., Dubey, A., Malla, M. A., & Kumar, A. (2023). Pesticide pestilence: Global scenario and recent advances in detection and degradation methods. *Journal of Environmental Management, 338*, 117680.

Raj, A., Kumar, A., & Vaghasiya, T. P. (2021). Multidisciplinary approaches to clean up heavy metal contaminated soil: A review. *Bulletin of Environment, Pharmacology and Life Sciences, 10*(6), 53–65.

Raman, R. K., Talukder, A., Mahanty, A., Sarkar, D. J., Das, B. K., Bhowmick, S., ... Mohanty, B. P. (2022). Arsenic bioaccumulation and identification of low-arsenic-accumulating food fishes for aquaculture in arsenic-contaminated ponds and associated aquatic ecosystems. *Biological Trace Element Research, 200*(6), 2923–2936.

Rehman, Q., Rehman, K., & Akash, M. S. H. (2021). Heavy metals and neurological disorders: From exposure to preventive interventions. In Environmental contaminants and neurological disorders. Cham: Springer International Publishing, 69–87.

Saini, L., Kumar, P., & Singh, P. J. (2024). Current status of wastewater in india/other countries/regions. In *Bioelectrochemical oxidation processes for wastewater treatment* (pp. 152–165).

Sakowski, S. A., Koubek, E. J., Chen, K. S., Goutman, S. A., & Feldman, E. L. (2024). Role of the exposome in neurodegenerative disease: Recent insights and future directions. *Annals of Neurology, 95*(4), 635–652.

Shabir, S., Sehgal, A., Dutta, J., Devgon, I., Singh, S. K., Alsanie, W. F., ... Singh, M. P. (2023). Therapeutic potential of green-engineered ZnO nanoparticles on rotenone-exposed *D. melanogaster* (Oregon R +): Unveiling ameliorated biochemical, cellular, and behavioral parameters. *Antioxidants, 12*(9), 1679.

Sighencea, M. G., Popescu, R.Ș., & Trifu, S. C. (2024). From fundamentals to innovation in Alzheimer's disease: Molecular findings and revolutionary therapies. *International Journal of Molecular Sciences, 25*(22), 12311.

Steinacker, P., Barschke, P., & Otto, M. (2019). Biomarkers for diseases with TDP-43 pathology. *Molecular and Cellular Neuroscience, 97*, 43–59.

Sun, Q., Li, Y., Shi, L., Hussain, R., Mehmood, K., Tang, Z., & Zhang, H. (2022). Heavy metals induced mitochondrial dysfunction in animals: Molecular mechanism of toxicity. *Toxicology, 469*, 153136.

Tandon, A., Singh, S. J., & Chaturvedi, R. K. (2021). Nanomedicine against Alzheimer's and Parkinson's disease. *Current Pharmaceutical Design, 27*(12), 1507–1545.

Thakur, M., Rachamalla, M., Niyogi, S., Datusalia, A. K., & Flora, S. J. S. (2021). Molecular mechanism of arsenic-induced neurotoxicity including neuronal dysfunctions. *International Journal of Molecular Sciences, 22*(18), 10077.

Toader, C., Dobrin, N., Brehar, F. M., Popa, C., Covache-Busuioc, R. A., Glavan, L. A., ... Ciurea, A. V. (2023). From recognition to remedy: The significance of biomarkers in neurodegenerative disease pathology. *International Journal of Molecular Sciences, 24*(22), 16119.

Vasta, R., Callegaro, S., Sgambetterra, S., Cabras, S., Di Pede, F., De Mattei, F., ... Calvo, A. (2023). The presymptomatic geographical distribution of ALS patients suggests the involvement of environmental factors in the disease pathogenesis. *Journal of Neurology, 270*(11), 5475–5482.

Volonté, C., Apolloni, S., Parisi, C., & Amadio, S. (2016). Purinergic contribution to amyotrophic lateral sclerosis. *Neuropharmacology, 104*, 180–193.

Wu, H., Liao, Q., Chillrud, S. N., Yang, Q., Huang, L., Bi, J., & Yan, B. (2016). Environmental exposure to cadmium: Health risk assessment and its associations with hypertension and impaired kidney function. *Scientific Reports, 6*(1), 29989.

Yang, J., Ran, M., Li, H., Lin, Y., Ma, K., Yang, Y., ... Yang, S. (2022). New insight into neurological degeneration: Inflammatory cytokines and blood–brain barrier. *Frontiers in Molecular Neuroscience, 15*, 1013933.

Yang, T., Chen, M. L., & Wang, J. H. (2015). Genetic and chemical modification of cells for selective separation and analysis of heavy metals of biological or environmental significance. *TrAC Trends in Analytical Chemistry, 66*, 90–102.

Zhang, Y., Xu, Y., & Zheng, L. (2020). Disease ionomics: Understanding the role of ions in complex disease. *International Journal of Molecular Sciences, 21*(22), 8646.

Chapter 16

Neurodevelopmental disorders linked to heavy metal exposure

Mastu Patel, and Prasann Kumar
Department of Agronomy, School of Agriculture, Lovely Professional University, Phagwara, Punjab, India

16.1 Introduction

Neurodevelopmental disorders are a group of disorders of the brain that can occur in children and may impair brain growth and function. Such disorders are known to develop in early childhood and may continue to affect productivity, social interaction, and mental health in the later years. Neurodevelopmental disorders consist of autism spectrum disorder (ASD), ADHD, intellectual disabilities (IDs), communication disorders, specific learning disorders, and others. The etiology of these diseases is complex because genes, environments, and epigenomes that distort the classic ontogeny of the brain are involved (Larrigan, Shah, Fernandes, & Mattar, 2021). Neurodevelopmental disorders affect children in unique ways, given that these are aspects that are central to life for children. The following are common symptoms of cognitive disability: attention problems, memory, problem-solving or learning disability, and academic achievement and information processing (Scandurra et al., 2019). Even social development may be affected, as learning disabilities may cause problems with interpreting body language signals, speech and language, or relationship development may make a child feel lonely or misunderstood by other kids and adults. Other impacted domains include emotional regulation—-many children become more anxious, moody, or frustrated due to the various difficulties they encounter in the environment. Therefore effective early and timely diagnosis plays a vital role in reducing the effects of neurodevelopmental disorders. Although there is no cure for the upper limb "nerves" dysfunction, therapeutic strategies include behavioral interventions, speech and occupational therapy, and educational programs and can significantly improve children's abilities (Nshimiyimana, Uwihoreye, Muhigirwa, & Niyonsega, 2023). Finally, family and community support were also identified as an effective factor that helps create a protective environment for the child's development. Neurodevelopmental disorders are critical to grasp and handle appropriately to allow each child to live a meaningful life (Guinchat, Cravero, Lefèvre-Utile, & Cohen, 2020).

There has been increasing concern in the last few years connecting the settings, specifically toxic heavy metals, to neurodevelopmental disorders. Metals causing more or less obvious toxicity include lead, mercury, arsenic, and cadmium, and it is well known that metals can cause damage to the nervous system (Kumar, Goud, Devi, Dey, & Dwivedi, 2022). Since these toxicants can penetrate the placental barrier and concentrate themselves in the brain, they are a special danger to pregnancies and young children at times when their brain development is most vulnerable. It may interfere with some crucial developmental processes, such as neurogenesis, synaptogenesis, and myelination, all affected by heavy metals. From a molecular standpoint, these metals cause toxicity through oxidative stress, disrupt impartial function, and influence neurotransmitter transmission. These disruptions are associated with learning difficulties, language delay, disruptive behaviors, learning disabilities, and the enhanced possibility of neurodevelopmental disorders, including ASD, ADHD, and IDs (Rubin, Coles, Barnhill, Coles, & Barnhill, 2023). Common ways people get exposed to these heavy metals encompass industrial discharge, water, soil, and food. Since neurodevelopmental disorders have become increasingly common, and these diseases have many years of social and economic consequences for patients and their families, there is an urgent need to classify them and their environmental factors. More focused research is needed to determine how these factors influence each other and to develop a valuable

Heavy Metal Toxicity and Neurodegeneration. DOI: https://doi.org/10.1016/B978-0-443-36575-1.00027-3

understanding of implementing preventive measures that reduce risk factors that affect children's neurodevelopment.

Sensitive periods of brain development are well-defined intervals wherein the brain experiences significant anatomic growth, connectivity, and plasticity and is thus vulnerable to positive or negative influences. The first ones are flexible or critical periods that may last for months or sometimes for the first year of life and are accompanied by "experience-dependent" processes such as synaptogenesis, synaptic pruning, and myelination. On the one hand, plasticity allows the brain to change and learn, but, on the other hand, the vulnerable developing brain is especially susceptible to toxicants. In the periods of early childhood and prenatal development, the brain is subjected to definite architectural constructions that determine its future function; the key sections of the brain that get critical development at this moment are the hippocampus, the prefrontal cortex, and the cerebellum (Keunen, Counsell, & Benders, 2017). The best examples are embryonic and fetal development; exposure to toxic substances such as heavy metals, air pollutants, and endocrine disruptors at these stages has severe consequences. For example, prenatal lead exposure, wire head damage, declined IQ, and attention disorder, prenatal mercury exposure causes motor skills and language problems (Parithathvi, Choudhari, & Dsouza, 2024). This increased susceptibility is due to the relative inadequacy of the blood–brain barrier during early development in blocking the toxic effects of a substance. Recognizing the importance of these sensitive periods emphasizes the need to avoid environmental toxicants when brain growth is most vulnerable, thus enhancing neurological and psychological well-being in children and through adulthood.

16.2 Mechanisms of action: heavy metal neurotoxicity in the developing brain

16.2.1 Disruption of neurotransmitter systems

Lead (Pb), mercury (Hg), arsenic (As), and cadmium (Cd) interfere with neurotransmitter systems necessary for the brain's orderly development and functioning (Huang, Liu, Duan, Yin, & Cao, 2024). These disruptions interfere with the basic synaptic transmission that goes a long way toward shaping cognitive, emotional, and behavioral functions.

- **Lead (Pb):** Lead inhibits synaptic transmission by modulating the activity of calcium-dependent processes in neurotransmitter release. It mimics the action of calcium ions and alters the release of synaptic vesicles containing essential neurotransmitters such as glutamate, dopamine, and GABA (Rahman & Guillemin, 2021). They do this by creating an imbalance that, when left unchecked, leads to excitotoxicity, which reduces synaptic plasticity, impairing learning and memory.
- **Mercury (Hg):** Methylmercury accumulates in astrocytes and neurons, raising extracellular glutamate levels through the blockade of various glutamate transporters. Such excitotoxicity causes neuronal death and contributes to hippocampus and cerebral cortex apoptosis. Mercury also inhibits GABAergic inhibitory transmission, thus increasing the excitability of neural networks (Penticoff & Fortin, 2023).
- **Cadmium (Cd):** Cadmium reduces dopamine production by downregulating tyrosine hydroxylase, a dopaminergic enzyme. It also decreases serotonin levels, weakening their ability to regulate mood and causing behavioral problems (Kupnicka et al., 2020).
- **Arsenic (As):** Arsenic interferes with the enzyme acetylcholinesterase, the causal agent of synaptic accumulation of Ach. Self-administration of nicotine leads to cholinergic overactivity, resulting in distortions of cognitive functions, including attention learning and memory processes (Albuquerque, Burke, Rao Gullapallib, & Mamczarza, 2017). These disruptions result in experiencing persistent engrams of synaptic transmission, learning, memory, and regulatory dysfunctions. Given that the neurotransmitter system matures progressively and, particularly, undergoes tremendous plasticity during early life, the alterations that occur due to the impact of various factors can be arbitrarily deleterious on the formation of the brain in development.

16.2.2 Interference with cell signaling pathways

Heavy metals disrupt basic molecular processes through which nervous system cells differentiate, form synapses, and divide. These disruptions often affect signaling molecules within the cell, transcription factors, and growth factors important in the brain's development (Jiang & Nardelli, 2016).

- **Calcium signaling dysregulation:** Lead and mercury disrupt calcium homeostasis by interacting with voltage-gated calcium channels, calcium pumps, and calcium-binding proteins. Abnormal calcium signaling interferes with developmental events such as neuronal migration, differentiation, or synaptic plasticity (Flora & Agrawal, 2017).
- **Disruption of neurotrophic factor signaling:** Mercury and cadmium interfere with the signaling of brain-derived neurotrophic factor (BDNF) and nerve growth factor (NGF). This results in less neuronal survival, differentiation, and lengthy process extension or axonal growth, ultimately paralleled by decreased overall synaptic density and connectivity (Sighencea, Popescu, & Trifu, 2024).
- **Oxidative stress-mediated signaling alterations:** To our knowledge, heavy metals cause oxidative stress, activating mitogen-activated protein kinase (MAPK),

which is involved in apoptosis and inflammatory responses. For example, p38 MAPK activation by cadmium leads to neuronal death, and arsenic blocks PI3K-Akt signaling, resulting in decreased cell survival and neurogenesis in the SH-SY5Y cells (Gupta, Ambasta, & Kumar, 2021).

- **Epigenetic modifications:** Tox metals have the potential to change DNA methylation genes, histone, and microRNA (miRNA) that control the gene expression of newly developed brains. This early life lead exposure also alters epigenetic marks such as DNA methylation, reducing synaptic plasticity memory and learning (Yu et al., 2024). Arsenic modifies histone acetylation, thus affecting chromosomal arrangements and gene expression. Heavy metals' disruption of cell signaling pathways causes proneurodevelopmental problems, including decreased synapse density, impaired neurogenesis, and increased neuronal death. These chronic molecular disruptions have consequences for cognitive, behavioral, and emotional well-being.

16.2.3 Induction of oxidative stress and alteration of brain structure

These changes are facilitated by the induction of oxidative stress and the alteration of brain structure. Transition metals are scientifically proven to provoke oxidative stress, which tends to cause cell damage and morphological changes in growing brains (Ijomone, Ifenatuoha, Aluko, Ijomone, & Aschner, 2020). Besides, the brain receives a high amount of oxygen, and its tissues contain high lipids prone to oxidation.

16.2.3.1 Mechanisms of oxidative stress induction

- **Lead (Pb):** Lead causes the production of free radicals and oxygen radicals and inhibits both total and catalase antioxidant enzymes while reducing glutathione levels. This oxidative disturbance affects both mitochondrial and neuronal membrane hostility and compromises energy metabolism and synaptic communication.
- **Mercury (Hg):** This, in turn, reacts with mercury, which has a predilection for cysteine thiol groups. This inactivates antioxidant enzymes such as glutathione peroxidase, producing heightened reactive oxygen species (ROS). This leads to the generation of ROS, which cause lipid peroxidation, protein oxidation, and DNA damage in the neuronal cells.
- **Cadmium (Cd):** Cadmium induces a Fenton-like reaction and produces hydroxyl radicals that cause oxidative damage to DNA and lipids. It also affects conventional mitochondrial function and hence increases ROS generation even further.
- **Arsenic (As):** Arsenic enhances ROS and RNS levels in cells and leads to scanning of their components and apoptosis. It also interferes with antioxidant enzyme function, compromising the brain's ability to protect against oxidative stress.

16.2.3.2 Structural alterations in the brain

- **Gray matter volume reduction:** Lead and mercury toxicity have reduced the gray matter voxels in the dorsolateral prefrontal cortex, hippocampus, and cerebellum. These findings decrease in parallel with executive control, memory, and motor planning impairments.
- **White matter integrity loss:** There is evidence that arsenic and cadmium cause demyelination and reduced signal conduction speed along white matter fibers.
- **Synaptic degeneration:** Heavy metals interfere with dendritic branching, spine density, and synaptic protein, affecting synaptic connections.

16.2.3.3 Behavioral and cognitive consequences of structural damage

These structural changes lead to learning and memory loss, attention problems, and other neuropsychological development disorders such as ASD and attention-deficient hyperkinetic disorder (ADHD) (Table 16.1).

16.3 Vulnerable populations: risks, disparities, and interventions

16.3.1 Vulnerable populations at increased risk of heavy metal exposure

Some populations are especially susceptible to heavy metal exposure during their critical period of brain development. Biological, environmental, and socioeconomic factors enhancing the level of this toxicant and its toxic impact on developing neural systems combine to put these groups particularly at risk due to disproportionate exposure levels.

- **Children living in poverty:** An essential source of lead exposure for children is in their homes with lead-based paint in older, deteriorating ones. They often live where heavy environmental contamination can come from mining, smelting, or improper waste disposal. Such settings are common sites for exposure to arsenic in contaminated drinking water or cadmium in industrial emissions. Poor nutrition makes you more vulnerable by increasing the body's uptake of heavy metals. For example, lead absorption in the gut is increased when you have iron or calcium deficiencies (Naja & Volesky, 2017).
- **Infants and young children:** The blood–brain barrier still develops during infancy and childhood, making the brain especially vulnerable to heavy metals. Yet breastfeeding exposes the baby to other germ exposures. Breast milk can accumulate heavy metals, such as mercury and arsenic, exposing babies to neurotoxic agents at a critical

TABLE 16.1 The neurotoxic mechanisms, impacts on brain development, and implications for heavy metal exposure, emphasizing their role in cognitive and behavioral impairments.

Heavy metal	Mechanism of neurotoxic action	Impact on brain development	Implications	References
Lead	Disrupts neurotransmitter release and uptake, induces oxidative stress	Impaired synaptic development, reduced cognitive abilities	Linked to ADHD, learning disabilities, and lower IQ	Rahman and Guillemin (2021)
Mercury	Binds to thiol groups in proteins, disrupting cell signaling and mitochondrial function	Neural cell apoptosis, impaired brain connectivity	Associated with autism spectrum disorders and cognitive decline	Penticoff and Fortin (2023)
Arsenic	Interferes with ATP production, induces oxidative damage	Reduced neurogenesis, disrupted neural differentiation	Behavioral problems, reduced memory, and learning capacity	Kupnicka et al. (2020)
Cadmium	Disrupts calcium signaling and induces oxidative stress	Altered neurotransmitter release, delayed brain development	Linked to impaired cognitive function and motor deficits	Albuquerque et al. (2017)
Manganese	Alters dopamine metabolism interferes with mitochondrial activity	Impaired motor control, emotional dysregulation	Linked to attention deficits and mood disorders	Jiang and Nardelli (2016)
Aluminum	Accumulates in the brain, induces inflammation and oxidative stress	Disrupted synaptic plasticity, memory impairment	Potential contributor to neurodegenerative conditions	Flora and Agrawal (2017)
Chromium	Interferes with neurotransmitter metabolism, causes DNA damage	Reduced synaptic plasticity	Linked to learning and memory deficits	Gupta et al. (2021)
Nickel	Induces oxidative stress, inhibits neurotransmitter uptake	Reduced neural differentiation	Associated with impaired motor and cognitive development	Yu et al. (2024)
Zinc (excess exposure)	Disrupts homeostasis of essential metals and interferes with synaptic function.	Impaired synaptic signaling	Affects memory, learning, and attention span	Ijomone et al. (2020)
Iron (overload)	Generates ROS, induces oxidative stress	Neural degeneration reduces cognitive abilities	Associated with behavioral and cognitive deficits	
Copper (excess exposure)	Disrupts mitochondrial function, promotes ROS formation	Altered neurogenesis and synapse function	Linked to neurodevelopmental disorders	Rahman and Guillemin (2021)
Selenium (excess exposure)	Interferes with redox homeostasis, induces DNA damage	Impaired cognitive development	Affects memory and emotional regulation	Penticoff and Fortin (2023)
Disruption of neurotransmitters	Heavy metals disrupt neurotransmitter synthesis, release, and reuptake	Imbalanced neural communication	This leads to mood, attention, and behavioral issues	Kupnicka et al. (2020)
Oxidative stress	Heavy metals induce oxidative stress and damage DNA, lipids, and proteins	Impaired cellular repair mechanisms	Increases susceptibility to neurodevelopmental disorders	(Albuquerque et al. (2017))
Mitochondrial dysfunction	Heavy metals impair energy production in neurons	Energy deficits, reduced neural plasticity	It affects learning and memory processes	Jiang and Nardelli (2016)
Epigenetic modifications	Heavy metals alter gene expression through DNA methylation	Dysregulated neural development	Long-term effects on brain structure and function	Flora and Agrawal (2017)

Alteration of calcium signaling	Heavy metals interfere with calcium-dependent signaling pathways	Impaired synaptic transmission	It affects memory formation and learning	Gupta et al. (2021)
Induction of neuroinflammation	Heavy metals activate microglia and astrocytes, leading to inflammation	Disrupted neural network formation	Contributes to cognitive and behavioral impairments	Yu et al. (2024)
Structural brain changes	Heavy metals affect gray and white matter volume	Reduced brain volume, impaired connectivity	Linked to cognitive delays and developmental disorders	Ijomone et al. (2020)
Behavioral impacts	Heavy metals disrupt brain regions regulating behavior	Emotional dysregulation, impulsivity	Increases risk for ADHD and mood disorders	

ROS, Reactive oxygen species.

time of their development (Samiee, Vahidinia, Javad, & Leili, 2019). Consequently, exposure risk in young children is elevated due to frequent hand-to-mouth activities and play in contaminated soils.

- **Residents of contaminated areas:** They live near mining sites, industrial facilities, and hazardous waste sites with high environmental contamination levels. In places such as South Asia, where millions are exposed to levels far higher than acceptable, arsenic in groundwater is a serious problem. Because indigenous groups and coastal populations rely on these resources as a source of sustenance, fish and other aquatic life are particularly contaminated with mercury and disproportionately burden these communities (Landrigan et al., 2020). During sensitive periods of brain development, these populations are at heightened risk of long-lasting neurological and cognitive consequences (Tables 16.1 and 16.2).

16.3.2 Social and economic factors contributing to disparities in exposure

Heavy metal environmental exposures are not evenly distributed throughout the population. Socioeconomic, political, and systemic factors combine to create disparities, making the risks difficult for vulnerable groups.

- **Poverty and lack of resources:** Poor families tend to live around industrial sites, landfills, or polluted waterways (also may have contaminated housing), which are also more polluted. Early detection and treatment of heavy metal toxicity are prevented by inadequate access to healthcare. Heavy metals have highly toxic effects that are exacerbated by malnutrition, which degrades what little detoxification mechanisms are available and increases metal absorption (Kumar et al., 2024).
- **Environmental injustice and systemic inequalities:** Ethnic minorities and indigenous populations are disproportionately affected by environmental contamination (Hoover, 2018). The core population of these communities lives near these places because of protective gaps in the regulatory framework and weak political representation; these gaps include failing to enforce laws to monitor air, soil, and water quality; such regulatory lapses do not protect these populations.
- **Educational and awareness gaps:** Only a limited understanding of the risk of heavy metal exposure can protect individuals and communities. In vulnerable populations, low literacy levels and barriers to language can hinder access to information on safe practices, including avoiding contaminated water or reducing occupational exposure.
- **Occupational exposure:** Many adults from vulnerable communities working in industry are at a high risk of heavy metal exposure in areas such as mining, metal recycling, and construction. Often, the family is secondarily contaminated with these exposures through exposures in the home and clothes. This requires a holistic understanding of how socioeconomic and systemic factors contribute to increased exposure to heavy metals and how it disproportionately affects vulnerable populations (Kordas et al., 2018).

16.3.3 Targeted interventions and public health policies

Consequently, targeted interventions and robust public health policies are necessary to mitigate the adverse effects of heavy metal exposure on vulnerable populations. Action on these measures needs to target not only the sources of exposure but also the socioeconomic inequalities that worsen risks.

- **Regulation and remediation of contaminated environments:** Governments and regulatory bodies must enforce stricter controls on industrial emissions and waste disposal to prevent environmental contamination. For high-burden areas, large-scale remediation efforts, like removing lead-based paint, treating arsenic-contaminated water, and cleaning polluted soils, are essential to lowering exposure risks. Heavily affected communities should be prioritized for policies that must remediate environmental injustice (Laidlaw et al., 2017).
- **Improving nutritional interventions:** Feeding nutritional programs for vulnerable populations can decrease heavy metal absorption and toxicity. For example, supplementation with iron, calcium, and zinc can also reduce the bioavailability of lead and cadmium. Heavy metals induce oxidative stress, which antioxidant-rich diets can counter. Another essential thing to ensure is regular monitoring of agricultural products for heavy metal contamination to ensure food safety (Kumar, Choudhury, Dey, & Sharma, 2024).
- **Community education and awareness campaigns:** Community education should be a public health campaign to educate communities about heavy metal exposure sources and the risks. Measures that promote safe practices, such as using filtered water, frequent hand washing, and proper waste disposal, can help reduce exposure. These initiatives can further reach and impact the community with collaboration from local leaders and community organizations (Shakir et al., 2017).
- **Healthcare and early intervention programs:** Putting good healthcare where people are at risk lets doctors identify and treat metal poisoning quickly and correctly. Routine public health screenings look for early exposure signs among children in places where they are at higher risk. We need wider access and lower prices for therapy methods backed by research.
- **Occupational safety measures:** Companies in risky work environments must provide workers with full safety gear and well-defined rules to keep them safe.

TABLE 16.2 The vulnerable populations, contributing factors, impacts of heavy metal exposure, and possible targeted interventions.

Vulnerable population	Contributing factors	Impacts of heavy metal exposure	Targeted interventions	References
Children living in poverty	Poor housing conditions, proximity to industrial areas, and inadequate nutrition	Cognitive delays, behavioral problems, lower academic performance	Implement stricter environmental regulations, and provide nutritional support	Naja and Volesky (2017)
Residents near industrial sites	Exposure to contaminated air, water, and soil	Increased risk of neurodevelopmental disorders, respiratory issues	Monitor and mitigate environmental contamination	Samiee et al. (2019)
Infants exposed to breastfeeding.	Accumulation of metals in the maternal body, inadequate healthcare access	Impaired brain development, weakened immune response	Educate mothers and improve maternal healthcare	Landrigan et al. (2020)
Indigenous communities	Reliance on locally sourced food and water from contaminated areas	Higher prevalence of neurodevelopmental and chronic diseases	Support sustainable resource use, monitor contamination	Kumar, Raj, and Kumar (2024)
Pregnant women	Increased vulnerability to transplacental heavy metal transfer	Impaired fetal brain development, preterm birth risks	Promote prenatal care, reduce maternal exposure	Hoover (2018)
Agricultural workers	Exposure to contaminated soil, water, and pesticides	Neurological effects, reproductive issues	Enforce workplace safety measures, provide personal protection	Kordas, Ravenscroft, Cao, and McLean (2018)
Urban slum dwellers	Exposure to polluted water, air, and lead-based paint	Cognitive and physical development delays	Improve housing, access to clean water, and sanitation	Laidlaw et al. (2017)
Refugees and displaced persons	Poor living conditions, proximity to contaminated environments	Heightened exposure to neurotoxicants, malnutrition	Establish safe and clean living spaces	Kumar et al. (2024)
Children in mining regions	Exposure to heavy metals in water, air, and locally sourced food	Reduced IQ, developmental delays, respiratory problems	Remediate contaminated areas, educate communities	Shakir et al. (2017)
Communities in developing countries	Limited regulation of industrial emissions, contaminated food, and water	Disproportionate burden of heavy metal–related diseases	International collaboration for stricter policies	Anyanwu, Ezejiofor, Igweze, and Orisakwe (2018)
Low-income communities in urban areas	Proximity to highways, factories, and industrial facilities	Higher lead and mercury exposure, cognitive impairments	Urban planning to minimize exposure zones	Naja and Volesky (2017)
Fishermen and coastal communities	Consumption of contaminated seafood	Mercury toxicity, neurological issues	Implement seafood monitoring programs	Samiee et al. (2019)
Infants and young children	Higher dietary intake per body weight, hand-to-mouth behavior	Developmental neurotoxicity, behavioral problems	Educate caregivers, improve testing, and exposure standards	Anyanwu et al. (2018)
Adolescents in industrial zones	Airborne and waterborne exposure to metals	Impaired cognitive development, hormonal disruptions	Promote health education, regulate industrial emissions	Landrigan et al. (2020)

(Continued)

TABLE 16.2 (Continued)

Vulnerable population	Contributing factors	Impacts of heavy metal exposure	Targeted interventions	References
Elderly in contaminated areas	Age-related vulnerability, bioaccumulation over time	Exacerbation of neurodegenerative diseases	Provide healthcare support and regular screening	Kumar et al. (2024)
Migrant workers	Poor workplace safety, exposure to contaminated food and water	Long-term health risks, developmental impacts on children	Enforce workplace regulations, improve housing conditions	Hoover (2018)
Communities near e-waste recycling sites	High exposure to heavy metals from improper recycling practices	Neurological damage, respiratory, and immune system issues	Promote safe recycling practices, community awareness	Kordas et al. (2018)
Pregnant teenagers	Limited knowledge of exposure risks, poor nutrition	Increased risk of congenital disabilities and neurodevelopmental issues	Provide education, prenatal care, and social support	Laidlaw et al. (2017)
Children in areas with poor waste management	Exposure to contaminated soil and water from waste dumping sites	Cognitive delays, stunted growth, and higher disease susceptibility	Improve waste management systems	Kumar et al. (2024)
Rural populations dependent on wells	Contaminated groundwater with arsenic and lead	Long-term neurological damage, cancer risk	Ensure access to clean water, test and treat water sources	Shakir et al. (2017)

Having workers checked regularly for health problems helps prevent them from long-term heavy metal exposure and keeps their families safe.

- **Global collaboration and policy alignment:** International groups such as the World Health Organization and the United Nations Environment Programme should help nations set safe exposure limits for heavy metals while working together to create plans for cleaning up polluted areas. Countries must work together to deal with pollution that moves across borders, especially mercury that travels great distances. Bundled aid packages from rich countries aid low-income countries in putting strong pollution-fighting measures into action. Taking specific steps and ensuring everyone gets equal health support will help keep these metals from hurting the health of those needing the most help (Anyanwu et al., 2018).

16.4 Specific neurodevelopmental disorders

16.4.1 Evidence linking heavy metal exposure to neurodevelopmental disorders

Many recent studies investigate whether toxins in heavy metals may affect how a child's brain develops, possibly creating learning and behavior problems. We know that heavy metals affect brain development during gestation and early infancy because studies in people, lab animals, and how things work have found this link.

- **ASD**: Kids with ASD usually show higher amounts of heavy metals in their blood or hair, according to several research studies. Scientific studies show that mercury from thimerosal in vaccines or fish pollution can change how the brain develops, affecting how a person with ASD talks with others and repeats actions (Morris, Puri, Frye, & Maes, 2018). Lead poisoning in the body seems to affect the same social abilities, language, and mental skills as ASD, since it changes how nerve cells talk to one another in the brain. When a baby's mother is exposed to metals like mercury and lead during pregnancy, researchers found that the brain parts that control social behavior may fail to develop fully.
- **Attention-deficit/hyperactivity disorder (ADHD):** Heavy metal exposure can make people more likely to develop ADHD. The brain's concentration of metals builds up when someone is affected by lead and specific pesticide types. Lead damages communication signals in our brains' frontal regions that help us stay focused and organize our thoughts. Too many metals harm children's control over their impulses, focus, and energy levels. New research shows that any amount of lead in a child's system makes that child more likely to develop ADHD. The study shows that mercury changes how our genes produce brain chemicals, harming a child's control systems for thinking and actions if they have ADHD (Xi & Wu, 2021).
- **ID:** Being exposed to heavy metals can cause ID, which makes it harder for a person to learn and function. Small amounts of lead lower IQ scores and cause issues with learning and memory ability. Lead impacts young brains by disrupting brain pathways that shape connections between neurons, affecting how nerve tissue forms protective insulating layers, and interfereing with growth-supporting nerve chemicals. Studies showed that long-term arsenic exposure causes decreased IQ scores, lower learning, and less memory in children. Researchers blame changes it causes in the brain's hippocampus during development (Bellinger, Matthews-Bellinger, & Kordas, 2016).

16.4.2 Mechanisms of heavy metal–induced neurodevelopmental disruptions

Heavy metals—lead, mercury, cadmium, and arsenic—harm a growing brain through multiple actions at the cell and molecular level, leading to harmful brain development. These heavy metals disrupt brain function and contribute to neurodevelopmental disorders using overlapping processes.

- **Oxidative stress and inflammation:** Heavy metals cause problems by creating harmful oxygen molecules called ROS that can destroy brain cells through cell damage from oxidized fats, burned proteins, and DNA damage. This reaction damages connections between brain cells and reduces cell growth, negatively affecting memory skills. Faithful to our earlier understanding, both mercury and lead have been found to target the blood—brain barrier by pumping up the production of damaging oxygen-based chemical reactions. Heavy metals trigger long-term inflammation in the body, damaging how immune signals work in the brain while making microglial cells that help sustain and refine neural connections work poorly (Karri et al., 2016).
- **Disruption of neurotransmitter systems:** Heavy metals cause problems by disrupting how our brain uses neurotransmitters for proper thinking and behavior. Heavy metals damage two major chemical processes in our brain—the dopamine and glutamate systems—that help us stay focused, remember new information, and decide what feels good. Children with ADHD have issues controlling their attention and impulsive behavior partly because their neurochemical pathways related to dopaminergic circuits are not working well. The way mercury disrupts communication between nerve cells that control gabapentin and glutamate activity in the mind probably causes the self-repetitive and social difficulties seen in ASD. When the body encounters cadmium or arsenic particles, they change how neurons

talk to each other using cholinergic messages—signals needed for memory, learning, and staying focused (Bell, Schwartz, Frye, Carter, & Standish, 2015).

- **Epigenetic modifications and developmental toxicity:** Heavy metals change how genes work in developing brains by turning specific genes on and off without affecting the original genetic information. How heavy metals alter our DNA affects how brain cells develop and work over time. Interactions between heavy metals and genes control how long metal exposure affects early brain growth and cell communication systems. Researchers found that when people are exposed to arsenic, key brain growth factors stop working correctly, which causes nerve cells to develop slowly and makes it harder for children to learn and understand (Grandjean, 2015).

16.4.3 Importance of early diagnosis and intervention for children with neurodevelopmental disorders

Finding and starting treatment immediately helps protect against long-term brain development delays from exposure to large amounts of heavy metals. Your brain stays very sensitive to outside influences during early childhood, and metals affect how it grows (Tizabi, Bennani, El Kouhen, Getachew, & Aschner, 2023). Timely diagnosis and supportive strategies in young children help lower their symptoms of developmental issues and enhance their entire lives from this point forward.

- **Early screening and monitoring:** Assessing small children who live near toxic areas or were exposed to toxins before birth is important because heavy metals can harm their brain development. Checking metal levels in children's blood and urine regularly helps us spot who's been exposed early on so we can start helping them right away. Today's medical methods conducting behavior tests and brain imaging—show how heavy metals damage thinking and behavior, making it more straightforward to jump in and help.
- **Environmental modifications and medical interventions:** After finding exposure to harmful metals, getting the person out of contaminated spaces and ending contact with toxic materials are key steps to restore health. Removing heavy metals from the body using chelation therapy helps reduce lead and mercury damage to your brain while keeping kids in their best mental state. Dedicated treatments to improve thinking, social, and behavior skills lead to better results for kids. Exceptional help for learning, behavior therapy, and getting early treatment helps children with these brain disorders better deal with their mental and social problems.
- **Long-term follow-up and support:** Parents must keep tracking a child's developmental progress after heavy metal exposure causes neurodevelopmental disorders and regularly adjust treatment to handle their ongoing health challenges. Keeping support going for children who have had toxic exposure will lessen the damage over time. Children with IDs need long-term help because they always need school and job support to live as independently as possible (Fig. 16.1).

16.5 Lead and mercury toxicity

16.5.1 Neurotoxic effects of lead and mercury on brain development

Lead and mercury are damaging substances linked to different problems with brain development. These metals damage the brain most during its growth and change early in life. These heavy metals mess with how brain cells work and communicate, overstress brain cells with harmful substances, and impact thinking and behavior growth (Kumar et al., 2022).

- **Lead toxicity:** Lead causes damage to the brain cells primarily by stopping how nerve cells release chemicals needed for communication, changing calcium signals throughout the nerve pathways, and affecting how brain connections adapt and grow. When babies and young children are exposed to lead, the metal stays in their brain tissues. It creates problems with the developing nervous system, damaging the frontal cortex and hippocampus, both vital for learning, thinking, and memory formation (Grandjean, 2015). Lead creates important neuron changes that diminish the way dendrites grow, and communications happen between cells. Since lead affects parts of our brain that control movement, attention, and behavior, it makes us more impulsive, energetic, and unable to focus. Because lead prevents normal brain barrier development, it worsens damage while moving through the body.
- **Mercury toxicity:** Methylmercury, mercury's natural chemical form, causes severe brain damage by disrupting brain cell motors, stopping neuron communication, changing how nerve cells send messages, and setting off brain inflammation. The body quickly takes in methylmercury from your food and lets it run through your blood to reach your brain, where it collects in large amounts. Mercury is attracted to proteins containing sulfur, including those needed for making and moving neurotransmitters, and disrupts nerve cells by mistiming how ions pass through their channels and communicate. Mercury also interferes with our brain's natural production of key growth and protection proteins, affecting how certain key brain regions develop and grow to control thoughts, movement skills, and senses (Kaur et al., 2021). Studies have found that mercury stops developing brain cells from linking correctly, leading to poor learning, memory, and behavior in affected children.

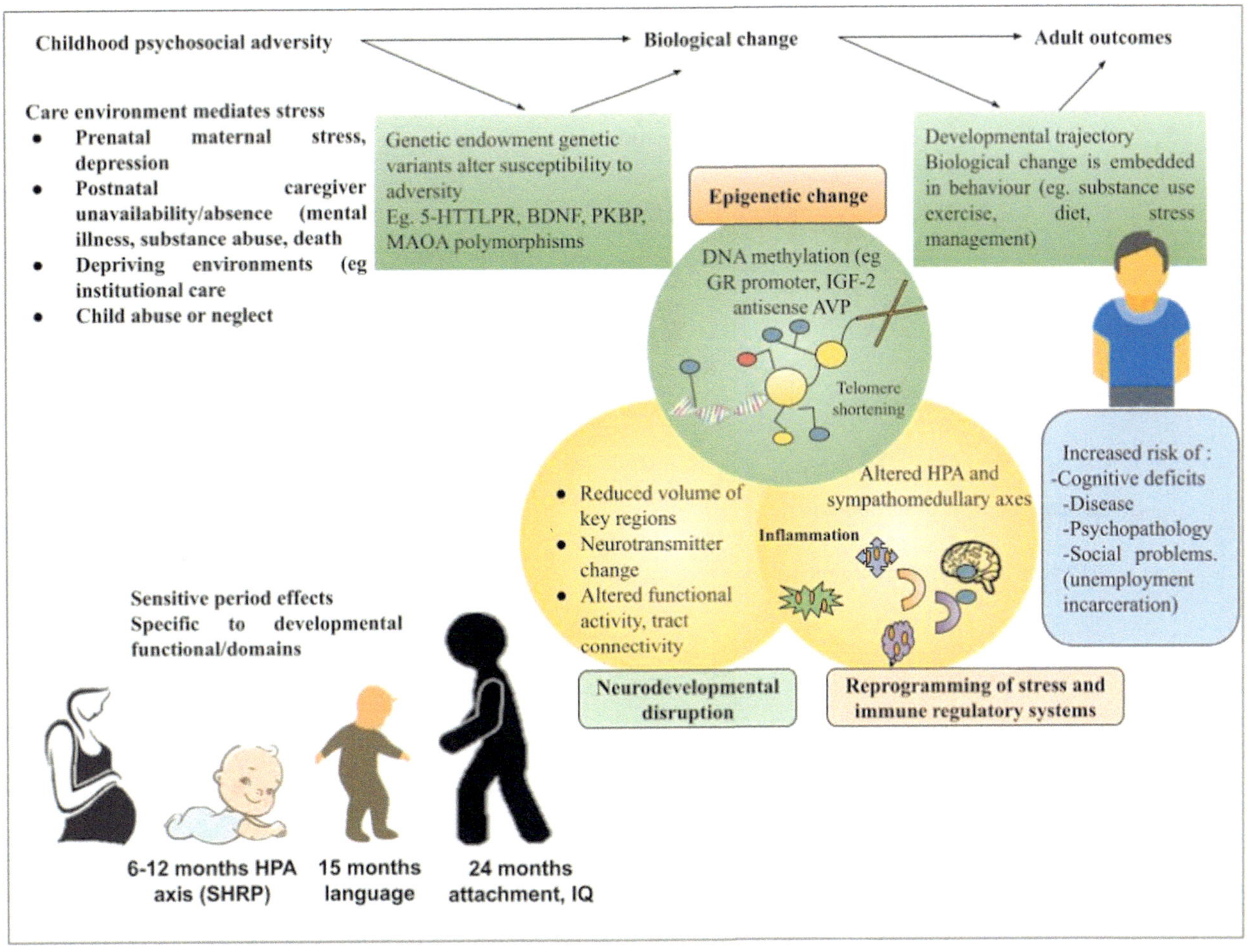

FIGURE 16.1 Several pathways link early life adversity to adult outcomes. Early exposure to adversity interacts with a child's genetic predisposition (e.g., variations in genetic polymorphisms), triggering a cascade of biological alterations across multiple levels.

Studies show mercury can change how the brain develops and kills nerve cells in the brain's main motor-controlling section (the cerebellum) and brain region linked to higher thinking (the cortex) (Fig. 16.2).

16.5.2 Historical and ongoing sources of lead and mercury exposure

Public health groups have been watching lead and mercury pollution for many years, which enters the environment through different sources everywhere. Governments have taken steps to control these metals, but they still harm people most at risk—young children and pregnant women—from getting overexposed to them.

- **Lead exposure sources**: Many people get lead in their bodies from old products like lead paint, gasoline chemicals, plumbing materials, and battery parts. Lead poisoning got a lot better after taking the lead out of gas and paint by the late 1900s, but some places still have problems with people getting dangerously high levels of lead in their bodies. Children living in older urban homes face lead poisoning risks because those properties often still have lead-based paint. Children can get hurt when they eat paint or breathe in dust. Many towns still have lead water systems, causing drinkable water to become contaminated by small amounts of lead. In poor and industrial neighborhoods with weaker environmental rules, people face ongoing risks from lead in their soil, industrial air pollution, and trash disposal. People who work in construction, battery factories, and paint jobs are still at risk of getting lead poisoning from their job exposure.
- **Mercury exposure sources:** Mercury, which used to be used back in the day for making thermometers and mining gold, is still around because it gets mixed in with our natural environment and comes out of factories. The primary way people get mercury in their bodies nowadays comes from eating fish contaminated with the mineral, which enters aquatic systems through industrial emissions and gets changed into easily absorbed molecules that rise the food chain. People who eat these big fish frequently face health dangers because mercury in

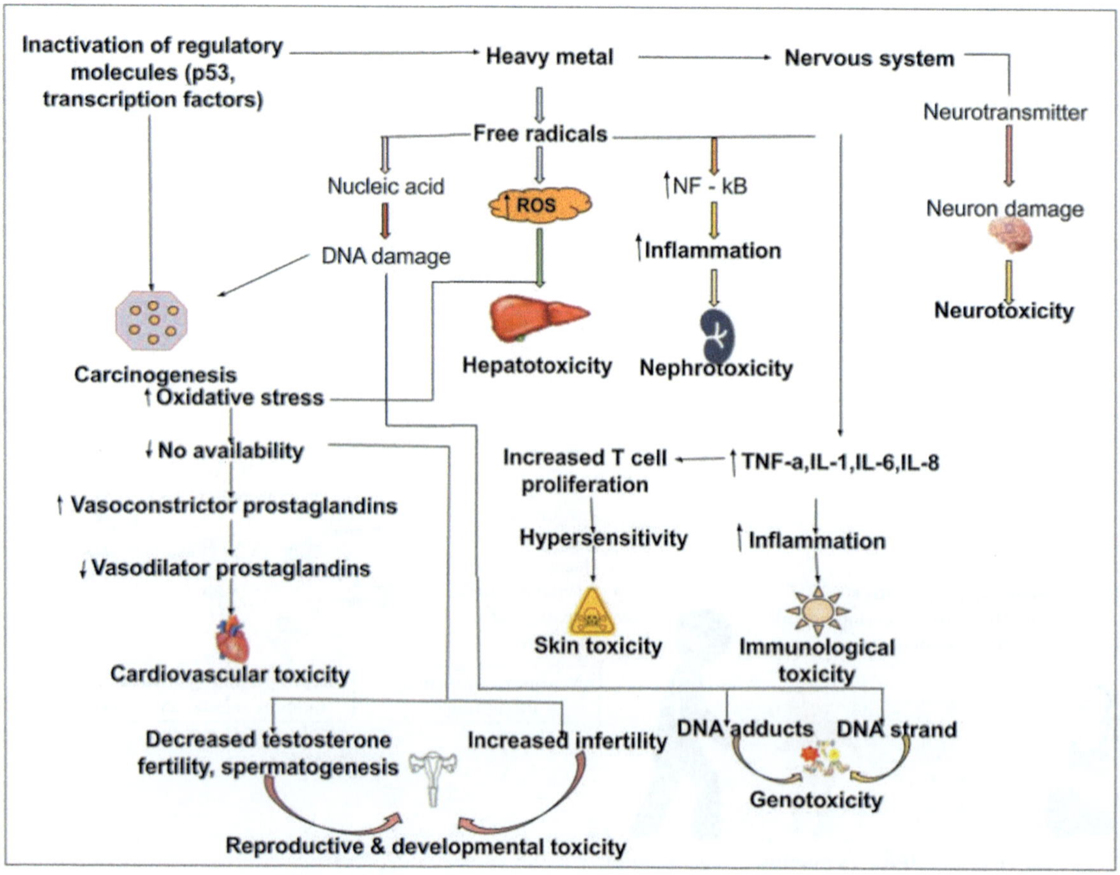

FIGURE 16.2 Heavy metals toxicological mechanisms and consequent health effects.

tuna, swordfish, and sharks builds up along the food chain. Mercury comes from three primary places: dental work, specific medical devices, and vaccines that used thimerosal (but this was fixed by taking thimerosal out of most vaccines).

16.5.3 Long-term consequences of lead and mercury exposure on brain development

Conducting brain development while young from lead and mercury levels leads to permanent thinking changes and behavior issues and raises the possibility of nervous system diseases as adults (Grandjean, 2015). These metals harm the brain most when they enter early on, when brain growth is at its most sensitive and flexible stage.

- **Cognitive deficits:** These two metals strongly show that we have permanent thinking and learning problems after contact with their particles. Lead harms a child's brain development, whether it is present in small amounts, and causes issues with IQ, school performance, memory, and attention. Also affecting brain development, lead exposure causes problems in how our brain organizes thoughts, plans tasks, and controls impulses. Children exposed to mercury have trouble learning, forget verbal instructions, and have difficulties finding their way in space. When mercury damages the brain's ability to make new nerve cells and form strong connections, it typically stops children from learning new things. They will struggle to learn in school and do everyday stuff throughout life (Ralston & Raymond., 2018).
- **Behavioral problems:** Kids who face lead and mercury exposure are more likely to develop behavioral changes that cause them to act too fast, recklessly, or too kind. Children poisoned by lead sometimes have problems learning and getting along with others, along with these behavior issues. The behavioral changes caused by lead toxicity occur because lead disrupts the functioning of chemical systems that regulate moods, control impulses, and utilize rewards. Wood's toxic chemicals cause people with mercury in their bodies to lose control of their emotions and struggle to behave well socially. Lead poisoning causes school struggles and situations with friends that harm children's mental health early in their lives.
- **Increased risk of neurodegenerative diseases:** Exposure to lead and mercury throughout life raises the risk of getting Alzheimer's, Parkinson's, and other brain diseases as people grow older. Being around lead for a

long time increases your chances of both faster memory loss and brain-wasting diseases down the road. Those who had lead exposure when they were young see more damage to their brain and nervous system take in middle age. Over time, people who breathe too much mercury can develop central nervous system disorders. Mercury slows down how brain proteins form tangles and plaques, which often appear in Alzheimer's patients, and harms energy centers in neurons, killing them. Both lead and mercury exposure create harmful situations inside your brain where free radicals damage cells and trigger harmful immune reactions, which push your brain's ageing issues along faster (Cariccio, Samà, Bramanti, & Mazzon, 2019).

- **Global and developmental impact**: During pregnancy, both lead and mercury poisoning can harm how a baby's brain develops, causing birth defects, smaller babies at birth, and late development. Heavy metal toxicities from lead and mercury add up to large medical bills to care for cognitive, behavioral, and other health issues that have affected us for many years. This also brings heavy costs from lost work productivity and increased care needs.

16.6 Environmental and public health implications

16.6.1 Broader environmental and public health implications of heavy metal exposure

Lead, mercury, arsenic, and cadmium entering the environment cause serious problems in air, water, and soil. These toxic metals damage the brain directly and indirectly through wide-ranging issues such as lower output at work, higher healthcare spending, and lasting developmental problems among groups of people (Penumantra, Shirbhate, Tiwari, & Narkhede, 2024). Heavy metals cause ongoing pollution that harms people and local areas today and could continue to damage future generations. Those with special needs and who live by factories or in poor communities are facing the worst effects of metal-contaminated environments.

- **Impact on vulnerable populations:** During their development, from before birth through their early years, kids are susceptible to harm done to their brains by toxic metals. Environmental heavy metals easily damage the young brain, and even small doses of metals can change how children's brains work and behave for their whole lives. High metal pollution and pesticide usage in areas with factories, poor living places, and farms push these people to receive more toxic metals. Health problems from heavy metals occur primarily in people struggling to afford proper healthcare, safe water, and housing. In poor neighborhoods and developing nations, people are more likely to face deadly levels of heavy metals because laws and garbage disposal aren't working well enough.
- **Ecosystem contamination and bioaccumulation:** Mercury and lead stay in the environment long, making everything around them dirty for months, years, and sometimes forever. These metals remain in nature for a long time and build up in the creatures' food, changing wildlife and ecosystems. Flowers show how mercury in water harms fish and other sea life, which hurts wildlife and long-term food production. People who mostly eat fish run a higher risk of getting heavy metals in their bodies when these metals move from animals at the bottom of the food chain to more enormous creatures. The heavy metals in soil can cause two major problems: lower farm productivity and unhealthy food due to higher trace metal content.
- **Chronic health issues:** Heavy metals that build up over time can cause various chronic health troubles—they hurt your heart, harm your kidneys, and raise cancer risks. These health problems require costly care over a lifetime, placing additional pressure on healthcare services. More people with prior heavy metal exposure are getting brain conditions like Alzheimer's and Parkinson's, making healthcare systems everywhere face more strain. These health problems continue to cost money, as patients need lifelong medical support. Public health risks from heavy metals will grow as children and older people become a more significant portion of our population (Dorsey, Sherer, Okun, & Bloem, 2020).

16.6.2 Advocacy for environmental regulations and public health interventions

Strong protection rules and medical help are needed to lower the effects of heavy metal pollution and keep it from hurting those who need it most. Good pollution controls should start with where metal pollution comes from, work to lower how metals affect the environment, and watch for and stop health risks before they happen through monitoring.

- **Environmental regulations:** Keeping heavy metals out of our environment follows two steps: governments control industrial pollution and ban dangerous chemicals from reaching us through products we buy and use. Rules keeping lead out of paint, gasoline, and plumbing equipment help lower lead exposure in all people, but most of all in children. Using better filters in coal plants and cleaner energy resources can help stop mercury from leaving factories as waste. Governments must establish and enforce rules requiring testing of air, water, and soil for heavy metals to monitor pollution, defining the actions to be taken when pollution is detected (Li, Liu, Lin, Liu, & Xie, 2019) vention on Mercury has helped lower global mercury pollution, we still need better enforcement of this treaty and new rules to protect our environment from long-term mercury damage.

- **Public health interventions:** Health officials and government agencies must allocate resources to programs that protect individuals at the highest risk. Children who live with painted houses containing lead or in contaminated soil areas need free blood lead checks, secure housing, and help cleaning their environment of dangerous chemicals. By teaching pregnant women and giving them support in polluted areas, public health can make sure fewer kids get exposed to mercury and other harmful metals during pregnancy. Doctors can use chelation therapy to lower heavy metal levels in people's bodies and decrease damage from acute poisoning (Aaseth, Skaug, Cao, & Andersen, 2015). Our public health programs need to teach people how to throw away safe household and industrial waste to keep toxic metals from getting into drinking water and soil. Local, national, and global groups working together are needed to find and carry out long-term ways to fight the harm heavy metals do to the environment and people's health.
- **Regulating agricultural practices**: Heavy metals find their way through agricultural food production when soil or water becomes polluted during farming. We need laws that stop farmers from using polluted water for their crops and from applying pesticides that include dangerous metals. By being more careful and regularly checking farm soil and water conditions, we can limit how heavy metals build up in the food grown for people and animals (Fig. 16.3). Organic farming and bioremediation on contaminated land help reduce farming's damage to the natural world. We must back factory farming methods that protect health while caring for nature (Sonone, Jadhav, Sankhla, & Kumar, 2020).

16.6.3 Importance of public education campaigns and preventive measures

People must learn about heavy metal risks through public education and information campaigns to protect themselves from exposure. Teaching people which sources and dangers cause heavy metals to harm us most helps us take detailed safety measures and get needed health checks early.

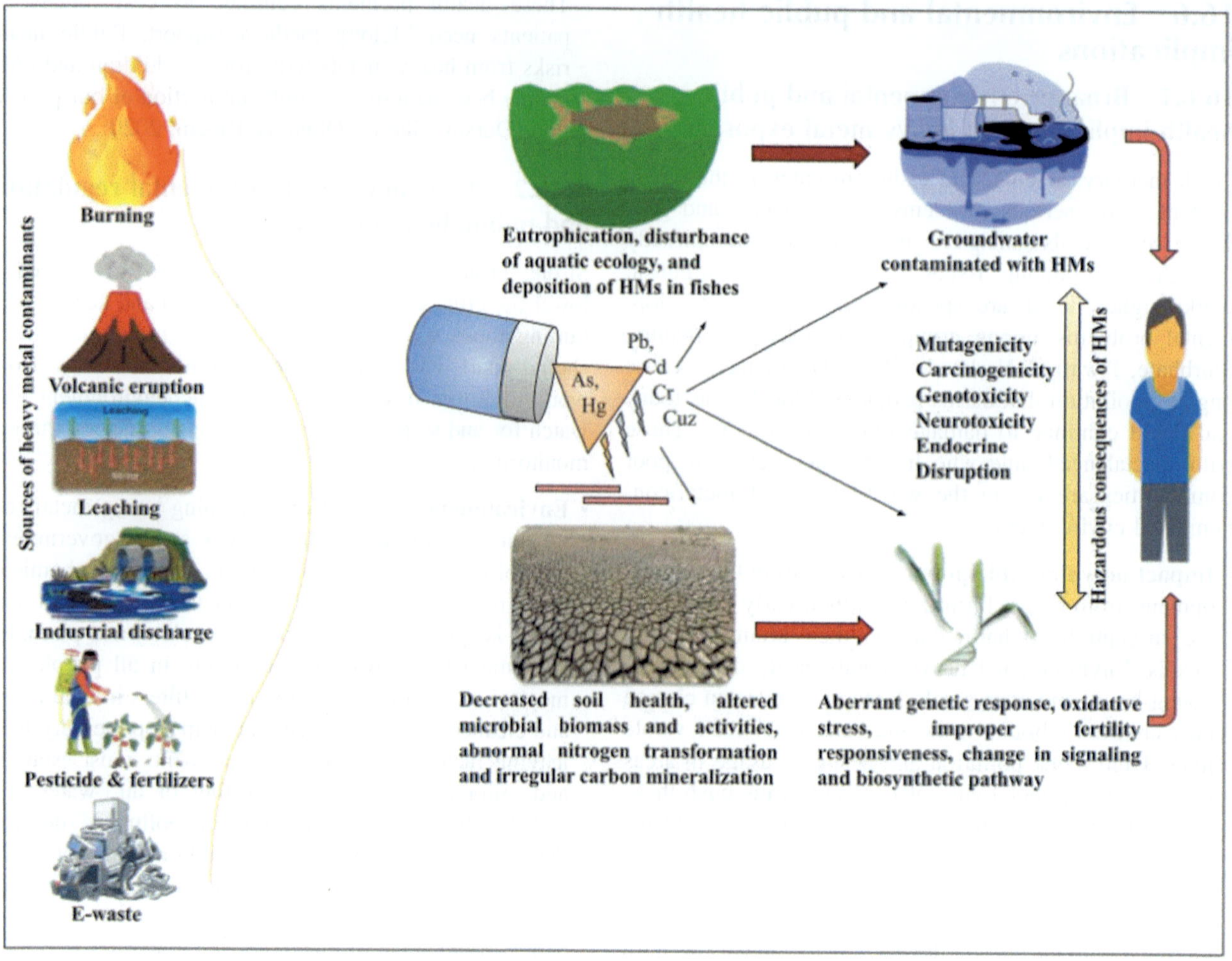

FIGURE 16.3 Microbe-mediated environmental remediation of heavy metals.

- **Raising awareness:** Our public education efforts must explain how metals can cause health problems and highlight the need for special protection for specific groups, including children and pregnant individuals. Knowing where metals come from in daily life—like painted house walls, tap water, and polluted fish—and how to stay safe goes into these awareness programs. Community activities and teams working with schools, medical centers, and eco groups spread the message to people who face the most significant risk from heavy metals. Educational programs should teach people and healthcare providers in locations with known metal exposure issues how to test for and handle this risk (Rehman, Fatima, Waheed, & Akash, 2018).
- **Promoting preventive measures:** The education programs should show people and neighborhoods the steps to reduce exposure risk. We show parents how their children can get sick from lead poisoning when they live in homes built before 1978 and tell them to clean regularly and fix any paint problems to lower their child's risk. People in places with lead pipes or polluted water must learn how using water filters helps keep lead exposure low. Groups responsible for public health can organize projects to show people which fish have high mercury levels and tell them what lower-mercury alternatives they can eat instead (Bostic, Sobal, & Bisogni, 2018). Part of our prevention work involves correctly removing mercury-bearing products like thermometers and batteries to stop any mercury from entering the environment. People can take fewer heavy metals in their bodies if they hear about and follow the steps these campaigns suggest.
- **Engaging communities in environmental monitoring:** Using community-based monitoring lets people in local neighborhoods take part in both learning about and stopping pollution from metal toxicity. Public health groups can teach residents how to test for heavy metal contamination in their area (such as water samples for lead and soil samples), which protects people from toxins better and makes residents feel they have a stake in their community. These programs help people stay healthy while showing them how to work together to solve problems in their area. When governments run short of money or industries create most of the pollution, community groups often do better work to remove pollutants.
- **Collaborating with healthcare providers:** Teachers and doctors need special training to detect and prevent heavy metal poisoning in their students and patients early. Regular screening of the people most at risk—children, pregnant mothers, and others—helps us find health problems early before they cause lasting damage.

16.7 Future directions and research needs

16.7.1 Key areas for future research on heavy metal exposure and neurodevelopment

Our understanding of how heavy metals affect brain development over time is very new, and researchers still need to find answers to essential questions about this topic. We must figure out how these heavy metals—lead, mercury, arsenic, and cadmium—damage developing brain cells to understand their impact on the brain's health (Karri, Schuhmacher, & Kumar, 2016). Future investigations must focus on key unfinished research areas to fill our current knowledge gaps and help minimize exposure damage.

- **Long-term effects of heavy metal exposure:** Research indicates that early exposure to heavy metals can impact how babies learn and behave, but scientists still lack a comprehensive understanding of how metal exposure affects children's developing brains. How people's brains change from test results measured as children to what happens when they become adults helps us understand if those toxins stay damaging or if the brain finds ways to recover independently. Researchers know that lead harms children's IQ, but they still do not have good data on how kids' thinking skills are affected later in life. We need researchers to explore how young people who come into contact with these toxins in their early years may develop Alzheimer's or Parkinson's diseases years later. By studying animals and people exposed to small heavy metal doses during critical growth stages, we can learn how these materials change how the brain looks and works long into the future.
- **Mechanisms of action:** Investigators must figure out the complex ways that various heavy metals damage brain development, working their way deeper into how cells and molecules react. Metals harm the developing brain by creating harmful oxidative stress, disrupting the way nerve cells connect and communicate with each other, and altering how neurons transmit their messages. When lead enters the system, it tricks calcium into jumping across cells and blocks signals sent by both dopamine and glutamate pathways—two key learning and attention drivers. Mercury damages brain cells by binding to sulfhydryl proteins and interfering with electrical signals and brain chemical messages. Studying how metals connect with brain molecules is key to designing treatments that fix and limit metal-based brain damage. Examining how different metals combine and work together when they are present simultaneously can help us better predict how they may lower brain development more severely than exposing people to just one of these metals.

- **Preventive and therapeutic strategies:** Our research must find ways to prevent brain development harm caused by heavy metal exposure and get better results when treating those currently affected. We must find ways to measure early mental exposure and develop methods to block exposure when developing brains are most vulnerable—both before birth and early in a child's life. Scientists must investigate how diet, chelation treatments, and brain protection medicines can help people who already carry high levels of heavy metals. Research indicates that antioxidants, such as N-acetylcysteine, protect against mercury-induced brain oxidative damage in animals. However, studies testing the same effect on people are still in the early stages. Scientists are working to create better tests (in blood or urine) that can find heavy metal toxicity early in life, making it easier to help affected children get the care they need.

16.7.2 Need for interdisciplinary collaborations to advance understanding

Understanding how heavy metals harm developing brains requires collaboration among multiple experts. Environmental scientists, toxicologists, and neurologists must work closely together to understand how ecological chemicals cause brain impairments in children (Landrigan, Landrigan, 2018). We need to bring together different types of experts to devise suitable treatments and preventions.

- **Environmental scientists and toxicologists:** Environmental scientists find and track where heavy metals enter the environment in air, water, soil, and food to give us essential information about contamination levels in different places. Toxicologists help us learn how each metal acts as a poison in the body's systems and what makes them dangerous. Scientists who study the environment must work with those who study poisons to improve our understanding of where people might come into contact with hazardous levels of metals and which groups of people are more likely to get sick. Examining mercury pollution in our nearest water supplies lets us locate and evaluate precisely what harms people in catchment areas might face. Learning how environmental elements influence people's reactions to heavy metals could teach us how to protect public health.
- **Neurologists and developmental psychologists:** Neurologists and developmental psychologists work together to show how heavy metal effects change children's thinking and acting. Their knowledge is required to spot three necessary developmental brain conditions in children caused by heavy metals—ASD, ADHD, and IDs. Combining neuroimaging tools like MRI and PET scans with group research will show how toxic metals impact brain structure and operation. By working together, experts can find how brain changes cause behavioral and cognitive problems, track potential markers for early illness detection, and develop new treatments. These varied study approaches can teach us why damage early in a child's brain growth causes long-term learning and behavioral problems.
- **Public health experts and epidemiologists:** With research projects, epidemiologists can follow heavy metal exposure changes and health issues among people, showing us what dangers need urgent public health action. Health professionals use research to develop prevention plans and practical public health policies. Collaboration helps build extensive studies of population groups to see how metal exposure affects a child's brain and mind growth over many years. The results of how well public health programs work at lowering exposure and risk can be determined with the help of experts who study how diseases spread through populations.

16.7.3 Translating research into public health interventions and environmental policies

Researching how heavy metals affect children's brain growth aims to turn scientific discoveries into real-world solutions to protect everyone's health. Our study findings must move out of the university and help make practical decisions to shield at-risk children from mental harm to their brain growth.

- **Public health interventions:** Our findings show what heavy metals do to children's brain growth, so we must use them to create health programs and community regulations that keep people safe. Our discoveries about lead harming brain growth inspired practical public health efforts to cut lead from gas, plumbing, and paint. We need the exact public health solutions to reduce people's exposure to other dangerous metals such as mercury and cadmium. Our findings can guide government health policy to switch from toxic chemicals, like lead in toys, to safer choices and limit mercury in dental fillings. Health experts can teach parents and guardians about places in their homes where children can get metal poisoning and how to do exposure tests and make their homes safe.
- **Environmental policies and regulations**: Environmental rules must reflect scientific research to reduce how widespread heavy metals make the planet dirty. More official rules must be expanded to restrict factories' dangerous metal releases while requiring cleaner production methods and better ways to eliminate waste. We must help nations follow their promises to adopt and apply the Mercury Minamata Convention. Based on regional research on heavy metal dangers, we must focus on cleaning up polluted areas while creating better and safer spaces for populations prone to suffering from metal poisoning. When we routinely check soil, air, and water

throughout industrial and farming regions, it lets us spot pollution early on, saving time to fix the problem.

- **Prevention and early intervention:** We need to study how heavy metals affect kids' brains, then build programs that test for these metals and help affected children before the problem becomes more serious. We need new public health rules to require quick lead and mercury tests for every child in areas with high levels of these substances. The government should enforce new policies that remove lead-based materials from houses and buildings while helping families pay to fix or test their plumbing and see if anyone is ill from lead. Earlier research shows that what works should guide medical professionals in creating further policies to protect children from toxic metal exposure. Researchers show us how taking action now based on scientific findings can protect against long-term brain and behavioral damage from heavy metal poisoning.

16.8 Conclusion

More research shows that harmful metals can cause neurodevelopmental problems, and many health experts now say this is a growing public health concern. If children get too much lead, mercury, arsenic, or cadmium in their bodies, it usually hurts their brain development. This causes them to have trouble thinking, changes their behavior, and makes them more likely to struggle with three common neurodevelopmental disorders: ASD, ADHD, and ID. Children and pregnant women who come into contact with these substances at life-changing brain development moments can struggle with lasting brain problems that stay through adulthood. While we are learning more about these dangers, there are still significant holes in what we know. We need more research to understand how these metals harm the brain and how exposure to multiple toxins combines. Heavy metals that enter early brain development have lasting effects on expected growth. In children exposed to heavy metals, their brain functions regularly get damaged, having trouble learning like normal children, struggling to focus, and experiencing memory failures. Most children exposed to heavy metals have behavioral issues like impulse control problems, extreme activity, and angry outbursts. While metal exposure while young may impact brain development, it raises our risk of developing neurological conditions during our older years, like Alzheimer's and Parkinson's disease. We must find exposure immediately and work on solutions to reduce these lasting brain effects. Specific groups of experts need to work together to fight the rise of brain disorders caused by environmental poisons. We need more research to understand the toxins' lasting damage to our brains, find out how they harm nerve cells, and develop treatments to prevent or reverse the harm. Working with scientists who study the environment, learn about poison effects, and work with brain specialists and health researchers is our best way to understand how toxic materials wreck our brains during growth and discover what we can do to make them matter less. Learning what causes health issues is only one step. We must understand how to apply this knowledge to close public health gaps and control pollutant levels in places with the highest risk. Public outreach efforts help people understand the dangers of heavy metals and show them how to protect themselves, from living in secure housing to changing their diets and monitoring their environment. According to research, finding and treating heavy metals in children early on can protect their brain growth and overall health. Together, continuing our work and taking specific steps can lower the number of disabilities caused by heavy metals affecting children's brains and keep future children safe from these brain risks.

References

Aaseth, J., Skaug, M. A., Cao, Y., & Andersen, O. (2015). Chelation in metal intoxication—Principles and paradigms. *Journal of Trace Elements in Medicine and Biology, 31*, 260–266.

Albuquerque, E., Burke, R., Rao Gullapallib, J., & Mamczarza, E. P. (2017). Developmental neurotoxicity of the organophosphorus pesticide chlorpyrifos: From animal behaviour to molecular mechanisms. *Journal of Neurochemistry, 142*, 188–225.

Anyanwu, B. O., Ezejiofor, A. N., Igweze, Z. N., & Orisakwe, O. E. (2018). Heavy metal mixture exposure and effects in developing nations: An update. *Toxics, 6*(4), 65.

Bell, I. R., Schwartz, G. E., Frye, J., Carter, B., & Standish, L. J. (2015). Extending the adaptive network nanomedicine model for homoeopathic medicines: Nanostructures as salient cell danger signals for adaptation. *Nanoscience and Technology, 2*(1), 1–22.

Bellinger, D. C., Matthews-Bellinger, J. A., & Kordas, K. (2016). A developmental perspective on early-life exposure to neurotoxicants. *Environment International, 94*, 103–112.

Bostic, S. M., Sobal, J., & Bisogni, C. A. (2018). Social representations of fish and seafood among midlife rural adults: Benefits, risks, and involvement. *Food Policy, 76*, 99–108.

Cariccio, V. L., Samà, A., Bramanti, P., & Mazzon, E. (2019). Mercury is involved in neuronal damage and neurodegenerative diseases. *Biological Trace Element Research, 187*(2), 341–356.

Dorsey, R., Sherer, T., Okun, M. S., & Bloem, B. R. (2020). *Ending Parkinson's disease: A prescription for action*. Hachette UK.

Flora, S. J., & Agrawal, S. (2017). Arsenic, cadmium, and lead. In Reproductive and developmental toxicology. Academic Press, 537–566.

Grandjean, P. (2015). Only one chance: How environmental pollution impairs brain development—and how to protect the next generation's brains. *Ecological Ethics and Science*.

Guinchat, V., Cravero, C., Lefèvre-Utile, J., & Cohen, D. (2020). Multidisciplinary treatment plan for challenging behaviors in neurodevelopmental disorders. In *Handbook of clinical neurology (Vol. 174)*. Elsevier, 301–321.

Gupta, R., Ambasta, R. K., & Kumar, P. (2021). Autophagy and apoptosis cascade: Which is more prominent in neuronal death? *Cellular and Molecular Life Sciences, 78*(24), 8001–8047.

Hoover, E. (2018). Environmental reproductive justice: Intersections in an American Indian community impacted by environmental contamination. *Environmental Sociology, 4*(1), 8–21.

Huang, M., Liu, Y., Duan, R., Yin, J., & Cao, S. (2024). Effects of continuous and pulse lead exposure on the swimming behaviour of tadpoles revealed by brain-gut axis analysis. *Journal of Hazardous Materials, 465*, 133267.

Ijomone, O. M., Ifenatuoha, C. W., Aluko, O. M., Ijomone, O. K., & Aschner, M. (2020). The ageing brain: Impact of heavy metal neurotoxicity. *Critical Reviews in Toxicology, 50*(9), 801–814.

Jiang, X., & Nardelli, J. (2016). Cellular and molecular introduction to brain development. *Neurobiology of Disease, 92*, 3–17.

Karri, V., Schuhmacher, M., & Kumar, V. (2016). Heavy metals (Pb, Cd, As and MeHg) as risk factors for cognitive dysfunction: A general review of metal mixture mechanism in the brain. *Environmental Toxicology and Pharmacology, 48*, 203–213.

Kaur, I., Behl, T., Aleya, L., Rahman, M. H., Kumar, A., Arora, S., & Akter, R. (2021). Role of metallic pollutants in neurodegeneration: Effects of aluminum, lead, mercury, and arsenic in mediating brain impairment events and autism spectrum disorder. *Environmental Science and Pollution Research, 28*, 8989–9001.

Keunen, K., Counsell, S. J., & Benders, M. J. (2017). The emergence of functional architecture during early brain development. *Neuroimage, 160*, 2–14.

Kordas, K., Ravenscroft, J., Cao, Y., & McLean, E. V. (2018). Lead exposure in low and middle-income countries: Perspectives and lessons on patterns, injustices, economics, and politics. *International Journal of Environmental Research and Public Health, 15*(11), 2351.

Kumar, P., Goud, E. L., Devi, P., Dey, S. R., & Dwivedi, P. (2022). Heavy metals: Transport in plants and their physiological and toxicological effects. In *Plant metal and metalloid transporters*. Singapore: Springer Nature Singapore, 23–54.

Kumar, P., Choudhury, D., Dey, S. R., & Sharma, M. (2024). Application of genetically modified crops against metallic contaminations. In *Bioremediation of emerging contaminants from soils*. Elsevier, 541–564.

Kumar, P., Raj, A., & Kumar, V. A. (2024). Approach to reduce agricultural waste via sustainable agricultural practices. In *Valorization of biomass wastes for environmental sustainability: Green practices for the rural circular economy*. Cham: Springer Nature Switzerland, 21–50.

Kupnicka, P., Kojder, K., Metryka, E., Kapczuk, P., Jeżewski, D., Gutowska, I., ... Baranowska-Bosiacka, I. (2020). Morphine-element interactions—The influence of selected chemical elements on neural pathways associated with addiction. *Journal of Trace Elements in Medicine and Biology, 60*, 126495.

Laidlaw, M. A., Filippelli, G. M., Brown, S., Paz-Ferreiro, J., Reichman, S. M., Netherway, P., ... Mielke, H. W. (2017). Case studies and evidence-based approaches to addressing urban soil lead contamination. *Applied Geochemistry, 83*, 14–30.

Landrigan, P. J., & Landrigan, M. M. (2018). *Children and environmental toxins: What everyone needs to know®*. Oxford University Press.

Landrigan, P. J., Stegeman, J. J., Fleming, L. E., Allemand, D., Anderson, D. M., Backer, L. C., ... Rampal, P. (2020). Human health and ocean pollution. *Annals of Global Health, 86*(1).

Larrigan, S., Shah, S., Fernandes, A., & Mattar, P. (2021). Chromatin remodeling in the brain—A NuRD evelopmental Odyssey. *International Journal of Molecular Sciences, 22*(9), 4768.

Li, T., Liu, Y., Lin, S., Liu, Y., & Xie, Y. (2019). Soil pollution management in China: A brief introduction. *Sustainability, 11*(3), 556.

Morris, G., Puri, B. K., Frye, R. E., & Maes, M. (2018). The putative role of environmental mercury in the pathogenesis and pathophysiology of autism spectrum disorders and subtypes. *Molecular Neurobiology, 55*, 4834–4856.

Naja, G. M., & Volesky, B. (2017). Toxicity and sources of Pb, Cd, Hg, Cr, As, and radionuclides in the environment. In *Handbook of advanced industrial and hazardous wastes management*. CRC Press, 855–903.

Nshimiyimana, J., Uwihoreye, P., Muhigirwa, J. C., & Niyonsega, T. (2023). Neurofunctional intervention approaches. In *Neurorehabilitation and physical therapy*. IntechOpen.

Parithathvi, A., Choudhari, N., & Dsouza, H. S. (2024). Prenatal and early life lead exposure induced neurotoxicity. *Human and Experimental Toxicology, 43* 09603271241285523.

Penticoff, H. B., & Fortin, J. S. (2023). Toxic/metabolic diseases of the nervous system. In *Neurobiology of brain disorders*. Academic Press, 379–401.

Penumantra, A., Shirbhate, N. S., Tiwari, P. S., & Narkhede, S. (2024). *The impact of environmental pollution on human health*. Academic Guru Publishing House.

Rahman, A., & Guillemin, G. J. (2021). Lead and excitotoxicity. In R. M. Kostrzewa (Ed.). *Handbook of neurotoxicity*. (pp. 1–39). Springer, Cham.

Ralston, N. V., & Raymond, L. J. (2018). Mercury's neurotoxicity is characterized by its disruption of selenium biochemistry. *Biochimica et Biophysica Acta (BBA)—General Subjects, 1862*(11), 2405–2416.

Rehman, K., Fatima, F., Waheed, I., & Akash, M. S. H. (2018). Prevalence of exposure of heavy metals and their impact on health consequences. *Journal of Cellular Biochemistry, 119*(1), 157–184.

Rubin, I. L., Coles, C. D., Barnhill, J., Coles, C. D., & Barnhill, J. (2023). Behavioral and mental health disorders (including attentional disorders). In *Neurodevelopmental pediatrics: Genetic and environmental influences*. Cham: Springer International Publishing, 655–674.

Samiee, F., Vahidinia, A., Javad, M. T., & Leili, M. (2019). Exposure to heavy metals released to the environment through breastfeeding: A probabilistic risk estimation. *Science of the Total Environment, 650*, 3075–3083.

Scandurra, V., Emberti Gialloreti, L., Barbanera, F., Scordo, M. R., Pierini, A., & Canitano, R. (2019). Neurodevelopmental disorders and adaptive functions: A study of children with autism spectrum disorders (ASD) and/or attention deficit and hyperactivity disorder (ADHD). *Frontiers in Psychiatry, 10*, 673.

Shakir, S. K., Azizullah, A., Murad, W., Daud, M. K., Nabeela, F., Rahman, H., ... Häder, D. P. (2017). Toxic metal pollution in Pakistan and its possible risks to public health. *Reviews of Environmental Contamination and Toxicology, 242*, 1–60.

Sighencea, M. G., Popescu, R., & Trifu, S. C. (2024). From fundamentals to innovation in Alzheimer's disease: Molecular findings and revolutionary therapies. *International Journal of Molecular Sciences, 25*(22), 12311.

Sonone, S. S., Jadhav, S., Sankhla, M. S., & Kumar, R. (2020). Water contamination by heavy metals and their toxic effect on aquaculture and human health through food Chain. *Letters in Applied NanoBioScience, 10*(2), 2148–2166.

Tizabi, Y., Bennani, S., El Kouhen, N., Getachew, B., & Aschner, M. (2023). Interaction of heavy metal lead with gut microbiota: Implications for autism spectrum disorder. *Biomolecules, 13*(10), 1549.

Xi, T., & Wu, J. (2021). A review on the mechanism between different factors and the occurrence of autism and ADHD. *Psychology Research and Behavior Management,* 393–403.

Yu, G., Wu, L., Su, Q., Ji, X., Zhou, J., Wu, S., ... Li, H. (2024). Neurotoxic effects of heavy metal pollutants in the environment: Focusing on epigenetic mechanisms. *Environmental Pollution,* 123563.

Chapter 17

Childhood exposure to heavy metals and long-term neurodevelopmental effects

Mastu Patel, and Prasann Kumar
Department of Agronomy, School of Agriculture, Lovely Professional University, Phagwara, Punjab, India

17.1 Introduction

The human brain, during the development process, is a growing organ. In its formation, specific periods play a crucial role in all other periods of human development. From perinatal to early childhood, these critical windows are associated with neuronal generation, migration, synaptogenesis, and myelination (Zeiss, 2021). At these times, the complexity of the nervous system is determined by genetic plans and external factors too. This plasticity, however, has its vice because the immature brain is sensitive to toxins in the environment, which can disrupt the developmental process and lead to developmental disorders with neurological consequences. Among the dangerous environmental pollutants, heavy metals are considered the most dangerous. Lead, mercury, cadmium, and manganese are common heavy metals in the environment resulting from natural and human activities (Kumar, Goud, Devi, Dey, & Dwivedi, 2022). In industrial application areas, the generation and lack of proper measures and control can cause the dispersion of these metals into the atmosphere, water, and the ground. In the case of the developing brain, the danger is further compounded by the fact that during this age, the blood–brain barrier ensures that unsuitable substances are not allowed into the head, and the spinal cord is not well developed. This semipermeable barrier enables heavy metals to pass through and affect the body, especially the central nervous system, which is essential for the healthy development of the brain (Mitra et al., 2022). The neonate's brain undergoes rapid postnatal development regarding cell division and differentiation and various signaling autophosphorylation reactions, critically dependent on the regulation of metal ions in the intracellular environment. Heavy metal ions can replace or compete with the required metal ions, affecting the enzymes, synthesis of neurotransmitters, and even signal transmission. For example, lead displaces calcium, which is involved in neuronal signaling and synaptic plasticity, among other things. This can negatively affect output, impacting synaptogenesis, the process of forming new connections due to learning and memory. Likewise, organomercury, especially methylmercury, can quickly pass through the placenta and settle in the fetal cerebral tissues. Methylmercury can also inhibit neuronal migration and the development of specific neural circuits, thus creating cognitive and motor deficits that may only be seen in development late in that course (Cariccio, Samà, Bramanti, & Mazzon, 2019).

Among these signs, one can discuss oxidative stress as one of the major pathways through which heavy metals induce neurotoxic effects. Oxidative stress is more likely in the developing neurological tissue, with high oxygen demand and low antioxidant potential. Heavy metals enhance the generation of reactive oxygen species (ROS), which in turn causes cellular damage to bingo lipids, proteins, and nucleic acids. This oxidative damage undermines the neuronal membrane integrity and affects the electron transport chain, thus proffering energy phases that hamper the development and survival of neurons (Chen et al., 2022). A new example of the effects of neurotoxicity during the development of the brain is another study done on the outskirts of the industrial city of Jamshedpur in India, where various lead smelting plants are located. This place is one of the areas where children experience high lead concentrations in their drinking water and the soil (Das, Ghosh, Mohanty, & Sukla, 2015). It focuses on longitudinal research exploring children from infancy to early school-aged children. They noticed that low concentrations, much lower than those known to cause definite toxic reactions in adults, were linked with changes

Heavy Metal Toxicity and Neurodegeneration. DOI: https://doi.org/10.1016/B978-0-443-36575-1.00016-9

in cognitive efficiency, which included severe compromise in such subtests as the ones that measure executive dysfunction and attention level. Neuroimaging studies of some of these children indicate that they have tiny brains associated with minor variations in some brain structures like the prefrontal cortex and the hippocampus, which play a significant role in higher cognitive functions (Gilmore, Knickmeyer, & Gao, 2018). These observations indicate that, indeed, chronic low-level toxic exposures at critical periods in the developmental process are not without far-reaching and irreversible consequences on the brain.

These findings for public health encompass numerous features and considerations, which are as follows: Such measures should start with controlling the spread of pollution in the environment, especially in industrial estates. This need calls for sound legislative measures that demand lower emissions, better waste disposal, and clean-up of affected areas. In targeting vulnerable populations, tests must be performed early before toxicity is experienced, especially during pregnancy or when children are affected (Ferguson, Penney, & Solo-Gabriele, 2017). In addition, there is an excellent need for developmental neuroscience, toxicology, and environmental epidemiology research to arrive at an integrated approach. This chapter is helpful as biomarker technology and neuroimaging have the potential to identify neurodevelopmental changes resulting from heavy metal exposure to aid early intervention. To facilitate this research, it is proposed to include genetic studies since genetic polymorphisms can increase some children's susceptibility to heavy metal neurotoxicity. The critical periods of brain development represent windows of extraordinary plasticity and profound vulnerability. Potential impacts of heavy metals on the still-developing nervous system include their capacity to inhibit key neurodevelopmental processes, given that young children's blood–brain barrier is not well developed and their antioxidant defense mechanisms are also poorly developed (Moretti et al., 2015). This real-life violent occurrence in Jamshedpur shows how heavy metals can affect the developing brain and why there is a need for proper environmental and safety measures, early detection, and practical research. In conclusion, it is essential to note that the only way to protect neurodevelopment is through referencing public health policy, continued and improved scientific identification, and community awareness on the means, ways, and opportunities toward improved development (Fig. 17.1).

17.2 Heavy metals: sources, exposure pathways, and bioavailability

The hazardous pollutants are lead, mercury, cadmium, and arsenic, which are released into the environment through industrial processes, mining activities, agriculture, and the disposal of wastes. These metals are dispersed in the ecosystems through the dispersion of toxic gases in the air and contamination of water, soil, and food (Gautam, Gautam, Banerjee, Chattopadhyaya, & Pandey, 2016). The major routes are inhalation, ingestion, and dermal contact; these include inhaled particles, ingestion of contaminated water and food items, and skin contact with contaminated soil and other products. Metals can dissolve in the environment, be adsorbed by soil and its particles, or be ingested by plants and other life forms (Kumar, Goud, Devi, & Koul, 2022). The level of bioavailability is determined by pH level, organic material, and the type of soil, which in turn define the extent of risks humans and wildlife are exposed to. It starts from natural sources as well as human-influenced sources. Some are mentioned next: natural sources of heavy metals include the disintegration of metal auriferous rocks by geological weathering, volcanic eruptions, and forest fires, while anthropogenic, original from man's activities like mining, smelting of metals, burning of fossil fuels, and disposal of industrial wastes (Raj, Kumar, & Vaghasiya, 2021). Fate routes are inhalation of contaminated air, intake of infected water and food products, and skin contact with contaminated soil. It, therefore, depends on chemical form, the position of the chemical on soil pH, the amount of organic matter that the chemical reacts with, and the environmental factors affecting it; thus bioavailability determines the amount that gets absorbed, distributed, and toxic to the organisms. Therefore knowledge of heavy metal sources, pathways of exposure to humans and wildlife, and those that are most bioavailable are essential in evaluating the risks to the environment and humans and developing efficient remediation measures that are safe, feasible, and sustainable anywhere in the world.

17.2.1 Common sources of childhood heavy metal exposure

Infants and children are at a higher risk of being affected by toxic heavy metals, given the fact that they are in a developmental stage and have tendencies that make them in touch with the substances more often (Anyanwu, Ezejiofor, Igweze, & Orisakwe, 2018). Sources of heavy metal contamination (HC) among children involve contaminated environments through influences such as industrial waste left behind, weak infrastructure in urban centers, and farming. For instance, lead-containing paints formerly used in older buildings still present a significant threat as the paint peels off into dust and small particles. In most large cities, industries and vehicles release lead and cadmium that settle on the ground, covering the soil; hence, children are likely to ingest them at play. Diet is a crucial factor in influencing the overall levels of heavy metals consumed in an individual's body. For instance,

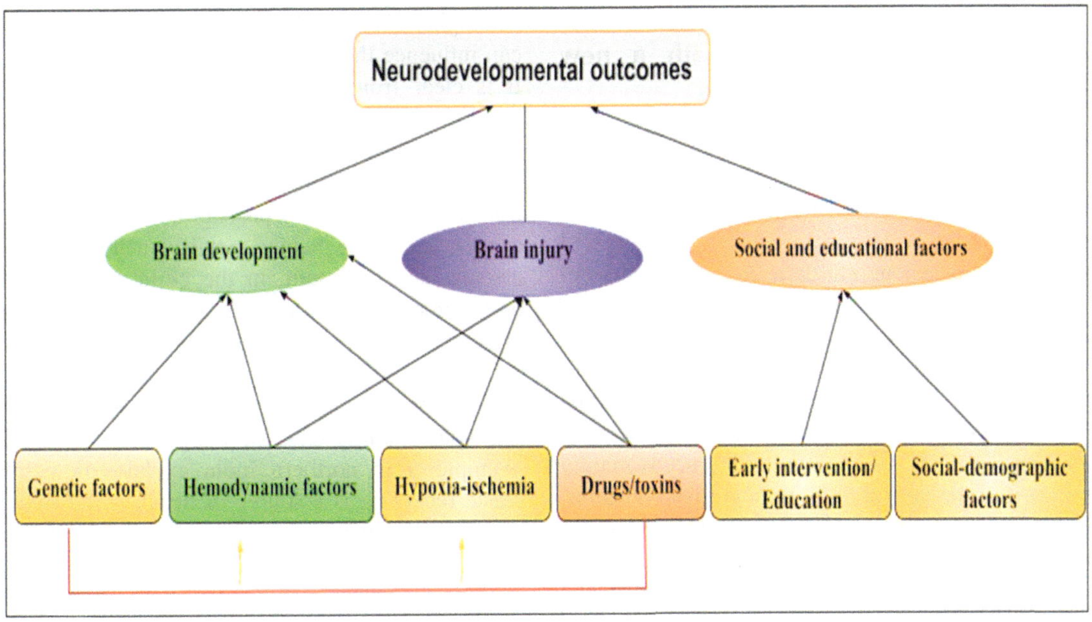

FIGURE 17.1 Factors affecting neurodevelopmental outcomes. Measured neurodevelopmental outcomes are directly influenced by how the brain.

children are most vulnerable to mercury, primarily through fish and seafood consumption, since methylmercury accumulates in the food chain, especially among fish and marine animals (Uddin, Khanom, & Islam, 2023). Likewise, when contaminated water is used to irrigate crops, children can also ingest inorganic arsenic via the food products they eat. Likewise, the crops grown in the home gardens within the urban communities may also be contaminated with toxic metals due to industrial residues. Therefore the study shows that environmental and dietary sources play a significant role in children's accumulation of heavy metals.

17.2.2 Routes of exposure and factors influencing bioavailability

A child can be exposed to heavy metals in three significant ways: ingestion, inhalation, and dermal contact. This is mainly attributed to the high ingestion rate because children often place their hands in their mouths. This behavior further exposes the soils, dust, and foods ready to be ingested, increasing the chances of contacting contaminated substances (Ungureanu & Mustatea, 2022). It is equally essential that the process of Inhalation is also important, especially to those who live in regions that have many people or industries, as they breathe in particulate matter filled with heavy metals. There is a beneficial partnership between PM2.5 and those tiny particles that enter the pathway leading to the lungs, allowing for cadmium absorption and lead in the body (Liang, Tian, Ho, Ip, & Mak, 2019). As with ingestion and inhalation, there are conditions where dermal exposure can be significant, namely, when children are repeatedly exposed to contaminated water or soil, and some chemical form of the metal is essentially capable of passing through the skin. The availability of heavy metals, according to researchers, is high, as the percentage that reaches the systemic circulation and makes its way into the bloodstream to act as a bioactive agent. They found that the chemical form of the metal is far more critical, where methylmercury, because of its fat solubility, is accumulated far in abundance than inorganic mercury. Inorganic arsenic is more soluble and bioavailable than organic arsenic compounds. As for its organic compounds, it is less soluble and less available to the human body than inorganic forms of As. The state of nutritional status is known to impact the absorption rates; calcium, iron, and zinc, for instance, have increased the absorption of heavy metals since they compete with them in the intestinal transport system (Einhorn, Haase, & Maares, 2024). For example, as a result, a low calcium diet increases the tendency to enhance lead uptake in the gastrointestinal system. Other factors are the stability of compounds and their solubility in the gastrointestinal tract: in conditions of low pH, certain metals have enhanced solubility; the composition of the intestinal microbiota can prevent absorption of these toxic elements or, on the contrary, can engorge them. All these factors influence how much of the heavy metal a child is exposed to and how much this exposure is transformed into a load on the body.

17.2.3 Focus on specific heavy metals: lead, mercury, arsenic, and cadmium with a new example

The standard heavy metals that are now deemed dangerous include lead, mercury, arsenic, and cadmium; these elements strongly affect the neurodevelopment of the human body. Lead is still one of the most frequently encountered environmental toxins among children, mainly because of its earlier use in paints and as a gasoline additive. The mechanism of action of lead involves the establishment of competition with calcium after it is ingested and swallowed, damage to the synapses, and disruption of neuronal signaling, resulting in cognitive and behavioral abnormalities (Gudadhe, Singh, & Ahsan, 2024). Mercury is mainly bioaccumulated in the human body through fish consumption, but the main toxicological form of mercury is methylmercury. Due to its high lipophilicity, it can pass through the blood–brain barrier. It is concentrated in the cells of the developmentally vulnerable brain, for example, due to assembled neuronal migration and synaptogenesis leading to long-term neurological impacts. Inorganic arsenic, which is often linked to contaminated drinking water or crops, can be absorbed well through the gastrointestinal tract and interferes with cell signaling pathways that are critical in the development of the brain (Thakur, Rachamalla, Niyogi, Datusalia, & Flora, 2021). This is because cadmium that is gotten from industrial processes and ingested by plants has a very long biological half-life; therefore a low chronic level of exposure to cadmium can accumulate in the body for a long time, causing continuous oxidative stress also disrupting the normal developmental process of the brain. A new example that fits into these dynamics concerns a study made in one of the semiurban communities in north India where informal e-waste recycling is prevalent. For instance, in this region, electronic devices such as televisions and computers are often recycled by scrapping and dismantled in the open air, releasing a mix of heavy metals into the local environment (Awasthi, Zeng, & Li, 2016). It is established that the sources of children's exposures were quantified by assessing blood, hair, and nail samples for lead, gadolinium, and cadmium, respectively. Specific evaluations pointed to pollution due to recycling and water irrigation coming from the surrounding areas. Diet studies also revealed that local fruits and vegetables that are part of the children's daily diet were severely contaminated with inorganic arsenic. The study also evaluated the mercury levels and realized that children who consume fish from the surrounding rivers had higher levels of methylmercury.

These were complemented by neurodevelopmental studies showing mild cognitive, attention, and motor deficits. More specifically, it was found that children with poor nutritional status, especially in terms of calcium and iron, had higher blood lead concentrations, proving that nutrition can influence the absorption and toxicity of heavy metals. It is clear from the comprehensive example above that several environmental factors, such as diet and child characteristics, enhance the risk and consequences of heavy metal exposure (Ferguson et al., 2017). Accumulation with early toxicity probability depends on conductive causes, including heavy metals, such as lead, mercury, arsenic, and cadmium, in the child's direct and indirect environment and diet. Different exposure pathways affecting ingestion, inhalation, and dermal penetration are influenced by the chemical form, nutritional status, gastric or intestinal pH, and microbiota, respectively, determining the bioavailability (Lu, Li, Yang, & Yu, 2021). The latest example from northern India explains how informal e-waste recycling is a source of heavy metal pollution with possible negative consequences for developing encephalopathy. Eradicating these obstacles necessitates broad public health approaches such as abatement of the environment, feeding modification, and community awareness to lower the young, impressionable mind's vulnerability (Fig. 17.2).

17.3 Neurotoxic mechanisms of heavy metal action in the developing brain

In infants and children's heads, heavy metals such as lead, mercury, and cadmium pose significant hazards to the developing brain, resulting in a chain of neurotoxic processes influencing cognitive, motor, and emotional learning. There are main pathways by which such metals cause toxicity, and one of them is oxidative stress, involving the ability of the metals to generate ROS beyond the body's antioxidant capacity, resulting in tissue damage (Adwas, Elsayed, Azab, & Quwaydir, 2019). This oxidative damage causes the neurons to malfunction, interferes with cell signaling, and destroys the cell membrane. Also, heavy metals can have toxic effects on the mitochondria, essential for energy production in the growing neurons. Disruptions in energy metabolism led to neuronal volatility, apoptosis, and developmental delays. Heavy metal also induces neurodegeneration in the developing brain and plays a crucial role in activating the microglia and astrocytes, which release inflammatory cytokines and ROS, leading to neuronal toxicity (Murumulla, Bandaru, & Challa, 2024). These molecular changes reinforce one another, leading to maldevelopment and dysfunction of the brain and raising the susceptibility to developmental disorders, learning disabilities, and neurobehavioral disturbances. Lead, mercury, cadmium, and other heavy metals, when ingested by pregnant women, can inflict severe damage to the brain of the fetus in many ways. The three metals at these developmental stages cause the impairment of synaptogenesis, neurogenesis, and myelination, which have long-term neurobehavioral implications.

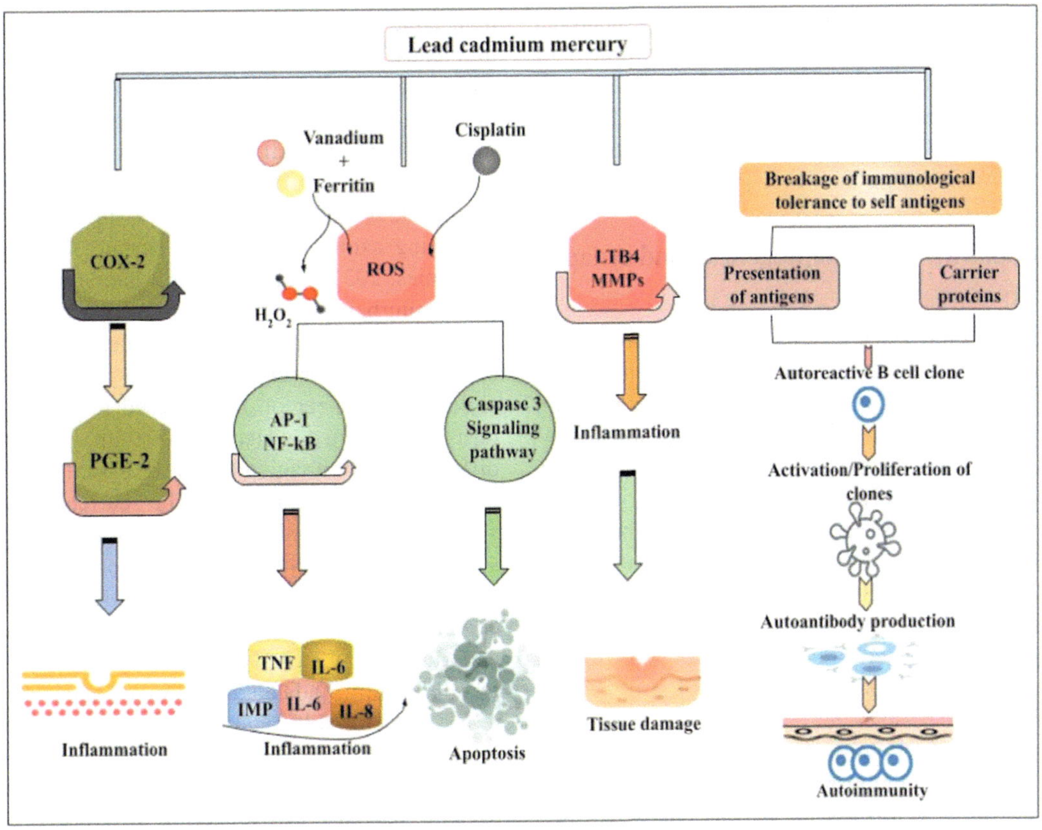

FIGURE 17.2 The effects of lead, cadmium, and mercury exposure on the induction of autoimmunity and inflammation.

They are also known to produce ROS that cause oxidative stress to the membranes, proteins, and DNA of cells. Additionally, heavy metals impair mitochondrial function, compromising energy production essential for neuronal growth and function (Bandaru, Ayyalasomayajula, Murumulla, & Challa, 2022). They also affect neurotransmitter systems, causing changes in calcium signals and synapse plasticity. Activation of the neuroinflammatory pathways leads to worse neural injury since microglial activation in treatment produces proinflammatory cytokines. In the same manner, this section shows that prenatal vulnerability, particularly to mercury, is associated with cognitive impairment and behavioral abnormalities. Altogether, these combined mechanisms arrest the organization of neural circuits, resulting in characterized deficits in cognition, motility, and sensation (Fagundes et al., 2022). It should also be noted that early adverse effects interfere with the future intact and further brain development.

17.3.1 Disruption of neuronal migration and differentiation

In the course of neural development, some of the principal events include neuronal migration and differentiation, which facilitate the proper formation of the networks of neuronal connections. During the developmental phase, neurons are produced at the ventricular zones and must move along the radial glial fibers toward the cortex (Ferent, Zaidi, & Francis, 2020). Some of these include lead and mercury, which could affect some processes. For instance, lead competes with calcium ions, which play a role in transmitting signaling pathways that help migrate neurons. If lead replaces calcium, it interferes with the normal signaling within cells and abnormal cell migrations and incorrect formation of the layers of the cortex are observed. Mercury, specifically methylmercury, has been demonstrated to impact migration by regulating adhesion molecules and disrupting the function of motor proteins needed for the cytoskeleton (Kang, Wang, Guo, & Yang, 2024). Thus neurons may develop abnormally, and the typical pattern of organization in the cortex may be disrupted, leading to cognitive and behavioral disturbances in later years. A cross-sectional survey conducted in an industrial area of an Eastern European country exposed children to high environmental lead from deteriorating urban facilities and infrastructure, which has been associated with developmental delays (Hazer & Gredeback, 2023). Brain scans of children affected by the drugs revealed that they had altered thickness of the cortex and conflicting neuronal organization, which is a manifestation of a problem of migration and differentiation. Hence, exploring such factors can play a significant role in illuminating the disastrous effects heavy metal elements exert on the structural project of the brain during the development period.

17.3.2 Impairment of synaptogenesis and synaptic plasticity

Synaptogenesis and synaptic plasticity are crucial in learning, memory, and cognition. Some key processes that affect heavy metals, such as arsenic and cadmium, are inhibited. For example, arsenic can affect the neurotrophic factors such as the brain-derived neurotrophic factor (BDNF) and its receptor tropomyosin receptor kinase B in the process of synapse development and maintenance (Nicoletti, Pajer, Calcagno, Pajenda, & Nógrádi, 2022). Lesser BDNF is the inhibition of the maturation of synapses and the consolidation of connections within the neural networks. The impact of cadmium on the synaptic proteins is related to oxidative stress that hinders the synaptic vesicles' release and recycling process. As a consequence of OP poisoning, there is less release of neurotransmitters and diminished activity of the receptors, which is the primary process of synaptic plasticity, the ability of the brain to cope with new information. There is evidence that when given subthreshold doses of cadmium during critical phases of development, animals have a reduced LTP in CA1 circuits in the hippocampus, which is associated with learning and memory (Avchalumov & Mandyam, 2021). In another study, children living close to the sites of agricultural use of arsenic pesticides in Southeast Asia were compared. These children not only had high blood arsenic concentrations but also proved to be inferior in tests performed on their memory and attention span. Studies using similar concentrations of arsenic on rodent models included a decrease in LTP in the hippocampus based on electrophysiological studies (Veisi, Khaleghzadeh-Ahangar, Fattahi, & Haghparast, 2023). Such data suggest that even if cells are exposed to heavy metals only during synaptogenesis, synaptic and cognitive ability will be affected.

17.3.3 Effects on myelination and white matter development

Myelination, the ability of oligodendrocytes to form the myelin sheath around the axons, is crucial for the fast conduction of electrical impulses and proper neuronal signaling. Abnormal development of myelination is likely to disrupt white matter and consequently impact motor skills and cognitive abilities (Bae, Kim, Suk, Jung, & Jo, 2020). It is also established that manganese and mercury hinder the function of oligodendrocytes. Manganese itself may cause, for instance, sedation and dyskinesias due to the manganese-induced abovementioned dysfunctionality of the cells and hinder the formation and maintenance of myelin. Experimental mercury toxicity to neurons was considered through decreased expression of myelin essential protein (MBP) and other myelin-related genes impacting axonal insulation (Chesnut, Hartung, Hogberg, & Pamies, 2021). A recent investigation based on a survey in a semiurban area of northern India, well-experienced in informal e-waste recycling, illustrates these effects. Lead, cadmium, and mercury are the heavy metal pollutants that children in this region are exposed to frequently through dust and polluted water. Modern imaging methods like diffusion tensor imaging (DTI) were used to measure white matter in these children. The current study showed significantly reduced FA in many tracts of white matter, such as the main body of the corpus callosum and the regions that are part of the frontal lobe in children of the contaminated area (Chen et al., 2023). These neuroimaging abnormalities matched with clinic-radiologic disability, which in turn indicated the functional impairments in terms of motor skills and reaction time. Since myelination inhibition requires heavy metal exposure, the present evidence demonstrates functional implications for neurotoxicity. Dietary investigations also revealed that most children had dietary deficiencies of protective minerals such as calcium and iron, which are known to have displacement properties for toxic heavy metals (Jomova et al., 2022). It is believed that high levels of heavy metals in the environment, jointly with malnutrition, contributed to higher levels of heavy metal accumulation and, thus, their more significant negative impact on myelination (Fig. 17.3).

17.4 Impact on neurotransmitter systems and cognitive function

There is evidence that exposure to heavy metals such as lead, mercury, and cadmium affects neurotransmitter functioning, which, in turn, affects the brain in the developed and developing fetus. These metals can inhibit the synthesis process, release neurotransmitters, and bind neurotransmitters to receptors such as dopamine, serotonin, and glutamate (Chen, Xu, & Chen, 2021). For instance, lead inhibits calcium's function, essential in releasing neurotransmitters and modifying synapses. In contrast, some effects of mercury impact specific enzymes associated with the synthesis of neurotransmitters and impair synaptic vesicles, causing imbalances in neural transmission at synapses. This interruption of neurotransmitter transmission contributes to impaired learning ability, reduced memory, and poor attention span. Scientific investigations that include animals and human persons

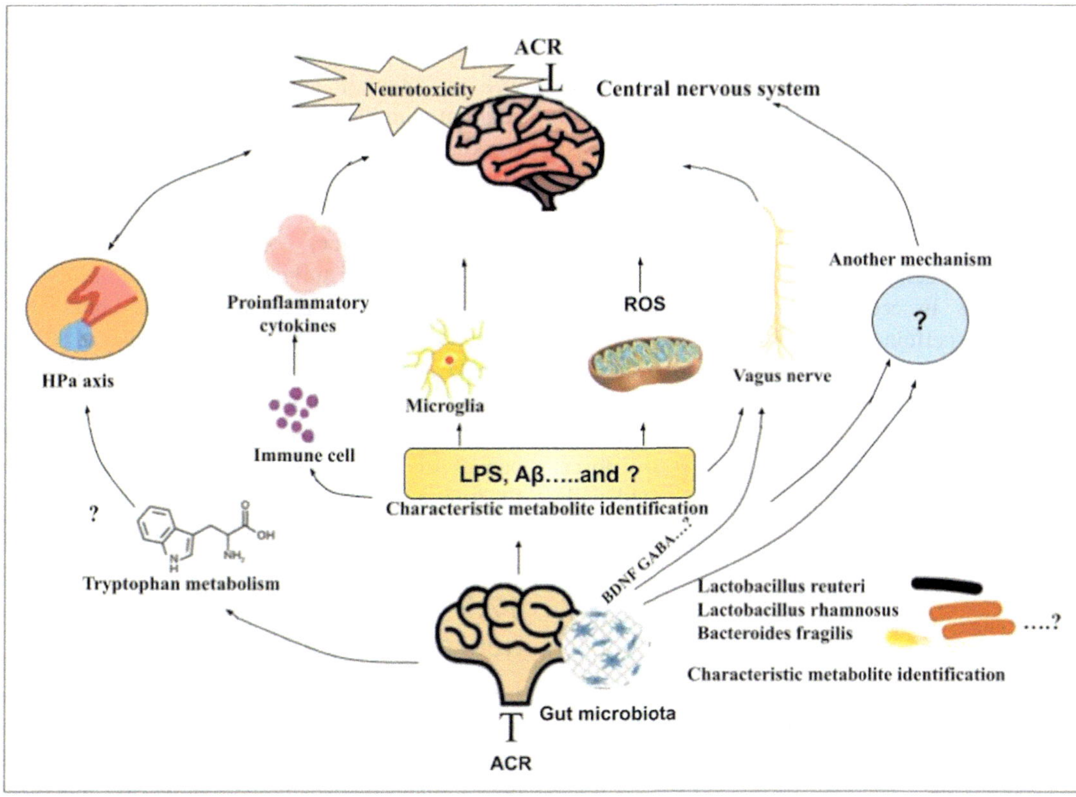

FIGURE 17.3 This process impairs the blood–brain barrier, ultimately leading to neural damage and deterioration.

exposed to lead and mercury have indicated that this causes long-term memory loss, poor performance in class, and behavior issues (Teleanu et al., 2022). These effects add up to the point where executives and other neurocognitive performances and motor skills deteriorate.

17.4.1 Alterations in neurotransmitter synthesis, release, and receptor function

For instance, lead, mercury, arsenic, and cadmium disrupt the chemical harmony of neurotransmitters in the developing brain, which is a form of neurotoxicity. Most significantly, the synthesis of neurotransmitters through enzymatic chemical reactions, which are very sensitive to specific temperatures of enzymes and the physical state of the body, though they can take place anywhere in the body. For instance, lead is known to interact with the tyrosine hydroxylase enzyme, which is central to dopamine synthesis since it is its rate-limiting enzyme (Monday, Younts, & Castillo, 2018). In this enzyme's active site, lead can substitute for calcium ions, thus hampering the synthesis of the required dopamine. Likewise, as already mentioned, all forms of mercury, particularly methylmercury, can interfere with the enzymes necessary to create the neurotransmitter acetylcholine, which is essential in learning and memory. In addition to the synthesis, heavy metals interfere with the release of transmitter substances at the synaptic cleft. For instance, cadmium can cause oxidative stress, affecting the synaptic vesicles, thus affecting the exocytosis process and the amount of glutamate and gamma-aminobutyric acid (GABA) released. Also, interference in one of the receptor functions is evident; the metals cause changes in the density and affinity of the receptors. Chronic exposure to arsenic has been associated with reducing the NMDA subtype of receptors needed for LTP and long-term memory (Liu, Feng, Wang, Xia, & Zheng, 2022). Damage to any of these areas, synapse formation, release of neurotransmitters, and receptor functionality, all disrupt the normal neurotransmission, affecting the coordination of neural network communication.

17.4.2 Consequences for cognitive development (attention, memory, executive function)

Neurotransmitter abnormality impacts cognitive development, which is a fact that the series of observations made here have established. Dopamine, acetylcholine, glutamate, and GABA constitute neurotransmitters that impact cognitive processes such as attention, working memory, and executive functions (Gasiorowska et al., 2021). These deficits can be caused by heavy metals that affect these systems, leading to impairments in these cognitive areas. For example, reduced dopamine production in the brain and changes in its receptors due to lead intake are linked to poor attention span and impulsiveness, which limit a child's ability to concentrate and regulate his conduct. Similarly, polarities in acetylcholine synthesis due to mercury entail forgetfulness or, more explicitly, an inability to consolidate information to be stored in the memory bank; hence, they cannot set up new information. The above effects are further supported by experimental data obtained using animal models. Exposure to low doses of cadmium during brain development is associated with significant deficits in spatial memory tasks and decreased performance of tasks involving executive functions (Soetrisno, Delgado-Saborit, 2020). These have been attributed to changes in synaptic plasticity and an imbalance in the concentration of neurotransmitters in the brain. Also, such changes are not temporary; early-life exposure to heavy metals induces neurodevelopmental changes that affect the remainder of a person's life and result in a decline in cognitive ability.

17.4.3 Potential links to specific neurodevelopmental disorders (attention-deficit/hyperactivity disorder, autism) with a new example

New evidence also indicates the contribution of heavy metals' neurotoxicity in the etiology of certain neurodevelopmental disorders like attention-deficit/hyperactivity disorder (ADHD) and autism spectrum disorder (ASD). Dysregulation of the dopamine and glutamate neurotransmitter systems has been implicated in causing the etiology of ADHD (Thapar, Cooper, & Rutter, 2017). Strong evidence from epidemiological research supports the rise in blood lead levels and its correlation with inattention, hyperactivity, and impulsivity in children with an ADHD-like phenotype. Likewise, in ASD, glutamatergic and GABAergic transmission has been implicated to be dysregulated. Heavy metals, which disrupt these neurotransmitter systems, can result in an imbalance between excitation and inhibition in the brain, a key determinant of neurodevelopment that can lead to the expression of autistic behavior (Homberg et al., 2016). In this research, researchers studied a child cohort who lived close to artisanal fishery communities where heavy metal contamination, mercury, and arsenic rose with the release of industrial effluent and the reuse of water for aquaculture activities. The research employed a multidisciplinary framework, using biomonitoring of blood mercury and arsenic levels, neuropsychological testing, and neuroimaging scans. The research findings established that children with higher mercury levels did not perform well on attention and executive function testing, while children with higher levels of arsenic performed poorly on working memory and processing speed. Of special note, this subgroup of children also displayed behavioral traits typical of ADHD, including increased impulsivity and decreased inhibitory control. Functional MRI neuroimaging showed disturbed patterns of activation in the prefrontal cortex—a region of the brain involved in executive function and already found to be disturbed in ADHD and ASD (Xi & Wu, 2021). Collectively, these findings suggest that heavy metal exposure in this population throughout several years is associated with disturbances of neurotransmitter systems regulating cognitive function as well as behavior regulation, and possibly with the etiology of neurodevelopmental disorders.

17.5 Assessing long-term neurodevelopmental effects

Long-term neurodevelopmental consequences of heavy metal exposure are dramatic and can impact brain function lifelong. The metals lead, mercury, and cadmium alter the development of the brain by impairing processes such as neurogenesis, synaptogenesis, and myelination, essential for cognitive and motor function. Repeated and long-term exposures, especially within windows of development, can contribute to long-standing neurobehavioral impairment (Wang et al., 2020). Such is the case for children with exposure to lead during early childhood who exhibit diminished IQ, attention, and executive function. Heavy metals also cause oxidative stress and inflammation, which destroy neuronal structures and impair cellular communication (Sun et al., 2022). Such continuous injuries can lead to neurodegenerative disorders in later years of life. Furthermore, heavy metal-induced changes in neurotransmitter systems also lead to long-term cognitive impairment and psychiatric illnesses, such as attention-deficit disorders or anxiety. Notably, such effects might not be fully realised until later in life, highlighting the importance of early detection, prevention, and ongoing monitoring to protect long-term neural health.

17.5.1 Challenges in measuring subtle neurodevelopmental deficits

Evaluation of long-term neurodevelopmental impact is tricky because cognitive and behavioral impairments that appear over time are subtle. Most neurodevelopmental alterations are not easily identified through standard clinical examination but occur as minute differences in functioning across multiple cognitive areas. These subtle impairments—impacting areas like attention, processing speed, working memory, and executive function—will not necessarily be overt until academic or social stressors arise later in childhood or adolescence (Astle, Holmes, Kievit, & Gathercole, 2022). Second, heterogeneity of developmental profiles implies that what is typical for one child may be a delay or impairment in another and that there may not be universal standards. Environmental exposures, genetic susceptibilities, and socioeconomic factors can also complicate measurements, requiring longitudinal designs that track people over many years. Practically, the challenge is to disentangle the subtle effects of early exposures like heavy metals, prenatal stress, or nutrient deficiencies against a background of continuous developmental changes. These problems require sensitive, valid measurement instruments that detect small neural functions and structural changes over time.

17.5.2 Use of neuroimaging techniques (magnetic resonance imaging, functional magnetic resonance imaging, electroencephalography)

Current neuroimaging methods offer unparalleled windows into the developing brain, providing objective biomarkers to augment behavioral evaluation. Structural magnetic resonance imaging (MRI) allows researchers to visualize brain structure, and using it, one can identify small changes in cortical thickness, white matter integrity, and brain volume that can arise from environmental exposure. Functional MRI (fMRI) takes these tools further by quantifying blood oxygen level–dependent signals, thus informing functional connectivity and activity patterns across brain networks engaged with cognitive function. Electroencephalography (EEG), on the other hand, provides high temporal resolution, registering quick changes in neural activity and enabling the detection of shifts in brain oscillations signaling disrupted neural processing or connectivity (Mele et al., 2019). Collectively, these methods can identify macroscopic and microscopic changes in brain structure and function and offer a complete view of neurodevelopmental pathways. For example, an investigation has been recently conducted on the effect of early-life exposure to low-level lead on child brain development in children from an urban community with documented environmental contamination. Minor but statistically significant cortical thinning was observed in regions linked to executive function and attention by structural MRI. fMRI analysis revealed changed connectivity patterns between the prefrontal cortex and other brain areas during cognitive tasks. In contrast, EEG recordings revealed deviations in the predicted power of specific neural oscillations associated with attention and memory (Bullock, Jackson, & Abbott, 2021). These convergent results highlight the sensitivity of neuroimaging techniques in picking up subtle neurodevelopmental changes that are not evident through clinical observation alone (Fig. 17.4).

17.5.3 Neuropsychological assessments and standardized cognitive tests

Together with neuroimaging, neuropsychological testing provides essential information on the functional significance of structural and neurophysiological alterations. Behavior and standardized cognitive tests are crafted to assess domains like attention, memory, speed of processing, and executive function. These allow researchers and clinicians to quantify cognition and detect deficits associated with previous environmental exposures or developmental insults. The use of extensive neuropsychological batteries, which are frequently given at more than one-time point, allows for the tracking of cognitive trajectories and the identification of early deviations from normal development (Annett, Patel, & Phipps, 2015). Researchers used a battery of cognitive tests to assess the long-term effects of exposure on children exposed to arsenic-contaminated drinking water in a rural community. Findings showed that children with greater arsenic exposure consistently did more poorly on tasks involving sustained attention and quick information processing relative to children in less contaminated communities. Neuroimaging results subsequently supported these behavioral impairments that showed decreased white matter integrity in key cognitive circuits. This multimethod approach, which paired neuropsychological assessment with neuroimaging, offered a strong framework for relating environmental exposure to tangible cognitive consequences (Vogt et al., 2017). A recent example from Southeast Asia further demonstrates the convergence of these approaches. School-age children were repeatedly assessed over several years in a coastal village exposed to heavy metals from industrial discharge and artisanal fishing. Investigators paired high-resolution MRI scans with standardized neuropsychological tests to determine the long-term neurodevelopmental effects of mercury and cadmium exposure (Vazquez Cervantes et al., 2023). Over time, the children had subtle impairments in executive function and attention tests, and neuroimaging measures of decreased frontal brain connectivity paralleled these. Moreover, EEG evaluations indicated disturbed patterns of brain activity in the context of cognitive tasks as markers of impaired neural processing. This broad assessment highlighted the sensitivity of newer neuroimaging and neuropsychological methods and the additive impacts of environmental poisons on the developing brain.

17.6 Genetic susceptibility and individual variability

Genetic susceptibility is also central in regulating individual responses to heavy metal exposures, affecting metabolism and toxicity. Specific genetic polymorphisms, for example, in metal transporter genes and detoxification

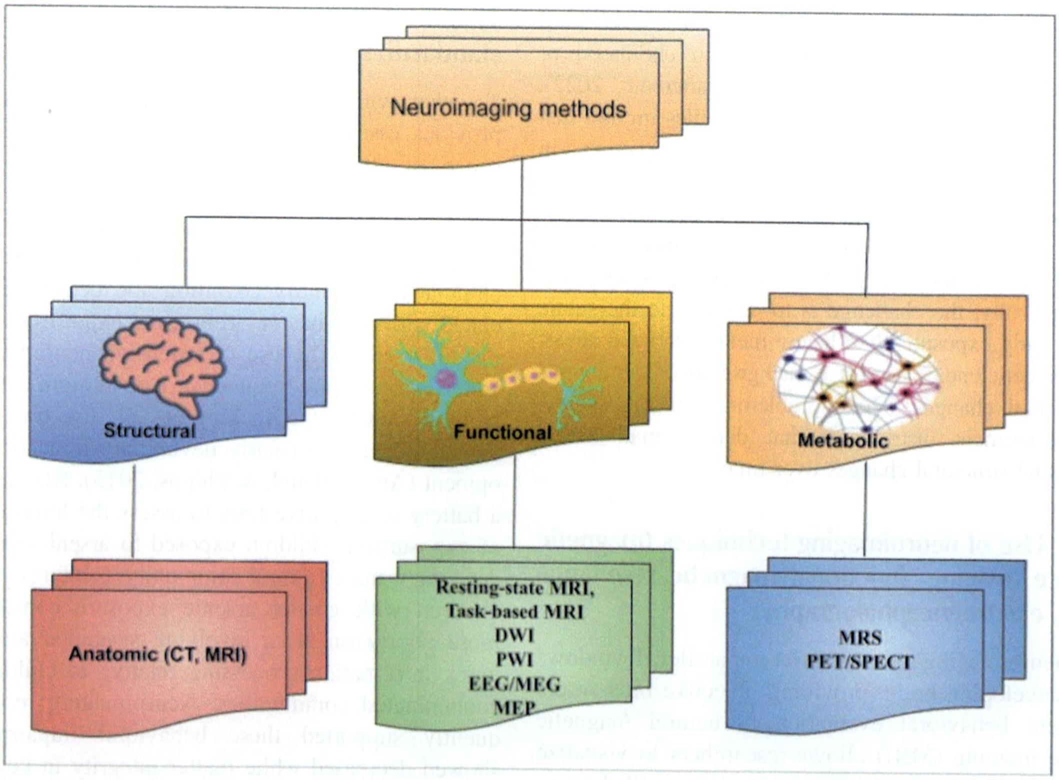

FIGURE 17.4 Classification of neuroimaging methods: CT (computer tomography), EEG, MEP (motor evoked potential), MRI (magnetic resonance imaging), MRS (magnetic spectroscopy), PET/SPECT (positron emission tomography/single-photon emission computed tomography).

enzymes, can influence the absorption, distribution, and excretion of metals such as lead, mercury, and cadmium (Andreoli & Sprovieri, 2017). These polymorphisms regulate enzyme functions that participate in oxidative stress responses, which could enhance neurotoxicity. Gene–environment interactions add to the complexity of neurodevelopmental outcomes since genetic susceptibility can enhance or suppress the effects of environmental exposures on brain development. For instance, children with specific variants in antioxidant defense genes are more susceptible to heavy metal–induced cognitive deficits. Personalized risk assessment, which combines genetic screening with exposure analysis, allows for targeted interventions and individualized preventive strategies (Bhandari, Paliwal, & Kuhad, 2020). These strategies have the potential to minimize neurodevelopmental risk through the identification of high-risk populations and the development of targeted interventions to reduce exposure and maximize detoxification. This new area of research enhances the capability to protect vulnerable individuals safely.

17.6.1 Influence of genetic polymorphisms on heavy metal metabolism and toxicity

Genetic polymorphisms are essential in modulating how people metabolize and react to heavy metal exposure. Polymorphisms in genes that code for metal transporters, detoxification enzymes, and antioxidant proteins can modify the absorption, distribution, and excretion of poisonous metals like lead, mercury, arsenic, and cadmium (Teodoro, Briguglio, Fenga, & Costa, 2019). For instance, polymorphisms within the gene encoding for divalent metal transporter 1 (DMT1, SLC11A2) determine the intake of essential metals such as iron and manganese and the toxic ones with similar imitative effects. Elevated expression or activity of DMT1 will increase intracellular concentrations of such metals and subsequently increase oxidative stress and cell injury (Cirovic, Cirovic, Yimthiang, Vesey, & Satarug, 2024). Genetic differences within detoxifying enzymes, for example, those belonging to the glutathione-*S*-transferase (GST) family, are similarly paramount. Null mutations in GSTT1 or GSTM1 decrease the capacity to conjugate and eliminate heavy metals, increasing the retention of these poisons. The high-risk alleles in these genes can thus make individuals more susceptible to increased neurotoxicity even at comparable environmental levels. For example, children with the GSTM1 null genotype have been found to retain more lead and exhibit more severe cognitive impairments than those possessing functional alleles (Rahbar et al., 2022). These polymorphisms highlight the biological heterogeneity of heavy metal toxicity, offering a molecular basis for why some people experience more serious neurodevelopmental consequences than others when exposed to the same environmental conditions.

17.6.2 Gene–environment interactions in neurodevelopmental outcomes

The interaction between environmental exposures and genetic background—termed gene–environment interactions—is an essential determinant of neurodevelopmental outcomes. In the case of heavy metals, the interaction between exposure level and genetic susceptibility is a powerful predictor of risk for cognitive and behavioral impairments (Flores-Dorantes, Díaz-López, & Gutiérrez-Aguilar, 2020). For instance, kids with polymorphisms that undermine detoxification processes (e.g., in GST or metallothionein genes) are less able to neutralize heavy metal-induced oxidative stress. When exposed to high doses of metals through contaminated water, soil, or food, the combined effect may become evident as impaired neuronal migration, synaptogenesis, and myelination—processes crucial for healthy brain development. Increasing evidence suggests that even slight differences in exposure can have contrasting effects based on genetic susceptibility. A good example is a study of an artisanal gold mining community in the Andes, where people are exposed to mercury emitted during gold processing. Scientists discovered that kids with a specific form of the ALAD gene, which is involved in the process of heme synthesis and known to have an affinity for lead and mercury, had more significant neurocognitive deficits than kids who do not carry this variant despite comparable exposure. The kids performed poorly on attention, working memory, and processing speed measures. Neuroimaging identified structural variations in the prefrontal cortex and hippocampus, essential for executive function and memory. This research illustrates how genetic influences can enhance the neurotoxic impact of heavy metals, thus raising the risk of neurodevelopmental disorders like ADHD and learning disabilities (Cheroni, Caporale, & Testa, 2020).

17.6.3 Personalized risk assessment and targeted interventions

With the diversity in genetic susceptibility and exposure patterns, individualized risk assessment holds great promise in preventing heavy metal neurotoxicity. Breakthroughs in genomic technologies currently allow for detecting high-risk genetic variants that lead to enhanced metal absorption and toxicity. Genetic screening incorporated into public health programs can enable the identification of children who are most susceptible to environmental toxins. This individual data can guide specific intervention strategies when paired with precise exposure measurements like blood, hair, or nail metal content biomonitoring (Marczylo, Jacobs, & Gant, 2016). For instance, in the abovedescribed Andean village, scientists instituted an intensive risk assessment program that combined routine mercury-level biomonitoring with genetic screening for ALAD variants. High-risk children were provided with nutritional counseling to increase their consumption of protective minerals like calcium and iron, which can competitively inhibit the absorption of toxic metals. Community education programs were also initiated to educate families on minimizing exposure through purified water and consuming low-mercury fish species. At a more significant level, policymakers employed these data to push for more stringent regulations on artisanal mining activities and to get funding approved for environmental remediation initiatives within the area. Combining personalized risk assessment with targeted interventions offers immediate payback by minimizing individual exposure and contributes to the long-term enhancement of community health. With our continuing advances in gene–environment interaction, additional genetic markers for susceptibility may be discovered, allowing even more targeted interventions. For example, future clinical trials could investigate the effectiveness of antioxidant therapies or chelation agents in the genetically high-risk subgroup, eventually translating into individualized treatment strategies that include genetic susceptibility and environmental history (Rauh & Margolis, 2016).

17.7 Environmental justice and health disparities

Environmental justice acknowledges that chemical exposure is not uniformly distributed. Disadvantaged groups, frequently marginalized by socioeconomic status (SES) and lack of political influence, disproportionately experience ecological burdens. They are likelier to live near industrial areas, dump sites, and contaminated waterways, resulting in excessive heavy metal and fluoride exposure (Kumari, Dhankhar, Abrol, & Yadav, 2024). These communities are compounded in risk by limited access to healthcare and fewer resources to counteract environmental threats. Socioeconomic conditions, including reduced income and fewer educational opportunities, enhance health disparities and lower community resilience. Environmental injustice must be addressed through targeted policies emphasizing reducing pollution, enhancing monitoring, and providing equal access to healthcare and remediation. Encouraging health equity requires the participation of communities, strong regulatory systems, and infrastructure investment in creating sustainable, healthy conditions for all, irrespective of location or socioeconomic position (Baciu, Negussie, Geller, Weinstein, & National Academies of Sciences, Engineering, & Medicine, & Committee on Community-Based Solutions to Promote

Health Equity in the United States, 2017). Such an inclusive strategy diminishes harmful exposures and empowers vulnerable populations to obtain cleaner, safer environments.

17.7.1 Unequal exposure patterns and disproportionate impact on vulnerable populations

Environmental poisons, such as heavy metals, are not distributed uniformly between groups. Exposure "hotspots" usually created around race, poverty, socioeconomic position, and neighborhoods seem to land with disproportionate burdens for vulnerable people. In the majority of circumstances, communities adjacent to industrial complexes, waste landfills, or locales with decrepit infrastructure more typically experience enhanced doses of poisonously high amounts of lead, mercury, arsenic, and cadmium. These exposures occur through numerous routes, such as contaminated air, soil, water, and diet (Kordas, Ravenscroft, Cao, & McLean, 2018). The developing brain is susceptible to these neurotoxins, placing children in these populations at heightened risk for permanent cognitive and motor impairments and potentially predisposing them to neurodegenerative disease in the future. For example, urban communities with a history of industrial contamination tend to have higher lead concentrations in the environment due to deteriorating lead-based paint and lingering industrial emissions. These children are at risk from hand-to-mouth activity that enhances ingestion of contaminated dust and soil. Additionally, in rural communities surrounding mining or smelting activity, people can be exposed continuously to several heavy metals simultaneously, which can interact synergistically to enhance neurotoxicity (Ruiz, Becerra, Jagai, Ard, & Sargis, 2018). The disproportionate exposure to environmental hazards, therefore, becomes a health inequity where already disadvantaged groups have a higher chance of developing disorders like Parkinson's disease.

17.7.2 Socioeconomic factors and access to healthcare resources

SES is an essential determinant of exposure levels and health effects. Low SES populations tend to have fewer resources to counteract environmental hazards. They might live in older, neglected housing near industrial plants or dumps, which raises exposure to heavy metals and decreases the ability to remediate (Said & Hernandez, 2015). Furthermore, communities with lower SES often experience lower access to quality healthcare, so early detection and treatment of neurodevelopmental and neurodegenerative disorders are less probable. This disparity in healthcare access is part of a cycle that allows environmental exposures to lead to long-term health effects that are not adequately addressed, further expanding health inequalities. Economic limitations also impact diet quality. Where fresh, nutrient-dense foods are not readily available, residents may subsist on cheap, locally produced fruits and vegetables, which are more likely to be contaminated with heavy metals from contaminated soil or water. Nutritional deficiencies can enhance the bioavailability of heavy metals because essential minerals such as calcium and iron, which are known to inhibit metal absorption competitively, are frequently absent. As a result, low SES children are not only exposed to higher lead levels but are also more susceptible biologically because they have poor nutritional protection. In some of the low-income city neighborhoods in the United States, research has found that children there have greater levels of blood lead than children who are better off. This inequality is exacerbated by the reality that low-income families usually have access barriers to regular healthcare services, thereby postponing the diagnosis and treatment of lead poisoning and related cognitive impairments (Heng et al., 2022). The overlap between socioeconomic disadvantage and environmental exposure is a situation where the most disadvantaged populations bear the most significant risk of developing severe health outcomes, including conditions that place them at risk for neurodegenerative diseases such as Parkinson's.

17.7.3 Addressing environmental injustice and promoting health equity: a new example

Tackling environmental injustice involves a holistic strategy that brings together community involvement, policy change, and targeted interventions to lower exposure and enhance health outcomes. Community-based researchers reported that the blood levels of lead and cadmium in local children were considerably higher than in regional averages. These results were complemented by environmental surveys indicating that residential neighborhoods are found near active manufacturing facilities and discontinued waste dumping areas. In reaction to these inequalities, a multifaceted public health intervention was launched. In association with environmental researchers and community-based organizations, local health departments created an integrated program focused on decreasing heavy metal exposure and buffering its health effects (Njoku, Sampson, 2023). The program had several key elements:

- **Environmental remediation and policy overhaul:** Regulatory bodies imposed tighter emissions controls on local industries and funded soil and water remediation initiatives. This involved equipping facilities with high-tech filtration systems and employing phytoremediation measures to leach heavy metals from polluted soils.
- **Community education and nutritional assistance:** Public health initiatives were implemented to inform

citizens of the hazards of heavy metal exposure and proven means of mitigating risk. Nutritional interventions were conducted to provide local families with foods high in calcium, iron, and other minerals that inhibit the absorption of toxic metals. Workshops were conducted in local schools and community centers to create awareness of safe practices and regular health screening.

- **Increased access to healthcare and early intervention:** Mobile health clinics were sent out to conduct regular biomonitoring and early neuropsychological testing for children. The clinics enabled early identification of exposure-related health problems, enabling timely interventions that could prevent long-term developmental effects.

17.8 Conclusion

The latest evidence emphasizes the devastating effect of exposure to heavy metals in childhood, for example, lead, mercury, arsenic, and cadmium, and neurodevelopmental outcomes in later life. There has been overwhelming evidence that these toxic substances disrupt fundamental processes like neuronal migration, synaptogenesis, and myelination and ultimately affect cognition and predispose to neurodevelopmental disorders. However, numerous issues remain unexplained about dose–response relations, the synergistic effects of chronic low-level exposures, and mechanisms of gene–environment interaction underlying individual susceptibility. Longitudinal studies following children over long periods, from early exposure into adolescence and adulthood, will be a research priority in the future. These studies are critical to defining the temporal course of neurodevelopmental impairments and identifying early biomarkers that predict subsequent cognitive and behavioral impairments. Advanced neuroimaging methods (e.g., MRI, fMRI, and DTI) and detailed neuropsychological testing provide sensitive tools to identify subtle alterations in brain structure and function that can be used as early markers of heavy metal–induced neurotoxicity. New research approaches, such as high-throughput genomics and exosomes, need to be incorporated into these longitudinal designs to elucidate the intricate interactions between genetic susceptibility and environmental exposures. New computational models and machine learning methods will assist in analyzing multifactorial information, allowing researchers to detect patterns that lead to unfavorable consequences. This holistic approach is essential to gain deeper insights into how heavy metals impact the developing brain and untangle the cumulative impact of prolonged exposure. Furthermore, research results must be translated into efficient prevention and intervention strategies. Translational research would need to address creating and validating targeted interventions—like nutritional supplements, environmental remediation, and public health education campaigns—that lower the bioavailability of heavy metals and offset their neurotoxic impacts. Interventions conducted in communities, based on solid epidemiological data and specially adapted to meet the unique requirements of vulnerable groups, are of key importance. For instance, regions with severe environmental pollution would be aided by water quality and soil clean-up programs, complemented by providing exposed communities with nutritional interventions that enhance natural resistance to heavy metal absorption. Combating childhood exposure to heavy metals and its eventual neurodevelopmental impact necessitates a multidisciplinary approach. By bringing basic science, clinical research, and practice together, we can work toward integrated approaches that heighten our comprehension of these interactions and give us tangible advantages in the real world. The direction for the future must stress requirements for precise and creative research methods and aggressive approaches to guard against the developing brain so that each child has an equal chance for maximum neurodevelopmental advantage.

References

Adwas, A. A., Elsayed, A., Azab, A. E., & Quwaydir, F. A. (2019). Oxidative stress and antioxidant mechanisms in human body. *Journal of Applied Biotechnology & Bioengineering, 6*(1), 43–47.

Andreoli, V., & Sprovieri, F. (2017). Genetic aspects of susceptibility to mercury toxicity: An overview. *International Journal of Environmental Research and Public Health, 14*(1), 93. Retrieved from https://doi.org/10.3390/ijerph14010093.

Annett, R. D., Patel, S. K., & Phipps, S. (2015). Monitoring and assessment of neuropsychological outcomes as a standard of care in pediatric oncology. *Pediatric Blood & Cancer, 62*(S5), S460–S513. Retrieved from https://doi.org/10.1002/pbc.25749.

Anyanwu, B. O., Ezejiofor, A. N., Igweze, Z. N., & Orisakwe, O. E. (2018). Heavy metal mixture exposure and effects in developing nations: An update. *Toxics, 6*(4), 65. Retrieved from https://doi.org/10.3390/toxics6040065.

Astle, D. E., Holmes, J., Kievit, R., & Gathercole, S. E. (2022). Annual research review: The transdiagnostic revolution in neurodevelopmental disorders. *Journal of Child Psychology and Psychiatry, 63*(4), 397–417. Retrieved from https://doi.org/10.1111/jcpp.13481.

Avchalumov, Y., & Mandyam, C. D. (2021). Plasticity in the hippocampus, neurogenesis and drugs of abuse. *Brain Sciences, 11*(3), 404. Retrieved from https://doi.org/10.3390/brainsci11030404.

Awasthi, A. K., Zeng, X., & Li, J. (2016). Environmental pollution of electronic waste recycling in India: A critical review. *Environmental Pollution, 211*, 259–270. Retrieved from https://doi.org/10.1016/j.envpol.2015.11.027.

Baciu, A., Negussie, Y., Geller, A., Weinstein, J. N., & National Academies of Sciences, Engineering, and Medicine, & Committee on Community-Based Solutions to Promote Health Equity in the United States (2017). Partners in promoting health equity in communities. In *Communities in action: Pathways to health equity*. National Academies Press (US).

Bae, H. G., Kim, T. K., Suk, H. Y., Jung, S., & Jo, D. G. (2020). White matter and neurological disorders. *Archives of Pharmacal Research, 43*, 920–931. Retrieved from https://doi.org/10.1007/s12272-020-01270-x.

Bandaru, L. J. M., Ayyalasomayajula, N., Murumulla, L., & Challa, S. (2022). Mechanisms associated with the dysregulation of mitochondrial function due to lead exposure and possible implications on the development of Alzheimer's disease. *Biometals an International Journal on the Role of Metal Ions in Biology, Biochemistry, and Medicine, 35*(1), 1–25. Retrieved from https://doi.org/10.1007/s10534-021-00360-7.

Bhandari, R., Paliwal, J. K., & Kuhad, A. (2020). Neuropsychopathology of autism spectrum disorder: Complex interplay of genetic, epigenetic, and environmental factors. In *Personalized food intervention and therapy for autism spectrum disorder management*. Springer, 97–141. Retrieved from https://doi.org/10.1007/978-3-030-30402-7_4.

Bullock, M., Jackson, G. D., & Abbott, D. F. (2021). Artifact reduction in simultaneous EEG-fMRI: A systematic review of methods and contemporary usage. *Frontiers in Neurology, 12*, 622719. Retrieved from https://doi.org/10.3389/fneur.2021.622719.

Cariccio, V. L., Samà, A., Bramanti, P., & Mazzon, E. (2019). Mercury involvement in neuronal damage and in neurodegenerative diseases. *Biological Trace Element Research, 187*(2), 341–356. Retrieved from https://doi.org/10.1007/s12011-018-1380-4.

Chen, Y., Xu, J., & Chen, Y. (2021). Regulation of neurotransmitters by the gut microbiota and effects on cognition in neurological disorders. *Nutrients, 13*(6), 2099. Retrieved from https://doi.org/10.3390/nu13062099.

Chen, Y., Hamidu, S., Yang, X., Yan, Y., Wang, Q., Li, L., ... Li, Y. (2022). Dietary supplements and natural products: An update on their clinical effectiveness and molecular mechanisms of action during accelerated biological aging. *Frontiers in Genetics, 13*, 880421. Retrieved from https://doi.org/10.3389/fgene.2022.880421.

Chen, Y., Wang, Y., Song, Z., Fan, Y., Gao, T., & Tang, X. (2023). Abnormal white matter changes in Alzheimer's disease based on diffusion tensor imaging: A systematic review. *Ageing Research Reviews, 87*, 101911. Retrieved from https://doi.org/10.1016/j.arr.2023.101911.

Cheroni, C., Caporale, N., & Testa, G. (2020). Autism spectrum disorder at the crossroad between genes and environment: Contributions, convergences, and interactions in ASD developmental pathophysiology. *Molecular Autism, 11*(1), 69. Retrieved from https://doi.org/10.1186/s13229-020-00370-1.

Chesnut, M., Hartung, T., Hogberg, H., & Pamies, D. (2021). Human oligodendrocytes and myelin in vitro to evaluate developmental neurotoxicity. *International Journal of Molecular Sciences, 22*(15), 7929. Retrieved from https://doi.org/10.3390/ijms22157929.

Cirovic, A., Cirovic, A., Yimthiang, S., Vesey, D. A., & Satarug, S. (2024). Modulation of adverse health effects of environmental cadmium exposure by zinc and its transporters. *Biomolecules, 14*(6), 650. Retrieved from https://doi.org/10.3390/biom14060650.

Das, A. P., Ghosh, S., Mohanty, S., & Sukla, L. B. (2015). Advances in manganese pollution and its bioremediation. *Environmental Microbial Biotechnology*, 313–328. Retrieved from https://doi.org/10.1007/978-3-319-19018-1_16.

Einhorn, V., Haase, H., & Maares, M. (2024). Interaction and competition for intestinal absorption by zinc, iron, copper, and manganese at the intestinal mucus layer. *Journal of Trace Elements in Medicine and Biology, 84*, 127459. Retrieved from https://doi.org/10.1016/j.jtemb.2024.127459.

Fagundes, B. H. F., Nascimento, P. C., Aragão, W. A. B., Chemelo, V. S., Bittencourt, L. O., Eiró-Quirino, L., ... Lima, R. R. (2022). Methylmercury exposure during prenatal and postnatal neurodevelopment promotes oxidative stress associated with motor and cognitive damages in rats: an environmental-experimental toxicology study. *Toxicology Reports, 9*, 563–574. Retrieved from https://doi.org/10.1016/j.toxrep.2022.02.014.

Ferent, J., Zaidi, D., & Francis, F. (2020). Extracellular control of radial glia proliferation and scaffolding during cortical development and pathology. *Frontiers in Cell and Developmental Biology, 8*, 578341. Retrieved from https://doi.org/10.3389/fcell.2020.578341.

Ferguson, A., Penney, R., & Solo-Gabriele, H. (2017). A review of the field on children's exposure to environmental contaminants: A risk assessment approach. *International Journal of Environmental Research and Public Health, 14*(3), 265. Retrieved from https://doi.org/10.3390/ijerph14030265.

Flores-Dorantes, M. T., Díaz-López, Y. E., & Gutiérrez-Aguilar, R. (2020). Environment and gene association with obesity and their impact on neurodegenerative and neurodevelopmental diseases. *Frontiers in Neuroscience, 14*, 863. Retrieved from https://doi.org/10.3389/fnins.2020.00863.

Gasiorowska, A., Wydrych, M., Drapich, P., Zadrozny, M., Steczkowska, M., Niewiadomski, W., & Niewiadomska, G. (2021). The biology and pathobiology of glutamatergic, cholinergic, and dopaminergic signaling in the aging brain. *Frontiers in Aging Neuroscience, 13*, 654931. Retrieved from https://doi.org/10.3389/fnagi.2021.654931.

Gautam, P. K., Gautam, R. K., Banerjee, S., Chattopadhyaya, M. C., & Pandey, J. D. (2016). Heavy metals in the environment: Fate, transport, toxicity and remediation technologies. *Nova Science Publishers, 60*, 101–130.

Gilmore, J. H., Knickmeyer, R. C., & Gao, W. (2018). Imaging structural and functional brain development in early childhood. *Nature Reviews Neuroscience, 19*(3), 123–137. Retrieved from https://doi.org/10.1038/nrn.2018.1.

Gudadhe, S., Singh, S. K., & Ahsan, J. (2024). Cellular and neurological effects of lead (Pb) toxicity. In *Lead toxicity mitigation: Sustainable nexus approaches*. Cham: Springer Nature Switzerland, 125–145. Retrieved from https://doi.org/10.1007/978-3-031-46146-0_6.

Hazer, L., & Gredeback, G. (2023). The effects of war, displacement, and trauma on child development. *Humanities and Social Sciences Communications, 10*(1), 1–19. Retrieved from https://doi.org/10.1057/s41599-023-02438-8.

Heng, Y. Y., Asad, I., Coleman, B., Menard, L., Benki-Nugent, S., Hussein Were, F., ... McHenry, M. S. (2022). Heavy metals and neurodevelopment of children in low and middle-income countries: A systematic review. *PLoS One, 17*(3), e0265536. Retrieved from https://doi.org/10.1371/journal.pone.0265536.

Homberg, J. R., Kyzar, E. J., Nguyen, M., Norton, W. H., Pittman, J., Poudel, M. K., ... Kalueff, A. V. (2016). Understanding autism and other neurodevelopmental disorders through experimental translational neurobehavioral models. *Neuroscience & Biobehavioral Reviews, 65*, 292–312. Retrieved from https://doi.org/10.1016/j.neubiorev.2016.03.013.

Jomova, K., Makova, M., Alomar, S. Y., Alwasel, S. H., Nepovimova, E., Kuca, K., ... Valko, M. (2022). Essential metals in health and disease. *Chemico-Biological Interactions, 367*, 110173. Retrieved from https://doi.org/10.1016/j.cbi.2022.110173.

Kang, B., Wang, J., Guo, S., & Yang, L. (2024). Mercury-induced toxicity: Mechanisms, molecular pathways, and gene regulation. *Science of the Total Environment*, 173577. Retrieved from https://doi.org/10.1016/j.scitotenv.2024.173577.

Kordas, K., Ravenscroft, J., Cao, Y., & McLean, E. V. (2018). Lead exposure in low and middle-income countries: Perspectives and lessons on patterns, injustices, economics, and politics. *International Journal of Environmental Research and Public Health, 15*(11), 2351. Retrieved from https://doi.org/10.3390/ijerph15112351.

Kumar, P., Goud, E. L., Devi, P., & Koul, B. (2022). 12 Metal pollutants in the environment. In M. Shah (Ed.). *Environmental microbiology: Emerging technologies* (pp. 291–324). Berlin, Boston: De Gruyter. Retrieved from https://doi.org/10.1515/9783110727227-012.

Kumar, P., Goud, E. L., Devi, P., Dey, S. R., & Dwivedi, P. (2022). Heavy metals: Transport in plants and their physiological and toxicological effects. In *Plant metal and metalloid transporters*. Singapore: Springer Nature Singapore, 23–54. Retrieved from https://doi.org/10.1007/978-981-19-6103-8_2.

Kumari, S., Dhankhar, H., Abrol, V., & Yadav, A. K. (2024). Effect of fluoride-contaminated water on the living being and their surroundings. In *Advanced treatment technologies for fluoride removal in water: Water purification*. Cham: Springer Nature Switzerland, 215–231. Retrieved from https://doi.org/10.1007/978-3-031-38845-3_12.

Liang, Y., Tian, L., Ho, K. F., Ip, M. S. M., & Mak, J. C. W. (2019). Mitochondrial toxicity of fine particulate matter (PM2. 5) collected in Hong Kong in human airway epithelial cells. In *The 24th medical research conference*. Hong Kong Academy of Medicine Press. Retrieved from http://hdl.handle.net/10722/274555.

Liu, J., Feng, X., Wang, Y., Xia, X., & Zheng, J. C. (2022). Astrocytes: GABAceptive and GABAergic cells in the brain. *Frontiers in Cellular Neuroscience, 16*, 892497. Retrieved from https://doi.org/10.3389/fncel.2022.892497.

Lu, M., Li, G., Yang, Y., & Yu, Y. (2021). A review on in-vitro oral bioaccessibility of organic pollutants and its application in human exposure assessment. *Science of the Total Environment, 752*, 142001. Retrieved from https://doi.org/10.1016/j.scitotenv.2020.142001.

Marczylo, E. L., Jacobs, M. N., & Gant, T. W. (2016). Environmentally induced epigenetic toxicity: Potential public health concerns. *Critical Reviews in Toxicology, 46*(8), 676–700. Retrieved from https://doi.org/10.1080/10408444.2016.1175417.

Mele, G., Cavaliere, C., Alfano, V., Orsini, M., Salvatore, M., & Aiello, M. (2019). Simultaneous EEG-fMRI for functional neurological assessment. *Frontiers in Neurology, 10*, 848. Retrieved from https://doi.org/10.3389/fneur.2019.00848.

Mitra, S., Chakraborty, A. J., Tareq, A. M., Emran, T. B., Nainu, F., Khusro, A., ... Simal-Gandara, J. (2022). Impact of heavy metals on the environment and human health: Novel therapeutic insights to counter the toxicity. *Journal of King Saud University-Science, 34*(3), 101865. Retrieved from https://doi.org/10.1016/j.jksus.2022.101865.

Monday, H. R., Younts, T. J., & Castillo, P. E. (2018). Long-term plasticity of neurotransmitter release: Emerging mechanisms and contributions to brain function and disease. *Annual Review of Neuroscience, 41*(1), 299–322. Retrieved from https://doi.org/10.1146/annurev-neuro-080317-062155.

Moretti, R., Pansiot, J., Bettati, D., Strazielle, N., Ghersi-Egea, J. F., Damante, G., ... Gressens, P. (2015). Blood-brain barrier dysfunction in disorders of the developing brain. *Frontiers in Neuroscience, 9*, 40. Retrieved from https://doi.org/10.3389/fnins.2015.00040.

Murumulla, L., Bandaru, L. J. M., & Challa, S. (2024). Heavy metal mediated progressive degeneration and its noxious effects on brain microenvironment. *Biological Trace Element Research, 202*(4), 1411–1427. Retrieved from https://doi.org/10.1007/s12011-023-03778-x.

Nicoletti, V. G., Pajer, K., Calcagno, D., Pajenda, G., & Nógrádi, A. (2022). The role of metals in the neuroregenerative action of BDNF, GDNF, NGF and other neurotrophic factors. *Biomolecules, 12*(8), 1015. Retrieved from https://doi.org/10.3390/biom12081015.

Njoku, A. U., & Sampson, N. R. (2023). Environment injustice and public health. In *Handbook of social sciences and global public health*. Cham: Springer International Publishing, 1987–2006. Retrieved from https://doi.org/10.1007/978-3-031-25110-8_37.

Rahbar, M. H., Samms-Vaughan, M., Kim, S., Saroukhani, S., Bressler, J., Hessabi, M., ... Loveland, K. A. (2022). Detoxification role of metabolic glutathione *S*-transferase (GST) genes in blood lead concentrations of Jamaican children with and without autism spectrum disorder. *Genes, 13*(6), 975. Retrieved from https://doi.org/10.3390/genes13060975.

Raj, A., Kumar, A., & Vaghasiya, T. P. (2021). Multidisciplinary approaches to clean up heavy metal contaminated soil: A review. *Bulletin of Environment, Pharmacology and Life Sciences, 10*(6), 53–65.

Rauh, V. A., & Margolis, A. E. (2016). Research review: Environmental exposures, neurodevelopment, and child mental health–new paradigms for the study of brain and behavioral effects. *Journal of Child Psychology and Psychiatry, 57*(7), 775–793. Retrieved from https://doi.org/10.1111/jcpp.12537.

Ruiz, D., Becerra, M., Jagai, J. S., Ard, K., & Sargis, R. M. (2018). Disparities in environmental exposures to endocrine-disrupting chemicals and diabetes risk in vulnerable populations. *Diabetes Care, 41*(1), 193–205. Retrieved from https://doi.org/10.2337/dc16-2765.

Said, S., & Hernandez, G. T. (2015). Environmental exposures, socioeconomics, disparities, and the kidneys. *Advances in Chronic Kidney Disease, 22*(1), 39–45. Retrieved from https://doi.org/10.1053/j.ackd.2014.09.003.

Soetrisno, F. N., & Delgado-Saborit, J. M. (2020). Chronic exposure to heavy metals from informal e-waste recycling plants and children's attention, executive function and academic performance. *Science of the Total Environment, 717*, 137099. Retrieved from https://doi.org/10.1016/j.scitotenv.2020.137099.

Sun, Q., Li, Y., Shi, L., Hussain, R., Mehmood, K., Tang, Z., & Zhang, H. (2022). Heavy metals induced mitochondrial dysfunction in animals: Molecular mechanism of toxicity. *Toxicology, 469*, 153136. Retrieved from https://doi.org/10.1016/j.tox.2022.153136.

Teleanu, R. I., Niculescu, A. G., Roza, E., Vladacenco, O., Grumezescu, A. M., & Teleanu, D. M. (2022). Neurotransmitters—Key factors in neurological and neurodegenerative disorders of the central nervous system. *International Journal of Molecular Sciences, 23*(11), 5954. Retrieved from https://doi.org/10.3390/ijms23115954.

Teodoro, M., Briguglio, G., Fenga, C., & Costa, C. (2019). Genetic polymorphisms as determinants of pesticide toxicity: Recent advances. *Toxicology Reports, 6*, 564–570. Retrieved from https://doi.org/10.1016/j.toxrep.2019.06.004.

Thakur, M., Rachamalla, M., Niyogi, S., Datusalia, A. K., & Flora, S. J. S. (2021). Molecular mechanism of arsenic-induced neurotoxicity including neuronal dysfunctions. *International Journal of Molecular Sciences, 22*(18), 10077. Retrieved from https://doi.org/10.3390/ijms221810077.

Thapar, A., Cooper, M., & Rutter, M. (2017). Neurodevelopmental disorders. *The Lancet Psychiatry, 4*(4), 339–346. Retrieved from https://doi.org/10.1016/S2215-0366(16)30376-5.

Uddin, S., Khanom, S., & Islam, M. R. (2023). Mercury contamination in food—An overview. In *Mercury toxicity: Challenges and solutions*33–70. Retrieved from https://doi.org/10.1007/978-981-99-7719-2_2.

Ungureanu, E. L., & Mustatea, G. (2022). Toxicity of heavy metals. In H. M. Saleh, & A. I. Hassan (Eds.). *Environmental impact and remediation of heavy metals*IntechOpenRetrieved from 10.5772/intechopen.102441.

Vazquez Cervantes, G. I., González Esquivel, D. F., Ramírez Ortega, D., Blanco Ayala, T., Ramos Chávez, L. A., López-López, H. E., ... Pérez de la Cruz, V. (2023). Mechanisms associated with cognitive and behavioral impairment induced by arsenic exposure. *Cells, 12*(21), 2537. Retrieved from https://doi.org/10.3390/cells12212537.

Veisi, A., Khaleghzadeh-Ahangar, H., Fattahi, M., & Haghparast, A. (2023). The role of Orexin-1 receptors within the hippocampal CA1 area in the extinction and reinstatement of methamphetamine-seeking behaviors. *Neurochemical Research, 48*(2), 671–680. Retrieved from https://doi.org/10.1007/s11064-022-03793-9.

Vogt, V. L., Aikia, M., Del Barrio, A., Boon, P., Borbely, C., Bran, E., & E-PILEPSY consortium (2017). Current standards of neuropsychological assessment in epilepsy surgery centers across Europe. *Epilepsia, 58*(3), 343–355. Retrieved from https://doi.org/10.1111/epi.13646.

Wang, L., Yin, Y. L., Liu, X. Z., Shen, P., Zheng, Y. G., Lan, X. R., ... Wang, J. Z. (2020). Current understanding of metal ions in the pathogenesis of Alzheimer's disease. *Translational Neurodegeneration, 9*, 1–13. Retrieved from https://doi.org/10.1186/s40035-020-00189-z.

Xi, T., & Wu, J. (2021). A review on the mechanism between different factors and the occurrence of autism and ADHD. *Psychology Research and Behavior Management, 14*, 393–403.

Zeiss, C. J. (2021). Comparative milestones in rodent and human postnatal central nervous system development. *Toxicologic Pathology, 49*(8), 1368–1373. Retrieved from https://doi.org/10.1177/01926233211046933.

Chapter 18

Environmental sources of heavy metals and their impacts on human health

Monika Sharma, and Prasann Kumar
Department of Agronomy, School of Agriculture, Lovely Professional University, Phagwara, Punjab, India

18.1 Introduction

Heavy metals are naturally found substances characterized by high atomic weight and density at least five times larger than water. They encompass essential and nonessential metals, with some necessary in trace quantities for biological processes, while others demonstrate significant toxicity even at low doses (Jaishankar, Tseten, Anbalagan, Mathew, & Beeregowda, 2014). The primary heavy metals of concern for environmental and human health effects are lead (Pb), mercury (Hg), cadmium (Cd), arsenic (As), and chromium (Cr). These metals endure in the environment owing to their nonbiodegradable characteristics and can bioaccumulate in organisms, resulting in chronic toxicity (Tchounwou, Yedjou, Patlolla, & Sutton, 2012). Heavy metals are often characterized by their density, atomic number, or toxicity. Although a globally recognized description is lacking, they are usually defined by their high density (>5 g/cm^3) and probable toxicity. Heavy metals are characterized as metallic elements with a density significantly greater than water. Assuming a correlation between heaviness and toxicity, heavy metals encompass metalloids, such as arsenic, which can generate toxicity at low exposure levels. In recent years, there has been a growing ecological and worldwide public health concern over environmental poisoning by these metals. Moreover, human exposure has escalated significantly due to the rapid growth in their use across many industrial, agricultural, home, and technology applications. Identified sources of heavy metals in the environment encompass geogenic, industrial, agricultural, pharmaceutical, residential effluents, and atmospheric origins. Environmental contamination is particularly significant in point source locations such as mining, foundries, smelters, and other metal-based industrial activities. Heavy metals are classified as trace elements due to their occurrence in minute amounts (ppb range to less than 10 ppm) throughout numerous environmental matrices. Their bioavailability is affected by physical variables like temperature, phase association, adsorption, and sequestration biological variables, including species traits, trophic relationships, and biochemical or physiological adaptations, significantly influence. Certain heavy metals, including iron (Fe), zinc (Zn), and copper (Cu), are vital micronutrients necessary for biological functions. Nonetheless, in excessive amounts, even necessary heavy metals can exhibit toxicity. Nonessential heavy metals such as mercury (Hg), cadmium (Cd), and lead (Pb) lack any reorganization role and exhibit substantial toxicity even at minimal quantities (Alloway, 2013). Toxicity and carcinogenicity generated by heavy metals encompass several mechanistic facets, some of which remain inadequately characterized or comprehended. Each metal possesses distinct characteristics and physicochemical qualities that contribute to its distinctive toxicological modes of action. This study analyzes the environmental prevalence, production and utilization, possible human exposure, and molecular mechanisms of toxicity, genotoxicity, and carcinogenicity of arsenic, cadmium, chromium, lead, and mercury.

Many treatment strategies have been devised to mitigate the toxicity of heavy metals. Natural products are effectively harmful (Tchounwou et al., 2012). Medicinal plants and natural items for the treatment of diverse treating diverse ailments have existed for the primary advantage of traditional or plant-based medicine is its perceived efficacy, together with a low incidence of severe, severe adverse effects. Tally-produced heavy metal toxicity in laboratory animals was markedly reduced with the use of several therapeutic herbs and natural products. Nanotechnological methods are garnering interest for their potential advantages in eliminating the detrimental effects of heavy

Heavy Metal Toxicity and Neurodegeneration. DOI: https://doi.org/10.1016/B978-0-443-36575-1.00030-3

metals. Nanotechnology is the interdisciplinary use of technologies to create nanoscale materials or devices, including principles from physics, chemistry, engineering, and biology. The emergence of nanotechnology has led to the meticulous development and production of nanomaterials for the detection and removal of heavy metals offering several and detecting and removing This chapter seeks to elucidate the environmental origins of heavy metals, their exposure pathways, and their effects on human health while emphasizing efforts to decrease associated risks and promote while emphasizing inability.

18.2 Environmental sources of heavy metals

Ecologically enduring environmental pollutants because they cannot naturally dissolve or disintegrate. In contrast to organic contaminants, which may be decomposed by microbial or chemical means, heavy metals persist in the environment eternally, accumulating in soil, water, and air. Their permanence is significantly affected by environmental conditions like pH, redox potential, temperature, and the availability of organic materials. In soils, heavy metals, including lead (Pb), cadmium (Cd), and arsenic (As), here to clay minerals and organic matter, diminishing their mobility while extending their persistence for decades or even centuries (Kabata-Pendias, 2011). In acidic circumstances, some metals exhibit increased solubility and leach into groundwater, causing pollution concerns. In aquatic ecosystems, heavy metals accumulate in sediments, which can be reintroduced into the water column owing to pH or microbial activity alteration stemming from persistent water contamination (Mason, Reinfelder, & Morel, 1996). The atmosphere functions as a conduit for metal transport, with emissions from industrial sources contributing to global deposition via precipitation and airborne particulate matter (Pacyna, Pacyna, & Aas, 2010).

18.3 Bioaccumulation of heavy metals in organisms

Bioaccumulation denotes the process by which organisms assimilate heavy metals from their surroundings at a pace above their capacity for excretion. This buildup transpires by direct exposure to polluted air, water, soil, and food consumption. In aquatic environments, metals like mercury (Hg) and lead (Pb) are absorbed by fish and shellfish via their gills or by the ingestion of polluted prey. Mercury bioaccumulates in fish muscle tissues in its methylated form, rendering it highly hazardous to aquatic forms and human consumers (Clarkson & Magos, 2006). Terrestrial plants uptake heavy metals from polluted soil, with crops including rice and leafy vegetables recognized for accumulating cadmium, subsequently entering the food chain (Meharg & Zhao, 2012). In animals, heavy metals accumulate in organs like the liver, kidneys, and bones, resulting in chronic toxicity and organ damage. The bioaccumulation rate includes metal speciation, environmental circumstances, and the organism's metabolic ability to detoxify or excrete the metals.

18.4 Biomagnification of heavy metals in trophic chains

Biomagnification is the phenomenon wherein concentrations of heavy metals escalate at successive trophic levels within the food chain. In contrast to bioaccumulation, which occurs inside a single organism, biomagnification denotes the increasing concentration of metals as they ascend via trophic levels. This process is especially alarming for mercury (Hg), lead (Pb), and cadmium (Cd), which accumulate in organisms and are not readily eliminated. In aquatic ecosystems, primary producers like algae absorb heavy metals from water, supporting herbivorous zooplankton, small fish, and ultimately, large predatory fish such as tuna and swordfish (Karimi, Fitzgerald, & Fisher, 2012). Consequently, apex predators and people who consume seafood are subjected to the most extraordinary levels of most significant in terrestrial ecosystems, herbivorous animals that plants collect heavy metals, which are then transferred to carnivorous predators, including humans (Miranda, López-Alonso, Castillo, Hernández, & Benedito, 2005). Biomagnification presents significant ecological and health hazards since heavy metals disrupt biological activities, resulting in neurological, developmental, and immune system diseases in higher organisms (Pain, Mateo, & Green, 2009).

18.5 Effects on ecosystems and human health

The persistence and bioaccumulation of heavy metals pose significant risks to ecosystems and human health. Heavy metal pollution in aquatic habitats diminishes biodiversity by impairing fish and amphibians, reproduction, behavior' survival, reproduction, and behavior. Elevated mercury levels, for instance, affect fish's neurological system and reproductive health, decreasing population (UNEP, 2013). Heavy metal contamination modifies soil microbial populations in terrestrial ecosystems, diminishing soil fertility and agricultural yield (Ali & Khan, 2018). Chronic exposure to bioaccumulated heavy metals via food, water, and air in humans is associated with severe health issues, including renal illness, cardiovascular diseases, neurotoxicity, and cancer (Tchounwou et al., 2012). Pregnant women and children are more susceptible to lead and mercury poisoning, which can lead to cognitive deficits and developmental abnormalities (Lanphear, Rauch, Auinger,

Allen, & Hornung, 2018). Considering these hazards, it is imperative to establish effective pollution control measures, enforce stringent regulatory limitations on industrial emissions, and advocate for sustainable farming practices to mitigate heavy metal exposure and its detrimental impacts on ecosystems and human populations.

18.6 Industrial contributions to heavy metal pollution

Industries are significant sources of heavy metal pollution, discharging harmful substances, including lead (Pb), mercury (Hg), cadmium (Cd), chromium (Cr), and arsenic (As), into the environment. Industrial activities, such as mining, metal smelting, coal combustion, electronic waste processing, and manufacturing, release heavy metals into the air, water, and soil. Mining operations produce substantial amounts of tailings laden with heavy metals, which, if inadequately handled, seep into adjacent water bodies and groundwater, polluting potable water sources (Jiang, Chen, & Wu, 2014; Jiang, Liu, Wang, & Li, 2014). Likewise, smelting and metal processing facilities release lead, cadmium, and arsenic into the atmosphere, precipitating terrestrial and aquatic environments via atmospheric deposition, resulting in extensive pollution (Pacyna et al., 2010). Effluents from electroplating, battery manufacturing, and textile sectors frequently possess elevated concentrations of heavy metals, which, if inadequately managed, accumulate in aquatic environments, jeopardizing human health (He, Yang, & Stoffella, 2019). The enduring presence of heavy metals in industrial waste sites complicates rehabilitation efforts, as these metals are nonbiodegradable and can retain their toxicity for decades.

18.7 Sources of heavy metal pollution in agriculture

Agricultural practices significantly contribute to heavy metal pollution, mainly due to the overuse of fertilizers, pesticides, wastewater irrigation, and livestock feed additives. Chemical fertilizers, especially those based on phosphate, contain minute quantities of cadmium and lead, which progressively build in the soil with continuous use (Nagajyoti, Lee, & Sreekanth, 2010). Historically, pesticides, including arsenic, copper, and mercury, were utilized in agriculture. Despite legislative limitations in several countries, residual contamination remains in soil and water sources (Nicholson, Smith, Alloway, Carlton-Smith, & Chambers, 2003). Using of untreated or inadequately treated industrial and municipal wastewater for irrigation delivers heavy metals, including chromium, nickel, and lead, into agricultural fields, hence elevating metal concentrations in crops. These toxins are subsequently transmitted to the human population via food intake, resulting in chronic health complications such as renal impairment, neurological problems, and carcinogenic consequences (Sharma & Agrawal, 2005). Moreover, polluted soils diminish agricultural output by modifying soil microbial activity, disrupting nutrient cycling, and resulting in prolonged soil deterioration.

18.8 Accumulation of heavy metals in crops and food chains

A significant consequence of industrial and agricultural heavy metal contamination is its accumulation in food crops. Plants uptake heavy metals from polluted soil and water via their root systems, resulting in bioaccumulation in consumable components, including leaves, grains, and fruits (Meharg & Zhao, 2012). Specific crops, such as rice, spinach, and lettuce, are recognized for their capacity to accumulate elevated concentrations of arsenic and cadmium, presenting considerable health hazards to human populations dependent on these staples (Zhao, Ma, Zhu, Tang, & McGrath, 2015). Livestock grazing in polluted areas or ingesting feed with heavy metals collect these harmful elements in their subsequent exposure to humans who consume meat, dairy, and eggs (Miranda et al., 2005). Chronic exposure to heavy metals through food results in developmental delays in children, renal and hepatic damage, and an elevated cancer risk. The issue is intensified in locations with inadequate regulatory control, where industrial and agricultural runoff directly pollute food production sequences.

18.9 Environmental and human health

The synergistic impacts of industrial and agricultural heavy metal contamination have catastrophic repercussions for ecosystems and human health. Metal pollution in aquatic habitats results in biodiversity loss by impairing the reproductive success and survival of fish, amphibians, and invertebrates (Mason et al., 1996). Soil pollution in terrestrial ecosystems diminishes crop yields and disturbs microbial populations, resulting in a decline in agricultural output over time (Ali & Khan, 2018). Heavy metals, including lead and mercury, are especially detrimental to human health, with prolonged exposure associated with neurological problems, cognitive deficits, and cardiovascular illnesses (Lanphear et al., 2018). In underdeveloped nations, where environmental restrictions tend to be less rigorous, industrial effluents and inadequate agricultural practices threaten human health. Urgent implementation of effective mitigation techniques, such as enhanced pollution regulations, sustainable farming practices, and bioremediation technology, is essential to diminish heavy metal contamination and its related hazards.

18.9.1 Sources and pathways of heavy metal exposure

1. **Natural sources of heavy metals:** Heavy metals are naturally present in the Earth's crust and are emitted into the environment via geological and weathering processes. Volcanic eruptions, rock erosion, and wildfires contribute to the natural baseline concentrations of heavy metals in soil, water, and air. The weathering of metal-rich rocks and minerals results in the leaching of elements, including arsenic (As), lead (Pb), and mercury (Hg), into groundwater and surface water, potentially endangering drinking water sources (Smedley & Kinniburgh, 2002). Furthermore, geothermal operations emit metals such as mercury and arsenic into the atmosphere and aquatic environments (Ball, Nordstrom, & Schoonen, 2010). Natural sources contribute to the environment by containing heavy metals; nevertheless, anthropogenic activities have markedly elevated their quantities, resulting in extensive pollution and public health issues.
2. **Emissions from industrial and urban sources:** Industrial operations are the primary cause of heavy metal pollution. Mining, metal smelting, industry, coal combustion, and waste incineration emit substantial amounts of lead, cadmium, chromium, and mercury into the environment (Pacyna et al., 2010). Smelting contributes to airborne heavy metals that deposit in soil and aquatic environments via atmospheric deposition (Nriagu & Pacyna, 1988). Moreover, burning in power plants and automobiles releases lead and mercury, which can disperse over considerable distances before depositing in ecosystems (UNEP, 2013). Urban areas are significant sources of heavy metal pollution due to elevated traffic emissions, industrial waste disposal, and the degradation of lead-containing infrastructure, including pipes and paints (Hu et al., 2016). These sources facilitate persistent exposure via inhalation, ingestion, and cutaneous contact. Moreover, e-waste recycling and the inappropriate disposal of electronic items substantially contribute to heavy metal pollution since discarded batteries and circuit boards contain elevated amounts of lead, cadmium, and mercury (Gupta, Yadav, Kumar, & Krishnan, 2019). Mining operations, such as open-pit mining, gold extraction, and coal mining, are major contributors to heavy metal contamination (Table 18.1). Mining activities leach heavy metals from ores into adjacent water bodies and soils, polluting agricultural fields (Ali, Damdimopoulou, Stenius, Adamsson, & Mäkelä, 2019; Ali, Khan, & Ilahi, 2019; Ali, Penttinen-Damdimopoulou, Stenius, & Sandalova, 2019). Acid mine drainage, prevalent in abandoned and operational mines, arrow the oxidation of sulfide minerals, producing acidic water metals like arsenic, cadmium, and lead. Mercury contamination is notably acute in artisanal and small-scale gold mining, where mercury is employed to extract gold from ore, resulting in significant pollution of aquatic life 2013.

18.10 Agricultural methods and heavy metal accumulation

Agricultural practices substantially contribute to heavy metal exposure by applying chemical fertilizers, such as fertilizers, wastewater irrigation, and animal feed. Phosphate-based fertilizers fertilizers include cadmium and lead, which build in soils over time and are absorbed by crops. Historically, insecticides comprising arsenic, copper, and mercury were extensively utilized, resulting in persistent pollution of soil and water sources (Gong, Zhao, & Wang, 2020). Furthermore, irrigation in fields with untreated industrial or municipal wastewater adds heavy metals, including chromium, nickel, and lead,

TABLE 18.1 Presence of toxic metals in industrial effluents.

Toxic metal	Industries releasing the metal
Arsenic	Fertilizer production, metal treatment, paint manufacturing, textile industry
Cadmium	Phosphate-based fertilizers, electronics manufacturing, pigment production, paint industry
Chromium	Metal finishing, leather tanning, rubber processing, photography
Copper	Electroplating, rayon manufacturing, electrical industry
Lead	Paint production, battery manufacturing
Nickel	Electroplating, iron, and steel industry
Zinc	Galvanization, iron, and steel plating
Mercury	Chlor-alkali industry, scientific instrument manufacturing, chemical production

polluting food crops (Sharma & Agrawal, 2005). Livestock husbandry adds to heavy metal pollution since animal feed additives may include copper and zinc, which are excreted into the environment and accumulate in soil and aquatic systems (Khan, Cao, Zheng, Huang, & Zhu, 2017). Livestock fed polluted feed and water collect heavy metals in their tissues, which can be transmitted to people by consuming products (Miranda et al., 2005). The buildup of heavy metals in food chains presents considerable health hazards, especially in areas with inadequate regulatory supervision.

18.11 Water pollution and human exposure

Water is a principal conduit for heavy metal exposure and drinking water sources. Industrial effluents, mining runoff, and landfill leaching introduce heavy metals, including arsenic, lead, and mercury, into aquatic systems such as rivers, lakes, and groundwater (Jiang, Chen, et al., 2014; Jiang, Liu, et al., 2014). Arsenic pollution in groundwater is a significant health concern in nations like Bangladesh and India, where millions depend on wells contaminated with arsenic for potable water (Meharg & Zhao, 2012). Heavy metals in water can bioaccumulate in aquatic species, resulting in exposure by ingesting shellfish (Karimi et al., 2012). Mercury pollution in aquatic environments leads to the development of methylmercury, a highly toxic substance that bioaccumulates in fish and presents significant neurological hazards to humans who ingest seafood (Clarkson & Magos, 2006). Heavy metal pollution in aquatic habitats arises from industrial effluents, oil spills, and inadequate waste management. Industrial effluents, including mercury, lead, mercury, lead, chromium, and coastal waterways, impact aquatic creatures and fisheries (Karimi et al., 2012). Mercury contamination in seas is a significant issue since it converts into methylmercury, a poisonous variant that bioaccumulates in fish and shellfish (Selin, 2009). Discharge of wastewater from tanneries and textile industries introduces chromium and other hazardous metals into aquatic ecosystems, jeopardizing biodiversity and affecting human populations dependent on polluted water supplies (Tchounwou et al., 2012). Plastic pollution in marine ecosystems exacerbates heavy metal contamination since microplastics adsorb and convey metals like cadmium and lead, elevating toxicity across marine food webs (Brennecke, Duarte, Paiva, Caçador, & Canning-Clode, 2016).

18.12 Atmospheric deposition and aerial exposure

Heavy metals are introduced into the environment via air channels and release delicate particulate matter from industrial sources, vehicular emissions, and combustion activities. Lead, mercury, cadmium, and arsenic are prevalent airborne contaminants that can be deposited in soil and water, resulting in prolonged pollution (Pacyna et al., 2010). Vacuum vapors can persist in the atmosphere for months, facilitating worldwide discrimination through rainfall or dry deposition (Selin, 2009). Urban populations have increased the risk of airborne heavy metal exposure via traffic emissions, industrial operations, and trash incineration (Hu et al., 2016). Prolonged exposure to airborne heavy metals is associated with respiratory ailments, cardiovascular problems, and neurological deficits (Lanphear et al., 2018).

18.13 Heavy metal contamination via trophic levels

Heavy metals infiltrate the food chain via soil, water, and air, producing crops, cattle, and seafood bioaccumulation. Plants absorb heavy metals from polluted soil and water via their root systems, leading to the buildup of harmful elements in consumable portions (Zhao et al., 2015). Rice, leafy vegetables, and wheat are especially vulnerable to heavy metal absorption, presenting nutritional hazards to human populations (Meharg & Zhao, 2012). Fish and shellfish acquire mercury, cadmium, and lead from contaminated environments, resulting in biomagnification in predatory species, including tuna and swordfish (Karimi et al., 2012). Individuals ingesting contaminated food face the danger of chronic heavy metal poisoning, resulting in health complications, including renal illness, cognitive deficits, and developmental abnormalities.

18.14 Occupational exposure, urbanization, and vehicular emissions of heavy metals

Employees in mining, metallurgy, battery fabrication, and electronics manufacturing have an elevated risk of heavy metal exposure from direct contact with hazardous substances. Occupational exposure transpires via inhalation of metal-laden dust, skin absorption, and ingestion of congestion and drinks in the workplace (Jaishankar et al., 2014). Lead poisoning poses a considerable risk to employees in the battery recycling and welding industries. Protective protocols for employees elevate exposure hazards (Needleman, 2004). Prolonged exposure to heavy metals in working environments has been associated with neurological problems, pulmonary ailments, and carcinogenic consequences (Tchounwou et al., 2012). Enforcing stringent occupational safety protocols, such as adequate ventilation, personal protective equipment, and routine health assessments, is crucial to mitigate the hazards of heavy metal exposure in the lace. Urbanization and transportation are major contributors to heavy metal pollution, especially in highly populated urban areas. Vehicle emissions, brake wear, and tire abrasion emit heavy metals like lead, cadmium, and zinc into the atmosphere and road dust, which humans can resuspend and breathe (Hu et al., 2016). Leaded GA although mostly discontinued in many nations, leaded gasoline led to persistent lead pollution in urban soils, particularly in historic city centers. Construction and demolition operations contribute to heavy metal pollution by emitting asbestos, lead-based paint particles, and cement dust containing chromium and arsenic into the environment.

18.15 Refuse management and sanitary landfills

Inadequate waste disposal, encompassing landfill leachate and waste incineration, significantly contributes to heavy metal pollution in soil and groundwater. Landfills frequently harbor batteries, electronic trash, and industrial by-products, which leak heavy metals into the adjacent environment, including lead, cadmium, and mercury (Gupta et al., 2019). The open incineration of waste items emits harmful heavy metal vapors into the atmosphere, heightening human exposure by inhalation. Moreover, the incorrect disposal of medical waste, which includes mercury-based thermometers and lead-lined radiation shields, exacerbates localized soil and water pollution.

18.16 Routes of human exposure to heavy metals

Heavy metals pose a considerable risk to human health via several exposure routes, including ingestion, inhalation, skin contact, and occupational exposure. These metals are present in the environment due to industrial activity, agricultural practices, and natural processes, resulting in extensive air, water, food, and soil pollution. Comprehending the routes of exposure is essential for reducing health risks and executing effective preventative strategies.

18.16.1 Consumption via contaminated food and water

One of the main ways people are exposed to heavy metals is through contaminated food and drinking water. Plants absorb heavy metals from polluted soil and water, producing bioaccumulation in edible parts such as grains, vegetables, and fruits (Zhao et al., 2015). Rice, in particular, is a primary dietary source of arsenic exposure because it absorbs arsenic from flooded paddy fields (Meharg & Zhao, 2012). Similarly, large predatory fish like tuna and swordfish accumulate high levels of methylmercury, a toxic substance formed by the methylation of inorganic mercury in aquatic ecosystems (Karimi et al., 2012). Contaminated drinking water is another serious concern, especially in regions that rely on arsenic-contaminated groundwater. Long-term consumption can lead to chronic arsenic poisoning, which is associated with skin lesions, cognitive impairments, and an increased risk of cancer (Smedley & Kinniburgh, 2002). Lead contamination in water sources, often due to corroded lead pipes, has also been linked to severe developmental and cognitive issues in children (Lanphear et al., 2018).

18.16.2 Inhalation of aerial heavy metals

Suspended heavy metals provide a considerable health hazard, especially in metropolitan and industrial regions. Vehicle exhaust, coal combustion, metal smelting, and industrial processes emit heavy metals like lead (Pb), cadmium (Cd), mercury (Hg), and arsenic (As) into the atmosphere (Pacyna et al., 2010). These metals manifest as tiny particulate matter (PM2.5 and PM10), which, upon inhalation, infiltrate the respiratory system, resulting in pulmonary illnesses, cardiovascular conditions, and systemic toxicity (Hu et al., 2016). Mercury vapor, released by coal-fired power plants and artisanal gold mining, may persist in the atmosphere for prolonged durations before depositing onto terrestrial and aquatic environments (Selin, 2009). Occupational exposure to dust containing heavy metals is a significant problem for mining, construction, and welding workers. Extended inhalation may result in chronic respiratory illnesses and metal toxicity (Jaishankar et al., 2014).

18.16.3 Cutaneous exposure to polluted soil and water

Direct dermal contact with polluted soil, water, or industrial waste may lead to heavy metal exposure. Metals like arsenic and chromium can permeate the skin and infiltrate the circulation, resulting in dermatological disorders and systemic toxicity (Tchounwou et al., 2012). Agricultural workers, construction laborers, and miners are especially vulnerable owing to their regular interaction with polluted substances. Exposure to arsenic via irrigation water and polluted soil has been associated with hyperkeratosis, skin malignancies, and other dermatological problems in areas with elevated arsenic levels (Meharg & Zhao, 2012). Moreover, some populations have recognized cosmetic goods containing heavy metals, such as lead-contaminated eyeliner, as a source of cutaneous exposure (Al-Saleh, Al-Enazi, & Shinwari, 2009). While skin absorption is often less critical than ingesting or inhalation, prolonged exposure may lead to cumulative systemic toxicity.

18.16.4 Occupational exposure to heavy metals

Employees in industrial environments, such as mining, battery production, electroplating, and smelting, have a heightened risk of heavy metal exposure from regular interaction with metal dust, fumes, and contaminated substances (Jaishankar et al., 2014). Occupational exposure transpires via air intake containing metals, skin absorption, and inadvertent ingestion of contaminated materials. Lead exposure is a recognized occupational hazard in the battery recycling and welding sectors, posing neurological, hematological, and reproductive health concerns to workers (Needleman, 2004). Mercury exposure in gold miners employing amalgamation procedures has led to significant neurological deficits and renal damage (UNEP, 2013). Without adequate protective measures, workers may transport heavy metal pollutants on their clothing and their families to harmful metals (Khan et al., 2017). Workplace safety rules, protective equipment, and routine health assessments are crucial for mitigating the hazards linked to occupational heavy metal exposure.

18.16.5 Exposure during prenatal and early childhood stages

Exposure to heavy metals during fetal development and early infancy can result in significant long-term health repercussions. Heavy metals can traverse the placental barrier, producing chemicals like lead, mercury, and cadmium (Grandjean & Landrigan, 2014). Maternal exposure to lead is associated with preterm births, low birth weight, and developmental delays, whereas prenatal mercury exposure can result in cognitive deficits and neurobehavioral problems in children (Lanphear et al., 2018). Children are especially susceptible to heavy metal poisoning because of their elevated absorption rates, immature neural systems, and habitual hand-to-mouth activity (Goyer & Clarkson, 2001). Exposure to lead from polluted home dust, paint, and water sources has been linked to diminished IQ, attention difficulties, and learning problems in children (Needleman, 2004). Minimum exposure and executing early intervention techniques are essential for averting the long-term health consequences of early-life heavy metal exposure.

18.17 Health consequences of heavy metal exposure

Heavy metal exposure poses a significant public health risk owing to its prolonged harmful effects on several organ systems (Table 18.2). The enduring characteristics of heavy metals enable their accumulation in biological tissues, resulting in chronic health conditions. Exposure to metals like lead (Pb), mercury (Hg), cadmium (Cd), arsenic (As), and chromium (Cr VI) has been associated with neurological problems, renal and hepatic damage, cardiovascular diseases, cancer, reproductive dysfunction, respiratory ailments, and immune system impairment. The intensity of these health impacts is contingent upon parameters like dosage, exposure length, and individual vulnerability. The subsequent sections offer an in-depth analysis of the health hazards linked to heavy metal exposure.

1. **Neurological disorders:** The nervous system is especially susceptible to heavy metal poisoning, resulting in consequences that range from cognitive impairments to severe neurodegenerative disorders. Lead (Pb) is a recognized neurotoxin, especially detrimental to children since it disrupts neurotransmitter function, synaptic development, and myelination. Chronic lead exposure is associated with diminished IQ, learning difficulties, attention deficiencies, and behavioral problems, including autism spectrum disorder (Lanphear et al., 2018). Methylmercury (Hg) traverses the blood–brain barrier, resulting in neurotoxicity manifested as memory deficits, tremors, and impaired motor coordination. The Minamata illness in Japan was a catastrophic instance of acute mercury poisoning, resulting in congenital brain damage and neurological problems (Harada, 1995). Exposure to arsenic (As) from polluted drinking water has been linked to neurocognitive deterioration, memory impairments, and peripheral neuropathy in both children and adults (Naujokas, Anderson, & Ahsan, 2013).
2. **Renal and hepatic toxicity:** The kidneys and liver are essential for detoxifying and excreting heavy metals, rendering them particularly vulnerable to poisoning. Cadmium (Cd) is a powerful nephrotoxin that accumulates in the kidneys, resulting in chronic kidney disease, proteinuria, and renal tubular failure (Satarug, Garrett, Sens, & Sens, 2018). Occupational exposure to cadmium, especially in industrial environments like battery manufacture, markedly elevates the risk of renal damage. Lead (Pb) exposure is linked to renal impairment, diminishing the glomerular filtration rate and fostering kidney fibrosis. Chronic arsenic exposure has been associated with hepatic damage, encompassing liver fibrosis, cirrhosis, and hepatocellular cancer. In populations with significant exposure, arsenic-induced hepatic damage frequently advances to critical situations (Guha Mazumder, 2008).
3. **Cardiovascular disorders:** Heavy metals exacerbate cardiovascular disorders by provoking oxidative stress, endothelial dysfunction, and inflammation, resulting in hypertension and arterial damage. Exposure to lead (Pb), even at minimal concentrations, has been linked to heightened blood pressure, arterial rigidity, and an increased likelihood of cardiovascular disease and stroke (Navas-Acien, Guallar, Silbergeld, & Rothenberg, 2007). Cadmium (Cd) exposure exacerbates atherosclerosis by impairing vascular endothelial cells and elevating

TABLE 18.2 Health impacts of chronic toxic metal exposure.

Toxic metal	Affected organs	Major exposure sources	Associated health effects
Arsenic	Lungs, nervous system, and skin	Industrial dust, medicinal applications, and contaminated water	Nasal septum perforation, lung cancer, peripheral neuropathy, skin disorders, skin cancer
Cadmium	Kidneys, bones, lungs	Industrial dust and fumes, contaminated food and water	Proteinuria, glucosuria, osteocalcin, aminoaciduria, emphysema
Chromium	Lungs	Industrial dust and fumes, polluted food	Nasal septum ulcers and perforation, respiratory cancer
Manganese	Nervous system	Industrial dust and fumes	Central and peripheral nervous system disorders
Lead	Nervous system, blood-forming system, kidneys	Industrial dust, fumes, and contaminated food	Encephalopathy, peripheral neuropathy, central nervous disorders, anemia
Nickel	Lungs, skin	Industrial dust, airborne particles	Cancer, skin inflammation (dermatitis)
Tin	Nervous system, lungs	Medicinal applications, industrial dust	Central nervous system dysfunction, vision impairments, EEG abnormalities, pneumoconiosis
Mercury	Nervous system, kidneys	Industrial dust, fumes, contaminated food, and water	Proteinuria

inflammatory markers, heightening the risk of myocardial infarction (Tellez-Plaza, Jones, & Dominguez-Lucas, 2013). Chronic exposure to arsenic (As) has been associated with hypertension, peripheral vascular disease, and ischemic heart disease. Research in Bangladesh identified a strong link between prolonged arsenic exposure from polluted groundwater and increased mortality from cardiovascular problems (Moon, Guallar, & Navas-Acien, 2013).

4. **Carcinogenic effects:** Numerous heavy metals have been designated as human carcinogens owing to their capacity to provoke genetic mutations, epigenetic modifications, and oxidative stress, resulting in cancer progression. Arsenic (As) is a well-documented carcinogen associated with malignancies of the skin, lungs, bladder, liver, and kidneys (Smith, Marshall, & Yuan, 2009). Prolonged use of arsenic-laden water has led to a proliferation of cancer cases, especially in South Asia. Hexavalent chromium (Cr VI) is a significant carcinogen that induces lung cancer, especially in industrial workers exposed to chromium dust (Costa, 1997). Cadmium (Cd) is linked to a heightened risk of lung, prostate, and breast cancer. Its capacity to imitate estrogen is associated with hormone-related malignancies, including breast and endometrial cancer (Waalkes, 2003). The carcinogenic properties of these metals highlight the necessity for rigorous controls for industrial emissions and occupational exposure thresholds.
5. **Endocrine disruption and reproductive toxicity:** Heavy metals damage the endocrine system, impairing hormone synthesis, reproductive function, and fetal development. Lead (Pb) exposure has been demonstrated to disrupt the hypothalamic–pituitary–adrenal axis, resulting in monthly irregularities, diminished fertility, and adverse pregnancy outcomes, including preterm delivery and low birth weight (Jiang, Chen, et al., 2014; Jiang, Liu, et al., 2014). Mercury (Hg) is a reproductive toxin that interferes with sex hormone equilibrium, resulting in infertility and developmental anomalies in neonates (Freire, Koifman, & Koifman, 2010a; Freire, Koifman, & Koifman, 2010b). Cadmium (Cd), by its capacity to imitate estrogen, is associated with reproductive diseases, such as polycystic ovary syndrome and an elevated risk of breast cancer (Ali, Damdimopoulou, et al., 2019; Ali, Khan, et al., 2019; Ali, Penttinen-Damdimopoulou, et al., 2019). Pregnant women exposed to heavy metals have an increased risk of miscarriage, stillbirth, and congenital abnormalities in their children.
6. **Respiratory diseases:** The inhalation of heavy metal particles from industrial emissions, mining activities, and vehicle pollution is a major contributor to respiratory disorders. Exposure to chromium (Cr VI), especially among employees in the electroplating and welding sectors, has been associated with lung cancer, chronic bronchitis, and occupational asthma (ATSDR, 2012). Inhalation of cadmium (Cd) significantly contributes to the risk of emphysema, chronic obstructive pulmonary disease, and lung fibrosis, particularly among workers exposed to cadmium fumes in metal

refineries (Henderson, Durbin, & Perry, 2012). Inhalation of airborne arsenic particles can lead to respiratory discomfort, persistent cough, and a heightened risk of bronchial cancer. These findings underscore the necessity for enhanced air quality legislation and preventive measures for employees in high-risk sectors.

7. **Dysfunction of the immune system:** Exposure to heavy metals has been shown to compromise immunological function, heightening vulnerability to infections, autoimmune disorders, and chronic inflammation. Mercury (Hg) exposure inhibits immune cell function, resulting in heightened susceptibility to bacterial and viral infections and autoimmune diseases such as lupus and multiple sclerosis (Moszczynski, 1997). Exposure to arsenic (As) has been demonstrated to impair immunological responses, resulting in heightened occurrences of respiratory infections and reduced vaccination efficiency, especially in children (Rosenberg, McCurdy, Roberts, & Lee, 2019). Lead (Pb) is associated with immunological dysregulation, impairing the body's capacity to combat infections. The immunotoxin effects of heavy metals underscore the necessity of minimizing these hazardous substances to preserve general health and immunity.

18.18 Conclusion and future perspectives

Environmental pollution, food quality, safety, and public health are closely interconnected. In recent decades, heavy metal concentrations in the ecosystem have risen significantly. The sources of heavy metal contamination in food crops vary between developing and industrialized nations. Heavy metals are persistent environmental pollutants that seriously threaten human health and ecosystems. Their widespread presence in air, water, and soil and their ability to accumulate in living organisms necessitate continuous monitoring and control. Addressing heavy metal contamination requires a comprehensive approach integrating scientific research, policy measures, and public awareness to reduce risks and safeguard global health. The pathways of heavy metal exposure can differ, making it crucial to understand how these toxic elements enter the body, their toxicity levels, and their potential health impacts. Our assessment indicates that oral ingestion is a primary route of exposure, leading to harmful effects on various organs. High levels of heavy metal exposure can result in neurological disorders, respiratory diseases, and even cancer. Individuals should minimize exposure by avoiding contaminated environments and reducing reliance on industrial products that contain harmful metals to mitigate heavy metal toxicity. Additionally, the use of heavy metals in industrial processes should be carefully regulated. Public awareness plays a vital role in reducing exposure, as informed individuals can make conscious choices to limit their interaction with toxic substances. Prolonged low-dose exposure to heavy metals remains a significant public health concern, particularly in regions with widespread pollution. Understanding the underlying mechanisms of heavy metal interactions is essential for assessing health risks and developing effective management strategies. Therefore further research is needed to clarify the molecular pathways involved in human exposure to multiple hazardous metals and their broader public health implications.

References

Agency for Toxic Substances and Disease Registry (ATSDR). (2012). *Toxicological profile for chromium*. U.S. Department of Health and Human Services.

Ali, H., & Khan, E. (2018). What are heavy metals? Long-standing controversy over the scientific use of the term 'heavy metals'—Proposal of a comprehensive definition. *Toxicological & Environmental Chemistry, 100*(1), 6–19.

Ali, H., Khan, E., & Ilahi, I. (2019). Environmental chemistry and ecotoxicology of hazardous heavy metals: Environmental persistence, toxicity, and bioaccumulation. *Journal of Chemistry, 2019*, 6730305.

Ali, I., Damdimopoulou, P., Stenius, U., Adamsson, A., & Mäkelä, S. I. (2019). Cadmium as a possible endocrine disruptor: A review on recent findings and possible mechanisms. *Environmental Health Perspectives, 127*(12), 12001.

Ali, I., Penttinen-Damdimopoulou, P., Stenius, U., & Sandalova, T. (2019). Cadmium as an endocrine disruptor: Distinct mechanisms of action and implications for public health. *Environmental Research, 178*, 108763.

Alloway, B. J. (2013). *Heavy metals in soils: Trace metals and metalloids in soils and their bioavailability* (3rd ed.). Springer Science & Business Media.

Al-Saleh, I., Al-Enazi, S., & Shinwari, N. (2009). Assessment of lead in cosmetic products. *Regulatory Toxicology and Pharmacology, 54*(2), 105–113.

Ball, J. W., Nordstrom, D. K., & Schoonen, M. A. (2010). Geochemical processes and controls on metal mobility in groundwater. *Geochemistry of the Environment, 2*, 47–82.

Brennecke, D., Duarte, B., Paiva, F., Caçador, I., & Canning-Clode, J. (2016). Microplastics as a vector for heavy metal contamination. *Environmental Pollution, 218*, 356–361.

Clarkson, T. W., & Magos, L. (2006). The toxicology of mercury and its chemical compounds. *Critical Reviews in Toxicology, 36*(8), 609–662.

Costa, M. (1997). Toxicity and carcinogenicity of Cr(VI) in animal models and humans. *Critical Reviews in Toxicology, 27*(5), 431–442.

Freire, C., Koifman, R. J., & Koifman, S. (2010a). Mercury exposure and reproductive outcomes in humans: A systematic review. *Journal of Toxicology and Environmental Health, Part B, 13*(7-8), 493–533.

Freire, C., Koifman, R. J., & Koifman, S. (2010b). Thyroid hormone levels and exposure to endocrine-disrupting chemicals among pregnant women. *Environmental Health Perspectives, 118*(5), 711–717.

Gong, Y., Zhao, D., & Wang, Q. (2020). Application of phosphate fertilizers containing heavy metals. *Science of the Total Environment, 715*, 136904.

Goyer, R. A., & Clarkson, T. W. (2001). Toxic effects of metalsIn *Casarett & Doull's toxicology: The basic science of poisons* (6th ed.). McGraw-Hill, 811–867.

Grandjean, P., & Landrigan, P. J. (2014). Neurobehavioral effects of developmental toxicity. *The Lancet Neurology, 13*(3), 330–338.

Guha Mazumder, D. N. (2008). Chronic arsenic toxicity & human health. *Indian Journal of Medical Research, 128*(4), 436–447.

Gupta, N., Yadav, K. K., Kumar, V., & Krishnan, S. (2019). Environmental impact of electronic waste and management strategies. *Environmental Science and Pollution Research, 26*(18), 18124–18135.

Harada, M. (1995). Minamata disease: Methylmercury poisoning in Japan caused by environmental pollution. *Critical Reviews in Toxicology, 25*(1), 1–24.

He, Z. L., Yang, X. E., & Stoffella, P. J. (2019). Trace elements in agroecosystems and impacts on the environment. *Journal of Trace Elements in Medicine and Biology, 25*(2), 125–140.

Henderson, R. G., Durbin, P. W., & Perry, P. M. (2012). Cadmium: Toxicology and environmental health concerns. *Journal of Toxicology and Applied Pharmacology, 258*(3), 228–235.

Hu, X., Zhang, Y., Luo, J., Wang, T., Lian, H., & Ding, Z. (2016). Bioaccessibility and health risk of arsenic and heavy metals in urban street dust from a mega-city, Nanjing, China. *Environmental Pollution, 214*, 73–81.

Jaishankar, M., Tseten, T., Anbalagan, N., Mathew, B. B., & Beeregowda, K. N. (2014). Toxicity, mechanism and health effects of some heavy metals. *Interdisciplinary Toxicology, 7*(2), 60–72.

Jiang, L., Liu, G., Wang, R., & Li, L. (2014). Heavy metal pollution and health risk assessment in the vicinity of a lead-zinc mine in China. *Environmental Science and Pollution Research, 21*(5), 3745–3753.

Jiang, Y., Chen, J., Wu, B., et al. (2014). Effects of lead exposure on the reproductive system: A review. *Biological Trace Element Research, 160*(1), 97–107.

Kabata-Pendias, A. (2011). *Trace elements in soils and plants (4th ed.)*. CRC Press.

Karimi, R., Fitzgerald, T. P., & Fisher, N. S. (2012). A quantitative synthesis of mercury in commercial seafood and implications for exposure in the U.S. population. *Environmental Health Perspectives, 120*(11), 1512–1519.

Khan, S., Cao, Q., Zheng, Y. M., Huang, Y. Z., & Zhu, Y. G. (2017). Health risks of heavy metals in contaminated soils and food crops irrigated with wastewater in Beijing, China. *Environmental Pollution, 152*(3), 686–692.

Lanphear, B. P., Rauch, S., Auinger, P., Allen, R. W., & Hornung, R. W. (2018). Low-level lead exposure and mortality in US adults: A population-based cohort study. *The Lancet Public Health, 3*(4), e177–e184.

Mason, R. P., Reinfelder, J. R., & Morel, F. M. M. (1996). Uptake, toxicity, and trophic transfer of mercury in a coastal diatom. *Environmental Science & Technology, 30*(6), 1835–1845.

Meharg, A. A., & Zhao, F. J. (2012). *Arsenic & rice*. Springer Science & Business Media.

Miranda, M., López-Alonso, M., Castillo, C., Hernández, J., & Benedito, J. L. (2005). Effects of moderate pollution on toxic and trace metal levels in calves from a polluted area of northern Spain. *Environment International, 31*(4), 543–548.

Moon, K., Guallar, E., & Navas-Acien, A. (2013). Arsenic exposure and cardiovascular disease: An updated systematic review. *Current Atherosclerosis Reports, 15*(6), 393.

Moszczynski, P. (1997). Mercury compounds and the immune system: A review. *International Journal of Occupational Medicine and Environmental Health, 10*(3), 247–258.

Nagajyoti, P. C., Lee, K. D., & Sreekanth, T. V. M. (2010). Heavy metals, occurrence and toxicity for plants: A review. *Environmental Chemistry Letters, 8*(3), 199–216.

Naujokas, M. F., Anderson, B., Ahsan, H., et al. (2013). The broad scope of health effects from chronic arsenic exposure: Update on a worldwide public health problem. *Environmental Health Perspectives, 121*(3), 295–302.

Navas-Acien, A., Guallar, E., Silbergeld, E. K., & Rothenberg, S. J. (2007). Lead exposure and cardiovascular disease: A systematic review. *Environmental Health Perspectives, 115*(3), 472–482.

Needleman, H. (2004). Lead poisoning. *Annual Review of Medicine, 55*, 209–222.

Nicholson, F. A., Smith, S. R., Alloway, B. J., Carlton-Smith, C., & Chambers, B. J. (2003). An inventory of heavy metals inputs to agricultural soils in England and Wales. *Science of the Total Environment, 311*(1-3), 205–219.

Nriagu, J. O., & Pacyna, J. M. (1988). Quantitative assessment of worldwide contamination of air, water, and soils by trace metals. *Nature, 333*(6169), 134–139.

Pacyna, J. M., Pacyna, E. G., & Aas, W. (2010). Changes of emissions and atmospheric deposition of mercury, lead, and cadmium. *Atmospheric Environment, 44*(2), 248–256.

Pain, D. J., Mateo, R., & Green, R. E. (2009). Effects of lead from ammunition on birds and other wildlife: A review and update. *Ambio: A Journal of the Human Environment, 38*(3), 236–245.

Rosenberg, A. M., McCurdy, L. E., Roberts, E. M., & Lee, D. (2019). Immunotoxicity of environmental contaminants: Impacts on child health. *Pediatrics, 143*(6), e20182563.

Satarug, S., Garrett, S. H., Sens, M. A., & Sens, D. A. (2018). Cadmium, environmental exposure, and health outcomes. *Environmental Health Perspectives, 126*(5), 057008.

Selin, N. E. (2009). Global biogeochemical cycling of mercury: A review. *Annual Review of Environment and Resources, 34*, 43–63.

Sharma, R. K., & Agrawal, M. (2005). Biological effects of heavy metals: An overview. *Journal of Environmental Biology, 26*(2), 301–313.

Smedley, P. L., & Kinniburgh, D. G. (2002). A review of the source, behavior, and distribution of arsenic in natural waters. *Applied Geochemistry, 17*(5), 517–568.

Smith, A. H., Marshall, G., Yuan, Y., et al. (2009). Increased mortality from lung cancer and bronchiectasis in young adults after exposure to arsenic in utero and in early childhood. *Environmental Health Perspectives, 118*(8), 1042–1046.

Tchounwou, P. B., Yedjou, C. G., Patlolla, A. K., & Sutton, D. J. (2012). Heavy metal toxicity and the environment. *Molecular, Clinical, and Environmental Toxicology, 101*, 133–164.

Tellez-Plaza, M., Jones, M. R., Dominguez-Lucas, A., et al. (2013). Cadmium exposure and cardiovascular risk. *Current Atherosclerosis Reports, 15*(10), 356.

UNEP (United Nations Environment Programme) (2013). Global mercury assessment 2013: Sources, emissions, releases and environmental transport. Geneva, Switzerland: UNEP Chemicals Branch.

Waalkes, M. P. (2003). Cadmium carcinogenesis. *Mutation Research, 533*(1–2), 107–120.

Zhao, F. J., Ma, Y., Zhu, Y. G., Tang, Z., & McGrath, S. P. (2015). Soil contamination in China: Current status and mitigation strategies. *Environmental Science & Technology, 49*(2), 750–759.

Chapter 19

Biomarkers of heavy metal–induced neurotoxicity: insights into mechanisms and disease impact

Krupanshi Bharadava[1], Sanjay Kumar[2], and Radhey Shyam Kaushal[1,3]

[1]*Biophysics & Structural Biology, Research & Development Cell, Parul University, Vadodara, Gujarat, India,* [2]*Structural and Molecular Biology, Center for Discovery and Innovation Building, The City College of New York, New York, NY, United States,* [3]*Department of Life Sciences—Biotechnology, Parul Institute of Applied Sciences, Parul University, Vadodara, Gujarat, India*

19.1 Introduction

Heavy metal toxicity has become a growing concern due to its widespread presence in the environment, impacting air, water, and food, with particular risks for individuals living in industrial or chemically contaminated areas. While metals are necessary for various physiological functions, their levels must be tightly regulated to prevent harmful effects. Neurodegenerative disorders like Alzheimer's disease (AD), Parkinson's disease (PD), and amyotrophic lateral sclerosis (ALS) have been related to disturbances in metal homeostasis, including excessive buildup or deficiency (Mitra et al., 2022). The mechanisms behind metal-induced toxicity are complex, but oxidative stress plays a central role, particularly with metals like iron (Fe) that catalyze Fenton reactions, generating reactive oxygen species (ROS) that cause cellular damage. This oxidative damage disrupts the cellular antioxidant defense (Aranda-Rivera, Cruz-Gregorio, Arancibia-Hernández, Hernández-Cruz, & Pedraza-Chaverri, 2022), such as the glutathione and thioredoxin systems, which protect neurons from stress-induced cell death. When these protective mechanisms are compromised, neuronal death accelerates, contributing to the progression of neurodegenerative disorders (Aranda-Rivera et al., 2022; Ren et al., 2017).

Furthermore, heavy metal exposure can cause epigenetic changes in DNA methylation, histone modification, and chromatin remodeling, all of which can affect the expression of genes linked to neurodegeneration. These epigenetic changes may exacerbate disease progression by activating genes contributing to neuronal dysfunction. As a result, biomarkers for heavy metal toxicity are gaining attention as valuable tools for understanding and diagnosing neurodegenerative diseases (Ijomone, Ifenatuoha, Aluko, Ijomone, & Aschner, 2020). These biomarkers help to identify those at risk, guide the creation of more focused treatment plans, and provide knowledge of the toxicity processes. This chapter thoroughly overviews heavy metal function as toxicity biomarkers, particularly in ALS, AD, and PD. It explores the molecular channels by which heavy metals cause neurotoxicity and emphasizes their part in the evolution of various diseases (Cheslow, Snook, & Waldman, 2024).

19.1.1 Cadmium (Cd)

The 48th element on the periodic table, cadmium (Cd), shows physicochemical characteristics similar to those of mercury and zinc. Mostly existing in the +2-oxidation state, its electronic configuration is [Kr] $4d^{10}5\ s^2$, reflecting its most stable and common valence state (Friberg, 2018). Low aqueous solubility marks the soft, silvery-white, combustible metal cadmium. Unlike many transition metals, cadmium demonstrates limited sensitivity to corrosion, making it a preferred choice for protective coatings on more reactive metals. Annually, global cadmium production surpasses 13,000 tons, primarily consumed in Ni–Cd batteries, metal coatings, color pigments, chemical stabilizers, and alloys (Hayat, Nauman, Nazir, Ali, & Bangash, 2019). Geochemically, cadmium is a relatively rare metal in the Earth's crust, mainly obtained from weathering cadmium-containing rocks and volcanic

Heavy Metal Toxicity and Neurodegeneration. DOI: https://doi.org/10.1016/B978-0-443-36575-1.00015-7

eruptions. Anthropogenic sources of cadmium pollution primarily arise from the mining and smelting of sulfide ores linked with zinc, copper, and lead. Industrial processes, including ore refinement, fossil fuel burning, and waste incineration, contribute significantly to cadmium emission into the environment, with estimates ranging between 5000 and 13,000 tons annually. Wind erosion of contaminated soils and industrial sites can also distribute cadmium particles into the atmosphere (Genchi, Sinicropi, Lauria, Carocci, & Catalano, 2020).

Cadmium exposure in humans primarily occurs through inhalation and, to a lesser extent, via ingestion. Once absorbed, cadmium exhibits a prolonged biological half-life of approximately 15–20 years (Rahimzadeh, Rahimzadeh, Kazemi, & Moghadamnia, 2017), leading to its bioaccumulation in tissues, particularly in the kidneys and liver. In aquatic environments, cadmium contamination arises from industrial effluents, mining runoff, landfill leachates, battery recycling sites, and excessive fertilizer application. Aquatic organisms readily assimilate cadmium, leading to its trophic transfer and bioaccumulation in the food web (Kar & Patra, 2021). Additionally, high cadmium concentrations in phosphate-based fertilizers have been implicated in reduced seed germination and impaired crop growth, highlighting the detrimental impact of excessive anthropogenic cadmium inputs on agroecosystems. Consequently, intensified human activities, particularly the overuse of chemical fertilizers, exacerbate cadmium contamination in soil and water systems, posing significant ecological and public health concerns (Alengebawy, Abdelkhalek, Qureshi, & Wang, 2021).

19.1.2 Arsenic

Arsenic (As), the 33rd element in the periodic table, typically appears in oxidation states of +5, +3, 0, and −3. Its electronic configuration is [Ar] $3d^{10}4 s^2 4p^3$. Arsenic is a metalloid because of its intermediate qualities between metals and nonmetals. It appears in three allotropic forms: α (yellow), β (black), and γ (gray). It occurs in organic and inorganic forms, with the latter being of greater toxicological significance (Chen & Costa, 2021). From an environmental and toxicological perspective, arsenic is categorized as a heavy metal due to its widespread distribution in soil and water, posing significant risks to human health and ecosystems. Arsenic is predominantly found in sedimentary deposits, metal ores, and igneous rocks, often associated with sulfur and oxygen. Arsenopyrite (FeAsS), a major arsenic-bearing mineral, is a primary source of arsenic release into the environment. The most prevalent atmospheric form of arsenic is arsenic trioxide (As_2O_3), whereas in aquatic and terrestrial environments, arsenic predominantly exists as inorganic arsenates (iAs^{5+}, $[AsO_4]^{3-}$) and arsenites (iAs^{3+}, $[AsO_3]^{3-}$) (Ahmed et al., 2022; Raju, 2022). Anthropogenic sources of arsenic contamination include industrial activities such as glass manufacturing, textile processing, ammunition production, paper manufacturing, and arsenic-based wood preservatives. Additionally, the cosmetic industry utilizes arsenic-containing color pigments in products such as eyeshadows, where percutaneous absorption poses a potential health hazard. To mitigate exposure risks, regulatory guidelines limit arsenic impurities in cosmetic formulations to below 5 ppm (Bhat, Ravi, Tian, & Singh, 2024).

Dietary exposure to arsenic occurs primarily through the consumption of seafood, mushrooms, rice, and poultry, which can accumulate organic arsenic species. Arsenic pollution is a serious public health issue in Bangladesh and West Bengal (India), where over 40 million people rely on arsenic-contaminated groundwater for drinking (Cubadda, Jackson, Cottingham, Van Horne, & Kurzius-Spencer, 2017). The World Health Organisation (WHO) sets a maximum arsenic limit of 10 μg/L in drinking water. Furthermore, traditional Chinese medicinal products are often enriched with arsenic for therapeutic purposes, inadvertently leading to cases of arsenic poisoning worldwide (Khosravi-Darani, Rehman, Katsoyiannis, Kokkinos, & Zouboulis, 2022). Mass spectrometric analysis of herbal formulations sold in the Dutch market revealed that heavy metal concentrations, including arsenic (As), mercury (Hg), and lead (Pb), exceeded safety thresholds by up to 20%. Chronic exposure to arsenic through skin contact poses a cancer risk, making strict monitoring and regulation of arsenic levels in food, water, and consumer products essential to protect public health (Jomova, Alomar, Nepovimova, Kuca, & Valko, 2024).

19.1.3 Mercury

Mercury (Hg) is the 80th element on the periodic table, with the electronic configuration [Xe] $4 f^{14}5d^{10}6 s^2$. The Earth's crust contains elemental (metallic) mercury, inorganic mercury compounds, and organic mercury species like methylmercury, among other forms of mercury that are found naturally (Gworek, Dmuchowski, & Baczewska-Dąbrowska, 2020). Mercury is a liquid at room temperature with minimal volatility, with a partial vapor pressure of 0.261 Pa at 25°C. However, its volatility increases dramatically at higher temperatures (Rudyk, Spirov, Samuel, & Joshi, 2017). Mercury exists typically in two oxidation states: +1 and +2, with the latter being the more stable. Mercury(II) compounds, like HgO, develop at temperatures above 350°C, but Hg_2^{2+} ions are less prevalent. Mercury(II) chloride ($HgCl_2$) is widely used in agriculture as a fungicide, in industrial operations as a catalyst in vinyl chloride manufacturing, and as a precursor for other mercury-based compounds. In contrast, mercury(II) sulfide (HgS) is utilized as a pigment in paints, rubber, and plastics (Kwon et al., 2020).

Environmental mercury exposure can occur through elemental, inorganic, and organic mercury compounds. Elemental mercury (Hg^0) is a silvery liquid at ambient conditions, which gradually evaporates into toxic, colorless, and odorless vapors. Although the use of mercury in thermometers and fluorescent bulbs has diminished, it is still found in some products, including dental amalgams, commonly referred to as "silver fillings" (Ha et al., 2017). Inorganic mercury is prevalent in the environment, particularly in minerals and geological formations. Mercury salts, including those linked with sulfur or chlorine, are present in water, soil, and airborne particles. The primary sources of elemental and inorganic mercury emissions caused by humans are industrial facilities and coal-burning power plants, which release effluents that are contaminated with mercury. Additionally, mercury compounds have been identified in certain cosmetic products, such as skin-lightening creams, manufactured in regions with weak regulatory standards, posing health hazards (mercury undergoes various chemical transformations in the environment, interacting with organic substances to form organomercury compounds). Methylmercury (MeHg), a highly toxic compound, is produced when microorganisms methylate inorganic mercury. Methylmercury bioaccumulates and biomagnifies in aquatic food webs, reaching harmful concentrations in fish and shellfish, which are the primary sources of human exposure. Due to its potent neurotoxic effects, methylmercury contamination remains a serious concern for public health and the environment (Alex, 2018).

19.1.4 Lead

Lead (Pb), a soft gray-blue metal with atomic number 82, has the electrical configuration [Xe] 4 $f^{14}5d^{10}6\ s^26p^2$. Galena (PbS), the principal mineral form of lead, contains around 87% lead by weight. Lead has strong electrical conductivity and is commonly found in +2 and +4 oxidation states, with Pb^{2+} being the more stable ion. Pb^{+} compounds can form complexes with diverse ligands, including carboxylates, thiolates, nitrogen, phosphorus, and halides, with coordination numbers ranging from 1 to 12 (Bjørklund, Dadar, Mutter, & Aaseth, 2017). Pb^{4+} compounds frequently adopt tetrahedral coordination with bigger ligands. Lead is a hazardous element that can drastically impact human health, wildlife, and ecosystems. Lead accumulates mainly in living organisms' bones, liver, and kidneys, with children, adolescents, and pregnant women being more sensitive due to their increased absorption rates (Leff, Stemmer, Tyrrell, & Jog, 2018). Environmental contamination with lead originates from industrial emissions, mining, wastewater discharge, and atmospheric deposition. Once in the environment, lead can persist, accumulating in soils and sediments over time. Ecotoxicological research has shown that lead negatively impacts plants, terrestrial organisms, and aquatic species by impairing growth, reproduction, and metabolic processes (Mousavi, Brodie, Payghamzadeh, Raiesi, & Strivastava, 2022).

Regulatory actions over the past several decades, such as removing lead from gasoline, have significantly decreased atmospheric lead concentrations. However, lead remains present in various household and industrial products, including paints, plumbing materials, pipes, glassware, batteries, cosmetics, and ammunition. The primary sources of human lead exposure are fossil fuel combustion, metal smelting, and various industrial activities (Kumar et al., 2020). Lead emissions into the atmosphere occur primarily as lead sulfate, oxides, and carbonates, with airborne particles depositing slowly. Consequently, inhalation is a primary route of exposure, particularly in industrial and urban environments. Lead exposure is particularly hazardous to children due to their greater absorption efficiency and increased risk of neurotoxic effects (Rahman & Singh, 2019). Prenatal exposure to lead is especially concerning as it can interfere with fetal development. To address these concerns, the US Centers for Disease Control and Prevention has established a blood lead reference level of 3.5 μg/dL for children aged 1–5 years, highlighting the need for continued monitoring and reduction of lead exposure risks (Lanphear et al., 2019).

19.2 Biomarkers of heavy metal neurotoxicity

19.2.1 Reactive oxygen species as biomarkers

Heavy metal exposure impairs mitochondrial function by affecting the electron transport chain (ETC), resulting in electron leakage and the subsequent univalent reduction of molecular oxygen, generating superoxide anions (O_2^-). These superoxide radicals undergo further conversion into hydrogen peroxide (H_2O_2) and hydroxyl radicals (•OH) by the Fenton and Haber–Weiss reactions, worsening oxidative stress and leading to cellular damage. The excessive formation of ROS produces oxidative changes in critical macromolecules, including DNA, proteins, and lipids, ultimately inducing neuronal malfunction and apoptosis (Zorov, Juhaszova, & Sollott, 2014). Oxidative stress indicators like malondialdehyde (MDA), 4-hydroxynonenal (4-HNE), and protein carbonylation are routinely found under these conditions. Concurrently, there is frequently a compromise in the antioxidant defense system, which includes enzymes like glutathione peroxidase (GPx), catalase (CAT), and superoxide dismutase (SOD). This impairs the cell's capacity to neutralize ROS and exacerbates oxidative damage. Long-term exposure to heavy metals causes chronic oxidative stress, which is a significant contributing factor to the etiology of neurodegenerative diseases like ALS, PD, and AD. Prolonged ROS buildup triggers proinflammatory signaling pathways, including mitogen-activated protein kinases (MAPKs) and nuclear factor kappa B (NF-κB), exacerbating neuronal degeneration and promoting inflammation. Heavy metals are environmental risk factors that contribute to the course of neurodegenerative diseases, as evidenced by the intricate interactions among oxidative stress, mitochondrial dysfunction, and inflammatory reactions (Ayala, Muñoz, & Argüelles, 2014; Dash et al., 2024; Jomova et al., 2023).

19.2.2 Lipid peroxidation as a biomarker

Lipid peroxidation, mediated by ROS, is crucial in neuronal membrane degradation during heavy metal-induced neurotoxicity. Neuronal membranes are especially sensitive due to their high amount of polyunsaturated fatty acids (PUFAs), which are highly prone to oxidative destruction. These PUFAs are the specific target of ROS, which sets off a chain reaction that produces harmful lipid peroxidation byproducts such as 4-HNE and MDA (Su et al., 2019). These reactive aldehydes harm proteins, nucleic acids, and other cellular constituents by forming adducts with them, which also causes apoptosis. Because they throw off the equilibrium between the prooxidant and antioxidant systems, heavy metals like mercury (Hg) and cadmium (Cd) make lipid peroxidation worse. The thiol (–SH) groups of antioxidant molecules are bound by these metals, which causes glutathione (GSH), an essential cellular antioxidant, to be depleted. Furthermore, heavy metals exacerbate oxidative damage by blocking the action of necessary antioxidant enzymes such as glutathione peroxidase (GPx), catalase (CAT), and superoxide dismutase (SOD). Consequently, the intensity of lipid peroxidation increases, causing alterations in ion homeostasis, decreased membrane fluidity, and increased membrane permeability (Barrera et al., 2018). This sequence of events compromises mitochondrial function, inducing the release of proapoptotic proteins, activation of caspases, and, finally, neuronal cell death. Lipid peroxidation is a crucial component of the pathogenesis of neurological disorders. Elevated levels of MDA and 4-HNE have been discovered in the brain, cerebrospinal fluid (CSF), and plasma of persons with AD and PD, highlighting their importance as indicators of oxidative stress. These aldehydes cause protein misfolding and aggregation, typical aspects of neurodegeneration. Alpha-synuclein (α-syn) aggregates more readily into Lewy bodies in PD, whereas 4-HNE promotes the development of amyloid-beta (Aβ) plaques in AD. These pathogenic aggregates lead to neuronal dysfunction and exacerbate disease progression (Bassil et al., 2020; Zheng, Sun, Luo, Li, & Huang, 2024).

19.3 Neuroinflammatory markers in neurodegenerative disease pathogenesis

19.3.1 Tumor necrosis factor-α

A key component of the brain's immune response, neurones and glial cells, including microglia and astrocytes, release the proinflammatory cytokine tumor necrosis factor-alpha (TNF-α). TNF-α is vital in controlling acute-phase inflammation, starting cytokine signaling pathways fundamental to the inflammatory process. Beyond its involvement in inflammation and injury, TNF-α has diverse physiological functions, including regulating immune responses, haematopoiesis, cardiovascular health-modulating dendritic cell maturation, synaptic plasticity, and sensory stimuli processing (Olmos & Lladó, 2014).

Moreover, TNF-α regulates α-amino-3-hydroxy-5-methyl-4-isoxazole propionic acid (AMPA) receptor signaling, essential for synaptic plasticity and function. By increasing the surface expression of AMPA receptors, TNF-α increases synaptic efficiency. Disruption of TNF-α-mediated signaling results in diminished synaptic activity, revealing the dual role of TNF-α in neuronal function. In the context of injury, TNF-α initially protects the brain by counteracting infections, neuronal damage, and toxicity (Freudenberg, Celikel, & Reif, 2015). But persistent inflammation, gliosis, synaptic loss, and glutamate toxicity are caused by chronic or dysregulated TNF-α production, which also contributes to HIV-associated dementia, PD, AD, and multiple sclerosis (MS). These disorders advance more quickly when TNF-α levels are elevated, making it a possible target for therapeutic therapies meant to lessen neuroinflammation and neuronal damage (Jung, Tweedie, Scerba, & Greig, 2019).

19.3.2 Interleukin-1β

An important inflammatory cytokine implicated in the pathophysiology of neurodegenerative illnesses such as MS is interleukin-1β (IL-1β). IL-1β is produced by neutrophils and monocyte-derived macrophages in models of experimental autoimmune encephalomyelitis, promoting their migration into the central nervous system (CNS) and intensifying the inflammatory response. Microglia and astrocytes, the key glial cells implicated in neuroinflammation, are highly responsive to IL-1β, which induces subsequent inflammatory cascades in the brain (Mendiola & Cardona, 2018).

IL-1β stimulates the production of cytokines, chemokines, adhesion molecules, and matrix metalloproteinases in astrocytes, contributing to the escalation of neuroinflammation. Additionally, IL-1β enhances IL-6 production and induces inducible nitric oxide synthase (iNOS) activity, further driving neuroinflammation. IL-1β also activates acetylcholinesterase (AChE) activity and amplifies microglial activation, worsening neuroinflammatory processes. As IL-1β levels rise, they promote a cycle of astrocyte activation, marked by the expression of the S100β protein, a biomarker for astrocyte activation. This perpetuates neuroinflammation, with IL-6 playing a critical role in enhancing astrogliosis, microglial activation, and the production of acute-phase proteins, all of which contribute to the progression of neurodegenerative diseases (Chen et al., 2024).

19.3.3 Nitric oxide (NO)

Nitric oxide (NO) is a tiny, bioactive molecule created by the enzyme nitric oxide synthase (NOS), which catalyzes L-arginine conversion into L-citrulline. Endothelial NOS (eNOS), neuronal NOS (nNOS), and inducible NOS (iNOS) are the three different kinds of NOS. While eNOS and nNOS control vasodilation and CNS functioning, iNOS is usually activated in response to inflammatory stimuli. iNOS is expressed in numerous cells, including microglia, astrocytes, neurons, and macrophages. Upon activation, iNOS creates higher quantities of NO and its reactive byproduct, peroxynitrite, which is highly destructive to cellular structures. These reactive species cause oxidative stress, impairing neuronal cells, blood–brain barrier (BBB) integrity, and synaptic function, all of which contribute to neurodegenerative illnesses such as MS, AD, and PD (Förstermann & Sessa, 2012; Król & Kepinska, 2020).

19.3.4 Cyclooxygenases (COX)

Cyclooxygenases (COX) convert arachidonic acid into prostanoids, which regulate inflammation and other physiological functions. Among the COX isoforms, COX-1 and COX-2 are particularly implicated in the pathology of neurodegenerative diseases. COX-2, in particular, is highly responsive to proinflammatory signals like TNF-α, IL-1β, and IL-2. COX-2 expression is induced in certain brain regions, particularly in specific neuronal populations, during the early stages of neurodegenerative diseases, suggesting its role in initiating or exacerbating neuroinflammation and neuronal damage (Juan, Pérez de la Lastra, Plou, & Pérez-Lebeña, 2021; Wang, Wu, Chen, et al., 2021; Wang, Wu, Bai, et al., 2021).

19.4 Metal homeostasis-associated proteins

19.4.1 Metallothioneins

Metallothioneins (MTs) are tiny, cysteine-rich proteins that perform a crucial function in detoxifying heavy metals and reducing oxidative stress. These proteins are essential in neurodegenerative illnesses such as PD and AD, where heavy metal poisoning causes neuronal death. MTs aid detoxification by adhering to hazardous metals such as cadmium, lead, and mercury, lowering their bioavailability and minimizing cell damage (Ruttkay-Nedecky et al., 2013). This relationship reduces inflammation, mitochondrial dysfunction, and oxidative stress. Additionally, MTs operate as free radical scavengers, limit lipid peroxidation, and regulate microglial activation, all diminishing neuroinflammation and preventing neuronal death. Additionally, MTs aid in regulating vital metals such as copper and zinc, which are necessary for preserving synaptic function and promoting neuroprotection (Miyazaki & Asanuma, 2023). Heavy metal exposure has been related to neurological disease development due to their capacity to produce ROS and increase protein aggregation. For instance, PD's signature α-synuclein aggregation is linked to elevated iron and zinc levels in the substantia nigra. MTs have shown potential as therapeutic targets in neurodegenerative disorders because of their ability to counteract the toxicity of α-synuclein and prevent metal-induced neurodegeneration. Studies in MT-null mice have revealed heightened vulnerability to oxidative stress from heavy metals, with increased neurotoxicity, lipid peroxidation, and neuroinflammation compared to wild-type controls (Pyatha, Kim, Lee, & Kim, 2022). Furthermore, MTs are known to promote neuronal survival through the activation of anti-apoptotic pathways, suggesting that targeting MTs could offer therapeutic benefits for mitigating heavy metal toxicity in neurodegenerative diseases. Despite their protective role, the exact function of MTs in the nervous system remains incompletely understood. Their expression and regulation vary across different brain regions, indicating a more complex role in neuroprotection. Future research should aim to clarify the mechanisms through which MTs interact with heavy metals and influence neurodegenerative pathways, potentially leading to the development of novel therapeutic strategies to combat metal-induced neurotoxicity (Chen, Miah, & Aschner, 2016).

19.4.2 Divalent metal transporter 1

Divalent metal transporter 1 (DMT1), also known as natural resistance-associated macrophage protein 2 (NRAMP2) or solute carrier family 11 member 2 (SLC11A2), is a key transporter involved in the uptake and regulation of divalent metal ions such as iron (Fe^{2+}), manganese (Mn^{2+}), and harmful heavy metals like lead (Pb^{2+}), cadmium (Cd^{2+}), and mercury (Hg^{2+}). DMT1 has a key function in metal homeostasis, specifically in the absorption of dietary iron in the duodenum, recycling of iron from red blood cells in macrophages, and metal transport in neuronal cells. Although DMT1 is primarily known for its role in enterocyte iron transport, it is also crucial to intake of hazardous metals. Studies have demonstrated that DMT1 expression is elevated in response to iron shortage, leading to a larger intake of dangerous metals—and worsening their neurotoxic effects (Aizenman & Mastroberardino, 2015). Specifically, DMT1 has been related to the buildup of cadmium and lead in the CNS, contributing to oxidative stress, mitochondrial dysfunction, and neurodegeneration. Elevated expression of DMT1 has been seen in brain regions affected by neurodegenerative illnesses such as PD and AD. In PD, DMT1-mediated iron absorption leads to iron dysregulation in the substantia nigra, leading in the destruction of dopaminergic neurons. Furthermore, iron overload caused by increased DMT1 expression has been shown to enhance α-synuclein aggregation, which is characteristic of PD pathology. Higher iron levels in the hippocampus and cortex are associated with amyloid-β plaque formation and tau hyperphosphorylation in AD, suggesting DMT1 involvement in disease progression (Ingrassia, Garavaglia, & Memo, 2019; Venkatesan et al., 2024).

The expression of DMT1 is tightly regulated through the iron regulatory protein/iron responsive element system, ensuring that DMT1 is upregulated when iron is low and downregulated when iron is sufficient to prevent excess accumulation. However, in pathological conditions, such as chronic exposure to heavy metals or neurodegeneration, this regulatory mechanism may become disrupted, leading to persistent DMT1 upregulation and metal toxicity (Skjørringe, Burkhart, Johnsen, & Moos, 2015). Due to its involvement in metal toxicity and neurodegeneration, DMT1 is being considered as a potential therapeutic target. Strategies like small-molecule inhibitors, RNA interference (RNAi), and gene-editing approaches are being explored to regulate DMT1 activity and mitigate metal-induced oxidative stress. Additionally, iron chelation therapies, such as deferoxamine (DFO) and deferiprone (DFP), have shown potential in reducing iron overload and alleviating neurodegeneration in PD and AD models (Shawki, Knight, Maliken, Niespodzany, & Mackenzie, 2012) (Table 19.1).

19.5 Impact of heavy metal toxicity on key neurodegenerative disorders

19.5.1 Alzheimer's disease

AD is a progressive neurodegenerative ailment that mostly affects adults aged 65 and older, constituting the most prevalent neurological disorder globally. It presently effects over 50 million individuals, with forecasts showing a rise to 152 million by 2050 (Richetti et al., 2011). AD is characterized by gradual cognitive decline, memory difficulties, and impairments in daily functioning, substantially affecting the quality of life (Singh, Day, Abdella, & Garg, 2024). The pathophysiology of AD is complicated and diverse, involving hereditary components; however, environmental impacts, such as exposure to heavy metals, solvents, and pesticides, have attracted increasing attention in recent studies. Heavy metals, such as iron, copper, and zinc, are among these environmental risk factors that have been linked to the development and course of AD through processes like metal dyshomeostasis and oxidative stress induction. The "amyloid hypothesis" posits that metal-induced disruption of metal homeostasis accelerates the aggregation of beta-amyloid plaques, a defining pathological feature of AD (Bakulski, Hu, & Park, 2020). These plaques, alongside neurofibrillary tangles (NFTs), contribute to neuronal injury and cognitive deterioration. Oxidative stress, driven by ROS generated from metal dysregulation, exacerbates neuronal damage. Furthermore, the dysregulated metabolism of metals leads to cellular dysfunction and neuronal apoptosis, further accelerating disease progression. Recent reviews emphasize the role of environmental exposures, such as air pollution and heavy metals, in triggering or aggravating AD through various molecular pathways. While early-onset AD is typically hereditary, the majority of cases occur in the elderly. Despite extensive research, effective treatments to halt the progression of AD remain elusive. A deeper understanding of the pathogenic mechanisms underlying metal-induced AD, particularly in relation to environmental factors such as metal exposure, is crucial. Identifying reliable biomarkers associated with metal toxicity could offer valuable insights into AD's pathogenesis and present novel opportunities for early disease detection and monitoring (Kabir et al., 2021).

19.5.1.1 Cadmium and Alzheimer's disease

Amyloid-beta (Aβ) aggregation is a critical characteristic in the development of AD. The combination of cadmium (Cd) with Aβ has been connected with an increased risk of AD, with specific metals leading to the production of NFTs (Xu et al., 2018). Specifically, cadmium has been shown to interact with Aβ1–42, a critical component of senile plaques (SPs) that significantly accelerates AD progression. This interaction occurs when cadmium binds to specific tyrosine and histidine residues on the N-terminal region of Aβ1–42, potentially blocking the ion channel activity of the peptide. Moreover, cadmium exposure can alter the expression of various proteins, further raising the risk of AD. For example, cadmium inhibits M1 receptors, resulting in reduced AChE-R levels and an overproduction of AChE-S. In the basal forebrain, cadmium exposure can trigger cell death in cholinergic neurons, mimicking the neurodegenerative processes observed in AD (Bakulski et al., 2020).

These disruptions are accompanied by increased levels of Aβ, glycogen synthase kinase 3 beta (GSK3β), and tau

TABLE 19.1 Biomarkers of heavy metal–induced neurotoxicity: detection and role in neurodegenerative diseases.

Category	Biomarker	Biological role	Associated heavy metals	Related diseases	Detection method	References
Oxidative stress and lipid peroxidation	ROS	Causes oxidative damage by disrupting mitochondrial function, leading to neuronal apoptosis	Pb, Hg, Cd, As	AD, PD, ALS	Fluorescence assays, ESR spectroscopy	Singh, Kukreti, Saso, and Kukreti (2019)
	MDA	Lipid peroxidation end-product, marker of oxidative stress and membrane damage	Pb, Hg, Cd, As	AD, PD, ALS	TBARS assay, HPLC	Singh et al. (2019)
	4-HNE	Marker of lipid peroxidation, disrupts protein function and induces neuroinflammation	Pb, Hg, Cd	AD, PD	ELISA, Western blot	Korkmaz, Butuner, and Roggenbuck (2018) and Zhong and Yin (2015)
	8-OHdG	Marker of DNA oxidative damage, associated with neurodegeneration	Pb, Hg, Cd	AD, PD	ELISA, HPLC–MS	Korkmaz et al. (2018)
	SOD	Antioxidant enzyme that neutralizes superoxide radicals, altered by metal toxicity	Pb, Hg, Cd, As	AD, PD, ALS	Spectrophotometric assays, Western blot	Zhong and Yin (2015) and Zitka et al. (2012)
	Glutathione (GSH/GSSG ratio)	Key antioxidant defense, decreased in oxidative stress	Pb, Hg, Cd	AD, PD	HPLC, fluorescent probes	Younus (2018) and Zitka et al. (2012)
Neuroinflammation and immune activation	IL-1β	Proinflammatory cytokine, involved in chronic neuroinflammation and microglial activation	Pb, Hg, Cd, As	AD, PD, ALS	ELISA, RT-qPCR	Matousek et al. (2012)
	TNF-α	Regulates apoptosis and neuroinflammation, elevated in heavy metal exposure	Pb, Hg, Cd	AD, PD	ELISA, immunohistochemistry	Matousek et al. (2012) and Plantone et al. (2023)
	IL-6	Induces neuroinflammation and BBB disruption	Pb, Hg, Cd	AD, PD	ELISA, RT-qPCR	Plantone et al. (2023)
	Iba1 (Ionized calcium-binding adapter molecule 1)	Microglial activation marker, indicative of neuroinflammation	Pb, Hg, Cd	AD, PD	Immunofluorescence, Western blot	Zhang et al. (2021)
	GFAP	Astrocyte activation marker, reflects neuroinflammatory and glial reactivity	Pb, Hg, Cd	AD, PD	Immunohistochemistry, Western blot	Yang and Wang (2015)

(Continued)

TABLE 19.1 (Continued)

Category	Biomarker	Biological role	Associated heavy metals	Related diseases	Detection method	References
Metal homeostasis and transport	MTs	Metal-binding proteins involved in detoxification and homeostasis	Pb, Hg, Cd, As	AD, PD, ALS	ELISA, Western blot	Li et al. (2017)
	DMT1	Facilitates metal ion transport, altered in heavy metal exposure	Pb, Cd	AD, PD	Western blot, immunohistochemistry	Li et al. (2017) and Xu et al. (2024)
	Ferroportin (FPN)	Iron exporter, dysregulated in neurodegenerative diseases	Pb, Hg, Cd	AD, PD	ELISA, RT-qPCR	Raha et al. (2022)
	Ferritin	Iron storage protein, increased levels indicate metal accumulation	Pb, Hg, Cd	AD, PD	ELISA, MRI	Kell and Pretorius (2014)
	Transferrin	Iron transport protein, altered in neurotoxicity and neurodegeneration	Pb, Hg	AD, PD	ELISA, Western blot	Bloom (2014)
Neuronal damage and protein aggregation	Amyloid-β (Aβ)	Forms plaques in AD, promotes oxidative stress and neuronal toxicity	Pb, Hg	AD	ELISA, immunohistochemistry	Bloom (2014)
	Tau protein	Hyperphosphorylated tau forms neurofibrillary tangles, linked to heavy metal exposure	Pb, Hg	AD	ELISA, Western blot	Bloom (2014)
	α-Synuclein	Aggregates in PD, its oligomers are neurotoxic and metal-sensitive	Pb, Hg	PD	ELISA, immunohistochemistry	Bloom (2014) and Tamagno, Guglielmotto, Vasciaveo, and Tabaton (2021)
	UPS proteins	Impaired protein degradation due to metal-induced proteotoxic stress	Pb, Hg	AD, PD, ALS	Western blot, immunoprecipitation	Kandel, Jung, and Neal (2024)
Glutamate excitotoxicity and neurotransmitter dysfunction	NMDA receptor subunits (NR1, NR2A, NR2B)	Glutamate receptor subunits, affected by metal toxicity, leading to excitotoxicity	Pb, Hg	AD, PD	Western blot, Patch-clamp electrophysiology	Nicosia, Giovenzana, Misztak, Mingardi, and Musazzi (2024)
	DAT	Regulates dopamine reuptake, inhibited in heavy metal exposure, leading to PD pathology	Pb, Hg	PD	Western blot, PET imaging	Richetti et al. (2011)
	AChE	nzyme regulating acetylcholine levels, inhibited by heavy metals, affecting cognition	Pb, Hg	AD	Enzyme activity assays	Richetti et al. (2011)

4-HNE, 4-Hydroxynonenal; *8-OHdG*, 8-hydroxy-2′-deoxyguanosine; *AChE*, acetylcholinesterase; *AD*, Alzheimer's disease; *ALS*, amyotrophic lateral sclerosis; *BBB*, blood–brain barrier; *DAT*, dopamine transporter; *DMT1*, divalent metal transporter 1; *ESR*, electron spin resonance; *GFAP*, glial fibrillary acidic protein; *IL-1β*, interleukin-1β; *MDA*, malondialdehyde; *MTs*, metallothioneins; *PD*, Parkinson's disease; *ROS*, reactive oxygen species; *SOD*, superoxide dismutase; *TBARS*, thiobarbituric acid reactive substances; *TNF-α*, tumor necrosis factor-α; *UPS*, ubiquitin-proteasome system.

filament formation—hallmarks of AD pathology. GSK3β, a critical component in the formation of tau paired helical filaments found in NFTs, serves as a marker for neuronal degeneration in AD. Studies show that cadmium exposure is linked to elevated metal levels in plasma, liver, and brain tissues of AD patients, as well as in hair and blood samples (Bakulski et al., 2020). Furthermore, workers exposed to cadmium often report cognitive impairments, including memory deficits, reduced psychomotor speed, and attention issues. In animal models, such as APP/PS1 transgenic mice, cadmium exposure induces memory and learning deficits, characteristic of AD. These findings suggest that cadmium contributes to the pathogenesis of AD by promoting the aggregation of Aβ and the formation of NFTs (Amaral et al., 2021; Reddy, 2013).

19.5.1.2 Mercury and Alzheimer's disease

Tubulin proteins polymerize into long chains or filaments, forming the structural framework of microtubules (Siblerud, Mutter, Moore, Naumann, & Walach, 2019). Mercury (Hg) disrupts this process by binding to tubulin with high affinity, leading to structural destabilization and impaired polymerization. This inhibition prevents microtubule formation, contributing to the development of NFTs and SPs, both of which are hallmark neuropathological features of AD (Azar, Yousef, El-Fawal, & Abdelnaser, 2021). Experimental studies on neuronal cells have demonstrated that mercury exposure induces axonal degeneration and promotes NFT formation. Similarly, investigations using stem cells revealed that mercury contributes to neuronal apoptosis and inhibits tubulin activity. Furthermore, mercury has been implicated in tau hyperphosphorylation, disrupting microtubule stabilization within neurons (Zhang, Wang, Ziober, Zhang, & Bagg, 2021). Mercury-induced oxidative stress (OS) further exacerbates tau phosphorylation, increasing its pathological burden. Given the interplay between tau and amyloid-β (Aβ), mercury exposure may exacerbate both Aβ pathology and tau-Aβ interactions, thereby accelerating AD progression. Moreover, Aβ accumulation has been shown to induce tau hyperphosphorylation, suggesting that Aβ deposition may activate intracellular signaling pathways that promote tau dysfunction. Dysregulation of tau is increasingly recognized as a critical factor in AD pathogenesis (Stancu, Vasconcelos, Terwel, & Dewachter, 2014). Additionally, mercury has been reported to influence the expression of the amyloid precursor protein (APP) gene, further implicating its role in Aβ pathology. The investigated effects of mercury on Aβ accumulation demonstrated that both mercury and methylmercury upregulated APP expression, thereby enhancing Aβ production. Concurrently, neprilysin (NEP), a key Aβ-degrading enzyme, exhibited reduced expression in PC12 cells following exposure to mercury and methylmercury, indicating that mercury facilitates Aβ accumulation by increasing APP levels while suppressing Aβ clearance (Wallin et al., 2020).

Methylmercury-induced astrogliosis has also been observed in AD pathology, with glial activation contributing to increased APP expression. Elevated mercury concentrations have been detected in the blood and brain tissue of AD patients, with additional findings indicating increased mercury levels in hair samples. Notably, individuals with neurodegenerative disorders exhibit significantly higher mercury concentrations compared to control subjects. Mercury exposure within the nervous system has been linked to cognitive impairments, including dementia, attention deficits, and memory loss. In a study by Haut et al., workers with occupational exposure to mercury vapor exhibited elevated blood mercury levels and cognitive dysfunction, further supporting the neurotoxic role of mercury (Haut et al., 1999).

19.5.1.3 Lead and Alzheimer's disease

Numerous studies have explored the molecular link between lead (Pb) exposure and AD, with oxidative DNA damage emerging as a significant contributing factor. As a hallmark of aging, oxidative DNA damage is also implicated in AD pathogenesis. Exposure to lead triggers oxidative stress (OS), which in turn elevates amyloid-beta (Aβ) levels, exacerbating neuronal oxidative damage and accelerating neurodegeneration. Additionally, OS-driven apoptotic mechanisms contribute to Aβ accumulation, further aggravating neuronal dysfunction (Bakulski, Rozek, Dolinoy, D., L. Paulson, & Hu, 2012). Lead-induced oxidative DNA damage has been associated with an imbalance between the accumulation of 8-hydroxy-2′-deoxyguanosine (oxo8dG) and the activity of 8-oxoguanine DNA glycosylase 1 (Ogg1), an enzyme responsible for repairing oxidative DNA lesions. Bolin et al. observed that oxo8dG levels initially increased during early postnatal development (postnatal day 5) and surged significantly two decades after the cessation of lead exposure, despite Ogg1 activity remaining unchanged. Interestingly, this study also reported an age-related decline in the inverse correlation between oxo8dG accumulation and Ogg1 function. In older animals, lead exposure did not significantly alter oxo8dG levels, implying that oxidative DNA damage in aging brains may be influenced by early-life environmental insults. Since oxo8dG accumulation is widely regarded as a biomarker for oxidative DNA damage, its presence may play a pivotal role in AD progression (Bolin et al., 2006).

Additionally, it has been demonstrated that exposure to lead changes the expression of genes associated with AD pathogenesis, specifically APP and β-site APP-cleaving enzyme 1 (BACE1), which are essential for Aβ production.

Proteolytic cleavage of APP produces Aβ peptides that aggregate into plaques, a key pathological feature of AD. Bolin et al. (2006) demonstrated that lead exposure during early development leads to epigenetic modifications, specifically hypomethylation of the APP gene, which enhances its transcriptional activity. This epigenetic deregulation results in excessive APP production, leading to an intracellular buildup of the protein. Elevated APP levels subsequently activate specificity protein 1 (Sp1), a transcription factor that regulates genes involved in AD pathology, ultimately facilitating Aβ aggregation and plaque formation in the brain (Bakulski et al., 2012). In hippocampal cell cultures, chronic exposure to low doses of lead was also found to upregulate APP expression. Similarly, Wu et al. provided evidence that lead exposure induces AD-like pathology in primates, as immunohistochemical analysis revealed intracellular Aβ accumulation and dense-core plaque formation, further supporting the neurotoxic role of lead in AD progression (Shreenivasa, Velayudhannair, Arockiasamy, Shama, & Sivakumar, 2023). Chronic occupational lead exposure remains a critical concern, particularly in industries such as battery manufacturing, where workers are subjected to significantly higher lead levels compared to the general population. Research indicates that these workers frequently experience neuropsychological impairments, including headaches, dizziness, memory deficits, and paresthesia, symptoms that closely resemble those observed in AD patients. Additionally, elevated blood lead levels have been documented in these individuals. Rahman et al. (2021) reported that lead-exposed industrial workers exhibited structural brain abnormalities associated with schizophrenia, along with functional alterations in key brain regions such as the frontal cortex and hippocampus areas critically affected in AD. These findings suggest that lead exposure may contribute to the development of neurocognitive disorders, including AD and schizophrenia, both of which are characterized by cognitive dysfunction and neurodegeneration.

19.5.1.4 *Arsenic and Alzheimer's disease*

The neurotoxic consequences of arsenic exposure remain incompletely explained at the molecular level; however, various ideas propose its participation in the etiology of neurodegenerative illnesses such as AD. One potential mechanism implicates arsenic-induced oxidative stress, where ROS damage cellular macromolecules, including lipids, proteins, and DNA, thereby affecting neural integrity. Studies in rodent models have demonstrated that arsenic exposure leads to oxidative DNA damage and impairs antioxidant defense mechanisms, particularly in the hippocampus, striatum, and cortex. This oxidative stress-mediated dysfunction includes the downregulation of ATP synthase and enhanced lipid peroxidation, contributing to neurodegenerative processes (Vázquez Cervantes et al., 2023). Epidemiological and experimental studies have found a link between arsenic exposure at levels ranging from 0.005 to 0.02 ppm in drinking water or 0.01 to 0.05 ppm in environmental settings and increased oxidative modifications to DNA and proteins, as well as increased neuroinflammatory responses in both humans and animal models. These findings support the notion that arsenic-induced oxidative stress may contribute to AD pathogenesis by boosting the buildup of oxidized metabolites, including damaged nucleic acids, lipids, and proteins, in the brains of affected patients (Rahman et al., 2021).

Moreover, arsenic exposure has been associated with cognitive impairments, including deficits in learning and memory, as demonstrated in animal studies. For instance, prenatal exposure to arsenic (0.75 mg/kg) resulted in neurobehavioral abnormalities in offspring (Zhou, Yi, Wang, Lim, & Tan, 2023). Similarly, exposure to inorganic arsenic (20 mg/L) during gestation and early postnatal development led to pronounced spatial memory deficits in murine models. These cognitive impairments may be attributed to arsenic-mediated disruptions in amyloidogenic pathways, which are implicated in AD-related neurodegeneration. Further research has revealed that exposure to organic arsenic compounds, such as dimethylarsinic acid (DMA) (5–10 μM for 12–24 hours), increases amyloid-beta (Aβ) levels in cholinergic SN56.B5.G4 cells, indicating that arsenic may contribute to Aβ accumulation within the brain (Zhou et al., 2023). Similar effects have been observed in the Tg2576 mouse model, which overexpresses a mutant APP, a widely used model for AD research. The elevation in Aβ levels following DMA exposure is believed to be associated with increased APP expression, while Aβ degradation mechanisms appear largely unaffected. Although the precise molecular pathways underlying arsenic-induced Aβ accumulation remain to be fully elucidated, oxidative stress and neuroinflammation likely play key roles, further supporting the hypothesis that arsenic exposure contributes to the molecular pathogenesis of AD (Vázquez Cervantes et al., 2023).

19.5.2 Parkinson's disease

PD is a progressive neurological condition largely impairing motor function due to the degeneration of dopaminergic neurons in the substantia nigra. This neuronal degradation results in typical symptoms, including tremors, bradykinesia (slowness of movement), muscle rigidity, and balance deficits. The prevalence of PD increases with age, affecting roughly 1%–2% of those over 60 and rising to 3%–8% among those aged 85–89. Epidemiological studies reveal a higher frequency in men, with a male-to-female ratio of roughly 1.5:1

(Channer et al., 2023). Under normal physiological conditions, dopaminergic neurons facilitate communication between brain regions by transmitting signals through dopamine release at the synapse. Dopamine binds to specific receptors on the postsynaptic neuron, ensuring proper neural signaling. Once this process is complete, dopamine is either reabsorbed by the presynaptic neuron or enzymatically degraded by monoamine oxidase B and catechol-*O*-methyltransferase to maintain synaptic homeostasis (Ramesh & Arachchige, 2023). In PD, the progressive loss of dopamine-producing neurons disrupts motor regulation, leading to a significant depletion of dopamine levels. This reduction impairs neural communication between the substantia nigra, basal ganglia, and other motor-associated regions. Once dopamine levels fall below approximately 20% of their normal concentration, clinical symptoms such as tremors, rigidity, and bradykinesia become evident (Yoshida et al., 2022).

The basal ganglia and cerebellum are critical structures involved in coordinating smooth and controlled movement. In a healthy brain, neural signals are efficiently transmitted from the CNS to the muscles. However, in PD, the deficiency of dopamine leads to hyperactivity in specific regions of the basal ganglia, causing inhibition of the thalamus and subsequent motor impairments, such as tremors. Additionally, the dopaminergic deficit triggers compensatory mechanisms, including increased glutamate signaling, which further exacerbates neurodegeneration and contributes to disease progression (Wu et al., 2024).

19.5.2.1 Mercury and Parkinson's disease

Mercury (Hg) is a highly toxic transition metal that exists in multiple forms, including elemental mercury (Hg^0), organic mercury (methylmercury, MeHg), and inorganic mercury (Hg^{2+} and Hg^{+}). While all mercury species pose significant health hazards, MeHg has been particularly associated with neurodegenerative disorders, including PD. Environmental exposure to Hg arises from both natural sources, such as volcanic eruptions and atmospheric deposition, and anthropogenic activities, including emissions from coal combustion, industrial waste incineration, and artisanal gold mining (Novo et al., 2021). Among the various forms of mercury, inorganic mercury is especially neurotoxic and is found in high concentrations in aquatic ecosystems. It readily penetrates the CNS by crossing the BBB through an L-cysteine-mediated transport mechanism, utilizing the neutral L-type amino acid carrier system. Once inside the brain, Hg disrupts protein synthesis by impairing ribosomal function and interfering with endoplasmic reticulum activity. Furthermore, MeHg is capable of transplacental transfer, with experimental evidence demonstrating substantial mercury accumulation in fetal mouse brains following maternal exposure, underscoring its potential role as a developmental neurotoxicant (Ke et al., 2019).

The neurotoxic effects of Hg are mediated by numerous cellular pathways, including oxidative stress, mitochondrial malfunction, and free radical production. Hg exposure depletes glutathione (GSH), a key antioxidant, consequently affecting cellular redox equilibrium. This deficiency leads to the increased production of ROS, which contribute to oxidative damage in neurons. Additionally, Hg interferes with critical metalloproteins by displacing zinc and copper ions, inhibiting normal sulfur binding and further worsening neurotoxicity. The inflammatory response generated by Hg includes elevation of proinflammatory cytokines such as TNF-α, which increases apoptosis and neuroinflammation, both of which are linked in PD pathogenesis. Epidemiological studies reveal a connection between Hg exposure and an increased risk of PD (Torrey & Simmons, 2023). Individuals with prolonged exposure to Hg vapors, such as those with occupational exposure in industrial settings or dental amalgam fillings, have been reported to exhibit PD-like symptoms, including tremors, bradykinesia, cognitive impairments, and cortical atrophy. Notably, research indicates that PD patients have significantly elevated Hg concentrations in their blood up to six times higher than in healthy controls. This observation supports the hypothesis that Hg acts as an environmental risk factor contributing to PD pathogenesis. At the molecular level, Hg-induced neurotoxicity is associated with disturbances in glutamate homeostasis. MeHg exposure inhibits astrocytic glutamate transporters, resulting in excessive extracellular glutamate buildup, which overstimulates *N*-methyl-D-aspartate (NMDA) receptors. This hyperactivation results in higher intracellular calcium levels, causing nitric oxide generation, mitochondrial malfunction, and oxidative damage. The oxidative changes of neuronal proteins further increase cellular dysfunction, contributing to the gradual degradation of dopaminergic neurons, a characteristic of PD (Nag & Cummins, 2022).

19.5.2.2 Lead and Parkinson's disease

Lead (Pb) is a very poisonous metal that has become a serious environmental problem due to its widespread use in many industries and goods. Although Pb is not necessary for the human body, it causes major health concerns, particularly to the neurological system, kidneys, bones, and heart. Research has demonstrated that continuous exposure to Pb might dramatically raise the chance of acquiring neurodegenerative disorders such as PD. Studies have indicated that those with higher levels of Pb exposure had a two- to threefold increased risk of PD compared to those with lower levels of exposure. For instance, a research in the United States found that persons with the greatest levels of lifetime Pb exposure were twice as likely to develop PD as those with the least exposure (Ortega et al., 2021). Children are more prone to Pb toxicity due to their developing BBB, which makes it easier for Pb to enter the brain. Studies have revealed that Pb exposure in early life can lead to long-term neurological, behavioral, and genetic alterations, enhancing the probability of developing neurodegenerative disorders as individuals grow older. Pb interferes with calcium, an essential element for various cellular processes, including neuronal signaling, myelination, and the functioning of glial cells. Pb disrupts these processes by mimicking calcium and entering neurons and glial cells through calcium channels (Rizzi & Tan, 2017).

Once Pb enters the brain, it can disrupt the release and reuptake of key neurotransmitters, such as dopamine and acetylcholine. These neurotransmitters are critical for normal brain function and movement. Pb's disruption of dopamine pathways is especially harmful in PD, as it leads to an imbalance of dopamine levels in the brain, impairing motor control. This disruption damages dopaminergic neurons and affects dopamine transport, contributing to the neurodegenerative changes seen in PD (Clemente-Suárez et al., 2023). Pb exposure also interacts with various other neurotransmitter systems, including those that employ acetylcholine, norepinephrine, and GABA. This interference adds to oxidative stress, mitochondrial malfunction, and damage to the cells that support neurons, such as astrocytes. Pb decreases mitochondrial energy production by inhibiting NMDA receptors and activating protein kinase C (PKC), which reduces the transfer of calcium within mitochondria. This leads to the production of ROS, which damage mitochondria and contribute to neuronal cell death (Reuben, 2018).

Additionally, Pb exposure can affect gene expression, a critical process implicated in the evolution of neurodegenerative disorders. Early Pb exposure has been demonstrated to change gene expression patterns, notably those related to tau protein buildup, a hallmark of various neurodegenerative disorders, including AD and PD. Pb suppresses the function of DNA methyltransferases, enzymes that regulate gene expression, leading to alterations in the expression of roughly 150 genes related to protein buildup and neurodegeneration (Zhang et al., 2012). Furthermore, elevated tau hyperphosphorylation and α-synuclein levels in the hippocampus—both of which are implicated in neurodegenerative processes—have been connected to Pb exposure. This results in apoptosis (cell death) and autophagy activation, both of which exacerbate neurological symptoms. Pb's effects extend to iron regulation in the brain, particularly in the presence of APPs, further exacerbating the damage to dopaminergic neurons. These effects, combined with oxidative stress and PKC activation, create a cycle of neurotoxicity that accelerates the development of PD (Branca et al., 2020; Reuben, 2018).

19.5.2.3 Cadmium and Parkinson's disease

Cadmium (Cd), a neurotoxic heavy metal, has a crucial role in worsening oxidative stress (OS) and progressing neurodegenerative illnesses, including PD. Cd promotes oxidative stress by activating critical redox-sensitive transcription factors, including NF-κB, activating protein-1 (AP-1), mammalian target of rapamycin, extracellular signal-regulated kinases (Erk1/2), and c-Jun N-terminal kinase (JNK). These pathways contribute to the progressive degradation of dopaminergic neurons, culminating in the characteristic PD symptoms, including tremors, bradykinesia, and rigidity. Moreover, Cd disrupts the BBB integrity by activating caspase-3-dependent signaling pathways, leading to the sustained opening of pannexin-1 channels. These channels facilitate excessive ATP leakage, disrupting the neurovascular unit's homeostasis, which causes neuronal damage and death (Wang & Du, 2013). This disruption in BBB permeability heightens the brain's susceptibility to Cd-induced neurotoxic effects. Cd also disturbs calcium homeostasis, particularly affecting the release of acetylcholine, a neurotransmitter vital for cognitive functions and motor control. The interference with acetylcholine release contributes to cognitive impairments, motor dysfunction, and other neurological deficits. Prolonged exposure to Cd has been associated with various neurotoxic effects, including learning disabilities, olfactory dysfunction, peripheral neuropathy, and motor dysfunction (Medda, De, & Maiti, 2021).

19.5.2.4 Arsenic and Parkinson's disease

Both organic and inorganic forms of arsenic are metalloids that cause neurotoxicity through a variety of molecular routes, such as the activation of apoptotic pathways in neurones, particularly through c-Jun N-terminal kinase 3 (JNK3) signaling and P38 MAPK. A main method by which arsenic generates neurotoxic consequences is the development of oxidative stress, which leads to DNA damage. The generation of ROS and lipid peroxides, coupled lower levels of antioxidants such as superoxide dismutase (SOD) and reduced glutathione (GSH), leads to severe cellular damage, notably in the brain. The metabolism of important neurotransmitters, such as glutamate, gamma-aminobutyric acid (GABA), acetylcholine, and dopamine, is also disturbed by prolonged exposure to arsenic (Pandey, 2017). Reduced monoamine levels from prolonged exposure, particularly in areas of the brain like the hippocampus, corpus striatum, and frontal cortex, contribute to the cognitive and motor deficits linked to arsenic toxicity. In addition to neurotransmitter problems, arsenic exposure changes calcium signaling, leading to mitochondrial malfunction and a reduction in ATP generation. The perturbation of mitochondrial activity and variations in membrane potential result in neuronal cell death. These findings underline the diverse neurotoxic effects of arsenic and its potential role in the etiology of neurodegenerative illnesses like PD (Medda et al., 2021).

19.5.3 Amyotrophic lateral sclerosis

ALS is a progressive neurological illness with an uncertain cause, despite intensive inquiry. About 90% of ALS cases are sporadic, with no known genetic basis, while 10% are familial, with an autosomal dominant inheritance pattern. Although external causes do not directly cause ALS, environmental and occupational exposures to chemical agents, metals, pesticides, electromagnetic fields, smoking, and excessive physical activity have all been associated with its development. These factors presumably combine with innate genetic predisposition, contributing to the start of ALS (Mathis, Goizet, Soulages, Vallat, & Le Masson, 2019). Over 30 genetic variants have been found in familial ALS, including those in important genes including chromosome 9 open-reading frame 72 (C9orf72), angiogenin (ANG), TAR-DNA binding protein (TARDBP, or TDP-43), superoxide dismutase 1 (SOD1), and fused in sarcoma RNA binding protein (FUS). ALS is a multifactorial illness that disrupts cellular and molecular processes such as glutamatergic excitotoxicity, axonal transport, RNA and protein metabolism, mitochondrial malfunction, and oxidative stress. Mitochondria are crucial for energy production, calcium buffering, phospholipid synthesis, and apoptosis regulation, and their malfunction is particularly damaging to neurons, which have high energy demands. Mitochondrial dysfunction and OS are important factors in motor neuron degeneration in ALS, while the specific processes remain unclear. This review underscores the crucial function of OS in ALS pathogenesis and explores prospective therapeutic options targeted at lowering OS and protecting neurons in ALS patients (Bovio et al., 2021).

19.5.3.1 Cadmium and amyotrophic lateral sclerosis

Cadmium (Cd) plays a role in neurotoxicity and motor neuron degeneration by inhibiting Cu/Zn superoxide dismutase 1 (SOD1) activity, disrupting the BBB, and inducing glutamate toxicity. It achieves this by upregulating glutamate dehydrogenase and downregulating glutamate uptake in glial cells, which exacerbates glutamate excitotoxicity. Elevated cadmium levels have been found in the CNS, blood, white matter, and gray matter of ALS patients. Studies in rodents and humans indicate that cadmium exposure impairs motor neuron conduction and spinal motor neuron function, potentially through reduced SOD1 activity. High levels of MTs, proteins associated with heavy metal exposure, and MT-bound cadmium have been detected in the liver and kidneys of ALS patients. Experiments in *Escherichia coli* show that cadmium reduces SOD1 activity by misfolding the Cu/Zn SOD1 protein and increases MT expression. However, some studies have found no definitive relationship between cadmium exposure and ALS (Farace et al., 2022).

19.5.3.2 Lead and amyotrophic lateral sclerosis

Lead (Pb) has been identified as elevated in CSF, blood, and tibia in ALS patients. Systematic reviews and metaanalyses suggest a strong association between lead exposure and ALS. Animal studies show that Pb^{2+} ions can cross the BBB and accumulate in the cortex, where they induce the formation of insoluble TDP-43, a hallmark of ALS pathology. Occupational exposure to lead, such as from shooting ranges or working with lead-based materials, has been associated with oxidative damage and an increased risk of ALS. Research from Denmark and the United States indicates that lead exposure, particularly more than 10 years before ALS diagnosis, increases the risk of ALS. Activities such as casting lead bullets or using lead fishing sinkers also appear to contribute to the risk (Andrew et al., 2018).

19.5.3.3 Mercury and amyotrophic lateral sclerosis

Mercury (Hg) is another metal suspected of contributing to ALS risk. Elevated mercury levels have been detected in the brains of ALS patients, and several studies support a link between mercury exposure and ALS. Case reports of mercury poisoning demonstrate ALS-like symptoms, suggesting that mercury could be a risk factor for ALS. Both animal models and human studies show increased mercury accumulation in ALS patients. Specifically, in mutant SOD1 ALS mice, mercury accumulates in spinal neurons, and rats exposed to methylmercury exhibit ALS-like neurological effects. In vitro studies indicate that methylmercury exposure impairs the ETC, induces oxidative stress, and disrupts the ubiquitin-proteasome system, all of which are mechanisms implicated in ALS. Moreover, mercury exposure may contribute to glutamate dysregulation and excitotoxicity, which are also central to ALS development (Thakur, Rachamalla, Niyogi, Datusalia, & Flora, 2021).

19.5.3.4 Arsenic and amyotrophic lateral sclerosis

Arsenic, a metalloid, has been implicated in ALS neurotoxicity through multiple mechanisms, particularly oxidative stress and apoptosis. Arsenic exposure activates key signaling pathways, including P38 MAPK and c-Jun N-terminal kinase 3 (JNK3), that promote cell death in cerebral neurons, contributing to neuronal degeneration in ALS. Arsenic exposure increases ROS, lipid peroxides, and decreases critical antioxidants, such as superoxide dismutase (SOD) and glutathione (GSH), exacerbating oxidative stress. This imbalance in ROS further damages neurons, disrupting cellular function (Teleanu et al., 2022).

Arsenic exposure also alters neurotransmitter metabolism, particularly dopamine, acetylcholine, GABA, and glutamate. Chronic exposure has been shown to reduce monoamine levels in key brain regions involved in motor control, such as the frontal cortex, corpus striatum, and hippocampus. This disruption in neurotransmitter regulation may contribute to motor neuron degeneration seen in ALS. Furthermore, arsenic exposure has been linked to mitochondrial dysfunction, a significant factor in ALS pathogenesis. Impairing mitochondrial function, including ATP production and intracellular calcium ion regulation, leads to neuronal cell death, aggravating ALS development (Ayeni et al., 2022).

19.6 Neurodegenerative diseases and biomarkers for heavy metal toxicity: clinical and therapeutic implications

19.6.1 Current diagnostic approaches

Genetic, environmental, and behavioral variables interact intricately to cause neurodegenerative disorders like AD, PD, and ALS. Environmental contaminants, particularly heavy metals such as cadmium (Cd), lead (Pb), mercury (Hg), and arsenic, are increasingly implicated in the development and progression of various illnesses, according to emerging data. The accumulation of these hazardous metals in the CNS can result in oxidative stress, mitochondrial dysfunction, neuroinflammation, and neuronal cell death, all of which play important roles in the etiology of neurodegenerative disorders (Toader et al., 2023). Diagnosis of neurodegenerative illnesses often involves clinical evaluations, neuroimaging techniques, and the discovery of biomarkers that follow disease development. For example, in PD, the hallmark indications are the depletion of dopamine and the destruction of dopaminergic neurons in the substantia nigra. In ALS, diagnosis comprises electromyography, clinical observation of motor deficits, and genetic testing for abnormalities such as those in the SOD1 and TARDBP genes. However, early detection remains challenging, as these diseases often manifest with subtle symptoms that progress slowly, complicating differentiation from other conditions. Biomarkers for heavy metal exposure, including blood, urine, and tissue concentrations, have become valuable tools for assessing environmental risk factors in neurodegenerative diseases. For example, elevated levels of Cd, Pb, Hg, and arsenic in biological samples have been linked to neurological damage and are being increasingly recognized as potential indicators of neurotoxic exposure. Advanced methods, such as mass spectrometry and atomic absorption spectroscopy, are utilized to detect and quantify these metals in biological samples, providing an early indicator of environmental exposure (Kim, Harrison, Aschner, & Bowman, 2022). However, identifying biomarkers that correlate with specific stages of neurodegenerative diseases remains a challenge, and more research is needed to refine these diagnostic tools.

19.6.2 Challenges in biomarker validation

The use of biomarkers to detect heavy metal toxicity and its relationship with neurodegenerative diseases presents several challenges. One major obstacle is the lack of specificity in biomarkers. Heavy metals such as Pb, Cd, Hg, and arsenic are ubiquitous in the environment, making it difficult to attribute elevated metal levels to a particular disease. Moreover, the effects of heavy metals on neuronal function are complex and may involve multiple pathways, such as oxidative stress, mitochondrial dysfunction, and glutamate excitotoxicity, which overlap with the pathogenesis of various neurodegenerative diseases. Another significant challenge in biomarker validation is the variability in individual responses to heavy metal exposure. Genetic predispositions, such as mutations in genes involved in metal homeostasis (e.g., SOD1, FUS, C9orf72) or antioxidant defense mechanisms (e.g., glutathione peroxidase), may modulate the impact of metal toxicity, leading to differences in disease progression and severity. Furthermore, the long latency period between exposure and disease onset complicates the establishment of causal relationships between heavy metal exposure and neurodegenerative diseases (Althobaiti, 2024).

To overcome these challenges, it is essential to develop more sensitive and specific biomarkers that can accurately reflect both the presence of heavy metal toxicity and its contribution to disease development. A multiomics approach, combining genomics, proteomics, and metabolomics, holds great promise for identifying novel biomarkers that can provide a more comprehensive understanding of the molecular mechanisms underlying metal-induced neurotoxicity.

19.6.3 Emerging therapeutic strategies

Given the strong association between heavy metal exposure and neurodegenerative diseases, emerging therapeutic strategies are focused on mitigating the toxic effects of metals and promoting neuronal repair. Several approaches are being explored to address metal toxicity in the CNS (Haidar, Fatema, Shoily, & Sajib, 2023):

1. **Chelation therapy**: Chelators are molecules that bind to heavy metals and facilitate their excretion from the body. In the context of neurodegenerative diseases,

chelation therapy is being investigated as a potential treatment to reduce the accumulation of toxic metals in the brain. For example, agents such as dimercaprol (BAL), EDTA, and DMSA have been used to treat heavy metal poisoning and are being studied for their potential to reverse neurotoxic effects in diseases like PD and ALS (Ashok et al., 2022; Haidar et al., 2023).

2. **Antioxidant therapy**: Oxidative stress plays a central role in the neurotoxicity induced by heavy metals. Antioxidants, such as *N*-acetylcysteine, melatonin, and coenzyme Q10, are being explored for their potential to protect against metal-induced oxidative damage. These compounds work by neutralizing free radicals and restoring redox balance, thus protecting neurons from further damage (Ijomone et al., 2020).
3. **Gene therapy**: With the increasing understanding of genetic predispositions to metal-induced neurodegeneration, gene therapy offers a promising avenue for treating heavy metal-related neurodegenerative diseases. For instance, strategies aimed at enhancing the expression of metal-binding proteins, such as MTs, or boosting antioxidant defense pathways could provide neuroprotective effects. Gene-editing technologies, like CRISPR-Cas9, may also offer the potential for correcting mutations in metal-related genes (e.g., SOD1 in ALS) (Ijomone et al., 2020).
4. **Neuroprotective agents**: Several compounds with neuroprotective properties are being investigated for their ability to counteract the neurotoxic effects of heavy metals. For example, curcumin, a natural compound with antiinflammatory and antioxidant effects, has been shown to reduce metal-induced neurotoxicity in preclinical models. Other compounds, such as brain-derived neurotrophic factor (BDNF) enhancers, are being explored for their potential to promote neuronal survival and regeneration in the context of metal-induced damage (Tao & Farokhzad, 2022).
5. **Nanomedicine**: The use of nanoparticles as delivery vehicles for drugs and chelating agents offers a novel approach to target heavy metal toxicity in the brain. Nanoparticles can cross the BBB more effectively than conventional drugs, allowing for more precise targeting of metal-induced neurotoxicity. Nanoparticles can also be functionalized to deliver specific neuroprotective agents to the affected brain regions, enhancing their therapeutic potential (Gorini & Tonacci, 2024).

19.7 Future perspectives and research directions

The relationship between heavy metal exposure and neurodegenerative diseases has garnered significant attention in recent years, with mounting evidence supporting the role of metals like cadmium (Cd), lead (Pb), mercury (Hg), and arsenic in the pathogenesis of conditions such as PD, ALS, and AD. Despite this progress, there are several important areas that require further exploration to deepen our understanding and improve the clinical management of these diseases.

1. **Longitudinal studies and early detection:** A major challenge in studying the impact of heavy metals on neurodegenerative diseases is the long latency period between exposure and the onset of symptoms. Longitudinal cohort studies involving large, diverse populations are needed to track the cumulative effects of metal exposure over time and to identify early biomarkers that can predict the onset of neurodegenerative diseases before clinical symptoms become apparent. Additionally, advanced imaging techniques and fluid-based biomarkers may offer noninvasive methods for early detection and monitoring of metal-induced neurotoxicity (Cheng et al., 2021).
2. **Mechanistic insights and pathways:** While much has been discovered regarding the role of oxidative stress and mitochondrial dysfunction in metal-induced neurodegeneration, there are still gaps in our understanding of the molecular mechanisms through which heavy metals contribute to neurodegenerative diseases. Research should focus on elucidating the precise pathways involved, such as the interaction between metals and neuroinflammatory processes, the impact on protein misfolding and aggregation (e.g., TDP-43 in ALS), and the disruption of calcium homeostasis and synaptic function. Understanding these mechanisms will be essential for developing targeted interventions and personalized therapeutic strategies (Barnham & Bush, 2014).
3. **Metal-specific and disease-specific biomarkers:** Current biomarkers for heavy metal exposure are general and often lack specificity for different metals or diseases. The development of metal-specific biomarkers that can distinguish between exposure to Cd, Pb, Hg, and arsenic, as well as identify the associated neurodegenerative diseases, is a key research priority. Advances in multiomics approaches (genomics, proteomics, metabolomics) may provide novel insights into metal-induced toxicity pathways and lead to the identification of biomarkers that can be used for early diagnosis, disease progression tracking, and treatment monitoring (Athar, Al Balushi, & Khan, 2021; Smith et al., 2017).
4. **Therapeutic interventions and drug development:** Although chelation therapy, antioxidants, and neuroprotective agents have shown promise in preclinical models, their effectiveness in clinical settings remains uncertain. Future research should focus on optimizing

existing therapies and developing new, more effective treatments that specifically target the neurotoxic effects of heavy metals. This includes the exploration of novel chelators with better BBB penetration, as well as the development of drugs that can modulate metal homeostasis or enhance neuronal resilience to oxidative stress. Additionally, personalized medicine approaches, which take into account genetic predispositions and environmental exposures, will be crucial for improving treatment outcomes (Dutta, Gorain, Choudhury, Roychoudhury, & Sengupta, 2022).

5. **Environmental and occupational exposure:** While much of the research has focused on individual heavy metal toxicity, there is a need for studies that investigate the combined effects of multiple metals and other environmental toxins. Understanding the synergistic effects of various pollutants on neurodegeneration will help create comprehensive public health policies to reduce occupational and environmental exposures. Moreover, developing global initiatives to monitor and regulate metal pollution, particularly in high-risk industries, is crucial for preventing exposure and protecting vulnerable populations (Sahu et al., 2021).
6. **Nanomedicine and novel drug delivery systems:** Nanotechnology in medicine offers exciting possibilities for improving the delivery of therapeutic agents across the BBB. Nanomedicine could allow for more precise targeting of metal-induced neurotoxicity and the delivery of chelators, antioxidants, or neuroprotective agents directly to the brain. Further research is needed to explore the nanomedicine-based therapies' safety, efficacy, and long-term effects on neurodegenerative diseases (Sahu et al., 2021).

19.8 Conclusion

The emerging role of heavy metals in the pathogenesis of neurodegenerative diseases such as PD, ALS, and AD underscores the need for continued research into the complex interactions between environmental toxins and genetic susceptibility. Heavy metals, including cadmium, lead, mercury, and arsenic, contribute to neurotoxicity through mechanisms such as oxidative stress, mitochondrial dysfunction, neuroinflammation, and excitotoxicity, all of which are central to the progression of neurodegenerative diseases. While current diagnostic approaches rely on clinical assessments and neuroimaging, there is a pressing need for more specific and sensitive biomarkers to detect metal-induced neurotoxicity and early stages of disease. Advances in molecular diagnostics, including multiomics approaches, hold promise for identifying novel biomarkers that can improve early detection, monitoring, and treatment efficacy. In terms of therapeutic interventions, promising strategies such as chelation therapy, antioxidants, and neuroprotective agents offer hope, but their clinical efficacy remains fully established. Nanomedicine and gene therapy also represent cutting-edge avenues for targeted treatment of metal-induced neurodegeneration. As research progresses, it is crucial to adopt a multidisciplinary approach that combines insights from toxicology, genetics, and clinical medicine to develop better diagnostic tools and effective therapies. By addressing the environmental and occupational risks associated with heavy metal exposure, we can mitigate their impact on public health and pave the way for more effective management and treatment of neurodegenerative diseases.

References

Ahmed, S. F., Kumar, P. S., Rozbu, M. R., Chowdhury, A. T., Nuzhat, S., Rafa, N., & Mofijur, M. (2022). Heavy metal toxicity, sources, and remediation techniques for contaminated water and soil. *Environmental Technology & Innovation, 25*, 102114.

Aizenman, E., & Mastroberardino, P. G. (2015). Metals and neurodegeneration. *Neurobiology of Disease, 81*, 1.

Alengebawy, A., Abdelkhalek, S. T., Qureshi, S. R., & Wang, M. Q. (2021). Heavy metals and pesticides toxicity in agricultural soil and plants: Ecological risks and human health implications. *Toxics, 9*(3), 42.

Alex, S. (2018). Mercury and its associated impacts on environment and human health: A review. *Journal of Environment and Health Science, 4*(2), 37–43.

Althobaiti, N. A. (2024). Heavy metals exposure and Alzheimer's disease: Underlying mechanisms and advancing therapeutic approaches. *Behavioural Brain Research,* 115212.

Amaral, A. C., Perez-Nievas, B. G., Chong, M. S. T., Gonzalez-Martinez, A., Argente-Escrig, H., Rubio-Guerra, S., & Gomez-Isla, T. (2021). Isoform-selective decrease of glycogen synthase kinase-3-beta (GSK-3β) reduces synaptic tau phosphorylation, transcellular spreading, and aggregation. *Iscience, 24*(2).

Andrew, A. S., Chen, C. Y., Caller, T. A., Tandan, R., Henegan, P. L., Jackson, B. P., & Stommel, E. W. (2018). Toenail mercury Levels are associated with amyotrophic lateral sclerosis risk. *Muscle & Nerve, 58*(1), 36–41.

Aranda-Rivera, A. K., Cruz-Gregorio, A., Arancibia-Hernández, Y. L., Hernández-Cruz, E. Y., & Pedraza-Chaverri, J. (2022). RONS and oxidative stress: An overview of basic concepts. *Oxygen, 2*(4), 437–478.

Ashok, A., Andrabi, S. S., Mansoor, S., Kuang, Y., Kwon, B. K., & Labhasetwar, V. (2022). Antioxidant therapy in oxidative stress-induced neurodegenerative diseases: Role of nanoparticle-based drug delivery systems in clinical translation. *Antioxidants, 11*(2), 408.

Athar, T., Al Balushi, K., & Khan, S. A. (2021). Recent advances on drug development and emerging therapeutic agents for Alzheimer's disease. *Molecular Biology Reports, 48*(7), 5629–5645.

Ayala, A., Muñoz, M. F., & Argüelles, S. (2014). Lipid peroxidation: production, metabolism, and signaling mechanisms of malondialdehyde and 4-hydroxy-2-nonenal. *Oxidative medicine and cellular longevity, 2014*(1), 360438.

Ayeni, E. A., Aldossary, A. M., Ayejoto, D. A., Gbadegesin, L. A., Alshehri, A. A., Alfassam, H. A., & Tawfik, E. A. (2022). Neurodegenerative diseases: implications of environmental and climatic influences on neurotransmitters and neuronal hormones activities. *International Journal of Environmental Research and Public Health, 19*(19), 12495.

Azar, J., Yousef, M. H., El-Fawal, H. A., & Abdelnaser, A. (2021). Mercury and Alzheimer's disease: A look at the links and evidence. *Metabolic Brain Disease, 36*, 361–374.

Bakulski, K. M., Hu, H., & Park, S. K. (2020). Lead, cadmium and Alzheimer's disease. *Genetics, neurology, behavior, and diet in dementia*, 813–830.

Bakulski, K. M., Seo, Y. A., Hickman, R. C., Brandt, D., Vadari, H. S., Hu, H., & Park, S. K. (2020). Heavy metals exposure and Alzheimer's disease and related dementias. *Journal of Alzheimer's Disease, 76*(4), 1215–1242.

Bakulski, M., Rozek, K. S., Dolinoy, D., L. C., L. Paulson, H., & Hu, H. (2012). Alzheimer's disease and environmental exposure to lead: the epidemiologic evidence and potential role of epigenetics. *Current Alzheimer Research, 9*(5), 563–573.

Barnham, K. J., & Bush, A. I. (2014). Biological metals and metal-targeting compounds in major neurodegenerative diseases. *Chemical Society Reviews, 43*(19), 6727–6749.

Barrera, G., Pizzimenti, S., Daga, M., Dianzani, C., Arcaro, A., Cetrangolo, G. P., & Gentile, F. (2018). Lipid peroxidation-derived aldehydes, 4-hydroxynonenal and malondialdehyde in aging-related disorders. *Antioxidants, 7*(8), 102.

Bassil, F., Brown, H. J., Pattabhiraman, S., Iwasyk, J. E., Maghames, C. M., Meymand, E. S., & Lee, V. M. Y. (2020). Amyloid-beta (Aβ) plaques promote seeding and spreading of alpha-synuclein and tau in a mouse model of Lewy body disorders with Aβ pathology. *Neuron, 105*(2), 260–275.

Bhat, A., Ravi, K., Tian, F., & Singh, B. (2024). Arsenic Contamination Needs Serious Attention: An opinion and global scenario. *Pollutants, 4*(2), 196–211.

Bjørklund, G., Dadar, M., Mutter, J., & Aaseth, J. (2017). The toxicology of mercury: Current research and emerging trends. *Environmental Research, 159*, 545–554.

Bloom, G. S. (2014). Amyloid-β and tau: The trigger and bullet in Alzheimer disease pathogenesis. *JAMA Neurology, 71*(4), 505–508.

Bolin, C. M., Basha, R., Cox, D., Zawia, N. H., Maloney, B., Lahiri, D. K., & Cardozo-Pelaez, F. (2006). Exposure to lead (Pb) and the developmental origin of oxidative DNA damage in the aging brain. *The FASEB Journal, 20*(6), 788–790.

Bovio, F., Sciandrone, B., Urani, C., Fusi, P., Forcella, M., & Regonesi, M. E. (2021). Superoxide dismutase 1 (SOD1) and cadmium: A three models approach to the comprehension of its neurotoxic effects. *Neurotoxicology, 84*, 125–135.

Branca, J. J. V., Fiorillo, C., Carrino, D., Paternostro, F., Taddei, N., Gulisano, M., & Becatti, M. (2020). Cadmium-induced oxidative stress: Focus on the central nervous system. *Antioxidants (basel), 9*, 492.

Channer, B., Matt, S. M., Nickoloff-Bybel, E. A., Pappa, V., Agarwal, Y., Wickman, J., & Khoshbouei, H. (2023). Dopamine, immunity, and disease. *Pharmacological Reviews, 75*(1), 62–158.

Chen, P., Miah, M. R., & Aschner, M. (2016). Metals and neurodegeneration. *F1000Research*, 5.

Chen, Q. Y., & Costa, M. (2021). Arsenic: A global environmental challenge. *Annual Review of Pharmacology and Toxicology, 61*(1), 47–63.

Chen, T., Dai, Y., Hu, C., Lin, Z., Wang, S., Yang, J., & Li, W. (2024). Cellular and molecular mechanisms of the blood–brain barrier dysfunction in neurodegenerative diseases. *Fluids and Barriers of the CNS, 21*(1), 60.

Cheng, H., Yang, B., Ke, T., Li, S., Yang, X., Aschner, M., & Chen, P. (2021). Mechanisms of metal-induced mitochondrial dysfunction in neurological disorders. *Toxics, 9*(6), 142.

Cheslow, L., Snook, A. E., & Waldman, S. A. (2024). Biomarkers for managing neurodegenerative diseases. *Biomolecules, 14*(4), 398.

Clemente-Suárez, V. J., Redondo-Flórez, L., Beltrán-Velasco, A. I., Ramos-Campo, D. J., Belinchón-deMiguel, P., Martinez-Guardado, I., & Tornero-Aguilera, J. F. (2023). Mitochondria and brain disease: A comprehensive review of pathological mechanisms and therapeutic opportunities. *Biomedicines, 11*(9), 2488.

Cubadda, F., Jackson, B. P., Cottingham, K. L., Van Horne, Y. O., & Kurzius-Spencer, M. (2017). Human exposure to dietary inorganic arsenic and other arsenic species: State of knowledge, gaps and uncertainties. *Science of the Total Environment, 579*, 1228–1239.

Dash, U. C., Bhol, N. K., Swain, S. K., Samal, R. R., Nayak, P. K., Raina, V., & Jena, A. B. (2024). Oxidative stress and inflammation in the pathogenesis of neurological disorders: Mechanisms and implications. *Acta Pharmaceutica Sinica B, 15*(1), 15–34.

Dutta, S., Gorain, B., Choudhury, H., Roychoudhury, S., & Sengupta, P. (2022). Environmental and occupational exposure of metals and female reproductive health. *Environmental Science and Pollution Research, 29*(41), 62067–62092.

Farace, C., Fiorito, G., Pisano, A., Etzi, F., Sabalic, A., Fenu, G., & Madeddu, R. (2022). Human tissue lead (Pb) levels and amyotrophic lateral sclerosis: A systematic review and meta-analysis of case–control studies. *Neurological Sciences, 43*(10), 5851–5859.

Förstermann, U., & Sessa, W. C. (2012). Nitric oxide synthases: Regulation and function. *European Heart Journal, 33*(7), 829–837.

Freudenberg, F., Celikel, T., & Reif, A. (2015). The role of α-amino-3-hydroxy-5-methyl-4-isoxazolepropionic acid (AMPA) receptors in depression: Central mediators of pathophysiology and antidepressant activity? *Neuroscience & Biobehavioral Reviews, 52*, 193–206.

Friberg, L. (2018). *Cadmium in the environment*. CRC press.

Genchi, G., Sinicropi, M. S., Lauria, G., Carocci, A., & Catalano, A. (2020). The effects of cadmium toxicity. *International Journal of Environmental Research and Public Health, 17*(11), 3782.

Gorini, F., & Tonacci, A. (2024). Metal toxicity and dementia including frontotemporal dementia: Current state of knowledge. *Antioxidants, 13*(8), 938.

Gworek, B., Dmuchowski, W., & Baczewska-Dąbrowska, A. H. (2020). Mercury in the terrestrial environment: A review. *Environmental Sciences Europe, 32*(1), 128.

Ha, E., Basu, N., Bose-O'Reilly, S., Dórea, J. G., McSorley, E., Sakamoto, M., & Chan, H. M. (2017). Current progress on understanding the impact of mercury on human health. *Environmental Research, 152*, 419–433.

Haidar, Z., Fatema, K., Shoily, S. S., & Sajib, A. A. (2023). Disease-associated metabolic pathways affected by heavy metals and metalloid. *Toxicology Reports, 10*, 554–570.

Haut, M. W., Morrow, L. A., Pool, D., Callahan, T. S., Haut, J. S., & Franzen, M. D. (1999). Neurobehavioral effects of acute exposure to inorganic mercury vapor. *Applied Neuropsychology, 6*(4), 193–200.

Hayat, M. T., Nauman, M., Nazir, N., Ali, S., & Bangash, N. (2019). Environmental hazards of cadmium: Past, present, and future. In *Cadmium toxicity and tolerance in plants*. Academic Press, 163–183.

Ijomone, O. M., Ifenatuoha, C. W., Aluko, O. M., Ijomone, O. K., & Aschner, M. (2020). The aging brain: Impact of heavy metal neurotoxicity. *Critical Reviews in Toxicology, 50*(9), 801–814.

Ijomone, O. M., Ijomone, O. K., Iroegbu, J. D., Ifenatuoha, C. W., Olung, N. F., & Aschner, M. (2020). Epigenetic influence of environmentally neurotoxic metals. *Neurotoxicology, 81*, 51–65.

Ingrassia, R., Garavaglia, B., & Memo, M. (2019). DMT1 expression and iron levels at the crossroads between aging and neurodegeneration. *Frontiers in Neuroscience, 13*, 575.

Jomova, K., Alomar, S. Y., Nepovimova, E., Kuca, K., & Valko, M. (2024). Heavy metals: Toxicity and human health effects. *Archives of Toxicology*, 1–57.

Jomova, K., Raptova, R., Alomar, S. Y., Alwasel, S. H., Nepovimova, E., Kuca, K., & Valko, M. (2023). Reactive oxygen species, toxicity, oxidative stress, and antioxidants: Chronic diseases and aging. *Archives of Toxicology, 97*(10), 2499–2574.

Juan, C. A., Pérez de la Lastra, J. M., Plou, F. J., & Pérez-Lebeña, E. (2021). The chemistry of reactive oxygen species (ROS) revisited: Outlining their role in biological macromolecules (DNA, lipids and proteins) and induced pathologies. *International Journal of Molecular Sciences, 22*(9), 4642.

Jung, Y. J., Tweedie, D., Scerba, M. T., & Greig, N. H. (2019). Neuroinflammation as a factor of neurodegenerative disease: Thalidomide analogs as treatments. *Frontiers in Cell and Developmental Biology, 7*, 313.

Kabir, M. T., Uddin, M. S., Zaman, S., Begum, Y., Ashraf, G. M., Bin-Jumah, M. N., & Abdel-Daim, M. M. (2021). Molecular mechanisms of metal toxicity in the pathogenesis of Alzheimer's disease. *Molecular Neurobiology, 58*, 1–20.

Kandel, R., Jung, J., & Neal, S. (2024). Proteotoxic stress and the ubiquitin proteasome system. In *Seminars in cell & developmental biology (Vol. 156)*. Academic Press, 107–120.

Kar, I., & Patra, A. K. (2021). Tissue bioaccumulation and toxicopathological effects of cadmium and its dietary amelioration in poultry—A review. *Biological Trace Element Research, 199*(10), 3846–3868.

Ke, T., Gonçalves, F. M., Gonçalves, C. L., Dos Santos, A. A., Rocha, J. B., Farina, M., & Aschner, M. (2019). Post-translational modifications in MeHg-induced neurotoxicity. *Biochimica et Biophysica Acta (BBA)—Molecular Basis of Disease, 1865*(8), 2068–2081.

Kell, D. B., & Pretorius, E. (2014). Serum ferritin is an important inflammatory disease marker, as it is mainly a leakage product from damaged cells. *Metallomics, 6*(4), 748–773.

Khosravi-Darani, K., Rehman, Y., Katsoyiannis, I. A., Kokkinos, E., & Zouboulis, A. I. (2022). Arsenic exposure via contaminated water and food sources. *Water, 14*(12), 1884.

Kim, H., Harrison, F. E., Aschner, M., & Bowman, A. B. (2022). Exposing the role of metals in neurological disorders: A focus on manganese. *Trends in Molecular Medicine, 28*(7), 555–568.

Korkmaz, K. S., Butuner, B. D., & Roggenbuck, D. (2018). Detection of 8-OHdG as a diagnostic biomarker. *Journal of Laboratory and Precision Medicine, 3*.

Król, M., & Kepinska, M. (2020). Human nitric oxide synthase—Its functions, polymorphisms, and inhibitors in the context of inflammation, diabetes and cardiovascular diseases. *International Journal of Molecular Sciences, 22*(1), 56.

Kumar, A., Kumar, A., MMS, C. P., Chaturvedi, A. K., Shabnam, A. A., Subrahmanyam, G., & Yadav, K. K. (2020). Lead toxicity: Health hazards, influence on food chain, and sustainable remediation approaches. *International Journal of Environmental Research and Public Health, 17*(7), 2179.

Kwon, S. Y., Blum, J. D., Yin, R., Tsui, M. T. K., Yang, Y. H., & Choi, J. W. (2020). Mercury stable isotopes for monitoring the effectiveness of the Minamata Convention on Mercury. *Earth-Science Reviews, 203*, 103111.

Lanphear, B. P., Hornung, R., Khoury, J., Yolton, K., Baghurst, P., Bellinger, D. C., & Roberts, R. (2019). Erratum: "Low-level environmental lead exposure and children's intellectual function: an international pooled analysis". *Environmental Health Perspectives, 127*(9), 099001.

Leff, T., Stemmer, P., Tyrrell, J., & Jog, R. (2018). Diabetes and exposure to environmental lead (Pb). *Toxics, 6*(3), 54.

Li, Y., Jiao, Q., Xu, H., Du, X., Shi, L., Jia, F., & Jiang, H. (2017). Biometal dyshomeostasis and toxic metal accumulations in the development of Alzheimer's disease. *Frontiers in Molecular Neuroscience, 10*, 339.

Mathis, S., Goizet, C., Soulages, A., Vallat, J. M., & Le Masson, G. (2019). Genetics of amyotrophic lateral sclerosis: A review. *Journal of the Neurological Sciences, 399*, 217–226.

Matousek, S. B., Ghosh, S., Shaftel, S. S., Kyrkanides, S., Olschowka, J. A., & O'banion, M. K. (2012). Chronic IL-1β-mediated neuroinflammation mitigates amyloid pathology in a mouse model of Alzheimer's disease without inducing overt neurodegeneration. *Journal of Neuroimmune Pharmacology, 7*, 156–164.

Medda, N., De, S. K., & Maiti, S. (2021). Different mechanisms of arsenic related signaling in cellular proliferation, apoptosis and neoplastic transformation. *Ecotoxicology and Environmental Safety, 208*, 111752.

Mendiola, A. S., & Cardona, A. E. (2018). The IL-1β phenomena in neuroinflammatory diseases. *Journal of Neural Transmission, 125*, 781–795.

Mitra, S., Chakraborty, A. J., Tareq, A. M., Emran, T. B., Nainu, F., Khusro, A., & Simal-Gandara, J. (2022). Impact of heavy metals on the environment and human health: Novel therapeutic insights to counter the toxicity. *Journal of King Saud University-Science, 34*(3), 101865.

Miyazaki, I., & Asanuma, M. (2023). Multifunctional metallothioneins as a target for neuroprotection in Parkinson's disease. *Antioxidants, 12*(4), 894.

Mousavi, S. M., Brodie, G., Payghamzadeh, K., Raiesi, T., & Strivastava, A. K. (2022). Lead bioavailability in the environment: Its Exposure And And Effects. *Journal of Advances in Environmental Health Research, 10*(1), 1–14.

Nag, R., & Cummins, E. (2022). Human health risk assessment of lead (Pb) through the environmental-food pathway. *Science of the Total Environment, 810*, 151168.

Nicosia, N., Giovenzana, M., Misztak, P., Mingardi, J., & Musazzi, L. (2024). Glutamate-mediated excitotoxicity in the pathogenesis and treatment of neurodevelopmental and adult mental disorders. *International Journal of Molecular Sciences, 25*(12), 6521.

Novo, J. P., Martins, B., Raposo, R. S., Pereira, F. C., Oriá, R. B., Malva, J. O., & Fontes-Ribeiro, C. (2021). Cellular and molecular mechanisms mediating methylmercury neurotoxicity and neuroinflammation. *International Journal of Molecular Sciences, 22*(6), 3101.

Olmos, G., & Lladó, J. (2014). Tumor necrosis factor alpha: A link between neuroinflammation and excitotoxicity. *Mediators of Inflammation, 2014*(1), 861231.

Ortega, D. R., Esquivel, D. F. G., Ayala, T. B., Pineda, B., Manzo, S. G., Quino, J. M., & de la Cruz, V. P. (2021). Cognitive impairment induced by lead exposure during lifespan: Mechanisms of lead neurotoxicity. *Toxics, 9*(2), 23.

Pandey, A. K. (2017). Disruption of neurosynaptic physiology and neuron network dysfunction in brain disorders: An environmental and occupational health perspective. *Activitas Nervosa Superior, 59*(2), 61–77.

Plantone, D., Pardini, M., Righi, D., Manco, C., Colombo, B. M., & De Stefano, N. (2023). The role of TNF-α in Alzheimer's disease: A narrative review. *Cells, 13*(1), 54.

Pyatha, S., Kim, H., Lee, D., & Kim, K. (2022). Association between heavy metal exposure and Parkinson's disease: a review of the mechanisms related to oxidative stress. *Antioxidants, 11*(12), 2467.

Raha, A. A., Biswas, A., Henderson, J., Chakraborty, S., Holland, A., Friedland, R. P., & Raha-Chowdhury, R. (2022). Interplay of ferritin accumulation and ferroportin loss in ageing brain: Implication for protein aggregation in down syndrome dementia, Alzheimer's, and Parkinson's diseases. *International Journal of Molecular Sciences, 23*(3), 1060.

Rahimzadeh, M. R., Rahimzadeh, M. R., Kazemi, S., & Moghadamnia, A. A. (2017). Cadmium toxicity and treatment: An update. *Caspian Journal of Internal Medicine, 8*(3), 135.

Rahman, M. A., Hannan, M. A., Uddin, M. J., Rahman, M. S., Rashid, M. M., & Kim, B. (2021). Exposure to environmental arsenic and emerging risk of Alzheimer's disease: Perspective mechanisms, management strategy, and future directions. *Toxics, 9*(8), 188.

Rahman, Z., & Singh, V. P. (2019). The relative impact of toxic heavy metals (THMs)(arsenic (As), cadmium (Cd), chromium (Cr)(VI), mercury (Hg), and lead (Pb)) on the total environment: An overview. *Environmental Monitoring and Assessment, 191*, 1–21.

Raju, N. J. (2022). Arsenic in the geo-environment: A review of sources, geochemical processes, toxicity and removal technologies. *Environmental Research, 203*, 111782.

Ramesh, S., & Arachchige, A. S. P. M. (2023). Depletion of dopamine in Parkinson's disease and relevant therapeutic options: A review of the literature. *AIMS Neuroscience, 10*(3), 200.

Reddy, P. H. (2013). Amyloid beta-induced glycogen synthase kinase 3β phosphorylated VDAC1 in Alzheimer's disease: Implications for synaptic dysfunction and neuronal damage. *Biochimica et Biophysica Acta (BBA)—Molecular Basis of Disease, 1832*(12), 1913–1921.

Ren, X., Zou, L., Zhang, X., Branco, V., Wang, J., Carvalho, C., & Lu, J. (2017). Redox signaling mediated by thioredoxin and glutathione systems in the central nervous system. *Antioxidants & Redox Signaling, 27*(13), 989–1010.

Reuben, A. (2018). Childhood lead exposure and adult neurodegenerative disease. *Journal of Alzheimer's Disease, 64*(1), 17–42.

Richetti, S. K., Rosemberg, D. B., Ventura-Lima, J., Monserrat, J. M., Bogo, M. R., & Bonan, C. D. (2011). Acetylcholinesterase activity and antioxidant capacity of zebrafish brain is altered by heavy metal exposure. *Neurotoxicology, 32*(1), 116–122.

Rizzi, G., & Tan, K. R. (2017). Dopamine and acetylcholine, a circuit point of view in Parkinson's disease. *Frontiers in Neural Circuits, 11*, 110.

Rudyk, S., Spirov, P., Samuel, P., & Joshi, S. J. (2017). Vaporization of crude oil by supercritical CO_2 at different temperatures and pressures: Example from gorm field in the Danish north sea. *Energy & Fuels, 31*(6), 6274–6283.

Ruttkay-Nedecky, B., Nejdl, L., Gumulec, J., Zitka, O., Masarik, M., Eckschlager, T., & Kizek, R. (2013). The role of metallothionein in oxidative stress. *International Journal of Molecular Sciences, 14*(3), 6044–6066.

Sahu, T., Ratre, Y. K., Chauhan, S., Bhaskar, L. V. K. S., Nair, M. P., & Verma, H. K. (2021). Nanotechnology based drug delivery system: Current strategies and emerging therapeutic potential for medical science. *Journal of Drug Delivery Science and Technology, 63*, 102487.

Shawki, A., Knight, P. B., Maliken, B. D., Niespodzany, E. J., & Mackenzie, B. (2012). H+-coupled divalent metal-ion transporter-1: Functional properties, physiological roles and therapeutics. *Current Topics in Membranes, 70*, 169–214.

Shreenivasa, S. A., Velayudhannair, K., Arockiasamy, J. P., Shama, G., & Sivakumar, R. (2023). Determination of heavy metals in various tissues of locally reared (country) chicken in major districts of Karnataka, India: Assessment of potential health risks. *Archives of Veterinary Science, 28*(4), 1.

Siblerud, R., Mutter, J., Moore, E., Naumann, J., & Walach, H. (2019). A hypothesis and evidence that mercury may be an etiological factor in Alzheimer's disease. *International Journal of Environmental Research and Public Health, 16*(24), 5152.

Singh, A., Kukreti, R., Saso, L., & Kukreti, S. (2019). Oxidative stress: A key modulator in neurodegenerative diseases. *Molecules (Basel, Switzerland), 24*(8), 1583.

Singh, B., Day, C. M., Abdella, S., & Garg, S. (2024). Alzheimer's disease current therapies, novel drug delivery systems and future directions for better disease management. *Journal of Controlled Release, 367*, 402–424.

Skjørringe, T., Burkhart, A., Johnsen, K. B., & Moos, T. (2015). Divalent metal transporter 1 (DMT1) in the brain: Implications for a role in iron transport at the blood-brain barrier, and neuronal and glial pathology. *Frontiers in Molecular Neuroscience, 8*, 19.

Smith, E. E., Cieslak, A., Barber, P., Chen, J., Chen, Y. W., Donnini, I., & Hachinski, V. (2017). Therapeutic strategies and drug development for vascular cognitive impairment. *Journal of the American Heart Association, 6*(5), e005568.

Stancu, I. C., Vasconcelos, B., Terwel, D., & Dewachter, I. (2014). Models of β-amyloid induced Tau-pathology: The long and "folded" road to understand the mechanism. *Molecular Neurodegeneration, 9*, 1–14.

Su, L. J., Zhang, J. H., Gomez, H., Murugan, R., Hong, X., Xu, D., & Peng, Z. Y. (2019). Reactive oxygen species-induced lipid peroxidation in apoptosis, autophagy, and ferroptosis. *Oxidative Medicine and Cellular Longevity, 2019*(1), 5080843.

Tamagno, E., Guglielmotto, M., Vasciaveo, V., & Tabaton, M. (2021). Oxidative stress and beta amyloid in Alzheimer's disease. Which comes first: The chicken or the egg? *Antioxidants, 10*(9), 1479.

Tao, W., & Farokhzad, O. C. (2022). Theranostic nanomedicine in the NIR-II window: Classification, fabrication, and biomedical applications. *Chemical Reviews, 122*(6), 5405–5407.

Teleanu, R. I., Niculescu, A. G., Roza, E., Vladâcenco, O., Grumezescu, A. M., & Teleanu, D. M. (2022). Neurotransmitters—Key factors in neurological and neurodegenerative disorders of the central nervous system. *International Journal of Molecular Sciences, 23*(11), 5954.

Thakur, M., Rachamalla, M., Niyogi, S., Datusalia, A. K., & Flora, S. J. S. (2021). Molecular mechanism of arsenic-induced neurotoxicity including neuronal dysfunctions. *International Journal of Molecular Sciences, 22*(18), 10077.

Toader, C., Dobrin, N., Brehar, F. M., Popa, C., Covache-Busuioc, R. A., Glavan, L. A., & Ciurea, A. V. (2023). From recognition to remedy: The significance of biomarkers in neurodegenerative disease pathology. *International Journal of Molecular Sciences, 24*(22), 16119.

Torrey, E. F., & Simmons, W. (2023). Mercury and Parkinson's disease: Promising leads, but research is needed. *Parkinson's Disease, 2023*(1), 4709322.

Vázquez Cervantes, G. I., González Esquivel, D. F., Ramírez Ortega, D., Blanco Ayala, T., Ramos Chávez, L. A., López-López, H. E., & Pérez de la Cruz, V. (2023). Mechanisms associated with cognitive and behavioral impairment induced by arsenic exposure. *Cells, 12*(21), 2537.

Venkatesan, D., Muthukumar, S., Iyer, M., Babu, H. W. S., Gopalakrishnan, A. V., Yadav, M. K., & Vellingiri, B. (2024). Heavy metals toxicity on epigenetic modifications in the pathogenesis of Alzheimer's disease (AD). *Journal of Biochemical and Molecular Toxicology, 38*(6), e23741.

Wallin, C., Friedemann, M., Sholts, S. B., Noormägi, A., Svantesson, T., Jarvet, J., & Wärmländer, S. K. (2020). Mercury and Alzheimer's disease: Hg (II) ions display specific binding to the amyloid-β peptide and hinder its fibrillization. *Biomolecules, 10*(1), 44.

Wang, B., & Du, Y. (2013). Cadmium and its neurotoxic effects. *Oxidative Medicine and Cellular Longevity, 2013*(1), 898034.

Wang, B., Wu, L., Chen, J., Dong, L., Chen, C., Wen, Z., & Wang, D. W. (2021). Metabolism pathways of arachidonic acids: Mechanisms and potential therapeutic targets. *Signal Transduction and Targeted Therapy, 6*(1), 94.

Wang, R., Wu, Z., Bai, L., Liu, R., Ba, Y., Zhang, H., & Huang, H. (2021). Resveratrol improved hippocampal neurogenesis following lead exposure in rats through activation of SIRT1 signaling. *Environmental Toxicology, 36*(8), 1664–1673.

Wu, Y. S., Osman, A. I., Hosny, M., Elgarahy, A. M., Eltaweil, A. S., Rooney, D. W., & Yap, P. S. (2024). The toxicity of mercury and its chemical compounds: Molecular mechanisms and environmental and human health implications: A comprehensive review. *Acs Omega, 9*(5), 5100–5126.

Xu, L., Zhang, W., Liu, X., Zhang, C., Wang, P., & Zhao, X. (2018). Circulatory levels of toxic metals (aluminum, cadmium, mercury, lead) in patients with Alzheimer's disease: A quantitative meta-analysis and systematic review. *Journal of Alzheimer's Disease, 62*(1), 361–372.

Xu, Z., Yang, S., Li, C., Xie, M., He, Y., Chen, S., & Yang, G. (2024). Characterization of metallothionein genes from Broussonetia papyrifera: metal binding and heavy metal tolerance mechanisms. *BMC Genomics, 25*, 563.

Yang, Z., & Wang, K. K. (2015). Glial fibrillary acidic protein: From intermediate filament assembly and gliosis to neurobiomarker. *Trends in Neurosciences, 38*(6), 364–374.

Yoshida, J., Oñate, M., Khatami, L., Vera, J., Nadim, F., & Khodakhah, K. (2022). Cerebellar contributions to the basal ganglia influence motor coordination, reward processing, and movement vigor. *Journal of Neuroscience, 42*(45), 8406–8415.

Younus, H. (2018). Therapeutic potentials of superoxide dismutase. *International Journal of Health Sciences, 12*(3), 88.

Zhang, H., Wei, W., Zhao, M., Ma, L., Jiang, X., Pei, H., & Li, H. (2021). Interaction between Aβ and tau in the pathogenesis of Alzheimer's disease. *International Journal of Biological Sciences, 17*(9), 2181.

Zhang, J., Cai, T., Zhao, F., Yao, T., Chen, Y., Liu, X., & Chen, J. (2012). The role of α-synuclein and tau hyperphosphorylation-mediated autophagy and apoptosis in lead-induced learning and memory injury. *International Journal of Biological Sciences, 8*(7), 935.

Zhang, X., Wang, L. P., Ziober, A., Zhang, P. J., & Bagg, A. (2021). Ionized calcium binding adaptor molecule 1 (IBA1) a novel, highly sensitive, and specific marker for histiocytic, dendritic, and monocytic neoplasms. *American Journal of Clinical Pathology, 156*(1), 86–99.

Zheng, Y., Sun, J., Luo, Z., Li, Y., & Huang, Y. (2024). Emerging mechanisms of lipid peroxidation in regulated cell death and its physiological implications. *Cell Death & Disease, 15*(11), 859.

Zhong, H., & Yin, H. (2015). Role of lipid peroxidation derived 4-hydroxynonenal (4-HNE) in cancer: Focusing on mitochondria. *Redox Biology, 4*, 193–199.

Zhou, Z. D., Yi, L. X., Wang, D. Q., Lim, T. M., & Tan, E. K. (2023). Role of dopamine in the pathophysiology of Parkinson's disease. *Translational Neurodegeneration, 12*(1), 44.

Zitka, O., Skalickova, S., Gumulec, J., Masarik, M., Adam, V., Hubalek, J., & Kizek, R. (2012). Redox status expressed as GSH: GSSG ratio as a marker for oxidative stress in paediatric tumour patients. *Oncology Letters, 4*(6), 1247–1253.

Zorov, D. B., Juhaszova, M., & Sollott, S. J. (2014). Mitochondrial reactive oxygen species (ROS) and ROS-induced ROS release. *Physiological Reviews, 94*(3), 909–950.

Chapter 20

Diagnostic techniques for heavy metal induced neurotoxicity

Prasann Kumar
Department of Agronomy, School of Agriculture, Lovely Professional University, Phagwara, Punjab, India

20.1 Introduction

Heavy metal–induced neurotoxicity refers to the damaging effects upon the nervous system of toxic metal exposure, which often causes serious neurological damage. Heavy metals such as lead (Pb), arsenic (As), mercury (Hg), and cadmium (Cd) can pass through the blood–brain barrier (BBB) and interfere with important cellular processes in neuronal cells (Karri et al., 2018; Yu et al., 2024). Neurotoxicity due to heavy metal poisoning can show up in many different forms, with anything from slight mental loss to severe progressive nerve disease. Early and correct diagnosis is crucial for preventing long-term damage. Heavy metal poisoning can also bring in a wide variety of symptoms when it strikes the central nervous system (CNS) and peripheral nervous system. The degree and nature of neurotoxicity are related to several factors, notably the type of metal, duration of exposure, dose, individual susceptibility, and age. Due to their developing nervous systems, children are particularly susceptible, while adult labor exposure carries great danger. Lead exposure primarily affects cognitive function, memory, and behavior. Chronic exposure during childhood is associated with reduced IQ and other mental deficits. In adults, it can lead to symptoms of mood disorder, cognitive decline as well as more serious peripheral nerve problems like odor nerve weakness (Andrade, Mateus, Batoréu, Aschner, & Marreilha dos Santos, 2014; Famurewa et al., 2022). Mercury toxicity, particularly from methylmercury, can cause sensory disturbances in coordination, tremors, and psychiatric symptoms such as depression and anxiety. Severe cases of mercury poisoning, such as Minamata disease, result in profound neurological problems, including paralysis and unconsciousness. Chronic arsenic exposure damages cognitive functioning and causes peripheral neuropathy, weakness, and sensory loss. It has also been associated with neurodegenerative diseases such as Alzheimer's (AD) and Parkinson's (PD). The primary targets of cadmium are the mast cells. The drug can trigger the partitions of tissues and organs—neuroinflammation, oxidative stress, and motor disease. There is a growing vulnerability to PD associated with long-term exposure. High levels of manganese result in a disorder similar to PD called manganese, characterized by tremors, slowness in movement, and rigidity. It can also cause psychiatric symptoms. Heavy metal exposure may also cause neurotoxic diseases such as AD, PD, and amyotrophic lateral sclerosis. This is because heavy metals found in brain tissue will worsen neuroinflammation, oxidative stress, mitochondrial dysfunction, and protein misfolding, all of which play important roles in the progression of this disease. A diagnosis of heavy metal–induced neurotoxicity is critical to preventing irreversible damage to the nervous system. The earlier, the better. Everything falls into place early on in detecting and controlling the disease. It enables healthcare workers to implement chelation therapy, change their diet, modify hazards, and make other interventions. If neurotoxicity is not diagnosed early, the damage to neurons can increase and be irreversible; eventually, it causes permanent disability and results in a lower quality of life. There are several problems with the early diagnosis of heavy metal poisoning, including symptoms that are not specific and sometimes overlap with other diseases like neurodegenerative disorders, psychiatric disorders, or metabolic diseases. Moreover, doctors may not always consider heavy metal exposure as a possible cause of neurological symptoms (Florio, Marasco, & Manna, 2025; Harischandra et al., 2018). Consequently, it is incorrectly diagnosed. Since heavy metals accumulate in body tissues over time, it may be some years before symptoms emerge,

Heavy Metal Toxicity and Neurodegeneration. DOI: https://doi.org/10.1016/B978-0-443-36575-1.00018-2

making it tricky to pinpoint the source. Genetic disposition, previous diseases, and nutritional status impact the degree of neurotoxic effects, making diagnosis difficult. Considering these complexities, the adoption of comprehensive diagnostic approaches is crucial for confirming heavy metal–induced neurotoxicity and planning treatment strategies. One must use a multidisciplinary approach that combines clinical assessment with laboratory tests, neuroimaging, and neurophysiological tests to diagnose heavy metal toxicity accurately. First and foremost are the clinical history and physical examination, the initial steps in diagnosis. Key elements include a history of occupational and environmental exposure and identifying sources such as pollution, contaminated food or water, and industrial work. A neurological examination looks at areas of motor function and coordination lesions; cognitive function checks should be performed, as well as sensory deficits and thought disorders. For children, there needs to be a developmental check to look for learning obstacles, language delays, or behaviors distinct from those expected for their age. Clinical laboratory tests showing blood heavy metal levels can provide data on exposure. Biden indicates significant exposure if a child's blood lead level (BLL) is over 5 μg/dL. Hair and nail analyses are useful for long-term detection because inhibition increases over time. Biomarkers of oxidative stress and neuroinflammation, such as malondialdehyde (MDA), superoxide dismutase (SOD), and inflammatory cytokines, indicate metal-related damage in cells. A chelation challenge test can determine how much heavy metal causes total body stress. It entails an injection of a chelating agent followed by uric acid collection. Advanced imaging modalities help reveal structural and physiological changes in the brain due to heavy metal exposure. It detects brain atrophy and white matter lesions associated with manganese toxicity. Magnetic resonance imaging (MRI), positron emission tomography (PET), and single-photon emission computed tomography (SPECT) evaluate neurochemical changes in cases of heavy metal–induced neurotoxicity, bit by bit, for example, in people exposed to lead or mercury. X-ray fluorescence spectroscopy, also called X-ray microanalysis, measures the quantity of accumulated heavy metals in bone which is how children can become ill. Neurophysiological tests such as electroencephalography (EEG) can detect aberrations in cerebral electrical activity, as they can show changes from ordinary for heavy metal poisoning (Jang & Hoffman, 2011; Rajagopalan et al., 2023; Samal & Subudhi, 2024). Nerve conduction studies (NCS) and electromyography (EMG) are used to investigate damage sensory nerve in the musculoskeletal system, especially when the cause is arsenic. Neurological and neuropsychological testing assesses memory, attention, executive function, and emotional health, which may be affected by heavy metal exposure. Heavy metal–induced neurotoxicity

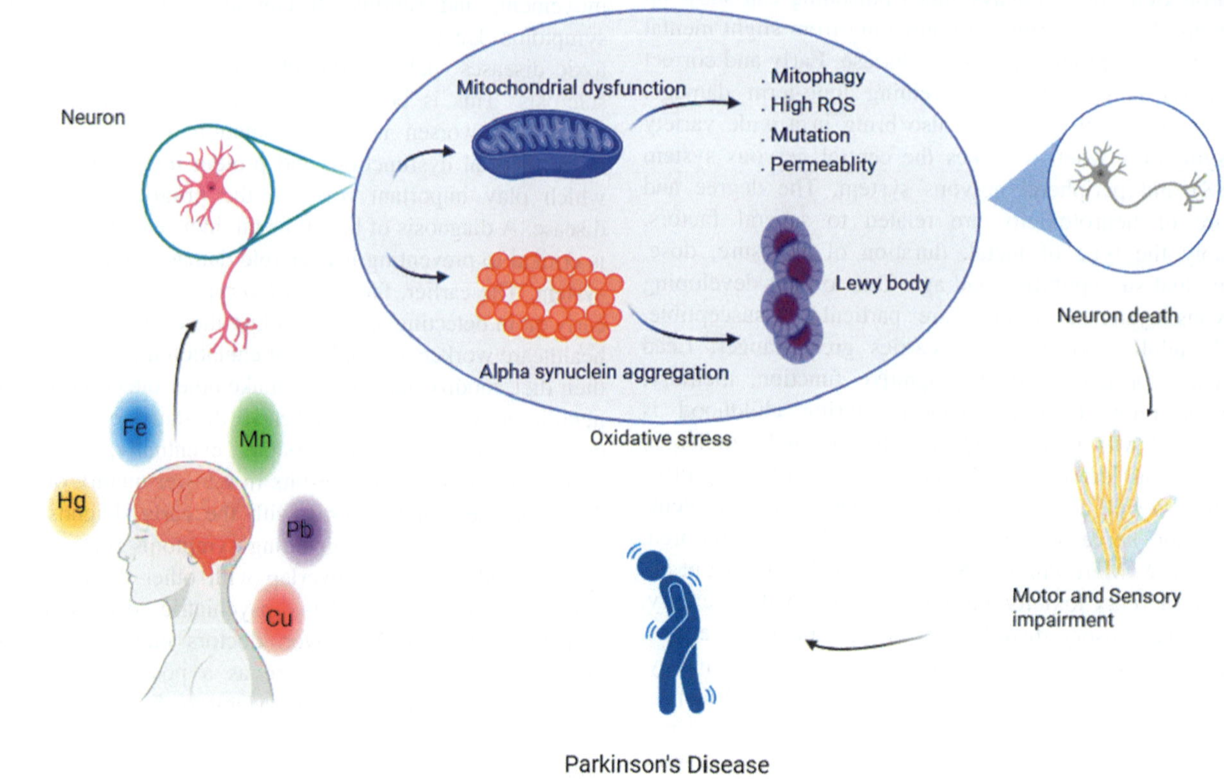

FIGURE 20.1 Heavy metals and Parkinson's disease.

presents a serious public health concern with an impact on cognitive and brain health. The clinical manifestations vary depending on the type of heavy metal, carrying with them not only cognitive impairment but also motor dysfunction up to full neurodegenerative disease. Early diagnosis is essential to prevent long-term neurological damage and improve patient prognosis. Comprehensive diagnostic approaches involving clinical history, laboratory tests, neuroimaging, and neurophysiological testing are necessary to effectively determine and manage heavy metal exposure. Future research should concentrate on advancing biomarkers and drug interventions to limit heavy metal–induced neurotoxicity and protect public health (Fig. 20.1).

20.2 Clinical assessment

A comprehensive medical history and neurological exam may be used to detect potential heavy metal exposure and related neurological symptoms. Which constitutes the groundwork for effective diagnosis. The medical history can yield invaluable insights into a patient's lifestyle, working environment, and possible exposure to resources. This is crucial in suspected heavy metal toxicity cases. This detailed history involves such factors as work environment, pastime activities, and eating and living habits. These things can reveal past or current exposure to toxic metals such as lead, mercury, arsenic, cadmium, and manganese (Gomes da Silva et al., 2024; Goutam Mukherjee, Ramesh Wanjari, Renu, Vellingiri, & Valsala Gopalakrishnan, 2022; Pomary et al., 2023). Occupational exposure in, for example, mining, welding, battery manufacturing, and paint making is a significant risk factor in enterprises where men are regularly exposed to excessively high levels of these metals. Similarly, chronic exposure to heavy metals via infected water, soil, and air can bring about the accumulation of toxic metal. Understanding these factors enables healthcare providers to judge the likelihood of exposure and pick out high-risk populations who may show symptoms of nervous system dysfunction—subtle or obvious. Neurological examination allows medical staff to match the patient's clinical history with identifiable signs of neurotoxicity. It is a step-by-step examination of thought, motor function, sensing, and reflecting skills. An unconsciousness due to heavy metal poisoning results. A hypoglycemic deficiency may cause any abnormality in the thinking process. For instance, impaired cognitive function, such as memory loss and attention deficit, may both indicate an underlying chronic lead or mercury exposure. Furthermore, muscle disturbances such as tremors, ataxia (incoordination), or rigidity may indicate mercury or manganese toxicity. Peripheral neuropathy, displaying symptoms of weakness, numbness, and tingling, is another common feature of chronic arsenic or cadmium exposure. In children, developmental delays in learning ability as well as behavior are often due to lead toxicity. Here, the emphasis placed on early neurological examination of vulnerable groups such as maple-asbestos workers and food processing workers thus becomes clear. It cannot be overstressed enough: tracing occupational and environmental exposure history is crucial to assessing an individual's risk of heavy metal–induced neurotoxicity. A detailed occupation history helps the healthcare provider point out certain industrial segments and job or work types with high risks, prompting his customisation of screenings and interventions. For example, battery recycling workers are at risk of lead poisoning sufferers, or in electronics waste processing jobs; people may come into contact with mercury or cadmium (Gade, Comfort, & Re, 2021; Paul, Samanta, Mallick, & Majumdar, 2024; Shanmugavel, Rene, Balakrishnan, Krishnakumar, & Jose, 2024). Hence, a thorough inquiry into environmental exposures is vital, for example, among residents bordering industrial sites or whose living water supplies have been polluted. Arsenic contamination is included as an example from the ground city all around. For example, the risk of long-term health effects from this draftsman's fine point strains your thoracic diaphragm. Therefore understanding these environmental factors makes it possible to assess the total risk of exposure and set up community-based interventions. Despite the importance of clinical history and neurological examination, these approaches have significant limitations that necessitate additional diagnostic tools. One major problem is that the symptoms associated with heavy metal toxicity are not specific to neurological disease alone but are often found in patients suffering from other neurodegenerative or psychiatric disorders. For instance, fatigue, headaches, and slight periods of memory loss can all be either heavy metal poisoning or multiple sclerosis. The reactions manager Both the time lag between exposure and symptoms manifesting—which may be years in some cases and causal relationships based solely on clinical evaluations make it difficult to establish direct links to poisonous heavy metal poisoning. As a result, diagnosis and treatment may be delayed for patients with heavy metal accumulation in the body tissues before these substances are finally revealed as having caused neurological symptoms. In addition, susceptibility to heavy metal toxicity varies on both a personal and individual level, influenced by genetic factors, nutritional status, age, and overall health status. This affects both the magnitude of symptoms and how quickly they progress. For example, children are more susceptible to lead neurotoxicity due to their developing nervous systems. Those with genetic shepherds, such as detoxification pathways, may be more susceptible to heavy metal–induced neurotoxicity. After many tests and diagnoses attempt to address these problems, such as BLLs that are taken from blood, urine samples collected on the spot, hair or nail measurements for heavy metal concentrations right next to the site of contact with living tissue—an area where all biological systems are interconnected; and blood samples processed in baggies biomarkers of oxidative stress and neuroinflammation are also useful for corroborating diagnoses. These may be detected in blood, urine or other bodily fluids samples (Boas, Remennik, & Reches, 2023; Caini et al., 2023; Hu, Wang, Niu, He, & Tang, 2024). Thus increased MDA levels and proinflammatory cytokines can indicate cell damage within the body. PET scans and SPECT are noninvasive imaging tools. They enable doctors to see what changes there are in the pattern of radioactivity emitted by a substance when it enters healthy tissue such as brain cells; if these have been disrupted by chemicals like lead or mercury, those effects can be seen on both a pet scan and a SPECT image MRI shows the manifestations of brain poisoning typical to manganese—like atrophy, lesions in white matter or malformations at the base of the ear. Meanwhile, PET and SPECT spotlight chemical changes that have been made to neurochemicals when hit with excess heavy metals such as lead or mercury, arthritis, diabetes, nerve damage, and carpal tunnel syndrome. EEG and NCS are vital tools for assessing the impact of heavy metals on brain electrical activity and peripheral nerve function, respectively. By studying EEG readings, we can detect abnormal patterns of brain waves associated with neurotoxicity. At the same time, the NCS gives us evidence that peripheral neuropathy is due to chronic exposure to arsenic. Early detection of heavy metal neurotoxicity is still difficult for all these advanced diagnostic methods. The sufferer's symptoms are intermittent and nonspecific (Table 20.1).

Diagnosis can be postponed or missed altogether in severe situations, necessitating a heightened awareness of the clinical part and regular screening of patients who are at high risk. Education and training programs should be established to help physicians recognize the signs and symptoms of heavy metal neurotoxicity as soon as possible, a process that must rely on public health initiatives to reduce environmental and occupational exposure in communities. Although extensive medical history and neurological examination are essential in identifying potential heavy metal exposure and any linked neurological symptoms, they have their limitations. It is necessary for accurate diagnosis and treatment to integrate clinical evaluation with laboratory tests, neuroimaging, and neurophysiological assessments (Bjørklund, Đorđević, Hamdan, Wallace, & Peana, 2024; Lewis, Azmoun, Shoieb, & Marquez, 2024; Okpoghono, Agbetuyi, Eyide, & Onyesom, 2024). What measures are being taken to reduce environmental exposure through workplace chemical safety management regulations, public education, and strengthened industrial pollution control? Indeed, what will be done with the control instruments operates no better than machinery for salt making. This chapter introduces the early detection and treatment of heavy metal neurotoxicity. As the field advances, continued research into new biomarkers and effective therapeutic strategies will be key in improving both early diagnosis and management of heavy metal–induced neurotoxicities. Understanding occupational and environmental exposure history is essential to evaluating risk factors and identifying targeted interventions.

20.3 Neuroimaging

A variety of neuroimaging techniques, including MRI and computed tomography (CT), have emerged as crucial tools for detecting structural abnormalities in the brain caused by heavy metal ion-induced neurotoxicity, which can be fatal (Naqvi, Kumar, & Flora, 2020; Nicolescu et al., 2010; Onyancha, Aigbe, Ukhurebor, & Muchiri, 2021). Neurotoxic effects such as neuronal degeneration, white matter damage, oxidative stress, and breakdown of the BBB can all be seen after chronic exposure to substances containing Pb, Hg, As, or Cd. Changes like these can be effectively seen through advanced neuroimaging modalities, shedding some light on how the brain comes to such a state. It provides direction for early diagnosis and treatment strategies. It is widely believed that MRI, with its superior soft tissue contrast resolution and ability to detect brain microstructure, is a good tool for identifying heavy metal–induced neurotoxicity. Various MRI sequences can give detailed insights into different aspects of neurotoxic damage. For example, by using T1-weighted and T2-weighted imaging, one can judge structural atrophy and white matter lesions. Meanwhile, diffusion-weighted imaging and diffusion tensor imaging provide a view into what microstructural changes are occurring, such as axonal damage and demyelination. Susceptibility-weighted imaging is beneficial for detecting heavy metal deposits because it is sensitive to paramagnetic substances such as iron and manganese, which accumulate in the brain after chronic metal exposure. Functional MRI (fMRI) then complements this approach by helping us observe changes in nerve activity due to heavy metal–induced neurotoxicity, showing altered brain function before any significant structural damage can be seen. The nature of CT imaging is such that it is less sensitive in picking up soft tissue abnormalities. However, its value in identifying early and late neurotoxic effects makes it indispensable and complementary to MRI. The speed with which we can acquire CT images and its ability to detect significant anatomical changes like cerebral atrophy, hydrocephalus, and calcifications in this setting makes imaging valuable for both acute emergencies and chronic measurements over time (Kim et al., 2025; Sears, Smith, Cai, & Zierold, 2024; Wellenberg et al., 2021). For example, in individuals suffering from lead poisoning, CT scans often show characteristic basal ganglia calcifications, an acknowledged sign

TABLE 20.1 Different neuroimaging techniques.

Neuroimaging techniques	Primary usage	Advantages	Limitations	Clinical applications	Key findings in neurotoxicity
CT	Brain structure imaging	Easily available and cost-effective	Limited soft tissue contrast and also have radiation exposure	Structural damage and acute metal exposure	Detects calcifications or structural changes due to metal deposition
MRI	Brain structural imaging	High-resolution imaging and no radiation	Costly and not suitable for patients with metal implants	Chronic metal exposure with subtle neurotoxic effects	White matter and gray matter changes along with edema or atrophy from toxicity
fMRI	Mapping of brain activity	Noninvasive and high spatial resolution	Limited temporal resolution with an indirect measure of activity	Functional impact of toxicity on cognition	Altered activation patterns in affected regions
DTI	White matter integrity	Sensitive to microstructural changes	Susceptible to motion artifacts	Chronic exposure; early detection of neurotoxicity	Detection of axonal damage or connectivity changes
PET	Molecular and metabolic imaging	High sensitivity to metabolic changes	It requires radioactive tracers and is also costly	Used for the detection of long-term effects of exposure	Hypo- or hypermetabolism in affected areas
SPECT	Mapping of blood flow and receptors	Less expensive than PET	It has lower resolution and also uses radioactive tracers	Acute and chronic perfusion changes	Abnormal blood flow in metal-affected regions

CT, Computed tomography; *DTI*, diffusion tensor imaging; *fMRI*, functional MRI; *MRI*, magnetic resonance imaging; *PET*, positron emission tomography; *SPECT*, single-photon emission computed tomography.

of chronic PB exposure. Also, CT imaging is beneficial in cases where metal-induced neurotoxicity mimics stroke symptoms, allowing one to differentiate between toxic effects and vascular pathology. In addition to traditional MRI and IPCT scanning, advanced neuroimaging techniques such as PET (PET1) and SPECT (SPECT2) offer insights into the functional effects of heavy metal–induced neurotoxicity. Focused on the brain, PET imaging with radioactive tracer labels like [18F] fluorodeoxyglucose allows one to judge metabolic changes in different brain regions. As a result, areas of hypometabolism consistent with heavy metal poisoning can often be seen. On the other hand, SPECT imaging gives information about regional cerebral blood flow abnormalities, frequently altered due to heavy metal–induced neurovascular damage (Askari et al., 2025; Nabavi, Safari, Kashkooli, Sadat Nabavizadeh, & Molavi Vardanjani, 2024; Xie et al., 2023). The complete information basis for effective research of heavy metal–induced neurotoxic effects lies in these different methods and emendations of neuroimaging, between structural (which is roughly accessible with today's technology) and functional metabolic analysis (still more of an art form). Although each has its strengths and limitations, different neuroimaging modalities can also measure the specific damage patterns in very different neurological damage. With its high resolution and versatility, MRI imaging is the best tool for revealing subtle changes in brain structure. However, its high cost, long acquisition time, and inducement of motion artifacts have restricted its widespread use, especially among pediatric patients and critically ill people. Rapid, readily available CT imaging cannot detect subtle neurotoxic damage early. It exposes patients to iodizing radiation, making it unsuitable for routine monitoring. Functional imaging techniques such as PET or SPECT are critical in obtaining metabolic data and perfusion. However, they carry high costs, have limited availability for many people, and require exposure to radioactive tracers, which may not be appropriate for specific groups of patients. This places them in inferior standing with other diagnostic tools in terms of convenience and safety. Heavy metal–induced neurotoxicity is also subject to a significant limitation in neuroimaging studies: distinguishing toxic effects from age-related or other neurodegenerative changes (Anchidin-Norocel, Gutt, Tătăranu, & Amariei, 2024; Dey et al., 2024; Sharma, Mehan, Khan, Gupta, & Narula, 2024; Tian et al., 2021). Chronic heavy metal exposure often leads to generalized brain atrophy and white matter lesions, findings similar to those of neurodegenerative disorders such as AD or PD. This overlap makes it necessary to combine neuroimaging conclusions with additional diagnostic modalities, including neuropsychological assessments, biomarker analysis, and electrophysiological studies, to arrive at a more actual diagnosis. There are specific chief trends in the way that neuroimaging deals with heavy metal–induced neurotoxicity. Together, these can give a more complete picture than any method alone. MRI data plus cerebrospinal fluid (CSF) analysis of biomarkers can help discriminate metal-induced neurotoxicity from other neurological disorders. Evidence such as elevated oxidative stress markers, inflammatory cytokines, and abnormal metal concentration in the CSF will support MRI findings. Similarly, complementing neuroimaging with blood-based biomarkers—forward-looking measures that, for example, can verify blood lead or mercury levels—can provide further evidence and gauge the severity of exposure.

Additional methods such as EEG and magnetoencephalography (MEG) complement neuroimaging and can be used to detect the functional changes in neural activity connected with heavy metal exposure. An abnormal pattern of EEG findings, including changes in theta power and decreased levels of alpha activity, has been identified among individuals with chronic lead and mercury exposure. This majorly disrupted neural connectivity. With its high available temporal resolution, MEG offers further elucidation of altered oscillations in neural networks at times of metal toxicity (Karthik, Karuna, Kumar, Saravanan, & Hemavathy, 2022; Kumar et al., 2025; Kwakye et al., 2019; Ma, Li, Wang, Wang, & He, 2024; Singh & Kostova, 2024; Zhou et al., 2024).

In addition, combining neuropsychological tests with neuroimaging allows investigators to examine how structural and functional imaging changes are related to cognitive and behavioral outcomes. Patients suffering from heavy metal–induced neurotoxicity typically exhibit cognitive disturbances such as memory impairment, executive dysfunction, and deficient concentration, quantifiable through standardised neuropsychological tests. When this cognitive profile is correlated with specific imaging markers—for instance, hippocampal atrophy seen on MRI or hypometabolism observed with PET—it makes the clinical significance of neuroimaging findings greater and aids in the prognosis and planning of treatment.

The diagnostic ability of neuroimaging for heavy metal toxicity is further bolstered by developing a new pipeline for neuroimaging analysis driven by artificial intelligence (AI). Machine learning algorithms, for example, can evaluate large datasets and identify subtle imaging cues associated with metal-induced neurotoxicity that are distinguishable from other neurological conditions. AI-based segmentation and quantification programs enhance the accuracy of discerning temporal changes in the volumes of brain regions affected by heavy metal exposure. This means they help diagnose diseases earlier and conduct long-term surveillance in a manner that is less burdensome on the patient than would otherwise be possible (Dey et al., 2024; Ma et al., 2024; Naqvi et al., 2020; Tian et al., 2021; Xie et al., 2023).

Heavy metal–induced brain damage can be visualized on MRI, CT, PET, and SPECT topographical images, thanks

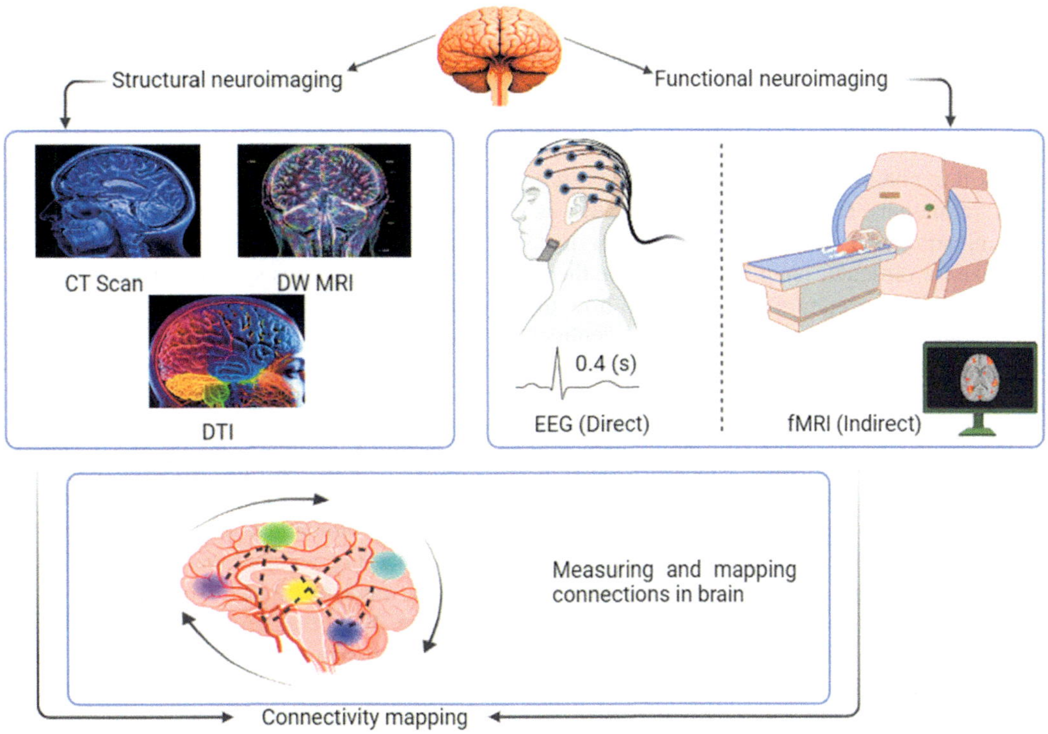

FIGURE 20.2 Different neuroimaging techniques.

in large part to the crucial role that these neuroimaging technologies play. Each technique has its advantages and disadvantages, but combined with other diagnostic methods—such as biomarker analysis, EEG, and neuropsychological assessments, the combined testing provides a comprehensive evaluation of metal-induced brain damage. As improvements in imaging technologies continue (including the application of AI algorithms in image analyses), the sensitivity and specificity of neuroimaging remain on the increase, giving way to better early diagnosis, intervention, and clinical outcomes for cases caused by heavy metal exposure (Fig. 20.2).

20.4 Biochemical and genetic testing

Detection of heavy metals and determination of potential biomarkers of neurotoxicity in blood, urine, hair, and CSF, as well as biological specimens, are essential assays. Blood analysis is the standard approach to identify recent exposure to heavy metals like lead, mercury, arsenic, and cadmium. Blood can correctly measure the systemic circulation levels of heavy metals, reflecting both recent intake from diet and release into it. For example, BLLs are well established as biomarkers of exposure and toxicity; over 5 μg/dL indicates potential neurodevelopmental problems along with physical symptoms such as headaches and anemia. Mercury in the blood suggests methylmercury exposure, commonly associated with fish eating. Arsenic exposure is seen by measuring total and speciated arsenic levels in blood, especially the distinction between inorganic arsenic and its toxic metabolites. These measures, however, have limits because of their dependence on blood levels of heavy metals or other substances circulating transiently in the body. Therefore additional methods are needed for comprehensive exposure assessment (Anchidin-Norocel et al., 2024; Ma et al., 2024; Sears et al., 2024; Tian et al., 2021; Xie et al., 2023).

Urine analysis is another important method to evaluate heavy metal exposure, especially for those elements mainly excreted by the kidneys. Heavy metals like cadmium and arsenic in the urine indicate chronic exposure and the cumulative burden that this puts on your body. Urinary cadmium is a reliable biomarker of long-term exposure, such as when workers inhale cadmium fumes over time while on the job. The urinary arsenic analysis likewise provides valuable information about your exposure to inorganic arsenic. The ratio of monomethyl arsenate to dimethyl arsenate indicates both metabolic efficiency and probable toxicity. Urine testing has the advantage of convenience and the ability to monitor changes in exposure over time. However, variability in urine dilution mandates normalization strategies such as creative correction to ensure accurate interpretations.

We often take hair analysis as a biomarker of lifelong or chronic heavy metal exposure, particularly mercury, arsenic, and lead. Hair provides a long-term exposure record because heavy metals are incorporated into the growing matrix and become part of it. For example, in hair, methylmercury is inclined to bind particularly well with keratin, so its concentration can be linked directly to the composition of a person's diet. The advantages of hair analysis are that it is noninvasive and can detect exposure levels over several months. However, confounders such as external contamination and individual variation in hair growth rates make for difficulties in interpreting data; there have to be rules (which are culture-specific) as to how to separate endogenous exposure from exogenous contamination (Dey et al., 2024; Karthik et al., 2022; Nabavi et al., 2024; Onyancha et al., 2021; Singh & Kostova, 2024).

For example, CSF analysis is critical in evaluating neurotoxic effects from heavy metal exposure since it provides direct insight into the involvement of the CNS. CSF biomarkers may indicate neuronal damage and neuroinflammation associated with heavy metal toxicity. These include neurofilament light chain, tau protein, and glial fibrillary acidic protein. Elevated CSF lead and mercury levels have been implicated in neurodegenerative diseases such as AD and PD. However, the invasive nature of lumbar puncture limits routine CSF analysis to research settings and severe cases of suspected neurotoxicity.

It is genetic testing that is key to identifying people with genetic polymorphisms that may increase their vulnerability to heavy metal neurotoxicity. For example, single-nucleotide polymorphisms (SNPs) in genes coding for metal-binding proteins, detoxification enzymes, and transporters are the source of individual variability in heavy metal metabolism and toxicity. For instance, polymorphisms in the ALAD (aminolevulinic acid dehydratase) gene influence lead metabolism and neurotoxicity risk, with the ALAD-2 variant being linked to more excellent lead retention and worse nervous system outcomes. Similarly, polymorphisms in GSTM1 and GSTT1 (glutathione-*S*-transferases) can change detoxification capacity, thereby bringing varying degrees of risk from cadmium or mercury poisoning. Genetic differences in MT1A (metallothionein) can shape how capable someone is of catching and holding on to heavy metals, thereby changing their neurotoxic effects. And, for yet another example, SNPs in genes associated with oxidative stress, such as SOD2 (superoxide dismutase) and NQO1 (NAD(P)H quinone dehydrogenase 1), make people respond differently to heavy metal-induced oxidative damage. Genetic screening to identify at-risk populations provides personalized risk assessment and targeted intervention strategies. Each biochemical and genetic test for diagnosing heavy metal exposure and monitoring its neurological effects has strengths and weaknesses that must be critically evaluated. Blood tests offer great accuracy in assessing recent exposure but may not reflect accumulation over the long term. Urine tests can help monitor continuing exposure, but adjustments need to be made for variability in dilution. Hair analysis records past exposure but can be contaminated by outside sources. CSF analysis provides direct evidence of neurotoxicity but perfect care for the patient and significant risk if even small amounts are taken from children during routine checks of their eyes, noses, or throats. Genetic testing provides insight into personal susceptibility but does not measure a person's exposure level or the effects heavy metals might have on their nervous system. An integrated approach that combines many biological materials plus the genetic profile maintains the reliability of risk prediction and assessment for heavy metal exposure (Askari et al., 2025; Dey et al., 2024; Karthik et al., 2022; Kim et al., 2025; Tian et al., 2021; Zhou et al., 2024).

Further improving the sensitivity and specificity of heavy metal toxicology is thus made possible by developing novel biomarkers and advanced analytical techniques. Emerging metabolomics and proteomics are used alongside modern biotechnology tools to prevent cancer tumors at both the genetic company level undergoing metabolic mapping to identify specific metabolic signatures change between healthy or carcinogen-exposed animals or cells and their families as species that are recurrently exposed to environmental pollutants have been found to exhibit higher levels of metallothionein in their liver and lower levels in their brain than rats living in a pollution-free environment. Advances in high-throughput sequencing and epigenetic analysis offer a better understanding of gene-environment interactions, telling us how heavy metals modulate DNA methylation patterns plus histone modifications. Such integrative methodologies will lay a solid foundation for understanding heavy metal neurotoxicity and provide the basis for adopting early detection, prevention, intervention, and treatment strategies (Anchidin-Norocel et al., 2024; Kim et al., 2025; Kumar et al., 2025; Naqvi et al., 2020; Singh & Kostova, 2024).

When testing for exposure to heavy metals, the diagnostic procedure involves a combination of blood, urine, hair, and CSF. Each method has its advantages and disadvantages. By analyzing the genetic differences between individuals susceptible to different heavy metals, it is possible to identify potential victims or carriers of cancer or Wilson's disease in advance. While biochemistry and genetic testing are valuable diagnostic and monitoring tools, merging multiple approaches makes such applications more accurate and robust than they would otherwise be. With rapidly developing omics technologies and the discovery of new biomarkers that will alter future directions in assessing heavy metal toxicity, no one knows what tomorrow holds.

20.5 Advanced analytical methods

Using such advanced analytical techniques as mass spectrometry (MS) and inductively coupled plasma mass spectroscopy (ICP-MS) has positively revolutionised bioanalytical chemistry. This is true, especially regarding the exact quantification of heavy metals in biological samples and their very high sensitivity levels. These methods enable us to detect trace elements at part-per-trillion (ppt) levels, thus accurately assessing heavy metal exposure and bioaccumulation. It is widely recognized for its exceptional sensitivity, selectivity, and ability to detect several heavy metals simultaneously. This includes everything from cadmium (Cd) to mercury (Hg), lead (Pb), and arsenic (As), among others. By utilizing such high-throughput analytical technologies, investigators can comprehensively evaluate the level of heavy metal contamination in biological matrices such as blood, urine, CSF, and tissues. This allows for a more profound understanding of toxicokinetics and bioavailability. Both enhances these methods and adds precision to quantifying heavy metals. By mitigating matrix effects and instrumental drift, isotope dilution mass spectroscopy forms a critical part of the scientific effort to improve heavy metal quantification.

Not only does elemental analysis tell us what elements are present, but also integrated metabolomics; it is now generally accepted that combining proteomics can be much more effective than either in elucidating the biochemical perturbations resulting from heavy metal–induced neurotoxicity. Since hundreds of small molecules exist in any biological system, metabolomics can help us understand how even a tiny molecular disorder may create havoc within the cell (Askari et al., 2025; Kumar et al., 2025; Naqvi et al., 2020; Sears et al., 2024; Sharma et al., 2024). Studies of this sort have now shown that lead causes concentrations leading to free radical production in the brain and increases oxidative stress throughout it: metabolism turns chaotic. Pioneering metabolomics studies have found that exposure to lead, for instance, disrupts the glutathione metabolic pathway so that free radical production within the brain rises and oxidative stress gets worse. Likewise, arsenic poisoning has been connected with malfunctions in cellular bioenergetics and abnormal amino acid profiles, suggesting the general impact of heavy metals on metabolism homeostasis at all levels, from cell to organism. This is important because analyzing specific metabolic biomarkers by MS with resolution MS should lead to an early diagnosis and intervention that is not just a possibility but a promising reality for tomorrow.

When proteins form posttranslational modifications on a large scale, it adds a new angle for us to study heavy metal toxicity at the molecular level at all times. For example, advances in tandem MS (MS/MS) and data-independent acquisition (DIA) proteomics allow us to identify and quantify variations of protein levels in response to heavy metal exposure. Neurotoxic heavy metals—all types—from mercury to cadmium—have caused protein misfolding in cells, leading to the endoplasmic reticulum stress and interruption of normal synaptic signaling pathways. Researchers who have carried out proteomic studies on neurons tissues and CSF have identified some key proteins such as heat shock proteins, metallothioneins and enzymes involved in neurotransmitter synthesis, as being differentially expressed following an attack by the awful metals Burden. They show that specific proteomic signatures might serve as biomarkers for heavy metal-induced neurotoxicity, providing new options for medical targets. In addition, embedding phosphoproteomics and redox proteomic methods into such a rich, phenomenal framework affords us a much better understanding of how heavy metals alter the acetylation and oxidation state of proteins, which are crucial factors in the normal function and survival of neurons (Kumar et al., 2025; Sharma et al., 2024; Singh & Kostova, 2024; Wellenberg et al., 2021; Xie et al., 2023).

Combining such advanced analytical techniques offers early diagnosis and personalized risk estimation from heavy metal poisoning. Downstream of these biological indicators, either specific metabolomic and/or proteomic biomarkers, make early diagnosis of heavy metal–induced CNS disorders possible and offer the chance to categorise patients by their susceptibility to such injuries. The perspective on personalized risk assessment can be refined further by integrating omics approaches that combine whole-genome genetic analysis and transcriptomics and measuring small molecules and their degradation products, peptides in cells or plasma (i.e., proteomics). It is possible to make a comprehensive molecular profile of participants exposed to heavy metals. Machine learning algorithms and AI-driven predictive modeling further increase our ability to correlate analytical findings with clinical outcomes. This then enhances risk assessment and decision-making in public health and occupational safety, reassuring about the future of heavy metal exposure (Bjørklund et al., 2024; Naqvi et al., 2020; Pomary et al., 2023; Sears et al., 2024; Shanmugavel et al., 2024).

Moreover, using advanced MS technologies to develop noninvasive and minimally invasive sampling techniques, such as dried blood spots, hair analysis, or saliva-based diagnostics, is one way in conjunction with these same strategies for biological specimen collection that significantly raises the likelihood of large-scale human monitoring programs being put into place. These improvements make it possible to take long-term measurements of chronic exposure and its impact on health. Also, adopting spatially resolved techniques like laser ablation ICP-MS (LA-ICP-MS) can permit careful mapping of heavy metal distribution to the tissue level, thereby offering better information about heavy metals' organ-specific toxicity and pathophysiological mechanisms in animals. In summary, this integration of MS-based heavy metal analysis with metabolomics and proteomics signifies a sea change

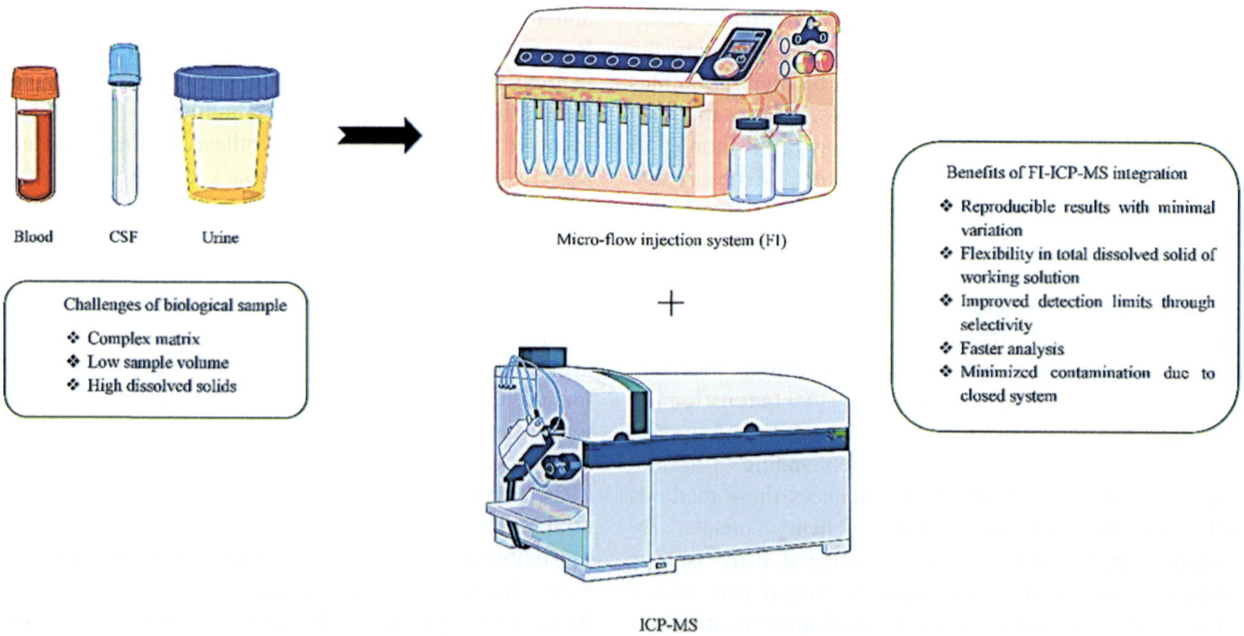

FIGURE 20.3 Integration of micro-flow injection system (FI) with ICP-MS to overcome challenges presented by biological.

in understanding how heavy metals impact living biological systems. It will provide robust, quantitative, and mechanistic insights that open up new possibilities for diagnosis, early intervention strategies tailored to suit each case individually, and better ways of preventing long-term injury from heavy metal exposure to people's health. This potential for early intervention strategies empowers us to be proactive about the future of heavy metal exposure (Fig. 20.3).

20.6 Challenges and future directions

The neuroscience community feels that heavy metal–induced neurotoxicity poses a significant threat to public health. This is mainly because these neurotoxins are present so foully in the environment: notably prevalent among toxic metals found would be mercury, for instance, lead, arsenic, and cadmium. Their effects on the brain have been widely documented. In the past several years, definitive studies have shown how toxic these metals can be. Neurobehavioral diseases, learning problems, and neurodegenerative disorders all follow similar patterns once afflicted with one type of heavy metal such as lead (Pb). Nonetheless, diagnosing heavy metal–induced neurotoxicity remains challenging due to the lack of specific biomarkers. Moreover, it is possible that even with current investigative tools and knowledge, we may never indeed arrive at such a diagnosis for several reasons, including individual differences in susceptibility to these toxins and limitations of current diagnostic techniques at our command. Thus, if heavy metal poisoning occurs only in some people but not all against whom you test this hypothesis, then, of course, under statistical analysis, an ordinary physician might assume a high false positive rate (Florio et al., 2025; Gade et al., 2021; Goutam Mukherjee et al., 2022; Karri et al., 2018). This would create conflicting information about what evidence to believe, making it difficult to map out any causal relationships (or even conclusively establish a relationship between exposure and effect). Finally, these biomarkers are, at best, single-point-in-time indicators of heavy metal exposure rather than continuous monitoring systems. Moreover, these systems can be less than associated with the severity of impaired functions, for example, liver damage from heavy metal toxicity may occur even after blood shows normal levels for some time and gradually get worse precisely because that is exactly what would happen to metal content in an inactive organ like one's brain tissue. The next major challenge is the interindividual variability in susceptibility to heavy metal–induced neurotoxicity. Participant's genetic factors, age, and nutritional status, such as coexposure to other environmental toxins, can all affect individual sensitivity. Take genetic polymorphisms in detoxification pathways, for example. Variations in metallothioneins or glutathione-related enzymes may either directly change the body's ability to detoxify heavy metals or indirectly increase the accumulation of toxic substances. At low exposure levels, children, pregnant women, and those with background neurologic diseases than the general population, susceptibility to neurotoxic effects is exceptionally high. These variations make it difficult for universal diagnostic thresholds to be established since a metal concentration in biological fluids that is toxic for one individual might be innocuous for another. The limitations of current diagnostic tools also make early detection and accurate assessment of heavy metal–induced neurotoxicity practically impossible (Boas et al., 2023; Caini et al., 2023; Gade et al., 2021; Pomary et al., 2023; Rajagopalan et al., 2023). Standard laboratory tests primarily measure heavy metal concentrations in blood, urine, hair, and nails but do not give a comprehensive picture of long-term neurotoxic effects. A more advanced method, such as neuroimaging, for example, MRI and PET scans, can show both the structural and functional changes in the brain that are associated with heavy metal exposure. However, these approaches are expensive and unavailable on a routine or standard basis in clinical settings. Electrophysiological tests (e.g., EEG, nerve conduction testing) may detect the presence of neurophysiological impairments. Still, they lack specificity in attributing what part of such impairment is due to heavy metals and what might otherwise come from a different neurological disease. The mainstay of neuropsychological assessment is still employed to discern cognitive and behavioral deficits induced by heavy metal exposure. However, these tests are highly vulnerable to subjective interpretation and influenced by confounding factors such as socioeconomic status, educational level, and concurrent mental illness. Furthermore, the amount of time it takes from the occurrence of exposure to the appearance of symptoms introduces yet another complication in diagnosis. Neurological damage might not manifest until years later when Someone is unaware that something terrible has happened (Famurewa et al., 2022; Gomes da Silva et al., 2024; Hu et al., 2024; Rajagopalan et al., 2023; Shanmugavel et al., 2024).

Future research is needed to work harder on developing more sensitive, specific, and accessible diagnostic techniques for the timely and accurate detection of heavy metal–induced neurotoxicity. It must find pathways of this kind to make real gains in industrial health. One promising development is that novel biomarkers reflecting ongoing neurotoxic effects rather than current levels of exposure can be picked up. Advances in omics technologies such as genomics, transcriptomics, proteomics, and metabolomics pave the way for discovering heavy metal–induced neurotoxicity molecular signatures. For example, differential gene expression profiles and protein biomarkers associated with oxidative stress, neuroinflammation, and mitochondrial dysfunction could be the first signs of neurotoxic damage. Liquid biopsy methods, including circulating microRNAs (miRNAs) and exosomal biomarkers, are now emerging as noninvasive means of detecting neurotoxic effects at the molecular level. miRNAs have critical functions in controlling the activity of nerve cells and have been implicated in heavy metal–induced neurotoxicity (Andrade et al., 2014; Bjørklund et al., 2024; Caini et al., 2023; Gomes da Silva et al., 2024; Jang & Hoffman, 2011; Paul, Samanta, Mallick, & Majumdar, 2024). The miRNA expression patterns in blood or CSF could thus be decisive steps forward in diagnosing neurotoxicity. Similarly, exosomes, small extracellular vesicles containing lipids, proteins, and nucleic acids, can mirror pathological changes in the brain, providing a window into neurotoxic processes. Neuroimaging techniques are increasingly headed to greater specificity for detecting the changes in the brain that heavy metals can produce. fMRI and diffusion tensor imaging can detect subtle changes in brain connectivity or white matter integrity accompanying neurotoxic exposure. New methods, such as MEG and neuroimaging-based metal mapping, can directly prove metal accumulation in brain tissue. Moreover, the combination of neuroimaging with machine learning algorithms can enhance diagnostic accuracy by recognizing patterns of brain abnormalities associated with specific heavy metal exposures.

Another frontier in diagnosing neurotoxicity caused by heavy metals is wearable biosensors and point-of-care testing devices. Advances in nanotechnology have led to the development of highly sensitive biosensors for detecting trace levels of heavy metals in body liquids. Portable electrochemical sensors and smartphone-based diagnostics tools could make heavy metal exposure monitoring immediate, especially among higher risk populations. In this way, it was not only children with enterovirus infection but also elderly persons with symptoms resulting from acute or chronic lead poisoning for whom continued surveillance and intervention were made.

Innovations like these could close the gap between laboratory-based diagnosis and field practice and provide timely interventions. Achieving these prospects requires a multidisciplinary and collaborative approach, including clinicians, researchers, public health authorities, and regulatory agencies. Given the complexity of heavy metal poisons and their broader implications, formulating new diagnostic capabilities and better patient outcomes calls for cross-disciplinary partnerships.

Clinicians are vital in identifying early signs and symptoms of neurotoxicity due to heavy metals and must incorporate comprehensive exposure assessments into routine medical history. Standardised screening procedures must be established for identifying people, especially at risk, such as employees with high occupational levels of heavy metal exposure or individuals who live near industrial sources of pollution. Exposure histories should be obtained through questionnaires, neurobehavioral tests, and biomarker tests as part of routine clinical practice to provide early diagnosis and treatment.

Researchers must pursue novel diagnostic strategies and validate budding biomarkers through large-scale epidemiological surveys and clinical trials. Longitudinal studies are vital for understanding the progression of heavy metal-induced toxic neuropathies and locating the first signs of as yet undiagnosed neurological disorders developing before an irreversible phase sets in. Moreover, efforts are needed to standardize biomarker research methods across different populations and exposure scenarios to enhance diagnostic reliability (Caini et al., 2023; Harischandra et al., 2018; Jang & Hoffman, 2011; Paul, Samanta, Mallick, & Majumdar, 2024; Yu et al., 2024).

Public health agencies have been pivotal in policies that reduce heavy metal exposure, generate new early detection strategies, and monitor related diseases. Programs to strengthen environmental monitoring, setting stricter exposure limits, and cutting-bled access to diagnostic testing are all imperative to reducing the burden of heavy metal–induced neurotoxicity. In addition, public health initiatives should be directed at educating communities about the hazards of heavy metal exposure and calling for preventative measures: more stringent industrial regulation, safe water supplies, and better nutrition intake. What we most urgently need now is a noninvasive method of diagnosis of cranial nerve micronuclei that can turn up definitive findings in every subject. Also, the clinical picture should be distinguishable from unrelated diseases with which it has often been confused. Diagnosing heavy metal–induced neurotoxicity remains a daunting task because there is no specific biomarker for the diagnosis, limitations of technology, and individual variances in people's vulnerability. Future research should prioritize developing sensitive and diagnostic techniques that are specific and widely accessible. This includes novel biomarkers, improved neuroimaging methods, and liquid biopsy approaches. Clinicians, researchers, and public health authorities should unite in collaborative efforts. Everyone's shared goal is a diagnosis capability already practiced and ready for action. This is crucial to catching heavy metal neurotoxicity early, which can then avoid fatal outcomes for patients in the long run.

20.7 Conclusion

Today, heavy metal–induced neurotoxicity is a serious health problem. People worldwide are confronted with toxic metals like lead, mercury, arsenic, and cadmium in the environment and on the job. When these heavy metals enter our bodies, as some are accumulated there, they cause permanent damage to our nervous systems. To help control or stop neurotoxicity, diagnostics need to be made promptly. Delayed detection usually leads to a progressive deterioration in the nervous system, cognitive deficits, and movement dysfunction, all of which negatively impact one's quality of life. The neurotoxic nature of heavy metals requires us to have a sound diagnostic framework: oftentimes, symptoms such as fatigue, memory loss, and depression may all be signs, but neither blood tests nor biopsies are sufficiently precise. Special diagnostic tests have been developed to assess exposure to heavy metals and their neurotoxic effects, each having merits and limitations. Traditional methods include blood, urine, and hair analysis. They can tell us the body's recent and chronic exposure to metals thus far. Despite extensive research, this method cannot reflect how an organ has become heavily infiltrated with metals. These include the brain, which, due to a person's lifetime accumulation, has often undergone cellular changes indicative of more than functional decline. Despite being a much more promising prospect than existing methods, neuroimaging technology is confined to the laboratory. It is expensive partly because of how much work it requires on each patient's case. As another way forward, new biochemical and genetic testing represents a breakthrough. Measures such as the level of certain enzymes in the blood from oxidative stress-indicative markers, epigenetic changes, or metallothionein gene expression offer an opportunity to observe individual makeups and potential pathways for neurotoxic preconditioning. This requires further acceptable and standardised guarantees before these methods become general practice. Because heavy metals cause so many different types of damage to the nervous system, a thorough diagnostic approach is required to draw accurate conclusions about the diagnosis. It both pinpoints the onset and indicates the direction treatment should follow. In evaluating the patient suspected of neurotoxicity from heavy metals, a full inquiry should be made about his exposure, as well as detailed physical and mental examinations paired with sensitive aptitude tests for cognition. However, neuroimaging equipment is expensive, and not every medical institution can afford it. More advanced medical tests, such as oxidative stress biomarkers, epigenetic changes, and metallothionein expression, can improve diagnostic precision by reflecting earlier pathophysiological changes. Investigations into the genes that set up someone's predisposition to strong reactions when metal homeostasis, transport, and detoxification pathways are disturbed. With new Omics technologies, including "metabolomics" and (proteomics), it is now possible to look for alternative early diagnostic indicators and assess therapeutic responses. However, these methods must be more widely validated before they can be universally accepted. Given the complex nature of heavy metal–induced neurotoxicity, comprehensive diagnoses require integrating various methods. It needs a panoramic view and a precise form check of heavy metal neurotoxic dust. However, even though considerable progress has been made in diagnosis, there are still several obstacles before it can take root and truly become universal. Among these challenges is the inability to find low-cost, decentralized means of assessing a person's biochemistry and metabolism. In the future, research should focus on creating cheap portable monitors.

Diagnostic instruments required in these locations will also be developed, based on machine learning analysis of individual cases. Public health efforts should concentrate on monitoring the environment, imposing safe practices in the workplace, and having communities take the initiative to monitor their health.

References

Anchidin-Norocel, L., Gutt, G., Tătăranu, E., & Amariei, S. (2024). Electrochemical sensors and biosensors: Effective tools for detecting heavy metals in water and food with possible implications for children's health. *International Journal of Electrochemical Science, 19*, 100643. Retrieved from https://doi.org/10.1016/j.ijoes.2024.100643.

Andrade, V., Mateus, M. L., Batoréu, M. C., Aschner, M., & Marreilha dos Santos, A. P. (2014). Changes in rat urinary porphyrin profiles predict the magnitude of the neurotoxic effects induced by a mixture of lead, arsenic and manganese. *Neurotoxicology, 45*, 168–177. Retrieved from https://doi.org/10.1016/j.neuro.2014.10.009.

Askari, V. R., Fadaei, M. S., Mostafazade, R., Dabbaghi, M. M., Fadaei, M. R., & Baradaran Rahimi, V. (2025). Chapter 27 – Biosensors used for detection of toxic and heavy metals in biological samples. In M. S. Hasnain, & A. K. Nayak (Eds.). *Aminabhavi TMBT-A of B in H* (pp. 629–647). Academic Press.

Bjørklund, G., Đorđević, A. B., Hamdan, H., Wallace, D. R., & Peana, M. (2024). Metal-induced autoimmunity in neurological disorders: A review of current understanding and future directions. *Autoimmunity Reviews, 23*, 103509. Retrieved from https://doi.org/10.1016/j.autrev.2023.103509.

Boas, D., Remennik, S., & Reches, M. (2023). Peptide-capped Au and Ag nanoparticles: Detection of heavy metals and photochemical core/shell formation. *Journal of Colloid and Interface Science, 631*, 66–76. Retrieved from https://doi.org/10.1016/j.jcis.2022.10.154.

Caini, S., Cozzolino, F., Saieva, C., Aprea, M. C., De Bonfioli Cavalcabo', N., Ermini, I., ... Masala, G. (2023). Serum heavy metals and breast cancer risk: A case-control study nested in the Florence cohort of the EPIC (European Prospective Investigation into Cancer and nutrition) study. *The Science of the Total Environment, 861*, 160568. Retrieved from https://doi.org/10.1016/j.scitotenv.2022.160568.

Dey, P., Kaur, M., Khajuria, A., Kaur, D., Singh, M., Alajangi, H. K., ... Barnwal, R. P. (2024). Heavy metal ion detection with nano-engineered materials: Scaling down for precision. *Microchemical Journal, Devoted to the Application of Microtechniques in all Branches of Science, 196*, 109672. Retrieved from https://doi.org/10.1016/j.microc.2023.109672.

Famurewa, A. C., Renu, K., Eladl, M. A., Chakraborty, R., Myakala, H., El-Sherbiny, M., ... Valsala Gopalakrishnan, A. (2022). Hesperidin and hesperetin against heavy metal toxicity: Insight on the molecular mechanism of mitigation. *Biomedicine & Pharmacotherapy=Biomedecine & Pharmacotherapie, 149*, 112914. Retrieved from https://doi.org/10.1016/j.biopha.2022.112914.

Florio, D., Marasco, D., & Manna, S. L. (2025). Approaches for developing peptide- and metal complexes- or chelators-based leads for anti-amyloid drugs. *Inorganica Chimica Acta, 577*, 122474. Retrieved from https://doi.org/10.1016/j.ica.2024.122474.

Gade, M., Comfort, N., & Re, D. B. (2021). Sex-specific neurotoxic effects of heavy metal pollutants: Epidemiological, experimental evidence and candidate mechanisms. *Environmental Research, 201*, 111558. Retrieved from https://doi.org/10.1016/j.envres.2021.111558.

Gomes da Silva, J., Chagas, C. A., Giselly dos Santos Souza, T., Chagas de Araújo, M., Alexandre de Araújo, L. C., Melo Santos, A. M., ... Betânia Melo de Oliveira, M. (2024). RETRACTED: Using structural equation modeling to assess the genotoxic and mutagenic effects of heavy metal contamination in the freshwater ecosystems: A study involving *Oreochromis niloticus* in an urban river. *The Science of the Total Environment, 913*, 169529. Retrieved from https://doi.org/10.1016/j.scitotenv.2023.169529.

Goutam Mukherjee, A., Ramesh Wanjari, U., Renu, K., Vellingiri, B., & Valsala Gopalakrishnan, A. (2022). Heavy metal and metalloid-induced reproductive toxicity. *Environmental Toxicology and Pharmacology, 92*, 103859. Retrieved from https://doi.org/10.1016/j.etap.2022.103859.

Harischandra, D. S., Ghaisas, S., Rokad, D., Zamanian, M., Jin, H., Anantharam, V., ... Kanthasamy, A. G. (2018). Environmental neurotoxicant manganese regulates exosome-mediated extracellular miRNAs in cell culture model of Parkinson's disease: Relevance to α-synuclein misfolding in metal neurotoxicity. *Neurotoxicology, 64*, 267–277. Retrieved from https://doi.org/10.1016/j.neuro.2017.04.007.

Hu, Y., Wang, X., Niu, Y., He, K., & Tang, M. (2024). Application of quantum dots in brain diseases and their neurotoxic mechanism. *Nanoscale Advances, 6*, 3733–3746. Retrieved from https://doi.org/10.1039/d4na00028e.

Jang, D. H., & Hoffman, R. S. (2011). Heavy metal chelation in neurotoxic exposures. *Neurologic Clinics, 29*, 607–622. Retrieved from https://doi.org/10.1016/j.ncl.2011.05.002.

Karri, V., Ramos, D., Martinez, J. B., Odena, A., Oliveira, E., Coort, S. L., ... Kumar, V. (2018). Differential protein expression of hippocampal cells associated with heavy metals (Pb, As, and MeHg) neurotoxicity: Deepening into the molecular mechanis of neurodegenerative diseases. *Journal of Proteomics, 187*, 106–125. Retrieved from https://doi.org/10.1016/j.jprot.2018.06.020.

Karthik, V., Karuna, B., Kumar, P. S., Saravanan, A., & Hemavathy, R. V. (2022). Development of lab-on-chip biosensor for the detection of toxic heavy metals: A review. *Chemosphere, 299*, 134427. Retrieved from https://doi.org/10.1016/j.chemosphere.2022.134427.

Kim, N.-E., Heo, M., Shin, H., Do, A. R., Kim, J., Kang, H.-G., ... Won, S. (2025). Heavy metal exposure and its effects on APOC3, CFAI, and ZA2G. *Journal of Hazardous Materials, 482*, 136574. Retrieved from https://doi.org/10.1016/j.jhazmat.2024.136574.

Kumar, A., Abhinav, S., Rachamalla, M., Ali, M., Navin, S., Kumar, D., & Ghosh, A. K. (2025). Chapter 29 – Treatment of heavy metal induced neurotoxicity via advanced drug delivery systems. In P. A. Chawla, R. Loebenberg, K. Dua, & V. Parikh (Eds.). *Chawla VBT-NDDS in the management of CNSD* (pp. 393–407). Academic Press.

Kwakye, G. F., Jiménez, J. A., Thomas, M. G., Kingsley, B. A., McIlvin, M., Saito, M. A., & Korley, E. M. (2019). Heterozygous huntingtin promotes cadmium neurotoxicity and neurodegeneration in striatal cells via altered metal transport and protein kinase C delta dependent oxidative stress and apoptosis signaling mechanisms. *Neurotoxicology, 70*, 48–61. Retrieved from https://doi.org/10.1016/j.neuro.2018.10.012.

Lewis, F., Azmoun, S., Shoieb, D., Marquez, R., & Lucchini, R. G. (2024). BT-RM in BS. In *Neurotoxicology of metals*. Elsevier.

Ma, C.-S., Li, D.-L., Wang, F., Wang, J.-P., & He, M.-T. (2024). Neurotoxicity from long-term exposure to 6-PPDQ: Recent advances. *Ecotoxicology and Environmental Safety, 282*, 116689. Retrieved from https://doi.org/10.1016/j.ecoenv.2024.116689.

Nabavi, A., Safari, F., Kashkooli, M., Sadat Nabavizadeh, S., & Molavi Vardanjani, H. (2024). Early prediction of cognitive impairment in adults aged 20 years and older using machine learning and biomarkers of heavy metal exposure. *Current Research in Toxicology, 7*, 100198. Retrieved from https://doi.org/10.1016/j.crtox.2024.100198.

Naqvi, S., Kumar, P., & Flora, S. J. S. (2020). Comparative efficacy of Nano and Bulk Monoisoamyl DMSA against arsenic-induced neurotoxicity in rats. *Biomedicine & Pharmacotherapy=Biomedecine & Pharmacotherapie, 132*, 110871. Retrieved from https://doi.org/10.1016/j.biopha.2020.110871.

Nicolescu, R., Petcu, C., Cordeanu, A., Fabritius, K., Schlumpf, M., Krebs, R., ... Winneke, G. (2010). Environmental exposure to lead, but not other neurotoxic metals, relates to core elements of ADHD in Romanian children: Performance and questionnaire data. *Environmental Research, 110*, 476–483. Retrieved from https://doi.org/10.1016/j.envres.2010.04.002.

Okpoghono, J., Agbetuyi, B. A., Eyide, T., & Onyesom, I. (2024). Influence of kaempferol against garri meal tainted with vacuum gas oil induced nephro- and neurotoxicity. *Food Chemistry Advances, 5*, 100794. Retrieved from https://doi.org/10.1016/j.focha.2024.100794.

Onyancha, R. B., Aigbe, U. O., Ukhurebor, K. E., & Muchiri, P. W. (2021). Facile synthesis and applications of carbon nanotubes in heavy-metal remediation and biomedical fields: A comprehensive review. *Journal of Molecular Structure, 1238*, 130462. Retrieved from https://doi.org/10.1016/j.molstruc.2021.130462.

Paul, P., Samanta, S., Mallick, A., & Majumdar, T. (2024). Searching the factual factor in emission tweaking of a reported bisindole based self-aggregation sensitive organic emitter: The missing link. *Chemical Physics Impact, 8*, 100526. Retrieved from https://doi.org/10.1016/j.chphi.2024.100526. (Accessed 24 May 2025).

Pomary, P. K., Eichau, S., Amigó, N., Barrios, L., Matesanz, F., García-Valdecasas, M., ... Garcia-Martin, M. L. (2023). Multifaceted analysis of cerebrospinal fluid and serum from progressive multiple sclerosis patients: Potential role of vitamin c and metal ion imbalance in the divergence of primary progressive multiple sclerosis and secondary progressive multiple sclerosis. *Journal of Proteome Research, 22*, 743–757. Retrieved from https://doi.org/10.1021/acs.jproteome.2c00460.

Rajagopalan, V., Venkataraman, S., Rajendran, D. S., Vinoth Kumar, V., Kumar, V. V., & Rangasamy, G. (2023). Acetylcholinesterase biosensors for electrochemical detection of neurotoxic pesticides and acetylcholine neurotransmitter: A literature review. *Environmental Research, 227*, 115724. Retrieved from https://doi.org/10.1016/j.envres.2023.115724.

Samal, R. R., & Subudhi, U. (2024). Modulation of antioxidant enzyme by light and heavy rare earth metals: A case study with catalase. *International Journal of Biological Macromolecules, 283*, 137820. Retrieved from https://doi.org/10.1016/j.ijbiomac.2024.137820.

Sears, L., Smith, M. J., Cai, L., & Zierold, K. M. (2024). The association of manganese overexposure and neurobehavioral function is moderated by arsenic: A metal mixture analysis of children living near coal ash storage sites. *Neurotoxicology, 103*, 78–86. Retrieved from https://doi.org/10.1016/j.neuro.2024.06.005.

Shanmugavel, A., Rene, E. R., Balakrishnan, S. P., Krishnakumar, N., & Jose, S. P. (2024). Heavy metal ion sensing strategies using fluorophores for environmental remediation. *Environmental Research, 260*, 119544. Retrieved from https://doi.org/10.1016/j.envres.2024.119544.

Sharma, S., Mehan, S., Khan, Z., Gupta, G. D., & Narula, A. S. (2024). Icariin prevents methylmercury-induced experimental neurotoxicity: Evidence from cerebrospinal fluid, blood plasma, brain samples, and in-silico investigations. *Heliyon, 10*, e24050. Retrieved from https://doi.org/10.1016/j.heliyon.2024.e24050.

Singh, A., & Kostova, I. (2024). Health effects of heavy metal contaminants Vis-à-Vis microbial response in their bioremediation. *Inorganica Chimica Acta, 568*, 122068. Retrieved from https://doi.org/10.1016/j.ica.2024.122068.

Tian, J., Hu, J., Liu, D., Yin, J., Chen, M., Zhou, L., & Yin, H. (2021). Cadmium chloride-induced transgenerational neurotoxicity in zebrafish development. *Environmental Toxicology and Pharmacology, 81*, 103545. Retrieved from https://doi.org/10.1016/j.etap.2020.103545.

Wellenberg, A., Brinkmann, V., Bornhorst, J., Ventura, N., Honnen, S., & Fritz, G. (2021). Cisplatin-induced neurotoxicity involves the disruption of serotonergic neurotransmission. *Pharmacological Research: The Official Journal of the Italian Pharmacological Society, 174*, 105921. Retrieved from https://doi.org/10.1016/j.phrs.2021.105921.

Xie, J., Zhou, F., Ouyang, L., Li, Q., Rao, S., Su, R., ... Fan, G. (2023). Insight into the effect of a heavy metal mixture on neurological damage in rats through combined serum metabolomic and brain proteomic analyses. *The Science of the Total Environment, 895*, 165009. Retrieved from https://doi.org/10.1016/j.scitotenv.2023.165009.

Yu, G., Wu, L., Su, Q., Ji, X., Zhou, J., Wu, S., ... Li, H. (2024). Neurotoxic effects of heavy metal pollutants in the environment: Focusing on epigenetic mechanisms. *Environmental Pollution (Barking, Essex: 1987), 345*, 123563. Retrieved from https://doi.org/10.1016/j.envpol.2024.123563.

Zhou, L., Li, R., Wang, F., Zhou, R., Xia, Y., Jiang, X., ... Chen, C. (2024). N6-methyladenosine demethylase FTO regulates neuronal oxidative stress via YTHDC1-ATF3 axis in arsenic-induced cognitive dysfunction. *Journal of Hazardous Materials, 480*, 135736. Retrieved from https://doi.org/10.1016/j.jhazmat.2024.135736.

Chapter 21

Prevention and mitigation strategies for heavy metal exposure

Makey Sri Chakra Naga Vardhan, Ashok S. Dambale, Prasann Kumar, and Nilabja Banerjee
Department of Agronomy, School of Agriculture, Lovely Professional University, Phagwara, Punjab, India

21.1 Introduction

Heavy metal contamination is one of the most pressing environmental challenges of the 21st century, with far-reaching consequences for ecosystems, human health, and economic stability. Heavy metals, including lead (Pb), mercury (Hg), cadmium (Cd), arsenic (As), chromium (Cr), and others, are naturally occurring elements with high atomic weights and densities that make them persistent environmental pollutants. Unlike organic pollutants, heavy metals do not degrade over time and can accumulate in the environment, leading to prolonged exposure risks (Kumar, Choudhury, Dey, & Sharma, 2024a; Sharma, Dey, Kumar, & Dwivedi, 2024). Their persistence is exacerbated by anthropogenic activities such as mining, industrialization, agriculture, and improper waste disposal. Despite their necessity in trace amounts for biological functions, such as zinc (Zn) and copper (Cu) being essential for enzymatic activities—excessive accumulation of these metals in biological systems can be toxic. The growing awareness of heavy metal pollution has led to the development of numerous remediation strategies; however, effective prevention remains the most sustainable solution for minimizing exposure risks. Heavy metal contamination originates from both natural and anthropogenic sources. Natural sources include volcanic eruptions, weathering of metal-bearing rocks, and forest fires, which release metals such as lead, mercury, and arsenic into the atmosphere, soil, and water bodies. However, anthropogenic activities are the primary contributors to heavy metal pollution, with industrial processes dominating their release into the environment. Industrial emissions from mining, smelting, electroplating, and manufacturing generate significant amounts of waste containing heavy metals, which leach into surrounding soil and water bodies, contaminating ecosystems. Similarly, metallurgical industries release mercury, lead, cadmium, and chromium through smelting and refining. Effluents from battery manufacturing, metal plating, and pigment production industries also contribute to severe heavy metal contamination in aquatic and terrestrial environments (Dey, Sharma, Kumar, & Dwivedi, 2024e; Kumar & Choudhury, 2024a; Kumar & Dwivedi, 2024). Modern agricultural practices, including pesticides, herbicides, and fertilizers, have significantly increased heavy metal accumulation in soil and water. Fertilizers, particularly phosphate-based ones, contain cadmium and arsenic, which gradually build up on farmlands. Additionally, sewage sludge and wastewater irrigation introduce heavy metals into farmlands, where they persist in the soil and enter the food chain through plant uptake, posing a direct risk to human and animal health. Heavy metals are widely present in municipal solid waste due to the disposal of batteries, electronic waste (e-waste), paints, plastics, and metal alloys. Inadequate waste management, such as open dumping and landfill leachate infiltration, releases lead, cadmium, and mercury into groundwater sources. Electronic waste, in particular, contains high concentrations of lead and cadmium from circuit boards, lithium batteries, and fluorescent lamps, making it a critical source of heavy metal pollution. The combustion of fossil fuels in transportation, power plants, and industries releases heavy metals into the atmosphere. Historically, leaded gasoline was a significant source of lead contamination, although its use has been phased out in many countries. However, emissions from diesel engines, coal combustion, and oil refining still contribute to mercury, arsenic, and chromium pollution, leading to widespread environmental contamination. Industrial discharge, improper wastewater treatment, and oil spills contribute to heavy metal accumulation in marine and freshwater ecosystems. Mercury pollution in

Heavy Metal Toxicity and Neurodegeneration. DOI: https://doi.org/10.1016/B978-0-443-36575-1.00022-4

water bodies, primarily from gold mining and industrial wastewater, leads to the formation of methylmercury, a highly toxic compound that bioaccumulates in fish and seafood. Human consumption of contaminated fish can lead to severe neurological and developmental disorders, particularly in infants and pregnant women (Dey, Kumar, & Singh, 2024; Kumar, Devi, & Panwar, 2024b; Kumar, Jose Anand, & Rajagopal, 2024e). Heavy metal exposure poses significant risks to human health, affecting multiple organ systems and leading to chronic diseases, neurological impairments, and even death. The severity of toxicity depends on factors such as the type of metal, exposure duration, concentration, and individuals' biological susceptibility. Heavy metals exert toxic effects by generating oxidative stress, disrupting enzyme function, and interfering with essential biological processes. Heavy metals such as lead, mercury, and arsenic are neurotoxic, causing cognitive impairments, developmental disorders, and neurodegenerative diseases. Lead exposure, particularly in children, decreases IQ, causes behavioral changes, and leads to learning disabilities. Chronic exposure to mercury, often through seafood consumption, leads to Minamata disease, characterized by tremors, memory loss, and motor dysfunction. Arsenic exposure has been linked to cognitive decline and an increased risk of Alzheimer's disease. The kidneys and liver are major detoxification organs, making them highly susceptible to heavy metal toxicity. Cadmium, in particular, accumulates in the kidneys, causing nephrotoxicity, renal dysfunction, and kidney failure. Chronic arsenic exposure leads to liver fibrosis, cirrhosis, and an increased risk of hepatocellular carcinoma. Long-term exposure to heavy metals has been associated with hypertension, atherosclerosis, and an increased risk of heart attacks and strokes. Lead and cadmium interfere with calcium channels, disrupting cardiovascular function. Inhalation of airborne heavy metal particulates, such as chromium and nickel, from industrial emissions increases the risk of lung cancer, chronic obstructive pulmonary disease, and respiratory inflammation. The International Agency for Research on Cancer classifies several heavy metals, including arsenic, cadmium, chromium, and nickel, as human carcinogens. These metals interfere with DNA repair mechanisms, induce genetic mutations, and promote tumorigenesis (Devi & Kumar, 2024a; Kumar & Choudhury, 2024b; Saini, Kumar, & Panwar, 2024). Chronic arsenic exposure has been linked to skin, lung, and bladder cancers, while chromium exposure is associated with lung cancer and nasal cavity tumors. Given heavy metals' persistent and bioaccumulating nature, prevention and mitigation strategies are crucial to minimizing exposure and reducing environmental contamination. Effective strategies must be multidimensional and integrate technological, regulatory, and public health approaches. Implementing strict environmental regulations, such as the Clean Water Act, Resource Conservation and Recovery Act, and international treaties like the Minamata Convention on Mercury, is essential to limit industrial emissions and waste disposal. Enforcing pollution control measures, such as installing air filters, wastewater treatment systems, and sustainable mining practices, can significantly reduce heavy metal contamination at its source. Reducing heavy metal–based fertilizers and pesticides and adopting organic farming techniques can minimize soil contamination. Crop rotation, biofertilisers, and using hyperaccumulator plants for phytoremediation are promising strategies for reducing crop metal uptake. Various remediation techniques, including bioremediation, phytoremediation, and adsorption-based methods using activated carbon and biochar, offer practical solutions for heavy metal removal from contaminated water and soil. Developing nanomaterial-based adsorbents and microbial consortia for heavy metal detoxification is an emerging area of research. Educating communities about the risks of heavy metal exposure and promoting dietary interventions—such as increased selenium, zinc, and antioxidants—can help mitigate toxic effects. Regular biomonitoring of at-risk populations and early detection of heavy metal poisoning through blood and urine tests are critical in preventing long-term health consequences (Devi, Kumar, & Singh, 2024a; Dey, Kumar, & Panwar, 2024; Sharma, Kumar, & Panwar, 2024). Heavy metal contamination remains a significant global challenge with severe environmental and health implications. Addressing this issue requires a comprehensive approach encompassing prevention, regulation, remediation, and public health initiatives. Integrating scientific advancements with policy-driven solutions can mitigate heavy metal exposure and protect human health and ecological sustainability (Table 21.1).

21.2 Primary prevention, controlling the sources of contamination

21.2.1 Industrial and agricultural regulation

Environmental regulations are critical in addressing global ecological challenges and balancing industrial growth and environmental protection. Regulatory frameworks such as those established by the Environmental Protection Agency (EPA) and the World Health Organization (WHO) are foundational pillars for enforcing pollution control, setting emission limits, and safeguarding public health. Governments worldwide have adopted stringent policies that regulate industrial waste disposal, air and water quality standards, and hazardous substance management. These regulations are vital in curbing environmental degradation, fostering corporate responsibility, and encouraging industries to invest in cleaner technologies (Devi, Kumar, & Singh, 2024b; Devi, Kumar, & Singh, 2024c; Kumar &

TABLE 21.1 Different sources of exposure to heavy metals and their impact.

Heavy metal	Sources of exposure	Health risks	Prevention strategies	Mitigation techniques	Technological interventions	Regulatory guidelines
Lead (Pb)	Industrial discharge, contaminated water, lead-based paints, and dust in urban areas	Short-term: headache, dizziness, nausea; long term: neurological damage, developmental delays in children, kidney damage, high blood pressure	Ban on lead-based paints, use of unleaded gasoline, regular monitoring of water quality, proper disposal of lead-containing waste	Phytoremediation using plants like mustard, lead-contaminated water filtration using activated carbon	Portable water testing kits, soil testing technologies, lead detection using portable XRF	WHO guidelines (lead in drinking water < 10 μg/L), EPA maximum allowable levels
Mercury (Hg)	Coal burning, industrial effluent, contaminated fish and seafood, waste incineration	Short-term: tremors, memory loss, insomnia; long term: kidney damage, neurological disorders, reproductive toxicity	Substituting mercury-containing devices with alternatives (e.g., digital thermometers), enforcing safe disposal of mercury waste, and using cleaner energy sources.	Activated carbon filtration, sulfur-based remediation in contaminated soil, use of bioremediation by microorganisms	Mercury detection using portable XRF, advanced air monitoring sensors, and satellite imagery for tracking mercury pollution	WHO (mercury in air: < 1.0 μg/m³), EPA water quality criteria for mercury
Cadmium (Cd)	Industrial emissions, phosphate fertilisers, contaminated food (e.g., shellfish), tobacco smoke	Short-term: nausea, vomiting, flu-like symptoms; long term: kidney damage, lung cancer, bone disease	Use of cadmium-free fertilisers, reducing industrial emissions, proper disposal of cadmium-containing batteries	Soil amendment with chelating agents (e.g., EDTA), phytoremediation using hyperaccumulators like Thlaspi	Cadmium detection using ion-selective electrodes, automated monitoring systems for industrial emissions	WHO guidelines (cadmium in drinking water < 3 μg/L), EPA's NAAQS
Arsenic (As)	Contaminated groundwater, industrial effluent, pesticides, wood preservatives	Short-term: skin lesions, nausea, vomiting; long term: skin cancer, lung cancer, cardiovascular disease, diabetes	Use of arsenic-free water sources, installation of filtration systems, banning arsenic-based pesticides	Reverse osmosis filtration, activated alumina adsorption, phytoremediation using arsenic-tolerant plants	Arsenic detection via portable field sensors, advanced groundwater monitoring technologies	WHO (arsenic in drinking water < 10 μg/L), EPA MCL for arsenic
Chromium (Cr)	Industrial discharge (tanning, metal plating), contaminated groundwater	Short-term: skin ulcers, gastrointestinal issues; long term: lung cancer, kidney damage, liver damage	Using chromium-free tanning processes, improving waste management in industries, regulating chromium emissions	Bioremediation using chromium-reducing bacteria, electrokinetic remediation of chromium-contaminated soils	Chromium detection using atomic absorption spectrometry, portable water and soil testing devices	WHO (chromium VI in drinking water < 50 μg/L), EPA guidelines on chromium emissions
Nickel (Ni)	Industrial emissions (refining, mining), contaminated food (e.g., nuts, chocolate), tobacco smoke	Short-term: allergic reactions, lung irritation; long term: cancer, kidney and liver damage, respiratory issues	Substituting nickel-containing alloys with alternatives, controlling industrial emissions, banning excessive nickel use in consumer products	Soil stabilization using biosorbents and adsorbents like activated carbon in water treatment	Nickel detection using XRF, LIBS	WHO (nickel in drinking water < 20 μg/L), OSHA occupational limits for nickel exposure

(Continued)

TABLE 21.1 (Continued)

Heavy metal	Sources of exposure	Health risks	Prevention strategies	Mitigation techniques	Technological interventions	Regulatory guidelines
Aluminum (Al)	Industrial emissions, contaminated water, food packaging, pharmaceuticals	Short-term: gastrointestinal issues, memory loss; long term: neurodegenerative diseases (e.g., Alzheimer's)	Limiting aluminum-based additives in food, monitoring aluminum levels in drinking water, improving waste disposal practices	Adsorption of aluminum from water using activated alumina, soil treatment with calcium sulfate	Detection using atomic absorption spectroscopy, field tests using colourimetric sensors	EPA regulations on aluminum levels in the water, WHO guidelines
Copper (Cu)	Agricultural runoff, industrial effluent, copper-based pesticides, drinking water pipes	Short-term: nausea, vomiting, stomach cramps; long term: liver and kidney damage, neurological disorders	Substituting copper-based fungicides, regulating industrial effluent, replacing old copper pipes	Phytoremediation uses copper-tolerant plants like Brassica and copper ion-exchange resins for water purification.	Copper detection using portable XRF, electrochemical sensors for real-time monitoring	WHO guidelines for copper in drinking water (2 mg/L), EPA copper standards for aquatic life
Zinc (Zn)	Industrial emissions, agricultural runoff, galvanized materials, sewage sludge	Short-term: nausea, vomiting, abdominal pain; long term: liver and kidney damage, interference with iron absorption	Improved industrial waste management, regulating zinc in agricultural runoff, and minimizing the use of zinc-containing pesticides	Electrochemical remediation, phytoremediation using zinc-hyperaccumulating plants like Arabidopsis	Zinc detection through ion-selective electrodes, electrochemical sensors for soil testing	WHO guidelines (zinc in drinking water < 5 mg/L), EPA regulations on zinc in aquatic ecosystems
Beryllium (Be)	Industrial processing (aerospace, electronics), contaminated air, soil, and water	Short-term: cough, shortness of breath, chest pain; Long-term: lung cancer, CBD	Enforcing strict industrial emissions controls, using protective equipment for workers, improving workplace safety	Soil stabilization using phosphates, bioremediation using plants like sunflower and Indian mustard	Beryllium detection via XRF, air monitoring using portable sensors	OSHA workplace limits for beryllium, EPA emission standards for beryllium

EPA, Environmental Protection Agency; *LIBS*, laser-induced breakdown spectroscopy; *MCL*, maximum contaminant level; *NAAQS*, National Ambient Air Quality Standards; *OSHA*, Occupational Safety and Health Administration; *WHO*, World Health Organization; *XRF*, X-ray fluorescence.
Source: Based on Kumar (2024a), Kumar, Singh, and Singh (2024l), and Sharma, Kumar, and Singh (2024).

Singh, 2024a). For example, the Clean Air Act and the Clean Water Act in the United States have significantly contributed to reducing air pollution and improving water quality by mandating industries to implement pollution control technologies and adhere to permissible emission limits. Similarly, the European Union's REACH (Registration, Evaluation, Authorization, and Restriction of Chemicals) regulation ensures that chemical substances used in manufacturing are assessed for their environmental and health impacts, thereby minimizing potential risks to humans and ecosystems.

In addition to regulatory frameworks, sustainable mining and industrial practices are essential for mitigating environmental impacts associated with resource extraction and manufacturing. Traditional mining methods have long been associated with land degradation, deforestation, and pollution of air and water bodies due to the release of toxic substances. To address these issues, sustainable mining practices emphasize reduced environmental footprint, enhanced resource efficiency, and rehabilitation of mining sites postextraction. Techniques such as biomining, which employs microorganisms to extract valuable metals from ores without generating toxic waste, and precision mining, which uses advanced sensors and artificial intelligence (AI) to optimize resource extraction while minimizing waste, are becoming increasingly prevalent. Moreover, many industries are adopting circular economy principles, where waste materials are recycled and repurposed instead of being disposed of in landfills or released into the environment (Kumar & Singh, 2024b, 2024c, 2024d). For instance, the steel and aluminum industries have significantly incorporated scrap metal recycling into their production processes, reducing the need for virgin ore extraction and minimizing energy consumption. Furthermore, transitioning to renewable energy sources in industrial operations is crucial for sustainability. Industries can drastically cut greenhouse gas emissions and reduce carbon footprint by replacing fossil fuels with solar, wind, and hydroelectric power.

Another critical aspect of environmental sustainability is the phasing out of heavy metal-containing products, which pose severe health and environmental risks. Heavy metals such as lead, mercury, and cadmium are known to be highly toxic and persistent in the environment, leading to bioaccumulation and severe health complications in humans and wildlife. One of the most significant milestones in reducing heavy metal pollution was the global phase-out of leaded gasoline. Leaded gasoline, once widely used in vehicles, was found to contribute significantly to air pollution and cause severe neurological damage, particularly in children (Devi & Kumar, 2024b; Kumar, Khokhar, & Dey, 2024g; Kumar, Saini, & Sharma, 2024h). Through concerted efforts by the EPA, WHO, and the United Nations Environment Programme (UNEP), countries worldwide have successfully transitioned to unleaded gasoline, improving air quality and reducing lead exposure. Similarly, eliminating mercury thermometers and other mercury-containing medical devices has been crucial in minimizing mercury pollution. Mercury exposure can lead to severe neurological disorders, kidney damage, and developmental issues in fetuses and young children. International agreements such as the Minamata Convention on Mercury, which aims to protect human health and the environment from mercury emissions, have led to the widespread ban of mercury in products and industrial processes. Additionally, the electronics industry has been making significant strides in reducing heavy metal usage by developing lead-free solders and implementing stringent regulations on e-waste disposal. For example, the Restriction of Hazardous Substances (RoHS) directive in the European Union restricts the use of specific hazardous materials in electrical and electronic products, encouraging manufacturers to adopt safer alternatives (Devi & Kumar, 2024c; Kumar, Karam, & Choudhury, 2024f; Kumar, Saurabh, & Sharma, 2024i).

Despite the progress in implementing environmental regulations, promoting sustainable industrial practices, and phasing out hazardous substances, several challenges remain. Compliance with environmental laws varies across different regions due to disparities in enforcement mechanisms, economic priorities, and technological capabilities. Developing countries, in particular, often struggle with inadequate infrastructure and financial resources to implement and enforce stringent environmental policies. Additionally, industries may resist adopting sustainable practices due to concerns about increased operational costs and potential disruptions to production processes. To address these challenges, fostering global cooperation, investing in capacity-building initiatives, and providing financial incentives for industries to transition toward greener alternatives are essential. Public awareness and advocacy also play a crucial role in driving environmental reforms. Educating consumers about hazardous substances' ecological and health impacts can encourage demand for eco-friendly products, compelling industries to adopt sustainable practices (Kumar & Choudhury, 2024c; Kumar et al., 2024a; Kumar, Srivastav, Chaudhary, van Hullebusch, & Busquets, 2024m).

Moreover, advancements in science and technology offer promising solutions for achieving environmental sustainability. Innovations such as green chemistry, which focuses on designing chemical products and processes that minimize the use and generation of hazardous substances, can significantly reduce the environmental footprint of industrial activities. Developing biodegradable materials and eco-friendly packaging alternatives reduces plastic pollution and promotes a circular economy. Additionally, monitoring and data analytics advancements enable more effective enforcement of environmental regulations. Remote sensing technologies, for instance, can detect illegal mining activities and deforestation in real time, allowing authorities to take swift action against violators. Blockchain technology can also enhance transparency and accountability in environmental compliance by providing a tamper-proof record of industrial activities and emissions data (Choudhury, Kumar, Zhimo, & Sahoo, 2024; Kumar, Sharma, & Singh, 2024j; Pandey, Kumar, & Dey, 2024).

The role of international collaboration cannot be overstated in addressing global environmental challenges. Agreements such as the Paris Agreement on Climate Change and the Basel Convention on Hazardous Waste Management emphasize the need for collective action to tackle pollution, reduce carbon emissions, and promote sustainable development. Cross-border partnerships between governments, industries, and research institutions facilitate knowledge-sharing, technology transfer, and the implementation of best practices in environmental management. Additionally, multinational corporations' corporate social responsibility (CSR) initiatives contribute to sustainability by investing in clean energy projects, supporting reforestation programs, and adopting environment-friendly supply chain practices (Devi kumar et al., 2024; Dey, Sharma, & Kumar, 2024c; Kumar, Sharma, & Singh, 2024k).

Implementing environmental regulations, sustainable mining and industrial practices, and phasing out heavy metal-containing products are crucial steps toward achieving a cleaner and healthier planet. While significant progress has been made in reducing pollution, promoting resource efficiency, and eliminating hazardous substances, ongoing efforts are necessary to address existing challenges and ensure long-term environmental sustainability. Strengthening regulatory frameworks, fostering technological innovation, enhancing international cooperation, and raising public awareness are key strategies for overcoming barriers and advancing toward a more sustainable future. By prioritizing environmental protection alongside economic development, societies can achieve a harmonious balance that benefits both present and future generations (Table 21.2).

21.2.2 Sustainable waste management

Proper disposal of electronic and industrial waste has become a critical issue worldwide due to the rapid increase in technological advancements and industrialization. As industries expand and the consumption of electronic products rises, the generation of waste containing hazardous materials such as lead, mercury, cadmium, and other toxic substances has surged, posing severe environmental and health risks. Addressing this issue requires a multifaceted approach, including strict regulations, public awareness, and sustainable waste management practices. One of the most effective methods to mitigate the negative impact of electronic and industrial waste is recycling and recovering heavy metals. Many electronic components and industrial byproducts contain valuable metals such as gold, silver, palladium, and copper, which can be extracted and reused in various applications. Advanced techniques such as hydrometallurgical and pyrometallurgical processes enable the recovery of these metals, reducing the need for virgin metal mining and minimizing environmental degradation (Devi, Kapoor, & Kumar, 2024; Kumar, 2024b, 2024c, 2024d; Kumar, Dwivedi, & Upadhyay, 2024d). Additionally, adopting a circular economy model where manufacturers design products with recyclability in mind can significantly enhance the efficiency of material recovery, thereby reducing the accumulation of hazardous waste. Furthermore, reducing landfill leachate contamination is essential in preventing the infiltration of toxic substances into groundwater and surrounding ecosystems. Landfills that receive improperly disposed electronic and industrial waste often generate leachate, a highly polluted liquid formed when rainwater percolates through waste materials, dissolving harmful chemicals and heavy metals. If not managed effectively, landfill leachate can contaminate soil, rivers, and drinking water sources, leading to serious ecological and public health consequences. Implementing engineered landfill liners, leachate treatment plants, and bioremediation techniques can help mitigate the risks of landfill leachate. Additionally, diverting hazardous waste from landfills through improved recycling programs and extended producer responsibility (EPR) initiatives can significantly reduce the volume of toxic leachate produced. Governments and industries must collaborate to develop stringent policies that mandate proper e-waste and industrial waste disposal, encouraging the adoption of sustainable recycling practices. Public awareness campaigns and incentives for responsible waste disposal can further reinforce environment-friendly behaviors among consumers and businesses (Devi, Dogra, Kumar, & Singh, 2024; Jagadesh et al., 2024; Kumar & Choudhury, 2024e; Raj, Goutam, Kumar, & Singh, 2024; Srivastav et al., 2024). Moreover, technological innovations in waste treatment, such as electrochemical recovery, solvent extraction, and bioleaching, offer promising solutions for extracting heavy metals in a more environment-friendly manner. Establishing dedicated e-waste recycling facilities equipped with state-of-the-art recovery technologies can also enhance efficiency and ensure that hazardous materials do not end up in landfills. Beyond technological solutions, a holistic approach to waste management involves integrating policies that promote resource efficiency, green manufacturing, and responsible consumption. Governments should implement stringent waste disposal and recycling regulations, ensuring compliance with

TABLE 21.2 Primary prevention, controlling the sources of contamination.

Contamination source	Prevention method	Applicable industry/field	Effectiveness	Environmental impact	Costs/investment	Challenges/limitations
Industrial pollution	Installation of pollution control devices (e.g., scrubbers, filters)	Manufacturing, power generation, chemical industry	High	Reduction in air and water pollutants, lower greenhouse gas emissions	High initial setup cost, maintenance expenses	Requires technological expertize, regulatory compliance
Agricultural runoff	Sustainable farming practices (e.g., no-till farming, crop rotation, cover crops)	Agriculture	Moderate	Reduced nutrient runoff, improved soil quality, and water retention	Moderate cost, ongoing monitoring	Farmers' resistance to new practices, initial yield loss
Air pollution (vehicle emissions)	Transition to electric vehicles, emission control technologies (e.g., catalytic converters)	Transportation, automotive	High	Reduced air pollution, lower carbon emissions, improved air quality	High upfront cost for infrastructure, incentives needed	The high initial cost of EVs, charging infrastructure development
Waste management	Zero-waste policies, recycling programs, composting, and waste-to-energy technologies	Waste management, municipalities	High	Reduced landfill waste, lower methane emissions, less waste incineration	High investment in infrastructure	Public awareness, logistical challenges in waste segregation
Water pollution (chemical spills)	Implementing spill containment systems, better chemical handling protocols, stricter regulations	Manufacturing, oil and gas, chemical industry	High	Prevention of water contamination, protection of aquatic ecosystems	High cost for containment systems, regulatory compliance	Need for stringent monitoring, public policy enforcement
Soil contamination (heavy metals)	Phytoremediation (using plants to absorb contaminants), bioremediation techniques	Agriculture, environmental management	Moderate	Restoration of soil health, reduction of heavy metal concentration in soil	Moderate-to-high, depending on the scale	A time-consuming process, not practical for all types of contamination
Marine pollution (plastic waste)	Bans on single-use plastics, promotion of biodegradable alternatives, improved waste collection systems	Marine industry, packaging, environmental NGOs	High	Reduction in plastic waste, protection of marine life and ecosystems	Medium cost for alternatives, public awareness campaigns	Resistance from plastic manufacturers, consumer habits
Industrial wastewater	On-site wastewater treatment systems (e.g., biological treatment, filtration)	Manufacturing, food processing, chemical industry	High	Improved water quality, reduced toxins entering waterways	High investment, ongoing maintenance	High operational costs, complex regulatory frameworks
Deforestation (air and soil pollution)	Reforestation programs, afforestation, agroforestry techniques	Forestry, agriculture	High	Increased carbon sequestration, improved biodiversity, better soil quality	Moderate-to-high, depending on the scale	Land availability, funding for long-term projects
		Mining, construction	Moderate			High initial cost, ongoing operational cost

(Continued)

TABLE 21.2 (Continued)

Contamination source	Prevention method	Applicable industry/field	Effectiveness	Environmental impact	Costs/investment	Challenges/limitations
Mining operations (heavy metals, dust)	Dust suppression technologies (e.g., water spraying, dust collectors), proper mine waste management			Reduced dust pollution and decreased water contamination from runoff	High setup costs, long-term maintenance	
Food contamination (pesticides, bacteria)	Organic farming practices, IPM, improved food safety protocols	Agriculture, food industry	High	Reduced pesticide residues in food, improved soil and water health	Moderate cost, training for farmers	Resistance from conventional farmers, lower yield potential in some cases
Construction waste	Material recycling, use of sustainable construction materials, waste minimization practices	Construction, building industry	Moderate	Reduced waste generation, conservation of natural resources	Medium cost for infrastructure and training	Limited recycling facilities, higher upfront material costs
Electromagnetic pollution	Innovative grid technologies, shielding of electrical equipment, regulating wireless signals	Telecommunications, electronics	Moderate	Reduced electromagnetic radiation exposure, improved public health	Moderate investment, technological development	Technological challenges, high equipment costs

IPM, Integrated Pest Management.
Source: *From* Devi, Dey, Sharma, Kumar, and Singh (2024), Devi, Kaur, and Kumar (2024), Kumar and Choudhury (2024d), Sharma and Kumar (2024).

international standards such as the Basel Convention on the Control of Transboundary Movements of Hazardous Wastes and their Disposal. Encouraging businesses to adopt sustainable production and disposal methods through tax incentives and subsidies can drive positive change within the industrial sector. Moreover, consumer education on the importance of proper disposal and the environmental impact of e-waste can foster responsible behavior, leading to increased participation in recycling programs. Adopting urban mining, where valuable materials are extracted from discarded electronic devices and industrial waste, presents sustainable solutions to mitigate resource depletion. Research into alternative materials and biodegradable electronic components can also be crucial in minimizing environmental impact (Kumar and Choudhury, 2024f; Kumar & Dey, 2024; Kumar & Singh, 2024e; Kumar, Dey, & Choudhury, 2024c). Companies should be encouraged to invest in product design that facilitates easy disassembly and component recovery, ensuring that electronic devices can be efficiently recycled at the end of their life cycle. In addition, developing standardised labeling for recyclable materials and components can simplify sorting processes and enhance the effectiveness of recycling facilities. Strengthening international collaboration on e-waste management and heavy metal recovery can further amplify efforts to create a sustainable global waste management system. Countries with advanced recycling infrastructure can assist developing nations in establishing effective waste management policies and technologies. Joint initiatives between governments, industries, and research institutions can accelerate innovation in waste treatment and resource recovery. Furthermore, fostering partnerships between the public and private sectors can lead to the development of large-scale recycling initiatives, creating job opportunities and economic growth while simultaneously addressing environmental concerns. The importance of landfill leachate management cannot be overstated, as the long-term environmental consequences of untreated leachate can be devastating. Implementing sustainable landfill designs that incorporate leachate collection systems, impermeable liners, and constructed wetlands for natural filtration can significantly reduce the risk of groundwater contamination (Jagadesh et al., 2024; Kumar & Choudhury, 2024g, 2024h; Kumar, 2024c, 2024d; Srivastav et al., 2024). Additionally, exploring alternative waste disposal methods such as incineration with energy recovery and plasma gasification can help minimize landfill dependency while generating useful byproducts such as syngas and slag, which can be repurposed in various industries. Implementing smart waste tracking systems utilizing blockchain and IoT (Internet of Things) can enhance transparency and accountability in waste disposal, ensuring that hazardous materials are properly handled and directed towards appropriate recycling or treatment facilities. Monitoring waste generation patterns through data analytics can help policymakers devise targeted waste reduction and resource optimization strategies. Moreover, integrating AI and machine learning (ML) in waste sorting processes can improve the efficiency of recycling plants by accurately identifying and separating recyclable materials from nonrecyclables. In conclusion, proper disposal of electronic and industrial waste, recycling and recovering heavy metals, and reducing landfill leachate contamination requires a comprehensive and coordinated effort among governments, industries, and consumers. By implementing sustainable waste management practices, investing in advanced recycling technologies, and enforcing stringent environmental regulations, we can mitigate the adverse effects of hazardous waste and move towards a more sustainable future. The transition to a circular economy, where materials are continuously repurposed, and waste generation is minimized, is key to addressing the growing challenges of electronic and industrial waste. Public participation, corporate responsibility, and government intervention are all crucial in ensuring that waste is managed environmentally, ultimately safeguarding human health and ecological integrity for future generations (Devi et al., 2024d; Kumar & Choudhury, 2024d; Kumar et al., 2024a, 2024j; Pandey et al., 2024).

21.2.3 Green chemistry approaches

The global demand for heavy metal-free alternatives is driven by increasing environmental and human health concerns related to heavy metal toxicity. Heavy metals, such as lead, cadmium, mercury, and chromium, are commonly used in electronics, pigments, batteries, and coatings. However, their persistence in the environment and bioaccumulate nature pose significant ecological and health risks, necessitating the development of sustainable and eco-friendly substitutes. Researchers and industries are actively exploring biocompatible, biodegradable, and nontoxic materials to replace heavy metals in various applications. One of the primary areas of heavy metal–free innovation is electronics and energy storage. Traditional lithium-ion batteries often contain cobalt and nickel, which pose environmental hazards due to mining and disposal challenges. Researchers are developing sodium-ion and lithium-sulfur batteries with sulfur-based cathodes and nonmetallic anodes to eliminate the need for toxic heavy metals. Additionally, graphene and carbon-based nanomaterials are being explored for supercapacitors, providing high efficiency without relying on heavy metal components (Devi et al., 2024c; Kumar & Singh 2024d, 2024b; Kumar, 2024a).

Perovskite solar cells have gained attention in semiconductors due to their high efficiency. However, conventional perovskites contain toxic lead, raising environmental contamination concerns. Scientists are actively developing lead-free perovskites using tin- and bismuth-based compounds, which show promising photovoltaic properties with reduced toxicity. To enhance sustainability, organic and dye-sensitized solar cells are also improved with nontoxic elements. Heavy metals have been widely used in pigments and dyes for coloring textiles, paints, plastics, and ceramics. Conventional pigments, such as cadmium-based reds and lead chromates, pose serious environmental risks due to their leaching potential. To address these concerns, researchers have developed biobased and mineral-based alternatives. Iron oxides, titanium dioxide, and cerium-based pigments provide vibrant colors without heavy metals. Natural pigments derived from plant extracts, such as anthocyanins, betalains, and chlorophyll derivatives, offer nontoxic and biodegradable options for sustainable colouration.

Microbial fermentation has emerged as a novel approach to producing eco-friendly pigments. Engineered bacteria, such as *Escherichia coli* and *Streptomyces* species, can synthesize natural dyes like indigo and carotenoids in controlled conditions. This method significantly reduces the environmental impact compared to synthetic dye production, which often involves toxic precursors and heavy metal catalysts. Traditional corrosion-resistant coatings usually contain heavy metals like hexavalent chromium, which is highly poisonous and carcinogenic. Green alternatives, such as graphene-based coatings, self-healing polymers, and biobased inhibitors, have been developed to enhance durability without environmental risks. Silica- and cerium-based coatings exhibit excellent corrosion resistance while minimizing toxicity. Zinc-based coatings, widely used for galvanisation, pose ecological concerns due to heavy metal leaching. Researchers have developed magnesium-aluminum alloys and polymer-based protective films as substitutes for zinc coatings. These alternatives provide high-performance corrosion resistance while reducing the ecological footprint of metal finishing processes (Dey, Sharma, Kumar, & Dwivedi, 2024d; Kumar & Choudhury, 2024b; Kumar & Dwivedi, 2024; Kumar et al., 2024b; Kumar, 2024a).

Conventional packaging materials, particularly plastics, contribute to severe environmental pollution. Heavy metals are often added as stabilizers in polyvinyl chloride and other polymers, raising toxicity concerns. Sustainable alternatives, such as biobased polymers derived from agricultural waste, seaweed, and microbial fermentation, are gaining traction. Polylactic acid, polyhydroxyalkanoates, and cellulose-based bioplastics offer excellent biodegradability while maintaining mechanical properties suitable for packaging applications. Nanocellulose and chitosan-based films are also being explored as biodegradable packaging materials. These materials provide barrier properties comparable to synthetic plastics derived from renewable resources. Edible coatings made from proteins and polysaccharides further enhance the sustainability of food packaging by extending shelf life and reducing plastic waste.

Catalysts play a crucial role in chemical manufacturing, yet many traditional catalysts contain heavy metals like platinum, palladium, and chromium. Researchers are developing nonmetallic and bioinspired catalysts to replace these toxic materials. Enzyme-based biocatalysts and metal-free carbon-based catalysts have shown remarkable efficiency in industrial processes, including pharmaceutical synthesis, fuel production, and wastewater treatment. Graphene oxide and nitrogen-doped carbon materials are effective catalysts for oxidation and reduction reactions, offering a sustainable alternative to metal-based catalysts (Devi et al., 2024c; Dey et al., 2024b; Kumar & Choudhury, 2024a, 2024b; Kumar, 2024a; Sharma Dey et al., 2024). Additionally, zeolite- and clay-based catalysts are optimized for applications in petrochemicals and polymerization processes, reducing dependency on toxic heavy metals. Eco-friendly production methods are essential for minimizing industrial processes' environmental impact. Sustainable manufacturing involves adopting cleaner production technologies, renewable energy sources, and waste valorization strategies. Life cycle assessment (LCA) methodologies evaluate materials' environmental footprints, guiding industries toward greener alternatives (Fig. 21.1).

The circular economy approach promotes resource efficiency by designing products with recyclability and reusability in mind. Recycling electronic waste (e-waste) to recover valuable materials while eliminating heavy metal contaminants is a key strategy. Researchers are exploring bioleaching techniques using bacteria and fungi to extract precious metals from e-waste without hazardous chemicals. Closed-loop recycling systems for plastics, metals, and composites further contribute to sustainable material management. The transition to heavy metal–free alternatives and sustainable materials faces several challenges, including cost, scalability, and performance optimization. Many eco-friendly materials require further research to achieve competitive properties compared to conventional heavy metal–based counterparts. However, nanotechnology, biotechnology, and material science advancements continue to drive innovations in this field. Government regulations and industry initiatives play a crucial role in promoting sustainable materials. Bans on hazardous substances, such as the European Union's REACH regulations and the RoHS directive, encourage industries to adopt safer alternatives. Financial incentives for green technologies, research funding, and consumer awareness initiatives further support the development of eco-friendly products (Devi & Kumar, 2024a; Devi et al., 2024c, 2024a; Dey

FIGURE 21.1 Different types of heavy metals and their harmful impacts.

Panwar et al., 2024; Kumar et al., 2024e; Sharma et al., 2024). As global sustainability goals emphasize pollution reduction and resource conservation, heavy metal–free alternatives and sustainable production methods will remain a priority for research and industry. Innovations in biopolymers, green chemistry, and renewable materials offer promising solutions for reducing environmental impacts while maintaining high-performance standards across various sectors. Integrating these technologies into mainstream applications will be instrumental in achieving a more sustainable and toxin-free future.

21.3 Secondary prevention: remediation strategies

21.3.1 Water remediation

Water remediation is a critical process for ensuring the availability of clean and safe water, particularly in the face of increasing pollution from industrial, agricultural, and domestic sources. One of the most effective techniques in water remediation is adsorption, which involves using materials such as activated carbon, biochar, zeolites, and nanomaterials to remove contaminants. Activated carbon is widely utilized due to its high surface area and strong adsorption capacity for organic pollutants, heavy metals, and chemical residues. Similarly, biochar, derived from biomass pyrolysis, is a sustainable and cost-effective alternative, efficiently removing toxins while contributing to carbon sequestration. Zeolites, microporous aluminosilicate minerals, play a crucial role in ion-exchange processes, effectively trapping heavy metals and ammonium ions from wastewater (Devi & Kumar, 2024a; Dey et al. 2024d, 2024a; Kumar et al., 2024e). In recent years, nanomaterials, including carbon nanotubes (CNTs), graphene oxide, and metal-organic frameworks, have emerged as advanced adsorbents with enhanced selectivity and adsorption capacity due to their unique physicochemical properties. In addition to adsorption, chemical precipitation and coagulation are widely employed to remove dissolved contaminants by converting them into insoluble forms that can be easily separated from water. Chemical precipitation involves the addition of reagents such as lime, ferric chloride, and aluminum sulfate to induce the formation of precipitates that trap heavy metals, phosphates, and other pollutants. Conversely, coagulation utilizes coagulants to destabilize and aggregate suspended particles, forming larger flocs that can be removed through sedimentation or filtration. This method is particularly effective in treating wastewater containing colloidal particles, organic matter, and heavy metals. Another crucial approach in water remediation is membrane filtration, which includes techniques such as reverse osmosis and nanofiltration. Reverse osmosis relies on a semipermeable membrane to separate contaminants from water by applying pressure, making it highly effective in desalination, removing dissolved salts, and eliminating organic compounds. Nanofiltration operates on a similar principle but allows selective separation of smaller molecules, making it suitable for removing divalent ions, pathogens, and organic pollutants while retaining essential minerals. Membrane filtration techniques are favored for producing high-purity water. Still, membrane fouling and energy consumption challenges necessitate ongoing advancements in membrane materials and operational efficiency. Additionally, electrochemical methods have gained prominence in water remediation due to their effectiveness in removing a wide range of pollutants. Electrocoagulation utilizes electrical currents to generate coagulants in situ, facilitating the aggregation and removal of contaminants such as heavy metals, suspended solids, and pathogens. This method offers advantages over conventional coagulation by reducing the need for chemical additives and generating less sludge (Kumar & Singh, 2024a; Kumar et al., 2024b; Sharma et al., 2024; 2024b). Electrodialysis, another electrochemical technique, employs ion-exchange membranes to selectively transport ions out of the water under the influence of an electric field, making it particularly useful for desalination and removal of charged contaminants. As water pollution challenges continue to escalate, integrating these techniques in a hybrid or sequential manner offers a promising approach to achieving comprehensive water remediation while minimizing environmental and economic costs.

21.3.2 Soil and sediment remediation

Due to its persistence, toxicity, and bioaccumulating nature, heavy metal contamination poses a significant environmental threat. Conventional remediation methods, such as soil excavation and chemical treatments, are often costly and environmentally disruptive. However, the emergence of eco-friendly bioremediation strategies, including phytoremediation and microbial remediation, offers hope. These sustainable and cost-effective alternatives mitigate heavy metal pollution and present a feasible solution for the future. Additionally, immobilization techniques using biochar, clay minerals, soil amendments, and soil washing and excavation contribute to a comprehensive remediation approach.

Phytoremediation is a biological approach that employs hyperaccumulator plants to extract, stabilize, or degrade heavy metals in contaminated soils. Hyperaccumulators, such as *Thlaspi caerulescens*, *Brassica juncea*, and *Pteris vittata*, can tolerate and sequester high concentrations of metals like cadmium (Cd), lead (Pb), arsenic (As), and zinc (Zn) in their biomass. These plants facilitate metal uptake through specialized root exudates that enhance bioavailability and transport metals into above-ground tissues. The harvested biomass containing accumulated metals can be processed through phytoextraction, reducing soil contamination. The potential of phytoremediation is further enhanced by advances in genetic engineering, which have improved the phytoremediation potential of certain species by enhancing metal uptake, transport efficiency, and tolerance mechanisms (Devi et al., 2024a; Dey et al. 2024e, 2024b; Kumar et al., 2024l).

Microbial remediation leverages the natural capacity of bacteria and fungi to immobilize or transform heavy metals into less toxic forms. Metal-resistant bacteria, such as *Pseudomonas*, *Bacillus*, and *Rhizobium* species, utilize biosorption, bioaccumulation, and biotransformation to detoxify contaminated environments. These microbes secrete extracellular polymeric substances that bind metal ions, reducing their bioavailability. Fungi, such as *Aspergillus* and *Penicillium*, play a crucial role in mycoremediation by producing chelating agents, organic acids, and oxidative enzymes that facilitate metal solubilization and uptake. The symbiotic relationship between mycorrhizal fungi and plant roots further enhances metal absorption by expanding the root surface area and improving nutrient uptake. Combining multiple metal-resistant strains, engineered microbial consortia have shown promising results in accelerating heavy metal removal while maintaining soil health.

Immobilization techniques involve applying materials that reduce metal mobility in contaminated soils, preventing leaching into groundwater and reducing bioavailability. Biochar, derived from the pyrolysis of organic biomass, is a highly porous adsorbent that binds heavy metals through ion exchange, precipitation, and surface complexation. Similarly, clay minerals, such as bentonite and kaolinite, exhibit high cation-exchange capacities, making them effective in sequestering metal contaminants. Soil amendments like zeolite, lime, and organic composts can alter soil pH and redox conditions, further enhancing metal immobilization. These techniques provide a long-term stabilization solution, improving soil fertility and microbial activity.

Soil washing is a physicochemical method that uses chelating agents, surfactants, or acids to extract heavy metals from contaminated soil. This technique enhances metal solubility, allowing them to be removed through liquid-phase separation. However, the effectiveness of soil washing depends on soil composition, metal speciation, and the choice of washing solution. Excavation, though highly effective, is often considered a last resort due to its high cost and environmental disruption. It involves physically removing contaminated soil and relocating it to secure landfills or treatment facilities (Devi et al., 2024a; Dey et al., 2024d; Kumar & Singh, 2024a; Kumar et al., 2024l; Saini et al., 2024).

By integrating phytoremediation, microbial remediation, immobilization techniques, and soil washing, a comprehensive and sustainable approach to heavy metal remediation can be achieved. However, the successful adoption of these strategies relies not only on scientific research and technological advancements but also on the support and regulation of policymakers. Future research focusing on genetic modifications, microbial engineering, eco-friendly amendments, and policy initiatives promoting the adoption of these strategies will further enhance the efficiency of these bioremediation strategies.

21.3.3 Air quality management

Airborne heavy metal emissions pose significant environmental and occupational health risks, as toxic metals such as lead (Pb), cadmium (Cd), mercury (Hg), arsenic (As), and chromium (Cr) are released from industrial activities, including metal smelting, fossil fuel combustion, mining, and waste incineration. These pollutants can accumulate in the atmosphere, enter the food chain, and cause severe health issues, such as respiratory disorders, neurological impairments, and carcinogenic effects. To mitigate such hazards, industries employ advanced air filtration and scrubbing technologies alongside strict monitoring protocols to control occupational exposure.

Air filtration and scrubbing systems are essential for capturing and neutralizing airborne heavy metal particles before they are released into the environment. Various industrial air filters, including high-efficiency particulate air (HEPA) filters, electrostatic precipitators, and activated carbon filters, are widely used to remove metal-laden aerosols. HEPA filters are highly effective in trapping delicate particulate matter, including metal oxides, with efficiencies exceeding 99.97% for particles as small as 0.3 μm. On the other hand, electromagnetic precipitators utilize an electrostatic charge to attract and collect metal particles from flue gases, making them suitable for high-temperature industrial emissions (Fig. 21.2).

In addition to dry filtration, wet scrubbers are crucial in removing heavy metal contaminants from exhaust streams. These systems introduce a liquid, usually water or an alkaline solution, into the gas flow, effectively dissolving and capturing metal pollutants through chemical reactions or absorption. Advanced scrubber technologies, such as packed beds and venturi scrubbers, enhance the removal efficiency of metal vapors like mercury and arsenic. Chemical reagents, such as sodium hydroxide (NaOH) or lime ($Ca(OH)_2$), are often added to neutralize acidic gases and convert metals into less volatile, stable compounds. Hybrid filtration systems that integrate both dry and wet technologies further enhance the overall efficiency of heavy metal emission control.

Industrial workers involved in metal refining, welding, battery manufacturing, and waste incineration are at a heightened risk of exposure to airborne heavy metals. Prolonged inhalation of metal fumes can result in chronic toxicity, neurodegenerative disorders, and pulmonary diseases. Therefore stringent monitoring and control strategies are necessary to ensure workplace safety. Real-time monitoring systems, such as inductively coupled plasma mass spectrometry and X-ray fluorescence analyzers, allow for detecting and quantifying airborne metal concentrations. Personal air samplers and biomonitoring techniques, including blood and urine analysis, also help assess individual exposure levels. To mitigate risks, industries implement engineering controls such as localized exhaust ventilation systems, which capture metal fumes at the source, and personal protective equipment (PPE), including respirators and protective clothing (Devi et al., 2024c; Dey et al., 2024b; Kumar & Choudhury, 2024a; Kumar et al., 2024l; Sharma Kumar et al., 2024).

Regulatory frameworks, such as those established by the Occupational Safety and Health Administration (OSHA) and the EPA, mandate permissible exposure limits for heavy metals, ensuring compliance with safety standards. Continuous air quality assessments and adopting best practices in industrial hygiene further contribute to minimizing occupational exposure. Reducing airborne heavy metal emissions requires a multifaceted approach, incorporating advanced filtration and scrubbing technologies, rigorous monitoring, and strict occupational safety measures. By implementing these strategies, industries can significantly reduce environmental contamination and safeguard human health from the detrimental effects of heavy metal exposure.

21.4 Tertiary prevention: reducing human health impacts

21.4.1 Medical interventions

Chelation therapy, a widely recognized medical approach for removing heavy metals from the body, primarily uses chelating agents such as ethylenediaminetetraacetic acid (EDTA) and meso-2,3-dimercaptosuccinic acid (DMSA). These agents bind to toxic metals like lead, mercury, arsenic, and cadmium, forming stable complexes excreted through the urine. EDTA, a synthetic amino acid, has a strong affinity for divalent and trivalent metal ions, making it particularly effective in cases of lead poisoning and cardiovascular conditions where calcified plaques contain heavy metals. DMSA, on the other hand, is an orally administered chelating agent with a high affinity for mercury and lead, often used in pediatric cases of lead poisoning due to its safety profile. Other chelating agents, such as dimercaprol (BAL) and deferoxamine, are used in specific arsenic and iron toxicity cases. The efficacy of chelation therapy depends on the chemical properties of the chelating agent, the route of administration, and the severity of metal exposure. However, the potential for side effects, including the depletion of essential minerals such as calcium, magnesium, and zinc, necessitates careful monitoring and supplementation to prevent deficiencies.

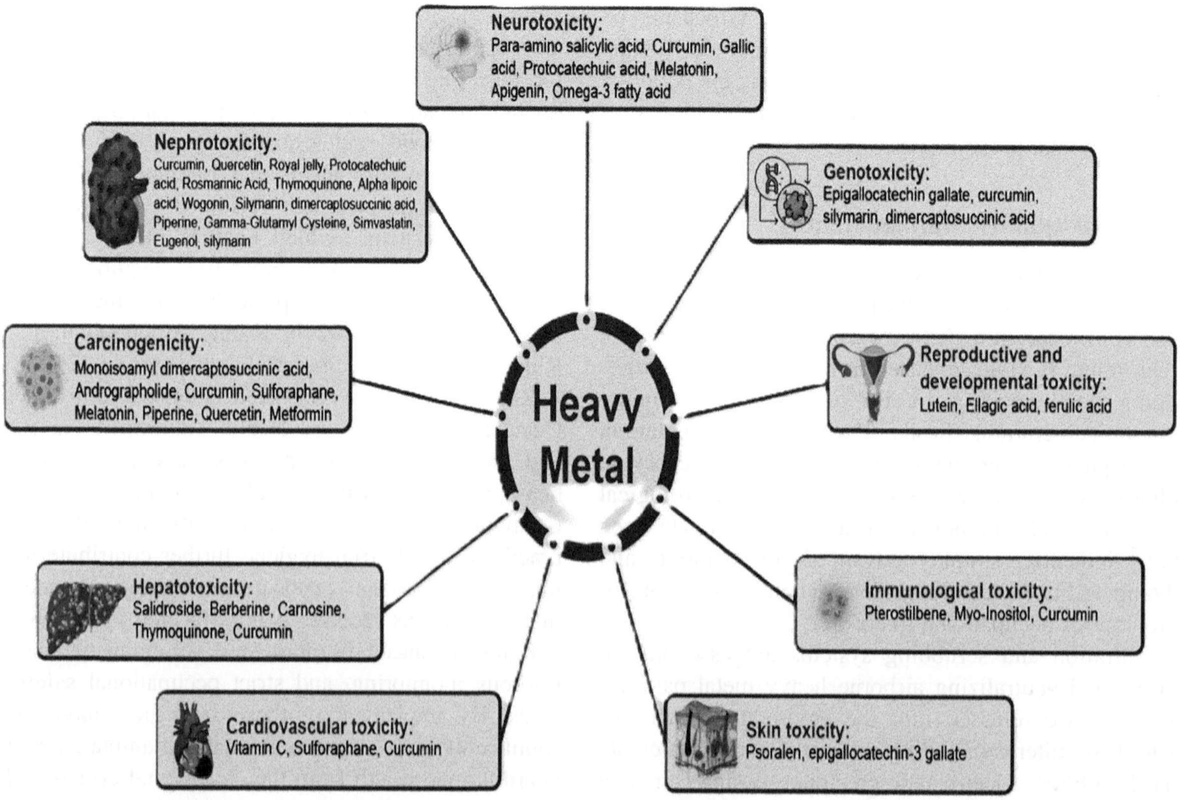

FIGURE 21.2 Toxicity of heavy metals on humans.

This careful monitoring is a crucial aspect of chelation therapy, ensuring the safety and well-being of the patient throughout the procedure (Devi & Kumar, 2024a; Devi et al., 2024a, 2024b; Dey et al., 2024e; Sharma et al., 2024).

Antioxidant supplementation, a critical detoxification component, plays a significant role in neutralizing reactive oxygen species and reducing oxidative stress induced by heavy metal toxicity. Vitamins such as vitamin C (ascorbic acid) and vitamin E (tocopherol) function as potent antioxidants, scavenging free radicals and preventing lipid peroxidation, a key mechanism of cellular damage caused by heavy metals. Vitamin C also enhances the excretion of certain metals, such as lead and cadmium, by increasing their solubility and facilitating renal clearance. Similarly, vitamin E protects cell membranes from oxidative damage and has been shown to mitigate the toxic effects of mercury and lead on neuronal tissues. Minerals such as selenium and zinc play a dual role in detoxification, not only as essential cofactors for antioxidant enzymes like glutathione peroxidase and superoxide dismutase but also by competing with toxic metals for binding sites, thereby reducing their bioavailability and toxicity. Selenium, in particular, forms stable complexes with mercury, decreasing its accumulation in tissues and mitigating its neurotoxic effects. Inducing metallothionein expression, zinc enhances the sequestration and excretion of heavy metals such as cadmium and lead, thereby reducing their toxic impact on organs such as the liver and kidneys.

Furthermore, glutathione, a tripeptide composed of glutamate, cysteine, and glycine, plays a central role in detoxification by directly conjugating with heavy metals to facilitate their elimination. *N*-Acetylcysteine, a glutathione precursor, is often used as a supplement to boost endogenous glutathione levels and enhance metal detoxification. Alpha-lipoic acid is another essential antioxidant that exhibits both water- and fat-soluble properties, allowing it to protect both cytosolic and membrane-bound cellular components from oxidative stress. It also acts as a metal chelator, binding to heavy metals and preventing their intracellular accumulation. Given the interplay between oxidative stress and heavy metal toxicity, combining chelation therapy with antioxidant supplementation represents a comprehensive approach to detoxification. However, careful consideration must be given to these interventions' dosage, duration, and potential interactions to optimize their benefits while minimizing adverse effects. Ongoing research continues to explore novel chelating agents and antioxidant strategies to improve heavy metal detoxification and mitigate associated health risks (Kumar & Dwivedi, 2024; Kumar et al., 2024l; Saini et al., 2024; Sharma et al., 2024; Sharma Dey et al., 2024; Sharma Kumar et al., 2024).

21.4.2 Dietary and nutritional strategies

Functional foods and dietary fibers are crucial in reducing heavy metal absorption in the human body by acting as natural chelators, binding toxic metals in the gastrointestinal tract, and preventing systemic circulation. Heavy metals such as lead, cadmium, mercury, and arsenic pose significant health risks due to their bioaccumulate nature, oxidative stress induction, and disruption of essential metabolic processes. Dietary fibers, particularly soluble fibers like pectin, beta-glucans, and inulin, form viscous gels in the intestine, reducing metal bioavailability by complexation and fecal excretion. Insoluble fibers, including cellulose and lignin, enhance intestinal motility and limit metal retention by reducing transit time. Functional foods enriched with bioactive compounds, such as polyphenols, flavonoids, and phytochemicals, exhibit metal-binding properties through hydroxyl and carboxyl functional groups, preventing heavy metal–induced oxidative damage. Spirulina, chlorella, and medicinal mushrooms have demonstrated significant heavy metal chelation effects due to their rich polysaccharide and protein content, which interact with toxic metals and facilitate their elimination. Fermented foods containing probiotic bacteria, such as *Lactobacillus* and *Bifidobacterium* species, contribute to metal detoxification by altering gut microbiota composition, enhancing metal excretion, and reducing intestinal absorption.

Additionally, consuming selenium, zinc, and sulfur-containing foods plays a fundamental role in detoxifying heavy metals by modulating antioxidant defense mechanisms and promoting metal excretion. Selenium, an essential trace element, enhances the activity of glutathione peroxidase, thioredoxin reductase, and selenoproteins, which mitigate oxidative damage caused by heavy metal toxicity. Selenium-containing foods such as Brazil nuts, garlic, seafood, and eggs have been shown to counteract mercury and cadmium toxicity by forming biologically inactive metal-selenide complexes, reducing their bioavailability. Zinc, a crucial micronutrient involved in metallothionein synthesis, competes with toxic metals for binding sites in the intestine, thereby reducing the absorption of lead and cadmium. Zinc-rich foods, including pumpkin seeds, chickpeas, and whole grains, fortify cellular defense mechanisms by maintaining redox balance and metal homeostasis. Sulfur-containing compounds, particularly those found in cruciferous vegetables like broccoli, Brussels sprouts, and kale, as well as in allium vegetables such as garlic and onions, play a vital role in detoxification pathways. These foods are rich in organosulfur compounds such as sulforaphane, allicin, and glutathione precursors, which enhance hepatic detoxification by upregulating phase II detoxifying enzymes, including glutathione-S-transferases. Sulfur-containing amino acids like cysteine and methionine facilitate the synthesis of glutathione, the body's primary intracellular antioxidant, which binds to heavy metals and supports their biliary and renal excretion (Kumar & Singh, 2024c; Kumar et al., 2024j, 2024m, 2024a, 2024g).

Furthermore, dietary interventions combining functional foods and targeted nutrient supplementation can significantly reduce heavy metal toxicity and associated health risks. Polyphenol-rich foods such as green tea, berries, and citrus fruits exhibit chelating properties by forming metal complexes and reducing their absorption. Probiotic and prebiotic combinations also support gut microbiota resilience, reducing intestinal inflammation and strengthening mucosal barriers against metal translocation. Integrating a diet rich in dietary fibers, functional foods, and essential minerals can effectively mitigate heavy metal toxicity, enhance detoxification processes, and protect against oxidative stress-related disorders. A comprehensive nutritional strategy, including selenium-, zinc-, and sulfur-containing foods, not only aids in heavy metal excretion but also supports immune function, neurological health, and metabolic balance, making it a crucial component of heavy metal detoxification protocols.

21.4.3 Public health policies and risk communication

Community awareness programs are pivotal in educating populations about environmental hazards, occupational risks, and preventive healthcare. These initiatives involve public campaigns, workshops, and interactive sessions to inform individuals about pollutants, hazardous chemicals, and health risks associated with industrial and agricultural activities. By integrating multimedia tools, mobile applications, and social media platforms, community engagement becomes more effective, allowing for the widespread dissemination of knowledge. Collaboration with healthcare professionals, policymakers, and local leaders enhances credibility and encourages community participation. For instance, educational campaigns on air and water pollution have significantly improved public hygiene practices, reduced exposure to contaminants, and influenced policy changes in many regions (Devi et al., 2024d; Kumar & Choudhury, 2024e; Kumar & Singh, 2024e; Kumar et al., 2024d; Raj et al., 2024).

Regular health screening and biomonitoring are crucial for early detection and management of diseases, especially among vulnerable populations such as children, the elderly, and individuals exposed to hazardous environments. These screenings help identify early biomarkers of exposure to heavy metals, pesticides, and industrial toxins, enabling timely medical interventions. Biomonitoring systematically measures chemicals and their metabolites in biological samples such as blood, urine, and hair, providing critical insights into the body's burden of toxic substances. This data is invaluable in assessing long-term exposure risks and developing public health strategies. For instance, lead biomonitoring programs have played a significant role in mitigating lead poisoning in children, leading to regulatory actions such as reducing lead in paints, gasoline, and plumbing materials. Additionally, biomonitoring of occupationally exposed workers, such as those in mining and chemical industries, helps identify early signs of toxicity and implement preventive measures before severe health effects manifest.

Occupational safety measures and worker protections are essential for safeguarding employees from workplace hazards, particularly in industries dealing with toxic chemicals, heavy machinery, and hazardous environments. PPE, such as respirators, gloves, and protective clothing, is the first defense against occupational exposures. Engineering controls, including proper ventilation systems, enclosed processing units, and automated monitoring, further minimize worker contact with harmful substances. Regulatory frameworks such as the OSHA standards and the International Labour Organization guidelines establish stringent safety protocols to protect workers from occupational diseases and injuries. Regular training programs ensure that workers are well-versed in safety procedures, emergency responses, and the correct use of protective equipment. Moreover, workplace risk assessments and safety audits help identify potential hazards and improve working conditions. In high-risk occupations such as construction, mining, and manufacturing, adherence to strict safety regulations has led to a significant reduction in workplace accidents and chronic illnesses. Policies such as paid sick leave, medical insurance, and compensation for occupational diseases further enhance worker protection, promoting physical and mental well-being (Devi et al., 2024c, 2024d; Kumar and Choudhury, 2024g, 2024f; Kumar et al., 2024d; Raj et al., 2024).

Integrating community awareness programs, regular health screening, and occupational safety measures foster a comprehensive approach to public health and worker safety. By raising awareness, facilitating early detection of health risks, and implementing stringent safety protocols; societies can reduce disease burdens, enhance workforce productivity, and promote sustainable development. Governments, industries, and communities must collaborate to implement these measures effectively, ultimately leading to healthier populations and safer working environments.

21.5 Case studies and real-world applications

One of the most notable environmental remediation projects has been the cleanup of arsenic-contaminated groundwater in Bangladesh. This country has faced one of the most severe arsenic poisoning crises in the world. Arsenic contamination in groundwater, primarily due to the natural dissolution of arsenic-bearing minerals, has led to widespread health issues, including skin lesions, cardiovascular diseases, and various cancers. The crisis, first identified in the 1990s, prompted international collaboration and research-driven intervention strategies to mitigate exposure.

Several successful approaches have been implemented to reduce arsenic exposure in Bangladesh. One of the primary strategies involved the installation of arsenic-removal filters at the household and community levels. Technologies such as the Kanchan Arsenic Filter, activated alumina filters, and iron-coated sand have shown efficacy in removing arsenic from drinking water. Additionally, deep tube wells, which access deeper, uncontaminated aquifers, have been widely adopted. Government-led initiatives, supported by organizations such as UNICEF and the WHO, have also promoted identifying and labeling safe wells and guiding communities toward arsenic-free water sources. Moreover, implementing rainwater harvesting and pond sand filtration systems has provided alternative safe drinking water sources, particularly in rural areas where contamination is most severe.

Phytoremediation techniques have also been explored as a sustainable approach to mitigate arsenic contamination in groundwater. Arsenic hyperaccumulator plants, such as *P. vittata* (Chinese brake fern), have been tested in constructed wetland systems to remove arsenic from soil and water. While phytoremediation offers a long-term and cost-effective strategy, its widespread application remains limited due to the slow accumulation rate and large land area requirements (Devi et al. 2024d, 2024a; Kumar et al., 2024c; Kumar, 2024c; Raj et al., 2024).

Despite these interventions, challenges remain, particularly in ensuring long-term sustainability and equitable access to arsenic-free water. Continued efforts in monitoring, policy implementation, and community education are essential to sustaining the progress made in arsenic mitigation in Bangladesh.

Removing lead from gasoline represents one of the most successful global policies for reducing heavy metal exposure and associated health risks. Leaded gasoline, widely used in the mid-20th century, was a significant source of environmental lead contamination, leading to severe public health consequences, particularly among children. The toxic effects of lead include neurodevelopmental impairment, cardiovascular diseases, and renal dysfunction.

Recognizing these dangers, policy-driven interventions led to a gradual phase-out of leaded gasoline worldwide.

The movement to eliminate leaded gasoline gained momentum in the 1970s when research revealed a direct correlation between atmospheric lead levels and blood lead concentrations in exposed populations. The United States pioneered regulatory actions through the Clean Air Act, which mandated the reduction of lead content in gasoline. This initiative, spearheaded by the US EPA, introduced unleaded gasoline and enforced stringent emission standards. Over time, the transition led to a dramatic decline in blood lead levels among American children, demonstrating the efficacy of policy-driven interventions.

Globally, the UNEP played a pivotal role in advocating for the elimination of leaded gasoline. By 2021, Algeria became the last country to phase out leaded fuel, marking a historic milestone in environmental health policy. The global ban on leaded gasoline has resulted in substantial public health benefits, including reduced cognitive impairment in children and lower cardiovascular mortality rates.

While removing lead from gasoline has been a significant success, lead exposure remains a concern in certain regions due to legacy contamination from industrial emissions, lead-based paints, and leaded plumbing systems. Ongoing policies targeting these sources, continued monitoring, and public awareness campaigns are essential to further minimizing lead-related health risks. Phytoremediation has emerged as a promising, eco-friendly, cost-effective approach for rehabilitating heavy metal-contaminated agricultural lands. Unlike traditional remediation methods such as soil excavation or chemical treatment, phytoremediation leverages plants' natural ability to accumulate, stabilize, or degrade toxic contaminants, restoring soil health and making land suitable for agricultural use (Devi et al., 2024e; Dey et al., 2024c; Kumar & Choudhury, 2024c; Kumar et al. 2024i, 2024a; Pandey et al., 2024).

One of the most successful phytoremediation applications has been in China's cadmium- and lead-contaminated agricultural soils. Due to extensive industrial activities and the use of phosphate fertilisers containing trace amounts of heavy metals, large tracts of farmland have become contaminated, posing serious risks to food safety and human health. To address this issue, researchers and policymakers have promoted the cultivation of hyperaccumulator plants such as Sedum alfredii and *B. juncea* (Indian mustard) in contaminated regions. These plants effectively extract heavy metals from the soil, and their harvested biomass can be safely disposed of or processed for metal recovery.

Similarly, sunflowers (*Helianthus annuus*) have been widely used in Eastern Europe to remediate agricultural lands contaminated with lead, arsenic, and uranium. Sunflowers' extensive root system enables them to absorb and concentrate toxic metals, which can be removed through periodic harvesting. This method has been particularly beneficial in rehabilitating farmland affected by industrial waste discharge and mining activities.

Recent advancements in phytoremediation have focused on enhancing plant metal uptake efficiency through biotechnological interventions. Genetic engineering has enabled the development of transgenic plants with improved metal hyperaccumulation capabilities. For instance, overexpression of metal transporter genes, such as those encoding ZIP (zinc-iron permease) and NRAMP (natural resistance-associated macrophage protein) family proteins, has been shown to increase metal uptake and sequestration in plant tissues. Additionally, soil amendments such as biochar, organic compost, and mycorrhizal fungi have been explored to enhance phytoremediation efficiency by improving soil structure and metal bioavailability.

Despite its advantages, phytoremediation faces several challenges, including the long duration required for effective remediation, potential biomass disposal issues, and the risk of metal translocation into the food chain. Researchers are investigating integrated approaches combining phytoremediation with microbial-assisted bioremediation, where heavy metal-resistant bacteria and fungi facilitate plant metal uptake and detoxification to mitigate these risks. Future research should also focus on optimizing plant selection for site-specific contamination scenarios and developing economically viable strategies for large-scale implementation. Government incentives and regulatory frameworks supporting phytoremediation-based land reclamation projects will be crucial in promoting this sustainable remediation technique.

The successful remediation of arsenic-contaminated groundwater in Bangladesh, the global removal of lead from gasoline, and the implementation of phytoremediation in contaminated agricultural lands illustrate the profound impact of science-driven environmental policies and innovative remediation strategies. While significant progress has been made, continued research, policy reinforcement, and technological advancements are necessary to address the remaining challenges in heavy metal contamination. Integrating multidisciplinary approaches and fostering international collaboration will ensure a cleaner and safer environment for future generations (Devi et al., 2024a, 2024b; Kumar & Singh, 2024c; Kumar, 2024a; Saini et al., 2024; Sharma et al., 2024).

21.6 Future directions and challenges

Nanotechnology has emerged as a transformative solution for heavy metal remediation in contaminated water and soil systems. Applying nanomaterials, such as metal oxides, carbon-based nanomaterials, and functionalised nanoparticles, has significantly improved the efficiency of heavy metal adsorption, precipitation, and catalytic degradation. Iron oxide nanoparticles (Fe_3O_4) have been widely studied due to their high surface area, magnetic properties, and strong affinity for metal ions. Moreover, graphene oxide (GO) and CNTs have demonstrated exceptional adsorption capabilities due to their large surface-to-volume ratio and functionalised surfaces that enhance metal binding.

The incorporation of nanocomposites has further enhanced the efficiency of remediation. For instance, due to its strong redox potential, nano zero-valent iron (nZVI) is extensively used to reduce toxic metals such as arsenic, chromium, and lead. Functionalized silica nanoparticles and polymer-coated nanomaterials have been engineered to enhance selectivity toward specific heavy metals while minimizing environmental toxicity. Another breakthrough in nanotechnology involves bioinspired nanomaterials, such as biogenic nanoparticles derived from plant extracts, which offer an eco-friendly and cost-effective alternative for remediation. Despite these advances, challenges such as nanoparticle aggregation, potential ecotoxicity, and the regeneration of nanomaterials remain areas of active research.

AI and the IoT have revolutionized environmental monitoring by enabling real-time data acquisition, analysis, and decision-making. IoT-based sensor networks are deployed in water bodies, industrial sites, and agricultural fields to continuously monitor heavy metal concentrations, pH levels, and other environmental parameters. These sensors transmit data to cloud platforms, where AI algorithms analyze patterns and predict contamination hotspots. ML models have been extensively used to optimize sensor placement, enhance detection accuracy, and forecast pollution trends.

Deep learning techniques, such as convolutional neural networks and recurrent neural networks, have been applied for anomaly detection in large datasets collected from environmental monitoring stations. AI-driven predictive modeling assists policymakers and industries in implementing proactive remediation strategies by identifying potential contamination risks before they escalate. Additionally, AI-powered robotic systems integrated with nanotechnology have been developed for autonomous heavy metal removal in water treatment plants. These innovative remediation systems optimize nanomaterial deployment, ensuring maximum efficiency while minimizing secondary pollution.

Despite AI and IoT's promising capabilities in environmental monitoring, challenges such as data security, sensor calibration, and the high cost of deployment hinder their widespread adoption. Moreover, integrating AI models with existing regulatory frameworks requires careful validation and standardization to ensure reliability and accuracy in real-world applications.

While numerous remediation technologies have demonstrated high efficiency at the laboratory and pilot scales, their large-scale implementation faces several challenges. One of the major obstacles is the heterogeneity of contaminated sites, where variations in soil composition, water chemistry, and pollutant distribution impact the effectiveness of remediation strategies. For instance, phytoremediation, which utilizes plants for heavy metal uptake, requires long-term monitoring and is influenced by site-specific factors such as climate, soil fertility, and microbial interactions.

The scalability of nanotechnology-based remediation is also limited by the cost of nanomaterial synthesis, stability, and potential environmental risks associated with nanoparticle leaching. Electrochemical and membrane-based remediation techniques, such as electrocoagulation and nanofiltration, require substantial energy inputs, making them less feasible for remote and resource-constrained regions. Moreover, the disposal of secondary waste generated during remediation, such as heavy metal-laden adsorbents, poses additional environmental concerns and requires proper management.

Infrastructure limitations, lack of skilled workforce, and inadequate stakeholder engagement impede large-scale remediation efforts. Transitioning from laboratory-scale research to field-scale implementation necessitates interdisciplinary collaboration among scientists, engineers, policymakers, and local communities. Innovative financing models, such as public-private partnerships and green investments, can facilitate the translation of novel remediation technologies into scalable solutions.

Regulative, economic, and institutional barriers often hinder the successful adoption of heavy metal mitigation strategies. One primary challenge is the lack of stringent enforcement of environmental policies in many regions, leading to continued industrial discharge and agricultural runoff containing heavy metals. Although international frameworks such as the Basel Convention and the Minamata Convention aim to regulate hazardous waste and mercury emissions, their implementation varies across countries due to differences in governance and economic priorities.

Economic constraints significantly limit the adoption of advanced remediation technologies. The high capital and operational costs associated with nanotechnology-based solutions, AI-driven monitoring systems, and large-scale remediation projects deter investment, particularly in developing nations. Additionally, the absence of financial incentives, such as tax benefits and subsidies for pollution control measures, discourages industries from adopting cleaner technologies. Market failures, including the undervaluation of ecosystem services and the lack of monetization of pollution reduction benefits, further exacerbate the issue.

Governments must integrate environmental sustainability into economic policies to overcome these barriers by introducing green credits, pollution taxes, and EPR schemes. International collaboration and technology transfer programs can also play a crucial role in bridging the gap between scientific advancements and practical implementation. Moreover, fostering public awareness and community participation in environmental conservation efforts can drive demand for sustainable remediation solutions and encourage CSR initiatives.

The convergence of nanotechnology, AI, and IoT holds immense potential for addressing heavy metal contamination in environmental systems. While significant progress has been made in developing high-efficiency remediation techniques, scalability, cost, and policy implementation challenges remain critical. Addressing these barriers requires a multidisciplinary approach that combines scientific innovation, regulatory support, economic incentives, and community engagement. By fostering collaboration between researchers, policymakers, and industry stakeholders, sustainable and scalable solutions for heavy metal mitigation can be achieved, ensuring a cleaner and healthier environment for future generations.

21.7 Conclusion

The research highlights the urgent need for effective measures to mitigate environmental heavy metal contamination. Key findings emphasize that industrial discharge, agricultural runoff, and improper waste disposal are major contributors to soil and water contamination. Heavy metals such as lead, cadmium, arsenic, and mercury pose significant health risks to humans, animals, and plants, leading to bioaccumulation and long-term ecological damage. The study also underscores the efficacy of various remediation strategies, including phytoremediation, biochar amendments, and microbial-assisted remediation. When combined with soil amendments like zeolite and biochar, engineered hyperaccumulator plants show promising results in extracting heavy metals while maintaining soil health. Additionally, innovative adsorbent materials and nanotechnology-based solutions have demonstrated high efficiency in water purification.

Furthermore, findings highlight that several environmental factors influence phytoremediation efficiency, such as soil pH, microbial diversity, and temperature variations. Integrating genetically engineered plants with specialized microbial consortia enhances heavy metal uptake, making it a viable solution for large-scale remediation projects. However, scalability and site-specific implementation remain challenges that require further research and investment. Another critical finding is the socioeconomic impact of heavy metal contamination, particularly in agricultural regions. Heavy metal uptake by crops can lead to food safety concerns, reduced agricultural productivity, and economic losses for farmers. Addressing this issue requires a holistic approach involving scientific advancements, policy interventions, and community engagement.

Tackling heavy metal contamination necessitates a multidisciplinary approach integrating environmental science, biotechnology, chemistry, engineering, and policymaking. Phytoremediation, for instance, requires expertize in plant genetics, soil chemistry, and microbial ecology to enhance metal uptake and degradation processes. Similarly, nanotechnology and material science advancements have led to the development of efficient adsorbents for heavy metal removal from water sources.

Collaboration between researchers, policymakers, and industry stakeholders is essential for developing cost-effective and sustainable remediation strategies. Engineers and material scientists can design improved filtration systems, while agronomists can explore soil amendments to reduce metal bioavailability. Data analytics and geographic information systems technology are crucial in mapping contamination hotspots and predicting long-term environmental impacts. Public health experts and toxicologists also assess human exposure risks and recommend safety thresholds. Understanding the pathways through which heavy metals enter the food chain helps design effective mitigation strategies. Legal experts and policymakers, on the other hand, ensure that regulatory frameworks align with scientific findings, promoting environmental safety and public health.

Education and community engagement are equally vital. Local communities and farmers need access to knowledge and resources to implement safe agricultural practices and reduce heavy metal exposure. Interdisciplinary collaboration ensures that scientific solutions translate into actionable policies that benefit the environment and society. Heavy metal contamination is a global challenge that requires collective action. Governments must implement stricter regulations on industrial emissions, waste disposal, and agricultural practices to minimize heavy metal pollution. Strengthening environmental laws, enforcing penalties for violations, and promoting sustainable technologies are crucial to preventing further contamination.

International cooperation is essential, as heavy metal pollution does not recognize borders. Transboundary pollution from industries and mining activities necessitates coordinated efforts between countries. Establishing global monitoring systems and data-sharing platforms can enhance early detection and response strategies. Investments in green technologies, such as phytoremediation and eco-friendly adsorbents, should be prioritized. Public-private partnerships can accelerate the development and deployment of innovative solutions, ensuring their practical application in contaminated areas. Additionally, incentives for industries to adopt cleaner production methods can significantly reduce heavy metal emissions. Sustainable policies must also protect vulnerable populations, particularly developing regions with limited access to clean water and safe agricultural practices. Governments should support initiatives that promote heavy metal testing in food and water supplies, ensuring public health and safety.

In conclusion, addressing heavy metal contamination requires a proactive, science-driven approach that integrates multiple disciplines and stakeholders. By fostering global cooperation, enforcing strict regulations, and investing in sustainable solutions, we can mitigate the risks associated with heavy metal exposure and create a healthier, more resilient environment for future generations. Acknowledgment: The author used AI for language improvement only; all final edits were made by the author (s). The human author(s) performed the editing.

References

Choudhury, D., Kumar, P., Zhimo, V. Y., & Sahoo, J. (2024). Crop rotation patterns and soil health management. In *Bioremediation of emerging contaminants from soils: Soil health conservation for improved ecology and food security*. Punjab, Jalandhar, India: Elsevier, Department of Plant Pathology, School of Agriculture, Lovely Professional University, 565–589. Retrieved from https://doi.org/10.1016/B978-0-443-13993-2.00025-6.

Devi, P., & Kumar, P. (2024a). Alleviation of different climatic conditions by foliar application of salicylic acid and sodium nitroprusside and their interactive effects on pigments and sugar content of maize under different sowing dates. *Nature Environment and Pollution Technology, 23*, 1161–1166. Retrieved from https://doi.org/10.46488/NEPT.2024.v23i02.050.

Devi, P., & Kumar, P. (2024b). Challenges in the storage of herbal medicine products and strategies for sustainable management. In *Reference series in phytochemistry*. Phagwara, India: Springer Science and Business Media B.V., Department of Agronomy, School of Agriculture, Lovely Professional University, 1739–1767. Retrieved from https://doi.org/10.1007/978-3-031-43199-9_37.

Devi, P., & Kumar, P. (2024c). Herbal medicine and pregnancy. In *Reference series in phytochemistry*. Punjab, Phagwara, India: Springer Science and Business Media B.V., Department of Agronomy, School of Agriculture, Lovely Professional University, 693–722. Retrieved from https://doi.org/10.1007/978-3-031-43199-9_25.

Devi, P., Dey, S. R., Sharma, K., Kumar, P., & Singh, J. (2024a). Waste management as an opportunity for the inclusion of vulnerable groups. In *Waste management in the circular economy*. Punjab, Phagwara: Springer International Publishing, Department of Agronomy, School of Agriculture, Lovely Professional University, 207–230. Retrieved from https://doi.org/10.1007/978-3-031-42426-7_11 144411, India.

Devi, P., Dogra, P., Kumar, P., & Singh, J. (2024). Global status of biomass energy programmes-challenges and roadmap. In *Biomass energy for sustainable development*. Phagwara, India: CRC Press, Lovely Professional University, 394–414. Retrieved from https://doi.org/10.1201/9781003406501-20.

Devi, P., Kapoor, P., & Kumar, P. (2024). Biomarkers for toxicity detection. In *Biomarkers in environmental and human health biomonitoring: An integrated perspective*. Punjab, Phagwara, India: Elsevier, Department of Agronomy, School of Agriculture, Lovely Professional University, 313–329. Retrieved from https://doi.org/10.1016/B978-0-443-13860-7.00004-5.

Devi, P., Kaur, A., & Kumar, P. (2024d). Phytoremediation of metal contaminated soil using energy crops: Soil health maintenance along with biofuel production. In *Bioremediation of emerging contaminants from soils: Soil health conservation for improved ecology and food security*. Punjab, Phagwara, India: Elsevier, Department of Agronomy, School of Agriculture, Lovely Professional University, 307–333. Retrieved from https://doi.org/10.1016/B978-0-443-13993-2.00015-3.

Devi, P., Kumar, P., Dey, S. R., Banik, D., Kumar, G., & Mehta, C. M. (2024e). Integrated omics approaches for nutrient stress management in plants. In *Current omics advancement in plant abiotic stress biology: A volume in developments in applied microbiology and biotechnology*. Punjab, Phagwara, India: Elsevier, School of Agriculture, Lovely Professional University, 93–117. Retrieved from https://doi.org/10.1016/B978-0-443-21625-1.00008-7.

Devi, P., Kumar, P., & Singh, J. (2024a). Sewage treatment and energy recovery by bioelectrochemical oxidation system. In *Bioelectrochemical oxidation processes for wastewater treatment*. Punjab, Phagwara, India: CRC Press, Lovely Professional University, 100–125. Retrieved from https://doi.org/10.1201/9781003368472-6.

Devi, P., Kumar, P., & Singh, J. (2024b). Application of potential microbes in bioremediation. In *Development in waste water treatment research and processes: role of environmental microbiology in industrial wastewater research*. Punjab, Phagwara, India: Elsevier, Department of Agronomy, School of Agriculture, Lovely Professional University, 201–225. Retrieved from https://doi.org/10.1016/B978-0-443-13609-2.00021-5.

Devi, P., Kumar, P., & Singh, J. (2024c). Biotechnological approaches as clean technology for the development of a sustainable environment. In *Development in waste water treatment research and processes: Role of environmental microbiology in industrial wastewater research*. Punjab, Phagwara, India: Elsevier, Department of Agronomy, School of Agriculture, Lovely Professional University, 381–403. Retrieved from https://doi.org/10.1016/B978-0-443-13609-2.00016-1.

Dey, S. R., Kumar, P., & Panwar, J. S. (2024a). Microbial desalination cell. In *Bioelectrochemical oxidation processes for wastewater treatment*. Punjab, Phagwara, India: CRC Press, Lovely Professional University, 189–217. Retrieved from https://doi.org/10.1201/9781003368472-10.

Dey, S. R., Kumar, P., & Singh, J. (2024b). Total soluble protein mediated morphological traits in mustard treated with thiourea and salicylic acid. *Nature Environment and Pollution Technology, 23*, 679–694. Retrieved from https://doi.org/10.46488/NEPT.2024.v23i02.007.

Dey, S. R., Sharma, M., & Kumar, P. (2024c). Phytoremediation of phenolic compounds from soil. In *Bioremediation of emerging contaminants from soils: soil health conservation for improved ecology and food security*. Punjab, Phagwara, India: Elsevier,

Department of Agronomy, School of Agriculture, Lovely Professional University, 335–358. Retrieved from https://doi.org/10.1016/B978-0-443-13993-2.00016-5.

Dey, S. R., Sharma, M., Kumar, P., & Dwivedi, P. (2024d). Genetic improvement of groundnut. In *Genetic engineering of crop plants for food and health security (Volume 2)*. Springer Nature, Department of Agronomy, School of Agriculture, Lovely Professional University, 281–306. Retrieved from https://doi.org/10.1007/978-981-97-3119-0_12.

Dey, S. R., Sharma, M., Kumar, P., & Dwivedi, P. (2024e). Genetic improvement of mustard. In *Genetic improvement of mustardgenetic engineering of crop plants for food and health security (volume 2)*. Phagwara, India: Springer Nature, Department of Agronomy, School of Agriculture, Lovely Professional University, 227–259. Retrieved from https://doi.org/10.1007/978-981-97-3119-0_10.

Jagadesh, M., Dash, M., Kumari, A., Singh, S. K., Verma, K. K., Kumar, P., ... Sharma, S. K. (2024). Revealing the hidden world of soil microbes: Metagenomic insights into plant, bacteria, and fungi interactions for sustainable agriculture and ecosystem restoration. *Microbiological Research, 285*. Retrieved from https://doi.org/10.1016/j.micres.2024.127764.

Kumar, P. (2024a). Epiphytes as a sustainable biomonitoring tool for environmental pollutants. In *Biomonitoring of pollutants in the Global South*. Punjab, Phagwara, India: Springer Nature, Department of Agronomy, School of Agriculture, Lovely Professional University, 359–390. Retrieved from https://doi.org/10.1007/978-981-97-1658-6_10.

Kumar, P. (2024b). Biochar production from marine algae and its application in the treatment of industrial effluent. In *Marine biomass: Biorefinery, bioproducts and environmental bioremediation*. Phagwara, Punjab: De Gruyter, Department of Agronomy, School of Agriculture, Lovely Professional University, 57–93. Retrieved from https://doi.org/10.1515/9783111353951-004 144411, India.

Kumar, P. (2024c). Application of marine biomass for carbon neutrality, sustainable environment, and economy. In *Marine biomass: Biorefinery, bioproducts and environmental bioremediation*. Phagwara, Punjab: De Gruyter, Department of Agronomy, School of Agriculture, Lovely Professional University, 95–128. Retrieved from https://doi.org/10.1515/9783111353951-005 144411, India.

Kumar, P. (2024d). Soil conservation for global sustainability. In *World sustainability series*. Punjab, Phagwara: Springer Science and Business Media Deutschland GmbH, Department of Agronomy, School of Agriculture, Lovely Professional University, 103–128. Retrieved from https://doi.org/10.1007/978-3-031-57456-6_6 144411, India.

Kumar, P., & Choudhury, D. (2024a). Pollutants from agricultural activities in the Global South. In *Biomonitoring of pollutants in theglobal south. Punjab*. Phagwara, India: Springer Nature, Department of Agronomy, School of Agriculture, Lovely Professional University, 89–126 https://doi.org/10.1007/978-981-97-1658-6_3.

Kumar, P., & Choudhury, D. (2024b). Seed Plants as a sustainable biomonitoring tool for environmental pollutants. In *Biomonitoring of pollutants in the Global South*. Punjab, Phagwara, India: Springer Nature, Department of Agronomy, Lovely Professional University, 321–357. Retrieved from https://doi.org/10.1007/978-981-97-1658-6_9.

Kumar, P., & Choudhury, D. (2024c). Optimized phytoremediation process for the sustainable management radionuclides. In *Bioremediation ofemerging contaminants from soils: soil health conservation for improved ecology and food security*. Punjab, Phagwara, India: Elsevier, 443–464. Retrieved from https://doi.org/10.1016/B978-0-443-13993-2.00020-7. Department of Agronomy, School of Agriculture, Lovely Professional University.

Kumar, P., & Choudhury, D. (2024d). Role of indigenous knowledge in agricultural soil reclamation without disturbing other ecosystems. In *Bioremediation of emerging contaminants from soils: Soil health conservation for improved ecology and food security*. Punjab, Phagwara, India: Elsevier, Department of Agronomy, School of Agriculture, Lovely Professional University, 465–488. Retrieved from https://doi.org/10.1016/B978-0-443-13993-2.00021-9.

Kumar, P., & Choudhury, D. (2024e). Microbial native soil bacteria against cadmium toxicity. In *Springer water*. Punjab, Phagwara: Springer Nature, Department of Agronomy, School of Agriculture, Lovely Professional University, 193–226. Retrieved from https://doi.org/10.1007/978-3-031-54005-9_9 144411, India.

Kumar, P., & Choudhury, D. (2024f). Toxicity of rhizospheric cadmium contaminated soil and its phytoremediation. In *Springer water*. Punjab, Phagwara: Springer Nature, Department of Agronomy, School of Agriculture, Lovely Professional University, 227–264. Retrieved from https://doi.org/10.1007/978-3-031-54005-9_10 144411, India.

Kumar, P., & Choudhury, D. (2024g). Hydroponic removal of organic contaminants from water. In *Springer water*. Punjab, Phagwara: Springer Nature, Department of Agronomy, School of Agriculture, Lovely Professional University, 143–164. Retrieved from https://doi.org/10.1007/978-3-031-53258-0_6 144411, India.

Kumar, P., & Choudhury, D. (2024h). Bioremediation of wastewater using hydroponics. In *Springer water*. Punjab, Phagwara: Springer Nature, Department of Agronomy, School of Agriculture, Lovely Professional University, 27–64. Retrieved from https://doi.org/10.1007/978-3-031-53258-0_2 144411, India.

Kumar, P., & Dey, S. R. (2024). Hydroponics phytoremediation: An overview. In *Springer water*. Punjab, Phagwara: Springer Nature, Department of Agronomy, School of Agriculture, Lovely Professional University, 361–396. Retrieved from https://doi.org/10.1007/978-3-031-53258-0_13 144411, India.

Kumar, P., & Dwivedi, P. (2024). Recent advances in the role of nitric oxide and cross-talk with signalling molecules under environmental stress in plants. In *Plant growth regulators: Resilience for sustainable agriculture*. Phagwara, India: Springer Nature, Lovely Professional University, 55–78. Retrieved from https://doi.org/10.1007/978-981-97-2918-0_4.

Kumar, P., & Singh, J. (2024a). Exploring enzyme biotechnology's role in green chemistry and advancements in pharmaceutical technologies. In *Enzyme biotechnology for environmental sustainability*. Punjab, Phagwara, India: Elsevier, Department of Agronomy, School of Agriculture, Lovely Professional University, 465–495. Retrieved from https://doi.org/10.1016/B978-0-443-22072-2.00015-2.

Kumar, P., & Singh, J. (2024b). Environmental and health risk assessment of chemical pollutants in drinking water and wastewaters. In *Trends in biological processes in industrial wastewater treatment*. Phagwara, Punjab, India: Institute of Physics Publishing, Department of Agronomy, School of Agriculture, School of Bioengineering and

Biosciences, Lovely Professional University, 5.1–5.30. Retrieved from https://doi.org/10.1088/978-0-7503-5678-7ch5.

Kumar, P., & Singh, J. (2024c). Bioprocesses in industrial wastewater treatment: Trends and prospects. In *Trends in biological processes in industrial wastewater treatment*. Phagwara, Punjab, India: Institute of Physics Publishing, Department of Agronomy, School of Agriculture, School of Bioengineering and Biosciences, Lovely Professional University, 15.1–15.50. Retrieved from https://doi.org/10.1088/978-0-7503-5678-7ch15.

Kumar, P., & Singh, J. (2024d). Emerging contaminants and ways to reduce them. In *Trends in biological processes in industrial wastewater treatment*. Phagwara, Punjab, India: Institute of Physics Publishing, Department of Agronomy, School of Agriculture, School of Bioengineering and Biosciences, Lovely Professional University, 4.1–4.34. Retrieved from https://doi.org/10.1088/978-0-7503-5678-7ch4.

Kumar, P., & Singh, J. (2024e). Hydroponic: An eco-friendly future. In *Springer water*. Punjab, Phagwara: Springer Nature, Department of Agronomy, School of Agriculture, Lovely Professional University, 223–267. Retrieved from https://doi.org/10.1007/978-3-031-53258-0_9 144411, India.

Kumar, P., Choudhury, D., Dey, S. R., & Sharma, M. (2024a). Application of genetically modified crops against metallic contaminations. In *Bioremediation of emerging contaminants from soils: Soil health conservation for improved ecology and food security*. Punjab, Phagwara, India: Elsevier, Department of Agronomy, School of Agriculture, Lovely Professional University, 541–564. Retrieved from https://doi.org/10.1016/B978-0-443-13993-2.00024-4.

Kumar, P., Devi, P., & Panwar, J. S. (2024b). Constructed wetland-bioelectrochemical oxidation systems: A hybrid system for wastewater treatment. In *Bioelectrochemical oxidation processes for wastewater treatment*. Punjab, Phagwara, India: CRC Press, Lovely Professional University, 126–151. Retrieved from https://doi.org/10.1201/9781003368472-7.

Kumar, P., Dey, S. R., & Choudhury, D. (2024c). Effectiveness of cadmium on biochemical shift of pea plant treated with mycorrhiza and putrescine. *Nature Environment and Pollution Technology, 23*, 275–285. Retrieved from https://doi.org/10.46488/NEPT.2024.v23i01.023.

Kumar, P., Dwivedi, P., & Upadhyay, S. K. (2024d). Optimization of polyamine and mycorrhiza in sorghum plant for removal of hazardous cadmium. *Plant Physiology and Biochemistry PPB/Societe Francaise de Physiologie Vegetale, 214*. Retrieved from https://doi.org/10.1016/j.plaphy.2024.108846.

Kumar, P., Jose Anand, A., & Rajagopal, A. R. (2024e). Sustainable use of marine resources using Ocean Technologies. In *Sustainable development goals: Technologies and opportunities*. Punjab, Phagwara: CRC Press, Department of Agronomy, School of Agriculture, Lovely Professional University, 249–274. Retrieved from https://doi.org/10.1201/9781003468257-14 144411, India.

Kumar, P., Karam, N., & Choudhury, D. (2024f). Assessment of the phytochemical constituents and metabolites in the medicinal plants and herbal medicine used in the treatment and management of skin diseases. In *Reference series in phytochemistry*. Punjab, Phagwara, India: Springer Science and Business Media B.V., Department of Agronomy, School of Agriculture, Lovely Professional University, 533–573. Retrieved from https://doi.org/10.1007/978-3-031-43199-9_21.

Kumar, P., Khokhar, A., & Dey, S. R. (2024g). Evidence from the use of herbal medicines in the management and prevention of common eye diseases. In *Reference series in phytochemistry*. Phagwara, India: Springer Science and Business Media B.V., Department of Agronomy, School of Agriculture, Lovely Professional University, 763–795. Retrieved from https://doi.org/10.1007/978-3-031-43199-9_27.

Kumar, P., Saini, L., & Sharma, M. (2024h). Assessment of the phytochemical constituents and metabolites in the medicinal plants and herbal medicine used in the treatment and management of respiratory diseases. In *Reference series in phytochemistry*. Punjab, Phagwara, India: Springer Science and Business Media B.V., Department of Agronomy, School of Agronomy, Lovely Professional University, 481–510. Retrieved from https://doi.org/10.1007/978-3-031-43199-9_19.

Kumar, P., Saurabh, S., & Sharma, K. (2024i). Assessment of the phytochemical constituents and metabolites in medicinal plants and herbal remedies used in the treatment and management of reproductive diseases: Polycystic ovary syndrome. In *Reference series in phytochemistry*. Punjab, Phagwara, India: Springer Science and Business Media B.V., Department of Agronomy, School of Agriculture, Lovely Professional University, 511–532. Retrieved from https://doi.org/10.1007/978-3-031-43199-9_20.

Kumar, P., Sharma, K., & Singh, J. (2024j). Analysis of complex microbial communities in soil and wastewater treatment processes. In *Microbial metagenomics in effluent treatment plant*. Punjab, Phagwara, India: Elsevier, Department of Agronomy, School of Agriculture, Lovely Professional University, 25–41. Retrieved from https://doi.org/10.1016/B978-0-443-13531-6.00002-1.

Kumar, P., Sharma, K., & Singh, J. (2024k). Response of microbial community to environment changes. In *Microbial metagenomics in effluent treatment plant*. Punjab, Phagwara, India: Elsevier, Department of Agronomy, School of Agriculture, Lovely Professional University, 43–61. Retrieved from https://doi.org/10.1016/B978-0-443-13531-6.00006-9.

Kumar, P., Singh, J., & Singh, G. (2024l). Renewable hydrogen opportunities and challenges. In *Renewable hydrogen: Opportunities and challenges in commercial success*. Punjab, Phagwara, India: Elsevier, Department of Agronomy, School of Agriculture, Lovely Professional University, 193–220. Retrieved from https://doi.org/10.1016/B978-0-323-95379-5.00010-9.

Kumar, P., Srivastav, A. L., Chaudhary, V., van Hullebusch, E. D., & Busquets, R. (2024m). *Bioremediation of emerging contaminants from soils: Soil health conservation for improved ecology and food security*. Punjab, Phagwara, India: Elsevier, Department of Agronomy, School of Agriculture, Lovely Professional University. Retrieved from https://doi.org/10.1016/C2022-0-02758-6.

Pandey, S. K., Kumar, P., & Dey, S. R. (2024). Xenobiotic stress management by microbial endophytes. In *Bioremediation of emerging contaminants from soils: Soil health conservation for improved ecology and food security*. Madhya Pradesh, Jabalpur, India: Elsevier, College of Agriculture, Powarkheda Narmadapuram, Jawaharlal Nehru Agriculture University, 241–259. Retrieved from https://doi.org/10.1016/B978-0-443-13993-2.00012-8.

Raj, A., Goutam, E., Kumar, P., & Singh, J. (2024). Polyphenol oxidase: An enzyme of plants. In *Polyphenol oxidases: Function, wastewater remediation, and biosensors*. Phagwara, Punjab: De Gruyter, Department of Agronomy, School of Agriculture, Lovely

Professional University, 105–128. Retrieved from https://doi.org/10.1515/9783111033525-004 144411, India.

Saini, L., Kumar, P., & Panwar, J. S. (2024). Current status of wastewater in India/Other Countries/Regions. In *Bioelectrochemical oxidation processes for wastewater treatment*. Punjab, Phagwara, India: CRC Press, Lovely Professional University, 152–165. Retrieved from https://doi.org/10.1201/9781003368472-8.

Sharma, K., & Kumar, P. (2024). Environmental threats posed by xenobiotics. In *Bioremediation of emerging contaminants from soils: Soil health conservation for improved ecology and food security*. Punjab, Phagwara, India: Elsevier, Department of Agronomy, School of Agriculture, Lovely Professional University, 183–201. Retrieved from https://doi.org/10.1016/B978-0-443-13993-2.00009-8.

Sharma, K., Kumar, P., & Singh, J. (2024). Ecology and diversity of microbial communities involved in the removal of priority contaminants and micropollutants in waste. In *Development in waste water treatment research and processes: Role of environmental microbiology in industrial wastewater research*. Punjab, Phagwara, India: Elsevier, Department of Agronomy, School of Agriculture, Lovely Professional University, 181–199. Retrieved from https://doi.org/10.1016/B978-0-443-13609-2.00014-8.

Sharma, M., Dey, S. R., Kumar, P., & Dwivedi, P. (2024). Insights into the genetic improvement of canola. In *Insights into the genetic improvement of canolain: Genetic engineering of crop plants for food and health security (Volume 2)*. Phagwara, India: Springer Nature, Department of Agronomy, School of Agriculture, Lovely Professional University, 205–226. Retrieved from https://doi.org/10.1007/978-981-97-3119-0_9.

Sharma, M., Kumar, P., & Panwar, J. S. (2024). Implementation of bioelectrochemical oxidation systems in existing wastewater treatment plant: Challenges in retrofitting. In *Bioelectrochemical oxidation processes for wastewater treatment*. Punjab, Phagwara, India: CRC Press, Lovely Professional University, 166–188. Retrieved from https://doi.org/10.1201/9781003368472-9.

Srivastav, A. L., Patel, N., Rani, L., Kumar, P., Dutt, I., Maddodi, B. S., & Chaudhary, V. K. (2024). Sustainable options for fertilizer management in agriculture to prevent water contamination: A review. *Environment, Development and Sustainability, 26*, 8303–8327. Retrieved from https://doi.org/10.1007/s10668-023-03117-z.

Chapter 22

Therapeutic approaches for heavy metal detoxification

Prasann Kumar
Department of Agronomy, School of Agriculture, Lovely Professional University, Phagwara, Punjab, India

22.1 Introduction

Monitoring and understanding metal concentrations in various environmental matrices have gained significant importance due to toxic metal elements' stability and bioaccumulation nature (Dey, Sharma, Kumar, & Dwivedi, 2024d; Sharma, Dey, Kumar, & Dwivedi, 2024a). Lead (Pb), mercury (Hg), cadmium (Cd), arsenic (As), and chromium (Cr) are naturally occurring elements in the environment, originating from both geogenic processes and anthropogenic activities. While trace amounts of metals such as zinc (Zn), copper (Cu), and iron (Fe) are essential for biological functions, excessive exposure to heavy metals poses severe health risks, including toxicity and chronic diseases. Heavy metals, characterized by their high atomic weight and density, persist in the environment without rapid degradation. Instead, they accumulate in ecosystems, water bodies, soils, and food chains, leading to widespread contamination. Industrialization, urbanization, and chemically intensive agricultural practices have exacerbated heavy metal pollution, increasing human exposure through multiple pathways. Unlike organic pollutants, heavy metals do not undergo biodegradation into less toxic forms, complicating their removal from the environment. Therefore effective detoxification strategies are urgently needed to mitigate their adverse impacts on human health and ecological systems (Dey, Sharma, Kumar, & Dwivedi, 2024e; Sharma, Kumar, & Panwar, 2024b). Heavy metals originate from multiple sources, including industrial, environmental, dietary, and occupational pathways. Industrial activities such as mining, smelting, battery production, and coal combustion significantly contribute to heavy metal emissions. The electroplating, leather tanning, and chemical industries also play substantial roles in heavy metal pollution. Industrial discharge into nearby water bodies contaminates drinking and irrigation water sources, severely threatening public health. Additionally, atmospheric emissions from industrial facilities can settle on crops, thereby infiltrating the food chain. Natural sources of heavy metals, such as forest fires, volcanic eruptions, and rock weathering, also contribute to environmental metal loads. However, anthropogenic activities far exceed natural contributions, necessitating stringent industrial regulations and pollution control measures to mitigate heavy metal exposure. Dietary exposure represents another significant pathway through which heavy metals enter the human body. Contaminated food and water are primary vectors for heavy metal ingestion, as crops absorb metals from polluted soil, irrigation water, and atmospheric deposition. Certain agricultural products, including rice, leafy vegetables, and root crops, exhibit a high propensity for metal accumulation due to their affinity for soil and waterborne contaminants. Mercury exposure, primarily from consuming contaminated seafood, occurs due to industrial mercury discharge into aquatic ecosystems, which bioaccumulates in fish species such as tuna and swordfish (Kumar & Dwivedi, 2024; Kumar, Devi, & Panwar, 2024b). Water contamination is a particularly pressing public health concern, as individuals consuming water contaminated with toxic metals such as arsenic, lead, and cadmium face serious health risks. This issue is especially prevalent in regions with inadequate water management and purification infrastructure. For instance, arsenic-contaminated groundwater remains a critical problem in Bangladesh, India, and parts of South America, where long-term exposure can lead to severe health complications. Heavy metals in food are further exacerbated by pesticide residues, chemical fertilizers, and wastewater irrigation, underscoring the need for stringent agricultural and food safety regulations. Occupational exposure to heavy metals is prevalent among workers in mining, welding, battery production, construction, and waste

Heavy Metal Toxicity and Neurodegeneration. DOI: https://doi.org/10.1016/B978-0-443-36575-1.00005-4

recycling industries. These occupations involve regular inhalation and dermal absorption of metal dust and fumes, leading to chronic toxicity. Workers involved in paint, ceramic, and plumbing material production frequently encounter lead poisoning, whereas those in battery and plastic manufacturing are at high risk of cadmium exposure. Prolonged exposure to heavy metals in occupational settings has been linked to severe health conditions, including lung diseases, kidney damage, neurological disorders, and an increased risk of cancer. Protective measures, such as personal protective equipment (PPE), improved ventilation, and exposure monitoring, are essential in mitigating occupational health risks. However, in many developing countries, inadequate enforcement of occupational health regulations exposes workers to metal poisoning hazards. Heavy metal toxicity can lead to both acute and chronic health complications, affecting multiple organ systems. The severity of toxicity depends on factors such as the type of metal, duration of exposure, route of entry, and individual susceptibility. Heavy metals exert toxic effects primarily through oxidative stress induction, disruption of cellular functions, and inhibition of enzymatic activity. Lead exposure is particularly detrimental to children, as it impairs cognitive development, reduces IQ scores, and increases the risk of behavioral disorders. In adults, chronic lead exposure is associated with hypertension, renal dysfunction, and reproductive health issues (Kumar, 2024a; Saini, Kumar, & Panwar, 2024). Mercury toxicity, particularly from methylmercury, adversely affects the nervous system, contributing to neurodegenerative diseases, memory impairment, and tremors. Prenatal mercury exposure poses severe developmental risks to infants, making it a critical concern for expectant mothers. Cadmium exposure has been linked to kidney damage, osteoporosis, and an elevated risk of lung cancer due to its carcinogenic properties. Arsenic exposure is associated with dermatological lesions, cardiovascular diseases, and several forms of cancer, including bladder and lung cancer. Chromium, particularly in its hexavalent form (Cr(VI)), is classified as a mutagen and carcinogen, contributing to respiratory disorders and increased cancer risk among exposed populations. Given the systemic toxicity of heavy metals, the development of effective detoxification strategies is imperative. As the global burden of heavy metal toxicity continues to rise, the need for effective detoxification strategies becomes increasingly urgent. Conventional detoxification methods, such as chelation therapy, involve the administration of chemical chelating agents, including ethylenediaminetetraacetic acid (EDTA), dimercaptosuccinic acid (DMSA), and 2,3-dimercapto-1-propanesulfonic acid (DMPS), which bind to heavy metals and facilitate their excretion. While chelation therapy has proven effective in treating acute metal poisoning, it presents risks such as depletion of essential minerals and adverse side effects. Consequently, researchers are exploring alternative detoxification approaches, including dietary and nutraceutical interventions. Antioxidants such as glutathione, vitamin C, and selenium are crucial in mitigating oxidative stress induced by heavy metals while enhancing the body's natural detoxification mechanisms. Additionally, certain functional foods and herbal remedies, including Spirulina, Chlorella, and Moringa, exhibit metal-binding properties, making them promising candidates for natural detoxification therapies. Microbiome modulation has emerged as another promising approach for heavy metal detoxification, utilizing probiotics and prebiotics to facilitate the binding and elimination of toxic metals in the gastrointestinal tract (Dey, Kumar, & Panwar, 2024a; Kumar, Jose Anand, & Rajagopal, 2024e). Studies have demonstrated that specific probiotic strains, such as Lactobacillus and Bifidobacterium, can sequester heavy metals, preventing their absorption into the bloodstream. Advances in nanotechnology-based detoxification strategies are also noteworthy, with engineered nanomaterials and bio-adsorbents being developed for targeted heavy metal removal. Additionally, environmental interventions such as phytoremediation, which employs metal-accumulating plants to remediate contaminated soils and water sources, offer sustainable solutions for reducing heavy metal exposure at its source. Despite the emergence of innovative detoxification strategies, several challenges remain, including the need for individualized detoxification protocols, the development of advanced diagnostic techniques for early heavy metal toxicity detection, and the implementation of stricter environmental regulations to prevent contamination. Addressing heavy metal toxicity requires a multifaceted approach encompassing clinical, nutritional, technological, and policy-driven solutions. Public awareness campaigns regarding heavy metal exposure risks, preventive measures such as dietary modifications, proper waste disposal, and improved healthcare interventions are essential to a comprehensive strategy. Future research should focus on developing safer, more effective detoxification methods, including novel chelating agents with minimal side effects, microbiome-based therapies, and nanotechnology-driven solutions. A concerted effort is required to implement comprehensive health and regulatory interventions to mitigate heavy metal exposure and safeguard public health for future generations (Table 22.1).

22.2 Mechanisms of heavy metal toxicity

Heavy metal exposure poses significant risks to biological systems due to its bioaccumulative nature, persistence in tissues, and capacity to interfere with fundamental biochemical and physiological processes. Bioaccumulation occurs when organisms absorb heavy metals at a rate faster than they can metabolize or excrete them, leading to progressively higher concentrations within tissues over

TABLE 22.1 Comparing key chelating agents used in heavy metal detoxification.

Chelating agent	Target metals	Mechanism of action	Clinical applications	Side effects	Contraindications
EDTA	Lead, cadmium, mercury, arsenic	Binds to metal ions and forms stable complexes excreted via urine	Lead poisoning, cardiovascular disease, calcium overload	Nephrotoxicity, hypocalcaemia, hypotension, headache	Kidney disease, hypocalcemia, pregnancy
DMSA	Lead, mercury, arsenic	Thiol groups bind to metals, increasing urinary excretion	Lead poisoning in children, mercury detox, arsenic poisoning	Gastrointestinal discomfort, rash, mild neutropenia	Liver disease, severe renal impairment
2,3-Dimercapto-1-propanesulfonic acid (DMPS)	Mercury, lead, arsenic, cadmium	Forms water-soluble metal complexes excreted via urine	Mercury poisoning, arsenic toxicity, industrial metal exposure	Allergic reactions, skin rash, nausea, kidney strain	Sulfur allergies, kidney disease, severe hepatic disease
Penicillamine	Copper, lead, mercury, arsenic	Binds to metals and promotes renal excretion	Wilson's disease, rheumatoid arthritis, heavy metal poisoning	Bone marrow suppression, nephrotoxicity, hypersensitivity	Penicillin allergy, renal insufficiency, pregnancy
DFO	Iron, aluminum	Chelates excess iron and aluminum, promotes renal excretion	Hemochromatosis, iron overload, aluminum toxicity (dialysis)	Hypotension, neurotoxicity, respiratory distress	Severe renal disease, pregnancy, hypersensitivity to the drug

DFO, Deferoxamine; *DMSA*, dimercaptosuccinic acid; *EDTA*, ethylenediaminetetraacetic acid.
Source: *Based on* Kumar and Choudhury (2024e) and Kumar and Singh (2024c).

time. This persistence is particularly concerning because heavy metals such as lead, mercury, cadmium, and arsenic do not readily degrade in biological environments, causing prolonged exposure and increasing toxicity risks. Once inside an organism, these metals interact with cellular components, often binding to proteins, enzymes, and nucleic acids, disrupting enzymatic and cellular functions. Many heavy metals have a high affinity for thiol (–SH) groups in proteins, leading to enzyme inhibition, protein misfolding, and disruption of critical metabolic pathways. For instance, mercury readily binds to cysteine residues, inhibiting vital enzymes involved in cellular respiration and antioxidant defense (Devi & Kumar, 2024a; Dey, Kumar, & Singh, 2024b; Kumar & Choudhury, 2024b). Similarly, lead interferes with calcium-dependent enzymatic processes by mimicking calcium ions, disrupting neurotransmitter release, and impairing various cellular functions. Such interference in enzymatic activity can cascade into broader metabolic disturbances, affecting energy production, DNA repair mechanisms, and membrane integrity. One of the most concerning outcomes of heavy metal toxicity is the induction of oxidative stress, a condition characterized by an imbalance between the production of reactive oxygen species (ROS) and the biological system's ability to detoxify them. Heavy metals catalyze the generation of ROS, including superoxide anions, hydrogen peroxide, and hydroxyl radicals, which cause extensive damage to lipids, proteins, and nucleic acids. This oxidative stress not only compromises cell viability but also triggers inflammation pathways. Inflammatory responses are mediated through the activation of transcription factors such as nuclear factor-kappa B (NF-κB) and activator protein-1 (AP-1), which upregulate proinflammatory cytokines like tumor necrosis factor-alpha (TNF-α), interleukins (IL-1β, IL-6), and chemokines. Chronic inflammation exacerbates tissue damage, disrupts cellular homeostasis, and contributes to various pathophysiological conditions. In particular, heavy metal–induced oxidative stress and inflammation have been implicated in neurotoxicity, nephrotoxicity, hepatotoxicity, and cardiovascular risks. Neurotoxicity is a significant concern with metals like lead, mercury, and arsenic, which cross the blood–brain barrier and interfere with neurotransmission, synaptic plasticity, and neuronal survival (Devi, Kumar, & Singh, 2024g; Kumar, Singh, & Singh, 2024l; Sharma, Kumar, & Singh, 2024). Lead exposure, for example, disrupts calcium-dependent signaling in neurons, leading to deficits in learning and memory, neurodevelopmental disorders, and neurodegenerative diseases. Mercury, particularly in its organic form (methylmercury), induces neuronal apoptosis by triggering mitochondrial dysfunction, increasing ROS production, and impairing antioxidant defenses. Similarly, arsenic affects dopaminergic and glutamatergic pathways, contributing to cognitive impairments and neurodegenerative conditions such as Parkinson's and Alzheimer's diseases. Nephrotoxicity, or kidney damage, is another major consequence of heavy metal exposure, as the kidneys serve as primary excretory organs for many toxic metals, making them highly susceptible to accumulation and damage. Cadmium, for instance, preferentially accumulates in the renal cortex, where it disrupts tubular reabsorption, induces apoptosis, and leads to chronic kidney disease. Mercury also damages the renal proximal tubules by binding to sulfhydryl groups in cellular proteins, impairing mitochondrial function, and causing glomerular dysfunction. Hepatotoxicity arises when heavy metals accumulate in the liver, disrupting hepatic metabolism, inducing oxidative stress, and promoting inflammatory cascades. Arsenic and cadmium are well-documented hepatotoxins that trigger liver fibrosis, hepatocyte apoptosis, and even carcinogenesis through epigenetic alterations and persistent inflammatory signaling. Cardiovascular risks associated with heavy metal exposure include hypertension, atherosclerosis, and myocardial dysfunction, primarily due to endothelial damage, increased oxidative stress, and altered lipid metabolism. Lead exposure, for instance, is strongly linked to hypertension as it interferes with nitric oxide (NO) signaling, leading to vasoconstriction and increased blood pressure. Cadmium contributes to cardiovascular disease by promoting endothelial cell apoptosis, enhancing oxidative stress, and inducing lipid peroxidation, accelerating atherosclerotic plaque formation. The genetic and epigenetic modifications induced by heavy metal exposure further complicate their toxicological impact (Devi, Kumar, & Singh, 2024h; Kumar & Singh, 2024a; Kumar & Singh, 2024b). Heavy metals can induce DNA damage through direct interactions with nucleotides or indirectly via oxidative stress-induced mutagenesis. Arsenic, for example, interferes with DNA repair enzymes, leading to genomic instability and increased mutation rates, which are associated with carcinogenesis. Epigenetic modifications, including DNA methylation, histone modification, and microRNA (miRNA) regulation, play a critical role in the long-term effects of heavy metal exposure. Cadmium exposure has been shown to alter DNA methylation patterns, leading to the silencing of tumor suppressor genes and activating oncogenes. Similarly, lead disrupts histone acetylation, affecting gene expression patterns related to neurodevelopment and cognitive function. These epigenetic changes can have transgenerational effects, meaning that exposure to heavy metals in one generation may influence disease susceptibility in subsequent generations. The cumulative impact of bioaccumulation, enzymatic disruption, oxidative stress, inflammation, organ toxicity, and genetic modifications underscores the severe health risks associated with heavy metal exposure. Understanding these mechanisms is crucial for developing therapeutic strategies to mitigate toxicity, including chelation therapy, antioxidant supplementation, and dietary interventions to reduce metal absorption and enhance detoxification pathways. Furthermore, regulatory measures to limit environmental contamination and occupational exposure to heavy metals remain vital in preventing long-term health effects on populations worldwide (Table 22.2).

22.3 Chelation therapy for heavy metal detoxification

Chelation therapy is a medical intervention to remove heavy metals from biological systems by employing chelating agents and specialized molecules that bind to metal ions and facilitate excretion. The mechanism of chelating agents is based on their ability to form stable, water-soluble complexes with metal ions, preventing them from participating in harmful biochemical interactions. Chelating agents typically contain functional groups such as carboxyl (–COOH), thiol (–SH), or amine ($–NH_2$) that coordinate metal ions through electrostatic and covalent interactions. Once bound, these complexes are transported to the kidneys or liver for elimination via urine or bile, reducing the bioavailability of toxic metals and mitigating their harmful effects on enzymatic functions, oxidative stress pathways, and cellular metabolism (Devi and Kumar, 2024b; Kumar, Karam, & Choudhury, 2024f; Kumar, Khokhar, & Dey, 2024g; Kumar, Saurabh, & Sharma, 2024i). The effectiveness of a chelating agent depends on its specificity for particular metal ions, ability to access intracellular compartments, and pharmacokinetic properties, including absorption, distribution, metabolism, and excretion. Some chelators are highly selective for specific heavy metals, while others have broad-spectrum activity and can target multiple toxic elements simultaneously. The clinical applications of chelating agents have been extensively explored, particularly in cases of lead, mercury, arsenic, and cadmium poisoning, as well as in conditions associated with excess iron or copper accumulation. One of the most commonly used chelating agents is EDTA, which has a high affinity for divalent and trivalent metal ions such as lead, cadmium, and iron. EDTA forms stable, soluble complexes with these metals, allowing their excretion through urine. It is primarily used in cases of acute and chronic lead poisoning and has also been investigated for its potential benefits in reducing vascular calcification in cardiovascular diseases. However, EDTA has limitations; it can also chelate essential minerals such as calcium, magnesium, and zinc, leading to electrolyte imbalances if not carefully monitored. Another clinically significant chelating agent is DMSA, which contains two thiol groups that enable strong binding to lead, mercury, and arsenic. DMSA is orally bioavailable and is widely used to treat lead poisoning in children, as it effectively reduces blood lead levels without significantly depleting essential minerals (Devi & Kumar, 2024c; Kumar & Choudhury, 2024c; Kumar, Choudhury, Dey, & Sharma, 2024a; Kumar, Srivastav, Chaudhary, van Hullebusch, & Busquets, 2024m). DMSA is also less toxic than some older chelators, making it a preferred option in pediatric and chronic heavy metal exposure cases. In addition to DMSA, DMPS is another sulfur-containing chelator that has demonstrated efficacy in treating mercury and arsenic poisoning. DMPS is more water-soluble than DMSA and can be administered orally and intravenously, providing flexibility in treatment. It enhances the renal excretion of mercury and arsenic while having a lower affinity for essential trace elements, making it a relatively safer alternative. Despite its advantages, due to regulatory restrictions, DMPS is not widely available in some regions. Penicillamine, a derivative of penicillin, is another important chelating agent used primarily for copper toxicity, particularly in Wilson's disease, a genetic disorder characterized by excessive copper accumulation in the liver, brain, and other tissues. Penicillamine binds to copper and promotes its excretion through urine, effectively lowering tissue copper levels and preventing the progression of hepatic and neurological symptoms. It is also used off-label for treating lead and mercury poisoning, though its side effects, including nephrotoxicity, dermatological reactions, and hematological complications, limit its widespread use. While chelation therapy provides critical benefits in metal detoxification, it also carries risks and limitations that must be carefully managed. One of the primary concerns is the nonspecific binding of chelating agents to essential minerals such as calcium, zinc, and magnesium, which can lead to deficiencies and metabolic disturbances. This is particularly relevant for EDTA, which can cause hypocalcaemia and cardiac arrhythmias if administered too rapidly. Additionally, chelation therapy can induce oxidative stress by mobilizing metal ions from tissues into circulation before excretion, potentially exacerbating cellular damage if not properly controlled. Side effects such as gastrointestinal distress, nephrotoxicity, hepatotoxicity, and allergic reactions are also common, varying depending on the specific chelating agent used. Moreover, the long-term efficacy of chelation therapy in chronic exposure cases remains a subject of debate, as heavy metals often accumulate in deep tissue compartments such as bone and brain, where chelators may have limited access. Another significant limitation is the potential redistribution of metals during therapy, in which mobilized metal ions may be temporarily deposited in sensitive organs before excreting, increasing transient toxicity risks (Choudhury, Kumar, Zhimo, & Sahoo, 2024; Dey, Sharma, & Kumar, 2024c; Kumar, Sharma, & Singh, 2024j; Pandey, Kumar, & Dey, 2024). Contraindications to chelation therapy

TABLE 22.2 Nutritional and dietary interventions for heavy metal detoxification.

Nutrient/compound	Source (food/supplement)	Mechanism of action	Heavy metals targeted	Recommended intake (general guidance)
Glutathione	Avocados, spinach, asparagus, whey protein	Antioxidant: binds to metals for excretion	Mercury, Lead, Cadmium, Arsenic	No established RDA; 250–1000 mg/day (supplemental)
N-Acetylcysteine	Eggs, garlic, onions, supplements	Precursor to glutathione; enhances detoxification pathways	Mercury, Lead, Cadmium	600–1800 mg/day (supplemental)
Vitamin C	Citrus fruits, bell peppers, berries	Antioxidant: enhances metal excretion and reduces oxidative stress	Lead, Cadmium, Mercury	75–90 mg/day (RDA); up to 2000 mg/day (upper limit)
Vitamin E	Nuts, seeds, vegetable oils, leafy greens	Protects against metal-induced oxidative damage	Lead, Cadmium, Mercury	15 mg/day (RDA); upper limit 1000 mg/day
Selenium	Brazil nuts, seafood, eggs	Supports glutathione function, reduces metal toxicity	Mercury, Arsenic, Cadmium	55 mcg/day (RDA); upper limit 400 mcg/day
Zinc	Meat, shellfish, legumes, nuts	Competes with metals for absorption, reduces toxicity	Lead, Cadmium, Arsenic	8–11 mg/day (RDA); upper limit 40 mg/day
Curcumin	Turmeric	Chelates metals reduces oxidative stress	Lead, Cadmium, Mercury, Arsenic	500–2000 mg/day (supplemental)
Quercetin	Onions, apples, grapes, berries	Binds to metals, enhances antioxidant defenses	Lead, Cadmium, Arsenic	500–1000 mg/day (supplemental)
Resveratrol	Red grapes, red wine, peanuts	Antioxidants modulate detoxification pathways	Mercury, Lead, Cadmium	5–500 mg/day (supplemental)
Chlorella	Green algae supplements	Binds to metals, enhances elimination via feces	Mercury, Lead, Cadmium, Arsenic	1–5 g/day (supplemental)
Spirulina	Blue-green algae supplements	Chelates heavy metals, supports liver detoxification	Lead, Mercury, Cadmium	1–5 g/day (supplemental)
Fiber (pectin, lignin, etc.)	Fruits, vegetables, whole grains	Binds to metals in the gut, reducing absorption	Lead, Mercury, Cadmium	25–38 g/day (RDA for fiber)

Source: *Based on* Kumar, Kumar and Singh (2024a), Kumar, Kumar and Singh (2024b), and Saini et al. (2024).

include renal insufficiency, as many chelators rely on renal excretion pathways and preexisting cardiovascular conditions that electrolyte imbalances may exacerbate. Furthermore, chelation therapy is not universally effective for all types of heavy metal poisoning; for example, it has limited efficacy in removing aluminum, which requires specific ligands for efficient clearance. Given these challenges, ongoing research is focused on developing emerging chelating agents with improved selectivity, reduced side effects, and enhanced penetration into deep tissue compartments. One promising area of research involves designing multifunctional chelators that bind heavy metals and possess antioxidant properties to mitigate oxidative stress during metal mobilization. For instance, novel thiol-based polyamine carboxylate chelators are being investigated for their ability to target intracellular metal pools while minimizing disruption to essential minerals. Nanotechnology-based chelators are also gaining attention, with nanoparticles engineered to selectively capture heavy metals and facilitate their excretion without affecting trace element homeostasis (Devi et al., 2024e; Kumar & Choudhury, 2024d; Kumar, Sharma, & Singh, 2024k; Sharma & Kumar, 2024). Additionally, hybrid chelators that combine chelating functions with antiinflammatory or neuroprotective properties are being explored to address the broader pathophysiological consequences of heavy metal toxicity. In drug development, efforts are being made to optimize existing chelators through structural modifications that enhance their pharmacokinetics and bioavailability. For example, lipophilic derivatives of DMSA and DMPS are being synthesized to improve their ability to cross the blood–brain barrier, making them more effective in treating neurological manifestations of heavy metal toxicity. Gene therapy approaches targeting metal transporters and detoxification enzymes are also under investigation, offering potential long-term solutions for managing heavy metal accumulation disorders. Furthermore, dietary and natural chelating agents, such as phytochelatins, polyphenols, and alginate-based compounds, are being explored as adjunctive therapies to conventional chelation. These naturally occurring chelators in foods such as garlic, cilantro, and pectin-rich fruits may offer a safer and more sustainable approach to mitigating chronic low-level heavy metal exposure. Integrating these emerging strategies with existing chelation protocols could revolutionize the management of heavy metal toxicity, providing safer, more effective, and personalized treatment options. Ultimately, while chelation therapy remains a cornerstone of heavy metal detoxification, continued research into novel chelating agents, advanced drug delivery systems, and complementary therapeutic approaches is essential to overcoming current limitations and improving clinical outcomes for affected individuals (Table 22.3).

22.4 Nutritional and nutraceutical approaches

Dietary modifications play a crucial role in reducing the bioavailability of heavy metals in the human body by influencing their absorption, distribution, and excretion. The strategic intake of specific nutrients, antioxidants, and metal-binding phytochemicals can mitigate the harmful effects of heavy metals and enhance the body's natural detoxification pathways. Several dietary components, including proteins, fibers, vitamins, and minerals, have been shown to interfere with heavy metal absorption in the gastrointestinal tract, thereby reducing their systemic toxicity. Consuming a diet rich in essential minerals such as calcium, iron, magnesium, and zinc can competitively inhibit the uptake of toxic metals like lead, cadmium, and mercury, as these metals often utilize similar transport pathways. For instance, lead absorption is significantly higher in individuals with calcium or iron deficiencies, highlighting the importance of adequate mineral intake in preventing heavy metal toxicity. Additionally, dietary fibers, particularly insoluble fibers like cellulose and lignin, can bind heavy metals in the gut and facilitate their excretion through feces, further limiting their bioavailability. Beyond macronutrient modulation, antioxidants and metal-binding phytochemicals play a pivotal role in neutralizing heavy metal–induced oxidative stress and promoting detoxification. Glutathione, a tripeptide composed of glutamine, cysteine, and glycine, is one of the most potent endogenous antioxidants in metal detoxification. It directly binds to toxic metals, forming glutathione-metal complexes excreted via the biliary and renal pathways (Devi, Dogra, Kumar, & Singh, 2024b; Kumar, Dwivedi, & Upadhyay, 2024d; Kumar, 2024c; Kumar, 2024d). However, heavy metal exposure depletes glutathione levels, making dietary and supplemental sources essential for maintaining adequate glutathione stores. *N*-Acetylcysteine (NAC), a precursor to glutathione, has been extensively studied for its ability to replenish intracellular glutathione levels and enhance metal excretion. NAC possesses free thiol groups that directly chelate metals like mercury, lead, and cadmium, reducing their toxicity while simultaneously acting as a powerful antioxidant to mitigate oxidative damage. Vitamins C and E are critical antioxidants in counteracting heavy metal–induced oxidative stress. Vitamin C, or ascorbic acid, reduces metal ions to less toxic forms and facilitates their excretion by chelating metals in circulation. Additionally, vitamin C enhances the bioavailability of iron, reducing competition between essential and toxic metals at absorption sites. Vitamin E, a lipid-soluble antioxidant, protects cell membranes from lipid peroxidation induced by heavy metal toxicity, thereby preventing cellular damage and inflammation. The synergistic action of vitamins C and E

TABLE 22.3 Probiotic and microbiome-based strategies for heavy metal detoxification.

Probiotic strain	Mechanism of action	Heavy metals affected	Clinical evidence	Potential benefits
Lactobacillus rhamnosus GG	Binds to heavy metals in the gut, prevents absorption	Lead, cadmium, mercury	Shown to reduce lead absorption in animal and human studies	Protects gut barrier, enhances metal excretion
Lactobacillus plantarum	Produces exopolysaccharides that bind metals	Lead, cadmium, arsenic	Demonstrated metal-binding abilities in vitro and animal models	Supports gut microbiome balance, reduces oxidative stress
Bifidobacterium bifidum	Enhances gut barrier, modulates immune response	Lead, mercury, arsenic	Preclinical studies suggest a reduction in metal toxicity	Protects intestinal lining, reduces inflammation
Bacillus subtilis	Forms biofilms that trap metals promote fecal excretion	Lead, mercury	Animal studies indicate improved heavy metal excretion	Improves gut detox pathways, enhances digestion
Saccharomyces boulardii	Absorbs heavy metals, prevents gut dysbiosis	Cadmium, lead, arsenic	Shown to protect against arsenic toxicity in animal studies	Reduces gut inflammation, supports liver detoxification
Lactobacillus casei	Enhances antioxidant activity, modulates gut microbiota	Lead, cadmium, arsenic	Evidence of reduced oxidative stress in heavy metal exposure models	Enhances antioxidant defenses, gut health improvement
Bifidobacterium longum	Chelates metals via bacterial cell wall interactions	Lead, mercury	Preclinical studies suggest potential heavy metal detoxification	Supports immune function, reduces toxic metal burden
Lactobacillus reuteri	Produces antimicrobial and detoxifying compounds	Lead, cadmium, arsenic	Studies show improved gut barrier integrity under toxicity	Supports gut microbiome balance, improves metabolic health
Clostridium butyricum	Produces short-chain fatty acids, reduces inflammation	Lead, arsenic	Research indicates metal detox support via gut microbiota	Strengthens gut lining, enhances detox enzyme activity
Escherichia coli Nissle 1917	Binds heavy metals, modulates gut immune response	Lead, cadmium, mercury	Human and animal trials suggest improved detoxification	Protects against heavy metal-induced gut damage

Source: Based on Devi, Kumar, and Singh (2024f) and Kumar (2024a).

has been shown to reduce the toxic effects of lead, mercury, and cadmium, particularly in individuals with chronic exposure. Selenium and zinc also play vital roles in mitigating heavy metal toxicity through their antioxidant and metal-binding properties. Selenium, an essential trace element, forms complexes with heavy metals such as mercury, preventing their interaction with cellular components (Jagadesh et al., 2024; Kumar & Choudhury, 2024e; Raj, Goutam, Kumar, & Singh, 2024; Srivastav et al., 2024). Selenoproteins, such as glutathione peroxidase, reduce oxidative stress induced by metal exposure and protect against DNA damage. Zinc competes with toxic metals for binding sites on proteins and transporters, thereby reducing the absorption and accumulation of lead and cadmium. Zinc also plays a role in synthesizing metallothioneins, small cysteine-rich proteins that sequester and detoxify heavy metals within cells. Polyphenols, a diverse group of plant-derived compounds, exhibit potent metal-chelating and antioxidant properties, making them valuable in dietary strategies for heavy metal detoxification. Curcumin, the active compound in turmeric, has been shown to chelate metals such as lead and cadmium, reducing their bioavailability and preventing oxidative damage. Additionally, curcumin enhances the expression of detoxification enzymes and supports liver function, further promoting metal elimination. Quercetin, a flavonoid found in fruits and vegetables, possesses strong chelating activity, particularly against iron and copper, and inhibits metal-induced oxidative damage by stabilizing free radicals. Resveratrol, another polyphenol found in grapes and red wine, has demonstrated neuroprotective effects against heavy metal toxicity by modulating inflammatory pathways and reducing metal-induced mitochondrial dysfunction. Functional foods and herbal remedies have gained significant attention for their role in detoxifying heavy metals from the body. Chlorella and spirulina, two nutrient-dense algae, have been widely studied for their metal-binding capabilities (Kumar & Choudhury, 2024f; Kumar & Dey, 2024; Kumar & Singh, 2024e; Kumar, Dey, & Choudhury, 2024c). These microalgae contain high levels of chlorophyll, which facilitates metal chelation and excretion through the gut. Studies have shown that chlorella supplementation significantly reduces blood mercury levels in individuals with chronic exposure. Cilantro (*Coriandrum sativum*) is another herbal remedy for mobilizing and eliminating heavy metals, particularly lead and mercury. It enhances the urinary excretion of metals and supports liver detoxification pathways. Garlic, rich in sulfur-containing compounds like allicin, has been shown to protect against lead and cadmium toxicity by enhancing glutathione synthesis and promoting metal excretion. Milk thistle (*Silybum marianum*), a hepatoprotective herb, supports liver detoxification by upregulating the production of detoxifying enzymes and reducing oxidative stress associated with heavy metal exposure. The gut microbiome plays a crucial role in heavy metal detoxification, and probiotics have emerged as a promising intervention in reducing heavy metal bioaccumulation. Beneficial gut bacteria, such as *Lactobacillus* and *Bifidobacterium* species, have demonstrated the ability to bind heavy metals in the gastrointestinal tract, preventing their absorption. These probiotic strains produce exopolysaccharides that chelate metals and promote their excretion (Devi, Kaur, & Kumar, 2024d; Kumar & Dey, 2024; Kumar, 2024c; Raj et al., 2024; Srivastav et al., 2024). Additionally, probiotics enhance gut barrier integrity, reducing the translocation of heavy metals into systemic circulation. Fermented foods, such as yogurt, kefir, and kimchi, provide natural probiotics and prebiotics that support a healthy gut microbiome and aid in metal detoxification. Modulating the gut microbiota through dietary interventions has been shown to reduce the toxic burden of heavy metals, particularly in individuals with compromised detoxification pathways. In summary, dietary modifications offer a multifaceted approach to reducing heavy metal bioavailability and toxicity. Essential minerals, antioxidants, and polyphenols compete with toxic metals for absorption, neutralizing oxidative damage, and enhancing detoxification. Functional foods and herbal remedies further support metal elimination by chelating metals, promoting liver detoxification, and modulating inflammatory pathways. The gut microbiome, through probiotic interventions, provides an additional mechanism for limiting heavy metal absorption and promoting excretion. A comprehensive dietary strategy that integrates these elements can significantly mitigate the health risks associated with heavy metal exposure, offering a natural and effective means of detoxification. Continued research in nutritional interventions and microbiome science will refine dietary strategies for managing heavy metal toxicity and improving overall health outcomes (Table 22.4).

22.5 Nanotechnology-based detoxification strategies

Engineered nanomaterials have emerged as powerful tools for the adsorption and removal of heavy metals from environmental and biological systems, leveraging their high surface area, tenable surface chemistry, and unique physicochemical properties to enhance metal-binding efficiency. These nanomaterials include carbon-based structures such as graphene oxide and carbon nanotubes, metal and metal oxide nanoparticles, polymeric nanostructures, and bio-inspired nanocomposites, each with distinct mechanisms for interacting with heavy metal ions. Functionalizing nanomaterials with specific ligands, such as thiols, amines, and carboxyl groups, enhances their selectivity and binding capacity for toxic metals like lead, mercury, cadmium, and arsenic. For instance, graphene

TABLE 22.4 Nanotechnology and enzyme-based detoxification approaches.

Nanomaterial/enzyme	Detoxification mechanism	Targeted metals	Safety considerations	Current Applications
Nano-selenium (SeNPs)	Binds and neutralizes heavy metals enhance antioxidant defense	Mercury, cadmium, lead	Potential cytotoxicity at high doses requires biocompatible formulations	Used in water purification, dietary supplementation for detox
Iron oxide nanoparticles (Fe_3O_4 NPs)	Adsorbs and removes heavy metals via magnetic separation	Arsenic, lead, cadmium	Concerns about bioaccumulation need surface modification	Wastewater treatment, biomedical detoxification strategies
Gold nanoparticles (AuNPs)	Chelate metals enhance enzymatic detox pathways	Mercury, arsenic	Possible oxidative stress, dose-dependent toxicity	Biosensors for metal detection, potential therapeutic applications
Chitosan-based nanomaterials	Binds heavy metals through electrostatic interactions	Lead, mercury, arsenic	Biodegradable but may have limited absorption efficiency	Water purification, dietary supplements
Carbon nanotubes (CNTs)	Adsorbs metals, provides a high surface area for detox	Cadmium, lead, chromium	Potential lung toxicity requires functionalisation	Industrial detox, environmental remediation
Metallothioneins (MTs)	Binds and sequesters metals in cells to prevent oxidative damage	Lead, cadmium, mercury	Low immunogenicity but potential overload risks	Used in engineered probiotics and bioremediation
Phytochelatins (PCs)	Chelate metals via thiol groups enhances excretion	Arsenic, cadmium, lead	Requires controlled expression in therapeutic models	Bioengineered plants and microbes for detox applications
Superoxide dismutase (SOD)	Reduces oxidative stress from metal toxicity	Cadmium, mercury	Can be degraded before reaching the target sites	Antioxidant therapy, metal-induced oxidative stress reduction
Laccase enzyme	Catalyses oxidation and removal of metal ions	Lead, chromium, arsenic	Stability issues require immobilization	Used in bioremediation and industrial effluent detox
Horseradish peroxidase (HRP)	Breaks down metal-organic complexes	Mercury, arsenic, chromium	Needs cofactors for optimal activity	Environmental cleanup, biosensor development
Nano-cerium oxide (CeO_2 NPs)	Scavenges reactive oxygen species (ROS), reduces metal-induced damage	Lead, cadmium, mercury	Potential toxicity at high doses requires surface modifications	Neuroprotection, environmental remediation
Silica nanoparticles (SiNPs)	Binds to metal ions, enhances cellular detoxification	Arsenic, lead, mercury	Needs biocompatible coatings for medical use	Drug delivery, heavy metal sensing, and removal

Source: Based on Choudhury et al. (2024), Devi et al. (2024), Devi, Kumar, Dey, et al. (2024), Devi and Kumar (2024c), and Sharma and Kumar (2024).

oxide nanoparticles have demonstrated exceptional adsorption potential due to their oxygen-containing functional groups, which facilitate strong electrostatic interactions with metal cations. Similarly, metal oxide nanoparticles, such as iron oxide and titanium dioxide, exhibit a high affinity for heavy metals, enabling their removal through adsorption, redox reactions, and precipitation mechanisms (Devi et al., 2024b; Devi, Kapoor, & Kumar, 2024c; Devi et al., 2024d; Kumar & Choudhury, 2024f; Kumar, 2024c). Functionalized polymeric nanostructures, such as dendrimers and nanogels, offer controlled metal chelation and release properties, allowing for targeted detoxification applications. These nanomaterials can be integrated into water treatment systems, soil remediation strategies, and biomedical applications to minimize heavy metal bioavailability and toxicity. In biomedical detoxification, the role of enzymes and nano chelators has gained significant attention, as they offer highly selective and efficient mechanisms for targeted heavy metal removal. Enzymes such as phytochelatins, metallothioneins, and oxidoreductases have evolved in nature to bind and neutralize heavy metals, and their incorporation into nanostructured carriers enhances their stability and bioactivity in therapeutic applications. Nanochelators, engineered with high specificity for heavy metal ions, provide a novel approach to detoxification by facilitating metal binding and excretion through renal and hepatic pathways. Functionalized nanoparticles with chelating ligands, such as EDTA, diethylenetriaminepentaacetic acid (DTPA), and thiol-based compounds, have demonstrated promising results in sequestering heavy metals in vivo, thereby reducing their systemic toxicity (Devi et al., 2024c; Jagadesh et al., 2024; Kumar & Choudhury, 2024g; Kumar et al., 2024d; Srivastav et al., 2024). Additionally, enzyme-mimicking nanomaterials, such as nanozymes, offer catalytic activity similar to natural enzymes, enabling the degradation and transformation of toxic metals into less harmful forms. The development of nanocarriers that encapsulate and deliver detoxifying agents to specific tissues affected by heavy metal exposure further enhances the efficacy of detoxification strategies. Despite the promising applications of nanotechnology in heavy metal detoxification, concerns regarding the biocompatibility and safety of engineered nanomaterials remain a significant challenge. The interactions of nanoparticles with biological systems can lead to unintended cytotoxicity, oxidative stress, immune responses, and potential bioaccumulation. The size, shape, surface charge, and composition of nanoparticles influence their biodistribution, cellular uptake, and clearance pathways, necessitating thorough biocompatibility assessments to ensure their safe application. For example, while effective in metal adsorption, some metal oxide nanoparticles can generate ROS that may exacerbate oxidative damage rather than mitigate it. Similarly, despite their high adsorption capacity, carbon-based nanomaterials pose risks related to pulmonary toxicity and inflammatory responses. Strategies to improve the safety of nanomaterials include surface modifications with biocompatible coatings, the use of biodegradable polymers, and the development of stimuli-responsive nanocarriers that release their cargo only upon specific biological triggers (Kumar & Choudhury, 2024f; Kumar et al., 2024c; Kumar et al., 2024d; Raj et al., 2024; Sharma & Kumar, 2024). Additionally, the long-term fate of nanoparticles in the human body and the environment must be carefully evaluated to prevent potential ecological and health risks associated with nanomaterial exposure. Advances in biosensors for heavy metal detection and detoxification monitoring have revolutionized the field of environmental and biomedical diagnostics, enabling real-time and highly sensitive detection of toxic metals in water, soil, food, and biological fluids. Nanotechnology-based biosensors leverage nanoparticles' unique optical, electrochemical, and magnetic properties to enhance detection sensitivity and specificity (Devi et al., 2024c; Jagadesh et al., 2024; Kumar & Choudhury, 2024f; Kumar et al., 2024c; Kumar, 2024d). For instance, gold and silver nanoparticles have been widely utilized in colourimetric sensors, where changes in their plasmonic properties indicate the presence of heavy metal ions. Similarly, quantum dots and fluorescent nanomaterials enable highly sensitive fluorescence-based detection of metals, providing rapid and precise measurements even at trace levels. Electrochemical biosensors, incorporating nanostructured electrodes and conductive polymers, offer portable and cost-effective solutions for in-field heavy metal analysis, making them valuable tools for environmental monitoring and point-of-care diagnostics. Beyond detection, biosensors are also being developed to monitor the efficacy of detoxification therapies, tracking metal levels in biological fluids to assess treatment progress and optimize therapeutic interventions. Integrating nanotechnology and bioelectronics, wearable biosensors and implantable monitoring devices represent the next frontier in real-time heavy metal exposure assessment and detoxification monitoring (Kumar and Singh, 2024d). These devices provide continuous data on metal bioaccumulation, allowing for personalized treatment strategies and early intervention in cases of acute or chronic heavy metal toxicity. Integrating artificial intelligence and machine learning algorithms with biosensor technology further enhances the predictive capabilities of heavy metal detection systems, enabling automated data analysis and risk assessment. Future research will enhance biosensors' selectivity, stability, and miniaturization, expanding their applicability to diverse environments and biological settings. Engineered nanomaterials present innovative solutions for heavy metal adsorption and removal, offering high efficiency and selectivity in environmental and biomedical applications. Enzymes and nano chelators in targeted detoxification strategies enhance metal binding and excretion while minimizing systemic toxicity. However, the safety and biocompatibility of nanomaterials must be carefully considered to prevent unintended health and environmental risks. Advances in biosensor technology provide powerful tools for heavy metal detection and detoxification monitoring, facilitating real-time assessment and personalized treatment approaches. Continued interdisciplinary research in nanotechnology, biomedicine, and environmental science will drive the development of safer and more effective nanomaterial-based solutions for mitigating heavy metal toxicity and improving public health (Devi et al., 2024b; Kumar & Choudhury, 2024e; Kumar & Singh, 2024e; Kumar et al., 2024c; Kumar et al., 2024d).

22.6 Microbiome and heavy metal detoxification

The gut microbiota plays a crucial role in metal metabolism and detoxification, acting as a natural defense system against the toxic effects of heavy metals by influencing their bioavailability, transformation, and excretion. The human gut is colonized by a diverse community of microorganisms, including bacteria, archaea, fungi, and viruses, that interact with dietary and environmental metals, modulating their absorption and systemic distribution. Certain gut microbes possess metal-binding proteins, enzymatic detoxification pathways, and redox capabilities that enable them to chelate, sequester, or biotransform toxic heavy metals such as lead, mercury, cadmium, and arsenic into less harmful forms. The gut microbiota achieves this through multiple mechanisms, including biosorption, biotransformation, methylation, and efflux transport systems that reduce heavy metal toxicity and promote their elimination via feces. However, exposure to heavy metals can also disrupt the gut microbial balance, leading to dysbiosis characterized by reduced microbial diversity, altered metabolic functions, and increased intestinal permeability, which can exacerbate systemic heavy metal toxicity by facilitating metal translocation into the bloodstream. Research has identified specific heavy metal-resistant probiotic strains, such as Lactobacillus and Bifidobacterium species, which exhibit strong metal-binding capacities and detoxification properties (Devi et al., 2024d; Kumar & Choudhury, 2024e; Kumar & Singh, 2024e; Kumar, 2024c; Raj et al., 2024). These beneficial microbes can reduce the bioavailability of toxic metals by trapping them within their cell walls, chelating them through surface-expressed proteins and exopolysaccharides, and altering metal speciation through enzymatic activity. *Lactobacillus rhamnosus* GG, for instance, has been shown to bind lead and cadmium effectively, preventing their absorption in the intestinal tract. At the same time, *Bifidobacterium breve* has demonstrated arsenic detoxification capabilities through methylation and sequestration mechanisms. These probiotic strains mitigate heavy metal toxicity and help restore gut microbial homeostasis, enhance intestinal barrier function, and reduce oxidative stress triggered by metal-induced inflammatory pathways. Furthermore, probiotic supplementation has been explored as a therapeutic strategy in populations with high environmental exposure to heavy metals, showing promising results in reducing metal accumulation and associated health risks. The potential of fecal microbiota transplantation (FMT) as a therapy for heavy metal toxicity is an emerging area of research that leverages the ability of a healthy donor's gut microbiota to restore microbial balance and improve detoxification pathways in affected individuals (Devi et al., 2024d; Jagadesh et al., 2024; Kumar & Dey, 2024; Kumar et al., 2024d; Kumar, 2024d; Srivastav et al., 2024). FMT involves the transfer of a complex microbial ecosystem from a healthy donor to a recipient with dysbiosis, facilitating the recolonization of beneficial microbes that can enhance metal metabolism and excretion. Studies suggest that FMT can reverse gut dysbiosis caused by heavy metal exposure, improve gut barrier integrity, and promote the proliferation of metal-resistant bacterial strains that support detoxification. In animal models, FMT has been shown to mitigate cadmium and lead-induced gut inflammation, enhance microbial-mediated metal sequestration, and reduce systemic oxidative stress, indicating its potential as a novel intervention for heavy metal poisoning. However, the clinical application of FMT for heavy metal detoxification requires further investigation to optimize donor selection, microbial composition, and administration protocols to ensure safety and efficacy. In addition to probiotics and FMT, using prebiotics and symbiotics in modulating gut–metal interactions presents a promising approach to enhancing microbial-mediated heavy metal detoxification. Prebiotics, which include non-digestible fibers such as inulin, fructooligosaccharides (FOS), and galactooligosaccharides (GOS), selectively stimulate the growth and activity of beneficial gut bacteria, improving their ability to bind and transform heavy metals. These dietary fibers create a favorable environment for probiotic strains like Lactobacillus and Bifidobacterium, enabling them to outcompete pathogenic bacteria that may exacerbate metal toxicity. Furthermore, prebiotics have been shown to reduce heavy metal bioavailability by altering gut pH, increasing mucus production, and enhancing the excretion of metal-bound microbial biomass through feces (Devi, Dey, Sharma, Kumar, & Singh, 2024a; Jagadesh et al., 2024; Kumar & Choudhury, 2024f; Kumar & Choudhury, 2024g; Kumar, 2024b; Kumar, 2024d). Combining probiotics and prebiotics offers an integrated approach to gut-mediated heavy metal detoxification by supplying beneficial bacteria and their preferred substrates for growth and activity. Clinical studies have demonstrated that synbiotic formulations containing Lactobacillus spp. and inulin can enhance mercury and arsenic excretion, reduce inflammation, and improve overall gut health in individuals with chronic heavy metal exposure. The synergistic effects of prebiotics and probiotics in modulating gut microbial metabolism and metal-binding capabilities underscore their potential as dietary interventions for mitigating heavy metal toxicity. While the role of gut microbiota in heavy metal metabolism and detoxification is increasingly recognized, further research is needed to fully elucidate the mechanisms underlying microbial-metal interactions and develop targeted microbiome-based therapies for heavy metal poisoning. Advances in metagenomic and metabolomic technologies enable deeper insights into gut microbes' functional capacities in metal detoxification, paving the way to identify novel probiotic strains, microbial consortia, and gut-derived metabolites with enhanced metal-binding properties. Developing personalized microbiome interventions, guided by individual gut microbial profiles and metal exposure levels, also holds promise for optimizing detoxification strategies tailored to specific populations at risk. In conclusion, the gut microbiota plays a pivotal role in modulating heavy metal bioavailability, metabolism, and detoxification through microbial-mediated metal binding, biotransformation, and excretion pathways. Heavy metal-resistant probiotic strains, such as *Lactobacillus* and *Bifidobacterium*, offer a natural and effective means of reducing metal toxicity while restoring gut microbial balance. Fecal microbiota transplantation represents a novel therapeutic approach with potential applications in reversing gut dysbiosis and enhancing microbial-mediated detoxification (Devi et al., 2024b; Devi et al., 2024c; Kumar & Choudhury, 2024g; Kumar & Singh, 2024e; Srivastav et al., 2024). Prebiotics and synbiotics further support gut–metal interactions by fostering the growth of beneficial microbes and enhancing their detoxification capabilities. As research advances, microbiome-based strategies hold significant potential for mitigating heavy metal toxicity and improving human health through targeted dietary and probiotic interventions (Table 22.5).

22.7 Phytoremediation and biodetoxification strategies

Medicinal plants with metal-absorbing properties have gained significant attention recently due to their ability to sequester and detoxify heavy metals from contaminated environments and biological systems. Plants such as *Moringa oleifera*, *Spirulina platensis*, and *Chlorella vulgaris* possess remarkable heavy metal-binding capabilities through various mechanisms, including biosorption, bioaccumulation, and enzymatic detoxification. *Moringa oleifera*, commonly known as the drumstick tree, has been extensively studied for its ability to remove toxic metals like lead, cadmium, arsenic, and chromium from water and soil. Moringa's leaves, seeds, and roots contain high concentrations of bioactive compounds such as polyphenols, flavonoids, and proteins with strong metal-chelating properties (Devi et al., 2024a; Kumar & Choudhury, 2024d; Kumar & Singh, 2024e; Kumar et al., 2024d; Raj et al., 2024). The anionic polyelectrolytes in *Moringa* seeds are particularly effective in coagulating metal ions, making them useful for water purification and environmental detoxification. Beyond environmental applications, *Moringa* has also demonstrated protective effects in biological systems by reducing heavy metal bioavailability, enhancing antioxidant defense mechanisms, and mitigating oxidative stress-induced damage caused by metal toxicity. *Spirulina platensis* and *Chlorella vulgaris* are microalgae with high affinities for heavy metal adsorption and bioaccumulation. These cyanobacteria are rich in metal-binding proteins, polysaccharides, and functional groups such as carboxyl, hydroxyl, and amino groups that facilitate the sequestration of toxic metals. *Spirulina* has been shown to effectively remove lead, mercury, and arsenic from aqueous solutions while providing hepatoprotective and neuroprotective benefits in cases of heavy metal poisoning. Similarly, *Chlorella* has demonstrated strong heavy metal chelation properties, enhancing metal excretion through feces while reducing systemic toxicity (Devi et al., 2024a; Jagadesh et al., 2024; Kumar & Choudhury, 2024f; Kumar & Choudhury, 2024g; Kumar, 2024b; Kumar, 2024d). The detoxification potential of these microalgae extends beyond passive metal binding; their bioactive compounds, including chlorophyll, carotenoids, and antioxidants, help counteract the oxidative damage induced by heavy metal exposure. Integrating medicinal plants and algae into dietary interventions and detoxification strategies provides a natural, cost-effective approach to mitigating heavy metal toxicity. In addition to plant-based detoxification mechanisms, fungi play a crucial role in mycoremediation, which uses fungi for heavy metal detoxification and environmental remediation. Certain fungal species, including *Pleurotus ostreatus* (oyster mushrooms), *Aspergillus niger*, and *Trametes versicolor*, exhibit strong heavy metal resistance and biosorption capabilities. These fungi produce extracellular enzymes, such as laccases, peroxidases, and hydrolases, that aid in metal transformation and sequestration. Mycorrhizal fungi, which form symbiotic associations with plant roots, enhance metal uptake and stabilization in soil by producing metal-chelating compounds like glomalin and organic acids. *Pleurotus ostreatus* has been extensively studied for its ability to bioremediate lead, cadmium, and mercury-contaminated environments through bioaccumulation and enzymatic transformation. The fungal cell wall, rich in chitin and

TABLE 22.5 Traditional and Alternative Medicine Approaches for Heavy Metal Detoxification:

Therapy/herbal remedy	Active compounds	Mechanism of detoxification	Heavy metals targeted	Traditional uses	Scientific evidence
Panchakarma (Ayurveda)	Triphala, guggul, turmeric	It enhances liver detox, and promotes excretion through sweat, urine, and feces	Lead, mercury, arsenic, cadmium	Ayurvedic cleansing therapy for toxin removal	Studies indicate improved metal excretion and antioxidant benefits
Triphala	Gallic acid, ellagic acid, chebulagic acid	Antioxidant activity enhances liver detox pathways, binds metals	Lead, arsenic, cadmium	Digestive aid, detoxification, rejuvenation	Research supports metal chelation and liver protection
Chlorella	Chlorophyll, beta-carotene, lutein	Binds heavy metals, stimulates glutathione production	Mercury, cadmium, lead	Detoxification, immune system support	Clinical studies show reduced metal levels in urine and blood
Spirulina	Phycocyanin, chlorophyll, B-vitamins	Chelates metals reduces oxidative stress, enhances excretion	Lead, arsenic, cadmium	Nutritional supplement, liver and kidney detoxification	Human and animal studies demonstrate metal chelation effects
Moringa	Quercetin, kaempferol, isothiocyanates	Binds heavy metals, promotes liver detox enzymes	Lead, cadmium, arsenic	Water purification, immune boosting, detox support	Evidence suggests protective effects against metal toxicity
Cilantro (coriander)	Linalool, quercetin, phenolic acids	Mobilizes and excretes heavy metals through urine	Mercury, lead, arsenic	Used in traditional medicine for detox and digestion	Studies indicate enhanced mercury excretion
Garlic (*Allium sativum*)	Allicin, sulfur compounds	Sulfur-containing compounds bind metals, reduce oxidative stress	Lead, cadmium, mercury	Blood purification, immunity boosting	Animal studies confirm reduced metal accumulation
Guggul (*Commiphora mukul*)	Guggulsterones, flavonoids	Induces liver detox enzymes, supports bile excretion of metals	Lead, arsenic, cadmium	Used in Ayurveda for purification and detox	Some studies show improved metal clearance rates
Turmeric (curcumin)	Curcumin, demethoxycurcumin	Reduces oxidative damage, binds metals, enhances liver detox	Lead, cadmium, mercury	Antiinflammatory, liver detox, immune modulation	Research supports neuroprotective and chelating effects
Milk thistle (*Silybum marianum*)	Silymarin, silibinin	Protects liver cells, enhances glutathione production	Lead, cadmium, arsenic	Used in liver detox and protection	Clinical trials support hepatoprotective effects
Ayurvedic Rasayanas	Various polyphenols, adaptogens	Enhances the body's natural detoxification pathways	Lead, mercury, arsenic	Rejuvenation therapy in Ayurveda	Some studies show improvements in heavy metal excretion
Homeopathy (*Mercurius solubilis, Arsenicum album*)	Highly diluted metal preparations	Stimulates the body's natural elimination pathways	Mercury, arsenic, lead	Traditional alternative medicine for detox	Limited scientific validation, primarily anecdotal evidence
Sauna therapy	Heat-induced sweating	Promotes heavy metal excretion through sweat	Lead, mercury, arsenic	Detoxification, cardiovascular health	Studies suggest increased metal excretion via sweating
Hyperbaric oxygen therapy	High-pressure oxygen	Reduces oxidative stress, enhances cellular detox processes	Lead, mercury	Used in alternative medicine for healing and detox	Some studies show potential benefits, but more research is needed

Source: *Based on* Devi et al., 2024d; Jagadesh et al., 2024; Kumar & Choudhury, 2024f; Kumar & Choudhury, 2024g; Kumar et al., 2024c; Kumar, 2024c.

glucans, provides a natural scaffold for metal binding, reducing bioavailability and toxicity. Furthermore, fungi such as *Trametes versicolor* have demonstrated the ability to degrade organic-metal complexes and convert toxic metal species into less bioavailable forms, enhancing the overall efficiency of metal detoxification. Using fungi with plants and microorganisms in bioremediation strategies holds great promise for sustainable heavy metal removal and ecosystem restoration. Enzyme-based detoxification mechanisms play a fundamental role in heavy metal sequestration, particularly through the actions of phytochelatins and metallothioneins—metal-binding peptides and proteins that facilitate intracellular metal homeostasis (Devi et al., 2024b; Kumar & Choudhury, 2024d; Kumar & Choudhury, 2024e; Kumar & Singh, 2024e; Kumar et al., 2024c; Kumar, 2024b; Raj et al., 2024). Phytochelatins, small cysteine-rich peptides synthesized by plants, algae, and some fungi, act as natural chelators that bind heavy metals such as cadmium, arsenic, and lead, forming stable complexes that are sequestered into vacuoles for detoxification. The enzymatic biosynthesis of phytochelatins is catalyzed by phytochelatin synthase, which responds to metal exposure by increasing phytochelatin production, thereby reducing metal toxicity at the cellular level. Metallothioneins, a class of low-molecular-weight cysteine-rich proteins found in plants, fungi, and animals, play a similar role in heavy metal detoxification by binding metal ions and regulating their intracellular concentrations. Metallothioneins protect cells from heavy metal–induced oxidative stress and participate in metal redistribution and homeostasis, facilitating metal tolerance and resistance in plants and microorganisms. The genetic engineering of plants with enhanced phytochelatin and metallothionein expression has been explored as a strategy to develop hyperaccumulator species capable of removing heavy metals from contaminated environments. By leveraging the natural detoxification mechanisms of these metal-binding proteins, researchers aim to enhance bioremediation efficiency and reduce the ecological impact of heavy metal pollution. Applying biochar and plant-based adsorbents for heavy metal sequestration is another promising environmental detoxification approach. Biochar, a carbon-rich material derived from the pyrolysis of organic biomass, has gained widespread attention for its ability to adsorb heavy metals from soil and water due to its high surface area, porous structure, and abundance of functional groups such as carboxyl and hydroxyl groups. The physicochemical properties of biochar can be tailored to optimize metal adsorption capacity by modifying feedstock composition, pyrolysis temperature, and posttreatment processes such as activation and functionalization (Devi et al., 2024c; Kumar & Choudhury, 2024h; Kumar & Singh, 2024e; Kumar, 2024c; Raj et al., 2024; Sharma & Kumar, 2024; Srivastav et al., 2024). Studies have demonstrated that biochar from agricultural residues, such as rice husks, coconut shells, and bamboo, effectively removes lead, arsenic, and cadmium from contaminated environments by immobilizing metals through surface complexation, ion exchange, and electrostatic interactions. Incorporating biochar into the soil reduces heavy metal bioavailability and improves soil fertility, microbial activity, and plant growth, making it a sustainable strategy for remediating polluted agricultural lands. In addition to biochar, plant-based adsorbents such as activated carbon derived from plant biomass, cellulose-based nanomaterials, and lignin-rich agricultural by-products have been explored for their heavy metal-binding capabilities. Using natural adsorbents derived from plant materials offers an eco-friendly, low-cost alternative to synthetic chelating agents, providing a sustainable means of mitigating heavy metal contamination. The synergistic application of biochar, medicinal plants, fungi, and enzymatic detoxification mechanisms presents a holistic approach to heavy metal sequestration and environmental restoration. As research in this field continues to evolve, the development of integrated bioremediation strategies that combine multiple detoxification pathways holds great promise for addressing the global challenge of heavy metal pollution. Future advancements in biotechnology, nanomaterials, and genetic engineering will further enhance the efficiency and scalability of plant-based and microbial detoxification systems, paving the way for innovative solutions in environmental and human health protection (Devi et al., 2024a; Kumar & Choudhury, 2024e; Kumar & Choudhury, 2024g; Kumar & Dey, 2024; Kumar, 2024b; Raj et al., 2024).

22.8 Emerging and alternative therapies

Ayurvedic and traditional medicine approaches have long been used for detoxification, with practices like Panchakarma, Triphala, and Ayurvedic detox herbs playing a crucial role in removing toxins, including heavy metals, from the body. Panchakarma is a five-step detoxification therapy in Ayurveda that involves procedures such as Vamana (therapeutic emesis), Virechana (purgation therapy), Basti (medicated enema), Nasya (nasal administration of herbal oils), and Raktamokshana (bloodletting). These techniques help cleanse different bodily systems, enhance liver function, and improve metabolic balance, thereby eliminating heavy metals stored in tissues. Panchakarma is often recommended for individuals exposed to environmental toxins, including industrial pollutants, pesticides, and heavy metals like lead, mercury, and arsenic(Devi et al., 2024b; Devi et al., 2024d; Kumar & Dey, 2024; Kumar et al., 2024c; Kumar et al., 2024d; Srivastav et al., 2024). Among the widely used Ayurvedic formulations, Triphala is a powerful detoxifying agent composed of three fruits: Amalaki (*Emblica officinalis*), Bibhitaki (*Terminalia bellirica*), and Haritaki

(*Terminalia chebula*). Triphala possesses strong antioxidant, antiinflammatory, and metal-binding properties, effectively neutralizing oxidative stress caused by heavy metals while promoting their excretion through the digestive system. The combination of bioactive compounds in Triphala helps regenerate liver cells, supports kidney function, and enhances bile secretion, all essential for detoxification. Several Ayurvedic herbs are also traditionally used to chelate and remove heavy metals, including Guduchi (*Tinospora cordifolia*), Manjistha (*Rubia cordifolia*), Neem (*Azadirachta indica*), and Turmeric (*Curcuma longa*). These herbs contain bioactive compounds such as curcumin, tannins, alkaloids, and flavonoids that enhance the body's natural detoxification processes, reduce heavy metal accumulation in tissues, and protect against oxidative damage. Curcumin, the active compound in turmeric, has been shown to chelate metals like cadmium and lead while reducing inflammation and oxidative stress in heavy metal–exposed individuals. Ayurvedic detox strategies focus on removing toxins, restoring balance to the body's doshas (Vata, Pitta, and Kapha), and strengthening overall vitality (Kumar & Dey, 2024; Kumar, 2024b; Kumar, 2024c; Raj et al., 2024; Sharma & Kumar, 2024; Srivastav et al., 2024). Homeopathic and alternative remedies have also gained attention as potential detoxification therapies, particularly in addressing heavy metal toxicity. Homeopathy operates on the principle of "like cures like" and uses highly diluted substances to stimulate the body's natural detoxification processes. Several homeopathic remedies, such as *Mercurius solubilis*, *Arsenicum album*, and *Plumbum metallicum*, are commonly prescribed for detoxifying mercury, arsenic, and lead, respectively. These remedies enhance the body's excretory functions, support liver and kidney health, and reduce the neurological and gastrointestinal symptoms of heavy metal poisoning. Another alternative detoxification approach involves herbal adaptogens such as Ashwagandha (*Withania somnifera*), Rhodiola (*Rhodiola rosea*), and Ginseng (*Panax ginseng*), which help modulate stress responses and improve resilience against environmental toxins. Traditional Chinese Medicine (TCM) also employs detoxification methods such as acupuncture, cupping therapy, and herbal formulations to support metal elimination. Herbs like *Schisandra chinensis*, *Gynostemma pentaphyllum*, and *Rehmannia glutinosa* are known for their hepatoprotective and chelating effects. These herbs enhance liver enzyme activity, boost antioxidant defenses, and aid in the biotransformation of toxic metals into excretable forms. In addition to herbal and homeopathic approaches, therapies like hyperbaric oxygen therapy (HBOT) and sauna therapy have gained prominence as modern detoxification modalities (Devi et al., 2024d; Jagadesh et al., 2024; Kumar & Choudhury, 2024g; Kumar & Singh, 2024e; Kumar et al., 2024d; Kumar, 2024d). HBOT involves breathing pure oxygen in a pressurized chamber, which increases oxygen delivery to tissues and promotes cellular detoxification. HBOT has been shown to reduce oxidative stress, enhance mitochondrial function, and support the body's natural ability to eliminate heavy metals. Research suggests that HBOT may accelerate the excretion of toxic metals by improving circulation, enhancing immune responses, and stimulating enzymatic pathways involved in detoxification. It is particularly beneficial for individuals suffering from mercury and lead toxicity, as these metals can impair oxygen transport and mitochondrial energy production. By increasing tissue oxygenation, HBOT helps counteract the cellular damage caused by heavy metals while supporting neuroprotection and cardiovascular health. Sauna therapy, particularly infrared sauna therapy, is another effective detoxification strategy that enhances heavy metal excretion through sweat. Studies have shown that metals such as arsenic, cadmium, and lead can be efficiently eliminated through perspiration. Infrared saunas penetrate deep into the tissues, promoting circulation, lymphatic drainage, and the mobilization of stored heavy metals from fat cells. Sweating has been recognized as one of the body's natural detoxification pathways, and sauna therapy can significantly reduce the body's toxic burden when combined with proper hydration and mineral replenishment. Many detox programs incorporate sauna therapy alongside nutritional support, such as replenishing essential minerals like magnesium, zinc, and selenium to prevent nutrient depletion caused by excessive sweating(Devi et al., 2024b; Kumar & Singh, 2024e; Kumar et al., 2024c; Kumar, 2024b; Srivastav et al., 2024). Integrating genetic and epigenetic therapies in detoxification enhancement represents a cutting-edge approach to personalized medicine. Genetic factors play a critical role in an individual's ability to detoxify heavy metals, as polymorphisms in detoxification-related genes can influence enzyme activity and metal metabolism. Key genes involved in heavy metal detoxification include *GSTM1*, *GSTT1* (glutathione-S-transferases), *MT1* (metallothioneins), *SOD2* (superoxide dismutase), and *NQO1* (NAD(P)H dehydrogenase). Genetic testing can help identify individuals with impaired detoxification capacity, allowing for targeted interventions to enhance metal clearance. Epigenetic modifications, such as DNA methylation and histone acetylation, can also influence detoxification pathways by regulating gene expression. Environmental toxins, including heavy metals, have been shown to alter epigenetic patterns, leading to dysregulated detoxification processes and increased susceptibility to toxicity-related diseases. Recent research has focused on developing epigenetic therapies that modulate detoxification-related genes to enhance the body's ability to eliminate metals (Kumar et al., 2024h). Nutrients such as methylfolate, betaine, and vitamin B12 support DNA methylation and histone modifications, improving detoxification efficiency. Emerging therapies, including small-molecule epigenetic modulators, aim to reverse heavy metal-induced epigenetic alterations and restore normal cellular function. The combination of genetic screening, personalized nutrition, and epigenetic interventions has the potential to revolutionize heavy metal detoxification strategies, making them more effective and individualized. In conclusion, integrating Ayurvedic and traditional medicine, homeopathic and alternative remedies, HBOT, sauna therapy, and genetic and epigenetic interventions provides a comprehensive and multifaceted approach to heavy metal detoxification. Each method offers unique benefits, from enhancing liver and kidney function to modulating gene expression and improving cellular resilience. As research in detoxification science continues to advance, the synergy between traditional wisdom and modern medical innovations holds great promise for developing more effective and personalized detoxification strategies for individuals exposed to heavy metals and environmental toxins (Kumar and Choudhury, 2024a).

22.9 Challenges and future perspectives

Existing detoxification methods for heavy metals, despite their effectiveness in certain cases, come with significant limitations that hinder their widespread application and efficacy. Conventional chelation therapy, which involves the administration of chelating agents such as EDTA, DMSA, or DMPS, is one of the primary clinical interventions for heavy metal poisoning. However, these therapies are not without risks. Chelating agents often exhibit nonselectivity, meaning they can bind to and deplete essential minerals such as calcium, magnesium, zinc, and iron, leading to potential nutritional deficiencies. Additionally, these agents may have toxic side effects, including kidney and liver damage, gastrointestinal distress, and allergic reactions. Chelation therapy is also primarily recommended for acute poisoning rather than chronic low-level exposure, which is a more widespread issue in environmental toxicity. Another major limitation of existing detoxification approaches is their inability to effectively target and remove metals stored in deep tissue compartments such as the brain, bones, and adipose tissues. Many heavy metals, including lead and mercury, accumulate in these tissues over time and are not easily mobilized by conventional chelation methods. Moreover, heavy metal detoxification is highly individualistic, as genetic variations among individuals influence their ability to metabolize and excrete toxicants. Polymorphisms in genes related to glutathione metabolism, metallothionein synthesis, and antioxidant defense mechanisms can significantly impact a person's susceptibility to heavy metal toxicity and their response to detoxification therapies. This underscores the need for personalized and precision detoxification strategies that consider genetic predispositions, individual toxic metal burdens, and metabolic variations to optimize treatment outcomes. Personalized detoxification involves genetic and metabolic profiling to tailor detox strategies to individual needs. Advances in pharmacogenomics have enabled the identification of genetic markers associated with detoxification pathways, allowing clinicians to develop customized treatment protocols. For example, individuals with impaired glutathione metabolism may benefit from targeted supplementation with NAC or glutathione precursors. At the same time, those with compromised metallothionein expression may require zinc and selenium support to enhance metal-binding capacity. Precision detoxification also integrates real-time biomonitoring to assess an individual's heavy metal exposure and detoxification progress. Advanced diagnostic tools, including high-resolution mass spectrometry and biosensors, enable the accurate detection of metal concentrations in blood, urine, and tissues, allowing for dynamic adjustments to detoxification protocols. Integrative approaches that combine multiple detoxification strategies offer a more holistic and effective means of addressing heavy metal toxicity. Since different detoxification pathways operate in synergy, a multimodal approach can enhance metal excretion and reduce systemic toxicity. For instance, combining chelation therapy with dietary interventions, antioxidant supplementation, and microbiome modulation may improve detoxification efficiency and minimize adverse effects. Nutritional strategies are crucial in integrative detoxification by supporting the body's endogenous metal-clearing mechanisms. Certain dietary components, such as polyphenols, fiber, and sulfur-containing compounds, have been shown to promote heavy metal excretion. Polyphenols like curcumin, quercetin, and resveratrol exhibit metal-chelating properties and help reduce oxidative stress caused by metal toxicity. Sulfur-containing foods such as garlic, onions, and cruciferous vegetables enhance glutathione synthesis, which is essential for detoxification. Fiber-rich foods facilitate the binding and elimination of metals through fecal excretion by preventing enterohepatic recirculation. The gut microbiome also plays a pivotal role in heavy metal detoxification, as certain probiotic strains can bind and neutralize toxic metals in the intestines, reducing their systemic absorption. Probiotic strains such as *Lactobacillus* and *Bifidobacterium* have been found to exhibit metal resistance and sequester heavy metals, thereby limiting their bioavailability. Prebiotic and synbiotic interventions that modulate gut microbial composition may further enhance detoxification capacity. Moreover, emerging biotechnological approaches, such as enzyme-based detoxification, offer promising avenues for targeted heavy metal removal. Enzymes like phytochelatins and metallothioneins naturally bind and sequester heavy metals, rendering them biologically inactive. Developing bioengineered enzymes and nanochelators for targeted metal detoxification could provide concrete and efficient alternatives to conventional chelation therapy. However, despite the potential of integrative and precision detoxification strategies, significant ethical, regulatory, and clinical challenges remain in the field of heavy metal detox research. One of the primary ethical concerns is the safety and long-term effects of novel detoxification interventions, particularly those involving nanotechnology, genetic modifications, and synthetic biological systems. Using engineered nanomaterials for heavy metal adsorption and removal raises questions about biocompatibility, unintended toxicity, and environmental impact. While effective in metal binding, nanoparticles may accumulate in biological systems and exert unknown effects on cellular processes. Additionally, the development of genetic and epigenetic therapies for detoxification enhancement necessitates rigorous ethical scrutiny, as interventions that alter gene expression could have unpredictable long-term consequences. Regulatory challenges also hinder the widespread adoption of innovative detoxification therapies. Many natural and alternative detoxification approaches, including herbal formulations, probiotics, and dietary supplements, fall under the category of nutraceuticals and are not subject to the same stringent regulatory approval processes as pharmaceutical drugs. This lack of standardized regulation leads to variability in product quality, efficacy, and safety, making it difficult for consumers and healthcare providers to assess the reliability of available detoxification products. Moreover, clinical research on heavy metal detoxification faces obstacles related to study design, funding, and ethical approval. Conducting large-scale clinical trials on detoxification interventions requires long-term monitoring, extensive biomarker analysis, and controlled exposure assessments, which can be logistically challenging and costly. Ethical considerations also arise in human studies involving heavy metal exposure, as deliberately exposing participants to toxic metals for research purposes is not feasible. As a result, much of the existing evidence on detoxification therapies comes from observational studies, animal models, or in vitro research, which may not fully capture the complexities of human detoxification processes. In light of these challenges, a multidisciplinary approach is needed to advance the field of heavy metal detoxification. Collaboration between toxicologists, clinicians, geneticists, microbiologists, and regulatory bodies can facilitate the development of standardized protocols, evidence-based guidelines, and safe therapeutic interventions. Public health initiatives that focus on reducing environmental heavy metal exposure, promoting preventive dietary strategies, and raising awareness about detoxification methods are also essential in mitigating the impact of heavy metal toxicity on global health. Ultimately, the future of heavy metal detoxification lies in the integration of personalized medicine, cutting-edge biotechnology, and holistic health approaches. By combining targeted therapies with natural detoxification mechanisms, leveraging advancements in diagnostic tools, and addressing ethical and regulatory challenges, it is possible to develop safer, more effective, and individualized detoxification solutions for individuals exposed to heavy metals.

22.10 Conclusion

Detoxification strategies for heavy metal exposure encompass various approaches, including chelation therapy, dietary modifications, microbiome interventions, nanotechnology-based solutions, and traditional medicine. The fundamental goal of these strategies is to reduce the bioavailability, systemic toxicity, and long-term health impacts of heavy metals by enhancing their excretion, neutralization, or sequestration in biologically inactive forms. Chelation therapy remains one of the most well-established clinical methods, with agents such as EDTA, DMSA, DMPS, and penicillamine used to bind and facilitate the removal of heavy metals through renal excretion. While effective in cases of acute metal poisoning, chelation therapy has several limitations, including potential depletion of essential minerals, nephrotoxicity, hepatotoxicity, and limited efficacy in addressing chronic, low-dose exposure scenarios. Consequently, complementary detoxification strategies have gained increasing attention to improve overall heavy metal clearance while mitigating the risks associated with conventional chelators. Dietary interventions play a critical role in detoxification by modulating metal absorption, enhancing endogenous antioxidant defenses, and promoting excretion through the gastrointestinal tract. Certain foods and nutrients, including glutathione, NAC, vitamin C, vitamin E, selenium, and zinc, have been demonstrated to possess metal-chelating properties and support enzymatic detoxification pathways. Polyphenolic compounds such as curcumin, quercetin, and resveratrol further exhibit antioxidative and antiinflammatory effects, reducing heavy metal–induced cellular damage. Functional foods, herbal remedies, and nutraceuticals—such as Moringa, Spirulina, and Chlorella—are increasingly being explored for their metal-absorbing properties, as they contain bioactive compounds facilitating heavy metal sequestration and detoxification. Furthermore, gut microbiota interventions represent a promising frontier in heavy metal detoxification. Specific probiotic strains, particularly those from the *Lactobacillus* and *Bifidobacterium* genera, have shown potential in binding and reducing the bioavailability of toxic metals. These probiotic strains produce exopolysaccharides that chelate metals, thereby preventing their absorption into the systemic circulation. Prebiotics and synbiotics, which enhance beneficial gut microbial populations, may further contribute to detoxification by modulating gut–metal interactions. Additionally, fecal microbiota transplantation (FMT) has been proposed as an experimental therapy for individuals with heavy metal–induced dysbiosis, although more clinical research is needed to validate its efficacy. Technological advancements in nanotechnology and enzyme-based detoxification have opened new avenues for heavy metal removal. Engineered nanomaterials, including metal-organic frameworks (MOFs), carbon-based adsorbents, and functionalised nanoparticles, have demonstrated high efficiency in adsorbing and capturing toxic metals. These materials can be applied in biomedical settings for detoxification or environmental remediation to prevent exposure at the source. Enzyme-based detoxification, leveraging metallothioneins and phytochelatins, provides another promising approach by selectively binding metals within biological systems, thus reducing their toxicity. However, concerns regarding the biocompatibility, long-term safety, and potential toxicity of nanotechnology-based interventions need to be carefully addressed through rigorous preclinical and clinical studies. Looking ahead, future research in heavy metal detoxification should focus on translational applications that integrate multidisciplinary approaches. Personalized detoxification strategies, informed by genetic, metabolic, and exposome profiling, can optimize treatment regimens based on an individual's detoxification capacity, dietary habits, and microbial composition. The development of precision medicine approaches in detoxification could enable targeted interventions that minimize side effects while maximizing heavy metal clearance. Additionally, applying advanced biosensors for real-time monitoring of heavy metal burden in biological systems could facilitate early detection and personalized treatment adjustments. Research should also explore novel chelating agents with improved safety profiles and higher specificity for toxic metals. Emerging compounds with enhanced bioavailability, reduced mineral depletion effects, and superior metal-binding affinity may provide viable alternatives to conventional chelators. Biotechnological innovations could offer microbiome-based solutions to heavy metal detoxification, including synthetic biology approaches to engineer probiotic strains with enhanced metal resistance and adsorption capabilities. Moreover, investigating epigenetic and gene-editing techniques to enhance endogenous detoxification pathways could open new therapeutic possibilities. Beyond scientific research, policy and public health interventions are essential to mitigate heavy metal exposure at the population level. Regulatory measures should focus on stricter monitoring and control of industrial emissions, agricultural contaminants, and water pollution to reduce environmental sources of heavy metal contamination. Establishing guidelines for permissible heavy metal levels in food, cosmetics, and consumer products can further safeguard public health. Governments and health agencies should also prioritize implementing community-based detoxification programs, particularly in high-risk regions with elevated heavy metal pollution. Public awareness campaigns should be widely promoted to educate individuals about dietary strategies, functional foods, and preventive measures for heavy metal toxicity. Encouraging lifestyle modifications, such as increasing fiber intake, consuming metal-binding phytochemicals, and incorporating probiotic-rich foods, can be accessible and cost-effective methods to enhance detoxification naturally. Investment in sustainable and eco-friendly remediation technologies is another crucial aspect of the heavy metal detoxification policy. Applying phytoremediation, mycoremediation, and biochar-based adsorbents in contaminated environments can help reduce heavy metal bioavailability in soil and water. Government incentives for industries adopting green technologies and pollution control measures can further reduce heavy metal exposure risks. Collaborative efforts between researchers, policymakers, healthcare practitioners, and environmental agencies are necessary to implement effective heavy metal detoxification strategies on a broader scale. Integrating scientific advancements with regulatory frameworks and public health initiatives can lead to more comprehensive and sustainable solutions for mitigating heavy metal toxicity. Ultimately, a multipronged approach that combines medical, nutritional, microbiome, technological, and policy-driven interventions will be required to address the complex challenge of heavy metal detoxification in clinical and environmental settings.

References

Choudhury, D., Kumar, P., Zhimo, V. Y., & Sahoo, J. (2024). Crop rotation patterns and soil health management. In *Bioremediation of emerging contaminants from soils: Soil health conservation for improved ecology and food security*. Punjab, Jalandhar, India: Elsevier, Department of Plant Pathology, School of Agriculture, Lovely Professional University, 565–589. Retrieved from https://doi.org/10.1016/B978-0-443-13993-2.00025-6.

Devi, P., Dey, S. R., Sharma, K., Kumar, P., & Singh, J. (2024a). Waste management as an opportunity for the inclusion of vulnerable groups. In *Waste management in the circular economy*. Punjab, Phagwara, India: Springer International Publishing, Department of Agronomy, School of Agriculture, Lovely Professional University, 207–230. Retrieved from https://doi.org/10.1007/978-3-031-42426-7_11.

Devi, P., Dogra, P., Kumar, P., & Singh, J. (2024b). Global status of biomass energy programmes-challenges and roadmap. In *Biomass energy for sustainable development*. Phagwara, India: CRC Press, Lovely Professional University, 394–414. Retrieved from https://doi.org/10.1201/9781003406501-20.

Devi, P., Kapoor, P., & Kumar, P. (2024c). Biomarkers for toxicity detection. In *Biomarkers in environmental and human health biomonitoring: an integrated perspective*. Punjab, Phagwara, India: Elsevier, Department of Agronomy, School of Agriculture, Lovely Professional University, 313–329. Retrieved from https://doi.org/10.1016/B978-0-443-13860-7.00004-5.

Devi, P., Kaur, A., & Kumar, P. (2024d). Phytoremediation of metal contaminated soil using energy crops: Soil health maintenance along with biofuel production. In *Bioremediation of emerging contaminants from soils: soil health conservation for improved ecology and food security*. Punjab, Phagwara, India: Elsevier, Department of Agronomy, School of Agriculture, Lovely Professional University, 307–333. Retrieved from https://doi.org/10.1016/B978-0-443-13993-2.00015-3.

Devi, P., & Kumar, P. (2024a). Alleviation of different climatic conditions by foliar application of salicylic acid and sodium nitroprusside and their interactive effects on pigments and sugar content of maize under different sowing dates. Nature Environment and Pollution Technology, *23*, 1161–1166. Retrieved from https://doi.org/10.46488/NEPT.2024.v23i02.050.

Devi, P., & Kumar, P. (2024b). Challenges in the storage of herbal medicine products and strategies for sustainable management. In *Reference series in phytochemistry*. Phagwara, India: Springer Science and Business Media B.V., Department of Agronomy, School of Agriculture, Lovely Professional University, 1739–1767. Retrieved from https://doi.org/10.1007/978-3-031-43199-9_37.

Devi, P., & Kumar, P. (2024c). Herbal medicine and pregnancy. In *Reference series in phytochemistry*. Punjab, Phagwara, India: Springer Science and Business Media B.V., Department of Agronomy, School of Agriculture, Lovely Professional University, 693–722. Retrieved from https://doi.org/10.1007/978-3-031-43199-9_25.

Devi, P., Kumar, P., Dey, S. R., Banik, D., Kumar, G., & Mehta, C. M. (2024e). Integrated omics approaches for nutrient stress management in plants. In *Current omics advancement in plant abiotic stress biology: A volume in developments in applied microbiology and biotechnology*. Punjab, Phagwara, India: Elsevier, School of Agriculture, Lovely Professional University, 93–117. Retrieved from https://doi.org/10.1016/B978-0-443-21625-1.00008-7.

Devi, P., Kumar, P., & Singh, J. (2024f). Sewage treatment and energy recovery by bioelectrochemical oxidation system. In *Bioelectrochemical oxidation processes for wastewater treatment*. Punjab, Phagwara, India: CRC Press, Lovely Professional University, 100–125. Retrieved from https://doi.org/10.1201/9781003368472-6.

Devi, P., Kumar, P., & Singh, J. (2024g). Biotechnological approaches as clean technology for the development of a sustainable environment. In *Development in waste water treatment research and processes: role of environmental microbiology in industrial wastewater research*. Punjab, Phagwara, India: Elsevier, Department of Agronomy, School of Agriculture, Lovely Professional University, 381–403. Retrieved from https://doi.org/10.1016/B978-0-443-13609-2.00016-1.

Devi, P., Kumar, P., & Singh, J. (2024h). Application of potential microbes in bioremediation. In *Development in waste water treatment research and processes: Role of environmental microbiology in industrial wastewater research*. Punjab, Phagwara, India: Elsevier, Department of Agronomy, School of Agriculture, Lovely Professional University, 201–225. Retrieved from https://doi.org/10.1016/B978-0-443-13609-2.00021-5.

Dey, S. R., Kumar, P., & Panwar, J. S. (2024a). Microbial desalination cell. In *Bioelectrochemical oxidation processes for wastewater treatment*. Punjab, Phagwara, India: CRC Press, Lovely Professional University, 189–217. Retrieved from https://doi.org/10.1201/9781003368472-10.

Dey, S. R., Kumar, P., & Singh, J. (2024b). Total soluble protein mediated morphological traits in mustard treated with thiourea and salicylic acid. *Nature Environment and Pollution Technology, 23*, 679–694. Retrieved from https://doi.org/10.46488/NEPT.2024.v23i02.007.

Dey, S. R., Sharma, M., & Kumar, P. (2024c). Phytoremediation of phenolic compounds from soil. In *Bioremediation of emerging contaminants from soils: Soil health conservation for improved ecology and food security*. Punjab, Phagwara, India: Elsevier, Department of Agronomy, School of Agriculture, Lovely Professional University, 335–358. Retrieved from https://doi.org/10.1016/B978-0-443-13993-2.00016-5.

Dey, S. R., Sharma, M., Kumar, P., & Dwivedi, P. (2024d). Genetic improvement of groundnut. In *Genetic engineering of crop plants for food and health security (Volume 2)*. Phagwara, India: Springer Nature, Department of Agronomy, School of Agriculture, Lovely Professional University, 281–306. Retrieved from https://doi.org/10.1007/978-981-97-3119-0_12.

Dey, S. R., Sharma, M., Kumar, P., & Dwivedi, P. (2024e). Genetic improvement of mustard. In *Genetic engineering of crop plants for food and health security (Volume 2)*. Phagwara, India: Springer Nature, Department of Agronomy, School of Agriculture, Lovely Professional University, 227–259. Retrieved from https://doi.org/10.1007/978-981-97-3119-0_10.

Jagadesh, M., Dash, M., Kumari, A., Singh, S. K., Verma, K. K., Kumar, P., ... Sharma, S. K. (2024). Revealing the hidden world of soil microbes: Metagenomic insights into plant, bacteria, and fungi interactions for sustainable agriculture and ecosystem restoration. *Microbiological Research, 285*. Retrieved from https://doi.org/10.1016/j.micres.2024.127764.

Kumar, P. (2024a). Epiphytes as a sustainable biomonitoring tool for environmental pollutants. In *Biomonitoring of pollutants in the global south*. Punjab, Phagwara, India: Springer Nature, Department of Agronomy, School of Agriculture, Lovely Professional University, 359–390. Retrieved from https://doi.org/10.1007/978-981-97-1658-6_10.

Kumar, P. (2024b). Biochar production from marine algae and its application in the treatment of industrial effluent. In *Marine biomass: Biorefinery, bioproducts and environmental bioremediation*. Phagwara, Punjab, India: De Gruyter, Department of Agronomy, School of Agriculture, Lovely Professional University, 57–93. Retrieved from https://doi.org/10.1515/9783111353951-004.

Kumar, P. (2024c). Application of marine biomass for carbon neutrality, sustainable environment, and economy. In *Marine biomass: Biorefinery, bioproducts and environmental bioremediation*. Phagwara, Punjab, India: De Gruyter, Department of Agronomy, School of Agriculture, Lovely Professional University, 95–128. Retrieved from https://doi.org/10.1515/9783111353951-005.

Kumar, P. (2024d). Soil conservation for global sustainability. In *World sustainability series*. Punjab, Phagwara, India: Springer Science and Business Media Deutschland GmbH, Department of Agronomy, School of Agriculture, Lovely Professional University, 103–128. Retrieved from https://doi.org/10.1007/978-3-031-57456-6_6.

Kumar, P., & Choudhury, D. (2024a). Pollutants from agricultural activities in the GlobaL South. In *Biomonitoring of pollutants in the Global South*. Punjab, Phagwara, India: Springer Nature, Department of Agronomy, School of Agriculture, Lovely Professional University, 89–126. Retrieved from https://doi.org/10.1007/978-981-97-1658-6_3.

Kumar, P., & Choudhury, D. (2024b). Seed plants as a sustainable biomonitoring tool for environmental pollutants. In *Biomonitoring of pollutants in the Global South*. Punjab, Phagwara, India: Springer Nature, Department of Agronomy, Lovely Professional University, 321–357. Retrieved from https://doi.org/10.1007/978-981-97-1658-6_9.

Kumar, P., & Choudhury, D. (2024c). Optimized phytoremediation process for the sustainable management radionuclides. In *Bioremediation of emerging contaminants from soils: soil health conservation for improved ecology and food security*. Punjab, Phagwara, India: Elsevier, Department of Agronomy, School of Agriculture, Lovely Professional University, 443–464. Retrieved from https://doi.org/10.1016/B978-0-443-13993-2.00020-7.

Kumar, P., & Choudhury, D. (2024d). Role of indigenous knowledge in agricultural soil reclamation without disturbing other ecosystems. In *Bioremediation of emerging contaminants from soils: Soil health conservation for improved ecology and food security*. Punjab, Phagwara, India: Elsevier, Department of Agronomy, School of Agriculture, Lovely Professional University, 465–488. Retrieved from https://doi.org/10.1016/B978-0-443-13993-2.00021-9.

Kumar, P., & Choudhury, D. (2024e). Microbial native soil bacteria against cadmium toxicity. In *Springer water*. Punjab, Phagwara, India: Springer Nature, Department of Agronomy, School of Agriculture, Lovely Professional University, 193–226. Retrieved from https://doi.org/10.1007/978-3-031-54005-9_9.

Kumar, P., & Choudhury, D. (2024f). Toxicity of rhizospheric cadmium contaminated soil and its phytoremediation. In *Springer water*. Punjab, Phagwara, India: Springer Nature, Department of Agronomy, School of Agriculture, Lovely Professional University, 227–264. Retrieved from https://doi.org/10.1007/978-3-031-54005-9_10.

Kumar, P., & Choudhury, D. (2024g). Bioremediation of wastewater using hydroponics. In *Springer water*. Punjab, Phagwara, India: Springer Nature, Department of Agronomy, School of Agriculture, Lovely Professional University, 27–64. Retrieved from https://doi.org/10.1007/978-3-031-53258-0_2.

Kumar, P., & Choudhury, D. (2024h). Hydroponic removal of organic contaminants from water. In *Springer water*. Punjab, Phagwara, India: Springer Nature, Department of Agronomy, School of Agriculture, Lovely Professional University, 143–164. Retrieved from https://doi.org/10.1007/978-3-031-53258-0_6.

Kumar, P., Choudhury, D., Dey, S. R., & Sharma, M. (2024a). Application of genetically modified crops against metallic contaminations. In *Bioremediation of emerging contaminants from soils: Soil health conservation for improved ecology and food security*. Punjab, Phagwara, India: Elsevier, Department of Agronomy, School of Agriculture, Lovely Professional University, 541–564. Retrieved from https://doi.org/10.1016/B978-0-443-13993-2.00024-4.

Kumar, P., Devi, P., & Panwar, J. S. (2024b). Constructed wetland-bioelectrochemical oxidation systems: A hybrid system for wastewater treatment. In *Bioelectrochemical oxidation processes for wastewater treatment*. Punjab, Phagwara, India: CRC Press, Lovely Professional University, 126–151. Retrieved from https://doi.org/10.1201/9781003368472-7.

Kumar, P., & Dey, S. R. (2024). Hydroponics phytoremediation: An overview. In *Springer water*. Punjab, Phagwara, India: Springer Nature, Department of Agronomy, School of Agriculture, Lovely Professional University, 361–396. Retrieved from https://doi.org/10.1007/978-3-031-53258-0_13.

Kumar, P., Dey, S. R., & Choudhury, D. (2024c). Effectiveness of cadmium on biochemical shift of pea plant treated with mycorrhiza and putrescine. *Nature Environment and Pollution Technology, 23*, 275–285. Retrieved from https://doi.org/10.46488/NEPT.2024.v23i01.023.

Kumar, P., & Dwivedi, P. (2024). Recent advances in the role of nitric oxide and cross-talk with signalling molecules under environmental stress in plants. In *Plant growth regulators: Resilience for sustainable agriculture*. Phagwara, India: Springer Nature, Lovely Professional University, 55–78. Retrieved from https://doi.org/10.1007/978-981-97-2918-0_4.

Kumar, P., Dwivedi, P., & Upadhyay, S. K. (2024d). Optimization of polyamine and mycorrhiza in sorghum plant for removal of hazardous cadmium. *Plant Physiology and Biochemistry: PPB/ Societe Francaise de Physiologie Vegetale, 214*. Retrieved from https://doi.org/10.1016/j.plaphy.2024.108846.

Kumar, P., Jose Anand, A., & Rajagopal, A. R. (2024e). Sustainable use of marine resources using ocean technologies. In *Sustainable development goals: Technologies and opportunities*. Punjab, Phagwara, India: CRC Press, Department of Agronomy, School of Agriculture, Lovely Professional University, 249–274. Retrieved from https://doi.org/10.1201/9781003468257-14.

Kumar, P., Karam, N., & Choudhury, D. (2024f). Assessment of the phytochemical constituents and metabolites in the medicinal plants and herbal medicine used in the treatment and management of skin diseases. In *Reference series in phytochemistry*. Punjab, Phagwara, India: Springer Science and Business Media B.V., Department of Agronomy, School of Agriculture, Lovely Professional University, 533–573. Retrieved from https://doi.org/10.1007/978-3-031-43199-9_21.

Kumar, P., Khokhar, A., & Dey, S. R. (2024g). Evidence from the use of herbal medicines in the management and prevention of common eye diseases. In *Reference series in phytochemistry*. Phagwara, India: Springer Science and Business Media B.V., Department of Agronomy, School of Agriculture, Lovely Professional University, 763–795. Retrieved from https://doi.org/10.1007/978-3-031-43199-9_27.

Kumar, P., Saini, L., & Sharma, M. (2024h). Assessment of the phytochemical constituents and metabolites in the medicinal plants and herbal medicine used in the treatment and management of respiratory diseases. In *Reference series in phytochemistry*. Punjab, Phagwara, India: Springer Science and Business Media B.V., Department of Agronomy, School of Agronomy, Lovely Professional University, 481–510. Retrieved from https://doi.org/10.1007/978-3-031-43199-9_19.

Kumar, P., Saurabh, S., & Sharma, K. (2024i). Assessment of the phytochemical constituents and metabolites in medicinal plants and herbal remedies used in the treatment and management of reproductive diseases: Polycystic ovary syndrome. In *Reference series in phytochemistry*. Punjab, Phagwara, India: Springer Science and Business Media B.V., Department of Agronomy, School of Agriculture, Lovely Professional University, 511–532. Retrieved from https://doi.org/10.1007/978-3-031-43199-9_20.

Kumar, P., Sharma, K., & Singh, J. (2024j). Analysis of complex microbial communities in soil and wastewater treatment processes.

In *Microbial metagenomics in effluent treatment plant*. Punjab, Phagwara, India: Elsevier, Department of Agronomy, School of Agriculture, Lovely Professional University, 25–41. Retrieved from https://doi.org/10.1016/B978-0-443-13531-6.00002-1.

Kumar, P., Sharma, K., & Singh, J. (2024k). Response of microbial community to environment changes. In *Microbial metagenomics in effluent treatment plant*. Punjab, Phagwara, India: Elsevier, Department of Agronomy, School of Agriculture, Lovely Professional University, 43–61. Retrieved from https://doi.org/10.1016/B978-0-443-13531-6.00006-9.

Kumar, P., & Singh, J. (2024a). Exploring enzyme biotechnology's role in green chemistry and advancements in pharmaceutical technologies. In *Enzyme biotechnology for environmental sustainability*. Punjab, Phagwara, India: Elsevier, Department of Agronomy, School of Agriculture, Lovely Professional University, 465–495. Retrieved from https://doi.org/10.1016/B978-0-443-22072-2.00015-2.

Kumar, P., & Singh, J. (2024b). Environmental and health risk assessment of chemical pollutants in drinking water and wastewaters. In *Trends in biological processes in industrial wastewater treatment*. Phagwara, Punjab, India: Institute of Physics Publishing, Department of Agronomy, School of Agriculture, School of Bioengineering and Biosciences, Lovely Professional University, 5.1–5.30. Retrieved from https://doi.org/10.1088/978-0-7503-5678-7ch5.

Kumar, P., & Singh, J. (2024c). Bioprocesses in industrial wastewater treatment: Trends and prospects. In *Trends in biological processes in industrial wastewater treatment*. Phagwara, Punjab, India: Institute of Physics Publishing, Department of Agronomy, School of Agriculture, School of Bioengineering and Biosciences, Lovely Professional University, 15.1–15.50. Retrieved from https://doi.org/10.1088/978-0-7503-5678-7ch15.

Kumar, P., & Singh, J. (2024d). Emerging contaminants and ways to reduce them. In *Trends in biological processes in industrial wastewater treatment*. Phagwara, Punjab, India: Institute of Physics Publishing, Department of Agronomy, School of Agriculture, School of Bioengineering and Biosciences, Lovely Professional University, 4.1–4.34. Retrieved from https://doi.org/10.1088/978-0-7503-5678-7ch4.

Kumar, P., & Singh, J. (2024e). Hydroponic: An eco-friendly future. In *Springer Water*. Punjab, Phagwara, India: Springer Nature, Department of Agronomy, School of Agriculture, Lovely Professional University, 223–267. Retrieved from https://doi.org/10.1007/978-3-031-53258-0_9.

Kumar, P., Singh, J., & Singh, G. (2024l). Renewable hydrogen opportunities and challenges. In *Renewable hydrogen: Opportunities and challenges in commercial success*. Punjab, Phagwara, India: Elsevier, Department of Agronomy, School of Agriculture, Lovely Professional University, 193–220. Retrieved from https://doi.org/10.1016/B978-0-323-95379-5.00010-9.

Kumar, P., Srivastav, A. L., Chaudhary, V., van Hullebusch, E. D., & Busquets, R. (2024m). Bioremediation of emerging contaminants from soils: soil health conservation for improved ecology and food security. In *Bioremediation of emerging contaminants from soils: Soil health conservation for improved ecology and food security*. Phagwara, Punjab, India: Elsevier, Department of Agronomy, School of Agriculture, Lovely Professional University. Retrieved from https://doi.org/10.1016/C2022-0-02758-6.

Pandey, S. K., Kumar, P., & Dey, S. R. (2024). Xenobiotic stress management by microbial endophytes. In *Bioremediation of emerging contaminants from soils: Soil health conservation for improved ecology and food security*. Jabalpur, India: Elsevier, College of Agriculture, Powarkheda Narmadapuram, Jawaharlal Nehru Agriculture University, Madhya Pradesh, 241–259. Retrieved from https://doi.org/10.1016/B978-0-443-13993-2.00012-8.

Raj, A., Goutam, E., Kumar, P., & Singh, J. (2024). Polyphenol oxidase: An enzyme of plants. In *Polyphenol oxidases: Function, wastewater remediation, and biosensors*. Phagwara, Punjab, India: De Gruyter, Department of Agronomy, School of Agriculture, Lovely Professional University, 105–128. Retrieved from https://doi.org/10.1515/9783111033525-004.

Saini, L., Kumar, P., & Panwar, J. S. (2024). Current status of wastewater in India/other countries/regions. In *Bioelectrochemical oxidation processes for wastewater treatment*. Punjab, Phagwara, India: CRC Press, Lovely Professional University, 152–165. Retrieved from https://doi.org/10.1201/9781003368472-8.

Sharma, K., & Kumar, P. (2024). Environmental threats posed by xenobiotics. In *Bioremediation of emerging contaminants from soils: Soil health conservation for improved ecology and food security*. Punjab, Phagwara, India: Elsevier, Department of Agronomy, School of Agriculture, Lovely Professional University, 183–201. Retrieved from https://doi.org/10.1016/B978-0-443-13993-2.00009-8.

Sharma, K., Kumar, P., & Singh, J. (2024). Ecology and diversity of microbial communities involved in the removal of priority contaminants and micropollutants in waste. In *Development in waste water treatment research and processes: Role of environmental microbiology in industrial wastewater research*. Punjab, Phagwara, India: Elsevier, Department of Agronomy, School of Agriculture, Lovely Professional University, 181–199. Retrieved from https://doi.org/10.1016/B978-0-443-13609-2.00014-8.

Sharma, M., Dey, S. R., Kumar, P., & Dwivedi, P. (2024a). Insights into the genetic improvement of canola. In *Genetic engineering of crop plants for food and health security: Volume 2*. Phagwara, India: Springer Nature, Department of Agronomy, School of Agriculture, Lovely Professional University, 205–226. Retrieved from https://doi.org/10.1007/978-981-97-3119-0_9.

Sharma, M., Kumar, P., & Panwar, J. S. (2024b). Implementation of bioelectrochemical oxidation systems in existing wastewater treatment plant: Challenges in retrofitting. In *Bioelectrochemical oxidation processes for wastewater treatment*. Punjab, Phagwara, India: CRC Press, Lovely Professional University, 166–188. Retrieved from https://doi.org/10.1201/9781003368472-9.

Srivastav, A. L., Patel, N., Rani, L., Kumar, P., Dutt, I., Maddodi, B. S., & Chaudhary, V. K. (2024). Sustainable options for fertilizer management in agriculture to prevent water contamination: A review. *Environment, Development and Sustainability, 26*, 8303–8327. Retrieved from https://doi.org/10.1007/s10668-023-03117-z.

Chapter 23

Dietary interventions for reducing heavy metal toxicity

Prasann Kumar[1], **and Shipa Rani Dey**[2]
[1]*Department of Agronomy, School of Agriculture, Lovely Professional University, Phagwara, Punjab, India,* [2]*Faculty of Agricultural Sciences, GLA University, Mathura, Uttar Pradesh, India*

23.1 Introduction

Heavy metal toxicity is a significant environmental and public health issue arising from the excessive accumulation of metals such as lead (Pb), mercury (Hg), cadmium (Cd), and arsenic (As) in biological systems. Unlike essential trace elements like iron (Fe) and zinc (Zn), which are required for normal physiological functions, heavy metals lack a known beneficial role in the human body (Dey, Sharma, Kumar, et al., 2024b; Sharma, Dey, et al., 2024). They can be highly toxic even at low concentrations. Their toxicity is primarily attributed to their ability to generate oxidative stress, disrupt enzyme function, interfere with essential metal homeostasis, and damage DNA, proteins, and lipids. Heavy metals persist in the environment due to their nondegradable nature, leading to bioaccumulation in organisms and biomagnification across the food chain. Exposure to these toxicants occurs through contaminated air, water, food, and occupational settings, with long-term accumulation in vital organs such as the liver, kidneys, brain, and bones. The toxic effects of heavy metals manifest in a range of health issues, including neurological disorders, carcinogenesis, renal dysfunction, and immune suppression. Their ability to mimic essential elements and compete for biochemical pathways further exacerbates their harmful effects, leading to cellular and systemic toxicity. Given the widespread contamination of ecosystems with heavy metals, understanding their sources, mechanisms of toxicity, and health implications is critical for developing effective preventive and therapeutic strategies (Dey, Sharma, Kumar, et al., 2024a; Saini et al., 2024). Heavy metal exposure occurs through multiple environmental and anthropogenic sources, including industrial discharge, agricultural activities, mining, electronic waste, and contaminated food and water. These sources contribute to the accumulation of toxic metals in the environment, increasing human and ecological risks. Industrial activities are among the primary contributors to heavy metal contamination. Metal mining, smelting, electroplating, battery manufacturing, and textile industries release significant amounts of lead, cadmium, and mercury into the environment. These metals contaminate the soil, water bodies, and air through emissions, effluents, and improper waste disposal. For instance, mining and ore-processing activities generate metal-laden tailings that leach into groundwater and surface water systems. Similarly, coal combustion in power plants releases mercury into the atmosphere, which later deposits in aquatic ecosystems through precipitation. Additionally, electronic waste recycling processes, especially in developing countries with inadequate regulations, expose workers and nearby communities to hazardous heavy metals. Using fertilisers, pesticides, and wastewater irrigation in agriculture significantly contributes to heavy metal accumulation in soils and crops. Phosphate fertilizers often contain cadmium as a contaminant, which, when applied to agricultural fields, accumulates in the soil and enters the food chain through plant uptake. Similarly, arsenic-based pesticides, historically used in cotton farming and fruit orchards, have contaminated soil and groundwater. Livestock rearing also contributes to heavy metal exposure, as animal feed may contain high levels of lead and cadmium, which bioaccumulate in meat and dairy products that humans consume. Heavy metal contamination of water resources poses a severe threat to public health. Natural sources, such as volcanic eruptions, geothermal activities, and rock weathering, contribute to the presence of heavy metals in groundwater (Dey, Kumar, & Singh, 2024;

Heavy Metal Toxicity and Neurodegeneration. DOI: https://doi.org/10.1016/B978-0-443-36575-1.00024-8

Sharma, Kumar et al., 2024). However, anthropogenic sources, including industrial effluents, sewage discharge, and landfill leachates, significantly amplify water pollution. Lead contamination in drinking water is a notable concern, as seen in cases where ageing lead pipes corrode, releasing lead into household water supplies. Mercury pollution in aquatic ecosystems, primarily from industrial discharge and artisanal gold mining, results in the bioaccumulation of methylmercury in fish, a primary source of human exposure. Additionally, arsenic contamination in groundwater, particularly in South and Southeast Asia, affects millions of people, leading to chronic arsenicosis. Heavy metals enter the human diet through contaminated crops, seafood, and processed foods. Plants grown in metal-polluted soil absorb toxic elements, accumulating in edible parts such as grains, fruits, and vegetables. Rice, for example, is particularly susceptible to arsenic accumulation due to its growth in flooded conditions that enhance arsenic bioavailability. Similarly, seafood, huge predatory fish such as tuna, swordfish, and shark, accumulate high levels of methylmercury, posing health risks to consumers. Processed and canned foods may also contribute to metal exposure due to contamination from packaging materials and food processing equipment (Dey, Kumar, & Panwar, 2024; Kumar & Dwivedi, 2024). Workers in metal refining, construction, and battery manufacturing industries are at high risk of inhaling heavy metal-laden dust and fumes. Chronic occupational exposure can lead to severe health complications, including lung disease, neurological disorders, and renal failure. Additionally, household products such as lead-based paints, cosmetics, and ceramics contribute to exposure, particularly in young children who ingest lead-contaminated dust or paint chips. The long-term accumulation of heavy metals in the body leads to various adverse health effects, affecting multiple organ systems. The severity of toxicity depends on factors such as the type of metal, route of exposure, dose, duration, and individual susceptibility. Heavy metals such as lead, mercury, and arsenic are potent neurotoxicants that impair cognitive function, disrupt neurotransmission, and cause neurodegenerative diseases. Lead exposure, particularly in children, results in reduced IQ, learning disabilities, behavioral disorders, and developmental delays. In its organic form (methylmercury), mercury crosses the blood–brain barrier and causes neurological impairments such as tremors, memory loss, and motor dysfunction. Chronic arsenic exposure is linked to neurobehavioral abnormalities and an increased risk of Alzheimer's and Parkinson's diseases. The kidneys and liver serve as primary detoxification organs, making them highly susceptible to heavy metal toxicity. Cadmium accumulates in the renal cortex, causing nephrotoxicity, proteinuria, and renal failure. Similarly, arsenic and mercury induce oxidative stress and inflammation in the liver, leading to hepatotoxicity, fibrosis, and increased risk of liver cancer (Kumar, Devi, et al., 2024; Kumar, Jose Anand, et al., 2024). Epidemiological studies have linked heavy metal exposure to cardiovascular diseases such as hypertension, atherosclerosis, and stroke. Lead and cadmium contribute to vascular damage, endothelial dysfunction, and oxidative stress, increasing the risk of hypertension and myocardial infarction. Arsenic exposure through drinking water has been associated with an increased prevalence of ischemic heart disease and peripheral vascular disorders. Several heavy metals, including arsenic, cadmium, and lead, are classified as carcinogens by the International Agency for Research on Cancer. These metals induce DNA damage, disrupt cell cycle regulation, and promote tumorigenesis through oxidative stress and epigenetic modifications. Arsenic exposure has been strongly correlated with skin, lung, and bladder cancers, while cadmium is implicated in lung and prostate cancer. Heavy metals interfere with hormone signaling pathways, leading to reproductive and endocrine dysfunctions. Lead and cadmium have been associated with reduced fertility, miscarriages, and congenital disabilities. Mercury disrupts thyroid function and affects fetal brain development, posing risks to pregnant women and infants. Chronic exposure to heavy metals weakens the immune system, increasing susceptibility to infections and autoimmune diseases. Arsenic and cadmium suppress immune cell function, reduce antibody production, and induce chronic inflammation, contributing to immune dysregulation (Kumar & Choudhury, 2024e, 2024g). Heavy metal toxicity remains a pressing global health concern due to widespread environmental contamination and its detrimental effects on human health. Understanding toxicity sources, pathways, and mechanisms is crucial for developing preventive strategies, including dietary interventions, improved environmental regulations, and safer industrial practices. Continued research is necessary to explore innovative approaches for heavy metal detoxification and minimize exposure risks in vulnerable populations (Table 23.1).

23.2 Mechanisms of heavy metal toxicity

Heavy metal toxicity exerts profound effects on biological systems by disrupting cellular and molecular pathways, leading to severe physiological and pathological consequences. Exposure to heavy metals such as lead (Pb), mercury (Hg), cadmium (Cd), and arsenic (As) initiates a cascade of cellular dysfunctions, primarily through oxidative stress, inflammation, enzymatic disruption, and DNA damage. These metals interact with biomolecules, alter signaling pathways, and impair critical cellular processes,

TABLE 23.1 Dietary interventions for reducing heavy metal toxicity.

Food type	Key bioactive compounds	Mechanism of action	Target heavy metals	Additional health benefits
Fruits	Pectin, polyphenols (flavonoids, anthocyanins)	Chelation, antioxidant protection	Lead (Pb), cadmium (Cd), mercury (Hg)	Antiinflammatory, cardiovascular health
Apples	Pectin, quercetin	Binds to metals, enhances excretion	Lead, mercury	Supports gut health
Citrus fruits	Vitamin C, hesperidin	Enhances glutathione synthesis	Cadmium, lead	Boosts immunity, reduces inflammation
Berries	Anthocyanins, ellagic acid	Chelation, free radical scavenging	Mercury, lead	Neuroprotective, antioxidant
Vegetables	Chlorophyll, sulfur compounds, fiber	Binds and removes heavy metals	Arsenic (As), lead, mercury	Supports detoxification pathways
Leafy greens	Chlorophyll, fiber	Prevents absorption in the intestines	Mercury, cadmium	Supports liver function
Cruciferous veg.	Sulforaphane, Indoles	Activates detoxification enzymes	Arsenic, lead	Anticancer properties
Herbs and spices	Curcumin, allicin, polyphenols	Chelation, antioxidant, liver support	Lead, mercury, cadmium	Antiinflammatory, gut health
Turmeric	Curcumin	Reduces oxidative stress, metal binding	Cadmium, lead	Brain and joint health
Garlic	Allicin, sulfur compounds	Enhances metal excretion	Lead, mercury	Cardiovascular benefits
Cilantro	Chlorophyll, flavonoids	Chelates heavy metals	Mercury, lead, arsenic	Detoxification, immune support
Fermented foods	Probiotics (*Lactobacillus*, *Bifidobacterium*)	Alters gut microbiota, prevents metal absorption	Cadmium, lead, mercury	Gut health, immune modulation
Yogurt	*Lactobacillus*, *Bifidobacterium*	Binds metals, reduces gut inflammation	Lead, arsenic	Supports digestion, immunity
Kimchi	Probiotics, vitamins	Enhances gut integrity, reduces toxicity	Cadmium, lead	Antioxidant, antiinflammatory
-Miso	Fermented soy compounds	Supports liver detox pathways	Mercury, lead	Supports metabolic health
Biofortified crops	Selenium, zinc, iron	Displacement of heavy metals, prevents absorption	Arsenic, lead, cadmium	Enhances nutrient deficiency prevention
Selenium-enriched rice	Selenium	Competes with arsenic absorption	Arsenic	Supports thyroid function
Zinc-enriched wheat	Zinc	Competes with cadmium absorption	Cadmium	Boosts immune function
Fibers and polysaccharides	Pectin, alginates	Binds heavy metals, prevents absorption	Lead, mercury, arsenic	Supports digestive health
Apples (pectin)	Pectin	Chelates and promotes excretion of metals	Lead, mercury	Gut microbiota support
Brown seaweed (alginates)	Alginates	Traps heavy metals in digestive tract	Arsenic, lead, cadmium	Supports thyroid health
Nutraceuticals	Chelators, antioxidants, sulfur compounds	Enhances detoxification enzymes, reduces oxidative stress	Mercury, lead, cadmium	Systemic health benefits
Spirulina	Chlorophyll, phycocyanin	Binds heavy metals, boosts liver detox	Arsenic, mercury	Antiinflammatory, immune boost
Chlorella	Chlorophyll, fiber	Chelates and enhances heavy metal excretion	Lead, mercury	Detoxification, metabolic support
Alpha-lipoic acid	Antioxidant, metal chelator	Enhances glutathione, supports detoxification	Arsenic, lead, mercury	Neuroprotective, antiaging

Source: Based on Devi et al. (2024), Devi and Kumar (2024a), and Kumar (2024).

ultimately resulting in toxicity and disease progression. Understanding the cellular and molecular mechanisms underlying heavy metal toxicity is crucial for developing effective therapeutic interventions and mitigation strategies (Devi, Kumar, & Singh, 2024b; Kumar, Singh, et al., 2024; Sharma, Kumar, et al., 2024). At the cellular level, heavy metals interfere with ion transport mechanisms, metal homeostasis, and redox balance, increasing intracellular metal accumulation and toxicity. Many heavy metals have a high affinity for sulfhydryl (–SH) groups, which are abundant in proteins, particularly enzymes and transcription factors. The binding of heavy metals to these groups disrupts protein conformation, leading to loss of function, misfolding, and aggregation. For instance, cadmium readily binds to metallothioneins, proteins responsible for metal detoxification, leading to their dysregulation and excessive cell accumulation. Similarly, mercury interacts with cysteine residues in cellular proteins, altering their structure and function, particularly in the nervous system, which impairs neurotransmitter function and synaptic signaling. A significant cellular consequence of heavy metal exposure is the generation of oxidative stress, a key driver of toxicity. Oxidative stress results from an imbalance between the production of reactive oxygen species (ROS) and the antioxidant defense mechanisms of the cell. Heavy metals catalyze the formation of ROS, including superoxide anions (O_2^-), hydrogen peroxide (H_2O_2), and hydroxyl radicals (•OH), which cause oxidative damage to cellular macromolecules. For example, arsenic generates ROS by inhibiting antioxidant enzymes such as glutathione peroxidase and superoxide dismutase, leading to lipid peroxidation, protein oxidation, and DNA strand breaks. Lead disrupts mitochondrial function by interfering with the electron transport chain, producing excessive ROS and dysfunction (Devi, Kumar, & Singh, 2024a; Kumar & Singh 2024c, 2024d). Mercury also contributes to oxidative stress by binding to thiol-containing antioxidants like glutathione, depleting cellular antioxidant reserves, and making cells more vulnerable to oxidative damage. Oxidative stress from heavy metals triggers a proinflammatory response, exacerbating cellular damage. Inflammation is a key pathological feature of heavy metal toxicity, driven by the activation of nuclear factor-kappa B (NF-κB). This transcription factor regulates the expression of proinflammatory cytokines such as tumor necrosis factor-alpha (TNF-α), interleukin-6 (IL-6), and interleukin-1 beta (IL-1β). Arsenic exposure activates NF-κB signaling, leading to chronic inflammation in tissues such as the liver, kidneys, and lungs. Cadmium induces inflammatory responses by stimulating macrophages to release cytokines and chemokines, further amplifying tissue damage. Lead exposure has been linked to neuroinflammation in the brain, where it activates microglia and astrocytes, leading to the release of inflammatory mediators that contribute to neurodegenerative disorders such as Alzheimer's and Parkinson's diseases (Kumar & Singh, 2024a, 2024b; Kumar, Saini, et al., 2024).

The disruption of enzymatic functions is another critical aspect of heavy metal toxicity. Many heavy metals act as enzyme inhibitors by binding to essential cofactors or catalytic sites, rendering enzymes inactive. Lead, for example, inhibits the activity of delta-aminolevulinic acid dehydratase, an enzyme involved in heme biosynthesis, leading to anemia and impaired oxygen transport. Cadmium disrupts calcium-dependent enzymes, interfering with cellular signaling and muscle contraction. Mercury inhibits thioredoxin reductase, an enzyme crucial for maintaining cellular redox balance, resulting in increased oxidative stress and cellular apoptosis. Arsenic disrupts pyruvate dehydrogenase, an essential enzyme in cellular respiration, leading to mitochondrial dysfunction and reduced ATP production. These disruptions in enzymatic activity have widespread consequences, affecting metabolic pathways, energy production, and cellular repair mechanisms. One of the most detrimental effects of heavy metal toxicity is DNA damage, which can lead to mutagenesis, carcinogenesis, and cell death. Heavy metals directly interact with DNA by forming adducts and crosslinks or indirectly by inducing oxidative stress that generates DNA-damaging free radicals. Cadmium, for instance, interferes with DNA repair mechanisms by inhibiting DNA repair enzymes such as poly(ADP-ribose) polymerase, leading to genomic instability and increased mutation rates. Arsenic exposure is associated with epigenetic modifications, including DNA methylation and histone alterations, which can lead to aberrant gene expression and cancer development. Lead has been shown to interfere with chromatin remodeling, affecting gene transcription and cellular differentiation. Mercury disrupts DNA replication by binding to nucleotides and impairing polymerase activity, leading to cell cycle arrest and apoptosis (Devi & Kumar, 2024b; Kumar, Khokhar, et al., 2024; Kumar, Saurabh, et al., 2024).

The molecular mechanisms underlying heavy metal-induced DNA damage involve complex interactions between oxidative stress, inflammation, and repair pathways. Activating p53, a tumor suppressor protein, in response to DNA damage is a critical protective mechanism, as it triggers cell cycle arrest and apoptosis in damaged cells. However, chronic exposure to heavy metals can lead to p53 dysfunction, allowing damaged cells to proliferate and potentially become cancerous. Heavy metals such as cadmium and arsenic also interfere with the base excision repair and nucleotide excision repair pathways, preventing cells from effectively repairing DNA lesions. This accumulation of DNA damage contributes to the initiation and progression of various cancers, including lung, bladder, and liver cancers. The effects of heavy metal toxicity are not limited to individual cells but extend to systemic toxicity, affecting multiple organ systems. In the nervous system, heavy metals impair neurotransmission, alter synaptic plasticity, and induce neuronal apoptosis, leading to cognitive decline and neurodevelopmental disorders. Mercury, for example, disrupts calcium signaling in neurons, affecting neurotransmitter release and synaptic function (Devi & Kumar, 2024c; Kumar, Choudhury, et al., 2024; Kumar, Karam, et al., 2024). Lead exposure in children results in reduced brain volume and impaired cognitive function due to its ability to cross the blood–brain barrier and interfere with neuronal differentiation. Cadmium exposure has been linked to neurodegenerative diseases through its effects on oxidative stress, inflammation, and protein misfolding. Heavy metals contribute to endothelial dysfunction, arterial stiffness, and hypertension in the cardiovascular system. Cadmium induces vascular inflammation and oxidative damage, increasing the risk of atherosclerosis and cardiovascular disease. Lead exposure is associated with increased blood pressure and myocardial infarction due to its effects on calcium signaling in vascular smooth muscle cells. Arsenic disrupts nitric oxide signaling, impairing vasodilation and promoting thrombosis, which can lead to stroke and other cardiovascular complications. The renal system is particularly vulnerable to heavy metal toxicity, as the kidneys play a central role in metal excretion and detoxification. Cadmium accumulates in renal tubular cells, causing nephrotoxicity and chronic kidney disease. Lead exposure leads to glomerular damage and proteinuria, contributing to renal insufficiency. Mercury-induced kidney damage is characterized by tubular necrosis, inflammation, and fibrosis, leading to progressive renal failure. The liver, a major detoxification organ, also suffers significant damage from heavy metal exposure. Arsenic and cadmium induce hepatic inflammation, fibrosis, and carcinogenesis through oxidative stress and DNA damage. Mercury disrupts bile acid metabolism, leading to cholestasis and liver dysfunction. The hepatotoxic effects of heavy metals compromise liver function, affecting metabolism, detoxification, and immune regulation (Choudhury et al., 2024; Kumar & Choudhury, 2024d; Kumar, Srivastav, et al., 2024; Pandey et al., 2024). The immune system is another target of heavy metal toxicity, with metals such as cadmium and arsenic impairing immune cell function and cytokine production. Chronic exposure to heavy metals suppresses immune responses, increasing susceptibility to infections and autoimmune diseases. Cadmium, for instance, reduces T-cell proliferation and antibody production, weakening adaptive immunity. Arsenic exposure is associated with increased inflammation and immune dysregulation, contributing to conditions such as asthma and rheumatoid arthritis. Overall, heavy metal toxicity exerts profound effects on cellular and molecular pathways, disrupting enzymatic functions, inducing oxidative stress, promoting inflammation, and causing DNA damage. These mechanisms collectively contribute to systemic toxicity, affecting multiple organ systems and increasing the risk of chronic diseases such as cancer, neurodegenerative disorders, cardiovascular disease, and kidney failure. Understanding these pathways is essential for developing targeted therapies and preventive strategies to mitigate the harmful effects of heavy metals. Continued research is needed to explore novel detoxification approaches, enhance metal chelation therapies, and implement effective regulatory policies to minimize heavy metal exposure and its associated health risks (Table 23.2).

23.3 Role of diet in heavy metal detoxification

Nutritional modulation is critical in regulating the absorption and metabolism of metals in the human body, influencing both essential and toxic metal levels. Metals such as iron, zinc, and copper are crucial for physiological functions, while heavy metals like lead, cadmium, and mercury pose significant health risks when accumulated in excessive amounts. The interplay between dietary components and metal metabolism can determine whether these elements serve beneficial or toxic roles. Nutrients such as vitamins, minerals, and antioxidants can mitigate the harmful effects of heavy metals by altering their absorption, distribution, and excretion. For instance, iron competes with lead for absorption through the divalent metal transporter 1 (DMT1) in the intestines, meaning sufficient dietary iron can reduce lead uptake. Similarly, zinc supplementation has been shown to lower cadmium accumulation by competing for common binding sites in cellular transport mechanisms. Likewise, calcium can inhibit lead absorption by blocking its uptake in the gut, reducing its systemic toxicity. Magnesium and selenium also act as protective agents by counteracting oxidative stress induced by heavy metals and supporting detoxification pathways (Devi et al., 2024; Kumar & Choudhury, 2024f; Kumar, Sharma, et al., 2024b; Sharma & Kumar, 2024). Dietary interventions are crucial in reducing metal toxicity, particularly in populations exposed to high environmental metal levels. A well-balanced diet rich in essential minerals can significantly reduce the bioavailability and retention of toxic metals in the body. For example, selenium, a vital trace element, forms complexes with mercury, reducing its bioaccumulation and promoting excretion. Additionally, sulfur-containing compounds in garlic and onions can chelate toxic metals, facilitating their removal. Phytochemicals in fruits and vegetables, such as polyphenols, flavonoids, and carotenoids, contribute to metal detoxification by enhancing

TABLE 23.2 Food-based chelation strategies for heavy metal detoxification.

Food category	Specific food items	Active chelating compounds	Target heavy metals	Chelation mechanism	Recommended intake	Supporting scientific evidence
Fruits	Apples, citrus fruits, berries	Pectin, vitamin C, flavonoids	Lead (Pb), mercury (Hg), cadmium (Cd)	Pectin binds heavy metals in the gut; Vitamin C enhances metal excretion through urine	1–2 servings per day	Studies show pectin reduces lead bioavailability in animal models
Vegetables	Leafy greens (spinach, kale), cruciferous vegetables (broccoli, cauliflower)	Chlorophyll, sulforaphane	Arsenic (As), lead (Pb)	Chlorophyll blocks absorption; sulforaphane boosts detoxification enzymes	2–3 servings per day	Chlorophyll-rich foods enhance mercury excretion
Herbs and spices	Turmeric, cilantro, garlic	Curcumin, allicin, flavonoids	Lead (Pb), arsenic (As), mercury (Hg)	Curcumin reduces oxidative stress; allicin binds to metals and aids excretion	1–2 teaspoons daily	Curcumin has demonstrated lead chelation properties
Fermented foods	Yogurt, kimchi, miso	Probiotics (*Lactobacillus*, *Bifidobacterium*)	Cadmium (Cd), lead (Pb)	Modifies gut microbiota to reduce metal absorption	1–2 servings per day	Probiotics have shown to bind heavy metals and prevent gut absorption
Biofortified crops	Selenium-enriched rice, zinc-enriched wheat	Selenium, zinc	Arsenic (As), cadmium (Cd)	Competes with toxic metals for absorption	Daily as staple food	Selenium reduces arsenic-induced toxicity
Seaweed and algae	Spirulina, chlorella, brown seaweed	Chlorophyll, alginates	Arsenic (As), lead (Pb), cadmium (Cd)	Binds to metals in intestines and promotes excretion	3–5 g of spirulina/chlorella daily	Studies confirm Spirulina aids mercury detoxification
Nuts and seeds	Flaxseeds, sunflower seeds, Brazil nuts	Fiber, selenium, alpha-linolenic acid	Lead (Pb), mercury (Hg)	Fiber binds metals in the gut; selenium neutralizes toxicity	1–2 tablespoons daily	Selenium supplementation reduces mercury accumulation
Fibers and polysaccharides	Apples, oats, brown rice	Pectin, beta-glucans	Lead (Pb), mercury (Hg)	Forms complexes with metals for elimination	25–30 g fiber per day	High-fiber diets linked to lower heavy metal burden
Nutraceuticals	Alpha-lipoic acid, N-acetylcysteine (NAC)	Antioxidants, sulfur compounds	Mercury (Hg), lead (Pb)	Enhances glutathione production and metal excretion	600–1200 mg NAC daily	NAC protects against mercury-induced oxidative stress

Source: Based on Devi, Kumar, Dey, et al. (2024), Dey, Sharma, and Kumar (2024), and Kumar, Sharma, et al. (2024a).

TABLE 23.3 Role of essential minerals and vitamins in heavy metal detoxification.

Nutrient	Natural food sources	Mechanism of action	Heavy metals targeted	Recommended daily intake	Additional health benefits
Zinc (Zn)	Oysters, pumpkin seeds, chickpeas, nuts	Competes with heavy metals for absorption, enhances metallothionein production	Cadmium (Cd), lead (Pb), arsenic (As)	8–11 mg (adults)	Supports immune function, wound healing, and enzyme activity
Selenium (Se)	Brazil nuts, fish, sunflower seeds, eggs	Neutralizes oxidative damage, forms insoluble complexes with heavy metals	Mercury (Hg), arsenic (As), lead (Pb)	55 μg (adults)	Protects against oxidative stress, enhances thyroid function
Magnesium (Mg)	Dark leafy greens, almonds, bananas, whole grains	Competes with heavy metals for binding sites, supports enzymatic detox pathways	Lead (Pb), aluminum (Al)	310–420 mg (adults)	Supports nerve function, muscle relaxation, and bone health
Calcium (Ca)	Dairy products, leafy greens, sardines	Reduces heavy metal absorption in the gut by competing for uptake	Lead (Pb), cadmium (Cd)	1000–1300 mg (adults)	Essential for bone health, muscle function, and nerve signaling
Iron (Fe)	Red meat, lentils, spinach, quinoa	Competes with heavy metals for transport proteins, reducing their absorption	Lead (Pb), cadmium (Cd)	8–18 mg (adults)	Prevents anemia, supports oxygen transport in blood
Sulfur (S)	Garlic, onions, cruciferous vegetables, eggs	Enhances glutathione production, binds to heavy metals for excretion	Mercury (Hg), lead (Pb), cadmium (Cd)	No specific RDI; adequate protein intake ensures supply	Supports liver detoxification and antioxidant defenses
Vitamin C (ascorbic acid)	Citrus fruits, bell peppers, strawberries	Enhances urinary excretion of heavy metals, reduces oxidative damage	Lead (Pb), mercury (Hg), arsenic (As)	75–90 mg (adults)	Strengthens immune system, promotes collagen synthesis
Vitamin E (tocopherol)	Nuts, seeds, spinach, avocado	Antioxidant protection against metal-induced oxidative stress	Lead (Pb), cadmium (Cd), mercury (Hg)	15 mg (adults)	Supports skin health, immune function, and cellular repair
Vitamin B6 (pyridoxine)	Bananas, poultry, potatoes, fortified cereals	Supports neurotransmitter function, aids in metal detox pathways	Lead (Pb), cadmium (Cd)	1.3–2.0 mg (adults)	Essential for brain health and amino acid metabolism
Vitamin B12 (cobalamin)	Fish, meat, dairy, fortified cereals	Supports red blood cell formation, reduces neurotoxicity from heavy metals	Lead (Pb), arsenic (As)	2.4 μg (adults)	Maintains nerve function and DNA synthesis
Alpha-lipoic acid	Spinach, broccoli, tomatoes, organ meat	Chelates heavy metals, regenerates antioxidants like vitamin C and E	Mercury (Hg), lead (Pb)	No specific RDI; 300–600 mg for therapeutic use	Enhances energy metabolism and neuroprotection

Source: Based on Devi et al. (2024, 2024), Kumar et al. (2024), and Kumar (2024b).

antioxidant defense systems and modulating metal-binding proteins like metallothioneins. Furthermore, dietary fiber can bind to heavy metals in the digestive tract, reducing their absorption and promoting fecal elimination. A diet deficient in essential nutrients can exacerbate heavy metal toxicity by increasing their uptake and retention. For example, iron-deficient individuals are more susceptible to lead poisoning due to increased intestinal absorption of lead. Similarly, low dietary zinc intake can enhance cadmium retention, leading to nephrotoxicity and bone demineralization. Hence, maintaining adequate levels of essential nutrients is crucial in preventing heavy metal toxicity. Besides nutrient supplementation, dietary strategies such as consuming probiotic-rich foods can also aid in detoxification. Certain probiotic strains, including Lactobacillus and Bifidobacterium species, have been found to bind heavy metals in the gut, reducing their absorption. Additionally, omega-3 fatty acids, commonly found in fish and flaxseeds, protect against heavy metal-induced oxidative damage by modulating inflammatory pathways. The role of diet in metal metabolism extends beyond individual nutrients, as dietary patterns influence long-term metal exposure outcomes. A Mediterranean-style diet rich in whole grains, lean proteins, healthy fats, and antioxidants has been associated with reduced heavy metal toxicity due to its ability to provide protective micronutrients and support detoxification mechanisms. In contrast, processed foods high in artificial additives and poor in essential nutrients can contribute to increased susceptibility to metal toxicity. Thus, dietary interventions represent a cost-effective and accessible strategy for mitigating metal toxicity, particularly in vulnerable populations with high environmental exposure. Adopting a nutrient-rich diet tailored to optimize metal metabolism and detoxification processes can significantly reduce health risks associated with heavy metal exposure while supporting overall well-being (Table 23.3).

23.4 Antioxidants and phytochemicals in detoxification

Polyphenols and flavonoids, abundant in foods such as green tea, berries, and citrus fruits, are crucial in promoting health due to their potent antioxidant, antiinflammatory, and detoxifying properties. These bioactive compounds scavenge free radicals, reduce oxidative stress, and modulate various signaling pathways to protect cells from damage. Green tea, rich in catechins such as epigallocatechin gallate, has been widely studied for its ability to reduce the risk of chronic diseases, including cardiovascular disorders, neurodegeneration, and certain cancers. The polyphenolic compounds in green tea exhibit potent antioxidant activity and enhance detoxification pathways by modulating phase I and phase II liver enzymes, aiding in the biotransformation and excretion of toxins. Similarly, berries such as blueberries, strawberries, and blackberries contain anthocyanins, quercetin, and ellagic acid, contributing to their vibrant colors and health-promoting effects. These flavonoids have been shown to improve endothelial function, regulate blood sugar levels, and provide neuroprotective benefits by reducing neuroinflammation and enhancing cognitive function (Devi et al., 2024; Jagadesh et al., 2024; Kumar, 2024a, 2024c; Raj et al., 2024). Citrus fruits, including oranges, lemons, and grapefruits, are excellent sources of hesperidin, naringenin, and vitamin C, which synergistically work to boost immune function, improve circulation, and facilitate collagen synthesis while neutralizing harmful free radicals that contribute to ageing and disease progression. In addition to polyphenols and flavonoids, curcumin, the primary bioactive compound in turmeric, has gained significant attention for its powerful antiinflammatory, antimicrobial, and anticancer properties. Curcumin modulates various molecular targets, including nuclear factor-kappa B (NF-κB), cyclooxygenase-2 (COX-2), and tumor necrosis factor-alpha (TNF-α), thereby exerting its protective effects against chronic inflammation and oxidative damage. When combined with piperine from black pepper, curcumin's bioavailability increases significantly, enhancing its therapeutic efficacy. Ginger, another potent rhizome, contains gingerol and shogaol, contributing to its antiinflammatory and digestive health benefits. Ginger has been shown to alleviate nausea, improve gastrointestinal motility, and reduce muscle pain and soreness, making it a valuable natural remedy for various ailments. Furthermore, chlorophyll-rich foods such as spirulina and chlorella have gained recognition for their detoxification potential, particularly in heavy metal chelation and liver health support. Spirulina, a blue-green algae, is packed with phycocyanin, which exhibits potent antioxidant and antiinflammatory properties, aiding in immune function and reducing the risk of oxidative stress-related diseases. Additionally, spirulina has been shown to enhance endurance, reduce fatigue, and support cardiovascular health by lowering cholesterol and blood pressure levels. Chlorella, another nutrient-dense algae, is mainly known for its ability to bind heavy metals such as lead, mercury, and cadmium, promoting their excretion from the body. The chlorophyll content in these algae also aids in detoxification by facilitating the elimination of harmful toxins and promoting optimal liver function. Incorporating polyphenol-rich foods, bioactive compounds like curcumin and ginger, and chlorophyll-dense superfoods into the diet provides a holistic approach to improving health, preventing chronic diseases, and enhancing detoxification processes, making them essential components of a well-balanced, nutrient-dense diet aimed at longevity and overall well-being (Table 23.4).

TABLE 23.4 Role of dietary fibers and polysaccharides in heavy metal detoxification.

Fiber type (soluble/insoluble)	Food sources	Binding mechanism	Target heavy metals	Additional detoxification benefits
Soluble fiber (pectin)	Apples, citrus fruits, carrots, berries	Forms gel-like matrix in the gut, binds heavy metals, enhances fecal excretion	Lead (Pb), cadmium (Cd), mercury (Hg)	Supports gut health, regulates blood sugar, lowers cholesterol
Soluble fiber (alginates)	Brown seaweed, kelp, wakame	Chelates heavy metals in the digestive tract, reduces intestinal absorption	Lead (Pb), mercury (Hg), arsenic (As)	Enhances thyroid health, supports immune function
Soluble fiber (beta-glucans)	Oats, barley, mushrooms	Binds to heavy metals, enhances immune function, supports gut microbiota	Lead (Pb), cadmium (Cd)	Reduces inflammation, lowers cholesterol levels
Insoluble fiber (cellulose)	Whole grains, nuts, seeds, vegetables	Increases fecal bulk, promotes metal excretion via stool	Lead (Pb), arsenic (As), aluminum (Al)	Supports bowel regularity, prevents constipation
Insoluble fiber (lignin)	Flaxseeds, whole wheat, fruits with seeds	Adsorbs heavy metals in the intestines, prevents reabsorption	Lead (Pb), cadmium (Cd)	Acts as a prebiotic, supports gut microbiota balance
Soluble fiber (gum and mucilage)	Psyllium husk, okra, chia seeds	Forms a viscous gel, slows metal absorption, promotes elimination	Mercury (Hg), arsenic (As)	Aids in digestion, supports cardiovascular health
Polysaccharides (chitin and chitosan)	Shellfish, mushrooms, fungi	Strong chelating properties, binds heavy metals, reduces bioavailability	Lead (Pb), mercury (Hg), chromium (Cr)	Supports weight management, enhances gut health
Polysaccharides (hemicellulose)	Whole grains, nuts, beans	Traps heavy metals in the gut, enhances excretion	Lead (Pb), cadmium (Cd), aluminum (Al)	Improves digestion, regulates blood sugar levels

Source: Based on Devi et al. (2024), Jagadesh et al. (2024), Kumar (2024a, 2024c), and Raj et al. (2024).

23.5 Dietary fibers and chelating agents

Dietary fiber is crucial in binding heavy metals and facilitating their excretion from the body. It is a natural detoxification agent that reduces metal bioavailability and prevents toxic accumulation. Both soluble and insoluble fibers contribute to this process in distinct ways. Soluble fibers, such as pectin and alginates, form gel-like structures in the digestive tract that trap heavy metals, preventing their absorption and promoting their elimination through feces. Insoluble fibers, on the other hand, add bulk to stool and accelerate gut motility, reducing the time toxic metals remain in the intestines and minimizing their reabsorption into the bloodstream. This dual action of dietary fiber is particularly beneficial for individuals exposed to environmental pollutants such as lead, cadmium, and mercury. Among soluble fibers, pectin, a naturally occurring polysaccharide found in apples, citrus fruits, and certain vegetables, has demonstrated significant metal-binding properties, particularly in reducing lead and mercury absorption (Kumar & Choudhury, 2024c, Kumar & Choudhury, 2024h; Kumar & Singh, 2024e; Srivastav et al., 2024). Pectin's gel-forming ability allows it to chelate metal ions in the gastrointestinal tract, effectively lowering their systemic bioavailability. Studies have shown that pectin supplementation can reduce heavy metal levels in individuals exposed to industrial pollution or contaminated food sources, making it a valuable dietary component for detoxification. Citrus pectin, in particular, has been studied for its ability to modulate intestinal absorption of toxic metals while exhibiting prebiotic effects that support gut microbiota, further enhancing detoxification pathways. Similarly, alginates, derived from brown seaweed, have gained attention for their ability to bind and promote the excretion of heavy metals, particularly strontium and radioactive isotopes. Alginates form a viscous gel in the intestines that selectively binds toxic metals, preventing their absorption while allowing essential minerals such as calcium and magnesium to remain bioavailable. This selective binding mechanism makes alginates an effective natural remedy for reducing heavy metal toxicity without causing mineral deficiencies. Research has shown that alginate supplementation significantly lowers strontium retention in the body, reducing its potential harmful effects on bone health and overall metabolism. Furthermore, alginates exhibit additional health benefits, including supporting gut health, reducing inflammation, and improving cholesterol metabolism. Combining soluble fibers like pectin and alginates with insoluble fibers from whole grains, vegetables, and seeds promotes metal detoxification, ensuring a well-balanced and efficient elimination process (Kumar & Choudhury,

TABLE 23.5 Probiotic strains and fermented foods for heavy metal detoxification.

Probiotic strain/fermented food	Heavy metals targeted	Detoxification mechanism
Lactobacillus rhamnosus GG	Lead (Pb), Cadmium (Cd)	Binds heavy metals in the gut and prevents absorption.
Lactobacillus plantarum	Lead (Pb), Mercury (Hg)	Chelate metals supports gut barrier function
Lactobacillus acidophilus	Arsenic (As), Cadmium (Cd)	Reduces metal bioavailability, promotes excretion
Bifidobacterium bifidum	Mercury (Hg), Lead (Pb)	Adsorbs heavy metals in the intestinal lumen enhances excretion
Bifidobacterium longum	Lead (Pb), Arsenic (As)	Reduces oxidative stress caused by metals, binds toxins
Saccharomyces boulardii	Lead (Pb), Cadmium (Cd)	Yeast-based chelation reduces metal absorption.
Fermented dairy (Yogurt, Kefir)	Lead (Pb), Mercury (Hg)	It contains probiotics that bind metals, support gut health
Fermented vegetables (Kimchi, Sauerkraut)	Cadmium (Cd), Arsenic (As)	Rich in probiotic strains that detoxify metals
Natto (fermented soybeans)	Mercury (Hg), Lead (Pb)	Contains Bacillus subtilis, enhances metal detoxification
Miso (fermented soy paste)	Lead (Pb), Arsenic (As)	Rich in beneficial microbes, it boosts antioxidant defenses
Kombucha (fermented tea)	Mercury (Hg), Cadmium (Cd)	Polyphenols and probiotics bind metals, reduce toxicity
Tempeh (fermented soybean cake)	Lead (Pb), Arsenic (As)	The fermentation process enhances the bioavailability of detox agents

Source: Based on Devi et al. (2024, 2024), Kumar and Choudhury (2024h), Kumar et al. (2024), and Kumar (2024c).

2024a, 2024b; Kumar & Dey, 2024; Kumar et al., 2024). Incorporating fiber-rich foods into the diet supports digestive health and provides a protective mechanism against toxic metal exposure, reinforcing the importance of dietary strategies in maintaining optimal health. A diet rich in apples, citrus fruits, seaweed, and fiber-dense whole foods enhances the body's ability to regulate heavy metal levels naturally, reducing the risk of toxicity and associated health complications. This underscores the significance of dietary interventions in mitigating heavy metal exposure, emphasizing that a well-structured, fiber-rich diet can be a powerful tool for detoxification, improving overall well-being and long-term health outcomes (Table 23.5).

23.6 Essential nutrients for heavy metal protection

Zinc, selenium, and magnesium play crucial roles in metal displacement by competing with toxic heavy metals for absorption, reducing their bioavailability, and mitigating their harmful effects on cellular function. Zinc, an essential trace element, competes with poisonous metals such as cadmium and lead for binding sites in proteins and transporters, effectively reducing their uptake and accumulation in the body. By maintaining adequate zinc levels, the body can prevent the incorporation of toxic metals into enzymes and structural proteins, thereby protecting cellular integrity and metabolic functions. Selenium, another vital micronutrient, exerts its protective effects by forming stable complexes with toxic metals such as mercury, preventing their interaction with critical biomolecules and enhancing their excretion. Selenium also plays a fundamental role in supporting antioxidant enzymes such as glutathione peroxidase, which helps neutralize oxidative damage caused by heavy metal exposure. Magnesium further contributes to metal displacement by acting as a natural antagonist to aluminum and lead, reducing their absorption and promoting detoxification through increased renal excretion (Devi et al., 2024, 2024; Kumar et al., 2024, 2024; Srivastav et al., 2024). The combined action of these essential minerals provides a powerful defense mechanism against heavy metal toxicity, emphasizing the importance of maintaining adequate dietary intake of zinc, selenium, and magnesium to counteract environmental and dietary metal exposure. In addition to mineral-based detoxification, sulfur-containing amino acids such as cysteine and methionine are crucial for heavy metal chelation and detoxification. These amino acids are precursors to glutathione, the body's primary antioxidant and detoxifying molecule. It binds to toxic metals such as arsenic, mercury, and lead, facilitating their removal through bile and urine. Cysteine, in particular, contains a thiol (–SH) group with a strong affinity for heavy metals, allowing it to form stable, nontoxic complexes that can be efficiently excreted. Methionine, an essential amino acid, supports methylation processes that enhance the biotransformation of toxic metals into forms that are more easily eliminated by the body. The presence of sulfur-containing amino acids in protein-rich foods such as eggs, garlic, onions, and legumes reinforces the body's ability to detoxify and protect against heavy metal–induced cellular damage. Furthermore, vitamins C and E are vital in reducing oxidative stress, a significant consequence of heavy metal toxicity. Heavy metals such as lead, cadmium, and mercury generate excessive ROS, leading to lipid peroxidation, DNA damage, and protein oxidation, contributing to chronic inflammation and disease development. Vitamin C, a potent water-soluble antioxidant, neutralizes free radicals by donating electrons, thereby reducing oxidative damage and enhancing immune function (Kumar & Dey, 2024; Kumar, 2024b; Raj et al., 2024; Sharma & Kumar, 2024; Srivastav et al., 2024). Additionally, vitamin C can chelate certain heavy metals, such as lead and arsenic, facilitating their excretion while protecting essential cellular components from oxidative harm. Vitamin E, a fat-soluble antioxidant, protects cell membranes from lipid peroxidation by scavenging ROS and stabilising polyunsaturated fatty acids. Vitamins C and E provide a synergistic protective effect against heavy metal-induced oxidative stress, reinforcing cellular defense mechanisms and improving overall resilience to environmental toxins. The combined action of essential minerals, sulfur-containing amino acids, and antioxidant vitamins highlights the importance of a nutrient-rich diet in mitigating heavy metal toxicity and supporting long-term health (Table 23.6).

23.7 Probiotics and gut microbiota in heavy metal detoxification

The gut microbiota plays a crucial role in modulating heavy metal bioavailability, detoxification, and overall health by influencing the absorption, sequestration, and excretion of toxic metals such as lead, cadmium, arsenic, and mercury. The trillions of microbes in the gut form a dynamic ecosystem that interacts with dietary components, environmental toxins, and host metabolism, significantly mitigating or exacerbating metal toxicity. The gut microbiota impacts heavy metal bioavailability by altering their chemical forms, reducing solubility, and preventing systemic absorption. Certain beneficial bacteria possess metal-chelating properties that bind heavy metals within the gut lumen, rendering them less bioavailable for absorption into the bloodstream and facilitating their elimination via feces. Additionally, microbial enzymes such as metallothioneins, glutathione-*S*-transferases, and phytochelatins can transform heavy metals into less toxic forms, reducing their detrimental effects on host cells. A well-balanced gut

TABLE 23.6 Biofortified crops for reducing heavy metal absorption.

Crop type	Fortified nutrient	Mechanism of action	Effectiveness in heavy metal detoxification
Zinc-biofortified wheat	Zinc (Zn)	Competes with heavy metals (Cd, Pb) for absorption, reducing bioavailability	High—lowers cadmium uptake and oxidative stress.
Iron-biofortified rice	Iron (Fe)	Reduces lead and arsenic uptake by competing for transporters	Moderate—enhances detoxification enzyme activity
Selenium-biofortified maize	Selenium (Se)	Enhances antioxidant defense, reduces mercury and arsenic toxicity	High—neutralizes reactive oxygen species
Iodine-biofortified potatoes	Iodine (I)	Supports thyroid detox functions, reduces heavy metal accumulation	Moderate—improves metal excretion through metabolism
Magnesium-biofortified millet	Magnesium (Mg)	Competes with lead and cadmium for absorption, promotes chelation	High—lowers metal-induced oxidative stress
Selenium-biofortified lentils	Selenium (Se)	Supports glutathione production, aids in mercury and cadmium detox	High—enhances metal excretion via kidneys
Calcium-biofortified beans	Calcium (Ca)	Reduces lead absorption by blocking uptake pathways	High—protects against lead accumulation in bones
Iron-biofortified pearl millet	Iron (Fe)	Displaces heavy metals in intestinal absorption sites	Moderate—reduces arsenic and cadmium bioavailability
ZinC-BIOFORTIFIED Rice	Zinc (Zn)	Reduces cadmium accumulation in grains, lowers oxidative stress	High—effective in arsenic and lead detoxification
Sulfur-biofortified garlic	Sulfur (S)	Boosts glutathione synthesis, enhances heavy metal excretion	High—effective in mercury and cadmium detox

Source: Based on Devi et al. (2024), Jagadesh et al. (2024), Kumar and Choudhury (2024a), Kumar and Singh (2024e), and Kumar (2024c).

microbiome also strengthens intestinal barrier integrity, preventing metals from crossing into systemic circulation and reducing overall toxicity. Conversely, dysbiosis, or an imbalance in gut microbial composition, can compromise the gut barrier, increasing heavy metal absorption and exacerbating their toxic effects. This highlights the importance of maintaining a diverse and healthy gut microbiota to mitigate heavy metal exposure and its associated health risks (Devi et al., 2024; Jagadesh et al., 2024; Kumar & Choudhury, 2024c, Kumar & Choudhury, 2024h; Kumar et al., 2024; Kumar, 2024a, 2024c; Raj et al., 2024; Srivastav et al., 2024). One of the most effective dietary strategies for supporting gut microbiota in heavy metal detoxification is consuming fermented foods and probiotics. Fermented foods such as yogurt, kefir, kimchi, sauerkraut, miso, and tempeh are rich sources of beneficial bacteria that contribute to gut health, enhance microbial diversity, and provide a natural means of reducing metal toxicity. Probiotics, specifically strains of Lactobacillus and Bifidobacterium, have been extensively studied for their ability to bind, sequester, and neutralize toxic heavy metals. These probiotic strains exhibit metal-chelating properties by producing exopolysaccharides, biofilms, and organic acids that trap metals within the gut, preventing their absorption and facilitating their excretion. Lactobacillus strains, particularly *Lactobacillus rhamnosus*, *Lactobacillus plantarum*, and *Lactobacillus casei*, have demonstrated strong heavy metal-binding capacities, effectively reducing lead, cadmium, and arsenic bioavailability. These bacteria can adhere to intestinal epithelial cells, forming a protective layer that prevents metal ions from interacting with and penetrating gut tissues. Additionally, *Lactobacillus* species produce metabolites such as lactic acid, hydrogen peroxide, and short-chain fatty acids that help neutralize oxidative stress induced by heavy metal exposure. *Lactobacillus rhamnosus* has been shown to reduce lead accumulation in animal studies by binding the metal within the gut and promoting its excretion through feces. Similarly, Bifidobacterium strains, including *Bifidobacterium breve*, *Bifidobacterium bifidum*, and *Bifidobacterium longum*, play a significant role in metal detoxification by reducing intestinal metal absorption and modulating immune responses. These bacteria produce extracellular polysaccharides with high metal-binding affinity, preventing the reabsorption of toxic metals and enhancing their clearance from the body. Bifidobacterium species also help maintain gut barrier integrity by increasing mucus production and strengthening tight junction proteins, which reduces metal-induced intestinal permeability. Additionally, probiotics contribute to heavy metal detoxification by modulating gut immune responses, reducing inflammation, and enhancing antioxidant enzyme activity. Heavy metal exposure often induces systemic inflammation and oxidative stress, leading to cellular damage and dysfunction. Probiotics help mitigate these effects by upregulating the production of antioxidant enzymes such as superoxide dismutase, glutathione peroxidase, and catalase, which counteract the oxidative damage caused by heavy metals. Furthermore, probiotics stimulate the production of antiinflammatory cytokines, reducing metal-induced inflammatory responses and improving overall gut health. Combining probiotics with prebiotics, known as synbiotics, further enhances heavy metal detoxification by promoting the growth of beneficial bacteria and increasing their detoxifying capacity. Prebiotics such as inulin, fructooligosaccharides, and galactooligosaccharides serve as food sources for probiotics, encouraging their proliferation and activity. This synergistic approach enhances microbial diversity, improves gut barrier function, and optimizes the gut microbiome's ability to neutralize and excrete toxic metals. In addition to direct heavy metal detoxification, probiotics also play a role in reducing the secondary health effects associated with metal toxicity. Heavy metal exposure has been linked to neurotoxicity, liver damage, kidney dysfunction, and immune suppression (Devi et al., 2024; Jagadesh et al., 2024; Kumar & Dey, 2024; Kumar et al., 2024; Kumar, 2024c). Probiotics help mitigate these effects by improving liver function, supporting detoxification pathways, and enhancing bile acid metabolism, which plays a crucial role in the excretion of metal-chelate complexes. Probiotic supplementation has also been shown to improve cognitive function and reduce neuroinflammation in cases of heavy metal–induced neurotoxicity, highlighting their potential as a therapeutic intervention for mitigating heavy metal–related health risks. The growing body of evidence supporting the role of gut microbiota in heavy metal detoxification underscores the importance of dietary and lifestyle interventions that support microbial balance. Incorporating probiotic-rich foods, consuming fiber-rich prebiotics, reducing processed foods, and minimizing exposure to antibiotics and environmental toxins all contribute to maintaining a healthy gut microbiome capable of effectively neutralizing heavy metals. Future research continues to explore the potential of engineered probiotic strains with enhanced metal-chelating capabilities, offering promising avenues for targeted detoxification therapies. As our understanding of the gut microbiota's role in heavy metal detoxification expands, it becomes increasingly clear that maintaining a diverse and resilient microbial ecosystem is fundamental to overall health and well-being. The interplay between probiotics, gut microbiota, and heavy metal detoxification represents a powerful, natural approach to mitigating toxicity, reducing oxidative stress, and enhancing overall resilience to environmental pollutants (Table 23.7).

23.8 Functional foods and nutraceutical approaches

The use of biofortified crops for heavy metal detoxification presents a promising approach to mitigating metal toxicity through dietary interventions, offering a natural and sustainable strategy to combat the adverse health effects associated with heavy metal exposure. Biofortification involves enhancing the nutritional composition of crops through conventional breeding, genetic engineering, or agronomic practices, resulting in plants with higher concentrations of essential nutrients such as zinc, selenium, and iron, which play a crucial role in displacing toxic metals from biological systems. Crops enriched with these beneficial elements help counteract heavy metal absorption by competing with toxic metals for cellular uptake and metabolic pathways, ultimately reducing their bioavailability and accumulation in tissues. For example, zinc-biofortified wheat and rice have been shown to lower cadmium and lead absorption by occupying the same transport pathways, thereby preventing toxic metal entry into cells. Similarly, selenium-biofortified crops such as garlic, onions, and lentils enhance the body's antioxidant defenses, reducing oxidative stress induced by heavy metal exposure and promoting the detoxification of mercury and arsenic (Devi et al., 2024; Kumar & Dey, 2024; Kumar et al., 2024; Raj et al., 2024; Srivastav et al., 2024). Additionally, iron-biofortified crops, such as iron-rich beans and pearl millet, help mitigate the effects of lead toxicity by restoring iron-dependent enzymatic functions disrupted by metal interference. Beyond essential minerals, biofortification strategies also aim to enhance the levels of natural chelators and detoxifying compounds within crops. For instance, plants rich in phytochelatins, metallothioneins, and polyphenols actively bind heavy metals in the gut, reducing their systemic absorption and facilitating excretion. Cruciferous vegetables, including broccoli, kale, and mustard greens, contain sulfur-rich compounds such as glucosinolates, which enhance heavy metal detoxification by upregulating phase II detoxification enzymes in the liver. Similarly, biofortified crops with increased levels of dietary fiber, including soluble fibers such as pectin and insoluble fibers such as cellulose, act as natural metal-binding agents that trap toxic metals in the digestive tract, preventing their reabsorption and enhancing fecal elimination. Integrating biofortified crops into daily diets offers a cost-effective and accessible means of reducing heavy metal toxicity, particularly in regions with high environmental contamination and limited healthcare resources.

TABLE 23.7 Biofortified crops for reducing heavy metal absorption.

Crop type	Fortified nutrient	Mechanism of action	Effectiveness in heavy metal detoxification	Regions where it is cultivated
Zinc-biofortified wheat	Zinc (Zn)	Competes with heavy metals like cadmium (Cd) for absorption, reducing bioavailability	High—reduces cadmium accumulation and oxidative stress	India, Pakistan, China, Mexico
Iron-biofortified rice	Iron (Fe)	Reduces arsenic (As) and lead (Pb) uptake by competing for transporters	Moderate—enhances detoxification enzyme activity	Bangladesh, Philippines, Brazil
Selenium-biofortified maize	Selenium (Se)	Boosts antioxidant defenses, reduces mercury (Hg) and arsenic toxicity	High—neutralizes oxidative damage from heavy metals	China, the United States, South Africa
Iodine-biofortified potatoes	Iodine (I)	Supports thyroid detox functions, reducing heavy metal accumulation	Moderate—improves excretion of metals via metabolism	Europe, Andean regions, India
Magnesium-biofortified millet	Magnesium (Mg)	Competes with lead (Pb) and cadmium (Cd) for absorption, enhances chelation	High—lowers metal-induced oxidative stress	Sub-Saharan Africa, India
Selenium-biofortified lentils	Selenium (Se)	Supports glutathione production, aids in mercury (Hg) and cadmium detox	High—enhances heavy metal excretion via kidneys	Canada, India, Australia
Calcium-biofortified beans	Calcium (Ca)	Reduces lead (Pb) absorption by blocking uptake pathways	High—protects against lead accumulation in bones	Latin America, Africa, India
Iron-biofortified pearl millet	Iron (Fe)	Displaces heavy metals in intestinal absorption sites	Moderate—reduces arsenic (As) and cadmium (Cd) bioavailability	India, Sahel region (Africa)
Zinc-biofortified rice	Zinc (Zn)	Reduces cadmium (Cd) accumulation in grains, lowers oxidative stress	High—effective in arsenic (As) and lead (Pb) detoxification	China, Vietnam, India
Sulfur-biofortified garlic	Sulfur (S)	Boosts glutathione synthesis, enhances heavy metal excretion	High—effective in mercury (Hg) and cadmium (Cd) detox	China, Korea, the United States

Source: Based on Devi et al. (2024), Jagadesh et al. (2024), Kumar and Choudhury (2024h), and Kumar (2024a).

Complementing biofortified crops, nutraceutical formulations have emerged as an effective dietary-based intervention for heavy metal chelation, leveraging bioactive compounds with strong metal-binding properties to facilitate detoxification. Nutraceuticals encompass functional foods, dietary supplements, and medicinal plant extracts and provide targeted support for heavy metal elimination while enhancing overall health and resilience against metal-induced oxidative stress. Key nutraceuticals include polyphenol-rich extracts from green tea, berries, and citrus fruits, which exhibit strong metal-chelating capabilities through their hydroxyl and carboxyl functional groups, allowing them to bind heavy metals such as cadmium and lead, neutralizing their toxicity. Curcumin, a bioactive compound derived from turmeric, has demonstrated exceptional potential in heavy metal chelation due to its ability to form stable complexes with mercury, lead, and arsenic, preventing their cellular uptake and reducing inflammation caused by metal-induced oxidative damage. Other powerful nutraceuticals include chlorophyll-rich formulations derived from spirulina and chlorella, which exhibit remarkable detoxifying effects by binding to toxic metals in the digestive tract, promoting excretion, and restoring antioxidant balance. Alginates, derived from brown seaweed, have gained significant attention for their ability to selectively bind heavy metals while preserving essential minerals, making them a valuable component in metal detoxification therapies (Devi et al., 2024; Kumar & Choudhury, 2024c; Kumar & Dey, 2024; Kumar, 2024a; Raj et al., 2024). Additionally, sulfur-containing amino acids such as cysteine and methionine are commonly incorporated into nutraceutical formulations to support glutathione synthesis, a crucial antioxidant and metal-binding molecule that plays a central role in detoxification. Combining these nutraceutical components in targeted formulations offers a promising avenue for enhancing heavy metal elimination while simultaneously addressing oxidative stress, inflammation, and cellular damage associated with metal exposure. Future trends in dietary-based interventions for heavy metal detoxification are expected to focus on personalized nutrition, precision nutraceuticals, and developing bioengineered functional foods with enhanced detoxifying properties. Advances in nutrigenomics and microbiome research pave the way for personalized detoxification strategies tailored to an individual's genetic makeup, gut microbiota composition, and specific heavy metal exposure risks. By leveraging big data analytics and AI-driven health assessments, future dietary interventions may be customized to optimize detoxification pathways, ensuring each individual's most effective combination of biofortified crops, probiotics, and nutraceuticals. Furthermore, the development of bioengineered functional foods designed to enhance heavy metal chelation is gaining traction, with researchers exploring genetic modifications and microbial fermentation techniques to produce crops with elevated levels of natural metal-binding compounds. Innovations in agricultural biotechnology, such as CRISPR-based genome editing, can potentially enhance the phytoremediation capabilities of food crops, enabling them to absorb and sequester heavy metals in nonedible plant tissues while maintaining high nutritional value in consumable portions. Additionally, advancements in nanotechnology are opening new possibilities for dietary-based metal detoxification, with nano-encapsulated chelators and bioactive compounds offering improved bioavailability, targeted delivery, and sustained release of detoxifying agents. These cutting-edge approaches have the potential to revolutionize dietary-based interventions, providing safe, effective, and scalable solutions for mitigating heavy metal toxicity on a global scale. Another key area of future research lies in integrating gut microbiota modulation with dietary detoxification strategies (Devi et al., 2024, 2024; Kumar 2024a, 2024b; Kumar & Choudhury, 2024c; Kumar & Singh, 2024e). The interplay between gut bacteria and heavy metal metabolism is becoming increasingly evident, with probiotic and prebiotic interventions showing promise in enhancing metal chelation, restoring gut barrier integrity, and reducing systemic toxicity. Future developments may involve the engineering of probiotic strains with enhanced metal-binding properties, capable of sequestering toxic metals directly within the gut and promoting their excretion through natural metabolic pathways. The application of synbiotics, combining probiotics with targeted prebiotics, is also expected to be crucial in optimizing gut microbiome composition for heavy metal detoxification. Additionally, emerging research on postbiotics—metabolites produced by beneficial bacteria—suggests that these bioactive compounds could offer novel approaches for reducing metal-induced oxidative stress and inflammation, further enhancing the efficacy of dietary-based interventions. In parallel with these advancements, policy initiatives and public health strategies to promote the adoption of dietary-based heavy metal detoxification measures will be essential in addressing global metal contamination challenges. Public awareness campaigns, educational programs, and regulatory frameworks supporting the production and consumption of biofortified crops, functional foods, and nutraceutical formulations will ensure widespread access to effective detoxification strategies. Collaborative efforts between governments, research institutions, and the food industry can facilitate the development of fortified food products, dietary guidelines, and sustainable agricultural practices prioritizing heavy metal mitigation. Moreover, investment in food safety monitoring, environmental pollution control, and risk assessment strategies will be vital in reducing heavy metal exposure at its source, complementing dietary interventions with proactive measures to minimize contamination in food and water supplies. Ultimately, the future of dietary-based heavy metal detoxification lies in an integrated, multidisciplinary approach that combines advances in biotechnology, nutrigenomics, microbiome science, and public health policy to develop effective, sustainable, and personalized strategies for protecting human health against metal toxicity. By harnessing the power of biofortified crops, nutraceutical innovations, and precision nutrition, the next generation of dietary interventions can transform heavy metal detoxification into a proactive, accessible, and scientifically driven solution for populations worldwide (Devi et al., 2024; Jagadesh et al., 2024; Kumar & Choudhury, 2024a; Kumar & Singh, 2024e; Kumar, 2024a, 2024b; Sharma & Kumar, 2024; Srivastav et al., 2024). As research continues to uncover new insights into the complex interactions between diet, gut microbiota, and heavy metal metabolism, the development of holistic, evidence-based dietary strategies will play a pivotal role in mitigating the long–term health impacts of heavy metal exposure, paving the way for a safer and healthier future for all.

23.9 Conclusion and future perspectives

The dietary strategies for heavy metal detoxification discussed in this work highlight the crucial role of biofortified crops, nutraceutical formulations, and microbiome interventions in mitigating metal toxicity. Key findings suggest that biofortified crops enriched with essential minerals such as zinc, selenium, and iron significantly reduce heavy metal bioavailability by competing with toxic metals for absorption and metabolic pathways. Functional foods and nutraceuticals, including polyphenols, curcumin, chlorophyll-rich compounds, alginates, and sulfur-containing amino acids, exhibit strong metal–chelating properties, promoting detoxification and reducing oxidative stress. The gut microbiota has emerged as a critical player in heavy metal metabolism, with probiotics such as Lactobacillus and Bifidobacterium strains demonstrating the ability to bind toxic metals, restore gut integrity, and enhance excretion. Integrating dietary fibers, including pectin-rich foods and alginates, further supports metal elimination by trapping digestive tract toxins, preventing systemic absorption. Despite these advancements, several research gaps remain in fully understanding and optimizing dietary strategies for heavy metal detoxification. One significant gap lies in the need for large-scale, long-term clinical trials to establish the efficacy of biofortified crops and nutraceutical formulations in human populations with varying levels of metal exposure. The bioavailability and optimal dosages of different metal-chelating compounds require further investigation to maximize their effectiveness without interfering with essential nutrient absorption. Additionally, the mechanisms by which gut microbiota influence metal detoxification remain incompletely understood, necessitating further exploration of microbiome–metabolite interactions and their potential for targeted therapeutic interventions. Future research should also focus on developing engineered probiotic strains with enhanced metal–binding properties and identifying postbiotic compounds capable of mitigating metal-induced oxidative stress. Personalized dietary interventions, informed by nutrigenomics and gut microbiome profiling, present a promising avenue for tailoring detoxification strategies to individual genetic predispositions and exposure risks. Furthermore, integrating nanotechnology in dietary-based interventions, such as nano–encapsulated chelators and controlled-release nutraceuticals, warrants further exploration to improve detoxifying agents' bioavailability and targeted action. From a policy perspective, there is an urgent need to implement dietary strategies for heavy metal detoxification in high-risk populations, particularly in regions with significant environmental contamination and limited access to healthcare. Governments and public health organizations should prioritize promoting biofortified crops through agricultural policies that support sustainable farming practices and fortification initiatives. Regulatory frameworks should encourage the development and accessibility of nutraceutical formulations with proven metal-chelating properties, ensuring their affordability and safety. Public awareness campaigns and educational programs should emphasize the role of diet in reducing heavy metal toxicity and empowering communities to make informed dietary choices. Additionally, policies aimed at monitoring and reducing heavy metal contamination in food and water supplies should be strengthened, integrating dietary-based interventions with broader environmental and public health measures. Collaborative efforts between researchers, policymakers, and the food industry will be essential in translating scientific advancements into practical, scalable solutions for mitigating heavy metal toxicity, ultimately improving public health outcomes and reducing the long-term burden of metal-related diseases.

Acknowledgement

The author used AI tools to improve language and readability, but all final editing was done by a human.

References

Choudhury, D., Kumar, P., Zhimo, V. Y., & Sahoo, J. (2024). Crop rotation patterns and soil health management. In *Bioremediation of emerging contaminants from soils: Soil health conservation for improved ecology and food security*. Elsevier, 565–589. Retrieved from https://doi.org/10.1016/B978-0-443-13993-2.00025-6.

Devi, P., & Kumar, P. (2024a). Alleviation of different climatic conditions by foliar application of salicylic acid and sodium nitroprusside and their interactive effects on pigments and sugar content of maize under different sowing dates. *Nature Environment and Pollution Technology, 23*(2), 1161–1166. Retrieved from https://doi.org/10.46488/NEPT.2024.v23i02.050.

Devi, P., & Kumar, P. (2024b). Challenges in the storage of herbal medicine products and strategies for sustainable management. In *Reference series in phytochemistry: Vol. Part F3164*. Springer Science and Business Media B.V, 1739–1767. Retrieved from https://doi.org/10.1007/978-3-031-43199-9_37.

Devi, P., & Kumar, P. (2024c). Herbal medicine and pregnancy. In *Reference series in phytochemistry: Vol. Part F3164*. Springer Science and Business Media B.V, 693–722. Retrieved from https://doi.org/10.1007/978-3-031-43199-9_25.

Devi, P., Kaur, A., & Kumar, P. (2024). Phytoremediation of metal contaminated soil using energy crops: Soil health maintenance along with biofuel production. In *Bioremediation of emerging contaminants from soils: Soil health conservation for improved ecology and food security*. Elsevier, 307–333. Retrieved from https://doi.org/10.1016/B978-0-443-13993-2.00015-3.

Devi, P., Dey, S. R., Sharma, K., Kumar, P., & Singh, J. (2024). Waste management as an opportunity for the inclusion of vulnerable groups. In *Waste management in the circular economy*. Springer International Publishing, 207–230. Retrieved from https://doi.org/10.1007/978-3-031-42426-7_11.

Devi, P., Dogra, P., Kumar, P., & Singh, J. (2024). Global status of biomass energy programmes-challenges and roadmap. In *Biomass energy for sustainable development*. CRC Press, 394–414. Retrieved from https://doi.org/10.1201/9781003406501-20.

Devi, P., Kapoor, P., & Kumar, P. (2024). Biomarkers for toxicity detection. In *Biomarkers in environmental and human health biomonitoring: An integrated perspective*. Elsevier, 313–329. Retrieved from https://doi.org/10.1016/B978-0-443-13860-7.00004-5.

Devi, P., Kumar, P., Dey, S. R., Banik, D., Kumar, G., & Mehta, C. M. (2024). Integrated omics approaches for nutrient stress management in plants. In *Current omics advancement in plant abiotic stress biology: A volume in developments in applied microbiology and biotechnology*. Elsevier, 93–117. Retrieved from https://doi.org/10.1016/B978-0-443-21625-1.00008-7.

Devi, P., Kumar, P., & Singh, J. (2024a). Application of potential microbes in bioremediation. In *Development in waste water treatment research and processes: Role of environmental microbiology in industrial wastewater research*. Elsevier, 201–225. Retrieved from https://doi.org/10.1016/B978-0-443-13609-2.00021-5.

Devi, P., Kumar, P., & Singh, J. (2024b). Biotechnological approaches as clean technology for the development of a sustainable environment. In *Development in waste water treatment research and processes: Role of environmental microbiology in industrial wastewater research*. Elsevier, 381–403. Retrieved from https://doi.org/10.1016/B978-0-443-13609-2.00016-1.

Dey, S. R., Kumar, P., & Panwar, J. S. (2024). Microbial desalination cell. In *Bioelectrochemical oxidation processes for wastewater treatment*. CRC Press, 189–217. Retrieved from https://doi.org/10.1201/9781003368472-10.

Dey, S. R., Kumar, P., & Singh, J. (2024). Total soluble protein mediated morphological traits in mustard treated with thiourea and salicylic acid. *Nature Environment and Pollution Technology, 23*(2), 679–694. Retrieved from https://doi.org/10.46488/NEPT.2024.v23i02.007.

Dey, S. R., Sharma, M., & Kumar, P. (2024). Phytoremediation of phenolic compounds from soil. In *Bioremediation of emerging contaminants from soils: Soil health conservation for improved ecology and food security*. Elsevier, 335–358. Retrieved from https://doi.org/10.1016/B978-0-443-13993-2.00016-5.

Dey, S. R., Sharma, M., Kumar, P., & Dwivedi, P. (2024a). Genetic improvement of groundnut. In Genetic engineering of crop plants for food and health security: Volume 2 *(Vol. 2*. Springer Nature, 281–306. Retrieved from https://doi.org/10.1007/978-981-97-3119-0_12.

Dey, S. R., Sharma, M., Kumar, P., & Dwivedi, P. (2024b). Genetic Improvement of Mustard. In Genetic engineering of crop plants for food and health security: Volume 2 *(Vol. 2*. Springer Nature, 227–259. Retrieved from https://doi.org/10.1007/978-981-97-3119-0_10.

Jagadesh, M., Dash, M., Kumari, A., Singh, S. K., Verma, K. K., Kumar, P., ... Sharma, S. K. (2024). Revealing the hidden world of soil microbes: Metagenomic insights into plant, bacteria, and fungi interactions for sustainable agriculture and ecosystem restoration. *Microbiological Research, 285*. Retrieved from https://doi.org/10.1016/j.micres.2024.127764.

Kumar, P. (2024a). Application of marine biomass for carbon neutrality, sustainable environment, and economy. In *Marine biomass: Biorefinery, bioproducts and environmental bioremediation*. De Gruyter, 95–128. Retrieved from https://doi.org/10.1515/9783111353951-005.

Kumar, P. (2024b). Biochar production from marine algae and its application in the treatment of industrial effluent. In *Marine biomass: Biorefinery, bioproducts and environmental bioremediation*. De Gruyter, 57–93. Retrieved from https://doi.org/10.1515/9783111353951-004.

Kumar, P. (2024c). Soil conservation for global sustainability. In *World sustainability series: Vol. Part F2887*. Springer Science and Business Media Deutschland GmbH, 103–128. Retrieved from https://doi.org/10.1007/978-3-031-57456-6_6.

Kumar, P., & Choudhury, D. (2024a). Bioremediation of wastewater using hydroponics. In Springer water: Vol. Part F2309. Springer Nature, 27–64. Retrieved from https://doi.org/10.1007/978-3-031-53258-0_2.

Kumar, P., & Choudhury, D. (2024b). Hydroponic removal of organic contaminants from water. In *Springer water: Vol. Part F2309*. Springer Nature, 143–164. Retrieved from https://doi.org/10.1007/978-3-031-53258-0_6.

Kumar, P., & Choudhury, D. (2024c). Microbial native soil bacteria against cadmium toxicity. In *Springer water: Vol. Part F2532*. Springer Nature, 193–226. Retrieved from https://doi.org/10.1007/978-3-031-54005-9_9.

Kumar, P., & Choudhury, D. (2024d). Optimized phytoremediation process for the sustainable management radionuclides. In *Bioremediation of emerging contaminants from soils: Soil health conservation for improved ecology and food security*. Elsevier, 443–464. Retrieved from https://doi.org/10.1016/B978-0-443-13993-2.00020-7.

Kumar, P., & Choudhury, D. (2024e). Pollutants from agricultural activities in the Global South. In *Biomonitoring of pollutants in the Global South*. Springer Nature, 89–126. Retrieved from https://doi.org/10.1007/978-981-97-1658-6_3.

Kumar, P., & Choudhury, D. (2024f). Role of indigenous knowledge in agricultural soil reclamation without disturbing other ecosystems. In *Bioremediation of emerging contaminants from soils: Soil health conservation for improved ecology and food security*. Elsevier, 465–488. Retrieved from https://doi.org/10.1016/B978-0-443-13993-2.00021-9.

Kumar, P., & Choudhury, D. (2024g). Seed plants as a sustainable biomonitoring tool for environmental pollutants. In *Biomonitoring of pollutants in the Global South*. Springer Nature, 321–357. Retrieved from https://doi.org/10.1007/978-981-97-1658-6_9.

Kumar, P., & Choudhury, D. (2024h). Toxicity of rhizospheric cadmium contaminated soil and its phytoremediation. In *Springer water: Vol. Part F2532*. Springer Nature, 227–264. Retrieved from https://doi.org/10.1007/978-3-031-54005-9_10.

Kumar, P., & Dey, S. R. (2024). Hydroponics phytoremediation: An overview. In *Springer water: Vol. Part F2309*. Springer Nature, 361–396. Retrieved from https://doi.org/10.1007/978-3-031-53258-0_13.

Kumar, P., & Dwivedi, P. (2024). Recent advances in the role of nitric oxide and cross-talk with signalling molecules under environmental stress in plants. In *Plant growth regulators: Resilience for sustainable agriculture*. Springer Nature, 55–78. Retrieved from https://doi.org/10.1007/978-981-97-2918-0_4.

Kumar, P., & Singh, J. (2024a). Bioprocesses in industrial wastewater treatment: Trends and prospects. In *Trends in biological processes in industrial wastewater treatment*. Institute of Physics Publishing, 15.1–15.50. Retrieved from https://doi.org/10.1088/978-0-7503-5678-7ch15.

Kumar, P., & Singh, J. (2024b). Emerging contaminants and ways to reduce them. In *Trends in biological processes in industrial wastewater treatment*. Institute of Physics Publishing, 4.1–4.34. Retrieved from https://doi.org/10.1088/978-0-7503-5678-7ch4.

Kumar, P., & Singh, J. (2024c). Environmental and health risk assessment of chemical pollutants in drinking water and wastewaters. In *Trends in biological processes in industrial wastewater treatment*. Institute of Physics Publishing, 5.1–5.30. Retrieved from https://doi.org/10.1088/978-0-7503-5678-7ch5.

Kumar, P., & Singh, J. (2024d). Exploring enzyme biotechnology's role in green chemistry and advancements in pharmaceutical technologies. In *Enzyme biotechnology for environmental sustainability*. Elsevier, 465–495. Retrieved from https://doi.org/10.1016/B978-0-443-22072-2.00015-2.

Kumar, P., & Singh, J. (2024e). Hydroponic: An eco-friendly future. In *Springer water: Vol. Part F2309*. Springer Nature, 223–267. Retrieved from https://doi.org/10.1007/978-3-031-53258-0_9.

Kumar, P., Karam, N., & Choudhury, D. (2024). Assessment of the phytochemical constituents and metabolites in the medicinal plants and herbal medicine used in the treatment and management of skin diseases. In *Reference series in phytochemistry: Vol. Part F3164*. Springer Science and Business Media B.V, 533–573. Retrieved from https://doi.org/10.1007/978-3-031-43199-9_21.

Kumar, P., Khokhar, A., & Dey, S. R. (2024). Evidence from the use of herbal medicines in the management and prevention of common eye diseases. In *Reference series in phytochemistry: Vol. Part F3164*. Springer Science and Business Media B.V, 763–795. Retrieved from https://doi.org/10.1007/978-3-031-43199-9_27.

Kumar, P., Saini, L., & Sharma, M. (2024). Assessment of the phytochemical constituents and metabolites in the medicinal plants and herbal medicine used in the treatment and management of respiratory diseases. In *Reference series in phytochemistry: Vol. Part F3164*. Springer Science and Business Media B.V, 481–510. Retrieved from https://doi.org/10.1007/978-3-031-43199-9_19.

Kumar, P., Saurabh, S., & Sharma, K. (2024). Assessment of the phytochemical constituents and metabolites in medicinal plants and herbal remedies used in the treatment and management of reproductive diseases: Polycystic ovary syndrome. In *Reference series in phytochemistry: Vol. Part F3164*. Springer Science and Business Media B.V, 511–532. Retrieved from https://doi.org/10.1007/978-3-031-43199-9_20.

Kumar, P., Choudhury, D., Dey, S. R., & Sharma, M. (2024). Application of genetically modified crops against metallic contaminations. In *Bioremediation of emerging contaminants from soils: Soil health conservation for improved ecology and food security*. Elsevier, 541–564. Retrieved from https://doi.org/10.1016/B978-0-443-13993-2.00024-4.

Kumar, P., Dey, S. R., & Choudhury, D. (2024). Effectiveness of cadmium on biochemical shift of pea plant treated with mycorrhiza and putrescine. *Nature Environment and Pollution Technology, 23*(1), 275–285. Retrieved from https://doi.org/10.46488/NEPT.2024.v23i01.023.

Kumar, P., Devi, P., & Panwar, J. S. (2024). Constructed wetland-bioelectrochemical oxidation systems: A hybrid system for wastewater treatment. In *Bioelectrochemical oxidation processes for wastewater treatment*. CRC Press, 126–151. Retrieved from https://doi.org/10.1201/9781003368472-7.

Kumar, P., Dwivedi, P., & Upadhyay, S. K. (2024). Optimization of polyamine and mycorrhiza in sorghum plant for removal of hazardous cadmium. *Plant Physiology and Biochemistry, 214*. Retrieved from https://doi.org/10.1016/j.plaphy.2024.108846.

Kumar, P., Jose Anand, A., & Rajagopal, A. R. (2024). Sustainable use of marine resources using ocean technologies. In *Sustainable development goals: Technologies and opportunities*. CRC Press, 249–274. Retrieved from https://doi.org/10.1201/9781003468257-14.

Kumar, P., Singh, J., & Singh, G. (2024). Renewable hydrogen opportunities and challenges. In *Renewable hydrogen: Opportunities and challenges in commercial success*. Elsevier, 193–220. Retrieved from https://doi.org/10.1016/B978-0-323-95379-5.00010-9.

Kumar, P., Srivastav, A. L., Chaudhary, V., van Hullebusch, E. D., & Busquets, R. (2024). *Bioremediation of emerging contaminants from soils: Soil health conservation for improved ecology and food security*. Elsevier. Retrieved from https://doi.org/10.1016/C2022-0-02758-6.

Kumar, P., Sharma, K., & Singh, J. (2024a). Analysis of complex microbial communities in soil and wastewater treatment processes. In *Microbial metagenomics in effluent treatment plant*. Elsevier, 25–41. Retrieved from https://doi.org/10.1016/B978-0-443-13531-6.00002-1.

Kumar, P., Sharma, K., & Singh, J. (2024b). Response of microbial community to environment changes. In *Microbial metagenomics in effluent treatment plant*. Elsevier, 43–61. Retrieved from https://doi.org/10.1016/B978-0-443-13531-6.00006-9.

Pandey, S. K., Kumar, P., & Dey, S. R. (2024). Xenobiotic stress management by microbial endophytes. In *Bioremediation of emerging contaminants from soils: soil health conservation for improved ecology and food security*. Elsevier, 241–259. Retrieved from https://doi.org/10.1016/B978-0-443-13993-2.00012-8.

Raj, A., Goutam, E., Kumar, P., & Singh, J. (2024). Polyphenol oxidase: An enzyme of plants. In *Polyphenol oxidases: Function, wastewater remediation, and biosensors*. De Gruyter, 105–128. Retrieved from https://doi.org/10.1515/9783111033525-004.

Saini, L., Kumar, P., & Panwar, J. S. (2024). Current status of wastewater in India/Other Countries/Regions. In *Bioelectrochemical oxidation processes for wastewater treatment*. CRC Press, 152–165. Retrieved from https://doi.org/10.1201/9781003368472-8.

Sharma, K., & Kumar, P. (2024). Environmental threats posed by xenobiotics. In *Bioremediation of emerging contaminants from soils: Soil health conservation for improved ecology and food security*. Elsevier, 183–201. Retrieved from https://doi.org/10.1016/B978-0-443-13993-2.00009-8.

Sharma, K., Kumar, P., & Singh, J. (2024). Ecology and diversity of microbial communities involved in the removal of priority contaminants and micropollutants in waste. In *Development in waste water treatment research and processes: Role of environmental microbiology in industrial wastewater research*. Elsevier, 181–199. Retrieved from https://doi.org/10.1016/B978-0-443-13609-2.00014-8.

Sharma, M., Dey, S. R., Kumar, P., & Dwivedi, P. (2024). Insights into the genetic improvement of canola. In Genetic engineering of crop plants for food and health security: Volume 2. Springer Nature, 205–226. Retrieved from https://doi.org/10.1007/978-981-97-3119-0_9.

Sharma, M., Kumar, P., & Panwar, J. S. (2024). Implementation of bioelectrochemical oxidation systems in existing wastewater treatment plant: Challenges in retrofitting. In *Bioelectrochemical oxidation processes for wastewater treatment*. CRC Press, 166–188. Retrieved from https://doi.org/10.1201/9781003368472-9.

Srivastav, A. L., Patel, N., Rani, L., Kumar, P., Dutt, I., Maddodi, B. S., & Chaudhary, V. K. (2024). Sustainable options for fertilizer management in agriculture to prevent water contamination: A review. *Environment, Development and Sustainability, 26*(4), 8303–8327. Retrieved from https://doi.org/10.1007/s10668-023-03117-z.

Chapter 24

Pharmacological treatments for neurodegeneration due to heavy metals

Prasann Kumar
Department of Agronomy, School of Agriculture, Lovely Professional University, Phagwara, Punjab, India

24.1 Introduction

Neurodegenerative diseases linked to heavy metal exposure represent a significant public health concern, given the increasing environmental contamination and the established neurotoxic effects of metals such as lead (Pb), mercury (Hg), cadmium (Cd), and arsenic (As). Heavy metals, due to their persistent nature and bioaccumulation potential, exert profound neurotoxic effects that contribute to the pathogenesis of various neurodegenerative disorders, including Alzheimer's disease (AD), Parkinson's disease (PD), amyotrophic lateral sclerosis (ALS), and multiple sclerosis (MS). Chronic exposure to these metals, whether through occupational settings, contaminated food and water, or air pollution, leads to the progressive deterioration of neuronal structures, impairing cognitive and motor functions (Hensley et al., 2022; Zhang, Li, Ren, Peng, & Luo, 2023). Studies have demonstrated that these metals can cross the blood–brain barrier (BBB) and accumulate in different brain regions, disrupting essential biochemical pathways and leading to neuronal death. Lead, for instance, is a well-documented neurotoxin with a high affinity for calcium-binding proteins, interfering with synaptic transmission and plasticity, which are crucial for learning and memory. Mercury, particularly methylmercury, is another potent neurotoxin that targets the central nervous system (CNS), impairing neurotransmission, promoting oxidative stress, and triggering neuronal apoptosis. Although primarily known for its nephrotoxicity, cadmium has been increasingly recognized for its role in neurodegeneration by inducing inflammation and interfering with cellular homeostasis (Fu, Li, Song, Cen, & Wu, 2023; Wankhede et al., 2025). Arsenic exposure, both organic and inorganic, has been implicated in cognitive impairments and neurobehavioral disorders, primarily through its effects on oxidative stress pathways and mitochondrial dysfunction.

The mechanisms underlying metal-induced neurodegeneration are multifaceted, with oxidative stress playing a central role. Heavy metals catalyze the generation of reactive oxygen species (ROS) and reactive nitrogen species (RNS), leading to lipid peroxidation, protein oxidation, and DNA damage, all contributing to neuronal injury. Due to its high oxygen consumption, lipid-rich environment, and relatively low antioxidant defenses, the brain is particularly vulnerable to oxidative damage. Lead, for example, disrupts the redox balance by inhibiting antioxidant enzymes such as superoxide dismutase (SOD) and catalase while simultaneously promoting lipid peroxidation and protein aggregation, hallmark features of neurodegenerative diseases (Arun et al., 2021; Moyano et al., 2022). Mercury exposure leads to mitochondrial dysfunction by impairing electron transport chain (ETC) activity, reducing adenosine triphosphate (ATP) production, and increasing ROS generation. This disruption in mitochondrial function results in the activation of apoptotic pathways, contributing to neuronal cell death. Similarly, cadmium-induced oxidative stress has been linked to the misfolding and aggregation of proteins such as α-synuclein and tau, which are key pathological markers of PD and AD, respectively. Arsenic's ability to interfere with cellular respiration and generate excessive free radicals further exacerbates oxidative stress, leading to neuronal apoptosis and synaptic dysfunction (Khan & Parvez, 2015; Stachowicz, 2023).

Neuroinflammation is another crucial mechanism through which heavy metals contribute to neurodegenerative processes. Chronic exposure to these metals results in sustained activation of microglia and astrocytes, leading to the release of proinflammatory cytokines such as tumor necrosis factor-alpha (TNF-α), interleukin-1β (IL-1β), and interleukin-6 (IL-6). This heightened inflammatory state exacerbates neuronal damage by increasing oxidative stress and impairing synaptic function. Lead exposure,

Heavy Metal Toxicity and Neurodegeneration. DOI: https://doi.org/10.1016/B978-0-443-36575-1.00021-2

for instance, has been shown to induce microglial activation and astrogliosis, contributing to neuroinflammatory cascades that accelerate neurodegeneration. Mercury induces neuroinflammation by activating toll-like receptor 4 (TLR4) signaling pathways, leading to excessive cytokine release and neuronal apoptosis (Anyanwu, Orish, Ezejiofor, Nwaogazie, & Orisakwe, 2020; Bhattacharya, Bose, & Munshi, 2024; Pota et al., 2024). Cadmium exposure disrupts calcium homeostasis, leading to excitotoxicity, further amplifying neuroinflammatory responses. Arsenic's impact on neuroinflammation is mediated through nuclear factor kappa B (NF-κB) activation, a key transcription factor in regulating inflammatory genes. Chronic inflammation induced by these metals not only contributes to neuronal loss but also affects neurogenesis and synaptic plasticity, critical processes for cognitive function and memory retention.

Mitochondrial dysfunction represents another key pathological mechanism in heavy metal–induced neurodegeneration. The mitochondria serve as the primary energy source for neurons, and any disruption in their function leads to cellular energy deficits, increased oxidative stress, and activation of apoptotic pathways. Lead exposure impairs mitochondrial bioenergetics by inhibiting key enzymes in ATP production, leading to energy depletion and neuronal stress. Mercury directly binds to mitochondrial proteins, disrupting their function and enhancing ROS production, further damaging mitochondrial DNA (mtDNA) and proteins. Cadmium exposure alters mitochondrial membrane potential and disrupts calcium homeostasis, increasing susceptibility to apoptosis. Arsenic compromises neuronal survival and function by interfering with oxidative phosphorylation and ATP synthesis. The interplay between mitochondrial dysfunction, oxidative stress, and neuroinflammation establishes a vicious cycle that accelerates neurodegenerative processes, ultimately leading to neuronal death (Aschner et al., 2024; Hussien et al., 2018; López-Granero, Aschner, & Sánchez-Santed, 2023).

The cumulative effect of these mechanisms highlights the complexity of heavy metal–induced neurodegeneration and its contribution to the pathogenesis of various neurodegenerative disorders. Given the persistent nature of heavy metals in biological systems and their ability to bioaccumulate over time, their role in neurotoxicity cannot be underestimated. The increasing evidence linking heavy metal exposure to AD, PD, ALS, and MS underscores the urgent need for effective mitigation strategies, including pharmacological interventions aimed at detoxification, antioxidant therapy, and neuroprotection. A better understanding of the molecular pathways involved in metal-induced neurodegeneration will aid in developing targeted therapeutic approaches to counteract the detrimental effects of these environmental toxins on brain health (Dong, Fu, & Jiang, 2024; Dooka et al., 2024; Lasisi-Sholola, Hammed, Ajike, Akhigbe, & Afolabi, 2024).

24.2 Pathophysiology of heavy metal–induced neurodegeneration

24.2.1 Disruption of neurotransmitter systems

The disruption of neurotransmitter systems is a key mechanism through which heavy metals contribute to neurodegeneration, leading to cognitive, motor, and behavioral impairments. Neurotransmitters, including acetylcholine, dopamine, serotonin, glutamate, and gamma-aminobutyric acid (GABA), play crucial roles in maintaining neuronal communication, synaptic plasticity, and overall brain function. However, heavy metals such as lead (Pb), mercury (Hg), cadmium (Cd), and arsenic (As) interfere with neurotransmitter synthesis, release, reuptake, and receptor function, leading to widespread neuronal dysfunction. Lead exposure, for instance, has been extensively studied for its neurotoxic effects on the developing brain, where it disrupts synaptic transmission by impairing calcium-dependent neurotransmitter release. Lead has a strong affinity for calcium channels and competes with calcium ions, disrupting synaptic vesicle exocytosis and reducing the release of acetylcholine and dopamine (Bjørklund, Đorđević, Hamdan, Wallace, & Peana, 2024; Famurewa et al., 2022; Kumar et al., 2025). This impairment in synaptic function contributes to learning deficits, attention disorders, and reduced cognitive abilities. Similarly, mercury, particularly methylmercury, interferes with neurotransmitter function by targeting glutamatergic and GABAergic systems. Mercury-induced excitotoxicity results from excessive glutamate release and impaired reuptake, leading to overstimulation of *N*-methyl-D-aspartate (NMDA) receptors, calcium influx, and subsequent neuronal death. This mechanism is closely linked to neurodegenerative diseases such as AD and PD, where glutamatergic dysfunction plays a pivotal role in disease progression. In addition to glutamatergic dysregulation, mercury exposure has been shown to alter dopamine signaling by reducing dopamine transporter expression, leading to dopamine accumulation in the synaptic cleft and increased oxidative stress.

Cadmium, another neurotoxic heavy metal, disrupts neurotransmitter systems by interfering with cholinergic and dopaminergic signaling pathways. Cadmium exposure has been associated with reduced acetylcholine synthesis and release, impairing cognitive function and memory processes. Additionally, cadmium disrupts dopamine metabolism by inhibiting tyrosine hydroxylase, the rate-limiting enzyme in dopamine biosynthesis. This disruption in dopamine homeostasis has been linked to Parkinsonian-like symptoms, as dopamine deficiency in the substantia nigra is a hallmark feature of PD (Grasso et al., 2017; Raj, Kaur, Gupta, & Singh, 2021; Wydrych et al., 2025). Furthermore, cadmium exposure has been found to impair serotoninergic neurotransmission by decreasing serotonin levels in the brain, which may contribute to mood disorders, depression, and anxiety-related behaviors. Arsenic, a widely distributed environmental toxicant, has also been implicated in neurotransmitter system disruption. Arsenic exposure affects cholinergic transmission by reducing acetylcholine levels and impairing muscarinic receptor function, leading to deficits in memory and learning. Arsenic-induced alterations in glutamatergic signaling contribute to excitotoxicity and neuronal apoptosis, while its impact on GABAergic inhibition results in increased neuronal excitability and susceptibility to seizures. Additionally, arsenic exposure has been shown to impair dopamine and serotonin metabolism, contributing to neurobehavioral disorders, including depression and cognitive decline. The cumulative effects of heavy metal–induced neurotransmitter system disruption highlight the complexity of their neurotoxicity and their role in neurodegenerative disease pathogenesis. Given the essential role of neurotransmitters in maintaining brain homeostasis, the dysregulation caused by heavy metal exposure underscores the need for targeted therapeutic interventions aimed at restoring neurotransmitter balance and mitigating the long-term neurological effects associated with heavy metal toxicity (Table 24.1).

24.2.2 Role of oxidative stress and free radicals

Oxidative stress and free radical generation play a central role in heavy metal–induced neurodegeneration, contributing to cellular damage, mitochondrial dysfunction, neuroinflammation, and neuronal apoptosis. The brain is particularly vulnerable to oxidative stress due to its high oxygen consumption, abundant polyunsaturated fatty acids (PUFAs), and relatively lower antioxidant defenses than other tissues. Heavy metals such as lead (Pb), mercury (Hg), cadmium (Cd), and arsenic (As) act as potent inducers of oxidative stress by generating ROS and RNS, which disrupt cellular homeostasis and trigger neurotoxic cascades (Khandayataray & Murthy, 2024; Pocivavsek, Schwarcz, & Erhardt, 2024; Reiter et al., 2024). These metals interfere with redox cycling, inhibit antioxidant defense systems, and promote the accumulation of oxidative damage to proteins, lipids, and nucleic acids. Lead exposure, for example, has been shown to increase lipid peroxidation while simultaneously inhibiting key antioxidant enzymes such as SOD, catalase (CAT), and glutathione peroxidase (GPx). This imbalance between ROS production and antioxidant capacity leads to neuronal injury, synaptic dysfunction, and cognitive decline, as observed in neurodegenerative conditions such as AD and PD. Mercury, particularly methylmercury, exacerbates oxidative stress by directly binding to thiol groups in proteins and depleting glutathione (GSH), a critical antioxidant responsible for neutralizing free radicals. Mercury-induced oxidative stress disrupts calcium homeostasis, leading to excitotoxicity and neuronal apoptosis through mitochondrial permeability transition pore (mPTP) activation.

Similarly, cadmium exposure produces excessive ROS, leading to mitochondrial dysfunction and oxidative damage to neural membranes. Cadmium disrupts the function of key mitochondrial enzymes in the ETC, leading to increased leakage of electrons and subsequent generation of superoxide anions (Abdolmaleki, Aliabadi, & Khaksar, 2025; Briffa, Sinagra, & Blundell, 2020; Green & Planchart, 2018; Song et al., 2020). This impairment in mitochondrial function further accelerates neuronal damage by reducing ATP production, promoting energy deficits, and triggering apoptotic pathways. Additionally, cadmium induces oxidative DNA damage, which compromises the genomic integrity of neural cells and increases susceptibility to neurodegenerative disorders. Arsenic exposure contributes to neurotoxicity through its ability to induce oxidative stress and disrupt cellular signaling pathways. Arsenic generates ROS by interfering with mitochondrial function and inhibiting antioxidant enzymes, leading to neuronal apoptosis and synaptic dysfunction. Arsenic-induced oxidative stress has been implicated in the misfolding and aggregation of proteins such as beta-amyloid (Aβ) and alpha-synuclein, which are hallmark features of AD and PD, respectively. Furthermore, arsenic exposure has been shown to activate NF-κB signaling, a key regulator of oxidative stress and inflammation, further exacerbating neuronal injury (Bi et al., 2022; Costa, 2017; Spiers, Cortina Chen, Barry, Bourgognon, & Steinert, 2022).

The interplay between oxidative stress and other neurotoxic mechanisms, such as neuroinflammation and mitochondrial dysfunction, creates a vicious cycle that accelerates neurodegeneration. Persistent oxidative stress leads to sustained activation of microglia and astrocytes, increasing the release of proinflammatory cytokines and exacerbating neuronal damage. Additionally, oxidative stress–induced mitochondrial dysfunction further amplifies ROS production, creating a self-perpetuating cycle of cellular injury. Given the critical role of oxidative stress in heavy metal–induced neurodegeneration, therapeutic strategies targeting antioxidant defense mechanisms hold promise in mitigating neurotoxic effects (Liu et al., 2023; Luan et al., 2024; Oteiza, Mackenzie, & Verstraeten, 2004). The use of natural and synthetic antioxidants, such as *N*-acetylcysteine (NAC), melatonin, resveratrol, and curcumin, has shown potential in reducing oxidative stress, restoring redox balance, and protecting against heavy metal–induced neuronal damage. Understanding the molecular mechanisms underlying oxidative stress in neurodegeneration is essential for developing targeted interventions

TABLE 24.1 Heavy metals implicated in neurodegeneration and their toxic effects.

Heavy metal	Primary sources	Mechanisms of neurotoxicity	Affected brain regions	Neurological disorders linked
Lead (Pb)	Lead-based paints, contaminated water, industrial emissions, batteries, ceramics, cosmetics	Oxidative stress, calcium dysregulation, inhibition of neurotransmitter release, apoptosis, BBB disruption	The hippocampus, prefrontal cortex, cerebellum	Alzheimer's disease, cognitive decline, developmental delays, ADHD
Mercury (Hg)	Seafood (methylmercury), dental amalgams, industrial waste, thermometers, vaccines (thimerosal)	Mitochondrial dysfunction, excitotoxicity, ROS generation, disruption of glutamate homeostasis	Cerebellum, motor cortex, visual cortex, basal ganglia	Parkinson's disease, ALS, ataxia, cognitive impairment
Cadmium (Cd)	Industrial pollution, cigarette smoke, contaminated food, batteries, fertilisers	Neuroinflammation, synaptic impairment, oxidative DNA damage, disruption of neurotransmitter systems	Basal ganglia, hippocampus, cerebral cortex	Memory impairment, motor dysfunction, neurodevelopmental disorders
Arsenic (As)	Contaminated drinking water, pesticides, mining activities, industrial processes	Epigenetic alterations, oxidative DNA damage, disruption of neuronal signaling pathways	Cerebral cortex, hippocampus, striatum	Cognitive decline, neurodevelopmental disorders, neurodegenerative diseases
Aluminum (Al)	Cookware, food packaging, drinking water, cosmetics, antiperspirants	Accumulation in neurons, oxidative stress, disruption of calcium homeostasis, protein aggregation	Hippocampus, neocortex, cerebellum	Alzheimer's disease, memory loss, cognitive impairment
Manganese (Mn)	Welding fumes, contaminated water, mining, fungicides, industrial emissions	Mitochondrial dysfunction, dopamine depletion, neuroinflammation, oxidative stress	Basal ganglia, substantia nigra, striatum	Parkinsonian syndromes, motor deficits, cognitive dysfunction
Iron (Fe)	Industrial pollution, dietary overload, blood transfusions	Excessive ROS production, ferroptosis, neuronal damage, protein aggregation	Substantia nigra, hippocampus, cortex	Alzheimer's disease, Parkinson's disease, neuroinflammation
Copper (Cu)	Pesticides, industrial waste, plumbing, excessive dietary intake	Imbalance in metal homeostasis, oxidative stress, beta-amyloid aggregation, mitochondrial toxicity	The hippocampus, basal ganglia, cerebral cortex	Wilson's disease, Alzheimer's disease, cognitive decline

ALS, Amyotrophic lateral sclerosis; *ROS*, reactive oxygen species.
Source: Based on Chakraborty and Aschner (2012), Husain and Leveckis (2024), and Tundo et al. (2023).

aimed at reducing the long-term neurological effects associated with heavy metal exposure (Carter & Blizard, 2016; Eddie-Amadi et al., 2022; Kciuk, Gielecińska, Kałuzińska-Kołat, Yahya, & Kontek, 2024).

24.2.3 Neuroinflammatory response and microglial activation

Neuroinflammatory response and microglial activation play a critical role in heavy metal–induced neurodegeneration, contributing to chronic neuronal damage, synaptic dysfunction, and the progression of neurodegenerative diseases such as AD, PD, and ALS. Microglia, the resident immune cells of the CNS, serve as the first line of defense against neurotoxic insults by detecting and responding to cellular damage, infections, and environmental toxins. However, prolonged or excessive microglia activation due to heavy metal exposure leads to a chronic inflammatory state that exacerbates neuronal injury through the sustained release of proinflammatory cytokines, ROS, and nitric oxide (NO). Heavy metals such as lead (Pb), mercury (Hg), cadmium (Cd), and arsenic (As) have been shown to activate microglia and astrocytes, triggering the overproduction of inflammatory mediators such as TNF-α, IL-1β, and IL-6. These cytokines amplify neuroinflammation by disrupting neuronal homeostasis, promoting oxidative stress, and inducing apoptotic pathways (Camilleri & Blundell, 2024; Panda et al., 2023; Tahir et al., 2024). Lead exposure, for instance, has been associated with increased levels of IL-1β and TNF-α in the brain, contributing to synaptic dysfunction and cognitive decline. Lead-induced microglial activation has also been shown to promote the production of NO and ROS, leading to neuronal damage through oxidative stress and excitotoxicity. Similarly, mercury exposure, particularly methylmercury, induces neuroinflammation by activating the NF-κB signaling pathway, a key regulator of inflammatory gene expression. NF-κB activation enhances the transcription of proinflammatory cytokines and chemokines, further exacerbating neuroinflammation and promoting neuronal apoptosis. Mercury-induced microglial activation has been implicated in synaptic loss and neurotransmitter imbalance, contributing to neurodegenerative processes (Lawes et al., 2020; Liu, Zhang, & Shi, 2025; Matés, Segura, Alonso, & Márquez, 2010).

Cadmium is another potent inducer of neuroinflammation, as it disrupts the BBB and facilitates the infiltration of peripheral immune cells into the CNS, intensifying the inflammatory response. Cadmium exposure has been shown to upregulate the expression of glial fibrillary acidic protein (GFAP), a marker of astrocyte activation, leading to increased production of inflammatory cytokines and oxidative stress–related molecules. Astrocyte activation further contributes to neuronal damage by impairing glutamate clearance, leading to excitotoxicity and synaptic dysfunction. Similarly, arsenic exposure triggers neuroinflammation by promoting microglial proliferation and increasing TNF-α, IL-6, and cyclooxygenase-2 (COX-2) levels in the brain. Arsenic-induced neuroinflammation has been linked to mitochondrial dysfunction and apoptosis, which contribute to cognitive deficits and motor impairments observed in neurodegenerative disorders. Chronic neuroinflammation also plays a crucial role in protein misfolding and aggregation, as seen in AD and PD. The sustained release of inflammatory mediators enhances the aggregation of amyloid-beta (Aβ) and alpha-synuclein, further activating microglia and astrocytes, perpetuating a vicious cycle of inflammation and neurodegeneration (Eddie-Amadi et al., 2022; Huat et al., 2019; Liu et al., 2023; Luan et al., 2024; Pan & Umapathy, 2024). Given the significant impact of neuroinflammation on heavy metal–induced neuronal damage, therapeutic strategies targeting microglial activation and inflammatory signaling pathways are being explored. Antiinflammatory agents, such as nonsteroidal antiinflammatory drugs (NSAIDs), minocycline, and polyphenolic compounds, have shown promise in attenuating microglial activation and reducing neuroinflammatory damage. Understanding the molecular mechanisms underlying microglial activation and neuroinflammation is essential for developing targeted interventions to mitigate the long-term neurological effects of heavy metal exposure.

24.2.4 DNA damage and epigenetic alterations

Understanding DNA damage and epigenetic alterations as critical components of the neurotoxic effects of heavy metals is of utmost importance. These effects contribute significantly to the pathogenesis of neurodegenerative diseases. Heavy metals such as lead (Pb), mercury (Hg), cadmium (Cd), and arsenic (As) can induce a range of DNA lesions, including base modifications, strand breaks, cross-links, and chromosomal aberrations. These compromise cellular integrity and promote neuronal dysfunction. The urgency lies in that DNA damage activates cellular responses that attempt to repair the lesions; however, prolonged exposure to heavy metals overwhelms these repair mechanisms, leading to genomic instability and the accumulation of mutations that drive neurodegeneration (Bi et al., 2022; Briffa et al., 2020; Carter & Blizard, 2016; Khandayataray & Murthy, 2024). One of the primary mechanisms by which heavy metals induce DNA damage is by generating ROS and RNS. These free radicals cause oxidative modifications to nucleic acids, disrupting the normal functioning of neurons. Lead exposure, for example, has been shown to increase DNA strand breaks and oxidative base damage by directly interacting with cellular components and disrupting redox homeostasis.

Similarly, mercury, especially in its methylated form, generates ROS that target the DNA molecule, leading to oxidative DNA damage and activating repair mechanisms such as base excision repair. However, prolonged mercury exposure can overwhelm the DNA repair capacity of neurons, leading to persistent DNA damage and cellular dysfunction. Cadmium exposure also induces DNA damage by increasing the formation of ROS and inhibiting DNA repair pathways. Cadmium has been shown to directly interact with the DNA repair enzyme poly(ADP-ribose) polymerase, leading to its depletion and impairing the repair of single-strand breaks. This inability to repair DNA damage accelerates the accumulation of genomic mutations and promotes neuronal cell death, characteristic of many neurodegenerative diseases (Costa, 2017; Husain & Leveckis, 2024; Kciuk et al., 2024; Reiter et al., 2024; Song et al., 2020).

In addition to direct DNA damage, heavy metals also induce epigenetic alterations that can influence gene expression, leading to persistent changes in neuronal function and contributing to the development of neurodegenerative disorders. Epigenetic modifications, such as DNA methylation, histone modifications, and noncoding RNA regulation, regulate gene expression without altering the underlying genetic code. Environmental factors, including exposure to toxic substances like heavy metals, can influence these modifications. DNA methylation, in particular, is crucial in controlling gene expression in neurons. Heavy metals such as arsenic and cadmium have been shown to alter DNA methylation patterns by inhibiting DNA methyltransferases (DNMTs), enzymes responsible for adding methyl groups to the DNA molecule (Chakraborty & Aschner, 2012; Green & Planchart, 2018; Pocivavsek et al., 2024). This disruption in DNA methylation leads to the activation of customarily silenced genes and the repression of essential neuronal genes, contributing to cellular dysfunction. For instance, arsenic exposure has been linked to the hypomethylation of the tumor suppressor gene p53, leading to its activation and increased susceptibility to apoptosis in neuronal cells. On the other hand, histone modifications, such as acetylation and methylation, also play a vital role in gene regulation. Heavy metals like lead and mercury interfere with histone acetyltransferases and histone deacetylases (HDACs), enzymes responsible for adding and removing acetyl groups to histones, respectively. These alterations in histone acetylation lead to changes in chromatin structure, making genes more or less accessible to transcription factors and ultimately influencing gene expression. For example, mercury exposure has been shown to alter the expression of genes involved in oxidative stress and inflammation by modifying histone acetylation, further exacerbating neuroinflammation and neuronal damage (Camilleri & Blundell, 2024; Eddie-Amadi et al., 2022; Oteiza et al., 2004).

Noncoding RNAs, including microRNAs (miRNAs), also play an essential role in regulating gene expression in response to heavy metal exposure. The audience's research is crucial in understanding how heavy metals can affect the expression of miRNAs, which in turn modulate the expression of target genes involved in neuronal survival, synaptic plasticity, and oxidative stress responses. For instance, cadmium exposure has been shown to increase the expression of miR-155, a proinflammatory miRNA, which contributes to microglial activation and promotes neuroinflammation. These epigenetic alterations induced by heavy metals impact gene expression and neuronal function, further contributing to the pathophysiology of neurodegenerative diseases (Luan et al., 2024; Matés et al., 2010; Spiers et al., 2022; Tahir et al., 2024). The cumulative effects of DNA damage and epigenetic modifications due to heavy metal exposure underscore the complexity of neurodegenerative processes and highlight the need for strategies to repair DNA, restore normal epigenetic regulation, and mitigate the long-term neurological effects of these toxicants. Your research is integral to understanding and developing these strategies (Table 24.2).

24.3 Pharmacological strategies for heavy metal detoxification

24.3.1 Chelation therapy (e.g., EDTA, DMSA, DMPS)

Chelation therapy is one of the primary medical interventions for heavy metal toxicity. It involves administering chelating agents such as ethylenediaminetetraacetic acid (EDTA), dimercaptosuccinic acid (DMSA), and 2,3-dimercapto-1-propanesulfonic acid (DMPS) to facilitate the removal of toxic metals from the body. Chelating agents form stable complexes with metal ions, allowing them to be excreted through urine or feces, thereby reducing their toxic burden and preventing further damage to biological systems, particularly within the nervous system. Chelation therapy has been widely utilized in cases of lead, mercury, arsenic, and cadmium poisoning, as these metals are known to accumulate in the brain, causing oxidative stress, neuroinflammation, mitochondrial dysfunction, and neurotransmitter imbalances. Among the most commonly used chelating agents, EDTA is highly effective at binding divalent and trivalent metal cations, making it particularly useful for lead poisoning. EDTA forms a stable complex with lead ions, which are then excreted primarily through the kidneys (Costa, 2017; Liu et al., 2025; Oteiza et al., 2004; Song et al., 2020). However, EDTA is not highly selective and can also chelate essential metals such as calcium, magnesium, and zinc, leading to potential side effects like electrolyte imbalances. Additionally, EDTA

TABLE 24.2 Pharmacological interventions for heavy metal–induced neurodegeneration.

Intervention	Mechanism of action	Key benefits	Current clinical status
Chelation therapy (EDTA, DMSA, DMPS)	Binds heavy metals, enhances renal excretion, reduces metal accumulation in the brain	Removes heavy metals from circulation, reduces oxidative stress, and neuroinflammation	FDA-approved for lead poisoning (DMSA), used off-label for neurotoxicity, ongoing clinical trials for expanded applications
Antioxidants (vitamin C, vitamin E, *N*-acetylcysteine, glutathione)	Neutralizes ROS, reduces oxidative damage, restores redox balance	Protects neurons from oxidative stress, prevents mitochondrial dysfunction, and improves cognitive function	Some are approved as dietary supplements, under investigation for neuroprotection
Antiinflammatory drugs (NSAIDs, corticosteroids)	Inhibits neuroinflammatory pathways (COX-2, NF-κB), reduces microglial activation	Mitigates inflammation-induced neuronal damage and delays disease progression	NSAIDs show mixed results in clinical trials; corticosteroids used in severe inflammatory conditions
Polyphenols and flavonoids (curcumin, resveratrol, quercetin, EGCG)	Scavenges free radicals inhibits metal-induced aggregation, and modulates inflammatory cytokines	Multifunctional neuroprotection reduces metal toxicity, enhances synaptic plasticity	Preclinical and early-phase clinical trials, limited bioavailability issues
NMDA receptor antagonists (memantine, ketamine)	Blocks excitotoxic glutamate activity, prevents calcium overload, and modulates synaptic transmission	Protects against excitotoxicity and neuronal death, improves cognition	Memantine approved for Alzheimer's, ketamine explored for neuroprotection
Mitochondrial-targeted therapies (coenzyme Q10, MitoQ, EPI-743)	Enhances mitochondrial function, reduces oxidative stress, and improves ATP production	Restores energy metabolism, delays neurodegeneration, supports cellular repair	CoQ10 is used as a supplement, MitoQ in clinical trials for neurodegenerative diseases
Nanomedicine-based detoxification (nanoparticles, liposomes, carbon nanotubes)	Selectively binds heavy metals, enhances the bioavailability of drugs, and delivers targeted therapy.	Improved metal clearance, reduced systemic toxicity, and brain-targeted drug delivery	Preclinical and early-phase trials are promising but need further validation.
Gene therapy and epigenetic modulation (CRISPR, siRNA, HDAC inhibitors)	Modifies gene expression, repairs metal-induced DNA damage, restores normal protein function	Long-term therapeutic potential targets disease at the molecular level	Experimental stage, ongoing research in neurodegenerative models
Multitarget approaches (combination therapies, personalized medicine)	Simultaneous targeting of oxidative stress, inflammation, excitotoxicity, and mitochondrial dysfunction	Enhanced efficacy, reduced side effects, and tailored treatment based on individual patient profiles	Emerging strategy under investigation for neurodegenerative diseases

DMPS, 2,3-Dimercapto-1-propanesulfonic acid; *DMSA*, dimercaptosuccinic acid; *EDTA*, ethylenediaminetetraacetic acid; *EGCG*, epigallocatechin gallate; *HDAC*, histone deacetylase; *NMDA*, *N*-methyl-D-aspartate; *NSAIDs*, nonsteroidal antiinflammatory drugs; *ROS*, reactive oxygen species.
Source: Based on Abdolmaleki et al. (2025), Costa (2017), and Luan et al. (2024).

cannot cross the BBB, making it less effective in directly reducing heavy metal concentrations in the CNS. Despite these limitations, EDTA remains a widely used chelating agent, especially in acute lead toxicity cases, and is often administered intravenously in clinical settings.

DMSA, a water-soluble dithiol compound, is another widely employed chelating agent that has demonstrated efficacy in removing lead, mercury, and arsenic. Unlike EDTA, DMSA has the advantage of being orally bioavailable and is less likely to cause depletion of essential minerals. DMSA can also cross the BBB to some extent, making it a valuable option for treating neurotoxicity associated with heavy metals. Studies have shown that DMSA effectively lowers blood lead levels and improves cognitive function in children with lead poisoning. Moreover, DMSA has demonstrated neuroprotective effects by reducing oxidative stress and inflammation, which are key contributors to heavy metal–induced neurodegeneration (Abdolmaleki et al., 2025; Carter & Blizard, 2016; Liu et al., 2025; Oteiza et al., 2004). Although DMSA is generally well tolerated, some patients may experience side effects such as gastrointestinal distress, skin rash, and mild elevations in liver enzymes. Long-term or improper use of DMSA may also lead to the redistribution of heavy metals, potentially exacerbating toxicity in specific organs if not carefully monitored.

DMPS, another dithiol-containing chelating agent, has shown high efficacy in mercury detoxification. DMPS is structurally similar to DMSA but possesses superior water solubility and enhanced mercury-binding capacity. It has been widely used for the treatment of mercury poisoning, including cases involving methylmercury exposure. DMPS can be administered orally and intravenously, allowing treatment regimens to be flexible. One of its key advantages over other chelators is its ability to form highly stable complexes with mercury, which prevents metal redistribution within the body. However, DMPS can bind to essential trace elements like other chelating agents, necessitating careful monitoring to avoid mineral deficiencies. Clinical studies have reported that DMPS significantly reduces mercury levels in the blood and tissues, improving neurological symptoms and cognitive function in affected individuals. Nevertheless, the use of DMPS remains controversial in certain medical circles due to potential side effects, including allergic reactions, gastrointestinal discomfort, and alterations in kidney function. Despite these concerns, chelation therapy remains a crucial approach in managing heavy metal toxicity, with ongoing research aimed at developing more selective and less toxic chelating agents that can efficiently remove heavy metals while preserving essential minerals (Camilleri & Blundell, 2024; Eddie-Amadi et al., 2022; Matés et al., 2010; Panda et al., 2023).

24.3.2 Limitations and risks of chelation therapy

Despite being widely used for removing toxic heavy metals from the body, Chelation therapy has several limitations and risks that must be carefully considered before its application. One of the primary limitations is the nonselectivity of most chelating agents, which bind to toxic metals and remove essential minerals such as calcium, magnesium, zinc, and iron. This can lead to mineral deficiencies, which may cause serious side effects such as hypocalcemia, electrolyte imbalances, and weakened immune function (Bi et al., 2022; Kciuk et al., 2024; Liu et al., 2023, 2025; Matés et al., 2010). For instance, EDTA, a commonly used chelator for lead poisoning, has a high affinity for divalent and trivalent metal cations, including calcium, and can potentially induce hypocalcemia, leading to cardiac arrhythmias, muscle cramps, and seizures in severe cases. Careful monitoring and supplementation of essential minerals are necessary to mitigate this risk during and after chelation therapy. Another major limitation of chelation therapy is its limited ability to cross the BBB, which restricts its effectiveness in directly removing heavy metals that have accumulated in the CNS. Many heavy metals, such as mercury and lead, preferentially deposit in brain tissue, contributing to neurodegeneration through oxidative stress, mitochondrial dysfunction, and neuroinflammation. Chelating agents such as EDTA and DMSA primarily act in the bloodstream and extracellular compartments, leading to a slow and often incomplete clearance of metals from the brain. While some chelators, such as DMSA and DMPS, exhibit some ability to penetrate the BBB, their effectiveness in fully detoxifying the CNS remains limited, necessitating prolonged treatment protocols that may increase the risk of adverse effects (Liu et al., 2025; Luan et al., 2024; Panda et al., 2023; Spiers et al., 2022).

Another risk associated with chelation therapy is the potential for redistribution of heavy metals within the body rather than complete elimination. Chelating agents work by forming stable complexes with metal ions. Still, in some cases, these complexes may dissociate before excretion, rereleasing toxic metals into different tissues. This redistribution can potentially increase toxicity in sensitive organs such as the kidneys, liver, and brain. For example, improper dosing or incomplete courses of DMSA treatment have been associated with increased mercury redistribution, which can exacerbate neurological symptoms rather than alleviate them (Abdolmaleki et al., 2025; Liu et al., 2025; Luan et al., 2024; Spiers et al., 2022; Tahir et al., 2024). Moreover, the excretion pathways of different chelating agents vary, with some primarily relying on renal clearance, posing an additional risk for individuals with preexisting kidney disease. EDTA, DMPS, and DMSA are excreted through the kidneys, meaning that patients with impaired renal function may experience a slower elimination of the chelator-metal complexes, leading to prolonged metal retention and an increased risk of nephrotoxicity. Therefore kidney function must be regularly monitored in patients undergoing chelation therapy, and dose adjustments should be made accordingly to prevent renal complications (Camilleri & Blundell, 2024; Carter & Blizard, 2016; Liu et al., 2023; Pan & Umapathy, 2024).

Beyond physiological risks, chelation therapy has been controversial due to its use in unapproved and nonevidence-based treatments. Some alternative medicine practitioners promote chelation therapy for conditions such as autism, cardiovascular disease, and neurodegenerative disorders despite a lack of scientific consensus on its efficacy for these applications. The misuse of chelation therapy in such contexts has led to cases of severe toxicity, including fatalities due to cardiac complications, hypocalcemia, or mismanagement of treatment regimens. Additionally, adverse effects such as gastrointestinal disturbances, headache, dizziness, fatigue, and skin rashes are commonly reported among patients undergoing chelation therapy, further complicating its safety profile. Given these

TABLE 24.3 Nanomedicine-based strategies for heavy metal detoxification and neuroprotection.

Type of NP	Function	Neuroprotective effects	Current research status
Metal NPs (gold, silver, platinum NPs)	Antioxidant properties neutralize ROS, reduce metal-induced toxicity	Protects against oxidative stress, prevents neuronal apoptosis, enhances neurodegenerations	Preclinical studies, potential applications in neurodegenerative diseases
Carbon-based nanomaterials (carbon nanotubes, graphene oxide, fullerenes)	Adsorbs heavy metals, enhances drug delivery, and protects neuronal structures	Reduces heavy metal bioavailability, supports neuronal repair, enhances drug efficacy	Preclinical studies, toxicity concerns under investigation
Lipid-based NPs (liposomes, solid lipid NPs, nanocapsules)	It encapsulates chelators and antioxidants and enhances BBB penetration	Facilitates targeted drug delivery to the brain, reduces systemic toxicity, and prolongs drug half-life	Early-phase clinical trials are promising for neurodegenerative therapy
Polymeric NPs (chitosan, PLGA, PEGylated NPs)	Controlled drug release, biodegradable, enhances bioavailability of neuroprotective agents	Improves delivery of chelators, protects neurons from heavy metal toxicity, and reduces inflammatory responses	Preclinical and early clinical trials, widely explored for neuroprotection
Magnetic NPs (iron oxide NPs, superparamagnetic NPs)	Binds and removes heavy metals, enables MRI tracking, enhances targeted therapy	Reduces metal burden, allows real-time monitoring, and improves BBB transport	Investigational stage, undergoing development for targeted detoxification
Ceramic and silica NPs (mesoporous silica NPs, zeolites)	Adsorbs toxic metals stabilizes neuroprotective compounds, prevents aggregation of poisonous proteins	Reduces neurotoxicity, protects against amyloid aggregation, and improves cognitive function	Experimental phase, promising in neurodegenerative models
Dendrimer-based nanocarriers	Encapsulates therapeutic agents, crosses the BBB, and provides multifunctional drug delivery	Enhances brain drug bioavailability and reduces heavy metal–induced oxidative damage	Ongoing research tested for neurodegenerative applications

BBB, blood–brain barrier; *NPs*, nanoparticles; *PLGA*, poly(lactic-*co*-glycolic acid); *ROS*, reactive oxygen species.
Source: Modified from Bi et al. (2022), Carter and Blizard (2016), Lawes et al. (2020), Luan et al. (2024), and Tahir et al. (2024).

limitations and risks, chelation therapy should only be used under strict medical supervision and in cases where the benefits outweigh the potential harms. Future research aims to develop more selective and less toxic chelating agents that minimize essential mineral loss, improve BBB penetration, and enhance heavy metal clearance, ultimately making chelation therapy a safer and more effective treatment for heavy metal toxicity (Table 24.3).

24.4 Antioxidant and antiinflammatory approaches

24.4.1 Role of antioxidants (vitamin C, vitamin E, *N*-acetylcysteine, glutathione)

Antioxidants are crucial in mitigating the toxic effects of heavy metal–induced oxidative stress, a key contributor to neurodegeneration. Heavy metals such as lead, mercury, cadmium, and arsenic generate excessive ROS and RNS, leading to lipid peroxidation, protein oxidation, mitochondrial dysfunction, and DNA damage. This oxidative stress disrupts neuronal function and promotes neuroinflammation, ultimately contributing to the pathogenesis of neurodegenerative diseases such as AD, PD, and ALS. Antioxidants counteract these damaging effects by neutralizing free radicals, enhancing the activity of endogenous antioxidant enzymes, and restoring redox balance (Abdolmaleki et al., 2025; Eddie-Amadi et al., 2022; Lawes et al., 2020; Oteiza et al., 2004; Spiers et al., 2022; Tahir et al., 2024). Among the most studied antioxidants, vitamin C (ascorbic acid), vitamin E (tocopherols), NAC, and glutathione (GSH) have demonstrated significant neuroprotective properties against heavy metal toxicity by reducing oxidative stress, modulating inflammatory pathways, and enhancing metal detoxification.

Vitamin C, a potent water-soluble antioxidant, scavenges free radicals and prevents lipid peroxidation in neuronal membranes, which are particularly vulnerable to oxidative damage due to their high lipid content. It also regenerates other antioxidants, such as vitamin E and glutathione, amplifying the overall antioxidant defense system. In cases of heavy metal exposure, vitamin C enhances metal chelation by forming stable complexes with lead, mercury, and cadmium, facilitating their excretion and reducing their accumulation in the brain. Studies have shown that vitamin C supplementation can reduce lead-induced cognitive impairment and improve behavioral deficits by decreasing oxidative stress and inflammation in the CNS. Vitamin C modulates neurotransmitter synthesis, particularly dopamine and serotonin, which are often disrupted by heavy metal toxicity, further supporting its neuroprotective role (Bi et al., 2022; Briffa et al., 2020; Carter & Blizard, 2016; Chakraborty & Aschner, 2012; Raj et al., 2021).

Vitamin E, a fat-soluble antioxidant, primarily protects cell membranes from lipid peroxidation caused by heavy metal–induced oxidative stress. It stabilizes neuronal membranes by preventing the oxidation of PUFAs, which are essential for maintaining synaptic connections' integrity and function. Vitamin E also modulates neuroinflammatory responses by inhibiting the activation of microglia and astrocytes, reducing the production of proinflammatory cytokines such as TNF-α and IL-6. Research has demonstrated that Vitamin E supplementation can mitigate the neurotoxic effects of mercury and cadmium, preserving cognitive function and preventing neuronal apoptosis. Moreover, vitamin E synergises with vitamin C in recycling oxidized antioxidants, ensuring sustained protection against oxidative stress.

NAC is a precursor to glutathione, the body's most abundant intracellular antioxidant, and plays a vital role in heavy metal detoxification. NAC directly scavenges ROS and RNS while also chelating heavy metals, reducing their bioavailability, and preventing interaction with cellular components. Additionally, NAC replenishes glutathione levels, often depleted in individuals exposed to heavy metals, restoring redox homeostasis and supporting mitochondrial function. Studies have indicated that NAC administration reduces neuroinflammation, improves cognitive deficits, and enhances motor coordination in models of heavy metal toxicity. It has effectively countered mercury-induced neurotoxicity by preventing mitochondrial damage and preserving neuronal viability (Abdolmaleki et al., 2025; Briffa et al., 2020; Husain & Leveckis, 2024; Reiter et al., 2024; Spiers et al., 2022).

Glutathione (GSH) is a tripeptide composed of glutamate, cysteine, and glycine and serves as the primary detoxifying agent for heavy metals. It neutralizes ROS, repairs oxidized proteins, and facilitates the excretion of metal-glutathione complexes through bile and urine. Heavy metals, especially mercury and cadmium, deplete cellular glutathione levels, increasing oxidative stress and neurodegeneration susceptibility. Supplementing with glutathione or its precursors, such as NAC, can restore antioxidant capacity, protect neuronal integrity, and enhance metal excretion. Overall, the combined use of antioxidants like vitamin C, vitamin E, NAC, and glutathione offers a promising strategy to counteract heavy metal–induced neurotoxicity by reducing oxidative damage, modulating inflammation, and promoting detoxification.

24.4.2 Antiinflammatory drugs (nonsteroidal antiinflammatory drugs, corticosteroids)

Antiinflammatory drugs, including NSAIDs and corticosteroids, play a crucial role in mitigating neuroinflammation induced by heavy metal exposure, which is a key contributor to neurodegenerative diseases. Heavy metals such as lead, mercury, cadmium, and arsenic trigger chronic neuroinflammatory responses by activating microglia and astrocytes, leading to the release of proinflammatory cytokines such as TNF-α, IL-6, and IL-1β. This inflammatory cascade exacerbates oxidative stress, disrupts neuronal function, and accelerates neuronal degeneration. By targeting inflammatory pathways, NSAIDs and corticosteroids offer potential therapeutic benefits in reducing the progression of neurodegenerative disorders linked to heavy metal toxicity (Bi et al., 2022; Briffa et al., 2020; Famurewa et al., 2022; Khandayataray & Murthy, 2024; Song et al., 2020; Wydrych et al., 2025).

NSAIDs, including ibuprofen, aspirin, and naproxen, exert their antiinflammatory effects primarily by inhibiting cyclooxygenase (COX) enzymes, particularly COX-1 and COX-2. These enzymes convert arachidonic acid into prostaglandins, which mediate inflammation and pain. In heavy metal–induced neurotoxicity, COX-2 expression is significantly upregulated, leading to excessive prostaglandin production, increased BBB permeability, and enhanced recruitment of inflammatory cells into the CNS. By blocking COX-2 activity, NSAIDs reduce prostaglandin synthesis, alleviating neuroinflammation and protecting neurons from inflammatory damage. Several studies have demonstrated that chronic NSAID use can reduce the risk of developing neurodegenerative diseases such as AD and PD, conditions that are exacerbated by heavy metal exposure (Grasso et al., 2017; Green & Planchart, 2018; Lasisi-Sholola et al., 2024; Reiter et al., 2024; Tundo et al., 2023). Additionally, NSAIDs have been shown to suppress microglial activation, limiting the release of toxic inflammatory mediators that contribute to neuronal dysfunction.

Despite their potential neuroprotective effects, NSAIDs have limitations and risks associated with long-term use. Chronic NSAID administration can lead to gastrointestinal ulcers, kidney damage, and an increased risk of cardiovascular events due to COX-1 inhibition, which affects protective prostaglandins in the gastric mucosa and renal system. Furthermore, NSAIDs may not fully reverse heavy metal–induced neurotoxicity since they primarily target inflammation without directly influencing metal accumulation or oxidative stress. Therefore while NSAIDs can be helpful as adjunctive therapy, they may need to be combined with other neuroprotective strategies to achieve optimal therapeutic benefits.

Corticosteroids, such as dexamethasone and prednisone, are potent antiinflammatory agents that suppress inflammation by inhibiting NF-κB signaling, a primary cytokine production and immune activation pathway. These drugs reduce the expression of proinflammatory cytokines, decrease microglial activation, and stabilize the BBB, preventing excessive infiltration of immune cells into the CNS. Corticosteroids also modulate the activity of reactive astrocytes, limiting the release of excitotoxic glutamate and contributing to neuronal damage in heavy metal toxicity. While corticosteroids are highly effective in controlling acute neuroinflammation, their long-term use is associated with significant adverse effects, including immunosuppression, osteoporosis, metabolic disturbances, and increased susceptibility to infections. Prolonged corticosteroid therapy can also lead to neuropsychiatric complications, such as anxiety, depression, and cognitive impairment, which may further exacerbate the neurological consequences of heavy metal exposure (Briffa et al., 2020; Khandayataray & Murthy, 2024; Kumar et al., 2025; Song et al., 2020; Wydrych et al., 2025).

Given these limitations, the use of antiinflammatory drugs in treating neurodegeneration due to heavy metals must be carefully monitored. While NSAIDs and corticosteroids offer symptomatic relief and can reduce inflammatory damage, they do not eliminate the underlying cause of toxicity. Future research should focus on developing targeted antiinflammatory therapies that modulate metal-induced inflammatory pathways while minimizing systemic side effects. Combining antiinflammatory drugs with antioxidants, chelators, and neuroprotective agents may offer a more comprehensive approach to mitigating heavy metal–induced neurodegeneration and improving long-term neurological outcomes.

24.4.3 Polyphenols and flavonoids as natural neuroprotectants

Polyphenols and flavonoids are natural bioactive compounds found abundantly in fruits, vegetables, tea, cocoa, and medicinal plants, and they have gained significant attention as potential neuroprotectants against heavy metal–induced neurodegeneration. These compounds exert multiple beneficial effects, including antioxidant, antiinflammatory, and metal-chelating properties. They are promising candidates for mitigating the neurological damage caused by toxic metals such as lead, mercury, cadmium, and arsenic. Heavy metal exposure disrupts neuronal homeostasis by generating oxidative stress, triggering chronic neuroinflammation, impairing mitochondrial function, and inducing apoptosis. Polyphenols and flavonoids counteract these mechanisms through their ability to scavenge ROS, modulate signaling pathways, and enhance endogenous detoxification systems. Their neuroprotective effects have been extensively studied in models of neurodegenerative diseases such as AD, PD, and Huntington's disease (HD), conditions that are exacerbated by environmental toxins, including heavy metals (Briffa et al., 2020; Chakraborty & Aschner, 2012; Dong et al., 2024; Kumar et al., 2025; Raj et al., 2021).

One of the key mechanisms by which polyphenols and flavonoids exert their neuroprotective effects is by reducing oxidative stress. Heavy metals catalyze the production of free radicals, leading to lipid peroxidation, DNA damage, and protein oxidation. Polyphenols such as resveratrol, quercetin, curcumin, and epigallocatechin gallate (EGCG) act as potent antioxidants by donating electrons to neutralize free radicals, thereby preventing cellular damage. Additionally, these compounds enhance the expression and activity of endogenous antioxidant enzymes such as SOD, catalase (CAT), and GPx, further strengthening the cellular defense against oxidative damage. Studies have shown that curcumin, a polyphenol derived from turmeric, can significantly reduce mercury-induced oxidative stress in neural tissues by upregulating antioxidant defenses and protecting mitochondria from dysfunction. Similarly, quercetin, a flavonoid found in onions, apples, and citrus fruits, has been demonstrated to protect against lead-induced neurotoxicity by reducing lipid peroxidation and restoring glutathione levels.

Another critical aspect of polyphenols and flavonoids in neuroprotection is their antiinflammatory properties. Heavy metal exposure activates microglia, the resident immune cells of the CNS, leading to the excessive production of proinflammatory cytokines such as TNF-α, IL-1β, and IL-6. Chronic neuroinflammation exacerbates neuronal damage and contributes to cognitive decline. Polyphenols such as resveratrol and EGCG inhibit neuroinflammatory pathways by suppressing NF-κB signaling, reducing the release of inflammatory mediators, and modulating microglial activation (Bjørklund et al., 2024; Chakraborty & Aschner, 2012; Green & Planchart, 2018; Reiter et al., 2024; Wydrych et al., 2025). Resveratrol, found in red grapes and berries, has been shown to reduce cadmium-induced neuroinflammation by inhibiting NF-κB and attenuating oxidative stress. Similarly, EGCG, the primary catechin in green tea, has demonstrated the ability to protect dopaminergic neurons from arsenic-induced neurotoxicity by reducing inflammation and preserving mitochondrial integrity.

Beyond their antioxidant and antiinflammatory properties, polyphenols and flavonoids also play a role in detoxification through their metal-chelating abilities. Certain polyphenols possess hydroxyl and catechol functional groups that allow them to bind heavy metals, forming stable complexes that facilitate metal excretion and reduce bioavailability in the CNS. Quercetin and curcumin, for example, have been shown to chelate lead and mercury, reducing their accumulation in neural tissues and mitigating their toxic effects. This chelating property makes polyphenols valuable adjuncts to conventional chelation therapy, offering a natural, safer alternative with fewer side effects.

Furthermore, polyphenols and flavonoids contribute to neurogenesis, synaptic plasticity, and cognitive enhancement. These compounds have been reported to activate signaling pathways such as brain-derived neurotrophic factor, which plays a crucial role in neuronal survival, differentiation, and synaptic connectivity. Flavonoids such as fisetin and luteolin have demonstrated the ability to promote neuroprotection and enhance memory function in models of heavy metal neurotoxicity. Their ability to restore cognitive function makes them promising candidates for preventing neurodegenerative diseases linked to environmental toxins (Briffa et al., 2020; Chakraborty & Aschner, 2012; Kumar et al., 2025; Lasisi-Sholola et al., 2024; Raj et al., 2021).

Polyphenols and flavonoids are potent natural neuroprotectants against heavy metal–induced toxicity through their antioxidant, antiinflammatory, and metal-chelating properties. Their ability to mitigate oxidative damage, suppress neuroinflammation, and enhance neuronal survival highlights their potential as therapeutic agents in managing heavy metal–related neurodegenerative conditions. Future research should optimize their bioavailability, explore synergistic effects with conventional treatments, and develop targeted delivery systems to maximize their neuroprotective benefits in clinical settings (Table 24.4).

24.5 Neuroprotective agents and emerging therapies

24.5.1 *N*-Methyl-D-aspartate receptor antagonists and neurotransmitter modulators

NMDA receptor antagonists and neurotransmitter modulators are crucial in mitigating the neurodegenerative effects of heavy metal exposure. Heavy metals such as lead, mercury, cadmium, and arsenic disrupt neurotransmission by interfering with glutamate signaling, which leads to excitotoxicity, oxidative stress, and neuronal damage. The NMDA receptor, a subtype of glutamate receptor, is particularly susceptible to dysfunction in heavy metals, as these toxic elements can induce excessive receptor activation, resulting in calcium overload, mitochondrial dysfunction, and neuronal apoptosis. NMDA receptor antagonists help counteract these effects by regulating glutamate-mediated excitotoxicity, thereby protecting neurons from further damage and preserving cognitive function. One of the most well-known NMDA receptor antagonists is memantine, an FDA-approved drug primarily used for treating AD. Memantine is a low-affinity, noncompetitive antagonist that blocks excessive NMDA receptor activity without inhibiting normal physiological function. This selective inhibition is crucial because NMDA receptors play an essential role in synaptic plasticity, learning, and memory (Chakraborty & Aschner, 2012; Dooka et al., 2024; Grasso et al., 2017; Reiter et al., 2024; Wydrych et al., 2025). In heavy metal neurotoxicity, memantine has been shown to mitigate lead-induced excitotoxicity by reducing intracellular calcium accumulation and preventing oxidative stress-mediated damage. Similarly, other NMDA receptor antagonists, such as ketamine and dextromethorphan, have demonstrated neuroprotective effects against heavy metal–induced neuronal dysfunction by reducing glutamate excitotoxicity and modulating downstream signaling pathways involved in cell survival. Apart from NMDA receptor antagonists, neurotransmitter modulators, including GABA agonists and dopamine stabilizers, are vital in counteracting heavy metal–induced neurotoxicity. Heavy metals disrupt the balance between excitatory and inhibitory neurotransmission, often leading to excessive neuronal firing and excitotoxicity. GABAergic neurotransmission, which provides inhibitory control in the CNS, is significantly impaired by heavy metal exposure, leading to hyperexcitability and neurodegeneration. Benzodiazepines, which enhance GABAergic activity by increasing the affinity of GABA for its receptors, can help restore inhibitory tone and reduce neuronal hyperactivity caused by heavy metals. Similarly, baclofen, a GABA-B receptor agonist, has shown potential in mitigating heavy metal–induced neurotoxicity by enhancing inhibitory neurotransmission and protecting against excitotoxic damage. Dopaminergic dysfunction is another key consequence of heavy metal exposure, particularly in cases of mercury and manganese toxicity, which are known to impair dopamine synthesis, transport, and receptor signaling (Dooka et al., 2024; Famurewa et al., 2022; Green & Planchart, 2018; Husain & Leveckis, 2024; Khandayataray & Murthy, 2024). This disruption contributes to motor deficits, cognitive impairment, and psychiatric symptoms observed in individuals exposed to high levels of these metals. Dopamine modulators, such as levodopa and dopamine agonizts, like pramipexole and ropinirole, can help restore dopaminergic function and alleviate symptoms associated with heavy metal–induced neurodegeneration. Additionally, serotonin and acetylcholine modulators, including selective serotonin reuptake inhibitors and acetylcholinesterase inhibitors, may offer neuroprotective benefits by modulating neurotransmitter systems affected by heavy metal toxicity. Overall, NMDA receptor antagonists and neurotransmitter modulators provide critical therapeutic

TABLE 24.4 Gene-based interventions for heavy metal–induced neurodegeneration.

Intervention type	Mechanism of action	Targeted genes/pathways	Therapeutic potential	Current research status
CRISPR-Cas9 gene editing	Directly modifies or repairs mutated genes involved in neurotoxicity and oxidative stress.	SOD1 (superoxide dismutase 1), PARK2 (Parkin), Nrf2 (antioxidant response regulator)	Corrects genetic susceptibility to metal-induced oxidative damage and enhances detoxification.	Yet it faces significant challenges in delivery and safety, underscoring the need for collaborative efforts and innovative solutions.
RNA interference (siRNA, miRNA therapy)	Silences toxic gene expression and modulates epigenetic responses to heavy metal exposure.	BACE1 (β-secretase in Alzheimer's), TNF-α (neuroinflammatory cytokine), APP (amyloid precursor protein)	Reduces neuroinflammation, prevents neurotoxic protein accumulation, protects against neuronal loss	These are underway, but the need for effective delivery systems is pressing, underscoring the urgency and importance of this issue.
Epigenetic modulators (HDAC inhibitors, DNA methylation modifiers)	Reverses heavy metal–induced epigenetic changes, restores gene expression balance	DNMT1 (DNA methyltransferase), HDAC (histone deacetylases), MeCP2 (methyl-CpG-binding protein)	Prevents aberrant gene silencing, enhances neuroprotection, promotes neuronal repair	Investigational phase, the potential for combination therapies
Gene therapy with viral vectors (AAV, lentivirus)	Delivers neuroprotective genes into affected neurons to restore function	GDNF (glial cell line-derived neurotrophic factor), BDNF (brain-derived neurotrophic factor), NGF (nerve growth factor)	Promotes neuronal survival, enhances regeneration, reduces neurodegeneration	Clinical trials ongoing for neurodegenerative diseases
Mitochondrial gene editing (mtDNA repair, mRNA therapy)	Repairs mitochondrial DNA mutations caused by heavy metals enhances energy metabolism	PGC-1α (mitochondrial biogenesis regulator), ND1 (mitochondrial NADH dehydrogenase)	Improves ATP production, reduces mitochondrial dysfunction, protects neurons	Early-stage research is promising but faces delivery challenges
Gene-activated scaffolds (biomaterial-based gene delivery)	Uses biomaterial platforms to deliver neuroprotective genes to affected brain regions	VEGF (vascular endothelial growth factor), IGF-1 (insulin-like growth factor)	Supports brain tissue repair, enhances neuronal survival, promotes angiogenesis	Experimental stage, potential for regenerative medicine

Source: Modified from Dong et al. (2024), Grasso et al. (2017), Kumar et al. (2025), Pocivavsek et al. (2024), and Tundo et al. (2023).

strategies for mitigating the neurological effects of heavy metal exposure. These agents offer potential neuroprotection and symptomatic relief in heavy metal–induced neurodegeneration by preventing excitotoxicity, restoring inhibitory balance, and stabilising neurotransmitter function. Future research should focus on optimizing these treatments, exploring their synergistic effects with antioxidants and chelators, and developing targeted drug delivery systems to enhance their efficacy and minimize side effects.

24.5.2 Mitochondrial-targeted therapies (coenzyme Q10, MitoQ)

Mitochondrial-targeted therapies, particularly those involving coenzyme Q10 (CoQ10) and its derivative MitoQ, have emerged as promising strategies for mitigating neurodegeneration caused by heavy metal exposure. Heavy metals such as lead, mercury, cadmium, and arsenic induce significant mitochondrial dysfunction, leading to impaired energy production, excessive ROS generation, and the activation of apoptotic pathways that contribute to neuronal damage (Briffa et al., 2020; Dong et al., 2024; Famurewa et al., 2022; Khandayataray & Murthy, 2024; Wydrych et al., 2025). Mitochondria play a crucial role in cellular metabolism, ATP production, and the maintenance of redox homeostasis, making them primary targets for toxicity. Since neurons are highly dependent on mitochondrial function due to their high energy demands, mitochondrial impairment caused by heavy metals accelerates neurodegenerative processes, resulting in cognitive deficits, motor dysfunction, and neuroinflammation. CoQ10 and MitoQ, as mitochondrial-targeted antioxidants, offer neuroprotection by restoring mitochondrial function, reducing oxidative stress, and preventing programmed cell death in affected neurons (Bjørklund et al., 2024; Briffa et al., 2020; Husain & Leveckis, 2024; Khandayataray & Murthy, 2024; Raj et al., 2021).

Coenzyme Q10, a naturally occurring lipid-soluble molecule in the inner mitochondrial membrane, plays an essential role in the ETC, facilitating electron transfer and ATP synthesis. Heavy metals disrupt this process by binding to mitochondrial complexes, leading to electron leakage and excessive superoxide production. CoQ10, as a potent antioxidant, helps neutralize these free radicals, thereby protecting mitochondria from oxidative damage. Studies have shown that CoQ10 supplementation can mitigate lead-induced mitochondrial dysfunction by preserving ATP levels, stabilizing mitochondrial membrane potential, and reducing lipid peroxidation in neural tissues. Additionally, CoQ10 has demonstrated the ability to protect against mercury toxicity by inhibiting mPTP opening, a key event in apoptosis initiation. By preventing mPTP opening, CoQ10 preserves mitochondrial integrity and prevents the release of proapoptotic factors such as cytochrome *c*, thereby reducing neuronal cell death (Kciuk et al., 2024; Matés et al., 2010; Oteiza et al., 2004; Panda et al., 2023; Tahir et al., 2024).

Despite its efficacy, CoQ10 has limitations regarding bioavailability and its ability to target mitochondria in neuronal tissues effectively. MitoQ, a mitochondria-targeted derivative of CoQ10, has been developed to overcome these limitations. MitoQ consists of a CoQ10 moiety linked to a lipophilic triphenylphosphonium cation, facilitating its selective accumulation within mitochondria. This targeted delivery enhances its antioxidant capacity, allowing it to more effectively neutralize ROS and prevent mitochondrial dysfunction. MitoQ has been shown to protect against arsenic-induced neurotoxicity by improving mitochondrial respiration, enhancing ATP production, and reducing oxidative stress-induced damage in neuronal cells (Costa, 2017; Eddie-Amadi et al., 2022; Liu et al., 2025; Pan & Umapathy, 2024; Panda et al., 2023). Furthermore, MitoQ's ability to mitigate cadmium-induced mitochondrial damage by preserving mitochondrial biogenesis and preventing excessive calcium influx highlights its potential as a neuroprotective agent.

Beyond their antioxidant properties, CoQ10 and MitoQ modulate key signaling pathways involved in mitochondrial health and neuroprotection. They activate nuclear factor erythroid 2-related factor 2 (Nrf2), a transcription factor that upregulates the expression of antioxidant enzymes such as SOD and GPx. Additionally, they modulate AMP-activated protein kinase and peroxisome proliferator-activated receptor gamma coactivator-1 alpha (PGC-1α), which regulate mitochondrial biogenesis and energy metabolism. By enhancing these pathways, mitochondrial-targeted therapies promote neuronal survival and functional recovery in heavy metal–induced neurodegeneration (Abdolmaleki et al., 2025; Camilleri & Blundell, 2024; Carter & Blizard, 2016; Lawes et al., 2020; Oteiza et al., 2004).

Overall, mitochondrial-targeted therapies such as CoQ10 and MitoQ offer a promising approach for counteracting the detrimental effects of heavy metals on neuronal health. Their ability to restore mitochondrial function, reduce oxidative stress, and inhibit apoptosis makes them valuable candidates for neuroprotection. Future research should focus on optimizing their formulations, exploring combination therapies with other neuroprotective agents, and conducting clinical trials to establish their efficacy in treating heavy metal–induced neurodegenerative disorders.

24.5.3 Nanomedicine approaches for heavy metal detoxification

Nanomedicine has emerged as a revolutionary approach for heavy metal detoxification, offering novel strategies for reducing toxicity and mitigating neurodegenerative effects associated with heavy metal exposure. Heavy metals such as lead, mercury, cadmium, and arsenic accumulate in the brain and other organs, causing oxidative stress, mitochondrial dysfunction, neuroinflammation, and DNA damage. Traditional detoxification strategies, including chelation therapy, have limitations such as low specificity, potential side effects, and incomplete removal of metals (Bi et al., 2022; Liu et al., 2025; Luan et al., 2024; Tahir et al., 2024). Nanomedicine, which utilizes engineered nanoparticles (NPs) with unique physicochemical properties, provides an innovative solution by enhancing the specificity, efficiency, and biocompatibility of metal detoxification processes. Different nanomaterials, including metal-based NPs, carbon-based nanostructures, polymeric NPs, and lipid-based nanocarriers, have been explored for their ability to bind, neutralize, and facilitate the excretion of heavy metals while minimizing toxicity.

One of the most promising approaches in nanomedicine for heavy metal detoxification involves using metal-based NPs such as iron oxide (Fe_3O_4) and gold (Au) NPs. Due to their strong magnetic properties and high surface reactivity, iron oxide NPs have been extensively investigated for their ability to chelate and remove heavy metals from biological systems. These NPs can be functionalised with chelating agents such as EDTA or diethylenetriaminepentaacetic acid, enhancing their ability to selectively bind to heavy metals and facilitate their excretion via the kidneys or liver. Similarly, gold NPs have been explored for their unique ability to interact with toxic metal ions through thiol and amine functionalisation, reducing the bioavailability of toxic metals and preventing their accumulation in the nervous system (Abdolmaleki et al., 2025; Chakraborty & Aschner, 2012; Grasso et al., 2017; Pocivavsek et al., 2024; Spiers et al., 2022).

Carbon-based nanostructures, including graphene oxide (GO) and carbon nanotubes, offer another promising avenue for heavy metal detoxification. These materials have a high surface area and excellent adsorption capacity, allowing them to bind heavy metal ions efficiently. Functionalized graphene oxide, in particular, has been shown to remove lead and cadmium from biological systems by forming stable complexes that can be excreted safely. Additionally, carbon-based nanostructures have intrinsic antioxidant properties, which help neutralize ROS and reduce oxidative damage associated with heavy metal toxicity (Bjørklund et al., 2024; Hussien et al., 2018; Lasisi-Sholola et al., 2024; Raj et al., 2021). The combination of heavy metal binding and antioxidative action makes these materials highly effective in mitigating neurodegenerative effects.

Polymeric NPs and lipid-based nanocarriers are also being explored for their potential in targeted heavy metal detoxification. Polymeric nanocarriers, such as chitosan and poly(lactic-*co*-glycolic acid) NPs, have been designed to encapsulate chelating agents or antioxidant compounds, improving their bioavailability and controlled release. These NPs enhance the efficacy of chelation therapy while reducing systemic toxicity. Lipid-based nanocarriers, including liposomes and solid lipid NPs, offer another promising strategy by encapsulating bioactive compounds such as glutathione, NAC, or curcumin, which counteract heavy metal–induced toxicity through chelation and antioxidative mechanisms.

Overall, nanomedicine-based approaches provide a highly efficient and targeted strategy for heavy metal detoxification. By improving specificity, bioavailability, and safety, nanotechnology holds great promise for developing next-generation therapeutics that minimize heavy metal–induced neurotoxicity. Future research should optimize nanomaterial formulations, evaluate their long-term biocompatibility, and conduct clinical trials to translate these promising approaches into viable treatments for heavy metal poisoning and associated neurodegenerative disorders.

24.5.4 Gene therapy and epigenetic interventions

Gene therapy and epigenetic interventions are emerging promising strategies for mitigating the neurotoxic effects of heavy metal exposure and preventing neurodegeneration. Heavy metals such as lead, mercury, cadmium, and arsenic disrupt regular gene expression, induce DNA damage, and alter epigenetic mechanisms such as DNA methylation, histone modifications, and microRNA (miRNA) regulation. These alterations contribute to neuronal dysfunction, oxidative stress, neuroinflammation, and apoptosis, ultimately leading to cognitive impairment and neurodegenerative diseases such as AD, PD, and ALS. Gene therapy and epigenetic interventions aim to restore normal gene function, counteract heavy metal–induced dysregulation, and promote neuronal survival by targeting specific molecular pathways associated with detoxification, antioxidant defense, and neuroprotection (Bjørklund et al., 2024; Lasisi-Sholola et al., 2024; López-Granero et al., 2023; Raj et al., 2021; Tundo et al., 2023).

Gene therapy approaches focus on delivering therapeutic genes or silencing toxic genes to restore normal cellular functions in neurons affected by heavy metal toxicity. One of the most promising strategies involves the use of viral and nonviral vectors to introduce genes encoding antioxidant enzymes such as SOD, catalase (CAT), and GPx. These enzymes play a critical role in scavenging ROS generated by heavy metals, thereby preventing oxidative damage to neuronal cells. Additionally, gene therapy targeting mitochondrial function has gained attention, with genes encoding peroxisome proliferator-activated receptor gamma coactivator-1 alpha (PGC-1α) and nuclear factor erythroid 2-related factor 2 (Nrf2) being introduced to enhance mitochondrial biogenesis and antioxidant responses. Another gene therapy strategy involves silencing toxic metal transporter genes that facilitate the entry of heavy metals into neurons. For instance, small interfering RNA or CRISPR-Cas9 gene-editing technology can be used to inhibit the expression of divalent metal transporter 1 (DMT1), a key transporter involved in lead and iron uptake, thereby reducing heavy metal accumulation in neural tissues (Dong et al., 2024; Husain & Leveckis, 2024; Khandayataray & Murthy, 2024; Kumar et al., 2025; Reiter et al., 2024).

On the other hand, epigenetic interventions focus on reversing heavy metal–induced changes in DNA methylation, histone modifications, and miRNA expression that contribute to neurodegeneration. Heavy metals interfere with DNA methylation patterns, leading to either hypermethylation of neuroprotective genes or hypomethylation of genes associated with inflammation and apoptosis. DNMT inhibitors and HDAC inhibitors are being explored as potential therapeutic agents to restore normal epigenetic regulation. HDAC inhibitors such as valproic acid and trichostatin A can upregulate gene expression in neuronal survival, synaptic plasticity, and detoxification pathways, counteracting heavy metal–induced neurotoxicity. Additionally, miRNA-based therapies have gained interest due to their ability to regulate multiple genes simultaneously. miRNAs such as miR-34a and miR-146a, which are dysregulated in response to heavy metal exposure, play crucial roles in inflammation and apoptosis. Restoring their regular expression using miRNA mimics or inhibitors can provide neuroprotection and enhance detoxification mechanisms (Grasso et al., 2017; Hussien et al., 2018; Kumar et al., 2025; Lasisi-Sholola et al., 2024; Tundo et al., 2023).

Overall, gene therapy and epigenetic interventions offer targeted and personalized treatment strategies for mitigating the effects of heavy metal toxicity on the nervous system. By restoring regular gene expression, enhancing antioxidant defenses, and modulating epigenetic pathways, these approaches hold great potential for preventing and treating serious metal-induced neurodegenerative diseases. Future research should focus on optimizing gene delivery methods, improving the safety and specificity of epigenetic drugs, and conducting clinical trials to validate their efficacy in human populations exposed to toxic metals (Table 24.5).

24.6 Future perspectives and challenges

Advances in personalized medicine have revolutionised the approach to neurodegenerative diseases, offering tailored therapeutic strategies based on an individual's genetic, molecular, and environmental factors. Neurodegenerative diseases such as AD, PD, ALS, and HD exhibit complex pathophysiology influenced by genetic predisposition, epigenetic modifications, environmental toxins, and lifestyle factors. With increasing knowledge of patient-specific biomarkers, genetic risk factors, and disease progression mechanisms, personalized medicine enables precision diagnostics, early interventions, and optimized treatments for individuals at risk or in the early stages of neurodegeneration. A breakthrough in personalized medicine for neurodegenerative diseases has been the advent of genomic sequencing and polygenic risk scoring, which help identify individuals susceptible to heavy metal–induced neurotoxicity and neurodegeneration. Whole-genome and whole-exome sequencing have provided insights into genetic mutations associated with disease susceptibility, including mutations in amyloid precursor protein, presenilin 1 (PSEN1), and presenilin 2 (PSEN2) in AD, or mutations in leucine-rich repeat kinase 2 (LRRK2) and glucocerebrosidase in PD. By understanding these genetic variations, clinicians can tailor interventions based on a patient's specific genetic profile, enabling a shift from a generalized treatment approach to a more targeted and effective strategy.

Another key component of personalized medicine is biomarker-based diagnostics, which allow for early detection and monitoring of neurodegenerative diseases. Biomarkers such as cerebrospinal fluid levels of amyloid-beta (Aβ), tau proteins, and neurofilament light chain (NfL) provide critical insights into disease progression. Advances in neuroimaging techniques, including positron emission tomography and functional magnetic resonance imaging, have further enhanced the ability to visualize disease-related changes at a molecular level. Additionally, liquid biopsy techniques using blood-based biomarkers, extracellular vesicles, and microRNAs (miRNAs) have emerged as promising tools for noninvasive monitoring of neurodegeneration. These advances allow for individualized treatment regimens based on a patient's disease progression stage, thus improving therapeutic outcomes. Personalized medicine also drives the development of novel therapeutic strategies such as gene therapy, RNA-based therapies, and patient-specific stem cell treatments. Gene therapy approaches targeting neurodegeneration aim to correct genetic mutations or modulate the expression of neuroprotective genes. For example, antisense oligonucleotides have been used to reduce the expression of toxic proteins in HD and ALS. CRISPR-Cas9 gene editing has also shown promise in correcting mutations associated with neurodegeneration. Additionally, induced pluripotent stem cells (iPSCs) derived from patients' cells are being explored as a potential regenerative therapy for replacing damaged neurons and restoring neural function.

Given the multifactorial nature of neurodegenerative diseases, multitarget therapies have gained increasing attention as a practical treatment approach. Traditional single-target drugs have shown limited success due to the complex interplay of molecular mechanisms underlying neurodegeneration, including oxidative stress, neuroinflammation,

TABLE 24.5 Future areas of exploration in heavy metal–induced neurodegeneration.

Research gap	Current limitations	Future areas of exploration
Mechanisms of metal-induced neurodegeneration	Limited understanding of metal-specific neurotoxic pathways and their interactions	Advanced multiomics approaches (proteomics, transcriptomics, metabolomics) to map neurotoxic pathways.
Long-term effects of low-dose metal exposure	Lack of longitudinal studies on chronic low-dose exposure and cumulative toxicity	Epidemiological studies tracking long-term neurodegenerative risks in exposed populations
Personalized medicine approaches	Current treatments follow a "one-size-fits-all" approach without considering genetic variability.	Development of biomarker-based precision medicine for individualized treatment strategies
Chelation therapy optimization	Risks of nonspecific metal removal, side effects, and BBB penetration challenges	Development of brain-targeted chelators with improved specificity and safety profiles
Role of gut microbiota in metal detoxification	Insufficient research on gut–brain–microbiome interactions in heavy metal toxicity	Investigation of microbiome-targeted therapies for neuroprotection and metal detoxification
Epigenetic and transgenerational effects	Limited studies on how heavy metal exposure alters epigenetic programming across generations	Exploration of heritable epigenetic changes and their impact on neurodegenerative disease susceptibility
Nanomedicine and targeted drug delivery	Challenges in nanoparticle safety, BBB penetration, and long-term toxicity	Development of biocompatible, targeted nanocarriers for drug delivery and metal sequestration
Gene and cell-based therapies	Ethical concerns, technical limitations, and delivery challenges in neurodegeneration	Advancements in CRISPR, stem cell therapy, and regenerative medicine for neuroprotection.
Synergistic multitarget therapies	Current treatments focus on single pathways rather than holistic disease mechanisms	Development of combination therapies integrating antioxidants, antiinflammatory agents, and neuroprotective drugs
Regulatory and policy frameworks	Inconsistent regulations on heavy metal exposure limits and lack of standardised treatment protocols	Establishment of global policies for metal exposure mitigation and neuroprotective guidelines

Source: Modified from Chakraborty and Aschner (2012), Dooka et al. (2024), Husain and Leveckis (2024), Raj et al. (2021), and Reiter et al. (2024).

mitochondrial dysfunction, and protein misfolding. Multitarget therapies, which involve a combination of drugs or multifunctional molecules, aim to simultaneously modulate multiple pathogenic pathways, thereby improving efficacy and reducing disease progression. One promising approach combines antioxidant therapy with antiinflammatory agents to address oxidative stress and chronic neuroinflammation, which play central roles in neurodegeneration. For instance, coenzyme Q10 (CoQ10) and NAC are being explored in combination with NSAIDs or natural antiinflammatory compounds such as curcumin and resveratrol. These combinations help reduce free radical damage while suppressing microglial activation and proinflammatory cytokine release.

Another multitarget approach focuses on mitochondrial-targeted therapies, which aim to restore mitochondrial function while simultaneously addressing neuroinflammation and oxidative stress. Mitochondrial dysfunction is a hallmark of many neurodegenerative diseases, and therapies such as MitoQ (a mitochondria-targeted antioxidant), nicotinamide riboside, and pyruvate supplementation have shown potential in improving mitochondrial energy production, reducing oxidative stress, and preventing neuronal death. Additionally, drugs targeting excitotoxicity and neurotransmitter imbalance, such as NMDA receptor antagonists (memantine) in combination with cholinesterase inhibitors (donepezil, rivastigmine), have been investigated for their synergistic neuroprotective effects in AD.

Despite significant advances in personalized medicine and multitarget therapies, there remain considerable gaps in research and areas for further exploration. One major challenge is the heterogeneity of neurodegenerative diseases, as different patients exhibit varying genetic, epigenetic, and environmental risk factors. This heterogeneity complicates the development of universal treatment strategies. It highlights the need for a more profound stratification of patient subgroups based on biomarker profiles, genetic risk factors, and disease phenotypes. Additionally, while genetic and epigenetic markers have provided valuable insights, the role of gut microbiota in neurodegenerative diseases remains underexplored. Emerging evidence suggests that gut microbiota dysbiosis contributes to neuroinflammation and neurodegeneration through the gut–brain axis, making microbiome-targeted interventions a promising avenue for future research.

Another key gap in research is the limited availability of effective drug-delivery systems for targeting neurodegeneration. The BBB presents a significant challenge in delivering therapeutic agents to the brain, as many drugs fail to cross this barrier efficiently. Advances in nanomedicine, including NP-based drug carriers and lipid-based nanocarriers, have shown promise in improving drug delivery to the CNS. However, further research is needed to optimize these delivery systems and evaluate their long-term safety and efficacy. Additionally, while gene and RNA-based therapies hold great potential, challenges such as immune responses, off-target effects, and long-term gene expression stability must be addressed to ensure clinical translation. Another pressing challenge is the need for large-scale, longitudinal studies to validate the efficacy of personalized medicine approaches. Many experimental therapies have shown promise in preclinical models but have failed to translate into successful clinical trials. This discrepancy highlights the need for improved preclinical models replicating human neurodegenerative disease pathology. Using patient-derived iPSC models and organoids offers a promising approach to studying disease mechanisms and testing personalized treatments in a more physiologically relevant context.

Furthermore, integrating artificial intelligence (AI) and machine learning (ML) into neurodegenerative disease research presents an exciting opportunity for advancing personalized medicine. AI-driven algorithms can analyze vast amounts of genomic, proteomic, and imaging data to identify novel biomarkers, predict disease progression, and optimize treatment regimens. AI-based drug discovery platforms can also accelerate the identification of multi-target therapeutic compounds by screening large chemical libraries and predicting drug interactions. However, challenges such as data standardization, ethical considerations, and the need for high-quality datasets must be addressed to fully harness AI's potential in neurodegenerative disease research.

Personalized medicine and multitarget therapies represent the future of neurodegenerative disease treatment, offering tailored and more effective interventions based on an individual's genetic and molecular profile. Advances in genomic sequencing, biomarker identification, and novel therapeutic strategies such as gene therapy, stem cell therapy, and mitochondrial-targeted interventions have significantly improved our understanding of neurodegeneration and opened new avenues for treatment. However, significant gaps remain in research, including the need for better patient stratification, improved drug delivery systems, microbiome-targeted therapies, and large-scale clinical validation of experimental treatments. Addressing these challenges requires a multidisciplinary approach combining molecular biology, pharmacology, AI, and clinical research to develop effective and personalized treatment strategies for neurodegenerative diseases. With continued advancements in precision medicine, neurodegenerative diseases may one day become manageable conditions, significantly improving the quality of life for affected individuals and reducing the global burden of these devastating disorders.

24.7 Conclusion

The key findings from this comprehensive exploration of neurodegenerative diseases linked to heavy metal exposure underscore the significant role of environmental toxins, oxidative stress, neuroinflammation, mitochondrial dysfunction, and genetic susceptibility in the pathogenesis of conditions such as AD, PD, ALS, and HD. Heavy metals, including lead, mercury, cadmium, and arsenic, contribute to neuronal damage through multiple mechanisms, including increased ROS production, disruption of neurotransmitter systems, DNA damage, and epigenetic modifications. The interplay between neuroinflammation and microglial activation further exacerbates disease progression, while mitochondrial dysfunction impairs cellular energy production, ultimately leading to neurodegeneration. Given the complexity of these processes, a multi-targeted therapeutic approach is necessary, integrating chelation therapy, antioxidant supplementation, antiinflammatory drugs, mitochondrial-targeted therapies, and gene-based interventions to mitigate the harmful effects of heavy metal–induced neurotoxicity. Additionally, emerging fields such as nanomedicine and personalized medicine offer promising avenues for more effective treatment strategies by enhancing drug delivery, improving patient-specific treatment regimens, and addressing the genetic and epigenetic variability observed in neurodegenerative disorders.

The implications for clinical practice and policy are far-reaching, emphasizing the urgent need for stricter environmental regulations to reduce heavy metal exposure, particularly in industrial areas and regions with high pollution levels. Public health policies should focus on regularly screening and monitoring heavy metal levels in at-risk populations, especially in occupational settings where workers are exposed to toxic metals. In clinical practice, early detection of neurodegenerative diseases through advanced biomarker-based diagnostics, neuroimaging techniques, and liquid biopsies could enable timely intervention and improve patient outcomes. Personalized medicine approaches, including genetic testing and biomarker profiling, should be integrated into routine clinical care to guide treatment decisions and enhance therapeutic efficacy. Healthcare practitioners should also be educated on the risks of heavy metal exposure and the potential benefits and limitations of interventions such as chelation therapy, antioxidant supplementation, and neuroprotective strategies. Furthermore, multidisciplinary collaborations between neurologists, toxicologists, geneticists, and environmental scientists are essential for developing comprehensive prevention and treatment strategies tailored to individual patient needs. At the policy level, investment in research funding is critical for advancing our understanding of neurodegenerative diseases' molecular mechanisms and accelerating the development of innovative therapeutic approaches.

Future research directions should focus on addressing the gaps in our current understanding of heavy metal–induced neurotoxicity and its long-term effects on brain health. While substantial progress has been made in elucidating the molecular pathways involved in neurodegeneration, further studies are needed to identify novel biomarkers to improve early disease detection and prognostic evaluation. Additionally, research on the gut–brain axis and the role of microbiota in neurodegenerative diseases presents an exciting frontier, as emerging evidence suggests that gut dysbiosis contributes to neuroinflammation and neuronal dysfunction. Advances in gene therapy, including CRISPR-based interventions, hold promise for correcting genetic mutations associated with neurodegeneration. Still, challenges such as off-target effects and immune responses must be overcome to ensure clinical applicability. Developing more effective drug delivery systems, including NP-based carriers and BBB-penetrating molecules, requires further exploration. Moreover, large-scale longitudinal studies are necessary to validate the efficacy of multitarget therapeutic strategies and personalized medicine approaches in diverse patient populations. As AI and ML continue to advance, their integration into neurodegenerative disease research could significantly enhance drug discovery, biomarker identification, and treatment optimization. Ultimately, a multidisciplinary, patient-centric approach combining cutting-edge scientific innovations with public health initiatives will be key to developing effective prevention and treatment strategies, improving patient quality of life, and reducing the global burden of neurodegenerative diseases.

Acknowledgment

The author used AI to enhance language and readability, with final editing done by a human.

References

Abdolmaleki, S., Aliabadi, A., & Khaksar, S. (2025). Bridging the gap between theory and treatment: Transition metal complexes as successful candidates in medicine. *Coordination Chemistry Reviews, 531*, 216477. Retrieved from https://doi.org/10.1016/j.ccr.2025.216477.

Anyanwu, B. O., Orish, C. N., Ezejiofor, A. N., Nwaogazie, I. L., & Orisakwe, O. E. (2020). Neuroprotective effect of Costus afer on low dose heavy metal mixture (lead, cadmium and mercury) induced neurotoxicity via antioxidant, anti-inflammatory activities. *Toxicology Reports, 7*, 1032–1038. Retrieved from https://doi.org/10.1016/j.toxrep.2020.08.008.

Arun, K. B., Madhavan, A., Sindhu, R., Emmanual, S., Binod, P., Pugazhendhi, A., ... Pandey, A. (2021). Probiotics and gut microbiome – Prospects and challenges in remediating heavy metal toxicity. *Journal of Hazardous Materials, 420*, 126676. Retrieved from https://doi.org/10.1016/j.jhazmat.2021.126676.

Aschner, M., Skalny, A. V., Paoliello, M. M. B., Tinkova, M. N., Martins, A. C., Santamaria, A., ... Tinkov, A. A. (2024). Retinal toxicity of heavy metals and its involvement in retinal pathology. *Food and Chemical Toxicology, 188*, 114685. Retrieved from https://doi.org/10.1016/j.fct.2024.114685.

Bhattacharya, S., Bose, M., & Munshi, C. (2024). *Paradigms in heavy metal induced autophagy: A toxicological approach*. Elsevier. Retrieved from https://doi.org/10.1016/B978-0-323-95488-4.00136-4.

Bi, S.-S., Talukder, M., Jin, H.-T., Lv, M.-W., Ge, J., Zhang, C., & Li, J.-L. (2022). Nano-selenium alleviates cadmium-induced cerebellar injury by activating metal regulatory transcription factor 1 mediated metal response. *Animal Nutrition, 11*, 402–412. Retrieved from https://doi.org/10.1016/j.aninu.2022.06.021.

Bjørklund, G., Đorđević, A. B., Hamdan, H., Wallace, D. R., & Peana, M. (2024). Metal-induced autoimmunity in neurological disorders: A review of current understanding and future directions. *Autoimmunity Reviews, 23*(3), 103509. Retrieved from https://doi.org/10.1016/j.autrev.2023.103509.

Briffa, J., Sinagra, E., & Blundell, R. (2020). Heavy metal pollution in the environment and their toxicological effects on humans. *Heliyon, 6*(9), e04691. Retrieved from https://doi.org/10.1016/j.heliyon.2020.e04691.

Camilleri, E., & Blundell, R. (2024). A comprehensive review of the phytochemicals, health benefits, pharmacological safety and medicinal prospects of Moringa oleifera. *Heliyon, 10*(6), e27807. Retrieved from https://doi.org/10.1016/j.heliyon.2024.e27807.

Carter, C. J., & Blizard, R. A. (2016). Autism genes are selectively targeted by environmental pollutants including pesticides, heavy metals, bisphenol A, phthalates and many others in food, cosmetics or household products. *Neurochemistry International, 101*, 83–109. Retrieved from https://doi.org/10.1016/j.neuint.2016.10.011.

Chakraborty, S., & Aschner, M. (2012). Altered manganese homeostasis: Implications for BLI-3-dependent dopaminergic neurodegeneration and SKN-1 protection in *C. elegans*. *Journal of Trace Elements in Medicine and Biology, 26*(2), 183–187. Retrieved from https://doi.org/10.1016/j.jtemb.2012.03.011.

Costa, L. G. (2017). Chapter 12 – Developmental exposure to metals and its contribution to age-related neurodegeneration. In A. R. White, M. Aschner, L. G. Costa, & A. I. Bush (Eds.). *Biometals in neurodegenerative diseases: Mechanisms and therapeutics* (pp. 217–229). Academic Press. Retrieved from https://doi.org/10.1016/B978-0-12-804562-6.00012-9.

Dong, Q., Fu, H., & Jiang, H. (2024). The role of exosome-shuttled miRNAs in heavy metal-induced peripheral tissues and neuroinflammation in Alzheimer's disease. *Biomedicine &*

Pharmacotherapy, 176, 116880. Retrieved from https://doi.org/10.1016/j.biopha.2024.116880.

Dooka, D. B., Anyachor, C. P., Orish, C. N., Ezejiofor, A. N., Obasi, C. N., Frazzoli, C., & Orisakwe, O. E. (2024). Understanding the impact of plant-derived food grade substances PDFGS on metal-induced neurodegeneration: A systematic review of preclinical evidence. *Journal of Trace Elements and Minerals, 7*, 100110. Retrieved from https://doi.org/10.1016/j.jtemin.2023.100110.

Eddie-Amadi, B. F., Ezejiofor, A. N., Orish, C. N., Rovira, J., Allison, T. A., & Orisakwe, O. E. (2022). Banana peel ameliorated hepato-renal damage and exerted anti-inflammatory and anti-apoptotic effects in metal mixture mediated hepatic nephropathy by activation of Nrf2/Hmox-1 and inhibition of Nfkb pathway. *Food and Chemical Toxicology, 170*, 113471. Retrieved from https://doi.org/10.1016/j.fct.2022.113471.

Famurewa, A. C., Renu, K., Eladl, M. A., Chakraborty, R., Myakala, H., El-Sherbiny, M., ... Valsala Gopalakrishnan, A. (2022). Hesperidin and hesperetin against heavy metal toxicity: Insight on the molecular mechanism of mitigation. *Biomedicine & Pharmacotherapy, 149*, 112914. Retrieved from https://doi.org/10.1016/j.biopha.2022.112914.

Fu, X., Li, H., Song, L., Cen, M., & Wu, J. (2023). Association of urinary heavy metals co-exposure and adult depression: Modification of physical activity. *Neurotoxicology, 95*, 117–126. Retrieved from https://doi.org/10.1016/j.neuro.2023.01.008.

Grasso, G., Santoro, A. M., Lanza, V., Sbardella, D., Tundo, G. R., Ciaccio, C., ... Milardi, D. (2017). The double faced role of copper in Aβ homeostasis: A survey on the interrelationship between metal dyshomeostasis, UPS functioning and autophagy in neurodegeneration. *Coordination Chemistry Reviews, 347*, 1–22. Retrieved from https://doi.org/10.1016/j.ccr.2017.06.004.

Green, A. J., & Planchart, A. (2018). The neurological toxicity of heavy metals: A fish perspective. *Comparative Biochemistry and Physiology Part C: Toxicology & Pharmacology, 208*, 12–19. Retrieved from https://doi.org/10.1016/j.cbpc.2017.11.008.

Hensley, K., Danekas, A., Farrell, W., Garcia, T., Mehboob, W., & White, M. (2022). At the intersection of sulfur redox chemistry, cellular signal transduction and proteostasis: A useful perspective from which to understand and treat neurodegeneration. *Free Radical Biology and Medicine, 178*, 161–173. Retrieved from https://doi.org/10.1016/j.freeradbiomed.2021.11.039.

Huat, T. J., Camats-Perna, J., Newcombe, E. A., Valmas, N., Kitazawa, M., & Medeiros, R. (2019). Metal toxicity links to Alzheimer's disease and neuroinflammation. *Journal of Molecular Biology, 431*(9), 1843–1868. Retrieved from https://doi.org/10.1016/j.jmb.2019.01.018.

Husain, S., & Leveckis, R. (2024). Pharmacological regulation of HIF-1α, RGC death, and glaucoma. *Current Opinion in Pharmacology, 77*, 102467. Retrieved from https://doi.org/10.1016/j.coph.2024.102467.

Hussien, H. M., Abd-Elmegied, A., Ghareeb, D. A., Hafez, H. S., Ahmed, H. E. A., & El-moneam, N. A. (2018). Neuroprotective effect of berberine against environmental heavy metals-induced neurotoxicity and Alzheimer's-like disease in rats. *Food and Chemical Toxicology, 111*, 432–444. Retrieved from https://doi.org/10.1016/j.fct.2017.11.025.

Kciuk, M., Gielecińska, A., Kałuzińska-Kołat, Ż., Yahya, E. B., & Kontek, R. (2024). Ferroptosis and cuproptosis: Metal-dependent cell death pathways activated in response to classical chemotherapy – Significance for cancer treatment? *Biochimica et Biophysica Acta (BBA) – Reviews on Cancer, 1879*(4), 189124. Retrieved from https://doi.org/10.1016/j.bbcan.2024.189124.

Khan, M. H. A., & Parvez, S. (2015). Hesperidin ameliorates heavy metal induced toxicity mediated by oxidative stress in brain of Wistar rats. *Journal of Trace Elements in Medicine and Biology, 31*, 53–60. Retrieved from https://doi.org/10.1016/j.jtemb.2015.03.002.

Khandayataray, P., & Murthy, M. K. (2024). Dietary interventions in mitigating the impact of environmental pollutants on Alzheimer's disease – A review. *Neuroscience, 563*, 148–166. Retrieved from https://doi.org/10.1016/j.neuroscience.2024.11.020.

Kumar, A., Abhinav, S., Rachamalla, M., Ali, M., Navin, S., Kumar, D., & Ghosh, A. K. (2025). Treatment of heavy metal induced neurotoxicity via advanced drug delivery systems. In P. A. Chawla, R. Loebenberg, K. Dua, V. Parikh, & V. Chawla (Eds.). *Novel drug delivery Systems in the management of CNS disorders* (pp. 393–407). Academic Press. Retrieved from https://doi.org/10.1016/B978-0-443-13474-6.00025-1.

Lasisi-Sholola, A. S., Hammed, S. O., Ajike, R. A., Akhigbe, R. E., & Afolabi, O. A. (2024). Estrogen replacement therapy reverses spatial memory loss and pyramidal cell neurodegeneration in the prefrontal cortex of lead-exposed ovariectomized Wistar rats. *Current Research in Toxicology, 7*, 100200. Retrieved from https://doi.org/10.1016/j.crtox.2024.100200.

Lawes, M., Pinkas, A., Frohlich, B. A., Iroegbu, J. D., Ijomone, O. M., & Aschner, M. (2020). Metal-induced neurotoxicity in a RAGE-expressing *C. elegans* model. *Neurotoxicology, 80*, 71–75. Retrieved from https://doi.org/10.1016/j.neuro.2020.06.013.

Liu, J., Li, T., Zhong, G., Pan, Y., Gao, M., Su, S., ... Shi, Q. (2023). Exploring the therapeutic potential of natural compounds for Alzheimer's disease: Mechanisms of action and pharmacological properties. *Biomedicine & Pharmacotherapy, 166*, 115406. Retrieved from https://doi.org/10.1016/j.biopha.2023.115406.

Liu, J., Zhang, F., & Shi, X. (2025). The role of metal nanocarriers, liposomes and chitosan-based nanoparticles in diabetic retinopathy treatment: A review study. *International Journal of Biological Macromolecules, 291*, 139017. Retrieved from https://doi.org/10.1016/j.ijbiomac.2024.139017.

Luan, M., Feng, Z., Zhu, W., Xing, Y., Ma, X., Zhu, J., ... Jia, Y. (2024). Mechanism of metal ion-induced cell death in gastrointestinal cancer. *Biomedicine & Pharmacotherapy, 174*, 116574. Retrieved from https://doi.org/10.1016/j.biopha.2024.116574.

López-Granero, C., Aschner, M., & Sánchez-Santed, F. (2023). Chapter 22 – Environmental toxicants (OPs and heavy metals) in the diet: What are their repercussions on behavioral/neurological systems? In C. R. Martin, V. B. Patel, & V. R. Preedy (Eds.). Academic Press. Retrieved from https://doi.org/10.1016/B978-0-323-89834-8.00038-6.

Matés, J. M., Segura, J. A., Alonso, F. J., & Márquez, J. (2010). Roles of dioxins and heavy metals in cancer and neurological diseases using ROS-mediated mechanisms. *Free Radical Biology and Medicine, 49*(9), 1328–1341. Retrieved from https://doi.org/10.1016/j.freeradbiomed.2010.07.028.

Moyano, P., Vicente-Zurdo, D., Blázquez-Barbadillo, C., Menéndez, J. C., González, J. F., Rosales-Conrado, N., & Pino, J.del (2022). Neuroprotective mechanisms of multitarget 7-aminophenanthridin-6(5*H*)-one derivatives against metal-induced amyloid proteins generation and aggregation. *Food and Chemical Toxicology, 167*, 113264. Retrieved from https://doi.org/10.1016/j.fct.2022.113264.

Oteiza, P. I., Mackenzie, G. G., & Verstraeten, S. V. (2004). Metals in neurodegeneration: Involvement of oxidants and oxidant-sensitive transcription factors. *Molecular Aspects of Medicine, 25*(1), 103–115. Retrieved from https://doi.org/10.1016/j.mam.2004.02.012.

Pan, I., & Umapathy, S. (2024). Probiotics an emerging therapeutic approach towards gut-brain-axis oriented chronic health issues induced by microplastics: A comprehensive review. *Heliyon, 10*(11), e32004. Retrieved from https://doi.org/10.1016/j.heliyon.2024.e32004.

Panda, S. R., Chaudhari, V. B., Ahmed, S., Kwatra, M., Jala, A., Ponneganti, S., ... Naidu, V. G. M. (2023). Ambient particulate matter (PM2.5) exposure contributes to neurodegeneration through the microbiome-gut-brain axis: Therapeutic role of melatonin. *Environmental Toxicology and Pharmacology, 101*, 104183. Retrieved from https://doi.org/10.1016/j.etap.2023.104183.

Pocivavsek, A., Schwarcz, R., & Erhardt, S. (2024). Neuroactive kynurenines as pharmacological targets: New experimental tools and exciting therapeutic opportunities. *Pharmacological Reviews, 76*(6), 978–1008. Retrieved from https://doi.org/10.1124/pharmrev.124.000239.

Pota, K., Nilewar, S., Mantsorov, C., Zumwalt, L., Nguyen, N., Bowers, C. J., ... Green, K. N. (2024). Impact of indole inclusion in the design of multi-tactical metal-binding tetra-aza macrocycles that target the molecular features of neurodegeneration. *ACS Chemical Neuroscience*. Retrieved from https://doi.org/10.1021/acschemneuro.4c00530.

Raj, K., Kaur, P., Gupta, G. D., & Singh, S. (2021). Metals associated neurodegeneration in Parkinson's disease: Insight to physiological, pathological mechanisms and management. *Neuroscience Letters, 753*, 135873. Retrieved from https://doi.org/10.1016/j.neulet.2021.135873.

Reiter, R. J., Sharma, R. N., Manucha, W., Rosales-Corral, S., Almieda Chuffa, L. G., de, Loh, D., ... Govitrapong, P. (2024). Dysfunctional mitochondria in age-related neurodegeneration: Utility of melatonin as an antioxidant treatment. *Ageing Research Reviews, 101*, 102480. Retrieved from https://doi.org/10.1016/j.arr.2024.102480.

Song, J., He, Y., Luo, C., Feng, B., Ran, F., Xu, H., ... Zhang, D. (2020). New progress in the pharmacology of protocatechuic acid: A compound ingested in daily foods and herbs frequently and heavily. *Pharmacological Research, 161*, 105109. Retrieved from https://doi.org/10.1016/j.phrs.2020.105109.

Spiers, J. G., Cortina Chen, H.-J., Barry, T. L., Bourgognon, J.-M., & Steinert, J. R. (2022). Redox stress and metal dys-homeostasis appear as hallmarks of early prion disease pathogenesis in mice. *Free Radical Biology and Medicine, 192*, 182–190. Retrieved from https://doi.org/10.1016/j.freeradbiomed.2022.09.025.

Stachowicz, K. (2023). Regulation of COX-2 expression by selected trace elements and heavy metals: Health implications, and changes in neuronal plasticity. A review. *Journal of Trace Elements in Medicine and Biology, 79*, 127226. Retrieved from https://doi.org/10.1016/j.jtemb.2023.127226.

Tahir, T., Javed, M., Ahmed, W., Iahtisham-Ul-Haq, Wang, Q., Khan, M. I., & Huang, Z. (2024). Therapeutic uses and pharmacological properties of the traditional South Asian medicinal plant *Withania coagulans* (Stocks) Dunal. *Journal of Herbal Medicine, 47*, 100926. Retrieved from https://doi.org/10.1016/j.hermed.2024.100926.

Tundo, G. R., Cascio, P., Milardi, D., Santoro, A. M., Graziani, G., Lacal, P. M., ... Sbardella, D. (2023). Targeting immunoproteasome in neurodegeneration: A glance to the future. *Pharmacology & Therapeutics, 241*, 108329. Retrieved from https://doi.org/10.1016/j.pharmthera.2022.108329.

Wankhede, N. L., Kale, M. B., Shende, P., Ratne, N., Taksande, B. G., Upaganlawar, A. B., & Magrulkar, S. V. (2025). Chapter 1 – Etiology and treatment challenges for neurodegenerative disorders. In V. Patel, M. S. Rajput, J. S. Shah, & T. Mehta (Eds.). Academic Press. Retrieved from https://doi.org/10.1016/B978-0-443-23841-3.00001-7.

Wydrych, A., Pakuła, B., Janikiewicz, J., Dobosz, A. M., Jakubek-Olszewska, P., Skowrońska, M., ... Więckowski, M. R. (2025). Metabolic impairments in neurodegeneration with brain iron accumulation. *Biochimica et Biophysica Acta (BBA) – Bioenergetics, 1866*(1), 149517. Retrieved from https://doi.org/10.1016/j.bbabio.2024.149517.

Zhang, Y.-Y., Li, X.-S., Ren, K.-D., Peng, J., & Luo, X.-J. (2023). Restoration of metal homeostasis: A potential strategy against neurodegenerative diseases. *Ageing Research Reviews, 87*, 101931. Retrieved from https://doi.org/10.1016/j.arr.2023.101931.

Chapter 25

The role of chelation therapy in heavy metal detoxification

Venkata Suneel Kumar Velugolu, and Prasann Kumar
Department of Agronomy, School of Agriculture, Lovely Professional University, Phagwara, Punjab, India

25.1 Introduction

Any toxic or poisonous metal at lower concentrations is known as a heavy metal. These are inorganic and metallic. They have a density of more than 5 g/cm^3. These heavy metals include chromium (Cr), lead (Pb), cadmium (Cd), mercury (Hg), copper (Cu), zinc (Zn), arsenic (As), iron (Fe), cobalt (Co), manganese (Mn), and nickel (Ni). These heavy metals are classified into two groups based on their toxicity level. Essential heavy metals: these are harmless at low concentrations (such as Cu, Fe, Zn, and Co). Nonessential heavy metals: These are highly toxic and dangerous at lower concentrations (such as Cd, Hg, As, and Cr).

Heavy metals such as copper, zinc, and iron will help in various biological processes in the human body when they are in trace amounts. Still, the same metals become toxic in excess amounts. Moreover, they induce oxidative stress by producing free radicals. They also reduce the antioxidant levels in the body. These metals will also inhibit the protein and DNA confirmation. These heavy metals will not degrade or be destroyed. They enter the body through industrial exposure, contaminated food or water, and sometimes specific medical treatments. These metals predominantly target the liver, heart, and other metabolically active tissues, contributing to the development of serious chronic diseases like cardiovascular disease, neurodegenerative disorders, and cancer. Heavy metal toxicity has become the most significant problem currently at the global level. So, finding an effective way to detox these metals from the body is substantial. Chelation therapy has become a widely used treatment for removing these heavy metals from the body (Table 25.1).

25.2 Major heavy metals and their sources

Heavy metal contamination is a significant environmental and health concern due to its persistent and toxic nature. These metals originate from natural and human-made sources, affecting air, water, soil, and food chains. One of the most hazardous heavy metals, lead (Pb), primarily comes from industrial emissions, old infrastructure, and natural geological deposits. Lead-based paints and plumbing materials used in older buildings pose serious risks as they deteriorate over time, releasing lead dust into the environment. Additionally, lead-acid batteries and improper disposal of electronic waste contribute to contamination. Industrial processes like mining and smelting release lead into the air and soil, while aviation fuel used in small aircraft still contains lead, adding to airborne pollution. Another highly toxic metal, mercury (Hg), is released into the environment through coal combustion, small-scale gold mining, and industrial waste. Coal-fired power plants emit mercury into the atmosphere, which later settles into water bodies, transforming into methylmercury, a highly toxic form that accumulates in fish and seafood. Artisanal gold miners use mercury to extract gold, leading to severe water and soil contamination. Mercury is also present in industrial effluents from chlor-alkali plants and cement production facilities, further polluting ecosystems. Consumption of mercury-contaminated fish, such as tuna and swordfish, is a significant source of human exposure, leading to neurological disorders such as tremors, vision changes, and numbness, and developmental issues like delays in cognitive development and learning disabilities.

Heavy Metal Toxicity and Neurodegeneration. DOI: https://doi.org/10.1016/B978-0-443-36575-1.00011-X

TABLE 25.1 Heavy metals and their effects.

Heavy metal	Common sources	Primary routes of exposure	Affected organs/system	Health effects	Chelation/detox treatment
Lead (Pb)	Old paints, batteries, plumbing pipes, industrial emissions, and contaminated soil	Inhalation (dust, fumes), ingestion (water, food), dermal (rare)	The brain, kidneys, nervous system, and cardiovascular system	Cognitive impairment, anemia, kidney dysfunction, high blood pressure	EDTA, DMSA, succimer
Mercury (Hg)	Contaminated fish, industrial waste, dental fillings (amalgam), thermometers, fluorescent bulbs	Inhalation (vapor), ingestion (fish, water), dermal (rare)	Nervous system, kidneys, immune system	Memory loss, tremors, coordination issues, kidney damage	DMSA, dimercaprol (BAL), selenium supplementation
Arsenic (As)	Pesticides, groundwater contamination, mining, industrial waste, rice, seafood	Ingestion (contaminated water, food), inhalation (industrial fumes)	Skin, liver, lungs, cardiovascular system	Skin lesions, increased cancer risk, heart disease, neuropathy	DMSA, dimercaprol, hydration therapy
Cadmium (Cd)	Cigarette smoke, industrial emissions, batteries, fertilizers, and contaminated food	Inhalation (smoke, industrial fumes), ingestion (food, water)	Kidneys, lungs, bones	Kidney failure, lung damage, osteoporosis, cancer	No specific chelator; supportive therapy
Chromium (Cr) (hexavalent chromium)	Leather tanning, stainless steel manufacturing, dyes, pigments	Inhalation (dust, fumes), ingestion (contaminated water, food)	Lungs, kidneys, skin	Lung cancer, dermatitis, kidney damage	No FDA-approved chelator; vitamin C therapy
Nickel (Ni)	Jewelry, batteries, stainless steel, industrial waste, tobacco smoke	Inhalation (dust, smoke), dermal (jewelry, coins), ingestion (food, water)	Skin, respiratory system, immune system	Allergic reactions, lung fibrosis, respiratory issues	No specific chelator; supportive therapy
Aluminum (Al)	Cookware, antiperspirants, processed foods, drinking water	Ingestion (food, water), inhalation (dust)	Brain, bones	Alzheimer's disease (controversial link), bone disorders	No effective chelation; citrate supplementation
Copper (Cu)	Pesticides, plumbing pipes, contaminated water, and industrial exposure	Ingestion (contaminated water, food), dermal (handling copper)	Liver, brain, and gastrointestinal tract	Wilson's disease (copper accumulation), liver failure	Penicillamine, trientine
Zinc (Zn) (excessive exposure)	Supplements, industrial fumes, brass materials	Inhalation (fumes), ingestion (supplements, contaminated water)	Gastrointestinal tract, immune system	Nausea, immune suppression, metal fume fever	No specific chelator; zinc intake reduction
Manganese (Mn)	Welding fumes, pesticides, contaminated water, steel industry	Inhalation (industrial exposure), ingestion (food, water)	Brain, nervous system	Parkinson-like symptoms, motor dysfunction	No FDA-approved chelation; supportive therapy
Iron (Fe) (excessive exposure)	Supplements, iron-rich foods, and blood transfusions	Ingestion (supplements, diet), intravenous (transfusions)	Liver, heart, pancreas	Hemochromatosis, diabetes, and heart failure	Deferoxamine, deferasirox

Another dangerous heavy metal, arsenic (As), naturally occurs in groundwater and is exacerbated by industrial activities. Millions of people worldwide, especially in regions like Bangladesh and parts of India, suffer from chronic arsenic poisoning due to contaminated drinking water. Arsenic is also found in pesticides and herbicides, historically used in agriculture, leading to soil and crop contamination. Industrial waste from mining and smelting operations contributes to arsenic pollution, while pressure-treated wood containing arsenic compounds releases toxic residues into the environment. **Cadmium (Cd)**, another highly toxic heavy metal, is mainly associated with industrial pollution and agricultural sources. Cigarette smoke is one of the leading sources of cadmium exposure, as tobacco plants absorb this metal from the soil. Additionally, nickel-cadmium batteries, though less common today, still contribute to environmental contamination when disposed of improperly. Using phosphate fertilizers in agriculture leads to cadmium accumulation in soil and food crops, further increasing human exposure. Industrial processes such as metal plating, smelting, and plastic manufacturing also release cadmium into the air, soil, and water, posing severe health risks, particularly to the kidneys and bones. Chromium (Cr), particularly in its hexavalent form (Cr^{6+}), is highly toxic and originates from leather tanning, stainless steel production, and electroplating industries. The leather tanning industry heavily relies on chromium-based compounds, resulting in wastewater contamination that pollutes nearby water sources. Industrial effluents from textile and metal industries discharge chromium into rivers and lakes, while improper disposal of industrial waste causes chromium to leach into soil and groundwater. Exposure to hexavalent chromium is particularly concerning, as it is a known carcinogen linked to lung cancer, skin irritation, and respiratory disorders. These heavy metals pose serious environmental and health risks, but the power to change this is in our hands. By advocating for stricter monitoring, regulation, and remediation efforts, we can help minimize exposure and mitigate their harmful effects. Addressing heavy metal pollution demands a combination of more stringent industrial regulations, improved waste management, and the adoption of cleaner technologies to reduce emissions and contamination. Let's use our collective voice to make a difference.

- **Lead**: Present in old paint, plumbing, and industrial emissions.
- **Mercury**: Present in certain fish, dental fillings, and industrial processes.
- **Cadmium**: Released from cigarette smoke, industrial pollution, and some batteries.
- **Arsenic**: Found in contaminated water, certain pesticides, and rice.
- **Other metals**: Antimony, gadolinium, and uranium, each with their sources and uses.

25.3 Chelation background

Chelation is a process where an organic molecule, called a chelating agent, binds to a metal ion (like lead or mercury) to form a stable complex. The chelating agent typically has atoms like sulfur, nitrogen, or oxygen that help to attach the metal. The strength of this bond depends on the types of atoms involved and how they are arranged.

Different metal ions inside the body, such as calcium, magnesium, and zinc, can compete with toxic metals for binding with the chelator. The effectiveness of the chelating agent depends on how easily it can reach the tissues, how strongly the poisonous metal is already bound to the body, how strongly the chelator binds to the metal, and the relative amounts of different metals present.

Chelators help remove metals from tissues and carry them to the kidneys for excretion through urine or to the liver for removal in bile. However, concerns exist that the body could reabsorb some metals through the kidney and liver processes.

The solubility of the chelate agent also plays a significant role in this therapy. Water-soluble chelators can easily be transported into the bloodstream and excreted through the kidneys. Other fat-soluble chelators, which can pass through cell membranes, including those in the brain, and remove metals from the cells, are also mainly excreted through bile (Table 25.2).

25.3.1 Types of chelating agents

- **Ethylenediaminetetraacetic acid (EDTA)**: This acid is commonly used to detox lead, mercury, and other heavy metals. It is usually injected into the body directly into the blood through veins.
- **Dimercaptosuccinic acid (DMSA)**: This acid is very effective in detoxing lead and is taken orally into the body.
- **Dimercaptopropane sulfonate (DMPS)**: This drug is used in mercury and arsenic poisoning cases and is available in oral and injectable forms.
- **Penicillamine**: This is primarily used for copper poisoning but can also chelate other heavy metals.

25.4 Mechanisms of chelation therapy

Chelation therapy uses unique chemical compounds called chelating agents, which form complexes with metal ions to form stable compounds that are nontoxic to humans. Water-soluble complexes can be excreted through urine or feces from the body (Table 25.3). The fundamental mechanism involves the chelating agent's ability to "claw" onto metal ions, preventing them from interacting with cellular structures or enzymes, thus reducing their toxic effects (Meyer, 2016). The mechanism of chelation therapy involves three stages: **binding**, **transport**, and **excretion** (Sharma, Kumar, & Singh, 2017).

TABLE 25.2 Chelation therapy and corresponding heavy metals.

Chelating agent	Target heavy metals	Mechanism of action	Route of administration	Potential side effects
EDTA	Lead (Pb), cadmium (Cd), chromium (Cr), manganese (Mn), zinc (Zn), iron (Fe)	Forms stable water-soluble complexes with metal ions, facilitating excretion via urine	IV	Kidney damage, hypocalcemia, hypotension, nausea
DMSA (Succimer)	Lead (Pb), mercury (Hg), arsenic (As), cadmium (Cd)	Binds with metal ions in blood and tissues, forming nontoxic, excretable complexes	Oral (capsules)	Gastrointestinal discomfort, skin rash, elevated liver enzymes
Dimercaprol (BAL)	Arsenic (As), mercury (Hg), lead (Pb), gold (Au)	Forms stable chelates with heavy metals, enhancing excretion through urine	IM	Hypertension, nausea, fever, pain at injection site
Penicillamine (D-penicillamine)	Lead (Pb), mercury (Hg), copper (Cu), zinc (Zn)	Forms stable water-soluble metal–chelate complexes, excreted via urine	Oral	Hypersensitivity reactions, nephrotoxicity, blood disorders
DFO	Iron (Fe), aluminum (Al)	Binds excess iron or aluminum, facilitating removal via urine and feces	IV/IM	Hypotension, allergic reactions, visual and auditory disturbances
DFX	Iron (Fe)	Selectively binds iron and excretes it through feces	Oral	Liver and kidney toxicity, gastrointestinal discomfort
DFP	Iron (Fe)	Chelates iron from cells and plasma, excreted via urine	Oral	Neutropenia, agranulocytosis, gastrointestinal issues
Unithiol (DMPS)	Mercury (Hg), arsenic (As), lead (Pb)	Binds with heavy metals in the bloodstream and enhances renal excretion	Oral/IV	Skin rashes, gastrointestinal discomfort, blood pressure fluctuations
Trientine	Copper (Cu)	Forms soluble complexes with copper, excreted via urine	Oral	Iron deficiency anemia, gastric irritation
NAC	Mercury (Hg), arsenic (As), lead (Pb)	Acts as an antioxidant and promotes metal detoxification via glutathione pathways	Oral/IV	Nausea, dizziness, allergic reactions

BAL, British anti-Lewisite; *DFO*, deferoxamine; *DFP*, deferiprone; *DFX*, deferasirox; *DMPS*, 2,3-dimercapto-1-propanesulfonic acid; *DMSA*, dimercaptosuccinic acid; *EDTA*, ethylenediaminetetraacetic acid; *IM*, intramuscular; *IV*, intravenous; *NAC*, *N*-acetylcysteine.

TABLE 25.3 Risks and benefits of chelation therapy.

Category	Details
Benefits of chelation therapy	**Explanation**
Heavy metal detoxification	It removes toxic heavy metals (lead, mercury, arsenic, cadmium, etc.), reducing their harmful effects on organs like the brain, kidneys, and liver.
Treatment for lead poisoning	Proven effective in treating acute and chronic lead poisoning, especially in children exposed to lead-based paints and contaminated water.
Iron overload management	It is used in conditions like **thalassemia**, where excess iron from frequent blood transfusions accumulates in the body, leading to organ damage.
Reduces cardiovascular risk	Some studies suggest that **EDTA chelation** may reduce arterial plaque, lower calcium buildup in arteries, and improve blood vessel function.
Alternative therapy for autism and neurodegenerative diseases	Some practitioners claim benefits in removing neurotoxic metals that may contribute to autism spectrum disorder and neurodegenerative diseases like Alzheimer's, though the evidence remains inconclusive.
Enhanced kidney and liver function	Chelation therapy may help improve kidney and liver function in individuals exposed to toxic metals by reducing the burden of heavy metals.
Potential benefits for diabetes	Some research suggests that removing toxic metals may improve insulin sensitivity and metabolic function.
Reduction of free radical damage	Chelating agents may reduce oxidative stress and inflammation, lowering cellular damage caused by toxic metals.
Potential risks and side effects	**Explanation**
Risk of kidney damage	Chelation therapy can place stress on the kidneys, increasing the risk of renal dysfunction, particularly in individuals with preexisting kidney disease.
Electrolyte imbalance and mineral deficiency	Chelating agents bind toxic metals and essential minerals (calcium, magnesium, zinc), leading to deficiencies that can cause muscle cramps, weakness, and bone demineralization.
Hypocalcemia (low blood calcium levels)	EDTA chelation can rapidly lower calcium levels, leading to serious complications like heart arrhythmias and muscle spasms.
Liver toxicity	Overusing chelating agents may cause liver strain, resulting in elevated liver enzymes or liver dysfunction in some patients.
Gastrointestinal issues	Common side effects include nausea, vomiting, diarrhea, and stomach pain, particularly with oral chelating agents like DMSA and Penicillamine.
Blood disorders	It can lead to anemia, neutropenia, or agranulocytosis, particularly with drugs like Deferiprone, which affects bone marrow function.
Neurological side effects	Due to the excessive removal of essential trace elements, high doses may cause dizziness, headaches, cognitive impairment, or neuropathy.
Potential for overuse and misuse	Some alternative health practitioners promote chelation therapy for unproven conditions, leading to unnecessary risks without medical supervision.
Allergic reactions	Some patients may experience **skin rashes, itching, fever, or anaphylactic reactions**, particularly with dimercaprol (BAL) and EDTA.
Requires medical supervision	Chelation therapy should only be administered under professional medical guidance, as improper use can lead to life-threatening complications.

BAL, British anti-Lewisite; *DMSA*, dimercaptosuccinic acid; *EDTA*, ethylenediaminetetraacetic acid.

1. **Binding**: Chelating agents contain functional groups, such as carboxyl, amine, or thiol groups, capable of forming bonds with the metal ions. These bonding forms a complex that stabilises the toxic metal into nontoxic and prevents it from interacting with other biological molecules like proteins or enzymes present

inside the body (Caldwell, Jones, & Dhingra, 2019). For example, EDTA has many carboxyl groups that can bind to divalent metals like lead, calcium, and iron (Martin & Wray, 2014).

2. **Transport**: After the chelator binds to the toxic metal, the complex is transported through the bloodstream to the kidneys or gastrointestinal tract (Meyer, 2016). This makes it easier for the body to eliminate the metal. The chelator–metal complex is often distributed to organs and tissues that facilitate excretion, most commonly the kidneys for renal excretion or the gastrointestinal tract for fecal excretion (Miller, Thompson, & Reyes, 2021).
3. **Excretion**: The last step is the excretion of the metal–chelator complex. Once the kidneys filter the complex, it is excreted in the urine. In gastrointestinal excretion cases, the complex is expelled in feces (Sharma, Kumar, & Singh, 2017). The effectiveness of chelation therapy depends on the chelating agent's affinity for specific metals, ensuring that the complex does not detach prematurely before elimination from the body (Martin & Wray, 2014).

25.5 Common chelating agents

Chelating agents are essential in treating various types of metal poisoning, each agent having a unique affinity for certain metals. Below are some of the most commonly used chelating agents in clinical medicine, along with their specific applications:

1. **Dimercaprol [British anti-Lewisite (BAL)]**: Dimercaprol, also known as British anti-Lewisite (BAL), is a sulfur-containing chelating agent traditionally used to treat poisoning caused by arsenic, mercury, and lead. Its molecular structure includes thiol groups (–SH) that can form stable complexes with these metals, rendering them water-soluble and facilitating their excretion through the urine. BAL is often injected intramuscularly into the body, and it is particularly effective for acute arsenic poisoning and acute mercury toxicity. However, it is not without side effects, including pain at the injection site and allergic reactions, which limit its use in some cases.
2. **EDTA**: EDTA is one of the most commonly used chelating agents in treating lead poisoning. It binds to divalent metal ions, such as lead and calcium, and forms a stable water-soluble compound that the kidneys can excrete. EDTA is also injected intravenously into the body. Prolonged use of this EDTA agent will cause the risk of nephrotoxicity. EDTA effectively detoxifies the blood lead.
3. **DMSA** is a water-soluble chelating agent taken orally. This chelator is mainly effective in treating lead and mercury poisoning. DMSA works by binding to heavy metals and forming stable complexes that can be eliminated through the urine. It is safer compared to dimercaprol and EDTA, making it suitable for outpatient treatment, especially for chronic low-level metal exposure. DMSA is easy to administer and has minimal side effects, such as mild gastrointestinal discomfort or rashes (Miller, Thompson, & Reyes, 2021).
4. **Penicillamine**: Penicillamine is mainly used to treat copper toxicity, particularly in patients with Wilson's disease, a genetic disorder that leads to copper accumulation in the brain and liver. Penicillamine chelates with copper metal and eliminates it through urine. It also treats mercury poisoning. Penicillamine is effective, but it also has a few side effects, including renal impairment, skin rashes, and hematologic abnormalities.
5. **Prussian blue**: Prussian blue is a unique chelator for cesium and thallium poisoning. Unlike other chelators, Prussian blue does not bind to metals in the bloodstream but instead works by trapping them in the gastrointestinal tract, where they can be excreted through feces. This makes it particularly useful for radiation emergencies and thallium poisoning. Prussian blue is given orally and is relatively safe, though its use is primarily restricted to specific forms of heavy metal toxicity.

25.6 Clinical applications of chelation therapy

Chelation therapy plays a vital role in treating various types of metal poisoning, including lead, mercury, arsenic, and, in some cases, cadmium. Each of these metals presents unique health risks, and using chelating agents helps reduce the toxic burden in the body and prevent long-term organ damage.

Heavy metal poisoning: The primary purpose of chelation therapy is to treat both acute and chronic heavy metal poisoning.

Heart disease and other conditions: Some specialists use chelation therapy to treat heart disease, but its effectiveness for these purposes is not well-established and remains a topic of debate.

- **Lead poisoning**: Continuous exposure to Lead, mainly in the case of children, has significant health issues due to its neurotoxic effects. Even low levels of lead exposure can impair growth and development, which leads to behavioral problems, while high levels lead to coma and death. Chelation therapy is commonly used to lower blood lead levels and mitigate the damage caused by lead accumulation. The chelating agents EDTA and DMSA are frequently used for lead poisoning (Caldwell, Jones, & Dhingra, 2019). Chelation therapy is essential in preventing long-term neurological and cognitive deficits in children exposed to lead (Martin & Wray, 2014).

- **Mercury poisoning**: Mercury exposure, particularly in industrial settings or from contaminated fish (such as methylmercury), poses significant health risks. Chronic mercury toxicity is associated with neurological damage, kidney dysfunction, and other systemic effects. BAL and DMSA are the main chelators used to treat mercury poisoning. Dimercaprol is particularly effective in treating inorganic mercury toxicity, while DMSA is used for both inorganic and organic mercury exposure. Treatment with these chelating agents helps to reduce mercury concentrations in the body and prevent further organ damage. However, careful monitoring is necessary, as chelation therapy itself can have adverse effects, including kidney damage, particularly if not adequately dosed. The timing and method of chelation therapy are crucial in optimizing outcomes for mercury poisoning (Meyer, 2016).
- **Arsenic poisoning**: Arsenic exposure, mainly through contaminated water, can lead to various health problems, like skin lesions, neuropathy, cancer, and cardiovascular diseases. Chronic arsenic toxicity is a known risk factor for lung, skin, and bladder cancers. Dimercaprol and DMSA are both used to treat acute arsenic poisoning. Dimercaprol is particularly useful for treating acute arsenic poisoning due to its ability to bind arsenic and facilitate its excretion via the urine. DMSA is also used for arsenic detoxification, especially in cases of chronic exposure, although it is generally considered less effective for acute cases.
- **Cadmium and other metal toxicities**: Cadmium toxicity, mainly due to occupational exposure (e.g., in battery manufacturing or mining), leads to kidney damage, osteoporosis, and high blood pressure. Although chelation therapy is not routinely used for cadmium poisoning, some studies suggest that EDTA might help remove cadmium from the kidneys and other tissues. However, its effectiveness is still debated (Miller, Thompson, & Reyes, 2021). Cadmium is primarily excreted through the urine. Prussian blue may also be employed in some instances of cesium and thallium poisoning, though its application for cadmium toxicity is limited.

Chelation therapy remains an essential tool in the management of heavy metal poisoning. The choice of chelating agent and the timing of administration depend on the type of metal involved, the severity of the poisoning, and the patient's clinical condition. Chelation therapy can significantly reduce the body's burden of toxic metals (Table 25.4).

25.7 Benefits of chelation therapy

Chelation therapy has several benefits, making it a valuable tool in the treatment of heavy metal poisoning. By using chelating agents to bind and eliminate toxic metals from the body, this therapy helps alleviate symptoms, prevent organ damage, and improve overall health outcomes in individuals exposed to harmful metals like lead, mercury, arsenic, and others. The primary benefits of chelation therapy include:

1. **Reduction of toxic metal burden**: One of the most essential benefits of chelation therapy is its ability to lower the concentration of toxic metals in the bloodstream and tissues. When metals like lead, mercury, or arsenic accumulate in the body, they can harm various systems, including the nervous, cardiovascular, and renal systems. Chelating agents bind to these metals and form stable, water-soluble complexes that are then excreted through the kidneys or gastrointestinal tract (Martin & Wray, 2014). This process effectively reduces the overall toxic metal burden, helping to alleviate poisoning symptoms such as fatigue, headaches, abdominal pain, and cognitive dysfunction (Meyer, 2016).
2. **Prevention of organ damage**: Chronic exposure to heavy metals can cause irreversible damage to many organs, including the kidneys, liver, and central nervous system. Lead poisoning can lead to renal failure and neurological deficits, while mercury exposure may damage the kidneys. Chelation therapy plays a key role in preventing further damage by removing these metals from organs and tissues. In this way, chelation therapy can act as a protective measure to maintain organ function and prevent the onset of chronic conditions associated with metal toxicity.
3. **Elimination of symptoms**: In cases of high exposure to heavy metals, chelation therapy can lead to noticeable improvements in symptoms, particularly those related to neurological and cognitive dysfunction. In chronic lead exposure, children are associated with developmental delays, learning difficulties, and behavioral problems. Chelation therapy, particularly with agents like **DMSA**, can improve cognitive function and reduce behavioral disturbances in children with lead poisoning. Similarly, in adults with mercury poisoning, chelation therapy can help alleviate symptoms such as tremors, memory loss, and mood swings. These benefits underscore the importance of chelation therapy in managing heavy metal poisoning, not only in removing toxic metals from the body but also in improving the overall health and well-being of individuals affected by metal toxicity (Table 25.5).

TABLE 25.4 Comparison of heavy metal accumulation and detoxification.

Heavy metal	Sources of exposure	Affected organs	Symptoms of toxicity	Biological half-life	Primary detoxification methods
Lead (Pb)	Lead-based paints, contaminated water, industrial emissions, old plumbing, batteries	Brain, kidneys, liver, bones	Cognitive impairment, anemia, developmental delays, kidney damage, hypertension	10–30 years (bones), weeks (blood)	Chelation therapy (EDTA, DMSA), calcium and iron supplementation, dietary fiber
Mercury (Hg)	Contaminated fish (methylmercury), dental amalgams, industrial waste, thermometers	Brain, kidneys, nervous system	Tremors, memory loss, mood swings, kidney damage, peripheral neuropathy	50 days (methylmercury), 90 days (inorganic mercury)	Chelation therapy (DMSA, DMPS), glutathione support, selenium supplementation
Arsenic (As)	Contaminated water, pesticides, industrial smelting, seafood, rice	Skin, liver, lungs, kidneys	Skin lesions, cancer, peripheral neuropathy, cardiovascular disease	1–2 days (blood), years (nails, hair, bones)	Chelation therapy (DMSA, DMPS), antioxidants (vitamin C, NAC), sulfur-containing foods
Cadmium (Cd)	Cigarette smoke, industrial waste, contaminated soil and water, batteries	Kidneys, bones, lungs	Osteoporosis, kidney failure, lung disease, and hypertension	7–30 years (kidneys, bones)	Chelation therapy (EDTA), zinc and selenium supplementation, activated charcoal
Aluminum (Al)	Processed foods, cookware, deodorants, vaccines, industrial exposure	Brain, bones	Neurotoxicity, Alzheimer's risk, bone disorders, anemia	10–50 years (bones), 24 hours (blood)	Silica-rich water, dietary fiber, antioxidants (curcumin, NAC), chelation (deferoxamine)
Chromium (Cr VI)	Industrial emissions, leather tanning, contaminated water	Lungs, liver, kidneys	Respiratory distress, DNA damage, liver & kidney failure, cancer	40 hours (blood), years (tissues)	Vitamin C, chelation therapy (EDTA, NAC), dietary detox foods
Nickel (Ni)	Stainless steel cookware, tobacco smoke, industrial waste	Skin, lungs, kidneys	Allergic dermatitis, lung fibrosis, DNA damage, and respiratory issues	10–30 hours (blood), years (bones)	Vitamin C, zinc supplementation, and chelation therapy (DMSA)
Copper (Cu)	Copper pipes, contaminated water, excess supplements, Wilson's disease	Liver, brain	Liver damage, neurological issues, mood swings, anemia	10–15 days (blood), lifetime (liver in Wilson's disease)	Chelation therapy (penicillamine, trientine), zinc supplementation, and molybdenum support
Iron (Fe) (overload cases)	Blood transfusions, excessive supplements, iron-rich diets, hemochromatosis	Liver, heart, pancreas	Liver cirrhosis, diabetes, heart failure, joint pain	6 hours (blood), years (liver, heart)	Chelation therapy (deferoxamine, deferasirox), phlebotomy, dietary adjustments

DMPS, 2,3-Dimercapto-1-propanesulfonic acid; *DMSA*, dimercaptosuccinic acid; *EDTA*, ethylenediaminetetraacetic acid; *NAC*, *N*-acetylcysteine.

TABLE 25.5 Sources, health risks, and detox methods of heavy metals.

Heavy metal	Common sources	Primary health risks	Affected organs	Detoxification methods
Lead (Pb)	Lead-based paints, contaminated water (old plumbing), industrial emissions, batteries, cosmetics	Cognitive impairment, anemia, developmental delays, kidney failure, hypertension, reproductive issues	Brain, kidneys, liver, bones	Chelation therapy (EDTA, DMSA), calcium and iron supplementation, dietary fiber, vitamin C and antioxidants
Mercury (Hg)	Contaminated seafood (methylmercury), dental fillings (amalgam), industrial waste, thermometers, coal-burning	Tremors, memory loss, mood disorders, kidney damage, peripheral neuropathy	Brain, kidneys, nervous system	Chelation therapy (DMSA, DMPS), glutathione support, selenium supplementation, omega-3 fatty acids
Arsenic (As)	Contaminated groundwater, pesticides, rice, industrial smelting, seafood	Skin lesions, lung/liver/kidney cancer, cardiovascular disease, diabetes, peripheral neuropathy	Skin, liver, lungs, kidneys	Chelation therapy (DMSA, DMPS), antioxidants (vitamin C, NAC), sulfur-rich foods (garlic, onions), activated charcoal
Cadmium (Cd)	Cigarette smoke, industrial pollution, contaminated soil & water, batteries, fertilizers	Osteoporosis, kidney failure, lung disease, hypertension, DNA damage	Kidneys, bones, lungs	Chelation therapy (EDTA), zinc and selenium supplementation, activated charcoal, high-fiber diet
Aluminum (Al)	Processed foods, cookware, deodorants, vaccines, and industrial exposure	Neurotoxicity, Alzheimer's risk, anemia, bone disorders, DNA damage	Brain, bones	Silica-rich water, dietary fiber, antioxidants (curcumin, NAC), chelation (deferoxamine), magnesium-rich diet
Chromium (Cr VI)	Industrial waste, leather tanning, contaminated water, stainless steel welding	Respiratory distress, DNA mutations, lung/liver/kidney cancer	Lungs, liver, kidneys	Vitamin C, chelation therapy (EDTA, NAC), antioxidants, dietary detox foods
Nickel (Ni)	Stainless steel cookware, tobacco smoke, industrial waste, and batteries	Allergic dermatitis, lung fibrosis, DNA damage, and respiratory problems	Skin, lungs, kidneys	Vitamin C, zinc supplementation, chelation therapy (DMSA), avoidance of nickel-based products
Copper (Cu)	Copper pipes, contaminated water, excess supplements, Wilson's disease	Liver damage, neurological issues, mood swings, anemia	Liver, brain	Chelation therapy (penicillamine, trientine), zinc supplementation, and molybdenum-rich foods
Iron (Fe) (overload cases)	Blood transfusions, excessive supplements, iron-rich diets, and hemochromatosis	Liver cirrhosis, diabetes, heart failure, joint pain	Liver, heart, pancreas	Chelation therapy (deferoxamine, deferasirox), phlebotomy, dietary adjustments (avoiding iron-rich foods, vitamin C restriction)

DMPS, 2,3-Dimercapto-1-propanesulfonic acid; *DMSA*, dimercaptosuccinic acid; *EDTA*, ethylenediaminetetraacetic acid; *NAC*, *N*-acetylcysteine.

25.7.1 Limitations and risks of chelation therapy

While chelation therapy can be beneficial, it has some limitations and risks:

1. **Side effects**: Chelating agents can cause side effects such as allergic reactions, gastrointestinal issues, and kidney damage, particularly with prolonged or improper use.
2. **Overuse and misuse**: Chelation therapy should only be done under the guidance of a medical professional. Misusing chelators (for example, for detox purposes) can lead to serious complications.
3. **Inability to treat all forms of toxicity**: Chelation therapy is impractical for all types of metal poisoning, and chelating agents may not effectively remove some metals.

25.8 Conclusion

Heavy metal toxicity is one of the significant health concerns globally. Metals such as lead, mercury, arsenic, and cadmium cause considerable health problems, including organ damage and neurological issues. Chelation therapy offers an effective way to manage both acute and chronic heavy metal poisoning. This therapy uses specific chelating agents like EDTA, DMSA, and dimercaprol to safely remove toxic metals from the body, helping to alleviate symptoms, prevent organ damage, and improve overall health. In summary, chelation therapy is a vital tool in the fight against heavy metal poisoning. Ongoing research and advancements in this field will likely continue to improve its efficacy and safety, ensuring it remains a crucial component of modern medical treatment for heavy metal detoxification.

References

Caldwell, K. L., Jones, R. L., & Dhingra, R. (2019). Chelation in the management of metal toxicity: Modern perspectives. *Clinical Toxicology, 57*, 356–367.

Martin, D., & Wray, J. (2014). EDTA and its clinical implications in metal chelation therapy. *Journal of Environmental Health Science, 12*, 112–118.

Meyer, J. (2016). Mechanisms of chelation: From molecular interactions to clinical applications. *Toxicol Reviews, 35*, 89–101.

Miller, A., Thompson, C., & Reyes, M. (2021). Advances in chelation therapy for toxic metal exposure. *Current Medicinal Chemistry, 28*, 1342–1354.

Sharma, A., Kumar, S., & Singh, R. (2017). Chelation therapy in heavy metal poisoning: A review. *International Journal of Medical Toxicology and Legal Medicine, 20*, 1–8.

Further reading

Andersen, O., & Aaseth, J. (2002). Molecular mechanisms of in vivo metal chelation: implications for clinical treatment of metal intoxications. *Environmental Health Perspectives, 110*(Suppl 5), 887–890.

Aremu, D. A., Madejczyk, M. S., & Ballatori, N. (2008). *N*-Acetylcysteine as a potential antidote and biomonitoring agent of methylmercury exposure. *Environmental Health Perspectives, 116*(1), 26–31.

Atsdr, U. S. (1997). Agency for toxic substances and disease registry. *Case Studies in Environmental Medicine*. Retrieved from http://www.atsdr.Cdc.Gov/HEC/CSEM/csem.html.

Awata, H., Linder, S., Mitchell, L. E., & Delclos, G. L. (2017). Association of dietary intake and biomarker levels of arsenic, cadmium, lead, and mercury among Asian populations in the United States: NHANES 2011–2012. *Environmental Health Perspectives, 125*(3), 314–323.

Bernhoft, R. A. (2012). Mercury toxicity and treatment: A review of the literature. *Journal of Environmental and Public Health, 2012*(1), 460508.

Ciesielski, T., Weuve, J., Bellinger, D. C., et al. (2012). Cadmium exposure and neurodevelopmental outcomes in U.S. children. *Environmental Health Perspectives, 120*, 758–763.

Delangle, P., & Mintz, E. (2012). Chelation therapy in Wilson's disease: From D-penicillamine to the design of selective bioinspired intracellular Cu (I) chelators. *Dalton Transactions, 41*(21), 6359–6370.

Flora, G., Gupta, D., & Tiwari, A. (2012). Toxicity of lead: A review with recent updates. *Interdisciplinary Toxicology, 5*(2), 47–58.

Flora, S. J. S. (2009). Metal poisoning: Threat and management. *Al Ameen Journal of Medical Sciences, 2*(2), 4–26.

Flora, S. J. S., Bhadauria, S., Kannan, G. M., & Singh, N. (2007). Arsenic induced oxidative stress and the role of antioxidant supplementation during chelation: A review. *Journal of Environmental Biology, 28*(2), 333.

Flora, S. J. S., Mittal, M., & Mehta, A. (2008). Heavy metal induced oxidative stress & its possible reversal by chelation therapy. *Indian Journal of Medical Research, 128*(4), 501–523.

Flora, S. J., & Pachauri, V. (2010). Chelation in metal intoxication. *International Journal of Environmental Research and Public Health, 7*(7), 2745–2788.

Fowler, B. A. (2009). Monitoring of human populations for early markers of cadmium toxicity: A review. *Toxicology and Applied Pharmacology, 238*(3), 294–300.

Fujihara, J., & Nishimoto, N. (2024). Thallium-poisoner's poison: An overview and review of current knowledge on the toxicological effects and mechanisms. *Current Research in Toxicology, 6*, 100157.

Genchi, G., Carocci, A., Lauria, G., Sinicropi, M. S., & Catalano, A. (2021). Thallium use, toxicity, and detoxification therapy: An overview. *Applied Sciences, 11*(18), 8322.

Hall, J. Á. (2002). Cellular mechanisms for heavy metal detoxification and tolerance. *Journal of Experimental Botany, 53*(366), 1–11.

Ito, Y., Niiya, Y., Otani, M., Sarai, S., & Shima, S. (1987). Effect of food intake on blood lead concentration in workers occupationally exposed to lead. *Toxicology Letters, 37*(2), 105–114.

Joshi, D., Mittal, D., Shrivastav, S., Shukla, S., & Srivastav, A. K. (2011). Combined effect of *N*-acetyl cysteine, zinc, and selenium against chronic dimethylmercury-induced oxidative stress: A biochemical and histopathological approach. *Archives of Environmental Contamination and Toxicology, 61*, 558–567.

Kim, J. J., Kim, Y. S., & Kumar, V. (2019). Heavy metal toxicity: An update of chelating therapeutic strategies. *Journal of Trace Elements in Medicine and Biology, 54*, 226–231.

Lanphear, B. P., Hornung, R., Khoury, J., Yolton, K., Baghurst, P., Bellinger, D. C., ... Roberts, R. (2019). Erratum: "Low-level environmental lead exposure and children's intellectual function: An international pooled analysis". *Environmental Health Perspectives, 127*(9), 099001.

Lee, Y. R., Kang, M. H., & Park, H. M. (2016). Treatment of zinc toxicosis in a dog with chelation using d-penicillamine. *Journal of Veterinary Emergency and Critical Care, 26*(6), 825–830.

Matović, V., Buha, A., Bulat, Z., & Đukić-Ćosić, D. (2011). Cadmium toxicity revisited: Focus on oxidative stress induction and interactions with zinc and magnesium. *Arhiv za Higijenu Rada i Toksikologiju, 62*(1), 65–75.

National Research Council. (2000). *Toxicological effects of methylmercury*. Washington, DC: National Academy Press. Retrieved from https://doi.org/10.17226/9899.

National Toxicology Program. (2012). NTP monograph on health effects of low-level lead. *NTP Monograph*(1) xiii-148.

Rooney, J. P. (2007). The role of thiols, dithiols, nutritional factors and interacting ligands in the toxicology of mercury. *Toxicology, 234*(3), 145–156.

Sakamoto, M. I. N. E. S. H. I., Murata, K., Kakita, A., & Sasaki, M. (2012). A review of mercury toxicity with special reference to methylmercury. *Environmental Chemistry and Toxicology of Mercury,* 501–516.

Sears, M. E., & Genuis, S. J. (2012). Environmental determinants of chronic disease and medical approaches: Recognition, avoidance, supportive therapy, and detoxification. *Journal of environmental and public health, 2012*(1), 356798.

Sears, M. E., Kerr, K. J., & Bray, R. I. (2012). Arsenic, cadmium, lead, and mercury in sweat: A systematic review. *Journal of Environmental and Public Health, 2012*(1), 184745.

Sherameti, I., & Varma, A. (Vol. Eds.), (2011). *Detoxification of heavy metals: 30*, Springer Science & Business Media.

Van Barneveld, A. A., & Van den Hamer, C. J. (1985). Influence of Ca and Mg on the uptake and deposition of Pb and Cd in mice. *Toxicology and Applied Pharmacology, 79*(1), 1–10.

Wang, X., & Cheng, L. (2020). Multifunctional Prussian blue-based nanomaterials: Preparation, modification, and theranostic applications. *Coordination Chemistry Reviews, 419*, 213393.

Chapter 26

Impact of heavy metals on neuroplasticity

Prasann Kumar[1], **and Shipa Rani Dey**[2]
[1]*Department of Agronomy, School of Agriculture, Lovely Professional University, Phagwara, Punjab, India,* [2]*Faculty of Agricultural Sciences, GLA University, Mathura, Uttar Pradesh, India*

26.1 Introduction

Neuroplasticity, also known as brain plasticity or neural plasticity, refers to the brain's remarkable ability to reorganize itself by forming new neural connections throughout life. This dynamic process is essential for learning, memory formation, recovery from brain injuries, and adaptation to environmental changes (Bai et al., 2022; Dooka et al., 2024; Mattson & Leak, 2024). Neuroplasticity operates through several mechanisms, including synaptic plasticity, which involves the strengthening or weakening of synapses in response to neuronal activity, and structural plasticity, which encompasses the formation of new neurons (neurogenesis) and changes in dendritic branching. The brain's ability to rewire allows individuals to acquire new skills, adapt to new experiences, and recover from damage such as strokes or traumatic brain injuries. Moreover, neuroplasticity plays a crucial role in cognitive development, emotional regulation, and overall mental health, enabling individuals to adapt to different life stages and challenges. However, despite its adaptability, neuroplasticity is highly sensitive to external factors, including environmental toxins such as heavy metals, which can interfere with these critical processes and lead to long-term cognitive and neurological impairments.

Environmental neurotoxicants adversely affect the nervous system, disrupting normal neural function. Among these, heavy metals such as lead (Pb), mercury (Hg), arsenic (As), cadmium (Cd), and manganese (Mn) are particularly concerning due to their ability to cross the blood–brain barrier (BBB) and accumulate in neural tissues. Unlike organic toxins, heavy metals persist in the environment and can bioaccumulate in living organisms, leading to prolonged exposure and heightened neurotoxic effects. These metals can interfere with neurotransmission, induce oxidative stress, promote neuroinflammation, and alter epigenetic regulation, all compromising neuroplasticity. Lead, for example, is a well-documented neurotoxin that disrupts calcium homeostasis, impairing synaptic function and reducing cognitive abilities. Mercury, particularly in its organic form (methylmercury), interferes with glutamate uptake and increases excitotoxicity, causing neuronal damage and impairing plasticity mechanisms. Arsenic exposure has been linked to neurodevelopmental deficits and cognitive impairments by inducing apoptosis in neural cells and altering synaptic connectivity. Cadmium, often found in cigarette smoke and industrial waste, disrupts dopamine regulation and impairs learning and memory processes (Azmitia, 2007; Thapa et al., 2024; Thilakaratne et al., 2024). Manganese, an essential trace element at low levels, becomes neurotoxic when present in excess, leading to disruptions in motor function and cognitive abilities. These heavy metals exert their neurotoxic effects through multiple pathways, including generating reactive oxygen species (ROS), mitochondrial dysfunction, DNA damage, and dysregulation of key neurotrophic factors such as brain-derived neurotrophic factor (BDNF). The impact of these metals on neuroplasticity is particularly concerning during critical developmental periods, such as infancy and childhood, when the brain is highly plastic and vulnerable to external insults. Prenatal and early-life exposure to heavy metals has been associated with neurodevelopmental disorders such as autism spectrum disorder (ASD) and attention deficit hyperactivity disorder (ADHD), highlighting the profound consequences of environmental neurotoxicants on brain function.

The rising concern about heavy metal exposure is primarily attributed to rapid industrialization, urbanization, and environmental pollution. Over the past century, human activities such as mining, smelting, fossil fuel combustion, agricultural practices, and improper waste disposal have

Heavy Metal Toxicity and Neurodegeneration. DOI: https://doi.org/10.1016/B978-0-443-36575-1.00029-7

significantly increased the levels of heavy metals in the environment. Industrial effluents, vehicular emissions, and electronic waste contribute to widespread air, water, and soil contamination, increasing human exposure. Heavy metals can enter the human body through inhalation, ingestion, or dermal absorption, accumulating over time in vital organs, including the brain. One of the most pressing concerns is lead contamination in drinking water, as evidenced by high-profile cases such as the Flint water crisis in the United States, where aging lead pipes led to widespread neurological and developmental issues in children. Similarly, mercury contamination in seafood from industrial discharge into water bodies poses significant risks to populations with high fish consumption. Arsenic contamination in groundwater, particularly in regions such as Bangladesh and parts of India, has been linked to severe neurological and cognitive impairments in exposed populations. Industrial workers, miners, and populations living near hazardous waste sites are at an elevated risk of heavy metal exposure, as occupational exposure to these toxicants often exceeds safe levels (Ciano Albanese et al., 2025; Fasone, 2021; Qu et al., 2025).

The impact of heavy metal exposure is not limited to direct neurological effects but also extends to broader societal and economic consequences. Neurodevelopmental deficits caused by heavy metal toxicity can lead to reduced educational attainment, lower productivity, and increased healthcare costs. The burden of neurological disorders linked to heavy metal exposure places a strain on public health systems, emphasizing the need for stringent environmental regulations and preventive measures. Despite the growing body of evidence highlighting the neurotoxic effects of heavy metals, regulatory measures and mitigation strategies remain insufficient in many parts of the world. While developed nations have implemented policies to reduce lead in gasoline, restrict mercury emissions, and regulate industrial waste disposal, many developing countries struggle with inadequate environmental policies and enforcement. Furthermore, climate change and extreme weather events exacerbate heavy metal pollution by increasing the mobilization of contaminants in water bodies and soil. The continued industrialization of emerging economies, coupled with insufficient regulatory frameworks, raises concerns about the long-term impact of heavy metal exposure on global brain health (Laaroussi et al., 2025; Mishra, Ghosh, Singh, & Sinha, 2025; Wani, Hammad Ahmad Shadab, & Afzal, 2021).

Addressing the challenges posed by heavy metal neurotoxicity requires a multifaceted approach that includes policy interventions, public awareness, and scientific advancements in neuroprotection. Governments and regulatory agencies must enforce stricter pollution control measures, improve waste management practices, and implement monitoring systems to track heavy metal levels in the environment. Public health initiatives that reduce exposure, such as replacing lead pipes, promoting safe agricultural practices, and educating communities about the dangers of contaminated water and food sources, are crucial in mitigating risks. From a scientific perspective, research into chelation therapy, dietary interventions, and neuroprotective compounds offers promising avenues for reducing the toxic effects of heavy metals on the brain. Natural antioxidants, polyphenols, and essential micronutrients such as zinc and selenium have been shown to counteract oxidative stress and support neuroplasticity, providing potential therapeutic strategies for individuals exposed to heavy metals. Neuroplasticity is fundamental to cognitive function, learning, and brain resilience, yet it is highly vulnerable to environmental toxins and hefty metals. The increasing prevalence of heavy metal contamination due to industrialization and pollution has raised significant concerns about its impact on neuroplasticity and overall brain health. Exposure to lead, mercury, arsenic, cadmium, and manganese disrupts synaptic function, induces oxidative stress, and impairs neurodevelopment, leading to cognitive deficits and neurological disorders (Luca et al., 2025; Mazuryk, Klepacka, Kutner, & Sharma, 2024; Tang, Bai, Ma, & Ou, 2023). The widespread contamination of air, water, and soil with heavy metals, coupled with inadequate regulatory measures in many parts of the world, underscores the urgency of addressing this growing environmental health crisis. Practical strategies to mitigate heavy metal–induced neurotoxicity include policy enforcement, public education, and scientific research into neuroprotective interventions. As industrial activities expand, proactive measures are essential to safeguard brain health and prevent the long-term consequences of heavy metal exposure on neuroplasticity and cognitive function.

26.2 Heavy metals and their sources of exposure

Lead (Pb) is among the best-studied neurotoxic heavy metals, with exposure occurring mainly through lead-containing paint, battery use, industrial waste, and contaminated water. Since regulatory efforts to eliminate lead-containing products have been instituted to varying degrees worldwide, legacy contamination represents a critical health risk, particularly in older buildings with decaying lead paint and soils and water supplies contaminated with lead. Lead exposure is hazardous for children, whose still-developing nervous systems are more sensitive, and even low exposure levels have been associated with cognitive deficits, learning disabilities, and behavioral issues. Calcium mimicry is one of the main mechanisms involved in lead neurotoxicity. Lead competes with calcium ions to permeate neurons through calcium channels, displacing calcium, a vital signal for various neuronal processes such as neurotransmitter release, synaptic plasticity, and general neuronal functioning. This disruption reduces long-term potentiation (LTP), an essential mechanism of learning and memory. Moreover, lead can provide oxidative stress by creating ROS and draining stores of antioxidants, including glutathione, resulting in lipid peroxidation, genomic damage, and mitochondrial failure. Such oxidative stress leads to neuronal apoptosis or enhances neuroinflammation, making cognitive deterioration more acute. Lead exposure also profoundly affects synaptic function since lead disrupts synaptogenesis and the release of key neurotransmitters like glutamate and dopamine. This disruption causes deficits in in attention, executive function, and emotional regulation. Moreover, lead exposure has been linked to epigenetic changes that alter neurodevelopmental gene expression patterns, leading to elevated risk of neurodevelopmental disorders, such as ADHD and lower intelligence quotient (IQ) in exposed populations. Lead also persists in the environment and bioaccumulates in bones, contributing to its long-term neurotoxic effects and making it a high-priority public health challenge (Alaerts, Swinnen, & Wenderoth, 2010; Ramaswamy et al., 2024; Schmidt & Schmidt, 2025).

Another well-known neurotoxic element is mercury (Hg), whose primary sources of exposure include dietary intake, dental amalgams, and industrial emissions. There are three common forms of mercury: elemental mercury, inorganic mercury, and organic mercury (e.g., methylmercury). It has the highest bioavailability of any form of mercury and the ability to cross the BBB, making it the most neurotoxic form. The main route of human exposure to methylmercury is eating contaminated seafood and large predatory fish like tuna, swordfish, and shark, which accumulate mercury through biomagnification. Inhibition of glutamate uptake induces excitotoxicity as a mechanism of mercury-induced neurotoxicity. Mercury impairs astrocytic functioning, whereby astrocytes are known to clear excess glutamate from the synaptic clefts. Chronic glutamate accumulation causes overactivation of *N*-methyl-D-aspartate (NMDA) receptors with excessive calcium influx, cell swelling, and ultimately cell death. Aside from excitotoxicity, mercury induces mitochondrial dysfunction by adversely affecting mitochondrial membrane potential, hindering ATP generation, and facilitating ROS production. This mitochondrial compromise leads to energy deficits in neuronal cells and, as a result, synaptic plasticity and cognitive impairment. Mercury exposure has been associated with neurodevelopmental disorders, including ASD and intellectual disabilities, especially in populations exposed to environmental mercury in utero. Methylmercury is toxic to pregnant women, and they risk passing the toxin onto their developing fetus, resulting in impaired brain development, motor dysfunction, and cognitive deficits. Moreover, mercury has been demonstrated to modify neurotransmitter systems, including dopaminergic and serotonergic pathways, which can lead to mood disorders, depression, and anxiety. Because mercury is ubiquitous in the environment and bioaccumulative in biological tissues, it remains a pivotal neurotoxicant deserving of strict regulatory control and public health intervention (Amooei, Meshkati, Nasiri, & Dakhili, 2021; Athanasopoulos, Karagiannis, & Tsolaki, 2016; El-Fawal, 2011a).

Another highly toxic heavy metal with potent neurotoxic effects is arsenic (As), which mainly comes from contaminated groundwater, pesticides, and industrial activities (e.g., mining or smelting). Chronic exposure to arsenic is a serious public health problem in areas around the world where groundwater is contaminated, such as parts of Bangladesh, India, and China, where millions of people depend on arsenic-tainted water for drinking and irrigation. The most persistent neurotoxic effects of arsenic exposure appear to be the interference with cognition and memory, with evidence mounting for a link between arsenic exposure and neurodegenerative diseases such as Alzheimer's disease (AD) and Parkinson's disease (PD). Epigenetic modifications are one of the main pathways through which arsenic exerts toxicity on the nervous system. Genotoxicity of arsenic induces ANS, which modifies bioavailable DNA methylation and concomitant histone off-sets, disturbed gene expression patterns impairing neurodevelopment, synaptic plasticity, and neuronal survival. These epigenetic modifications collectively lead to sustained cognitive deficits and a heightened risk for neurodevelopmental disorders in offspring with early lifetime exposure to arsenic. Moreover, arsenic-induced apoptosis triggers oxidation processes and impairs mitochondrial respiration. Neuronal damage, synaptic loss, and learning and memory deficits are caused by the generation of ROS and depletion of antioxidant defenses. Research has demonstrated that exposure to arsenic results in hippocampal atrophy and lowered levels of critical neurotrophic factors (including the BDNF, which is vitally involved in the processes of synaptic plasticity and long-lasting potentiation necessary for the development of memory). Moreover, arsenic has been associated with alterations in cholinergic and dopaminergic neurotransmission and cognitive decline, as well as mood disorders. As a result, its toxicity is still an ongoing public health concern that needs urgent remediation efforts (Resendiz, Watkins, Öztürk, & Zhou, 2023; Scattolin, Resegue, & Rosário, 2022; Womersley et al., 2024).

Cadmium (Cd) is another toxic heavy metal that exerts significant neurotoxic effects and is ubiquitous in cigarette smoke, industrial waste, and contaminated food sources, including rice and leafy vegetables grown in contaminated soil. Cadmium differs from other heavy metals in that it has a long biological half-life, leading to the accumulation of cadmium in the tissues, including the brain, over time. Cadmium exposure poses a significant threat to health, especially synaptic plasticity and neurodevelopmental disorders; cadmium also disturbs calcium homeostasis in neurons, which affects synaptic transmission and LTP, both of which are required for learning and memory. In addition, cadmium disrupts vesicular transport and decreases the expression of synaptic proteins, thereby inhibiting neurotransmitter release by leading to cognitive deficits and increased risk of neurodevelopmental disorders, including ADHD and learning disabilities, among children exposed to cadmium during these critical periods of neurodevelopment. Moreover, cadmium can cause the formation of ROS and initiate proinflammatory factors that contribute to brain injury and apoptosis. Cadmium exposure activates microglial cells, leading to the release of proinflammatory cytokines, thus amplifying neurotoxicity (Di Pino, Guglielmelli, & Rossini, 2009; Guarnieri et al., 2024; Snoeijs et al., 2005). Cadmium has also been implicated in dopaminergic toxin exposure, which has been associated with the pathogenesis of PD. Reducing humans' cadmium impacts through better industrial regulation and dietary changes is essential to preventing these neurotoxic effects due to the ubiquity of cadmium in the environment and its ability to bioaccumulate.

Manganese (Mn), an essential trace element, is necessary for diverse enzymatic processes; however, excessive exposure results in neurotoxicity, particularly altering dopaminergic function and movement. Familiar sources of manganese exposure include welding fumes, drinking water, pesticides, and occupations like mining and metalworking. Chronic exposure to this toxic metal has been linked to a neurological disorder named manganism, which is an illness that bears no little resemblance to Parkinson's: motor impairments, tremors, and bradykinesia. However, the primary way that manganese can poison the brain is by interfering with dopamine. Manganese deposits are typically found in the basal ganglia, a brain region implicated in motor control, where manganese disrupts dopamine production, release, and reuptake. This disruption leads to dopamine depletion, motor dysfunction, and cognitive dysfunction. Manganese also causes oxidative stress and mitochondrial dysfunction, possibly resulting in neuronal apoptosis and synaptic loss. Manganese exposure decreases executive function, working memory, and attention, and children exposed to excess levels of manganese via drinking water have consistently been shown to suffer cognitive deficits (PDF). Moreover, manganese has been associated with heightened neuroinflammation, where higher levels of proinflammatory cytokines lead to neurodegeneration. The extensive use of manganese in various industrial processes and the risk of environmental contamination mean that manganese exposure is still a significant health concern, particularly among occupational workers and at-risk groups. Lead, mercury, arsenic, cadmium, and manganese are the most neurotoxic metals, and three are used in making different metals. It confers its effects through various mechanisms, such as calcium mimicry, oxidative stress, mitochondrial dysfunction, and neurotransmitter dysregulation, resulting in cognitive dysfunction and neurodevelopmental and neurodegenerative diseases. Given the increasing environmental pollution of these metals due to industrialization, pollution, and limited regulatory control, there is a pressing need for preventative public health initiatives, regulatory policies, and further investigation into neuroprotective strategies. The role of heavy metal neurotoxicity is critical in protecting the central nervous system (CNS), especially among susceptible populations, which include children, pregnant women, and workers with high exposure to these toxicants (Table 26.1).

26.3 Mechanismss of heavy metal–induced neuroplasticity disruptions

26.3.1 Synaptic dysfunction and neurotransmitter alterations

Three crucial neurotransmitter systems are impacted by the neurotoxicity of heavy metals, namely, the glutamatergic, dopaminergic, and cholinergic pathways that are relevant for cognition, memory, and motor control. Heavy metals [such as lead (Pb), mercury (Hg), arsenic (As), cadmium (Cd), and manganese (Mn)] are particularly disruptive to these systems, interfering with neurotransmitter secretion, receptor activity, and synaptic plasticity. The glutamatergic system is especially susceptible to heavy metal–induced toxicity, the predominant excitatory neurotransmitter system in the brain. Glutamate is crucial for synaptic transmission, learning, and memory; glutamate transporters and astrocytes tightly control its homeostasis. Heavy metals like mercury prevent glutamate uptake and lead to the impairment of astrocytic activities, causing excessive accumulation in the synaptic cleft. This causes excitotoxicity, during which excessive NMDA receptor activation causes overwhelming calcium influx into neurons, which results in oxidative stress, mitochondrial dysfunction, and neuronal apoptosis. Due to this disruption, impaired synaptic plasticity contributes to the cognitive deficits observed in heavy metal–exposed individuals. Heavy metal toxicity can also affect the dopaminergic system, which is involved in reward, motivation, motor control, and executive function. Manganese exposure, for instance, interferes with dopamine synthesis and transport in the basal ganglia, causing motor deficits akin to PD. At the neurotransmission level, lead exposure has been shown to reduce dopamine receptor density, impair dopamine signaling, and lead to deficits in attention, learning, and emotional regulation (Gao et al., 2025; Peleg et al., 2022; Shahlaei et al., 2024). Likewise, cadmium and arsenic disrupt the homeostasis of dopamine and increase neuroinflammatory processes and oxidative stress within dopaminergic neurons. There is excellent disruption in

TABLE 26.1 Heavy metals and their impact on neuroplasticity.

Heavy metal	Sources	Primary neurotoxic effects	Mechanism of neuroplasticity disruption	Associated neurological disorders
Lead (Pb)	Paint, water pipes, industrial emissions, batteries, contaminated soil	Cognitive deficits, reduced synaptic function, behavioral issues	Inhibition of NMDA receptors, oxidative stress, disruption of calcium homeostasis, apoptosis	ADHD, learning disabilities, dementia, reduced IQ
Mercury (Hg)	Fish (methylmercury), dental amalgams, industrial waste, thermometers, vaccines (historically, thimerosal)	Motor dysfunction, memory loss, tremors, speech impairment	Mitochondrial dysfunction, neuroinflammation, neurotransmitter imbalance (glutamate and GABA), impaired synaptogenesis	Autism spectrum disorder, Alzheimer's disease, Parkinson's disease, neuropathy
Arsenic (As)	Contaminated water, pesticides, industrial byproducts, mining, rice, and seafood	Cognitive impairment, behavioral changes, peripheral neuropathy	Disruption of synaptic signaling, epigenetic alterations, oxidative stress, DNA damage	Neurodevelopmental disorders, intellectual disability, neurodegeneration
Cadmium (Cd)	Batteries, cigarette smoke, contaminated soil, industrial waste, and shellfish	Neurodegeneration, oxidative stress, learning, and memory impairment	Disrupts calcium signaling, induces neuroinflammation, impairs blood–brain barrier function	Parkinson's disease, neurocognitive decline, peripheral neuropathy
Aluminum (Al)	Cookware, food packaging, antiperspirants, contaminated water	Memory loss, motor impairment, cognitive dysfunction	Disrupts synaptic plasticity, promotes amyloid-beta accumulation, and induces oxidative stress	Alzheimer's disease, cognitive impairment, neuroinflammation
Manganese (Mn)	Welding fumes, contaminated water, pesticides, and mining	Motor dysfunction, tremors, reduced dopamine levels	Alters dopamine metabolism, oxidative stress, neuroinflammation, mitochondrial dysfunction	Parkinsonism, cognitive decline, psychiatric disorders
Chromium (Cr-VI)	Industrial waste, contaminated drinking water, and stainless steel production	Memory impairment, behavioral disturbances, neurotoxicity	Induces oxidative damage, disrupts neurotransmitter systems, and inhibits neuronal differentiation	Neurocognitive disorders, motor dysfunction, and psychiatric effects
Nickel (Ni)	Industrial emissions, cigarette smoke, batteries, contaminated food	Cognitive impairment reduces neuronal viability	Induces DNA damage, alters synaptic transmission, and promotes neuroinflammation	Neurodegenerative diseases, memory deficits, mood disorders

ADHD, Attention deficit hyperactivity disorder; *GABA*, gamma-aminobutyric acid; *IQ*, intelligence quotient; *NMDA*, *N*-methyl-D-aspartate.
Source: Anyachor et al. (2023), Kamiński et al. (2024), Padhan, Simran, Kumar, and Verma (2025), and Shehryar (2025).

dopamine functioning, leading to neurodevelopmental disorders such as ADHD and an increase in susceptibility to neurodegenerative diseases. Heavy metals also negatively impact the cholinergic system, essential for memory, learning, and attention. Moreover, the inhibition of choline acetyltransferase (ChAT) (the synthetic enzyme of acetylcholine) by arsenic exposure has been proposed as a mechanism leading to reduced cholinergic system function. However, less acetylcholine is available as an excitatory element, disrupting cognitive function, especially across tasks where working memory and attention are necessary. Acute liver failure due to the inhibition of acetylcholinesterase (AChE), which is responsible for breaking down acetylcholine in the synaptic cleft, has also been associated with mercury toxicity and cholinergic dysfunction. This perturbation causes excessive stimulation of cholinergic receptors, resulting in excitotoxicity of the neurons, which in turn causes cognitive decline (Rahmatinia et al., 2024; Scher, Ludington, Smith, Klemming, & Pilon, 2024; Skalny et al., 2021). Besides interfering with the neurotransmitter systems, heavy metals also hinder synaptic plasticity by affecting LTP and long-term depression (LTD) and essential processes in learning and memory formation. LTP is a process that increases the strength of synaptic connections due to repeated stimuli, amplifying signal transmission between neurons and reinforcing memory consolidation. In contrast, LTD reduces synaptic strength, strengthening synaptic pruning and memory flexibility. Heavy metals like lead and cadmium impair LTP by stunting calcium-mediated signaling that downregulates the expression of synaptic proteins fundamental to the physical strengthening of synaptic connections. For example, lead prevents the function of the NMDA receptor, which leads to synaptic modification required for memory encoding. Likewise, mercury-induced oxidative stress and mitochondrial dysfunction impair LTP by flawed neuronal energy metabolism and low synaptic efficacy. In addition, heavy metals also alter LTD through loss of balance between excitatory and inhibitory signaling in neural circuits, leading to synaptically excessive weakening and cognitive flexibility deficits. These disturbances lead to learning disabilities, memory deficits, and higher predisposition to neurodegenerative illnesses like AD and PD. Considering the critical role of LTP and LTD in neural plasticity, heavy metal–induced impairments of these processes demonstrate the significant consequences environmental neurotoxicant exposure can have on cognitive and neurological health.

26.3.2 Oxidative stress and mitochondrial dysfunction

Neurotoxic effects of heavy metals are induced by inflammation of oxidative stress and mitochondrial disorder in the brain, resulting in neuron harm and cognitive deficit. All of these adverse conditions: lead (Pb), mercury (Hg), arsenic (As), cadmium (Cd), and especially manganese (Mn) can disturb the dynamic balance between oxidants and antioxidants in cells. They augment ROS such as superoxides, hydroxyl radicals, and hydrogen peroxide, simultaneously impairing mitochondrial cytocontrol. Under natural conditions, cells keep the production of ROS (or reactive oxygen species) in equilibrium with antioxidant defense mechanisms such as SOD (superoxide dismutase), glutathione peroxidase and catalaze all functioning faithfully. Exposure to heavy metals may diminish these natural protections, resulting in the accumulation of too many ROS, which can damage cellular components of all kinds or lead to neuron cell apoptosis. Increased production of ROS also damages lipid membranes, causing lipid peroxidation. The cell's phospholipids deteriorate for neuronal cell membranes and myelin sheaths, which are indispensable for signal conduction (El-Fawal, 2011b; Haase, Prasad, Cole, Baggett-Strehlau, & Wyatt, 2011; Lee et al., 2024; Spero et al., 2024). Runs on the membrane disturb neurotransmission, cause synaptic failure, and increase neuronal excitability. All of these mean that neurodegenerative processes are encouraged. Moreover, the allocation by ROS of synaptic proteins, essential ionic channels, and the elements that make up the cytoskeleton are other factors contributing to neuronal malfunction. Oxidative stress fatigues DNA, which is a further consequence of producing ROS. It can cause base modifications or strand breaks and chromosomal aberrations. These all lead to a state of genomic disarray, which can turn on proapoptotic pathways within neurons. Neurons are particularly susceptible because they have a high metabolic rate and limited capacity to regenerate, so heavy metal–induced neurotoxicity is characterized primarily by oxidative stress. In addition to increasing ROS production, heavy metals impede the bioenergetics of mitochondria. Mitochondria generate ATP using oxidative phosphorylation. Nonetheless, that does not prevent heavy metals from targeting components such as respiratory chain complexes in mitochondria, mitochondrial DNA (mtDNA), and antioxidant systems. This leads to mitochondrial disorder. For example, mercury and cadmium block mitochondrial respiratory chain complexes. This lowers the production of ATP and increases electron leakage to induce ROS further. Heavy metal lead disrupts calcium homeostasis, which can cause mitochondria to swell up and lead to the release of cytochrome *c*—which starts an apoptotic cascade. Arsenic affects the workings of mitochondrial enzymes involved in oxidative phosphorylation. This lowers the energy supply for neural cells, resulting in death; manganese accumulates within the mitochondria. This causes failure of most mitochondrial functions, including ATP production, and also changes in membrane character that, in turn, lead to swelling (Azargoonjahromi, 2023; Liang et al., 2023; Müller, Shevchenko, Gerhardt, Kiefer, & Vollstädt-Klein, 2021; Safhi et al., 2024). The results are devastating: apoptosis throughout the brain microvasculature; increased production of internuclear oxygen free radicals, which interfered with AIS recovered when not immediately analyzed after surgery; hindered release agents for centrally acting neurotransmitters such as arginine-rich peptides; shockingly, we did not see any significant results from our efforts The decline in antioxidant capacity and the reduced availability of ATP incite the brain's cellular destruction. This further weakens its oxygen uptake. We might even expect infection again, which would be catastrophic for a weakened immune system. Neuronal metabolisms are energy-dependent processes such as synaptic transmission, transmitter' release at nerve endings, and, even more importantly, neuron differentiation. These functions are essential for maintaining neural plasticity, that is, our ability to learn from new information and harmonize different types of data processing into a coherent whole structure (as currently enabled through 'intermediate memory circuits). The combination of oxidative stress and mitochondrial dysfunction adds insult to injury as neurodegenerative processes take their course, making it all but inevitable that diseases such as AD, PD, and amyotrophic lateral sclerosis (ALS) will eventually turn up in the brain. Because mitochondria are vital for neurons to remain alive and kicking, the heavy metal–induced damage to mitochondria is an essential pathological factor underlying neurotoxicity. Strategies now being developed to ease the body's load on animal organizational resources hold great promise for mitigating heavy metal–dependent oxidative stress and remedying mitochondrial dysfunctions. Antioxidant therapies, dietary strategies, and chelation therapy are needed if those toxic effects on the brain must be reversed or contained.

26.3.3 Neuroinflammation and microglial activation

Chronic inflammation and cytokine dysregulation play a significant role in heavy metal–induced neurotoxicity, contributing to long-term neuronal damage and cognitive impairments. Heavy metals such as lead (Pb), mercury (Hg), arsenic (As), cadmium (Cd), and manganese (Mn) act as potent neuroinflammatory agents, triggering the activation of microglia and astrocytes, the brain's primary immune cells. Under normal conditions, microglia maintain homeostasis by clearing cellular debris and regulating immune responses. However, prolonged exposure to heavy metals leads to excessive microglial activation, resulting in the overproduction of proinflammatory cytokines such as tumor necrosis factor-alpha (TNF-α), interleukin-1 beta (IL-1β), and interleukin-6 (IL-6). This dysregulation amplifies neuroinflammation, disrupts synaptic function, and increases oxidative stress, exacerbating neuronal damage. In parallel, astrocytes, which provide metabolic and structural support to neurons, become reactive in response to heavy metal exposure, further contributing to the release of inflammatory mediators and the breakdown of neuronal homeostasis (Grace, Duran, Ann Ottinger, & Maness, 2024; Kumar, Gupta, & Pratihar, 2025; Massie, Kantak, Narayanan, & Wittenberg, 2015). The sustained release of proinflammatory cytokines damages neurons and alters neurotransmitter balance, leading to cognitive function, learning, and memory impairments. Additionally, chronic inflammation interferes with neurogenesis, reducing the brain's ability to repair and regenerate damaged neuronal circuits. The BBB, a highly selective structure that protects the brain from harmful substances, is crucial in regulating neuroinflammation. However, heavy metal exposure compromises BBB integrity, allowing neurotoxic substances, immune cells, and inflammatory mediators to infiltrate the CNS. Lead, for example, disrupts tight junction proteins such as claudin-5 and occludin, increasing BBB permeability and facilitating the entry of circulating inflammatory cytokines and toxins into the brain. Mercury accumulates in endothelial cells lining the BBB, inducing oxidative stress and endothelial dysfunction, further weakening the barrier and promoting neuroinflammation. Arsenic and cadmium similarly induce endothelial cell apoptosis, leading to vascular leakage and increased susceptibility to CNS inflammation. Once the BBB is compromised, systemic inflammatory responses triggered by heavy metals can more easily influence the brain, exacerbating neurodegenerative processes. The breakdown of the BBB also exposes neurons to peripheral immune cells, triggering additional inflammatory cascades that accelerate neuronal degeneration. The cumulative effects of BBB disruption, chronic neuroinflammation, and cytokine dysregulation create a vicious cycle where persistent immune activation leads to sustained neuronal damage and cognitive decline. This process is particularly concerning in neurodevelopmental disorders such as ASD and ADHD, as well as in neurodegenerative diseases like AD and PD, where inflammation is a key pathological feature. Given the critical role of inflammation in heavy metal neurotoxicity, therapeutic strategies aimed at reducing neuroinflammation and preserving BBB integrity, such as anti-inflammatory drugs, antioxidants, and dietary interventions, are essential for mitigating the long-term neurological effects of heavy metal exposure. Strengthening the BBB through pharmacological or nutritional approaches may help limit the infiltration of inflammatory mediators and protect against further neuronal damage, offering potential avenues for preventing heavy metal–induced neurotoxicity and cognitive decline (Table 26.2).

26.3.4 Epigenetic modifications and gene expression changes

Epigenetic modifications and gene expression changes are critical mechanisms through which heavy metals exert long-term neurotoxic effects, influencing neurodevelopmental and neurodegenerative diseases. Heavy metals such

TABLE 26.2 Neuroplasticity-related markers and their alterations under heavy metal toxicity.

Marker	Function in neuroplasticity	Alterations under heavy metal toxicity	Affected heavy metals	Affected brain regions
BDNF	Supports neuronal survival, synaptic plasticity, learning, and memory	Decreased expression, impaired neurogenesis, reduced synaptic plasticity	Lead (Pb), mercury (Hg), arsenic (As), cadmium (Cd), aluminum (Al)	Hippocampus, prefrontal cortex, striatum
SYP	Presynaptic vesicle protein critical for synaptic transmission	Reduced expression, synaptic dysfunction, impaired neurotransmitter release	Lead (Pb), mercury (Hg), cadmium (Cd)	Hippocampus, cortex
PSD-95	Scaffolding protein for synapse stabilization and plasticity	Decreased levels, loss of dendritic spine density, impaired synaptic connectivity	Lead (Pb), mercury (Hg), arsenic (As)	Hippocampus, cortex
CREB	Regulates memory formation, neuronal survival, and synaptic plasticity	Downregulated expression, impaired LTP, and memory formation	Lead (Pb), mercury (Hg), cadmium (Cd), aluminum (Al)	Hippocampus, prefrontal cortex
GAP-43	Axonal growth, synaptic remodeling, neuronal regeneration	Decreased expression, impaired axonal regeneration, and plasticity	Lead (Pb), mercury (Hg), cadmium (Cd)	Hippocampus, cortex
NF-κB	Regulates neuronal survival, inflammation, and synaptic plasticity	Hyperactivation, neuroinflammation, neuronal apoptosis	Mercury (Hg), arsenic (As), cadmium (Cd), manganese (Mn)	Hippocampus, amygdala
DCX	Marker of neurogenesis and immature neurons	Reduced expression, impaired neuronal migration, and differentiation	Lead (Pb), mercury (Hg), arsenic (As)	Hippocampus, subventricular zone
α-Synuclein	Involved in synaptic vesicle regulation and neurotransmission	Accumulation, neurotoxicity, aggregation into Lewy bodies	Manganese (Mn), aluminum (Al)	Substantia nigra, basal ganglia
NR2B	Critical for synaptic plasticity and LTP	Downregulated expression, impaired excitatory synaptic transmission	Lead (Pb), mercury (Hg), arsenic (As)	Hippocampus, cortex
GFAP	Marker of astrocyte activation and neuroinflammation	Increased expression, gliosis, neuroinflammation	Mercury (Hg), arsenic (As), cadmium (Cd), manganese (Mn)	Cortex, hippocampus, basal ganglia

BDNF, Brain-derived neurotrophic factor; *CREB*, cAMP response element-binding protein; *DCX*, doublecortin; *GAP-43*, growth-associated protein-43; *GFAP*, glial fibrillary acidic protein; *LTP*, long-term potentiation; *NF-κB*, nuclear factor kappa B; *NR2B*, NMDA receptor subunit 2B; *PSD-95*, postsynaptic density protein-95; *SYP*, synaptophysin.
Source: Based on Bottini and MacDougall-Shackleton (2023), Scher et al. (2024), Shahlaei et al. (2024), Skalny et al. (2021), and Zhang et al. (2020).

as lead (Pb), mercury (Hg), arsenic (As), cadmium (Cd), and manganese (Mn) can alter the epigenetic landscape of neurons by disrupting DNA methylation, histone modifications, and noncoding RNA expression, ultimately affecting gene regulation, synaptic plasticity, and neuronal survival. DNA methylation, one of the most well-characterized epigenetic modifications, involves adding a methyl group to cytosine residues in CpG islands, typically leading to transcriptional repression. Heavy metal exposure can cause global hypomethylation or locus-specific hypermethylation, leading to aberrant gene expression patterns that impair neuronal function (Grace et al., 2024; Kumar et al., 2025; Lee et al., 2024; Liang et al., 2023). For instance, lead exposure during early development has been associated with altered methylation of genes involved in synaptic plasticity, such as BDNF, resulting in long-lasting cognitive deficits and increased susceptibility to neurodevelopmental disorders like ASD and ADHD. Similarly, arsenic exposure has been shown to induce hypermethylation of tumor suppressor genes and hypomethylation of proto-oncogenes in the brain, disrupting cellular homeostasis and promoting neurodegeneration. In addition to DNA methylation, histone modifications, including acetylation, methylation, phosphorylation, and ubiquitination, play a crucial role in regulating chromatin structure and gene expression. Heavy metals interfere with these processes by altering the activity of histone-modifying enzymes such as histone deacetylases (HDACs) and histone acetyltransferases, leading to transcriptional dysregulation. For example, cadmium exposure has been linked to increased histone deacetylation, resulting in the repression of genes involved in neuronal differentiation and synaptic function. Conversely, mercury disrupts histone methylation patterns, impairing the expression of genes critical for mitochondrial function and oxidative stress response. These histone modifications contribute to neuroinflammation, oxidative damage, and synaptic dysfunction, which are hallmark features of neurodegenerative diseases such as AD and PD. Noncoding RNAs, particularly microRNAs (miRNAs) and long noncoding RNAs (lncRNAs), further mediate the effects of heavy metal exposure on gene expression by posttranscriptionally regulating target mRNAs. Dysregulation of miRNAs due to heavy metal toxicity can lead to the overexpression or suppression of genes involved in neuronal survival, apoptosis, and synaptic plasticity. For instance, lead exposure has been shown to upregulate miR-132, a microRNA associated with synaptic dysfunction and cognitive impairment. In contrast, arsenic exposure alters the expression of miR-124, which plays a critical role in neurogenesis and synaptic plasticity. The combined impact of DNA methylation changes, histone modifications, and miRNA dysregulation creates a cascade of molecular alterations that disrupt neuronal homeostasis, increase vulnerability to neurodevelopmental disorders, and accelerate the progression of neurodegenerative diseases. Given the reversible nature of epigenetic modifications, targeting these pathways through epigenetic therapies, including HDAC inhibitors, DNA methylation modulators, and miRNA-based interventions, presents a promising strategy for mitigating the adverse effects of heavy metal–induced neurotoxicity and promoting neuroprotection (Haase et al., 2011; Rahmatinia et al., 2024; Shahlaei et al., 2024).

26.3.5 Gut–brain axis and heavy metal neurotoxicity

The gut–brain axis plays a crucial role in maintaining neurological health, and disruptions in this bidirectional communication system due to heavy metal exposure can significantly contribute to neurotoxicity. Heavy metals such as lead (Pb), mercury (Hg), arsenic (As), cadmium (Cd), and manganese (Mn) can induce gut dysbiosis, alter microbiome composition, increase gut permeability, and trigger systemic inflammation, all of which have profound effects on brain function. The gut microbiome is a complex community of bacteria, fungi, and other microorganisms that regulate immune responses, produce neurotransmitters, and influence brain function through metabolic pathways. However, exposure to heavy metals disrupts the delicate balance of gut microbiota, leading to an overgrowth of pathogenic bacteria while reducing beneficial species such as *Lactobacillus* and *Bifidobacterium*. This dysbiosis alters the production of short-chain fatty acids like butyrate, acetate, and propionate, which are essential for maintaining the integrity of the gut lining, regulating neuroinflammation, and modulating neurotransmitter synthesis. For instance, butyrate deficiency due to microbiome disruption impairs the production of BDNF, a key protein involved in synaptic plasticity, learning, and memory. Additionally, certain gut bacteria play a role in detoxifying heavy metals by sequestering them or modifying their bioavailability. Their depletion due to dysbiosis exacerbates metal toxicity, allowing for excellent absorption into the systemic circulation and ultimately crossing the BBB.

Heavy metal exposure also compromises gut barrier integrity, leading to a condition known as “leaky gut,” where tight junction proteins such as occludin and claudin become disrupted, increasing intestinal permeability. As a result, toxic metal ions, bacterial endotoxins like lipopolysaccharides, and inflammatory cytokines enter the bloodstream, triggering systemic inflammation that extends to the brain. This systemic inflammation exacerbates microglial activation, leading to the overproduction of proinflammatory cytokines such as TNF-α, IL-6, and IL-1β, which contribute to neuronal damage and cognitive deficits. Mercury, for example, has been shown to increase gut permeability while inducing neuroinflammation, leading to impairments in cognition, mood, and behavior. Similarly, arsenic disrupts intestinal tight junctions and alters microbial metabolites, promoting neurotoxic effects such as mitochondrial dysfunction and oxidative stress in the brain (Kumar et al., 2025; Müller et al., 2021; Spero et al., 2024).

Furthermore, the gut microbiota influences neurotransmitter production, and heavy metal–induced dysbiosis affects key neurotransmitter systems, including serotonin, dopamine, and gamma-aminobutyric acid. For instance, lead and cadmium exposure reduce serotonin-producing bacteria, leading to mood disorders such as depression and anxiety. Manganese toxicity has been linked to excessive accumulation in the basal ganglia, disrupting dopamine homeostasis and contributing to motor deficits resembling PD. Additionally, gut microbiome imbalances influence the hypothalamic–pituitary–adrenal axis, increasing stress hormone levels, which further aggravate neurotoxicity and cognitive dysfunction. The cumulative effects of dysbiosis, increased gut permeability, and systemic inflammation create a vicious cycle that perpetuates neurodegeneration and impairs cognitive function. Addressing gut–brain axis disruptions through probiotics, prebiotics, dietary interventions, and microbiome-targeted therapies presents a promising approach to mitigating heavy metal–induced neurotoxicity and restoring neurological health (Table 26.3).

26.4 Heavy metals and neurodevelopmental and neurodegenerative disorders

26.4.1 Autism spectrum disorder

ASD is a complex neurodevelopmental condition characterized by deficits in social interaction, communication difficulties, and repetitive behaviors, with a rising prevalence worldwide. While genetic factors play a crucial role in ASD, emerging research highlights the significance of environmental influences and hefty metal exposure in early-life neurodevelopmental alterations. Heavy metals such as lead (Pb), mercury (Hg), arsenic (As), cadmium (Cd), and manganese (Mn) are potent neurotoxicants that can interfere with critical developmental processes in the fetal and early postnatal brain. These metals can cross the placenta, disrupt neuronal differentiation, alter neurotransmitter systems, induce oxidative stress, and provoke neuroinflammatory responses, all of which are implicated in the pathogenesis of ASD. Prenatal and early childhood exposure to heavy metals can have lasting consequences on brain development, contributing to structural and functional abnormalities associated with ASD-like behaviors. Lead, for example, mimics calcium ions and disrupts synaptic function, leading to impaired neural connectivity and cognitive dysfunction. Studies have shown that even low levels of lead exposure during early childhood are associated with increased ASD risk, reduced IQ, attentional deficits, and altered social behavior. Mercury, particularly in the form of methylmercury from contaminated seafood or ethylmercury from certain medical products, can interfere with glutamate homeostasis, disrupt mitochondrial function, and cause neuronal apoptosis, mechanisms that align with the neuropathology observed in ASD.

Arsenic exposure, primarily from contaminated drinking water and industrial pollution, has been linked to epigenetic modifications that alter gene expression patterns critical for brain development. Research suggests that arsenic-induced DNA methylation changes may influence neurodevelopmental genes, leading to deficits in synaptic plasticity, memory formation, and behavioral regulation, all of which are relevant to ASD. Cadmium, another heavy metal commonly found in cigarette smoke and industrial waste, can impair calcium-dependent neurotransmission, induce oxidative stress, and trigger inflammatory responses in the developing brain, potentially contributing to the neurodevelopmental abnormalities seen in ASD. Elevated cadmium levels have been detected in children with ASD, further supporting the hypothesis that environmental heavy metal exposure plays a role in the disorder's etiology. Manganese, an essential trace element required for enzymatic activity and neurotransmitter metabolism, becomes neurotoxic when present in excessive amounts. Excessive manganese exposure, often from contaminated drinking water or occupational exposure, has been associated with impaired dopamine regulation, motor deficits, and ASD-like behavioral phenotypes, highlighting its potential role in neurodevelopmental disruptions (Azargoonjahromi, 2023; Lee et al., 2024; Massie et al., 2015; Shahlaei et al., 2024; Skalny et al., 2021).

Beyond direct neurotoxicity, heavy metals can exacerbate ASD risk through immune system dysregulation and chronic neuroinflammation. Elevated levels of proinflammatory cytokines, such as IL-6 and TNF-α, are commonly observed in children with ASD, suggesting an inflammatory component in the disorder's pathogenesis. Heavy metals can activate microglial cells, leading to sustained neuroinflammatory responses that disrupt synaptic pruning, neuronal migration, and myelination. Given the growing evidence linking heavy metal exposure to ASD, it is imperative to implement strategies for reducing environmental contamination, enhancing public health policies, and developing potential therapeutic interventions that target oxidative stress, inflammation, and epigenetic dysregulation to mitigate the impact of heavy metals on neurodevelopment.

26.4.2 Attention-deficit hyperactivity disorder

Dopamine is a critical neurotransmitter involved in regulating executive function, motivation, reward processing,

TABLE 26.3 Protective strategies against heavy metal–induced neurotoxicity.

Protective strategy	Mechanism of action	Examples	Targeted heavy metals	Neuroprotective effects
Dietary interventions	It enhances antioxidant defense, reduces oxidative stress, and inhibits metal absorption.	Curcumin, quercetin, vitamin C, vitamin E, omega-3 fatty acids, selenium, zinc	Lead (Pb), mercury (Hg), cadmium (Cd), arsenic (As), aluminum (Al)	Reduces oxidative stress, enhances synaptic plasticity, and promotes neuronal survival
Chelation therapy	Binds and removes heavy metals from the body	EDTA, DMSA, DMPS, BAL	Lead (Pb), mercury (Hg), arsenic (As), cadmium (Cd)	Lowers heavy metal burden, improves cognitive function, and reduces neuroinflammation
Pharmacological treatments	Modulates neurotransmitter balance, prevents excitotoxicity, and enhances neuroprotection.	NAC, melatonin, lipoic acid, resveratrol	Lead (Pb), mercury (Hg), cadmium (Cd), arsenic (As)	Protects against mitochondrial dysfunction, reduces apoptosis, and prevents cognitive decline
Neuroprotective agents	Prevents neuronal degeneration, supports neurogenesis, and improves synaptic plasticity	BDNF enhancers, CREB activators, glutathione precursors, antiinflammatory agents (ibuprofen, minocycline)	Lead (Pb), mercury (Hg), cadmium (Cd), arsenic (As), manganese (Mn)	Restores neuroplasticity, reduces neuroinflammation, enhances memory and learning
Gut microbiome modulation	Reduces metal absorption, detoxifies heavy metals, and prevents gut dysbiosis	Probiotics (*Lactobacillus*, *Bifidobacterium*), prebiotics, fermented foods	Lead (Pb), mercury (Hg), cadmium (Cd), arsenic (As)	Prevents neuroinflammation, enhances gut–brain axis communication, reduces heavy metal bioavailability
Lifestyle modifications	Reduces exposure, enhances detoxification, and promotes overall brain health	Regular exercise, sauna therapy, hydration, and avoiding contaminated water and food	Lead (Pb), mercury (Hg), cadmium (Cd), arsenic (As)	Enhances neuroplasticity, improves blood circulation, supports toxin elimination

BAL, British anti-Lewisite; *BDNF*, brain-derived neurotrophic factor; *CREB*, cAMP response element-binding protein; *DMPS*, dimercapto-propane sulfonate; *DMSA*, dimercaptosuccinic acid; *EDTA*, ethylenediaminetetraacetic acid; *NAC*, *N*-acetylcysteine.
Source: Based on Bottini and MacDougall-Shackleton (2023), Müller et al. (2021), Rahmatinia et al. (2024), and Safhi et al. (2024).

and motor control, and disruptions in dopamine signaling have been implicated in numerous neurological and neuropsychiatric disorders. Exposure to heavy metals such as lead (Pb), mercury (Hg), arsenic (As), cadmium (Cd), and manganese (Mn) has been shown to interfere with dopamine homeostasis, leading to cognitive impairments, behavioral abnormalities, and deficits in executive function. Dopaminergic neurons, primarily located in the substantia nigra and ventral tegmental area, project to key brain regions such as the prefrontal cortex and striatum, areas that govern decision-making, working memory, attention, and impulse control. Heavy metals exert neurotoxic effects on these pathways by altering dopamine synthesis, transport, receptor function, and degradation, ultimately contributing to executive function and motor control dysfunctions. Lead exposure, for example, has been shown to reduce dopamine receptor density and impair dopamine transporter (DAT) function, leading to altered synaptic dopamine levels and cognitive deficits resembling ADHD. Studies have linked early-life lead exposure to reduced cognitive flexibility, impaired inhibitory control, and deficits in working memory, all of which are hallmarks of executive dysfunction. Mercury, particularly methylmercury, disrupts dopamine signaling by inhibiting dopamine uptake and causing oxidative stress-induced neurodegeneration in dopaminergic neurons. This disruption has been associated with mood disorders, impulsivity, and deficits in cognitive performance, reinforcing the link between heavy metal exposure and impaired executive function (Bottini & MacDougall-Shackleton, 2023; Liang et al., 2023; Rahmatinia et al., 2024; Shahlaei et al., 2024; Spero et al., 2024).

Arsenic exposure has also been found to alter dopamine metabolism, leading to reduced dopamine levels and an imbalance in neurotransmitter systems essential for cognitive function. Chronic arsenic toxicity has been associated with learning and memory deficits, reduced attention span, and impaired problem-solving abilities, further supporting its role in executive dysfunction. Similarly, cadmium has been shown to interfere with DAT function and induce neuroinflammation, contributing to changes in behavior and cognitive performance. Studies in animal models have demonstrated that cadmium exposure reduces dopamine turnover and impairs synaptic transmission, which may underlie the observed deficits in attention, decision-making, and goal-directed behavior. Manganese, while an essential trace element, becomes neurotoxic when present in excessive amounts, leading to a condition known as manganism, characterized by motor deficits resembling PD. Excessive manganese accumulation in the basal ganglia disrupts dopamine release and receptor function, impairing executive functions such as impulse control, task-switching, and cognitive flexibility. Additionally, manganese-induced oxidative stress and neuroinflammation further exacerbate dopaminergic dysfunction, leading to long-term neurocognitive deficits.

Given dopamine's essential role in regulating higher order cognitive functions, disruptions caused by heavy metal exposure have profound implications for neurodevelopment and behavior. Individuals exposed to high levels of these neurotoxic metals are at increased risk of developing attention deficits, poor impulse control, and impaired decision-making abilities, which can persist into adulthood. Understanding the mechanisms by which heavy metals disrupt dopamine signaling is crucial for developing strategies to mitigate their effects, including dietary interventions, chelation therapies, and policies aimed at reducing environmental exposure to these harmful neurotoxicants (Haase et al., 2011; Liang et al., 2023; Massie et al., 2015; Skalny et al., 2021; Spero et al., 2024).

26.4.3 Alzheimer's disease

The role of heavy metals as risk factors in amyloid-beta (Aβ) aggregation has gained significant attention due to their potential contribution to neurodegenerative diseases, particularly AD. Heavy metals such as lead (Pb), mercury (Hg), arsenic (As), cadmium (Cd), and manganese (Mn) have been shown to disrupt protein homeostasis, promote oxidative stress, and interfere with cellular mechanisms that regulate amyloid precursor protein (APP) processing, leading to increased Aβ deposition. Aβ aggregation is a hallmark of AD pathology, contributing to synaptic dysfunction, neuronal loss, and cognitive decline. Heavy metals can accelerate Aβ oligomerisation and plaque formation through direct metal-protein interactions and indirect mechanisms involving oxidative stress, mitochondrial dysfunction, and inflammation. Lead exposure, particularly during early life, has been linked to an increased risk of AD due to its ability to enhance Aβ production and aggregation. Studies have shown that lead can disrupt APP processing by altering the activity of secretases, the enzymes responsible for cleaving APP into Aβ peptides. Lead exposure also induces epigenetic modifications that result in long-term changes in gene expression related to Aβ metabolism, further increasing susceptibility to AD in later life. Additionally, lead can enhance Aβ aggregation by binding directly to Aβ peptides, altering their structural properties, and promoting the formation of toxic oligomers and fibrils.

Mercury is another heavy metal implicated in Aβ aggregation. Mercury exposure, particularly in the form of methylmercury, induces oxidative stress and mitochondrial dysfunction, which contribute to Aβ accumulation. Mercury can also interact directly with Aβ peptides, altering their conformation and increasing their propensity to form aggregates. Furthermore, mercury has been shown to impair proteasomal and lysosomal degradation pathways, leading to reduced clearance of Aβ and an increased burden of toxic amyloid deposits. Chronic mercury exposure has been associated with cognitive impairments and neuroinflammatory responses that exacerbate Aβ pathology. Arsenic exposure has also been linked to increased Aβ aggregation through mechanisms involving oxidative stress and epigenetic modifications. Arsenic disrupts mitochondrial function, increasing ROS that promote protein misfolding and aggregation (El-Fawal, 2011b; Haase et al., 2011; Kumar et al., 2025; Lee et al., 2024; Müller et al., 2021). Additionally, arsenic-induced epigenetic changes can upregulate genes involved in Aβ production while downregulating those responsible for Aβ clearance, further contributing to amyloid pathology. Similarly, cadmium exposure has been shown to accelerate Aβ aggregation by promoting oxidative stress and interfering with metal ion homeostasis, particularly zinc and copper, which play crucial roles in regulating Aβ aggregation and clearance.

Manganese is an essential trace element, but too much can be toxic to the brain. Manganese exposure has been linked to increased Aβ aggregation through mechanisms involving oxidative stress, neuroinflammation, and impaired proteolytic degradation of Aβ peptides. Studies have shown that manganese can bind to Aβ peptides and alter their aggregation dynamics, forming neurotoxic amyloid fibrils. Furthermore, manganese-induced neuroinflammation can exacerbate Aβ pathology by activating microglia and astrocytes, which release proinflammatory cytokines that further promote amyloid deposition. Overall, heavy metals play a significant role in Aβ aggregation and the progression of neurodegenerative diseases like AD. By inducing oxidative stress, disrupting metal ion homeostasis, impairing protein clearance mechanisms, and directly interacting with Aβ peptides, heavy metals contribute to the formation of toxic amyloid aggregates. Understanding these mechanisms is crucial for developing therapeutic interventions that target heavy metal–induced amyloid pathology and mitigate the risk of AD and other neurodegenerative disorders.

26.4.4 Parkinson's disease

Manganese (Mn) toxicity has been extensively studied for its detrimental effects on the CNS, particularly its role in dopaminergic neuron degeneration. While manganese is an essential trace element required for enzymatic functions, mitochondrial activity, and antioxidant defense, excessive exposure can lead to neurotoxicity, primarily affecting the basal ganglia and dopaminergic pathways. Chronic manganese exposure has been associated with manganism, a neurological disorder with symptoms similar to PD, including bradykinesia, rigidity, tremors, and postural instability. Unlike idiopathic PD, manganism does not typically respond to dopamine replacement therapies, indicating distinct underlying neurotoxic mechanisms. The primary mode of manganese-induced neurotoxicity involves oxidative stress, mitochondrial dysfunction, neuroinflammation, and impaired dopamine metabolism, leading to progressive degeneration of dopaminergic neurons in the substantia nigra and striatum (Lee et al., 2024; Liang et al., 2023; Massie et al., 2015; Shahlaei et al., 2024; Skalny et al., 2021).

Manganese readily crosses the BBB via active transport mechanisms, accumulating in brain regions with high dopaminergic activity, particularly the globus pallidus and substantia nigra. Once inside the brain, manganese disrupts iron homeostasis, producing excessive ROS and oxidative stress. Dopaminergic neurons are highly susceptible to oxidative damage due to the inherent reactivity of dopamine, which can undergo autooxidation, forming neurotoxic quinones and hydrogen peroxide. Manganese further exacerbates this oxidative burden by impairing mitochondrial function, reducing ATP production, and increasing mitochondrial permeability, leading to neuronal apoptosis. Studies have shown that manganese exposure decreases the activity of the electron transport chains (ETCs) complex I and complex IV, leading to energy deficits and increased ROS generation, which contribute to dopaminergic neurodegeneration.

Another key mechanism of manganese-induced dopaminergic toxicity is its interference with dopamine metabolism and synaptic transmission. Manganese disrupts vesicular monoamine transporter 2, which is responsible for packaging dopamine into synaptic vesicles for regulated release. This disruption leads to an accumulation of free cytosolic dopamine, which undergoes oxidation, further increasing oxidative stress and neuronal damage. Additionally, manganese exposure alters the expression of DATs, leading to reduced dopamine reuptake and imbalances in extracellular dopamine levels. These changes contribute to synaptic dysfunction, impairing dopaminergic signaling and motor control. Neuroinflammation is another major contributor to manganese-induced dopaminergic neurodegeneration. Manganese activates microglia and astrocytes, releasing proinflammatory cytokines such as TNF-α, IL-1β, and IL-6. Chronic neuroinflammation exacerbates oxidative stress and promotes neuronal apoptosis, accelerating dopaminergic cell loss. Furthermore, manganese has been shown to impair glutamate homeostasis by disrupting astrocytic glutamate transporters, leading to excitotoxicity, a process that further damages dopaminergic neurons. Prolonged manganese exposure has also been linked to epigenetic modifications influencing gene expression patterns associated with dopaminergic function and neurodegeneration. Changes in DNA methylation, histone modifications, and noncoding RNA expression contribute to the progressive decline of dopaminergic neurons, leading to long-term neurotoxic effects. The cumulative impact of oxidative stress, mitochondrial dysfunction, dopamine dysregulation, neuroinflammation, and epigenetic changes highlights the multifaceted nature of manganese toxicity in dopaminergic neuron degeneration. Understanding these mechanisms is essential for developing therapeutic strategies to mitigate manganese-induced neurotoxicity and prevent related neurological disorders (Azargoonjahromi, 2023; El-Fawal, 2011b; Massie et al., 2015; Shahlaei et al., 2024; Skalny et al., 2021).

26.4.5 Amyotrophic lateral sclerosis and other motor disorders

ALS and other motor disorders have been increasingly linked to environmental factors, particularly exposure to heavy metals, contributing to neurodegeneration in motor neurons. ALS is a progressive neurodegenerative disease characterized by the selective loss of motor neurons in the spinal cord, brainstem, and motor cortex, leading to muscle weakness, paralysis, and, ultimately, respiratory failure. While genetic mutations play a significant role in ALS pathogenesis, environmental factors such as heavy metal exposure have been identified as major contributors to disease onset and progression. Heavy metals, including lead (Pb), mercury (Hg), arsenic (As), cadmium (Cd), and manganese (Mn), are known to accumulate in the nervous system, triggering oxidative stress, mitochondrial dysfunction, neuroinflammation, and impaired protein homeostasis, all of which contribute to motor neuron degeneration. One of the primary mechanisms through which heavy metals contribute to ALS and motor neuron disorders is oxidative stress. Motor neurons are highly susceptible to oxidative damage due to their high metabolic rate, long axonal projections, and limited capacity for antioxidant defense. Heavy metals, notably lead, mercury, and arsenic, promote the generation of ROS, which damage cellular components such as lipids, proteins, and DNA. Mercury, for example, disrupts thiol homeostasis by binding to cysteine residues in antioxidant enzymes such as glutathione peroxidase, reducing the cell's ability to neutralize oxidative stress. Similarly, arsenic exposure has been shown to inhibit superoxide dismutase (SOD) and catalaze activity, further exacerbating oxidative damage in motor neurons. Mitochondrial dysfunction is another critical factor in heavy metal–induced motor neuron degeneration. Mitochondria play a vital role in energy production, calcium homeostasis, and apoptosis regulation, and their impairment has been strongly associated with ALS pathogenesis. Heavy metals interfere with mitochondrial bioenergetics by inhibiting ETC complexes, reducing ATP production, and increasing ROS generation. Cadmium, for example, disrupts mitochondrial membrane potential, causing cytochrome *c* release and activation of apoptotic pathways (Bottini & MacDougall-Shackleton, 2023; Grace et al., 2024; Haase et al., 2011; Rahmatinia et al., 2024; Shahlaei et al., 2024; Spero et al., 2024). Additionally, lead and manganese have been shown to impair mitochondrial calcium buffering, leading to excitotoxicity and neuronal death. The cumulative impact of mitochondrial dysfunction in motor neurons results in energy deficits, increased oxidative burden, and eventual neurodegeneration. Neuroinflammation also plays a crucial role in heavy metal–induced motor neuron degeneration. Chronic exposure to heavy metals triggers microglial activation, resulting in the release of proinflammatory cytokines such as TNF-α, IL-6, and IL-1β. This persistent inflammatory state exacerbates oxidative stress and mitochondrial dysfunction, creating a toxic environment that accelerates motor neuron loss. Manganese, in particular, has been linked to increased neuroinflammation in ALS patients, with elevated levels of inflammatory markers detected in cerebrospinal fluid.

Furthermore, heavy metals have been shown to interfere with protein homeostasis, a key factor in ALS pathology. Protein misfolding and aggregation are hallmarks of ALS, with toxic inclusions of TDP-43, SOD1, and FUS proteins observed in affected motor neurons. Heavy metals such as lead and mercury exacerbate protein misfolding by disrupting the ubiquitin–proteasome system and autophagy pathways, accumulating neurotoxic aggregates. The failure to clear misfolded proteins results in endoplasmic reticulum stress and neuronal apoptosis, further contributing to disease progression. In addition to ALS, heavy metal exposure has been implicated in other motor disorders, including PD, multiple sclerosis, and manganism. The disruption of dopaminergic and glutamatergic signaling caused by heavy metals can lead to motor dysfunction, tremors, rigidity, and spasticity, symptoms commonly observed in these disorders. For instance, manganese exposure has been closely linked to Parkinsonian symptoms, as it selectively accumulates in the basal ganglia, disrupting dopamine metabolism and inducing oxidative damage in dopaminergic neurons.

Given the growing evidence of heavy metals as risk factors for ALS and other motor disorders, efforts to mitigate exposure are crucial. Occupational safety measures, environmental regulations, and public health interventions to reduce heavy metal contamination in air, water, and food sources are essential for preventing neurodegeneration and motor dysfunction. Additionally, developing therapeutic strategies targeting oxidative stress, mitochondrial dysfunction, neuroinflammation, and protein misfolding may help in slowing the progression of ALS and related disorders. Understanding the complex interplay between heavy metal exposure and motor neuron degeneration is critical for developing practical preventive and therapeutic approaches in neurodegenerative diseases (Lee et al., 2024; Massie et al., 2015; Müller et al., 2021; Safhi et al., 2024; Skalny et al., 2021; Spero et al., 2024).

26.5 Strategies for mitigating heavy metal–induced neurotoxicity

26.5.1 Chelation therapy and metal detoxification

Chelation therapy is a widely studied approach for removing toxic heavy metals from the body, offering potential relief from heavy metal–induced neurotoxicity and systemic damage. This therapy involves the administration of chelating agents—compounds that bind to metal ions to form stable complexes that are excreted through urine or feces. Chelating agents like ethylenediaminetetraacetic acid (EDTA), meso-2,3-dimercaptosuccinic acid (DMSA), and 2,3-dimercapto-1-propanesulfonic acid (DMPS) have been used to treat metal poisoning from lead (Pb), mercury (Hg), arsenic (As), and cadmium (Cd). While chelation therapy can be highly effective in reducing heavy metal burdens, its use comes with potential risks and limitations, especially concerning long-term neurological effects. Understanding the balance between therapeutic efficacy and adverse effects is crucial for determining the appropriate clinical applications of chelation therapy in heavy metal detoxification.

EDTA, one of the earliest and most commonly used chelating agents, is particularly effective for treating lead poisoning. It binds to divalent and trivalent metal cations, forming stable, water-soluble complexes that are eliminated through the kidneys. EDTA has been shown to lower blood lead levels and reduce some symptoms of lead toxicity, including cognitive impairments and cardiovascular dysfunction. However, EDTA is nonspecific and can bind to essential minerals like calcium, zinc, and magnesium, leading to potential deficiencies. In some cases, rapid administration or high doses of EDTA can cause hypocalcemia, arrhythmias, and renal toxicity, highlighting the need for careful monitoring during treatment. Additionally, EDTA is less effective at removing metals tightly bound to tissues, such as lead stored in bones or mercury accumulated in the brain. It limits its utility for chronic, low-level exposures that result in deep tissue accumulation (El-Fawal, 2011b; Gao et al., 2025; Haase et al., 2011; Müller et al., 2021; Skalny et al., 2021).

DMSA is a water-soluble, orally bioavailable chelating agent that is particularly effective for lead and mercury detoxification. DMSA has a high affinity for sulfhydryl groups, allowing it to bind to heavy metals and facilitate their excretion via the kidneys. Clinical studies have demonstrated that DMSA can improve cognitive function and reduce behavioral symptoms in children with lead poisoning while also showing promise for reducing mercury levels in cases of dental amalgam exposure or fish consumption. Unlike EDTA, DMSA has a lower risk of depleting essential minerals, although some studies have reported mild gastrointestinal side effects, rashes, and liver enzyme elevations. One limitation of DMSA is its reduced ability to cross the BBB, making it less effective for clearing heavy metals that accumulate in the brain, which are often the most damaging in neurotoxicity. DMPS is another chelating agent with a strong affinity for mercury, arsenic, and lead, and it can be administered orally or intravenously. DMPS has been shown to lower blood and urinary mercury levels in individuals exposed to environmental or occupational mercury, and it has been used to treat arsenic poisoning in industrial settings. DMPS is generally well-tolerated, but like DMSA, it may cause side effects such as skin reactions, nausea, and mild renal stress. Some studies have also raised concerns that DMPS may redistribute heavy metals rather than eliminate them, potentially leading to transient increases in metal levels in sensitive organs like the brain. Despite their potential to lower heavy metal burdens, chelating agents do not directly reverse neuronal damage caused by oxidative stress, mitochondrial dysfunction, or neuroinflammation. While reducing metal loads may slow disease progression, adjunct therapies targeting these underlying mechanisms are likely necessary for comprehensive neuroprotection. Research into combination therapies pairing chelating agents with antioxidants, antiinflammatory compounds, or neurotrophic factors may offer a more holistic approach to mitigating heavy metal–induced neurotoxicity. Ultimately, while chelation therapy remains a valuable tool for treating acute metal poisoning, its application for chronic low-level exposure and long-term neurological health requires further investigation to optimize safety, efficacy, and long-term outcomes for patients affected by heavy metal toxicity (Lee et al., 2024; Liang et al., 2023; Massie et al., 2015; Rahmatinia et al., 2024; Shahlaei et al., 2024).

26.5.2 Dietary and nutritional interventions

Dietary and nutritional interventions are crucial in mitigating the neurotoxic effects of heavy metal exposure by enhancing the body's natural detoxification pathways, reducing oxidative stress, and promoting neuronal resilience. Vitamin C, E, and glutathione antioxidants are the most well-studied dietary compounds. These help neutralize ROS generated by heavy metals, thereby preventing neuronal damage. Heavy metals such as lead (Pb), mercury (Hg), cadmium (Cd), and arsenic (As) induce oxidative stress by disrupting cellular redox balance and impairing mitochondrial function, leading to lipid peroxidation, protein misfolding, and DNA damage. Dietary antioxidants protect against neurodegeneration associated with heavy metal toxicity by scavenging free radicals and reinforcing antioxidant defenses. Vitamin C, or ascorbic acid, is a water-soluble antioxidant that directly neutralizes free radicals and enhances the regeneration of other antioxidants, such as vitamin E. It also reduces metal toxicity by chelating certain metals, notably lead, and promoting their excretion. Studies have demonstrated that vitamin C supplementation reduces oxidative stress markers and improves cognitive function in individuals with chronic lead exposure, underscoring its neuroprotective potential. Similarly, vitamin E, a fat-soluble antioxidant, protects neuronal membranes from lipid peroxidation, which is exacerbated by heavy metal–induced oxidative stress. Heavy metals, particularly mercury and cadmium, target the lipid-rich brain tissue, leading to oxidative damage that compromises synaptic function and neuronal viability. By incorporating into cell membranes, vitamin E helps maintain membrane integrity and prevents the chain reactions of lipid peroxidation that contribute to neurodegenerative processes. Research has shown that vitamin E supplementation alleviates mercury-induced neurotoxicity by reducing oxidative damage and improving neurotransmitter balance, suggesting its potential role in dietary interventions for individuals at risk of heavy metal exposure (Gao et al., 2025; Kumar et al., 2025; Liang et al., 2023; Safhi et al., 2024; Skalny et al., 2021; Spero et al., 2024).

Glutathione, a tripeptide composed of glutamate, cysteine, and glycine, is one of the body's most important endogenous antioxidants and plays a central role in detoxification. It facilitates the conjugation and elimination of heavy metals through the liver and provides cellular protection against metal–induced oxidative stress. Many heavy metals, including arsenic and cadmium, deplete intracellular glutathione levels, exacerbating oxidative damage and increasing vulnerability to neurodegeneration. Dietary strategies aimed at boosting glutathione levels—such as consuming sulfur-rich foods (garlic, onions, cruciferous vegetables) or supplementing with precursors like *N*-acetylcysteine—have been shown to support detoxification pathways and enhance neuronal resilience in cases of heavy metal exposure. In addition to antioxidants, essential trace minerals such as zinc and selenium play a pivotal role in metal detoxification by competing with toxic metals for binding sites and supporting enzymatic processes involved in detoxification. Zinc, an essential cofactor for numerous enzymes, competes with heavy metals such as lead and cadmium for absorption and cellular binding sites, reducing their bioavailability and toxicity. Zinc also supports the function of metallothioneins, proteins that sequester heavy metals and facilitate their excretion. Selenium, another essential trace element, is known for its role in detoxification, as it binds to mercury to form biologically inert complexes that are safely excreted. Selenium-dependent enzymes, such as glutathione peroxidases, also help mitigate oxidative stress caused by heavy metal exposure. Dietary intake of selenium-rich foods, including Brazil nuts, fish, and whole grains, has been associated with reduced heavy metal toxicity and improved neurological outcomes. Overall, dietary and nutritional interventions provide a promising, noninvasive strategy for counteracting heavy metal toxicity and its neurotoxic effects. By incorporating antioxidants, essential trace minerals, and dietary strategies that enhance detoxification pathways, individuals can reduce their vulnerability to heavy metal–induced neurodegeneration and support overall brain health. Further research into optimal dietary interventions and their synergistic effects with other detoxification strategies may offer new avenues for mitigating the impact of heavy metals on neuroplasticity and cognitive function.

26.5.3 Neuroprotective compounds and natural remedies

Neuroprotective compounds and natural remedies, particularly flavonoids, polyphenols, and herbal extracts, have garnered significant attention for their potential to mitigate heavy metal–induced neurotoxicity. These bioactive compounds exhibit potent antioxidant, antiinflammatory, and metal-chelating properties that help protect neurons from the detrimental effects of heavy metals such as lead (Pb), mercury (Hg), cadmium (Cd), arsenic (As), and manganese (Mn). Heavy metal exposure induces oxidative stress, neuroinflammation, mitochondrial dysfunction, and synaptic impairments, all of which contribute to neurodegenerative diseases and cognitive deficits. By targeting these pathological mechanisms, flavonoids, polyphenols, and herbal extracts provide a promising avenue for neuroprotection and detoxification. Flavonoids, a diverse group of plant-derived polyphenolic compounds, have been extensively studied for their neuroprotective effects. They act as powerful antioxidants, neutralizing ROS from heavy metal toxicity and reducing oxidative damage to neuronal cells. Flavonoids such as quercetin, epicatechin, and kaempferol enhance synaptic plasticity, modulate neurotransmitter activity, and promote neurogenesis, all of which counteract the neurotoxic effects of heavy metals. Quercetin, found in apples, onions, and berries, has been shown to chelate heavy metals, reducing their bioavailability and preventing their accumulation in the brain. Studies have demonstrated that quercetin alleviates lead-induced neurotoxicity by restoring antioxidant enzyme levels and reducing neuronal apoptosis. Similarly abundant in green tea and cocoa, epicatechin has been reported to enhance cognitive function and protect against mercury-induced neuronal damage through its potent antioxidant properties. Polyphenols, another class of plant-based bioactive compounds, play a crucial role in combating heavy metal–induced neurodegeneration. Resveratrol, a well-known polyphenol found in grapes, red wine, and peanuts, exerts neuroprotective effects by modulating inflammatory pathways, enhancing mitochondrial function, and reducing amyloid-beta accumulation, which is relevant in conditions such as AD (Gao et al., 2025; Kumar et al., 2025; Lee et al., 2024; Müller et al., 2021; Shahlaei et al., 2024). Resveratrol has also been shown to mitigate cadmium-induced oxidative stress and cognitive dysfunction by upregulating antioxidant defenses and reducing neuroinflammation. Curcumin, the active component of turmeric, is another potent polyphenol with metal-chelating, antiinflammatory, and neuroprotective properties. It has been found to reduce the accumulation of heavy metals in the brain, particularly lead and arsenic, by enhancing their excretion and protecting neurons from oxidative stress-induced apoptosis. Herbal extracts, used in traditional medicine for centuries, offer promising neuroprotective benefits against heavy metal toxicity. Ginkgo biloba, one of the most well-studied medicinal herbs, has been shown to improve cognitive function, reduce oxidative stress, and enhance cerebral blood flow. Studies have indicated that Ginkgo biloba extract protects against manganese-induced neurotoxicity by modulating dopamine regulation and reducing mitochondrial dysfunction. Bacopa monnieri, an adaptogenic herb widely used in Ayurvedic medicine, is known for its cognitive-enhancing and neuroprotective effects. It has been reported to protect against lead-induced cognitive deficits by restoring cholinergic function and reducing oxidative stress. Ashwagandha, another adaptogenic herb, exhibits potent antioxidant and antiinflammatory properties, making it a potential therapeutic agent for heavy metal detoxification and neuroprotection. Flavonoids, polyphenols, and herbal extracts provide a natural and effective means of combating heavy metal–induced neurotoxicity. Their ability to reduce oxidative stress, modulate inflammatory pathways, chelate heavy metals, and enhance synaptic function makes them valuable neuroprotective agents. Incorporating these natural compounds into dietary and therapeutic strategies may help mitigate the long-term effects of heavy metal exposure on brain health and cognitive fun.

26.5.4 Lifestyle and behavioral interventions

Lifestyle and behavioral interventions play a crucial role in mitigating the adverse effects of heavy metal neurotoxicity by focusing on reducing exposure, improving diet, and engaging in cognitive training. Heavy metals such as lead, mercury, cadmium, arsenic, and manganese are prevalent in the environment due to industrial pollution, contaminated water, pesticides, and certain consumer products; taking proactive steps to limit exposure is essential for preserving neurological health. One of the most effective strategies to reduce heavy metal exposure is to ensure clean, uncontaminated water consumption, as arsenic and lead contamination in groundwater remains a significant public health concern. Installing water filtration systems that specifically target heavy metals can help minimize intake. Similarly, avoiding high-mercury seafood, such as swordfish, shark, and king mackerel, can significantly reduce the risk of mercury-induced neurotoxicity. For individuals exposed to occupational hazards, such as welders or battery factory workers, using personal protective equipment, improving workplace ventilation, and adhering to safety regulations can help minimize metal inhalation and absorption. In addition to reducing exposure, adopting a nutrient-rich diet can help counteract the toxic effects of heavy metals by enhancing detoxification pathways and supporting neurological function. A diet rich in antioxidants, essential minerals, and antiinflammatory foods is critical in neutralizing oxidative stress caused by heavy metal accumulation (Haase et al., 2011; Kumar et al., 2025; Liang et al., 2023; Rahmatinia et al., 2024; Spero et al., 2024). Consuming foods high in vitamin C, such as citrus fruits, bell peppers, and strawberries, enhances metal chelation by promoting glutathione production, a key antioxidant involved in detoxification. Similarly, vitamin E, found in nuts, seeds, and green leafy vegetables, protects neuronal cells from oxidative damage by stabilising cell membranes and reducing lipid peroxidation. Minerals such as zinc and selenium also play vital roles in heavy metal detoxification by competing with toxic metals for absorption sites and promoting their excretion. Zinc, found in foods like pumpkin seeds, chickpeas, and beef, helps reduce lead and cadmium accumulation in the brain. At the same time, selenium, abundant in Brazil nuts, fish, and eggs, enhances mercury detoxification and protects against neuroinflammation. A high-fiber diet that includes whole grains, legumes, and vegetables supports gut health and facilitates the excretion of heavy metals through bile and feces, reducing their systemic bioavailability. Cognitive training and mental stimulation are essential lifestyle interventions that can help counteract the cognitive deficits associated with heavy metal neurotoxicity. Engaging in activities that challenge the brain, such as puzzles, memory exercises, learning new skills, and playing musical instruments, promotes neuroplasticity and strengthens neural connections. Studies have shown that cognitive training can improve executive function, working memory, and attention in individuals exposed to neurotoxicants. Mindfulness practices such as meditation and yoga have also been found to reduce stress-related inflammation, which may help protect against heavy metal–induced neurodegeneration. Regular physical exercise further supports brain health by enhancing blood circulation, promoting neurogenesis, and stimulating the release of neurotrophic factors such as BDNF, which aids neuronal repair and resilience. Incorporating these lifestyle and behavioral interventions—reducing exposure, improving diet, and engaging in cognitive training—can significantly mitigate the harmful effects of heavy metal exposure on neurological health. By adopting a proactive and holistic approach, individuals can enhance their cognitive function, reduce neurotoxic risks, and promote long-term brain resilience (Azargoonjahromi, 2023; Bottini & MacDougall-Shackleton, 2023; Grace et al., 2024; Lee et al., 2024; Massie et al., 2015; Shahlaei et al., 2024).

26.6 Future directions and research perspectives

The growing concerns over heavy metal neurotoxicity underscore the urgent need for further epidemiological studies to understand the full scope of exposure, its long-term effects, and the interplay between genetic and environmental factors in determining susceptibility. Heavy metals such as lead, mercury, cadmium, arsenic, and manganese are pervasive in the environment, originating from industrial pollution, contaminated water sources, food, and occupational exposure. While research has established a strong link between heavy metal toxicity and various neurodevelopmental and neurodegenerative disorders, significant gaps remain in understanding dose–response relationships, critical exposure windows, and individual variability in metal accumulation and toxicity. Large-scale epidemiological studies can help identify populations at the highest risk, including children, pregnant women, and individuals with genetic predispositions to neurotoxicity. Prospective cohort studies tracking heavy metal exposure over time with comprehensive neurological assessments can provide valuable insights into how early-life exposure translates into cognitive and motor impairments later in life. Furthermore, integrating environmental monitoring data with epidemiological research can help establish regional risk assessments, guiding policy interventions to mitigate exposure sources.

Advancements in biomarker research hold immense promise for the early detection of heavy metal–induced neurotoxicity, which is critical for implementing timely interventions. Traditional biomarkers, such as blood and urine metal concentrations, offer a snapshot of recent exposure but fail to reflect chronic accumulation and target organ toxicity. Emerging biomarkers, including hair and toenail metal analysis, provide a longer term exposure profile, while advanced neuroimaging techniques, such as magnetic resonance imaging and positron emission tomography, are helping to visualize structural and functional changes in the brain associated with heavy metal accumulation. In addition, molecular biomarkers such as oxidative stress markers, inflammatory cytokines, and DNA damage indicators are gaining recognition for their potential to signal early neuronal damage before clinical symptoms manifest. Proteomic and metabolomic studies further uncover novel biomarkers associated with heavy metal–induced metabolic disruptions, paving the way for more precise risk assessments. Integrating these biomarkers into routine screening programs could enable early diagnosis, allowing for interventions that prevent irreversible neurological damage. Moreover, developing noninvasive biomarkers, such as those detectable in saliva or exhaled breath, could facilitate large-scale screening efforts, particularly in vulnerable populations such as children and industrial workers.

The advent of precision medicine is revolutionising neuroprotection against heavy metal toxicity by tailoring interventions based on an individual's genetic profile, environmental exposures, and metabolic responses. Precision medicine acknowledges that individuals metabolize and detoxify heavy metals differently due to genetic variations in metal transporters, detoxification enzymes, and antioxidant defense mechanisms. Genome-wide association studies and epigenetic analyses have identified key genetic polymorphisms that influence susceptibility to metal-induced neurotoxicity, offering potential avenues for personalized interventions. For example, individuals with mutations in genes involved in glutathione metabolism may benefit from targeted antioxidant supplementation to enhance detoxification. Pharmacogenomic approaches are also being explored to optimize chelation therapy, as genetic factors influence an individual's response to metal-binding agents such as EDTA, DMSA, and DMPS. Additionally, microbiome-based precision medicine strategies are being investigated, given the role of gut bacteria in modulating heavy metal absorption and excretion. Tailoring probiotic and prebiotic interventions to an individual's gut microbiome composition could offer a novel approach to reducing systemic heavy metal burden while supporting overall brain health.

In the search for effective neuroprotective interventions, identifying potential therapeutic targets for restoring neuroplasticity is a critical area of research. Heavy metal exposure is known to impair synaptic plasticity, reduce neurogenesis, and alter neurotransmitter balance, leading to cognitive dysfunction and motor impairments. One promising therapeutic approach involves targeting neurotrophic factors such as BDNF, which plays a crucial role in synaptic plasticity and neuronal survival. Enhancing BDNF signaling through pharmacological agents, dietary interventions, or physical exercise has shown potential in mitigating heavy metal–induced synaptic dysfunction. Another key target is the NMDA receptor, which is essential for learning and memory but is often disrupted by heavy metal toxicity. Modulating NMDA receptor function through selective agonists or antagonists could help restore excitatory neurotransmission and improve cognitive outcomes. Furthermore, the role of mitochondrial function in neuroplasticity cannot be overlooked, as heavy metals impair mitochondrial bioenergetics and increase oxidative stress. Mitochondrial-targeted antioxidants, such as coenzyme Q10 and alpha-lipoic acid, are being investigated for their potential to protect against heavy metal–induced neuronal damage.

Epigenetic modifications, including DNA methylation, histone acetylation, and noncoding RNA regulation, are promising therapeutic targets for counteracting heavy metal–induced neurotoxicity. Heavy metals have been shown to induce epigenetic changes that alter gene expression patterns critical for brain function. Developing epigenetic modulators, such as HDAC inhibitors, may help reverse detrimental epigenetic alterations and restore normal neurodevelopmental processes. Recent studies suggest that certain dietary compounds, such as polyphenols and sulforaphane, can exert epigenetic regulatory effects, offering a natural approach to mitigating metal-induced neurotoxicity. Understanding how heavy metals influence the epigenome will be crucial for designing targeted interventions to reverse neurodevelopmental and neurodegenerative effects.

26.7 Conclusion

Heavy metal neurotoxicity poses a significant and escalating threat to public health, with exposure linked to cognitive decline, neurodevelopmental disorders, and a range of neurological diseases. Key findings from extensive research highlight that heavy metals such as lead, mercury, cadmium, and arsenic can infiltrate the nervous system, crossing the BBB and inducing oxidative stress, mitochondrial dysfunction, and inflammation, all contributing to neuronal damage. Chronic exposure, even at low levels, has been associated with neurodegenerative conditions such as AD, PD, and ALS, as well as developmental disorders in children, including reduced IQ, attention deficits, and behavioral abnormalities. Emerging studies further indicate that prenatal and early-life exposure to these toxic metals can cause irreversible neurodevelopmental damage, emphasizing the critical window of vulnerability in infants and young children. The urgency of mitigating heavy metal neurotoxicity has never been more pronounced, as industrialization, mining, electronic waste, and agricultural activities continue contributing to environmental contamination, leading to widespread human exposure through food, water, air, and consumer products. The disproportionate impact on marginalized communities, particularly those in low-income and developing regions, underscores an environmental justice issue where vulnerable populations face the highest exposure levels and the least access to healthcare and mitigation resources. Additionally, climate change-induced environmental changes are exacerbating the mobilization of heavy metals in ecosystems, increasing bioavailability and subsequent human ingestion. The severity of the neurotoxic effects necessitates immediate intervention through stringent policy regulations, proactive public health initiatives, and technological advancements in heavy metal detection and removal. Governments worldwide must enforce stricter rules on industrial emissions, mining activities, and the use of heavy metals in consumer products while also enhancing monitoring systems to detect contamination in food, water, and soil. Investing in green technologies for waste management, safer alternatives for industrial processes, and remediation techniques such as phytoremediation and biochar-infused filtration systems can significantly reduce environmental heavy metal loads. Public health initiatives should focus on large-scale awareness campaigns about the dangers of heavy metal exposure, mainly targeting high-risk communities. Routine biomonitoring programs, nutritional interventions using metal-chelating agents, and the development of neuroprotective dietary strategies should be integrated into healthcare policies to mitigate long-term neurotoxic effects. International collaborations are also essential, as heavy metal pollution is a transboundary issue requiring coordinated global efforts to reduce emissions and establish universal safety standards. Future research should focus on developing advanced biomarkers for early detection of neurotoxicity, exploring genetic susceptibilities to heavy metal exposure, and innovating cost-effective remediation strategies to remove contaminants from the environment efficiently. With the growing recognition of heavy metal neurotoxicity as a public health crisis, a multidisciplinary approach involving scientists, policymakers, industry leaders, and healthcare professionals is crucial to implementing sustainable solutions that protect current and future generations from the debilitating effects of heavy metal exposure.

Acknowledgment

The Author(s) used AI for language; human performed all final editing.

References

Alaerts, K., Swinnen, S. P., & Wenderoth, N. (2010). Observing how others lift light or heavy objects: Which visual cues mediate the encoding of muscular force in the primary motor cortex? *Neuropsychologia, 48*(7), 2082–2090. Retrieved from https://doi.org/10.1016/j.neuropsychologia.2010.03.029.

Amooei, M., Meshkati, Z., Nasiri, R., & Dakhili, A. B. (2021). Cognitive decline prevention in offspring of Pb^{+2} exposed mice by maternal aerobic training and Cur/$CaCO_3$@Cur supplementations: In vitro and in vivo studies. *Ecotoxicology and Environmental Safety, 209*, 111785. Retrieved from https://doi.org/10.1016/j.ecoenv.2020.111785.

Anyachor, C. P., Orish, C. N., Ezejiofor, A. N., Cirovic, A., Cirovic, A., Ezealisiji, K. M., ... Orisakwe, O. E. (2023). Nickel and aluminium mixture elicit memory impairment by activation of oxidative stress, COX-2, and diminution of AChE, BDNF and NGF levels in cerebral cortex and hippocampus of male albino rats. *Current Research in Toxicology, 5*, 100129. Retrieved from https://doi.org/10.1016/j.crtox.2023.100129.

Athanasopoulos, D., Karagiannis, G., & Tsolaki, M. (2016). Recent findings in Alzheimer disease and Nutrition Focusing on Epigenetics.

Advances in Nutrition, 7(5), 917–927. Retrieved from https://doi.org/10.3945/an.116.012229.

Azargoonjahromi, A. (2023). Dual role of nitric oxide in Alzheimer's disease. *Nitric Oxide: Biology and Chemistry/Official Journal of the Nitric Oxide Society, 134–135*, 23–37. Retrieved from https://doi.org/10.1016/j.niox.2023.03.003.

Azmitia, E. C. (2007). Serotonin and brain: Evolution, neuroplasticity, and homeostasis. In *The pharmacology of neurogenesis and neuroenhancement (vol. 77)*. Academic Press, 31–56. Retrieved from https://doi.org/10.1016/S0074-7742(06)77002-7.

Bai, L., Wu, Y., Wang, R., Liu, R., Liu, M., Li, Q., ... Huang, H. (2022). Prepubertal exposure to Pb alters autophagy in the brain of aging mice: A time-series based model. *Brain Research Bulletin, 189*, 22–33. Retrieved from https://doi.org/10.1016/j.brainresbull.2022.08.013.

Bottini, C. L. J., & MacDougall-Shackleton, S. A. (2023). Methylmercury effects on avian brains. *Neurotoxicology, 96*, 140–153. Retrieved from https://doi.org/10.1016/j.neuro.2023.04.004.

Ciano Albanese, N., Poggini, S., Reccagni, A., Barezzi, C., Salciccia, C., Poleggi, A., & Branchi, I. (2025). Adolescent social isolation induces sex-specific behavioral and neural alterations. *Psychoneuroendocrinology, 172*, 107264. Retrieved from https://doi.org/10.1016/j.psyneuen.2024.107264.

Di Pino, G., Guglielmelli, E., & Rossini, P. M. (2009). Neuroplasticity in amputees: Main implications on bidirectional interfacing of cybernetic hand prostheses. *Progress in Neurobiology, 88*(2), 114–126. Retrieved from https://doi.org/10.1016/j.pneurobio.2009.03.001.

Dooka, D. B., Anyachor, C. P., Orish, C. N., Ezejiofor, A. N., Obasi, C. N., Frazzoli, C., & Orisakwe, O. E. (2024). Understanding the impact of Plant-derived Food Grade Substances PDFGS on metal-induced neurodegeneration: A systematic review of preclinical evidence. *Journal of Trace Elements and Minerals, 7*, 100110. Retrieved from https://doi.org/10.1016/j.jtemin.2023.100110.

El-Fawal, H. A. N. (2011a). Neurotoxicology. Elsevier, 87–106. Retrieved from https://doi.org/10.1016/B978-0-444-52272-6.00297-X.

El-Fawal, H. A. N. (2011b). In E. Nriagu (Ed.). Neurotoxicology (pp. 614–633). Elsevier. Retrieved from https://doi.org/10.1016/B978-0-444-63951-6.00297-7.

Fasone, J. (2021). Using the Trager® approach to reduce the impact of foot drop: Case report. *EXPLORE, 17*(3), 252–254. Retrieved from https://doi.org/10.1016/j.explore.2020.06.001.

Gao, W., Jing, S., He, C., Saberi, H., Sharma, H. S., Han, F., & Chen, L. (2025). Advancements in neurodegenerative diseases: Pathogenesis and novel neurorestorative interventions. *Journal of Neurorestoratology, 13*(2), 100176. Retrieved from https://doi.org/10.1016/j.jnrt.2024.100176.

Grace, J., Duran, E., Ann Ottinger, M., & Maness, T. (2024). Sublethal effects of early-life exposure to common and emerging contaminants in birds. *Current Research in Toxicology, 7*, 100190. Retrieved from https://doi.org/10.1016/j.crtox.2024.100190.

Guarnieri, L., Bosco, F., Leo, A., Citraro, R., Palma, E., De Sarro, G., & Mollace, V. (2024). Impact of micronutrients and nutraceuticals on cognitive function and performance in Alzheimer's disease. *Ageing Research Reviews, 95*, 102210. Retrieved from https://doi.org/10.1016/j.arr.2024.102210.

Haase, G. M., Prasad, K. N., Cole, W. C., Baggett-Strehlau, J. M., & Wyatt, S. E. (2011). Antioxidant micronutrient impact on hearing disorders: Concept, rationale, and evidence. *American Journal of Otolaryngology, 32*(1), 55–61. Retrieved from https://doi.org/10.1016/j.amjoto.2009.09.002.

Kamiński, P., Lorek, M., Baszyński, J., Tadrowski, T., Gorzelańczyk, E. J., Feit, J., ... Kurhaluk, N. (2024). Role of antioxidants in the neurobiology of drug addiction: An update. *Biomedicine & Pharmacotherapy, 175*, 116604. Retrieved from https://doi.org/10.1016/j.biopha.2024.116604.

Kumar, V., Gupta, P., & Pratihar, D. K. (2025). Chapter 4 – Advancing ankle–foot orthosis design through biomechanics, robotics, and additive manufacturing: A review. In F. Iqbal, P. Gupta, V. Kumar, & D. K. Pratihar (Eds.). *Medical robots and devices: New developments and advances* (pp. 65–84). Academic Press. Retrieved from https://doi.org/10.1016/B978-0-443-22206-1.00006-1.

Laaroussi, M., Boukholda, K., Essaidi, O., Berroug, L., Malqui, H., Anarghou, H., ... Chigr, F. (2025). Multigenerational impact of chronic exposure to mercury chloride on maternal care, puberty, fertility, and hypothalamic function in female mice. *Environmental Research, 264*, 120396. Retrieved from https://doi.org/10.1016/j.envres.2024.120396.

Lee, B. H., Cevizci, M., Lieblich, S. E., Ibrahim, M., Wen, Y., Eid, R. S., ... Galea, L. A. M. (2024). Exploring the parity paradox: Differential effects on neuroplasticity and inflammation by APOEe4 genotype at middle age. *Brain, Behavior, and Immunity, 120*, 54–70. Retrieved from https://doi.org/10.1016/j.bbi.2024.05.019.

Liang, Y., Wang, H., Wu, B., Peng, N., Yu, D., Wu, X., & Zhong, X. (2023). The emerging role of N6-methyladenine RNA methylation in metal ion metabolism and metal-induced carcinogenesis. *Environmental Pollution, 331*, 121897. Retrieved from https://doi.org/10.1016/j.envpol.2023.121897.

Luca, B. G.de, Almeida, P. P., Junior, R. R., Soares, D. J. S., Frantz, E. D. C., Miranda-Alves, L., ... Magliano, D. C. (2025). Environmental contamination by bisphenols: From plastic production to modulation of the intestinal morphophysiology in experimental models. *Food and Chemical Toxicology, 197*, 115280. Retrieved from https://doi.org/10.1016/j.fct.2025.115280.

Massie, C. L., Kantak, S. S., Narayanan, P., & Wittenberg, G. F. (2015). Timing of motor cortical stimulation during planar robotic training differentially impacts neuroplasticity in older adults. *Clinical Neurophysiology, 126*(5), 1024–1032. Retrieved from https://doi.org/10.1016/j.clinph.2014.06.053.

Mattson, M. P., & Leak, R. K. (2024). The hormesis principle of neuroplasticity and neuroprotection. *Cell Metabolism, 36*(2), 315–337. Retrieved from https://doi.org/10.1016/j.cmet.2023.12.022.

Mazuryk, J., Klepacka, K., Kutner, W., & Sharma, P. S. (2024). Glyphosate: Impact on the microbiota-gut-brain axis and the immune-nervous system, and clinical cases of multiorgan toxicity. *Ecotoxicology and Environmental Safety, 271*, 115965. Retrieved from https://doi.org/10.1016/j.ecoenv.2024.115965.

Mishra, P. K., Ghosh, S., Singh, K. K., & Sinha, J. K. (2025). Chapter 11 – Physical exercise and lifestyle management: Preventive measures for Alzheimer's disease. In S. K. Singh, & R. B. Maccioni (Eds.). *A New Era in Alzheimer'S Research* (pp. 169–182). Academic Press. Retrieved from https://doi.org/10.1016/B978-0-443-15540-6.00003-3. (Accessed 26 May 2025).

Müller, T., Shevchenko, Y., Gerhardt, S., Kiefer, F., & Vollstädt-Klein, S. (2021). The influence of perceived stress and self-control on efficacy of repeated transcranial direct current stimulation in non-treatment-seeking smokers. *Drug and Alcohol Dependence, 226*, 108861. Retrieved from https://doi.org/10.1016/j.drugalcdep.2021.108861.

Padhan, P., Simran, Kumar, N., & Verma, S. (2025). Glutathione S-transferase: A keystone in Parkinson's disease pathogenesis and therapy. *Molecular and Cellular Neuroscience, 132*, 103981. Retrieved from https://doi.org/10.1016/j.mcn.2024.103981.

Peleg, L. C., Rabinovitch, D., Lavie, Y., Rabbie, D. M., Horowitz, I., Fruchter, E., & Gruenwald, I. (2022). Post-SSRI sexual dysfunction (PSSD): Biological plausibility, symptoms, diagnosis, and presumed risk factors. *Sexual Medicine Reviews, 10*(1), 91–98. Retrieved from https://doi.org/10.1016/j.sxmr.2021.07.001.

Qu, T., Sun, Q., Tan, B., Wei, H., Qiu, X., Xu, X., ... Zhang, S. (2025). Integration of network toxicology and transcriptomics reveals the novel neurotoxic mechanisms of 2, 2′, 4, 4′-tetrabromodiphenyl ether. *Journal of Hazardous Materials, 486*, 136999. Retrieved from https://doi.org/10.1016/j.jhazmat.2024.136999.

Rahmatinia, M., Mohseni-Bandpei, A., Khodagholi, F., Abdollahifar, M.-A., Amouei Torkmahalleh, M., Hassani Moghaddam, M., ... Shahsavani, A. (2024). Exposure to different PM2.5 extracts induces gliosis and changes behavior in male rats similar to autism spectrum disorders features. *Environmental Pollution, 340*, 122804. Retrieved from https://doi.org/10.1016/j.envpol.2023.122804.

Ramaswamy, M., Philip, J. L., Priya, V., Priyadarshini, S., Ramasamy, M., Jeevitha, G. C., ... Ahmad, F. (2024). Therapeutic use of music in neurological disorders: A concise narrative review. *Heliyon, 10*(16), e35564. Retrieved from https://doi.org/10.1016/j.heliyon.2024.e35564.

Resendiz, M., Watkins, D. S., Öztürk, N. C., & Zhou, F. C. (2023). Chapter 32 - Environmental Influence on Epigenetics. In E. Tollefsbol (Ed.). Academic Press. Retrieved from https://doi.org/10.1016/B978-0-323-91909-8.00036-0.

Safhi, A. Y., Albariqi, A. H., Sabei, F. Y., Alsalhi, A., Khalil, F. M. A., Waheed, A., ... Alissa, M. (2024). Journey into tomorrow: cardiovascular wellbeing transformed by nano-scale innovations. *Current Problems in Cardiology, 49*(4), 102428. Retrieved from https://doi.org/10.1016/j.cpcardiol.2024.102428.

Scattolin, M. A.de A., Resegue, R. M., & Rosário, M. C.do (2022). The impact of the environment on neurodevelopmental disorders in early childhood. *Jornal de Pediatria, 98*, S66–S72. Retrieved from https://doi.org/10.1016/j.jped.2021.11.002.

Scher, M. S., Ludington, S., Smith, V., Klemming, S., & Pilon, B. (2024). Brain care bundles applied over each and successive generations. *Seminars in Fetal and Neonatal Medicine, 29*(4), 101558. Retrieved from https://doi.org/10.1016/j.siny.2024.101558.

Schmidt, M. A., & Schmidt, J. C. (2025). Chapter 17 – Using functionally characterized networks in the application of precision medicine in spaceflight. In M. A. Schmidt, & M. J Legato (Eds.). Academic Press. Retrieved from https://doi.org/10.1016/B978-0-443-13850-8.00018-X.

Shahlaei, M., Asl, S. M., Derakhshani, A., Kurek, L., Karges, J., Macgregor, R., ... Saboury, A. A. (2024). Platinum-based drugs in cancer treatment: Expanding horizons and overcoming resistance. *Journal of Molecular Structure, 1301*, 137366. Retrieved from https://doi.org/10.1016/j.molstruc.2023.137366.

Shehryar, Z. (2025). Chapter 3 – The genetic face of brain diseases. In T. S. Koduru, R. A. M. Osmani, E. Singh, & S. Dutta (Eds.). Academic Press. Retrieved from https://doi.org/10.1016/B978-0-443-28822-7.00012-X.

Skalny, A. V., Aschner, M., Bobrovnitsky, I. P., Chen, P., Tsatsakis, A., Paoliello, M. M. B., ... Tinkov, A. A. (2021). Environmental and health hazards of military metal pollution. *Environmental Research, 201*, 111568. Retrieved from https://doi.org/10.1016/j.envres.2021.111568.

Snoeijs, T., Dauwe, T., Pinxten, R., Darras, V. M., Arckens, L., & Eens, M. (2005). The combined effect of lead exposure and high or low dietary calcium on health and immunocompetence in the zebra finch (*Taeniopygia guttata*). *Environmental Pollution, 134*(1), 123–132. Retrieved from https://doi.org/10.1016/j.envpol.2004.07.009.

Spero, V., Scherma, M., D'Amelio, S., Collu, R., Dedoni, S., Camoglio, C., ... Fadda, P. (2024). Activity-based anorexia (ABA) model: Effects on brain neuroinflammation, redox balance and neuroplasticity during the acute phase. *Neurochemistry International, 180*, 105842. Retrieved from https://doi.org/10.1016/j.neuint.2024.105842.

Tang, R., Bai, Q., Ma, S., & Ou, J. (2023). Materials, workflows and applications of IMAC for phosphoproteome profiling in the recent decade: A review. *TrAC Trends in Analytical Chemistry, 158*, 116862. Retrieved from https://doi.org/10.1016/j.trac.2022.116862.

Thapa, R., Moglad, E., Afzal, M., Gupta, G., Bhat, A. A., almalki, W. H., ... Dua, K. (2024). ncRNAs and their impact on dopaminergic neurons: Autophagy pathways in Parkinson's disease. *Ageing Research Reviews, 98*, 102327. Retrieved from https://doi.org/10.1016/j.arr.2024.102327.

Thilakaratne, R., Lin, P.-I. D., Rifas-Shiman, S. L., Landero, J., Wright, R. O., Bellinger, D., ... Cardenas, A. (2024). Cross-sectional and prospective associations of early childhood circulating metals with early and mid-childhood cognition in the Project Viva cohort. *Environmental Research, 246*, 118068. Retrieved from https://doi.org/10.1016/j.envres.2023.118068.

Wani, A. L., Hammad Ahmad Shadab, G. G., & Afzal, M. (2021). Lead and zinc interactions – An influence of zinc over lead related toxic manifestations. *Journal of Trace Elements in Medicine and Biology, 64*, 126702. Retrieved from https://doi.org/10.1016/j.jtemb.2020.126702.

Womersley, J. S., Obellianne, C., Padula, A. E., Lopez, M. F., Griffin, W. C., Ball, L. E., ... Mulholland, P. J. (2024). Adaptations in glutathione-based redox protein signaling pathways and alcohol drinking across species. *Biomedicine & Pharmacotherapy, 180*, 117514. Retrieved from https://doi.org/10.1016/j.biopha.2024.117514.

Zhang, H.-Y., Yang, H.-X., Liu, Q., Xie, M.-J., Zhang, J., Liu, X., ... Wang, M.-Q. (2020). Injury responses of Sprague-Dawley rat jaw muscles to an experimental unilateral anterior crossbite prosthesis. *Archives of Oral Biology, 109*, 104588. Retrieved from https://doi.org/10.1016/j.archoralbio.2019.104588.

Chapter 27

Heavy metal toxicity and psychiatric disorders

Debjani Choudhury[1], **and Prasann Kumar**[2]

[1]*Department of Plant Pathology, School of Agriculture, Lovely Professional University, Phagwara, Punjab, India,* [2]*Department of Agronomy, School of Agriculture, Lovely Professional University, Phagwara, Punjab, India*

27.1 Introduction

Heavy metals are naturally occurring elements with high atomic weights and densities that can pose significant health risks even at low concentrations. While some, such as zinc (Zn), copper (Cu), and iron (Fe), are essential for biological functions, others, including lead (Pb), mercury (Hg), cadmium (Cd), and arsenic (As), have no known physiological roles and can be highly toxic (de Godoi Pereira, dos Santos, Geris, & Malta, 2023; Gade, Comfort, & Re, 2021; Raj, Kaur, Gupta, & Singh, 2021). Heavy metal exposure occurs through various environmental and occupational sources, including contaminated air, water, food, and soil, industrial emissions, mining, and the improper disposal of electronic waste. Due to their persistence in the environment and bioaccumulate nature, heavy metals can accumulate in biological tissues over time, leading to chronic toxicity. The toxic effects of heavy metals are primarily mediated through oxidative stress, inflammation, and disruption of essential cellular processes. Once inside the body, heavy metals can interfere with enzymatic reactions, disrupt mitochondrial function, and alter the balance of important minerals, leading to cellular damage. The nervous system, liver, kidneys, and cardiovascular system are particularly vulnerable to heavy metal toxicity. Neurological effects of heavy metal exposure include cognitive impairment, developmental delays, and neurodegenerative disorders, while systemic effects may manifest as renal dysfunction, cardiovascular disease, and endocrine disruptions. Recent research has established strong links between heavy metal toxicity and psychiatric disorders, with evidence suggesting that heavy metals play a crucial role in neurodevelopmental and neuropsychiatric conditions. The central nervous system (CNS) is highly susceptible to heavy metal–induced damage due to its high metabolic activity, reliance on neurotransmitter signaling, and limited regenerative capacity. Heavy metals such as lead, mercury, cadmium, and arsenic can cross the blood—brain barrier (BBB) and exert neurotoxic effects by inducing oxidative stress, impairing neurotransmitter function, and triggering neuroinflammation. One of the most concerning aspects of heavy metal exposure is its impact on early brain development (Jang & Hoffman, 2011; Nivetha et al., 2023; Xie et al., 2023). Lead (Pb), for instance, has been extensively studied for its role in children's cognitive deficits and behavioral abnormalities. Even low levels of lead exposure during critical periods of neurodevelopment have been associated with reduced intelligence quotient (IQ), attention-deficit/hyperactivity disorder (ADHD), and increased risk of aggressive and antisocial behavior. The neurotoxic effects of lead are mediated through its ability to disrupt calcium signaling, impair synaptic plasticity, and alter gene expression patterns involved in neuronal growth and differentiation. Similarly, mercury (Hg), particularly in its organic form as methylmercury (MeHg), is a potent neurotoxin that has been implicated in psychiatric disorders. Methylmercury exposure, primarily through contaminated seafood consumption, has been linked to autism spectrum disorder (ASD), depression, and cognitive dysfunction. The neurotoxic effects of mercury are attributed to its ability to bind to sulfhydryl groups in proteins, disrupt mitochondrial function, and promote excitotoxicity through excessive glutamate release. Cadmium (Cd) and arsenic (As) have also been associated with adverse mental health outcomes. Cadmium exposure, mainly from cigarette smoke, contaminated food, and industrial pollution, has been linked to increased anxiety, depression, and neurodegenerative

Heavy Metal Toxicity and Neurodegeneration. DOI: https://doi.org/10.1016/B978-0-443-36575-1.00004-2

disorders. Cadmium disrupts dopaminergic and serotonergic signaling, leading to mood disturbances and cognitive impairment. Arsenic, commonly found in contaminated groundwater, has been implicated in neurobehavioral changes, memory deficits, and increased susceptibility to psychiatric disorders. Chronic arsenic exposure has been shown to alter gene expression in the brain, disrupt neurotransmitter homeostasis, and induce inflammatory pathways that contribute to mental health disorders (Erkman et al., 2025; Shen et al., 2024; Singh & Kostova, 2024). The growing concern over the role of heavy metals in psychiatric disorders is further supported by epidemiological studies that have demonstrated a correlation between environmental metal exposure and the prevalence of mental health conditions. For instance, populations living in areas with high ecological lead contamination have shown higher rates of anxiety, depression, and cognitive decline. Similarly, studies examining prenatal and early childhood exposure to heavy metals have highlighted their long-term effects on mental health, underscoring the need for preventive measures. This chapter aims to provide a comprehensive overview of the relationship between heavy metal toxicity and psychiatric disorders, focusing on the underlying mechanisms, epidemiological evidence, and potential mitigation strategies. By addressing these key objectives, this chapter will contribute to the knowledge of the neuropsychiatric effects of heavy metal toxicity and provide a foundation for future research and clinical interventions. Given the increasing industrialization, environmental pollution, and widespread exposure to heavy metals, understanding their impact on mental health is crucial for developing effective strategies to protect vulnerable populations. The scope of this chapter encompasses a multidisciplinary approach, integrating insights from toxicology, neuroscience, psychiatry, environmental health, and public policy. The chapter will synthesize current research findings, clinical case studies, and policy frameworks to present a holistic perspective. Additionally, it will highlight the importance of early detection, prevention, and management strategies to mitigate the effects of heavy metal toxicity on mental health. The relationship between environmental pollutants and psychiatric disorders is an area of growing concern, and heavy metals represent one of the most significant neurotoxic threats in modern society (Fu, Li, Song, Cen, & Wu, 2023; Kim, Ha, Hwang, & Kim, 2025; Krishna Mitra & Chakraborty, 2025; Nigam, Sharma, Tripathy, & Kumar, 2021). As research continues to uncover the complex interactions between heavy metals and the nervous system, there is an urgent need for increased awareness, policy interventions, and targeted therapies to address this pressing public health challenge. This chapter aims to serve as a comprehensive resource for researchers, clinicians, and policymakers seeking to understand and mitigate the impact of heavy metal toxicity on mental health. The chapter will emphasize the importance of a proactive approach to addressing heavy metal toxicity and its psychiatric implications. By integrating scientific research with practical solutions, it is possible to reduce the burden of heavy metal–induced psychiatric disorders and improve public health outcomes. Future research should focus on developing advanced diagnostic tools, exploring novel therapeutic targets, and implementing stricter environmental regulations to minimize heavy metal exposure. Through a collaborative effort involving scientists, healthcare professionals, and policymakers, it is possible to mitigate the neuropsychiatric impact of heavy metals and promote mental well-being in affected populations (Table 27.1 and Fig. 27.1).

27.2 Common heavy metals associated with psychiatric disorders

Heavy metal exposure has been increasingly recognized as a significant factor contributing to psychiatric disorders, highlighting the need for interdisciplinary collaboration between environmental health and psychiatry. Among the various metals implicated in neuropsychiatric dysfunction, lead (Pb), mercury (Hg), cadmium (Cd), and arsenic (As) are among the most extensively studied due to their well-documented neurotoxic effects. Other metals, such as aluminum (Al) and manganese (Mn), have also been linked to mental health conditions, particularly in cases of chronic environmental or occupational exposure (He et al., 2024; Huang, Pan, Chin, Hsu, & Guo, 2023; Menon, Chang, & Kim, 2016; Yan & Saxena, 2022). These metals exert their effects through various mechanisms, including oxidative stress, neuroinflammation, neurotransmitter imbalances, and epigenetic modifications, which collectively contribute to the development of psychiatric disorders such as depression, anxiety, schizophrenia, and neurodevelopmental conditions like ASD and ADHD.

Lead (Pb) is one of the most widely studied neurotoxic metals and is strongly associated with neurodevelopmental and psychiatric disorders. Lead exposure primarily occurs through contaminated water, air, soil, and deteriorating lead-based paints. It is particularly concerning in children due to its profound effects on brain development. Lead can readily cross the BBB and accumulate in the CNS, where it disrupts neuronal signaling, impairs synaptic plasticity, and interferes with calcium-dependent neurotransmitter release. Lead-induced neurotoxicity is primarily attributed to its ability to induce oxidative stress and inflammation, leading to neuronal apoptosis and white matter abnormalities. Epidemiological studies have linked early childhood lead exposure to reduced IQ, learning disabilities, impulsivity, and increased risks of ADHD. In adults, chronic lead exposure has been associated with cognitive decline,

TABLE 27.1 Common heavy metals and their neurotoxic effects on psychiatric health.

Heavy metal	Primary sources of exposure	Mechanisms of neurotoxicity	Affected brain regions	Associated psychiatric disorders	Potential biomarkers for detection
Lead (Pb)	Old paints, contaminated water, industrial emissions, batteries, and soil	Oxidative stress, calcium dysregulation, mitochondrial damage, disruption of neurotransmitters (dopamine, glutamate)	The prefrontal cortex, hippocampus, and basal ganglia	ADHD, cognitive impairment, depression, schizophrenia, anxiety	Blood lead levels, bone lead, and hair analysis
Mercury (Hg)	Contaminated seafood (methylmercury), dental amalgams, industrial waste, vaccines (thimerosal)	Neuroinflammation, mitochondrial dysfunction, impairment of GABAergic and dopaminergic systems	Cerebellum, cortical neurons, limbic system	Anxiety, mood disorders, autism spectrum disorder, and memory impairment	Hair mercury levels, urine mercury, and blood mercury
Cadmium (Cd)	Tobacco smoke, contaminated food (rice, vegetables), industrial exposure (battery manufacturing, smelting)	Oxidative stress, inhibition of antioxidant enzymes, and apoptosis of neural cells	Hippocampus, cortex, striatum	Depression, anxiety, cognitive deficits, neurodegeneration	Urine cadmium, blood cadmium, and hair analysis
Arsenic (As)	Contaminated groundwater, pesticides, industrial processes, and rice consumption	Disruption of neurotransmitter signaling, DNA methylation changes, increased oxidative stress	Cerebral cortex, hippocampus	Depression, mood instability, cognitive dysfunction, neurodevelopmental disorders	Urine arsenic, hair arsenic, blood arsenic
Aluminum (Al)	Cookware, food additives, vaccines, contaminated water, and occupational exposure	Neuroinflammation, impaired synaptic plasticity, and disruption of iron homeostasis	Hippocampus, cerebral cortex	Alzheimer's disease, cognitive decline, and mood disorders	Serum aluminum, urine aluminum, hair analysis
Manganese (Mn)	Drinking water, occupational exposure (welding, mining), contaminated food, fertilisers	Dopaminergic system dysfunction, oxidative stress, impaired mitochondrial function	Basal ganglia, substantia nigra, striatum	Parkinsonism, depression, cognitive impairment, behavioral changes	Blood manganese, urine manganese, and hair analysis

ADHD, Attention-deficit/hyperactivity disorder.
Source: Based on Abdelmonem, Kamal, Elbaz, Khalifa, and Abdelnaser (2025), Amadi et al. (2022), Balali-Mood et al. (2025).

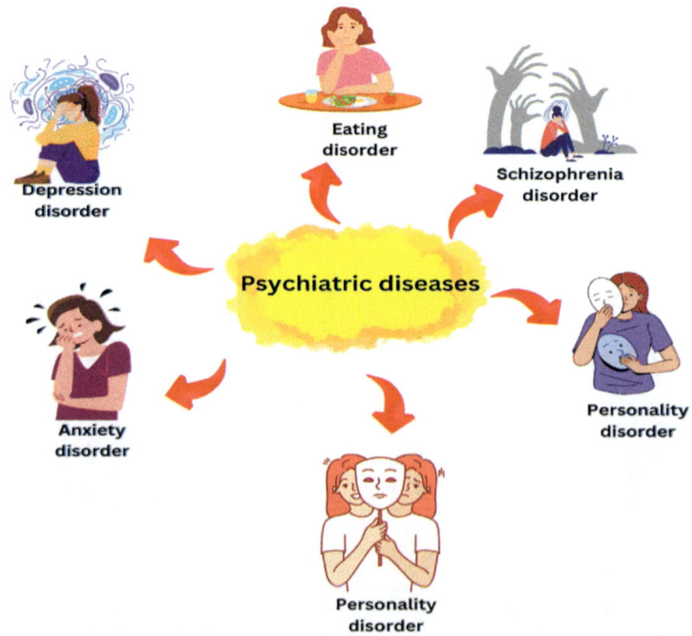

FIGURE 27.1 Heavy metal toxicity and psychiatric disease.

depression, and an elevated risk of developing schizophrenia. One of the key mechanisms through which lead contributes to psychiatric disorders is its impact on the dopaminergic and glutamatergic systems, both of which play crucial roles in mood regulation and cognitive function. Disruptions in these neurotransmitter systems can result in symptoms characteristic of schizophrenia, bipolar disorder, and major depressive disorder (MDD). Additionally, lead-induced epigenetic modifications have been observed, including DNA methylation changes in genes involved in neurodevelopment and synaptic plasticity, which may have long-term consequences on mental health (Ekholm, Ljunggren, & Helmfrid, 2025; Kumkar et al., 2023; Liu et al., 2014; Vunain, Mishra, & Mamba, 2016).

Mercury (Hg) is another potent neurotoxin implicated in psychiatric and neurodevelopmental disorders. Mercury exposure occurs through multiple sources, including industrial emissions, dental amalgams, and consuming contaminated seafood, which contains methylmercury (MeHg), the most neurotoxic form of mercury. Mercury has a strong affinity for sulfhydryl (–SH) protein groups, allowing it to disrupt enzymatic activity, interfere with antioxidant defenses, and impair mitochondrial function. The neurotoxic effects of mercury primarily stem from its ability to promote excitotoxicity by increasing glutamate levels, leading to excessive neuronal firing and oxidative damage. Mercury exposure has been linked to ASD, with studies showing higher levels of mercury in children with autism compared to neurotypical children. The proposed mechanisms for mercury-induced ASD include disruption of glutathione (GSH) metabolism, increased neuroinflammation, and altered synaptic transmission. Mercury exposure has also been associated with depression, anxiety, and cognitive impairments. Chronic exposure to low levels of mercury can result in mood disturbances, irritability, and memory deficits, a phenomenon historically known as "mad hatter syndrome" due to its prevalence among 19th-century hat makers exposed to mercury vapors. The long-term psychiatric effects of mercury exposure highlight the need for stringent regulations on mercury emissions and dietary guidelines to reduce mercury intake from seafood consumption (Aljumaili, El-Waseif, AbdulJabbar Suleiman, & El-Dein A. Ewais, 2022; Ellwanger & Chies, 2023; Gu et al., 2019; Maqbool et al., 2023).

Given the potential for preventive measures and policy changes to mitigate heavy metal exposure, this is an area ripe for further research and public health intervention. Cadmium (Cd) is a heavy metal with well-documented toxic effects on multiple organ systems, including the brain. Cadmium exposure primarily occurs through cigarette smoke, contaminated food, industrial emissions, and occupational exposure. Unlike lead and mercury, which have well-characterized effects on the nervous system, cadmium's neurotoxicity has only recently begun to receive significant attention in psychiatric research. Cadmium accumulates in the brain over time, disrupting calcium signaling, increasing oxidative stress, and altering neurotransmitter release. Studies have linked cadmium exposure to increased risks of depression, anxiety, and neurodegenerative disorders such as Alzheimer's and Parkinson's disease (PD). The impact of cadmium on mental health is thought to be mediated through its ability to disrupt the dopaminergic system, leading to alterations in mood regulation and cognitive function. Furthermore, cadmium-induced neuroinflammation and BBB dysfunction can contribute to neuropsychiatric disorders by promoting neurodegenerative processes and impairing neuronal communication. The role of cadmium in psychiatric disorders remains an emerging area of research. Still, growing evidence suggests that reducing cadmium exposure through dietary and environmental interventions may have significant mental health benefits (Bjørklund et al., 2018; Castro-González & Méndez-Armenta, 2008; Mukherjee et al., 2021; Rashaid et al., 2021).

Arsenic (As) is a well-known environmental toxin that has been linked to various psychiatric and cognitive disorders. Arsenic exposure occurs primarily through contaminated drinking water, industrial pollution, and certain pesticides. Chronic arsenic exposure has been associated with memory deficits, mood disturbances, and an increased risk of neurodevelopmental disorders. The neurotoxic effects of arsenic are mediated through its ability to generate reactive oxygen species (ROS), disrupt mitochondrial function, and impair DNA repair mechanisms. One of the most concerning aspects of arsenic exposure is its impact on early brain development. Prenatal and early childhood exposure to arsenic has been linked to lower IQ scores, attention deficits, and an increased risk of ASD and ADHD. In adults, chronic arsenic exposure has been associated with an elevated risk of depression and cognitive decline. Arsenic has also been shown to interfere with neurotransmitter systems, particularly the cholinergic and serotonergic pathways, which play crucial roles in learning, memory, and mood regulation. Additionally, arsenic-induced epigenetic modifications have been observed in genes involved in neurodevelopment and neuroplasticity, suggesting that arsenic exposure may have long-lasting effects on brain function. Given the widespread contamination of drinking water with arsenic in many parts of the world, addressing arsenic exposure remains a critical public health priority for reducing the burden of mental health disorders (Emurotu et al., 2025; Muñoz et al., 2020; Yousef et al., 2013; Zebbiche et al., 2024). Other metals, such as aluminum (Al) and manganese (Mn), have also been implicated in psychiatric and neurodegenerative disorders. Aluminum exposure occurs through contaminated food, drinking water, certain medications, and industrial exposure. Although the role of aluminum in mental health disorders remains controversial, some studies have suggested a link between aluminum accumulation in the brain and an increased risk of Alzheimer's disease (AD) and cognitive decline. Aluminum is believed to contribute to neurotoxicity by inducing oxidative stress, promoting beta-amyloid aggregation, and impairing neurotransmitter function. Manganese, an essential trace element required for normal brain function, can become neurotoxic at high concentrations. Chronic manganese exposure, primarily through occupational sources such as welding, mining, and contaminated drinking water, has been associated with mood disturbances, cognitive impairment, and Parkinsonian-like symptoms. Manganese toxicity is believed to be mediated by its impact on the dopaminergic system, leading to mood regulation and motor function alterations. The neurotoxic effects of manganese have been well-documented in cases of manganese-induced Parkinsonism. Still, emerging research suggests that manganese exposure may also contribute to psychiatric disorders such as depression and anxiety. Heavy metal exposure represents a significant and often overlooked factor in the etiology of psychiatric disorders. Lead, mercury, cadmium, and arsenic are among the most well-documented neurotoxic metals, with strong associations with cognitive deficits, mood disorders, and neurodevelopmental conditions. Other metals, such as aluminum and manganese, also contribute to mental health disturbances, particularly in cases of chronic exposure. The mechanisms through which heavy metals affect the brain involve oxidative stress, neuroinflammation, neurotransmitter imbalances, and epigenetic modifications (Channegowda, 2020; Liu, Ai, Xia, & Wang, 2024; Saravanan et al., 2024). Given the increasing prevalence of mental health disorders and the widespread environmental contamination with heavy metals, there is an urgent need for further research to better understand the role of heavy metals in psychiatric conditions and to develop effective prevention and intervention strategies. Reducing heavy metal exposure through environmental regulations, dietary modifications, and chelation therapies may offer promising approaches for mitigating the neuropsychiatric effects of heavy metal toxicity and improving public mental health outcomes (Table 27.2, Fig. 27.2).

27.3 Mechanisms of neurotoxicity and psychiatric impact

Heavy metal exposure contributes to neurotoxicity and psychiatric disorders through multiple overlapping mechanisms, including oxidative stress, neuroinflammation, neurotransmitter system disruptions, BBB dysfunction, epigenetic modifications, mitochondrial dysfunction, and neuronal apoptosis. These toxic effects can lead to neurodevelopmental delays, cognitive impairments, mood disorders, and neurodegenerative diseases. The interplay of these factors exacerbates mental health conditions by impairing neural signaling, synaptic plasticity, and overall brain homeostasis. Oxidative stress and neuroinflammation are among the most critical pathways through which heavy metals exert neurotoxic effects. Heavy metals such as lead (Pb), mercury (Hg), cadmium (Cd), and arsenic (As)

TABLE 27.2 Heavy metals, oxidative stress, and neuroinflammation.

Heavy metal	Oxidative stress markers	Inflammatory markers	Potential protective antioxidants
Lead (Pb)	↑ ROS, ↑ Lipid peroxidation (MDA, 4-HNE), ↓ GSH	↑ TNF-α, ↑ IL-6, ↑ NF-κB	Vitamin C, vitamin E, NAC, selenium
Mercury (Hg)	↑ ROS, ↑ mitochondrial dysfunction, ↓ SOD	↑ IL-1β, ↑ IL-6, ↑ COX-2	Alpha-lipoic acid, glutathione, curcumin
Cadmium (Cd)	↑ Lipid peroxidation, ↑ NO, ↓ CAT	↑ TNF-α, ↑ IL-8, ↑ CRP	Zinc, resveratrol, polyphenols
Arsenic (As)	↑ DNA oxidation (8-OHdG), ↑ Protein carbonylation, ↓ GSH	↑ IL-1β, ↑ IL-18, ↑ NF-κB	SAMe, omega-3 fatty acids, melatonin
Aluminum (Al)	↑ Hydrogen peroxide (H_2O_2), ↑ MDA, ↓ GPx	↑ IL-6, ↑ TNF-α, ↑ Microglial activation	Quercetin, flavonoids, astaxanthin
Manganese (Mn)	↑ ROS, ↑ mitochondrial dysfunction, ↓ SOD	↑ IL-1β, ↑ TNF-α, ↑ TGF-β	Magnesium, CoQ10, green tea catechins

CAT, Catalase; *CoQ10*, coenzyme Q10; *CRP*, C-reactive protein; *GPx*, glutathione peroxidase; *GSH*, glutathione; *IL-6*, interleukin-6; *IL-1β*, interleukin-1 beta; *MDA*, malondialdehyde; *NAC*, *N*-acetylcysteine; *NO*, nitric oxide; *ROS*, reactive oxygen species; *SAMe*, *S*-adenosylmethionine; *SOD*, superoxide dismutase; *TNF-α*, tumor necrosis factor-alpha.
Source: Based on Iqbal et al. (2018), Sun, Gao, Zhang, Cheng, and Hu (2024), Vänskä et al. (2019).

promote the excessive generation of ROS and reactive nitrogen species (RNS), which damage cellular macromolecules, including lipids, proteins, and DNA. The brain is highly susceptible to oxidative damage due to its high lipid content, oxygen consumption, and relatively low antioxidant defense capacity. Heavy metal–induced oxidative stress results in lipid peroxidation, damaging membranes and impairing neuronal function. Simultaneously, metals such as mercury and arsenic deplete GSH, a crucial antioxidant in the brain, further exacerbating oxidative damage. The activation of microglia and astrocytes in response to oxidative stress results in chronic neuroinflammation, characterized by the release of proinflammatory cytokines such as tumor necrosis factor-alpha (TNF-α), interleukin-6 (IL-6), and interleukin-1 beta (IL-1β). This persistent inflammatory state disrupts everyday neuronal communication and contributes to synaptic dysfunction, a hallmark of psychiatric disorders such as schizophrenia, depression, and ASD. Chronic inflammation can also promote neurodegeneration by triggering excitotoxicity and apoptotic cell death, further compounding the adverse effects of oxidative stress. The disruption of neurotransmitter systems is another fundamental mechanism by which heavy metals contribute to psychiatric disorders (Habib et al., 2016; Jiang & Zhao, 2024; Porru, Esplugues, Llop, & Delgado-Saborit, 2024; Woimant & Trocello, 2014). Neurotransmitters such as dopamine (DA), serotonin, and glutamate play crucial roles in mood regulation, cognition, and behavior. Heavy metals interfere with these systems by altering neurotransmitter synthesis, release, reuptake, and receptor function. Lead exposure, for example, has been shown to disrupt the dopaminergic system by reducing DA transporter (DAT) expression and impairing DA release in key brain regions such as the prefrontal cortex and striatum. These alterations are linked to cognitive deficits, impulsivity, and hyperactivity, characteristic of ADHD. Similarly, mercury exposure affects the serotonergic system by depleting serotonin (5-HT) levels and altering serotonin receptor function, contributing to anxiety, depression, and mood instability. Cadmium has been reported to interfere with glutamatergic signaling by increasing extracellular glutamate levels, leading to excitotoxicity, neuronal damage, and increased susceptibility to neuropsychiatric disorders such as schizophrenia. Arsenic exposure has been shown to impair gamma-aminobutyric acid (GABA)ergic neurotransmission, leading to an imbalance between excitatory and inhibitory signaling in the brain, which is implicated in anxiety disorders, epilepsy, and ASD. The disruption of neurotransmitter systems by heavy metals not only affects cognitive and emotional stability but also contributes to behavioral abnormalities observed in various psychiatric conditions. BBB dysfunction is another critical pathway through which heavy metals contribute to neurotoxicity and psychiatric disorders. The BBB is a selectively permeable membrane that protects the brain from harmful substances, allowing essential nutrients to enter. However, heavy metals such as lead, mercury, and cadmium can breach the BBB by directly damaging endothelial cells or increasing permeability through inflammatory signaling. Lead, for instance, has been shown to disrupt tight junction proteins such as claudins and occludins, leading to

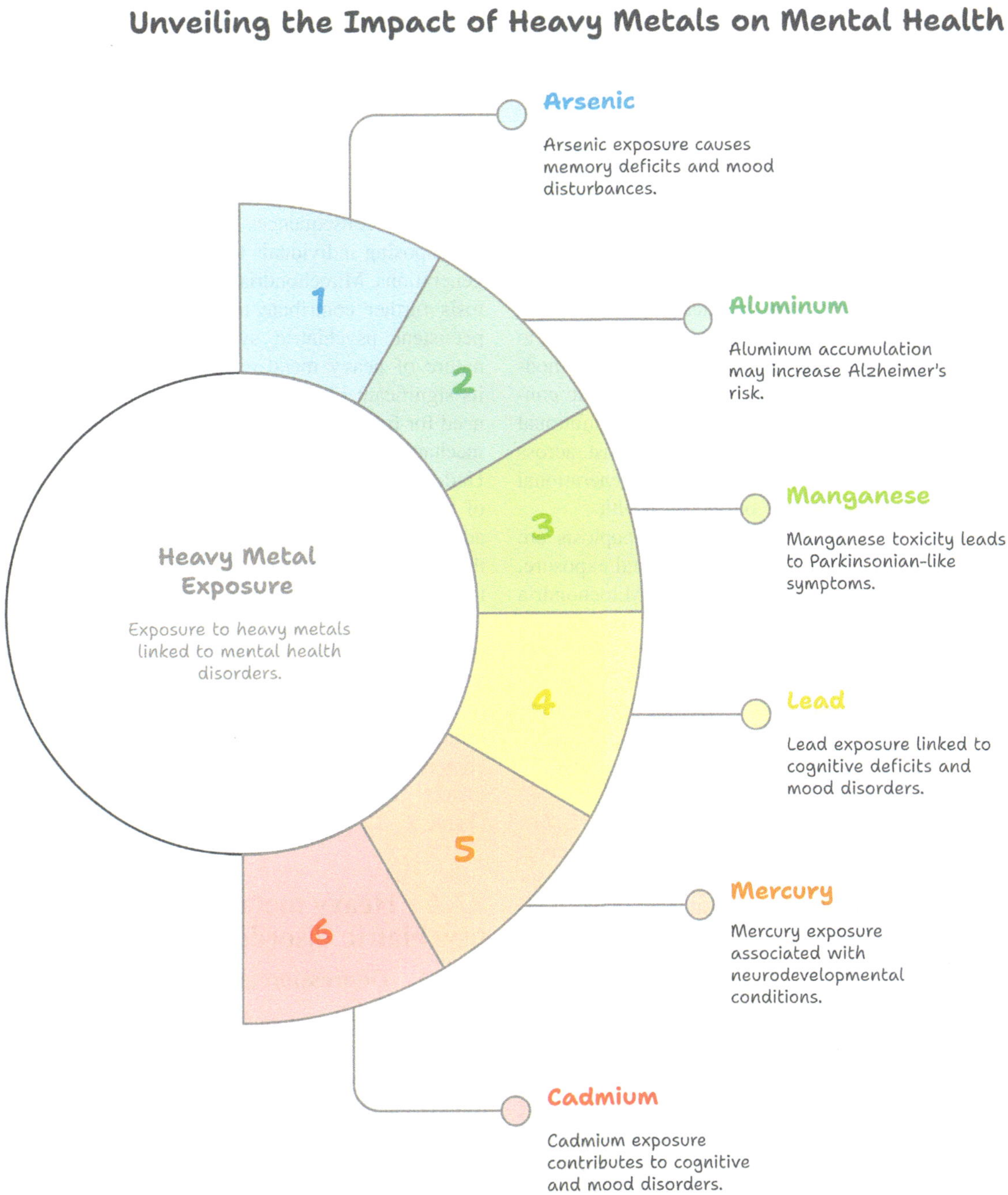

FIGURE 27.2 Unveiling the impact of heavy metals on mental health.

increased BBB permeability and allowing neurotoxicants and inflammatory mediators to enter the brain. Mercury accumulates in endothelial cells and induces oxidative stress, further weakening the integrity of the BBB. Cadmium exposure triggers inflammation-induced BBB dysfunction, facilitating the infiltration of peripheral immune cells into the brain, thereby exacerbating neuroinflammation and neuronal damage. The compromised BBB not only increases the brain's vulnerability to further toxic insults but also contributes to chronic neuroinflammation, which is a key factor in the pathophysiology of psychiatric disorders such as depression, schizophrenia, and bipolar disorder (Ayuso-Álvarez et al., 2019; Leite et al., 2022; Lodge et al., 2022; Saghazadeh & Rezaei, 2017).

Epigenetic modifications and neurodevelopmental toxicity are additional mechanisms by which heavy metals influence mental health. Epigenetic changes involve modifications to DNA and histones that regulate gene expression without altering the underlying genetic code. Heavy metals such as lead, arsenic, and mercury have been shown to induce DNA methylation, histone modifications, and microRNA (miRNA) dysregulation, leading to long-term alterations in gene expression related to brain development and function. Lead exposure during early life has been linked to hypermethylation of genes involved in synaptic plasticity and neurodevelopment, which may contribute to cognitive deficits and an increased risk of neurodevelopmental disorders. Arsenic exposure has been shown to alter DNA methylation patterns in genes associated with mood regulation, increasing susceptibility to depression and anxiety. Mercury-induced epigenetic modifications affect neuronal differentiation and synaptic connectivity, critical for proper cognitive and emotional function. These epigenetic alterations can persist across generations, raising concerns about the transgenerational effects of heavy metal exposure on mental health.

Mitochondrial dysfunction and neuronal apoptosis are key cellular processes disrupted by heavy metal exposure, further contributing to psychiatric disorders. Mitochondria are essential for energy production, calcium homeostasis, and apoptotic regulation, all crucial for maintaining neuronal function. Heavy metals such as cadmium, arsenic, and mercury impair mitochondrial function by inhibiting oxidative phosphorylation, reducing adenosine triphosphate (ATP) production, and increasing mitochondrial ROS generation (Ayuso-Álvarez et al., 2019; Habib et al., 2016; Muñoz et al., 2020; Saravanan et al., 2024). Mercury, for example, binds to mitochondrial thiol groups, disrupting the electron transport chain (ETC) function and leading to energy deficits and oxidative damage. Cadmium-induced mitochondrial dysfunction results in calcium overload, triggering apoptotic pathways that lead to neuronal loss. Arsenic disrupts mitochondrial dynamics by impairing mitophagy, which removes damaged mitochondria, leading to an accumulation of dysfunctional mitochondria and increased neuronal vulnerability to stress. The resulting energy deficits impair synaptic function, reduce neuroplasticity, and contribute to cognitive decline and mood disorders. The activation of apoptotic pathways further exacerbates neuronal loss, leading to structural brain changes associated with psychiatric conditions such as MDD and schizophrenia (Jiang & Zhao, 2024; Mukherjee et al., 2021; Porru et al., 2024; Sun et al., 2024; Zebbiche et al., 2024).

Heavy metal exposure exerts profound neurotoxic effects through multiple interconnected mechanisms, including oxidative stress, neuroinflammation, neurotransmitter system disruptions, BBB dysfunction, epigenetic modifications, mitochondrial dysfunction, and neuronal apoptosis. These processes collectively contribute to cognitive deficits, emotional instability, and an increased risk of neuropsychiatric disorders. Oxidative stress and neuroinflammation are central mediators of metal–induced neurotoxicity, triggering events that disrupt neuronal signaling and impair brain homeostasis. The disruption of neurotransmitter systems further exacerbates mood disorders and cognitive impairments, while BBB dysfunction increases the brain's vulnerability to neurotoxic insults. Epigenetic modifications induced by heavy metals have long-term consequences on brain function, potentially predisposing individuals to mental health disorders across generations. Mitochondrial dysfunction and neuronal apoptosis further contribute to neurodegeneration, leading to persistent psychiatric symptoms. Given the pervasive nature of heavy metal exposure in the environment and its significant impact on mental health, there is an urgent need for further research to elucidate the precise molecular mechanisms underlying metal-induced neurotoxicity. Understanding these pathways can aid in the development of targeted therapeutic strategies aimed at mitigating the adverse effects of heavy metals on the brain and reducing the burden of psychiatric disorders. Public health interventions, including stricter environmental regulations, improved dietary guidelines, and the development of metal chelation therapies, may provide effective means to protect against heavy metal–induced neurotoxicity and promote mental well-being. Addressing heavy metal exposure as a critical factor in psychiatric disorders can lead to more comprehensive prevention and treatment strategies, ultimately improving mental health outcomes worldwide (Table 27.3).

27.4 Heavy metal exposure and specific psychiatric disorders

27.4.1 Depression and anxiety disorders

Depression and anxiety disorders are among the most prevalent and debilitating mental health conditions worldwide, with growing evidence linking their pathophysiology to environmental toxins, including heavy metal exposure. These psychiatric disorders are characterized by persistent emotional disturbances, cognitive impairments, and physiological dysregulation, often exacerbated by oxidative stress, neuroinflammation, neurotransmitter imbalances, mitochondrial dysfunction, and epigenetic modifications. The complex interplay between genetic predisposition, environmental factors, and neurobiological mechanisms underscores the need to explore how heavy metals such as lead (Pb), mercury (Hg), cadmium (Cd), and arsenic (As) contribute to the onset and progression of depression and anxiety disorders. One of the primary mechanisms through which heavy metals induce depressive and anxious behaviors is oxidative stress. The brain is highly vulnerable to oxidative damage due to its high oxygen consumption, lipid-rich composition, and limited antioxidant defense

TABLE 27.3 Impact of heavy metals on neurotransmitter systems and psychiatric disorders.

Heavy metal	Affected neurotransmitter (s)	Receptor alterations	Associated psychiatric conditions	Potential pharmacological interventions
Lead (Pb)	↓ DA, ↓ glutamate, ↓ GABA	↓ D1 and D2 receptor density, ↑ NMDA receptor excitotoxicity	ADHD, schizophrenia, cognitive deficits, depression	Dopamine agonists (L-Dopa), NMDA receptor antagonists (memantine)
Mercury (Hg)	↓ Serotonin (5-HT), ↓ dopamine, ↓ GABA	↓ 5-HT1A/2A receptors, ↑ NMDA excitotoxicity, ↓ GABA-A receptor function	Depression, anxiety, ASD	SSRIs (fluoxetine), GABA modulators (benzodiazepines)
Cadmium (Cd)	↓ Dopamine, ↓ glutamate, ↓ ACh	↓ D2 receptor function, ↓ NMDA receptor density, ↓ cholinergic receptor activity	Mood disorders, cognitive impairment, Parkinson's-like symptoms	Dopamine precursors (L-Dopa), cholinesterase inhibitors (donepezil)
Arsenic (As)	↓ Dopamine, ↓ serotonin, ↑ glutamate toxicity	↓ D1/D2 receptor sensitivity, ↑ NMDA-mediated excitotoxicity	Depression, psychosis, neurodegeneration	Antioxidants (*N*-acetylcysteine), NMDA antagonists (ketamine)
Aluminum (Al)	↓ Acetylcholine, ↓ dopamine, ↑ glutamate	↓ Nicotinic and muscarinic receptor function, ↑ Glutamate toxicity	Alzheimer's disease, cognitive decline, anxiety disorders	Cholinesterase inhibitors (galantamine), memantine (NMDA antagonist)
Manganese (Mn)	↓ Dopamine, ↓ GABA, ↑ glutamate	↓ D2 receptor density, ↓ GABAergic inhibition, ↑ glutamate excitotoxicity	Parkinson's-like symptoms, psychosis, irritability	Dopamine agonists (pramipexole), GABA enhancers (gabapentin)

ACh, Acetylcholine; *ADHD*, attention-deficit/hyperactivity disorder; *ASD*, autism spectrum disorders; *DA*, dopamine; *GABA*, gamma-aminobutyric acid; *SSRIs*, selective serotonin reuptake inhibitors; *NMDA*, N-methyl-D-aspartate; *PTSD*, post-traumatic stress disorder.
Source: Based on Ayuso-Álvarez et al. (2019), Habib et al. (2016), Saghazadeh and Rezaei (2017), Saravanan et al. (2024), Woimant and Trocello (2014).

mechanisms. Heavy metals disrupt redox homeostasis by generating excessive ROS and depleting essential antioxidants such as GSH. The resulting oxidative stress damages neuronal membranes, alters synaptic function, and impairs neurotransmission, contributing to mood disturbances. Mercury and cadmium, for instance, have been shown to induce oxidative stress-mediated neurotoxicity, leading to depressive-like behaviors in animal models. Chronic oxidative damage also triggers inflammatory cascades, activating microglia and astrocytes, which release proinflammatory cytokines such as IL-6, TNF-α, and IL-1β. These inflammatory mediators interfere with neurogenesis, synaptic plasticity, critical emotional regulation, and cognitive resilience processes. Persistent neuroinflammation has been implicated in MDD and generalized anxiety disorder, with elevated levels of proinflammatory markers observed in patients suffering from these conditions. Neurotransmitter dysregulation is another key factor in heavy metal–induced depression and anxiety disorders. Neurotransmitters such as serotonin (5-HT), DA, and GABA play critical roles in regulating mood, stress responses, and emotional stability. Heavy metals disrupt these neurotransmitter systems by interfering with their synthesis, release, receptor binding, and reuptake mechanisms. Lead exposure, for example, has been associated with reduced serotonin levels in the hippocampus, a brain region essential for mood regulation. Serotonin deficits are strongly linked to depressive symptoms, including anhedonia, low energy, and feelings of hopelessness. Mercury and arsenic have been shown to impair DA signaling, contributing to motivational deficits, anhedonia, and cognitive dysfunction. Cadmium exposure alters GABAergic neurotransmission, leading to increased excitability in brain regions associated with fear and anxiety responses (Habib et al., 2016; Jiang & Zhao, 2024; Liu et al., 2024; Muñoz et al., 2020; Sun et al., 2024). These disruptions collectively contribute to the emotional instability and heightened stress sensitivity observed in individuals suffering from depression and anxiety disorders. BBB dysfunction further exacerbates the neurotoxic effects of heavy metals, increasing susceptibility to mood disorders. The BBB serves as a protective barrier, regulating the entry of substances into the CNS. Heavy metals such as lead and mercury compromise BBB integrity by inducing endothelial cell damage, increasing permeability, and facilitating the infiltration of inflammatory mediators into the brain. This breakdown of the BBB allows neurotoxic agents to accumulate, further propagating neuroinflammation and oxidative damage. Increased permeability of the BBB has been observed in patients with depression and anxiety, suggesting that environmental toxins, including heavy metals, play a role in the pathogenesis of these disorders. Epigenetic modifications induced by heavy metal exposure also contribute to long-term susceptibility to depression and anxiety disorders. Epigenetic mechanisms, including DNA methylation, histone modifications, and miRNA regulation, influence gene expression without altering the genetic code. Heavy metals such as arsenic and cadmium have been shown to induce hypermethylation of genes associated with neurogenesis and synaptic plasticity, leading to impaired neuronal function. Additionally, metal-induced alterations in miRNA expression can disrupt stress response pathways, increasing vulnerability to mood disorders. These epigenetic changes may have transgenerational effects, raising concerns about the heritable impact of heavy metal exposure on mental health. Mitochondrial dysfunction and neuronal apoptosis further contribute to the pathophysiology of depression and anxiety. Mitochondria play a crucial role in energy production, calcium homeostasis, and neuronal survival. Heavy metals such as mercury and cadmium disrupt mitochondrial function by impairing oxidative phosphorylation, reducing ATP production, and increasing mitochondrial ROS generation. Energy deficits in the brain lead to impaired synaptic function, reduced neuroplasticity, and increased susceptibility to stress-related disorders. Additionally, mitochondrial dysfunction triggers apoptotic pathways, leading to neuronal loss in brain regions involved in mood regulation, such as the prefrontal cortex and hippocampus. The loss of neuronal integrity further contributes to cognitive deficits, emotional dysregulation, and heightened stress sensitivity, all of which are hallmarks of depression and anxiety disorders. Heavy metal exposure plays a significant role in the pathogenesis of depression and anxiety disorders through multiple interrelated mechanisms, including oxidative stress, neuroinflammation, neurotransmitter dysregulation, BBB dysfunction, epigenetic modifications, mitochondrial impairment, and neuronal apoptosis. The cumulative impact of these factors disrupts brain homeostasis, leading to persistent mood disturbances and heightened vulnerability to stress-related disorders. Given the increasing prevalence of environmental heavy metal contamination, there is an urgent need for further research to elucidate the precise molecular mechanisms underlying metal-induced neurotoxicity and its impact on mental health. Understanding these pathways can inform the development of targeted therapeutic interventions, including antioxidant-based treatments, antiinflammatory strategies, and epigenetic therapies, to mitigate the adverse effects of heavy metal exposure on mood disorders. Additionally, public health initiatives aimed at reducing heavy metal contamination in food, water, and air, along with improved screening and early intervention programs, may help prevent the onset and progression of depression and anxiety disorders linked to environmental toxicants (Ayuso-Álvarez et al., 2019; Iqbal et al., 2018; Leite et al., 2022; Porru et al., 2024; Saravanan et al., 2024; Yousef et al., 2013).

27.4.2 Schizophrenia and psychotic disorders

Schizophrenia and other psychotic disorders are severe mental illnesses characterized by hallucinations, delusions, cognitive deficits, and emotional dysregulation (Fig. 27.2). While the exact etiology of schizophrenia remains complex and multifactorial, growing evidence suggests that environmental factors, including heavy metal exposure, play a significant role in its onset and progression. Heavy metals such as lead (Pb), mercury (Hg), cadmium (Cd), and arsenic (As) have been implicated in neurodevelopmental disruptions, neurotransmitter imbalances, oxidative stress, neuroinflammation, and epigenetic modifications, all of which contribute to the pathophysiology of schizophrenia and psychotic disorders. The neurotoxic effects of these metals, especially during critical periods of brain development, can lead to structural and functional abnormalities in key brain regions such as the prefrontal cortex, hippocampus, and striatum, which are associated with psychosis. One of the primary mechanisms through which heavy metals contribute to schizophrenia is oxidative stress. The brain's high metabolic activity and oxygen consumption make it particularly vulnerable to oxidative damage, and heavy metals exacerbate this vulnerability by generating excessive ROS and impairing antioxidant defense systems. Lead and mercury have been shown to increase lipid peroxidation and deplete essential antioxidants such as GSH, leading to neuronal damage and synaptic dysfunction. The increased oxidative burden disrupts neural connectivity and contributes to the cognitive and perceptual disturbances characteristic of schizophrenia. Furthermore, oxidative stress has been linked to the dopaminergic dysfunction observed in schizophrenia, where excessive DA activity in the mesolimbic pathway contributes to positive symptoms such as hallucinations and delusions. In contrast, reduced DA function in the prefrontal cortex leads to cognitive and negative symptoms such as social withdrawal and anhedonia. Neuroinflammation also plays a crucial role in heavy metal–induced schizophrenia and psychotic disorders. Chronic exposure to metals such as cadmium and arsenic activates microglia and astrocytes, triggering the release of proinflammatory cytokines such as IL-6, TNF-α, and IL-1β. This persistent neuroinflammatory state disrupts neuronal signaling, impairs synaptic plasticity, and contributes to brain volume reductions commonly observed in schizophrenia patients. Studies have shown that individuals with schizophrenia have elevated levels of neuroinflammatory markers, suggesting a strong link between immune system dysfunction and psychosis. Additionally, inflammation-induced changes in glutamatergic neurotransmission contribute to excitotoxicity, further exacerbating neuronal damage and cognitive impairment. Heavy metals also interfere with the balance of key neurotransmitter systems, particularly DA, glutamate, and GABA, which are heavily implicated in schizophrenia. Lead exposure has been associated with increased dopaminergic activity, mimicking the hyperdopaminergic state observed in schizophrenia patients. Mercury and cadmium disrupt glutamate signaling, leading to excitotoxic damage and impaired cognitive function. Furthermore, arsenic exposure alters GABAergic transmission, reducing inhibitory control over excitatory pathways and contributing to psychotic symptoms (Emurotu et al., 2025; Iqbal et al., 2018; Jiang & Zhao, 2024; Porru et al., 2024; Saghazadeh & Rezaei, 2017). These disruptions in neurotransmitter function collectively contribute to the disorganized thinking, perceptual abnormalities, and cognitive deficits characteristic of schizophrenia. BBB dysfunction further exacerbates the neurotoxic effects of heavy metals in schizophrenia. The BBB serves as a critical protective barrier, regulating the transport of substances into the brain and preventing neurotoxic agents from entering the CNS. Heavy metals such as lead and mercury compromise BBB integrity by increasing endothelial cell permeability and accumulating neurotoxic substances in the brain. Studies have shown that schizophrenia patients exhibit increased BBB permeability, suggesting that environmental toxins, including heavy metals, may contribute to disease progression by facilitating neurotoxic infiltration and inflammation (Fig. 27.3). Epigenetic modifications induced by heavy metal exposure further influence the risk of schizophrenia and psychotic disorders. Heavy metals such as cadmium and arsenic have been shown to alter DNA methylation patterns, histone modifications, and miRNA expression, leading to dysregulation of genes involved in neurodevelopment and synaptic plasticity. These epigenetic changes can have long-term consequences, increasing susceptibility to psychotic disorders across generations. Research has shown that schizophrenia patients exhibit widespread epigenetic alterations, particularly in genes related to neurotransmission and neuroinflammation, suggesting that environmental factors such as heavy metal exposure may play a role in disease onset through epigenetic reprogramming. Mitochondrial dysfunction and neuronal apoptosis also contribute to the pathophysiology of schizophrenia and heavy metal toxicity. Mitochondria are essential for energy production, neuronal survival, and synaptic function, and their impairment has been strongly linked to schizophrenia. Heavy metals such as mercury and cadmium disrupt mitochondrial function by impairing oxidative phosphorylation, reducing ATP production, and increasing mitochondrial ROS generation. The resulting energy deficits impair synaptic transmission, leading to cognitive dysfunction and negative symptoms commonly observed in schizophrenia. Additionally, mitochondrial dysfunction triggers apoptotic pathways, leading to neuronal loss in critical brain regions such as the prefrontal cortex and hippocampus. The loss of neural integrity

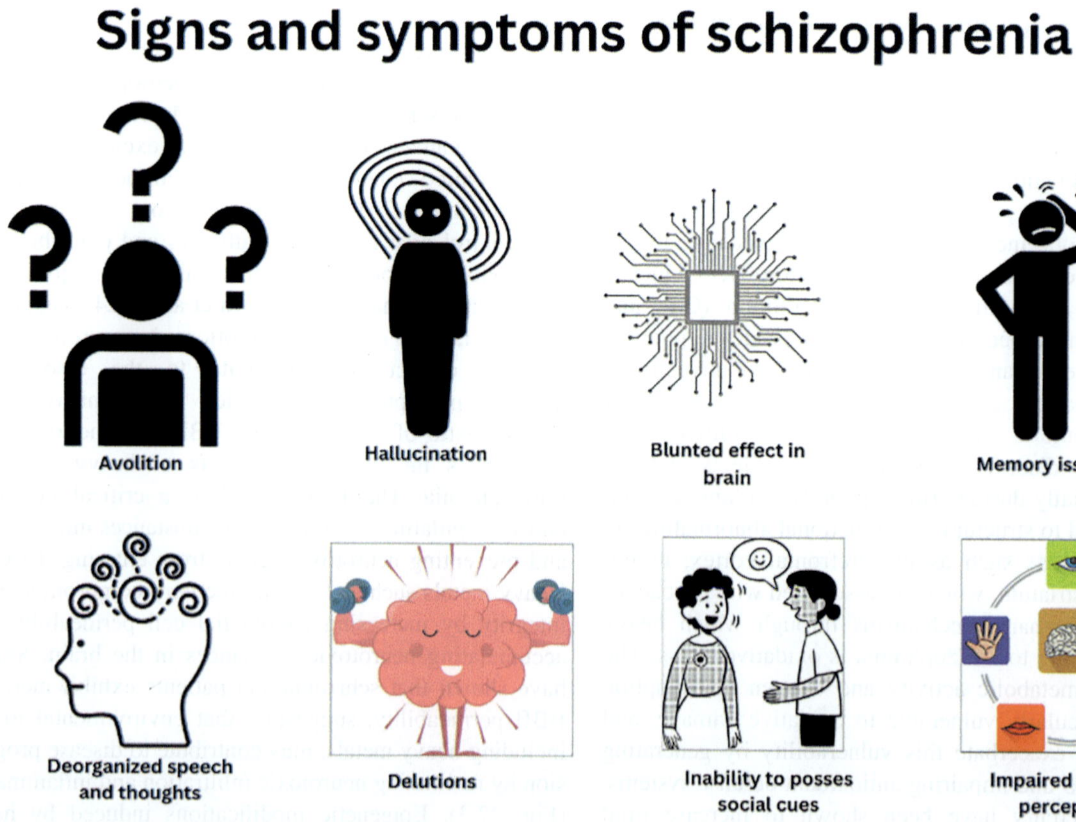

FIGURE 27.3 Signs and symptoms of schizophrenia.

further contributes to the cognitive, emotional, and perceptual disturbances characteristic of schizophrenia. Heavy metal exposure plays a significant role in the etiology and progression of schizophrenia and psychotic disorders through multiple interrelated mechanisms, including oxidative stress, neuroinflammation, neurotransmitter imbalances, BBB dysfunction, epigenetic modifications, mitochondrial impairment, and neuronal apoptosis. These neurotoxic effects disrupt brain homeostasis, leading to the characteristic symptoms of schizophrenia, including hallucinations, delusions, cognitive deficits, and emotional dysregulation. Given the increasing environmental burden of heavy metal contamination, further research is needed to elucidate the precise molecular mechanisms underlying metal-induced neurotoxicity and its impact on schizophrenia. Additionally, targeted interventions, such as antioxidant therapies, antiinflammatory strategies, and epigenetic treatments, may offer novel therapeutic approaches to mitigate the effects of heavy metal exposure on psychotic disorders. Public health measures aimed at reducing heavy metal contamination in food, water, and air, along with improved screening and early intervention programs, may help prevent the onset and progression of schizophrenia and related psychotic disorders linked to environmental toxicants (Channegowda, 2020; Jiang & Zhao, 2024; Muñoz et al., 2020; Vänskä et al., 2019).

27.4.3 Bipolar disorder and mood instability

Bipolar disorder is a chronic psychiatric illness characterized by extreme mood fluctuations between manic, hypomanic, and depressive states. The precise etiology of bipolar disorder remains multifactorial, involving genetic predisposition, neurobiological dysfunction, and environmental influences, including heavy metal exposure. Emerging evidence suggests that heavy metals such as lead (Pb), mercury (Hg), cadmium (Cd), and arsenic (As) contribute to the pathophysiology of bipolar disorder through mechanisms such as oxidative stress, neuroinflammation, neurotransmitter disruption, mitochondrial dysfunction, and epigenetic modifications. Given the critical role of metal homeostasis in brain function, even low-level exposure to these neurotoxic metals can significantly impact mood regulation, emotional stability, and cognitive function, increasing the risk of bipolar disorder and related affective disturbances. One of the primary mechanisms through which heavy metals influence bipolar disorder is oxidative stress. The brain's high oxygen consumption and lipid-rich environment make it particularly susceptible to oxidative damage, which has been implicated in the pathogenesis of bipolar disorder. Heavy metals such as lead and mercury increase the generation of ROS while simultaneously depleting antioxidant defenses, including GSH and superoxide dismutase (SOD). This oxidative imbalance leads to neuronal damage, synaptic dysfunction, and impaired neuroplasticity, which are all hallmarks of bipolar disorder. Studies have shown that bipolar patients exhibit elevated oxidative stress markers, suggesting that environmental factors such as heavy metal exposure may exacerbate the neurodegenerative processes underlying mood instability. Neuroinflammation is another critical factor linking heavy metal exposure to bipolar disorder. Chronic inflammation is increasingly recognized as a key component of mood disorders, with elevated levels of proinflammatory cytokines such as IL-6, TNF-α, and IL-1β observed in bipolar patients. Heavy metals such as cadmium and arsenic activate microglia and astrocytes, triggering a sustained inflammatory response in the CNS. This persistent neuroinflammation disrupts the delicate balance of excitatory and inhibitory neurotransmission, impairing emotional regulation and increasing susceptibility to mood swings. Inflammatory signaling also affects the hypothalamic–pituitary–adrenal axis, leading to dysregulated stress responses that contribute to manic and depressive episodes. Neurotransmitter dysregulation plays a pivotal role in bipolar disorder, and heavy metals interfere with key neurotransmitter systems, including DA, serotonin, glutamate, and GABA. Lead exposure, for example, disrupts DA homeostasis, contributing to the hyperdopaminergic state observed during manic episodes and the reduced dopaminergic activity associated with depressive states. Mercury and cadmium affect serotonin synthesis and receptor function, leading to mood instability, irritability, and emotional dysregulation. Arsenic disrupts glutamatergic and GABAergic signaling, impairing synaptic plasticity and increasing the risk of excitotoxicity, which has been linked to mood disorders. The cumulative effect of these neurotransmitter imbalances is a dysregulated mood state that characterises bipolar disorder. Mitochondrial dysfunction is another major consequence of heavy metal exposure, with significant implications for bipolar disorder. Mitochondria play a crucial role in energy metabolism, synaptic transmission, and neuroplasticity, and their impairment leads to neuronal instability and mood dysregulation. Heavy metals such as mercury and cadmium impair mitochondrial function by disrupting oxidative phosphorylation, reducing ATP production, and increasing mitochondrial ROS levels (Channegowda, 2020; Emurotu et al., 2025; Jiang & Zhao, 2024; Lodge et al., 2022; Vänskä et al., 2019). The resulting energy deficits affect high-demand brain regions such as the prefrontal cortex and limbic system, which are involved in mood regulation and executive function. Mitochondrial dysfunction has been strongly linked to bipolar disorder, with studies showing altered mitochondrial gene expression and bioenergetic deficits in bipolar patients. Heavy metal exposure may, therefore, exacerbate these mitochondrial abnormalities, contributing to the neurobiological basis of mood instability. Epigenetic modifications induced by heavy metals further influence the risk of bipolar disorder. Environmental toxins such as lead and arsenic have been shown to alter DNA methylation, histone modifications, and miRNA expression, leading to long-term changes in gene expression related to mood regulation and emotional processing. These epigenetic changes may underlie the heritable component of bipolar disorder, as studies have found altered DNA methylation patterns in genes related to neurotransmitter function, neuroinflammation, and mitochondrial activity in bipolar patients. Additionally, epigenetic modifications induced by heavy metal exposure may increase susceptibility to bipolar disorder across generations, highlighting the importance of environmental factors in disease risk. BBB dysfunction is another critical mechanism through which heavy metals contribute to bipolar disorder. The BBB serves as a protective barrier that regulates the entry of neurotoxic substances into the brain. Heavy metals such as lead and mercury disrupt BBB integrity by increasing endothelial permeability and altering tight junction proteins. This compromised barrier allows neurotoxic substances to infiltrate the CNS, triggering neuroinflammatory responses and exacerbating the neurochemical imbalances associated with bipolar disorder. Studies have shown that bipolar patients exhibit abnormalities in BBB function, suggesting that environmental toxins may play a role in the development and progression of mood disorders. Heavy metal exposure contributes significantly to the pathogenesis of bipolar disorder and mood instability through multiple interrelated mechanisms, including oxidative stress, neuroinflammation, neurotransmitter disruption, mitochondrial dysfunction, epigenetic modifications, and BBB impairment. These neurotoxic effects disrupt the delicate balance of mood regulation, increasing the risk of manic and depressive episodes characteristic of bipolar disorder. Given the rising environmental burden of heavy metal contamination, further research is needed to elucidate the precise molecular mechanisms underlying metal-induced neurotoxicity and its impact on bipolar disorder. Targeted interventions, such as antioxidant therapies, antiinflammatory treatments, and epigenetic modulation, may offer novel therapeutic strategies to mitigate the effects of heavy metal exposure on mood disorders. Additionally, public health initiatives aimed at reducing heavy metal contamination in food, water, and air, along with early detection and prevention strategies, may help decrease the incidence and severity of bipolar disorder linked to environmental neurotoxicants (Ayuso-Álvarez et al., 2019; Emurotu et al., 2025; Iqbal et al., 2018; Saravanan et al., 2024; Woimant & Trocello, 2014; Yousef et al., 2013).

27.4.4 Neurodevelopmental disorders

Neurodevelopmental disorders, including ADHD and ASDs, are complex conditions characterized by impairments in cognitive, social, and behavioral functions (Fig. 27.3). These disorders have a multifactorial etiology involving genetic susceptibility, environmental factors, and epigenetic modifications that influence brain development and function. Growing evidence suggests that heavy metal exposure plays a significant role in the pathogenesis of neurodevelopmental disorders by disrupting key neurobiological processes, including oxidative stress, neuroinflammation, neurotransmitter dysregulation, mitochondrial dysfunction, and epigenetic alterations. Prenatal and early-life exposure to neurotoxic metals such as lead (Pb), mercury (Hg), cadmium (Cd), and arsenic (As), as well as other relevant metals like aluminum (Al) and manganese (Mn), has been linked to an increased risk of ADHD and ASD, underscoring the critical impact of environmental pollutants on neurodevelopmental trajectories. One of the primary mechanisms by which heavy metals contribute to ADHD and ASD is oxidative stress. The developing brain is particularly vulnerable to oxidative damage due to its high metabolic rate, extensive synaptogenesis, and immature antioxidant defense systems. Heavy metals such as lead and mercury generate excessive ROS, leading to lipid peroxidation, DNA damage, and neuronal apoptosis. Studies have shown that children with ASD and ADHD exhibit elevated biomarkers of oxidative stress, including malondialdehyde and reduced GSH levels. This oxidative imbalance disrupts normal neuronal function and connectivity, impairing cognitive flexibility, attention regulation, and social interaction, all hallmark features of neurodevelopmental disorders. Neuroinflammation is another critical factor linking heavy metal exposure to neurodevelopmental disorders (Fig. 27.4). Prolonged activation of microglia and astrocytes in response to heavy metals leads to the overproduction of proinflammatory cytokines such as IL-6, TNF-α, and IL-1β.

Persistent neuroinflammation interferes with synaptic pruning, neural circuit formation, and myelination processes essential for normal cognitive and social development. Postmortem studies of ASD brains have revealed increased microglial activation and elevated inflammatory markers, suggesting that neuroinflammation plays a key role in ASD pathophysiology. Similarly, ADHD has been associated with chronic neuroinflammatory states that disrupt prefrontal cortex function, leading to attentional deficits and hyperactivity (Habib et al., 2016; Iqbal et al., 2018; Lodge et al., 2022; Mukherjee et al., 2021; Woimant & Trocello, 2014). Heavy metals also disrupt neurotransmitter systems that regulate attention, impulse control, and social behaviors. Lead exposure, for example, is known to interfere with DA homeostasis, contributing to the DA dysregulation observed in ADHD. Mercury and cadmium impair serotoninergic and glutamatergic signaling, which is critical for mood regulation, learning, and social interactions. Arsenic exposure has been linked to GABAergic dysfunction, affecting inhibitory control and increasing hyperactivity and impulsivity. These neurotransmitter imbalances align with the neurochemical abnormalities observed in ADHD and ASD, providing a mechanistic link between heavy metal toxicity and neurodevelopmental impairments. Mitochondrial dysfunction is another significant consequence of heavy metal exposure that contributes to ADHD and ASD. The brain relies on mitochondrial energy production to sustain neuronal activity, synaptic plasticity, and neurodevelopmental processes. Heavy metals such as mercury, lead, and cadmium impair mitochondrial function by inhibiting ETC complexes, reducing ATP production, and increasing mitochondrial ROS levels. Mitochondrial dysfunction has been consistently reported in ASD, with studies showing reduced oxidative phosphorylation capacity and altered mitochondrial gene expression in affected individuals. Similarly, ADHD has been associated with energy metabolism deficits that impair prefrontal cortex function, leading to attentional and behavioral dysregulation. Epigenetic modifications induced by heavy metal exposure further exacerbate neurodevelopmental vulnerabilities. Prenatal and early-life exposure to lead, mercury, and arsenic has been shown to alter DNA methylation, histone modifications, and miRNA expression patterns in genes involved in neurodevelopment. These epigenetic changes may have long-lasting effects on brain structure and function, increasing the risk of ASD and ADHD. For example, lead exposure has been linked to hypermethylation of DA-related genes, affecting DA receptor expression and contributing to attentional deficits in ADHD. Arsenic-induced epigenetic changes have been implicated in synaptic plasticity alterations associated with ASD. BBB dysfunction is another key mechanism by which heavy metals influence neurodevelopmental disorders. The BBB plays a crucial role in protecting the developing brain from toxic insults by regulating the entry of harmful substances. Heavy metals such as lead, mercury, and manganese disrupt BBB integrity by increasing endothelial permeability and altering tight junction proteins. This compromised barrier allows neurotoxic substances to infiltrate the brain, exacerbating neuroinflammatory responses and further disrupting neural circuit formation. Studies have shown that children with ASD exhibit abnormalities in BBB function, suggesting that environmental toxins may contribute to the development of the disorder. Heavy metal exposure plays a significant role in the etiology of neurodevelopmental disorders such as ADHD and ASD through multiple interrelated mechanisms, including oxidative stress, neuroinflammation, neurotransmitter

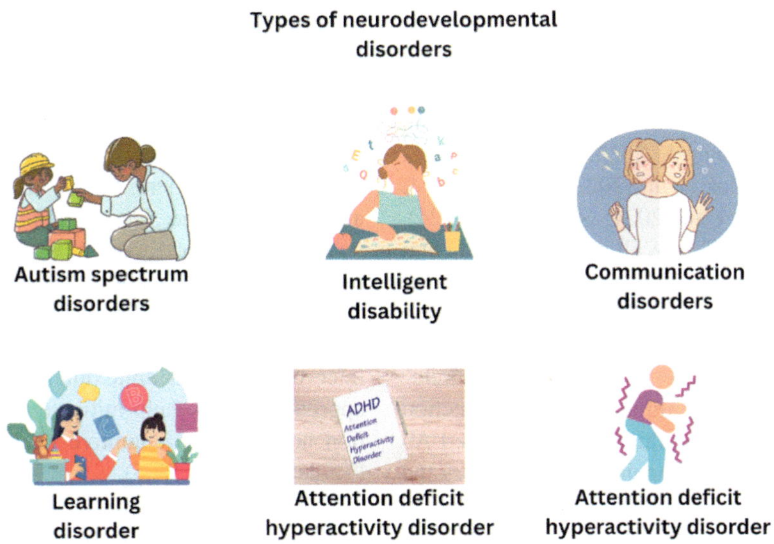

FIGURE 27.4 Types of neurodevelopmental disorders.

disruption, mitochondrial dysfunction, epigenetic modifications, and BBB impairment (Emurotu et al., 2025; Iqbal et al., 2018; Jiang & Zhao, 2024; Saghazadeh & Rezaei, 2017). Given the increasing environmental burden of heavy metal contamination, further research is needed to elucidate the precise molecular pathways underlying metal-induced neurotoxicity and its impact on neurodevelopment. Targeted interventions, such as antioxidant therapies, antiinflammatory treatments, and epigenetic modulation, may offer novel therapeutic strategies to mitigate the effects of heavy metal exposure on neurodevelopmental disorders. Additionally, public health initiatives aimed at reducing heavy metal contamination in food, water, and air, along with early detection and prevention strategies, may help decrease the incidence and severity of ADHD and ASD linked to environmental neurotoxicants.

27.4.5 Cognitive decline and neurodegenerative diseases

Heavy metal exposure has been increasingly recognized as a significant environmental risk factor for cognitive decline and neurodegenerative diseases, including AD and PD (Fig. 27.4). These progressive disorders are characterized by neuronal loss, synaptic dysfunction, and accumulation of pathological proteins, leading to cognitive impairments and motor dysfunction. While ageing and genetic predisposition play key roles in neurodegeneration, environmental toxins such as lead (Pb), mercury (Hg), cadmium (Cd), arsenic (As), aluminum (Al), and manganese (Mn) contribute to disease progression through oxidative stress, neuroinflammation, mitochondrial dysfunction, and impaired protein homeostasis. Chronic exposure to these metals accelerates neuronal damage, disrupts neurotransmitter balance, and induces epigenetic modifications, thereby increasing the risk of developing AD and PD. One of the primary mechanisms linking heavy metals to neurodegeneration is oxidative stress. The brain is highly susceptible to oxidative damage due to its high oxygen consumption, lipid-rich composition, and limited antioxidant defenses. Heavy metals generate excessive ROS and RNS, leading to lipid peroxidation, protein oxidation, and DNA damage. In AD, oxidative stress exacerbates the aggregation of amyloid-beta (Aβ) plaques and hyperphosphorylated tau, which are hallmark pathological features of the disease. Studies have shown that lead and aluminum exposure increase Aβ production and tau phosphorylation by activating stress kinases and impairing proteasomal degradation. Similarly, in PD, oxidative stress contributes to the degeneration of dopaminergic neurons in the substantia nigra, a key pathological event in the disorder. Manganese and mercury have been implicated in DA depletion by promoting oxidative damage to dopaminergic neurons, accelerating the onset of PD symptoms. Neuroinflammation is another major factor in heavy metal–induced neurodegeneration. Chronic exposure to neurotoxic metals activates microglia and astrocytes, leading to excessive production of proinflammatory cytokines such as IL-6, TNF-α, and IL-1β. Persistent neuroinflammation contributes to neuronal death, disrupts synaptic plasticity, and enhances amyloid plaque deposition in AD. In PD, microglial activation exacerbates dopaminergic neuron loss and promotes the misfolding of α-synuclein, a key protein implicated in Lewy body formation. Lead, cadmium, and arsenic have been shown to induce neuroinflammatory responses in experimental models, suggesting a strong link between heavy metal toxicity and the inflammatory processes underlying neurodegenerative diseases. Mitochondrial dysfunction is another critical consequence of heavy metal exposure that contributes to cognitive decline and neurodegeneration (Emurotu et al., 2025; Iqbal et al., 2018; Jiang & Zhao, 2024; Muñoz et al., 2020; Saghazadeh & Rezaei, 2017; Saravanan et al., 2024; Sun et al., 2024). Mitochondria play a vital role in ATP production, calcium homeostasis, and apoptosis regulation, all essential for neuronal survival. Heavy metals such as mercury, lead, and cadmium disrupt mitochondrial function by impairing ETC complexes, reducing ATP synthesis, and increasing mitochondrial ROS levels. In AD, mitochondrial dysfunction leads to energy deficits, synaptic failure, and increased Aβ toxicity. In PD, mitochondrial impairment accelerates the degeneration of dopaminergic neurons, contributing to motor deficits. Studies have demonstrated that individuals with high heavy metal burdens exhibit altered mitochondrial dynamics, further highlighting the role of environmental toxins in neurodegenerative diseases. Heavy metals also interfere with neurotransmitter systems critical for cognitive and motor functions. Lead, mercury, and arsenic disrupt cholinergic signaling, essential for learning, memory, and attention. In AD, reduced acetylcholine levels due to heavy metal–induced cholinergic dysfunction contribute to progressive cognitive impairment. In PD, manganese, mercury, and cadmium affect DA synthesis and receptor function, leading to movement disorders, rigidity, and tremors. Additionally, glutamate excitotoxicity induced by heavy metals exacerbates neuronal death in both AD and PD by overstimulating NMDA receptors and increasing intracellular calcium levels. Another significant mechanism linking heavy metals to neurodegenerative diseases is epigenetic modification. Heavy metals alter DNA methylation, histone modifications, and miRNA expression, leading to dysregulation of genes involved in neuroprotection, inflammation, and synaptic plasticity. Lead exposure has been associated with hypermethylation of genes related to neurodevelopment, while arsenic and cadmium alter histone acetylation patterns, contributing to neuronal dysfunction. These epigenetic changes may have long-lasting effects on cognitive health and increase susceptibility to AD and PD in later life. BBB dysfunction further exacerbates the effects of heavy metal exposure on neurodegeneration. The BBB protects the brain from toxic insults by regulating the entry of harmful substances. Still, heavy metals such as lead, mercury, and manganese disrupt their integrity by increasing endothelial permeability and altering tight junction proteins. This compromised barrier allows neurotoxic metals to accumulate in the brain, promoting neuroinflammation and accelerating disease pathology. Studies have shown that individuals with AD exhibit increased BBB permeability, suggesting that environmental toxins may play a role in disease onset and progression. Heavy metal exposure significantly contributes to cognitive decline and neurodegenerative diseases through multiple interrelated mechanisms, including oxidative stress, neuroinflammation, mitochondrial dysfunction, neurotransmitter disruption, epigenetic modifications, and BBB impairment. Given the rising prevalence of AD and PD and the increasing burden of environmental pollution, further research is needed to understand the precise molecular pathways underlying heavy metal–induced neurotoxicity. Preventive strategies, including reducing environmental heavy metal contamination, dietary interventions with antioxidants, and targeted therapeutic approaches, may help mitigate the risk of neurodegeneration and improve cognitive health in ageing populations (Table 27.4, Fig. 27.5).

27.5 Epidemiological and clinical evidence

The growing body of epidemiological and clinical research has provided compelling evidence linking heavy metal exposure to psychiatric disorders across different populations. Numerous population-based studies have explored the association between chronic exposure to metals such as lead (Pb), mercury (Hg), cadmium (Cd), arsenic (As), aluminum (Al), and manganese (Mn) with a range of psychiatric disorders, including depression, anxiety, schizophrenia, bipolar disorder, and neurodevelopmental disorders. These studies highlight the detrimental impact of metal toxicity on mental health by examining large cohorts, environmental exposure levels, and biomarkers of heavy metal burden. The findings suggest that prolonged exposure to neurotoxic metals, often due to industrial pollution, contaminated drinking water, occupational hazards, and dietary intake, contributes to the onset and progression of psychiatric conditions. Furthermore, variations in exposure levels based on geographic location, socioeconomic status, and genetic predisposition play a crucial role in determining the extent of mental health outcomes among affected populations. Several large-scale epidemiological studies have demonstrated a clear association between lead exposure and an increased risk of psychiatric disorders. Lead is a well-established neurotoxin that accumulates in the brain and disrupts synaptic function, neurotransmitter balance, and neurodevelopmental processes. Longitudinal cohort studies, such as the Cincinnati Lead Study and the Dunedin Multidisciplinary Health and Development Study, have shown that early-life lead exposure is significantly associated with cognitive impairments, attention deficits, aggressive behavior, and an increased risk of schizophrenia-like psychosis in adulthood. Blood lead levels (BLLs) measured in childhood have been correlated with an elevated prevalence of mood disorders, impulsivity, and antisocial behavior later in life. Furthermore, metaanalyses

TABLE 27.4 Comparison of biomonitoring techniques for assessing heavy metal exposure.

Biomonitoring technique	Sample type	Advantages	Limitations	Recommended use cases
Blood analysis	Blood	Measures recent exposure, reliable for toxic levels, standardized testing methods	Short detection window, invasive sample collection, fluctuations due to homeostasis	Acute and recent exposure assessment, occupational health monitoring
Urine analysis	Urine	Noninvasive, helpful in detecting excreted metals, effective for cadmium, arsenic, and mercury	Variable excretion rates, influenced by hydration status, may not reflect the total body burden	Chronic and recent exposure evaluation, environmental and occupational studies
Hair analysis	Hair	Reflects long-term exposure, noninvasive, useful for mercury and arsenic detection	Contamination risk from external sources, variability in hair growth, and metabolism	Retrospective exposure analysis, epidemiological studies, and forensic toxicology
Tissue analysis	Organ/ tissue biopsy	Measures long-term accumulation, most accurate for assessing metal storage in organs	Highly invasive, ethical concerns, limited availability for routine testing	Research-based exposure assessment, postmortem toxicology, and chronic bioaccumulation studies

Source: Based on Emurotu et al. (2025), Iqbal et al. (2018), Jiang and Zhao (2024), Muñoz et al. (2020), Saghazadeh and Rezaei (2017), Saravanan et al. (2024), Sun et al. (2024).

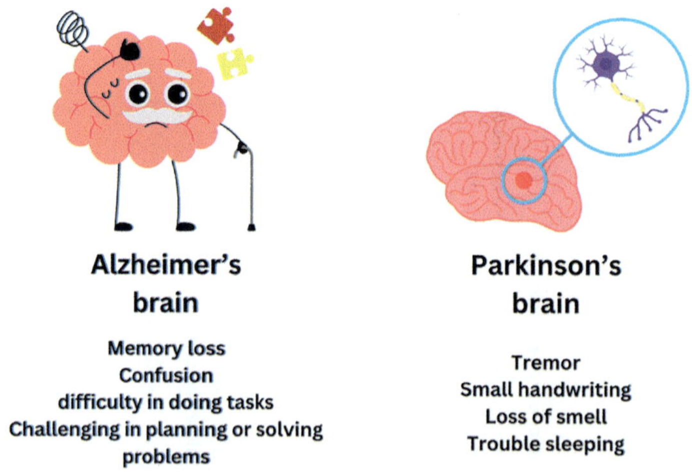

FIGURE 27.5 Alzheimer's and Parkinson's brain.

of multiple studies have indicated that populations with chronic lead exposure have a higher incidence of depression and anxiety, supporting the neurotoxic role of lead in mental health disorders. Mercury exposure has also been extensively studied for psychiatric disorders, particularly in populations exposed to methylmercury through seafood consumption and occupational settings. Epidemiological investigations, such as those conducted in the Seychelles Child Development Study and the Faroe Islands cohort, have demonstrated that prenatal and early-life exposure to mercury can lead to neurobehavioral disturbances, reduced IQ, and increased susceptibility to ASD and ADHD. Mercury's ability to disrupt neurotransmitter function, mainly by impairing DA and serotonin pathways, has been implicated in mood instability, anxiety, and depressive symptoms. Studies analyzing populations residing near artisanal gold mining areas, where mercury is frequently used for gold extraction, have reported increased rates of psychiatric distress, cognitive dysfunction, and suicidal tendencies, further linking chronic mercury exposure to mental health deterioration. Cadmium and arsenic exposure have been increasingly recognized as risk factors for psychiatric disorders in population-based studies. Cadmium, a heavy metal commonly found in cigarette smoke, industrial emissions, and contaminated food, has been linked to oxidative stress and neuroinflammation, both of which contribute to depression and cognitive decline. Studies using National Health and Nutrition Examination Survey data have shown that individuals with higher urinary cadmium levels exhibit greater odds of developing depression and anxiety (Ayuso-Álvarez et al., 2019; Channegowda, 2020; Emurotu et al., 2025; Leite et al., 2022; Porru et al., 2024). Similarly, arsenic exposure, particularly in regions with contaminated groundwater, such as Bangladesh, India, and parts of the United States, has been associated with an increased prevalence of neurodevelopmental disorders, mood instability, and cognitive impairments. Studies examining children exposed to high arsenic levels in drinking water have reported deficits in attention, memory, and executive function, along with an elevated risk of autism and ADHD. Regional and demographic variations in heavy metal exposure significantly influence mental health outcomes, with specific populations disproportionately affected due to geographic location, socioeconomic factors, and occupational hazards. Industrialized regions with high pollution levels, such as parts of China, India, and Latin America, have reported more significant burdens of heavy metal–related psychiatric disorders compared to areas with lower environmental contamination. Urban populations often experience higher exposure to lead from deteriorating infrastructure, industrial emissions, and vehicular pollution. In contrast, rural communities may face more significant risks from arsenic-contaminated groundwater, pesticide-laden agricultural products, and mercury exposure from small-scale mining activities. These regional disparities in mental exposure contribute to varying patterns of psychiatric illness across different populations. Socioeconomic status plays a critical role in determining susceptibility to heavy metal toxicity and its psychiatric consequences. Lower income populations often reside in areas with higher environmental contamination, limited access to clean water, and inadequate healthcare resources, exacerbating their vulnerability to heavy metal–induced mental health disorders. Children in impoverished communities are more likely to be exposed to lead from lead-based paint, contaminated soil, and substandard housing, increasing their risk of neurodevelopmental disorders such as ADHD and learning disabilities. Similarly, workers in hazardous occupations, such as battery manufacturing, welding, and mining, are at heightened risk of chronic heavy metal exposure, leading to cumulative neurotoxic effects that manifest as anxiety, depression, and cognitive impairment over time. Studies examining occupational cohorts have consistently demonstrated higher psychiatric morbidity in individuals exposed to neurotoxic metals compared to the general population. Genetic predisposition and individual susceptibility factors also contribute to variations in mental health outcomes following heavy metal exposure. Polymorphisms in genes involved in metal metabolism, detoxification pathways, and neurotransmitter regulation can influence an individual's ability to tolerate or excrete toxic metals. For example, genetic variations in metallothioneins, GSH-related enzymes, and DATs have been implicated in differential vulnerability to lead and mercury neurotoxicity. Epigenetic modifications induced by heavy metal exposure further complicate the interplay between genetics and environmental factors, leading to transgenerational effects on mental health. Studies have shown that prenatal exposure to heavy metals can result in heritable changes in DNA methylation patterns, potentially predisposing offspring to psychiatric disorders even in the absence of direct exposure (Ayuso-Álvarez et al., 2019; Channegowda, 2020; Vänskä et al., 2019; Woimant & Trocello, 2014; Yousef et al., 2013). Additionally, age and sex differences contribute to the heterogeneous effects of heavy metal exposure on psychiatric disorders. Children and pregnant women are particularly susceptible to heavy metal–induced neurotoxicity due to ongoing brain development and increased metal absorption rates. In utero exposure to heavy metals can lead to long-lasting alterations in brain structure and function, increasing the risk of autism, ADHD, and intellectual disabilities. Elderly individuals, on the other hand, may experience accelerated cognitive decline and a higher incidence of neurodegenerative diseases when exposed to heavy metals due to age-related reductions in detoxification capacity and cumulative lifetime exposure. Sex differences in heavy metal metabolism, hormonal regulation, and neuroimmune responses further influence psychiatric outcomes, with some studies suggesting that males may be more susceptible to lead-induced aggression and hyperactivity. At the same time, females may exhibit greater vulnerability to cadmium-induced depression and anxiety. Population-based studies have provided robust evidence linking heavy metal exposure to a broad spectrum of psychiatric disorders, with significant variations observed based on regional, demographic, and genetic factors. The complex interplay between environmental pollution, socioeconomic status, genetic predisposition, and individual susceptibility underscores the need for targeted public health interventions to mitigate heavy metal toxicity and its psychiatric consequences. Strategies such as stricter environmental regulations, improved water filtration systems, occupational safety measures, and early screening programs for metal toxicity could help reduce the burden of heavy metal–related mental health disorders. Future research should focus on longitudinal studies to elucidate causal relationships between specific metals and psychiatric conditions, as well as explore potential therapeutic approaches to counteract heavy metal–induced neurotoxicity. Understanding the intricate connections between environmental toxins and mental health is essential for developing effective policies and interventions aimed at safeguarding vulnerable populations from the adverse effects of heavy metal exposure on brain function and psychiatric well-being (Table 27.5, Fig. 27.5).

27.6 Diagnostic and biomarker approaches

The assessment of heavy metal toxicity and its impact on psychiatric health requires robust biomonitoring techniques, neuroimaging modalities, and advanced diagnostic tools for early detection. Biomonitoring provides essential data on metal exposure levels, bioaccumulation, and physiological impacts, while neuroimaging and psychiatric assessments help elucidate the underlying neuropathological mechanisms linking heavy metal exposure to psychiatric disorders. Recent advances in analytical chemistry, imaging technology, and molecular diagnostics have improved the ability to detect and quantify heavy metal–induced neurotoxicity at earlier stages, paving the way for preventive interventions and targeted treatments. Biomonitoring is a critical tool for assessing the levels of heavy metals in biological matrices such as blood, urine, hair, and tissues. These techniques allow researchers and clinicians to determine exposure levels, identify at-risk populations, and establish correlations between heavy metal burden and psychiatric disorders (Ayuso-Álvarez et al., 2019; Channegowda, 2020; Emurotu et al., 2025; Iqbal et al., 2018; Porru et al., 2024).

27.6.1 Blood analysis

Blood testing is one of the most widely used methods for assessing acute and chronic heavy metal exposure. Blood samples provide real-time information about circulating metal concentrations, making them particularly useful for detecting recent exposure. For example, BLLs are a well-established biomarker for lead toxicity, with thresholds set by organizations like the Centers for Disease Control and Prevention to assess neurotoxic risk. Similarly, blood mercury levels can indicate methylmercury exposure from seafood consumption, while blood cadmium levels reflect environmental and occupational exposure. However, since metals such as lead and cadmium accumulate in bone and soft tissues over time, blood measurements may not always reflect long-term exposure.

TABLE 27.5 Detoxification methods, nutritional strategies, and pharmacological interventions for mitigating heavy metal–induced neurotoxicity.

Intervention type	Method/agent	Mechanism of action	Key clinical applications
Detoxification methods	Chelation therapy (e.g., EDTA, DMSA, DMPS)	Binds to heavy metals in blood and tissues, facilitating excretion through urine or bile	Lead poisoning, mercury toxicity, and arsenic detoxification
	Activated charcoal	Adsorbs ingested heavy metals, reducing gastrointestinal absorption	Acute heavy metal poisoning, emergency detoxification
	Hemodialysis and hemoperfusion	Removes metals from circulation in cases of severe toxicity	Severe metal poisoning, renal impairment cases
Nutritional strategies	Antioxidants (vitamin C, E, *N*-acetylcysteine)	Scavenge free radicals, reduce oxidative stress, and prevent neuronal damage	General neuroprotection, heavy metal exposure management
	Essential minerals (zinc, selenium, magnesium, calcium)	Compete with toxic metals for absorption and binding sites, reducing bioavailability and toxicity	Lead and cadmium toxicity prevention, neurodegenerative disease risk reduction
	Phytochemicals (curcumin, quercetin, resveratrol)	Modulate metal chelation, reduce inflammation, enhance antioxidant response	Heavy metal–induced neuroinflammation, cognitive decline prevention
Pharmacological interventions	NMDA receptor antagonists (memantine, ketamine)	Prevent glutamate excitotoxicity induced by heavy metal exposure	Neuroprotection in Alzheimer's, heavy metal–induced cognitive impairment
	Dopamine agonists (levodopa, pramipexole)	Restore dopamine function impaired by metals like manganese and lead	Parkinson's-like symptoms, neurotoxicity-related motor dysfunction
	GABA modulators (benzodiazepines, gabapentin)	Enhance inhibitory neurotransmission to counteract excitotoxicity from metal exposure	Anxiety, seizure prevention in metal toxicity cases
	Cholinesterase inhibitors (donepezil, galantamine)	Improve acetylcholine signaling disrupted by aluminum and other metals	Alzheimer's disease, cognitive decline management

DMPS, Dimercaptopropane sulfonate; *DMSA*, dimercaptosuccinic acid; *EDTA*, ethylenediaminetetraacetic acid; *GABA*, gamma-aminobutyric acid.
Source: Based on Ayuso-Álvarez et al. (2019), Channegowda (2020), Vänskä et al. (2019), Woimant and Trocello (2014), Yousef et al. (2013).

27.6.2 Urine analysis

Urinary excretion of heavy metals provides insights into cumulative exposure and the body's detoxification processes. Urine analysis is commonly used for detecting arsenic, cadmium, and mercury exposure, as these metals are primarily eliminated through the kidneys. For example, increased urinary arsenic levels have been associated with neurodevelopmental disorders and cognitive impairment in children. Chelation challenge tests, in which chelating agents are administered to enhance metal excretion, are often used to assess body burden in cases of chronic exposure. While urine analysis is valuable for monitoring metal clearance, it is influenced by kidney function and hydration status, necessitating standardized protocols for accurate interpretation.

27.6.3 Hair analysis

Hair mineral analysis is a noninvasive biomonitoring technique that provides a long-term record of heavy metal exposure. Unlike blood or urine, which reflect recent exposure, hair analysis can indicate chronic accumulation of metals such as lead, mercury, aluminum, and manganese over weeks or months. Studies have shown that elevated hair mercury levels correlate with neurodevelopmental deficits, including lower IQ scores and increased risk of ASDs. However, external contamination and variability in hair growth rates can influence results, requiring careful sample preparation and standardized testing methodologies.

27.6.4 Tissue analysis

Heavy metal deposition in bone, liver, and brain tissues provides critical insights into long-term toxicity. Bone lead analysis, performed using X-ray fluorescence, is particularly useful for assessing lifetime lead exposure, as lead is known to accumulate in skeletal structures. Postmortem brain tissue analysis has revealed significant accumulations of aluminum and mercury in individuals with neurodegenerative diseases such as Alzheimer's and Parkinson's, further supporting the role of heavy metals in neurotoxicity. While tissue analysis offers valuable retrospective data, its invasive nature limits its application in clinical practice.

Neuroimaging techniques have revolutionized the understanding of heavy metal–induced psychiatric dysfunction by providing detailed visualizations of structural and functional changes in the brain. These imaging modalities help correlate metal exposure with alterations in neural pathways, neurotransmitter systems, and cognitive function, offering valuable insights into the mechanisms underlying psychiatric disorders.

27.6.5 Magnetic resonance imaging

Magnetic resonance imaging (MRI) is a noninvasive imaging tool to examine brain structure, connectivity, and tissue integrity. Studies utilizing MRI have demonstrated that individuals exposed to high levels of lead or mercury exhibit reduced gray matter volume, particularly in the prefrontal cortex and hippocampus—regions implicated in mood regulation, memory, and executive function. Diffusion tensor imaging (DTI), a specialized MRI technique, has revealed white matter abnormalities in children with lead exposure, suggesting disruptions in neuronal connectivity that may contribute to attention deficits and impulsivity.

27.6.6 Functional magnetic resonance imaging

Functional magnetic resonance imaging (fMRI) measures brain activity by detecting changes in blood flow and oxygenation. Research on heavy metal neurotoxicity has shown altered activation patterns in brain regions associated with psychiatric disorders. For instance, fMRI studies of lead-exposed individuals have demonstrated impaired prefrontal cortex function, which is linked to increased impulsivity, aggression, and risk of schizophrenia. Similarly, mercury exposure has been associated with dysregulated activity in the amygdala and limbic system, which may contribute to heightened anxiety and mood instability.

27.6.7 Positron emission tomography

Positron emission tomography (PET) imaging assesses neurotransmitter activity and metabolic changes in the brain. Heavy metals such as lead and mercury are known to disrupt DA, serotonin, and glutamate systems, contributing to psychiatric symptoms. PET studies have shown that lead exposure reduces DAT availability in the striatum, similar to findings observed in patients with schizophrenia. Mercury has been linked to alterations in serotonin receptors, which may underlie its association with depression and anxiety. These findings highlight the utility of PET imaging in identifying neurochemical imbalances associated with metal-induced psychiatric disorders.

27.6.8 Electroencephalography

Electroencephalography (EEG) is a valuable tool for assessing brain wave activity and detecting abnormalities in neural oscillations. Studies have reported that children exposed to lead and manganese exhibit altered EEG patterns, including increased slow-wave activity and decreased coherence between brain regions. These disruptions in neural synchronization have been linked to cognitive deficits, ADHD, and ASDs. EEG-based biomarkers are being explored for early diagnosis and monitoring of heavy metal–induced neurological dysfunction.

Recent technological advancements have improved the ability to detect and predict psychiatric dysfunction resulting from heavy metal exposure at earlier stages. These developments include high-throughput biomonitoring techniques, machine learning-based predictive models, and precision medicine approaches to identify vulnerable individuals.

27.6.9 Omics technologies

Genomic, transcriptomic, proteomic, and metabolomic analyses have provided new insights into the molecular pathways disrupted by heavy metal toxicity. Epigenetic studies have shown that lead and cadmium exposure can alter DNA methylation patterns, influencing gene expression related to neurodevelopment and psychiatric disorders. Metabolomics has identified distinct metabolic signatures associated with heavy metal–induced oxidative stress and neurotransmitter dysregulation, offering potential biomarkers for early detection.

27.6.10 Machine learning and artificial intelligence

Integrating machine learning algorithms with neuroimaging and biomonitoring data has enhanced the accuracy of predicting psychiatric risks associated with heavy metal exposure. AI-driven models can analyze complex datasets to identify patterns of neural dysfunction, enabling early diagnosis and targeted interventions. These approaches are particularly valuable for differentiating heavy metal–induced psychiatric symptoms from other neurological and psychiatric conditions.

27.6.11 Wearable biosensors and point-of-care diagnostics

The development of portable biosensors capable of detecting heavy metal levels in biological fluids offers a promising approach for real-time exposure monitoring. Advances in nanotechnology have enabled the creation of highly sensitive point-of-care devices that can measure trace metal concentrations with minimal sample volumes. These innovations facilitate early screening and personalized risk assessment, particularly in high-risk populations.

27.6.12 Preventive and therapeutic strategies

Early detection of heavy metal–induced psychiatric dysfunction allows for timely interventions aimed at reducing exposure and mitigating neurotoxic effects. Chelation therapy, antioxidant supplementation, and dietary modifications are being explored as potential strategies to counteract heavy metal toxicity. Additionally, neuroprotective compounds such as *N*-acetylcysteine (NAC), curcumin, and melatonin have shown promise in preclinical studies for attenuating heavy metal–induced oxidative stress and neuroinflammation. Advancements in biomonitoring, neuroimaging, and diagnostic technologies have significantly improved the understanding and early detection of heavy metal–induced psychiatric disorders. Integrating these tools into clinical practice and public health initiatives will be essential for identifying at-risk individuals, preventing long-term neurological damage, and developing effective therapeutic strategies to mitigate the impact of heavy metal toxicity on mental health. Future research should focus on refining biomarkers, enhancing imaging modalities, and leveraging AI-driven predictive models to advance early detection and intervention strategies further (Table 27.6).

27.7 Prevention and management strategies

The increasing recognition of heavy metals as neurotoxic agents linked to psychiatric disorders has led to the development of various detoxification, nutritional, pharmacological, and environmental strategies aimed at reducing exposure and mitigating their harmful effects. Heavy metals such as lead, mercury, cadmium, and arsenic disrupt neuronal function through oxidative stress, neuroinflammation, neurotransmitter dysregulation, and mitochondrial impairment. Consequently, therapeutic interventions focus on detoxification, neuroprotection, and policy-driven environmental management to prevent accumulation and associated neuropsychiatric complications. Advancements in chelation therapy, dietary interventions, pharmacological treatments, and regulatory policies provide a multifaceted approach to combat heavy metal toxicity, ensuring individual and public health protection. Detoxification of heavy metals is a critical step in mitigating neurotoxicity and its psychiatric manifestations. Chelation therapy, which involves the administration of chelating agents to bind and facilitate the excretion of heavy metals, is a widely used method for removing toxic metal burdens. Several chelating agents have effectively mobilized specific metals from tissues and reduced systemic toxicity.

27.7.1 Chelation therapy

The choice of chelating agent depends on the metal in question and its bioavailability in tissues. For lead poisoning, **dimercaptosuccinic acid (DMSA)** and **ethylenediaminetetraacetic acid** are used to enhance urinary excretion. DMSA is particularly effective in reducing BLLs and has been shown to improve cognitive and behavioral deficits in children with lead exposure. **Dimercaprol (BAL)** and **dimercaptopropane sulfonate** are chelators used for arsenic and mercury detoxification, acting by forming stable complexes that are excreted through urine. In cases of cadmium toxicity, **D-penicillamine** is sometimes employed, although its use is limited due to potential side effects such as nephrotoxicity. Despite the effectiveness of chelation therapy, prolonged or excessive use may lead to the depletion of essential trace minerals such as zinc and magnesium, necessitating careful monitoring and supplementation.

27.7.2 Dietary interventions

Nutritional strategies play a complementary role in enhancing natural detoxification processes. Specific dietary components can facilitate heavy metal excretion while preventing absorption and accumulation. **Dietary fiber** in whole grains, fruits, and vegetables binds to heavy metals in the gastrointestinal tract, reducing systemic uptake. **Sulfhydryl-containing foods** like garlic and onions have chelating properties that help neutralize metal ions. **Chlorella and spirulina**, both microalgae, have demonstrated metal-binding capacity and antioxidant effects, making them promising dietary detoxifiers. Additionally, **pectin**, a natural polysaccharide found in citrus fruits, has been shown to enhance lead excretion and reduce blood metal levels.

Heavy metal exposure induces oxidative stress and neuroinflammation, necessitating the use of antioxidants and essential minerals to mitigate neuronal damage. Nutritional strategies focus on replenishing antioxidant defenses and ensuring adequate levels of essential trace elements that compete with toxic metals for cellular uptake.

27.7.3 Antioxidant therapy

Oxidative stress is a key mechanism of heavy metal–induced neurotoxicity, leading to lipid peroxidation, DNA

TABLE 27.6 Genetic factors influencing individual susceptibility to heavy metal–induced psychiatric disorders.

Gene	Associated heavy metals	Affected pathways	Psychiatric implications
APOE	Lead (Pb), mercury (Hg), aluminum (Al)	Lipid metabolism, blood–brain barrier integrity	Increased risk of Alzheimer's, cognitive decline, and neuroinflammation
DRD2	Lead (Pb), manganese (Mn)	Dopaminergic neurotransmission	Schizophrenia, psychosis, mood instability
COMT	Lead (Pb), mercury (Hg)	Dopamine metabolism, stress response	Anxiety, bipolar disorder, and schizophrenia susceptibility
SLC6A4 (serotonin transporter, 5-HTT)	Lead (Pb), cadmium (Cd)	Serotonin reuptake and regulation	Depression, anxiety, and emotional dysregulation
MT1A/MT2A (metallothioneins)	Cadmium (Cd), arsenic (As), mercury (Hg)	Heavy metal detoxification, oxidative stress regulation	Increased neurotoxicity, cognitive impairment, and psychiatric vulnerability
NR3C1 (glucocorticoid receptor gene)	Lead (Pb), arsenic (As)	HPA axis regulation, stress response	Depression, PTSD, impaired stress resilience
GRIN2A/GRIN2B (glutamate receptor subunits)	Lead (Pb), mercury (Hg), manganese (Mn)	NMDA receptor function, synaptic plasticity	Schizophrenia, neurodevelopmental disorders, and excitotoxicity
ALAD	Lead (Pb)	Heme biosynthesis, neuroinflammation	Lead poisoning susceptibility, cognitive decline, and mood disorders
BDNF	Mercury (Hg), lead (Pb)	Neuronal growth, synaptic plasticity	Depression, anxiety, and neurodegenerative disease vulnerability
PON1	Arsenic (As), lead (Pb), mercury (Hg)	Detoxification of organophosphates and metals, oxidative stress regulation	Autism spectrum disorders, cognitive deficits, neuroinflammation

ALAD, Delta-aminolevulinic acid dehydratase; *APOE*, apolipoprotein E; *BDNF*, brain-derived neurotrophic factor; *COMT*, catechol-*O*-methyltransferase; *DRD2*, dopamine receptor D2; *HPA*, hypothalamic–pituitary–adrenal; *PON1*, paraoxonase 1.
Source: Based on Channegowda (2020), Jiang and Zhao (2024), Mukherjee et al. (2021), Vänskä et al. (2019), Woimant and Trocello (2014).

damage, and mitochondrial dysfunction. Several antioxidants have been investigated for their protective roles against metal toxicity.

- **Vitamin C (ascorbic acid):** A potent antioxidant that scavenges free radicals and enhances metal excretion. Studies have shown that vitamin C supplementation reduces BLLs and improves cognitive function in lead-exposed individuals.
- **Vitamin E (α-tocopherol):** Protects against lipid peroxidation in neuronal membranes, reducing oxidative damage from metals such as mercury and cadmium.
- **GSH:** A critical intracellular antioxidant involved in detoxification. Heavy metals deplete GSH reserves, leading to increased oxidative stress. **NAC**, a GSH precursor, has been shown to restore antioxidant defenses and reduce neuroinflammation.
- **Alpha-lipoic acid (ALA):** A metal chelator with antioxidant properties that crosses the BBB, offering neuroprotection against mercury-induced cognitive decline.
- **Polyphenols (curcumin, resveratrol):** Found in turmeric and grapes, these compounds have antiinflammatory and neuroprotective effects that counteract metal-induced neuronal damage.

27.7.4 Essential minerals

Certain essential minerals protect against toxic accumulation by competing with heavy metals for absorption.

- **Zinc:** A key competitor of lead and cadmium in cellular transport systems, reducing their bioavailability and toxicity. Zinc supplementation has been shown to protect against lead-induced neurotoxicity.

- **Selenium:** Forms stable complexes with mercury, mitigating its neurotoxic effects and reducing oxidative stress. Selenium also enhances GSH peroxidase activity, improving antioxidant defenses.
- **Magnesium:** Protects against aluminum and lead toxicity by preventing their entry into neuronal cells and stabilizing mitochondrial function.
- **Iron:** Adequate iron levels reduce lead absorption, as both metals compete for the exact transport mechanisms in the gut.

In addition to detoxification and nutritional strategies, pharmacological interventions are being explored to mitigate the neuropsychiatric consequences of heavy metal exposure. Certain drugs and neuroprotective agents target specific pathways involved in metal-induced toxicity.

- **Chelating agents with neuroprotective properties:** While traditional chelators focus on metal excretion, newer formulations are developed with added neuroprotective benefits. **Deferoxamine**, an iron chelator, has effectively reduced aluminum-induced cognitive deficits.
- **Antiinflammatory drugs:** Nonsteroidal antiinflammatory drugs and corticosteroids have been explored for their ability to reduce neuroinflammation caused by metal exposure.
- **NMDA receptor modulators:** Heavy metals disrupt glutamatergic neurotransmission, leading to excitotoxicity. **Memantine**, an NMDA receptor antagonist used in AD, has been investigated for its potential to mitigate lead and mercury-induced neuronal damage.
- **Dopaminergic and serotonergic agents:** Given the role of heavy metals in disrupting DA and serotonin systems, medications such as **L-DOPA** and selective serotonin reuptake inhibitors may be beneficial for treating metal-induced psychiatric disorders.

Reducing heavy metal exposure at a population level requires stringent environmental regulations, public health policies, and community-based interventions. Governments and regulatory bodies have implemented several measures to limit metal contamination in air, water, soil, and consumer products.

- **Regulations on industrial emissions:** Limiting heavy metal discharge from factories, mining operations, and coal-fired power plants is crucial in reducing environmental contamination. Stricter emission standards and improved waste management practices help minimize metal release into ecosystems.
- **Water filtration and remediation:** Lead contamination in drinking water remains a significant concern. Implementing water filtration technologies such as reverse osmosis and activated carbon adsorption can effectively remove heavy metals. Phytoremediation, which uses plants to extract metals from contaminated soil and water, is also being explored as a sustainable solution.
- **Eliminating lead-based products:** Phasing out leaded gasoline, lead-based paints, and specific electronic components has significantly reduced lead exposure.
- **Workplace safety measures:** Occupational exposure to heavy metals in industries such as welding, battery manufacturing, and mining necessitates protective equipment, regular health screenings, and strict safety protocols.
- **Educational and public awareness campaigns:** Informing communities about sources of heavy metal exposure and preventive strategies enhances public health initiatives. Awareness programs on dietary interventions, proper disposal of e-waste, and safer consumer choices contribute to reducing exposure risks.

A comprehensive approach incorporating detoxification strategies, nutritional supplementation, pharmacological interventions, and environmental policies is essential for mitigating heavy metal–induced neurotoxicity. Advances in chelation therapy, antioxidant therapies, and regulatory frameworks offer promising avenues for reducing exposure and protecting mental health. Future research should focus on refining detoxification protocols, developing targeted neuroprotective therapies, and enhancing environmental policies to minimize the long-term impact of heavy metal toxicity on neuropsychiatric health.

27.8 Future directions and research perspectives

Neurotoxicology and psychiatry are increasingly converging as research continues to uncover the profound effects of environmental toxins and hefty metals on mental health. Heavy metals such as lead, mercury, cadmium, and arsenic have been shown to disrupt neurodevelopmental processes, alter neurotransmitter function, and induce oxidative stress, contributing to a spectrum of psychiatric disorders, including depression, anxiety, schizophrenia, and neurodevelopmental disorders such as autism and ADHD. Emerging trends in neurotoxicology emphasize the need for a multidisciplinary approach, integrating genetics, environmental health, and psychiatry to develop targeted therapeutic interventions. Genetic predisposition and individual susceptibility play a critical role in determining the severity of neurotoxic effects, highlighting the importance of personalized medicine in psychiatric care. Integrating environmental health strategies into psychiatric care offers a novel approach to mitigating the burden of toxin-induced mental health disorders. Moreover, identifying potential therapeutic targets, including neuroprotective agents, antioxidant therapies, and epigenetic modulators, could pave the way for effective treatments against heavy metal–induced neuropsychiatric dysfunction. Recent advancements in neurotoxicology and psychiatry have provided more profound insights into the complex mechanisms by which heavy metals influence brain function and contribute to psychiatric disorders. Traditional neurotoxicological studies focused primarily on acute poisoning and severe neurological impairment; however, emerging research is now highlighting the effects of chronic low-level exposure on mental health and cognition. With the advent of high-throughput screening methods, omics technologies (genomics, proteomics, metabolomics), and advanced neuroimaging techniques, scientists can now examine the subclinical and long-term effects of heavy metal exposure on neural circuits, synaptic plasticity, and psychiatric outcomes. One key emerging trend is the recognition of epigenetic modifications as a critical mechanism linking environmental toxins to psychiatric disorders. Heavy metals such as lead and cadmium have been shown to alter DNA methylation patterns, histone modifications, and miRNA expression, leading to dysregulated gene expression in neurotransmitter signaling, neuroinflammation, and neurodevelopment pathways. Epigenetic alterations may underlie the long-lasting effects of prenatal and early-life exposure to heavy metals, increasing the risk of psychiatric conditions later in life. Another significant advancement is using neuroimaging and biomonitoring tools to detect early signs of neurotoxic damage. fMRI, PET, and DTI have been instrumental in identifying structural and functional changes in the brain associated with heavy metal exposure. These technologies help visualize alterations in brain regions implicated in mood regulation, executive function, and emotional processing, such as the prefrontal cortex, hippocampus, and amygdala. Emerging research is also investigating the role of the gut–brain axis in mediating heavy metal toxicity, as alterations in gut microbiota composition have been linked to both neurotoxic effects and psychiatric disorders. Not all individuals exposed to heavy metals develop psychiatric symptoms, suggesting that genetic factors play a crucial role in determining susceptibility to neurotoxic impacts. Variability in genes involved in metal metabolism, oxidative stress response, neurotransmitter function, and neuroinflammation can influence an individual's resilience or vulnerability to heavy metal–induced psychiatric disorders. Polymorphisms in metal transport and detoxification genes are particularly relevant. For instance, genetic variations in SLC39A8, a zinc and manganese transporter, have been associated with altered metal homeostasis and increased susceptibility to schizophrenia. Similarly, polymorphisms in MT1A and MT2A, encoding metallothioneins responsible for heavy metal detoxification, may modulate the extent of metal accumulation in the brain and its neurotoxic effects. Oxidative stress is a primary mechanism through which heavy metals exert neurotoxicity, and genetic variations in antioxidant defense genes, such as SOD2 (SOD 2) and GPX1 (GSH peroxidase 1), may influence an individual's ability to counteract metal-induced oxidative damage. Individuals with genetic polymorphisms that reduce antioxidant enzyme activity may be more prone to neuroinflammation and subsequent psychiatric dysfunction. Genes involved in neurotransmitter metabolism and synaptic plasticity also contribute to differential susceptibility. COMT (catechol-*O*-methyltransferase) polymorphisms, which affect DA metabolism, have been linked to variations in cognitive function and psychiatric risk following lead exposure. Variants in BDNF (brain-derived neurotrophic factor), a critical factor for neuronal survival and synaptic plasticity, may modulate the neurodevelopmental impact of mercury and lead exposure, influencing the risk of mood disorders and cognitive impairment. Understanding genetic susceptibility can aid in developing personalized interventions, allowing for targeted detoxification strategies, dietary modifications, and pharmacological treatments tailored to an individual's genetic profile.

The traditional approach to psychiatric care has primarily focused on genetic and psychological factors, often neglecting environmental influences such as heavy metal exposure. Integrating environmental health into psychiatric care represents a paradigm shift, acknowledging the role of neurotoxic agents in mental health disorders and promoting preventive strategies alongside conventional psychiatric treatments. One crucial aspect of this integration is routine biomonitoring of heavy metals in at-risk populations, particularly in regions with high industrial pollution, contaminated water sources, or occupational exposure risks. Measuring metal levels in blood, urine, hair, and tissue samples can help identify individuals with elevated toxic burdens and facilitate early interventions before significant neuropsychiatric symptoms emerge. Environmental health education should be incorporated into psychiatric practice, empowering clinicians to assess potential sources of exposure in their patients' environments. This may include evaluating occupational history, dietary habits, and residential proximity to industrial sites. Implementing nutritional interventions, such as promoting diets rich in antioxidants and essential minerals, can serve as an adjunct to psychiatric treatment, helping to mitigate metal-induced oxidative stress and neurotransmitter dysregulation. Incorporating environmental health policies into public mental health initiatives is also crucial. This includes advocating for stricter regulations on industrial emissions, improving water filtration systems, reducing the use of toxic metals in consumer products, and implementing safer waste disposal practices to minimize environmental contamination. Given the complex interplay between heavy metals and neuropsychiatric dysfunction, therapeutic strategies must target multiple pathways to mitigate neurotoxic effects effectively. Several promising therapeutic targets are emerging, ranging from traditional detoxification agents to novel pharmacological interventions.

- Neuroprotective agents: Drugs that enhance neuroprotection and synaptic plasticity may counteract metal-induced neuronal damage. Memantine, an NMDA receptor antagonist, has shown potential in reducing excitotoxicity caused by heavy metal exposure. Selegiline, a monoamine oxidase inhibitor, may protect dopaminergic neurons from mercury-induced neurodegeneration.
- Epigenetic modulators: Given the role of epigenetic modifications in metal-induced neurotoxicity, drugs targeting DNA methylation and histone modifications could offer novel therapeutic avenues. Histone deacetylase inhibitors, such as valproic acid, have shown promise in reversing epigenetic alterations linked to psychiatric disorders.
- Antioxidant therapies: Targeting oxidative stress with antioxidants such as NAC, ALA, and polyphenols (curcumin, resveratrol) may reduce neuroinflammation and protect against heavy metal–induced neuronal damage.
- Microbiota-based therapies: Emerging research suggests that gut microbiota play a role in modulating heavy metal toxicity. Probiotic supplementation may help restore gut microbiome balance and reduce systemic inflammation linked to neuropsychiatric symptoms.

Integrating insights from neurotoxicology, genetics, and environmental health into psychiatric care is essential for addressing the growing burden of heavy metal–induced mental health disorders. Advances in biomonitoring, personalized medicine, and targeted therapies offer promising avenues for early detection and intervention. Future research should focus on refining therapeutic strategies, exploring novel neuroprotective compounds, and implementing comprehensive environmental policies to reduce heavy metal exposure and its psychiatric consequences.

27.9 Conclusion

Research on heavy metal toxicity and its impact on mental health has revealed significant associations between chronic exposure to neurotoxic metals such as lead, mercury, arsenic, and cadmium and the development of psychiatric disorders, cognitive impairments, and neurodevelopmental conditions. Population-based studies have demonstrated that regions with high environmental contamination exhibit an increased prevalence of mental health disorders, particularly among vulnerable groups such as children, pregnant women, and occupational workers. Biomonitoring techniques, including blood, urine, hair, and tissue analysis, have provided critical insights into individual toxic burdens. At the same time, neuroimaging and psychiatric assessments have elucidated the structural and functional brain changes associated with metal exposure. Advances in early detection strategies, including epigenetic profiling and neurotoxicological biomarkers, offer promising avenues for identifying at-risk individuals before significant neurological damage occurs. Therapeutic interventions such as chelation therapy, dietary modifications, and pharmacological approaches targeting oxidative stress and neurotransmitter imbalances have shown potential in mitigating heavy metal–induced psychiatric dysfunction. Additionally, emerging research highlights the role of genetics and individual susceptibility in determining the severity of neurotoxic effects, underscoring the need for personalized treatment strategies.

Despite growing awareness of the psychiatric consequences of heavy metal exposure, addressing this issue requires an interdisciplinary approach that integrates environmental health, psychiatry, toxicology, public policy, and precision medicine. There is an urgent need for collaborative research that bridges the gap between neurotoxicology and mental health sciences, enabling a more comprehensive understanding of how chronic exposure to heavy metals influences brain function across different demographic groups. Future studies should focus on longitudinal analyses, examining how metal exposure during critical developmental periods translates into psychiatric outcomes. Additionally, genome-wide association studies and epigenetic research can help identify genetic markers of susceptibility, allowing for targeted preventive interventions. Expanding access to neuroimaging and biomonitoring programs in high-risk populations will also aid in early diagnosis and intervention, reducing the long-term burden of heavy metal–induced psychiatric disorders.

From a policy perspective, governments and regulatory agencies must implement stricter environmental regulations to limit industrial emissions, improve water quality, and reduce heavy metal contamination in food sources. Public health campaigns should emphasize the importance of regular heavy metal screening, especially in areas with known contamination risks. Integrating environmental health assessments into routine psychiatric evaluations can help clinicians identify ecological factors contributing to mental health disorders, enabling more holistic and effective treatment approaches. Furthermore, investment in cleaner technologies, safer waste disposal practices, and alternative materials to replace toxic metals in industrial processes can significantly reduce exposure risks at a population level.

In conclusion, mitigating the impact of heavy metal toxicity on mental health requires a multifaceted strategy combining scientific research, medical innovation, and proactive policy interventions. Addressing this issue is crucial for improving mental health outcomes and promoting overall public health and environmental sustainability. By fostering collaboration between scientists, healthcare providers, policymakers, and environmental advocates, we can develop effective strategies to protect future generations from the neurotoxic effects of heavy metals and reduce the burden of psychiatric disorders linked to ecological contamination.

References

Abdelmonem, B. H., Kamal, L. T., Elbaz, R. M., Khalifa, M. R., & Abdelnaser, A. (2025). From contamination to detection: The growing threat of heavy metals. *Heliyon, 11*(1), e41713. Retrieved from https://doi.org/10.1016/j.heliyon.2025.e41713.

Aljumaili, O. I., El-Waseif, A. A., AbdulJabbar Suleiman, A., & El-Dein A. Ewais, E. (2022). Assessment of hair aluminium, cobalt, and mercury in a specimen of autistic Iraqi patients: Environmental risk factors of heavy metals in autism. *Materials Today: Proceedings, 57*, 350–353. Retrieved from https://doi.org/10.1016/j.matpr.2021.09.235.

Amadi, C. N., Orish, C. N., Frazzoli, C., & Orisakwe, O. E. (2022). Association of autism with toxic metals: A systematic review of case-control studies. *Pharmacology, Biochemistry, and Behavior, 212*, 173313. Retrieved from https://doi.org/10.1016/j.pbb.2021.173313.

Ayuso-Álvarez, A., Simón, L., Nuñez, O., Rodríguez-Blázquez, C., Martín-Méndez, I., Bel-lán, A., ... Galán, I. (2019). Association between heavy metals and metalloids in topsoil and mental health in the adult population of Spain. *Environmental Research, 179*, 108784. Retrieved from https://doi.org/10.1016/j.envres.2019.108784.

Balali-Mood, M., Eizadi-Mood, N., Hassanian-Moghaddam, H., Etemad, L., Moshiri, M., Vahabzadeh, M., & Sadeghi, M. (2025). Recent advances in the clinical management of intoxication by five heavy metals: Mercury, lead, chromium, cadmium and arsenic. *Heliyon, 11*(4), e42696. Retrieved from https://doi.org/10.1016/j.heliyon.2025.e42696.

Bjørklund, G., Skalny, A. V., Rahman, M. M., Dadar, M., Yassa, H. A., Aaseth, J., ... Tinkov, A. A. (2018). Toxic metal (loid)-based pollutants and their possible role in autism spectrum disorder. *Environmental Research, 166*, 234–250. Retrieved from https://doi.org/10.1016/j.envres.2018.05.020.

Castro-González, M. I., & Méndez-Armenta, M. (2008). Heavy metals: Implications associated to fish consumption. *Environmental Toxicology and Pharmacology, 26*(3), 263–271. Retrieved from https://doi.org/10.1016/j.etap.2008.06.001.

Channegowda, M. (2020). Recent advances in environmentally benign hierarchical inorganic nano-adsorbents for the removal of poisonous metal ions in water: A review with mechanistic insight into toxicity and adsorption. *Nanoscale Advances, 2*(12), 5529–5554. Retrieved from https://doi.org/10.1039/d0na00650e.

de Godoi Pereira, M., dos Santos, A. V., Geris, R., & Malta, M. (2023). Chapter 2—Advanced fungal bio-based materials for remediation of toxic metals in aquatic ecosystems. In D. A. Giannakoudakis, L. Meili, & I. Anastopoulos (Eds.). *Novel materials for environmental remediation applications* (pp. 35–62). Elsevier. Retrieved from https://doi.org/10.1016/B978-0-323-91894-7.00007-4.

Ekholm, L., Ljunggren, S., & Helmfrid, I. (2025). Historical exposure to metals in contaminated areas and its impact on cardio- and cerebrovascular health. *International Journal of Hygiene and Environmental Health, 263*, 114461. Retrieved from https://doi.org/10.1016/j.ijheh.2024.114461.

Ellwanger, J. H., & Chies, J. A. B. (2023). Brazil's heavy metal pollution harms humans and ecosystems. *Science in One Health, 2*, 100019. Retrieved from https://doi.org/10.1016/j.soh.2023.100019.

Emurotu, J. E., Abubakar, T. A., Chukwu, L., Umudi, Q. E., Imumorin, V. I., Paul, G. D., & Oboni, G. U. (2025). Concentration and health risk assessment of toxic metals in giblets of free-range chickens in Lokoja, Nigeria. *Journal of Trace Elements and Minerals, 11*, 100209. Retrieved from https://doi.org/10.1016/j.jtemin.2024.100209.

Erkman, F. T., Yıldırım, M. Ş., Sabuncuoğlu, S., Demir, C., Değirmenci, B., Girgin, G., ... Akcan, R. (2025). Heavy metal levels in postmortem blood samples: Unraveling links to suicide and neurological impacts. *Forensic Science International, 367*, 112390. Retrieved from https://doi.org/10.1016/j.forsciint.2025.112390.

Fu, X., Li, H., Song, L., Cen, M., & Wu, J. (2023). Association of urinary heavy metals co-exposure and adult depression: Modification of physical activity. *NeuroToxicology, 95*, 117–126. Retrieved from https://doi.org/10.1016/j.neuro.2023.01.008.

Gade, M., Comfort, N., & Re, D. B. (2021). Sex-specific neurotoxic effects of heavy metal pollutants: Epidemiological, experimental evidence and candidate mechanisms. *Environmental Research, 201*, 111558. Retrieved from https://doi.org/10.1016/j.envres.2021.111558.

Gu, Q., Yu, T., Yang, Z., Ji, J., Hou, Q., Wang, L., ... Zhang, Q. (2019). Prediction and risk assessment of five heavy metals in maize and peanut: A case study of Guangxi, China. *Environmental Toxicology and Pharmacology, 70*, 103199. Retrieved from https://doi.org/10.1016/j.etap.2019.103199.

Habib, M. R., Mohamed, A. H., Osman, G. Y., Mossalem, H. S., Sharaf El-Din, A. T., & Croll, R. P. (2016). *Biomphalaria alexandrina* as a bioindicator of metal toxicity. *Chemosphere, 157*, 97–106. Retrieved from https://doi.org/10.1016/j.chemosphere.2016.05.012.

He, S., Jiang, T., Zhang, D., Li, M., Yu, T., Zhai, M., ... Liang, C. (2024). Association of exposure to multiple heavy metals during pregnancy with the risk of gestational diabetes mellitus and insulin secretion phase after glucose stimulation. *Environmental Research, 248*, 118237. Retrieved from https://doi.org/10.1016/j.envres.2024.118237.

Huang, C., Pan, S., Chin, W., Hsu, J., & Guo, Y. L. (2023). Urinary heavy metals and attention-deficit/hyperactivity symptoms of preschool children: A mixed-exposure analysis. *Ecotoxicology and Environmental Safety, 268*, 115714. Retrieved from https://doi.org/10.1016/j.ecoenv.2023.115714.

Iqbal, G., Zada, W., Mannan, A., & Ahmed, T. (2018). Elevated heavy metals levels in cognitively impaired patients from Pakistan. *Environmental Toxicology and Pharmacology, 60*, 100–109. Retrieved from https://doi.org/10.1016/j.etap.2018.04.011.

Jang, D. H., & Hoffman, R. S. (2011). Heavy metal chelation in neurotoxic exposures. *Neurologic Clinics, 29*(3), 607–622. Retrieved from https://doi.org/10.1016/j.ncl.2011.05.002.

Jiang, M., & Zhao, H. (2024). Joint association of heavy metals and polycyclic aromatic hydrocarbons exposure with depression in adults. *Environmental Research, 242*, 117807. Retrieved from https://doi.org/10.1016/j.envres.2023.117807.

Kim, B., Ha, Y., Hwang, J., & Kim, H.-J. (2025). Association between chronic ambient heavy metal exposure and mental health in Korean adult patients with asthma and the general population. *Chemosphere, 370*, 144002. Retrieved from https://doi.org/10.1016/j.chemosphere.2024.144002.

Krishna Mitra, A., & Chakraborty, N. (2025). Advancing luminescent sensing: Harnessing metal-organic frameworks for enhanced detection and analysis. *Inorganica Chimica Acta, 578*, 122520. Retrieved from https://doi.org/10.1016/j.ica.2024.122520.

Kumkar, P., Verma, C. R., Hýsek, Š., Pise, M., Źółtowska, S., Gosavi, S. M., ... Kalous, L. (2023). Contaminants and their ecological risk assessment in beach sediments and water along the Maharashtra coast of India: A comprehensive approach using microplastics, heavy metal (loid)s, pharmaceuticals, personal care products and plasticisers. *Science of The Total Environment, 892*, 164712. Retrieved from https://doi.org/10.1016/j.scitotenv.2023.164712.

Leite, F. D., Carvalho, P. M. S., Oliveira, R. G., Lopes, M. C., Domingues, I., Correia, P. M. M., ... Silva, A. L. M. (2022). Analysis of Zebrafish contamination with heavy metals using a FF-XRF imaging system based on a MPGD. *Spectrochimica Acta Part B: Atomic Spectroscopy, 198*, 106545. Retrieved from https://doi.org/10.1016/j.sab.2022.106545.

Liu, W., Huo, X., Liu, D., Zeng, X., Zhang, Y., & Xu, X. (2014). S100β in heavy metal-related child attention-deficit hyperactivity disorder in an informal e-waste recycling area. *NeuroToxicology, 45*, 185–191. Retrieved from https://doi.org/10.1016/j.neuro.2014.10.013.

Liu, Z.-H., Ai, S., Xia, Y., & Wang, H.-L. (2024). Intestinal toxicity of Pb: Structural and functional damages, effects on distal organs and preventive strategies. *Science of The Total Environment, 931*, 172781. Retrieved from https://doi.org/10.1016/j.scitotenv.2024.172781.

Lodge, E. K., Guseh, N. S., Martin, C. L., Fry, R. C., White, A. J., Ward-Caviness, C. K., ... Aiello, A. E. (2022). The effect of residential proximity to brownfields, highways, and heavy traffic on serum metal levels in the Detroit Neighborhood Health Study. *Environmental Advances, 9*, 100278. Retrieved from https://doi.org/10.1016/j.envadv.2022.100278.

Maqbool, A., Shahid, A., Jahan, Z., Bilal Khan Niazi, M., Ali Inam, M., Tawfeek, A. M., ... Saeed Akhtar, M. (2023). Development of ZnO-GO-NiO membrane for removal of lead and cadmium heavy metal ions from wastewater. *Chemosphere, 338*, 139622. Retrieved from https://doi.org/10.1016/j.chemosphere.2023.139622.

Menon, A. V., Chang, J., & Kim, J. (2016). Mechanisms of divalent metal toxicity in affective disorders. *Toxicology, 339*, 58–72. Retrieved from https://doi.org/10.1016/j.tox.2015.11.001.

Mukherjee, S., Bhattacharyya, S., Ghosh, K., Pal, S., Halder, A., Naseri, M., ... Bhattacharyya, N. (2021). Sensory development for heavy metal detection: A review on translation from conventional analysis to field-portable sensor. *Trends in Food Science & Technology, 109*, 674–689. Retrieved from https://doi.org/10.1016/j.tifs.2021.01.062.

Muñoz, M. P., Rubilar, P., Valdés, M., Muñoz-Quezada, M. T., Gómez, A., Saavedra, M., & Iglesias, V. (2020). Attention deficit hyperactivity disorder and its association with heavy metals in children from northern Chile. *International Journal of Hygiene and Environmental Health, 226*, 113483. Retrieved from https://doi.org/10.1016/j.ijheh.2020.113483.

Nigam, A., Sharma, N., Tripathy, S., & Kumar, M. (2021). Development of semiconductor based heavy metal ion sensors for water analysis: A review. *Sensors and Actuators A: Physical, 330*, 112879. Retrieved from https://doi.org/10.1016/j.sna.2021.112879.

Nivetha, N., Srivarshine, B., Sowmya, B., Rajendiran, M., Saravanan, P., Rajeshkannan, R., ... Dragoi, E.-N. (2023). A comprehensive review on bio-stimulation and bio-enhancement towards remediation of heavy metals degeneration. *Chemosphere, 312*, 137099. Retrieved from https://doi.org/10.1016/j.chemosphere.2022.137099.

Porru, S., Esplugues, A., Llop, S., & Delgado-Saborit, J. M. (2024). The effects of heavy metal exposure on brain and gut microbiota: A systematic review of animal studies. *Environmental Pollution, 348*, 123732. Retrieved from https://doi.org/10.1016/j.envpol.2024.123732.

Raj, K., Kaur, P., Gupta, G. D., & Singh, S. (2021). Metals associated neurodegeneration in Parkinson's disease: Insight to physiological, pathological mechanisms and management. *Neuroscience Letters, 753*, 135873. Retrieved from https://doi.org/10.1016/j.neulet.2021.135873.

Rashaid, A. H. B., Nusair, S. D., Alqhazo, M. T., Adams, J. B., Abu-Dalo, M. A., & Bashtawi, M. A. (2021). Heavy metals and trace elements in scalp hair samples of children with severe autism spectrum disorder: A case-control study on Jordanian children. *Journal of Trace Elements in Medicine and Biology, 67*, 126790. Retrieved from https://doi.org/10.1016/j.jtemb.2021.126790.

Saghazadeh, A., & Rezaei, N. (2017). Systematic review and meta-analysis links autism and toxic metals and highlights the impact of country development status: Higher blood and erythrocyte levels for mercury and lead, and higher hair antimony, cadmium, lead, and mercury. *Progress in Neuro-Psychopharmacology and Biological Psychiatry, 79*, 340–368. Retrieved from https://doi.org/10.1016/j.pnpbp.2017.07.011.

Saravanan, P., Saravanan, V., Rajeshkannan, R., Arnica, G., Rajasimman, M., Baskar, G., & Pugazhendhi, A. (2024). Comprehensive review on toxic heavy metals in the aquatic system: Sources, identification, treatment strategies, and health risk assessment. *Environmental Research, 258*, 119440. Retrieved from https://doi.org/10.1016/j.envres.2024.119440.

Shen, B., Lu, R., Lv, M., Chen, J., Li, J., Long, J., ... Gong, Z. (2024). Association between the levels of toxic heavy metals and schizophrenia in the population of Guangxi, China: A case-control study. *Environmental Pollution, 363*, 125179. Retrieved from https://doi.org/10.1016/j.envpol.2024.125179.

Singh, A., & Kostova, I. (2024). Health effects of heavy metal contaminants Vis-à-Vis microbial response in their bioremediation. *Inorganica Chimica Acta, 568*, 122068. Retrieved from https://doi.org/10.1016/j.ica.2024.122068.

Sun, R., Gao, S., Zhang, K., Cheng, W.-T., & Hu, G. (2024). Recent advances in alginate-based composite gel spheres for removal of heavy metals. *International Journal of Biological Macromolecules, 268*, 131853. Retrieved from https://doi.org/10.1016/j.ijbiomac.2024.131853.

Vänskä, M., Diab, S. Y., Perko, K., Quota, S. R., Albarqouni, N. M. A., Myöhänen, A., ... Manduca, P. (2019). Toxic environment of war: Maternal prenatal heavy metal load predicts infant emotional development. *Infant Behavior and Development, 55*, 1–9. Retrieved from https://doi.org/10.1016/j.infbeh.2019.01.002.

Vunain, E., Mishra, A. K., & Mamba, B. B. (2016). Dendrimers, mesoporous silicas and chitosan-based nanosorbents for the removal of heavy-metal ions: A review. *International Journal of Biological Macromolecules, 86*, 570–586. Retrieved from https://doi.org/10.1016/j.ijbiomac.2016.02.005.

Woimant, F., & Trocello, J.-M. (2014). Chapter 57—Disorders of heavy metals. In J. Biller, & J. M. Ferro (Eds.). *Handbook of clinical neurology. Neurologic aspects of systemic disease part II* (pp. 851–864). Elsevier. Retrieved from https://doi.org/10.1016/B978-0-7020-4087-0.00057-7 Vol. 120.

Xie, J., Zhou, F., Ouyang, L., Li, Q., Rao, S., Su, R., ... Fan, G. (2023). Insight into the effect of a heavy metal mixture on neurological damage in rats through combined serum metabolomic and brain proteomic analyses. *Science of The Total Environment, 895*, 165009. Retrieved from https://doi.org/10.1016/j.scitotenv.2023.165009.

Yan, W., & Saxena, D. (2022). Chapter 8—Interaction of toxic metals with the gut microbiome. In G. F. Nordberg, & M. Costa (Eds.). *Handbook on the toxicology of metals* (pp. 209–216). Academic Press. Retrieved from https://doi.org/10.1016/B978-0-12-823292-7.00006-1 5th ed.

Yousef, S., Eapen, V., Zoubeidi, T., Kosanovic, M., Mabrouk, A. A., & Adem, A. (2013). Learning disorder and blood concentration of heavy metals in the United Arab Emirates. *Asian Journal of Psychiatry, 6*(5), 394–400. Retrieved from https://doi.org/10.1016/j.ajp.2013.04.005.

Zebbiche, Y., Stambouli, B., Kaddour-Benkada, I., Amziane, A., Chebli, A. I., Achouri, M. Y., ... Alamir, B. (2024). Trace element levels and autism spectrum disorder in a sample of Algerian children: A case-control study investigation. *Research in Autism Spectrum Disorders, 110*, 102287. Retrieved from https://doi.org/10.1016/j.rasd.2023.102287.

Chapter 28

Epidemiological studies on heavy metal exposure and neurodegeneration

Debjani Choudhury[1], and Prasann Kumar[2]

[1]Department of Plant Pathology, School of Agriculture, Lovely Professional University, Phagwara, Punjab, India, [2]Department of Agronomy, School of Agriculture, Lovely Professional University, Phagwara, Punjab, India

28.1 Introduction

Globally, neurodegenerative disease (ND) is a prevalent and expanding cause of death and morbidity, especially in the elderly. Despite frequently shared characteristics, the clinical manifestations and underlying physiology of the various NDs vary widely. Accurate diagnosis is essential because it permits more precise prognostication and often directs particular management and treatment. Parkinson's disease (PD), Alzheimer's disease (AD), amyotrophic lateral sclerosis (ALS), multiple sclerosis (MS), and Huntington's disease (HD) are just a few of the many diseases that fall under the umbrella of neurodegenerative processes showing various symptoms (Table 28.1) even causing the death of patients. Because essential metals are involved in many physiological processes, they are vital to preserving human health. Both an abundance and a shortage of these metals can cause neurotoxicity through neuroinflammation and astrocyte and microglia activation. Astrocytes and microglia are essential modulators of neuroinflammation and prolong their overactivation in the central nervous system (CNS). The CNS is a unique system with a controlled immune response compared to the peripheral immune system. According to research, immune cells from the peripheral system may infiltrate the CNS as a result of a strong inflammatory response in the peripheral system brought on by things like systemic exposure to lipopolysaccharides or viral infections (Sharma et al., 2021).

Inflammation brought on by metals has lately been recognized as a key pathophysiological mechanism underlying neurological disorders. Nevertheless, the precise pathophysiological mechanism of metals remains unclear. This chapter gives a synopsis of a few of the most prevalent neurodegenerative illnesses. Because essential metals are involved in many physiological processes, they are vital to preserving human health. However, both too much and too little of these metals can cause neurotoxicity through neuroinflammation and astrocyte and microglia activation. When these cells are activated, proinflammatory cytokines may be released (Table 28.1).

28.2 Neurodegenerative diseases

28.2.1 Epidemiology of Alzheimer's disease

One of the most NDs, AD is typified by visuospatial deficits, language deterioration, amnesic-type memory, dementia, and cognitive and behavioral impairment. An estimated 24 million people worldwide are thought to be affected by AD, the most prevalent type of dementia, which accounts for 60%–80% of all dementia cases (Sosa-Ortiz, Acosta-Castillo, & Prince, 2012). It mainly affects older people, though it can also strike younger people. As people age, the prevalence of AD significantly rises by 15 times between the ages of 65 and 85. There seems to be worldwide variation in the disease burden despite the possibility that these reported differences result from methodological differences (Sosa-Ortiz et al., 2012). AD rarely manifests before age 65; those cases are referred to as "early-onset" AD. The incidence rate also rises with age, and the annual risk varies from 0.5% in people aged 65–69 to 6% over 85 (Mayeux & Stern, 2012). According to another report, more than 50 million people worldwide had dementia in 2018, according to Alzheimer's Disease International estimates, 2019.

According to the World Alzheimer's Report, this number is anticipated to rise to 82 million by 2030

Heavy Metal Toxicity and Neurodegeneration. DOI: https://doi.org/10.1016/B978-0-443-36575-1.00001-7

TABLE 28.1 Symptoms of neurodegenerative disease.

Neurodegenerative disease	Symptoms
Alzheimer's disease	Memory loss affects daily life, personality, and mood swings. Withdrawal from work and social activities, difficulty with work like speaking or writing, confusion, poor judgment
Parkinson's disease	Motor symptoms like shaking, bradykinesia, stiffness, balance issues, sudden inability to move, nonmotor symptoms like anosmia, sleep problems, facial masking, mood swings, dizziness, micrographia, soft voice
Huntington's disease	Subtle mood, difficulty in remembering things, chorea, clumsiness, difficulty in speaking or swallowing, bradykinesia, trouble walking, Cognitive symptoms like memory problems, lack of awareness, lack of understanding, Emotional symptoms like depression and anxiety, social withdrawal, irritability or aggression
Amyotrophic lateral sclerosis	Muscle weakness, cramping, clumsiness, dysarthria, difficulty in walking, dysphagia
Multiple sclerosis	Vision problems, Numbness, fatigue, balance and coordination issues, muscle weakness, muscle stiffness, bowel problems, cognitive issues, pain and sensory changes, tremors, heat sensitivity, dysarthria, tremors, hearing loss, seizures, muscle atrophy, breathing difficulty, uncontrolled laughing or crying

progressively and 152 million by 2050, with AD accounting for 60%–80% of all dementia cases (Lane, Hardy, & Schott, 2018). There are more than 20 risk factors for AD, including ageing, genetic risk factors, acquired factors like type 2 diabetes, obesity, hypertension, and cerebrovascular disease, and unhealthy lifestyle choices like smoking, body clock disturbance, depression, and environmental pollution (Crous-Bou, Minguillón, Gramunt, & Molinuevo, 2017; Ferreira Silva et al., 2019). Two primary neuropathological features of AD are the accumulation of amyloid beta (Aβ) and neurofibrillary tangles (NFTs) of hyperphosphorylated tau (Laurent, Buée, & Blum, 2018). The idea that neuroinflammation plays a role in the pathophysiology of AD is strongly supported by mounting evidence (Barron, Gartlon, Dawson, Atkinson, & Pardon, 2017). Likewise, earlier research has demonstrated that a significant cause of mild cognitive impairment is AD in chronic inflammation (Qiu et al., 2021). Studies conducted in vivo showed that prolonged inflammation increased amyloid deposits (Guo, Yu, Grass, de Beer, & Kindy, 2002). Additionally, the pathophysiology of AD involves C-reactive protein, a marker of chronic inflammation, according to the results of a Mendelian randomization analysis (Zhang, Wang, & Liu, 2022).

According to the Mayo Clinic Study on Aging (MCAS), the only sizable epidemiological study that tackles all of these intricate issues (Jack et al., 2019). About 1500–5000 participants in the MACS, between the ages of 60 and 89, had their biomarkers for tau and amyloid evaluated using PET imaging. At the age of 70, the prevalence of AD and clinically defined AD was very low, but it steadily increased with age; by the time people were 80, the prevalence of AD, as determined by biomarkers, was roughly three times that of clinically defined probable AD. The study's findings are highly significant. The intriguing finding of the MCAS study is that when taking into account the whole clinical spectrum (including dementia and MCI), the vast majority of participants with biologically defined AD do not exhibit clinical symptoms until they are 85 years old. In a study in Goteborg, Sweden (Kern et al., 2018), the prevalence of amyloid pathology was 22.8%, t-tau pathology was 33.2%, and p-tau pathology was 6.9% among a small sample of roughly 300 subjects who had a CDR report of 0 (normal individuals). Thus nearly half of the 70 years old with CDR score of 0 had AD pathology. Thus the findings of two population-based studies are comparable: in healthy older adults, tau and amyloid pathologies are relatively common (Keshavan et al., 2021; Vos et al., 2013). As a result, many individuals over 65 exhibit signs of the distinctive neuropathological accumulation of AD even though they do not yet exhibit symptoms.

28.3 Heavy metals and Alzheimer's disease

28.3.1 Alzheimer's disease and heavy metals

Impairment in at least one cognitive domain, such as memory, language, perception, attention, social cognition, and executive function, is a hallmark of dementia (Bassil and Mollaei, 2012; Sachdev et al., 2014). A diverse group of illnesses collectively referred to as dementia includes Lewy body dementia, vascular dementia, and frontotemporal dementia. About 70% of dementia is caused by AD, a common type of dementia (Brookmeyer et al., 2011). Pharmacological trials for AD have been unsuccessful despite the pressing need and enormous research efforts (Lane et al., 2018). Neuropathologic alterations seen years before symptoms appear make treatment more difficult (Swerdlow, 2007). Characterizing risk factors is crucial, especially when treatment is difficult (Egan et al., 2019). Both environmental and genetic factors increase the risk of AD and related dementias (Gatz, Chandra, Fratiglioni, & Nancy, 2006).

The identification of modifiable environmental risk factors can significantly impact the prevention and treatment of AD and related dementias. Numerous environmental chemicals have long been known to be neurotoxic (Miodovnik, 2011), especially in human neurodevelopment and laboratory models. Studies on environmental AD and related dementia frequently examine exposures at the onset of clinical symptoms. Still, pertinent exposures may have happened years, decades earlier, or even early in life.

With their extensive exposure in the population, the role of heavy metals like lead, cadmium, and manganese stands out among environmental factors. Even with low exposure levels in the general population, lead and cadmium are noteworthy metals for their neurotoxic effects. Although manganese is a necessary metal for many bodily processes, including the health of neurons, excessive or low concentrations can be harmful, it is essential to comprehend how these heavy metals contribute to the development of AD and associated dementias.

Lead causes about 1% of the world's disease burden, which includes long-term impacts on children's IQ and behavioral issues (Grandjean & Bellanger, 2017). However, there is evidence that this is understated (Shaffer et al., 2019). Exposure to lead is a historical and contemporary issue. Lead toxicity was noted as early as 370 BCE (Tong, von Schirnding, & Prapamontol, 2000) and more recently in the Flint, Michigan community (Hanna-Attisha, LaChance, Sadler, & Champney Schnepp, 2016). Although lead's persistence in soil, dust, and built environments makes its removal from our lives and environments challenging (Needleman, Schell, Bellinger, Leviton, & Allred, 1990), removing lead from paint and gasoline is a significant public health success (Dissanayake & Erickson, 2012). Lead is still utilized in many industrial applications, such as lead–acid storage batteries for automobiles, despite US legislative efforts to reduce lead exposure (Ericson et al., 2016). Lead mining, smelting, and electronic waste recycling are all linked to elevated lead levels worldwide (Papanikolaou, Eleftheria, Belivanis, & Tsatsakis, 2005). Endogenous exposure to lead may be the main cause of exposure in older adults. Lead accumulates frequently and is expected rather slowly and is a known neurotoxicant, causing nonspecific brain disruption (Fig. 28.1). First, lead is a redox-inactive metal that causes oxidative stress by depleting thiols and damaging the antioxidant defense system (Sanders, Liu, Buchner, & Tchounwou, 2009). Excessive oxidative stress causes mitochondrial damage, stress in the endoplasmic reticulum, and eventually neuronal apoptosis (Ercal, Gurer-Orhan, & Aykin-Burns, 2001). Lead exposure causes excitotoxic damage to neurons due to calcium overactivation. Lead modifies normal metal signaling and disturbs the homeostatic levels of essential metals.

According to the International Agency for Research on Cancer, Cd is a group I carcinogen and does not have any vital physiological role in humans (Straif et al., 2009). Chronic exposure to low levels of Cd raises the risk of diabetes, osteoporosis, kidney damage, hypertension, and reduced lung infection (Satarug, Scott, Mary Ann Sens, & Donald, 2011). Despite the paucity of human evidence, cadmium has recently been identified as a neurotoxicant. Because low iron increases cadmium absorption, women typically have higher cadmium levels than men, and smokers have higher cadmium levels than nonsmokers (Vacchi-Suzzi, Kruse, Harrington, Levine, & Meliker, 2016; Wong, Roberts, & Saab, 2022).

Exposure to cadmium through ingestion and inhalation interacts with the lung and gastrointestinal tract (Fig. 28.2). These tissues absorb cadmium, which enters the bloodstream. Only trace amounts of cadmium can pass through the bloodstream in healthy adults (Takeda, 2000). The choroid plexus in the brain prevents blood toxicants from entering the cerebrospinal fluid and preserves the homeostatic environment inside the CNS (Takeda, Takefuta, Ijiro, Okada, & Oku, 1999). By preventing adult neurogenesis in the hippocampus and olfactory bulb, cadmium enters the CNS directly through the olfactory system and causes long-lasting irreversible harm. The biological processes by which cadmium causes neurotoxicity are supported by toxicological research. By altering the blood–brain barrier (BBB)'s permeability and interacting with other neurotoxicants, cadmium may also cause neurotoxicity by causing tau NFT formation and Aβ aggregation. After exposure to Cd, pathogenic processes lead to AD pathology and cognitive impairment

Manganese is the 12th most abundant element on the Earth and the fifth most abundant metal in the environment (Cai et al., 2021). It is a vital trace metal needed for healthy development and physiological functions, including blood coagulation, bone formation, immunological response, metabolism of carbohydrates, and brain activity (O'Neal & Zheng, 2015). Manganese is essential for human health, but too much of it can be neurotoxic and build up in the brain, leading to manganism, an irreversible Parkinsonian syndrome (Roth, Ponzoni, & Aschner, 2013). High levels of manganese exposure impair cognitive function and play a role in the etiology of AD (Li et al., 2017). Manganese levels in the whole body are strictly regulated by

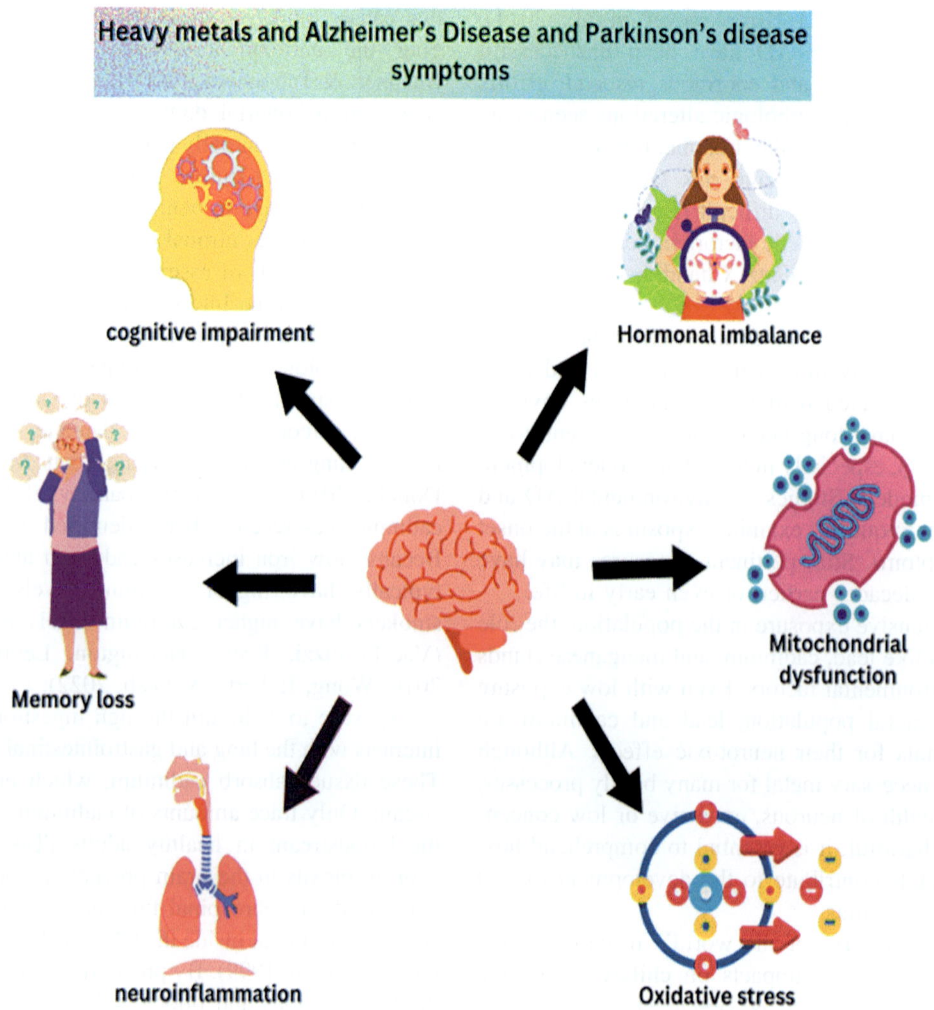

FIGURE 28.1 Effect of heavy metals on neurodegenerative disease (Alzheimer's and Parkinson's disease).

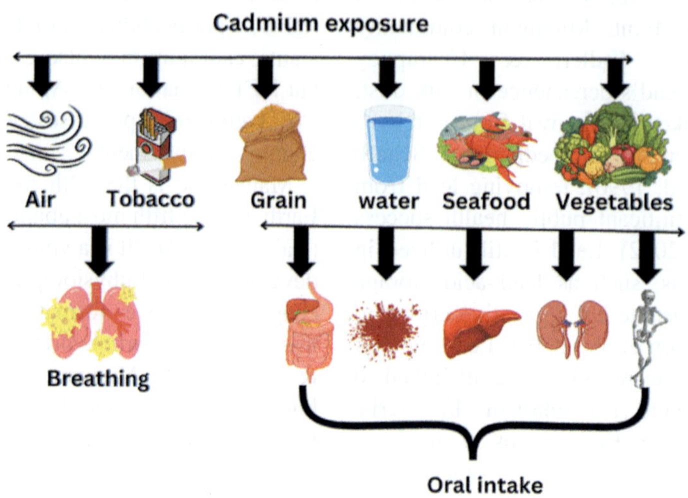

FIGURE 28.2 Exposure to cadmium through oral intake and breathing.

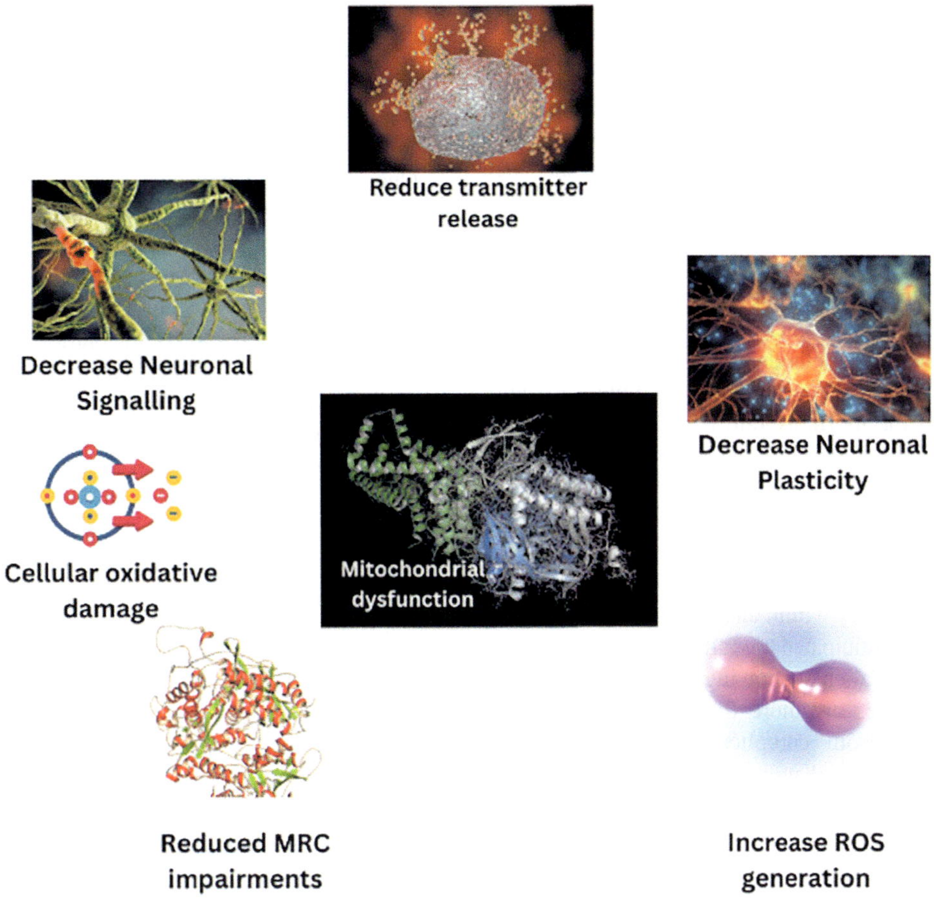

FIGURE 28.3 Effects of mitochondrial dysfunction.

controlling intestinal absorption and excretion of the metal through homeostatic mechanisms because manganese serves as both a necessary nutrient and a potent toxicant (Aschner and Aschner, 2005; Myers et al., 2003).

Traditionally, high exposure levels in drinking water or air cause manganese toxicity. Manganese finds extensive application in both commercial and industrial processes. The most common industries with excessive occupational manganese exposure are mining, welding, dry battery manufacturing, and the use of organochemical fungicides. In patients undergoing total parenteral nutrition therapy (Aschner et al., 2015; Nagatomo et al., 1999), patients with hepatic encephalopathy (Janocha-Litwin, Marianska, Serafinska, & Simon, 2015), and ephedra abusers (Guilarte, 2013), manganese toxicity can also result from a compromised or underdeveloped excretion system.

Neurotoxicity caused by manganese has been thoroughly investigated. Oxidative stress, mitochondrial dysfunction (Fig. 28.3), autophagy dysregulation, intracellular toxic metabolite accumulation, and apoptosis are some of the underlying mechanisms (Butterworth, Spahr, Fontaine, & Layrargues, 1995; Peres et al., 2016). Through the generation of mitochondrial reactive oxygen species, mitochondrial dysfunction contributes to the pathophysiology of AD (Selfridge, Lezi, Lu, & Swerdlow, 2013; Zhao and Zhao, 2013). Despite the extremely slow efflux, manganese builds up in the brain's mitochondria (Gavin, Gunter, & Gunter, 1990; Liccione, Mahin, 1988). The mitochondria contain the powerful antioxidant enzyme MnSOD. Overdosing on manganese can reduce MnSOD activity, which raises the production of reactive oxygen species and ultimately causes mitochondrial dysfunction (Wei and Lee, 2002). Both rodent brains and cultured neurons experience Aβ-related neurotoxicity when exposed to high manganese levels (Tong et al., 2014). Moreover, increased manganese levels resulted in Aβ-related cognitive impairments in the mouse model of AD (Tong et al., 2014).

Furthermore, tau hyperphosphorylation (Cai et al., 2011) brought on by manganese exposure may result in the development of NFTs, one of the leading clinical structural alterations in AD. Because manganese has an affinity for Aβ, high manganese exposure may hasten the buildup of Aβ in the brain, increasing Aβ neurotoxicity and hastening the course of the disease.

28.4 Epidemiology of Parkinson's disease

An increasing number of people have PD, which some have dubbed a pandemic, and this number will rise as people live longer (Dorsey, Sherer, Okun, & Bloem, 2018). According to a systematic analysis of data from the 2016 Global Burden of Disease study, the age-standardized prevalence increase, considering demographics, was less pronounced at 22% between 1990 and 2016. In contrast, the crude prevalence increased by about 74% (Dorsey, Elbaz, et al., 2018). The lack of high-quality, country-specific data makes the methods used to estimate mortality and prevalence globally and nationally complex. Because of worries about the mortality rate from PD being under-estimated, the highest mortality-to-prevalence ratios from three wealthy nations were extrapolated to all other nations, which may not have been broadly applicable.

While most regions demonstrated an increase, there is a notable variation in the observed temporal changes in prevalence and mortality within and between regions. Most of the increase is anticipated to fall in low and middle-income countries. Patients showing PD symptoms and other chronic diseases are more likely to live with their condition and die from complications as their illness worsens (Tunstall-Pedoe, Kuulasmaa, Mähönen, Tolonen, & Ruokokoski, 1999).

Given the lengthy latency period between exposure and disease diagnosis, it is significant that many of the toxicant chemicals linked to PD are persistent in the environment. For instance, trichloroethylene is found in almost one-third of drinking water supplies in the United States and in soil, air, food, and human breastmilk. Air pollution and occupational exposure to welding or specific metals are additional toxicant exposures linked to PD (Murata, Barnhill, & Bronstein, 2022). The possibility that these associations have causal effects is strengthened by the biochemical effects linked to PD, such as paraquat, rotenone, 2,4-D, and several organochlorines. These effects include inflammation, mitochondrial dysfunction, epigenetic methylation, and changes in the microbiome, all of which are believed to be significant to PD (Goldman, 2014; Tanner, Goldman, Ross, & Grate, 2014).

28.5 Heavy metals in Parkinson's disease

The role of metals in the cause and development of Parkinson's disease (Fig. 28.1). The etiology and pathological process of PD are greatly influenced by exposure to toxic metals and the body's depletion of essential metals. Fe deficiency causes PD with restless legs syndrome by impairing the peripheral or CNS and lowering dopamine and serotonin levels in the brain (Piao et al., 2017). In addition, the accumulation of Fe in the gray matter nuclei of the basal ganglia and midbrain has been suggested to play a role in the etiology of PD (Xuan et al., 2017). Bradykinesia, rigidity, postural instability, and slowly developing asymmetric tremors are the hallmarks of PD, the second most common neurodegenerative illness (Schneider, Iourinets, & Richard, 2017).

One of the most prevalent metals, iron, is required as a cofactor for many biological processes, such as the synthesis of neurotransmitters, DNA, mitochondrial respiration, and oxygen transport. The SN and globus pallidus of the healthy brain have greater Fe concentrations (Götz, Double, Gerlach, Youdim, & Riederere, 2004). In contrast, the concentration of Fe is lower in the white matter, cortical gray matter, cerebellum, and midbrain (Ward, Zucca, Duyn, Crichton, & Zecca, 2014); Fe concentrations seem to be higher in the parts of the brain related to motor functions than in the parts unrelated to motor functions, this clarifies the connection between motor-associated disease and Fe loading (Cass et al., 2007). In the substantia nigra of the brain, Fe accumulation is exceptionally high (Dexter et al., 1989; Griffiths, Dobson, Jones, & Clarke, 1999). Although the high concentration of iron in the SN is neurotoxic (Mezzaroba, Alfieri, Colado Simão, & Vissoci Reiche, 2019), iron is also essential for synthesizing dopamine because it is a cofactor for the enzyme tyrosine hydroxylase, which restricts the synthesis of chemical transmitters (Garza-Lombó, Yanahi, Liliana, María, & Franco, 2018). Therefore too much or too little iron can harm dopaminergic neurons. Additionally, glial cells and dopamine neurons play a role in iron metabolism as PD progresses. Neuroprotection in the early stages of PD may be provided by myelin synthesis and neurotransmitter regulation. However, in the late stages of PD, elevated Fe levels may result in neurodegeneration by raising ROS levels (Shi et al., 2019). Thus an imbalance in metal levels, particularly those of iron, may pose a risk to dopaminergic neurons.

The pathophysiology and development of PD are influenced by the activation of microglia and astrocytes, the two main glial cell types (Moreno, Streifel, Sullivan, Hanneman, & Tjalkens, 2011). Additionally, astrocytes also play a significant role in the development and etiology of PD (Lee et al., 2010). Massive α-synuclein immunoreactive astrocytes were concentrated in prosencephalic sites in the middle and advanced stages of PD (Braak, Sastre, & Tredici, 2007). With functions in immune defense, bone formation, cellular energy balance maintenance, and the inhibition of free radical formation, Mn is a crucial micronutrient for normal growth and development (Erikson, Syversen, Aschner, & Aschner, 2005). Mn is used as a cofactor by several enzymes, including glutamate synthetase, Mn superoxide dismutase (SOD), arginase, serine/threonine protein phosphatase I, and pyruvate decarboxylase, to control a variety of biochemical and cellular processes essential for the synthesis of neurotransmitters and the functions of neurons and glial cells (Fitsanakis, Au, Erikson, & Aschner, 2006).

The Mn causes synuclein aggregation, which increases the production of free radicals, and impairs mitochondrial electron transport (Martinez-Finley, Gavin, Aschner, & Gunter, 2013). Progressive neuronal degeneration results from Mn accumulation in neurons, primarily in the globus pallidus of the basal ganglia, SN, and subthalamic nucleus, followed by the striatum (Bouabid, Delaville, De Deurwaerdère, Lakhdar-Ghazal, & Benazzouz, 2014; Klos, Chandler, Kumar, Ahlskog, & Josephs, 2006). In PD, the function of the basal ganglia is disrupted. Motor impairment manifests as widespread rigidity, cock-walk gait, tremor, and generalized bradykinesia, resembling PD (Calne, Chu, Huang, Lu, & Olanow, 1994). Mn toxicity is typified by sensory disturbance. Numerous have looked into the potential link between chronic Mn exposure and PD, and they have found a direct correlation between occupational Mn exposure and PD, especially for workers who have been exposed to Mn for more than 30 years (Lai, Marion, Teschke, & Tsui, 2002). Several studies have shown a connection between elevated Mn levels in the brain and compromised dopaminergic neurotransmission. However, a precise mechanism for how Mn causes dopaminergic neurodegeneration has not yet been established. Furthermore, there is proof that Mn exposure causes dopamine levels to rise, fall, or both.

Mercury is a transition metal produced by artificial and natural processes, including pollution from incinerators and coal-fired power plants, volcanic eruptions, and evaporation. Elemental mercury (Hg0), organic mercury (MeHg), and inorganic mercury (Hg^{2+} and Hg^{+}) are among the different chemical forms of mercury. From different research, it has been observed that mercury exposure in any form is harmful to living beings. The most hazardous form of mercury, inorganic mercury, is present in high concentrations in aquatic environments (Rice et al., 2014) and has a high rate of entry into the CNS (Aschner, Erikson, Hernández, & Tjalkens, 2009), making it a serious neurotoxicant. The CNS is a primary target organ for mercury poisoning because of its effective delivery into the brain. Affecting ribosomal activity and endoplasmic reticulum function interferes with protein synthesis and biological components in the developing brain (Jan et al., 2015). High Hg levels in the brain of fetal mice exposed directly to MeHg during pregnancy were found, indicating both high transport into the developing brain and effective transplacental transfer of MeHg (Watanabe, Yoshida, Kasanuma, Kun, & Satoh, 1999).

Numerous in vitro and in vivo investigations have documented that exposure to MeHg causes GSH to be depleted, which is crucial for maintaining redox homeostasis (Ni et al., 2011). Numerous facets of neurotoxicity brought on by MeHg exposure have been linked to decreased GSH (Farina, Aschner, & Rocha, 2011). Reduced GSH levels have also been observed in the cerebellum and cerebral region of animals exposed to low doses of MeHg in the micromolar range (Franco et al., 2009), suggesting that both MeHg–GSH interactions and the production of ROS through GSH-dependent mechanisms cause MeHg-induced GSH oxidation (Franco et al., 2007).

PCu is the third most common essential metal in the human body, with the liver and brain having the highest concentration. The environment contains a lot of Cu, an essential cofactor for several enzymes in the CNS (Scheiber and Dringen, 2013). Thus balanced Cu metabolism is crucial for preserving the equilibrium of cell survival by controlling biochemical pathways such as oxidase activity, electron transfer, iron metabolism, and mitochondrial ROS production. Based on a default intake of 2 L/day and a body weight of 70 kg, the Environmental Protection Agency recommended that the Cu content in drinking water not exceed 1.3 mg/L (Taylor, Tsuji, McArdle, Adams, & Goodfellow, 2023). Significant lipid peroxidation reactions in the brain may result from excess Cu. However, neurological disorders can result from both an excess of Cu and a deficiency (Kaler, 2011).

28.6 Parkinson's disease and heavy metals (paper)

The Cu is essential for various biological processes because it is a cofactor or structural element of several cuproproteins. By interacting with ceruloplasmin, which exhibits Cu-based oxidase activity, Cu also contributes to Fe homeostasis. Furthermore, Cu interacts with synaptic proteins and neurotransmitter receptors to play an important role in synapses. By affecting synaptic activity, signaling cascades, and excitotoxic cell death brought on by neurotrophic substances, Cu is involved in a variety of neuronal cell activities (D'Ambrosi & Rossi, 2015). As a catalyst in Fenton chemistry, Cu is a transition metal that contributed to the creation of free radicals (Wang et al., 2012) and hydrogen peroxide. The metabolism of monoamine oxidase produces hydrogen peroxide, which is particularly important in parts of the brain that process biological amines like dopamine. The millimolar concentration of various metal ions, including Cu, has been shown in vitro to stimulate the formation of the partially folded amyloidogenic form of amyloid-protein, which is more prone to aggregate (Binolfi et al., 2006). Compared to healthy individuals of the same age, PD patients have lower blood ceruloplasmin levels, oxidase activities, and the number of Cu atoms per ceruloplasmin molecule (Kim et al., 2018). By binding to ceruloplasmin, Cu increases ferroxidase activity and promotes Fe homeostasis, as a result, it may be involved in indirect toxicity caused by variations in Fe concentration (Patel et al., 2002).

One of the most dangerous heavy metals for human health is lead (Pb), considered unnecessary (Csavina et al., 2012). Because lead has so many applications in both industry and daily life, exposure to lead has been rising. Pb is poisonous to many bodily systems and organs, such as the heart, intestines, kidneys, reproductive organs, bones, and nervous system. Epidemiological research has shown a positive correlation between Pb exposure. The Pb is poisonous to many bodily systems and organs, such as the heart, intestines, kidneys, reproductive organs, bones, and nervous system. Epidemiological research has shown a positive correlation between Pb exposure and PD, with a 2–3-fold increased risk for PD linked to Pb exposure (Weisskopf et al., 2010).

Numerous studies have demonstrated that neuroinflammation plays a significant role in the etiology of PD (230). According to epidemiological research, the white blood cell count rises by 1.7 and for every 1% increase in blood lead level among occupationally exposed Pb workers (He et al., 2021). According to a case study conducted in the United States, people with lifetime Pb exposures in the highest quartile were twice as likely to develop PD as people in the lowest quartile (Coon et al., 2006). The immature BBB of children makes it easier for Pb to pass through, making their brains more susceptible to exposure than those of adults (Weisskopf et al., 2010). According to several experimental studies, Pb exposure at an early age may have long-term neurological, behavioral, and epigenetic effects. This suggests that children exposed to Pb are significantly more likely to develop NDs as they get older.

Pb is a hazardous metal that can replace the function of Ca in the body and enter neurons and glial cells after passing through calcium channels in the BBB. Many neurotransmission processes in the brain that rely on Ca ions are impacted by Pb, which alters the release of and uptake of neurotransmitters like dopamine and acetylcholine (Loikkanen, Naarala, & Savolainen, 1998). It has been proposed that Pb influences how gene regulation is altered in populations exposed to Pb (M. Bakulski, Rozek, Dolinoy, Paulson, & Hu, 2012). Pb exposure during early life alters gene expression and regulation, which are linked to the pathophysiology of tau-related NDs (Rahman, Brew, & Guillemin,2011).

28.7 Epidemiology of Huntington's disease

HD is characterized by chorea and uncontrolled movements, but cognitive and psychological abnormalities are also prevalent and can negatively impact the quality of life. Overall, patient populations exhibit remarkably diverse HD symptoms and disease progression. It is a ND (Fig. 28.4) that is genetic and progressive. It presents with psychiatric disorders, chorea, and a progressive decline in motor and

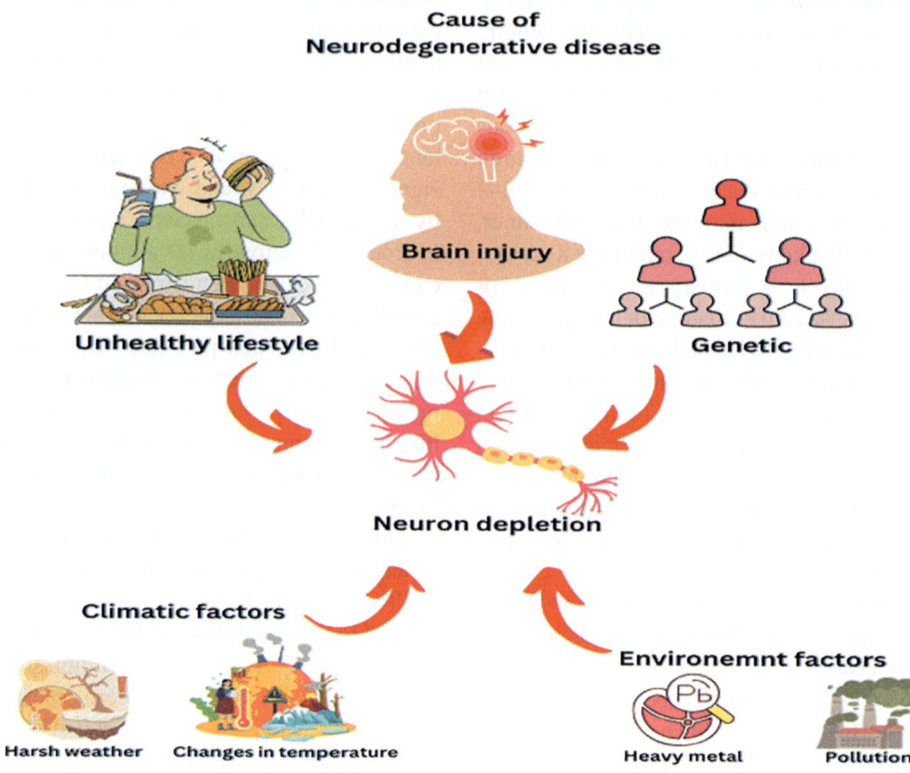

FIGURE 28.4 Cause of neurodegenerative disease.

cognitive dysfunctions (51). A mutation in the huntingtin (HTT) gene results in HD), a deadly hereditary or autosomal dominant neurodegenerative illness (Ananbeh, Vodicka, & Kupcova Skalnikova, 2021). Abnormalities in the handling or deposition of copper (Cu), iron (Fe), calcium (Ca), and manganese (Mn) in the tissue or cells, as well as in their neuronal functions, have been identified in HD. These abnormalities can either be the cause of pathology or the outcome of disease processes.

HD is an ND that is comparatively uncommon; its global prevalence is 2.71 cases per 100,000 people (Pringsheim et al., 2012). At 8.87 cases per 100,000 people, North America has the highest rate, followed by Australia at 8.4 and Europe at 6.37 (Papanna, Lazzari, & Rabottini, 2024; Stanisławska-Sachadyn et al., 2024). Oman also reported a comparatively high rate of 7.36 (Squitieri et al., 2020). South Korea and Russia report 2.22 and 1.91 cases per 100,000, respectively, while Africa reports 0.25 (low prevalence), 1.75 in Indian immigrants of the United Kingdom, 2.39 prevalence in East Asia, added to Middle Eastern (25–28).

A genetic mutation in the HTT gene is the leading cause of HD. This mutation arises from an atypical extension of the trinucleotide CAG repeat inside the HTT gene, which causes an enlarged poly-Q tract in the mutant huntingtin protein (Tandon, Aggarwal, & Sarkar, 2024). A genetic mutation in the HTT gene is the leading cause of HD. This mutation arises from an atypical extension of the trinucleotide CAG repeat inside the HTT gene, which causes an enlarged poly-Q tract in the mutant huntingtin protein (Tandon et al., 2024). Because of its strong link to immunoinflammation and capacity to induce neuronal death, mHTT aggregation triggers the inflammatory response and raises the concentration of inflammatory mediators centrally. As a result, this mutation is crucial to the neurodegeneration linked to HDs (Saba et al., 2022). Investigations have shown that the main pathological processes linked to HD include dopaminergic neuronal dysfunction, mitochondrial dysfunction, oxidative stress, irregular autophagy, aberrant protein accumulation, disrupted gene transcription, decreased trophic support, and NMDAr-mediated and non-NMDAr-mediated excitotoxicity (Murrough, Abdallah, & Mathew, 2017). These abnormalities are thought to be caused by a combination of the unfavorable gain-of-function effects of mHTT and the dysfunctional effects of wild-type huntingtin in the HD-affected brain.

It is still unknown if HD pathology results from toxic gain of function from the mutant HTT protein, loss of function from the normal HTT protein, or a combination of the two (Bichell, Timothy, Tipps, & Bowman, 2017). HD symptoms are still present in transgenic mouse models with one mutant allele and two functional wild-type alleles, indicating that the toxicity of the mutation is the underlying cause of the disease (Zuccato, Valenza, & Cattaneo, 2010). Conditional mouse models that express lower levels of mutant HTT in the cortex and striatum lessen neurodegeneration in these brain regions, and no behavioral abnormalities are observed.

28.8 Huntington's disease and heavy metals

Although there have been established links between heavy metals and NDs (Cohen et al., 2013; Matés, Segura, Alonso, & Márquez, 2010), little is known about how metal exposure affects HD development and symptoms, Changes in metallostasis can impact the functions of transition metals like Cu, Fe, Zn, Ca, and Mn as cofactors for enzymes, catalysts in reduction–oxidation reactions, and metabolic components—all of which are critical for neurological function (Wright and Baccarelli, 2007). The pathophysiology of HD involves each of the transition metals.

Metals and ions are essential for many biological processes, including oxygen transport, cell homeostasis regulation, and maintenance of cellular function. They also aid in cellular communication, maintain electrical charges and osmotic pressure during the electron transfer chain, and regulate nucleotide base pairing, stacking, stability, and DNA transcription, all of which restrict control of their intercellular concentration (Choi, Aid, & Bosetti, 2009). Dysregulated redox-active metals, including iron, copper, cadmium, and manganese, can raise oxidative stress and free radical production, which can result in cellular dysfunction and ultimately cell death (Paithankar, Saini, Dwivedi, Sharma, & Chowdhuri, 2021).

Furthermore, both in vitro and in vivo interactions between mutant huntingtin, insecticides, and heavy metals have been noted in HD models. For instance, in an HD model, the expression of mutant huntingtin decreases neurotoxicity and accumulation, indicating modifications in the ataxia telangiectasia mutant tumor protein p53 signaling pathway. The disruption in intracellular iron levels, crucial for oxidative energy metabolism and primarily controlled by the Huntingtin protein, is also connected to the etiology of HDs. Early-stage alterations in brain iron metabolism linked to HDs have been identified in MRI studies.

28.9 Epidemiology of amyotrophic lateral sclerosis

ALS, a progressive ND, impacts neurons in the brainstem, spinal cord, and motor cortex and causes muscle weakness, paralysis, and fasciculation, which are the hallmarks of ALS's clinical manifestations (Hulisz, 2018). Although survival varies, the majority of patients pass away 3–4 years after the onset of symptoms (van Es et al., 2017). With an estimated incidence of 2.6 to 3 per 100,000 person/year, ALS is the most prevalent motor neuron disease with an adult onset. WQS results indicated that when combined in late childhood and adolescence, barium, Cu, Pb, strontium, and zinc are the strongest drivers linked to ALS, confirmed by the ALS mouse model (Figueroa-Romero et al., 2020). Additionally, a prospective cohort study suggested that Zn, Pb, and Cd might be involved in the genesis of ALS (Peters et al., 2021). With few available treatments, ALS is a progressive and deadly illness that typically prolongs life by less than 6 months. According to a study, the number of ALS cases increased across all age groups, genders, and races between 2022 and 2030. Due to an anticipated decline in the population, it was predicted that the 51–65 age group would see a decrease in cases. We predict that there will be 32,893 cases of ALS in 2022, or an incidence of 9.9 cases per 100,000 people (95% CI 9.8, 10.0). Through 2030, this will gradually rise to 36,308 cases and a prevalence of 10.5 cases per 100,000 people (95% CI 10.4, 10.6). This represents a prevalence increase of roughly 0.1 cases annually. Throughout these years, the prevalence of ALS has continued to be higher in men (from 17,273 to 17,855) than in women (from 14,696 to 15,235). The white population will continue to account for most cases, which are expected to consistently fall between 22,000 and 23,000 cases. It is anticipated that the number of black and African cases will consistently fall between 1900 and 2000 (Mehta et al., 2025) (Table 28.2).

TABLE 28.2 Effect of heavy metals on neurodegenerative diseases.

Sl no.	**Parkinson's disease**	
	Metals	Their effects
1.	Manganese (Mn)	Chronic exposure leads to "manganism" which causes motor impairment by influencing the basal ganglia's dopamine-producing neurons. Raises inflammation and oxidative stress in the brain
2.	Lead (Pb)	Promotes neuroinflammation, interferes with dopamine signaling, and damages mitochondria
3.	Mercury (Hg)	Increases oxidative stress and disturbs calcium homeostasis, which harms neurons. Contribute to dopaminergic neurodegeneration, a major characteristic of PD
4.	Cadmium (Cd)	Interferes with mitochondrial and dopamine metabolism. May cause α-synuclein aggregation, a key pathogenic characteristic of PD
5.	Arsenic (As)	Causes dopaminergic neurons to undergo oxidative stress and death connected to a higher incidence of PD and neurodegeneration
	Alzheimer's disease	
1.	Lead (Pb)	Lead exposure throughout early life with a higher chance of developing AD later in life
		A key potential indicator of AD, β-amyloid aggregation is increased
		Causes mitochondrial malfunction and neuroinflammation, which results in the death of neurons
2.	Mercury (Hg)	Neurotoxic effects impair memory and brain function
		Boosts β-amyloid buildup and oxidative stress.
		Cholinergic neurotransmission, which is essential for cognitive function, is inhibited
3.	Cadmium (Cd)	It causes neurofibrillary tangles, promotes tau hyperphosphorylation, causes synaptic dysfunction, and triggers neuronal apoptosis. It also penetrates the BBB and builds up in the brain.
4.	Arsenic (As)	Long-term exposure to tainted water raises the risk of AD
		This leads to neuroinflammation and oxidative DNA damage
		Causes memory loss by interfering with synaptic plasticity
5.	Aluminum	Contributes to neurodegeneration by upsetting iron homeostasis linked to increased β-amyloid plaque formation
		Encourages cognitive deterioration and neuroinflammation

(Continued)

TABLE 28.2 (Continued)

Sl no.	Parkinson's disease	
	Huntington's disease	
1.	Manganese (Mn)	Overexposure causes neurotoxicity and oxidative stress, which especially affects the basal ganglia
		Promotes protein aggregation, possibly impacting mutant huntingtin (mHTT) toxicity.
2.	Lead (Pb)	Alters dopamine and calcium homeostasis, impacts brain signaling pathways essential for movement and cognition, and increases oxidative stress
		Exposure during pregnancy or early life may exacerbate the onset and severity of HD
3.	Mercury (Hg)	Builds up in the central nervous system, resulting in neuroinflammation and mitochondrial dysfunction, causing excitotoxicity, reducing neuronal survival
4.	Cadmium (Cd)	It prevents neurotoxic factors necessary for neuron survival, increases oxidative damage, and causes the misfolding of mutant huntingtin.
5.	Arsenic (As)	Induces neuroinflammation and oxidative stress, interrupts cellular energy production, motor dysfunction
	Amyotrophic lateral sclerosis	
1.	Lead (Pb)	Inhibits neurotransmission-related enzymes and causes the death of motor neurons, causing oxidative stress in the brain
2.	Mercury (Hg)	Causes excitotoxicity which is caused by the disruption of glutamate transporters, increases neuronal death by interfering with mitochondrial function.
3.	Arsenic (As)	This leads to oxidative stress by increasing ROS and modifies gene expression, which may impact the survival of motor neurons, chronic exposure may lead to neurological dysfunction.
4.	Cadmium (Cd)	By inducing oxidative damage and protein misfolding, it induces neuronal apoptosis and disrupts calcium signaling which is necessary for healthy nerve functioning
5.	Manganese (Mn)	Overexposure causes oxidative stress, mitochondrial dysfunction, and neuroinflammation, essential in small amounts, toxic at high levels
	Multiple Sclerosis	
1.	Lead (Pb)	Disrupts BBB, which allows immune cells to attack myelin, and induces chronic inflammation
2.	Mercury (Hg)	causes mitochondrial dysfunction leading to fatigue and neurodegeneration, causes axonal damage and demyelination, increases autoimmune activity, and modifies immune function
3.	Iron (Fe)	Neurodegeneration and oxidative stress are caused by iron overload, excessive iron supplements, and red meat increase the production of inflammatory cytokines
4.	Cadmium (Cd)	Contributes to immunological dysregulation and oxidative stress, damages neurons and myelin, and impacts Ca signaling, causing nerve dysfunction
5.	Aluminum (Al)	Encourages persistent neuroinflammation, and may affect the function of the BBB, vaccines, cookware, and food packaging are all associated with cognitive decline in MS patients

AD, Alzheimer's disease; *BBB*, blood–brain barrier; *HD*, Huntington's disease; *MS*, multiple sclerosis; *PD*, Parkinson's disease.
Source: From Roth et al. (2013), Mehta et al. (2025), Milo, Korczyn, Manouchehri, and Stüve (2020), and Murata et al. (2022).

28.10 Amyotrophic lateral sclerosis and heavy metals

For many years, metal exposure has been proposed as a potential risk factor for ALS (Sutedja et al., 2009; van Es et al., 2017). Exposures can happen in various work environments, such as manufacturing, plumbing, welding, metal smelting, and refining. However, they can also happen in the general public through tainted food and drinking water, air pollution, cigarette smoking, prescription drugs, or dietary supplements (Sutedja et al., 2009). A first prospective study examined the relationship between blood metal levels and ALS risk. It was observed that Lead and cadmium may be linked to an increased risk of ALS, with evidence of an inverse association for zinc, according to metal concentrations in blood from ALS patients before the disease onset and controls (Peters et al., 2021).

It has been demonstrated that misfold superoxide dismutase-1 (SOD 1), a pathological hallmark of Familial ALS activates astrocytes and microglia. (Kiernan et al., 2011; Wang et al., 2022). The pathophysiology of ALS involves the activation of peripherally derived inflammatory cells and glial cells (Thompson, Turner, 2019). In the cerebrospinal fluid of ALS patients, a case-control study discovered dysregulated proinflammatory cytokines like IL-6, IL-10, and IFN-γ (Mitchell et al., 2009). Furthermore, it has been noted that ALS patients exhibit changes in the populations of circulating immune cells and cytokines (Thompson, Turner, 2019; Mitchell et al., 2009). Neuroinflammation was detected in the CNS and spinal cords of ALS patients and mouse models (McGeer and McGeer, 2002), characterized by the activation of microglia and astrocytes, accompanied by an increase in inflammatory mediators and cytokines in the cerebrospinal fluid and brain, considered the hallmark of ALS (Mishra et al., 2017).

28.11 Epidemiology of multiple sclerosis

It is one of the most prevalent conditions, characterized by an immune-mediated illness brought on by an autoimmune attack on the CNS and progressive demyelination due to persistent inflammation (Dehghanifiroozabadi et al., 2019). In developed nations, this illness is the leading cause of disability among young and middle-aged individuals (Kotelnikova et al., 2017). Over 2.5 million people worldwide are afflicted with MS each year. Most patients have a relapsing-remitting form of the illness (Giacoppo et al., 2014; Nicoletti et al., 2016). Although the precise cause of MS is unknown, many factors, including immunological, genetic, environmental, and infectious ones, as well as dietary choices, lifestyle choices, and habits, all play a role in its development. Neuritis, ophthalmoplegia, pyramidal tract syndrome, psychiatric symptoms, and neuroaxonal degeneration are among the primary pathological characteristics of MS (Lee et al., 2019). The pathophysiology of MS is mediated by inflammation in the brain and spinal cord (Milo et al., 2020). It has been demonstrated that reducing neuroinflammation slows the progression of MS (Gharagozloo et al., 2018). According to case–control study results, the serum IL-1β levels of MS patients were four times greater than CSF levels (Reale et al., 2015).

A study conducted in which the relationship between MS patients and control subjects in terms of their circulating levels of arsenic, lead, mercury, and cadmium was evaluated in a metaanalysis showed that the circulatory As and Cd levels were noticeably higher in MS patients than in controls. However, circulating levels of Pb and Hg in MS patients and controls did not differ significantly (Sarihi et al., 2021).

28.12 Heavy metals and multiple sclerosis

According to studies, magnesium is one of the most significant minerals in MS. The body needs adequate Mg for bone health, muscle contraction, and nerve function (10). Additionally, cells require Mg to move Ca and K from the cell wall. Furthermore, cells are shielded by Mg for nickel, beryllium, lead, mercury, and aluminum. Oxidative stress in the kidneys and brain may result from iron and cadmium buildup. Aluminum and lead are brought on by Mg deficiency. In a study conducted in Japan in 1990, the Mg levels of four patients and five controls in Japan were observed. According to findings, MS patients had reduced Mg levels in all visceral organs and central nervous tissues, which implies that patients with MS should have their Mg levels checked (Yasui et al., 1990). A study of 101 patients with relapsing-remitting MS at the Neurological Research Center, Tehran, revealed that Mg and Folate intake are the only nutrients that affect fatigue reduction. The researchers linked extreme fatigue in MS patients to a diet lower in Mg and folate. This research has indicated that women with MS exhibit reduced folate, Mg, Vit E, and other nutrients with antiinflammatory or antioxidant qualities compared to healthy controls. (Bitarafan et al., 2014)

About 35 years ago, research on how zinc affects MS patients started. Males had higher zinc levels than females in the control and MS patient groups (M. Grabrucker, Rowan, & Garner, 2011). The important finding is that, in patients under 45, there was a significant decrease in zinc and copper levels compared to the control group (males and females). However, no changes in these levels were seen in patients over 45 (Bowling, 2006). According to studies, between 20% and50% of MS patients experience depression while they are ill. The severity of the disease, the location of lesions, neurological symptoms, and a patient's disability are not associated with depression symptoms. Zn and Mg levels are lower in MS patients than in the control group (Mousavi, Habibbollahi, & Mahmoudian, 2006; Salari, Khomand, Arasteh, Yousefzamani, & Hassanzadeh, 2015).

Research has indicated that the pathophysiology of degenerative nervous diseases is significantly influenced by compromised Fe and Cu metabolism (Johnson, 2000). Like zinc, Cu is a redox-active metal crucial for the catalysis of free radical formation and active antioxidant enzymes. Numerous animal species may experience decreased myelination due to Cu and Zn deficiencies, and it has been demonstrated that these deficiencies are crucial for the development and progression of neurological disorders (Johnson, 2000).

Johnson (2000) examined the pathophysiology of MS patients in Washington by examining the buildup of Cu, Cd, Fe, Zn, Mg, vitamin B6, and Vit D deficiency. These findings indicate that too much cu inhibits the synthesis of uric acid and reduces the absorption of Zn. Consequently, a higher concentration of cu results in a higher production of superoxide (Johnson, 2000). Additionally, observations of 60 patients in Italy revealed that MS patients had low cu levels because cu is involved in transmitting the pulse in the nervous and muscular systems. An essential element, copper, plays a role in the metabolism of the neurotransmitter dopamine (Forte et al., 2005)

In this context, a study was conducted in Kashan, Iran, where 35 patients had typical MS symptoms. The researchers looked into the difference between the serum levels of Mg, Cu, and Zn in MS patients and healthy controls and found that the MS patients had lower levels of these elements than the control group. Additionally, it was noted that women had lower mean serum Mg levels but higher mean serum copper and zinc levels in both MS patients and control groups compared to men (Wingerchuk, 2010).

One of the most harmful nonradioactive elements frequently absorbed is mercury, a potent neurotoxin that affects essential cell functions. Mercury toxicity is caused by the generation of free radicals and possible alterations of the cell redox reaction. Additionally, mercury prevents and degrades the cytoskeleton's development in nerve cells. According to studies on the prevalence of apoptosis among various rat brain cultures, Mercury increases spontaneous apoptosis in immature cultures, particularly astrocytes (Issa et al., 2003). The idea that elevated mercury levels in the serum may raise the incidence of MS disease is expressed by higher mercury levels in MS patients compared to healthy subjects (Table 28.3).

A thorough investigation of 1718 MS patients in Isfahan, Iran, by several nephrologists showed that the epidemiological risk of MS in the area was relatively high and was rising. In 2012 some studies investigated the possible connection between serum mercury levels and the development of MS in Isfahan due to the high rate of MS and industrial pollution in the city. There were notable differences between the serum levels of Hg in MS patients and healthy subjects. A difference in the level of Hg was observed in both males and females where MS patients have abnormally high mercury levels (Zapadniuk, 1992)

As the second leading cause of death worldwide in 2015, NDs pose a threat to the world's aging population (Feigin et al., 2017). Only 40% of patients with AD will survive for more than 5 years after diagnosis, 80% of patients will only live for an additional year, and 55% will live for three more years. (Akushevich, Yashkin, Kravchenko, & Yashin, 2021). However, according to research, the average survival time for a person with PD is 9.1 years after diagnosis (de Lau, Schipper, Hofman, Koudstaal, & Breteler, 2005). According

TABLE 28.3 Descriptive epidemiology and population-based research in neurodegenerative diseases: present and future (Logroscino et al., 2022).

Epidemiology	Present	Future
Disease diagnosis	Based on clinical criteria	Based on both biological and clinical criteria
Intervention strategies	On the idea that illness is a single biological entity. Reduced amyloid in AD and decreased synuclein in PD.	Customized based on various disease phenotypes that have a similar biological mechanism. The use of precision medicine is an approach.
The gold standard for valid disease diagnosis	Primarily clinical. In certain instances, pathological verification	Large population-based ageing cohorts with deep phenotyping will make analysis based on biological signals from outliers easier.
Definition of disease in projections	Disease diagnosis by clinical means (AD, PD, HD, etc.)	Disease preclinical diagnosis aided by biomarker utilization. The idea of prodromal AD is sound and has already been applied in population-based research.
Models for implementing	Biomarkers for clinical pathological model: Biomarkers are validated after clinical phenotypes are established	System's biology model: Biomarkers are found through larger, phenotype-agnostic ageing studies.
Biomarker types utilized in population-based research.	Costly PET scans; biomarkers based on CSF are thought to be intrusive	Advanced radiotracer-free neuroimaging biomarkers are less expensive and more widely available than blood-based biomarkers. A multimodel strategy would increase precision.

AD, Alzheimer's disease; *HD*, Huntington's disease; *PD*, Parkinson's disease.
Source: From Logroscino, G., Urso, D., & Savica, R. (2022). Descriptive epidemiology of neurodegenerative diseases: What are the critical questions?. Neuroepidemiology, 56(5), 309–318.

to another research, 75% of patients with ALS pass away within 5 years of being diagnosed. These findings highlight the need for early evaluation of NDD risk and prevalence to provide adequate individual and population-level interventions. However, at this time, there are no noninvasive liquid biopsy tests available to diagnose sporadic NDD cases. Therefore it is imperative to look for new noninvasive techniques to evaluate patient populations proactively before common pathologies manifest to increase life expectancy and quality of life after diagnosis (Newell et al., 2024). Identification and analysis of the variables influencing the onset and progression of NDDs may advance knowledge of the pathophysiology of NDDs and enhance population health in general. Developing research involved associating environmental chemical exposures with neurological diseases (Schmidt, 2022). Heavy metals associated with NDDs include the metalloid As and transition metals Cd, Hg, and Mn in descending order of importance (Manivannan, Manivannan, Samuel, Beach, & Halden, 2019; Newell, Adhikari, & Halden, 2022; Nielsen et al., 2021).

The symptoms of an NDD may not appear for years or even decades after exposure to a dangerous environmental trigger. Furthermore, for environmental pollutants of interest to human populations, the effects of high-dose, acute exposure as opposed to low-dose, chronic exposure are frequently unknown (Nabi and Tabassum, 2022). Most experimental research examining environmental pollutants' duration and dosage effects has been restricted to animal models, which are not always suitable for human use (Nabi and Tabassum, 2022). Various biometrics show a toxicant's variable metabolic duration before its excretion from the human body, complicating the assessment of human exposure. Because humans excrete toxic metals and metabolites at different rates, discrete measurements are not as helpful in determining lifetime exposure (Newell et al., 2024).

28.13 Conclusion

Improving the lifespan and quality of prospective patients requires an early diagnosis of NDDs. Epidemiologists can monitor areas more vulnerable to the development of NDDs by quantifying the levels of toxic metals and metalloids in environmental sources and providing proof of exposure. The symptoms of an NDD may not appear for years or decades after exposure to a dangerous environmental trigger. Furthermore, for environmental pollutants of interest to human populations, the effect of high-dose, acute exposure as opposed to low-dose, chronic exposure is frequently unknown; the majority of experimental research examining the duration and dosage effects of environmental pollutants has been restricted to animal models that are not always suitable for human use.

References

Akushevich, I., Yashkin, A. P., Kravchenko, J., & Yashin, A. I. (2021). Analysis of time trends in Alzheimer's disease and related dementias using partitioning approach. *Journal of Alzheimer's Disease, 82*(3), 1277–1289.

Ananbeh, H., Vodicka, P., & Kupcova Skalnikova, H. (2021). Emerging roles of exosomes in Huntington's disease. *International Journal of Molecular Sciences, 22*(8), 4085.

Aschner, J. L., & Aschner, M. (2005). Nutritional aspects of manganese homeostasis. *Molecular Aspects of Medicine, 26*(4–5), 353–362.

Aschner, J. L., Anderson, A., Slaughter, J. C., Aschner, M., Steele, S., Beller, A., ... Maitre, N. L. (2015). Neuroimaging identifies increased manganese deposition in infants receiving parenteral nutrition. *The American Journal of Clinical Nutrition, 102*(6), 1482–1489.

Aschner, M., Erikson, K. M., Hernández, E. H., & Tjalkens, R. (2009). Manganese and Its role in Parkinson's disease: From transport to neuropathology. *Neuromolecular Medicine, 11*, 252–266.

Bakulski, M., Rozek, L. S., Dolinoy, D. C., Paulson, H. L., & Hu, H. (2012). Alzheimer's disease and environmental exposure to lead: The epidemiologic evidence and potential role of epigenetics. *Current Alzheimer Research, 9*(5), 563–573.

Barron, M., Gartlon, J., Dawson, L. A., Atkinson, P. J., & Pardon, M.-C. (2017). A state of delirium: Deciphering the effect of inflammation on tau pathology in Alzheimer's disease. *Experimental Gerontology, 94*, 103–107.

Bassil, N., & Mollaei, C. (2012). *Alzheimer's dementia: A brief review*, 192–199.

Bichell, T. J. V., Timothy, C. H., Tipps, K. G., & Bowman, A. B. (2017). Metal biology associated with Huntington's disease. In *Biometals in neurodegenerative diseases*. Elsevier, 231–263.

Binolfi, A., Rasia, R. M., Bertoncini, C. W., Ceolin, M., Zweckstetter, M., Griesinger, C., ... Fernández, C. O. (2006). Interaction of α-synuclein with divalent metal ions reveals key differences: A link between structure, binding specificity and fibrillation enhancement. *Journal of the American Chemical Society, 128*(30), 9893–9901.

Bitarafan, S., Harirchian, M.-H., Nafissi, S., Sahraian, M.-A., Togha, M., Siassi, F., Saedisomeolia, A., et al. (2014). Dietary intake of nutrients and its correlation with fatigue in multiple sclerosis patients. *Iranian Journal of Neurology, 13*(1), 28.

Bouabid, S., Delaville, C., De Deurwaerdère, P., Lakhdar-Ghazal, N., & Benazzouz, A. (2014). Manganese-induced atypical parkinsonism is associated with altered basal ganglia activity and changes in tissue levels of monoamines in the rat. *PLoS One, 9*(6), e98952.

Bowling, A. C. (2006). *Complementary and alternative medicine and multiple sclerosis*. Demos Medical Publishing.

Braak, H., Sastre, M., & Tredici, K. D. (2007). Development of α-synuclein immunoreactive astrocytes in the forebrain parallels stages of intraneuronal pathology in sporadic Parkinson's disease. *Acta Neuropathologica, 114*, 231–241.

Brookmeyer, R., Denis, A. E., Hebert, Liesi, Langa, K. M., Heeringa, S. G., Plassman, B. L., & Walter, A. K. (2011). National estimates of the prevalence of Alzheimer's disease in the United States. *Alzheimer's & Dementia, 7*(1), 61–73.

Butterworth, R. F., Spahr, L., Fontaine, S., & Layrargues, G. P. (1995). Manganese toxicity, dopaminergic dysfunction and hepatic encephalopathy. *Metabolic Brain Disease, 10*, 259–267.

Cai, J., Guan, H., Jiao, X., Yang, J., Chen, X., Zhang, H., Zheng, Y., et al. (2021). NLRP3 inflammasome mediated pyroptosis is involved in cadmium exposure-induced neuroinflammation through the IL-1β/IkB-α-NF-KB-NLRP3 feedback loop in swine. *Toxicology, 453*, 152720.

Cai, T., Che, H., Yao, T., Chen, Y., Huang, C., Zhang, W., Du, K., et al. (2011). Manganese induces tau hyperphosphorylation through the activation of ERK MAPK pathway in PC12 cells. *Toxicological Sciences, 119*(1), 169–177.

Calne, D. B., Chu, N.-S., Huang, C.-C., Lu, C.-S., & Olanow, W. (1994). Manganism and idiopathic parkinsonism: Similarities and differences. *Neurology, 44*(9), 1583.

Cass, W. A., Grondin, R., Andersen, A. H., Zhang, Z., Hardy, P. A., Hussey-Andersen, L. K., ... Gash, D. M. (2007). Iron accumulation in the striatum predicts aging-related decline in motor function in rhesus monkeys. *Neurobiology of Aging, 28*(2), 258–271.

Choi, S.-H., Aid, S., & Bosetti, F. (2009). The distinct roles of cyclooxygenase-1 and -2 in neuroinflammation: Implications for translational research. *Trends in Pharmacological Sciences, 30*(4), 174–181.

Cohen, Y., Megyeri, M., Chen, O. C. W., Condomitti, G., Riezman, I., Loizides-Mangold, U., Abdul-Sada, A., et al. (2013). The yeast P5 type ATPase, Spf1, regulates manganese transport into the endoplasmic reticulum. *PLoS One, 8*(12), e85519.

Coon, S., Stark, A., Peterson, E., Gloi, A., Kortsha, G., Pounds, J., ... Gorell, J. (2006). Whole-body lifetime occupational lead exposure and risk of Parkinson's disease. *Environmental Health Perspectives, 114*(12), 1872–1876.

Crous-Bou, M., Minguillón, C., Gramunt, N., & Molinuevo, J. L. (2017). Alzheimer's disease prevention: From risk factors to early intervention. *Alzheimer's research & therapy, 9*, 1–9.

Csavina, J., Field, J., Taylor, M. P., Gao, S., Landázuri, A., Betterton, E. A., & Sáez, A. E. (2012). A review on the importance of metals and metalloids in atmospheric dust and aerosol from mining operations. *Science of the Total Environment, 433*, 58–73.

Dehghanifiroozabadi, M., Noferesti, P., Amirabadizadeh, A., Nakhaee, S., Aaseth, J., Noorbakhsh, F., & Mehrpour, O. (2019). Blood lead levels and multiple sclerosis: A case-control study. *Multiple Sclerosis and Related Disorders, 27*, 151–155.

Dexter, D. T., Carter, C. J., Wells, F. R., Javoy-Agid, F., Agid, Y., Lees, A., ... Marsden, C. D. (1989). Basal lipid peroxidation in substantia nigra is increased in Parkinson's disease. *Journal of Neurochemistry, 52*(2), 381–389.

Dissanayake, V., & Erickson, T. B. (2012). Ball and chain: The global burden of lead poisoning. *Clinical Toxicology, 50*(6), 528–531.

Dorsey, E. R., Elbaz, A., Nichols, E., Abbasi, N., Abd-Allah, F., Abdelalim, A., Adsuar, J. C., et al. (2018). Global, regional, and national burden of Parkinson's disease, 1990–2016: A systematic analysis for the global burden of disease study 2016. *The Lancet Neurology, 17*(11), 939–953.

Dorsey, E. R., Sherer, T., Okun, M. S., & Bloem, B. R. (2018). The emerging evidence of the Parkinson pandemic. *Journal of Parkinson's Disease, 8*(s1), S3–S8.

D'Ambrosi, N., & Rossi, L. (2015). Copper at synapse: Release, binding and modulation of neurotransmission. *Neurochemistry International, 90*, 36–45.

Egan, M. F., Kost, J., Voss, T., Mukai, Y., Aisen, P. S., Cummings, J. L., Tariot, P. N., et al. (2019). Randomized trial of verubecestat for prodromal Alzheimer's disease. *New England Journal of Medicine, 380*(15), 1408–1420.

Ercal, N., Gurer-Orhan, H., & Aykin-Burns, N. (2001). Toxic metals and oxidative stress Part I: Mechanisms involved in metal-induced oxidative damage. *Current Topics in Medicinal Chemistry, 1*(6), 529–539.

Ericson, B., Landrigan, P., Taylor, M. P., Frostad, J., Caravanos, J., John, K., & Richard, F. (2016). The global burden of lead toxicity attributable to informal used lead-acid battery sites. *Annals of Global Health, 82*(5), 686–699.

Erikson, K. M., Syversen, T., Aschner, J. L., & Aschner, M. (2005). Interactions between excessive manganese exposures and dietary iron-deficiency in neurodegeneration. *Environmental Toxicology and Pharmacology, 19*(3), 415–421.

van Es, M. A., Hardiman, O., Chio, A., Al-Chalabi, A., Pasterkamp, R. J., & Veldink, J. H. (2017). "Van Den Berg 376 LH." Amyotrophic lateral sclerosis. *Lancet, 390*(10107), 2084–2098.

Farina, M., Aschner, M., & Rocha, J. B. T. (2011). Oxidative stress in MeHg-induced neurotoxicity. *Toxicology and Applied Pharmacology, 256*(3), 405–417.

Feigin, V. L., Abajobir, A. A., Abate, K. H., Abd-Allah, F., Abdulle, A. M., Abera, S. F., Abyu, G. Y., et al. (2017). Global, regional, and national burden of neurological disorders during 1990–2015: A systematic analysis for the global burden of disease study 2015. *The Lancet Neurology, 16*(11), 877–897.

Figueroa-Romero, C., Mikhail, K. A., Gennings, C., Curtin, P., Bello, G. A., Botero, T. M., Goutman, S. A., et al. (2020). Early life metal dysregulation in amyotrophic lateral sclerosis. *Annals of Clinical and Translational Neurology, 7*(6), 872–882.

Fitsanakis, V. A., Au, C., Erikson, K. M., & Aschner, M. (2006). The effects of manganese on glutamate, dopamine and γ-aminobutyric acid regulation. *Neurochemistry International, 48*(6–7), 426–433.

Forte, G., Andrea, V., Simone, S., Anna, G., Forte Lorenzo, F.-T., Stefania, C., Beatrice, B., et al. (2005). Quantification of chemical elements in blood of patients affected by multiple sclerosis. *Annali dell'Istituto superiore di sanita, 41*(2), 213–216.

Franco, J. L., Braga, H. C., Stringari, J., Missau, F. C., Posser, T., Mendes, B. G., Leal, R. B., et al. (2007). Mercurial-induced hydrogen peroxide generation in mouse brain mitochondria: Protective effects of quercetin. *Chemical Research in Toxicology, 20*(12), 1919–1926.

Franco, J. L., Posser, T., Dunkley, P. R., Dickson, P. W., Mattos, J. J., Martins, R., Bainy, A. C. D., et al. (2009). Methylmercury neurotoxicity is associated with inhibition of the antioxidant enzyme glutathione peroxidase. *Free Radical Biology and Medicine, 47*(4), 449–457.

Garza-Lombó, C., Yanahi, P., Liliana, Q., María, E. G., & Franco, R. (2018). Neurotoxicity linked to dysfunctional metal ion homeostasis and xenobiotic metal exposure: Redox signaling and oxidative stress. *Antioxidants & Redox Signaling, 28*(18), 1669–1703.

Gatz, M., Chandra, A. R., Fratiglioni, L., ... Nancy, L. P. (2006). Role of genes and environments for explaining Alzheimer disease. *Archives of General Psychiatry, 63*(2), 168–174.

Gavin, C. E., Gunter, K. K., & Gunter, T. E. (1990). Manganese and calcium efflux kinetics in brain mitochondria. relevance to manganese toxicity. *Biochemical Journal, 266*(2), 329–334.

Gharagozloo, M., Gris, K. V., Mahvelati, T., Amrani, A., Lukens, J. R., & Gris, D. (2018). NLR-dependent regulation of inflammation in multiple sclerosis. *Frontiers in Immunology, 8*, 2012.

Giacoppo, S., Galuppo, M., Calabrò, R. S., D'Aleo, G., Marra, A., Sessa, E., Bua, D. G., et al. (2014). Heavy metals and neurodegenerative diseases: An observational study. *Biological Trace Element Research, 161*, 151–160.

Goldman, S. M. (2014). Environmental toxins and Parkinson's disease. *Annual Review of Pharmacology and Toxicology, 54*(1), 141–164.

Grabrucker, M., Rowan, M., & Garner, C. C. (2011). Brain-delivery of zinc-ions as potential treatment for neurological diseases: Mini review. *Drug Delivery Letters, 1*(1), 13–23.

Grandjean, P., & Bellanger, M. (2017). Calculation of the disease burden associated with environmental chemical exposures: Application of toxicological information in health economic estimation. *Environmental Health, 16*, 1–13.

Griffiths, P. D., Dobson, B. R., Jones, G. R., & Clarke, D. T. (1999). Iron in the basal ganglia in Parkinson's disease: An in vitro study using extended X-ray absorption fine structure and cryo-electron microscopy. *Brain, 122*(4), 667–673.

Guilarte, T. R. (2013). Manganese neurotoxicity: New perspectives from behavioral, neuroimaging, and neuropathological studies in humans and non-human primates. *Frontiers in Aging Neuroscience, 5*, 23.

Guo, J.-tao, Yu, J., Grass, D., de Beer, F. C., & Kindy, M. S. (2002). Inflammation-dependent cerebral deposition of serum amyloid a protein in a mouse model of amyloidosis. *Journal of Neuroscience, 22*(14), 5900–5909.

Götz, M. E., Double, K. A. Y., Gerlach, M., Youdim, M. B. H., & Riederere, Peter (2004). The relevance of Iron in the pathogenesis of Parkinson's disease. *Annals of the New York Academy of Sciences, 1012*(1), 193–208.

Hanna-Attisha, M., LaChance, J., Sadler, R. C., & Champney Schnepp, A. (2016). Elevated blood lead levels in children associated with the flint drinking water crisis: A spatial analysis of risk and public health response. *American Journal of Public Health, 106*(2), 283–290.

He, P., Yang, C., He, D., Zhao, S., Xie, Y., Wang, H., & Ma, J. (2021). Blood lead, systemic inflammation, and blood pressure: Exploring associations and mediation effects in workers exposed to lead. *Biological Trace Element Research, 199*(7), 2573–2581.

Hulisz, D. (2018). Disease state overview. *The American Journal of Managed Care, 24*(15), S320–S326.

Issa, Y., Watts, D. C., Duxbury, A. J., Brunton, P. A., Watson, M. B., & Waters, C. M. (2003). Mercuric chloride: Toxicity and apoptosis in a human oligodendroglial cell line MO3. 13. *Biomaterials, 24*(6), 981–987.

Jack, C. R., Therneau, T. M., Weigand, S. D., Wiste, H. J., Knopman, D. S., Vemuri, P., Lowe, V. J., et al. (2019). Prevalence of biologically vs clinically defined Alzheimer spectrum entities using the national institute on aging–Alzheimer's Association Research Framework. *JAMA Neurology, 76*(10), 1174–1183.

Jan, A. T., Azam, M., Siddiqui, K., Ali, A., Choi, I., Mohd, Q., & Haq, R. (2015). Heavy metals and human health: Mechanistic insight into toxicity and counter defense system of antioxidants. *International Journal of Molecular Sciences, 16*(12), 29592–29630.

Janocha-Litwin, J., Marianska, K., Serafinska, S., & Simon, K. (2015). Manganese encephalopathy among ephedron abusers-case reports. *Journal of Neuroimaging, 25*(5), 832–835.

Johnson, S. (2000). The possible role of gradual accumulation of copper, cadmium, lead and iron and gradual depletion of zinc, magnesium, selenium, vitamins B2, B6, D, and E and essential fatty acids in multiple sclerosis. *Medical Hypotheses, 55*(3), 239–241.

Kaler, S. G. (2011). ATP7A-related copper transport diseases—Emerging concepts and future trends. *Nature Reviews Neurology, 7*(1), 15–29.

Kern, S., Zetterberg, H., Kern, J., Zettergren, A., Waern, M., Höglund, K., Andreasson, U., et al. (2018). Prevalence of preclinical Alzheimer disease: Comparison of current classification systems. *Neurology, 90*(19), e1682–e1691.

Keshavan, A., Pannee, J., Karikari, T. K., Rodriguez, J. L., Ashton, N. J., Nicholas, J. M., Cash, D. M., et al. (2021). Population-based blood screening for preclinical Alzheimer's disease in a British birth cohort at age 70. *Brain, 144*(2), 434–449.

Kiernan, M. C., Vucic, S., Cheah, B. C., Turner, M. R., Eisen, A., Hardiman, O., ... Zoing, M. C. (2011). Amyotrophic lateral sclerosis. *The Lancet, 377*(9769), 942–955.

Kim, M.-J., Oh, S.-B., Kim, J., Kim, K., Ryu, H.-S., Kim, M. S., Ayton, S., et al. (2018). Association of metals with the risk and clinical characteristics of Parkinson's disease. *Parkinsonism & Related Disorders, 55*, 117–121.

Klos, K. J., Chandler, M., Kumar, N., Ahlskog, J. E., & Josephs, K. A. (2006). Neuropsychological profiles of manganese neurotoxicity. *European Journal of Neurology, 13*(10), 1139–1141.

Kotelnikova, E., Kiani, N. A., Abad, E., Martinez-Lapiscina, E. H., Andorra, M., Zubizarreta, I., Pulido-Valdeolivas, I., et al. (2017). Dynamics and heterogeneity of brain damage in multiple sclerosis. *PLoS Computational Biology, 13*(10), e1005757.

Lai, B. C. L., Marion, S. A., Teschke, K., & Tsui, J. K. C. (2002). Occupational and environmental risk factors for Parkinson's disease. *Parkinsonism & Related Disorders, 8*(5), 297–309.

Lane, C. A., Hardy, J., & Schott, J. M. (2018). Alzheimer's disease. *European Journal of Neurology, 25*(1), 59–70.

de Lau, L. M. L., Schipper, C. M. A., Hofman, A., Koudstaal, P. J., & Breteler, M. M. B. (2005). Prognosis of Parkinson disease: Risk of dementia and mortality: The Rotterdam study. *Archives of Neurology, 62*(8), 1265–1269.

Laurent, C., Buée, L., & Blum, D. (2018). Tau and neuroinflammation: What impact for Alzheimer's disease and tauopathies? *Biomedical Journal, 41*(1), 21–33.

Lee, E., Hwang, I., Park, S., Hong, S., Hwang, B., Cho, Y., ... Yu, J.-W. (2019). MPTP-driven NLRP3 inflammasome activation in microglia plays a central role in dopaminergic neurodegeneration. *Cell Death & Differentiation, 26*(2), 213–228.

Lee, H.-J., Suk, J.-E., Patrick, C., Bae, E.-J., Cho, J.-H., Rho, S., ... Lee, S.-J. (2010). Direct transfer of α-synuclein from neuron to astroglia causes inflammatory responses in synucleinopathies. *Journal of Biological Chemistry, 285*(12), 9262–9272.

Li, Y., Jiao, Q., Xu, H., Du, X., Shi, L., Jia, F., & Jiang, H. (2017). Biometal dyshomeostasis and toxic metal accumulations in the development of Alzheimer's disease. *Frontiers in Molecular Neuroscience, 10*, 339.

Liccione, J. O. H. N. J., & Mahin, D. M. (1988). Selective vulnerability of glutathione metabolism and cellular defense mechanisms in rat striatum to manganese. *Journal of Pharmacology and Experimental Therapeutics, 247*(1), 156–161.

Logroscino, G., Urso, D., & Savica, R. (2022). Descriptive epidemiology of neurodegenerative diseases: What are the critical questions? *Neuroepidemiology, 56*(5), 309–318.

Loikkanen, J. J., Naarala, J., & Savolainen, K. M. (1998). Modification of glutamate-induced oxidative stress by lead: The role of extracellular calcium. *Free Radical Biology and Medicine, 24*(2), 377–384.

Manivannan, B., Manivannan, Y., Samuel, S., Beach, T. G., & Halden, R. U. (2019). Assessment of persistent, bioaccumulative and toxic organic environmental pollutants in liver and adipose tissue of Alzheimer's disease patients and age-matched controls. *Current Alzheimer Research, 16*(11), 1039–1049.

Martinez-Finley, E. J., Gavin, C. E., Aschner, M., & Gunter, T. E. (2013). Manganese neurotoxicity and the role of reactive oxygen species. *Free Radical Biology and Medicine, 62*, 65–75.

Matés, J. M., Segura, J. A., Alonso, F. J., & Márquez, J. (2010). Roles of dioxins and heavy metals in cancer and neurological diseases using ROS-mediated mechanisms. *Free Radical Biology and Medicine, 49*(9), 1328–1341.

Mayeux, R., & Stern, Y. (2012). Epidemiology of Alzheimer disease. *Cold Spring Harbor Perspectives in Medicine, 2*(8), a006239.

McGeer, P. L., & McGeer, E. G.12362410 (2002). Inflammatory processes in amyotrophic lateral sclerosis. *Muscle & Nerve: Official Journal of the American Association of Electrodiagnostic Medicine, 26*(4), 459–470.

Mehta, P., Raymond, J., Nair, T., Han, M., Berry, J., Punjani, R., ... Horton, D. K. (2025). Amyotrophic lateral sclerosis estimated prevalence cases from 2022 to 2030, data from the national ALS registry. *Amyotrophic Lateral Sclerosis and Frontotemporal Degeneration*, 1–6.

Mezzaroba, L., Alfieri, D. F., Colado Simão, A. N., & Vissoci Reiche, E. M. (2019). The role of zinc, copper, manganese and iron in neurodegenerative diseases. *Neurotoxicology, 74*, 230–241.

Milo, R., Korczyn, A. D., Manouchehri, N., & Stüve, O. (2020). The temporal and causal relationship between inflammation and neurodegeneration in multiple sclerosis. *Multiple Sclerosis Journal, 26*(8), 876–886.

Miodovnik, A. (2011). Environmental neurotoxicants and developing brain. *Mount Sinai Journal of Medicine: A Journal of Translational and Personalized Medicine, 78*(1), 58–77.

Mishra, P.-S., Vijayalakshmi, K., Nalini, A., Sathyaprabha, T. N., Kramer, B. W., Alladi, P. A., & Raju, T. R. (2017). Etiogenic factors present in the cerebrospinal fluid from amyotrophic lateral sclerosis patients induce predominantly pro-inflammatory responses in microglia. *Journal of Neuroinflammation, 14*, 1–18.

Mitchell, R. M., Freeman, W. M., Randazzo, W. T., Stephens, H. E., Beard, J. L., Simmons, Z., & Connor, J. R. (2009). A CSF biomarker panel for identification of patients with amyotrophic lateral sclerosis. *Neurology, 72*(1), 14–19.

Moreno, J. A., Streifel, K. M., Sullivan, K. A., Hanneman, W. H., & Tjalkens, R. B. (2011). Manganese-induced NF-KB activation and nitrosative stress is decreased by estrogen in juvenile mice. *Toxicological Sciences, 122*(1), 121–133.

Mousavi, S. A., Habibbollahi, H., & Mahmoudian, F. (2006). *Low serum zinc level in dpression..*

Murata, H., Barnhill, L. M., & Bronstein, J. M. (2022). Air pollution and the risk of Parkinson's disease: A review. *Movement Disorders, 37*(5), 894–904.

MurroughJ.W.AbdallahC.G.MathewS.J.2017Targeting glutamate signalling in depression: Progress and prospects. Nature Reviews. Drug Discovery, 16(7), 472–486..

Myers, J. E., Fourie, M., Zogoe, H. B. A., Naik, I., Theodorou, P., Tassel, H., ... Thompson, M. L. (2003). Nervous system effects of occupational manganese exposure on South African manganese mineworkers. *Neurotoxicology, 24*(4–5), 649–656.

Nabi, M., & Tabassum, N. (2022). Role of environmental toxicants on neurodegenerative disorders. *Frontiers in Toxicology, 4*, 837579.

Nagatomo, S., Umehara, F., Hanada, K., Nobuhara, Y., Takenaga, S., Arimura, K., & Osame, M. (1999). Manganese intoxication during total parenteral nutrition: Report of two cases and review of the literature. *Journal of the Neurological Sciences, 162*(1), 102–105.

Needleman, H. L., Schell, A., Bellinger, D., Leviton, A., & Allred, E. N. (1990). The long-term effects of exposure to low doses of lead in childhood: An 11-year follow-up report. *New England Journal of Medicine, 322*(2), 83–88.

Newell, M. E., Adhikari, S., & Halden, R. U. (2022). Systematic and state-of the science review of the role of environmental factors in amyotrophic lateral sclerosis (ALS) or lou Gehrig's disease. *Science of the Total Environment, 817*, 152504.

Newell, M. E., Babbrah, A., Aravindan, A., Rathnam, R., Kiernan, R., Driver, E. M., ... Halden, R. U. (2024). Prevalence rates of neurodegenerative diseases versus human exposures to heavy metals across the United States. *Science of the Total Environment, 928*, 172260.

Ni, M., Li, X., Yin, Z., Sidoryk-Węgrzynowicz, M., Jiang, H., Farina, M., ... Aschner, M. (2011). Comparative study on the response of rat primary astrocytes and microglia to methylmercury toxicity. *Glia, 59*(5), 810–820.

Nicoletti, A., Messina, S., Bruno, E., Mostile, G., Quattrocchi, G., Raciti, L., Dibilio, V., et al. (2016). Risk factors in multiple sclerosis: A population-based case–control study in Sicily. Background and Methods. *Neurological Sciences, 37*, 1931–1937.

Nielsen, S., Searles, M. N., Warden, M., Sallmén, M., Sainio, S., Uuksulainen, H., ... Brad, A. R. (2021). Solvent exposed occupations and risk of Parkinson disease in Finland. *Clinical Parkinsonism & Related Disorders, 4*, 100092.

O'Neal, S. L., & Zheng, W. (2015). Manganese toxicity upon overexposure: A decade in review. *Current Environmental Health Reports, 2*, 315–328.

Paithankar, J. G., Saini, S., Dwivedi, S., Sharma, A., & Chowdhuri, D. K. (2021). Heavy metal associated health hazards: An interplay of oxidative stress and signal transduction. *Chemosphere, 262*, 128350.

Papanikolaou, N. C., Eleftheria, G. H., Belivanis, S., Tzanakakis, G. N., ... Tsatsakis, A. M. (2005). Lead toxicity update. A brief review. *Medical Science Monitor, 11*(10), RA329–RA336.

Papanna, B., Lazzari, C., & Rabottini, M. (2024). Huntington's disease prevalence in Asia: A systematic review and meta-analysis. *Rivista di Psichiatria, 59*(1), 4–12.

Patel, B. N., Dunn, R. J., Jeong, S. Y., Zhu, Q., Julien, J.-P., & David, S. (2002). Ceruloplasmin regulates iron levels in the CNS and prevents free radical injury. *Journal of Neuroscience, 22*(15), 6578–6586.

Peres, T. V., Schettinger, M. R. C., Chen, P., Carvalho, F., Avila, D. S., Bowman, A. B., & Aschner, M. (2016). Manganese-induced neurotoxicity: A review of its behavioral consequences and neuroprotective strategies. *BMC Pharmacology and Toxicology, 17*, 1–20.

Peters, S., Broberg, K., Gallo, V., Levi, M., Kippler, M., Vineis, P., Veldink, J., et al. (2021). Blood metal levels and amyotrophic lateral sclerosis risk: A prospective cohort. *Annals of Neurology, 89*(1), 125–133.

Piao, Y.-S., Lian, T.-H., Hu, Y., Zuo, L.-J., Guo, P., Yu, S.-Y., Liu, L., et al. (2017). Restless legs syndrome in Parkinson disease: Clinical characteristics, abnormal iron metabolism and altered neurotransmitters. *Scientific Reports, 7*(1), 10547.

Pringsheim, T., Wiltshire, K., Day, L., Dykeman, J., Steeves, T., & Jette, N. (2012). The incidence and prevalence of Huntington's disease: A systematic review and meta-analysis. *Movement Disorders, 27*(9), 1083–1091.

Qiu, S., Palavicini, J. P., Wang, J., Gonzalez, N. S., He, S., Dustin, E., Zou, C., et al. (2021). Adult-onset CNS myelin sulfatide deficiency is sufficient to cause Alzheimer's disease-like neuroinflammation and cognitive impairment. *Molecular Neurodegeneration, 16*, 1–24.

Rahman, A., Brew, B. J., & Guillemin, G. J. (2011). Lead dysregulates serine/threonine protein phosphatases in human neurons. *Neurochemical Research, 36*, 195–204.

Reale, M., Di Bari, M., Di Nicola, M., D'Angelo, C., De Angelis, F., Velluto, L., & Tata, A. M. (2015). Nicotinic receptor activation negatively modulates pro-inflammatory cytokine production in multiple sclerosis patients. *International Immunopharmacology, 29*(1), 152–157.

Rice, K. M., Walker Jr, E. M., Wu, M., Gillette, C., & Blough, E. R. (2014). Environmental mercury and its toxic effects. *Journal of Preventive Medicine and Public Health, 47*(2), 74.

Roth, J., Ponzoni, S., & Aschner, M. (2013). Manganese homeostasis and transport. *Metallomics and the Cell*, 169–201.

Saba, J., Couselo, F. L., Bruno, J., Carniglia, L., Durand, D., Lasaga, M., & Caruso, C. (2022). Neuroinflammation in Huntington's disease: A starring role for astrocyte and microglia. *Current Neuropharmacology, 20*(6), 1116.

Sachdev, P. S., Blacker, D., Blazer, D. G., Ganguli, M., Jeste, D. V., Paulsen, J. S., & Petersen, R. C. (2014). Classifying neurocognitive disorders: The DSM-5 approach. *Nature Reviews Neurology, 10*(11), 634–642.

Salari, S., Khomand, P., Arasteh, M., Yousefzamani, B., & Hassanzadeh, K. (2015). Zinc sulphate: A reasonable choice for depression management in patients with multiple sclerosis: A randomized, double-blind, placebo-controlled clinical trial. *Pharmacological Reports, 67*, 606–609.

Sanders, T., Liu, Y., Buchner, V., & Tchounwou, P. B. (2009). Neurotoxic effects and biomarkers of lead exposure: A review. *Reviews on Environmental Health, 24*(1), 15–46.

Sarihi, S., Niknam, M., Mahjour, S., Hosseini-Bensenjan, M., Moazzen, F., Soltanabadi, S., & Akbari, H. (2021). Toxic heavy metal concentrations in multiple sclerosis patients: A systematic review and meta-analysis. *EXCLI Journal, 20*, 1571.

Satarug, S., Scott, H. G., Mary, A. S., & Donald, A. S. (2011). Cadmium, environmental exposure, and health outcomes. *Ciencia & Saude Coletiva, 16*(5), 2587–2602.

Scheiber, I. F., & Dringen, R. (2013). Astrocyte functions in the copper homeostasis of the brain. *Neurochemistry International, 62*(5), 556–565.

Schmidt, C. W. (2022). *Hemispheres of influence: Bridging the disconnect between environmental neurotoxicology and clinical practice*.

Schneider, R. B., Iourinets, J., & Richard, I. H. (2017). Parkinson's disease psychosis: Presentation, diagnosis and management. *Neurodegenerative Disease Management, 7*(6), 365–376.

Selfridge, J. E., Lezi, E., Lu, J., & Swerdlow, R. H. (2013). Role of mitochondrial homeostasis and dynamics in Alzheimer's disease. *Neurobiology of Disease, 51*, 3–12.

Shaffer, R. M., Sellers, S. P., Baker, M. G., de Buen Kalman, R., Frostad, J., Suter, M. K., Anenberg, S. C., et al. (2019). Improving and expanding estimates of the global burden of disease due to environmental health risk factors. *Environmental Health Perspectives, 127*(10), 105001.

Sharma, R., Zamani, A., Dill, L. K., Sun, M., Chu, E., Robinson, M. J., ... Semple, B. D. (2021). A systemic immune challenge to model hospital-acquired infections independently regulates immune responses after pediatric traumatic brain injury. *Journal of Neuroinflammation, 18*, 1–23.

Shi, L., Huang, C., Luo, Q., Rogers, E., Xia, Y., Liu, W., Ma, W., et al. (2019). The association of iron and the pathologies of Parkinson's diseases in MPTP/MPP+-induced neuronal degeneration in non-human primates and in cell culture. *Frontiers in Aging Neuroscience, 11*, 215.

Silva, F., Vinicius, M., Loures, C.de M. G., Alves, L. C. V., Souza, L. C.de, Borges, K. B. G., & Carvalho, M.das G. (2019). Alzheimer's disease: Risk factors and potentially protective measures. *Journal of Biomedical Science*, 26.

Sosa-Ortiz, A. L., Acosta-Castillo, I., & Prince, M. J. (2012). Epidemiology of dementias and Alzheimer's disease. *Archives of Medical Research, 43*(8), 600–608.

Squitieri, F., Maffi, S., Al Harasi, S., Salmi, Q. A., Alessio, B. D., Capelli, G., & Mazza, T. (2020). Incidence and prevalence of Huntington disease (HD) in the Sultanate of Oman: The first Middle East post-HTT service-based study. *Journal of Neurology, Neurosurgery & Psychiatry, 91*(12), 1359–1360.

Stanisławska-Sachadyn, A., Krzemiński, M., Zielonka, D., Krygier, M., Ziętkiewicz, E., Sławek, J., ... REGISTRY Investigators of the European Huntington's Disease Network (EHDN (2024). Sex Contribution to average age at onset of Huntington's disease depends on the number of (CAG) n repeats. *Scientific Reports, 14*(1), 15729.

Straif, K., Benbrahim-Tallaa, L., Baan, R., Grosse, Y., Secretan, B., El Ghissassi, F., Bouvard, V., et al. (2009). A review of human carcinogens—Part C: Metals, arsenic, dusts, and fibres. *The Lancet Oncology, 10*(5), 453–454.

Sutedja, N. A., Veldink, J. H., Fischer, K., Kromhout, H., Heederik, D., Huisman, M. H. B., ... van den Berg, L. H. (2009). Exposure to chemicals and metals and risk of amyotrophic lateral sclerosis: A systematic review. *Amyotrophic Lateral Sclerosis, 10*(5–6), 302–309.

Swerdlow, R. H. (2007). Is aging part of Alzheimer's disease, or is Alzheimer's disease part of aging? *Neurobiology of Aging, 28*(10), 1465–1480.

Takeda, A. (2000). Movement of zinc and its functional significance in the brain. *Brain Research Reviews, 34*(3), 137–148.

Takeda, A., Takefuta, S., Ijiro, H., Okada, S., & Oku, N. (1999). 109Cd transport in rat brain. *Brain Research Bulletin, 49*(6), 453–457.

Tandon, S., Aggarwal, P., & Sarkar, S. (2024). Polyglutamine disorders: Pathogenesis and potential drug interventions. *Life Sciences*122562.

Tanner, C. M., Goldman, S. M., Ross, G. W., & Grate, S. J. (2014). The disease intersection of susceptibility and exposure: Chemical exposures and neurodegenerative disease risk. *Alzheimer's & Dementia, 10*, S213–S225.

Taylor, A. A., Tsuji, J. S., McArdle, M. E., Adams, W. J., & Goodfellow, W. L., Jr. (2023). Recommended reference values for risk assessment of oral exposure to copper. *Risk Analysis, 43*(2), 211–218.

Thompson, A. G., & Turner, M. R. (2019). Untangling neuroinflammation in amyotrophic lateral sclerosis. *Journal of Neurology, Neurosurgery & Psychiatry, 90*(12), 1303–1304.

Tong, S., von Schirnding, Y. E., & Prapamontol, T. (2000). Environmental lead exposure: A public health problem of global dimensions. *Bulletin of the World Health Organization, 78*(9), 1068–1077.

Tong, Y., Yang, H., Tian, X., Wang, H., Zhou, T., Zhang, S., Yu, J., et al. (2014). High manganese, a risk for Alzheimer's disease: High manganese induces amyloid-β related cognitive impairment. *Journal of Alzheimer's Disease, 42*(3), 865–878.

Tunstall-Pedoe, H., Kuulasmaa, K., Mähönen, M., Tolonen, H., & Ruokokoski, E. (1999). Contribution of trends in survival and coronar Y-event rates to changes in coronary heart disease mortality: 10-Year results from 37 WHO MONICA project populations. *The Lancet, 353*(9164), 1547–1557.

Vacchi-Suzzi, C., Kruse, D., Harrington, J., Levine, K., & Meliker, J. R. (2016). Is urinary cadmium a biomarker of long-term exposure in humans? A review. *Current Environmental Health Reports, 3*, 450–458.

Vos, S. J. B., Xiong, C., Visser, P. J., Jasielec, M. S., Hassenstab, J., Grant, E. A., Cairns, N. J., et al. (2013). Preclinical Alzheimer's disease and its outcome: A longitudinal cohort study. *The Lancet Neurology, 12*(10), 957–965.

Wang, H., Wang, M., Wang, B., Li, M., Chen, H., Yu, X., ... Chai, Z. (2012). The distribution profile and oxidation states of biometals in APP transgenic mouse brain: Dyshomeostasis with age and as a function of the development of Alzheimer's disease. *Metallomics, 4*(3), 289–296.

Wang, T., Tomas, D., Perera, N. D., Cuic, B., Luikinga, S., Viden, A., Barton, S. K., et al. (2022). Ferroptosis mediates selective motor neuron death in amyotrophic lateral sclerosis. *Cell Death & Differentiation, 29*(6), 1187–1198.

Ward, R. J., Zucca, R. F. A., Duyn, J. H., Crichton, R. R., & Zecca, L. (2014). The role of iron in brain ageing and neurodegenerative disorders. *The Lancet Neurology, 13*(10), 1045–1060.

Watanabe, C., Yoshida, K., Kasanuma, Y., Kun, Y., & Satoh, H. (1999). In utero methylmercury exposure differentially affects the activities of selenoenzymes in the fetal mouse brain. *Environmental Research, 80*(3), 208–214.

Wei, Y.-H., & Lee, H.-C. (2002). Oxidative stress, mitochondrial DNA mutation, and impairment of antioxidant enzymes in aging. *Experimental Biology and Medicine, 227*(9), 671–682.

Weisskopf, M. G., Weuve, J., Nie, H., Saint-Hilaire, M.-H., Sudarsky, L., Simon, D. K., Hersh, B., et al. (2010). Association of cumulative lead exposure with Parkinson's disease. *Environmental Health Perspectives, 118*(11), 1609–1613.

Wingerchuk, D. M. (2010). *Supplementing our understanding of vitamin D and multiple sclerosis.*.

Wong, C., Roberts, S. M., & Saab, I. N. (2022). Review of regulatory reference values and background levels for heavy metals in the human diet. *Regulatory Toxicology and Pharmacology, 130*, 105122.

Wright, R. O., & Baccarelli, A. (2007). Metals and neurotoxicology. *The Journal of Nutrition, 137*(12), 2809–2813.

Xuan, M., Guan, X., Gu, Q., Shen, Z., Yu, X., Qiu, T., Luo, X., et al. (2017). Different iron deposition patterns in early- and middle-late-onset Parkinson's disease. *Parkinsonism & Related Disorders, 44*, 23–27.

Yasui, M., Yase, Y., Ando, K., Adachi, K., Mukoyama, M., & Ohsugi, K. (1990). Magnesium concentration in brains from multiple sclerosis patients. *Acta Neurologica Scandinavica, 81*(3), 197–200.

Zapadniuk, B. V. (1992). The incidence of multiple sclerosis and the content of cobalt, boron, zinc, manganese and molybdenum in the arable soils of different climatic zones of ukraine. *Likars' ka Sprava, 1*, 111–113.

Zhang, Z., Wang, M., & Liu, X. (2022). C-reactive protein and risk of Alzheimer's disease. *Neurobiology of Aging, 109*, 259–263.

Zhao, Y., & Zhao, B. (2013). Oxidative stress and the pathogenesis of Alzheimer' s disease. *Oxidative Medicine and Cellular Longevity 2013,* (1), 316523.

Zuccato, C., Valenza, M., & Cattaneo, E. (2010). Molecular mechanisms and potential therapeutical targets in Huntington's Disease. *Physiological Reviews, 90*(3), 905–981.

Chapter 29

Regulatory standards and guidelines for heavy metal exposure

B. Gopika, and Prasann Kumar
Department of Agronomy, School of Agriculture, Lovely Professional University, Phagwara, Punjab, India

29.1 Introduction

To a greater extent, environmental pollution is caused by industrialization and mining activities, which result in ecological health deterioration. Heavy metals (HM) are the most important contributors to environmental pollution compared to all kinds of pollution (Nedelkoska & Doran, 2000). The United States Environmental Protection Agency (UEPA) listed HMs as a priority pollutant. Over 70,000 chemicals are currently used worldwide (Cairns, Smith, & Orvos, 1988). Among them, the most hazardous are lead, mercury, arsenic, and cadmium, which are on the list of the US Agency for Toxic Substances and Disease Registry (ATSDR) and are primarily found in toxic wastes. Potentially poisonous elements are commonly categorised as "trace metals" or "heavy metals." Trace metals occur naturally in soils at very low concentrations, considered 1 mol/m³. HMs are a poorly defined group of elements showing metallic properties with an atomic mass more significant than 20 (except alkali metals) and a specific gravity greater than 5 (Rascio & Navari-Izzo, 2011). According to Park et al. (2011), HMs are defined as elements with an atomic density greater than 6 cm^{-3} (except for A, B, and Se).

HMs are categorised into essential (Mo, Mn Cu, Ni, Fe, Zn) and nonessential (Cd, Ni, As, Hg, Pb). Essential elements, like Cu, are vital for metabolic processes in the human body, which are necessary for hemoglobin production and the metabolism of carbohydrates (Vasudevaraju, Govindaraju, Palanisamy, Sambamurti, & Rao, 2008). Similarly, essential HMs play a significant role in plant growth as they serve as enzyme cofactors, activate biochemical reactions, and enhance ductility, conductivity, and cation stability (Stohs & Bagchi, 1995). However, if these elements are found beyond their optimal levels, they can lead to toxicity in plants and humans.

Unlike essential metals, nonessential HMs are toxic even at low concentrations. They are not metabolically transformed or easily degraded in the environment, resulting in long-term persistence. Various human activities such as industrial, domestic, agricultural, medicinal, technological applications, or natural events (volcanic eruptions and rock weathering) are responsible for releasing HMs into the environment. Because of their nonbiodegradable and hazardous nature, these metals accumulate in the atmosphere for longer, showing significant threats to plants, animals, and humans.

Various anthropogenic activities, such as urbanization, smelting, sludge disposal, military operations, mining, waste dumping, and excessive pesticide and insecticide use, affect soil quality and plant effectiveness (Chowdhary et al., 2018; Rehman et al., 2021; Saleem, Ali et al., 2020; Saleem, Fahad, et al., 2020; Shahid et al., 2015). Due to HM contamination, plants inhibit growth, induce toxic symptoms, and pose serious risks to human health by entering the food chain (Afzal et al., 2021; Saleem, Ali et al., 2020; Saleem, Fahad, et al., 2020).

HMs disrupt humans' metabolic functions, and their toxicity damages vital organs and glands such as the heart, brain, kidneys, bones, and liver. These metals replace crucial essential nutrients and disrupt their biological functions. They are introduced into the body by consuming foods and beverages, skin exposure, and inhaling air. Chronic exposure to different HMs, even at low concentrations, leads to an increase in the risk of cancer.

Park et al. (2004) reported an increased lifetime risk of lung cancer due to occupational exposure to dust and mist, which contains hexavalent chromium. Hiroaki et al. (2014) found a potential increase in postmenopausal breast cancer resulting from the consumption of cadmium-contaminated rice and other foods. Additionally, acute and chronic exposure to arsenic can also cause various health issues,

Heavy Metal Toxicity and Neurodegeneration. DOI: https://doi.org/10.1016/B978-0-443-36575-1.00003-0

including dermal, respiratory, cardiovascular, gastrointestinal, hematological, hepatic, renal, neurological, developmental, reproductive, and immunological disorders, as well as genotoxic, mutagenic, and carcinogenic effects, such as liver cancer (Lin, Sung, Chen, & Guo, 2013).

The toxicity of mercury leads to Minamata disease, while cadmium poisoning causes Itai–Itai disease. Among other health issues, HM toxicity can also induce organ-specific toxicity, including nephrotoxicity, neurotoxicity, hepatotoxicity, skin toxicity, and cardiovascular toxicity.

The US Food and Drug Administration (FDA) declared dietary exposure to HMs a public concern. The FDA targets minimizing the contamination of food, nutritional supplements, and cosmetics with HMs.

Regulations such as the Federal Food, Drug, and Cosmetic Act and the Clean Water Act (CWA) set quality standards for Food and Water Consumption (Schultz & Schultz, 1981). Additionally, in response to concerns about HM exposure in infants and young children, the FDA launched the "Closer to Zero" action plan to minimize their risk.

This chapter reviews policies in which regulatory bodies collaborate with scientific communities to implement a legal framework for efficiently removing and monitoring HMs to mitigate health issues and support global environmental sustainability.

29.2 Heavy metal of concern

More than 20 million hectares of land are contaminated with HMs such as zinc (Zn), lead (Pb), nickel (Ni), arsenic (As), mercury (Hg), copper (Cu), cadmium (Cd), and chromium (Cr). According to the ATSDR, mercury, lead, cadmium, and arsenic are highly hazardous to plants and humans.

The World Health Organization (WHO) has also raised concerns about ecosystem degradation in many countries due to HM contamination in drinking water, particularly arsenic, lead, cadmium, and chromium.

29.2.1 Mercury

Mercury (Hg) is a highly hazardous HM that severe risks to marine life and human health. Mercury naturally occurs in its metallic form, a shiny silver-white, odorless liquid that turns into colorless, odorless gas when heated. Mercury is very toxic and highly bioaccumulative.

29.2.2 Occurrence and source

Mercury is a highly hazardous HM present in the ecosystem. Sources of mercury include geothermal activity, gold mining, coal combustion, and anthropogenic emissions. It may also be found in medicinal waste, plastics, and scientific instruments such as thermometers, thermostats, and dental amalgams (Cifuentes, Lindemann, & Barton, 1996).

Mercury is a unique metal in various environmental forms, such as HgS, Hg^{2+}, and methyl-Hg. Among these, Hg^{2} is the predominant form in agricultural soil. In its ionic state, Hg^{2+} is primarily absorbed by soil and accumulated in aquatic plants (Park & Zheng, 2012).

29.2.3 Impact on health and environment

Mercury transforms into highly toxic methylmercury when it reacts with aquatic sediments (Gworek, Dmuchowski, & Baczewska-Dąbrowska, 2020). Methylmercury penetrates the human body through the food chain, such as contaminated fish, seafood, and wildlife that ingest toxic microorganisms. Once absorbed, it spreads throughout the body and results in various neurological disorders (Rice, Walker, Wu, Gillette, & Blough, 2014).

Long-term exposure to elemental mercury (Hg) leads to blurred vision, unsteady walking, and poor concentration. High-level exposure to Hg can cause severe neurological damage and even lead to death. Low-level exposure causes depression, skin rashes, memory loss in adults, and tremors (shaking), while in the case of children, it results in peeling of the hands and feet along with redness (Rice et al., 2014). Organic Hg poisoning enters humans' bodies through the consumption of food and has led to Minamata disease in Japan and Iraq. Although dimethyl mercury is more toxic than methyl mercury (just a few mL spilt on the skin causes death) (Chen et al., 2006).

29.2.4 Lead

Lead is a highly toxic metal, and its widespread use is due to significant environmental contamination and health issues globally. Lead appears to be a bright silvery metal with a slight bluish color in a dry atmosphere. When lead is exposed to air, it begins to tarnish, forming a complex mixture of compounds depending on environmental conditions (Jaishankar, Tseten, Anbalagan, Mathew, & Beeregowda, 2014).

29.2.5 Occurrence and sources

Lead is primarily released into the environment through mining, ore processing, and the recycling of lead-acid batteries in battery manufacturing industries. It also presents lead-based paint, plumbing pipes, automobile exhaust, and industrial waste. Natural sources, including sea salt spray and volcanic eruptions, also enhance the environmental lead levels (Bharti & Sharma, 2022).

29.2.6 Impact on health and environment

Lead is dangerous to marine species and fish, first in the aquatic food chain, because they accumulate a high lead level in their gills and livers (Nair, Jayalakshmy, Balachandran, & Joseph,2006). Humans consume fish as part of their everyday diet. Lead enters the human body after absorption, is transported into circulation, and accumulates in body tissues (Castro-González & Méndez-Armenta, 2008).

High exposure to lead (Pb) in humans causes hypertension and cardiovascular diseases, in addition to peripheral vascular, cerebrovascular, and ischemic heart diseases. Lead also affects the reproductive system in both men and women. In males, it reduces libido, causes chromosomal damage to sperm, alters testosterone levels, disrupts spermatogenesis, and leads to infertility. In females, it leads to miscarriage, premature membrane rupture, preterm delivery, and infertility.

29.2.7 Arsenic

Arsenic is a member of the Nitrogen family, which exhibits both metallic and nonmetallic properties. Arsenic is most commonly found in soil, water, and living organisms. This belongs to a group of substances known as trace elements (Smedley & Kinniburgh, 2002). Periodically, it is known as "the king of poisons" and "the poison of kings" (Vahidnia, Van der Voet, & De Wolff, 2007). Arsenic reacts with sulfur, oxygen, chloride, and other substances in the air to form inorganic arsenic compounds (Odstrcil, Carino, Ricci, & Mandalunis,2010).

Arsenic possesses a semimetal characteristic that is highly toxic and carcinogenic and mainly present in the form of oxides, sulfides, or salts with elements of iron, sodium, calcium, and copper (Singh, Kumar, & Sahu, 2007). Arsenic is the 20th most abundant element found on Earth, and its inorganic forms, including arsenate and arsenate compounds, threaten the environment and all living beings.

29.2.8 Occurrence and source

Being a naturally occurring element with metalloid properties, arsenic (As) is characterized as widely spread and predominantly cycled through water in the environment. Arsenic is often detected in various substances due to organic and inorganic arsenic compounds in the atmosphere. Generally, arsenic is found in natural water sources, basin-fill deposits, volcanic materials, geothermal emissions, landfills, and mining sites. It is also emitted from sources like paints and wood preservatives. Trivalent arsenic (As(III)) and pentavalent arsenic (As(V)) are the most frequently occurring oxidation states of arsenic (Rathi et al., 2024).

29.2.8.1 Impact on health and environment

Inorganic arsenic is characterized as a carcinogen. Low-to-moderate level exposure of (As) results in diabetes, hepatic and renal dysfunction, as well as neurological complications. The skin is susceptible to arsenic exposure, which results in dermatitis. Mainly, women are more vulnerable to arsenic-induced skin disorders compared to men (Vahidnia et al., 2007). Nearly 50% of arsenic is excreted through urine, while the remaining portion is released through face, skin, hair, nails, and lungs (Das, 1995). Arsenic concentration can be detected in the human body's urine, face, skin, hair, nails, and lungs.

Drinking water that contains 50 µg/L of arsenic increases the risk of lung cancer, bladder cancer, and arsenic-related skin lesions. However, it is also implicated in several ailments, including conjunctivitis, hyperpigmentation, hyperkeratosis, peripheral vascular and nervous system disorders, skin cancer, cardiovascular diseases, gangrene, leucomelanosis, nonpitting swellings, splenomegaly, and hepatomegaly. According to the International Agency for Research on Cancer (IARC), when arsenic concentration exceeds the threshold for chronic poisoning in humans, it is classified as a Group 1 carcinogen (Rathi et al., 2024).

29.2.9 Cadmium

According to the ATSDR ranking, cadmium is the seventh most important toxic element among HMs. It is a by-product of zinc production to which humans and animals may be exposed. Cadmium (Cd), a soft, white, silvery metal with an atomic weight of 112.4 and atomic number 48, belongs to Group XII of the periodic table—cadmium shares similar physical and chemical properties with zinc and mercury (Jaishankar et al., 2014).

29.2.9.1 Occurrence and source

Cadmium (Cd) is a naturally occurring element found in small amounts in rocks, soils, and the Earth's crust (Fleischer et al., 1974; Nagajyoti, Lee, & Sreekanth, 2010). Moreover, human activities such as smelting, mining, waste incineration, burning fossil fuels, and producing Cd-Ni batteries (Genchi, Carocci, Lauria, Sinicropi, & Catalano, 2020) are the chief contributors to Cd accumulation in the soil ecosystem. During Cd-Ni battery manufacturing, Cadmium is also exposed to the atmosphere, as well as through forest fires, wind-blown dust, and sea spray (Uraguchi & Fujiwara, 2012; Hogervorst et al., 2007). Due to HM contamination of rivers and drinking water with Cd is a significant concern, especially in areas near industrial facilities or mining sites (Järup & Åkesson, 2009).

29.2.9.2 Impact on health and environment

According to the IARC and the United States National Toxicology Program, cadmium (Cd) is classified as a toxic human carcinogen. Cd contamination with cereals, rice, and shellfish has been identified, with individuals typically ingesting between 8 and 25 µg of Cd per day through food consumption. Cd exposure has been related to acute kidney injury linked to diabetes, as well as elevated blood pressure, nerve damage, and osteoporosis. In severe cases, it has also been associated with cancer, Alzheimer's disease, and Parkinson's disease due to its carcinogenic and mutagenic properties (Rathi et al., 2024).

29.2.10 Chromium

With an average concentration of approximately 40 mg/kg in US soils, chromium (Cr) is the 21st most abundant element in the Earth's crust. It is commonly found in ferrous and other metal oxides, frequently occurring along with manganese and vanadium. Its applications are primarily attributed to its properties as a coloring pigment. Crocoite, a rare and stable chromium mineral, is one of its naturally occurring forms. Chromium is beneficial due to its properties as a coloring pigment. Crocoite, a rare and stable chromium mineral, is the most naturally occurring form in the environment (Rathi et al., 2024).

29.2.11 Occurrence and source

Chromium enters the atmosphere in oxidation states ranging from Cr^{2+} to Cr^{6+} (Rodríguez et al., 2007). Trivalent chromium (Cr^{3+}) and hexavalent chromium (Cr^{6+}) are the most common forms of Cr, being toxic to animals, humans, and plants (Monalisa Mohanty, Patra, 2013).

Chromium typically occurs in various valence states in the environment, including Cr(III) and Cr(VI) (Escudero-Castejon, Sanchez-Segado, Parirenyatwa, & Jha, 2016; Ma, Xu, Chen, Li, & Feng, 2019; Ripley, Li, 2018; Tumolo et al., 2020), which are most significant because of their unique characteristics. Chromium is released into the environment through air, water, soil, natural processes, and human activities, such as industrial operations.

Chromium naturally occurs in the environment from oil and coal combustion, petroleum, ferrochromate refractory materials, pigment oxidants, catalysts, chromium steel, fertilizers, oil well drilling, and metal plating tanneries. Chromium is released into the environment through anthropogenic activities, including sewage and fertilizer applications (Ghani & Ghani, 2011).

29.2.12 Impact on health and environment

Chromium (Cr) is a significant pollutant in groundwater and is identified as one of the most toxic trace metals (Shanker, 2019; Wu et al., 2017; Yan & Chen, 2019). Due to industrial activities and urban sources, Cr(VI) accumulation in sediment and water is closely linked (Gorny et al., 2016). Cr(VI) can enter the human body through inhalation, food, or water consumption. Once ingested, it can severely affect the liver, kidneys, gut, immune system, and blood. Moreover, Cr(VI) toxicity is mitigated by specific biological mechanisms that involve reduction processes in body fluids (De Flora, Serra, Camoirano, & Zanacchi, 1989).

Chronic exposure to chromium (Cr), even at low concentrations, such as in occupational settings, can affect the skin, blood, eyes, respiratory system, and immune system (Zhang et al., 2014). Studies revealed that Cr levels often exceed the threshold linked with human-induced pollution, which ranges from 70 to 90 μg/L. Skin exposure to hexavalent Cr results in dermatitis, ulceration, and increased sensitivity (Rathi et al., 2024).

29.2.13 Nickel

Nickel is considered a minor essential metal for plant growth and development, but it becomes toxic when exposed to higher concentrations and is classified as an HM.

29.2.14 Occurrence and sources

Nickel enters the atmosphere from natural sources such as weathering of rocks, wildfires, and wind-blown. Several anthropogenic activities, such as Industrial dust, electroplating, production of iron and steel, food processing industries, chemical industries, incineration of waste, fertilizers, industrial aerosols, mining and metallurgy, battery, and combustion of coal, are also sources of nickel contamination. Ni is also released from tobacco and stainless-steel kitchen utensils (Bader, Gonzalez, Goodell, Ali, & Pillai, 1999).

29.2.15 Impact on health and environment

The presence of Ni in the ecosystem affects plant growth and development. Plants show toxicity when exposed to more than 100 mg dm^3 of Ni (Bharti & Sharma, 2022). Ni is primarily known as a carcinogen, mainly in the lungs, leading to chronic bronchitis and reduced lung function. The kidney is the most affected organ by nickel exposure. Acute nickel toxicity can result in frank hematuria and kidney injury (Schaumlöffel, 2012). It also affects inflammation in bronchitis, luminal congestion, alveolar cell hyperplasia, alveolar congestion, carcinogenesis, chromosomal damage, and mutagenesis. Additionally, Ni^{2+} inhibits natural killer cell activity, inducing tumor development.

29.3 Global regulatory framework

29.3.1 World Health Organization

The WHO helps monitor, assess, and regulate HM exposure to protect the ecosystem. It sets limits on permissible exposure levels of toxic HMs such as lead (Pb), mercury (Hg), cadmium (Cd), and arsenic (As) through air, water, or food and occupational exposure. International organizations, such as FAO (Food and Agriculture Organization) and UNEP (United Nations Environment Programme), collaborate with

TABLE 29.1 Standard threshold level for toxic heavy metals and their effect on human health, as recommended by World Health Organization (WHO) (Shrestha et al., 2021).

S. no.	Heavy metal	Recommended safe limits by WHO in drinking water	Recommended safe limits by WHO in wastewater	Toxic effects
1.	Mercury	< 0.006 mg/L	0.05 mg/L	"Oral ingestion creates problems in the gastrointestinal tract; kidney damage, increases the incidence of some benign tumors."
2.	Nickel	0.02–0.07 mg/L	0.02 mg/L	"Irritability, nausea, vomiting, difficulty sleeping."
3.	Arsenic	< 0.01 mg/L	N/A	"Chronic arsenic, including dermal lesions such as hyper-and hypopigmentation, peripheral neuropathy, skin cancer, bladder and lung cancers, and peripheral vascular disease."
4.	Cadmium	0.003–0.005 mg/ L	0.003 mg/L	"Carcinogenic by the inhalation route, kidney problems due to accumulation in the kidneys."
5.	Chromium	< 0.05 mg/L	0.05 mg/L	"Chromium (VI) is carcinogenic to humans, causing lung cancer."
6.	Lead	< 0.01 mg/L	0.01 mg/L	"Accumulates in the skeleton and creates bone problems, adverse effects on central and peripheral nervous systems."
7.	Selenium	< 0.01 mg/L	N/A	"Long-term exposure is manifested in nails, hair, and liver, affecting the synthesis of a liver protein."
8.	Copper	< 2 mg/L	1 mg/L	"Gastrointestinal effects, carriers of the gene for Wilson disease, a metabolic disorder of copper homeostasis"
9.	Zinc	< 3 mg/L	2–5 mg/L	"Skin irritations, anemia, nausea and vomiting."
10.	Manganese	< 0.12 mg/L	< 0.2 mg/L	"Neurological effects following inhalation exposure, psychological symptoms (irritability, emotional lability)."

WHO to develop safety standards, risk assessments, and policy recommendations to prevent HM contamination. WHO emphasizes through its reports and guidelines the adverse health effects of exposure to HMs, including neurotoxicity, organ damage, carcinogenicity, and developmental disorders, particularly in vulnerable populations (Table 29.1).

29.3.2 Food and Agriculture Organization/Codex Alimentarius

Based on the main principles recommended by the Codex Alimentarius, this standard is a reliable guide to dealing with contaminants and toxins in food and feed. It provides the maximum levels and associated sampling plans of contaminants and natural toxicants in food and feed that the Codex Alimentarius Commission recommends be applied to commodities moving in international trade.

This standard, which includes only maximum levels of contaminants and natural toxicants in feed where the contaminant can be transferred to food of animal origin, is designed to be highly protective of public health.

29.3.3 Arsenic

29.3.3.1 Toxicological guidance value

The 72nd meeting of the Joint FAO/WHO Expert Committee on Food Additives (JECFA) (2010) analyzed the lower limit on the benchmark dose for a 0.5% increase in the incidence of lung cancer (BMDL 0.5) linked with exposure to inorganic arsenic was evaluated from epidemiological studies to be 3.0 μg/kg bw/day (2–7 μg/kg bw/day based on the range of estimated total dietary exposure) using a various of assumptions to estimate overall dietary exposure to inorganic arsenic from both drinking water and food. The Joint FAO/WHO Expert Committee on Food Additives (JECFA) concluded that the provisional tolerable weekly intake (PTWI) of 15 μg/kg bw (corresponding to 2.1 μg/kg bw/day) set in 1994 was too close to the BMDL 0.5 and was no longer considered safe. As a result, the former PTWI was withdrawn (Table 29.2).

29.3.4 Cadmium

29.3.4.1 Toxicological guidance value

With the view of cadmium's long half-life, daily consumption of cadmium through food has minimal or negligible impact on overall exposure. Dietary intake should be monitored over several months to evaluate the health risks associated with exposure to cadmium, and the tolerable intake should be considered for at least 1 month. By this perspective, JECFA, at its 73rd meeting in 2010, agreed to express tolerable intake as a monthly value in the form of the provisional tolerable monthly intake and set it at 25 μg/kg body weight (Table 29.3).

29.3.5 Lead

29.3.5.1 Toxicological guidance value

At the 73rd meeting in 2010, JECFA determined previously established PTWI of 25 μg/kg bw based on dose–response analyses. The evaluation showed that this level of exposure could lead to a reduction of at least 3 IQ points in children and an increase in systolic blood pressure of approximately 3 mmHg (0.4 kPa) in adults. While such effects may appear minor individually, they are significant when considering population-wide shifts in IQ and blood pressure. Consequently, JECFA considered that the PTWI was no longer sufficiently protective of health and withdrew it (Table 29.4).

29.3.6 Mercury

29.3.6.1 Toxicological guidance value

At the 72nd meeting in 2010, JECFA implemented a PTWI of 4 μg/kg bw for inorganic mercury, replacing the previous PTWI of 5 μg/kg bw for total mercury, which

TABLE 29.2 Threshold limits and guidance levels for Arsenic in food products.

S. no.	Commodity/product name	ML (mg/kg)	Portion of the commodity/product to which the ML applies
1.	Natural mineral waters	0.01	–
2.	Rice, husked	0.35	Whole commodity
3.	Fat spreads and blended spreads	0.1	–
4.	Edible fats and oils	0.1	Whole commodity
5.	Rice, polished	0.2	Whole commodity
6.	Salt, food grade	0.5	–

ML, Maximum level.

TABLE 29.3 Threshold limits and guidance levels for cadmium in food products.

S. no.	Commodity/product name	ML (mg/kg)	Portion of the commodity/product to which the ML applies
1.	Fruiting vegetables	0.05	Whole commodity after removal of stems. Sweet corn and fresh corn: kernels plus cob without husk.
2.	Brassica vegetables	0.05	Head cabbages and kohlrabi: whole commodity as marketed, after removal of obviously decomposed or withered leaves. Cauliflower and broccoli: flower heads (immature inflorescence only). Brussels sprouts: "buttons" only.
3.	Leafy vegetables	0.2	Whole commodity as usually marketed, after removal of obviously decomposed or withered leaves.
4.	Bulb vegetables	0.05	Bulb/dry onions and garlic: whole commodity after removal of roots and adhering soil and whatever parchment skin is easily detached
5.	Legume vegetables	0.1	Whole commodity as consumed. The succulent forms may be consumed as whole pods or shelled products.
6.	Stalk and stem vegetables	0.1	Whole commodity as marketed after removal of obviously decomposed or withered leaves. Rhubarb: leaf stems only. Globe artichoke: flower head only. Celery and asparagus: Remove adhering soil.
7.	Root and tuber vegetables	0.1	After removing the tops, wash the whole commodity and remove adhering soil (e.g., by rinsing in running water or gently brushing off the dry commodity). Potato: peeled potato.
8.	Cereal grains	0.1	Whole commodity
9.	Pulses	0.1	Whole commodity
10.	Rice, polished	0.4	Whole commodity
11.	Wheat	0.2	Whole commodity
12.	Natural mineral waters	0.003	
13.	Salt, food grade	0.5	
14.	Marine bivalve molluscs	2	The whole commodity after the removal of the shell
15.	Chocolates containing or declaring < 30% total cocoa solids on a dry matter basis	0.3	Whole commodity as prepared for wholesale or retail distribution
16.	Chocolate containing or declaring ≥30% to < 50% total cocoa solids on a dry matter basis	0.7	Whole commodity as prepared for wholesale or retail distribution
17.	Chocolate containing or declaring ≥ 50% to < 70% total cocoa solids on a dry matter basis	0.8	Whole commodity as prepared for wholesale or retail distribution
18.	Chocolate containing or declaring ≥ 70% total cocoa solids on a dry matter basis	0.9	Whole commodity as prepared for wholesale or retail distribution
19.	Cocoa powder (100% total cocoa solids on a dry matter basis) ready for consumption	2.0	Whole commodity as prepared for wholesale or retail distribution

ML, Maximum level.

TABLE 29.4 Threshold limits and guidance levels for Lead in food products.

S. no.	Commodity/product name	ML (mg/kg)	Portion of the commodity/product to which the ML applies
1.	Fruits	0.1	Whole commodity. Berries and other small fruits: the whole commodity after removal of caps and stems. Pome fruits: whole commodity after removal of stems. Stone fruits, dates, and olives: whole commodity after removal of stems and stones, but the level calculated and expressed on the entire commodity without stem. Pineapple: whole commodity after removal of crown. Avocados, mangoes, and similar fruits with hard seeds are whole commodities after stone removal, but they are calculated based on whole fruit.
2.	Berries and other small fruits	0.1	The whole commodity after removal of caps and stems.
3.	Brassica vegetables	0.1	Head cabbages and kohlrabi: whole commodity as marketed, after removal of obviously decomposed or withered leaves. Cauliflower and broccoli: flower heads (immature inflorescence only). Brussels sprouts: "buttons" only.
4.	Bulb vegetables	0.1	Bulb/dry onions and garlic are the whole commodity after the roots, and adhering soil is removed, and whatever parchment skin is easily detached.
5.	Fruiting vegetables	0.05	The whole commodity after the removal of stems Sweet corn and fresh corn: kernels plus cob without husk.
6.	Legume vegetables	0.1	Whole commodity as consumed. The succulent forms may be consumed as whole pods or shelled products.
7.	Leafy vegetables	0.3	Whole commodity as usually marketed, after removal of obviously decomposed or withered leaves
8.	Fresh farmed mushrooms (common mushrooms [*Agaricus bisporus*], shiitake mushrooms [*Lentinula edodes*], and oyster mushrooms [*Pleurotus ostreatus*])	0.3	Whole commodity
9.	Root and tuber vegetables	0.1	After removing the tops, wash the whole commodity and remove adhering soil (e.g., by rinsing in running water or gently brushing off the dry commodity). Potato: peeled potato
10.	Cereal grains	0.2	Whole commodity

11.	Pulses	0.1	Whole commodity
12.	Cereal-based foods for infants and young children	0.02	Whole commodity as sold; not reconstituted or otherwise prepared for consumption
13.	Canned fruits	0.1	The ML applies to the product as consumed.
14.	Jams, jellies, and marmalades	0.4	
15.	Fruit juices	0.03	Whole commodity (not concentrated) or commodity reconstituted to the original juice concentration, ready to drink. The ML also applies to nectars, ready-to-drink
16.	Fish	0.3	Whole commodity (in general, after removing the digestive tract)
17.	Meat of cattle, pigs, and sheep	0.1	Whole commodity (without bones)
18.	Meat and fat of poultry	0.1	Whole commodity (without bones)
19.	Edible fats and oils	0.08	Whole commodity as prepared for wholesale or retail distribution
20.	Milk	0.02	Whole commodity
21.	Natural mineral waters	0.01	
22.	Salt, food grade	1	Whole commodity as prepared for wholesale or retail distribution
23.	White and refined sugar, corn, and maple sirups, honey	0.1	Whole commodity
24.	Wine	0.1	Whole commodity
25.	Fortified/liqueur wine	0.15	Whole commodity

ML, Maximum level.

had been determined at the sixteenth meeting. The newly established PTWI for inorganic mercury applies specifically to dietary exposure from sources other than fish and shellfish. For mercury exposure from fish and shellfish, the previously established PTWI for methylmercury remains applicable (Table 29.5).

29.3.7 Methylmercury

Toxicological guidance value: PTWI 0.0016 mg/kg bw (2003, confirmed in 2006) (Table 29.6).

29.3.8 International legal standards

The regulatory framework for HMs is governed by international legal standards, which promote an essential role in managing HMs by implementing standards for environmental protection, human health, and sustainable resource utilization. Organizations such as the UNEP and the International Council on Mining and Metals have developed various international treaties and agreements to minimize the adverse effects of HM contamination.

29.3.8.1 United Nations Environment Programme and heavy metals

UNEP, founded in 1972, is a leading global environmental protection authority. It helps in minimizing pollution and industrial emissions while promoting sustainable resource use. The main target of UNEP is to minimize environmental and health risks associated with the contamination of air, water, and soil with HMs through various initiatives and global agreements. They formulate policies and capacity-building programs and conduct scientific assessments to safely manage HMs globally.

29.3.8.2 Basel convention and heavy metal waste management

UNEP's most important contribution was the Basel Convention on the Control of Transboundary Movements of Hazardous Wastes and Their Disposal, adopted in 1989. The treaty targets to regulate the management, transport, and disposal of hazardous waste, including mining by-products contaminated with HMs. They focus primarily on metal mining, which draws special attention and, more specifically, on the problem of mining by-products and tailings, which also contain hazardous substances, including toxic chemicals and HMs. They established a legal basis for restricting the international movement of dangerous waste, avoiding illegal dumping and minimizing pollution. The Basel Convention assists in reducing the adverse effects of HM contamination from mining activities by formulating national policies and regulations.

TABLE 29.5 Threshold limits and guidance levels for mercury in food products.

S. no.	Commodity/product name	ML (mg/kg)	Portion of the commodity/product to which the ML applies
1.	Natural mineral waters	0.001	
2.	Salt food grade	0.1	

ML, Maximum level.

TABLE 29.6 Threshold limits and guidance levels for methyl mercury in food products.

S. no.	Commodity/ product name	ML (mg/kg)	Portion of the commodity/product to which the ML applies
1.	Alfonsino	1.5	Whole commodity, fresh or frozen (in general, after removing the digestive tract)
2.	Marlin	1.7	
3.	Orange roughy	0.8	
4.	Pink cusk-eel	1.0	
5.	Shark	1.6	
6.	Tuna	1.2	

ML, Maximum level.

29.3.8.3 Minamata Convention

Another key contribution from UNEP is the Minamata Convention on Mercury, established in 2017. This treaty is significant to Artisanal and Small-scale Gold Mining, in which mercury is widely used in gold extraction, which leads to serious environmental and health risks. The convention aims to reduce mercury emissions and releases, phase out its use in specific products and processes, and promote safer technologies. It established a framework for responsible mercury management, emission reduction, and community protection for mining operations. By setting international standards, the Minamata Convention is vital in evaluating the environmental and health risks of mercury in metal mining (Kulshrestha, Acharya, Mathur, & Pandey, 2024).

29.4 National regulations

29.4.1 United States

29.4.1.1 Safe Drinking Water Act

The Safe Drinking Water Act (SDWA), passed by Congress in 1947, aims to protect public health by regulating the nation's public drinking water supply. This legislation mainly addresses all water sources for human consumption, irrespective of surface or groundwater. The SDWA empowers the US Environmental Protection Agency to set national health-based standards to protect against naturally occurring and artificial contaminants in drinking water.

29.4.1.2 Roles and responsibilities

The US EPA acts as the primary enforcement authority, ensures that tests for water system contamination evaluate plans to upgrade systems, conducts on-site inspections and sanitary surveys, provides training and technical assistance, and takes legal action against drinking water systems that are not meeting standards. To ensure safe drinking water, this law sets various protections against contamination or provides ways to remove contaminated water. Public water systems ensure that contaminant levels are relatively safe within the limits to keep them under regulation.

29.4.1.3 National Primary Drinking Water Regulations

The National Primary Drinking Water Regulations includes legally enforceable primary standards and treatment techniques that public water systems must follow. Primary standards and treatment techniques help protect public health by limiting contaminants in drinking water.

The US Environmental Protection Agency (EPA) has set maximum contaminant levels (MCLs) for various HMs under the SDWA to ensure the safety of drinking water (Table 29.7).

Lead is managed by a Treatment Technique that requires systems to control the corrosiveness of their water. Water systems must take additional steps if more than 10% of tap water samples exceed the action level. For lead, the action level is 0.015 mg/L.

29.4.2 Clean Water Act

The CWA establishes the basic structure for regulating discharges of pollutants into the waters of the United States and regulating quality standards for surface waters. The basis of the CWA was enacted in 1948 and was known as the Federal Water Pollution Control Act, but the Act was significantly reorganized and expanded in 1972.

Under the CWA, the EPA has established pollution control programs, including setting industry wastewater

TABLE 29.7 Specific heavy metals and their respective maximum contaminant levels (MCLs) by the EPA.

S. no.	Contaminant	MCL	Health effects
1.	Arsenic	0.010 mg/L (10 ppb)	Skin damage, circulatory system problems, increased risk of cancer.
2.	Cadmium	0.005 mg/L (5 ppb)	Kidney damage.
3.	Chromium	0.1 mg/L (100 ppb)	Allergic dermatitis, potential carcinogenic effects.
4.	Lead	Treatment technique: action level at 0.015 mg/L (15 ppb)	Delays in physical or mental development in children; kidney problems and high blood pressure in adults.
5.	Mercury	0.002 mg/L (2 ppb)	Kidney damage.
6.	Selenium	0.05 mg/L (50 ppb)	Hair or fingernail loss, numbness in fingers or toes, circulatory problems.

standards. EPA also developed the national water quality criteria recommendations for pollutants in surface waters

29.4.3 Food and Drug Administration

The FDA (FDA or USFDA) is an agency of the United States Department of Health and Human Services. The FDA supervises and guarantees the safety of various food categories, such as foods, tobacco products, dietary supplements, prescription and over-the-counter medications, vaccines, biopharmaceuticals, blood transfusions, medical devices, electromagnetic radiation emitting devices, veterinary products, and cosmetics. The FDA also notifies other laws, especially Section 361 of the Public Health Service Act and the corresponding regulations.

The US FDA monitors and regulates HM contamination, such as arsenic, lead, cadmium, and mercury, in foods and dietary supplements to ensure public safety. These toxic elements can enter the food chain through environmental contamination, including soil, water, air, and industrial processes. The FDA's works are mainly focused on protecting vulnerable populations, such as infants and young children, who are more susceptible to these contaminants' potential adverse health effects.

The FDA monitors levels of contaminants in the food supply, sets limits, and provides guidance to food manufacturers on meeting their legal requirements to implement preventive controls to minimize or prevent chemical hazards in foods. The FDA collaborates with other US and international regulatory agencies on contaminant monitoring and regulatory activities.

The FDA informs policy decisions. This includes setting contaminant limits, providing industry guidance on best practices, and educating consumers on minimizing exposure. For instance, as part of its Closer to Zero initiative, the FDA has set action levels for lead in foods for babies and young children (Table 29.8).

29.4.4 India

29.4.4.1 Bureau of Indian Standards

The BIS is India's national standard body that ensures the defined quality and safety standards for food, water, soil, and industrial products. It sets maximum permissible limits for HMs to protect public health and ensure regulatory compliance. These standards are based on international standards like WHO, FAO, and CODEX Alimentarius but are specifically designed for Indian environmental conditions and practices (Table 29.9).

29.4.5 Food Safety and Standards Authority of India

The Food Safety and Standards Authority of India (FSSAI) is the regulatory body that defines food safety and quality standards in India. Established by the Food Safety and Standards Act of 2006 to regulate and monitor food safety, FSSAI implements the maximum limits for contaminants like HMs in food products to ensure public health.

FSSAI's guidelines are based on global standards like WHO, FAO, and CODEX Alimentarius and have also been adapted to India's unique dietary plans, environmental conditions, and food production systems. The HM limits are enforced through the Food Safety and Standards (Contaminants, Toxins, and Residues) Regulations, 2011, which specify maximum allowable concentrations to reduce the risk of toxic exposure through food consumption (Table 29.10).

29.4.6 Provisions Under Food Safety and Standards Act, 2006

FSSAI set maximum limits for metal contaminants under the Food Safety and Standards (Contaminants, Toxins and Residues) Regulations, 2011 (FSSR).

29.4.7 EU REACH

REACH (Registration, Evaluation, Authorization and Restriction of Chemicals) is a broad (European Union) regulation established in 2007. The main goal is to protect human health and the environment from the threat that chemicals can pose. REACH also aims to develop innovation and competitiveness in the EU chemicals industry. REACH demands that manufacturers and importers register substances they manufacture or import at one tonne or more quantities per year.

Certain hazardous substances, like HMs, can also be authorized or restricted under REACH, minimizing their product use. The EU has established standards for drinking water quality to protect public health, especially from HM contaminants. These limits are implemented under the Drinking Water Directive, revised in December 2020 (PDF) to implement new scientific knowledge and focus on emerging pollutants (Table 29.11).

29.4.8 Heavy metal legislation

The effects of HMs have become known, especially on human health, and their use has become more restricted. For example, mercury and lead were banned except for limited uses, and their emissions are continuously monitored. Government authorities have established limits for HM concentration in various media to reduce ecosystems' exposure to these toxic elements, especially in water. According to WHO, Safe Drinking Water is referred to as water that does not pose significant health risks over a lifetime of consumption (Vareda, Valente, & Durães, 2019).

JECFA (2017) recommends PTWI for these substances. Most of those values were eliminated because it was

TABLE 29.8 The action level of heavy metals in foodstuffs by the Food and Drug Administration.

S. no	Heavy metal	Food product	FDA limit	References
1.	Arsenic	Bottled water	10 parts per billion (ppb)	https://www.fda.gov/food/environmenta
		Apple juice	10 ppb	https://www.fda.gov/regulatory
		Infant rice cereals	100 ppb	https://www.fda.gov/regulatory
2.	Lead	Bottled water	5 ppb	https://www.fda.gov/food/environmenta
3.	Lead action level in processed food for babies and young children	1. For fruits, vegetables, yoghurts, custards/puddings, and single-ingredient meats	10 ppb	https://www.cov.com/en/news
		2. For single-ingredient root vegetables;	20 ppb	https://www.cov.com/en/news
		3. For dry infant cereals.	20 ppb	https://www.cov.com/en/news-
4.	Cadmium	Bottled water	5 ppb	FDA Environmental Contaminants in Food
5.	Mercury	Bottled water	2 ppb	https://www.fda.gov/food/environmental
		Wheat seeds (pink kernels)	1 ppm	https://www.fda.gov/regulatory
6.	Methyl mercury	Fish and shellfish (edible portion only)	1 part per million (ppm)	https://www.fda.gov/regulatory

TABLE 29.9 Bureau of Indian Standards guidelines for heavy metals in drinking water.

S. no	Heavy metal	Maximum permissible limit (mg/L)	Reference
1.	Lead (Pb)	0.01	https://cpcb.nic.in/wqm/BIS_Drinking_Water_Specification.pdf
2.	Cadmium (Cd)	0.003	
3.	Arsenic (As)	0.01	
4.	Mercury (Hg)	0.001	
5.	Chromium (Cr)	0.05	

TABLE 29.10 Threshold levels of Heavy metals in a wide range of foodstuffs.

S. no.	Heavy metal	Article of food	Maximum limit (parts per million by weight)	References
1.	Lead (Pb)	Brassica vegetables, excluding kale, leafy vegetables (excluding spinach)	0.3	https://www.fssai.gov.in/
		Bulb vegetables, fruiting vegetables, cucurbits and Root and tuber vegetables	0.1	
		Cereal grains, except for buckwheat	0.2	
		Fish	0.3	
		Salt	2.0	
2.	Cadmium (Cd)	Rice polished	0.4	
		Brassica, bulb and fruiting vegetables, cucurbits	0.05	
		Legume, root, and tuber vegetables	0.1	
3.	Arsenic (As)	Milk	0.1	
		Vegetables	1.1	
4.	Mercury (Hg)	Fish	0.5	
		Predatory fish	1.0	
5.	Chromium (Cr)	Vegetables	1.0	
		All fishery products	12	

determined that those concentrations did not protect human health. The proposed values all have very low concentrations, indicating these elements' bioaccumulation and acute toxicity effects in humans. The median lethal dose values for mice or rats also show acute toxicity of HMs exposure (Table 29.12).

A few countries also impose maximum concentration limits for these pollutants in irrigation water (Table 29.13), which shows the necessity of regulating HM levels in food.

In Canada, values for irrigation water are comparatively stricter than those of FAO, and chromium, copper, lead, and zinc concentrations are lower than those defined by. The lead limit suggested by FAO is very high because lead is a highly toxic element. This relates to the fact that lead tends to bind readily with organic and colloidal matter within the soil and is readily soluble with only a minute portion, thus making it bioavailable (Table 29.13). Moreover, Plants' uptake leads through roots and their uptake and translocation are reduced in aerial parts (Pourrut, Shahid, Dumat, Winterton, & Pinelli, 2011).

In most countries, land legislation is lacking for other contaminated media, such as soils and sediments. However,

TABLE 29.11 Standard limits of heavy metals on drinking water for human consumption.

S. no.	Heavy metal	Maximum permissible limit (mg/L or μg/L)	References
1.	Lead (Pb)	5 μg/L (previously 10 μg/L)	https://eur-lex.europa.eu/legal-content/EN/TXT/?uri=CELEX%3A32020L2184
2.	Arsenic (As)	10 μg/L	
3.	Cadmium (Cd)	5 μg/L	
4.	Mercury (Hg)	1 μg/L	
5.	Nickel (Ni)	20 μg/L	
6.	Chromium (Cr)	25 μg/L	
7.	Antimony (Sb)	10 μg/L	
8.	Selenium (Se)	20 μg/L	
9.	Boron (B)	1.5 mg/L	

the EU and Canada have laws regulating contaminants in agricultural soils. The maximum limits for HMs in agricultural soils are intended to ensure that these can be used for cultivation and that plants do not absorb excessive amounts of these pollutants, which remain edible (Tables 29.14 and 29.15).

29.5 Policy responses and regulations on heavy metals exposure

As per the UNEP, many International and regional conventions are implemented to protect the environment against persistent pollutants, including HMs. They form the legal and regulatory framework for the government to implement the necessary approach programs for assessing and monitoring HMs and evaluating their impact on the ecosystem. Many international conventions have been established to control HMs' hazardous threats in real-world scenarios, including the Rotterdam Convention on the Prior Informed Consent Procedure for Certain Hazardous Chemicals and Pesticides in International Trade (2004). This convention mainly targets facilitating the responsibilities of importing and utilizing hazardous substances (Abdelmonem, Kamal, Elbaz, Khalifa, & Abdelnaser, 2025).

Additionally, another convention on Long-Range Transboundary Air Pollutants (LRTAP) Protocol on Heavy Metals (2003) focused on preventing and reducing potential sources of air pollution, specifically on Hg, Pb, and Cd contamination. Besides, similar regulations, such as the Basel Convention on the Control of Transboundary Movements of Hazardous Wastes and their Disposal (1992) (Chen & Ding, 2023), are designed to regulate and avoid hazardous health problems due to waste disposal.

Furthermore, based on the United Nations (UN) agenda to adopt the Sustainable development goals and more international legislations and policies, the European region has already succeeded in minimizing industrial emissions until the year 2021, particularly in HM pollution where Pb emissions, Hg emissions and Cd emissions reduced by 42%, 47%, and 37%, respectively, in the 27 EU countries. Moreover, the EU still managed to uplift industrial growth and achieved its economic value by 16% over the same period (Abdelmonem et al., 2025).

While the regulatory authorities enforce global norms and laws for HMs control, the research experts have also joined in the regulatory goal and directed the research field in a similar direction. Lastly, establishing Global legislation and protocols and sharing unique documents from international organizations, such as the WHO and UN, by monitoring and reducing HM pollution will be significant for developing countries to achieve environmental sustainability goals.

29.6 Conclusion

HMs are a serious hazard to human health and the environment due to their persistence, toxicity, bioaccumulation in ecosystems, poisoning, and eventually adverse health effects in animals and humans through the food chain. HMs such as arsenic, lead, mercury, and cadmium can lead to varying health problems, from increased chance of organ damage to developing systemic diseases. HMs can

TABLE 29.12 The maximum concentration level (MCL) for heavy metals in drinking water and their toxicity (Vareda et al., 2019).

		MCL in drinking water/μg/L					Toxicity/mg/kg bw	bw[a]
S. no	Heavy metal	WHO[b]	EU[c]	United States[d]	Canada[e]	China[f]	PTWI[g]	LD50[h]
1.	As	10	10	10	10	10	Withdrawn (0.0021 daily)	41.0[l]
2.	Cd	3	5	5	5	5	0.025 monthly	5.2[m] 88.0[l]
3.	Cr	50	50	100	50	50[j]		–
4.	Cu	2000	2000	1300	–	1000	0.500 daily[k]	5.6[m] 584.0[l]
5.	Pb	10	10	15	10	10	Withdrawn (0.025[i])	–
6.	Hg	6	1	2	1	–	0.004	6.6[m] 1.0[l]
7.	Ni	7	20	–	–	20	–	–
8.	Zn	–				1000	1.0 daily[k]	20.2[m]

[a] *Bodyweight.*
[b] *World Health Organization (2017).*
[c] *Council of the European Union (1998).*
[d] *US Environmental Protection Agency (2018).*
[e] *Water and Air Quality Bureau - Healthy Environments and Consumer Safety Branch (2017).*
[f] *Ministry of Health of China (2006).*
[g] *Provisional tolerable weekly intake for humans JECFA (2017).*
[h] *Median lethal dose values obtained in mice or rats.*
[i] *Former value.*
[j] *Reported as hexavalent chromium.*
[k] *Maximum tolerable intake.*
[l] *Hunt et al. (2012).*
[m] *Jones et al. (1979).*

TABLE 29.13 The recommended maximum concentration level for heavy metals in irrigation water (Vareda et al., 2019).

S. no.	Heavy metals	Recommended Maximum Concentration level (μg/L)		
		FAO (Pescod, 1992)	Portugal (Do Ambiente, 1998)	Canada (Canadian Council of Ministers of the Environment, 1999)
1.	As	100	100	100
2.	Cd	10	10	5
3.	Cr	100	100	8 Cr(VI) 5 Cr(III)
4.	Cu	200	200	200[a]
5.	Pb	5000	5000	200
6.	Hg	–	–	–
7.	Ni	200	500	200
8.	Zn	2000	2000	1000 pH < 6.5

[a] *Value defined for cereals; 1.0 mg/L for tolerant crops.*

TABLE 29.14 Permissible thresholds for heavy metal concentrations in agricultural soils.

S. no	Heavy metals	Limit concentration/mg/kg dw[a]				
		Agricultural soils			Sediments	
		EU, current value[b] (soil pH: 6 ≤ pH < 7)	EU, proposal[c] (soil pH: 6 ≤ pH < 7) <	Canada[d]	US guidelines[e,f]	Canada[g,h]
1.	As	–	–	12	33.0	5.9
2.	Cd	1–3	1	1.4	4.98	0.6
3.	Cr	–	75	64	111	37.3
4.	Cu	50–140	50	63	149	35.7
5.	Pb	50–300	70	70	128	35.0
6.	Hg	1–1.5	0.5	6.6	1.06	0.170
7.	Ni	30–75	50	45	48.6	–
8.	Zn	150–300	150	200	459	123.0

[a] *Dry weight.*
[b] *Council of the European Communities (1986)*
[c] *Gawlik and Bidoglio (2006)*
[d] *Canadian Council of Ministers of the Environment (1999a)*
[e] *Ingersoll et al. (2000); National Oceanic and Atmospheric Administration's Office of Response and Restoration (2008).*
[f] *Consensus-based probable effect level.*
[g] *Canadian Council of Ministers of the Environment (2001).*
[h] *Canadian Interim Sediment Quality Guideline.*

also be absorbed by plants, which can impact plant growth and consumers alike. Controlling pollution by monitoring HM levels is crucial. HM exposure leads to potentially adverse effects to pose nervous and cardiovascular health hazards. These metals can accumulate in the body and disrupt normal physiological processes by entering the food chain. HMs can be detected and quantified using various analytical techniques such as AAS, ICP-MS, ICP-AES, and AFS. Assessing HM pollution in soil and water is crucial for protecting the environment and public health.

TABLE 29.15 Toxicological profiles of heavy metals by various international organizations.

S. no.	Heavy metals	Compartment	Agency	Description	Limits	References
1.	Chromium	Air	WHO	"Air Quality Guidelines Chromium (VI)"	1 μg/m^3 for lifetime risk of 4×10^{-2}	www.atsdr.cdc.gov
		Workplace	OSHA	"Legal limit over an 8-hour workday of Cr(VI)."	0.005 mg/m^3	
			OSHA	"Legal limit over an 8-hour workday of Cr(III)."	0.5 mg/m^3	
			OSHA	"Legal limit over an 8-hour workday of Cr(0)."	1 mg/m^3	
			NIOSH	"Legal limit over a workday of 8 h for Cr(0), Cr(II) and Cr(III)"	0.5 mg/m^3	
			NIOSH	"Exposure limit over a workday of 10 h for Cr(VI)."	0.001 mg/m^3	
		Drinking water	WHO	"Drinking-Water Quality Guidelines for Total Chromium."	0.05 mg/L	
			EPA	"Maximum contaminant level for total chromium"	0.1 mg/L	
		Bottled Water	FDA	"Not to exceed the total chromium concentration."	0.1 mg/L	
2.	Arsenic	Air	WHO	"Air Quality Guidelines"	1.5×10^{-3} μg/m^3	
			WHO	"Air Quality Guidelines for estimation of cancer risk for a lifetime exposure."	1 μg/m^3	
		Workplace Air	OSHA	"Legal limit over an 8-hour workday of airborne arsenic in places that use inorganic arsenic."	10 μg/m^3	
		Drinking water	WHO	"Drinking water threshold"	0.01 mg/L	
			EPA	"Drinking water threshold"	0.05 mg/L	
3.	Cadmium	Air	WHO	"Air Quality Guidelines"	5 ng/m^3	
		Workplace	OSHA	"Legal limit over an 8-hour the workday of cadmium in air."	5 μg/m3	
		Drinking water	EPA	"Drinking water at this concentration for 10 days will not cause adverse effects in youths."	0.04 mg/L	
			EPA	"Drinking water at this concentration for a lifetime exposure dose not cause adverse effects."	0.005 mg/L	
			WHO	"Drinking Water Quality Guidelines"	0.003 mg/L	
		Bottled water	FDA	"Bottled water threshold"	0.005 mg/L	
4.	Mercury	Intake	WHO	"Permissible weekly intake"	5 μg/kg total 3.3 μg/kg methylmercury	
		Workplace Air	OSHA	"Legal limit over an 8-hour workday of organic mercury in the air."	0.1 mg/m^3	
			OSHA	"Legal limit over an 8-hour workday of organic mercury in the air."	0.05 mg/m^3	
			NIOSH	"Legal limit over a 10-hour workday of organic mercury in the air."	0.05 mg/m^3	
		Drinking water	EPA	"Drinking water inorganic mercury threshold"	0.002 mg/L	
			WHO	"Drinking Water Quality Guidelines for health-related organics regarding all mercury forms."	0.001 mg/L	
		SeaFood	FDA	"Max legal limit of methylmercury"	1 ppm	

5.	Nickel	Air	WHO	"Air Quality Guidelines"	3.8×10^{-4} $(\mu g/m^3)^{-1}$	
		Workplace Air	OSHA	"Legal limit over an 8-hour workday of Ni in the air."	1 mg/m^3	
		Drinking water	WHO	"Drinking-Water Quality Guidelines"	0.07 mg/L	
			EPA	"Drinking water threshold"	0.1 mg/L	
6.	Lead	Air	WHO	Air Quality Guidelines	No data	
		Workplace Air	OSHA	"Legal limit over an 8-hour workday in general industry for elemental, organic and inorganic lead."	0.5 mg/m^3	
			OSHA	"Legal limit over an 8-hour workday in general industry for tetraethyl lead and tetramethyl lead."	0.075 mg/m^3	
			OSHA	"Legal limit over an 8-hour workday in shipyards and construction sites for elemental, organic and inorganic lead."	50 $\mu g/m^3$	
			OSHA	"Legal limit over an 8-hour workday in shipyards and construction sites for tetraethyl lead."	0.1 mg/m^3	
			OSHA	"Legal limit over an 8-hour workday in shipyards and construction sites for tetramethyl lead"	0.15 mg/m^3	
			OSHA	"Action level over an 8-hour workday for industry and construction for elemental, organic and inorganic lead."	30 $\mu g/m^3$	
			OSHA	"Medical removal protection for general industry—Temporary removal blood lead level."	≥ 60 $\mu g/m^3$	
			OSHA	"Medical removal protection for general industry—Return to work blood lead level."	≤ 40 $\mu g/m^3$	
			OSHA	"Medical removal protection for construction and shipyards—Temporary blood lead level removal."	≥ 50 $\mu g/m^3$	
			OSHA	"Medical removal protection for construction and shipyards—Return to work blood lead level."	< 40 $\mu g/m^3$	
			NIOSH	"Recommended exposure limit of lead and its compounds."	0.05 mg/m^3	
			NIOSH	"Recommended exposure limit of tetraethyl lead."	0.075 mg/m^3	
			NIOSH	"Immediately dangerous to life or health concentration of lead and its compounds."	100 mg/m^3	
			NIOSH	"Immediately dangerous to life or health concentration of tetraethyl lead."	40 mg/m^3	
		Drinking water	EPA	"Drinking water action level"	0.015 mg/L	
			WHO	"' Drinking Water Quality Guidelines"	0.01 mg/L	
		Bottled water	FDA	"Legal limit"	No data	

Recent advances in remediation technologies, both traditional and collective nanotechnology, can potentially address the challenge of HM exposure. International policymakers have developed protocols to prevent HM pollution from damaging the environment and human health. Additionally, implementing legal frameworks and public awareness are essential for establishing pollution control strategies. By integrating these approaches, the mining industry can evolve toward a model that meets economic and resource needs and contributes to environmental sustainability and social equity.

References

Abdelmonem, B. H., Kamal, L. T., Elbaz, R. M., Khalifa, M., & Abdelnaser, A. (2025). From contamination to detection: The growing threat of heavy metals. *Heliyon, 11*(1), e41713. Retrieved from https://doi.org/10.1016/j.heliyon.2025.e41713.

Afzal, J., Wang, X., Saleem, M. H., Sun, X., Hussain, S., Khan, I., ... Hu, C. (2021). Applying ferrous sulfate alleviates cadmium's negative impact in rice (*Oryza sativa* L.). *Biocell, 45*(6), 1631.

Bader, J. L., Gonzalez, G., Goodell, P. C., Ali, A. M. S., & Pillai, S. D. (1999). Aerobic reduction of hexavalent chromium in soil by indigenous microorganisms. *Bioremediation Journal, 3*(3), 201–211.

Bharti, R., & Sharma, R. (2022). Effect of heavy metals: An overview. *Materials Today: Proceedings, 51*, 880–885.

Cairns, J., Smith, E. P., & Orvos, D. (1988). The problem of validating simulation of hazardous exposure in natural systems. In *Proceedings of the 1988 Summer computer conference* (pp. 448–454). The Society for Computer Simulation International, San Diego.

Canadian Council of Ministers of the Environment. (1999). *Canadian water quality*.

Castro-González, M. I., & Méndez-Armenta, M. (2008). Heavy metals: Implications associated to fish consumption. *Environmental Toxicology and Pharmacology, 26*(3), 263–271.

Chen, S., & Ding, Y. (2023). Tackling heavy metal pollution: Evaluating governance models and frameworks. *Sustainability, 15*(22), 15863.

Chen, Y. W., Huang, C. F., Tsai, K. S., Yang, R. S., Yen, C. C., Yang, C. Y., ... Liu, S. H. (2006). Methylmercury induces pancreatic β-cell apoptosis and dysfunction. *Chemical Research in Toxicology, 19*(8), 1080–1085.

Chowdhary, P., Yadav, A., Singh, R., Chandra, R., Singh, D. P., Raj, A., & Bharagava, R. N. (2018). Stress response of *Triticum aestivum* L. and *Brassica juncea* L. against heavy metals growing at distillery and tannery wastewater contaminated site. *Chemosphere, 206*, 122–131.

Cifuentes, F. R., Lindemann, W. C., & Barton, L. L. (1996). Chromium sorption and reduction in soil with implications to bioremediation. *Soil Science, 161*(4), 233–241.

Das, D. (1995). *Arsenic species along with other metal/metalloid present and responsible for arsenic episode in groundwater of West Bengal and a cheap technique to remove arsenic, thus making the groundwater suitable for drinking and cooking* (Thesis submitted for the degree of Ph. D., Science). Jadavpur University, 260.

De Flora, S., Serra, D., Camoirano, A., & Zanacchi, P. (1989). Metabolic reduction of chromium, as related to its carcinogenic properties. *Biological Trace Element Research, 21*, 179–187.

Do Ambiente, M. (1998). *Decreto-lei no 236/98 de 1 de agosto. Diário da República: I série A, (176)*.

Escudero-Castejon, L., Sanchez-Segado, S., Parirenyatwa, S., & Jha, A. (2016). Formation of chromium-containing molten salt phase during roasting of chromite ore with sodium and potassium hydroxides. *Journal for Manufacturing Science and Production, 16*(4), 215–225.

Fleischer, M., Sarofim, A. F., Fassett, D. W., Hammond, P., Shacklette, H. T., Nisbet, I. C., & Epstein, S. (1974). Environmental impact of cadmium: A review by the Panel on Hazardous Trace Substances. *Environmental Health Perspectives, 7*, 253–323.

Gawlik, B. M., & Bidoglio, G. (2006). *Background values in European soils and sewage sludges*. European Commission.

Genchi, G., Carocci, A., Lauria, G., Sinicropi, M. S., & Catalano, A. (2020). Nickel: Human health and environmental toxicology. *International Journal of Environmental Research and Public Health, 17*(3), 679.

Ghani, A., & Ghani, A. (2011). Effect of chromium toxicity on growth, chlorophyll and some mineral nutrients of *Brassica juncea* L. *Egyptian Academic Journal of Biological Sciences, H. Botany, 2*(1), 9–15.

Gorny, J., Billon, G., Noiriel, C., Dumoulin, D., Lesven, L., & Madé, B. (2016). Chromium behavior in aquatic environments: A review. *Environmental Reviews, 24*(4), 503–516.

Gworek, B., Dmuchowski, W., & Baczewska-Dąbrowska, A. H. (2020). Mercury in the terrestrial environment: A review. *Environmental Sciences Europe; Revue Litteraire Mensuelle, 32*(1), 128.

Hogervorst, J., Plusquin, M., Vangronsveld, J., Nawrot, T., Cuypers, A., Van Hecke, E., & Staessen, J. A. (2007). House dust as possible route of environmental exposure to cadmium and lead in the adult general population. *Environmental Research, 103*(1), 30–37.

Hunt, P. R., Olejnik, N., & Sprando, R. L. (2012). Toxicity ranking of heavy metals with screening method using adult *Caenorhabditis elegans* and propidium iodide replicates toxicity ranking in rat. *Food and Chemical Toxicology, 50*(9), 3280–3290.

Ingersoll, C. G., MacDonald, D. D., Wang, N., Crane, J. L., Field, L. J., Haverland, P. S., & Smorong, D. E. (2001). Predictions of sediment toxicity using consensus-based freshwater sediment quality guidelines. *Archives of Environmental Contamination and Toxicology, 41*, 8–21.

Jaishankar, M., Tseten, T., Anbalagan, N., Mathew, B. B., & Beeregowda, K. N. (2014). Toxicity, mechanism and health effects of some heavy metals. *Interdisciplinary Toxicology, 7*(2), 60–72.

Järup, L., & Åkesson, A. (2009). Current status of cadmium as an environmental health problem. *Toxicology and Applied Pharmacology, 238*(3), 201–208.

Jones, M. M., Schoenheit, J. E., & Weaver, A. D. (1979). Pretreatment and heavy metal LD50 values. *Toxicology and Applied Pharmacology, 49*(1), 41–44.

Kulshrestha, S., Acharya, N., Mathur, K., & Pandey, T. (2024). Legal framework and regulatory compliance in metal mining—An analysis of environmental and operational standards. *Journal of Mines, Metals and Fuels, 72*(10), 1035–1047.

Lin, H. J., Sung, T. I., Chen, C. Y., & Guo, H. R. (2013). Arsenic levels in drinking water and mortality of liver cancer in Taiwan. *Journal of Hazardous Materials, 262*, 1132–1138.

Ma, L., Xu, J., Chen, N., Li, M., & Feng, C. (2019). Microbial reduction fate of chromium (Cr) in aqueous solution by mixed bacterial consortium. *Ecotoxicology and Environmental Safety, 170*, 763–770.

Monalisa Mohanty, M. M., & Patra, H. K. (2013). *Effect of ionic and chelate assisted hexavalent chromium on mung bean seedlings (Vigna radiata L. wilczek. var k-851) during seedling growth*.

Nagajyoti, P. C., Lee, K. D., & Sreekanth, T. V. M. (2010). Heavy metals, occurrence and toxicity for plants: A review. *Environmental Chemistry Letters, 8*, 199–216.

Nair, M., Jayalakshmy, K. V., Balachandran, K. K., & Joseph, T. (2006). Bioaccumulation of toxic metals by fish in a semi-enclosed tropical ecosystem. *Environmental Forensics, 7*(3), 197–206.

Nedelkoska, T. V., & Doran, P. M. (2000). Hyperaccumulation of cadmium by hairy roots of *Thlaspi caerulescens*. *Biotechnology and Bioengineering, 67*(5), 607–615.

Odstrcil, A. D. C. A., Carino, S. N., Ricci, J. C. D., & Mandalunis, P. M. (2010). Effect of arsenic in endochondral ossification of experimental animals. *Experimental and Toxicologic Pathology, 62*(3), 243–249.

Park, J. D., & Zheng, W. (2012). Human exposure and health effects of inorganic and elemental mercury. *Journal of Preventive Medicine and Public Health, 45*(6), 344.

Park, J. H., Lamb, D., Paneerselvam, P., Choppala, G., Bolan, N., & Chung, J. W. (2011). Role of organic amendments on enhanced bioremediation of heavy metal (loid) contaminated soils. *Journal of Hazardous Materials, 185*(2-3), 549–574.

Park, R. M., Bena, J. F., Stayner, L. T., Smith, R. J., Gibb, H. J., & Lees, P. S. (2004). Hexavalent chromium and lung cancer in the chromate industry: A quantitative risk assessment. *Risk Analysis: An International Journal, 24*(5), 1099–1108.

Pescod, M. B. (1992). *Wastewater treatment and use in agriculture-FAO irrigation and drainage paper 47*. Food and Agriculture Organization of the United Nations.

Pourrut, B., Shahid, M., Dumat, C., Winterton, P., & Pinelli, E. (2011). Lead uptake, toxicity, and detoxification in plants. *Reviews of Environmental Contamination and Toxicology, volume 213*, 113–136.

Rascio, N., & Navari-Izzo, F. (2011). Heavy metal hyperaccumulating plants: How and why do they do it? And what makes them so interesting? *Plant Science, 180*(2), 169–181.

Rathi, B. S., Kumar, P. S., Susana, J. N. I., Virgin, J. F., Dharani, R., Sanjay, S., & Rangasamy, G. (2024). Recent research progress on the removal of heavy metals from wastewater using modified zeolites: A critical review. *Desalination and Water Treatment, 100573*.

Rehman, M., Saleem, M. H., Fahad, S., Maqbool, Z., Peng, D., Deng, G., & Liu, L. (2021). Medium nitrogen optimized *Boehmeria nivea* L. growth in copper contaminated soil. *Chemosphere, 266*, 128972.

Rice, K. M., Walker, E. M., Jr, Wu, M., Gillette, C., & Blough, E. R. (2014). Environmental mercury and its toxic effects. *Journal of Preventive Medicine and Public Health, 47*(2), 74.

Ripley, E. M., & Li, C. (2018). Metallic ore deposits associated with mafic to ultramafic igneous rocks. In Sisir K. Mondal, & William L. Griffin (Eds.). *Processes and ore deposits of ultramafic-mafic magmas through space and time* (pp. 79–111). Elsevier. Retrieved from https://doi.org/10.1016/B978-0-12-811159-8.00004-4.

Rodríguez, M. C., Barsanti, L., Passarelli, V., Evangelista, V., Conforti, V., & Gualtieri, P. (2007). Effects of chromium on photosynthetic and photoreceptive apparatus of the alga *Chlamydomonas reinhardtii*. *Environmental Research, 105*(2), 234–239.

Saleem, M. H., Ali, S., Rehman, M., Rana, M. S., Rizwan, M., Kamran, M., & Liu, L. (2020). Influence of phosphorus on copper phytoextraction via modulating cellular organelles in two jute (*Corchorus capsularis* L.) varieties grown in a copper mining soil of Hubei Province, China. *Chemosphere, 248*, 126032.

Saleem, M. H., Fahad, S., Khan, S. U., Din, M., Ullah, A., Sabagh, A. E., & Liu, L. (2020). Copper-induced oxidative stress, initiation of antioxidants and phytoremediation potential of flax (*Linum usitatissimum* L.) seedlings grown under the mixing of two different soils of China. *Environmental Science and Pollution Research, 27*, 5211–5221.

Schaumlöffel, D. (2012). Nickel species: Analysis and toxic effects. *Journal of Trace Elements in Medicine and Biology, 26*(1), 1–6.

Schultz, H. W., & Schultz, H. W. (1981). Federal Food, Drug, and Cosmetic Act: Sections 381, 391 and 392 imports and exports; miscellaneous. In H. W. Schultz (Ed.). *Food law handbook* (pp. 621–625). Dordrecht: Springer. Retrieved from https://doi.org/10.1007/978-94-011-7373-5_51.

Shahid, M., Khalid, S., Abbas, G., Shahid, N., Nadeem, M., Sabir, M., & Dumat, C. (2015). Heavy metal stress and crop productivity. In K. Hakeem (Ed.). *Crop production and global environmental issues* (pp. 1–25). Cham: Springer. Retrieved from https://doi.org/10.1007/978-3-319-23162-4_1.

Shanker, A. K. (2019). *Chromium: Environmental pollution, health effects and mode of action*.

Shrestha, R., Ban, S., Devkota, S., Sharma, S., Joshi, R., Tiwari, A. P., & Joshi, M. K. (2021). Technological trends in heavy metals removal from industrial wastewater: A review. *Journal of Environmental Chemical Engineering, 9*(4), 105688.

Singh N., Kumar D., & Sahu A. (2007). *Arsenic in the environment: Eff ects on human*.

Smedley, P. L., & Kinniburgh, D. G. (2002). A review of the source, behaviour and distribution of arsenic in natural waters. *Applied Geochemistry, 17*(5), 517–568.

Stohs, S. J., & Bagchi, D. (1995). Oxidative mechanisms in the toxicity of metal ions. *Free Radical Biology and Medicine, 18*(2), 321–336.

Tumolo, M., Ancona, V., De Paola, D., Losacco, D., Campanale, C., Massarelli, C., & Uricchio, V. F. (2020). Chromium pollution in European water, sources, health risk, and remediation strategies: An overview. *International Journal of Environmental Research and Public Health, 17*(15), 5438.

Uraguchi, S., & Fujiwara, T. (2012). Cadmium transport and tolerance in rice: Perspectives for reducing grain cadmium accumulation. *Rice, 5*, 1–8.

Vahidnia, A., Van der Voet, G. B., & De Wolff, F. A. (2007). Arsenic neurotoxicity—A review. *Human & Experimental Toxicology, 26*(10), 823–832.

Vareda, J. P., Valente, A. J., & Durães, L. (2019). Assessment of heavy metal pollution from anthropogenic activities and remediation strategies: A review. *Journal of Environmental Management, 246*, 101–118.

Vasudevaraju, P., Govindaraju, M., Palanisamy, A. P., Sambamurti, K., & Rao, K. S. J. (2008). Molecular toxicity of aluminium in relation to neurodegeneration. *Indian Journal of Medical Research, 128*(4), 545–556.

Wu, M., Li, G., Jiang, X., Xiao, Q., Niu, M., Wang, Z., & Wang, Y. (2017). Non-biological reduction of Cr(VI) by reacting with humic acids composted from cattle manure. *RSC Advances, 7*(43), 26903–26911.

Yan, B. Z., & Chen, Z. F. (2019). Influence of pH on Cr(VI) reduction by organic reducing substances from sugarcane molasses. *Applied Water Science, 9*, 1–8.

Zhang, R., Xiang, Y., Ran, Q., Deng, X., Xiao, Y., Xiang, L., & Li, Z. (2014). Involvement of calcium, reactive oxygen species, and ATP in hexavalent chromium-induced damage in red blood cells. *Cellular Physiology and Biochemistry, 34*(5), 1780–1791.

Index

Note: Page numbers followed by "*f*" and "*t*" refer to figures and tables, respectively.

C

H

I

N

O

9780443365751